CAMBRIDGE LIBRARY COLLECTION

Books of enduring scholarly value

Botany and Horticulture

Until the nineteenth century, the investigation of natural phenomena, plants and animals was considered either the preserve of elite scholars or a pastime for the leisured upper classes. As increasing academic rigour and systematisation was brought to the study of 'natural history', its subdisciplines were adopted into university curricula, and learned societies (such as the Royal Horticultural Society, founded in 1804) were established to support research in these areas. A related development was strong enthusiasm for exotic garden plants, which resulted in plant collecting expeditions to every corner of the globe, sometimes with tragic consequences. This series includes accounts of some of those expeditions, detailed reference works on the flora of different regions, and practical advice for amateur and professional gardeners.

A Dictionary of the Economic Products of India

A Scottish doctor and botanist, George Watt (1851–1930) had studied the flora of India for more than a decade before he took on the task of compiling this monumental work. Assisted by numerous contributors, he set about organising vast amounts of information on India's commercial plants and produce, including scientific and vernacular names, properties, domestic and medical uses, trade statistics, and published sources. Watt hoped that the dictionary, 'though not a strictly scientific publication', would be found 'sufficiently accurate in its scientific details for all practical and commercial purposes'. First published in six volumes between 1889 and 1893, with an index volume completed in 1896, the whole work is now reissued in nine separate parts. Volume 4 (1890) contains entries from *Gossypium* (the cotton genus) to *Linociera intermedia* (a species of small tree, used for timber).

A Dictionary of the Economic Products of India

Volume 4: Gossypium to Linociera

George Watt

Cambridge University Press

CAMBRIDGE
UNIVERSITY PRESS

University Printing House, Cambridge, CB2 8BS, United Kingdom

Published in the United States of America by Cambridge University Press, New York

Cambridge University Press is part of the University of Cambridge.
It furthers the University's mission by disseminating knowledge in the pursuit of education, learning and research at the highest international levels of excellence.

www.cambridge.org
Information on this title: www.cambridge.org/9781108068765

This edition first published 1890
This digitally printed version 2014

ISBN 978-1-108-06876-5 Paperback

ERRATUM.—VOLUMES III-V.

Attention has been drawn to the fact that an error runs through Volumes III, IV, and V of the Dictionary, wherein the consecutive numbers of letters G and L, in passing from Volume III-IV and from Volume IV-V respectively, have been partially duplicated. The numbers in those volumes, which had not been issued from the Press when the error was pointed out, have been corrected, and will, as corrected, form the future reference numbers. It is suggested, that those who possess uncorrected volumes should adopt a similar course, making the first reference number in Volume IV G 381 and in Volume V L 379.

EDGAR THURSTON,

Offg. Reporter on Economic Products
to the Government of India.

A

DICTIONARY.

OF

THE ECONOMIC PRODUCTS OF INDIA.

BY

GEORGE WATT, M.B., C.M., C.I.E.

REPORTER ON ECONOMIC PRODUCTS WITH THE GOVERNMENT OF INDIA.
OFFICIER D'ACADEMIE; FELLOW OF THE LINNEAN SOCIETY; CORRESPONDING MEMBER OF THE ROYAL HORTICULTURAL SOCIETY, &C., &C.

(ASSISTED BY NUMEROUS CONTRIBUTORS.)
IN SIX VOLUMES.

VOLUME IV.,
Gossypium to Linociera.

Published under the Authority of the Government of India,
Department of Revenue and Agriculture.

LONDON:
W. H. ALLEN & Co., 13 WATERLOO PLACE, S.W., PUBLISHERS TO THE INDIA OFFICE.

CALCUTTA:
OFFICE OF THE SUPERINTENDENT OF GOVERNMENT PRINTING, INDIA,
8, HASTINGS STREET.

1890.

CALCUTTA:
GOVERNMENT OF INDIA CENTRAL PRINTING OFFICE,
8, HASTINGS STREET.

DICTIONARY
OF
THE ECONOMIC PRODUCTS OF INDIA.

GOSSYPIUM, *Linn.; Gen. Pl., I., 209.* 381

A genus of the Natural Order MALVACEÆ, the species of which are widely distributed on both sides of the Equator, and in both hemispheres. On the North they extend, under cultivation, as far as the shores of Southern Europe, and on the South to the Cape of Good Hope.

The limitation of the forms of these very important cultivated plants, to definite species, varieties, hybrids, and races, has much perplexed writers on the subject. The vast importance of the floss, obtained from the seeds, has induced an extensive cultivation, and that, too, under almost every condition of climate and soil. The ease with which climatic conditions originate local forms, and the rapidity with which hybrids are produced, have *both* combined to bring about a degree of complexity that perhaps exists with no other agricultural crop. The departures from original specific conditions are in fact so great as to render it almost impossible to determine whether the cotton-yielding plants have been derived from three or more species, or are all mere developments from a protean ancestor which possessed indigenous Asiatic, African, and American representatives. This being so, the authors have found it undesirable, at present, to attempt more than a brief review of the botanical literature of the subject. They are conscious that many ambiguities and even errors may be thereby reproduced, but unavoidably so, for, until a thorough and original investigation has been conducted with the living plants, nothing definite can be published regarding the cultivated cottons of India The classification which follows may, however, be accepted as denoting some of the chief forms recognised by us, but the accuracy of the relegation of vernacular names and of the restriction of special properties to definite forms, is open to the gravest doubt. This is due to the authors having been practically limited to a compilation from published works and official correspondence. The primary object of the present effort may in fact be viewed as an attempt to suggest some of the main lines on which a scientific classification of the Indian indigenous and exotic cottons, and of their hybrids, might in the future be conducted, rather than to produce a treatise professing originality. It must, however, be admitted that no real good can be accomplished until an original investigation has been conducted, since everything depends on knowing whether or not a form recommended for experimental cultivation is suited to a particular district.

This knowledge could doubtless be arrived at after the different forms of cotton had been scientifically worked up, and the degree of their hybridisation and the nature of *that* hybridisation clearly established. Indeed, it is at present impossible to write with any degree of confidence regarding this—one of the most important of Indian crops,—since it is often difficult to know what forms are referred to by writers who employ local and vernacular names only, when discussing their properties. Thus, for example, a District Officer writing of *Nurma* cotton may be understood to be speaking of some of the forms of **G. arboreum** or of some hybrid between that species and **G. herbaceum**—*Deshi* cotton with purple instead of yellow flowers. Indeed, it seems highly improbable that any form of Indian cotton can be said to belong, strictly speaking, to the Linnean species, **G. herbaceum.** On the other hand, nearly every writer on Indian cotton has denied the existence of **G. arboreum** as a field crop, whereas the chief cotton of Eastern and Northern India seems undoubtedly to be a form of that species. In fact, in this distinction lies the chief difference between the Indian cottons of the Eastern and Western peninsulas, the latter being mainly derived from a form allied to, but quite distinct from, **G. herbaceum.**

Much has been done to discover the soils suited for cotton generally, and under orders of Government, many valuable experiments have been made, with exotic forms. What would appear to be the foremost and most essential enquiry has, apparently, however, been entirely neglected, namely, a scientific and exhaustive investigation into the existing conditions of Indian cotton. Until this has been done experiment can be but blind groping in the dark, which by chance may now and then fall upon facts of importance, but only after needless expenditure of time and money. To effect improvement in the indigenous cottons would of necessity confer a more lasting benefit to the country than even success in acclimatisation of exotic forms. To accomplish this, the endemic cottons must first be worked up. The climatic and other causes which tend to preserve or destroy good or bad properties must be thoroughly established. The influences of hybridisation must be worked out on a scientific basis, by testing the strength of strain, from this species or that, best suited to the environment. The tendency of retrogression from a prized hybrid to either ancestor should not be a matter of periodic deprecation; it should be understood, and, if possible, prevented or the exhausted stock promptly replenished. When experiment and study have furnished the required data on which to act, it will almost for certain be found, that to preserve the cotton of a province or district up to a required standard, specially grown and artificially hybridised stock will be imperatively necessary. Indeed, the establishment of nurseries or seed farms, for this purpose, on a more rational basis than heretofore attempted, would, in all probability, prove the greatest reform possible for the cotton industry of India. When the quality of stock has been improved, attention might then be turned to the modes of cultivation and methods of cleaning the fibre, but these are less defective than the present evil of a degeneration of stock against which the poverty and indebtedness of the cultivator render him helpless. It is significant that in the country that once supplied Europe with its raw cotton and cotton goods, the latter of a quality far superior to the manufactures of the present day, there should exist neglected forms of **Gossypium** which have been lost or allowed to decline and become unknown, in less than a century, while exotic forms and hybrids often, it is feared, of inferior quality, have taken their places.

It will be observed by a comparison with what **Dr. Maxwell T. Masters** has written that, in discussing the forms of **Gossypium,** the authors have on the whole followed botanically (as is customary in this work) the classification of the *Flora of British India.* **Dr. Masters** recognises four species:—**G. Stocksii** (an indigenous plant met with on the limestone rocks of the Coast of Sind); **G. herbaceum** (the chief source of the so-called indigenous or *deshi* cultivated cottons); **G. barbadense** (the American cotton); and **G. arboreum** (the tree cotton, which, according, to **Masters,** is probably a native of Africa). **Dr. Masters,** and following him many other authors, incline to the view that **G. Stocksii** is the wild form, from which the cultivated **G. herbaceum** has been derived. As opposed, however, to this it may be pointed out that, while the description of the leaf of **G. Stocksii** (as given by **Masters**) would in part support that view, its naked seed and laciniated bracteoles point to an affinity to **G. barbadense.** This so-called wild species is only found, however, near the coast, and hence it may prove but an acclimatised form or possibly one of the numerous hybrids that undoubtedly exist in India. **Todaro** seems, however, to have advanced a new theory regarding the Indian cottons. The vast majority of the important growths, he informs us, are hybrids, not true species, hence under altered conditions, as, for example, on being removed from one district to another, they manifest a strong tendency to depart from their recognised and valued properties. He has further described many forms as species between which his hybrids have been produced, *and he practically excludes* **G. herbaceum** *from being an Indian species.* Of his species he urges, for example, that

G. Wightianum has a stronger claim to being recognised as a truly indigenous Indian plant than **G. herbaceum**. A variety of **G. herbaceum**, *viz.*, **microcarpum**, he informs us, is, however, frequently met with, as also a second variety, **G. obtusifolium**, but the typical form of **G. herbaceum**, according to **Todaro**, does not occur in India. His variety, **microcarpum**, he affirms, is the *Dhollera* cotton of merchants, a plant which, like **G. neglectum**, *Tod.* (referred by most botanists to **G. arboreum**), readily hybridises with **G. Wightianum**. These hybrids constitute the better class of Indian cottons, such as the Hingánghat, Oomraoti, &c. **Dr. Masters** follows the usual acceptation of the species **G. herbaceum**; and, doubtless according to him, **G. Wightianum** would be regarded as a variety or hybrid between that species and one or other of the forms of **G. arboreum**. In *Indian Herbaria* **G. Wightianum**, *Tod.*, is often named **G. herbaceum**, *var.* **hirsutum**; this is the case, for example, with **Kurz's** specimens from Pegu in the Calcutta Herbarium. The plant is certainly more densely hairy than **G. herbaceum**, and the hairs are stellate, which gives it a velvety and ashy appearance. It seems, in fact, probable that much of the confusion regarding **Todaro's** species is due to its having been spoken of ultimately as **G. hirsutum**, *Linn.*, a plant to which it has little or no affinity. There is probably no doubt that, whether recognised as a hybrid or species, **G. Wightianum** is of the greatest importance to India, even much more so than either **G. herbaceum** or **G. arboreum**, between which in many of its characteristics it is intermediate. **Todaro's** experiments would, however, support the theory of **G. Wightianum** being an independent species. The hybrids produced with it from **G. neglectum** or **G. herbaceum**, *var.* **microcarpum**, were fertile and afforded seed which, in some cases, yielded cotton of a superior quality, but in a year or two these hybrids gradually lost their merit (if not carefully cultivated), and reverted, in the majority of cases, to **G. neglectum**, more rarely to **G. Wightianum**. Experimenting with **G. herbaceum** and **G. Wightianum** by growing them in sterile soils and without irrigation, **Todaro** found the latter species to preserve its characteristics even more persistently than the former. Further, he noted that the good qualities of a hybrid were longer maintained when it was grown at a distance from either of its ancestors.

Todaro has, in a like manner, suggested a line of enquiry with **G. arboreum**, of the greatest practical importance. He isolates a condition which he designates **G. neglectum** from **G. arboreum**, *Linn.* According to **Todaro**, **G. neglectum** has played a far more important part in the Indian cotton question than the true **G. arboreum**. He cites the following synonyms and plates for his new species:—

G. neglectum, *Tod.*, with the synonyms **herbaceum** (*var.* China cotton of Roxburgh and of Royle, non-*Linn.*); **G. herbaceum**, *Wight, Ic.*, *t.* *11*; **G. arboreum**, *Parlatore, in part*; **Cudapariti**, *Rheede.*

Under his species he also forms a variety with the name **Roxburghianum**, and mentions the following synonyms for that:—**G. herbaceum** (*var.* Dacca Cotton, of Roxburgh); and **G. indicum** (*var.* Dacca Cotton, of Royle).

G. neglectum, thus isolated from **G. arboreum**, differs chiefly in being sub-herbaceous, in having broader, less pointed segments to the leaves, and in possessing an acutely pointed ovary. The type form of **G. arboreum**, *Linn.*, is more arborescent, has the segments of the leaves (**Todaro** affirms), bristle-tipped, with the fruit rounded. But in **Parlatore's** picture, cited as in part **G. neglectum**, the flowers are purple, whereas in *Royle's Ill. Him. Bot.*, *table 23, fig. 1* (which is *Wight's Ic., t. 11.*, quoted above), the flowers are shown as yellow with a purple centre. **Todaro** states that while **G. arboreum**, *Linn.*, occurs only near temples and in gardens, **G. neglectum** is extensively cultivated as a field crop, and is in fact the chief source of the so-called Bengal cottons of commerce. It is said to have always yellow flowers with a purple centre, and in this respect resembles **G. Wightianum** and other Indian cottons. Hence it would appear probable that all the purple-flowered field cottons of

India are hybrids with a strong strain of **G. arboreum**, thus preserving the characteristic flowers of that species. Of this nature very probably was **Hove's** red-flowered cotton, which he saw in 1787, extensively cultivated in Bombay. Indeed, similar evidence exists, which goes some way towards proving, that a red-flowered cotton was at one time largely grown before the present greater popularity of yellow-flowered forms.

We have thus briefly indicated some of **Todaro's** discoveries and spoken of a few of his species by name before the reader has been made fully acquainted with these, but we need only add, in this connection, that, while **Todaro** has placed the cotton industries under a heavy obligation for his careful experiments and study, we think he was tempted to form too many species. It seems, in fact, very probable that future research will confirm his opinions alone regarding **G. neglectum** and **G. Wightianum.** But, indeed, while these forms are likely to be regarded as the main types to which all the Indian growths are traceable, it is by no means certain that even these may be retained as species. Be that as it may, in various strains of hybridisation or as seminal sports, *they* have, by selection and cultivation, given origin apparently to the chief Indian races. Of this nature we regard many of the other species described by **Todaro** such as **G. cernum,** *Tod.*; **G. indicum,** *Lamk.*; **G. intermedium,** *Tod.*; **G. nanking,** *Meyer*; and **G. roseum,** *Tod.* Indeed, it would not be difficult to show that all the forms of Indian cotton are very probably but cultivated races and hybrids from two species. If we assume, for this purpose, the existence in India of some form of **G. herbaceum** (or an allied extinct species) the taint of other specific influence in the hybrids would be traceable to **G. arboreum,** *Linn.* The red-flowered field crop (alluded to above) would have been one of the earliest hybrids from these species, which, by selection and possible further crossing, might be assumed to have originated **G. neglectum,** and, by a continuation of the process, to have ultimately yielded **G. Wightianum** and all the other known races. **Todaro** selects certain more prevalent types and raises these to specific value, but pronounces the connecting forms as hybrids. The transition from one extreme to the other is, however, so complete that it seems almost arbitrary to isolate botanically any form. In an agricultural point of view, the isolation is, however, of vital importance, and we are bound to admit that **Todaro's** position is supported by two facts, *viz.*, (1) his own experiments in hybridisation and in testing the stability of character; (2) the fact that the only truly naturalised or, shall we say wild, cotton which we have had the opportunity of examining, possesses all the characters of his **G. Wightianum.** As already remarked, **G. Stocksii** (which by-the-bye, **Todaro** reduces as a synonym of **G. herbaceum,** *Linn.*) would seem from its description (given in the *Flora of British India*) to possess many of the characteristics of the American series of cottons. We have not seen a specimen of that plant, and cannot therefore express a definite opinion regarding it; but the so called wild cotton of Rajputana and of some parts of the Panjáb is almost identically that figured by **Todaro,** as his **G. Wightianum.** Further more, like the wild cotton discovered by **Sir J. Kirk** in Africa, it manifests a tendency to become scandent.

With a panorama before us, of forms no less puzzling than might be prepared from the allied genus **Althæa** (were the triumphs of the gardeners' art in the production of new kinds of holly-hock to receive specific recognition), it need not be wondered at, that we have failed completely to establish the nature of the races of Indian cotton. It must of necessity be difficult, if not impossible, to determine to what species or hybrid, popular writers allude in the works from which the present article has had to be compiled. The authors have been forced, therefore, to accept the simpler nomenclature of referring the Indian cottons to **G. herbaceum, G. arboreum,** and exotic species. Where they have been in a position to say the cotton of a province or district is **G. Wightianum, G. neglectum,** or a hybrid, this has been done, but while, through necessity, compelled to accept the older nomenclature, they feel satisfied that, on the lines established by **Todaro,** though perhaps somewhat simplified, most of the obscure features of Indian cotton are likely to be worked out.

Before proceeding to discuss the species of **Gossypium** met with in India, the authors take this opportunity of thanking their numerous correspondents who have obligingly allowed them to consult a large series of botanical specimens. Among these may be specially mentioned **Dr. G. King** and **Dr. D. Prain**, for sending a selection of the Calcutta Herbarium sets, accompanied with valuable notes and suggestions; **Mr. J. F. Duthie**, for not only contributing the Saharanpore sets, but for assisting the authors in the determination of the cottons obtained from the Provinces of India; **Mr. Edgar Thurston**, for his kindness in forwarding the collections of the Madras Herbarium. These Madras botanical collections have, in one or two instances, amplified the information derived from the samples received from Madras District Officers. **Mr. Ozanne**, Director of Land Records and Agriculture, Bombay, has also contributed a few specimens, and **Mr. Hare**, Settlement Officer, Berar, furnished interesting notes regarding Hyderabad cottons. From these sources of information, it has been possible to form some idea of the Indian cottons; but unfortunately botanical specimens, being but rarely associated with vernacular or commercial names, much remains still to be done in assigning the specific or hybrid designations for the commercial growths of Indian cotton.

I. Gossypium arboreum, *Linn.; Fl. Br. Ind.*, *I.*, *347*; MALVACEÆ. 2 (382)

Syn.—This species is frequently, though incorrectly, called, by Indian writers, **G. religiosum**: see *Dalz. & Gibs., Bomb. Fl. Supp., 8; Graham, Cat. Bomb. Pl., 15; also* **G. nigrum**, *Ham.; W. & A. Prod., 55.*

DIAGNOSTIC CHARACTERS.—*Leaves*, more or less hairy and gland, dotted, ⅔ segmented, or almost cut to the base into 5 or 7 lobes, mostly 5. *Segments*, contracted at the base, narrow, ovate, linear, acuminate, or ovate lanceolate, *not ½ as broad as long, central lobe often having a small, supplementary segment, or tooth, in the deep-rounded, lateral sinus. Bracteoles, ovate, cordate acute, toothed or entire. Flowers*, purple or purple with yellow centre, rarely white. *Seeds*, free from each other, covered with white cotton overlying a dense green or blackish down; *cotton*, not readily separable from the seed.

The supplementary tooth on one or both sides of the middle lobe of the leaf forms a most peculiar character, and in many cases a ready eye-mark in the separation of **G. arboreum** from **G. herbaceum.** It will be found in the remarks under habitat that, according to **Todaro** (the most recent author on the cultivated cottons of the world), three species are included under the above definition, *viz.*, **G. arboreum**, *L.*, **G. neglectum**, *Tod.*, and **G. sanguineum**, *Hasskarl.* It is very doubtful, however, whether these are separable, and the numerous hybrids that undoubtedly exists between **G. arboreum** and the Indian states of **G. herbaceum**, render this subject extremely obscure. **Todaro** affirms that these hybrids are mostly between his **G. neglectum** and the form of **G. herbaceum** which he has isolated under the name **G. Wightianum.** There seems no doubt that many of the cultivated cottons, especially on the higher lands, are either **G. neglectum**, *Tod.*, or hybrids with a strong **G. arboreum** or **G. neglectum** strain. The pure **G. arboreum** exists now, as it did in **Roxburgh's** time, near temples or "in the gardens of the curious over most parts of India, where it is in flower the greater part of the year." **Mr. Duncan**, for example, describes a large, bushy form of cotton (very probably a hybrid) as found in the Benares district, which yields cotton for five or six years and is there known as *Nurma.* This is also reported to be cultivated at Malwa and at Calpee in the gardens belonging to the Rajah of Jalaun. According to most writers, the pugris, worn by certain Hindus, are made from *Nurma* cotton. It was this fact apparently that led **Linnæus** to distinguish a **Gossypium** with the name **religiosum.** It has since transpired, however, that the plant so designated was not the source of the cotton alluded to, but that it was, in fact, derived from **G. arboreum.** The following passage may be given in passing, regarding this supposed sacred character of **G. arboreum**:—

In the Bengal Asiatic Society's Journal (*Vol. XI., 290*), there occurs an article on **G. arboreum** by **Dr. A. Burn.** "The *Nurma* cotton," he there wrote, "is a perennial plant, lasting for four or five years or more. It grows in every kind of soil; but attains perfection only in the light, sandy, *gorát* lands. The wool is fine, silky, of considerable strength, and fully an inch long. Hedgerows, gardens, and groves of trees about the abodes of devotees and temples, are the places where the plant is found. Muslins and turbans are made from it. Since the introduction of European cloth, the culture of this cotton has almost entirely ceased. Its yield per acre is estimated at one hundred pounds of clean cotton in the first year, and in the second at from three to four hundred pounds. The great hindrance to its cultivation is the fact that it requires protection throughout the year. The price of this cotton in the market at Broach is always double that of the common country article. But there are never more than a few pounds procurable." In the *Broach Gazetteer* it is added to the above that "*Nurma* or *deva kapas* **(Gossypium religiosum)**, would seem to be grown only to a small extent, chiefly near temples and the dwellings of ascetics. It is used in making the caste thread (*Jandi*)." Before passing from this subject it may be pointed out that the above opinion as to the value of *Nurma* cotton, is most probably greatly over-estimated.

A far more serious error has, however, been made in regarding one of the conditions of this species as a form of **G. herbaceum.** This has been cleared up by **Todaro,** who has isolated as species many forms hitherto viewed as but cultivated races. Some of these he has associated with **G. arboreum,** others with **G. herbaceum.** While viewing **Todaro** as having materially advanced the study of the species of **Gossypium,** we are not prepared to agree entirely with him that **G. Wightianum** and **G. neglectum** are good species. They are perhaps at most varietios, if, indeed, they are anything more than hybrids, derived from the two originally Asiatic (or Old World) species, represented by **G. arboreum** and **G. herbaceum.** But, whether species, varieties, or hybrids. they are, however, of such importance as to necessitate separate recognition, and the following passages, compiled from all available sources, are intended to allow the reader to form his own opinion as to the desirability of their retention or separation from the Linnean types.

Todaro's separation of **G. arboreum,** *L.*, into two forms may be exhibited as follows:—

(383) 3 *a* **G. arboreum,** *proper.*

Conf. with pp. 8-14; 63; 102; 103; 119; 134; 142.

Diagnostic Characters.—Arborescent, cultivated in gardens only or near temples; not a regular field crop. *Leaves*, 5-7 lobed with an extra tooth in the left side sinus (or both sides) of the central lobe; tnick, leathery, sub-glabrous or with short, abortively stellate hairs on the leaves and longer hairs on the petioles and young stems: lobes bristle-tipped: petiole long, rigid: *stipules* falcate. *Inflorescence* axillary, generally one-flowered; peduncle about ¾ the length of the petiole, but jointed above the middle, bearing a small leaf and two stipules at the joint. *Flowers* small, purple, a little less than twice the length of the bracteoles; *bracteoles* medium-sized, base with round ears, apex ovate, acute toothed or sub-entire. Ovary rounded; seeds with greyish black velvet underneath the floss.

Todaro figures this plant, as well as **G. sanguineum,** *Hasskarl*, on Table I.: we regard both these as representing the Linnean plant. The differences in **sanguineum** are very slight:—purplish stems, petioles, and bracteoles; lobes of the leaves scarcely constricted at the base, and lower pair much smaller than the others; bracteoles, as a rule, more deeply toothed. **Wight's** *Ic., t. 10* (taken from *Royle, Ill. Him. Bot., t. 23, f. 2*), is a very indifferent representation of **G. arboreum**; in fact in hairiness and shape of bracteoles it more nearly resembles the form indicated above as **G. sanguineum,** *Hasskarl.* **G. arboreum** is the *Nurma* or *Deo* cotton of most writers.

β **G. neglectum**, *Tod.* 4 (384)

Diagnostic Characters.—A large bush, chiefly grown as a field crop, and sometimes seen not more than 18 inches high. *Leaves*, 3, 5, 7-lobed, but extra tooth in sinus less distinct and lobes more linear and simply acute, rarely bristle-tipped: densely coated with long spreading hairs, and, if hybridised with **G. Wightianum,** having stellate hairs as well: *stipules* broad, linear. *Inflorescence* on short lateral branches, 2-4-flowered. *Flowers* yellow with purple centre or yellow with purple tinge (**Todaro** doubts the flowers being ever pure purple, as in **G. arboreum,** so that the so-called purple-flowered field crops, of indigenous plants, are either **G. arboreum** or hybrids of that species): *bracteoles* very large with greatly elongate lateral ears but in shape above ovate acute, toothed. *Ovary* pointed, boll of cotton sometimes elongating from the capsule so as to resemble a kidney-cotton: seed with green velvet below the floss.

Conf. with pp. 8-14; 98; 102; 111; 118; 129; 134; 142; 147.

An excellent plate of this plant will be found in "The Field and Garden Crops" under the name of **G. herbaceum. Parlatore's** plate of **G. arboreum** represents, on the whole, the Garo long-bolled condition of **G. neglectum** except in the colour of the flowers, and **Wight's** *Ic., t. 11* (taken from *Royle, l.c. t. 23, f. 1*), is a good illustration, except in the fact of the inflorescence being shown solitary flowered, which it rarely is in typical conditions of this species—the *radhia* and *manúa* cotton of Upper India. **Todaro** says that, having cultivated this plant for fifteen years, he never saw it with red flowers but always yellow and of a brighter tint than in **G. herbaceum** or **G. maritimum**—more resembling in fact the flowers of **G. Wightianum.** But **Roxburgh's** Dacca cotton, which **Todaro** refers to this species, has the petals tinged with red. **G. neglectum** is extensively cultivated in Bengal, the North-West Provinces, and the Panjáb, constituting to a large extent the so-called "Bengals" of commerce.

Conf. with p. 60.

For probable hybrids of this species, *see* the remarks under **G. herbaceum.**

I.

(*For II. see p. 16.*)

THE ARBOREUM SERIES OF COTTONS. 5 (385)

Vern.—*Nurma, deo kapas,* Hind., (*Borailli, tangori?* Dacca) Beng.; *Budi kaskom, bhoga kuskom,* Santal; *Bogali, nurma,* Bundel.; *Manua, radhia, nurma,* N.-W. P.; *Kapas,* Pb.; *Mannua, deo,* C. P.; *Deo kapas,* Bomb.; *Deva kápúsa* (according to **Sakharam Arjun**), Mar.; *Deo kurpas* (God's cotton), Mysore; *Shem paratie* (according to **Ainslie**), *semparuthi,* Tam.; *Patti,* Tel.; *Nu-wa* (according to **Mason**), Burm.; *Kárpásamu* (according to **Elliot**), Sans.

References.—*Roxb., Fl. Ind., Ed. C.B.C., 519; Dalz. & Gibs., Bomb. Fl. Supp., 8; DC. Origin Cult. Pl., 405; Graham, Cat. Bomb. Pl., 15; Sir W. Elliot, Fl. Andh., 84, 146; Rheede, Hort. Mal.; Rumph. Amb.; Burm. Fl. Ind.; Mason, Burma & Its People, 518, 756; Hove, Journal of Tour in Bombay made in 1787; Ainslie, Mat. Ind., II., 284; Dymock, Mat. Med. W. Ind., 2nd Ed., 110; Pharmacog. Ind., 225; Bent. & Trim., Med. Pl., 37; S. Arjun, Bomb. Drugs, 17; Lisboa, U. Pl. Bomb., 228; Royle, Cult. & Com. of Cotton in India, 144-145; Prod. Res., 221. In the Jours. & Trans. of the Agri-Hort. Soc. of India, frequent mention is made of Nurma cotton: of these the following may be specially cited:— Trans. V. (Proc.), 65; VII. (Proc.), 17; Jour. (Old Series), I., 278; II., 437; IV., 106, 108; XI., 295, et seq.; (New Series) II. (Proc. 1870), XLVI.*

Habitat.—The typical condition of this species is a low tree or shrub, met with in garden cultivation throughout India. Except, perhaps, in a state of hybridisation it rarely occurs as a field crop, but, as explained above, it seems probable the purple-flowered indigenous cotton crop, mentioned by many writers, is of this nature. The chief commercial form of this species is, however, the plant isolated by **Todaro** under the name of **G. neglectum.**

The Indian Arboreous Cottons.

CULTIVATION. 6 (386)

CULTIVATION.

G. neglectum is extensively cultivated as a field crop. It has bright yellow flowers, and deeply palmately segmented leaves, which in shape are scarcely distinguishable from **G. arboreum** proper, except in that they are more herbaceous and very much more hairy. How far **G. neglectum**, *Tod.*, may, by future observers, be confirmed as of specific value, we are not at present prepared to affirm. It seems, indeed, quite likely that it may prove but a hybrid between **G. arboreum** and one of the plants belonging to the series which we prefer to continue designating **G. herbaceum.** But whether species or hybrid, there is nothing gained, but, on the contrary, everything lost, by swamping so important a crop—the Bengals of Commerce—with either **G. arboreum** or **G. herbaceum.** To this species **Todaro** refers all the China and Dacca cottons described by **Roxburgh** as forms of **G. herbaceum**, but while some of the Dacca cottons seem to be of this nature, others most certainly are not. There is, however, a much stronger taint of **G. neglectum** in the cottons of Eastern and Northern than in Western India, where **G. Wightianum** becomes the prevailing form. The long-bolled Garo hill cotton agrees admirably with Todaro's description of **G. neglectum**, and the cottons of the Santal country are very similar, though they differ, not only from the Garo cottons but from each other. The **Rev. A. Campbell**, of the Santal Mission, who has kindly furnished admirable botanical specimens of the two cotton crops of Manbhum, informs us that they are grown at different seasons, and are quite distinct, being known as *Bhoga-kaskom* and *Budhi-kaskom* respectively. The latter (*Budhi*), he says, is generally grown in gardens or on the homestead lands. It flowers in November and yields a larger crop than the former (*Bhoga*), but, as it requires a richer soil, is much less cultivated. The *Bhoga* is the staple crop of Manbhum. It is sown in June-July and flowers in October; in rich soils it attains a height of 4 feet but is more frequently met with from 2 to 3 feet in height. **Mr. Campbell** also alludes to a third form of cotton known as *Jurgundi*. As he has not furnished a sample of this we are unable to say what it may be: it is, however, sown on high lands in June-July and flowers after the rains. The floss is slightly red coloured but of good quality and well thought of for making cloth. It has a stronger fibre than *Bhoga*, but the yield is considerably less, so that it is by no means so popular. The difference between *Bhoga* and *Budhi* as seen in the herbarium is, however, so slight that it is impossible to give a description that would isolate the one from the other. They are both forms of, or at least are allied to, **G. neglectum**, *Tod.*, and it would thus appear that that species has been sufficiently cultivated to give origin to races that are grown at different seasons of the year. If after all we accept **G. neglectum** as but a hybrid, some of its forms approach **G. arboreum**, while others have a stronger strain of **G. herbaceum.**

On many parts of the Himálaya cotton cultivation is prosecuted to a limited extent, ascending in Kumaon, Garwahl, and Murree, to altitudes close on 5,000 feet above the sea. The plant grown in the Simla district is apparently a hybrid stock, having a strong taint of **G. arboreum**, and much resembles the *Semparuthi* cotton of South India. **Mr. Vaupell** mentions a shrub met with in the neighbourhood of large towns in the Eastern districts of Gujarat, in spots most favourable for irrigation. Its wool, he states, is the finest of any, of a beautiful silky staple, upwards of an inch in length, and only used in the manufacture of the finest muslins. It is, he adds, but sparingly cultivated. *Nurma* cotton is also specially mentioned as a very fine cotton sent from Khorassan, and, according to several writers, is well known in Malwa. **Dalzell** and **Gibson** fell into the error,

CULTIVATION.

frequently made some twenty or thirty years ago, of assigning to this species the name **G. religiosum,** *Roxb.*, but they call it by its most general vernacular name *Nurma* and speak of it as "arboreous, slender, hardy, having the habit of a tree; cultivated rather extensively in the North-West Gujarat as a triennial (?), also in Sind. It derives its name '*Deo kupas*' from being most extensively used for the sacred thread of the Banians, *Moonj*." There would appear to be little doubt that the plant referred to above was a form of **G. arboreum.** With reference to the remarks below, on **Hove's** red-flowered cotton of Bombay, it is interesting to notice that **Dalzell** and **Gibson's** Gujarat field crop of **G. arboreum** is very probably one and the same as that described by **Hove** in 1787.

The Transactions and Journals of the Agri-Horticultural Society of India teem with papers and reports on cotton, but only from about the year 1850 do these deal at any length with the indigenous growths. The following passages may, however, be given in this place as presumably alluding to **G. arboreum.** So long ago as 1837 a **Mr. D. F. McLeod** of Seoní (Central Provinces) furnished the Agri-Horticultural Society with two samples of tree-cotton, both of which appear to have been **G. arboreum,** *L.*, as defined in the *Flora of British India.* These were then known as *Munnooa* and *Deo.* The former was a green and the latter a black seeded condition. They were perennials and "yielded their cotton in the hot weather, and not as the common country annuals, at the close of the rains." "These varieties are planted," says **Mr. McLeod,** "by the natives near their dwellings, with a view to shelter, and the produce is chiefly used for making Brahminical threads. The *Munnooa* is also cultivated extensively in fields on the ranges east of Mirzapore." It may be surmised that the *Munnooa* cotton of the above passage was **G. neglectum,** *Tod.*, and the *Deo*, **G. arboreum,** *L.*

Conf. with p. 129.

In 1842, an interesting enquiry was instituted into the subject of the **Chanderí** (Gwalior) cotton, from which the once famous muslins were made. This was at first attributed to **G. arboreum,** but **Mr. C. Fraser** and other writers assigned it to a peculiar form (*Bararía*), a Berar cotton, grown on a limited number of fields around two villages, or simply to imported Berar cotton. Incidentally, however, **Mr. Fraser** also deals with the subject of *Nurma* cotton, remarking that it was found all over Hindustán, but in Chanderí was "never employed, nor grown in large quantities. The plant stands about seven feet high and is bushy; solitary plants are met with occasionally in private gardens, and the produce is worked up into Brahminical threads for the higher classes of society." (*Jour. Agri-Hort. Soc., Old Series, Vol. I., 278.*) The Chamber of Commerce, Bombay, in their Annual Report for 1842-43 refers to the two forms of *Nurma* cotton, grown in Broach, by **Dr. Burn.** One of these, **Dr. Burn** wrote, was "a very fine, soft, strong, long, and clear-coloured staple, superior to the best Broach, and which would compete with the very best American short-staple cottons, could it be produced in sufficient quantity." The Committee in their report on the samples supported this opinion, adding that, "it was extremely advisable that steps should be taken to extend the cultivation of so choice an article of Indian produce." Nothing was, however, apparently done in this direction, and the cotton referred to is now probably completely lost. But it would, perhaps, be unsafe to assign, to any known form of **G. arboreum,** so flattering an opinion, the more so when the identification of the plant referred to is dependent on the mere assertion that it was a form of *Nurma.* Long before the date of the above report, numerous experiments had been made in acclimatising most of the American cottons, and it is, therefore, possible that the fibre may have been got from one of the larger growths of American which often assume the size of the

The Indian Arboreous Cottons.

CULTIVATION.

Indian tree cotton. Peruvian cotton, for example, is, like **G. arboreum,** frequently met with near temples and grows into a large bush or tree. **Mr. Blount**, Government Cotton Planter, in a report (1845) on the cottons which he found the natives cultivating in Goruckpore near Benares, mentioned, however, what appears to have been the same two forms of **G. arboreum.** These were known as *Bogalí* and *Nurma.* The former, he wrote, was grown as a field crop, being sown at the commencement of the rains, and the latter raised near huts only. Both these plants produced their fibre in February and March. The ordinary cotton of the district—that most extensively cultivated—he contrasted with these, stating that it was known as *Kuklí.* It was planted in February and gathered in June and July. There would seem little doubt but that, like the *Munnaoa* cotton of the Central Provinces, the *Bogalí* must have been a field crop which nearly, if not absolutely, answers to **Todaro's G. neglectum.** Many subsequent writers dwell on 'the tree cotton,' 'the red-flowered cotton,' 'the *Nurma* cotton' of the Benares Division of the North-West Provinces—a peculiar and perfectly distinct crop from the ordinary cotton of these provinces.

Conf. with p. 11.

Turning now to Bengal: in a long and most instructive report on the indigenous cottons of Eastern Bengal, written by **Mr. T. Allan Wise** in June 1860, much light may be assumed to be thrown on the subject of **Todaro's G. neglectum.** From that paper, there would appear to be considerable likelihood that two or three of the Dacca cottons are forms of **G. arboreum,** one most probably **G. neglectum,** *Tod.* Whether that form was actually (as **Todaro** asserts) the reddish-tinged plant, with pointed leaves (described by **Roxburgh** as a variety of **G. herbaceum**), may be open to doubt. **Mr. Wise** specialises eight forms of cotton as found in Eastern Bengal, *viz.*, (1) Tipperah Hills cotton: (2) *Sheraj* cotton: (3) *Bogga* cotton: (4) *Borailli* cotton: (5) Dacca cotton: (6) Dacca *Tangari* cotton: (7) Common Bengal cotton: and (8) Foreign cottons lately introduced. We shall revert to the subject of most of these cottons under **G. herbaceum**; but it is desirable to deal here with those that appear to belong to the **G. arboreum** series. The *Sheraj* cotton, **Mr. Wise** states, "very probably comes from the Garo hills. It is considered, after *Borailli,* the second best cotton found in the bazárs of Mymensing, and from it the cloth worn by the better class of natives is made. It is bought by the ryots in the bazárs for their wives and daughters to spin into thread, which they sell to the professional weavers to make into cloth for the markets." The *Bogga* was a very inferior sort of cotton, used in making the *do-sutti* or "two-thread" cloth employed for sails. It was said to come from the mountains of Assam. The *Borailli* "is the finest kind of cotton procurable in these marts, and from it is made the very fine thin cloths which the landed proprietors and wealthy natives are fond of wearing. It is the largest cotton plant I know, reaching the height of some eight or nine feet, with beautiful drooping branches, which, if erect, would measure more, and, as it is a perennial, must be very profitable; but it only grows in high village lands quite clear of inundation. It bears pods every month in the year for three or four years in succession, and being in every way such a different plant from any of the Dacca kinds, I am inclined to think it is peculiar to Mymensing district, or, more probably, is a foreign kind imported here by some of the early Portuguese settlers who had large villages in the district." The cotton plant of Dacca, from which the famous muslins were made, was, **Mr. Wise** informs us, in his time, an annual crop, found on the low-lying, rich, alluval lands, which were periodically inundated. It was sown in September or October, in lines, between which chillies were grown. In connection with the account given above of Santal cottons, it has been stated that, although the two crops there grown differ materially from an agricultural point

Conf. with p. 29.

CULTIVATION.

of view, they are botanically almost identical. To what extent all the Dacca cottons may, in the same way, be but forms of **G. neglectum** we cannot discover. That that species (or hybrid) actually does occur in Eastern Bengal we have no doubt, but we have seen specimens of what we take to be **G. herbaceum,** *var.* **obtusifolium,** from Dacca, as also others that appear to be hybrids of **G. neglectum** with **G. Wightianum.** It is thus impossible to determine to what species, varieties, or hybrids **Mr. Wise's** eight Dacca cottons belong, more than to discover which of them may have been the Dacca cottons examined by **Roxburgh** and which **Todaro** consigns collectively to his **G. neglectum.**

It seems highly probable that the *Sheraj* and *Borailli* cottons, at least, are forms or hybrids of **G. arboreum,** while the Dacca *Tangori* cotton, the cultivation of which is now practically abandoned, may have been **G. neglectum.** Speaking of the Dacca *Tangori* cotton **Mr. Wise** remarks that it was grown on high lands and appeared a species peculiar to the district. "It grows on the high red clay lands to the north of Dacca and attains the height of five feet. It is sown in July and the crop reaped in February. It bears a light crop for two or three years, and is probably the same kind that formed the attraction to Cospassia and the rest cf these jungles which once teemed with a wealthy people who had brick houses, tanks, &c., all now in ruins, but which seem to have been gradually relinquished. Thus only a few years ago at Sodapore, a village to the north-west of Dacca, large fields of this cotton were sown, but now not a single patch is to be found there The rayats say it is too troublesome to cultivate, but I doubt rather they do not manage it properly, do not allow the land to lie fallow every fourth year as was formerly done, and the crop is consequently abandoned." (*Agri.-Hort. Soc. Jour., Old Series, Vol. XI., l.c.*)

In a more recent report (written by **Mr. H. J. S. Cotton,** 1876) "on Cotton Cultivation in the Interior of Bengal," the recognition of two forms of arboreous indigenous **Gossypium** is placed beyond doubt. Of the northern parts of Rajshahye he wrote: "The varieties are locally known under three names, *viz., Boya-bonga, Chéngta-bonga,* and *Bura-muri.* **Mr. Fasson,** the Joint Magistrate who drew up the report for the district, says:—'There are certainly two strongly-marked varieties, both apparently **G. arboreum,** though neither answer the description exactly. The first of these is an erect, ligneous, slender-looking plant, growing six or eight feet high, with three or generally five-lobed palmate leaves; lobes acute, elongated, and lanceolate. The flowers large and showy, reddish pink in colour, with purple claws. The bracts of the outer calyx are cordate-ovate, either entire or obscurely tridentate at the apex; capsules large and three-celled; seeds black, slightly covered with greyish fuzz. The cotton is fine and silky, but of extremely short staple. This variety is called *Chengta-bonga* or *Bon kapas.*'"

Before proceeding to give **Mr. Fasson's** account of the second form, it may be here remarked that the above plant would seem to be unmistakeably a form of **G. arboreum.** The fact of its having leaves sometimes three-lobed, and still further of its being a field crop, would lead to the supposition, that it may have been but a hybrid of that species with one of the **G. herbaceum** series, and there would seem no room for doubt, that this is the red-flowered form of **G. herbaceum** alluded to by many popular writers. But in addition to the colour of the flower, we have the character of the seed and the deeply-cut leaves, with acute elongate, and lanceolate lobes—characters that clearly point to its having been either a cultivated state of **G. arboreum** or some hybrid with a powerful strain from that species. Whether this form still exists we have no means of saying; but the question of a possible improvement of Indian stock would

CULTIVATION.

seem to hinge upon the possession of all the early forms, and this is one well worthy of being followed up.

We turn now to **Mr. Fasson's** second variety. This is described as having had broader leaves, sub-acute lobes, and more deeply toothed bracteoles, the fruit, 5-celled, the staple less plentiful though longer, and fuzz of a grey colour. This is the *Boya-bonga*, and by **Mr. Fasson** was thought to be **Roxburgh's G. obtusifolium**: it was, however, more probably **G. neglectum.**

Mr. Cotton's valuable report, from which the above passages have been abstracted, contains much of great interest regarding the cottons of Bengal; enough to raise the hope of possible future good work, but not enough to solve all the obscure problems that involve the subject. In Bhagulpore the tree-cotton grown near the villages is said to have been known as *gajer*, a perennial which yielded a crop for three or four years.

The long-bolled condition of the Garo cotton has led to its being often spoken of as a kidney-cotton. It is, however, quite different from **G. acuminatum,** *Roxb.*, **G. vitifolium,** *Lamk.*, **G. brasiliense,** *Macf.*, and **G. racemosum,** *Poir*, the kidney-cottons. It has in fact a much closer resemblance to **G. peruvianum,** for in both the seeds have an under-coat of velvet. The long-bolled condition is, moreover, not due to a mechanical union of the seeds, as in the kidney-cottons, but to a firm intermixture of the wool which holds the contents of the capsule together in a mass, often in the case of the Garo cottons, protruding to a length of three to six inches. The matted character of the Garo cotton suits it, in an eminent degree, to one of the purposes for which it is employed. In weaving a sort of quilt, the tuft of wool from a seed is said to be placed, by the hand, within the weft of the texture and thus firmly fixed, on the upper surface, forming rows which in a dense mass, imitate a thick and crude plush.

We have incidentally alluded above to **Dr. Hove**, but it may perhaps be as well to discuss in this place more fully the inferences to be drawn from that indefatigable traveller's labours. He was sent to India apparently by the British Government, in 1787, to report on Indian cotton and cotton goods. From the treatment he received in Bombay it may be inferred that, for some unaccountable reason, the East India Company did not look with approbation on the object of his mission. He lived for two years among the people, the guest in most cases of the Native Princes and rich proprietors, and devoted himself to the study of cotton. He was presumably an expert on the subject and therefore probably familiar with all the American cottons then to be found in the European markets. On one point his silence is significant; he nowhere came across any cotton which suggested a resemblance to American forms. **G. hirsutum** (the New Orleans and Georgia cottons) grown, at the present day, to a considerable extent, in many of the districts where he explored, were not found by him, and, indeed, the first record of their introduction was not until 1811. In the face of this negative evidence we have **Roxburgh's** description of **G. hirsutum, G. religiosum,** and **G. acuminatum.** Of the two former, the father of Indian botany wrote that they had only recently been introduced into India, but of the latter he remarked, "Said to be a native of the mountains to the north and westward of Bengal. I do not find that this species is ever cultivated. It is readily distinguished by its superior size and large black seeds, which adhere firmly to each other." The three species named most undoubtedly belong to the American series of cottons, with large broad leaves and deeply laciniated bracteoles. **Royle**, in his account of **G. acuminatum,** *Roxb.*, largely removed the hesitation with which **Roxburgh** spoke of its habitat, for he affirmed that it was a native of the mountain tracts in question. **Todaro** seems to have accepted this

CULTIVATION.

assertion, and he gives no other region as the habitat of the species. Since the time of **Roxburgh**, however, no collector has recorded **G. acuminatum** as found anywhere except in those garden and farms where efforts have been made to acclimatise exotic cottons. The Calcutta Herbarium possesses specimens dried from plants grown in the gardens. Indeed, **G. acuminatum** is much less frequently cultivated than **G. peruvianum**, to which species **Royle** incorrectly referred **Roxburgh's** plant. It may, in fact, be said that Peruvian cotton is to some extent cultivated in the Madras Presidency. It is the large-leaved and conspicously flowered species with immense bracteoles, of which specimens have been contributed by several officers in South India. As already remarked **G. peruvianum** has velvety seeds free from each other, but in foliage, flowers, and bracteoles it much resembles the kidney-cottons.

But **Roxburgh**, in addition to the three species named, also described **G. barbadense** (Bourbon cotton), remarking that it had then been only recently introduced. Seeing that **Roxburgh** practically began to study the cottons of India about the time of **Hove's** visit, these distinguished botanical pioneers may be said to have found the cotton crops, at the close of the last century, to consist of forms of the plants generally known as **G. arboreum** and **G. herbaceum**. Indeed, **Roxburgh** questioned the former species being ever grown as a field crop, and, as already stated, he appears to have classed the yellow-flowered form (**G. neglectum**, *Tod.*) under **G. herbaceum**. He was, however, in considerable doubt regarding **G. arboreum**, and probably rightly conjectured that the *Cuduparti* of **Rheede** (*Hort. Mal. I., p. 55, t. 31*) might prove a distinct species. This **Todaro** refers to his **G. neglectum**, but as opposed to this arrangement it may be pointed out that, according to **Roxburgh**, the most characteristic feature of the *Cuduparti* is its round fruit, a character which would place it with **G. arboreum** proper. This confusion may, however, be accepted as indicating a recognition, on the part of **Roxburgh**, that there were two forms of **G arboreum**. He does not seem to have entertained the idea that the field crops with "petals tinged with red," and those with yellow flowers but with the foliage of **arboreum**, might be hybrids of that species, and probably had not seen the red-flowered growths of which **Hove** wrote. **Hove**, in fact, found three forms of cotton in Western India, but as his report was not published till 1851, **Roxburgh** appears to have remained ignorant of **Hove's** discoveries. The three cottons seen by **Hove** were—(1) a yellow-flowered form which yielded the best staple and was always grown on the lower-lying lands: (2) a red-flowered kind, cultivated on the higher sandy soils which afforded a short but fine staple: and (3) a form peculiar to Kathiawar, which he came across accidentally in a garden. There are many features of **Hove's** red-flowered cotton that justify the suspicion that it was a hybrid, with a pronounced strain of **G. arboreum**. He wrote of this cotton in Cambay on the 6th November as then in full bloom with scarlet flowers, and quite *another species from the yellow-flowered bush grown at Diroll in Broach.* "On my journey," he continues, "to Kerwan in Cambay, for the space of 16 miles, wherever I cast my eye, I could see nothing else but cotton plantations. Where the soil consisted of a heavy clay those districts were planted with the yellow sort, and those which consisted of sand, or were situated higher from the adjacent ground, were planted with the red species." He then goes on to say that in the second year the red-flowered bushes grow to a height of seven feet, but in order to make them bushy and to cause them to yield a large crop, they have to be twice pruned,—once when they are only three feet high, the shoots are then cut down by a foot, and the second pruning takes place after the first crop has been collected. At Desberah in Broach, he was

GOSSYPIUM arboreum.

The Indian Arboreous Cottons.

CULTIVATION.

told, the red cotton was known as *Dyva Nerma Capass.* As seen at Sabermatty it had blossoms of a prodigiously large size, not unlike "the **Gossypium arboreum** on the coast of Guinea." At Sunerwara the red-flowered species was called *Semul.* One of the most remarkable features of **Hove's** account of the Bombay cotton cultivation, one hundred years ago, is the stress he lays upon the necessity for free irrigation, with yellow-flowered cotton, which in this respect differs, he adds, materially from the red.

The third form of cotton seen and described by **Hove** was a form from Kathiawar which he saw accidentally. At Hanley village, close to the river Dahder, he came across a solitary plant of this cotton, which, he was told, had come from Junaghar in Kathiawar. The cultivation of this had been abandoned at Hanley because the plant formed so dense a bush as to kill the grains grown between the rows of the cotton plants. It was large podded and a more luxuriant plant than either of the former two sorts. **Hove** expressed astonishment that it had not taken the place of the yellow and red-flowered ordinary sorts. This was explained, however, for it not only suffocated the grains grown with it in the same field, but "possessed such a sweetness by nature that the insects destroyed the greatest part before it came to full perfection." It was apparently not so destroyed, however, in Kathiawar, for **Hove** was told that to the merits of this cotton the people of that region owed their reputation of being the most prosperous in all Guzerat. What cotton **Hove** here alludes to seems open to grave doubt, but it could not have been an American exotic, or he would most probably have recognised it as such. In the remarks under Bombay it will be found Kathiawar is shown to be, even today, famous for many kinds of cotton and some of great merit.

In another passage **Hove** alludes to the roots of red cotton being much more bitter than those of the yellow, and states that in consequence they were used as a medicine for fevers. He also came across a *khaki*-coloured cotton, but his description is too imperfect to allow of any opinion as to whether or not it was Nankin, the produce of **G. religiosum,** which **Roxburgh** shortly after described as occurring in India. It should be here added that all the wild cottons are reputed to have yellow flosses, and that several of the forms of **G. herbaceum** and one or two of **G. arboreum** have also yellow-coloured flosses, while the cotton from **G. religiosum** is by no means always yellow-coloured.

This brief notice of the Indian forms of **G. arboreum** and **G. neglectum** may now be concluded by the remark that, according to **DeCandolle, G. arboreum** came originally from Africa, where it has been seen in a wild state. From this opinion **Masters,** however, appears to dissent (see the remarks under **G. barbadense**), for he states that the wild plants found at the present day in Africa are more nearly allied to **G. barbadense.**

MEDICINE.

MEDICINE.

Root. 7 (387)

In the above remarks it may have been observed that **Dr. Hove** is quoted as the authority for the statement that in Bombay the ROOT was at his time used in the treatment of fever, as it had the reputation of being more bitter than that of **G. herbaceum. Dr. Dymock** says that "in the Konkan the root of the *Deokapás* (fairy or sacred cotton bush), rubbed to a paste with the juice of patchouli leaves, has a reputation as a promoter of granulation in wounds, and the JUICE of the LEAVES, made into a paste with the seeds of **Vernonia anthelmintica,** is applied to eruptions of the skin following fever. In Pudukota the leaves, ground and mixed with milk, are given for strangury." In the United States Dispensatory the bark of the root of **G. herbaceum** and other species is officinal.

Juice of the Leaves. 8 (388)

MEDICINE, Petals, 9 (389)

SPECIAL OPINION.—§ "The PETALS of the red cotton flower, squeezed and soaked in human or cow's milk, is used as a soothing and effective application for conjunctivitis of infants." (*Surgeon-Major Thomson, M.D., C.I.E., Madras*).

10 (390)

11. Gossypium barbadense, *Linn.; Fl. Br. Ind., I., 347.*

THE AMERICAN COTTONS.

For the purpose of the present article, all the American cottons are treated of collectively, but those only in any degree of detail, that have been successfully acclimatised in India. The New World cottons are recognised by most authors as the forms with broad (only half-segmented) leaves and ovate obtuse, deeply laciniated bracteoles. The leaves in the great majority are sub-glabrous, those of typical Asiatic origin being very hairy. **Todaro** puts little faith in the specific value of the under-coat of tomentum, which clothes the seed (in many forms of **Gossypium**), but it seems highly probable that that peculiarity was originally confined to Asiatic cottons. The further peculiarity of a series of cottons having the seeds mechanically united into a kidney-like mass would, on the whole, appear also to belong to certain American cottons. The species here viewed as of more immediate American origin, were, at the time of the discovery of America, found in a state of cultivation from the West Indies to Peru and from Mexico to Brazil.

Distribution. 11 (391)

The geographical distribution of the series, according to **Humboldt**, is from 0° to 34° of latitude, or where the annual temperature ranges from 82° to 68°. The several forms, races, or hybrids, of this assemblage, vary greatly in external characters and commercial value, though botanically it may be doubted how far it is desirable to isolate as species the forms recognised as of pure (not hybrid) origin. Generally speaking, the staple improves with proximity to the sea, but it cannot be said that there is any fixed ratio of variability to climate or position within the area of distribution. Thus the cotton of Demarara (14° south) is fairly good, that from Jamaica (20° north) bad, while the cotton of Georgia and Carolina, nearly at the most northern limit of the area, is very superior. Commenting on this variability **Royle** says, "It would appear, therefore, that not only is temperature necessary to be considered, but also the due balance between the supply of moisture to the roots and its escape by the leaves, as well as all the varied processes of a judicious culture in addition to the choice of species or variety to be cultivated in any particular locality."

Dr. Maxwell T Masters (*Jour. Linn. Soc., XIX., 213*) has given an entirely new turn to the enquiry into the so-called American cottons. A manifestation apparently of this type is, he says, the prevailing one in tropical Africa. It has been found, he affirms, by African travellers, in a state of cultivation near lake Nyassa, along the Zambesi, Shire, and the Rovuma valleys, near lake Tsad, in Abyssinia, and on the Somali coast. On the western side of the continent and adjacent islands it is also the species principally cultivated. Along the Nile valley, however, extending as far as Khartum, **G. herbaceum, Dr. Masters** states, is the form usually grown. He further describes a new species of **Gossypium** (**G. Kirkii**) which, there would appear to be no doubt, is a truly indigenous wild African form. It is interesting to notice that **Dr. Masters** regards **G. Kirkii** and **G. anomalum** as the only truly wild African forms and makes no mention of **G. arboreum.** On the other hand, **DeCandolle** (*Origin Cult. Pl., 405*) writes of **G. arboreum**:—"It is a native of tropical Africa and has been seen wild in Upper Guinea, in Abyssinia, Sennaar, and Upper Egypt. So great a number of collectors have brought it from these countries that there is no room for doubt, but cultivation has so diffused

CULTIVATION. Distribution.

and mixed this species with others that it has been described under several names in works on Southern Asia." **DeCandolle** appears to favour the idea that the cotton of Egypt, referred to by ancient writers, was probably derived from **G. arboreum**, grown in Upper Egypt. But he adds, "The Arabs and, afterwards, Europeans preferred and transported into different countries the herbaceous cotton rather than the tree-cotton which yields a poorer product and requires more heat."

It may be here repeated that authors agree in thinking all the truly wild species of **Gossypium** possess a yellow or Nankín colour. The description of **G. Stocksii**, as given in the *Flora of British India*, may also be here repeated; it recalls, far more a hybrid or acclimatised condition of **G. barbadense** than suggests the idea of its being the original wild state of **G. herbaceum**. We have, however, not seen an authentic specimen of **G. Stocksii**, and do not, therefore, advance the above explanation authoritatively. How far early cultivation and hybridisation may have to account for the mingling of types is open to grave doubt. Indeed, it is by no means proved that a hybrid between Asiatic and American cotton is possible or impossible: authors are very nearly divided in their opinion. The alternative explanation of velvety-seeded American cottons, a theory favoured by the African wild species and by **G. Stocksii**, would be the existence of a series of species which, like **G. hirsutum**, possess certain of the peculiarities of both American and Asiatic cottons—a series of **Gossypium** in that case indigenous to both the New and the Old Worlds.

The most characteristic features of the American cottons may be said to be in their more entire leaves. The upper leaves are in fact often only angled. The lower ones are 3- to 5- lobed, the lobes broad and often suddenly acuminate, the sinus acute or at least not rounded, and never possessed of supplementary teeth. The bracteoles in outline are almost obtuse, instead of acute, and are deeply gashed. **Royle** kept the kidney or Peruvian cottons distinct from the Sea Island, Georgian, and Bourbon. From an economic point of view it would seem the more natural course to do so, for, in addition to their leaves being more deeply segmented, the seeds have the peculiar character of cohering together in a kidney-shaped mass, hence their popular name of Kidney Cotton.

II.

(*For I., see p. 7 and for III., see p. 26*).

CLASSIFICATION. 12 (392)

THE AMERICAN SERIES OF COTTONS.

The following arrangement is practically that given by **Todaro, but,** as repeatedly stated, we consider some of the forms to be hybrids rather than species. That being so we prefer to retain the Roxburghian form of **G. religiosum** to any attempt at discovering what was probably the Linnean plant of that name. **Todaro** places the majority of the American forms in his 5th sub-section which he designates as **Magnibracteata** or those with a large bracteole, cordate, and deeply laciniated, and in which the seeds are free from each other. The others he assigns to his 6th section, **Synspermia,** or those with adherent seeds.

I. MAGNIBRACTEATA. (*For Section II., see p. 21.*)

A. Capsules small.

Under this classification Todaro discusses several species, two, or perhaps three, only being of interest. These are:—

13 (393) (1) **G. religiosum**, *Roxb.* (*? Linn.*)

DIAGNOSTIC CHARACTERS.—*Leaves* very hairy, with 3 to 5 triangular, acute or acuminate, lobes; stipules cordate acuminate. *Bracteoles* deeply laciniate; petals pure yellow or yellowish white. *Seeds* covered with firmly adherent short tawny down and long wool, which is either of the same colour as the down or white.

This is generally known as Nankin cotton. (*Conf. with pages 29, 31, 79, 112, 130, 134-135.*) According to Roxburgh it is closely allied to **G. hirsutum,** and by Todaro the Indian Nankin plant is placed under that species and apart from **G. religiosum,** *Linn.* Roxburgh appears to regard it as having come to India from China. Its peculiarities may be said to be those of a cross between the typical American and Asiatic forms, probably a hybrid which took its birth in India or China. (*Conf.* with **G. nanking,** see p. 29.) CLASSIFICATION.

Interest has been taken periodically in the subject of Nankin cotton, but it is feared the existence of *khaki*-coloured forms of all the Indian species and hybrids, has tended to destroy the advantages that might have been expected to accrue from the careful cultivation of this plant. It seems at least probable that any brown or reddish-brown cotton has been accepted and cultivated as Nankin and hence called **G. religiosum,** whereas the selection and development of white qualities of **G. religiosum,** *Roxb.*, might prove of great value, but of course apart entirely from the craze for "nature-dyed *khaki* cotton."

(2) **Gossypium mexicanum,** *Tod., Cult. dei Cot. 193, t. VI.* 14 (394)

MEXICAN COTTON.

DIAGNOSTIC CHARACTERS.—Shrubby, hairy, ultimately glabrous. *Leaves,* deeply cordate, sub-rotund, one-half cut into 3-5 lobes, lower pair very obscure; lobes oblong, triangular, acute or acuminate; stipules deciduous. *Flowers,* small, yellow or with a purple flush; petals obliquely truncate, often with a very small circular purple spot; bracteoles sub-equal with the corolla, broadly ovate, base rounded or slightly cordate, deeply laciniated. *Capsule,* small, rotund, slightly acute, 3-4 celled: seeds small, truncate, velvety underneath the floss; fuzz short, deeply rusty-coloured, floss long, pale reddish white.

It seems probable that this may be the species with small pale-coloured flowers alluded to by certain writers on Indian cotton. Dr. Wight contributed to the Agri-Horticultural Society of India two long papers on Mexican cotton. These will be found in the *Journals (Old Series) VI., 189-196*; in his second paper he gave the results of experiments with this cotton in the various provinces (*Jour., Old Series, VII., 194-215*). In a still earlier issue (*Journal, Old Series, Vol. I., 25-34*), Dr. Spry gave it as his opinion that Mexican cotton was admirably suited to Behar.

B. Capsule large, leaves also large and palmately lobed.

There are many more forms described by Todaro under this series than need be here dealt with. The following are the more important—

(3) **G. hirsutum,** *L.*

SAW-GINNED DHARWAR COTTON: NEW ORLEANS COTTON: UPLAND GEORGIAN: BAMIA, &c., &c. Conf. with p. 130.

References.—*Roxb., Fl. Ind., Ed. C.B.C., 521; Royle, Ill. Him. Bot., I., 100; Parl. Sp. Cot., p. 41, tab. V.; Gossypium album Ham. in Linn. Trans., XII., p. 482; Wight & Arn., Prod., Fl. Ind. Or., I., 54 (ex-parte); G. barbadense, var. Upland Georgian, Royle, Cot. in India, p. 148, pl. III., f. 4; G. barbadense, var. C., Wight, Ill. Ind. Bot., 84, tab. 28; G. herbaceum, var. 2, hirsutum, Masters, in Fl. Br. Ind., I., 347; Agri.-Hort. Soc., Ind., Trans.:—II., 61-63, 150, 151, 300, 403; III., 7, 133, 131, 138-180, 212; V., 50, 183; VI., 107, 110, 113, 230, 246; VII., App., 13, 14, 22; VIII., 226, 244; Journals:—(Old Series), I., 187; II., Sel., 2, 321, 485; III., Proc., 67, Sel. 262; VI., 119; X., 211; XI., 588; XII., 178, 179, 215, 220, 223, 225, 234, 243, 255-260, 263; Proc., 42; XIII., 89, 155, 157 (Pro. 1864), 30, 38.* 15 (395)

DIAGNOSTIC CHARACTERS.—The description of the Indian form of this plant, as given in Roxburgh's *Flora India* and the *Flora of British India*, may be expressed as:—shrubby, tender parts and young leaves softly hairy and of a pale moss green. *Leaves,* 3-5 triangularly lobed, being cut to one-half their depth. *Flowers,* large, yellow or yellowish white without a purple spot; bracteoles deeply laciniated, obtuse. *Capsule* rounded ovate, 4-celled, apex rounded then terminated with a short beak, seeds free, down, firmly adhering, green or grey, cotton, white, fine, and long.

CLASSIFICATION.

This species varies greatly in the thickness and colour of its velvety coat (on the seeds) and also in the length and fineness of the staple. **Roxburgh** wrote of it: "This green-seeded variety has only been of late introduced into India, where the cotton is much admired by the natives." During the past fifty years, however, the cultivation of this cotton has gradually concentrated in the Deccan. It is of course also met with in other parts of India, but success has chiefly attended its acclimatisation in Dharwar and the neighbouring districts; the cotton is known in the trade as Saw-ginned Dharwar. This has come by many writers to be spoken of as a form of **G. herbaceum**, and indeed in the *Flora of British India*, it appears as a variety of that species. In foliage, flowers, bracteoles, and cotton, however, it clearly belongs to the American series. In the velvety seeds and strong tendency to wooly stellate pubescence on the leaves it manifests an approximation to the Indian cottons, which may be due to hybridisation. Indeed, in the Sutlej Valley, near Suní, a form of **G. hirsutum** occurs which appears a hybrid. It is there known as *Nurma*, and its flowers are often quite pink. This is almost identical with the Sahet kaskom cotton of Santal country.

Egyptian cotton, at least the plant so called in India, is apparently a peculiar form of **G. hirsutum**, which in some districts has shown itself fairly suited to this country. Bamia cotton (once, though erroneously, said to have been a hybrid between ordinary Egyptian cotton and **Hibiscus esculentus**), is, according to **Todaro**, a hybrid between the Egyptian form of **G. hirsutum** and **G. maritimum**—Sea Island cotton. By cultivation in Italy **Todaro** found that it rapidly reverted to the latter species, never to the former. In Egypt, Sea Island cotton is perhaps grown to an equal extent with New Orleans. The Bami race, which appeared spontaneously, establishes beyond all doubt, therefore, that the American series of cottons hybridise amongst each other. Further, if **G. hirsutum** be itself accepted as a hybrid, it might even be inferred, that it originated from **G. maritimum** with some plant belonging to the **G. herbaceum** series, since Bamia reverted to the admitted specific ancestor, **G. maritimum**, never to the doubtful specific form, **G. hirsutum**. Indeed, on this line of reasoning, it might still further be assumed that the failures recorded in the effort to fertilize New Orleans with Indian cotton, were due to the **G. herbaceum** ancestor of New Orleans being distinct from the Indian cotton, or to **G. maritimum** being the stronger element of its hybridisation, towards which it would readily pass by further crossing or by continued cultivation.

16 (396) **(4) Gossypium maritimum,** *Tod., Cult. dei Cot., t. VII., p. 225.* [Egyptian).

SEA ISLAND COTTON; (Bamia, a hybrid between Sea Island and

References.—*G. barbadense, var. Sea Island, Royle, Cotton in India, 146, tab. III., fig. 3; Wight, Ill., Pl. 28 B.; Gossypium barbadense, Parl. in Sp. dei Cot., 83, tab. III.; G. barbadense, Masters, Fl. Br. Ind., I., 347 (non-Linn.); Bent. & Trim., Med. Pl. (an excellent plate), G. barbadense, No. 37; Agri-Hort. Soc. Ind. Trans., II., 148, 152, 397, 399, 403; III., 150; V., 183, 202; VIII., 226, 244; Jour. (Old Series), I., 32; IX., Pro., 70; X., 211; XI., 588; XIII., Sel. 88.*

DIAGNOSTIC CHARACTERS.—A glabrous annual, stem erect, much branched. *Leaves*, rotund ovate, sub-cordate, less than one-third cut into 3-5 lobes, or only angled, the upper ones entire; lobes ovate to lanceolate, oblong, acute, sinus almost rounded. *Inflorescence* solitary flowered, extra axillary; bracteoles broadly ovate, cordate, not united together, at the base deeply laciniated; flowers yellow, often almost white, rarely fully expanding. *Capsule* ovate-conic, acute, 3-4 locular, 6-9 seeded: seeds black, glabrous below the floss.

According to **Todaro** there are three varieties or rather hybrids of this species; *viz.*, **jumelianum, degeneratum,** and **polycarpum.** Only the last-mentioned need be discussed in this place, but in passing it may be said

CLASSIFICATION.

that Bourbon cotton, which we are told was introduced into India in 1804, belongs mainly to one or other of the forms of this species, or according to some writers, to **G. barbadense.** The term Bourbon is in fact almost generic in its significance—at least it is not specific, but includes many widely different flosses. For many years '**Heath's** cotton,' a form of Bourbon, found a ready market. It was seemingly the same plant which yields the staple, known up to the present day, as 'Salems.' (The reader is referred to the provincial section of this article which deals with South Indian cotton, pp. 102-104.)

G. maritimum *var.* **polycarpum,** *Tod., t. VIII.*, is the hybrid already alluded to which affords the Egyptian cotton known as Bamia. This, as already stated, is viewed by Todaro (*Cult., dei Cot., pp. 36-37*) as a hybrid between ordinary Egyptian cotton (one of the forms of **G. hirsutum**) and Sea Island cotton—**G. maritimum.** While admitting, however, that this is the nature of the plant in question, **Todaro** assigns it the position of a variety. He defines it as stem rigid, erect, rarely branched; flowering peduncles 1-3 in each axil, the lower one bearing one to two flowers. In his introductory chapter **Todaro** contends that Bamia cotton is, agriculturally speaking, very interesting, because, not being branched, many plants can be grown on a given space, and since more than one flower is formed at each node, the yield is very great. Cultivated in Italy, on poor soil, it lost, however, all these advantages and assumed the ordinary form of Sea Island cotton. In a like manner the condition which **Todaro** designates as variety **degeneratum** was found naturally produced in a field containing New Orleans and Sea Island, but the seeds were not sufficiently fertile to allow of its being cultivated for many years.

Before dismissing the subject of Bamia cotton it may be remarked that an extensive Indian official correspondence was conducted, from about the year 1877, on that subject. Large supplies of seed were obtained and the plant was experimentally cultivated in every province of India, but in a few years all interest ceased, and it is not known to what extent (if at all) Bamia cotton is now grown in India (see Proceedings, Revenue and Agricultural Department—Papers on the Experimental Cultivation of Bamia cotton in India—July 1880). In the Kew Report for 1887, p. 26, a short note occurs on this subject, in which the idea of Bamia being a hybrid between **Hibiscus esculentus** and Egyptian cotton is repudiated. **Sir J. D. Hooker** says of this cotton: "I doubt that it is anything more than a well-marked seminal sport, with a fastigiate habit, from some kind of Egyptian cotton, the bulk of which belongs to the Sea Island variety of **Gossypium barbadense,** to which species the Bamia cotton must also be referred. The cultivation of the Bamia cotton is said in Egypt to require more irrigation than the ordinary kinds."

In the *Journal of the Agri-Horticultural Society of India* (*New Series*), *Vol. VI., pp. 72-73 and 177-193,* in the last-quoted passage, **Mr. C. L. Tupper** (then Under-Secretary in the Revenue and Agricultural Department), gives a review of all the experiments conducted in India with the view of acclimatising Bamia cotton.

(5) **Gossypium barbadense,** *Linn.; Todaro, Cult. dei Cot., 234.* 17 (397)

A fairly good representation of this plant is the unpublished engraving in the India Office Museum collection, bearing the name **G. barbadense.** This form may possibly be the chief source of the so-called Bourbon Cotton of India.

References.—*Agri.-Hort. Soc. Ind., Trans. II., III., 132; V., 183; VI., 23, 238; Jour. (Old Series), I., 187.*

DIAGNOSTIC CHARACTERS.—**Todaro,** while insisting upon the separation of this from **G. maritimum,** simply defines it as "foliis trilobis, integerrimis." These

Conf. with pp. 102-104.

CLASSIFICATION.

were the words used by **Linnæus** in his descriptive designation of **G. barbadense,** *Linn.*, which very probably included **G. maritimum. Todaro** contends, however, that the confusion regarding this species is mainly traceable to **Swartz,** who described cultivated plants and says that in Barbadoes, there is a **Gossypium** with a three-lobed leaf corresponding closely to the Linnean type, and which could not be included with **G. maritimum.** He therefore regards the Barbadoes plant as very possibly **G. barbadense,** *Linn.* This argument is strengthened, he thinks, by the fact that Sea Island cotton (**G. maritimum**) was not cultivated in the West Indies at the time **Swartz** wrote.

After a critical inspection of all the plates cited, and a careful consideration of the meagre literature of these two plants as separated by **Todaro,** we do not think that they are likely, in the future, to be retained as distinct botanically. They are doubtless, however, worthy of separate recognition, since they yield commercial forms of cotton, and thus, both agriculturally and commercially, are of interest. **G. maritimum,** *Tod.*, is an annual plant, largely cultivated, while **G. barbadense** is perennial. We possess samples from various districts in Western and Southern India, that appear to be the **G. barbadense,** *Linn.*, accepted as such by **Todaro. Dr. Prain** has also kindly allowed us to inspect a specimen of it which he collected in the Andaman Islands on Mount Harriet. It was there found in an arboreous condition and apparently quite naturalised.

Conf. with p. 152.

It seems highly probable that this is, to a large extent, the **G. barbadense,** *Roxb., Fl. Ind., p. 521,* which, he says, had been introduced into India about the beginning of the century from the Island of Bourbon—hence its commercial name of Bourbon cotton. His remark regarding its introduction into Bourbon, from the West Indies, is interesting, since it somewhat confirms **Todaro's** statement of its having been originally a native of these islands. There are, however, one or two difficult points regarding **Roxburgh's** Bourbon cotton. It is described as having "leaves smooth, with five acute, short, broad lobes;" "colour of the corolla uniformly yellow." These characters would ally the Roxburghian plant more nearly to **G. hirsutum,** but, on the other hand, **Roxburgh** adds, "Seeds free, black, and without any other pubescence than the long, fine, white wool." From this it would appear that the plant should be assigned a place with **G barbadense,** *Linn.*, as popularly accepted, namely, the combined forms discussed above under that name and **G. maritimum,** *Tod.* It thus seems probable that **Roxburgh** did not separate these plants, and that his Bourbon may have been one of the numerous races of Sea Island cotton. **G. barbadense,** *Linn.*, and **G. maritimum,** *Tod.*, may thus be said to manifest a parallelism with **G. arboreum,** *L.*, and **G. neglectum,** *Tod.*, the latter in both cases being the chief field plant, while the former is more arborescent and much less frequently cultivated, but on that account the more entitled to recognition botanically.

Wight's *Illustration, 28 a* (reproduced in a reduced form by **Royle,** *Cult. Cotton in Ind., Pl. III., f. 2*) under the name of Bourbon cotton, has the broader, less segmented leaf of **G. barbadense,** but the 5-lobed condition of **G. maritimum.**

(398) 18 (6) **Gossypium peruvianum,** *Cav.; Tod., Cult. dei Cot., 240.*

PERUVIAN COTTON.

This is said by **Todaro** to be a native of Peru, and to be cultivated in Spain. We have seen one or two samples of it from various parts of India—notably a good sheet from the Madras Herbarium. This was apparently cultivated at Triplicane by **Dr. George Bidie, C.I.E.**

In foliage and bracteoles it much resembles **G. acuminatum,** *Roxb.*, but the stipules are very large, being fully an inch and a half long and the leaves thicker, smaller, and the segments more constricted at the base than in G.

CLASSIFICATION.

vitifolium and **G. brasiliense**—the kidney cottons—with which it was confused by **Royle** and, following him, by many Indian popular writers. This same mistake has also been made by **Masters** (*Flora of British India, I., 347*) where **G. peruvianum,** *Cav.*, is quoted under **G. acuminatum,** *Roxb.*, a plant which, at the same time, is reduced to a variety of **G. barbadense.**

Peruvian cotton has the seeds free from each other and coated with a thick velvet below the floss. It is thus either a distinct species (and in that case a native of Peru) or a Peruvian hybrid between one of the kidney cottons and some of the forms of **G. herbaceum.** The very large flowers and immense deeply gashed bracteoles are its most ready eye-marks when taken in conjunction with the fact of the seeds being free from each other.

Mention is made of this cotton in several of the reports, &c., which deal with the introduction of American cottons into India, notably in connection with Orissa. (See *Trans. Agri.-Hort. Soc. Ind., II., 51.*)

II. SYNSPERMIA. (*For Section I, see p. 16.*) 19 (399)

Seeds adherent into a mass—hence the name kidney-cottons. **Todaro** describes eight species under this section, and three of these are met with in India. There are several very obscure points regarding these plants, and, so far as the forms met with in India are concerned, it seems doubtful if they deserve a higher rank than that of varieties, if, indeed, they should not rather be accepted as cultivated races of one species. **Roxburgh** was the first to describe the plant now known under the name **G. acuminatum.** Regarding the habitat of that form he wrote," said o be a native of the mountains to the north and westward of Bengal." But the exact location of the region there referred to is extremely difficult, if not impossible. Nepal would perhaps be on the north-west of Bengal as now understood, but if Behar be excluded from Bengal, **Roxburgh's** sentence would then indicate Bundelkhand or thereabouts. But, it should be observed, he gives no exact locality, his words being "said to be a native &c." It thus seems highly probable that **Roxburgh** described his species from plants raised in the Botanic Gardens, the information given above being the report furnished to him by his native gardener, who, perhaps, procured the seed from some of his friends as that of an interesting new plant. It is not like **Roxburgh's** work generally to leave so obscure a passage, if it had been in his power to be more precise. **Royle** (*Illustrations of Himálayan Botany, 98*) republished **Roxburgh's** description and habitat of this species, but apparently had not seen an example of the plant. He, however, described it as distinct from **G. vitifolium** and from **G. peruvianum.** In his later work (*Cult. and Com. of Cotton in India, 149*) he reduced all the kidney-cottons, including **G. acuminatum,** *Roxb.*, to **G. peruvianum,** *Cav.*—a reduction due very probably to the misconception of viewing **G. peruvianum** as a kidney-cotton. While we cannot accept this part of **Royle's** work, as remarked above, we consider **G. acuminatum,** *Roxb.*, **G. vitifolium,** *Lam.* (*non Roxb.*), and **G. brasiliense,** *Macf.*, as doubtfully worthy of separate specific positions Apparently no modern botanist has found **G. acuminatum,** *Roxb., except in garden cultivation*, and there is absolutely nothing to justify the statement that it is a native of any part of India. It is significant that **Todaro** should give no other locality for it than that originally published by **Roxburgh**, and, indeed, he only refers to the Indian authors mentioned above. Among the samples of **Gossypium** seen by us occur one or two specimens, named **G. acuminatum,** *Roxb.*, by the authorities in charge of the Indian herbaria, but in each case the samples are shown to have been obtained from plants grown in Botanic Gardens. From Madras we have received specimens of kidney-cottons,

CLASSIFICATION. but are not prepared to say whether these should be referred to **G. acuminatum**, *Roxb.*, to **G. vitifolium**, *Lam.*, or **G. brasiliense**, *Macf.* **Mr. J. F. Duthie** has a specimen of kidney-cotton (very possibly **G. acuminatum**, *Roxb.*) which he recently collected in Merwara but in a state of cultivation —the Saharanpur herbarium has no sample of wild **G. acuminatum.**

(400) 20 (7) **G. acuminatum**, *Roxb., Fl. Ind., Ed. C.B.C., 520.*

References.—*Todaro, Cult., dei Cot., 249; Wight, Ill., t. 27; Royle, Ill. Him. Bot., 98; Cult. & Com. Cotton in India (in part), IV., 149; G. barbadense, var. acuminatum, Mast., Fl. Br. Ind., I., 347 (ex. certain syn.); Trans. Agri.-Hort. Soc. Ind., II., 149.*

DIAGNOSTIC CHARACTERS.—Sub-arboreous. *Leaves*, 3-5 lobed, lobes divergent, ovate, acute the middle one slightly the longest. *Flowers* large, bracteoles broadly ovate, deeply cordate and glandular at the base. *Capsule* ovate, acuminate, cells 8-10 seeded, black glabrous except the long white wool.

In a paper on **Gossypium acuminatum**, *Roxb.*, written by **Dr. N. Wallich**, dated April 3rd, 1834, considerable light is thrown on the report, originally made by **Roxburgh**, of this species being possibly a native of India. **Wallich** wrote:—"The accompanying seeds and samples of cotton were communicated to me in January and February last by **Major L. R. Stacy**, from Nusseerabad, where he first discovered the tree producing it in a *fakeer's* garden, and afterwards met with several others in the same vicinity."

"**Dr. Roxburgh** calls this species **G. acuminatum**, and says, it is reported to be a native of the mountains in the north-eastern parts of Bengal and some other parts of Hindustan. I have heard it asserted confidently, that it originally came from Surinam." "Be this as it may, the sort is remarkable, inasmuch as it grows to be a tree 12-16 feet, and is very productive. The seeds adhere together in a very peculiar manner; the cotton is easily and completely separable, milk-white, and of a long staple, but it is harsh and woolly to the touch."

In the remarks below, under **G. brasiliense**, *Macf.*, it will be found that **Mr. C. B. Clarke**, like **Dr. Wallich** in the passage just quoted, attributes to **Roxburgh** the assertion that his **G. acuminatum** was a native of the mountains in the north-*eastern* parts of Bengal. **Wallich** further states that, according to **Roxburgh**, it occurs in "some other parts of Hindustan." What **Roxburgh**, however, did remark was, that it was "said to be a native of the mountains to the north and *westward* of Bengal." **Wallich's** suggestion that it may have originally come from Surinam in Guiana —we feel strongly disposed to add—has been confirmed by all subsequent evidence. In other words, we are very nearly satisfied that all the kidney-cottons are of New World origin. In that case **G. acuminatum** would probably have to be referred to one or other of the western types instead of being retained as an Asiatic species.

(401) 21 (8) **G. vitifolium**, *Lam.* (*non Roxb.*)

Syn.—G. LATIFOLIUM, *Rumph., Amb. IV., 37, t. 13.*

References.—*Todaro, Cult. dei Cot., 251, tab. XII., fig. 1.*

Habitat.—Cultivated in the Island of France and, to some extent, in India and other parts of the world. This species is, by many writers, confused with Sea Island Cotton, though the two plants have nothing in common.

A shrubby glabrous species, with the leaves hairy below; the upper leaves 3-lobed, the lower 5; lobes ovate-lanceolate, acuminate.

The vine-leaved kidney-cotton is repeatedly mentioned in the various reports (*Agri.-Horti. Soc., India*) which appeared on the introduction of American cottons into India. Valuations are given of the cotton produced,

CLASSIFICATION.

but no detailed paper can be found on the subject, so that the extent to which it may now exist in India cannot be discovered. But, from the fact of its being figured by **Rumphius** in 1750, we may infer that it was, perhaps, one of the earliest forms of kidney-cotton which found its way to Asia.

(9) **Gossypium brasiliense,** *Macf.; Todaro, Cult. dei Cot.*, 265. 22 (402)

PERNAMBUCO OR BRAZILIAN COTTON.

References.—*G. religiosum, Parl. (in part, non Linn.); Royle, Cult. & Com. Cotton in India, 152; Agri.-Hort. Soc., Ind., Trans., II., 55, 124, 126, 146, 147; III., 39; V., 51, 54-56, Proc., 39; VI., 100, 116, 131; VII., App. 14; Journals (Old Series), I., 187; IV., 207; XIII., Sel. 89.*

DIAGNOSTIC CHARACTERS.—An extensively branched bush. *Leaves* very large, 5-7 lobed, with the sinus thrown up as a fold between the lobes. *Flowers* large, yellow (or, according to some writers, with a purple base), convolute, protruding considerably from the bracteoles, which are very large and deeply laciniated. *Capsule* ovate acuminate, shorter than the bracteoles, cells 7-9 seeded; seeds naked except for the long wool.

It seems very probable that this is the chief form of kidney-cotton met with in India. One or other form occurs very frequently in gardens and near temples, taking the place of **G. arboreum** of former times. In an official correspondence conducted in 1871, regarding the Nicobar Islands, mention is made of a very superior cotton. **Colonel Man** wrote that "it was found growing in an almost wild state, and, therefore, the specimens could only be taken as a slight evidence of what the island might produce, if skilled labour and care were devoted to the cultivation of the plant." The samples furnished were forwarded to the Chamber of Commerce at Bombay, and the report on them was to the effect that "the cotton is much superior to any grown in India. It is also superior to American cotton, except the fine Sea Island description. It greatly resembles Brazilian cotton both in colour and in staple." Later on we are told of other samples of Nicobar cottons grown from Sea Island seed, which had been furnished by the Secretary of State for India, as well as further samples of what is called "the indigenous cotton." The report on these samples gave a valuation of 20 to 24*d.* per pound for the Sea Island and 8 to 8½*d.* for the indigenous. The latter was pronounced to be Pernambuco stock of "good useful quality." **Mr. C. B. Clarke** was apparently asked, however, to examine the plant, since a report exists in the official correspondence (No. 329, dated 21st February 1871) from the pen of that distinguished botanist in which he says it is probably **Roxburgh's "G. acuminatum,** a plant indigenous in the hills of north and east Bengal." It is not known upon what this statement is based, but, as already stated, we have no proof at present of any kidney-cotton being indigenous to India. **Mr. Clarke** further makes, on the authority of the late **Mr. Scott**, the contradictory statement that it "cannot be Pernambuco or closely allied thereto, because in Pernambuco the seeds (while similarly adherent to each other) are covered with grey wool." The species resembling **G. acuminatum** with velvety seeds is **G. peruvianum,** a plant with free seeds. Thus while **Roxburgh's** remarks regarding **G. acuminatum** have not been confirmed by fresh collections, each succeeding writer has distorted the original statements regarding it. It seems probable that **Roxburgh's G. acuminatum** may be but a smaller-leaved condition of this species—one with thicker more rotundly palmate leaves, and which has become practically acclimatised in India. Under the provincial notices will be found occasional mention of kidney-cotons, and the reader might, in this connection specially, consult what has been said under Burma.

CONCLUDING REMARKS. 23 (403)

Concluding Remarks regarding the American Cottons.

To give even the briefest review of the numerous experiments which have been performed, to indicate the purposes for which large sums of money have been spent, or to exhibit the true force of the disappointing reaction, consequent on the futile efforts to acclimatise these cottons in India, would occupy many pages. Suffice it to say that neither money nor labour have been spared, but nearly everywhere the same conclusion has been arrived at, *viz.*, failure. In one or two localities, however, fairly good results have been obtained, and the Saw-ginned or New Orleans cotton of Dharwar may be said to be the practical outcome of a half century's energies, on the part of the public, the Agri-Horticultural Society of India, and the Government. Scattered throughout India may now be found the depauperated survivors of all these attempts at acclimatisation, some practically naturalised but others preserved as mere garden curiosities. It may be stated that the final verdict has been pronounced, *viz.*, that, if seed could be annually procured and of good quality, many of the American cottons might be cultivated in India, but that they all rapidly lose their special merits when grown from Indian-produced seed. Even the Dharwar exotic, though in some respects a superior cotton to many of the indigenous growths, is a sadly degenerated stock from the New Orleans of America and Egypt. What influence, if any, hybridisation of Indian with American or American with Indian may have exercised in this degeneration it is impossible to tell, but, as repeatedly urged, it would seem to us that the natural direction of future experiments should be in the hybridisation and selection of superior qualities from indigenous or naturalised stocks rather than further efforts at acclimatisation. The first step towards this end is, of necessity, the acquirement of detailed scientific knowledge as to the forms of cotton that now exist in India; the second, the establishment of the properties and characteristics of each species, variety, and hybrid, and the ascertainment of the lines on which hybridisation and selection should in future be conducted.

The reader will find full particulars regarding the Bombay experiments which have been conducted with a view to improving the cottons of that presidency in the chapter devoted to Bombay. But numerous papers have appeared from time to time in the Journals of the Agri.-Horticultural Society of India, chiefly from the pens of **Drs. Wight & Spry**, and in the reports of the Government Cotton Farms, and these may be also consulted. The better papers of this nature will be found in the following volumes:—

Agri-Horticultural Soc. of India: Transactions III., 98, 144; VII., App. 1; Proc., 164, 195; VIII., 1; Journals (Old Series), I, 60, 114 (farm reports); II., 53, 237, App. 2, 158, 281, 321; App. 85 (farm reports); III., App. 126 (Wight); IV., 135; V., 37, 83; VI., 118, 189 (Wight); VII., 20, 211, (Wight); VIII., App. 27 (Wight).

Dr. Forbes Watson writes on this subject:—"With regard to improvements in selection of seed and in cotton cultivation generally, no perceptible progress has been made, in spite of incessant efforts during the past fifty years. The system of Government cotton farms adopted for years seemed as if specially devised to render abortive the services of some of the zealous agents employed at them. The plan under which a certain number of acres was for a season taken over from the rayat and appropriated for experimental purposes, the locality of the experiments being shifted from year to year, made it perfectly impossible to attempt any consecutive experiments as to improved methods of cultivation, rotation of crops, or improvement of the plant by selection, such as could be carried out at a permanent agricultural experimental station, of the kind to be found in hundreds on the Continent and in the United States, and in

CONCLUDING REMARKS.

which the cultivation of cotton could be combined with experiments on the general agriculture of the country, with the progress of which the improvement of the cotton staple is indissolubly connected. There appears to be a natural adaptation of the Indian cotton plant to the dry climate of India, which enables it, by means of its long tap-root, to draw sustenance and moisture from greater depths of the soil than is the case with the American variety, with its number of lateral roots spreading near the surface. Hence it is quite as important to proceed with experiments in the improvement of the native staple by selection, as to attempt the introduction of foreign varieties which, whatever their advantages may be, are not so well adapted to sustain the frequent seasons of extreme drought occurring in India.

"It may be remarked that there is less reason now than appeared some time ago to apprehend a serious diminution of the Indian cotton trade through the increasing competition of American cotton. The Indian cotton is not only finding a new and rapidly increasing market in India itself, but it also has taken a firm hold of the continental markets, and it is possible that the appreciation of the merits and the extensive use of Indian cotton by the continental manufacturers may be one of the causes which enable them to compete so successfully as they have done of late with the cotton manufacturers of England, among whom the use of the Indian cotton has declined in a very marked manner. It is also due to this continental competition for Indian cotton, that its market value has of late approached more nearly to that of the American than at any previous period.

"Finally, the competition of the American cotton is not the only circumstance which regulates the cultivation of cotton in India. Unless the prices should fall much below even their present level, cotton will continue to be largely grown in India, as over a large part of the country it enters into the regular rotation of crops, and being at all times readily convertible into money, is relied on by the rayat to produce a large portion of the cash required to meet the Government rent and other payments. So long as no convenient substitute is discovered which can replace it in this respect, the cultivation of cotton is in no danger of falling much below its present extent. It may be remarked, too, that a period of low prices is much more favourable for the introduction of reforms in the cultivation and preparation of the cotton, than a period of high prices, in which there is a good demand for the most inferior descriptions which can be brought into the market." (*Report on Cotton Gins, Pt. I., 30-31.*)

III. Gossypium herbaceum, *Linn.; Fl. Br. Ind., I., 346.* 24 (404)

Vern.—*Rui, kupás* (the floss), HIND.; *Kapás, tula,* BENG.; *Rui,* (the floss), PB.; *Vaum,* SIND.; *Kapas, rui,* BOMB.; *Ru, kapás,* GUZ.; *Kapás,* DECCAN; *Vun-paratie, parutti,* TAM.; *Pauttie, edudi, paratti, paritt,* TEL.; *Wah, wa,* BURM.; *Karpasi, karpas,* SANS.; *Kurtam ussul,* ARAB.; *Pambah,* PERS. NOTE.—It seems probable that most, if not all, of the above names are those given to the wool, not the plant. Some racial names for the various cotton crops will be found in the provincial chapters and in the remarks under **Gossypium arboreum,** pp. 8—14.

References.—*Roxb., Fl. Ind., Ed. C.B.C., 519; DC., Orig. Cult. Pl., 402-405; Sir W. Elliot, Fl. Andh., 157, 146; Irvine, Mat. Med. Patna, 88; Medical Topog., Oudh, 6; Mat. Med., S. Ind. (in MSS.), 40; U. C. Dutt, Mat. Med. Hindus, 303; Dymock, Mat Med. W. Ind. 2nd Ed., 109-112; Birdwood, Bomb. Prod., 315; Baden Powell, Pb. Pr., 477-496; Drury, U. Pl. Ind., 229 234; Atkinson, Him. Dist. (Vol. X., N.-W.P. Gaz.), 738; Useful Pl. Bom. (Vol. XXV., Bomb. Gaz.), 215, 228; Royle, Prod. Res., 312-355: Fibrous Pl., 264: Cult. and Com. Cotton in India, many passages; Manual and Guide, Saidapet Farm, Madras, 17, 59-61; Liotard, Mem. Paper-making Mat., 24; Shortt, Man. Ind. Agri., 1-92; Linschoten, Voyage to East Indies (Ed. Bur-*

CLASSIFICATION. *nell, Tiele and Yule), Vol. I., 55, 60, 95, 129 ; Hove, Jour. of Tour in Bombay in 1787, pp. 26-27, 29, 30, 34-35, 36, 37-38, 39, 40-41, 42, 45, 58, 69, 76, 81, 106-107, 109, 127, 131-132, 148-149, 155-156, 158, 161, 167, 169 ; Man. Madras Adm., Vol. I., 289-290, 334, 335, 339 ; Nicholson, Man. Coimbatore, 232-235 ; Gazetteers and Settlement Reports too numerous to be quoted ; Indian Forester, XII., app. 7 ; XIV., 109.*

To prevent any possible misunderstanding, it may again be explained that we use the name **G. herbaceum,** as the most convenient for a series of forms, one of which (**G. Wightianum,** *Tod.*) may be of specific interest, but the others, while entitled to lower positions botanically, are nevertheless equally valuable as races of cultivated cotton. **Todaro** appears to have established one or two important facts regarding these cottons, and these may be briefly summarised :—(1) **G. herbaceum** *proper* does not occur in India, and indeed the attempts at introducing it have been less successful than were those with the American cottons : (2) The bulk of the commercial cottons of India are hybrids : (3) The species of greatest importance to India is **G. Wightianum,** *Tod.* : hybrids from that species constitute the better cottons of Western and Southern India : (4) **G. neglectum,** *Tod.* (a form of **G. arboreum,** according to many authors, and of **G. herbaceum** by others), also occurs in India, and hybrids from that species constitute the bulk of the cottons of the Eastern and Northern divisions of India : (5) Two varieties of **G. herbaceum** are found in India, ***viz.*, obtusifolium** and **microcarpum.** The former is met with chiefly in Bengal and Madras, and the latter is largely the source of the *Dhollera* cotton.

As already remarked under **G. arboreum** we regard many of **Todaro's** forms of this species as mere hybrids, but it may serve a useful purpose to give here the names of the species or hybrids to which we allude and to suggest their possible nature.

III.

(*For I. see page 7, and for II., page 16.*)

THE HERBACEUM OR ASIATIC SERIES OF COTTONS.

25 (405) (1) **Gossypium herbaceum,** *Linn.*

Habitat.—This form, **Todaro** says, is a native of Asia, and perhaps also of Egypt. Accepting **G. Stocksii** as a wild form of it, he gives also the rocks of Sind as one of its habitats, thus extending the area to India. As far as we can judge, the original home of **G. herbaceum** may have been to the North-West of India proper, corresponding more nearly with the region of its present cultivation, *viz.*, Egypt, Asia Minor, Northern Africa, and Southern Europe. Although now extensively cultivated in the United States, it was probably introduced there from Europe. It was a new cultivation, **DeCandolle** remarks, one hundred years ago, since in 1774 a bale from North America was confiscated at Liverpool on the plea that the cotton plant did not grow there, but there is probably little or nothing to support **DeCandolle's** appropriation of that circumstance as a historic fact specially connected with **G. herbaceum.** It would seem simply to prove that America up to that date was not known to produce cotton. We know, however, that the **G. barbadense** series of cottons had been grown there from time immemorial by the aboriginal tribes long anterior to any record of the introduction of **G. herbaceum.** It would appear also that the incident alluded to by **DeCandolle** took place in 1784 not 1774.

Todaro, however, admits two varieties as met with under cultivation in India, and if our suggestion proves correct, *viz.*, that even **G. Wightianum** may be but a hybrid or sport, or at most a variety, of **G. herbaceum,** *Linn.*, India, as commonly reported, might then be accepted as one of the chief habitats of the species. If, on the other hand, **Todaro's** views

CLASSIFICATION.

regarding **G. Wightianum** be confirmed by more extended study of the Indian cottons, that plant is probably the original condition from which all the better cottons of India have been derived. Under **G. Wightianum** mention will be found of a wild form of cotton, found in Rajputana and the Panjáb, which appears to be that species, a fact which powerfully supports **Todaro's** position.

DIAGNOSTIC CHARACTERS.—Perennial and bushy in the warmer areas, annual where killed by winter. *Stems* erect, sparsely hairy, branches spreading, the bush sub-pyramidal. *Leaves* pale green, thick, leathery, and larger and broader than in **G. Wightianum,** half segmented into 3-5, often 7 lobes: lobes ovate-rotund, constricted below, the middle one acuminate, all shortly bristle-tipped; upper surface reticulate, pubescent, with longer hairs on the veins and felted below when young with bifid or trifid (not stellate) hairs, margins ciliate. *Stipules* linear lanceolate, those nearer the peduncles shorter and broader. *Flowers* longer than the bracteoles, petals obovate, unequally cuneate, yellow, with a large purple patch, slightly united together below; *peduncles* about half the length of the petioles; *bracteoles* ovate sub-obtuse, cordate, inciso-dentate, the entire portion sub-rotund, united to each other below, in texture thinner and more numerously veined than in Indian cottons. *Capsule* ovate-rotund 4-5-celled, cells 6-7 seeded, seeds ovate, beaked at the hilum; undervelvet or fuzz as it is often called short, firmly adhering, whitish-grey; wool fairly long and white.

The chief features of this plant, which distinguish it from the ordinary Indian cottons, are its broader leaves, more rounded lobes, the absence of stellate felted hairs, and the shape of the bracteoles, which approach much nearer to those of the American than to any of the Indian cottons. The remark that the seeds are beaked is of considerable interest, since, if specific, this would indicate the inferior quality of cotton, with a spine on the end of the seed, alluded to by **Mr. Hare** (see remarks below in connection with Berar), as one of the forms of **G. herbaceum,** *Linn.*

We have seen only one or two examples of what we regard as true **G. herbaceum** in Indian collections. Of these may be specially mentioned the Gilgit specimens issued as No. 250 from Kew, and a specimen collected by the Editor in a garden on the Nilghiri Hills.

But the following are forms which we recognise as possibly only varieties of **G. herbaceum,** and the forms of the species which alone exist in India as field crops:—

Var. microcarpum, *Tod.* 26 (406)

This is described as possessing much smaller fruits than the type, and of an ovate, sub-rotund shape, with the apex deeply depressed.

Conf. with p. 57.

According to **Todaro** this corresponds to much of the *Dhollera* cotton, a form which is grown in Kathiawar, Cutch, Kaira, and Cambay, &c.

Var. obtusifolium. 27 (407)

Todaro regards this as a species, but we can see very little to separate it. Indeed, like *var.* **microcarpum,** it is a form that connects in some respects **G. Wightianum** with **G. herbaceum.** The description originally given by **Roxburgh** appears to have been simply reproduced by **Todaro,** also by **Masters** in the *Flora of British India.* It is not apparent whether or not these authors have specially studied the plant.

Conf. with p. 102.

DIAGNOSTIC CHARACTERS OF OBTUSIFOLIUM.—Shrubby, much branched. *Leaves* small, 3, rarely 5, lobed, lobes ovate, obtuse, entire; *stipules* falcate; *bracteoles* ovate, obtuse entire or nearly so. *Capsule* ovate, 3-celled, with 3 seeds in each cell; seeds free, coated with closely adhering green velvet below the ashy-white wool. **Todaro** remarks that it differs from all the other species in the entire bracteoles and in having only three seeds in each cell.

Roxburgh supposed this to be a native of Ceylon, but **Thwaites** and other more recent writers have failed to confirm this statement. The plant, as recognised above, has, however, been collected by **Griffith** in Ava, and the Editor came across it in a semi-wild condition on the Naga

CLASSIFICATION.

Hills and in Manipur, found on deserted lands near ruined villages. Samples have also been obtained from Madras of a plant with small and mostly three-lobed leaves, which appears to be the obtuse-leaved cotton of most writers, and we have grounds for believing that the plant we allude to, at least, is fairly extensively cultivated throughout India, especially in the warmer areas and on poorer soils. Indeed, it seems probable that it is the inferior cotton with beaked seeds referred to above.

(408) 28 (2) **Gossypium Wightianum**, *Tod., Cult. dei Cot., 141.*

This, according to **Todaro**, is the chief cultivated cotton plant of India, being mainly met with, in its purer forms, on the Western side of the Peninsula. It is apparently the **G. indicum** of **Ure's** *Cotton Manufacture, p. 126.*

Conf. with pp. 64, 85, 88, 98, 102, 111, 118, 129, 132, 142, 147, 150.

DIAGNOSTIC CHARACTERS.—*Stems* erect, somewhat hairy, branches spreading and ascending. *Leaves* when young densely matted, with short, thick, stellate hairs, which fall off in patches leaving a few scattered all over both surfaces; ovate-rotund, almost obsoletely cordate, 3-5, rarely 7, lobed, lobes ovate, oblong, acute, constricted at the base into the rounded sinus which in the young leaves rises up as a fold: *stipules* on the peduncles almost ovate, the others linear-lanceolate acuminate. *Flowers* yellow with a deep-purple patch at the base, becoming reddish on the outside on passing maturity; *bracteoles* small, slightly united at the base, ovate cordate, acute, shortly toothed: *peduncles* erect in flower, recurved in fruit, one-fourth the length of the petioles. *Capsule* small, ovate acute, 4-celled, with 8 seeds in each cell; seeds small, ovate, sub-rotund, velvet very short and firmly adhering, wool white or whitish-red.

Syn.—**Todaro** cites the following references to this form:—G. HERBACEUM, *var.* OBTUSIFOLIUM, *Wight, Ic., t. 9;* G. ALBUM, *W. & A.;* G. HERBACEUM, *Mast. in Hook., Fl. Br. Ind., ex parte.*

The peculiarities of the leaves, bracteoles, and capsules of this species are contrasted with those of **G. herbaceum**, *Linn.* in **Todaro's** plate IV. The leaves of the latter are larger, broader, more rotund and much more deeply cordate, with the lobes more rounded, and acute-apiculate, instead of ovate, oblong, acute as in **G. Wightianum.** The bracteole is also broader, much thinner in texture, more deeply cordate and often gashed, somewhat resembling the bracteoles of the **G. barbadense** series. We have very carefully examined a large assortment of specimens of **G. Wightianum** and have come to put a considerable degree of faith in the value of the character of the hairs in isolating that plant not only from **G. herbaceum,** but also from **G. neglectum.** The latter has long spreading hairs and is only stellately tomentose when hybridised with **G. Wightianum** just as that species shows a strain of **G. neglectum** by the presence on the veins, of a few of the long hairs.

We have had the pleasure of examining in **Mr. J. F. Duthie's** collections of **Gossypium,** now deposited in the Saharunpur herbarium, one or two specimens bearing on their labels the remarks "the wild cotton of Merwara," "apparently quite wild in the forests west of Bár in Merwara." **Mr. A. E. Lowrie** has also contributed a sample of the same plant found in fruit in Ajmir on the 30th October 1883. **Mr. Lowrie** remarks of this, "the wild cotton I send you was found as a creeper over other trees." These samples were kindly examined (in consultation) by **Mr. Drummond,** Deputy Commissioner, Karnal, who thought he could identify them as the wild cotton alluded to by him in his remarks below, in connection with the Province of the Panjáb. **Mr. Drummond's** description of the Karnal 'wild plant' would lead to the supposition of its being **G. neglectum,** but we have seen no specimen of the plant referred to. We have already incidentally alluded to these samples, and need only add that they appear to be the wild, or at least semi-wild, condition of **G. Wightianum.**

Conf. with p. 119.

(409) 29 HYBRIDS OF THE HERBACEUM SERIES.

There are numerous hybrids from **G. Wightianum,** but all we have examined are hybrids with **G. neglectum,** more rarely with **G. arboreum.** These hybrids have more deeply segmented leaves than **G. Wightianum,** and possess the extra tooth in the sinus which we regard as characteristic of the **G. arboreum** series. They are either very hairy, if hybridised with **G. neglectum** (having both kinds of hairs) or almost sub-glabrous, if with **G. arboreum.** Of this latter class may be mentioned the common Himálayan crop of which **Dr. J. E. T. Aitchison's** No. 462 from Kuram Valley may be mentioned as a good type. We in fact regard Todaro as having isolated as species some of these hybrids, for example, the following, taken up in alphabetical order :— CLASSIFICATION.

(1) **G. cernum,** *Tod., Cult. dei Cot., 160.* 30

This is said to be cultivated in India, its chief characteristic feature being the reflexed upper portion of the bracts. The description of the leaf as given by **Todaro** would point to its being a hybrid from **G. neglectum** with pale sulphur-coloured flowers with a purple spot. The capsule is said to be fairly large, cylindric, acute, 4- rarely 3- celled, cells 7-11 seeded, wool short, white or reddish.

(2) **G. indicum,** *Lamk. ; Tod., Cult. dei Cot., 150.* 31

This, according to **Todaro,** is the *Capas* of **Rumphius** (*Amb. IV., 33, t. 12*). The Linnean species, **G. herbaceum,** is, he thinks, a combination of the characters of **G. Wightianum** with the plant *he* isolates as **G. herbaceum,** *Linn.*, proper, and **G. indicum,** *Lamk.* The present restricted species is characterised by having 3-lobed leaves which are eglandular and shortly pointed, and small bracteoles, sub-entire or slightly toothed on the apex. But that description recalls **G. obtusifolium,** of which it may be but a form with more pointed leaves.

(3) **G. intermedium,** *Tod., Cult. dei Cot., 153.* 32

This, **Todaro,** on the authority of **Royle** (*Cotton in India, 142*), says, is cultivated in India, chiefly in the Deccan, and is distinguished by the free re-curved segments of the style. It would, however, appear to be one of the hybrids between **G. Wightianum** and **G. neglectum,** in which the segments of the leaves are usually broad as well as the stipules. The flower is large and opens wide in flowering. Capsule ovate acute, rough on the surface, 3-celled, each cell 8-seeded : wool reddish white.

(4) **G. nanking,** *Meyen ; Tod, Cult. dei Cot., 147.* 33

This, **Todaro** says, is the true Nankin cotton—a Chinese plant often grown in India under the name of **G. religiosum** ; the Nankin or reddish cotton of India being, he affirms, perfectly distinct and a form of **G. hirsutum.** The description given by **Todaro,** however, as also his figure III., suggests its being possibly a hybrid between **G. Wightianum,** *Tod.*, and **G. herbaceum,** *Linn.* The lower leaves are sub-rotund, obsoletely cordate, 5-7 lobed, the lobes oblong obtuse, with the base only slightly constricted ; upper leaves 3-5 lobed, lobes ovate obtuse. *Flowers* pale yellow with a purple spot ; *bracteoles* exceptionally small, ovate sub-rotund ; *peduncles* recurved in fruit, much shorter than the petioles. *Capsules* ovate acute (in the figure shown as sub-rotund) ; 3-celled, cells mostly 5-seeded, wool reddish yellow. Conf. with p. 17.

(5) **G. roseum,** *Tod., Cult. dei Cot., 164.* 34

This is also said to be cultivated in India and is described as being erect with slender spreading branches. In the description of the leaf it approaches most to **G. neglectum.** *Flowers* white or pale yellow with a purple tinge on the outside, petals remaining convolutely tubular, thus never completely expanding, also united together below : *bracteoles* rotund-ovate, dentate more or less united below. *Calyx* constricted at the base. *Capsule* very small, ovate acuminate, red-coloured, 3-celled, cells 5-6 seeded ; velvet ashy-coloured as also the wool. Conf. with pp. 10-11.

This would appear to answer to the description of some of the Cottons of Eastern Bengal and Assam.

Variability of the Capsule and Seed in Gossypium. 35

It will be seen from the above notes on some of **Todaro's** species that they would appear to possess few, if any, characters that could be accepted as absolutely separating them. They differ mainly in the shape of the cap-

 Variability of the Capsule and Seed.

CLASSIFICATION.

sule, the number of cells and seeds, and in the minor characters of the flower. Seeing that **Gossypium** is solely cultivated for its wool, it is not to be wondered at that every possible form and condition of fruit should be met with. Some years ago considerable interest was awakened (mainly by **Dr. Forbes Watson**) in the nature of the plants obtained from 3-4-5-celled growths of the so-called **G. herbaceum** of India. In the Annual Reports of the Government Experimental Farms, in the Bombay Presidency, numerous returns are published on this subject, which would seem to break down completely all botanical arguments based on these characters. Thus in Hyderabad (Sind) various experiments were performed, both with American and Sindí cottons, the results proving that the former show a most pronounced tendency to produce 4-celled and the latter 3-celled capsules, regardless of whether the seed cultivated had been obtained from 3-4-5-celled pods.

AMERICAN SEED.

Kinds used as Seed.	Total number of Pods collected.	PERCENTAGE OF PODS OF EACH KIND OBTAINED.		
		3-celled.	4-celled.	5-celled.
3-celled	4,633	6·6	78·4	14·9
4-celled	6,354	2·7	80·3	17·0
5-celled	4,026	2·0	69·0	29·0
Not selected	5,814	5·1	76·7	18·2

SINDI SEED.

Kinds used as Seed.	Total number of Pods collected.	PERCENTAGE OF PODS OF EACH KIND OBTAINED.		
		3-celled.	4-celled.	5-celled.
3-celled	17,893	66·8	33·1	0·1
4-celled	23,146	66·2	33·6	0·2
5-celled	3,280	34·1	63·5	2·4
Not selected	21,251	57·2	42·3	0·5

Similar results were obtained by experiments continued in the following year (*Annual Report, 1887-88, pp. 5-7*), and commenting on the experiments with the American seed it is stated, "The greatest percentage of pods are in every case pods with 4-cells, the normal number in this class of cottons. The seed from 5-celled pods has, however, produced the greatest percentage of similar celled capsules." The remark regarding the Sindí cottons may also be reproduced :—" In this, as in the American, the percentage of capsules having the larger number of cells is lower than last year, but the tendency seems to be in the same direction,—that is to say, the seeds from pods with the greatest number of cells give the highest proportion of similar celled pods." We are not told the particular growths of either American or Sindí cottons selected for these experiments, but the inference seems justifiable that all the former shewed a strong tendency to become 4-celled, while the latter became 3-celled. From these facts we may at least assume that these are the chief specific peculiarities of the fruit in the two great series of **Gossypium**, and learn the futility of making species out of races or hybrids of Indian Cotton according to the size, shape or divisions of the capsule.

(415)

DYE FROM GOSSYPIUM AND THE DYEING OF COTTON GOODS.

DYE. 36 (416)

The colouring matter seen in the wild cottons, and also in Nankín cotton, has been found, by **Schunck** (*Memoirs of the Literary and Philosophical Society of Manchester*), to be present, to a limited extent, in all forms of cotton. **Schunck's** paper (*l.c.*) 'On some Constituents of the Cotton Fibre,' investigated the substances contained in or attached to the framework of cellulose which constitutes the staple. He found that these substances were almost insoluble in water but soluble in hot alkaline lye. The action of concentrated aqueous solutions of caustic soda and potash had, however, been some time before observed by **M. Alkan** and turned to practical use by **Mr. John Mercer**, who patented a process now known as mercerising, by which the fibre was prepared for certain dye re-actions. Mercerising may be expressed as an extreme and powerful process of hydrolysis or bleaching, accompanied with structural changes in the fibre. Not only is cellulose separated from non-cellulose, but the shape or contour of the ultimate fibre-cells is altered. **Schunck**, however, confined himself to the examination of the substances separated in solution by alkaline lye, and which are afterwards precipitated by acids. The acid he employed was an excess of sulphuric, and the copious brown flocculent precipitate when thrown down from the alkaline solution was found to be 0·337 per cent. of the weight of '*Dhollera*' yarn employed. Similar yarn of Middling Orleans yielded 0·48 per cent. of precipitate.

On examining the material thus obtained, he found it to contain (1) Cotton wax: (2) Margaric acid; (3) A colouring matter easily soluble in alcohol: (4) A colouring matter sparingly soluble in alcohol: (5) Pectic acid: and (6) Albuminous matter.

Of these substances the pectic acid far exceeds the others in quantity, then follow the colouring matters. But it is well known that, on bleaching cotton, it sustains a loss of 5 per cent. of its weight, whereas the total weight of the substances removed by **Schunck** and afterwards precipitated by acid, amounted to only about ½ per cent. It is, therefore, assumed that the difference in loss on bleaching, was either removed by the water or not precipitated from the alkaline solution by the acids used by **Schunck**. It is known, for example, that certain derivatives of pectine are not precipitated by acids Hence it is not improbable, as contended by writers on this subject, that the substance that escaped observation in **Schunck's** experiments was parapectic acid, since that acid is readily soluble in pure water. A portion also of the pectic acid originally removed is assumed to be possibly converted, through the action of the alkali, into metapectic acid—a substance very soluble in water, and not precipitable by acids.

Conf. with p. 79.

These discoveries in the nature of the substances removed from cotton by bleaching and mercerising, have a direct bearing on many of the textile and dyeing industries. They are interesting also from a purely botanical consideration, since we learn that the colouring matter, in cotton, is a normal constituent, which yields to bleaching. The interest in Nankín or *khakí* cottons, as supplying a nature-dyed textile of a desired colour, was, therefore, in the light thrown on the subject through **Schunck's** discoveries, very probably mis-directed, since the colour must of necessity be removable by the ordinary methods of washing with soap or other alkalis.

Very little is known regarding a tinctorial property possessed by the species of **Gossypium.** Recently, through the kindness of **Mr. Beheram Shah**, Forest Officer, Palanpur State, Bombay, we have had the pleasure of receiving a piece of cotton cloth said to have been dyed with a decoction of the ROOTS of **G. herbaceum** mordanted with alum. A soft delicate colour has been thereby produced which, if permanent, might be turned to

Roots. 37 (417)

DYE. much account. It is remarkable that no other Indian writer appears to have observed the Natives of this country using the root; few even mention Flowers. 38 the FLOWERS as being so used, which, like those of **Thespesia populnea**, afford a yellow dye. **Mr. Liotard**, in his memorandum on the Dyes of India, incidentally mentions the flowers as being employed for dyeing purposes in the Mainpuri district.

Mr. T. Wardle, in his report on the "*Dyes of India*," states that the Collectors of Jacobabad and of Hyderabad (Sind) had furnished him with samples of the flowers. **Mr. Wardle** obtained from these, according to the mordants used, various shades, which ranged from bright yellow, greenish yellow, and fawn colour, to orange, reddish, and brownish yellow. Commenting on this subject **Mr. Wardle** wrote, "Cotton flowers are very rich in yellow colouring matter. They will give bright yellow colours, and, as a dye-stuff, have much the same properties as **Butea** flowers."

It may be here added that the above facts of a knowledge of the dye-property of cotton roots and flowers, being practically confined to Sind and Palanpur, might be accepted as denoting a more ancient acquaintance with the plant. If this inference be justifiable, it might perhaps be viewed as confirming the accuracy of the reports regarding a wild **Gossypium** in Sind, in certain parts of the Panjáb, Rajputana, and the hotter localities of the North-West Provinces.

Blue Dye. 39 In **Gmelin's** *Handbook* the subject of a BLUE DYE obtained from
Seeds. 40 cotton SEED is discussed. **Dr. Warden**, in a note on this subject, says, "Cotton seeds contain a principle which, by the action of acids, yields a blue colouring matter—cotton seed blue. This blue is not altered by reducing agents, but is at once decomposed by oxidising ones. It is insoluble in water. An alcoholic solution gives a fine blue colour to fabrics either mordanted or unmordanted with alum; the colour, however, fades rapidly."

The subject of DYEING and CALICO-PRINTING is too extensive to be dealt with collectively in this article. The reader is, therefore, referred to technical works such as *Buck's Dyes of the North-West Provinces; McCann's Dyes and Tans of Bengal; Liotard's Memorandum on the Dyes of India; Crooke's Handbook of Dyeing: Hummel, The Dyeing of Textile Fabrics, &c.* Also to the separate articles in this work on the more important dye materials such as INDIGO, SAFFLOWER, LAC, MADDER, MORINDA, &c., &c.

FIBRE. 41

FIBRE.

This will be treated of in detail under the section of this article designated COTTON. It may be as well, however, to give here a brief account of the microscopical and chemical features of the floss. The cotton staple consists of elongated simple cells, developed around the seed, and is not therefore a bundle of cells or vessels as in the case of Flax, Jute, and other commercial fibres. Under the microscope cotton is seen to be an elongated cell or fibre, ribbon-shaped, with corded edges, thickest at the base (the end of attachment to the seed), and continued upwards without material change for three-fourths of its length, after which it tapers to a point. Each fibre is twisted many times on itself and on transverse section it is seen to be a collapsed cylinder, the walls, compared with the calibre, being of a considerable thickness. This is the appearance of the perfect fibre, but there are often found many imperfectly developed or immature ones, which, though twisted like the perfect ones, are thin and brittle, and owing to the deficiency of cellulose do not manifest the double outline of margin. The peculiar twistings give strength to the fibre and facilitate the process of spinning, the natural twistings becoming interlocked. These twistings do not exist on the fibre before the dehiscence of the pod. On ripening nourishment from the seed is gradually discontinued, a vacuum is thus left in the interior, first, at the apex, afterwards along the length to the base. The atmospheric pressure occurring on these hollow tubes causes them to twist and collapse. FIBRE.

The mean length of the cells (popularly the staple) in Sea Island cotton is 1·65 inch; in Egyptian 1·50 inch; in Pernambuco 1·25 inch: in American 1·10 inch; in Indian 0·90 to 0·65 inches. Cotton floss consists of nearly pure cellulose, together with only 1 to 1⅓ per cent. of inorganic matter. The specific gravity of New Orleans is said to be 1·165, and Bourbon 1·110. Cotton is tasteless, devoid of smell, insoluble in water, alcohol, ether, fixed and volatile oils, or vegetable acids. It is soluble in strong alkaline solutions and decomposed by concentrated mineral acids. The action of nitric acid, &c., is referred to in the section under Medicine in the formation of Gun-cotton and Collodion. The colouring matter present in the fibre has already been dealt with and need not be again discussed in this place (*Conf.* with *Dye, p. 32*). Cotton exhibits a strong affinity for alumina, hence the use of alum as a mordant in dyeing. Iron stains it a yellow colour which, unless recent, cannot be removed by alkalis or soap. Oxide of tin, which has an affinity for the staple, is frequently used as a mordant in dyeing the fibre. Nitric acid and heat decompose cotton wool and form oxalic acid. Sulphuric acid chars it, chlorine bleaches it. It is extremely combustible, burning with a clear bright flame.

The STEM yields a good FIBRE, which may be separated by retting. Stem. 42
Several writers have alluded to this subject and recommended its utilisation, but apparently the people of India are not aware of this fact, since no Fibre. 43
mention is made of their putting it to any useful purpose. Most malvaceous plants are rich in fibre, and doubtless the cotton stems are not in this respect different from those of **Hibiscus, Urena, Pavonia,** &c. It is probable that in the cultivation for the floss a branched stem is produced which might not be conveniently treated for its fibre. If it could be, the supply would naturally be very great and afford a new source of income to the people.

COTTON-SEED, OIL, AND CAKE. 44

References.—*Hawke, Report on Oils of S. India, 39; Cooke, Oils and Oil-seeds of India, 45; Useful Pl. Bomb., 215; Simmonds, Tropical Agriculture, 404-405; Jour. Soc. Arts, 27th July 1883; Indian Agriculturist, Oct. 12th, & Nov. 2nd, 1889; Edinb. Quart. Jour. Agri., 1854; Spons, Encyclopædia; Balfour, Cyclopædia, &c., &c.*

After the floss has been removed, by the process technically known as ginning, the SEEDS are utilised either as cattle-food or as a source of oil. SEEDS. 45
Prior to the American war the seeds were viewed as useless, the difficulty of preparing them as a manure being such as to prevent, to a large extent, their utilisation even as a fertilizer. A writer, in the Report for 1874, issued by the Commissioner of Agriculture, Washington, says: "The seed in its natural state, or when converted into meal, has greater manurial value than any other single substance." He supports this statement by giving an analysis of the ash of cotton-seed as follows:—

	Per cent.		Per cent.
Potash	35·440	Sulphuric acid . . .	3·222
Soda	0·810	Oxide of iron (or alumina) .	1·075
Magnesia	15·067	Chlorine	0·490
Lime	4·450	Carbonic acid . . .	3·465
Phosphoric acid	30·016	Sand and charcoal . . .	5·965
			100

In spite of being so valuable a substance immense quantities were in America annually destroyed as useless; to some extent this is the

SEED & OIL.

case in India to-day. Shortly after the war of secession attention was earnestly directed, however, towards the utilisation of the immense annual production of cotton-seed. The OIL was prepared and found to be of superior quality, suitable for many industrial—and even, when purified, for culinary—purposes. The OIL-CAKE was also discovered to be one of the best muscle-forming articles of cattle diet, and as a manure it was seen to be practically unsurpassed. These facts were no sooner fully understood than a large and important trade sprang into existence, which has greatly enhanced the value of American cotton cultivation. In the Report for 1888, the Commissioner of Agriculture of the United States publishes a long and interesting report on the subject of cotton-seed oil as an adulterant for lard. From that valuable paper the following facts may be abstracted. Cotton-seed oil being a liquid at ordinary temperatures, its specific gravity can be easily taken. For the purpose of comparison the rate of variation in the specific gravity of the oil can be determined, and its specific gravity at any given temperature calculated: or its specific gravity can be directly determined at 35°, 40° or 100° as may be desired. In the samples examined the specific gravities of the oils at 35° varied from 0·9132 to 0·9154—the mean of 19 samples having been 0·9142.

OIL. (426) 46

OIL-CAKE. (427) 47

The colour produced in cotton oil by sulphuric and nitric acids is a characteristic mark of the greatest value. This varies from a deep reddish brown to almost black. The refractive index of cotton oil is distinctly higher than that of lard. The variation in the index of refraction is inversely as the temperature. For a temperature of 25° the mean refractive index of the samples examined was 1·4674.

The rise of temperature which cotton-oil sustains with sulphuric acid is a very prominent diagnostic sign. In the samples examined the lowest increment of temperature noted was 80·4° and the highest 90·2°. Cotton oil also possesses, in a much higher degree than lard, the property of absorbing iodine, the mean of results obtained having been 109·2. But an even more characteristic property of cotton oil is its reduction of silver to the metallic state.

Uses. (428) 48

USES OF COTTON-SEED OIL AND CAKE.—"The crude oil has 28 to 30 times the viscosity of water. At 20°C. it has a specific gravity of 0·9283 and at 15°C. of 0·9306. It congeals at 1·9°C. to 2·7°C. In taste and odour it resembles linseed oil, and in other properties it is intermediate between a drying and a non-drying oil. The refined oil has a specific gravity of 0·9264 at 15°C. and congeals at ·0°C. to 1·1°C." (*Pharmacog. Ind.*). "Refined cotton oil has a pleasant taste, is almost odourless, and possesses a faint yellow colour. It resembles olive oil so closely that it is naturally an excellent substitute. But it has a slight drying quality which renders it unfit for lubricating machinery. It will never, therefore, take the place of sweet oil for that purpose." (*Rept., Comm. Agri., U. S. A., l.c.*). "Chemically cotton-seed oil consists of palmatin and olein, and its ultimate percentage composition is Carbon 76·40, Hydrogen 11·40, Oxygen 12·20. (*Braunt.*) Cotton-seed oil is not suitable for pharmaceutical purposes" (*Pharmacog. Ind.*). Cotton-seed oil "is extensively mixed with olive-oil and often replaces it altogether; as a preventive measure, the Italian Government has levied a heavy duty on its importation. Large quantities of it are consumed by soap-makers, in combination with other oils and fats. For lubricating purposes, some manufacturers prepare a 'winter oil' from it, which does not thicken in cold weather, by precipitating and removing the stearine; but its genuineness must limit its application. Its value in New York is about 18—20*d.* a gallon. The cake remaining after the expression of the oil is invaluable as a cattle-food and as a fertiliser, and is an important article of commerce." (*Spons' Encycl.*).

SEED & OIL.

"The prejudice against the use of the oil for domestic purposes is fast disappearing, and refiners now sell it largely for cooking purposes. In the New England States, the cotton-seed meal is used as cattle-food, the cattle being penned and the manure utilised as a fertiliser; while in the south it is in most instances applied direct to the ground, or in combination with other fertilisers." (*Jour. Soc. Arts, l.c.*). Many writers deal with the value of cotton-seed cake as an article of diet for cattle, especially milch-cows. Cattle do not, it is stated, seem to take to it willingly at first, but eventually get to like it and then thrive remarkably. About 27 imperial stones of cake are got from 4 cwts. of seed. Long ago **Ainslie** wrote of South India that the seeds were fattening food for cattle, so that for many generations this appears to have been the chief, if not the only, use of the immense quantity of seed annually produced. Mr. **H. J. S. Cotton**, however, in a report published in 1876 on the Cotton Cultivation and Trade of the interior of Bengal, wrote:—"There is no trade in cotton-seed, either export or import, the cotton-grower from year to year preserving only sufficient seed of his own plant for future sowing." It is thus highly probable that in many parts of India the value of the seed, even as a cattle-food, is and has been neglected. The value of the cake is however fully understood here and there all over Asia. In the report of the Yarkand Mission, for example, it is stated that the seeds are pressed for oil and the cake given to fatten cattle.

Oil Expression. 49 (429)

EXPRESSION OF OIL.—"The seed coming from the ginning operation still has some fibre adhering to it, and has a tendency to accumulate in masses. These are thrown into a machine containing a screw-knife revolving in a trough, which divides the materials into particles fit for the screening operation. This is conducted, first, in a sieve with meshes that allow the sand and dirt to pass, while retaining the seed, and then in one through which the seed can escape, but not husks and coarse foreign matters. The cleaned seed is next passed through a special gin for removing all remaining fibre (useful for paper-making and other purposes), and finally through a hulling-machine or decorticator, consisting of fixed and revolving knives set so close as to sever the seeds. The huller made by **Mr. D. Kalmweiler**, 120, Center Street, New York, was favourably noticed at the Centennial Exhibition. Thus treated the seeds are taken through one or more separators, which pass the kernels but retain the shells. The kernels are pressed into cakes between iron rolls, and are then placed in steam-jacketed iron tanks, 4 feet wide and 15 inches deep, where by constant stirring, and the action of dry heat obtained from injecting steam at 35℔ a square inch into the jacket, the oil is liberated from the cells in the course of about five minutes. The heated mass is then filled into sacks and subjected to repeated hydraulic pressure, till most of the oil is extracted. By this American plan, the yield from 1,000℔ of seed averages 490℔ of husks, 10℔ of cotton, 365℔ of cake, and 135℔ of oil." (*Spons' Encycl.*). There are other American methods, also specially patented processes in England, but the above conveys the main ideas, *viz.*, great pressure on a cake of cotton meal prepared by baking, so to speak, the decorticated seed. According to the reports published in America the usual yield appears to be 35 gallons of crude oil (or 315℔), 22℔ of cotton floss obtained from the process of cleaning the seeds, and 750℔ of cake which does not contain more than 10 per cent. of oil. The husk, at the same time, furnishes more than sufficient fuel for the machinery. In *Spons' Encyclopædia* the following passage occurs regarding the yield:—"The yield of oil varies with the season and the locality in which the seed is produced. In the United States, it is reckoned that for each 1℔ of ginned cotton there are 3℔ of seed, or a total approaching 4,000 million ℔, half of which only

GOSSYPIUM herbaceum. **The Cotton-Seed, Oil, and Cake.**

SEED & OIL. is required for sowing. One authority estimates that 100℔ of seed give 2 gallons of oil, 48℔ of oil-cake, and 6℔ refuse fit for soap-making; another says that one ton of American seed gives 20 gallons of oil; a third, that the oil-product is 37 per cent. of the weight of the kernels; a fourth that 2 gallons of oil and 96℔ of cake are afforded by 1 cwt. of seeds, a fifth that 1 ton of cotton material yields at the rate of some 35 gallons of oil."

Indian Uses. 50 (430) INDIAN USES.—In India cotton-seed is chiefly used as a cattle-food, not as an oil-yielder. Indeed, in this case, as in many other features of cotton cultivation and utilisation, the natives of India do not manifest that intimate knowledge which characterises their acquaintance with most of the products of their country. While they grow many plants purely as sources of oil, for food, and lighting purposes, they seem to be practically ignorant of the fact that cotton-seed affords one of the most wholesome of oils. In *Balfour's Cyclopædia of India* it is stated that "when obtained by pressure its colour, owing to the presence of a resinous substance, is of a very dark red, and 10 to 15 per cent. is lost in bleaching it. When prepared by steaming the seeds, and collecting the oil by skimming it from the surface of the water, it has a bland, light-coloured appearance. (*Faulkner.*)" In *Cooke's Oils and Oil-seeds in India* it is stated that cotton-seed is "in some localities considered a better food for the working bullock than grain. It produces, under the action of the native oil mill, 25 per cent. of a good oil, which, by being purified, might grow into an extensive article of commerce. **Mr. Bingham** does not think it would answer to export the seed, as, owing to the fibre adhering to it, and perhaps other causes, it is very liable to heat and deteriorate in bulk. The oil is said to be a very useful one, more so than most others for machinery purposes." (*Agri.-Hort. Soc. Ind. Jour., XII., 343*). The above passage is practically a reprint of **Mr. Bingham's** remarks, in his account of the oil-seeds of the Sasseram district (published in 1862), and it has been repeated in most works on this subject without being in any way amplified or confirmed. The singular silence, on the uses of cotton-seed, in the Gazetteers is practically a proof of the total neglect that prevails throughout India. The *kapas* or cotton with its seed is sold to the dealers who clean it and dispose of the seed as they please, returning a certain amount to the cultivators to be sown in the succeeding season. Were it possible to induce the cultivators to clean their own cotton, the prospects of a large cotton-seed trade might be entertained. Whatever profit the dealers obtain from the seeds seems to be kept a secret, and all reports on cotton terminate with the sale on the part of the cultivator to the dealer. On this subject **Mr. J. E. O'Conor** (in his Reviews of Trade) has on more occasions than one urged the importance of placing the traffic on a more sound basis. In 1886-87 he wrote: "It is curious that cotton-seed, of which thousands of tons are crushed in Europe every year for its oil, which is used as a substitute for olive-oil, is as yet almost entirely neglected in India. Attention has been directed to the possibility of a large trade in this seed more than once, but as yet without effect, and the seed still goes mostly to waste." Again in 1888-89 he wrote: "Cotton-seed is specified separately in the returns for the first time this year. The exports amounted to 135,581 cwt., worth over three lakhs of rupees. The quantity is probably not a tenth of that available for export, and it may be observed here that the United Kingdom imported in 1888 more than five million cwts. of cotton-seed (valued at £1,657,432), of which only 81,300 cwt. came from India, the bulk being imported from Egypt and Brazil. India should take a large share in this trade." But is the Indian cultivator in no way to participate, in this new source of revenue, from the Indian cotton crop?

SEED & OIL.

It is admitted by many American writers that the income derived from the seed has made all the difference between profitable and unprofitable cotton cultivation. If the dealers in the future are to continue to pay the cultivators the same average amount for cotton with its seed, as they did in the past (when the seed was worthless), an undue profit will fall into the hands that least deserve it. In this respect it would seem clearly the duty of those in authority to keep before the eyes of the cultivators the progress made in the cotton-seed traffic. The best incentive to cleaning their own cotton, preserving their own seed, and improving the cultivation generally, would doubtless be secured through personal gain.

Some idea of the magnitude of this possible future trade may be learned from the fact that since India produced last year a little over 9 million cwt. of cleaned cotton, that amount must have been obtained from 27 million cwt. of seed. Allowing half to be required for home consumption and seed for next crop, over 600,000 tons should have been available for export, or twice the quantity imported by the United Kingdom from all sources.

SPECIAL OPINIONS.—§ "The seeds are given to cattle to increase and improve their supply of milk. The seeds furnish an oil which is used for lighting purposes." (*Narain Misser, Kathe Bazar Dispensary, Hoshangabad, Central Provinces*). "The seeds are largely used in various parts of India in feeding cows, to increase the quantity and improve the quality of their milk." (*Assistant Surgeon Nunda Lal Ghose, Bankipur*).

MEDICINE.

MEDICINE.

Herbaceous part. 51

According to the United States Dispensatory "the HERBACEOUS PART of this plant contains much mucilage, and has been used as a demulcent. The seeds yield by expression a FIXED OIL of the drying kind, which is employed for making soap and other purposes. The BARK of the root has been supposed to possess medicinal virtues, and is now recognised in the *U. S. Pharmacopœia.*" "Cotton SEEDS have been employed in the Southern States with great asserted success in the treatment of intermittents." "A pint of the seeds is boiled in a quart of water to a pint, and a tea-cupful of the decoction is given to the patient in bed, an hour or two before the expected return of the chill." Ainslie says that the seeds of **G. barbadense,** in the form of an emulsion, are given in dysentery and are supposed to be pectoral. They yield by expression an oil which is much used, and is considered to have, in a peculiar manner, the virtues, when externally applied, of clearing the skin of spots and freckles. A tea made of the young LEAVES is recommended in lax habits, and for preparing a vapour bath for the anus, in cases of tenesmus. Of the root of **G. herbaceum,** Ainslie remarks that he suspects it possesses little real virtue. The Tamil doctors are, however, he adds, in the habit of prescribing a decoction of it in strangury and gravel. Dr. Dymock (*Pharm. Ind., I., 225*) says: "Eastern physicians consider all parts of the cotton plant to be hot and moist; a syrup of the FLOWERS is prescribed in hypochondriasis on account of its stimulating and exhilarant effect; a poultice of them is applied to burns and scalds. Cotton CLOTH or mixed fabrics of cotton with wool or silk are recommended as the most healthy for wear. Burnt COTTON is applied to sores and wounds to promote healthy granulation; dropsical or paralysed limbs are wrapped in cotton after the application of ginger or zedoary *lép* (plaster); pounded cotton seed, mixed with ginger and water, is applied in orchitis. Cotton is also used as a moxa, and the seeds as a laxative, expectorant, and aphrodisiac. The juice of the leaves is considered a good remedy in dysentery, and the leaves with oil are applied as a plaster to gouty joints; a hip bath of the young leaves and roots is recom-

Fixed Oil. 52

Bark. 53

Seeds. 54

Leaves. 55

Flowers. 56

Cloth. 57

Burnt Cotton. 58

MEDICINE. mended in uterine colic." In the *Gazetteer of Bulandshahr* cotton-seeds,
binaula, are included in a list of the more commonly employed indigenous
drugs: the properties assigned to these seeds are not however stated. Atten-
Root-bark. tion appears to have been first drawn to the emmenagogue property of the
59 cotton ROOT-BARK through the observation made by **Dr. Bouchelle**, of Mis-
sissippi, that it was used by Negro females to produce abortion. On this
(439) subject **Dr. Dymock** writes,—"There appears to be little doubt that it acts
like ergot upon the uterus, and is useful in dysmenorrhœa and suppression
of the menses when produced by cold, a decoction of four ounces of the bark
in two pints of water boiled down to one pint, may be used in doses of two
ounces every 20 to 30 minutes, or the fluid extract may be prescribed in
doses of from 30 to 60 minims." The above passage is almost word for
word as originally published by the discoverer, so that the knowledge of
this property is apparently confined to America. This fact may have a
historic significance, for although the people of India seem to have had
a peculiar aptitude in discovering emmenagogues and aphrodisiacs, they
failed to discover any such properties in this case, and the cultivation of cotton might therefore be inferred to be less ancient than is generally supposed.

Preparation of Cotton. A special preparation of COTTON is made, which removes its greasiness
60 and renders it absorbent. This absorbent cotton may be further medicated
with carbolic acid, salicylic acid, or boracic acid, and in this form used as
(440) an application or dressing to wounds. Its effects are chiefly due to the
absorption of effused fluids and the protection it affords from the air. It
is similarly used to arrest hœmorrhage, its action being mechanical, but for
Styptic. this purpose the special preparation known as 'STYPTIC COTTON' is gene-
(441) 61 rally employed. GUN-COTTON or PYROXYLIN is prepared by dipping the
Gun-cotton. floss in a mixture of equal parts of nitric and sulphuric acids, and by after-
62 wards washing freely with water and drying. This explosive substance is
(442) readily soluble in a mixture of ether and rectified spirit, forming the officinal
COLLODIUM or COLLODION.

CHEMISTRY. CHEMISTRY OF GOSSYPIUM BARK.—In the United States Dispensatory,
63 taken for the most part from the American Journal of Pharmacy, will be found
the chief facts regarding the chemistry of this substance. These opinions
(443) are briefly reviewed in the *Pharmacographia Indica* as follows:—"The
bark contains *starch*, and when fresh, according to **W. A. Taylor** (1876), a
chromogen, which dissolves in alcohol with a fresh yellow, gradually changing to a bright brownish red. The same change takes place on keeping
the bark for some time, when it yields a red tincture with alcohol. This
substance was examined by **Professor Wayne** (1872) and **W. C. Staehle**
(1878), who regard it as of a resinous nature. The latter obtained about
8 per cent. of this substance, which is soluble in 14 parts of alcohol, 15 of
chloroform, 23 of ether, and 122 of benzol; also in alkalis, from which
solution it is again precipitated by acids. The potassa solution diluted
with water is of a sage green tint. Glucose was likewise observed, and
the aqueous solution of the alcoholic extract contained a principle which
gave a purplish-black precipitate with ferric chloride. **C. C. Drueding**
(1877) obtained also a yellow resin soluble in petroleum benzine, a fixed
oil, a little tannin, and 6 per cent. of ash."

SPECIAL OPINIONS—§ "Cotton-wool is a filterer of atmospheric germs
preventing their access to wounds, cavities of the nose, ear, &c., and for
this purpose is very useful after operations, including those on the eye."
(*Brigade Surgeon G. A. Watson, Allahabad*). "The seeds largely increase
the flow of milk, especially in cows, for whom *bandla* (sic? *banauli*—ED.) is
one of the best articles of diet. When given to women the *bandla* is beaten
up in a mortar with a little water; the mass is then strained and the juice
mixed with rice and cooked into *khír*." (*Surgeon-Major C. W. Calthrop*,

Morar). "The seeds are considered to be a nervine tonic and are given for headaches and brain affections." (*Surgeon-Major Robb, Civil Surgeon, Ahmedabad*). "Best application in burns and blisters." (*Assistant Surgeon Shib Chunder Bhuttacharji, Chanda, Central Provinces*). "The seeds, deprived of their outer coat, powdered and given in milk (in doses of 2 drachms) are used as a tonic." (*1st Class Hospital Assistant Lal Mahomed, Main Dispensary, Hoshangabad, Central Provinces.*) "The young fruit is given to check dysentery." (*V. Ummegudien, Mettapollian, Madras.*) "The oil from the seeds is used as a liniment in rheumatic affections." (*Honorary Surgeon P. Kinsley, Chicacole, Ganjam, Madras.*) **MEDICINE.**

DOMESTIC AND SACRED.

The sacred uses of cotton have already been alluded to. Except among one or two aboriginal tribes the plant as such cannot be said to be sacred, but the thread prepared from the so-called *Nurma* cottons is that alone prescribed to be worn as the Brahminical string. The STEMS of the cotton bush are largely used for agricultural purposes, being made into baskets in Bombay and other parts of India. Throughout the Panjáb they are used in making the wattle sides of carts, being supposed better suited to such purposes than for basket-work. One of the most universal usages of the dry cotton stems is as fuel: this may be seen in all the cotton districts. **SACRED.** 64 **DOMESTIC. Stems.** 65

In *U. C. Dutt's Materia Medica of the Hindus* (under the Diamond) occurs the following curious passage:—"Diamond is purified for medical purposes by being enclosed within a lemon and boiled in the juice of the leaves of **Sesbania grandiflora** (*vaka*). It is reduced to a powder in the following manner: A piece of the root of a Cotton plant is beaten to a paste with the juice of some betel leaves. Both these vegetables should not be less than three years old. Diamond is enclosed within this paste and roasted in a pit of fire. The process is repeated seven times, when the stone is easily reduced to a fine powder."

IV. G. Stocksii, *Mast.; Fl. Br. Ind., I., 346.* 66

References.—*Tod., Cult. dei Cot., 132; Dymock, Warden, and Hooper, Pharm. Ind., 224-228; Murray, Pl. and Drugs, Sind, 64.*

DIAGNOSTIC CHARACTERS.—*Leaves* glabrous, palmately 3-5-lobed, segments roundish or oblong obtuse. *Bracteoles* deeply laciniate, ssgments linear lanceolate. *Flowers* small, yellow. *Capsule* ovoid. *Seeds* devoid of felted down, covered with yellow adherent cotton not easily separable.

Habitat.—A shrub found on the lime-stone rocks of the coasts of Sind.

Dr. Masters and, following him, **Todaro** and **DeCandolle**, regard this as the wild condition of **G. herbaceum.** It may, however, be pointed out that the glabrous leaves, deeply laciniated bracteoles, and naked seeds ally it more strongly to some of the forms of **G. barbadense.** The interest in this observation lies in the fact that the most recently discovered wild species in Africa is viewed by **Masters** as very nearly allied to **G. barbadense.** Doubt may thus be admissible as to the desirability of a hard-and-fast isolation of the forms of **Gossypium** into Old and New World types, represented by **G. herbaceum** in the former and **G. barbadense** in the latter. **Stocks** himself, while describing his wild plant, in the *Gazetteer of Sind,* calls it **G. obtusifolium,** *Roxb.*; and thus did not view it as distinct.

V. Hybrids and Hybridisation.

If the character of velvety seeds (below the floss) be accepted as an Asiatic peculiarity, the absence of velvet would have to be added to the collective characters given above for the indigenous American species of **Gossypium.** And if this view receive more support than it has hitherto **HYBRIDS.** 67

HYBRIDS.

obtained, then the forms of cotton with broad leaves, laciniated bracteoles and velvety seeds (such as **G. hirsutum**), would have to be regarded as hybrids. As opposed, however, to this explanation of the velvety-seeded American cottons, there are one or two considerations of importance:—Certain writers affirm that it is impossible to hybridise the American naked seeded with the Asiatic velvety forms, or, indeed, the Asiatic with any American. It is further contended that the American velvety seeded cottons were known and even held to be indigenous Western species prior to the recorded introduction of Asiatic cottons in America, which probably took place towards the end of the last century. So again **G. hirsutum** and other velvety-seeded American forms show little tendency to revert to Asiatic types, though in India they degenerate to almost valueless states. But, on the other hand, the ease, in certain cases, with which the velvet is inherited or lost under simple cultivation, would point to its being an acquired and not an originally specific peculiarity, at least in the debateable species, the more so since, in the pure types of naked-seeded American or velvety Asiatic, the velvet is neither acquired nor lost.

HYBRIDISATION. 68 (448)

It would thus seem desirable to have the question of hybridisation more carefully investigated. Not having the opportunity of performing such experiments we are compelled to content ourselves with a review of what has been written. That the Indian cottons have been freely hybridised, naturally and artificially, goes without saying. That the same is the case with the American is equally true: the prized New World cottons are in fact mostly sports or hybrids. The debatable point is, can a hybrid be produced, or does such exist, between Indian and American pure cottons? We are bound to admit that there seems, even in the face of all that has been written to the contrary, no good reason for supposing that American and Asiatic species cannot be hybridised. But it may be as well to give here the opinions of writers on the subject. **Major Trevor Clarke** (*Proceedings, Madras Agri-Hort. Soc., June 1871*), in a letter to **Mr. H. Rivett-Carnac**, for example, made a direct statement opposed to the possibility of such hybrids:—"I strongly doubt," he wrote, "the possibility of any cross between exotic and Indian sorts, and fear you will be disappointed in this respect when your plants come to perfection. The Persian, which is an American exotic or occidental sort, will cross with Sea Island, Orleans, or any other cotton of the Western World, but not with the true Indian plant."

Mr. W. Walton (*in Cotton, its Culture, Trade, and Manufacture*, Bombay, 1880) devotes a chapter to the subject of Hybridisation, He seems to have been influenced by so liberal an admiration of **Dr. Wight's** botanical merits as to accept unreservedly the dictum laid down by that author that hybridisation of *any two species of cotton* "would be contrary to the laws of nature." To do so, wrote **Dr. Wight**, "it would be necessary to open the flowers before natural blooming, to cut out their stamens before shedding their pollen, and then to apply the pollen of the intended male parent to the virgin stigma; if its own pollen had once been applied, the other will not take effect." As opposed to this emphatic statement it need only be said that in India at least the difficulty is to find a cultivated cotton that is not a hybrid. Many authors describe, in great detail, their various methods of hybridisation, so that, in addition to the fact of most of the favourite growths of cotton being botanically hybrids, no such difficulty exists as stated by **Wight**. The failures which **Mr. Walton** cites, in support of his concurrence with **Wight**, were failures to naturally fertilise New Orleans with "native Indian cotton." But the difficulty arises, what is native Indian cotton, and even still more so, what is New Orleans? We are, for example, strongly disposed to think the latter a hybrid, as

HYBRIDISATION.

most of the Indian cottons certainly are; hence **Dr. Forbes, Mr. A. Elphinston,** and other writers on this subject very probably tried to produce a cross between two hybrids—a matter which might have been anticipated to prove difficult. We are therefore unable to support **Mr. Walton's** contention, that these failures in one line of experiment prove the impossibility of any hybridisation. On the contrary, we believe that few genera are more easily hybridised artificially than, or have been so extensively hybridised in nature as, the species of **Gossypium.** This idea is borne out by **Todaro's** experiments in the hybridisation of cottons. That careful observer not only crossed many species artificially, but he gives it as his experience that it is almost impossible to check the natural hybridisation of Asiatic cottons. For example, he observed that **G. Wightianum** may be crossed with **G. neglectum** or with **G. herbaceum,** *var.* **microcarpum,** and that the hybrids take intermediate forms which connect the ancestral types. By continued cultivation he found that these hybrids were all fertile, but that they gradually approached their specific types if grown near each other, though preserved if they were kept apart. On the other hand, he noted that **G. herbaceum,** *var.* **microcarpum,** fertilised **G. Wightianum** and **G. neglectum,** but was never fertilised by them. By continued cultivation of **G. herbaceum,** *var.* **microcarpum,** on sterile soils, **Todaro** noted that that form was far more liable to change its characters than was **G. Wightianum** when similarly treated. He then adds that in these hybrids it is nearly always **G. Wightianum** that fertilises **G. neglectum,** the reversion going to **G. neglectum,** rarely to **G. Wightianum.** These are facts of great importance, and doubtless there are many others of a similar nature which should imperatively be understood, before good results can be obtained in hybridisation. It would appear to be essentially necessary, for example, first of all to thoroughly establish the characters of the species and to possess stock of the purest strain. Having done so, experiments might next be conducted with the object of discovering the direction of successful and of injurious hybridisation, so as to be able to foster the one and avoid or check the other. Whether the hairy-seeded American cottons are species or only hybrids with an Asiatic strain we are not prepared to state emphatically at present, but we repeat that nothing has as yet transpired that proves a cross impossible between American and Asiatic cottons.

In the Annual Reports of the Experimental Farms in the Bombay Presidency numerous reports occur on the subject of Hybrids and Hybridisation. An interesting paper on this subject will also be found in the Report for 1872 (North-West Provinces and Oudh) from the pen of **Mr. J. Simpson,** Superintendent of the Cotton Farm at Bulandshahr. In the Journals of the Agri-Horticultural Society of India numerous papers also appear on this subject. Thus, for example, *Volume I. (Old Series), 183-193,* **Dr. J. V. Thompson** of Sydney describes his process of hybridisation in detail, and gives particulars of his favourite hybrid between Georgian and Maltese brown cottons, to which he gave the name of *Georgia tinta.* This was probably a cross between **G. barbadense** and **G. herbaceum,** or possibly **G. religiosum,** *Roxb.* In *Volume VI. (Old Series), 149-153,* **Mr. H. Hamilton Bell** discusses certain Indian hybrids. In *Volume I. (New Series), Proceedings 1867, x.,* **Major Trevor Clarke** furnishes particulars of his having been working hard at the subject of getting the arboreous sorts *to seed and cross.* Speaking of the hybrids he *had* produced, he wrote, "Many were mere scientific experiments, some were the produce of bad sorts, some no hybrids at all, and so on." In the same volume (*Proceedings for 1869*), **Mr. S. H. Robinson** furnishes the Agri-Horticultural Society with cotton grown at Goosree from seed of a hybrid cotton produced

HYBRIDISATION. by Major Trevor Clarke, a cross between Sea Island and Bourbon (*Conf.* with the remarks regarding **Bamia Cotton**, p. 18 and p. 19.)

It is perhaps needless to quote other passages in support of the contention that hybridisation is possible, or that many hybrids exist. Suffice it to add that we are strongly disposed to regard all American velvety-seeded cottons as hybrids between the American and Asiatic series of forms. When, or how these were produced is quite another matter, but that in India they should show no sign to revert to the **G. herbaceum** series establishes one peculiarity of their hybridisation, but does not disprove their origin as here suggested. There are many examples of hybrids produced in a particular manner, which invariably, in retrogression, produce one and not the other strain. To arrive at a more precise conception of these supposed American-produced hybrids, it would be necessary to know whether, in the home of their birth, they show a tendency to revert to the **G. barbadense** series of characters. Their behaviour in India is of less consideration, since all American cottons degenerate rapidly in this country and often assume perfectly new, though inferior, forms. The acclimatised New Orleans cotton of India has lost, for example, many of its original properties, but is as distinct as ever from either of its suggested ancestors.

COTTON.

References.—*J. H. Van Linschoten, Voyage to the East Indies (1598), I., 55, 129; Dunlop, the Cotton Trade, 1862; Arnold, History of the Cotton Famine, 1865; Watts, the Cotton Supply Association, 1871; Annual Reports of the Cotton Supply Association; Donnell, History of Cotton, 1872; Royle, Culture and Commerce of Cotton in India, 1851; Prod. Res., 12, 42, 67, 75, 78, 84 214, 221; Aytoun, Origin and Distrib. of Cotton Soils of India; Forbes Watson, Rep. on Cotton Gins, and Cleaning and Quality of Indian Cotton; Rivett-Carnac, Minute on Cultivation of Cotton; Medlicott, Cotton Handbook for Bengal, 1862; Moore, East Indian Cotton, its Adulterations, and Acts passed to suppress the same, 1872; Schöfield, Note on Ind. Cot., 1888; Reports of Proc. of East Ind. Co., on Cot., 1836; Numerous Official Repts., and Cor., Rev. & Agri. Dept., Govt. of India; Parly. Papers:—On Cotton exported from India, 1847; On Measures taken to Improve Cotton Cultivation in India, 1857; Cotton Cultivation in India and Prospect of Supplies to England, 1863; The Cotton Supply Reporter; Ure, Cotton Manufactures, 1861.*

HISTORY.

HISTORY. 69 (449)

Perhaps no more remarkable example of a sudden development exists in the whole history of Economic Products, than in the case of cotton. The enormous importance, at the present day, of the fibre in the agriculture, trade, manufactures, and daily wants of the people of nearly the whole world, render it difficult to believe that only a few hundred years ago cotton and its products were practically unknown to the civilised nations of the West.

But it is perhaps even more singular that a fibre which has for centuries been the staple article of clothing in India, and elsewhere in the East, should scarcely find a place in the early classical literature of these countries. Nearly all the beautiful and the useful plants of India, if not of Asia, have their properties extolled by the Sanskrit poets, and indeed are frequently dedicated to the gods, but cotton—the plant above all others which might be expected to have formed a theme for nature worship—is passed by, with only the most incidental allusions to its properties. Indeed, it is only in modern works that the fibre is unmistakeably referred to. Evidences exist, however, amongst certain of the aboriginal tribes of Ind a of what is perhaps a much more ancient knowledge of the plant. For example, the Khonds grow a cotton-bush in the place selected for a new

HISTORY.

settlement, and in after years the village plant is tended as sacred and carefully watered. This custom is probably of very ancient origin, and may denote a superstitious regard for the plant, probably derived from the knowledge and appreciation of its valuable properties.

Cotton in Sanskrit Literature.

The Sanskrit word translated "cotton" is perhaps first mentioned in the Institutes of Manu, where, in *Book II., No. 44,* it is stated that, "the sacrificial thread of a Brahman must be made of cotton, so as to be put on over his head in three strings." The word used in that passage and translated cotton is कार्पासी (*kárpásí*) from which has been derived, according to most writers, the modern Hindustani *kapás,* and even the Hebrew *karpás* (the "green hangings" of the Book of Esther). The frequent confusion in ancient classical terms for textile fibres, which will be further commented on below, has led many to believe that the *kárpásí* of the Institutes of Manu may not necessarily have been cotton. The probability, however, is that it really was so. Not only does the term in general use for cotton at the present day evidently owe its origin to that word, but the brahminical threads are, up to the present date, made almost, if not quite, exclusively from the fibre of **G. arboreum**, a plant, in many parts, considered the exclusive property of the Brahmans, and generally planted near temples, or the abodes of *fakírs.*

Be this as it may, it is certain that at the time of the Institutes of Manu spinning and weaving were not only known, but the art of starching or weighting a textile was also practised, and the admissible amount prescribed. In *Book VIII., No. 236,* it is enjoined that for "stealing thread, raw cotton, materials for making spirituous liquors, &c.," the fine should be thrice the value of the article stolen. Again, at *No. 397,* "Let a weaver, who has received ten *palas* of cotton thread, give them back increased to eleven by the rice water, and the like used in weaving; he who does otherwise shall pay a fine of twelve panas." A still earlier notice of the process of starching probably occurs in the *Rig Veda, Hymn 105, v. 8,* where the following is to be found: "Cares consume me Satakratu, although thy worshipper, as a rat gnaws a weaver's thread." Commentators have explained that the rat's temptation was probably the starch, in which the threads had been steeped to improve their tenacity and strengthen the fabric.

The picture thus raised of the character of the textile industries of India, some two thousand years ago, meets its parallel only in the injunctions laid down for domestic life. *Book VIII., No. 396, of the Institutes of Manu* deals with the regulations concerning washermen. "Let a washerman wash the clothes of his employers by little and little, or piece by piece, and not hastily, on a smooth board of *Sálmali*-wood: let him never mix the clothes of one person with the clothes of another, nor suffer any but the owner to wear them." (For *Sálmali*-wood *cf.* with **Bombax** and **Eriodendron.**) We are thus led to believe that not only were spinning and weaving fully understood, but that a civilisation existed in that early age, in which the social habits of the people differed little from those of the present day. Accepting as correct the interpretation of the early *kárpásí* as cotton, we have no alternative but to believe that the arts of spinning and weaving were perfectly understood in the East, at a time when the textile industries of the West, through ignorance of cotton, were in a much more backward condition.

Cotton in early classical Literature.

It is extremely difficult, if not quite impossible, to determine the earliest mention of cotton in Persian, Arabic, and the Western classics, owing to the confusion of names for different fibres. Thus *katán* كتان is an old Arabic word for flax, but may also have been used for cotton, since it apparently gave origin to the European words now in use for that fibre, such as Cotton, Cotone (*Ital.*), Coton (*Fr.*), &c.; on the other hand, the Ara-

HISTORY.

bic *Kurpas* (probably derived from the Sanskrit), the modern *kutn* قطن and *panbah* پنبه of the Persians are all now translated cotton, though their original meaning is doubtful.

In Greek the word *karpason* is generally translated "cotton," but has often been rendered "flax." The *βύσσος* of the early Greeks was the fine linen used for making mummy cloths, but by later writers it was applied to cotton. The words λίνον and *linum* have sometimes been translated "cotton," just as *karpason* has often been rendered "flax." **Yates** (in his *Textrinum Antiquorum*) says that some poets assume a license in the use of the word *carbasus*, and **Dr. Robertson** points out that doubt may be admissible as to whether cotton was even known to the early Romans, since garments made of that fibre are nowhere specified in the laws *de Publicanis et Vectigalibus*, while, on the other hand, imported spices and precious stones find a recognised place. It may also be viewed as significant that, with writers at the beginning of the Christian era, it was customary to speak of cotton and cotton goods by names not in use prior to that date. Thus **Pliny** refers to the *Gossypinos* or *Gossypion* of the island of Tylos, and a misapplication of a term derived from silk became general, *e.g.*, *Bambacinus* (made of cotton), *Bambocarius* (cotton dealer), &c. *Bombax* is the generic name given by botanists to the silk-cotton tree.

It is thus evident that no reliance can be placed on any of the terms above mentioned as necessarily meaning "cotton," and that it is only possible to identify a reference to it by the context.

Herodotus is perhaps the first who *seems* to refer to the fibre. In his account of India he writes, "The wild trees of that country bear fleeces as their fruit, surpassing those of sheep in beauty and excellence; and the Indians use cloth made from these trees." He further states that the cuirass sent by Amasis, King of Egypt, to Sparta, was "adorned with gold and with fleeces from trees." **Ctisias**, the contemporary of **Herodotus**, was also acquainted with cloth made from tree-wool.

But perhaps the first unmistakeable reference to cotton is by **Theophrastus**, subsequent to the expedition of Alexander into India. He describes the trees from which the Indians made cloth, as "having a leaf like that of a black mulberry, the whole plant resembling the dog-rose. They set them in the plains arranged in rows, so as to look like vines in a distance." The plant thus described is unmistakeably cotton, and the passage indicates that not only was the fibre known, but that the plant was cultivated, at least in North-Western India, at that early date. **Nearchus** described the garments made from this tree-wool, stating that the natives wore a cloth "which reached to the middle of the leg, a sheet folded about the shoulders, and a turban rolled round the head," a description which accurately describes the clothing of the people of many parts of the country at the present day.

Cotton as an article of Commerce.

The first mention of cotton, considered as an article of commerce, is found in "the Periplus of the Erythræan Sea," by **Arrian**, who probably lived towards the end of the first century. That writer, who was at the same time a merchant and traveller, sailed over the whole of the Arabian Sea, and described in detail the articles of trade of many of the coast towns of India. According to **Arrian** it appears that, in his time, the Arabs exported Indian cotton to Aduli, a port on the Red Sea, and that a trade was established with Patiala, Ariake, and Barygaza—the modern Broach. From the latter a great quantity of cotton cloth, made in the neighbourhood and also obtained from the remote parts of the country, was exported. Masulia, the modern Masulipatam, was also then, as it is to this day, famous for the manufacture of excellent cotton stuffs. The muslins of

HISTORY.

Dacca seem also to have been known. They are described as superior to all others, and are said to have been called *gangitiki* by the Greeks, a name indicative of their manufacture on the banks of the Ganges.

Spread of Cotton—to Europe,

The knowledge of the manufacture of cotton appears to have extended into Arabia and Persia about the beginning of the Christian era. From these countries cloth locally manufactured, or imported thence from India, may have found its way to Rome and the richer cities of the Greeks. But such supplies must have been very rarely obtained, and the silence of classical authors on the subject is a proof of the want of regular imports of the material. Even after regular commercial intercourse had been established between Rome and the East, cotton, considered as a material capable of being woven, appears to have excited little interest. It may in fact be inferred from the silence of writers of the middle ages that cotton was even then but little known. While they write in full detail of the ordinary articles of clothing, of wool, linen, and silk, they make no mention of cotton.

From India, there is no doubt, the cultivation of cotton spread into Persia, Arabia, and Egypt, from whence it probably extended into Central and Western Africa. From Persia the culture migrated into Syria and Asia Minor, also into Turkey in Europe, and thence into other parts of Southern Europe. It is doubtful at what period the plant first crossed the Mediterranean, but the earliest account of its cultivation on the European shores, is to be found in the works of Eben el Awan of Seville, who lived in the twelfth century. He describes the method employed in Sicily and Spain, to which the plant had doubtless been brought by the Arabs. About the same period cotton stuffs, manufactured in Turkestan and by the Armenians, became an important article of commerce between Asia and the Crimea.

On its arrival in Europe, cotton-wool was first employed in the manufacture of paper, and for padding quilts, cushions, &c. The art of spinning its short staple was only learnt in Europe after the delicate manufactures of the East had permeated the whole world. Thus, long before any mention is made of cotton manufacture in Europe, records exist of paper made of cotton being commonly used at Cambridge.

to China.

But more astonishing still than the slow progress of cotton to Europe is the fact that China, which, from very remote times, held intimate commercial intercourse with India, should have known nothing of the arts of culture and manufacture of cotton until the thirteenth century, when they were introduced into that empire by the Mongol-Tartar dynasty. Two Arabian travellers who visited China in the ninth century observed, "The Chinese dress, not in cotton as the Arabians do, but in silk." In the sixth century records exist that the Emperor Ou-ti possessed a robe of cotton, which had probably been presented by some foreign ambassador, and was held in great esteem. Towards the end of the seventh century it is known that the plant was grown in gardens for the beauty of its flowers. At first, as in England, there was great opposition to the new staple, but the cultivation gradually increased, until it reached its present importance, being now in China as in India a valued crop. The quantity produced annually has been estimated at as much as 12,000,000 bales. Whether that be a correct estimate of Chinese production or not, the outturn must be enormous, since the lower orders (now-a-days) are all clothed in cotton.

Cotton in the New World.

Turning to the New World, we find that cotton has probably been used from the earliest times. It is connected with some of the most ancient beliefs of the aboriginal peoples of South America. Thus, M. Agassiz, in his work on Brazil, recounts a strange legend respecting the cotton plant, according to which Caro Sacarbu, the first of men and a demi-god, pulled

HISTORY.

the ancestors of the human race from the bowels of the earth by means of a cotton rope. **Columbus** found cotton in use among the natives of Hispaniola, but only in the most primitive way. **Cortez** found the manufacture in a much more advanced condition in Mexico. The Spanish historian informs us that "The Mexicans made large webs, as delicate and fine as those of Holland. They wove their cloths of different figures and colours, representing different animals and flowers. Of feathers interwoven with cotton they made mantles and bed curtains, carpets, gowns, and other things, not less soft than beautiful. With cotton also they interwove the finest hair of the belly of rabbits and hares, after having spun it into thread; of this they made most beautiful cloths, and in particular winter waistcoats, for their lords."

Spread of the Cotton Industries.

With the rise of Muhammadanism and the conquests of the Saracens, the knowledge of spinning and weaving cotton probably spread to Europe, consequently we find the delicate cotton fabrics of the East first imitated in Italy and Spain, during the twelfth or thirteenth century. From these localities the art of cotton manufacture became diffused and extensively established all along the Southern shores of the Mediterranean, but it was confined to that area till the sixteenth century. The Dutch then succeeded in depriving the Portuguese of their Eastern colonies, and not only extended their own imports of cotton goods, but, towards the end of the century, began to manufacture at home.

From the Low Countries the industry passed over to England in the seventeenth century. On the arrival of cotton goods at the Mediterranean ports, these fabrics became known and valued in England, and gave rise to imitation. Thus **Camden**, writing of Manchester in 1590, says, "This town excels the towns immediately around it in handsomeness, populousness, woollen manufacture, market place, church, and college, but did much more excel them in the last age, as well as by the glory of its woollen cloths, which they call, 'Manchester cottons.'" In 1641 **Roberts**, in his "Treasure of Traffic," writes, "The town of Manchester buys cotton wool from London, that comes from Cyprus and Smyrna and works the same into fustians, vermillions, and dimities." About this time, however, calicoes began to be imported in greater quantities from Holland, Portugal, and also from India direct, with the result that Indian chintzes became so generally worn (to the detriment of the woollen and flaxen manufactures at home) as to excite popular feeling against them. Disputes arose as to the question of duty on these goods. Thus, in 1664, **Pepys** writes in his diary, "**Sir Martin Noel** told us the dispute between him, as framer of the additional duty, and the East India Company, whether calico be linen or no, which he says it is, having ever been returned so. They say it is made of cotton wool and grows upon trees." In 1678 complaints from those interested in woollen manufactures became louder and a pamphlet was issued, called "The Ancient Trades Decayed and Repaired again," lamenting the detrimental effect of imported cottons. This dissatisfaction steadily increased, till, in 1700, Acts were passed which prohibited the introduction of printed calicoes for domestic use, either as apparel or furniture, under a penalty of £200 on the wearer or seller. But this did not prevent the continued use of cotton goods, quantities of which were smuggled into the country.

Towards the end of the seventeenth century cotton printing was established in England, and English prints began to combine with Indian in ousting woollen goods. Thus in 1708 the *Weekly Review* of **Daniel De Foe** contained the statement that "above half of the woollen manufacture was entirely lost, half of the people scattered and ruined, and all this by the interference of the East India Trade." Dissatisfaction increased, till in

G. 69

(449)

HISTORY.

1720, an Act was passed prohibiting altogether "the *use* or *wear* in Great Britain, in any garment or apparel whatsoever, of any printed, painted, stained, or dyed calico, under the penalty of forfeiting to the informer the sum of £5." By the same Act the use of printed or dyed calicoes, "in or about any bed, chair, cushion, window curtain, or any other sort of household stuff or furniture," was forbidden under a penalty of £20, and the same attached to the seller of the article. This Act, however, though confirmed in 1721, appears to have had little permanent effect, since in 1728 we find another author writing of the prevalent evil of the consumption of Indian cotton manufactures, which he ascribed to the "passion of ladies for their fashion."

In 1736 the law, since it struck at the existence of the now rising cotton manufactures of England, was modified so far as to permit the manufacture of calicoes "provided the warp thereof is entirely made of linen thread."

During the succeeding thirty years the mechanical achievements of **Wyatt Kay, Hargreaves,** and **Arkwright,** by reducing the cost of manufacture, and contributing to the production of more perfect and cheaper fabrics, placed the English manufacturer in a position to dispense with a considerable proportion of the protective legislation which previously existed. Accordingly, a law was passed in 1774, sanctioning the manufacture of purely cotton goods, rendering English calicoes subject to a duty of 3*d.* per square yard, but still prohibiting the importation of such goods. The numerous inventions perfected about the end of the eighteenth century, together with the withdrawal of restriction in manufacture, led to so wonderful a development of the cotton trade, that it became not only necessary to seek new markets for English manufactures, but also for more abundant supplies of the raw material. In 1697 the amount of cotton imported into England was 1,976,359 ℔, up to 1749 it did not materially increase, in 1751 it was 2,976,610℔, the average from 1771 to 1775 was 4,764,589℔; but from 1796 to 1800 it increased to the annual amount of 37,350,276℔. At this period the West Indies appear to have become the best and largest source of supply, and prices seem to have ranged from 20*d.* to 30*d.* per ℔, though in 1799 the average price of West India and Berbice cotton was as high as 37*d.* per ℔.

Up to the commencement of the present century America contributed but a very insignificant portion of the cotton consumed. It is stated that "in 1784 an American ship, which had imported eight bags of cotton into Liverpool, was seized, on the plea that so much cotton could not be the produce of the United States." The imports from America in fact continued to be very small up to the end of the century. In 1790, for example, Europe received 18 bags; in 1791 the total exports of cotton from the United States were 189,316℔, but in 1794 they amounted to 1,601,700℔, and by the year 1800 they had increased to nearly 18,000,000℔.

At the close of the war in 1815, when prospects of foreign trade began to brighten and to lead to an increased demand, the production of cotton in America received a fresh impetus, and the subsequent progress was rapid and continuous. The lead then taken by the Southern States was steadily maintained for nearly half a century, and a practical monopoly was eventually established. With increased production came lower prices; cotton worth 22*d.* per ℔ two years before fell in 1819 to 1*s.*, in 1848 to 4½*d.* per ℔.

This great increase in the importance of America, as a cotton-growing country, early led the Directors of the Honourable East India Company to commence operations for the purpose of improving the quality and increasing the quantity of Indian cotton for export to England. Full information

G. 69

(449)

HISTORY.

regarding cultivation and trade was called for as early as 1788. In 1790 numerous reports were received which showed the chief districts of the several provinces in which cotton cultivation was then carried on, and recorded many interesting facts regarding the methods of cultivation. In a "Report on the Proceedings of the East India Company in regard to the Production of Cotton Wool" the following passage occurs:—"From this period great attention was bestowed with a view to the production of a larger and better supply of cotton in the provinces under the Company's Government, as well for the advantage of the native cultivators and manufacturers, as for the benefit of Great Britain. Endeavours were made to prevent adulteration in the cotton purchased from the native merchants; and in order to reduce the expense of freight, screws for compression were brought into successful use, and subsequently carried to great perfection."

The imports into Great Britain from India during the years 1800 to 1809 averaged 12,700 bales per annum; and owing to the influence on the cotton trade of the American embargo in 1808, the quantity brought over in 1810 amounted to 79,000 bales (27,783,700lb). "In 1818 the export to the United Kingdom amounted to 247,300 bales (86,555,000lb)—the largest quantity exported from India up to 1833." The report continues, "In the import of every year there was generally among the consignments some good clean cotton; but the greater proportion was not suitable for spinning in this country, and was therefore re-exported to the Continent." The Company persevered in their energetic measures for the improvement of Indian cotton, but without much success.

The final result of the contest between America and India in cotton supply is too well known to require any recapitulation in this place. From 1818 onwards the respective quantities of Indian and American cotton in the English market never again approached each other till about the time of the American war. During the first fifty years of the present century, Indian cotton exports were characterised by the most extraordinary fluctuations. Thus in 1821 the imports from India had fallen to 20,000 bales; in 1841 they rose to 278,000 bales; in 1848 they again sank to 49,000 bales. America, in short, had gained command of the market, and India was considered only as a supplementary source of supply, resorted to mainly in the event of a short crop from the West. It was not till the blockade of the Southern Ports, during the American civil war, that India again became one of the main sources of supply to the English cotton market. The effect of that war is not only shown by the increase of Indian exports but by the extraordinary rise in value. Thus, previous to war, Indian cotton sold for between 2 and 3 annas per pound, and during the crisis attained the high value of 11 annas 5 pie. With the recovery of American trade, Indian cotton again decreased to its normal value and has since fluctuated between 4 pence and six pence a pound. The following analytical tables of the imports into Great Britain, and statement of prices, will perhaps, however, exhibit, in the shortest way, the striking fluctuations of the American and Indian trades during the ten years previous to and following the War of Secession:—

G. 69
(449)

HISTORY.

Imports of Raw Cotton into the United Kingdom, in Bales, during the ten years previous to the American War, viz., from 1850 to 1859, inclusive.

	1850.	1851.	1852.	1853.	1854.	1855.	1856.	1857.	1858.	1859.
America	1,182,970	1,397,112	1,788,685	1,532,063	1,666,479	1,623,478	1,758,295	1,481,715	1,863,147	2,086,124
India	309,168	326,474	212,361	485,527	308,293	396,014	463,932	680,466	360,980	509,695
Brazil	171,359	108,670	144,214	132,443	107,393	134,762	122,411	168,340	106,127	124,930
Egypt	79,376	63,833	189,865	105,398	81,085	115,018	112,911	75,598	105,603	101,427
Smyrna	157	3,630	20	...	...	...	...	...	500	1,700
Peru	...	47	90	2,879	1,637	1,441	4,300	3,788	1,812	1,496
West Indies, &c. . . .	5,107	4,799	12,043	5,960	7,710	7,505	7,020	7,679	4,460	3,528
TOTAL .	1,748,137	1,904,565	2,357,278	2,264,270	2,172,597	2,278,218	2,468,869	2,417,586	2,442,629	2,828,900
TOTAL IN BALES OF 400 lb EACH.	1,714,000	1,900,250	2,313,000	2,255,750	2,216,500	2,252,750	2,552,500	2,440,250	2,563,750	2,977,000
PRICES OF INDIAN COTTON IN ANNAS PER lb	...	...	...	...	2—3	Not given.	2—3	2—8	3	2—7

HISTORY.

Imports of Raw Cotton into the United Kingdom, in Bales, for the ten years ending 1869, together with those of 1870.

	1860.	1861.	1862.	1863.	1864.	1865.	1866.	1867.	1868.	1869.	1870.
America . .	2,580,980	1,841,643	72,369	132,028	197,776	460,606	1,162,745	1,225,686	1,269,060	1,039,641	1,664,010
India . . .	562,738	986,290	1,071,768	1,229,984	1,399,514	1,266,513	1,847,759	1,508,754	1,451,979	1,496,426	1,063,540
Brazil . . .	103,084	99,224	133,807	137,142	212,192	340,261	407,646	437,208	636,897	514,200	402,760
Egypt . . .	110,009	97,759	135,420	205,788	257,102	333,575	167,451	181,173	188,689	185,670	219,920
Smyrna . .	100	957	14,851	42,282	61,793	79,803	32,770	16,995	12,758	40,957	...
Peru . . .	2,515	2,667	5,400	15,521	27,059	71,794	49,081	64,423	58,881	62,228	...
China and Japan .	...	...	...	160,807	399,074	141,610	18,844	1,942	93	...	...
West Indies, &c. .	7,259	7,188	11,436	8,610	32,586	61,159	62,745	64,593	41,770	43,042	112,100
TOTAL .	3,366,685	3,035,728	1,445,051	1,932,152	2,587,096	2,755,321	3,749,041	3,500,774	3,660,127	3,382,164	3,462,330
TOTAL IN BALES OF 400lb EACH.	3,589,500	3,153,500	1,332,750	1,729,500	2,240,250	2,416,000	3,384,500	3,156,340	3,320,210	3,053,067	3,289,214
PRICES OF INDIAN COTTON IN ANNAS PER lb.	3—7	4—2	6—4	10—5	11—5	7—1	6—2	5—3	4—7	5—6	5—5

G. 69
(449)

HISTORY.

Causes of decline in English Imports of Indian Cotton.

The causes of the decline of India's hold on the European market, during the first half of this century, are abundantly shown in every report on the subject. Imperfections in methods of picking, cleaning, and packing, together with extensive adulteration, rendered many of the Indian consignments practically valueless, and largely increased the demand for the clean American cottons, which thus rapidly obtained a stronger hold on the market. The great development in machinery, which coincidently took place, resulted in improvements being directed towards the utilisation of the clean, long-stapled, high quality American cottons. As a consequence, the short-stapled Indian flosses soon became entirely unsuited for English machinery, and thus their market price rapidly fell. With diminished prices, and a bad name, inducement to improve the quality of consignments became less and less, till such a statement as the following became only too true:—"An inferior and dirty article is by far the most profitable article of native trade, even so to European merchants."

Notwithstanding many endeavours on the part of the East India Company, but little improvement was effected. In 1829 a Regulation (III. of that year) was introduced for the suppression of the adulteration of cotton, but with no practical result. Attempts were therefore directed towards the improvement of Indian cotton by the introduction of American seed—attempts which were energetically aided by the Agri.-Horticultural Society of India, and by many private individuals. Thus, in 1830, the Society started a farm of 500 *bighas* at Akra, near Calcutta, for the experimental cultivation of cotton. This endeavour was prosecuted, and acclimatised seed freely distributed till June 1833, when the lease of the farm having expired, the Society did not deem it necessary to keep up the establishment. In 1837, £200 of the Society's money was set apart for procuring seed, and an annual grant of R1,000 voted for the same purpose. And the whole of that amount was ordered to be expended on Upland Georgian. The earlier publications of the Society in fact teem with accounts of similar patriotic efforts.

In 1838, ten experienced planters were procured by Government, from the Southern States of America, for the purpose of establishing the cultivation of New Orleans cotton. They were furnished with seeds, ploughs, hoes, and gins, and began operations about 1840-41. Three were stationed in Bombay, chiefly at Broach; three in Madras, at Tinnevelly and afterwards at Coimbatore; and the remaining four, sent to Bengal, commenced experimental cultivation in the Doab and Bandelkhand. With varying success these cotton farmers continued their efforts, until it became manifest that permanent improvements were not likely to be attained. Besides many natural difficulties, much opposition from the natives was encountered. This was overcome in Dharwar only, the one great instance of success, by the personal influence of the Collector and his assistants. For a full account of these experiments the reader is referred to the publications of the Agri.-Horticultural Society of India.

Notwithstanding the want of success which attended these endeavours, a firm conviction was still entertained that the speediest and surest way to the improvement of Indian cotton was by the extensive use of exotic seed. Large quantities were accordingly sent to India by the Cotton Supply Association, and were also distributed by the Agri.-Horticultural Society of India, but with the old result of general failure. Up to the date of the American war, and the consequent cotton famine, it appeared as if Indian cotton would be gradually driven out of the market. At that time, however, merchants in Europe naturally turned to this country for a great part of their supply, and energetic efforts were renewed by Government to meet the demand and improve the quality. These endeavours were materially

HISTORY.

For an account of the working of that Act, Conf with pp. 72—78.

aided by private enterprise; notably so by the English Cotton Supply Association. In 1863 a Cotton Commissioner was appointed in the Bombay Presidency; in 1864 a similar appointment was made for Berar and the Central Provinces. A Cotton Frauds Bill was submitted in 1862-63, and became law as the Bombay Cotton Frauds Act, IX. of 1863.

In conclusion it may be said that experiments on the introduction of American seed have been steadily kept up at the Government Farms but with little success, and of late years attention has been consequently more and more directed towards improving the Indian staple, by careful selection of seed and adoption of the best methods of cultivation. Since the re-establishment of the native staple in European markets, consequent on the cotton famine, it has been found that a good market does exist. The exports have remained fairly steady, and should the staple not further deteriorate, they are likely to continue regular.

The reader is referred to the article on "TRADE" for information as to the present position of India as a cotton-exporting country. The remarkable increase since 1862, and the present immense importance of the exports, render it manifest that every effort ought to be made to retain and expand a trade of so much value to this country.

CULTIVATION OF COTTON IN INDIA.

REVIEW OF AREA, OUTTURN, AND CONSUMPTIDN.

CULTIVATION.

Area and Outturn.

(450) 70

The statistics of area of agricultural crops in India are year by year being brought to a state of greater completion. In the case of cotton special attention has been paid to the subject, with the result that the figures now given* are believed to be fairly accurate. This remark applies to most of the provinces and Native States in India, with the exception of Bengal, a province regarding which the available information as to area and outturn is of the most meagre character. An attempt was, however, made, some few years ago, to determine the Bengal area and outturn of cotton, and the figures then published have had to be accepted in the present statement, for the year 1888-89, though everything points to their being under rather than over the mark. In a "Note" on cotton cultivation issued by the Government of India, in 1888, the average area for the five years preceding 1885-86, was put down at 14,051,800 acres, and the outturn at 6,775,000 cwts. of cleaned cotton. For most provinces of India a certain reduction (as far as possible allowed for in this article) has to be made, owing to the fact that a considerable percentage of the cotton lands is regularly occupied with mixed crops, *viz.*, one or two rows of cotton alternating with a row of one or other of the bushy pulses or some such crop which can conveniently be grown on the same field with cotton. The estimate of yield has, in many provinces, been arrived at, after the conduction of careful crop experiments, and these, on the whole, have demonstrated that in former returns the yield was understated. Moreover, for the races and hybrids of indigenous cotton (**Gossypium herbaceum** and **G. Wightianum**), it is now generally accepted that 30 : 100 is a safer estimate of cleaned from uncleaned cotton than 25 : 100 as previously allowed. Accepting these opinions as correct, the area under cotton in 1888-89 was 13,998,639 acres, and the outturn 9,219,493 cwt. This exhibits a decrease in the area but an increase in the outturn of approximately 2,500,000 cwt. of cleaned cotton over that given in the "Note" above alluded to. It is believed that these results indicate a departure of a more superficial than actual character, however, being the adjustment

* The year 1888-89 has been chosen (though later figures of foreign trade are available), because it is the last for which corresponding Agricultural and Internal Trade Statistics are obtainable.

CULTIVATION.

Area & Outturn.

of the error arising from mixed crops, which would restrict the area actually occupied by the crop but raise the acreage outturn. Indeed, it seems probable that, instead of the cultivation of cotton having declined, as the figures given above might be supposed to indicate, it is actually on the increase, and greatly owing to the favour shown for mixed crop cultivation. For example, if the standard of least error be taken in determining this point, *viz.*, the total consumption of Indian cotton in the home and foreign markets, we learn that the exports (as a direct result of the American War) suddenly expanded. In 1865-66, they attained the magnitude of 7,170,986 cwt., valued at R35,58,73,890. In the year previous, though the quantity exported was only 4,687,972 cwt., the greatest recorded value was then obtained, *viz.*, R37,57,36,370. From the date of the War of Secession the exports declined, till in 1869-70 they amounted to only 4,953,379 cwt. They again, however, recovered, and attained, in 1871-72, their maximum quantity, namely, 7,225,411 cwt.; but it is noteworthy that this remarkable production of Indian cotton decreased rather than increased the value, for the exports were then declared at only R21,27,24,297. From 1871-72, the exports declined once more and reached their lowest recorded level in 1878-79, a year of famine, when they were 2,966,060 cwt., valued at R7,91,30,458. Since that date they have steadily improved, having been, in 1885-86, 4,189,718 cwt.; in 1886-87, 5,432,648 cwt.; in 1887-88, 5,374,542 cwt.; in 1888-89, 5,331,904 cwt.; and in 1889-90, 6,328,898 cwt. It may thus be admitted that the foreign exports justify the opinion of an expansion in the production of cotton, the more so since the consumption by Indian mills has manifested during the past five years,—indeed during the past thirty years—a corresponding steady increase. The statistics of these mills showed, during the five years ending 1866-67, an average annual consumption of only 210,000 cwt., while last year they took 3,111,024 cwt., being an increase in twenty-two years of 1,381½ per cent. These facts may thus be admitted to afford a tangible conception of the cultivation of cotton in India, which can be elaborated by a review of the figures of internal and foreign trade. Working back from the returns of marine and rail-borne traffic, for example, a powerful confirmation is obtained of the accuracy of the agricultural statistics (in the case of surveyed provinces), and accepting the data thereby arrived at, it becomes possible to fill up approximately accurate figures for such provinces as Bengal, which publish only generalisations regarding their area and outturn of crops. Some years ago an attempt was made to register the road traffic to and from the larger cities of India. This was found both troublesome and expensive and had accordingly to be abandoned. But the more important railways having fallen into the hands of Government, an arrangement has been made by which each province is divided into certain blocks. The interchange of produce between these blocks is carefully recorded and the traffic published annually. A similar system of registration on rivers and canals having been organised, it has thus, during the past few years at least, become possible to form a fairly clear idea of the major portion of the internal trade of the country. Had it been possible to carry out a registration of road trade, an even more complete conception of outturn would have been derivable from the movements of produce from district to district and province to province. But by bearing in mind the error involved through want of road statistics we are, nevertheless, in a position to arrive at a fairly accurate idea of the intra- and inter-provincial transactions, from the annual reports of the rail, river, and canal traffic. A provincial balance sheet of imports and exports, doubtless, ignores at the same time local consumption, since a product, such as cotton, may be conveyed a few

CULTIVATION.

Area & Outturn.

miles from the field to the loom without being carried over any possible registration route. The amounts thus locally consumed would, therefore, escape registration, however carefully road traffic was recorded, but by various analytical methods, figures have been derived, which are believed to approximately express local consumption. The accuracy of these estimates is, indeed, confirmed by an analysis of all trade transactions. The net exports over imports being deducted from outturn, the balance shown would be available for local consumption, and that balance, in most cases, represents the actual figure stipulated for by official estimates as the amount requisite. It need only be here added, in this connection, that the most serious error, involved in all tabular statements of the production and trade in cotton, is due to the season of cultivation and harvest not corresponding to the official year in trade statistics. It does not follow, for example, that the area shown as under cotton, in any one year, afforded the whole supply for the trade of the same period. Indeed, an annual statement for twelve months ending 30th June (the cotton commercial year), would be more correct than a year ending 31st March—the official year. By such an arrangement, however, the relation of the cotton production to the imports and exports of raw cotton and piece goods would be lost sight of. The error here indicated is, at the same time, one that admits of correction by the average of a series of official years being compared with a corresponding average of commercial returns. As this involves, however, considerable labour, scarcely commensurate with the advantages gained, the authors have in the present review of the Indian Cotton Trade preferred to adhere to the official year, even though aware that the transactions in rail, river, canal, and marine returns, for the first few months of each year, refer mostly to the previous year's crop. As explained above, this is due to the official year closing in March, whereas the whole of the crop is not in the market before the end of June. Allowing for this error and for the want of road statistics, it is believed that the internal transactions in cotton under rail, river, and canal traffic, confirm the agricultural statistics of area, outturn, and local consumption, just as the returns of internal trade are again confirmed by the foreign exports from India.

Conf. with p. 161.

In 1869, **Mr. H. Rivett-Carnac** (then Cotton Commissioner for the Central Provinces and Berar) estimated the Indian consumption of cotton at 600,000 bales (of $3\frac{1}{2}$ cwt.). There were in all only seventeen mills at work in that year, and these used 90,000 bales, so that the local consumption for hand-looms and all other purposes, would on that estimate have been 510,000 bales, or 1,785,000 cwt. The author of the valuable work "Indian Cotton Statistics" has proved satisfactorily that the annual average production of cotton in India at the present day is a little over 2,600,000 bales, or 9,100,000 cwt. There are now at work 124 mills in India, and last commercial year these worked up 888,654 bales. The foreign exports from India, during the official year (which in part only corresponds to the commercial year), were 1,523,401 bales. If it be admissible to add these two quantities together and deduct the total from the estimated average annual production, there would have been left for local consumption in all India, during the year 1888-89, 187,945 bales or 657,807 cwt., a quantity which, there are grounds for believing, would have been considerably under the actual requirements of the country. But the above estimate does not take into account the foreign and trans-frontier imports, and thus deals with actual foreign exports, not net exports, from the country. The outturn, according to the most recent official returns, was, besides, about 120,000 cwt., in excess of the estimated average, so that it seems safe to add that the actual supply was

CULTIVATION.

Area & Outturn.

100,000 bales (or 350,000 cwt.) in excess of the average annual outturn. If this assumption be correct, India consumed, other than by mills, during the year under notice, close on 1,000,000 cwt., the total traffic being as follows :—

	Cwt.
Used by Indian Mills	3,110,289
Exported to Foreign Countries	5,331,904
Consumed in India	1,000,000
	9,442,193

It will be observed that, according to **Mr. H. Rivett-Carnac**, the hand-loom consumption of cotton in 1869-70 was considerably in excess of the present estimated consumption, but this is doubtless a direct consequence of the extension of mills in India and of the increasing favour shown for foreign piece goods by certain classes. The estimate of 1,000,000 cwt. of cotton, as used up per annum in India, excludes from consideration the amount of Indian mill piece goods and yarns, also of foreign piece goods and yarns brought into the country. That amount expressed, however, to the population of 250 millions, would be about 0·45℔ per head. **Royle** (*Culture and Commerce of Cotton in India, p. 19*) publishes an estimate showing that 5℔ of cotton cloth or 32½ yards is about the amount of clothing worn in India by a man and his wife, or 2½℔ per head of adult population. Accepting the lesser requirements of the juveniles and infants as balancing the greater demands of the well-to-do, a population of 250,000,000, consuming 2½℔ per head, would require 5,580,357 cwt. of cloth, or, correcting this to raw cotton by adding 3 per cent. for loss in spinning and weaving, they would annually use up 5,747,767 cwt. of cotton. A calculation of total consumption has been made by more recent investigators with practically the same result. It has, for example, been determined that on the average five yards of cloth go to the pound of cotton. A man and woman with two children would require 40 yards of cloth or 8℔ of cotton. The average of these two calculations points to the people of India using a little over 3,200,000,000 yards of cotton cloth. Adding to that, the minor uses of cotton (upholstery, lamps, &c.), they would require in all 6,000,000 cwt. of raw cotton, and this is very probably a liberal estimate of annual consumption.

Annual requirements of India in cotton cloth.

71 (451)

The following abstract statement for the year 1888-89 will be found to bring together the main facts regarding the raw cotton trade of the whole of India. It shows an outturn of 9,219,493 cwt., with a net export from the producing areas of 6,792,503 cwt., thus leaving a balance of 2,426,980 for local hand-loom and mill consumption in the provinces. Of these provincial exports nearly the whole goes to the port towns and is accordingly shown as imports to the ports. These imports constitute the stock from which the foreign exports are drawn, the amounts used by the mills in the port towns forming a balance which largely returns again as yarns or piece goods to the provinces, and, together with the foreign goods, constitutes the supply of cotton used by the people of India over hand-loom manufactures. It will be observed the net exports from the provinces are less than the imports by the towns; but this is due to two considerations: (1) the supplies in April, May, and June were the last portions of the crop of 1887-88. They were thus not drawn from the area cropped during the official year of 1888-89, and the difference between the yield in these months in the two years has to be added or deducted as the case may be; (2) the absence of road statistics doubtless brings into the port towns an amount which appears in the sea-coast traffic, but has not been given in the exports from the provinces :—

G. 71 (451)

Abstract Statement of the Production of, and Trade in, Raw Cotton for the year ending 31st March 1889.

Presidencies, Provinces, and Native States that produce the Indian Cotton.	References to pages where fuller details occur.	I. Population (including Ports).	II. Area under Cotton in	III. Outturn of cleaned Cotton in	IV. Total Imports.	V. Total Exports.	VI. Net Exports or Net Imports, drawn from production.	VII. Amount available for local consumption, worked out from VI., III., or V.
			Acres.	Cwts.	Cwts.	Cwts.	Cwts.	Cwts.
1. Bombay Presidency and Native States (excluding Bombay Port).	60	13,267,400	5,198,400	3,563,700	457,470	3,427,184	2,969,714	593,986
2. Sind (excluding Karáchi)	84	2,413,800	96,400	189,600	77,079	224,957	147,878	41,650
3. Berar	87	2,672,673	1,991,551	1,276,061	8,704	1,333,186	1,324,482	48,448
4. Rajputana and Central India	97	19,530,299	*886,419	*877,607	3,780	600,411	596,631	280,976
5. Madras (excluding Ports)	99	28,871,376	1,794,510	801,120	98,039	592,865	494,826	309,343
6. North-West Provinces and Oudh	109	43,268,599	1,399,388	706,344	122,006	825,734	703,728	2,616
7. Panjáb	117	18,799,319	759,465	931,824	1,889	295,976	294,087	637,737
8. Central Provinces	128	9,838,791	613,348	351,923	8,756	170,130	161,374	190,549
9. Nizám's Dominions	132	9,845,594	*1,016,565	*307,002	687	82,529	81,842	225,160
10. Bengal	133	69,586,861	†162,000	†139,000	119,755	123,317	3,562	135,438
11. Assam	139	4,932,046	40,588	54,359	3,671	29,435	25,764	28,593
12. Mysore	147	4,186,188	29,814	11,459	3,159	1,120	2,039	13,498
13. Burma (Lower)	148	6,736,771	10,191	9,494	55,295	48,897	6,398	15,892
Provincial Totals	...	...	13,998,639	9,219,493	960,290	7,755,741	6,812,325	2,426,980
Outturn according to Mr. Beaufort	...	...	...	9,625,000				
Chief Ports.								
14. Bombay Port					6,002,539	4,385,435	*Nil.*	+1,617,104
15. Karáchi ditto					268,552	269,039	*Nil.*	—487
16. Calcutta ditto					690,722	403,740	*Nil.*	+286,982
17. Madras Ports					620,734	672,240	*Nil.*	—51,506
Port Town Totals					7,582,547	5,730,454		1,852,093
Total Foreign Transactions (Imports of Foreign Cotton; Exports of Indian)					64,627	5,331,904		
Total Indian Trade (Imports by the Port Towns and Exports to the Provinces)					7,517,920	398,550		

* Averages of several years, not actual returns for 1888-89.
† Estimate made in the year 1878.

CULTIVATION. **Area & Outturn.**

Below will be found brief accounts of cotton cultivation in each province, the attempt having been made to bring together all available information. In the case of provinces possessing port towns, the provincial transactions have, as far as possible, been kept distinct, so that production might not be confused with markets of supply from which the foreign exports take place. With reference to the port towns, the areas from which supplies are drawn have been exhibited, so that it is possible to judge approximately what provincial cotton goes to each special foreign market; but before proceeding with provincial cultivation it may be as well to say something of:—

COMMERCIAL FORMS. 72 (452)

THE COMMERCIAL FORMS OF INDIAN COTTON.

The Indian cotton of commerce is divided into certain well-defined forms, designated according to the localities in which each is chiefly cultivated, and distinguished by certain peculiarities of staple. For an account of the botanical nature of these forms the reader is referred to the descriptions of the various species in the chapters on Botany (pp. 1—39), also to the provincial accounts, in which it has been attempted, as far as possible, with the limited material available, to refer each to its proper botanical position. It is, however, almost impossible to give accurate botanical definitions of these forms. Though the commercial terms may at one time have signified special forms of cotton in addition to the port of shipment or the place of production, in the present day they merely denote different standards of quality irrespective of race. Thus **Mr. Ozanne**, Director of Land Records and Agriculture, Bombay, in a note kindly furnished to the editor, writes, "A bale of cotton classed under any of these standards may contain the produce of different races, provided the staples blend together and are not easily detected. Thus the newly-created standard '*belati*' (see *Oomras*) has been found to contain staples of four distinct kinds, in certain proportions, which often determine the place of a mixture under a standard, and consequently its price."

In this paragraph a short description of the area under each form (as recognised commercially), the progress made in its cultivation, and the ports supplied, will be attempted, but the authors are indebted for much of the information given regarding area, locality, &c., of the commercial growths of cotton, to **Mr. Beaufort's** valuable compilation of Indian cotton statistics; they have also to acknowledge useful information received from **Mr. Ozanne**.

Dholleras. 73 (453) Conf. with pp. 27, 62—63.

"*Dholleras.*"—Is a race of cotton (called after Dhollera, till lately, the chief port of shipment on the Gulf of Cambay), the cultivation of which has, of recent years, shown a very marked expansion. It is principally produced in the Native States of Kathiawar, and includes the produce of the two annual varieties *lalia* and *vagadia*, and the two perennials *jeria*, and *rogi*. **Mr. Beaufort** writes:—"Excluding the State of Baroda, a portion of which till 1885-86 was not included in the Kathiawar Native States, the acreage has risen from a mean of 1,408,000, acres to 1,996,000 acres. In the Baroda State, in the last four seasons, it has averaged 168,000 acres, making a total average area devoted to this 'growth' of 2,547,000 acres. Of this, 1,760,000 acres are in Kathiawar, 360,000 in Ahmedabad, 1,790 in Cutch, 168,000, as stated, in Baroda (districts of Kadi and Amreli), 70,000 in Palanpur, and the remaining 14,000 acres in Kaira, Mahi Kantha, and Cambay." In 1888-89 the acreage was somewhat lower than the average above cited, *viz.*, 2,469,000 acres. **Mr. Ozanne** informs us that *Dholleras* are now sold in Europe under the names of *Bhownagger, Mowa, Wadwhair, Veerumgaum*, and *Veraval Cutch*. The land is prepared for this crop in May to June, the seed is sown in June, and picking takes place from February to April, or March to May. The average outturn is 1,600,000 cwt. of cleaned cotton, of which the average receipts in Bombay are 1,407,000 cwt.

COMMERCIAL FORMS.

Bengals.

(454) 74

Conf. with pp. 7—14; 96—99; 109—1 114; 118—16; 134—135.

"*Bengals.*"—A race of cotton grown chiefly in the Panjáb, the North-West Provinces and Oudh, Bengal, Rajputana, and Central India. The area under the crop is the largest returned for any 'growth' of cotton, the average during the past five years having been 3,492,000 acres. Of this the Panjáb grew 860,oco, the North-West Provinces 1,536,000, Oudh 79,600, Bengal 162,000, and Rajputana and Central India 840,000 acres. The land is prepared in May-June, the seed is sown in June, picking takes place from October to January, and the average yield from the total area is said to be 2,730,000 cwt. "*Bengals*" from the Panjáb go to Karáchi, Calcutta, and Bombay for shipment; those from the North-West Provinces and Oudh to Calcutta and Bombay; those from Bengal to Calcutta; and those from Rajputana and Central India to Bombay.

Oomras.

(455) 75

Conf. with pp. 62, 68; 88—92.

"*Oomras*" or "*Oomrawuttee cotton.*"—Is divided in Liverpool into three separate growths—*Oomras, Khandesh*, and *Bilatee**; but in the Bombay market the names of the principal producing districts or provinces are the leading heads under which statistics are compiled. "*Khandesh*" is produced in the district of that name, also to a small extent in Nassik. The area under the crop in these districts during the past five years averaged 887,400 acres. "*Barsee*" and "*Naggar*" are named after the towns of Barsee and Ahmednaggar, through which the bulk of the cotton passes, but, according to **Mr. Beaufort**, the greater portion is actually grown in the Nizam's Territory. The area in the Bombay districts during the past five years averaged 63,600 acres. According to **Mr. Ozanne**, three forms of cotton are sold under the name of *Oomras*:—a good-stapled form, *Akola*; a medium staple from Indore and Oojain, Jalgaum and Dhulia, and third, the short-stapled *Bilatee** from Khandesh and Oomrawatti, Shegaum, and Akote. It is practically, however, a mixture of four distinct kinds, *banni*, *jarivaradi*, acclimatised American, and the ordinary variety of the Deccan. But *Oomras* proper is *banni* and *jari* only.

The finest description of "*Oomras*" is produced in the Berars, but the area under the crop in that province has shown a decided falling-off at least during the past twenty years. Thus, in 1878-79, as much as 2,207,889 acres were planted with it, while the average for the past five years has only reached 1,957,200 acres. It is grown in all the districts, the average in each for the same period having been—Amraoti 480,000 acres, Akola 479,000 acres, Buldana 300,000 acres, Wun 231,000 acres, Bassein 233,000 acres, and Ellichpur 231,000 acres, making a total of nearly 1,960,000 acres. The total average area under "*Oomras*" for all India, during the past five years, was, as shown above, about 2,910,000 acres. The average crop for the same period was 2,007,000 cwt., of which 1,836,000 cwt. was sent to Bombay for shipment, the rest being retained for local consumption.

The land is prepared for the "*Oomras*" crop in May-June, the seed is sown in June, and picking is carried on from November to March.

Westerns.

(456) 76

Conf. with pp. 101—104; 132.

"*Westerns.*"—This race, as its name implies, is one of the chief cottons of Madras, the Nizam's Dominions, and Western India generally. In the Madras Presidency, Bellary is the principal district, with 250,000 acres, followed by Kurnul with 205,000, Anantapore with 110,000, and Cadappah with 100,000 acres, making a total of 665,000 acres. The average area under the crop during the three years preceding 1887-88 in the Nizam's Dominions was 973,600 acres, and in Sholapore (Bombay) 25,000, making an average grand total under "*Westerns*" in India, of about 1,660,000 acres. The average outturn for the past five years was 735,000 cwt., of which 112,000 cwt. was sent to Bombay for shipment, the bulk of the remain-

* *Vilayati-Khandesh or Varadi*. This while not, strictly speaking, a commercial form, is now so largely employed for adulteration that it may be viewed as of commercial importance. *Conf.* with pp. 62, 75, 89—91.

der going to Madras sea-ports. The land is prepared for the crop in July-August, the seed is sown in August-September, and the picking, which commences in March-April, is completed by May or early in June.

COMMERCIAL FORMS.

Dharwars. 77 (457) Conf. with pp. 17—18; 62—63; 67.

"*Dharwars.*"—This is an important Bombay cotton, grown principally in Dharwar, Bijapore, and the other Southern Mahratta States. The average area covered by the crop during the last three years was 277,300 acres, of which Dharwar contributed 191,000, Bijapore 25,000, and the remaining Southern Mahratta States, 61,000 acres. The average yield for the same period was 122,000 cwt., of which the greater part went for shipment to Bombay. *Dharwar* cotton proper is the produce of the American or New Orleans variety, introduced into these districts about half a century ago. The periods of sowing and picking correspond with those of "*Westerns.*"

Coomptas. 78 (458) Conf. with pp. 62; 68.

"*Coomptas.*"—This is another Bombay cotton of some importance, grown under the same conditions and in the same seasons as *Dharwars.* During the past three years the average acreage was 968,300, a considerable diminution on the figures for the four years ending 1883-84, in which the average was 1,118,250 acres. Bijapore, with 347,000 acres, is the chief district of its cultivation, followed by Dharwar with 287,000 Belgaum with 172,000, Kolhapur with 31,000, and the remaining Southern Mahratta States with 132,000 acres.

The average outturn for the same period was 420,000 cwt., of which the great bulk went to Bombay for shipment. It is supposed to be the native race of the Bombay Karnatak, and is named after one of the principal ports from which cotton of that district used to be shipped. In Kanarese (the prevailing language of the country) it is called *javári-hatti*=" country cotton," to distinguish it from the acclimatised American form.

Broach. 79 (459) Conf. with pp. 62—63.

"*Broach.*"—This is the cotton of Baroda, Broach, and Surat. The total area taken up by the crop in the four years ending 1888-89, was 597,000 acres, of which Broach contributed 241,000, Baroda 227,000, and Surat 129,000 acres. The average outturn for the same years was 525,000 cwt., of which 133,000 cwt. was despatched to Bombay for export. Broach cotton consists for the most part of the *lalia* form, and is locally known as ***kahnam***, *i.e.*, "black soil" cotton, to distinguish it from the ***rogi*** perennial cotton of the light soils of Gujarat. As in the case of most of the other Bombay cottons, sowing is conducted in June, the picking commencing in February and finishing in April.

Coconadas. 80 (460) Conf. with pp. 27; 101—102.

"*Coconadas.*"—A red cotton, principally produced in the Kistna district of Madras, where on an average about 200,000 acres are taken up by it. Nellore also contains about 25,000 acres under the crop, and Godaveri 20,000, making a total of 245,000 acres. The average crop is about 80,500 cwt., of which nearly the whole is sent to Guntoor to be pressed, and is despatched from thence chiefly to Coconada for export. A small quantity, however, is transported by the Buckingham Canal to Madras. This, like the other Madras cottons, is grown during the north-east monsoon, being sown in September-October, and picked in February to June.

Tinnevellys. 81 (461) Conf. with pp. 101—102.

"*Tinnevellys.*"—Are entirely shipped from Tuticorin, from whence nearly the whole is despatched coastwise to Bombay for transhipment to ports in Europe. The average area is estimated by **Mr. Beaufort** at 742,000 acres, distributed as follows:—Tinnevelly, 320,000; Coimbatore, 250,000; Madura, 120,000; Trichinopoly, 26,000; Salem, 17,000; South Arcot, 6,000; and Tanjore, 3,000 acres. The annual average crop is about 450,000 cwt. It is the latest of the cottons, being sown in October-November, and picked from March to April.

Salems. 82 (462) Conf. with pp. 19; 102—104.

In passing, it may be remarked that the cotton known as "*Salems*" or "*Coimbatores*" of older trade returns, was a race grown from Bourbon seed, which was at one time cultivated to some extent in Coimbatore and

COMMERCIAL FORMS.

Salem, and found a ready sale in China. Of late years, however, the stock has greatly deteriorated, and, according to Mr. Beaufort, the variety has, so far as export trade is concerned, virtually ceased to exist.

Hinganghat.

" *Hinganghát* "—A race of cotton peculiar to the Central Provinces,
83 named after the town Hinganghát. The cultivation of this cotton appears to have greatly fallen off during the last fifteen years. Thus, in the four years ending 1878-79, the area averaged 800,200 acres, while, in the last five, it has amounted to only 608,800 acres, probably owing to other crops, such as wheat and linseed, having proved more remunerative. Mr. Beaufort gives the following distribution of the area of cultivation at the present time: "Of the four divisions, into which this province is divided for administrative purposes, the Nagpore one produces most cotton, more than half of the total area being in this division. Nearly a third or, say, 220,000 acres, is in the Wardha district, while the Nagpore district has an average over 120,000 acres, and Chanda about 20,000 acres. In the Nerbudda division the districts of Newar and Chindwárá, have each about 45,000 acres under cotton, Narsinghpur about 30,000, Raipur, Bilaspur, and Sambulpur, in the Chhatisgarh division, together devote about 55,000 acres annually, the remaining 75,000 acres being cultivated in Saugor, Jubbul-

Conf. with p. 129.

Sind.

84 pur, and Damoh of the Jubbulpur division."

Conf. with p. 85.

The average crop is estimated at 226,000 cwt., of which the greater part appears to be retained for local consumption, or transported by road, rail or river to some other province, only 42,000 cwt. being sent to Bombay for purposes of export.

Assam

Hinganghát is one of the earliest of all cottons to come into the
85 market, being sown in May-June and picked from November to March.

Conf. with pp. 7, 141—142.

" *Sind.*"—Cotton cultivation in Sind has shown a steady increase during the past twenty years—the average annual area having risen from 57,900 acres in the decade ending 1879-80, to 73,000 acres in the past five years. Hyderabad, with an average area of 53,000 acres, is the principal producing district. The average crop for the past five years was 105,000

CULTIVATION in BOMBAY.

86 cwt., of which the greater part was transported to Karáchi for shipment. The crop resembles most other north-west monsoon cottons in its season of growth, being sown in June-July and picked in November to January.

Key to Provinces. Bom. pp. 60 to 84. Sind pp. 84—86. Berar pp. 86—96. Raj. pp. 96—99. Mad. pp. 99—109. N. W. P. pp. 109—117. Pan. pp. 117, 128. C.P. pp. 128—133. Niz.'s Domns. pp. 132—133. Ben. pp. 133—139. Assam pp. 139—146. Mysore pp. 147—148. Bur. pp. 148—152. Andamans p. 152.

" *Assam.*"—Assam cotton is produced over an average area of 40,000 acres, on which the outturn is calculated to be as high as 150℔ per acre. The average crop is estimated at 53,400 cwt., of which the greater portion is transported by river to Calcutta for export.

I. BOMBAY.

References.—*Hove, Tour in Bombay, (1787), 26-30, 34-45, 58, 69, 76, 81, 106-107, 109, 127, 155-158, 161, 167, 169; Walton, History of Cotton, 1880; Forbes, Gin Factory at Dharwar, 1862-63; Rep. on Admn. of Cotton Dept., 1868-69; Annual Repts., Bomb. Cotton Dept.; Cotton Repts., Agri. Dept.; Rept. from Director of Land Records & Agri., 1889; Admn. Repts. in many passages; Gazetteers: II., 63, 178, 392-404, 411, 461; III., 50, 75; IV., 55; V., 106, 125, 245, 294, 371; VII., 79, 87, 97; VIII., 185-186, 248-253, 262; X., 150; XII., 153-162; XIII., 290; XVI., 101, 167; also many other scattered passages in these and other volumes; Agri. Hort. Soc. of India, Jours.—I., 194; II. (Pt. II.), 439; V., 37-78; XIII., 133-146.*

REVIEW OF AREA, OUTTURN, AND CONSUMPTION OF COTTON.—In the Table p. 56, the area of cotton in Bombay and its Native States during the year ending 1888-89, is given as having been 5,198,400 acres. According to official returns, the Bombay cotton area during the past twelve years has

Area.

87 fluctuated slightly, while progressing forwards from 2,863,306 acres in 1877-78, to its present extent. The outturn has similarly shown a progression: in 1884-85, it was 3,432,600 cwt.; in 1885-86, 4,166,900 cwt.; in

CULTIVATION in BOMBAY.

1886-87, 3,873,800 cwt.; in 1887-88, 3,227,100 cwt.; and in 1888-89, 3,563,700 cwt. These figures would give an average yield, for the past five years, of 3,652,800 cwt. from an average acreage of 5,161,300—a result which would exhibit 79·2lb of cleaned cotton as the yield per acre, the yield of cleaned to seed-cotton having been taken as 30 : 100. It has been customary to speak of the acreage yield of Bombay as 50·5lb, but the figure now advanced has been confirmed in various ways, for example, in dealing both with commercial and agricultural statistics (see pp. 52-57, also 62), and is believed to be very nearly correct.

Yield. 88 (468)

But the Presidency of Bombay imported and exported cotton from and to other provinces, and, at the same time, consigned large quantities to its port town, which were destined mainly for the foreign traffic from Bombay. The balance of all these transactions on outturn, resulted in a net surplus remaining in the presidency of 593,986 cwt.—the amount available for local consumption. Similarly, all transactions to and from the port town showed a net balance for the local market (chiefly the power-mills) of 1,617,104 cwt. The total balance left in Bombay Presidency and its port town, therefore, amounted to 2,211,090 cwt.

Conf. with p. 161.

It must be here urged that the error due to the overlapping of outturn and traffic for the months of April, May, and June, of any cotton season with the official statistics of trade of the year ending 31st March, is a constant factor which prevents actual comparison between commercial and official statistics. But in dealing with the possible sources of error in Bombay cotton statistics, the Director of Land Records and Agriculture drew attention in 1889 to an even still more serious error :—"It must be noted, that in the estimates of outturn, the produce of over 100,000 acres, representing the area of unsurveyed villages, is not taken into account. If the produce of this unsurveyed area be added, the estimated outturn will leave a considerable margin, for the excess demand of the well-to-do classes and other miscellaneous uses, after meeting the full demand for local consumption."

Consumption. 89 (469)

In Bombay there are working, at the present moment, 55 mills in the island, and up-country 21. The former consumed, during the year ending 30th June, 1,973,055 cwt., and the latter 363,888 cwt. The total of these figures would give a greater consumption, were the official and commercial figures to be arbitrarily compared, than the available balance shown above. The imperfection of road statistics has, in most provinces, however, a serious influence. Thus large quantities are carried by road, across provincial frontiers, to collecting centres, particularly near the railways, and these appear in statistics of rail-borne trade without having been shown as exports from their actual areas of production. Thus, for example, large amounts of cotton from the Nizam's Dominions arrive at Sholapur and Ahmednagar by road, and are then conveyed by rail to Bombay. The totals of the entire cotton transactions of India, however, when carried over a number of years, so completely agree with production and trade, as to justify the inference that while errors exist in this province, and that, more serious in one case than in another, the figures here given may be accepted as fairly correct. This recommendation of accuracy would appear more especially applicable to Bombay, since the relation of area, outturn, and yield, bears a close approximation to all available statistics of the sales and movements of cotton to and from the presidency.

In a recent official correspondence regarding the cotton trade of Bombay and its Native States, it has been estimated that 20,000 cwt. are consumed for wicks, about 80,000 cwt. for beds and cushions, and about 619,400 cwt. for cloth of local manufacture, thus making a total of 719,000 cwt. Were a similar statement to that in the table, p. 56, to be worked but

CULTIVATION in BOMBAY.

for the past five years instead of one, the annual average left for local consumption in the presidency (excluding its port) would be found to have been 702,200 cwt. As already remarked, the up-country mills in Bombay Presidency consumed, in the year 1888-89 (ending 30th June), 363,888 cwt., and their annual average for the past five years, is found to have been 316,780 cwt. Deducting that amount from the total estimated annual average we learn that the Bombay Presidency, excluding its mills, consumes 387,420 cwt. a year. By a careful comparison, of outturn and total estimated consumption, *of Bombay-grown cotton*, the Director of Land Records and Agriculture has arrived at the conclusion that the average consumption of purely local cotton, per head of population, is 3·62lb, which, on the assumption of its all being made into cloth of the Bombay texture and weight (*viz.*, $4\frac{1}{3}$ yards to the pound of cotton), would be equal to 15 yards. This would allow for loss on manufacture, but deducting the equivalent for consumption of cotton in minor usages, an average of 13 yards is arrived at as probably more nearly the correct consumption. On this further factor the Bombay statistics may again be confirmed: hence it may in fact be safely said that about four parts are exported, and one part remains to meet the local consumption.

Chief Districts.
(470) 90

CHIEF DISTRICTS OF COTTON CULTIVATION AND RACES GROWN.—These are indicated by the following figures for the year 1888-89:—

	Chief Races grown.	Average acreage.	Outturn.
			Cwt.
Khandesh	Oomrawattee	921,202	519,389
Dharwar	Dharwar Kumta	475,952	156,333
Bijapur	Coompta (Kumta)	336,160	116,293
Ahmedabad	Dhollera	327,238	278,561
Broach	Broach	259,469	194,183
Belgaum	Coompta (Kumta)	146,687	77,406
Surat	Broach	89,626	57,040
Ahmednagar	Oomrawattee	23,862	1,562

In the other districts of Bombay proper, the cotton area is very small.

The following passages, extracted from the *Gazetteers*, give particulars regarding the cotton grown in the more important of the above regions. While not an exhaustive selection, it is believed the facts narrated in these passages convey the main features of the growths or races of **Gossypium** met with in Bombay.

Speaking of the cotton grown in Khándesh, for example, it is stated that, in 1878-79, there was a tillage area under *kapás* of 590,703 acres. Cotton was even then regarded as the chief crop in Khándesh. The local form, known as *varhádi* or *berár*,* was thought to have come to the district through Málwa. The author of the *Gazetteer* says: "It is short-stapled, harsh, and brittle, and has lately been largely supplanted by two foreign varieties—Hinganghát of two kinds, *banni* and *jeri*, from the Central Provinces, and Dharwar or acclimatised New Orleans from Dhárwár." In a foot-note, the *banni* is said to be an earlier variety with good staple, but leafy: the *jeri*, a later form (coming into the market a month or six weeks later), is whiter and freer from leaf, but has a poorer staple. The writer goes on to say that the Dhárwár is slightly longer in staple but much weaker than the Hinganghát, which, if well picked and cleaned, fetches a higher price. Dhárwár cotton, with larger and fewer pods, is the more easily picked. Being close-podded it can also be picked cleaner than Hinganghát, but from its larger and more clinging seeds, it is more apt to be

* Conf with Valayatí-Khandesh, pp. 58, 75.

CULTIVATION in BOMBAY.

Chief Districts.

stained in ginning. Of Dhárwár, it is remarked, there are three kinds of cotton grown, *viz.*, **Gossypium arboreum**, *dev kápas* (God's Cotton), used in making sacred threads and temple lamp wicks; **G. indicum**, *jvári-hatti*, that is, country cotton; and **G. barbadense**, *viláyati-hatti* or American cotton. Of these three forms, **G. arboreum**, a perennial bush, 8 to 12 feet high, is grown occasionally all over the district—in gardens, beside wells and streams, and near temples. "It is much like the Brazilian or Peruvian cotton plant, and, though this is unlikely, it is often said to be an American exotic." The *jvári-hatti*, generally known in the Bombay market as *kumta* (Coompta) cotton, is largely planted all over the black-soil plain. The *viláyati-hatti*, commonly known as SAW-GINNED DHARWAR, which was introduced by the Government in 1842, has thriven well and has come to occupy about a quarter of the district cotton area. Among the cotton-producing districts of the Bombay Karnatak, Dhárwár stands first, and both its American and its local cotton are highly esteemed. All evidence goes to show that with fair treatment in preparing them for the market, the two varieties grown in Dhárwár will rank among the best cottons of India. In Bijapur, the same forms of cotton appear to be grown as described above in Dhárwár.

Of Broach, it is remarked, there are two forms of cotton—the annual of black soils, *lália*, and the triennial of light soils, *jária*. These are the chief forms, but, according to the *Gazetteer*, two others are less frequently met with, *viz.*, *roji* and *nurma*. The latter appears to be **Gossypium arboreum**, which will be found described in various passages throughout this article, more particularly at pages 5—15. The *roji* is an inferior cotton, grown mainly in Baroda territory and brought into Broach district, chiefly with the object of being mixed with the regular Broach cottons. Of Ahmedabad, it is remarked that no foreign cotton is grown. "The local varieties are, in the Viramgaum sub-division, *jatvaria*; in Dholka, *bhália* and *vagadía*; in Dhandhuka, *lália* and *vagadía*; and, in small quantities near the city of Ahmedabad, *jaria*. Except *jária* these are all varieties. The *jária* is allowed to grow for four seasons." The *jária* is grown along with millets; the other kinds are never treated as mixed crops. The *vagadía* is a hard-shelled cotton, which must be picked with the pods, the other kinds ought, in picking, to be separated from the pods. But this is nowhere the practice, and, in consequence, the shell, broken and mixed with the wool, greatly lowers its value. Mr. Ozanne informs the writers, in a recent note, kindly furnished on the subject, that the Sanitary Commissioner attributes the high death rate at Viramgaum and other places where the *vagadiá* is cleaned by hand, to the diseases which result from the practice.

Vagria or vagadia cotton.

91 (471)

The Native State of Kathiawar is a most important producing area in the cotton supply of Bombay. Within the past few years, the area under the crop has been remarkably expanded, and the amount of *Dhollera* cotton brought to market accordingly greatly increased. Kathiawar, with Baroda, Ahmedabad, Cutch, Palanpur, Kaira, Mahi Kantha, and Cambay constitutes the region from which *Dhollera* cotton is derived, and during 1888-89, 2,547,000 acres were under that crop. In the *Gazetteer of Kathiawar* it is stated that two kinds of cotton are grown, *viz.*, *vágria* or *dhánknia*, and *pumadia* or *lália*. *Vágria* is almost solely grown in the north and east districts in Jhálávád and Dhandhuka, and *lália* throughout the rest of the province. The pod of the *vágria* does not open when ripe so as to allow the seed-cotton to be picked. It only slightly bursts and continues to clasp the cotton tightly, so that the pod and all has to be picked. The *pumadia*, when ripe, bursts open, and the cotton is easily picked from the pod as it hangs from the tree. The *vágria* cotton, though

Cotton pod which does not open.

92 (472)

CULTIVATION in BOMBAY.

Chief Districts.

inferior to it, can, by careful handling, be prepared so as to look very much like *pumadia*. *Vágria* has the advantage that the pods can be stored during the rains without injury.

Vágria cotton stored and cleaned in December is called *navlodh*. It is whiter than *pumadia*, but, in other respects, poorer. The chief varieties of exported cotton are locally called after the division from which they come, as Hálár and Jhálávád. The best cotton comes from the Pálitána State. The whole crop of Pálitána cotton is only about 2,100 tons, but it is by far the best prepared for the market, being well cleaned and free from adulteration. It is bulky, of a good bright creamy white with a yellowish tinge, and the staple is even, fine, and fairly strong and long. It classes from good to fine. The cotton grown in the Bhavnagar districts is called *desán*. It closely resembles Pálitána but is not so well prepared and is occasionally damped to add to its weight, or is mixed with seed, especially when prices rise beyond the rates at which the cotton is usually sold. The Bhavnagar crop averages about 9,800 tons, and most of it classes in Bombay and Europe as good. With care it could easily be made as fine as Pálitána. The cotton grown in Wadhwán, Limbdi, Halvad, Chuda, and Gujarvedi is known as *vágria Jhálávád*, and is grown from a peculiar seed. When properly picked and cleaned, it is reckoned good cotton, being a bright white which does not readily fade. Its staple too is strong, fairly long, and even, and has much body and bulk. Spinners readily buy it to mix with other kinds of cotton which lack these qualities. It is seldom properly prepared, most of it being full of fine-powdered leaf, which, after a certain stage, it is almost impossible to remove.

Jatvad cotton, grown between Wadhwán and Viramgaum, finds its way into the Kathiawar markets; it is of a very low quality. *Mahuva* cotton, grown chiefly in the Mahuva, Lilia, Amreli, Kundla, and Talaja districts, closely resembles Pálitána cotton, and, like Pálitána, classes as good and fine. What specially recommends it to buyers is its purity and cleanness. "In Bhál and the neighbouring district of Dhandhuka, a better cotton is known as *pumadia* or *lália*." "The bulk of the crop from these districts, as well as that from Limbdi, Ranpur, Borsad, Cambay, and Dholka, is marred by leaf and is known as *sakalia*. A weak, flimsy cotton finds its way into Kathiawar from these districts and is known in Bombay as *sagar*. It is harsh, short-stapled, and of a dirty white, much like wool in appearance and touch, and mixed with salt and clay. It is known at once by its smell. The cotton of Sorath and Hálár is inferior to that grown in other parts of Kathiawar, and is known in Bombay and Europe by the names of its place of export, Veraval and Mangrol. It is soft and flimsy, of a bluish white, and though the staple is fairly long and silky, it is weak and much mixed with dark broken leaf and seed. Here, as elsewhere, except in Mahuva and Pálitána, adulteration is common and most difficult to stop."

The above passages, taken almost at random from the *Gazetteers*, indicate the complex nature of the subject of the Bombay cottons. It seems probable that most of the forms mentioned are local cultivated races that owe their peculiarities to selection and methods of cultivation, continued over many generations, if not centuries. For an account of the botany of the indigenous races of Bombay cottons, the reader is referred to the special Chapter on that subject (pages 9, 12—14, 17, 28.). It need be only here added that by far the most prevalent form is the plant now known as **G. Wightianum.** (*Conf. with pp. 88, 129*).

It is also perhaps needless in this place to allude to the forms of exotic cottons that have been tried in Bombay. The history of the experiments with exotic cottons will be found detailed below, and it is probably enough to add that two or three forms have assumed acclimatised, though dege-

CULTIVATION in BOMBAY.

Conf. with pp. 20, 22.

Chief Districts.

nerated, conditions. These are **G. hirsutum**, *Roxb.*—New Orleans cotton, for the most part called saw-ginned Dharwar cotton: **G. acuminatum**, *Roxb.*—the kidney-cotton, most generally met with in the presidency—and **G. peruvianum**, *Cav.* The two last-mentioned exist mainly as garden crops; they grow, for the most part, into trees, and have thus come to bear the vernacular names of **G. arboreum**, the floss indeed being used in the preparation of the sacred Brahminical thread. To some extent, also, the remains of the cultivation of other exotic cottons exist, the Georgian, Sea Island, and other forms being occasionally met with in degenerated and neglected conditions.

The difficulty that exists, in accounting for the special properties of each of the Indian races of cotton, or of establishing the causation of inferiority or superiority, must remain in the obscurity that involves many features of the cotton industry, until definite experiments have been made with the view of ascertaining the features of each species botanically, and the taints of character, injurious or otherwise, that have passed into the races or hybrids now grown in the country. Such experiments, if conducted coincidently with chemical and meteorological investigations, would afford the key by which many obscure problems might be solved, and a tangible advancement made towards the improvement of the cotton supply. It would almost seem as if this much-to-be desired object had been retarded, through experiments at acclimatisation taking the place of selection and improvement of existing forms. The fact that so many races of cotton occur in one district, as have been mentioned above, in connection with Kathiawar, seems to show that India need not be dependent on America for new and improved forms of cotton. These remarks are not, however, intended to throw doubt on the propriety of attempts at acclimatisation, but rather to deprecate the false impression that too frequently exists in India, *viz.*, that no other methods of improvement are open to the agricultural reformer.

Method. 93 (473)

CULTIVATION.—SOILS: ROTATION OF CROPS: PLOUGHING: IRRIGATION: &c.—One of the earliest and, at the same time, one of the best detailed accounts of Bombay cotton is that written by **Dr. Hove** in 1787. **Linschoten**, and also **Rheede**, 200 years before the date of **Hove's** visit, briefly described the cottons of India. The former traveller made special mention of the muslins of Cambay and of Dacca, and of the cotton of Coromandel and the calicoes of various places in the Madras Presidency. The distinguished Polish traveller (**Hove**) came, however, to India for the express purpose of studying the cotton question, and his report is replete with interest. The botanical results of his study have already been briefly discussed under **G. arboreum** (to which the reader is referred, p. 12). **Hove** found, in Guzerat and other parts of Bombay, cotton cultivation relatively in as flourishing a condition as it is to-day. For miles and miles, as far as the eye could see, plantation after plantation of cotton stretched, across Cambay. On the lower-lying lands a form of **G. herbaceum**—the yellow-flowered cotton as he calls it—occurred, and on the higher sandy soils **G. arboreum**—the tree cotton with red flowers—was grown. One of the most striking features of **Hove's** account is the fact that **G. herbaceum** was apparently often grown on inundated land or even in brackish swamps. The fields were banked up to retain the rain-water, and wells were dug, in their vicinity, from which water was drawn by crude Persian wheels or leathern buckets elevated by bullocks. His description of the methods of well irrigation is, in fact, identically that pursued with other crops, though, at the present day, only rarely resorted to with cotton in Bombay. In 1789, it would thus appear, the cultivators regarded a high state of irrigation as an essential to successful cotton cultivation. Within the past few years this idea seems to have been returning to Bombay; but, for at least half a century, it has been

CULTIVATION in BOMBAY.

Method.

the habit to grow cotton only on soils possessed of deep-seated moisture, or in districts subject to rain at the season necessary for cotton-growing. The degree of inundation, which **Hove** seems to have witnessed, is abundantly exemplified by his statement that with yellow cotton, rice was often grown between the lines of bushes, whereas, with the red cotton, *bajra, juar*, or gram were grown. The yellow cotton was sown, he remarks, in the end of the rains, so that it might not be destroyed by too great inundation. It consequently "ripens a month later than the red cotton, grown on the higher lands and having the advantage of the powerful dews, which refresh the pods and promote vegetation, the wool gets not only compacter and finer, but likewise produces a greater crop. But the higher grounds, on which cotton ripens in the cold season, are not without some advantages. The dews falling on the wool that discovers itself by the elasticity of the capsula bleaches it, by the assistance of the sun, and contributes, in some measure, to its fineness. But where they have not the conveniency of water to support it in the period when the pod begins to swell, the crop is but poor and inferior in quality to those that have either a succour by freshes or are situated lower where the rains may have more influence." The idea of the necessity of artificial irrigation at the time the fruit is forming, has, within the past few years, begun to be felt, and many writers hold that the defect of harshness complained of against Indian cotton as a whole, would be remedied by a return to the system of cultivation which prevailed in the last century, where water, if not obtainable naturally, was supplied freely by artificial means.

Dr. **Hove** gives an interesting account of the method of ploughing and sowing cotton, as witnessed by him at Gorga, where, he says, the produce was the finest. As soon as the rain ceased, they ploughed all the high lands in the following manner: "A man measures the lines out by walking at the distance of 10 yards before the plough, drawing a line by a hooked piece of wood that scratches the earth which the ploughman follows, and indents a furrow at the depth of six inches by an iron cutter, which is in the shape of an elephant's tooth mounted into a solid piece of wood. The car at the bottom is hollowed on both sides into a beading by which the furrow is widened, and the earth both pressed on the sides and prevented from falling back and filling it. The beam the oxen or buffaloes are fastened to, may be placed agreeably to their size, and the reins are passed through a hole in the upper part of the plough tail. The plough is so light, the cattle so well trained, and the earth so very brittle and mild, that it is more a play-work than labour, both to the farmer and cattle." "They do not cover the seeds but leave it to the first shower of rain which fills the furrows with the turned out earth, and the plantation is as level as a table."

In the *Bombay Gazetteers* and *Agricultural Department Annual Reports* much valuable information will be found regarding the method of cultivation pursued at the present day. The limited space at our disposal prevents us from even reviewing all the available information, but the following passages, selected under certain important sectional headings, will be found to exhibit the main features of this important brarch of Bombay agriculture.

In the *Khandesh Gazetteer* it is stated that "cotton grows both in black and light soil. It is seldom sown in the same field oftener than once in three years, the intermediate crops being wheat and millet. With a moderate rainfall the black-soil crop, and with a heavy rain-fall the light-soil crop, is the better. There is no special ploughing of the field for cotton. After the first or second rainfall, the heavy hoe, *vakhar*, is passed over the field to loosen and clean it. Manure is seldom laid down imme-

CULTIVATION in BOMBAY.

Method.

diately before sowing, as the natives hold that it should be in the ground a year before the seed is sown." In the account of the Bijapur district it is stated, that "No crop takes more out of the soil than cotton. It never thrives in the same field for two successive years, but must be rotated with Indian millet, wheat, or gram. The cotton fields are enriched with the ordinary manure. Fresh manure is believed to heat the soil and, therefore, applied to the land the year before the cotton is sown. Prior to sowing partly by the hand and partly with the hoe, the field is cleared of the stumps of the previous crop, and if the field is overgrown with the *karige* grass, it is ploughed with the larger plough or *neglai*. After the ground is cleared, the clods are broken by a heavy wooden beam. In the latter part of August the land becomes fit for sowing. The seeds are rubbed in fresh bullock dung and water, and are then dropped through the hollow tubes of the seed drill or *kurgi*. The seed drill is immediately followed by the hoe which closes the drills. The seed leaves show in six to eight days, and in about a month the plants are three or four inches high. The farmer then works the grubber between the rows of seedlings, rooting out young weeds and grass, the surface is turned, and the soil is heaped at the roots of the young plants. Weeds are also removed by labourers with a sickle. The crop is ready for picking late in February or early in March. A good crop yields five and sometimes six, pickings; a poor crop not more than three or four." Of Ahmedabad it is said that "of the three varieties of soil, mixed, *besar*, is the best, black, *káli*, the next, and light, *gorádu*, the least suited to the growth of cotton." "Manure is used in light, but not in mixed, soils. Even in light soils it improves the crop only after a good rainfall. In July, when the land is ready, the seed cleaned by rubbing with earth and ashes in a crude frame, is, at the rate of ten pounds the acre, sown from a drill-plough with three or four tubes or feeders. The watering of cotton, common a hundred years ago (December 1787), but apparently in 1850 gone out of use, has again come into practice in Viramgam Dholka, and Dhundhuka. Watering much increases the outturn, but is open to the objection that it makes the crop more apt to take harm from frost. When full grown the plants stand from 3½ to 5 feet high. Fields sown in July, flower in September and October, and pod in December. In light soil, before the time of flowering, cloudy weather, or even slight rain, though it somewhat keeps back the picking, does good. Much rain increases the size of the seed and lessens the outturn; and after the pods are formed and when near bursting, rain harms the crop."

In the *Dhárwár Gazetteer* an interesting account is given of acclimatised American cotton, from which the following passage may be given regarding the soil and climate of the district: "The upland plain of Dhárwár enjoys the unusual advantage of two rainy seasons, the south-west between June and October, and the north-east or Madras between October and December. The north-east rains give the country a fresh supply of moisture in October, and often again in November, and a small degree still later on. This moisture, with the cool November nights, has had a large share in successfully acclimatising New Orleans cotton. In the Dhárwár cotton plains, the yearly rainfall ranges from 25 to 30 inches. During the cotton-growing months, that is, from September to February, the returns for the five years ending 1882, show a greatest heat of 97° in February, and a least heat of 58° in December. For cotton to thrive, the soil should be loose and open enough to allow the air and sun to pass below the surface, and still more to let excessive and untimely rain drain from the roots. These qualities the crumbling gaping soil of Dhárwár has in an unusual degree. The Dhárwár husbandmen describe

CULTIVATION in BOMBAY.

Method.

their cotton lands as of two kinds: *huluk-yeri*, which is a mixture of black and red soil, and *yeri*, pure black soil. Both local and American cottons are planted in both these soils, but *huluk-yeri*, or black and red, is generally considered best for New Orleans cotton, and *yeri* or pure black for local cotton. The black and red is considered the richer of the two, but in a bad season, blight and other diseases show themselves sooner, and to a greater extent, in black and red than in pure black. The great merit of these two soils is the surprising length of time during which the under-soil keeps moist. It is this underground dampness that enables the cotton plant, especially the American form, to mature as late as March and April. When the surface of the field is baked and gaping with the heat the cotton bushes are still green, because their tap-roots are down in the cool moist under-soil. Cotton is seldom grown in red soil; the outturn is too small to pay at ordinary prices." In a foot-note to the above passage the chemical analysis of the Dhárwár cotton soil is given as follows:—"An analysis of the best cotton soil showed in 4,500 grains 3,324 grains of very fine soil, 936 grains of impalpable powder, and 240 grains of coarse pebbles like jasper, with pieces, like burnt tiles, strongly retentive of moisture. The impalpable portion consisted of 18·000 grains of water, 0·450 of organic matter, 0·083 of chloride of sodium, 0·007 of sulphate of lime, 0·027 of phosphate of lime, 0·0450 of carbonate of lime, 0·013 of carbonate of magnesia, 15·200 of peroxide of iron, 16·500 of alumina, 0·085 of potash, 48·000 of silica combined and free as sand, and 1·185 loss; total 100·000."

Seasons. 94 (474)

SEASONS OF SOWING AND PICKING BOMBAY COTTONS.—The *Oomrawuttee, Dhollera, and Broach* growths are sown in June, the land having been prepared during May-June. The first pickings commence in December and continue till April and May.

The *Oomrawuttee* is the earliest in maturing, the pickings being complete in March. This is followed by the *Dhollera*, the pickings of which commence in February and are completed in April. The *Broach* begins to yield in March, the harvest being finished in May. It will thus be seen that a large portion of the Bombay cottons grown on the acreage of one year's statistics are included in the trade returns of the following year owing to the official year ending 31st March.

The *Coompta and Dharwar* are much later still. The land is prepared in July and August, the seed sown in August, the first pickings made in March, and the crop completely gathered in May. The whole of the yield of this class of cotton, from the area of one year's returns, therefore, appears in the trade transactions of the succeeding year.

Of Ahmedabad it is stated, that "beginning in early years in January, but oftener in February, the picking season lasts till the end of March and sometimes till April. Except the Dhandhuka *vágadia* or hard-shelled cotton, which must be picked with the pods, the other kinds ought, in picking, to be separated from the pod. But this is nowhere the practice and, in consequence, the shell, broken and mixed with the wool, greatly lowers its value. The one-year plant yields two and, sometimes, three pickings, the three-year plant always three. The picked cotton is gathered in heaps and at the end of the day carried to the yard near the cultivator's house. Here the pods are broken and the wool drawn out by the hand, and, to loosen and free it from leaf, it is laid on a rope framework or on the ground, and beaten with sticks." In the *Broach Gazetteer* it is remarked: "The time when a crop of cotton ripens varies according to the season. After a light fall of rain the cotton harvest is early, after a heavy fall it is late. If the rainfall has been light, picking begins about the 20th of December, and is over by the 10th of February. If the rainfall

CULTIVATION in BOMBAY.

Seasons.

has been heavy, the pods do not burst till the middle of February, and the cotton is not all housed till the last week in April. In an average season the picking begins about the close of January and ends in March. Before all the cotton is secured, the field has generally been thrice picked, with a fortnight's interval between each picking. The average proportion in weight of seed (*kapásia*), to cleaned cotton (*rúi*) is one-third of cleaned cotton to two-thirds of seed." "Before the time of picking, some of the cotton grows over-ripe, and, falling to the ground, takes up dust and leaf. In this way a portion of the cotton is damaged before it leaves the growers' hands. But the intentional mixture of dirt and earth to add weight to the cotton is not the work of the cultivator but of the dealer, *wakhário*. Cleaned cotton is divided into two classes, the better, *tumel*, and the inferior, *rasi*, or poor."

Cost & Profit. 95 (475)

COST OF CULTIVATION AND PROFIT.—The estimated cost of cultivation per acre on an *average field in a good season* is stated in the *Khándesh Gazetteer* (*1880*) to be as follows for two of the chief cotton-producing districts:—

	Khandesh. £ s. d.	Broach. £ s. d.
Labour	0 9 0	0 12 0
Manure	seldom used.	
Seed	0 0 7	0 0 4½
Rental	0 2 0	0 6 0
TOTAL .	0 11 7	0 18 4½

Conf. with prices, pp. 50 & 51.

In the former district (with the prices then current) the cultivation left a net profit of 14*s*. 8*d*. an acre, in the latter 16*s*. 7½*d*.

In the *Gazetteer for Dhárwár* it is stated (*page 285*), that "the cost of growing cotton is difficult to determine. Much depends on the grower, the number of cattle he owns, the area of land he holds, the number of persons in his house, and many other conditions which more or less affect his actual cash outlay in growing cotton. Roughly, the acre cost of growing American and local cotton is 11*s*. 4½*d*. (R5-11-0). As the value of the American crop may be set down at £1-10*s*. (R15) and the value of the local crop at £1-4*s*.-9*d*. (R12-6-0), the American leaves a net profit of 18*s*. 7½*d*. (R9-5-0) and the local of 13*s*. 4½*d*. (R6-11-0). To the net profit on the country cotton a small amount may be added, as in many cases the husbandman's family themselves clean the cotton."

In the *Khándesh Gazetteer* it is stated that the cotton crop is usually mortgaged to the money-lender, who receives it in the raw or unginned state, and gives back to the cultivator such seed as he may want for feeding his cattle and for sowing. This system prevails more or less throughout the presidency. The local dealers get loans from the larger buyers; from these they, in their turn, make advances to the growers, so that a state of indebtedness, dependent on the crop, prevails more or less in every grade of the cotton transactions.

Mixed Crop. 96 (476)

MIXED CROP CULTIVATION.—On this subject in 1886-87 the Director of Land Records and Agriculture wrote as follows:—"Respecting the cultivation of cotton and the merit of the local expedient of intersowing occasional rows of *jowari* or *túr*, experiments made in the season under report prove that the cultivator, even in a good cotton season, secures a larger profit from the mixed sowings than from cotton alone, and that in a year unfavourable to cotton he ensures a fair return. But, on the other hand, the experiments do not show a similar merit in the local practice of ploughing up poor cotton and following with a late crop. It is better to intersow gram or linseed between the rows of poor cotton." Many writers

CULTIVATION in BOMBAY.

Mixed Crop.

allude in similar terms to these advantages as justifying the growing tendency to extend the cultivation of cotton as a mixed crop. The habit is certainly less injurious to the cotton industry than the reckless admixture in the same field of two qualities of cotton by which means adulteration of good and bad qualities becomes unavoidable. It would seem, however, that even this pernicious practice is rarely a wilful action on the part of the cultivator, but is due, either intentionally or accidentally, to the buyer of the staple returning seed to the cultivator from a mixture obtained by ginning the cottons purchased from many growers. It would seem that this mixing of seed is only too frequently an intentional action, the dealer being thus saved the trouble of mixing bad cotton with good.

Selection of seed. 97 (477)

SELECTION OF SEED.—**Forbes Watson** expressed the opinion that the seed from four and five-celled pods would produce a better yield and a better staple than that from the ordinary three-celled condition. Experiments have been made, at the Hyderabad Sind Farm, to test this theory, and the results of the trials may be seen by reference to the tables on p. 30.

Whatever merit can be attached to the information therein tabulated there would appear to be no doubt that much greater success would be attendant on careful selection of seed than on experiments in acclimatising foreign prize varieties. If there be a demand for cheap cottons, as seems the case (both in India and Europe) care should be taken that the cultivation and trade in such staples be kept distinct from the good growths. The mixing of the seed of two qualities is not only direct adulteration but the surest way to secure, through hybridisation, the degeneration of the better stocks of Indian cotton.

Diseases. 98 (478)

DISEASES TO WHICH BOMBAY COTTON IS LIABLE.—In Dhárwár it is stated that the cotton crop is liable to two classes of disease. *Benithgi rog* is brought on by continued hard, cutting, easterly winds. The leaves turn red and become blighted. The flowers and pods fall off without maturing and the plant slowly dies. *Karaghi rog* is brought on by easterly winds accompanied with morning dews and fogs: it disappears if a westerly wind sets in before the disease has gained too strong a hold. Similar diseases are alluded to in most district reports.

The following account of the diseases to which the crop is liable may be abstracted from **Mr. Walton's** *History of Cotton in Bombay*:—Cotton is naturally a very hardy plant, and stands some wonderful vicissitudes of climate and weather, but, nothwithstanding this, there are some sudden atmospheric and other changes, which seriously affect it. The most trying consequences are produced by untimely, sudden, and heavy falls of rain, frequent changes of, and trying, winds, and cloudy weather; frost also injuriously affects the plant, but this seldom happens in the Kaládgi and Belgaum country.

The rayats often say their cotton plants are struck with *rog* (disease), when nothing of the kind has really occurred. One set of circumstances which causes this mistake, is when unusual heat comes on with an excessively dry atmosphere; and the roots of the plants have not reached down to a moist sub-soil. When this happens the branches and leaves first droop, then dry up, and eventually turn brown, getting the appearance of having been burned up. It is this brown appearance that the farmers sometimes mistake for blight, when the phenomenon has in reality occurred from the above causes only.

Cotton is essentially a sun plant, and if the sub-soil of the fields, where it is sown, is in proper condition for imparting the required sustenance through the tap-roots, no more rain is needed, and at most times, especially if heavy, any rain that falls after the bushes are about a foot high is injurious. If the fall is heavy, even before that time, it almost

CULTIVATION in BOMBAY. Diseases.

always does serious harm. Anything and everything that happens to cotton, and makes the plant look unhealthy, if even for a time only, is called by the native cultivators *rog*.

According to the local farmers, there are some five or six of these diseases to which cotton plants are liable, and they are given below in the order in which they may affect the staple in its gradual stages of growth as follows:—

Banti roga. 99 (479)

(1) *Banti roga*, or 'yellow disease.' This is caused by untimely and excessive rain flooding the fields. The effects are worst where drainage is defective. The stems and branches become a dirty yellow colour, the leaves red, and the bush droops; and if the disease is not arrested by the excessive moisture getting away in time, the plant dies.

Banji roga. 100 (480)

(2) *Banji roga*, or 'barren disease.' This is brought about by hard east winds blowing night and day, together with cloudy weather. The progress of the plant seems to be arrested; it ceases almost entirely to show fresh leaves, and the development of flowers and bolls quite stops.

Gugari roga. 101 (481)

(3) *Gugari roga:* this name literally means "half-cooked grain," and implies that the state of the soil and atmosphere has had the effect of partially boiling or cooking the plants. It is brought about by excessive moisture, and dull oppressive weather, with heavy clouds overhead, and but little and variable winds. The leaves fold and dry up.

Shidi hayu. 102 (482)

(4) *Shidi hayu.*—This *rog* is caused by long-continued harsh north-east winds, and is characterised by the leaves drooping; it is seldom fatal to the plant.

Kari jigi roga. 103 (483)

(5) *Kari jigi roga*, or the 'black sticky disease.'—This is quite the worst of all the ailments that the cotton plant is liable to, since, when it has once taken hold of a field, and the plants are far advanced, they hardly ever recover. It is produced by long continued dews and unceasing easterly winds, and shows itself by a thick, dark, gum-like substance covering the plants. This so entirely covers and closes up the leaves, that they, the flowers, and the half-developed cotton-bolls, all die and drop off, and, in a short time, a field, that had the appearance of strong, green, healthy bushes, shows nothing but some charred-looking dirty sticks.

Both descriptions of cotton are liable to all these ailments, but the acclimatised American may be considered the more susceptible of the two. The plants show their wonderful vitality and innate hardiness by rapidly recovering from disease, when the causes, which have produced it, are removed. All the diseases cease when a favourable change occurs, and genial, seasonable weather ensues for a sufficiently long period. If any of the ailments have attacked the plants in the earlier stages of their growth, and they recover, no eventual harm is done. The case is, however, different when it happens at a late stage of their cultivation,—the yield of cotton is then affected both in quantity and quality, the fibre is almost sure to be short and weak, and often dull in colour. (*History of Cotton, 113-114.*)

In 1867-68, a great part of the cotton-crop of Bombay suffered from blight, the effects of which were investigated and fully described by Dr. Forbes. The cause, he believed, to be exceptional climatic conditions similar to those detailed above; the effect was drooping, withering, and final death of the plants. Dr. Forbes sums up his report as follows:—

"This wind-blight is the bane of the southern districts. I have seen seasons pass without the slightest appearance of it, but of late, its occurrence has been more frequent, and its influence progressively more severely felt * * *. It may be described as the effects of a hot wind, more injurious, however, from its peculiar dryness, than from the heat which

CULTIVATION in BOMBAY.

accompanies it; its action upon the cotton plant is direct and speedy, and no amount of moisture in the soil will avert it."

Cotton picking. 104 (484)

COTTON PICKING—In Bombay, the picking of indigenous cotton usually commences in December and is completed in May. In the Presidency, as a whole, it would seem that the above dates are correct, although slightly different periods are recorded, in works treating of this subject. Thus, cotton picking in Khándesh is said to extend from the middle of October to the middle of January (*Gaz., 154*). According to Mr. Beaufort (*Indian Cotton Statistics, 24*) picking commences in December and is usually completed in March. The difference between these statements may be due to the introduction, since the date of the *Gazetteer*, of forms of cotton which mature later. "The average proportion of clean to seed-cotton is one to three. Seed-cotton, fallen on the ground, contains a certain amount of dirt, which is partially removed by beating it on the *jhanji* or *thátri*, a bamboo or cotton stalk wicker-work frame." (*Gazetteer*). In various passages, quoted above, the picking and cleaning of cotton has been incidentally dealt with, and the habit of criminal adulteration laid mainly to the charge of the dealer, not the cultivator. In the paragraph below, on the *Improvement of Bombay Cotton*, the legislative measures formerly taken, and since repealed, to check adulteration, will be found briefly reviewed. The final conclusion arrived at by most writers seems to have been that careful selection of seed on the part of the cultivator, and stringent measures by the wholesale dealers to bar systematic adulteration from the market, would effect a complete reformation in the Indian cotton industry. It has been contended that the matter rests more with the merchant than the Government, and that what impurities are due to the primitive methods of collection would soon be naturally reformed if the necessity for such reform were made to be felt not only by the primary dealers but by the cultivators. Pressure brought to bear on the local dealers, by those at the centres of baling, would go a long way towards giving a new life to Indian cotton cultivation.

Improvement. 105 (485)

IMPROVEMENT OF BOMBAY COTTON.—LEGISLATION TO CHECK ADULTERATION.—In the "Note" on Cotton, published by the Revenue and Agricultural Department, the following passages occur regarding the history of the legislative measures taken to prevent the injuries effected on the cotton industry through the practice of adulteration, that for some years prevailed to an alarming extent:—

"The subject of the improvement of the cotton culture and trade of the Bombay Presidency has occupied the attention of Government and of the commercial public for more than half a century. The introduction of American and other forms commenced seventy years ago. But the evil existed then, as it does even at the present day, of mixing indigenous with exotic, thus bringing discredit on the industry. It was not till after 1860 that measures were planned for endeavouring to stop this evil. In that year the practice of mixing local and New Orleans brought the Indian-grown form of the latter into disrepute. Shortly after, Government, however, appointed a Commission to enquire into the subject. The Commissioners found that—'during the season in which the enquiries were made, little or no local or American cotton had been shipped clean or unmixed. Besides the mixing of different varieties of cotton the dealers admitted that their cotton was mixed with seeds and other rubbish, and that it compared badly with the exports of other years. Many of the local dealers were anxious that the trade should be regulated by law and placed under inspection. In their report the Commissioners stated that the evils of the Dhárwár cotton trade were beyond usual remedies, and affected not only local but general interests. Nothing but the energetic action of Government could check so widespread an evil. Existing laws were insufficient; a fresh Act was required. With their report they submitted the draft of a Cotton Frauds Bill. This measure became law in July 1863 as the Bombay Cotton Frauds Act, IX. of 1863. The first Cotton Inspector appointed for Dhárwár was Captain (now Colonel) R. Hassard of the Bombay Staff Corps, who had already received charge of the Dhárwár factory from Dr. Forbes on his appointment as Cotton Commissioner." (*Dhárwár Gazetteer, 294—295*).

CULTIVATION in BOMBAY.

Improvement.

"The Act proved more or less inoperative. The Inspector found it very difficult to establish the mixing, at the gins, as fraudulent under its provisions, and its operation thus brought to light a defect in the provisions regarding fraudulent mixing. To be fraudulent, mixing must take place in cotton either offered for sale or offered for pressing. There were no presses in Dhárwár and the cotton was sold, not in Dhárwár but in Bombay. So the Inspector might see in a ginning yard a heap of local, a heap of American, and a third heap of seed to be added as a make-weight, and yet fail to secure a conviction." (*Dhárwár Gazetteer, 295*). The result was that out of many prosecutions, all of which were aggravated cases, clearly within the spirit of the law, very few convictions were procured. In 1873, the European merchants of the district as well as the local officers consequently pressed for a more stringent Act. In 1874, Government appointed a special Commission to enquire into the necessity for continuing the law to suppress cotton frauds. 'The majority of the Commission after collecting a large amount of evidence were of opinion that though it was not advisable to annul the Act it was preferable to place it in abeyance for a time. When the matter was referred to the Secretary of State, the Bombay Government were directed to prepare a fresh Act with the object of remedying the defects of the existing measure." (*Dhárwár Gazetteer, 299.*)

Legislation. 106 (486)

"Act IX. of 1863 was repealed by Act VII. of 1878. These Acts appear to have been extended to Berar but not to the Central Provinces (*Cotton Commissioner's Report for 1868-69, 215*). In September 1879 the Government of India recommended that all special legislation for the suppression of cotton frauds should cease. The Secretary of State did not then agree with the view held by the Government of India, though in March 1880 he sanctioned its proposals and desired the Bombay Government to do away with the special cotton fraud prevention establishment. Act VII. of 1878 was repealed by Act I. of 1882. 'According to Mr. Walton (Cotton Inspector) the opinion of the local European Agents and Native merchants was opposed to the giving up of Government efforts to check frauds. According to Mr. P. Chrystal, a Bombay merchant who is well acquainted with the Belgaum and Dhárwár cotton trade, the Bombay dealers and merchants in American Dhárwár and Coompta cotton think (1883) that the Cotton Frauds Act failed to stop adulteration in the Bombay-Karnatak. Mr. Chrystal thinks that since the Act has been stopped, there has been no noticeable increase in the adulteration" (*Dhárwár Gazetteer, 301*).

Admixture. 107 (487)

CAUSES OF ADMIXTURE.—The Collector of Khandesh, when referring to this admixture in 1878-79, pointed out that admixture begins with the sowing. The Khándesh ryots do not keep a portion of their crop for seed, as is customary in the Deccan, but depend for seed on the *bania*. The *bania* receives all kinds of cotton. That purchased from the pickers, who receive their wages in kind, must often be mixed. He sells this and what he has received direct from the cultivators to the large dealer, who has it ginned and pressed. The admixture so far may be done in ignorance, but on Mr. Robertson's authority I say that wilful admixture takes place at the hands of the large dealers and of the agents of the large ginning and pressing factories. At no single stage of this complicated business would honest and wise conduct succeed in preventing the evil. If the rayats sow pure seed, and keep their seed pure, the *banias* will mix it. If the *banias* were to guarantee good seed and to sell only such, they would have to charge a higher price for it, and the rayats would probably prefer using cheaper stuff which they would manage to find. If the large firms were to reject mixed cotton, their factories would be idle. Combined action is essential, but as unlikely as it is essential."

"The Director of Land Records and Agriculture, Bombay, in his annual reports (the main facts from which may be here reproduced) has repeatedly discussed the subject of adulteration and its possible prevention:—

"**1884-85.**—The subject of adulteration is of such importance that I feel it my duty here to state carefully the whole case before Government, for only a full statement can explain *the final deliberate decision that practically nothing at all can be done.*

"The Secretary of State forwarded a communication received from the Liverpool Chamber of Commerce calling attention to the increased adulteration of East India cotton—a subject of very great gravity—a growing evil which seriously imperils the foremost staple of India, at the same time affecting and disturbing most grievously the cotton trade. In due course the opinion of the Bombay Cotton Trade Association was solicited. This body informed Government that their attention had already been directed to the matter in such a way that they had been induced to appoint a special Committee to consider the question of mixing and adulteration in all its bearings. The conclusion arrived at was then announced—

"Having thoroughly considered previous legislative enactments on cotton frauds,

CULTIVATION in BOMBAY.

and the discussions connected therewith, by the light of more recent experience, and having further before them, the resolutions come to by the Liverpool Chamber of Commerce, they had, nevertheless, come to the conclusion that special legislation was not advisable. The Association then formulated its advice to Government, *viz.* :—

Admixture.

"(1) that Government should discourage the growing practice of ginning different growths together, thus producing a mixed and deteriorated seed;

(2) that Government should point out to the cultivators the benefit of persevering with the cultivation of the finer descriptions of the staple; and

(3) that Government should aid them in procuring improved seed for that purpose.

"With this advice it was suggested that the action initiated by Mr. A. T. Crawford, Commissioner, S. D., in Dhárwár, was a step in the right direction, and might be imitated in other districts with advantage.

"The proceedings of the meeting held to consider the report of the Special Committee were at the same time forwarded to Government.

"The Chairman premised his remarks by stating that some of the members of the Special Committee, after perusal of the pamphlet issued by the Liverpool Chamber, had entered on their deliberations with strong views in favour of special legislation, but that when the effect on the trade caused by the late Cotton Frauds Act was discussed, and when the manner in which the trade has been conducted since the withdrawal of special legislation was considered, the feeling became universal that further legislation was not advisable.

"It was admitted that 'a certain amount of admixture' had become prevalent since the abolition of the Cotton Frauds Act, and that this admixture was a decided evil, hurtful to the trade, and a source of loss to merchants. (It could not be expected that this meeting should concern themselves with the effect of admixture on the producers. That was not their province.)

"The Chairman gave his opinion that mixing had during the past season been of a less consistent character, and that professional mixers had not been so successful. He charged the native dealers with the origination of the plan of mixing, turning thereby their special knowledge of cotton to account. Mixed bales passed muster, and they became bolder. But the growth of the practice soon brought matters to a crisis, and a set of rules was formulated to combat the evil. The penalty clauses were, however, soon found unworkable. Contracts were made by the native dealers *inter se* with the omission of these clauses, and the European merchants were compelled to make them optional. But even with this relaxation of the stand taken against mixing, the Chairman thought that an improvement had set in.

"The deliberations of the Committee were then described as to the form which legislation, if decided on, should take. Fatal objections appeared to arise against every suggested penal provision. In fine, it seemed clear that a mild Act would be unworkable, and that any other form of legislation would entail a completely equipped staff of Inspectors, armed with extensive powers, which would simply be the re-institution of the Cotton Frauds Department with all the features detrimental to trade which had brought about its abolition. Therefore it was decided to ask Government for aid.

"It was prominently—but no emphasis could be too strong—stated that deterioration of seed was the great evil to be feared, and the grievous fact was noticed that during the previous year large quantities of *dhollera* cotton had been brought into Broach, and ginned with the fine Broach *kapás*. This, of course, meant irremediable mixture of seed, which would surely show itself in the coming crop.

"Next, the improvement effected by Mr. Ashburner, by substituting Hinganghát seed for the indigenous growth of the Khándesh district, was referred to, and it was admitted that, *during the time in which the evil of admixture had risen and thriven*, the deterioration of Khándesh cotton had gone on. The deduction was justly drawn, that it was this admixture by the trade which fostered the deterioration,—in other words, which fostered the cultivation of the short-stapled *varadi*.

"The Chairman, in proposing the form of letter in which the assistance of Government should be invoked, inserted a clause to the following effect—that it had been found from actual experience impossible, owing to the constitution of their own body, to establish and adhere to a set of trade rules which would effectually put a stop to the practice of mixing.

"This clause was omitted by a majority, on the ground that the Association was quite strong enough to make a rigid set of rules, which, if loyally adhered to, would go a long way to putting an end to the system of mixing.

"It remains to be seen whether this much-desired result can be effected by the Association. If the trade can and will stop the mixing, then Government's decision to do nothing at all will be fully justified. If, on the other hand, the trade cannot or

G. 107

(487)

CULTIVATION in BOMBAY.

Admixture.

will not stop mixing, it will, in my humble opinion, be incumbent on Government to seek to stop it without permitting interference with the trade or the obstruction caused by Inspectors, to weigh against the demands of the country and the welfare of the rayat. But I am anticipating.

"The Collector of Khándesh, Mr. W. H. Propert, was the first to lay before Government his estimate of the practicability of the advice offered by the Association.

"He pointed out that no effort had been relaxed by the Revenue officers—(1) to prevent the extension of *varadi**, for he had oftentimes gone near overstepping the bounds of legal authority to check its spread; (2) to encourage the cultivation of improved varieties, for at the Maheji Exhibition handsome prizes had been yearly awarded for the best samples of exotic cotton; and (3) to supply good seed, for the Government Farm had for its main object the production and distribution of as much good seed as possible; but that in 1882 the rayats got out of hand for two reasons. First, they had become fully aware that no restraining authority remained in the hands of the Collector, and, secondly, the price given for *varadi* had reached or nearly reached that accorded by the trade to the better staples of the Dhárwári (acclimatized American) and Hinganghâti.

"He showed that the producer has very little voice in the question as to what crop should be sown. He gets his seed from the middleman, and his crop is virtually disposed of, seed and all, before it is grown.

"He admitted that the ryot prefers to grow the indigenous variety, which is reared with far more ease and yields a heavier crop. When the demand rose for this commodity, it was very easy indeed for the middleman to extend its growth at will. And he asserted forcibly that if for one season (1883) the merchants had refused to buy *varadi*, or had at any time, within the past three or four years, assisted him in his efforts to sustain the good name ot Khándesh cotton, the *varadi* cotton would have been virtually stamped out.

"Next, turning to the question of the distribution of good seed, he indicated his reluctance to look again to the Berars, because there was evidence that the system of admixture had worked harm there as well as in Khándesh, and there was little prospect of advantage from the re-importation of this Hinganghâti variety, which is not so valuable or generally suitable as the Dhárwári. And so with regard to Dhárwár; the fact of Mr. Crawford's movement to improve the cotton of that district shows that the staple has deteriorated there also. Therefore there is risk in applying for seed thence. But Mr. Propert argued that in Khándesh itself it is possible to collect a large quantity of good seed by separately picking the bolls of the exotic cotton grown in and among the indigenous crop, relying on the fact that the former ripens earlier than the latter.

"I may parenthetically remark that good cotton can be procured from Dhárwár. Experience has shown that seed acclimatized there succeeds better in Khándesh than freshly-imported seed; and though the Dhárwár seed has deteriorated, its transfer to Khándesh will almost certainly give fair results. The plan recommended by Mr. Propert of collecting seed in Khándesh, too, is feasible. But, as is clearly argued, there is no possibility, under present adverse circumstances, of getting good seed sown. By the kind offices of one of the leading merchants in Bombay I was informed that his up-country agents could not, if they would, at present induce cultivators to sow the exotic cotton. Perhaps this opinion should be received with caution; for, as all the reports tend to show, the rayat is not the arbiter in this matter.

"Mr. Propert sums up with his advice that dishonest mixing should be made penal. He holds out the alternative that the Bombay merchants for one year should refuse to buy the mixed cotton. But the reply to this is that the merchants cannot afford to consider any such scheme.

"The Commissioner Mr. E. P. Robertson—after, I think very wisely, insisting that the coercive measures adopted by Mr. Ashburner some years ago were altogether over-estimated as regards their good results—says that Government have no right to interfere by dictating to the people what cotton they should sow, but fully supports the demand for a law to punish adulteration—very pertinently remarking that the merchants can hardly plead for the re-imposition of the law which their vehement opposition caused to be abolished, however much they may be convinced that that law is wanted.

"Independently I represented my view of the question. The object of my letter was to show that the advice of the Cotton Trade Association is futile, based on a misconception of the position of the rayat. I tried also to give reasons why the method of improvement initiated by Mr. Crawford in Dhárwár would do no good in Khandesh.

* *Varadi* in Khandesh = *Vilayati-Khandesh* in Berar Conf. with pp. 58; 89-91.

CULTIVATION in BOMBAY.

Admixture

I called attention to the late discussions in the Berars, and to the late Mr. Ridsdale's able review of the whole matter, in which he showed that interference with the rayat is not justifiable on either administrative or economic grounds. This review was called forth by the resolution of a meeting that the action of Mr. Ashburner in 1873 had secured wonderful results in Khandesh, but had been detrimental to the Berars, in that it caused an influx into that Province of the seed of the short-stapled variety ousted from Khandesh (which in the Berars is called Khándeshi, and which in Khándesh is called Berari or *varadi*), thereby working much harm, and that the imitation of the policy of Mr. Ashburner would rehabilitate the cotton of the province. The Government of India, Revenue and Agricultural Department, prevented any action being taken by the plain statement of the plain fact that Mr. Ashburner's policy in Khandesh had not secured lasting results. I endeavoured to make it clear that though the Dháiwári cotton does well in some parts and some soils in Khándesh, it is not suitable to others—that it would be an evil to attempt to stamp out *varadi* (or Khándeshi cotton), inasmuch as that variety is the only paying one in its peculiar soils. Further, that the production of *varadi* has its recognised and justifiable value, though the cotton is useless for export, except for purposes of admixture.

"I ventured to suggest that the earnest desire of the Cotton Trade Association to evolve a workable method of preventing fraudulent admixture, and the no less earnest anxiety of the District Officers to assist in the movement, would be best fostered by the appointment of a Joint Commission of merchants and revenue officers to enquire into and discuss the whole matter. On the one hand, the merchants would see how unworkable the advice offered by them really is, and, on the other hand, the Revenue Officers would learn much as to the most feasible method of meeting the evil without disturbance of the trade.

"With well-sifted information on the various points noticed, the Commission, I thought, would be in a position to recommend a feasible working-plan. The merchants would then be entitled to call on District Officers for co-operation, and the District Officers would know what support they could expect from the trade. At present in their eyes, in spite of the enlightened and well-intentioned declarations and warnings of the organised mercantile associations, the trade is doing everything it can to checkmate their efforts at every step.

"I casually hinted at the feasibility of establishing warehouses near the ginning and pressing factories where cotton received should be sorted and cleaned. I need only say that this plan is universally declared to be unpractical, and I do not wish to pretend that I am able to elaborate a scheme in connection with it.

"All these papers were considered by the Chamber of Commerce and the Cotton Trade Association. They were practically a challenge to the trade, and the trade could not take up the challenge. The former body explicitly stated that the admixture is due to a demand in Europe for inferior stuff.

"So long as there exists a demand in the consuming market of Europe for the inferior descriptions of cotton, so long will merchants continue to purchase, middlemen to mix, and growers to cultivate it. No good end could be served by attempting to interfere with the natural course of the trade.

"The Chamber went on to urge that there are signs of a re-action against mixed cotton, and advised Government to take energetic steps to distribute good seed throughout the districts, presumably to be ready to meet an improved demand.

"The Cotton Trade Association, in disposing of the reference, contended that there are buyers in Europe of adulterated and mixed stuff on its merits, and that merchants are bound to deal in it. It is, therefore, presumable that much of the adulteration which now goes on is not fraudulent. Certainly the former proceedings of the Association give me, at any rate, a different idea of the matter. It seems to me to be merely removing the fraud one stage further on, for these buyers in Europe of adulterated and mixed stuff on its merits, paying the high price they do for the chief ingredients (worthless on *its* merits) of the mixture, cannot be credited with honest dealing.

"The orders of Government were that no steps to stop the supply of a commodity which is grown to meet a demand by consumers can be taken. As regards the latest advice of the Chamber, it was pointed out that, till the improved demand for good cotton does arise, it would be premature to distribute good seed.

"It must be drawn from these orders that Government do not approve of any legislation to check fraudulent mixing.

"The course of events must be watched; but whatever is done in the matter must be done without consultation with the trade. It is, in my humble opinion, hardly fair to ask the merchants to show how the rayats can be protected from by-injury resulting from the course of the trade. On the other hand, the perusal of the

CULTIVATION in BOMBAY.

Admixture.

deliberations of the Cotton Trade Association gives the fullest confidence that whatever can be done on their part will be done.

"The Commissioner, N. D., Mr. G. F. Sheppard, brought to the notice of Government a specific instance of the importation into Broach of 179 tons of *dhollera* cotton to be ginned. It may be presumed that this was for the purposes of admixture. Clean cotton cannot be successfully mixed. Mr. Sheppard also noticed that annually a very inferior variety of cotton is exported to Dhollera from the Kaira district. He enlarged on the greater danger from deterioration of the seed than from mere admixture of staples, and urged that the subject demands consideration in the interests, *not* of the trader, but of Government itself.

"He contended further that it is easier to prevent admixture of seed at the beginning, than to restore the high character which a crop has once lost. He advised that the Collectors of Broach, Surat, and Ahmedabad should be directed to watch importations of inferior *kapás* and authorised to buy up the seed either pure or mixed. There can be no objection to such a plan. The seed would be useful for cattle food and the loss nothing, because it may be presumed that a lot of seed, stigmatized as mixed, would not knowingly be used for sowing. I trust that the Commissioner will carry his suggestion into practice.

"Mr. Stormont, of the Bhadgaon Farm, made a short tour chiefly to ascertain the extent of admixture *in the field*. The following may be quoted from his report:—

"*Tour in Cotton Districts.*—About the beginning of February the Collector directed me to make a short tour in the eastern talukas with the object mainly of finding out to what extent mixed cotton seed was sown by the cultivators; also whether, as alleged, merchants were paying preference prices for the short-stapled *varadi*. My inspection embraced a circle of districts including Páchora, Bhusaval, Savda, Chopda, and Erandol. The results of the analysis of numerous samples are arranged in the margin.

Table VI.—Analysis of Cotton.

Place.	Average Mixture.		
	Varadi.	Hingan-ghâti.	American.
	Per cent.	Per cent.	Per cent.
Páchora	25	30	45
Yaval . . .	58	40	6
Arrawad . . .	70	30	...
Chopda . . .	75	25	...
Dharangaon . .	50	20	30
Erandol . . .	58	29	13

"A glance at the table will show that the crop of Khándesh still contains a fair percentage of good cotton. At Páchora it is nearly half American; the short-staple being only a quarter of the whole. This is admittedly due to the spare farm seed being annually sold in the neighbourhood. At Bhusaval no reliable sample was obtainable, but about Yaval the American element diminishes to o per cent.; at Arrawad and onward to Chopda it altogether disappears, until the vicinity of Dharangaon is reached, where with the Hinganghâti it makes up half the crop. Coming southward by Erandol the *varadi* again predominates, being just outside the circle under the influence of the farm.

"*Opinion of Cotton Merchants.*—At Dharangaon a number of merchants and others met me to discuss the question of cotton improvement. The smaller class of local dealers were strongly opposed to any measures of restraint, declaring that although the price of *varadi* was comparatively low, still the cultivators liked to sow a certain mixture, it being the more reliable poor man's crop.

"The Bombay buyers, on the other hand, were unanimous in opinion that only the operation of an Act could ever restore the reputation of Khándesh cotton and the consequent prosperity which its producers enjoyed for so many years. All, however, were agreed as to the beneficial results which must necessarily follow the distribution of large quantities of good seed.

"It seems to me that Mr. Stormont has missed one point, deducible from his enquiry,—that is, that the further he went from the railway and from cotton centres (*e.g.*, Dharangaon) the less the percentage of good staple. This fact would seem to show that the suitability of the indigenous variety, apart from the support accorded by the demand for admixture, has a strong bearing on the increase of *varadi*. The deduction may be wrong, but if it is correct, it is a warning that a place for this variety must be recognised as necessary (see paragraph above).

1885-86.—"I have very little to place on record respecting this important staple. No further movement has taken place in the direction of check against admixture,

CULTIVATION in BOMBAY.

Admixture.

and I have no district agency to record the damage done by carting seed-cotton from districts, where the staple is bad, but the colour good and the cotton cheap, to be ginned with superior cotton. That this admixture goes on I have no doubt, and it is certain that the mixed seed is sown in the localities where the ginning is done, to the great detriment of the crop. One circumstance, however, is worthy of note. The Liverpool Chamber of Commerce has established a new standard which will be known as *biláti*. It will include the short-stapled *omrawattee*, and will apparently be a direct check against the export of the short-stapled *varadi* as it is called in Khándesh, or Khándeshi as the Berar people chose to denote it, under the standard of *omrawattee*. It remains to be seen from the quotations and trade returns the price which the new standard will command and the proportion it holds to the whole crop."

In concluding this series of quotations from the best sources regarding adulteration and the measures that have been taken to repress the practice, it may be added that the modern demand for Indian cotton, consequent upon the great trade created by the American war, was not the primary cause of these measures. From the earliest records of the Indian export traffic (somewhere during the last decade of the eighteenth century) up to the present time, complaints have been rife. Thus, Dr. Hove in 1787 wrote, "I observed to-day, when they were embaling the cotton, if they" (the dealers at Synapoor) "adulterate it with any ingredients as it is reported in Bombay. But in justice to the inhabitants, I must say that they paid a particular attention that nothing impure was communicated, in which state they deliver it to Broach, where the cotton *is* adulterated with their own inferior sorts and that of Baonegar, Bodra, and Dolea to make it resemble that of Ahmood, which is first mixed together, as I am informed, and then passed again through the cylinders, by means of which it is so intermixed that the ablest connoisseurs mistake it for the original. To this adulteration the merchant has nothing to say and is very glad to obtain it in that manner. But the same on coming to Surat is adulterated by Europeans in such a manner that all this which was sent from here of late on our Company's account to China lies unsold and for which they paid the contractor a most exorbitant price who gained unlawfully, by the information I had here, no less than £25,000 sterling. As the method of adulteration was not secretly enough performed, it was of course immediately communicated by the agents to their correspondents here, that they might also adulterate it and send it in that manner to Broach. The method which the well-known contractor for that cotton practised was as follows: he purchased a large quantity of old cotton seeds which is sold publicly in the bazárs as food for cattle, then opening the bales he intermixed a certain quantity of it with some other rubbish, and closed it in the first state in which he delivered the contracted quantity to the respective commanders on the Company's account for China."

INTRODUCTION OF BETTER QUALITIES.

In referring to the introduction of exotic cotton into the presidency, the Director of Land Records and Agriculture says in his report for 1884-85:—

Exotic Cottons. 108 (488)

"(*a*) *American*.—The valuable staple acclimatized from American seed is well established. Its cultivation is liable to fluctuations with the rise and fall of the demand for good stapled cotton, but it will not disappear. Its hold is especially strong on the three talukas of Bankapur, Karajgi, and Ranebennur in Dhárwár, and the large areas of the surrounding Native States. It does not do well in the deep black-soil talukas. Its advantages are:—

"(1) that it can be put into the market sooner It ripens earlier;
"(2) its outturn becomes sooner certain, and advances can, therefore, be earlier secured on the crop;
"(3) cleaned by the saw-gin it economizes manual labour, and this at the time when this labour is in great demand;

CULTIVATION in BOMBAY.

Exotic Cottons.

"(4) in good years its yield is very high;

"(5) though more liable to blight than the indigenous variety, it possesses greater power of recovery.

"The question of introducing this variety into neighbouring districts, *e.g.*, Bijapur, is very important."

The experiments with exotic seed, however, appear to have proved unsatisfactory in 1884-85 and 1885-86, but further trial was subsequently made with fresh seed imported from America. This was tried in Khándesh also, but the Director, in his report for 1886-87, states that it had been clearly demonstrated that it would be unwise to force exotic American cotton on the Khándesh cultivator as the soil and climate of the district was unsuited to its growth.

(*b*) ***Egyptian.***—The Egyptian black-seeded cotton was tried in Dhárwár in 1884-85 but failed. It has apparently been also tried in several districts without success.

Conf. with p. 17.

(*c*) ***Nankin.***—In June 1884, the Government of India supplied the Director with Nankin cotton seed to test the suitability of the Nankin or *Khaki*-coloured cotton which, it was contemplated, should be used for the summer clothing of the troops in place of dyed cloth. A guarantee was given that the produce would be purchased at 4 annas per ℔ of cleaned cotton. The Director wrote in 1884-85 regarding this experiment:—"The cultivation of Nankin cotton in this presidency has not got beyond the experimental stage, and even experimental cultivation is in its infancy. The only instance known of a farmer growing this cotton on his own account is in the Shevgaon Taluka of the Ahmednagar district, where one Narayan Raghunath, a Brahman holder of some means, cultivates it yearly as a garden crop to the extent of five *gunthas*. The seed is said to have been imported from Delhi about twelve years ago. The cultivator manufactures the produce into cloth for his own use.

"On the whole, although the outturn under experiment was smaller than that of the local variety, it is not to be doubted that as large an area as may be desired can be grown with this cotton if the Government of India continues its offer of 4 annas per ℔ of clean lint; but without such a stimulus it is not at all likely to be grown at all. I am very much averse to any attempt at forcing this variety on the country. It may injure our staples by hybridisation. Its seed may get mixed with the other varieties, and a most harmful, undesirable mixture will result, which, once established, will take long to remedy. No alacrity is shown by the Military Department to utilise the natural *khaki* cotton in lieu of the dyed material, and it is not unlikely that a mordant may soon be discovered which will prevent the now unsatisfactory variations in the colour of dyed *khaki*. Again, experiment has yet to be carried on to decide whether the colour of the Nankin cotton can be relied on. I have instituted experiments on this point at the Hyderabad Farm. The Nankin cotton has no merit as to staple or yield; and my advice on the matter, as it stands at present, is that all attempts to force its growth should be abandoned. I earnestly ask that pressure may not be brought to bear on me to do more than experiment with this variety.

"The following extract from a letter of the Chamber of Commerce—to whom a small sample of the Shevgaon cotton which was decidedly superior to that grown on the Bhadgaon Farm, was sent—also supports my view:—

"'The staple is short, irregular, and wasty, and would be most unprofitable stuff to work. Under the circumstances, the Committee would strongly advise that the cultivation of this description be confined to those places where it is indigenous, as there would be great danger of its hybri-

CULTIVATION in BOMBAY.

Exotic Cottons.

dising with the local varieties to their detriment if introduced largely into this presidency.'"

The experiments regarding the possibility of securing a uniform colour in Nankin cotton by selecting the seeds from bolls of the desired tint were concluded in 1886-87, when the Director stated that—"The result of adequate trial is that it is very difficult to obtain clean cotton of a uniform colour, though it is possible to do so with great care in selection of seed; but as the variety of cotton is poor in yield and undesirable in every respect, the conclusion is not of great importance. The Military Department has now withdrawn its special encouragement of this variety by experimental results."

Perhaps the first person who saw khaki-coloured cotton in India was the late **Dr. Hove**, who, in 1781, came across fields of yellow-coloured cotton. At first, on being shown the cotton, he thought it had been dyed, but afterwards he had abundant proof that this was not so. In the attempts narrated above, history seems only to have repeated itself, for, if unknown in 1884, the knowledge in the cotton had only died out. It was perfectly understood a century ago.

Continued Experiments. 109 (489)

CONTINUED EXPERIMENTS WITH AMERICAN COTTONS.—The Director found that the exotic cotton seed (acclimatized American) had degenerated, and he therefore determined on importing fresh seed from America. The following extracts from his reports refer to the subject:—

1884-85.—"I noticed in my report last year that there was a spontaneous demand for fresh seed in Bankapur of the *white-seeded American variety*, the only one of very numerous kinds which has really shown itself worthy of support. But I found that no further action had been taken. The seed has degenerated. It is harder and smaller than it was. Though a good year seems to check this degeneracy, there is, no doubt, need for new importation. The demand for new seed is not confined to Dhárwár. From Kolhapur I have received a large indent. I have placed myself in communication with the Chamber of Commerce, and fresh seed will be imported.

"But I have decided on adopting a slightly different plan from that followed heretofore. Fresh seed is very expensive. It never does well the first year. I propose therefore to sow the seed in that part of the district (*e.g.*, Bankapur) where the variety succeeds best (I may notice that in the deep black-soil talukas of Ron and Navalgund it gives a poor crop in comparison with the yield of the lighter black districts of Bankapur, Karajgi, and Ranebennur) and to distribute the produce at a moderate charge. This plan I shall recommend to the Kolhapur State. Specific proposals will be made to Government for sanction to hire land for one year for the purpose."

On the subject of the soil best suited to American cotton, the Director wrote: "**Dr. Royle** pointed out (1840) one result of experiment which will, I believe, by all be recognised as correct, *viz.*, that the black soil is not so well suited for American cotton as the poorer soils. I cannot but think that attempts to introduce the improved varieties without regard to this fact were so far misguided."

This opinion should be compared with that arrived at regarding the soils of the Central Provinces found best suited for American cotton.

1886-87.—"I described last year the new importation of seed and its distribution. I am now able to give a very satisfactory account of the results. In Dhárwár in six villages of selected talukas, *i.e.*, talukas where the exotic cotton has held its own without the aid of official pressure, sowings were made by Mr. Price. A little delay arose owing to an untoward break late in August 1886. The season was, on the whole, favourable. The average yield was 158lb per acre of seed-cotton. The proportion of

CULTIVATION in BOMBAY.

Continued Experiments.

clean cotton to seed was found to be 1 to 2½, a high proportion when compared with the produce of the old acclimatised seed, which is generally about 1 to 3. Mr. Baines, while Acting Director, inspected the fields and placed on record the following observations :—

"(1) The higher light black soil showed better results than the deep rich plain.

"(2) The better the growth of the crop, the less the injury by blight and discolouration.

"(3) The pods were larger, though fewer, than on the crop raised from old acclimatised seed.

"(4) The plants were stouter and lower in growth than the old cotton.

"It must be remembered that these results are for the first year of introduction, while the process of acclimatisation was only commencing. The smaller growth of the plants seems to point to some hybridisation with indigenous cottons in the case of the old cotton.

"The crop was considerably better in some places than in others, but, on the whole, it was everywhere fairly good. The net cost to Government of the importation, experimental sowings, and distribution has been only R233.

"*Sowings of new American seed cotton in Dhárwár, 1886.*

Name of Village.		Area sown.	Amount of seed.		Date of sowing.	Outturn in seed-cotton.	
			Per acre.	Total.		Acre Rate.	Total.
		A. G.	℔	℔		℔	℔
Hubli	Kusugal .	2 39	10	29½	29th September .	80	237
	Halihal .	14 35	7	102½	29th August .	128	1,908
Banakpur .	Ingalgi .	25 10	5	125	26th " .	158½	4,006
	Budihal .	8 7	6½	53	26th " .	173	1,414
	Kotgiri .	19 5	6½	125	7th September .	240½	4,603
Karajgi .	Devgiri .	16 9	7½	125	14th " .	97	1,570
Total .		86 25	6½	560		158½	13,738

"I have, under the head, farms, shown the result of experimental sowings at Bhadgaon. A large portion of the new seed was taken over by the Kolhapur Agency. In some States the season was unfavourable. The report received from Miraj Junior states : 'The cotton plant raised from the new American seed is 3 feet high as compared with 2¾ feet in the case of the old cotton. The pods are fewer but larger, about in the proportion of 2 to 3, but the weight is greater, ten pods of the new weighed 10½ tolas, while the same number of the old weighed only 8½.' These notes correspond with Mr. Baines' observations.

"In the Panch Mahals Mr. B. A. Dalal sowed the new cotton. It gave a poor yield, but he likes the quality of the sample."

GINNING 110

Conf. with pp. 152-157.

(490)

Cleaning or Ginning of Cotton in Bombay.—Mr. Beaufort writes under this subject :—"The *Broach* and *Dhárwár* growths are the only ones which are entirely cleaned by steam or hand saw-gins, but the quantity of *oomras* and *dholleras,* thus treated, is yearly increasing. All other descriptions of Indian cotton are hand-cleaned by the native *churka* or foot-roller." The following extracts from the annual reports issued by the Director of Land Records and Agriculture give the most recent informa-

GINNING in BOMBAY.

tion on the subject of saw-gins and the extent to which they are at present used in Bombay Presidency :—

1884-85.—" One of the most important questions in Dhárwár is with respect to the saw-gin. A reference to the Gazetteer* will show the various steps which led to the introduction of this necessary concomitant of the American cotton, which has established itself in the district, and the reasons why the management of the saw-gin factories still remains in the hands of Government. Three years ago it was discussed whether Government could not safely retire from the direct manufacture, distribution, and repair of the gins; but it was resolved to await the advent of the railway, when it might reasonably be expected that steam saw-gins would be set up by private firms. The railway is opened, but there is no indication yet of any such enterprise. There are steam gins at Hubli and Gadag, but the roller pattern is used.

" I sought to ascertain whether the people could not now be left to themselves to arrange for the repair of their gins. There is a great deal of discontent among them. The factories are made self-supporting by a subscription of R6 per annum per gin from each owner. The tax may be evaded by the deposit of the saws and spindles at one of the factories. The owners are entitled to the services of trained hands on payment, over and above the general subscription, for wages and material. The complaint is that the trained workmen kept are too few, and that only those villagers who live close to the factory get any benefit. On the other hand, more labourers would entail a larger subscription, and apparently the people are not careful to have their gins repaired in the slack season, but wait till the ginning time comes, and then, finding something go wrong, all clamour for the factory-men. At Hubli and Ingalgi the rayats assert that they can make their own arrangements, provided Government will assist them in procuring new saws and certain working parts and new gins when required; but everywhere else there was a consensus of opinion that Government must keep the factories, and I believe this is a necessity. It can only be hoped that private steam saw-ginning factories may yet be started.

" There is one gin worked by bullock power at Halyal near Hubli. At Devgiri I saw another being set up. There is a third, now unused, also near Hubli, and at Ingalgi there is a prospect of another. This utilisation of bullock power must be encouraged. It will reduce the labour now falling on the factory workmen, for the rayat who works a gin with bullocks is in a position to employ his own skilled hand. It may be advisable to seek to introduce Gordon & Co.'s bullock power into this district. This machine has no place to fill in Khándesh, but there may be benefit in Dhárwár.

" A new pattern of saw-gin has lately been introduced into Dhárwár. It is a great improvement; but the people, I think naturally, complain that the improved grid might have been adapted to the old frame-work and other working parts instead of necessitating their purchasing an entirely new gin. They do not like the new grid. It is too good, and cleans the cotton too well to suit the degenerated demands of the trade. The larger outturn of the old grid is solely due to the large percentage of seed carried through. The same feeling shows itself also in respect to the foot-roller. This rude implement worked by women can compete, and does compete, successfully with the steam-rollers, because the trade is not particular, and prefers buying cheaply half-clean stuff to giving a remunerative price for well-cleaned cotton.

" I may mention that the foot-roller—for indigenous cotton only—is the implement used all over the Southern Mahratta Country, but that

* Dharwar Statistical Account, Vol. XXII., 366-373.

G. 110

(490)

GINNING in BOMBAY.

strange to say, the hand-*charka*, very similar to the Khándesh *charka*, is again found in the very south of the district."

1885-86.—"I am glad to report that the Haliyal saw-gins are now worked by steam instead of by bullock power. This is, however, the only place in the whole Presidency except the two Government farms, where the steam saw-gin is used to clean cotton. No change has taken place regarding the manufacture, repair, and distribution of hand saw-gins in Dhárwár, described in last report.

"At Bhadgaon, now that Mr. Stormont has a steam engine, he will work saw-gins to clean the farm cotton and the small amount of the Dhárwár-American still grown in the neighbourhood from farm-seed. At Hyderabad, an experiment with the saw-gin for cleaning the indigenous cotton has been most successful. The whole of the farm crop was ginned with the improved saw-gin sent from the Bhadgaon Farm driven by a six horse-power engine. Mr. Strachan reports that the cost was R1-3-1 per maund of 84lb as compared with R2½ to R2¾ for the native hand-*charka*. The clean cotton sold for R15 per maund, while the ordinary native was selling for R13-8. The seed could have been sold for R1-10 per maund, but the price of the clean cotton alone was more than would have been got for both cotton and seed had it been sold unginned. The outturn of the gin was 18 maunds per day of nine hours. The Karachi Chamber reported, 'saw-ginned Sind native cotton from selected seed appears to be a great success. The staple does not show any signs of having suffered from the process of ginning, and is valued at about R1½ to R2 more than the average native *tanda*.' The hand-ginned samples of selected Sind cotton, *i.e.*, grown from the selected seed, were valued at R1 to R1½ above the then existing average selection of Sind. It is noticed further by Mr. Strachan that the saw-gin does not crush the seed, and that the percentage of clean cotton is greater than that obtained by hand-cleaning."

1886-87.—"The Dhárwár Factory is languishing. Perhaps the degeneration of staple in the exotic cotton is the cause. The extra price does not encourage or repay the use of the rather expensive gins. Private enterprise has not, as was hoped, relieved Government of the necessity for maintaining the factory. It has been decided, however, that if the Local Boards will not take over the factory, they will be closed on 31st August 1889. By this time the newly-imported seed may alter the prospect of affairs.

"This gin is wonderfully appreciated in Sind for indigenous cotton. It is only used in Dhárwár for exotic. I have in former reports given details of the working in the Sind Farm. Saw-gins were exhibited at the Shikarpur Show and were much liked. But a further indication of approval must be recorded. A large cotton dealer at Hyderabad, Sind, has contracted with the Government farm to gin considerable quantities of cotton with the saw-gin driven by the farm engine. The result of this working is not yet known."

Bombay Cotton Trade. III (491)

ANALYSIS OF LAST YEAR'S TRADE.—The total imports into the Bombay Presidency in 1888-89 amounted to 457,470 cwt., of which 119,949 cwt. came by rail, 287,553 cwt. by road, and 53 cwt. by coast, from external blocks; 47,900 cwt. by rail and 1,925 cwt. by coast, from Bombay port. The largest amounts received by rail were from the Central Provinces, Rajputana, Central India, and the North-West Provinces and Oudh; by road, from the Nizam's Territory.

The exports amounted to 3,427,184 cwt., of which 3,372,925 cwt. went to Bombay port, a few cwt. only going to Madras, the Madras seaports, the Central Provinces, Berar, and Bengal. Of the total exports by rail to Bombay port (2,130,897 cwt.), Khándesh, Násik and Ahmednagar contri-

Bombay Cotton Trade.

buted 974,186 cwt., Gujarát and Káthiawár 694,080 cwt., Poona and Sholápur, 208,112 cwt., the Southern Marátha Railway, 148,005 cwt., the region south of Narbada and below the Gháts, 102,883 cwt., and Marmagao Castle Rock, 3,631 cwt. The largest importing areas by rail were Gujarát and Káthiawár, the region south of Narbada, Khándesh, Násik, and Ahmednagar.

CULTIVATION in SIND. 112 (492)

2. SIND.

References.—*Stocks, Rep. on Sind; Gazetteer, 10, 13, 101-102, 169, 192, 426, 475, 520, 521, 559, 603, 654, 694, 746, 805, 851.*

Area. 113 (493)

REVIEW OF THE AREA, &c.—Sind shows a steady increase in the area of land under cotton cultivation, the annual average having risen from 57,900 acres in the ten years preceding to 65,600 during the past decade. For the last five years the average has been 73,000 acres. That this increase appears to be steady and continuous is shewn by the area for 1888-89, which was 96,400 acres. The total outturn indicates a corresponding increase, and in the year under consideration is returned as 184,200 cwt. for the British districts, and 5,400 for Native States, making a total of 189,600 cwt. These figures give an average outturn per acre of nearly 220℔, a very much larger figure than that returned for any other locality.

The net exports amounted to 147,950 cwt., leaving, when deducted from the outturn, 41,650 cwt. as available for local consumption. This, divided by the population, gives the amount locally consumed as 1·9℔ per head, equivalent to little under 9 yards of cloth. This figure cannot, however, be accepted as representing the actual consumption per head, since an average of the last four years gives only 15,000 cwt. as annually available for local consumption, an amount equivalent to 0·7℔ of local cotton per head.

The above figures, liable as they are to most of the errors pointed out in discussing the same question in other provinces, are also still further complicated by the very great deficiency in accuracy of the returns from Native States. It appears probable that, while the outturn may be approximately accurate, the area under the crop may be greatly under-estimated. If this be so, an explanation is at once afforded of the apparently excessive figure above derived for average area outturn.

District. 114 (494)

DISTRICTS WHERE GROWN.—By far the greater part of the cotton-crop is grown in Hyderabad, as will readily be seen from the following analysis of the area and outturn of the province:—

DISTRICT.	Area.	Outturn.
	Acres.	Cwt.
Karachi	1,764	3,473
Hyderabad	67,751	133,233
Shikarpur	11,275	22,172
Upper Sind Frontier	2,723	5,355
Thar and Parkar	10,171	20,002
Native States	2,700*	5,400*

* Estimated.

Races. 115 (495)

RACES OF COTTON IN SIND.—The whole of the cotton grown in the province is known, commercially, as "Sind." Experiments conducted at the Hyderabad Farm have shown that the fibre is good, and that the boll when cleaned furnishes the unusually large average of 36 per cent. of fibre. The staple, though strong, is short and frequently indifferently cleaned. Thus, in the *Gazetteer of Sind*, it is stated that "the great desiderata required to make Sind cotton more sought after, in the cotton marts of

CULTIVATION in SIND.

Races.

the world, would seem to be an increased length of staple and greater cleanliness. These improvements, it is believed, might be effected by more careful attention to the culture of the plant, and to the picking and cleaning of the wool."

Even so early as 1839 we find the cotton of Sind referred to by **Sir Edward Ryan** as a very flourishing plant, "cultivated in Upper Sind with great care by the aid of irrigation from the Indus."

Like most other commonly cultivated Indian cottons, that of Sind appears to be **G. Wightianum,** *Tod.*, or hybrid forms of that species. The hybrids are extremely variable, tending to revert to **G. arboreum,** and, in doing so, to deteriorate and give rise to local races with marked differences in quality of staple. Thus, the Superintendent of the Experimental Farm at Hyderabad writes:—"During the past season I have had a better opportunity, than ever previously, of comparing the cotton grown in different districts of Sind when it was brought in for ginning, and from no district did I see cotton so good as that grown on the farm from selected seed. This fact, I believe, is generally admitted. The Tanda cotton is best for length of staple; Hála comes close to it for length, and tops it for outturn of clean staple to seed; from the other districts the cotton is inferior to these. The poorest of all comes from Jhudda, in the Thar and Párkar districts. The seeds are smaller than those of the Sindi proper, but much more numerous, while the covering of cotton on them is very short and very thin. This year some of the farm-selected Sindi cotton has been sent for trial in the Thar and Párkar district, and I trust it will give such satisfaction as will induce the Deputy Commissioner to call for more of it next year." It may be noted in passing that the experiments at Hyderabad seem to have been attended with an unusual amount of success, most of which appears to have been due to the evident appreciation of native cotton on the part of the management of the farm. By careful selection of the best seeds, and attention to the best methods of cultivation and cleaning, a very exceptionally large outturn of a good fibre was obtained, which sold at R2 over the market rate, and which, at the same time, yielded good seed that could be sold for R2 per maund. There can be no doubt that the distribution of such seed, sold at only a little over the bazár rate, would greatly improve the cotton of the province generally.

Conjointly with these laudable endeavours to improve the local races of cotton, numerous experiments were conducted with the object of determining the suitability of foreign seeds to the climate and soil of Sind. The result of these has been to prove that no seed can in any way compete, under the local conditions of soil and climate, with the commonly cultivated Sindi race. Bourbon cotton, the exotic chiefly experimented with, gave a crop of only 316℔ in one case, 152℔ in the other, of seed or uncleaned cotton to the acre. Native cotton, on the other hand, yielded, in the same year (with similar care in cultivation), 1,775℔ of uncleaned cotton to the acre.

Method. 116 (496)

SOIL, MANURING, AND METHOD OF CULTIVATION.—The following account is extracted from the *Sind Gazetteer*: "Cotton is cultivated in two ways—*sailabi* and *bosi*. The first requires frequent watering after being planted, and the seed is sown on the sides of ridges after the surface has been inundated, the holes being made at a distance of about a foot and a half from each other.

"The second description is sown on the surface of the lands left by the inundation; no after-waterings are required, the dew, which falls heavily, affording sufficient moisture. The only care required is to keep the earth about the stems loose and free from weeds. Cotton is sown in Upper Sind at the end of February or beginning of March, sometimes in May or

CULTIVATION in SIND. Method. June, and picked in July and August, also in November and December. After picking, the cattle are turned in to graze, and the crops left for a second year. Cattle dung is used as a manure inthe proportion of about 12 maunds to a bighá. In other parts of Sind, cotton is not cultivated till the canals fill in June, and the crop, in consequence, is not picked till November or even December. The crop is liable to injury from bug" (? cotton-borer), "frost, and locusts." The rainfall of Sind, as is well known, is excessively small, so the greater part of the cotton, as of other crops, is cultivated by irrigation. Watering is effected by what is called '*charkhi*' or wheel irrigation.

Cultivation at the Hyderabad Experimental Farm has proved that, to ensure a good crop, (1) rotation ought to be kept up; (2) sowing ought to be broadcast, and should be carried out before the end of June. It was also proved that late cotton is very liable to the attacks of green fly in October.

Cleaning, 117 (497) CLEANING—Is, as a rule, performed by the ordinary native *charka*, which is similar to that described elsewhere. Recent experiments have proved that Sindi cotton is easily and more cheaply cleaned by the saw-gin than by hand; that the percentage of cleaned cotton so obtained is greater, and that the produce fetches R2 per maund more than the market rate.

It appears undoubted that the introduction of saw-gins, together with the efforts to improve the stock by careful selection of seed, may do much to improve the quality of the *Sindi* cotton of commerce.

Sind Trade. 118 (498) ANALYSIS OF LAST YEAR'S TRADE.—With the increase in importance of Karáchi as a commercial centre, the trade in cotton from the province has undergone a remarkable increase. Sind formerly imported the cotton it required, to the amount of thousands of cwt. annually, principally from Kachh and Gujarát, indeed, it was not until 1840 that the plant began to be at all extensively cultivated in the province itself. The fibre at first held a very poor position in trade, and in 1863 shewed a difference in value as compared with "Fair Dhollera" cotton of from 4*d*. to 6*d*. per pound. In 1870, it had so far improved as to reduce this difference to 1*d*. or 1½*d*. In the year under consideration (1888-89), the average value of the exports per cwt. amounted to over R20, an amount almost equal to that obtained for the average cwt. of Bombay cotton in the same year. The outturn has steadily increased, until last year, as already stated, it reached a total of 189,600 cwt.

The imports amounted to 77,079 cwt., of which nearly the whole, *viz.*, 74,012 cwt., was derived from the Panjáb, the small remainder came from the port town of Karáchi, and trans-frontier from Lus Bela, Khelát, and by the Sind-Pishin Railway. The exports amounted to 224,927 cwt., of which a small amount went to the Panjáb and trans-frontier to the localities above-named; all the rest, *viz.*, 224,330 cwt., was sent to Karáchi for shipment.

CULTIVAdION in BERAR. 119 (499)

3. BERAR.

References.—*Rivett-Carnac, Rep. on Operations of Cotton Dept., 1867-1869; Annual Reports, Cent. Provs. and Berar Cotton Dept.; Hume, Note on Cotton in Berar (1885); Official Corres. and Reports; Agri.-Hort. Society of India, Jours. (Old Series), XI., 456, 472; XII., 22-29; (New Series), I., Proc., lxi.*

Area. 120 (500) REVIEW OF THE AREA, &c.—This province, after Bombay, has the largestarea under cotton; but, in comparison to the total cropped area, the fibre is grown much more extensively in the latter than in the former tract. The cultivation of cotton in Berar reached its maximum in 1878-79, when as much as 2,207,889 acres were planted, but two years later a minimum,

CULTIVATION in BERAR.

Area.

for the last fourteen years, of 1,755,946 acres was recorded. Since that date it has again gradually increased, and, in the year under consideration, attained 1,991,551 acres.

Reference to the table, p. 56, will show that the total outturn in the same year was 1,276,061 cwt. of cleaned cotton, an average of 71·68℔ to the acre. Comparison with similar figures for 1887-88 shew a very marked increase in the production per acre, the total outturn in 1888-89 being nearly double that of the former year, while the area increased less than 100,000 acres. By a similar calculation to that followed in the case of Bombay and other regions, the total amount available for consumption in the year under consideration, is found to have been a minus quantity of 48,448 cwt. This fact is only to be explained by the absence of registration of road traffic, and by the error which necessarily arises from considering statistics for one year only. On this subject the Commissioner of the Hyderabad Districts writes :—

"The replies received from Deputy Commissioners shew that information as to imports and exports cannot possibly be given, as there are no statistics of road-borne trade in Berar. Information on consumption and outturn is included in the statement under submission, but the figures are only approximate. It is difficult to come to any conclusion as to the consumption of indigenous cotton per head of population, as no attempt has yet been made to ascertain what quantity of cotton is reserved by cultivators and others for local consumption. In the Amráoti district the consumption is estimated at one-tenth of the outturn, while in the Basein district the proportion, which estimated consumption bears to outturn, is 1 to 32. With such estimates arrived at by tahsildars, the figures must certainly be untrustworthy."

The Deputy Commissioner of Amráoti, commenting on the grave error which must arise from want of road trade returns, writes :—"The figures of yield and consumption given above are for Berar, but in addition cotton is poured into it on all sides from the Central Provinces and the Nizam's Territory. In Amráoti district alone the cotton markets of Amráoti, Dattapur, Karinja, and Murtizapur receive annually thousands of hundred-weights of cotton from the Nimar, Wardha, and Chanda districts of the Central Provinces, and from the Nizam's territories south of the Paingunga river, Rajur, Ealabad, Bela, and, perhaps, even more western taluks."

The exports from Berar, on the other hand, go principally to Bombay port for shipment, and are consequently nearly, if not quite entirely, represented in the rail-borne trade statistics.

With such an error in the available trade-returns of the province, it is obviously necessary to arrive at an estimate of the local consumption by some other means. This has been done in the recent note on cotton, supplied to the Government of India, from Berar, by accepting the estimates of tahsildars as approximately correct. These, though not as absolutely correct figures, may perhaps be accepted as fair approximations to the truth, and shew that the average in the year under consideration was 4℔ per head, indicating a total consumption of 103,725 cwt. This large amount, however, leaves only 1,172,337 cwt. to meet the net export of 1,324,482 cwt. The road imports alluded to must, therefore, have been over 150,000 cwt.

It appears probable, however, that the above estimate of 4℔ of cotton per head may be considerably in excess of the truth. The figure is certainly a very large one in comparison with that arrived at for other provinces, and is commented on, as follows, in the recent report above referred to : "The Deputy Commissioner of Ellichpur is of opinion that

CULTIVATION in BERAR. Area.

the difference" (the large excess of this figure over that of other provinces) "is probably due to the fact that the people of the valley of Berar are richer than their neighbours of the Central Provinces, &c., who have, on the whole, a less fertile country to live in, and consequently, do not consume cotton and other products of the soil to the same extent as the Berar *Kunbi* does; but the Resident feels bound to say that he fears the difference may be due to the figures being based on the merest guesses."

Chief Districts. 121 (501)

DISTRICTS WHERE GROWN.—The valley of the "Payanghát," where, it has been estimated, cotton occupies as much as 40 per cent. of the cultivated area, is the locality where the best cotton is grown. Here, deep rich black soil, sometimes extending to 30 or 40 feet in depth, is found mostly in the Akola and Amráoti districts (the former including the town of Khamgaon, the centre of the cotton trade), also in the northern portion of the Buldána district. The large trading towns of Amráoti, Khamgaon, Akot, Shegaon, &c., at which places there are presses, are situated in the same lowland country; but cotton trading is not limited to these large emporiums, as petty dealings take place, during the season, at every substantial village, weekly market, and even railway station, throughout the province. In the case of distant towns the rayats find ready sale for their produce, since the country possesses good roads, which act as feeders leading either to the nearest railway, or to one of the large marts.

The comparative share taken by each district during 1888-89 in this extensive cultivation will be seen from the following table:—

DISTRICT.	Area under Cotton.		Total out-turn of Cotton cleaned.
	Unmixed. Acres.	Mixed. Acres.	Cwts.
Amráoti	139,272	363,038	296,004
Akola	431,226	51,502	228,590
Ellichpur, excluding Melghát	194,397	12,833	101,152
Melghát	4,312	249	3,169
Buldána	258,340	55,199	348,126
Wun	131,942	118,569	204,280
Basim	230,672	...	94,740
TOTAL	1,390,161	601,390	1,276,061

Races. 122 (502)

RACES OF COTTON IN BERAR—The produce of Berar is known commercially as *Oomras* or *Amráoti* (better known as *Oomrawattee*) cotton. Like *Hinganghát*, however, it in reality consists of two distinct varieties, known by the same vernacular names of *bani* and *jari*. The former is cultivated, for the most part, towards the Southern Gháts, the latter in the deep black cotton soil of the low-lying districts. The average yield of both is pretty much the same, but since *jari* yields a larger proportion of fibre to seed, and has a stronger, though shorter and coarser, staple than *bani*, it is preferred by the cultivators. As already stated (*see pp. 64, 129*) both are forms of **G. Wightianum**, *Tod.* In 1842, **Mr. Mercer**, one of the experts sent to India to cultivate cotton, reported of Amráoti:—"The people are more nearly right in their method of cultivation than in any other part of India; the cotton is good and only requires more care in cleaning to make it a most desirable article."

Mr. Hare, Settlement Officer, Yeotmal, in a recent letter to the editor, gave the following interesting information regarding the cottons of his

CULTIVATION in BERAR.

Races.

district:—"Akote used to be famous for its *jari*. Then came the short-stapled *houri* which, being a heavy cropper, partially displaced the *jari*. In the last two or three years both have been replaced by a new variety called *kateli*, from a thorn on the seed*, which is much more profitable to sow than either of the others. This is probably a hybrid, and, if so, an attempt might be made to improve it still further, as the staple is not good. The cultivators have not the slightest idea where it came from." The remarks about a new variety threatening to displace the famous *jari* cotton of Berar is of interest, and it is to be feared that this *kateli*, from the poorness of its staple, may be productive of much harm. Specimens have not been received and consequently the authors are unable to determine the botanical nature of this race. It seems possible, however, that it may be either a form of *Vilayti-Khándesh*, which is known to be rapidly spreading in Berar, to the certain detriment of the once famous staple of that province, or the "Dhárwár" cotton referred to below. Dr. Hume, in his note on cotton (1885), describes three additional forms to *bani* and *jari*. The first of these is a *rabi* crop of *jari*, grown at the same time as wheat, and picked in March and April. It is sown on marshy and water-logged land, which, though quite uncultivable during the rains, retains water, and is thus available for cultivation in the cold weather. The soil is rich and deep, and supplies the cotton with water well into the hot weather. The crop, however, is not so heavy as that of *kharif jari*.

The second form is "Dhárwár," an acclimatised American cotton. Dr. Hume describes it as follows:—"The staple is rather shorter than that of the usual cotton. It is not grown as a separate crop, but plants of it appear at intervals in a cotton field. The natives call it '*gogli jari*,' the large *jari* or large-seeded *jari*. As a market cotton it has no separate existence, owing to the small quantity of it grown and the admixture of it with other cotton during the picking of the crops. Since the merchants have made such a determined stand against the Khándesh cotton, this American cotton is becoming more extensively cultivated than it ever was before."

Vilayati-Khandesh. **Conf. with pp. 58, 75.**

The third breed is *Vilayti-Khándesh*, which, according to Dr. Hume, came to Berar, principally because the Collector of Khándesh forbade its cultivation in his own district. He adds:—"The Khandesh variety is being grown largely to the ousting of the other varieties, and the most certain ruin of the Berar cotton trade. At present they obtain from *Vilayti-Khándesh*† an early crop, also a large one, getting three or four pickings instead of two or three as they get from the indigenous cotton. They receive R3 or R4 a bale less in price than for the indigenous cotton, but the greater bulk compensates, and much more, for this small loss. But this apparent prosperity will be short-lived, for it is only by mixing this *Vilayti-Khándesh* cotton with the indigenous cottons that merchants get it accepted in Bombay, and mixing cannot be carried to a greater extent than a third of the quantity with *bani* and a much smaller proportion with *jari*. When this Khándesh cotton has succeeded in ousting the *bani* and *jari*, there seems every prospect of extensive distress and ruin. The Khándesh variety is getting better than it was when first introduced, according to some people; but however good the Khándesh may become by hybridisation, it can never be as good as the old *jari* and *bani*; any hybrid of Khándesh must be a deterioration on them."

"It is a coarse hardy plant, that will grow almost anywhere that cotton will grow, and produces a large crop; but the staple is short and

* The seed of most forms of **G. herbaceum** proper is beaked *Conf.* with p. 27.

† It seems highly probable that *Vilayti-Khándesh* or *Varadi* is a hybrid form, very probably derived from **G. herbaceum** *var.* **microcarpum** on **G. neglectum**. The authors, however, have not seen an authentic botanical specimen.

CULTIVATION in BERAR. Races.

hard and is objected to as cotton in the home markets, and consequently in Bombay. It is a bad variety, and is doing a lot of injury, by its cultivation in Berar, to the Berar cotton reputation. The plant looks 'jungly;' the high split leaves show want of breeding and cultivation, which is borne out by the short-stapled fibre."

The following history of the attempts to extirpate the *Velayti-Khándesh* in Berar is of interest, and may be quoted in entirety from the Note on Cotton issued by the Revenue and Agricultural Department, Government of India:—

"This inferior cotton formed the subject of correspondence between the Resident, Hyderabad, and the Government of India in 1883-84. The Resident, **Mr. J. G. Cordery,** reported in August 1883, that vigorous measures were adopted in 1875-76 for the extirpation of this variety (*Vilayti*), which possesses no staple but is a very prolific crop, costs less to produce, and is clean and good to look at, and is, therefore, largely used for purposes of adulteration. It was added that the admixture had of late years been in some markets carried too far, and that it was apprehended that unless its growth could be discouraged, the reputation of Berar cotton would be much injured. The measures taken in 1875-76 were successful for the time, and proved, it was said, permanently successful in Khándesh. The cultivators who possessed adulterated or inferior seed were induced to receive in exchange for it seed of the same indigenous variety. Advances were made in Berar for this purpose amounting to R18,000. It does not appear that the agriculturists were compelled to accept the proposed exchange, but a special officer was deputed to the work and much attention and enquiry was needed for its successful execution. It was stated that the reason why the success was not permanent in Berar was probably the fact that, in January 1878, instructions were issued by the then Resident, **Sir R. Meade,** to Deputy Commissioners that no future attempts should be made to control cotton cultivation, and hence it was supposed by the growers that Government had become indifferent about the matter. The evil was said to have spread to the Amráoti district. **Mr. Cordery** was inclined to return to the policy formerly adopted, and he, therefore, asked sanction for another advance (R15,000) to purchase the seed required for the elimination of the *vilayti* seed from Berar."

To this application the Government of India replied in 1883 as follows:—

"I am directed to acknowledge the receipt of your letter on the present prospects of cotton in Berar and the increasing extent to which an inferior staple known as *vilayti,* and originally introduced from Khándesh, is being grown by cultivators. You refer to the measures adopted in 1875-76 in Berar, and previously in Khandésh, for the extirpation of this variety of cotton by the distribution of superior seed, by money advances, and by executive orders. And you consider that, though these measures produced no permanent results in Berar, and were abandoned in 1878, yet the large interests at stake, and the success which attended similar measures in Khándesh, justify further efforts in Berar. You accordingly solicit the sanction of the Government of India to the expenditure of R15,000 in advances for seed, of which sum you are disposed to think that not more than one-half might be recovered by Government.

"In reply, I am directed to say that, after giving the subject of your letter its careful attention, the Government of India has come to the conclusion that any attempt on the part of the State to make the Berar cultivator grow one class of cotton when his immediate interests prompt him to grow another, is almost certain to end in failure. The measures taken in

CULTIVATION in BERAR.

Races.

Khándesh have not been permanently successful any more than in Berar. The steps taken by **Mr. Ashburner** in 1876 succeeded at the time, but as soon as the vigorous measures to which he had to resort ceased with his departure from the district, the cultivators of Khándesh returned to the inferior variety of cotton. Writing in 1882, **Mr. Propert**, the Collector of Khándesh, was obliged to confess that all his endeavours during six years of office to discourage the cultivation of the inferior cotton had been practically fruitless.

"Under these circumstances, the Government of India, before sanctioning any expenditure on the purchase and distribution of seed in the Berars, would like to be assured that there is some probability of the experiment being productive of more lasting results than have been achieved hitherto. If there is any difficulty in procuring good seed, the Government of India will not object to the institution of any agency for obtaining and distributing it. But the Government of India is unwilling, in the face of past experience, to countenance a system of advances year after year on the mere chance of the cultivators of Berar being thereby induced to abandon the cultivation of a description of cotton for which they have shown a decided preference. Indeed, it appears to the Government of India that it has yet to be shown that the cultivators are wrong and do not know their own interests best. The short-stapled cotton appears to possess certain advantages in regard to soil and climate over the long-stapled variety: it is said to require less rain, and the facts seem to indicate the possibility of greater economy and profit in growing it. These are points which appear to the Government of India to deserve investigation, and I am to suggest that they might profitably form the subject of a series of experimental trials (which should from year to year be carefully watched and the results recorded) in a limited area or in a small number of villages. Proof cannot but eventually be forthcoming as to which cotton returns most profit, and it is by the test of the profits to be derived, rather than by any action on the part of Government that the cultivator will be influenced in his selection of staples."

Mr. Cordery in October 1884 again wrote as follows:—

"I have the honour to forward what appears to me to be a careful and able report from the Officiating Commissioner upon the manner in which the Province of Berar is affected by the large and increasing admixture of an inferior variety of seed with that which produces a longer staple of cotton.

"If, as **Mr. Ridsdale** asserts, the new crop has created a market of its own with a designation known in Liverpool and commanding a sufficient profit, there can be no need for any action whatever. The question has, of course, been represented in a very different light by other officers; but the conclusions drawn by the Officiating Commissioner would appear to be more in accordance with the probabilities deducible from the recent extension in the growth of this variety than those based on the statements of traders disappointed in not finding the better variety which they desire to purchase. In such cases the term adulteration is not truly applicable: there is no concealment on the part of the cultivator, and the buyer is not deceived. The substitution of an inferior staple for one which formerly enjoyed a better reputation and a larger price must, under these circumstances, be held to be due to natural causes, against which it is useless to contest.

"The suggestions, therefore, made in my letter No. 368, dated 17th August 1883, were based on the mistaken belief that the complaint lay against an adulterated article being brought to a market, which (if this course were persisted in) would ultimately be closed to Berar. And on the fuller information before me, I have no desire to press them further.

G. 122

(502)

CULTIVATION in BERAR. Races.

The Government of India replied that it agreed with the Resident that a case had not been made out for State interference.

"In the Land Revenue Administration Report on Berar during 1886-87 it is mentioned that *vilayti* cotton is fast driving the old *bani* cotton out of the fields to the great grief of all cotton merchants who like *bani* because of its long staple, which fetches a higher price than the shorter staple produced by the hardier plant. The large introduction of the latter around Amráoti is stated to have caused the present cotton known as *umras* to be looked on as very inferior to what it was a few years ago. It is asserted that the reasons for its superior popularity are, besides its extra productiveness, the fact that the plant is not so easily injured by either excessive wet or drought: it is an earlier crop, and does not come into bearing all at once. These facts have been proved from experiments carried out for some time now on the Amráoti Experimental Farm. The Deputy Commissioner of Amráoti considers it necessary to endeavour to substitute, if possible, a variety combining the hardihood of the *vilayti* with the long staple of the *bani*, and he suggested that Government should obtain the services of a scientific agriculturist to see if anything could be done towards this end. The Commissioner, in remarking that the suggestion was a good one, added that the value of the product was sufficient to warrant the expenditure of a large sum in making such experiments, and supported the proposal that an agricultural expert should be deputed to Berar to advise on the possibility of improving the cotton of the Province.

"These proposals met with the approval of the Government of India, the advice of **Mr. Ozanne**, Director of the Agricultural Department, Bombay, was solicited, and seed was obtained from that department for issue to the *rayats* in 1888.

"It is to be hoped that these efforts may be productive of good in exterminating the inferior exotic breeds which, during the past few years, has been steadily obtaining a stronger hold on the agriculture of the province, and at the same time ousting the really fine native plant."

Soil, and manuring. 123 (503)

CULTIVATION—SOIL AND MANURING.—The land on which the greater part of Berar cotton is produced is similar to that fully described under the Central Provinces (*see p. 130*). It is of great fertility, extends to some considerable depth, and has the advantage of sufficient undulation of surface to maintain a natural system of drainage.

The following remarks regarding the application of manure may be quoted from **Mr. Hobson's** valuable "Note on Cotton in Berar":—

"The advantage of manuring is understood and acknowledged by most of the cultivators in Berar, but their views are limited by their means and hereditary customs. High farming cannot be conducted without stock, and in this the *rayats* are wanting, though there is a vast improvement to be observed under this head compared with the state of matters twenty or even ten years ago. Want of fuel in the best cotton tracts necessitates a resort by the farmer to cow-dung, which might otherwise be put down in the fields as manure. Most villages, however, can now boast of a manure heap without the village site, which during the hot weather is utilised for their fields by the more industrious and far-sighted of the villagers, but numbers acting up to too much time-honoured customs, or more probably being too lazy, prefer to follow in their forefathers' steps, and content themselves with inferior crops. The *rayat*, however, in the valley of Berar, who has overcome many old prejudices by rubbing shoulders with his more experienced brother employed at the large cotton marts, has learnt what is to his advantage, and knowing that his crops greatly depend on his own exertions, will work and resort to experiments upon which his ancestors would have looked with wonderment and passed on. Knowledge gained

CULTIVATION in BERAR.

Soil.

by experience will doubtless filter through the mass of prejudice still to be found above the Gháts, and it is to be hoped that the day is not far distant when the use and advantage of thorough manuring will be generally understood by even the *rayats* in the more distant and uncivilized parts of Berar.

The manure generally used is from the refuse heaps collected outside the villages by daily contribution from each house. Various other manures, such as poudrette, guano, and bones, have been tried, and are even now employed in some exceptional cases, but, as a rule, resort is had to the village dust-bins. Even this, however, is employed only by a few for cotton soils, as the rayats are of the opinion (doubtless based on experience) that manure with water does very well, and with a good rainfall its effects are excellent, but, should rainfall be scant, it does more harm than good by exciting the plant and driving it to wood. Irrigation is very rarely practised."

Rotation. 124 (504)

ROTATION.—Mr. Hobson states, that the rotation generally followed, is 1st, wheat; 2nd, **Lathyrus sativus**; 3rd cotton; 4th, linseed (where practicable); 5th, *jowari;* 6th, wheat again. The same writer remarks that, except in the Buldána and Ellichpur districts, the outturn of cotton and *jowari* has decreased, having been replaced by linseed and wheat, to which the *rayat* has found it more profitable in every way to devote his ground. The cultivation of both last-mentioned crops, however, requires a considerable amount of manure, and since the soil is given rest by constant rotation, the cultivator finds it indirectly to his advantage to continue cultivating cotton.

Associated Crops. 125 (505)

ASSOCIATED CROPS.—*Toor* (**Cajanus indicus**) in the proportion of one-tenth is generally associated with cotton; but in the villages to the south and above the Gháts *bajri* and *til* are sometimes put down instead.

Sowing, &c. 126 (506)

SOWING, PLOUGHING, REAPING, &C.—Mr. Hobson writes, "The method of cultivation is rude and simple; deep ploughing is but rarely resorted to, the *rayat* considering (possibly with some reason) that the power of the sun would soon burn out all the nourishment that might be in the soil. Cotton, as before mentioned, is sown in rotation, and in the deep black soil of the valley generally succeeds **Lathyrus sativus.** Being an autumnal crop, the seed is sown in the early part of June as soon as possible after the first fall of heavy rain.

The *surta*, which is a rough description of drill plough, is chiefly used when sowing cotton. Ten pounds of seed is considered sufficient for an acre of land; some soils, less productive than the black, requires a larger quantity of seed. The cotton being sown, the rayat does not trouble himself further until the weeds have cropped up. The *dauri*, a kind of scarifier drawn by bullocks, is then used; it has the double advantage of rooting up the weeds and loosening the earth and throwing it up round the plants. Women are now sent in to gather up the roots which may happen to be on the surface of the ground. They also thin the young plants where they happen to be too thick. The number of times the field is weeded depends upon the resources of the cultivator, some having their fields weeded three or four times."

"Picking commences in November, and properly weeded and cared-for crops yield three gatherings; when picked it is carried to the threshing floors, or enclosures specially prepared and placed in long stacks. When the stack is completed, the owner spreads ashes longways and crossways over it, in order to detect any tampering. Weighing would be a tedious process, that which he follows is much simpler and has in the rayat's eyes the sanctity of being a time-honoured custom handed down from his ancestors. The cultivators have a superstitious predilection in favour of having their cotton picked by women. These women are remunerated by a share of the cotton."

CULTIVATION in BERAR.

The above remarks refer to the finer form of cotton known as *bani. Jari* is cultivated, in much the same way, but is sown some fifteen days later, and takes longer in coming to maturity, so that the first picking seldom comes off before the 15th of December.

Injuries. 127

INJURIES.—**Mr. Hobson** writes to the following effect: "Cotton requires rain shortly after sowing, otherwise it does not germinate; on the other hand, if there is too much rain, the plants rot. Should it rain during *Swatí Nakshatra, i.e.*, the dark quarter of October or the beginning of November, it is generally supposed that the crops will be seriously injured. The natives have a proverb, 'Padé swatí kapús na milé wati,' which, freely translated, means, 'should it rain during swatí, you will not be able even to get cotton for wicks.'

"The bolls are subject to attacks from insects (cotton-borers), and a promising crop early in the season thus not unfrequently ends in failure and disappointment. White-ants, too, sometimes attack the plants, but, as a rule, the cotton is a hardy plant, and, given a fair monsoon, there is little to fear."

Cleaning. 128

CLEANING.—The instrument employed in cleaning and separating the fibre from the seed is called *retsa.* It is similar to the *charka* elsewhere described (p. 115), consisting of two horizontal rollers, one of which is sometimes iron, but usually both are of wood. Generally, this is of such a size that it can be worked by one woman, but a larger kind is occasionally to be seen which requires a woman at each end to work it.

After being cleaned the cotton is scutched, This process is somewhat peculiar, and is described as follows by **Mr. Hobson**:—"The scutch, called *kamán*, is a wooden instrument about 3½ feet in length, embellished at each end with traditional signs and carvings. To the *kamán* is attached a bow-string made of sheep-gut, termed *tát*. Scutching is termed *dhúnkna*, the person who works the instrument, *pinjári*. Having fixed one end of the *kamán* to a beam with a string called *dhaní*, he seats himself with the loose cotton in front of him and the *tát* resting on it. He then passes his left arm through a roll of cloth called *háthá*, which goes over his elbow and gives him a purchase. With the left hand he holds a piece of stick called *dhúndúri*, which is tied to the *kamán* to steady the whole instrument; with his right hand he takes a kind of elongated dumb-bell called *dastara*, the end of which he catches in the bow-string, and constantly twangs the latter. This causes the *tát* to strike the cotton at each twang, thus separating the fibres. The floss, by this process, becomes loose enough to be subsequently twisted by women round pieces of wood, forming spools called *pelu*. These are afterwards worked into spinning thread."

Yield. 129

YIELD PER ACRE.—"As already stated, the average yield per acre for the year under consideration, as derived by dividing the total outturn by the total acreage, was 71·68℔."

The following figures, however, subject to the errors indicated at pp. 53—54 & 87, are given as the actual outturn in each district:—

District.	Outturn per acre of cleaned cotton.	
	Unmixed.	Mixed.
	℔	℔
Amráoti	66	58
Akola	49	26
Ellichpur	58	49
Melghát	80	40
Buldána	32	28
Wun	90	72
Basim	46	...

The outturn per acre shews a very large increase over that of the previous year, perhaps due, in part, to the increased cultivation of inferior but more productive races (*cff.* p. 92). **CULTIVATION in BERAR.**

COST AND PROFIT OF CULTIVATION.—Mr. Hobson writes, "Preparing, sowing, and weeding cost on an average R7 an acre, but much depends, of course, on the district, and whether sufficient hands are obtainable when wanted. Some small farmers sow a much larger surface than they can afford to use and clean properly; the crops naturally are poor, and the owner becomes disgusted and consequently careless of saving even a portion of his crop. Such men are said 'to try a large mouthful, having only a small mouth.' **Cost of, 130 (510)**

"Picking and Weeding are heavy expenses; the cotton-pickers are not paid in cash but in kind; the rates vary according to the market, the average may be said to be one-tenth of the first picking, one-sixth of the second, and one-third for the third, equal to from R3 to R4 an acre. Separating the fibre from the seed by passing it through the *retsa* is paid for, sometimes in kind, to the value of annas 5 to 7 per maund (28lb) of seed cleaned, which equals R2 to R3 an acre. The total cost may be taken at R13 to R15 an acre, depending greatly on the district, leaving the cultivator a fair margin of profit. This, of course, again depends on the demand.

"*Kapás* when cleaned gives 26 per cent. of cleaned cotton. Besides the cotton, the cultivator has the cotton-seed, *sarki*, to feed his bullocks upon, and cotton stalks, which are used for numerous purposes, such as roofing grain stores, &c., but cotton is not the crop the rayat now pins his faith on; it is being driven out of the field by the more profitable cereals."

PRESSING AND PACKING.—In 1866 there was not a single cotton press at work in Berar, at the present time there are said to be 31 such presses. (*Indian Cotton Statistics.*) The following description of these presses and their effect on the cotton trade of the province may be quoted from the *Berar Gazetteer*:— **Pressing & Packing. 131 (511)**

"But it was the introduction of pressing that promoted as much as, or more than, any other reform the safe and expeditious consignment of our inland cotton to the seaport. In 1866 there was not a single cotton-press at work in Berar, though it seems that as early as 1836 Messrs. Kikaji and Pestanji, merchants of Bombay and Hyderabad, had set up one screw at Khamgaon. During the year 1867, thirty-two half-presses and two full-presses were set up, and the subjoined statement details subsequent progress:—

YEARS.	Full-Presses.	Half-Presses.
1868	14	81
1869	19	125
1870	19	74

"During the season of 1869-70 the number of half-presses has very sensibly diminished, because the railway rates of carriage have been raised on half-pressed bales to an amount that renders full-pressing very much more advantageous. The effect of this change has been to throw most of the cotton into the hands of the merchants who buy for direct export to England, since the Bombay market does not like to invest in bales that cannot be opened out for examination of the cotton. Therefore most of

CULTIVATION in BERAR.

Pressing & Packing.

the cotton sent down in 1870 by railway from Berar has been full-pressed, as the following figures show :-

Full-pressed	122,932
Half-pressed	69,585
Dokras $\left(\frac{50,790}{3}\right)$	16,930
	209,447 bales of

3½ cwt., or equal to 104,723 Bombay kandis.

"In the market of Berar all the cotton is sold, between producer and dealer, by the *boja* of 280lb, or 266lb nett; about three of them go to the Bombay kandi. The word means generally a load, and in the Berar cotton trade it meant particularly the load of a pack-bullock.

"In the half-pressed bale the cotton is condensed to about 12lb the cubic foot, and in the full-pressed bale the density varies from 28lb to 32lb the cubic foot. A full-pressed bale contains generally 3¼ cwt., but Khamgaon presses now usually succeed in getting 3½ cwt. within the bale. The half-pressed bale sometimes contains as much as 5 cwt., but then it is nearly three times the bulk of the full-pressed. In these inland districts the dry weather makes cotton so elastic that the best presses find much difficulty in obtaining the density of a bale pressed on the sea-board.

"The full-pressed bale and half-pressed bale are equal to about half a Bombay kandi, and the *dokra* to one-third of a bale."

The advantages of full-pressing over half-pressing appear to have become gradually more and more appreciated, until now the former has completely ousted the latter process. Considerable quantities are still, however, sent by rail as *dokras,* to be pressed for shipment at the seaboard.

Berar Trade. 132

(512)

ANALYSIS OF LAST YEAR'S TRADE.—The imports into Berar by rail during the year under consideration, 1888-89, amounted to 8,704 cwt., the exports to 1,333,186 cwt. Of the imports nearly the whole were derived from the Central Provinces; but Bombay, the North-West Provinces and Oudh, Rajputana and Central India, and Bombay port contributed small amounts. As already remarked these rail-borne imports represent but a very small fraction of the total, thousands of cwt. being received by road at the great markets of the province from the Central Provinces and the Nizam's territories. Regarding the latter **Mr. Hobson** wrote in 1887: "In former years large quantities of cotton used to come across from that part of His Highness the Nizam's Dominion, bordering on Berar, but this of late years, possibly owing to bad seasons, has fallen off; however, should the proposed Akola-Hingoli line of railway be completed, it will tap that portion of His Highness' Dominions, and doubtless the cotton export may return to its former average."

Of the total rail-borne exports, 1,290,807 cwt. went to Bombay for shipment, leaving only 42,379 cwt., of which nearly all was exported to the North-West Provinces and Oudh. Very small quantities were received by the Central Provinces, Bombay, and Bengal.

CULTIVATION in RAJPUTANA.

(513) 133

4. RAJPUTANA AND CENTRAL INDIA.

References.—*Gazetteer of Rajputana, 98, 150, 227, 254, 269; Cotton Reports of Rev. and Agri. Dept.; Agri.-Hort. Soc. of Ind. Jour. (Old Series) :—I., 275, 278; II., (Pt. I.), 62; III., 257-274.*

Area. 134

(514)

REVIEW OF THE AREA, &c.—It is impossible to determine whether the area under cotton cultivation in Central India has increased or not, since it is only within the past four years that any attempts have been made to estimate it. The outturn and the exports have, however, largely increased, and now supply a great part of the 'Bengals' received for shipment at Bombay.

The average area under the crop during the two years 1885-86 and 1886-87, was returned as 886,419 acres, and the average outturn as 877,607 cwt. These figures show an average yield per acre of 110lb. Later figures are not available for Central India, but in Rajputana alone the area in 1888-89 is returned as 455,256 acres, the outturn as 493,206 cwt. The acreage during that year was less than the average, because the rains were late and less was sown. The yield was also much under the average. The net exports from Rajputana and Central India in 1888-89 amounted to 596,631 cwt., leaving (on the assumption that the crop for that year equalled the average given above) 280,976 cwt. for local consumption, or dividing this sum by the population, 1·6lb per head. This closely corresponds to the estimate of 255,000 cwt., given in the "Statistical Tables." The average outturn in Dholpur is stated by the Gazetteer to be 640lb of uncleaned cotton per acre. **CULTIVATION in RAJPUTANA. Area.**

DISTRICTS WHERE GROWN.—The relative importance of the various States, as cotton-growing regions, is shown in the following table, the figures of which are quoted from **Mr. Beaufort's** "Indian Cotton Statistics":— **Districts. 135** (515)

RAJPUTANA.		CENTRAL INDIA.	
State.	Area.	State.	Area.
	Acres.		Acres.
Jaipur	140,000	Gwalior	120,000
Meywar	120,000	Indore	51,000
Ulwar	60,000	Bundelkhand	50,000
Bhurtpore	65,000	Bhopal	25,000
Marwar	25,000	Bhopawar	25,000
Kotah	16,000	Western Malwa	14,000
Dholpur	30,000	Baghalkhand	5,000
Bundi	20,000		
Tonk	15,000		
Kishengarh	9,000		
Jhallawar	10,000		
Minor States	40,000		
TOTAL	550,000	TOTAL	290,000

The above totals amount to a slightly smaller area than the average derived from Government returns for the two years ending 1886-87, but correspond sufficiently to indicate the relative importance of each State.

RACES OF COTTON.—All the cottons exported from Rajputana and Central India are included commercially in the group of "Bengals," and are probably nearly all forms of **G. herbaceum** as that species has been described in the *Flora of British India.* Numerous references are to be found in the publications of the *Agri.-Horticultural Society of India,* to a particularly fine cotton called *barareea,* cultivated near Chanderi and employed in manufacturing the fine cloth called *mahmúdi.* **Major Sleeman** describes this fibre as coming from Indore to Chanderi, and also as being produced in a few fields in the villages of Herawal and Singwara in the district. He states that "if the seeds of the *barareea* cotton be sown in any other fields of that district, the cotton produced is found to be like the produce of the ordinary seed, and to lose all the peculiar qualities of the *barareea* cotton." In a more recent communication by **Mr. Hamilton Bell** it is stated, that the imported *barareea* floss is superior to that locally produced. In another article on the subject **Mr. Fraser** states, that the greater part of Chanderi **Races. 136** (516)

CULTIVATION in RAJPUTANA.

Races.

cotton is imported from Berar, and in yet another Dr. Irvine writes that a considerable portion is derived from the Saugor district.

These statements appear to be supported by the name of the cotton—*barareea*, and point to the fact that the floss in question was probably the produce of a fine quality of *bani*, *i.e.*, **G. Wightianum**, *Tod.*

Several writers allude to what appears to be still another kind of cotton, and very probably **G. neglectum**, *Tod.*, as cultivated in Malwa.

Soils. (577) 137

SOILS AND METHOD OF CULTIVATION.—The soil of Southern Jaipur, Tonk, and Ulwar is fertile though light; that of the extensive plains of the Meywar plateau is fertile when irrigated. In the latter region almost every village has its artificial lake or tank. Behind the retaining embankments, or in the beds, also wherever there are wells, large crops of cotton, wheat, opium, and sugarcane, are raised.

Localities which are free from inundation are usually selected. The seed is sown in the months of April and June; the first sowings are watered by wells and yield a spare crop; the later depend on rain. Picking commences in October and is completed by the end of the year. The seed is sown broadcast, the land is immediately afterwards lightly ploughed, and when the crop has reached some height the plough is again drawn in parallel lines. Land thus prepared is said to require little weeding, for the plough furrows deep and removes roots of grass, &c. Weeding which may be required is done by hand, and is again turned up by hand with the trowel or *koddlí*.

Maize and cotton are the only *kharíf* crops which are manured. About eight cart-loads of manure are applied to the *bigha*. Night-soil, cow, goat, and sheep dung, ashes, and all sorts of refuse are heaped together near villages, and afterwards utilised as manure. The best kind is the dung of goats and sheep. Fuel being so dear in Rajputana, cow-dung is only utilised in the rainy season when it cannot be made into cakes.

No definite information is available regarding rotation or associated crops in Rajputana and Central India as a whole. In Dholpur the crop is said to be generally mixed with *uríd*, *til*, *arhár*, and *san*.

The following account of the method of cultivating the fine Chanderi cotton may be quoted from Dr. Irvine's article on the subject (*Trans. Agri.-Hort. Soc. Ind., VIII., 64*). "Chanderi cotton is sown, as usual elsewhere, in June. After the first fall of rain the ground is ploughed, another shower is allowed to fall, when the ground is manured and worked in with the harrow, then again ploughed, after which the seed is sown broadcast and the ground again harrowed, which is nearly the course of the usual cultivation. The seed is then left to spring up under the influence of the rains, when at a moderate height the young plants are carefully hoed, and during the growth the hoeing is sometimes repeated until the matured plant is just about to flower.

"Should a cessation of the rains occur, then an artificial irrigation is twice or thrice resorted to in September and October. Each *bigha* requires four people to hoe and do other requisite labour. The result on the same land is by no means equal—some *khéts* are five times gathered and some seven times. The cotton fit for *mahmúdí* is never collected oftener than three times, the remainder is common."

Cleaning. (578) 138

CLEANING.—The ordinary *Bengals* cotton of Rajputana and Central India is cleaned by means of the native gin or *charka*. Chanderi cotton, however, when destined for the manufacture of the fine muslins of that district is said to be entirely cleaned by hand-picking.

CULTIVATION in RAJPUTANA.

COST AND PROFIT OF CULTIVATION.—The following figures, given in the *Rajputana Gazetteer* for Dholpur, may also be accepted as fairly representing the cost and profit in most of the neighbouring States:—

Cost. 139

COST—	R	a.	p.
Ploughing	1	8	0
Manure	0	11	0
Seeds	0	8	3
Weeding	3	12	0
Picking	6	6	0
Cleaning	0	11	0
Wear and Tear	0	3	6
Rent	11	4	0
TOTAL	25	0	0

The value of the cotton crop may be put at R26-10-9, the other grains R7-6-9, seed after cleaning R1-4-0, total R35-5-6, showing a profit of R10-5-6.

Provincial Trade. 140

ANALYSIS OF LAST YEAR'S TRADE.—The exports from Rajputana and Central India are increasing yearly; in 1885-86 they were only 399,285 cwt., while in 1888-89 they amounted to 600,411 cwt. The greater part is received by Bombay port, which derives much the largest proportion of its "Bengals" from Rajputana and Central India. The imports from these States into Bombay by rail during the five years ending 1886-87 averaged 387,000 cwt.; in the year under consideration they amounted to 497,311 cwt. Of the remaining exports, 23,136 cwt. went to Calcutta, 38,291 cwt. to the North-West Provinces and Oudh, 38,824 cwt. to Bombay Presidency, and small quantities to the Central Provinces, the Panjáb, Berar, and Bengal. As already stated, the road trade with neighbouring states and provinces is large.

The imports are unimportant. In 1888-89 they amounted to only 3,780 cwt., of which 129 cwt. was derived from Bombay Town, 2 cwt. from Calcutta, 2,328 cwt. from the North-West Provinces and Oudh, 955 cwt. from the Central Provinces, 133 cwt. from the Panjáb, and 122 cwt. from Bombay.

CULTIVATION in MADRAS. 141

5. MADRAS.

References.—*Buchanan-Hamilton, Journey from Madras through Mysore, Canara, & Malabar, I., 40, 203, 378, 411; II., 157, 198, 221, 253, 254, 263, 286, 290, 302, 313, 323, 326, 450, 520, 545, 562; III., 317, 323, 351; Shortt, Essay on Cotton in Manual on Indian Agri., 1889, 1-97; Mad. Man. of Admn., I., 289, 299, 334, 335, 359; District Manuals: Cuddapah, 201-204; Coimbatore, 232, 235; Trichinopoly, 74; Settl. Repts.: Chingleput, para. 60; A Selection of Papers showing the measures taken since 1847 to promote Cotton Cultivation in India (Parliamentary), 1858, Pt. II.; Mad. Board of Rev. Proc., June 1st, 1889, 266; Repts. by Agri. Dept.; Exper. Farms Repts.; Note by Dep. Land Rec. & Agri., 1888; Rept. by Collector of Bellary, 1888; by Collector of Kuddapah, 1888; Note by Mad. Chamber of Commerce, 1889; Extracts from the Proc. of the Govt. of Mad., Rev. & Agri. Dept., 15th Feb., 1890; Elliot, Fl. Andh., 141, 143, 147, 157; Wheeler, Handbook of Cotton in Mad.; Sel. from the Records of the Mad. Govt., 1856, 10, 12, 16, 23, 37, 60, 69, 75, 166; Agri.-Hort. Soc. of India, Journals (Old Series), II., Sel., 2, 437; III., Sel., 127; VI., 118-122, 189-196; VII., 20, 30; XIII., 176, 180; Pro. of Agri.-Hort. Soc. of Mad., 1862, Nov., 5; 1863, Feb., 1-2; June, 1-2; Aug., 3, 5; Sept., 4, 5; Nov., 2, 5; Dec., 8; 1864; Jan., 10; Feb., 1, 4; Aug., 2; Sept., 1-3; Nov., 2; 1865, Mar., 2-3; May, 2-3; Aug., 1; Oct., 1; 1866, Jan., 2; Mar., 3-4; April, 7-9; 1867, Mar., 22; 1868, 129-131; 1869, Nov., 3, 4; 1871, 40; (New Series), I., 284; IV., 50.*

Area. 142

REVIEW OF THE AREA, &C.—The total area under cotton in Madras during the year 1888-89 was returned as 1,794,510 acres. Of this, 1,119,337 acres

CULTIVATION in MADRAS.

Area.

were grown in Government villages, 419,249 acres in Inam villages, and 255,924 in zemindári villages. The area returned for the first may be accepted as fairly accurate, but the figures for the two last, more especially those for the zemindáris, are less reliable. The cultivation of cotton is steadily increasing in the presidency, the total area, for the year under consideration exceeding the average of the three years preceding it by 90,000 acres.

The total outturn in 1888-89 is calculated to have been 801,120 cwt. The net export deducted from this amount leaves 309,343 cwt. as available for local consumption, a sum which, when divided by the population, gives an average of only 1·2℔ per head. The Madras Board of Revenue, in a recent report to the Government of India, make the following statement:—"On the question of the amount of clothing required to supply the wants of the population, the Board has made careful enquiries, and finds, that as a rule, the people in the Telegu districts use more than those in the Tamil districts, and that, while the lower classes in all districts get on with very little clothing, the more well-to-do, middling, and upper classes consume a much larger quantity. Taking all the classes of the population into account, the Board thinks that about 10 yards of cloth for an adult male, and 16 yards for an adult female, would not be too liberal an allowance to make for the clothing required for a whole year. For juveniles, male and female, excluding children under five years who practically go without any kind of clothing, half the above allowance would be sufficient." Applying these averages the conclusion is arrived at that 302,534,321 yards of cloth, representing 60,506,864℔ of raw cotton, are consumed annually. By deducting the net imports of piece goods and twist from this amount, the total amount of raw cotton used in spinning locally is arrived at, and in the year under consideration, is found to have been 28,871,376℔ or 257,780 cwt. As already stated, 309,343 cwt. is found, from calculation based on the outturn, to have been left in the presidency as available for local consumption, so 257,780 cwt. appears to be well within the mark. The 50,000 cwt. of excess is probably partly exported to Madras seaports by road (not included in the trade returns), partly consumed locally for purposes of upholstery, padding, &c. Of the 257,780 cwt. locally consumed in the manufacture of cloth, 169,804 cwt. were worked up by the nine spinning and weaving mills in the presidency, the balance was used by the poorer classes for spinning into the thread employed in making the coarser cloths of the rural population.

Districts 143 (523)

DISTRICTS WHERE GROWN.—The principal cotton-producing districts are Bellary, Kistna, Kurnúl (which run along the Kistna and Tangabadra rivers), Anantapur (south-east of Bellary), Cuddapah (south of Kurnúl), Coimbatore, Madura, and Tinnevelly, the three latter districts being situated in the south of the presidency.

The proportion of the total area held by each of these in 1888-89 is shown in the following table:—

DISTRICTS.	Area under Cotton.
	Acres.
Kistna	162,159
Bellary	292,803
Kurnúl	216,151
Anantapur	126,547
Cuddapah	113,981
Coimbatore	250,635
Madura	148,297
Tinnevelly	378,072

CULTIVATION in MADRAS.

Races.

144 (524)

Races of Cotton in Madras.—As already stated in describing the commercial "growths" of cotton, there are four kinds grown in Madras, *viz.*, Tinnevelly, Westerns, Coconadas, and Salems. Of these, "Tinnevelly" is produced in the Tinnevelly district and the southern portion of Madura; "Salem" in Salem and the Coimbatore districts; "Westerns" in Bellary, Anantapur, portions of Kurnul and Cuddapah, and in the Nizam's Dominions near Raichore and Lingasagor; and "Coconadas" in the Kistna district.

According to a note by the Madras Chamber of Commerce, dated September 1889, "Tinnevelly" is the most valuable of these. It has a strong, though not particularly long staple, and a very pure white colour. It is, accordingly, in considerable demand among spinners at home for mixing with Americans, and is one of the few Indian cottons suitable for this purpose. "Salem" is inferior, and is considerably used in adulterating the more valuable "Tinnevelly." "Westerns" have been subdivided into "Westerns" and "Northerns," the latter being the chief cotton of Cuddapah and Kurnúl. The principal markets for "Northerns" are Tadputri, Prudatur, Jangalpille, and Koilkuntla; for "Westerns" Bellary, Adoni, and Raichore. The staple of Westerns is rough, but fairly long and strong and of a good white, but, owing to imperfections in cleaning, always contains a considerable proportion of broken leaf. "Northerns" much resembles "Westerns" in length and strength of staple, but is silkier, and were it not for the slightly reddish tinge which it commonly possesses, would be in greater demand. The staple of "Coconadas" is fairly long, strong, and silky, but, owing to its redness, this cotton, also, is used by comparatively few spinners. It is, however, employed in the manufacture of lace, and is said to take dyes more readily than other growths.

In the Provincial cotton report for 1882-83, the races found in Madras are described as follows:—

"There are two forms of indigenous cotton usually grown in this presidency—the one depending on the south-west monsoon, and the other on the north-east. The former is sown between May and July, and the latter between September and November. In Tinnevelly, however, both species are sown in the same season, *i.e.*, October to November. They are known in the Telugu districts as the white and the red species, and in the Tamil districts as *uppam* or *ukkam* and *ladam* or *nadam*. In the Ganjam and Vizagapatam districts, they are also known as the *punasa pratti* (*pratti* means cotton) and *paira* or *burada pratti*. The latter is grown on irrigated lands in the Ganjam district. In all other districts, the crop is grown on dry or unirrigated lands and is often sown with dry grains. The white or the *uppam* kind (short staple) is sown in black cotton soil, and its fibre is considered superior to that of the *yerra* or red cotton. The plant lasts from six to twelve months. The *ladam* species is grown generally on red or gravelly soils and lasts from three to four or five years in the Tinnevelly, Coimbatore, and Salem districts. The Collector of Tinnevelly reports that this crop yields ahout 62½℔ of uncleaned cotton per acre in the first year the tenth month after sowing, and about 125℔ per acre in each of the second and third years from the ninth month after the picking of the previous crop. The fibre is said to be thin and fine, and to be used in the manufacture of superior cloths. The plants grow to a height of from five to six feet, and in Salem yied two harvests in the second year—one in September, the other in Januay. In the third year, the plants are lopped at a height of one foot from the ground, and then only one crop is obtained."

CULTIVATION in MADRAS.

Races.

Royle describes the *uppam* race as a superior form of **G. herbaceum**, with a staple which more nearly resembles that of American Upland than does any other indigenous cotton. *Nadam*, on the other hand, he describes as a triennial plant yielding a much inferior staple, which is also more difficult to clean. The same writer states that plate 11 in **Wight's** *Icones* is an accurate representation of *uppam*. The plate in question, according to **Todaro**, is typical of his species **G. neglectum.**

Specimens of *uppam*, however, which the authors have had the pleasure of receiving from Madras, appear to be typical of **G. Wightianum**, *Tod.*, a species figured in **Wight's** *Icones, t. 9*.

In the absence of a complete set of specimens of the commercial forms of Madas cotton it is extremely difficult to determine the botanical nature of each, but, from information available, it would appear that the better class, "Tinnevellies" and "Westerns," are produced for the most part by *uppam*, that "Salems" are derived from a degenerated form of **G. barbadense** (*see p. 19*), and that the other inferior cottons of the presidency may be *nadam*. It appears impossible to correctly determine the latter. In the above description everything points to the plant being **G. neglectum,** *Tod.*, but on the other hand specimens seen by the authors, bearing the vernacular name of ***nadam***, are typical **G. herbaceum,** ***var.*** **obtusifolium.** Specimens of nearly typical **G. herbaceum,** and of **G. arboreum,** have also been obtained from different parts of the Madras Presidency, but whether they are regularly cultivated, or would be included under the general term of ***nadam***, the authors cannot say. Between **G. herbaceum, G. Wightianum,** and **G. neglectum** every intermediate condition seems to exist. Variations from **G. Wightianum** towards **G. neglectum** are much inferior to those towards the **G. herbaceum** series. Of the former may be mentioned a form known as ***karumkanni***, of the latter that known as ***villaikanni***.

The differences in value and quality of the several commercial forms may, after all, depend, as much on local variations of soil, climate, methods of cultivation, and processes adopted in cleaning, as on any racial distinction.

Besides these Indian races, small quantities of various exotic cottons are grown in certain localities, the results of numerous experiments which have been carried on in Madras for many years. Of these, the only one of commercial importance is "Bourbon cotton" **(G. barbadense)**, which at one time produced the commercial "growth" known as "Salems," and which is still grown to a considerable extent, though in a miserably deteriorated condition, in Coimbatore. The Collector of that district furnished an interesting report on the origin and progress of this exotic, from which the following is extracted :—

"The *Shem parutti* (? *Seemai paruthi*) is that known as Bourbon, and will last as long as five or six years; it grows only on red or gravel soils, especially the calcareous and quartzose; hence *one* reason why it is extensively grown in Erode Taluk, in which there is not an acre of black cotton soil.

"Bourbon is found in Erode and Dhárápuram Taluks, especially in the Perundurai and Kángyam Firkas. The following explanation has been given me of its spread in those taluks : about 1820, Mr. Heath bought the cotton godowns in this part from the East India Company and introduced the Bourbon cotton; he was succeeded by Mr. G. D. Fischer in 1824, and this firm, trading largely in Bourbon cotton, induced extensive cultivation. Gradually the firm's dealings fell off, and consequently, for reasons to be noted, the cultivation of Bourbon declined. The firm's godowns were chiefly at Erode, near Perundurai, Chennimalai, and in the Kángyam Firka. The whole course of the cultivation and trade is a striking com-

CULTIVATION in MADRAS.

Races.

ment on the importance of private enterprise and British capital in developing the country and improving cultivation."

In a subsequent report the Collector of Coimbatore wrote :—

"Messrs. Stanes & Co. also inform me that Bourbon, though of excellent quality and much in demand if properly cultivated, is now, as grown in Coimbatore, of miserable quality and hardly worth buying. The reason is to be found in the wretched cultivation, the lack of manure, the continual cropping of the same ground, the complete removal of the plant, stem, lint, and seed from the ground, and the deterioration of seed from want of renewal. It is understood that Messrs. Fischer & Co. used to obtain and distribute fresh seed, and the rayats, as regards their indigenous cotton, expressly state that seed costs more than the market price because they have to choose only good seed. Apparently the exotic cotton deteriorates unless *fresh* seed is supplied ; whether this should be the care of the Agricultural Department or be left to the enterprise of merchants is a question. But there can be no doubt that all cotton outturn in Coimbatore has much deteriorated, and that the quantity and quality of the lint and seed are poor.

"It may also be noted that the cotton going south is intimately mixed with Tinnevelly cotton and exported as such, and that the admixture has been detected by the home brokers. It is a pity that such risks should be run by the trade when every effort ought to be made to improve the name of Indian cotton."

The above report elicited the following remarks from the local Board of Revenue :—

"The Board regret to learn that the quality of the Bourbon cotton is declining in the Coimbatore district. From Mr. Nicholson's report its diminished quantity and deteriorated quality are plainly due to exhaustion of the soil resulting from the continued cropping of unmanured land. The superiority of Bourbon cotton to the indigenous variety has been clearly demonstrated to the rayats of Coimbatore, and no expenditure by Government on the importation and distribution of fresh seed would have any permanent effect, nor indeed be justifiable, so long as the rayats pursue their present exhausting methods of cultivation."

It may be of interest to note that records exist of Bourbon cotton in Coimbatore as early as 1804. In the Selections from the Records of the East India Company, dated 1812, a report occurs by the Collector of Coimbatore in which it is stated that there were three forms of cotton at that time in the district, known, as they are now, as ***nadam***, ***copum*** (*? oopum*), and ***shem parutti***. The first two are evidently identical with those bearing the same vernacular names. The last, described as a "large plant cultivated in gardens, and used only for spinning Brahminical threads," would appear to have been a form of **G. arboreum**. The application of the name at one time used for **G. arboreum**, to the exotic American plant of the present day, is interesting. The Collector then goes on to state that Bourbon cotton had been introduced eight years before, but he appears to have been very sceptical as to any good results ensuing.

The above-quoted remarks of the Board of Revenue completely sum up the history of many years of earnest endeavour on the part of Government and also of private enterprise. Good seed has been procured, has been found with careful cultivation to do well, and has been distributed, time after time, to the rayats. As far as can be gathered from available literature, the only result has been that the stock thus introduced has either completely died out in a few years, or, as in the case of Bourbon cotton in Coimbatore, has given rise to a race of cotton much inferior to that natural to the country. So inferior is the produce obtained, that, instead of fetch-

CULTIVATION in MADRAS.

Races.

ing a higher price in the home market, it can only be got rid of at all by using it to adulterate Tinnevelly cotton.

The following remarks on the subject of the introduction of new cottons by the Madras Chamber of Commerce are of interest :—"The Chamber is led to suppose that most varieties of seed have been tried from time to time in nearly all districts, and that the variety at present grown is the one that has been adopted owing to its ascertained suitability to the climate and soil of the particular districts. In the sowing time of 1886-87, a systematic attempt was made to introduce Broach cotton into the Bellary district; but the plants mostly withered before maturity, and generally the rayats sustained loss over it. Any attempt at the introduction of new seed on a large scale should be deprecated unless the seed has first been experimented with in the same climate and soil for more than one season. Failure generally results, and, even when the plants thrive, the introduction of a new variety is not always satisfactory, as it is some time before the proper market value can be determined by buyers and sellers, and the new variety is often so mixed with the old, that buyers will not pay the same price for the mixture as they would for the cottons were they sown separately. The fact of there being in one district two classes of cotton, one of which may command a better price than the other, leads to the extensive adulteration of the better quality by admixture of the inferior."

Notwithstanding these facts, numerous experiments were again made at the Saidapet farm, for some years preceding 1886, with all sorts of exotic seed. These once more abundantly proved that with due care and a judicious use of manure, it was possible to grow cotton of a finer quality than the produce of the indigenous races. But there appears to be little reason for believing that, if the Yca valley, New Orleans, or Egyptian cotton, which were all cultivated with great success on the farm, were distributed amongst the rayats, the result would be in the least more satisfactory than that obtained from the Bourbon in Coimbatore.

Soils.

(525) 145 METHOD OF CULTIVATION; SOILS, AND MANURES.—**Dr. Shortt** writes as follows in his "Essay on Cotton":—

"In the culture practised by the native planter, the soil selected is black, soft, porous, and composed of decomposed basalt. It is well known as the *regur* or 'black cotton soil.' Manure is seldom used, but in some districts the usual dung-heaps or wood-ashes are lightly scattered over the ground, and should cattle not be allowed to enter the fields, the leaves and twigs of the previous year are frequently permitted to remain on the ground."

Irrigation.

(526) 146 IRRIGATION.—As a general rule, cotton fields are not irrigated,—the rayat has, therefore, to depend for the successful growth of his crops, on the periodical rains.

Sowing.

(527) 147 SOWING.—The seed of his own growth or of the same district is sown over and over again, each successive year, and this appears to have continued for centuries. It is prepared for sowing by being steeped in a solution of cow-dung and dried in the sun to prevent sticking together, mixed with dry grain or pulse and sown broadcast; in some districts the seed is sown in parallel lines, alternately with pulse, by means of a drill.

Sowing takes place in some districts as early as June, but seldom later than September, depending, to a great extent, on the particular seasons.

The seedlings appear between the third and seventh day. When they are three weeks old, the plantation is hand-weeded, which process is repeated some two or three times during the growth of the plant. Frequently when the seedlings are but a fortnight old, a plough is run through the field to loosen the earth as well as to facilitate the removal of weeds.

The plant generally begins to flower about the fourth, and to ripen its

CULTIVATION in MADRAS.

fruit at about the sixth month of its growth, but in some rare instances, this does not take place until the eighth month."

Sowing.

In the *Coimbatore District Manual* it is stated that *uppam* is sown in October-November, and yields its harvest in March-April. Occasionally, if there is good rain in April, it is interploughed and a small crop taken again in July. It is usually sown with Bengal gram in lines, or mixed with *thenei* and coriander, and the whole sown broadcast together.

The *nadam* is sown either in July-August mixed with *kambu* and pulses, or in April-May with *cholam* and castor, or *ginjelly* and *dhal*; sometimes with gram in September. In any case the cereal or pulse crop is reaped in the usual way, and the young cotton plants which have grown up under their shelter are left. In the first and third cases, the first cotton crop is picked in the following July. In the second a crop is got in February, and then not again till the following February. Thenceforward the crops per year are obtained. The crop in the third year yields a small outturn, and is interploughed as often as possible, gram, *kambu*, and *sámei* being sown.

In a memorandum furnished by the Revenue Department, Madura district, it is stated that, "sowing begins in October and ends in December. Picking is carried on twice in the year from March to April, and from June to July. The land is prepared by ploughing it five times before sowing, each time after a shower, and is ploughed over after sowing. The seeds are sown in *kerisal*, black cotton soil, red soil, and *portal*. The ground should be manured well, ploughed thrice at least, and the seeds then sown. After the plants have grown to the height of nearly a span, weeding is performed thrice. Ten measures of the seed will sow an acre."

The naturally rude and careless manner in which the native is known to conduct all his agricultural operations is pre-eminently shown in the cultivation of cotton, a circumstance which at once accounts for its usual deficiency in quality. It is surprising to witness, even under these disadvantages, what a large amount is turned out in a tolerably fair season.

Picking. 148 (528)

PICKING.—Dr. Short continues, "The pods, as a rule, are not collected as they ripen, but are allowed to remain until the whole crop of the field is ready; indeed, so little importance is attached to the speedy collection of the harvest, that even then the rayat consults his own convenience rather than the importance of his duty. In many instances he is either unable to procure labour, or has not the means of doing so; often he is not permitted to gather his produce in small quantities as the fruit ripens, but has to await the pleasure of the merchant from whom he has received an advance. Frequently, also, the crop has to be assessed before he is permitted to gather it, during which delay, in most instances two-thirds of the produce fall to the ground where it becomes mixed with dust. Should it happen to rain during this interval, 50 per cent. of the produce is irretrievably lost. In some districts the rayat gathers the early pods, so as to be enabled to come early into market. It is then cleanly gathered. But when the cotton season has fully set in, the price fluctuates, and the rayat being in no hurry, to enter the market, thinks the produce may just as well lie in the field as in his house.

"When the cotton at last is gathered, no care is evinced in removing the dried leaves and other extraneous substances found attached to it. It is generally taken off with all these impurities still clinging to it, thrown into some receptacle, generally a large open bamboo basket, and carried to the stack yard, where it is heaped up amidst the dust which necessarily abounds there. I have often witnessed heaps of cotton, both cleaned and uncleaned, thus lying exposed to wind and weather.

Ginning. 149 (529)

CLEANING AND GINNING.—The produce thus collected is next freed from

CULTIVATION in MADRAS.

Ginning.

seed. This operation is, as in most other localities, conducted by means of an Indian *churka*. The cotton, as brought in from the field, is freed, to a certain extent, of its capsules, stalks, leaves, &c., and then submitted to the *churka*. In most districts the cotton is previously spread out on a common charpoy or country cot and beaten with switches, when a portion of the minute particles of earthy or other extraneous matter still adhering falls through; but from the slovenly mode of gathering the cotton, no effort, however persevering, can afterwards entirely free it from its impurities.

This beating or thrashing is more intended to loosen the wool from the seed (to which it adheres somewhat tenaciously), in order to facilitate the operation of ginning, rather than to clean the cotton.

The operation of ginning is generally performed by females, who hold the seed in the left hand to feed the cylinders, and with the right simultaneously turn the *churka*. A woman, by manual labour alone, can clean only about four pounds of cotton, whereas with the *churka*, she can free from about twenty-five to thirty pounds in a day of ten hours.

The Madras Chamber of Commerce, in their report referred to above, state that much of the Westerns cotton of commerce is deficient in cleanliness, owing to the common practice of cleaning the fibre by rolling it on stones with iron rods. This is called "stone cotton" in contradistinction to *churka*-cleaned or "jowari cotton." The remark is made that a large proportion of the Westerns crop, owing to this system of cleaning, contains too much seed, and is, accordingly, not readily marketable. It appears to be highly desirable that this laborious and ineffectual process of stone-cleaning should be replaced at least by the *churka*.

Packing, (530) 150

PACKING.—**Dr. Shortt** gives the following description of the operation of packing:—To a strong, perpendicular, wooden post, a cross piece like the balance of a scale is suspended, the post being partly buried in the ground. To the cross piece, a bag of *cumbly* or gunny is attached, intended as a covering to the bale. From five to six maunds of cotton are now thrown in and compressed by men who jump into the gunny and stamp down the cotton with their dirty feet. From six to ten men are sometimes thus employed, whilst others are engaged in bringing the sides of the covering together, and securing them by sewing. In some places the gunny bags are hung by three ropes from the roof of the house; a fourth also, attached to the roof, hangs free in the centre, to which the men stamping the cotton hold on. These bales vary in size and shape, according to the mode by whch they are to be conveyed to the nearest market for export, and which may be either pack bullocks or carts. From the time of picking to that of export, cotton is liable to adulteration on the part of every one of the many through whose hands it passes.

Diseases, (531) 151

DISEASES—The cotton crop in Madras appears, from **Dr. Shortt's** exhaustive essay, to be remarkably liable to the attacks of numerous insects, especially aphides and larvæ of moths, and to one or more fungous parasites. Space cannot be allowed for a full description of these in such an article as the present. To prevent the ravages of insects **Dr. Shortt** recommends smoking the plants with tobacco, sulphur, or damp straw, or dusting the infested parts with tobacco or black peppers. Rats, squirrels, mice, and other larger animals are all said to be extremely destructive and may have to be got rid of by trapping, poisoning, &c.

In the *Coimbatore District Manual*, the diseases which attack cotton are said to be known locally as *sunnámbu*, *pén* (*asugani* or *poriyan*), *murulei*, and *adakkei*. The first-named is distinguished by brownish-white fungoid spots on the leaves and bolls, and greatly diminishes the yield. The second and third are "due to excessive east winds." *Pén* shows itself by "black spots over the whole leaf and stem;" the

CULTIVATION in MADRAS.

murulei by sickly drooping, "roughness of the leaves and brownish-yellow spots." From the description they would seem to be due to fungoid pests, possibly resembling the cotton-rust and cotton-blight (**Cercospora gossypina**) of America. The fourth is said to be due to excessive rain, the plant runs to wood, and numerous hard bolls are formed, but no fibre.

Seasons. 152 (532)

Seasons of Sowing and Reaping.—The subjoined particulars as to the time cotton is sown and harvested in the chief producing districts has been taken from a report furnished by the Government of Madras in 1878 in reply to the enquiries made by the Famine Commission:—

Districts.	Sown.	Reaped.
Bellary . .	3rd July to 23rd September.	9th March to 19th June.
Kistna . .	August to October.	January to April.
Kurnúl . .	August.	March.
Anantapur . .	July to September.	March to June
Cuddapah . .	July.	3rd, 5th, and 7th months after sowing.
Coimbatore . .	15th July to 14th September.	14th June to 14th August.
Madura . .	October or November.	January to April.
Tinnevelly . .	September to January.	,, August.

Judging from the above table sowings take place from July to January, and the crop is harvested from January to August. In the Provincial cotton report for 1882-83, however, it is stated that *uppam* is sown between May and July, *nadam* between September and November, the former depending on the south-west monsoon, the latter on the north-east. On the other hand, the Collector of Coimbatore reports, that in his district *uppam* is sown in October-November, and reaped in April-May. From these conflicting statements it would appear that both varieties may be sown either during the north-east or the south-west monsoon.

Cost. 153 (533)

Cost and Profit of Cultivation.—The annexed statement shows the cost of cultivation per acre as given by Collectors in 1882-83 and the value of the average outturn of cleaned cotton at annas 3-6 per ℔:—

Districts.	Average Cost per Acre. Including Assessment.	Excluding Assessment.	Excluding Assessment and Cost of Manuring and Fencing.	Value of Outturn per acre at 3½ annas per ℔.
	R a. p.	R a. p.	R a. p.	R a. p.
North Arcot		Not given.		
Trichinopoly	9 15 0	8 14 4	5 13 6	3 15 0
South Arcot	10 6 9	8 10 8	6 6 3	4 13 0
Nellore	7 7 10	6 6 6	3 9 8	5 4 0
Anantapur	2 13 2	2 5 0	2 5 0	5 11 0
Bellary	3 6 10	2 7 1	2 2 3	5 14 6
Kurnúl	4 1 2	2 14 0	2 14 0	6 9 0
Kistna	8 5 0	6 5 0	3 5 0	6 12 0
Madura	8 13 5	7 5 9	5 2 2	8 15 6
Tanjore	10 1 2	8 13 9	7 1 3	9 6 6
Coimbatore	6 12 0	5 8 0	5 8 0	9 10 0
Ganjam	12 11 9	10 15 6	7 15 6	9 10 0
Godávari	7 9 11	5 11 7	5 6 4	9 10 0
Cuddapah	5 8 0	4 0 0	3 4 0	10 4 6
Salem	10 14 0	9 8 0	5 8 0	10 4 6
Vizagapatam	11 2 11	8 5 6	5 1 0	11 9 6
Tinnevelly	7 9 8	6 12 9	5 2 4	15 8 6
South Canara . . .	14 0 0	...	...	19 4 0

CULTIVATION in MADRAS, Cost.

The Government of Madras commented on the above table as follows:—

"Analysing the details of the items given in the district reports, it is observed that the value of seed varies from 2 to 3 annas in Kurnúl, Bellary, Anantapur, and Kistna, to more than double that rate in Cuddapah, South Arcot, Tanjore, Trichinopoly, Coimbatore, and Salem. The cost of ploughing is given at more than R3 per acre in South Arcot, as against less than half the amount in several of the other districts. There are similar disparities in the other items also. The total average cost for the Ganjam district is more than double that given in Fasli 1291 (1881-82). The explanation furnished by the Collector is that many of the items do not involve actual outlay. In Mr. Nicholson's report the cost of cultivation in black cotton soils is worked out at R7 per acre, which is probably nearer the truth." (*Paragraph 8, Cotton Report, 1882-83*).

If it be assumed that the average outturn be proportionately equal in value to the exports from the presidency by rail, the average value per acre in the year under consideration may be arrived at as follows. The average exports by rail were 433,925 cwt., value R1,06,31,102, therefore the value of 1 cwt. was approximately R22. The value of the average outturn (50lb) was therefore R9-13, an amount which, assuming the above cost of R7 pe acre, would leave R2-13 to cover the cost of cleaning, packing, and to give a small profit.

Madras Trade. 154 (534)

ANALYSIS OF THE TRADE.—The great proportion of the cotton produced in the Ceded Districts (Bellary, Anantapur, Cuddapah) and Kurnúl is sent to Madras by rail for export. Some portion is also sent to Bombay. The cotton produced in Godávari, Kistna, and Nellore is sent by canal and road to Cocanada, and since the opening of the Buckingham Canal a part goes by that route to Madras. The produce of the Southern districts is exported from Tuticorin, to which it is conveyed, partly by rail and partly by road.

The following statement, which has been extracted from the provincial cotton report for 1882-83, shows the mode in which the fibre is conveyed to the ports of shipment from the interior of the presidency:—

DISTRICTS.	To what Places sent.	Mode of Conveyance.	In what shape carried.
Anantapur	Madras	By rail	Mostly in pressed bales.
Bellary	Madras and Bombay	Do.	In pressed bales.
Canara, South	Mangalore and Barkur.	By road	In unpressed bales.
Cuddapah	Madras	By rail	In pressed bales.
Godávari	Cocanada	By canal and road	In pressed and half-pressed bales.
Kistna	Cocanada, Madras, and Masulipatam.	By road and canal	In pressed bales.
Kurnúl	Madras and Bombay	By road and rail and in boats.	In loose bales.
Madura	Tuticorin and Madras.	By road and rail	In unpressed bales.
Malabar	Beypore, Cannanore, and Calicut.	By rail and road	In pressed and unpressed bales.
Tinnevelly	Tuticorin	By road and rail	Do. do.
Trichinopoly	Madras and Tuticorin	By rail	......
Tanjore	Negapatam	Do.	In pressed bales.

CULTIVATION in MADRAS.

Trade.

It is impossible, from want of perfect statistics regarding road, river, and canal traffic to make an attempt at giving complete figures regarding the provincial trade of Madras. The following statements must, therefore, be recognised as referring to rail and sea traffic only. Probably the road transactions in cotton are comparatively small; but, on the other hand, the canal-borne traffic is large and important. Much of the cotton produced in the western part of the Presidency finds its way to Guntúr, whence it is despatched in great part by canal to Cocanada, Masulipatam, and Madras. The only figures available for this traffic are those in the Administration Report of the Public Works Department, Irrigation Branch, from which it would appear that 10,276 tons of raw cotton were registered in 1888-89 as passing up and down the Godávari, Kistna, Kurnúl, Cuddapah, and Buckingham Canals. Of this amount 7,947 tons passed down towards the seaports, of which 1,496 tons went down the Buckingham Canal towards Madras. Though it is impossible to give a complete analysis of this trade, its importance must be remembered in consulting the following figures, as it probably greatly increases the exports from the province to the seaports, and may, in part, explain the apparent deficiency in the amount available for consumption in the town of Madras.

The total registered exports by rail and sea in 1888-89 amounted to 592,865 cwt., of which 575,980 went to the seaports of the presidency for shipment. Of the remainder, 12,259 cwt. went to Bombay port, 3,043 cwt. to Mysore, 1,577 cwt. to Bombay Presidency, and 6 cwt. to the Nizam's territory. The exports by rail to Madras seaports amounted to 417,040 cwt., and were made up by 235,275 cwt. to Madras town, 155,832 cwt. to Tuticorin, 14,786 cwt. to Calicut, 7,102 cwt. to Pondicherry, and 4,045 cwt. to Negapatam.

The total imports amounted to 98,039 cwt., of which 52,966 cwt. were derived from the presidency seaports, 45,073 from other Presidencics and provinces. Of the latter amount 30,226 cwt. came from the Nizam's territory, 14,787 cwt. from Bombay, and 60 cwt. from Mysore.

6. NORTH-WEST PROVINCES AND OUDH.

CULTIVATION in N.-W. P. & OUDH. 155 (535)

References.—*Duthie & Fuller, Field & Garden Crops, II., 75, pl. xviii.; Atkinson, Him. Dist., 738; Bonavia, Notes on Foreign Cult. in Oudh (1863); Gazetteers:—I., 79; II., 167; III., 225, 375, 463; IV., lxviii.; Rep. Dir. Land Rec. and Agri., N.-W. P., Oct. 1889; Govt. Rep. on Cotton Crop of N.-W. P.; Settl. Reports:—Azimgarh, 123; Shajahanpur, xv.; Allahabad, 25-26; Bareilly, 82, 87; Selections from Records of Govt. of N.-W. P. (Second Series), IV. (ii.), 99; VI., 283, 568; A selection of papers shewing the measures taken since 1847 to improve Cotton cultivation in India (Parliamentary), 1857, Pt. I.; Agri-Hort. Soc. of India, Transactions:—III., 133, 177; V., 54, 55, 67; VII., Proc. 17; VIII., 64; Journals (Old Series):—I., 25, 37, 114, 184, 187; II., 337; IV., 106, 107, 108; VI., 149; XII., 17-21, 212-264.*

Area. 156 (536)

Review of Area, &c.—As will be seen from the table, p. 56, the total area under cotton in these Provinces in 1888-89, including mixed crops, amounted to 1,399,388 acres. As already stated in discussing the races of Indian cotton, the average area has shown a marked increase during the past twenty years. In the year under consideration, however, a considerable falling off is exhibited, the average during the five years preceding having been 1,615,600, and the maximum in 1886-87, no less than 1,885,487 acres.

The total outturn in 1888-89 was 706,344 cwt., giving the average of about 60lb per acre. A comparison of the total import, 112,006 cwt.,

CULTIVATION in N.-W. P. & OUDH.

Area.

with the total export, 825,734 cwt., shews the net export to have been 703,728 cwt., a quantity very nearly equal to the estimated total production. Thus, by deducting the net export from the total production, only 2,616 cwt. are shown to have been left for local consumption, a result which is obviously highly erroneous. In a report on cotton, dated 1889, by the Government of the North-West Provinces, it is explained that this error is due to the unregistered road traffic, a great deal of cotton being imported by road from the Panjáb and Rajputana to take the rail at Agra, Hathras, and other cotton-pressing districts. In 1877-78, a fairly comprehensive registration system for trade by road and river was enforced, with the result that the net import by these channels was found to average 200,000 cwt. a year. Assuming that this import continues, the figures above given, as available for local consumption, will be supplemented by a credit of 200,000 cwt. per annum in favour of the province. This will represent fully two-thirds of the estimated home consumption (300,000 cwt.), the remainder could be accounted for by very slightly raising the estimate of yield. The writer of the report above referred to, commenting on these facts, writes, "The weak point in this argument is, that possibly with improved railway communication with the Panjáb and Rajputana, road imports may have fallen off since 1879. As to this I have no figures. There is no doubt, however, that cotton still travels largely by road even where the railway competes directly with this mode of conveyance, and that all but a very small portion of the cotton exported from Agra city by rail finds its way into it by road. As the rail exports from Agra city average about one-third of the total rail exports of the province, this fact is significant." From these arguments it appears safe to assume a local consumption of approximately 300,000 cwt., which, when distributed over the population of 43,268,599, gives an average of a little less than ·8℔ per head, or, taking Indian cotton-cloth at 4·3 yards per ℔, about 3·44 yards to each individual. Commenting on the small average consumption of local cotton, the author of the above-quoted report writes, "Although the estimated home consumption of raw cotton is put at 300,000 cwt. a year, it is quite possible that, for a series of years, it may fall below this if the prices of cotton are high and the demand active in Calcutta and Bombay. This has been the case during the last few years. The native weaver then finds it cheaper to buy imported yarn, and the poorer classes economise in the matter of quilts and cotton-padded garments. These are the two chief uses to which raw cotton is put in the province, and the inagnitude of the imports into it of yarn and piece goods of all kinds show how readily a substitute for the demand of raw cotton can be found. I will give the leading figures of the yarn, twist, and piece goods trade. The net imports of piece goods into the North-West Provinces and Oudh represent an average value of 450 or 500 lakhs of rupees a year, the net imports of twist and yarn average 40 lakhs a year. As 300,000 cwt. of raw cotton are worth, at most, only 60 lakhs, it is obvious that the home consumption of raw cotton is a comparatively minor item in the total annual expenditure on clothing of the population of these provinces."

Of the 300,000 cwt. allowed above for local consumption, a considerable proportion is worked up by the five cotton mills which at present exist in the provinces. In 1888-89, these are returned as having consumed 94,766 cwt., of which, however, a certain amount may have been imported as yarn or twist.

Districts. (537) 157

Districts where Grown.—Almost the entire cotton crop of the United Provinces is grown within an irregular belt of country running down the Ganges-Jamna Doab, to which Rohilkhand and Bundelkhand

CULTIVATION in N.-W. P. & OUDH.

Area.

are appendages. The shares taken by each division in the general acreage and outturn of 1888-89 are shown in the following table:—

Division.	Area under Cotton.		Total outturn of cleaned Cotton in cwt.
	Mixed.	Unmixed.	
	Acres.	Acres.	
Meerut	207,540	165,110	287,235
Agra	382,001	61,653	250,238
Rohilkhand	134,232	31,074	62,009
Allahabad	276,884	5,440	69,912
Benares	6,810	2,710	2,495
Jhansi	60,206	2,898	14,632
Lucknow	24,982	2,535	8,386
Sitapur	23,802	3,803	7,711
Fyzabad	613	584	821
Rae Bareli	695	756	719
Tarai (dist.)	2,824	2,236	2,086
Total	1,120,589	278,799	706,344

Races.
158 (538)

Races of Cotton.—The only cotton of commercial importance grown in the North-West Provinces is the form known as '*Bengals,*' which probably consists of hybrid forms between **G. Wightianum** and **G. neglectum,** and is almost identical with many of the Panjáb forms with short white staple. In addition to this, however, purer forms of **G. neglectum,** *Tod.*, and **G. arboreum,** are sparsely cultivated in parts of Oudh and the more eastern districts of the North-West Provinces, where the latter is ordinarily known as *narma*, the former as *manua*. A good illustration of **G. neglectum,** *Tod.*, the *manua* cotton, is given in *Duthie and Fuller's Field and Garden Crops* under the name of **G. herbaceum.** A superior form of *manua* grown in the Allahabad district is called *radya* (*cff.* the account given under **G. arboreum** for Mirzapur and Goruckpur cottons, p. 10). Both these differ considerably from the ordinary cotton in the season of their growth, not bearing cotton till the hot weather months, instead of at the end of the rains. In a recent communication on *narma* and *manua* collectively, the Government of the North-West Provinces reports that "the area in a single holding rarely exceeds a quarter of an acre, and its total area in the United Provinces does not exceed 13,000 acres. In outturn and other details it does not differ much from the ordinary variety of the province."

Conf. with p. 7.

The hill cottons of Kumaon and Garhwal are, like those of the Panjáb-Himálaya, hybrids with a strong strain of **G. arboreum** on **G. Wightianum.**

Numerous attempts have been made by Government to introduce the finer varieties of American cotton into this, as into other, cotton-growing regions. The following exhaustive account of these experiments may be quoted in entirety from the Note on Cotton by the Revenue and Agricultural Department:—

"Experiments were undertaken at the Cawnpore Experimental Farm in 1881-82, in which trials of Hinganghát cotton and of the varieties of American cotton known as New Orleans and Upland Georgian, were made on two classes of soil, a light loam and a heavy loam, against two varieties of country cotton—one, grown in Bundelkhand and known as *kulpahar*, and the other, that commonly cultivated in the Cawnpore district. The results went to show (1) that for light soils the indigenous cotton is by far the most pro-

CULTIVATION in N. W. P. & OUDH.

Races.

fitable; (2) that on better class soils, and with careful cultivation, New Orleans cotton can be grown with success and profit.

The results of trials with Nankin cotton are thus summarised in the Cawnpore Experimental Farm Report for the *kharif* season of 1882 (page 6):—

"Of this variety it is necessary, in view of the interest lately taken in it, to make a separate note. To all appearance, in the field before the bolls burst, the plant is the same as that of New Orleans or Upland Georgian, and quite different, therefore, in growth and leaf to country cotton. It was grown at the farm some few years ago, but discontinued owing to the unfavourable reports on its staple received from Calcutta. This year some seed was received from the Saharanpur Botanical Gardens and from the Rawatpur estate, where it had been grown, as well as on other estates in the Cawnpore district, for several years past. On the plant the cotton is of a dirty fawn colour mixed with white; bolls of fawn-coloured lint and of pure white lint being sometimes found on the same plant. After ginning, spinning, and weaving, the result is a dark khaki-coloured cloth of uniform tint, admirably adapted for army clothing. The colour can, apparently, be bleached, judging from a coat shown to me at the Elgin Mills, but does not appear to fade with ordinary washing and wear. The mills object to its staple as being difficult to work with machinery, and the English operatives at the Elgin Mills say that on account of this difficulty it is known to the trade in England as 'rotten cotton'—a difficulty, however, which does not seem to be recognised by native weavers.

"'This cotton is cultivated in Central India and woven into cloth for the regiments of Central India Horse. **Lieutenant-Colonel Martin, C.B.**, in reply to inquiries, has kindly sent two samples of cloth of first and second quality. The Officer Commanding the 23rd Pioneers has sent samples of cloth and cotton from crops raised near Mian Mir by **Subadar-Major Natha Singh, A.D.C.**, and from the Government of Madras have been received specimens of the cotton, and of cloth manufactured from it at the Coimbatore Jail. From the Cotton Hand-book of Madras it appears that samples of Nankin cotton took prizes at the Madras Exhibition in 1859. The Secretary to the Muir Mills, Cawnpore, writes:—

"'This Company has grown for some years crops of this cotton and has manufactured drill from the same. I enclose a pattern of the drill for your examination. We have endeavoured to introduce the cotton among the cultivators but have not succeeded to any extent. We purchase any quantities of the cotton we may have offered to us, but at present very small quantities have been secured, and we are therefore unable to manufacture cloth for sale of the pattern sent you.

"'All the samples of Nankin cotton eceived appear to be of fairly uniform colour save that from the Muir Mills, which is brighter and lighter than the rest, though but a slight shade different to the first quality of cloth used by the Central India Horse.

"'In the report of the Mission to Yarkand, page 479, reference is made to a reddish-coloured cotton sold in the Yarkand bazár and known as *kara kiwaz*, which maintains the colour after washing. **Lieutenant-Colonel Gordon, C.B.**, to whom I sent specimens of the farm-grown cotton, speaking from recollection, seems to think Nankin identical with *kara kiwaz*, and has kindly promised to try and procure seed from Yarkand.

"'To sum up: Nankin cotton has long passed the experimental stage. It has been grown successfully, has been woven successfully, and is approved of by those who have had good opportunities of observing it in use, but of its value there are not at present sufficient data, as, owing to absence of demand, the market rate has been purely nominal.'

CULTIVATION in N.-W. P. & OUDH. Races.

In the *kharif* season of 1884 three varieties of Nankin cotton were tried on the farm, *viz.*:—(1) With seed procured from Yarkand. (2) With seed received from China through the Government of India. (3) With seed acclimatised on the farm.

In reporting the results the Director of Land Records and Agriculture wrote as follows :—"There is a marked difference in the habits of the three varieties. Nos. 1 and 2 appear to be of the Oriental variety, having few bunches and small deeply-indented leaves similar to the ordinary cotton of these provinces. No. 3 is distinctly occidental or similar in growth and habit to American cotton. Germination in case of the China seed was very irregular. Fields sown at the rate of 18 to 20 seers (36 to 40℔) of seed per acre had not as many plants as fields sown with 3 seers of the acclimatised variety. The plants were very stunted; most of them died before October, and those which survived bore scarcely any cotton. The fate of plants raised from the Yarkand seed was not much better. The fields sown with acclimatised seed fared comparatively well."

Samples of American and of country cotton cleaned in 1882-83 with the Burgess (Emery) saw-gin and with the ordinary country *charkhi* were submitted to the Manager, Elgin Mills, Cawnpore, for opinion. His report is as follows :—

"'After carefully examining the samples and comparing the cotton ginned with the native *charkhi* in the ordinary manner with that ginned with the Emery saw-gin, I have arrived at the conclusion that the staple has not been injuriously affected by the latter process, *i.e.*, cut, broken, or otherwise appreciably damaged; while the superiority of the saw-gin over the old primitive method is very marked, so far as opening out and cleaning the cotton is concerned, as applying to Upland Georgian and New Orleans. The effect of the saw-gin on the short-stapled country cotton, however, is that the seed is partially broken in the process, and small particles adhering to the fibre are passed through the machine, which materially reduces its value for spinning purposes. I am of opinion, therefore, that the Emery saw-gin, while doing its work admirably on the long-stapled varieties, is not so well suited for cotton grown in the North-West Provinces and usually known as 'Bengal.'"

The Director of Land Records and Agriculture adds :—

"'I may note that a sample of the country cotton cleaned by the saw-gin and sent to the Cawnpore market was rejected by the dealers, not on account of any supposed injury to the staple, but on account of its clean appearance, the dealers suspecting that it was old cotton cleaned with a Belna's bow.'

In the *kharif* season of 1887, experiments with cotton of two varieties, (1) Louisiana, and (2) Egyptian, were tried on the farm, before and after the rains. The early-sown plots gave more satisfactory results than those sown late. In each case the Egyptian variety is said to have proved more suitable to the climate and soil of the north-west than the other. In two plots country cotton, and in two other New Orleans, were sown. In both cases the yield with gypsum and kainite was greater than without them. New Orleans gave an outturn of 145℔ clean cotton per acre, and country cotton 121℔.

Nankin cotton was also tried; the points to be determined were—(*a*) for how many years the plants will continue to bear; (*b*) whether the yield is equal to the annually-sown crop or not; (*c*) whether there is any change of colour in the produce of the annual and perennial plants; and lastly (*d*) whether ratooning makes any difference? With regard to (*a*), it has been sufficiently proved that the Nankin cotton plant is perennial, and will yield

CULTIVATION in N.-W. P. & OUDH. Races.

and stand for several years. If kept long, it thrives well, becomes more bushy, and produces more pods. But compared with the fresh-sown cotton a greater number of pods do not yield fine cotton. It is also shown that with the age of the plant the seed becomes smaller and smaller, and therefore the percentage of clean cotton on seed increases. With regard to (*b*), the unfavourable nature of the season interfered much with the experiments. One of the plots sown this year yielded 172℔ cleaned cotton per acre, and the other only 40℔, while the produce of the plots sown three years ago was 85℔ per acre in each case. With regard to (*c*) and (*d*), there was apparently no difference between the colour of the produce of the two kinds of plants, nor was there any difference between the results of ratooned and unratooned plants. The foregoing trials have conclusively shown that early sowing of all *kharif* crops, especially cotton, is advantageous.

Soils. 159 SOILS AND MANURES.—**Messrs. Duthie and Fuller** write, "Cotton land may be either the very best or the very worst in a village. As a rule, cotton is grown on good land, a loam being preferred, and is either manured itself, or reaps some benefit from a manuring applied to a crop which preceded it. District returns show that about 32 per cent. of the cotton crop is grown on land manured specially for it, 39 per cent. on land manured in the previous year or two years, and 38 per cent. on land altogether unmanured. A very large proportion is grown with manure, but, on the other hand, it is a common crop on poor soils, such as the raviney calcareous tracts in the neighbourhood of great rivers, which is said to actually improve by the manure of the leaves which fall from it. When sown on high class soil it is generally grown alone, while on poor ground it is almost invariably mixed with a large proportion of pulses and oil-seeds. Hardly any of the Bundelkhand cotton, which is by far the best in the provinces, receives manure, nor does the black soil on which it is generally grown appear to require it."

Method. 160 METHOD OF CULTIVATION.—The following account of the method of cultivation pursued is extracted from **Messrs. Duthie and Fuller's** *Field and Garden Crops*:—

Seasons. 161 "SEASONS.—Cotton is a *kharif* crop, being the one first sown after the commencement of the rains" (with the exception of a large portion of the crop of the Meerut district, which is sown with the aid of artificial irrigation a month or so before the setting in of the monsoon), "and yielding its produce from October to January. This is with the ordinary variety, but *narma* and *radya* cotton do not bear a crop till the April and May following their sowing, and thus occupy the ground for at least eleven months. Cotton fills no place in any special rotation of crops, although it is reported generally to succeed sugar-cane in **Meerut**, and to intervene between two cereal crops in Bareilly, the deduction being merely that it is grown on good land which had at all events been manured in the preceding year. It is off the ground too late to admit of its being followed by a *rabi* crop in the same year, but an ingenious method of gaining a second crop off cotton fields is to sow the oil-seed *duán* (**Eruca sativa**) broadcast amidst the crop just before it is finally weeded. The seeds are buried in the operation of weeding, and the *duán* plants do not become tall enough to interfere with the cotton until the latter has finished bearing.

Mixtures. MIXTURES.—Cotton is comparatively seldom grown alone, but is, as a rule, associated with four or five subordinate crops, amongst which *arhar* is the chief. The *arhar* is generally sown in parallel lines, not broadcast, and it is said that the cotton plants find in its shelter some protection from cold winds and frost. The oil-seed *til*, or *gingelly*, occupies the first place amongst the remaining subordinate crops, which comprise the pulses

CULTIVATION in N.-W. P. & OUDH.

úrd or *múng* sown broadcast, and an edging of castor and of the fibre plant known as *patsan* (**Hibiscus cannabinus**).

Sowing. 162 (542)

TILLAGE AND SOWING.—The land is ploughed from four to six times on the first fall of rain, and the seed is sown broadcast at the rate of 4 to 6 seers per acre and ploughed in. The seed is generally rubbed with cow-dung before sowing, which prevents its clinging together in masses as it would otherwise do, and is also said to stimulate its growth. Irrigation is applied to about one field in seven, and this much only in Canal Districts, where watering will not cost more than from one to two rupees.

Narma cotton requires but little water, although it has the whole of the cold and part of the hot weather to stand before it produces its fibre. The *radya* variety, however, is said to require copious irrigation. It is essential to the proper growth of the plants that they be kept free from weeds, and the ground is, as a rule, carefully weeded by hand at least twice in the season, and often four times.

Picking. 163 (543)

PICKING.—The cotton bolls begin to open in October, and picking is in progress from then till the end of January, unless cut short sooner by frost,—the great enemy of the cotton plant. Good fields are picked every third or fourth day, but only between sunrise and mid-day, while the cotton remains damp with the night's dew and comes away easily. If force is necessary to separate it from the boll, bits of pod-shell come away with it, which are technically known as 'leaf' and greatly damage the commercial value of the produce. Cotton picking is generally done by women, who are remunerated by receiving $\frac{1}{8}$th to $\frac{1}{11}$th of the pickings.

Cleaning. 164 (544)

CLEANING.—For 'ginning' or separating the cotton fibres from the seed, a simple but ingenious machine is used (called a *charkhi*), consisting of two small rollers about a foot long (one of iron, the other of wood), each with one end turned into an endless screw, and so geared one into the other, that when one—the wooden one—is turned by a handle, the other also turns in the opposite direction. When cotton is applied to the rollers the fibres are drawn through, and are in this way parted from the seeds. With this instrument a woman can turn out from 4 to 5℔ of clean cotton fibre a day. The proportion of fibre and seed varies considerably, being in great measure dependent on the quality of cultivation. Occasionally it rises as high as $\frac{2}{5}$ths or falls as low as $\frac{1}{4}$th, but $\frac{1}{3}$rd is the general average. It is interesting to note that an instrument practically identical with the *charkhi* is used for cotton cleaning by the negroes of the Southern States of America.

Diseases. 165 (545)

INJURIES AND DISEASES.—Stagnant water is most harmful to the cotton plant, especially at the commencement of its growth, and fields selected for cotton are, as a rule, those in which water does not lodge. Rain, when the pods have begun to open, is also most damaging, since the fibre becomes discoloured and rotten. Early frosts may altogether termin ae the picking season a month or six weeks before it would otherwise have ended, and hence the eagerness shown to get the cotton seed into the ground as soon as possible. Caterpillars are often very destructive, sometimes stripping a field entirely of its leaves, and an immense deal of loss results from the ravages of a small white grub (called *súndi*) which lives within the pod."

Yield. 166 (546)

YIELD PER ACRE.—The question of yield per acre in the North-West Provinces, has formed the subject of considerable discussion, and has given rise to many conflicting statements. Thus, **Messrs. Duthie and Fuller** write: "There is no crop the outturn of which has been so systematically under-rated as that of cotton, and if we are to believe the district reports of the last three years, the provincial average is only 59·8℔ per acre, in which case it may be demonstrated that it would not pay to grow it at all.

CULTIVATION in N.-W. P. & OUDH.

Yield.

After consideration of the estimates arrived at by settlement officers, which exhibit, it must be said, the most astounding discrepancies, and utilising the experience of two years on the Cawnpore Farm, an all-round estimate of 170lb of clean cotton per acre of irrigated, and 150lb per acre of unirrigated land is the lowest which can be safely struck, except for Oudh and the Benares Division, where 100lb may be taken as sufficient. For cotton mixed with *arhar*, these outturns should be reduced by about 25 per cent."

On the other hand, the recent report by the Director of Land Records and Agriculture, dated 1889, contains the following:—"Considering that the American average is only 160lb per acre, the estimate given in *Field and Garden Crops* appears high for India, and the enquiries prosecuted by this Department confirm this suspicion. After considerable correspondence with agriculturists of all classes, the following standards of yield have finally been adopted by this Department as the full average yield for the several portions of the North-West Provinces and Oudh:—

Districts.	Outturn of Cleaned Cotton per acre.	
	Alone.	Mixed.
	lb	lb
Gangetic Doab	130	80
Rohilkhand and Terai	110	60
Bundelkhand	120	60
Benares Division and Jaunpur	100	50
Oudh	110	55

The latter figures being probably the more carefully and thoroughly estimated of the two, have been adopted by the authors in their statistics of total yield of the provinces. At the same time the argument based on the yield of American cotton does not appear to be altogether without flaw. Many Indian cottons, for example, the Sindi, produce a very much larger outturn than that laid down for the cotton of the North-West Provinces by **Messrs Duthie** and **Fuller**, and there seems no reason why the native cotton in this province should not produce equally largely.

Cost. (547) 167

Cost & Profit of Cultivation.—The cost of cultivation is estimated by **Messrs. Duthie** and **Fuller**, as follows:—

	R	a.	p.
Ploughing (four times)	3	0	0
Clod crushing (twice)	0	4	0
Seed (nominal)	0	2	0
Sowing	0	13	0
Weeding (twice)	3	0	0
Picking ($\frac{1}{10}$th produce on 200lb)	4	0	0
Cleaning (at $1\frac{1}{2}$ anna per 10lb)	1	14	0
Total	13	1	0
Manure (100 maunds)	3	0	0
Rent	6	8	0
Grand Total	22	9	0

It appears probable that the above cost is considerably over-estimated, the more so since comparison with similar statements regarding Bombay

CULTIVATION in N.-W. P. & OUDH. Cost.

and other provinces, shows that in the latter the cost is returned at only from one-fourth to half of the above grand total. If the above-quoted statistics as to outturn per acre be accepted as correct, R22-9, expended on the cultivation, is much too large to admit of the crop being grown at all. Thus, assuming the value of the total production to be proportionately equal to the value of the exports, we have, 825,734 cwt.: 706,344 cwt., : : R1,99,43,447 : R1,70,59,736, which last figure, when divided by the total area, gives an average value of R12 per acre. It may be noted that this figure very closely corresponds with the value obtained by a similar calculation for the acre crop of the Central Provinces.

Provincial Trade. 168

ANALYSIS OF LAST YEAR'S TRADE.—Of the total exports, 825,734 cwt., the largest proportion, *viz.*, 397,825 cwt., went to Bombay, while 348,233 cwt. went to Calcutta. Smaller quantities were consigned to Bengal and Bombay Presidencies, Rajputana and Central India, the Central Provinces, the Panjáb and Berar. The imports, which amounted to 122,006 cwt., were mainly derived from Rajputana and Central India, Berar, the Panjáb and the Central Provinces, while small quantities were obtained from Calcutta, Bombay port, Bombay, Bengal, and Karáchi.

A small trans-frontier trade exists with Tibet and Nepal, to the former of which 1 cwt. was exported, to the latter 649 cwt.

As already stated, the unregistered road traffic probably causes the imports to be considerably under-estimated.

CULTIVATION in the PANJAB. 169

7. PANJAB.

References.—*Baden Powell, Pb. Prod., 477-496; Cotton Repts. by Agr. Depts.; Rept. by Dir. Land Rec. and Agri., Panjáb, Dec. 1889; Rev. and Agri. Dept. Repts. in many passages; Settle. Repts.:—Hazara, 88; Kohat, 124; Bannu, 80; Dera Ismail Khan, 347, 348; Dera Ghazi Khan, 9; Jhang, 84, 91, 92; Montgomery, 107; Lahore, 8; Gujerat (App), xxxviii; Kangra, 25, 27, Rohtak, 192; Gazetteers:—Dera Ismail Khan, 127; Dera Ghazi Khan, 82; Jhang, 112; Muzaffargarh, 91; Multan, 97; Montgomery, 112; Lahore, 82, 97; Gujerat, 78; Gujranwala, 52; Shahpur, 65, 75; Gurdaspur, 57; Hoshiarpur, 92, 111; Ludhiana, 137; Gurgaon, 81; Karnal, 172; Agri. Hort. Soc. of India, Journals (Old Series), VIII., 137; XIII., Proc. xiii.; Select papers from the Agri. and Hort. Soc. of the Panjáb, I. (up to 1862), 48, 64, 123, 105-106, 158-160.*

Area. 170

REVIEW OF THE AREA, &c.—The total area under cotton in the Panjáb in 1888-89 was returned as 759,465 acres. In the year under consideration, a marked decrease in area is noticeable, but this is owing to exceptional climatic causes. On the whole, the cultivation is yearly increasing. Thus, during the three quinquennial periods of the past fifteen years, the following were the averages:—for the first 715,800 acres, for the second 837,600 acres, and for the third 859,800 acres.

The total outturn in the year under review was returned as 931,824 cwt., an exceptionally high outturn in comparison with the figures of acreage. Deducting the net export from that sum the total amount available for local consumption during the year is seen to have been 637,737 cwt. This figure, however, is liable to several sources of error. The returns of outturn per acre vary inexplicably in neighbouring districts, and are probably, in some cases at least, erroneous. Here, as in the North-West Provinces, the absence of a regular system of registration of road traffic may have given rise to some considerable error. While, in the North-West Provinces, the imports are probably under estimated, in this province on the other hand, the exports by road, if registered, would in all probability much augment the figure of net export. If, however, the amount available for local consumption be accepted at 637,737 cwt., the

CULTIVATION in the PANJAB.

Area.

amount of local raw cotton utilised during the year must have been 3·8℔ per head, equivalent to about 16 yards of cotton.

Perhaps this figure may not be much above the truth. The cold climate of the Panjáb necessitates a large amount of clothing; the cloth used is almost entirely cotton, and padded coats and quilts are very largely employed by all classes of the population.

Districts.

(551) 171 DISTRICTS WHERE GROWN.—The districts which grew the largest areas in 1888-89 may be most conveniently shewn in the following tabular statement:—

DISTRICT.	AREA UNDER COTTON.		Outturn of cleaned cotton.
	Unmixed.	Mixed.	
	Acres.	Acres.	Cwt.
Gurgaon	81,700	...	116,714
Rohtak	45,863	...	98,760
Karnal	...	31,222	61,329
Montgomery	28,158	...	65,367
Umballa	47,827	...	57,243
Multan	52,562	...	46,056
Jullundur	17,393	2,805	43,593
Sialkot	27.225	9,075	42,612
Lahore	41,982	...	41,232
Delhi	29,500	...	39,509
Shahpur	32,771	...	35,259
Dera Ghazi Khan	32,282	499	34,944
Rawal Pindi	25,466	...	27,285

Races.

(552) 172 RACES OF COTTON IN THE PANJAB.—Panjáb cotton, included commercially in the group of "Bengals," may be botanically considered as hybrids between **G. Wightianum**, *Tod.*, and **G. neglectum**, *Tod.*, or **G. arboreum**. Specimens, which the authors have had the privilege of examining, shew a much greater tendency towards **G. neglectum** or **G. arboreum** than do the Hinganghát and Berar cottons, which are nearly pure **G. Wightianum.**

Mr. Drummond, Deputy Commissioner, Karnal District, writes, "The standard name of the crop in the Panjáb proper is *karpas*, pronounced in most of the local dialects *karpa*. But in the tract between the Sutlej and the Jumna, except in certain minor portions which conform to the Panjáb rather than to Hindustan, or where there are (?) Sikh settlers (as in parts of the Hissar, Karnal, and Umballa districts), the name of the crop is *bari*. The seeds are known apparently everywhere in the cis-Indus tract as *banaula* or *banauli*. The ordinary name for the actual cotton is generally *ru* or *rui*, but this word is primarily applicable to fluff of any sort, whether of animal or vegetable origin. The wild **Gossypium** of the Thanésar country is known to the Brahmans and Hindú Rájputs as *nurma*, in allusion possibly to its ritual use, or to the purity of the snowy bolls."

Several forms occurring locally are known by distinct vernacular names. Among these may be mentioned the *nurma*, above noticed by **Mr. Drummond**, and frequent in other districts. It is said to have a red flower, and a greenish seed, both of which characters distinguish it from the ordinary field crop, of which the flower is yellow, and the seed grey. This so-called wild cotton, occasionally cultivated, is probably a form of **G. neglectum**, *Tod.*, though the Deputy Commissioner of the Jhang District believes it to be of Egyptian origin.

Conf. with pp. 7, 28.

The Panjáb-Himálayan forms appear to be hybrids between **G. arboreum** and **G. Wightianum.** Aitchison's 462, collected in the Kuram valley, is also of this nature, having a strong strain of the former. A specimen collected during the Gilgit expedition (No. 250) is, perhaps, the nearest approach to **G. herbaceum,** which the authors have seen from any region within the Indian botanical area.

CULTIVATION in the PANJAB. Races.

Mr **Drummond**, in an interesting note on the cotton of his district kindly furnished to the authors, writes: "The mass of the scrub of the Nardak is *Dhák* (**Butea frondosa**), but species belonging both to the Siwalik and the Deccan type of flora respectively occur, and among these is a species of **Gossypium,** which I am satisfied is indigenous, or at least spontaneously produced from a remote antiquity. This wild cotton is well known to the modern inhabitants of the Nardak, by whom it appears to be called in their vernacular '*narma*' or '*bun-narma,*' and the staple is used by preference for making up the sacred thread of the Kulchattar Brahmans." The plant referred to is, in all probability—like the "*narma*" already mentioned—**G. neglectum,** *Tod.*

Wild Cotton in the Panjab. 173 (553) Conf. with p. 28.

The same writer continues, "In the Karnál district the cultivators do not appear to distinguish with any precision special races or breeds of the cotton plant, but more than one race undoubtedly occurs. A red or pinkish flowered kind is not uncommon" (probably a cultivated form of **G. neglectum**) "in the Panipat Tahsil, and in parts of the Karnál sub-collectorate. The Nardak field cotton is barely distinguishable, except as regards its habit, from the wild **Gossypium** of the neighbouring thickets.

"The spontaneous form has an almost zig-zag stem, and a rather flexuous habit, which obviously contrasts with the rigid and twiggy appearance of the cultivated shrub. This habit is doubtless determined by the usual manner of growth of the wild cotton, which generally springs up inside a clump of *hiún* (**Capparis sepiara,** *Linn.*) or some other of the prevalent shrubs."

During the past twenty-five years experiments have been steadily conducted with exotic seed. It is unnecessary, in this place, to enter into an exhaustive consideration of these; for a full account the reader is referred to **Baden Powell's** *Panjáb Products, 488-496,* also to the Panjáb cotton reports. Suffice it to say, that Mexican and New Orleans seem to have promised well, while the Sea Island, Egyptian, and Nankin races did not succeed.

Exotic Cottons. 174 (554)

In 1875-76 attempts were made to introduce Hinganghát, but, apparently, with little success, for we find the Deputy Commissioner of Rawalpindi reporting, "The people do not care for this variety, as they consider the cultivation costly and the yield scanty."

Soils. 175 (555)

SOILS AND METHOD OF CULTIVATION.—The soil of the Panjáb appears generally to be well suited for the cultivation of cotton, if irrigated or low-lying and subject to inundation. Thus, Mr. **Berkely**, in an interesting report on experimental cotton cultivation near Delhi, writes, "The selection of the soil is by no means a difficulty; all varieties of equally productive soils have appeared to me equally well adapted for cotton. Thus, for instance, wherever sugar-cane, wheat, or gram grows luxuriantly, it may be expected that cotton will thrive equally well."

Methods. 176 (556)

In the report on the settlement of the Jhang district it is stated that the crop grows best on the *utár* wells in a strong loam, and that it does not succeed on *sailáb* lands. Perhaps one reason for this is that the mode of cultivation in the latter is more slovenly, but even on good wells in *sailáb* lands the crop is said to be always lighter than in the uplands.

Manure is frequently, but by no means always, applied. The extent

CULTIVATION in the PANJAB.
Methods.

to which irrigation was employed in cotton lands during the year 1888-89 may be most conveniently shown in the following table:—

IRRIGATED BY		Flooded lands.	Dry lands.
Canals.	Wells.		
Acres.	Acres.	Acres.	Acres.
231,900	186,900	21,600	315,900

The following extracts from Settlement Reports and Gazetteers may be quoted, as describing, with a fair amount of accuracy, the methods followed in the more important cotton-growing districts:—

In the Gazetteer of Gurgaon it is stated that :—" Cotton is sown on wells or where there are other means of irrigation, and on rain lands as soon as the first rain falls. The land is generally ploughed three or four times, the seed is sown broadcast, having been first rolled in cow-dung, so as to separate the individual seeds : about eight seers go to the acre. Cotton sown on wells has to be watered every fifteen or twenty days until rain falls ; weeding is required three or four times ; ten labourers will weed half a bigah a day for one rupee."

In the report on the settlement of the Jhang district, a long account of the methods pursued in the district occur, from which the following may be quoted :—" The cotton of Shorkot, grown on the *utár* soil, irrigated during the hot weather months from *jhallars*, or by inundation from canals, is very good. Land intended for cotton ought to be ploughed up once beforehand, after the cold weather rains. It is then manured ; all cotton land ought to be manured, but a good deal never is. The manure is spread and the first watering is given. If the *zamindar* is lazy, he sows the cotton seed smeared in cow-dung broadcast. The land is then ploughed twice and rolled. If the *zamindar* is industrious, he will plough the land twice to cover in the seed. The well beds and water channels are then made. In Chiniot cotton is sown much earlier than in the two southern tahsils. Sowings are made from the end of *Chait* to the middle of *Jeth* (April and May). About 32℔ of seed per acre are used. Early sown cotton is ready to pick in *Bhadron* (August-September) ; all *Bhadron* pickings belong to the tenant. The proprietor does not share in the pickings before the 1st *Assin* (September-November) and he takes nothing after the *Lohir* festival, the 1st *Magh* (January-February). There is not much left after the 10th January, but what there is the tenant takes. Very little *mudhi* cotton is grown in this district. There is not enough rain. Cotton is hardly ever grown alone. Melons, *jowar, mandua, kangnia, swank* are almost invariably found in the cotton fields. Melons are sown with the cotton. The other crops are sown later on and are used principally for fodder. *Jowar* is hardly ever allowed to ripen, and the writer cannot remember ever having seen an instance of the other three crops ripening, the reason being that they are sown where the soil is hard and saline and not well suited for cotton. Hence the cotton is light and the deficiency is made up by the associated crop. In this district the crop is not usually ploughed after the bushes have reached some height. The fields are hoed and weeded and the *jowar* or other seed is then scattered broadcast in between the cotton bushes. A watering is at once given and the seed usually germinates. Less *mandua, kangni,* and *swank* are grown in Chiniot than in the other tahsils. During the hot months cotton is watered every sixth day. In the early stages cotton is liable to be injured by drought and hot winds. Too much rain is also injurious to cotton. In 1878 the

CULTIVATION in the PANJAB.

Methods.

rains were very unfavourable and the cotton on well lands ran greatly to wood. The bushes were very fine, but the outturn was nothing more than average. The '*tela*' blight also attacks cotton. Early frosts do more damage than anything else. Two kinds of cotton are grown in this district, but the red-leaved plant is not often seen. The ordinary country plant is the most common."

The following description of the methods employed in Shahpur is also of interest, and may be quoted: "Cotton has always been very largely grown in this district. Few wells are without their patch of two or three acres of the plant. More than this cannot ordinarily be set apart for its culture, as it is a crop that requires constant attention in weeding and watering. The seed is put into the ground in March at the rate of eight seers (16 ℔) to the acre, and the pickings give an average of about one and a half maunds of cleaned cotton per acre. The same plants are often made to yield three crops, by cutting them down level with the ground each year after the cotton has been gathered; at the same time the soil is well ploughed up between the roots and manured."

Mr. Drummond, Deputy Commissioner of Karnál, has kindly furnished the following account of cotton cultivation in the Eastern Districts of the Panjáb:—

"In the Nardak, country cotton seems to have been cultivated from an ancient period. It is usually sown on uplying patches (ordinarily clearings, or paddy reclaimed), on the edges of the water hollows, and watered by lift from natural hollows, or excavated reservoirs, or from aqueducts led out of these. In some cases these are fed from branches of the Sarsuti and Chautang, on which numerous wide dams and similar works have been made to facilitate watering for different crops, but primarily rice, none of these being probably of high antiquity.

"Wells in the Nardak are rare, and confined for the most part to the environs of the village homestead. The cotton cultivation of the Nardak, in so far as it depends on these wells or on outlying wells, which are almost all of very modern origin, is an innovation. In fact, it is not uncommon to find the husbandman irrigating his cotton plots from the village tank, though they may grow within a few yards of a well, which is reserved for the tillage of manured wheat or tobacco.

"This is a very different condition of affairs from that which prevails in other portions of the district, where cotton is now grown. In the Khadar of the Jumna the cotton crop is mainly dependent on wells worked usually with the Persian wheel. In the belt again between the Jumna valley and the Nardak (comprised in the Bángar of the Karnál and Panipat tahsils), the cotton is almost entirely dependent on the Western Jumna canal

"Wells in the Nardak are, of course, restricted by the great depth of the water level, which slopes from 50 feet below the surface in the north of the Kaithal sub-collectorate, to over 150 feet in the south-west corner of the same sub-division. The Nardak cotton is, perhaps, consumed almost entirely on the spot: but the canal-irrigated and Jumna valley outturn goes to swell the *Khadar* cotton supply of the Delhi market.

"*Khadar* or *Khadir* cotton is a commercial term for the produce of the districts (in the Panjáb and North-West Provinces) on the Jumna from Umballa to Delhi, both inclusive.

"In the Eastern districts of the Panjáb, cotton is grown on all but the lightest soils, but it prefers the more tenacious loams. It requires copious watering and certain seasons, but unless the water be drained off naturally or by artificial means at the proper juncture, the flowers and, consequently, the pods, will suffer. Hence the Khadar is always more speculative than the result in the canal-irrigated uplands, or in the Nardak, notwithstanding

G. 176

CULTIVATION in the PANJAB.

Methods.

the fact that the cultivation in the last-named tract is so primitive in its character. On the other hand, too strong a sun when the buds are forming is most adverse to the field. The plants, if exposed at this stage to an excess of moisture, are drawn out and run to stalk. The blossom is then sparse and light, and of flowers produced, only a portion mature pods.

"The usual time for sowing in the Jumna Khadar and the neighbouring uplands is just at the first heavy burst of the periodical rains, or from about the 20th of June to the 10th of July. But on land commanded by wells the seed is sometimes put in earlier. The practice of sowing in March or even earlier, which is reported from some districts of the Panjáb seems to be unheard of in the Eastern districts. In these the wells and labour are not set free for cotton before April.

"Cotton is also raised as an unirrigated crop in parts of the Karnál and other Eastern districts, but nearly always in such a situation that the young plants can be watered, at a pinch, from a tank, or by catch-water drains leading from the waste. It is only in a specially favourable season that a cotton-patch can be left to the unaided rainfall. In the Eastern Panjáb the soil is not always manured for cotton. In the Khadar it is often sown on what is known as *sailába* or spongy land, too far from the village site for the transfer of manure. In the Nardák the grey loam does not require more manuring than it receives indirectly from collections of flocks and herds at particular stations; possibly the soil contains particles of lime washed down from the Himálayan formations near which the hill streams originate.

"The pods are picked from November to January. Near towns, and especially the large town of Panipat, the picking forms an occupation for the women of the more industrious classes of the cultivating lower orders. The hire is sometimes paid in cash; more frequently as a proportionate share of the day's gathering.

"The plan of sowing certain leguminous crops (chiefly species of **Melilotus** and **Trigonella**) with the cotton, as an after-crop to be used as fodder for well bullocks when the natural grass begins to fail, which is pretty general in the Central Panjáb, seems to be unknown in the Eastern districts. This fact is probably due to the comparative richness of the pasture and the greater area of waste and fallow which still prevails, especially in the Karnál and Hissar districts.

"Cotton mixed with other crops is commonest in the Khadar of the Jumna (*til,* melons, and certain autumn pulses are the crops most commonly sown with it), but throughout the Karnál district, and generally in the Eastern districts, *san* (**Hibiscus cannabinus,** *Linn.*) is, as a rule, sown in strips along the edges of the cotton pastures, or in alternate ribbons with the main staple. The Karnál cultivators assign several explanations, more or less reasonable, for this practice, but cotton when so treated seems undoubtedly to have a better chance than when unprotected. Possibly the tall stems of the **Hibiscus** help to moderate extremes of sun-light and wind, but the primary object, as in the parallel case of gram being edged with *kasumbha* (**Carthamus tinctorius**), is likely enough to keep off antelopes and cattle, which the traditions of the *Dharmdharti* (home of the law, or *Aryavartta*) do not permit to be kept off by more vigorous measures.

"In the eastern portions of the province cotton is one of those crops to which the system of agricultural association (*lána*) is almost universally applied. Indeed, in those tracts in which the wells are over 40 feet in depth (in such districts as Karnál and Hissar), the cultivation of cotton, except by the combination of individual resources in the way of oxen, implements, and labour, would scarcely be practicable."

CULTIVATION in the PANJAB.

Picking. 177 (557)

Picking and Cleaning.—The following is extracted from a monograph on cotton manufactures in the Panjáb, 1883-84:—"Cotton is picked about November and December. The pods do not all ripen at the same time, and it is therefore necessary to go over the field several times. This work is chiefly performed by women. They are usually paid by a portion, generally one-tenth, of the cotton they pick. The pods are exposed to the sun and beaten to make the husks (*doda*) separate from the fibre.

"The seed to which the fibre is attached is then extracted by a small hand machine which takes the place of the cotton gin. This consists of two rollers, one of which is generally of iron and smaller than the other, which is of wood; but in some districts both rollers are stated to be of wood. By means of a spiral turn at the end of the iron roller, which engages a projection of the wooden roller, the two are made to revolve in opposite directions when the handle attached to one of them is turned. The fibre (*rui*) is drawn through the rollers, while the seed (*binola* or *varema*), which cannot pass between them is left behind. This machine, which is called *belna*, is capable of ginning about 12 seers of cotton in a day. At Multán a steam cotton gin has been established . . . which employs 70 persons.

"In the Hazára district there are at Pakhli some *belnas* worked by water power. The *belna*-worker is commonly paid by receiving the cotton seed which he extracts, but sometimes receives cash. The operation is also very frequently performed by members of the household of the person to whom the cotton belongs. In the census tables 3,257 men and 1,345 women are recorded as cotton ginners, but these figures do not represent the numbers really so employed. The cleaned fibre usually weighs one-third, the seed two-thirds, of the whole contents of the pod, but with inferior cotton the fibre is sometimes only one-fourth of the whole."

Seasons. 178 (558)

Seasons of Sowing and Picking.—The period of sowing occurs at different times in different parts of the province, and is also largely regulated by the rainfall. In Gurgaon it is sown on wells, or where there are other means of irrigation, from March to May, and on rain lands in June-July, as soon as the first rain falls. The early-sown cotton begins to bear in August-September, the later in October-November, the plants continue to bear till they are killed off by frost. The early cotton has thus the great advantage of ripening a month earlier than the other, and of being generally able to bring out all its pods before the frost comes. If the rains or hot westerly winds are excessive, the pods are said not to ripen.

In the Shahpur district the seed is sown in March, but the process of gathering does not commence till October and lasts till the end of December or even later. In other districts of the Panjáb, sowings, as a rule, go on from the middle of February to the end of March, and in well-watered lands to April and May; on all rain lands they are much later. Picking ought always to be completed before January, owing to the danger of loss from frost. The crop is sent to market from January onwards, and is available for export, according to the means of transit, from January to May or June.

Diseases. 179 (559)

Diseases.—Dr. Johnstone, in a valuable note on the blights of cotton communicated to the Agri-Horticultural Society of the Panjáb, described the crop as liable to four special pests, *viz.*, (1) the larva of **Helicopis cupido**, which attacks the young plant; (2) the larva of **Deprescaria gossipiella** (weevil) which infests the seed in harvest and is known as the "*toka*" by the natives; (3) a large hairy caterpillar (*bhûngo*) which attacks the leaves and stem on the approach of the rainy season; and (4) an **Aphis** which frequently attacks the crop in great quantities, and has, on several

CULTIVATION in the PANJAB.

Diseases.

occasions, greatly diminished the harvest. The disease produced on the plant by its ravages is known to the Panjábis as *thela* and is characterised by withering of the leaves and finally by the death of the plant.

According to **Dr. Johnstone** the **Aphis** attacks crops on certain soils more readily than on others, hard, compact land being most favourable to the appearance of the insect, soft, damp, loamy soil less so, and sandy land least of all.

Dr. Johnstone states that the following methods of treatment are efficient cures for these blights:—

(1) The larva of **Helicopis cupido** may be put to flight by sprinkling ashes over the young plant.

(2) The weevil can be killed by scalding the seed with hot water.

(3) The large caterpillar, *bhúngo*, has many natural enemies, being preyed on by starlings and other birds. If, however, they attain large numbers and become very destructive, they must be picked off the plants by hand.

(4) Regarding the **Aphis Dr. Johnstone** makes the very interesting statement that "it does not exist in any cotton field grown in the proximity of *san* hemp, and consequently recommends that hemp should be planted here and there along with the cotton. The Deputy Commissioner of Gujerát, however, in forwarding **Dr. Johnstone's** note, states that in his opinion the action of *san* hemp described by the latter requires corroboration before being accepted as a proved fact.

It is worthy of note, in connection with **Dr. Johnstone's** theory, that in many parts of the province a row of *san* is sowed round the cotton fields (*cff.* p. 122).

Cotton is said in the Muzaffargarh district to be subject to the following diseases: (1), *múlá*, a blight that begins at the stem and spreads over the plant; (2), the soil becoming water-logged (*soma*); (3), a red worm that attacks the cotton in the pod.

Yield. (560) 180

YIELD PER ACRE.—The yield per acre appears to vary very largely in different parts of the province, being returned as 260lb to the acre in Montgomery, while in Jhelum it is shown as having been only 80lb. Dividing the total outturn for 1888-89 by the total area, an average outturn of 137lb per acre is obtained, an average a little lower than that of Sind and considerably above that returned for most other parts of India.

The following are given as the average outturns per acre (for 1888-89) of the chief cotton-growing districts of the province:—

DISTRICT.	YIELD PER ACRE.	
	Unmixed.	Mixed.
	lb	lb
Gurgaon	120	...
Rohtak	240	.
Karnál	...	220
Montgomery	260	...
Umballa	120	...
Multán	102	...
Jálandhar	248	210
Sialkot	100	32
Lahore	110	...
Delhi	150	...
Shahpur	120	116
Dera Ghazi Khan	120	80
Rawal Pindi	120	...

CULTIVATION in the PANJAB.

Yield.

Mr. Drummond has kindly furnished the following note in explanation of the variation in the figures of yield per acre: "This varies, throughout the province within wide limits, and even in the same district marked differences exist. A good Khadar plot, properly manured and steadily irrigated from a well, in a good season would probably produce as much as 400℔ to the acre. On the other hand, a Nardak patch, on which the same crop has been raised for some autumns in succession, will yield so little, should the reservoirs, from which it is usually watered fail, that the cultivator will not even expend, in picking it, the menial labour for which there may be a demand in some other direction. But the average outturn per acre in the eastern districts in an average season should probably not be placed below 200℔."

Cost. 181

COST OF CULTIVATION.—Details are given in the subjoined extracts from the Provincial Cotton Report for 1882-83, but, judging from the remarks thereon they do not appear to be very reliable. It is to be regretted that no more recent statistics are available:—

"The following tables show the cost of the cultivation of cotton in those districts from which detailed estimates have been furnished, distinguished, according as irrigation and manure are, or are not, together or separately applied; also a detail of the cost of each agricultural operation on that class of land in which the cultivation is most complete, that is to say, land which is both manured and irrigated:—

Statement showing the cost of cultivation of cotton in certain districts under the four heads A, B, C, and D, i.e., *irrigated and manured, irrigated but not manured, unirrigated but manured, unirrigated and unmanured.*

DISTRICT.	A.	B.	C.	D.
	R a. p.	R a. p.	R a. p.	R a. p.
Delhi	31 3 7	24 13 4	19 11 2	14 1 3
Gurgaon	20 13 0	17 11 0	11 15 0	8 12 0
Karnál	23 10 0	19 9 6	15 9 0	8 10 9
Hissar	17 0 0	...	...	8 8 0
Rohtak	14 3 7	11 3 10	8 11 1	6 8 7
Ambala	29 2 9	21 2 10	21 2 1	13 7 2
Ludhiana	23 0 10	17 15 5	...	10 3 5
Jálandhar	37 13 4	30 15 6	25 15 6	17 11 11
Amritsar	27 6 4	18 9 6	11 4 4	9 0 2
Gurdaspur	24 10 0	21 7 0	16 13 0	13 10 0
Lahore	20 6 0	12 4 0	9 4 1	5 5 0
Gujerat	17 1 0	15 7 0	10 5 0	8 5 0
Multán	16 11 10	10 6 5	...	7 12 10
Jhang	10 0 9	8 5 3	6 10 9	5 1 0
Montgomery	24 0 3	16 7 6	...	7 6 3
Dera Ghazi Khan	22 12 6	12 5 0	...	6 2 6

CULTIVATION in the PANJAB. Cost.

Estimated Cost of Cotton Cultivation under Class A (irrigated and manured) of each agricultural process in those districts of the Panjáb from which detailed estimates have been furnished.

	Delhi.	Gurgaon.	Karnal.	Hissar.	Rohtak.	Ambala.	Amritsar.	Gurdas-pur.	Lahore.	Gujerat.	Montgo-mery.	Dera Ghazi Khan.
	R a. p.	R a. p	R a. p.	R a. p.	R a. p.	R a. p.	R a. p.	R a. p.	R a. p.	R a. p.	R a. p.	R a. p.
1. Ploughing before sowing .	Detail is not given in the report.	Ditto.	6 0 0	3 0 0	2 10 0	4 4 9	2 12 0	4 0 0	2 0 0	2 0 0	1 4 0	1 8 0
2. Manuring before sowing .			2 8 0	2 8 0	2 3 6	5 3 2	6 1 4	1 0 0	1 0 0	0 14 0	1 3 9	3 2 0
3. Watering before sowing .			4 0 0	2 0 0	0 12 0	2 6 0	3 8 0	2 0 0	1 0 0	1 12 0	1 4 0	0 10 9
4. Seed and its preparation for sowing.			0 4 0	1 0 0	0 6 9	0 11 9	0 8 8	0 8 0	0 5 0	0 5 0	0 3 3	0 11 3
5. Sowing			...	0 8 0	1 4 3	0 10 0	0 7 0	0 2 0	1 0 0	0 10 0	1 4 0	0 15 9
6. Watering after sowing. .		Ditto	...	3 0 0	0 14 0	3 5 2	5 2 8	4 0 0	6 0 0	4 0 0	10 6 0	3 5 3
7. Hoeing and gleaning . .			1 12 0	2 0 0	1 8 2	4 0 7	2 0 8	3 0 0	2 1 0	1 0 0	1 2 0	1 9 6
8. Manuring after sowing .			...	1 0 0	1 3 5	0 8 0	...	1 0 0	2 0 0	0 8 0	0 3 0	3 0 0
9. Picking			2 11 6	1 0 0	1 12 0	3 13 7	3 6 0	5 4 0	2 0 0	3 0 0	2 13 10	3 0 0
10. Cleaning			6 6 6	1 0 0	1 9 6	4 3 9	3 8 0	3 12 0	3 0 0	3 0 0	4 4 5	4 14 0
TOTAL .	31 3 7	20 13 0	32 10 0	17 0 0	14 3 7	29 2 9	27 6 4	24 10 0	20 6 0	17 1 0	24 0 3	22 12 6
	M. S. C.	M. S. C.	M. S C.	M. S. C.	M. S. C.	M. S. C.	M. S. C.	M. S. C.	M. S. C.	M. S. C.	M. S. C.	M. S. C.
Produce per acre . .	6 0 0	3 0 0	4 20 0	3 20 0	2 0 0	6 0 0	4 10 0	5 0 0	4 0 0	2 0 0	4 20 0	4 0 0
	R a. p.	R a. p.	R a. p.	R a. p.	R a. p.	R a. p.	R a. p.	R a. p.	R a. p.	R a. p.	R a. p.	R a. p.
Cost per maund of uncleaned cotton.	5 3 3	6 15 0	5 4 0	4 14 0	7 1 9	4 13 9	6 7 1	4 14 10	5 1 6	8 8 6	5 5 4	5 11 1

(561)

CULTIVATION in the PANJAB.

Cost.

"The Deputy Commissioner of Muzaffargarh, Mr. Gladstone, finds that the tahsildars of his district, under the impression, apparently, that some return must be furnished for cost of cultivation under each of the heads A, B, C, and D, which the prescribed form contains, have not paid sufficient regard to the real state of the fact, which is that cotton is never grown upon unirrigated land in that district. It is not impossible that similar remarks might have been made of the returns of other localities. As a rule, cotton is one of the crops to which the zemindar devotes all the resources of labour and fertilising material at his command, and unless he possesses sufficient of these, he will not attempt the cultivation at all. To this rule there are some exceptions, and among them may be instanced the districts of the Rawal Pindi Division, where this crop is largely grown with little care in unirrigated land, three crops being often taken off the same plants.

"In details the estimates differ widely from one another, and those for the same district in this report, and in the former one do not always agree very closely. To take the Karnál estimate, for example, the operation of ploughing before sowing is now reckoned to cost only R6 an acre instead of R9 as before, and those of picking and of cleaning to cost together only R9-2 as against R12. Again, nothing is put down in the present estimate for either watering or manuring after sowing, for which R2-8 and R1-14 were charged in the former calculation, but the cost of the same operations *previous to* sowing is increased from R1-4 and R2-8 to R2-8 and R4 respectively. The district report contains no explanation of the differences; but since the cost in Karnál, as reported of the first and last operations, *viz.*, ploughing and cleaning, was, and still is, far above what it is said to be in any other district, it may be inferred that the present year's figures are a nearer approach to accuracy than those given before. The produce per acre of uncleaned cotton in Rohtak and in Gujerat is given as only two maunds. This is evidently an under-estimate, for the resulting cost of production of a maund of uncleaned cotton is thereby raised to a figure which would leave the producer a considerable loser, as a comparison with the prices per cwt. of uncleaned cotton in the same districts will show."

Mr Drummond has furnished the following note on the subject of cost of cultivation:—

"This in the Eastern Panjáb is peculiarly difficult to estimate, owing for one thing to the system of co-operation to which reference has been made under a previous head. The calculations usually furnished from the province to the Agricultural Department appear frequently to include too many items. For example, ploughing and manuring cannot be said to cost the bulk of the cultivating owners and fixed tenants of the Eastern districts any sum appreciable in currency.

"On the other hand, if the cost at which a capitalist could effect the same operations is to be taken, then there are several items which should be included in addition to the value of labour; notably a proportion of interest at the market rate on the original cost of well-sinking, and other items which the zemindar would recognise as included, such as the share given in kind to menials and artificers for their general contribution to his farming work. Where Persian wheels are used this is a material consideration.

"In the case of villages which enjoy canal irrigation from the Western Jumna Canal, the cost of irrigation can be pretty accurately estimated, and if to this important item, there be added an estimated figure for the value of labour and depreciation of stock, and a further percentage taken for miscellaneous expenditure, the cost per acre of raising uncleaned cotton

CULTIVATION in the PANJAB.

Cost.

fit for the Delhi market (and excluding, therefore, cost of delivery to the broker) cannot exceed R11 or R12 at the utmost. The market value of the manure, if added, would hardly bring the total estimated cost to more than R13 or R13-8.

Panjab Trade (562) 182

Analysis of Last Year's Trade.—The total exports in 1888-89 amounted to 295,976 cwt., of which 96,281 cwt. went to Bombay, 44,121 cwt. to Karáchi, 42,123 to Calcutta, and 108,988 cwt. to other provinces and presidencies. Of the last mentioned quantity much the greatest proportion went to Sind, smaller quantities to the North-Western Provinces and Oudh, Bengal, Bombay, and Rajputana. As already stated, it is highly probable that a considerable unregistered export road trade exists between the Panjáb and the North-West Provinces, more particularly Agra. In addition to the above, 1,304 cwt. were exported across the frontier to Kashmir, and 3,159 cwt. to Sewestan, Tirah, Kabul, Bajour, and Ladak. The total imports were very unimportant, amounting in all to only 1,889 cwt., to which the North-West Provinces and Oudh, Rajputana, and Central India, Sind, Karachi, Calcutta, and Bengal contributed small quantities, while the largest amount, *viz.*, 880 cwt., was imported across the frontier from Kashmir.

CULTIVATION in the CENTRAL PROVINCES. (563) 183

8. CENTRAL PROVINCES.

References.—*Rept. on Cotton by Chief Commissioner, C. P., Sept. 1889; Carnac, Rept. on operations of Cotton Dept., 1867-1869; Annual Reports of Cent. Prov. and Berar Cotton Dept.; Reports, Agri. Dept.; Cotton Reports, Rev. & Agri. Dept.; Exp. Farm. Reports; Administration Reports in many passages; Settlement Reports:—Jubbulpore, 87; Chanda, 81, 96 99, 113; Wardah, 63, 67, 75; Raepore, 56; Nursingpore, 53; Baitool, 77; Saugor, 3; Nimar, 1, 193-195; Nagpore (Supp.), 272; Upper Godavari, 36; Morris, Godavari Dist., 68; Gazetteer (1870), 6, 64, 115, 131, 136, 168, 204, 212, 327, 331, 365, 417, 516; Agri.-Hort. Soc. of India, Transactions, V., Proc., 65; Journals (New Series), I., Proc., lxi.*

Area. (564) 184

Review of the Area, Outturn, and Consumption.—The total area under cotton in these provinces during the year 1888-89, as will be seen from the table (p. 56) was 613,348 acres. As already stated in commenting on the facts regarding *Hinganghát* cotton, the area has markedly fallen off within the past twenty years. Thus, about fourteen years ago, the cultivation averaged 800,200 acres, five years ago it dropped to 688,400, while in the last five the average has only been 608,800 acres. The total outturn in the year under consideration was returned as 351,923 cwt. of cleaned cotton, giving an average yield to the acre of 64·2℔. A comparison of the total imports from all sources, 8,756 cwt., with the total export, 170,130, shews the net export to have been 161,374 cwt., which, when deducted from the above given total production, leaves 190,549 cwt. in the provinces for local consumption. When distributed over the population of 9,838,791, this amount gives an average per head of nearly 2·2℔, or, taking Indian cotton cloth at 4⅓ yards per ℔, about 9·5 yards of cloth to each individual. In giving the above figures calculated from available information for one year, however, it would be urged that in these provinces, as elsewhere, they are only approximations, and are open to several sources of grave error. Thus, in a communication from the Government of the Central Provinces, dated 1889, it is stated:—

1st—That the outturn cannot be precisely ascertained, as the statistics of area on which it is compiled are in most cases only estimates, and sometimes extremely rough ones.

2nd—That, granted the estimate of production is correct, the conclusions as to its consumption are justified only on the assumption that the exports and imports by road, of which no statistics are available, balance each other.

CULTIVATION in the CENTRAL PROVINCES.

3rd—That the years of production and of trade statistics do not correspond.

Area.

An interesting calculation, subject, however, to the above errors, is made in the same report, from which it would appear that the total consumption per head in these provinces, of local and imported cotton and cotton cloth, is approximately 3·18lb, while the production per head is 3·22lb per head. This shows that, on the whole, the provinces consume as much as they produce.

A considerable portion of the amount above calculated as available for local consumption is annually consumed by the cotton mills of the province, of which there are three. Thus, in 1889, these mills are stated to have worked up 91,021 cwt. of raw cotton, leaving approximately 100,000 cwt. for hand-loom manufactures, mattresses, wicks, &c.

Districts. 184 (564)

Districts where grown.—The cotton country of the Central Provinces lies on the left bank of the Wardha river. In the north, where the river debouches from the Satpura Hills, the cotton cultivation covers a rich but narrow strip along the bank. This strip widens as it proceeds southwards until it ultimately attains a width of some 50 miles. The well-known mart of Hinganghát may be said to be situated where the cotton field is widest; but the whole plain, though capable of producing cotton, is not entirely occupied by it, for wheat and *jowari* alternate with cotton fields. After reaching its greatest width, the cotton country again narrows, until at last, in the south of the Chanda district, it is lost in the encircling brushwood and jungle.

The districts in which the cotton crop is of most commercial importance, are Nimar, Wardha, Chhindwara, and Nagpur. The rainfall in these provinces is very much lighter on their western border than elsewhere, and it is in the western districts that cotton is most profitably grown. In Chanda, which supplies the mart of Hinganghát with what is reported to be its finest quality of cotton, the kind known as *jari*, is also produced. The following figures represent the average area under cotton, during the three years 1885-1888, in the above-mentioned chief cotton-growing districts:—

	Unmixed.	Mixed.
Wardha (acres)	7,813	137,978
Nagpur ,,	24,954	88,596
Chhindwara ,,	23,041	26,508
Nimar ,,	45,097	1,278

Races. 185 (565)

Races of Cotton.—As already stated the greater portion of cotton grown in these provinces belongs to the growth known commercially as *Hinganghát*; indeed, this race is the only one of importance from a trade point of view. It occurs in two different forms, one of which is known as *bani*, the other as *jari*. The former is longer in staple and commands a price equal to that quoted for any other Indian cotton; the latter is whiter and freer from leaf but of poorer staple. Botanically, both races are forms of **G. Wightianum**, *Tod.*, the inferior form shewing degeneration through hybridisation with **G. neglectum**. In the remarks under **G. arboreum** a passage has been quoted regarding Seoni district, from which it will be found that **G. neglectum** is cultivated under the name *manua*, and **G. arboreum** as *deo* (p. 9). Here, as elsewhere, however, numerous experiments have been made with the object of attempting to introduce cottons of a higher class, In 1882-83, *Broach* cotton was planted at the Nagpur Experimental Farm, but the results appear to have been unsatisfactory. It suffered much from a drought which prevailed in that year, and afforded no remunerative yield. In 1883-84 experiments were commenced at the Government Farm with New Orleans (**G. hirsutum**), and Upper Georgian cottons (**G. barba-**

Cotton with p. 9.

CULTIVATION in the CENTRAL PROVINCES.

Races.

dense), which have since been continued with fairly good results. Parallel experiments were made with the fibre obtained from these and with that of the *bani* and *jari* country or *Hinganghát* cotton, by the Managers of the Cotton-Mills at Nagpur and Hinganghát, with the following results :—

Comparative value of :—	New Orleans.		Upland Georgian.		Bani.		Jari.	
	Hand-ginned.	Saw-ginned.	Hand-ginned.	Saw-ginned.	Hand-ginned.	Saw-ginned.	Hand-ginned.	Saw-ginned.
	R	R	R	R	R	R	R	R
Nagpur Mills	75	55	75	78	73	70	55	56
Hinganghát Mills	70	68	78	78	78	75	61	60

The above-cited figures are of considerable interest, shewing, as they do, how markedly the New Orleans form of **G. hirsutum** deteriorated in value after being saw-ginned, while the Upland Georgian, on the other hand, was, if anything, improved by the process, and the Indian cottons were only slightly damaged. The high value of the better class of country cotton is also worthy of notice. A recent communication from the Manager of the Khándesh Experimental Farm contains the information that the short-stapled *jari* now fetches almost as good a price as the *bani*, owing to its containing a higher percentage of fibre to seed, and to its purer white colour. A continuation of the cultivation of American cottons demonstrated that New Orleans was unsuited for the climate and soil of the provinces, but that Upland Georgian was capable of yielding very satisfactory results. It was proved to give a fair crop,—173lb on manured, and from 71 to 111lb on unmanured land,—and to produce a strong, good staple superior to the best Hinganghát. In 1886, over 1,000lb of seed of the Upland Georgian cotton were distributed to the rayats in the Wardha district, and the resulting crop appears to have been highly esteemed both by the cultivators and cotton dealers.

Mention is made in several reports on the cotton of the Central Provinces of a *khaki* cotton which is grown in small quantities, and is consumed locally, none coming into the market. This is probably Nankin cotton (**G. religiosum.** Conf. with pp. 16-17).

Soils. 186 (566)

Cultivation :—Soils, Seasons, &c.—The following description of the "cotton soil" of the Central Provinces, extracted from the *Central Provinces Gazetteer*, may be of interest, since the soil seems to be one of the best in India for cotton cultivation, and, at the same time, appears to lend itself readily to the growth of one of the best races of American cottons :—

"The black soil or *regar*, or, as it is not uncommonly called, the 'cotton soil,' forms one of the most marked varieties in these provinces. It is the common soil of the Deccan, Malwa, Narbada valley, &c. It varies greatly in colour, in consistence, and, with these, in fertility, but throughout is marked by the constant character of being a highly argillaceous, somewhat calcareous, clay, very adhesive when wetted, and from its very absorbent nature expanding and contracting to a very remarkable extent, under the successive influence of moisture and dryness. It, therefore, becomes fissured in every direction by huge cracks in the hot weather. It also retains a good deal of moisture and requires therefore less irrigation than more sandy ground. The colour of this soil, often a deep and well-marked black, with every variation, from this to a brownish black, would appear to be solely due to an admixture of vegetable (organic) matter in a soil originally very clayey.

CULTIVATION in the CENTRAL PROVINCES.

Soils.

Thus, deposits of precisely the same character as this *regar* are being formed now at the bottom of every *jhil* in the country, and throughout the very area where the *regar* is best marked, it is by no means an uncommon thing to find the slopes of the small hills or undulations formed of more sandy reddish soil, while the hollows below consist solely of the finest *regar*. This appears to be due to the more argillaceous and finer portions of the decomposed rocks below being washed away by ordinary pluvial action from the slopes and accumulated in the hollows, where this finer mud forms a soil much more retentive of moisture, and which therefore rapidly becomes more impregnated with organic matter, and is often marshy. *Regar* can thus be formed, wherever a truly argillaceous soil is found, and its general, but by no means universal, absence over the metamorphic and other rocks is easily accounted for by the fact that these rocks for the most part yield sandy, not clayey, soils. It is never of any very great depth, and excepting when re-arranged by rivers in their recent deposits, is therefore never met with at any great distance below the surface.

"Obviously formed from the re-arranged wash of the older and more widely-extended soils we find large areas of very fertile soil, consisting of clays rather more sandy than the older alluvium, and not therefore so black or adhesive. Though rarely formed altogether of the true *regar* soil, it frequently contains a large proportion of this mixed with other clays and sands. Every intermediate form of soil occurs, and it would by no means be an easy task to distinguish them all. From an agricultural point of view, it is interesting to see how exactly the limits of certain kinds of cultivation coincide with the limits of these marked varieties of the alluvial deposits of the country." (*Central Provinces Gazetteer, 1870, xlvi., xlvii.*). An analysis of the soil shows its composition to be as follows: silica, 48·2; alumina, 20·3; carbonate of lime, 16; oxide of iron, 1; water and organic matter, 4·3 per cent.

Methods. 187 (567)

Methods.—The soil is prepared for sowing in May to June, the seed is sown in June, picking commences in November to December, and is completed from February to March.

The following account of the method followed is extracted from the *Central Provinces Gazetteer, 1870*: "Cotton is sown and reaped about the same time as *jowar*. As it does not early obtain any considerable height it requires more weeding and banking up with the *kolfa* (bullock hoe). The process of gathering the pods is a most negligent one, most of them being allowed to drop and get discoloured on the ground. It is the worst feature, perhaps, in the husbandry of Nimar, but the cultivators urge in excuse, that it would not pay to keep people running up and down the cotton fields every day for a month or so, when so little is grown and the return so small. The process of cleaning is the same as in other parts of the country."

It is encouraging to note, that though the total area under cotton in the Central Provinces has diminished of late years, improvements are being effected in the methods of cultivation and gathering. Weeding is more carefully attended to than it used to be, and many cultivators are said to have adopted a system of light, but careful, manuring, which, as noted above, greatly increases the crop.

Yield. 188 (568)

Yield per Acre.—The following figures are given by the Government of the Central Provinces as the average outturn of cleaned cotton per acre during the three years ending 1887-88:—

Jabalpur, 36·48lb; Narbada, 43·59lb, Nagpur, 61·33lb; Chhattisgarh, 32·19lb.

In 1870, Mr. J. G. Fraser of Gopalpur, Jaunpur, reported the results of cultivating Hinganghát cotton by transplanting from a broadcast sowing

CULTIVATION in the CENTRAL PROVINCES.

made in the rains. The yield was 376℔ per acre of cleaned cotton. The reports on the samples published by the Agri.-Horticultural Society pronounced the staple to be very good, silky, of good length and strength, measuring a little over 1 inch, while the average Hinganghát is 1·4 inch.

Cost.
COST AND PROFIT OF CULTIVATION.—It is difficult to arrive at any-
189 thing like an exact estimate of the total cost of cultivation, and of the profit derived per acre. Assuming the value of the total production, however, to be proportionately equal to the value of the exports, we have 170,130 cwt.: 351,923 cwt : R37,51,382 : R77,59,934: which last figure when divided by the total area gives an average value of R11 per acre. The profit derivable from this sum would appear to be in most cases considerably less than that from wheat, linseed, or *jowar*.

Provincial Trade.
ANALYSIS OF LAST YEAR'S TRADE.—Of the total production of 1888-89,
190 *viz.*, 351,923 cwt., 170,130 cwt. is returned as having been exported. Of this, much the largest proportion went to the port of Bombay for export, 57,237 cwt. is shewn as having gone to the province of Bombay excluding the port, and smaller quantities to the North-West Provinces and Oudh, Berar, Rájputana and Central India, Bengal and Calcutta. The imports are unimportant, amounting to only 8,756 cwt. The chief amount was obtained from Berar, the North-West Provinces and Oudh, and Bombay.

COTTON CULTIVATION in the NIZAM'S DOMINIONS.

9. NIZAM'S DOMINIONS.

REVIEW OF THE AREA, OUTTURN, &c.—The average area during the five years ending 1885-86 (the last period for which figures are available) was
191 1,016,565 acres, yielding 307,002 cwt. Statistics for road traffic not being
Area.
192 statistics of rail-borne trade by accepting the road exports and imports as balancing each other. On this assumption the net export in 1888-89 was 81,842 cwt. of cleaned cotton. Taking for granted that the production in that year was approximately equal to the average of the five years above stated, the amount left for local use would have been 225,160 cwt., a figure considerably in excess of the local consumption estimated from other sources of information, *viz.*, 131,000 cwt. (*Note on Cotton by Revenue and Agricultural Department*). It may be pointed out in passing, however, that the exports representing, as they do, those by rail only, are, necessarily, in a country such as Hyderabad, but a small proportion of the total, and that, in all probability, the exports by road may more than double the above amount. Indeed, in the "Note on Cotton" the total exports are said to average 300,000 cwt. With such imperfect data, therefore, any calculation as to the amount consumed per head would be useless.

At present two cotton mills exist in the Dominions, which are stated to consume approximately 37,522 cwt. of cleaned cotton annually.

Races.
RACES OF COTTON CULTIVATED.—In the northern division of the Niz-
193 am's Dominions the description of cotton known as *Barsee* and *Nugger* (*Oomras*) is produced, and, as previously stated (p. 83), finds its way into Bombay entirely by road, and is hence credited in the rail returns of the latter. In the rest of the Dominions the chief staple grown is "Westerns" (one of the forms of **G. Wightianum**, *Tod.*). A considerable portion of the produce of the Kunar Idlábád district finds its way to the Hinganghát market, is greatly valued, and fetches high prices.

Provincial Trade.
ANALYSIS OF LAST YEAR'S TRADE.—Rail-borne imports during the year
194 under consideration were very insignificant. Of the 687 cwt. imported, 677 cwt. came from Bombay, 6 cwt. from Madras, and 4 cwt. from Bombay port.

The rail-borne exports amounted to 82,529 cwt., of which 38,609 cwt. was despatched to Bombay port, and 7,429 cwt. to the seaports of Madras for shipment; 30,226 cwt. went to the presidency of Madras, and 6,265 cwt. to Bombay. Besides these registered exports, however, a large amount is exported by road to the Central Provinces, Bombay, Berar, and probably also to Madras. CULTIVATION in NIZAM'S DOMINIONS.

10. BENGAL.

CULTIVATION in BENGAL. 195 (575)

References.—*A Selection of Papers showing the measures taken since 1847 to promote Cotton Cultivation in India (Parliamentary), Pt. I; Hunter, Orissa, II., 45, 134, App. iv.; Medlicott, Hand-book on Cotton for Bengal (1862); Reports by Director of Land Records and Agriculture; Rept. by the Govt. of Bengal, 1876, 1878; Correspondence on Cult. of Cotton in the Sanderbands; Agri-Hort. Soc. of Ind., Trans., II., 111, 132, 150; V., 51, 52, 187; VI., 106, 239, 244; VII., Proc., 17; Journals (Old Series), III., 9, 69, Sel. 291; IV., 92-105, 135, 151-172, 241-248; V., 197-205; VII., 287-292; XI., 414-421, 434-440, 491-513, 534-550; Proc. xviii., xii., 29-76, 212-264, 273-286, 325-329.*

Area. 196 (576)

REVIEW OF THE AREA, &c.—In the absence of agricultural statistics, for the Lower Provinces, it is not known what is the exact area under cotton. It is, therefore, necessary to accept as approximations the following statistics for 1876-77, which have been extracted from a report furnished by the Government of Bengal in 1878:—" There are, as far as can be ascertained, 162,000 acres under cotton cultivation in Bengal, yielding 138,800 cwt. of cleaned cotton." Accepting the above figure of acreage as correct, and the total outturn as, approximately, 139,000 cwt., the average yield per acre would have been 96℔.

In the year 1888-89 the net export amounted to 3,562 cwt., which would have left from the accepted average outturn 135,438 cwt. as available for local consumption, or ·21℔ per head of population. This small figure may either be due to an under-estimate of outturn, or to the fact that the increasing use of foreign twist and cloth is gradually but surely driving locally-grown cotton out of the bazárs and is thus leaving a greater amount available for export.

Districts. 197 (577)

DISTRICTS WHERE GROWN.—The largest area under cotton cultivation is in the district of Sarun, in which 31,000 acres were estimated to be under the plant during 1876-77. In the same year the Chittagong Hill Tracts came next in importance with 28,000, then Cuttack with 20,000, Lohardugga with 15,000, Durbhanga with 12,000, Midnapore with 10,000, and Manbhoom with 10,000 acres. But, although Sarun is shown to have had the largest area under cultivation, the outturn was not sufficient even for local requirements. The largest outturn was in the Chittagong Hill Tracts, where it amounted to 50,000 cwt.; next in order came Midnapore, with an outturn of 12,800, Cuttack with 12,500, and Julpigorí with 10,300 cwt.

In the General Administration Report for 1882-83 the following additional particulars are given:—

" The cultivation of cotton is not of very great importance in any of the districts of Bengal, with the exception of the Chittagong and Tipperah Hill Tracts, Cuttack, and Julpigorí. In the plains of Bengal, which are so fertile in other produce, the production of cotton is an inconsiderable industry, and nothing is exported, while much is imported from the North-West Provinces. The cotton grown in Bengal is in fact insufficient for the requirements of the people and has to be supplemented by cotton brought by land or river from the west, and by an increasing importation of English piece-goods.

" Generally speaking, then, it may be said that the production of cotton does not form an important industry of the cultivators in Bengal. It is cultivated, not as an article of commerce but, for the sake of variety and for domestic use, only the agricultural classes preferring strong

CULTIVATION in BENGAL. Districts.

home-spun cloth to the less durable machine-made European piece-goods procurable in the bazár. There is no anxiety on the part of the cultivators to extend or improve the cultivation of cotton." (*General Administration Report, 1882-83.*)

The last paragraph of the above extract appears to take a rather too gloomy view of cotton culture in Bengal, for both the exports and cultivation appear to have increased during the past few years.

In his report for 1886-87 (page 8), the Director of Land Records and Agriculture writes as follows:—"Cotton was formerly grown extensively in the Dacca and Mymensingh districts, in a large tract of land, the soil as well as the physical aspect of which are very well suited to the cultivation of the plant. The cotton raised here, though rather short in staple, was the finest known in the world, and formed the material out of which the very delicate and extremely beautiful Dacca muslin was manufactured. Since the decline of that celebrated fabric, the cultivation of cotton has almost entirely ceased in this tract. Considering the high price that cotton fetches in most parts of Bengal and the improvement which has been effected in the yield by the introduction of some of the best American varieties, the cultivation of this important crop may be revived in the locality with great advantage."

Races. 198 (578)

RACES OF COTTON.—Though this province gives its name to the most extensively cultivated of all Indian cottons, "Bengals," it appears probable that much of the locally-grown cotton, most of which is used for home consumption, does not belong to **G. herbaceum** in any of its forms, as defined by the *Flora of British India*. It is needless in this place to again enter into a consideration of the plants which produce the finer staples of Mymensingh and Dacca, a full account of which will be found in the descriptions of **G. arboreum** (pp. 7-13) and **G. herbaceum** (pp. 25-29). Suffice it to say that, besides the ordinary hybrid forms of the **G. herbaceum** series, **G. neglectum**, *Tod.*, almost in its type form appears to be largely cultivated, especially in the eastern parts of the Lower Provinces.

Conf. with p. 24.

Numerous experiments have been made in Bengal, as elsewhere, with the object of introducing exotic seed. For a description of the earlier and more energetically carried out of these, the reader is referred to the article on **G. barbadense**, also to the exhaustive literature on the subject contained in the publications of the Agri.-Horticultural Society of India, above cited, and to **Mr. Medlicott's** *Cotton Hand-book for Bengal.*

The following extracts from provincial reports, relating to the attempts made in later years, may, however, be quoted. It will be seen that recent experiments have been chiefly directed towards the introduction of Nankin cotton:—

Conf. with p. 17.

"It having been brought to notice that a species of indigenous khaki-coloured cotton is grown in Durbhanga, Chittagong, and Hill Tipperah, it was thought that the cultivation of Nankin cotton would be most likely to succeed in these or in the neighbouring districts. Arrangements have accordingly been made for the experimental cultivation of Nankin cotton in Mozufferpore and Hill Tipperah. It has been found that this cotton can be grown with success in the Government Estate of Khoordah in the Orissa Division." (*Agricultural Report, Bengal, 1885-86.*)

At the Burdwan Ráj experimental station "eight different varieties of cotton were experimented upon. The plots were manured with bone-meal at the rate of 1,080lb per acre, with the following results:—

"Nankin gave a yield of $370\frac{1}{2}$lb; Deba Lonacinu, $97\frac{1}{2}$lb; Lonacinu, 468lb; Kapasia, 156lb; Upland, $136\frac{1}{2}$lb; New Orleans 156lb; Hill Tipperah and Gáro Hills failed to germinate.

"It will be noticed that, unlike previous experiments in Burdwan and

CULTIVATION in BENGAL.

Experiments with exotic forms. 199 (579)

unlike experiments made in other districts, very good results were obtained with Nankin cotton this year in the Burdwan farm. The plants looked quite healthy, and proved more hardy than any other varieties experimented upon. They have put forth new shoots and leaves, and a second crop is expected. Experiments on the cultivation of Nankin cotton were also tried in Mozufferpore and Hill Tipperah, but without success. Mr. Sen at first tried Nankin cotton as a winter crop, and, as such, it did not germinate; but planted in the beginning of the rainy season, in drills 30 inches apart, and manured with bone-meal, Nankin cotton, as the figures quoted above show, gave very good results. I do not, however, think it will ever be extensively grown in Bengal." (*Agricultural Report, Bengal, 1886-87.*)

It appears probable that the anticipation in the last paragraph will prove correct. The growth of Nankin cotton in several districts of Bengal is not unknown. Cotton of the natural *khaki* colour is grown extensively in some districts of the Patna Division. In 1883 khaki-coloured cotton seed was obtained from the Government Experimental Farm at Cawnpore and distributed to the Commissioners of Burdwan, Chittagong, Bhagulpore, Orissa, and Chutia Nagpur for experiment in their respective divisions. The results of the trials reported by them were not very promising. Thus in the report on the trials in Manbhoom it is stated that "Dr. H. W. Hill sowed broadcast the pound of seed received by him. Only a small portion germinated, and the cotton produced was of small quantity and of a brownish red colour. Dr. Hill thinks that the *khaki* cotton is a hybrid between **Gossypium herbaceum** and **G. barbadense.*** He says that a species of this plant, if not indigenous, is cultivated to some extent in parts of Patkom and Burrabhoom, chiefly by the Sonthals, who call it *khuruah kapas*, in contradistinction to *hurawah kapus*, or white cotton. The colour of this *khuruah kapas* is of a much lighter shade than that of the *khaki* cotton experimented on, and the Sonthals spin and weave it into a sort of coarse cloth. The species found in the district is considered by Dr. Hill to be a hybrid of **Gossypium herbaceum.**"† In the report from the Commissioner of the Orissa Division, it is stated that the khaki-coloured cotton is liable to hybridise, which, to a great extent, destroys the uniformity of the colour. In the agricultural report for 1887-88, Nankin is said to have yielded 594lb per acre, Lonacinu, 270lb, Upland 135, and New Orleans 162.

Soils. 200 (580)

SOILS AND METHOD OF CULTIVATION.—The following information is chiefly extracted from the report on cotton published by the Bengal Government in 1876:—

I. RAJSHAHYE AND KUCH BEHAR DIVISION.—The cotton plant is found within the following four districts of this division—Rajshahye, Bogra, Julpigori, and Darjeeling, but can be said to be cultivated only within the two last-named districts.

In the Darjeeling district, the cultivation is almost entirely confined to the Terai at the foot of the hills, where the cultivators are the Mechis and Dhimals, who inhabit the jungle tracts near Nuksurbari, Adulpore, Kyna-noka, and Champasari, and also the country west of the Teesta and north of the Julpigori district. Virgin soil, or, at the least, land for the most part with jungle growth, is selected, of sufficient elevation to secure natural drainage. The land is prepared from February to April,

* If he received the form generally known as **G. religiosum** he was probably not far from correct in that opinion.—*Ed.*

† It is one of the hybrids of **G. neglectum.**—*Ed.*

CULTIVATION in BENGAL.

Soils.

during which months the jungle is cut down and burned, this being the sole manure received by the plant. Sowing commences in April and extends to June, the crop being gathered in February and March. Occasionally cotton is sown in the Indian corn and *bhadoi dhan* fields, these crops being cleared off in succession before the cotton has ripened, but this plan is seldom resorted to, as the Mechis do not, as a rule, raise their own food supply, but purchase it with the proceeds of their cotton crops. As the cultivation at best is very rude, a second crop is scarcely ever raised on the same field: for this reason the cultivators select virgin or fallow soils, depending upon nature to make up for their laziness.

In the Julpigorí district cultivation is similarly carried on by the Mechis and Gáros, who grow it on high lands near the foot of the hills in the Mynagore tahsil, and in the Buxa sub-division. The land, naturally rich, is cleared of jungle, turned over a little with a *dao*, and the seed is sown. Sowing time lasts during May and June, and the crop is picked in February and March.

In the northern parts of the Rajshahye district the rayats grow a number of scattered cotton plants in their mulberry fields; on an average there are not more than, say, 50 or 60 plants in a *bigha*.

II. Dacca Division.—No cotton is now grown in this division for external trade, the very little that is cultivated being entirely used for home consumption. The imports are, however, considerable, chiefly from Chittagong, Assam, Hill Tipperah, Sylhet, and the North-West Provinces.

In the district of Mymensingh proper, and on the skirts of the Modhupur forest, on that laterite soil where fine qualities of cotton were formerly grown in sufficient quantities to make Dacca muslin world-famous, a languid and badly-tended cultivation is still maintained. The produce supplies the looms of a few weavers at Bazitpore and Polsia, in Attia.

A small amount of cultivation is also carried on in the Lallmye range of the Tipperah district. Patches of ground are cleared by burning the jungle, and turned up with the Tipperah *kodal* or hoe. The seed is sown in March and April, and the crop is gathered in December and January. The variety grown is known locally as *bhuta* or *bhunta*.

III. Chittagong Division.—Little or no cotton is grown in the Chittagong district itself, but in the Hill Tracts the crop is extremely important. It is grown both on lofty hills and low lands. The climate is damp, the soil varies with the situation of the *júm* on which the plant is reared Sowing takes place in May after the first heavy shower of rain. If, however, much rain falls immediately after the crop is sown, there is danger of the seed rotting in the ground, before germination takes place. Heavy showers in the latter part of September may also cause much damage to the pods. The description of cotton grown is termed *phulshuta*, the fibre of which is short but soft and very pure in colour. To what extent Roxburgh's **G. obtusifolium** is cultivated in the Eastern division of India the authors are not prepared to say, but samples of that variety of **G. herbaceum** have been seen by them from the Nága Hills, Chittagong, and Burma.

IV. Patna Division.—This division contains the largest cotton-growing district of Bengal, *viz.*, Sarun. In this district cotton is grown as a secondary crop, the seed being sown broadcast in fields prepared for other crops. Westerly winds with moderate rains are favourable to its growth; continued easterly winds are injurious and are supposed to favour the appearance of the boll-worm.

The chief cottons of Sarun are known locally as—1, *Bhagtha*; 2, *Bhochri*; 3, *Jathua*; and 4, *Kockti*. The first three forms are called collectively *Bysakhi*, being harvested in *Bysakh* (April), the fourth is called *Bhadoi*, being harvested in *Bhado* (August). The latter is sown in January and February, the first three in June and July. It seems probable that the crops reaped in the hot weather are approximately typical forms of **G. neglectum**, *Tod.*, the others conditions of the hybrid which yields the commercial "*Bengals.*" In Gya three kinds are distinguished, *viz.*, *braisa* or *bunga*, *rerhia*, and *bhochri*, of which the first is the most valuable, but the second most grown, since it yields a greater outturn. Two forms chiefly are grown in Mozufferpore known as *bhogla* and *bhochri*, the former of which is the better. An exotic called *belati* (possibly Peruvian cotton) is also grown, to some extent, round houses. In Durbhanga very little is grown. The crop recorded is of three kinds, *kockti*, *bhoera*, and *bhagla*. Of these, the first is very much prized, as the cloth prepared from its floss is very lasting. It is sown twice a year, first in October to November, second in February and March. The first crop is uncertain, the second seldom fails.

CULTIVATION in BENGAL.

Soils.

Conf. with pp. 10-12.

The authors find it impossible to relegate all the forms distinguished by the above vernacular names to their respective species, varieties or hybrids; they have, however, mentioned them in the hope that some or all may be collected and botanically identified.

V. Bhagulpore Division.—In this division cotton is generally grown either with other crops, or on the boundaries of fields, as a hedge. It is never irrigated. In the Sonthal Pergunnahs it is grown on second class upland soil in company with *serguja* or *til*. The samples from this division seen by us are **G. neglectum**, *Tod*.

VI. Orissa Division.—In the Cuttack district two sorts of cotton are grown, locally distinguished as *achua* and *haldia*. The former is preferred to the latter, as it contains a greater amount of fibre in proportion to the seeds, and the staple is finer, softer, and more glossy *achua* is grown on lower lying ground, a small part of the area being irrigated. *Haldia* is cultivated on higher lands near dwellings. The latter is probably **G. neglectum**, *Tod.*, the former a condition of **G. herbaceum**. Small quantities are annually raised in Púrí and Balasore, but neither of these districts is of sufficient importance to call for description.

VII. Chutia Nagpur Division.—Cotton cultivation is not carried on on an extensive scale in any part of this division. The people generally grow it in small patches in their *baree* or homestead lands, or in the jungles on variable areas only partially cleared. Many villages and tracts lying on the borders of rivers possess very fine alluvial soil which probably (with irrigation) would yield very fine crops. But the people do not care to grow cotton as a staple of trade, preferring those crops, such as cereals, which give them least trouble.

The principal cotton-growing district of the division is Palamow. In Chutia Nagpur proper, the most important cotton cultivating locality is Biru, where the staple forms one of the main sources of income to the people. In the Sili and Tamar pergunnahs to the south-east of the district, a large quantity of the better class of irrigated cotton is produced.

In Hazaribagh, Singhboom, and Manbhoom the outturn of cotton is small, and is entirely used for home consumption. Specimens seen by the authors from the last-mentioned district belong to **G. neglectum**, *Tod.* (*see p. 8*).

Seasons of Sowing and Reaping.—The information given below of the time when cotton is sown and picked is taken from the report by

Seasons. 201 (587)

CULTIVATION in BENGAL.

the Government of Bengal in November 1878, since later particulars are not available :—

Seasons.

DISTRICT.	When sown.	When picked.
Midnapore . .	May or June. October to November.	September to March. February to June.
Chittagong Hill Tracts	April and May.	November and December.
Durbhanga . .	October. May or June.	August and September. March and April.
Sarun . . .	June and July. January or February.	April and May. August and September.
Cuttack . . .	June and July. November and December.	October and November. May and June.
Lohardugga . .	October. June.	April and May. November to January.
Manbhoom . .	May to July. September to December.	October to December. February to April.

Yield.
(582) 202 **Yield per Acre.**—According to agricultural returns the average yield all over the presidency is 91℔ per acre. The returns for various districts, however, show great differences. Thus, in Mymensingh, the outturn is put as high as 476℔ per acre; in the Chittagong Hill Tracts it is said to be 200℔; in Midnapore, 144℔; in Manbhoom, 90℔; in Cuttack, 70℔; in Lohardugga, 56℔; in Durbhunga, 30℔; and in Sarun, only 20 to 24℔ per acre.

It is difficult to explain why such great variation in yield should exist in neighbouring districts, in which the method of cultivation and soil do not necessarily greatly differ. At the best, however, the figures are only approximations, and must at present be accepted for want of better.

For the yield of various races experimentally grown in the Burdwan Raj Experimental Station in 1886-87, see p. 134.

Cost.
(583) 203 **Cost and Profit of Cultivation.**—In the Bengal Government Report already referred to, the cost of cultivation, and selling price of cleaned cotton per cwt., is given for three districts as follows :—

DISTRICT.	Outturn per acre.	Cost of cultivation.	Average selling price per cwt.
	℔	R	R
Burdwan	120	17	18
Bancoorah	125 50	10 6	20
Julpigorí	120	16	25

The profit in Burdwan would, according to these figures, be a little over R2, in Bancoorah about R7-8, and in Julpigorí R10-12. These figures are the only ones available, but cannot be accepted as representing the state of matters in Bengal generally.

Bengal Trade.
(584) 204 **Analysis of the Trade.**—The total imports into the province, excluding Calcutta, amounted to 119,755 cwt., of which 19,492 cwt. came from Calcutta, 66,599 cwt. from the North-West Provinces and Oudh, 21,963 cwt. from Assam, 6,541 cwt. from the Panjáb, 986 cwt. from the Central Provinces, 29 cwt. from Berar, 94 cwt. from British Burma, and 12 cwt.

from Madras; 4,029 cwt. were also imported by trans-frontier routes from Nepal and Sikkim. **CULTIVATION in BENGAL. Trade.**

The total exports from the province were 123,317 cwt., of which 120,219 cwt. went to Calcutta, 152 cwt. to the North-West Provinces and Oudh, 89 cwt. to Assam, 7 cwt. to the Panjáb, 1 cwt. to the Central Provinces, 7 cwt. to British Burma, and 2,842 cwt. by frontier routes to Nepal and Sikkim.

The above figures have been derived from the rail and river statistics published by the Revenue and Agricultural Department, and from the Bengal statement of Sea-borne Trade and Navigation, and, consequently, do not include road traffic. The exports from Bengal to Calcutta are mostly from Dacca and the other parts of Eastern Bengal. Details of the proportions contributed by the principal districts for the past three years are given below:—

Whence exported.	1886-87.	1887-88.	1888-89.
Dacca	21,583	39,368	53,286
24-Pergunnahs	13,723	20,233	16,629
Chittagong	6,659	20,548	21,474
Furridpur	5,969	7,819	9,814
Noakhally	4,562	4,700	8,875
Rungpur	2,884	4,183	3,512
Hooghly	4,307	6,444	4,543
Nuddea	1,956	1,710	1,836
Muttra	*Nil.*	*Nil.*	6,489
Other districts in Bengal	2,482	1,661	1,995
Other districts in Behar	4,109	1,358	759

The above figures to some extent include road returns which have not been embraced by the foregoing statement. From the table it will be seen that exports from Bengal to its chief town are gradually increasing, the increase being most marked in the case of Dacca and Chittagong. It is difficult to determine how much of the increase in the case of the former is dependent on re-exports of cotton imported from other provinces, and railed to Calcutta. In the case of Chittagong the increase is probably due to an extension of cultivation and increase of outturn in the Chittagong Hill Tracts.

II. ASSAM.

References.—*Darrah, Note on Cotton in Assam, 1885; Agri-Horticultural Society of India, Journals: (Old Series), IV., 207; VII., 45-50; IX., 251, 336-342; (New Series), III., Sel. 35.* **CULTIVATION in ASSAM. 205** (585)

Review of the area, &c.—Owing to the absence of an establishment for registering agricultural statistics over the greater part of the province, it is almost impossible to frame an accurate estimate of the area under a particular crop. This difficulty is said by Mr. Darrah, the Director of Land Records and Agriculture in Assam, to be particularly intricate in the case of cotton, a crop which is almost unknown in the plains of the Assam Valley where alone the rudiments of a registering system are to be found. In the hill districts, on the other hand, where no such system of any kind exists, the crop is a very important one. The figures given below are, therefore, at the best only fair approximations to the actual state of affairs. The total area under cotton in Assam during the year 1888-89 was returned as 40,588 acres. The total outturn in the same year was 54,359 cwt., giving the high average yield of 150℔ per acre. Deducting the net export from the total outturn the amount available for local consumption is seen to have **Area. 206** (586)

CULTIVATION in ASSAM. Area.

been 28,593 cwt. This, divided by the population, gives the consumption of home-grown cleaned cotton at 0·67℔ per head. The Director of Land Records and Agriculture, commenting on this figure, remarks, "At first sight the figures, thus obtained, appear ludicrously wrong. But, though I do not contend that they can be trusted, or even that they possess any value, I would submit that they are not so absurd as they seem at first sight to be. A very large proportion of the population in Assam wears no indigenous cotton whatever—traders, officials of all ranks, tea planters, establishments, garden coolies, pleaders, priests and their disciples, well-to-do Assamese, and, in fact, all but the very poorest classes, probably never use indigenous cotton at all. Silk is largely worn by all respectable Assamese, and, if cotton cloths are required, they are usually of English make. Indigenous (Assam grown) cotton is probably in little demand, except amongst the hill tribes who mainly grow it." In the same report the average consumption per head in the Gáro Hills is stated to be 3·17℔, in Cachar 2·64, while in Nowgong it is returned as only 0·21℔.

Districts. (587) 207

DISTRICTS WHERE GROWN.—The extent of cultivation in the several districts of the province may be learned by the following extract from the Monograph on Cotton in Assam, written, in 1885, by the Director of Land Records and Agriculture:—

I. *Assam Valley*.—" In the plains patches of cotton are met with here and there. The Miris of Lakhimpur and Sibsagar grow a little for their own domestic consumption. In some *mauzas* of the northern frontier the cultivation of cotton on the lower slopes of the Bhután hills, and in the light soil at their foot, used once to be carried on extensively. It was known as *gari* cultivation, and the Chapaguri mauza in Kámrup was once famous for it. In the Goálpára district some cotton is grown in the hilly portions of the Sidli and Bijni parganas. In Kámrup *jhúming* (removing jungle for purposes of cultivation) is carried on for cotton to a limited extent on that portion of the Khásia hills which is included within the district. The only district in the Assam valley which grows cotton largely is Nowgong. The exception is due to the fact that the range of the Mikir hills is included within its boundary. The Mikirs grow cotton, not only for domestic consumption, but also for export, and boat-loads of it may be seen going down the Kopili and Kollong towards the close of the cold weather."

II. *Surma Valley*.—" Similarly, in the Surma valley, cotton is only occasionally cultivated in the plains districts. The Mikirs and Tipperas grow some by *jhúming* in pergunnah Mulagul in Sylhet. About 50 acres are supposed to be under cotton in Sunamganj.

" The area under cotton in Cachar is high, but that is because the district includes the sub-division known as the North Cachar hills, which are really a portion of the Central Assam range. Here cotton is largely grown, for export as well as for local consumption, by hill Cacharis, Nagas, and Kukis. The Sadr sub-division grows but little cotton."

III. *Cultivation in the plains*.—" Altogether, however, the cultivation of cotton in the plains districts of Assam is, with the exception of that in Nowgong and the North Cachar Hills, totally insignificant. In the regular plains it may be said never to be cultivated at all; but where the slopes of the adjacent hills are included within the borders of these districts cotton is grown to a limited extent."

IV. *Cultivation in the hills*.—" In the three hill districts of the province and the sub-division of North Cachar, however, the case is quite different. In the range of mountains which extends from the head of the Brahmaputra valley to the confines of Mymensingh cotton is nearly everywhere a staple crop. The one exception is the high plateau of the Khásia hills where the

CULTIVATION in ASSAM. Districts.

climate is too cold. The Gáros grow it in very large quantities. The inhabitants of the Khasia and Jaintia Hills carry down hundreds of maunds every year to Kámrup and Nowgong. The Nágás (Lhotas and Rengmas) export partly to Nowgong, but principally to Sibsagar."

As already stated the figures of outturn and area in the hill districts of Assam are based on calculations which are open to grave error. They have been arrived at as follows:—"The usual system has been to ascertain, by enquiries at the various submontane fairs, the quantity that has been exported in a given year. An addition is made for the amount taken out of the country by traders who do not visit the fairs, and also for the amount believed to be locally consumed. The result gives the probable quantity produced. This is divided by the assumed outturn per acre, and an estimate of the area is thus found."

By calculations carried out on this plan in the case of the hill districts, and by the imperfect system of registration in the case of the plains, the following figures have been arrived at for the year 1888-89:—

DISTRICT.	Area under Cotton.	Total outturn of cleaned Cotton.
	Acres.	Cwt.
Sylhet	2,714	3,634
Cachar	8,956	12,794
Goálpára	1,445	3,380
Kámrup	1,059	472
Darrang	96	128
Nowgong	2,089	932
Sibsagar	34	45
Lakhimpur	5	6
Khásia Hills	3,324	3,086
Gáro Hills	20,866	33,534
Nágá Hills	Not reported.	Not reported.
TOTAL	40,588	58,011

Races. 208 (588)

RACES OF COTTON.—Mr. Darrah, in the report above quoted, writes as follows:—"The varieties of cotton are not numerous, but the names by which the crop is known differ from district to district, and peculiarities of soil, climate, and method of cultivation, have no doubt produced divergences from the original type. It is not easy, therefore, to say exactly how many really different kinds there are. Roughly speaking, there are two well-marked forms:—

"(1) The large-bolled high growing cotton, known as *dhál* (white flowers) in Lakhimpur, as *bogakapáh* in the Majuli, as *khungi deva* in Cachar, as *kil* in the Gáro hills, and as *borkapáh* (*lit.*, large cotton) in Nowgong. This is probably also the same as the *bhugai* of Sylhet. In Nowgong, this species is grown on level ground, has a smaller number of seeds than the second variety (mentioned below), can be ginned more easily, can be plucked twice a year instead of once, and bears for three seasons. The *kil* of the Gáro hills is very nearly the same, except that the crop is annual, that it is grown everywhere on the hill-sides, and not confined to level ground, and that it can only be plucked once a year. The pods are very large, sometimes as much as eight inches in length, and when they burst the contents come out in a cataract of cotton which gives a field the appearance of being covered with snow. This variety is, however, not as much in request for ordinary purposes as the smaller kind. The fibre is

CULTIVATION in ASSAM. Races.

said by the trade to be harsh and to twist badly. It is better adapted for mixing with wool than for any other purpose.

"(2) The small, round-bolled species, known as *shet* (reddish flowers) in Lakhimpur, as *thumsa* in Cachar, as *ukynphád* in the Jaintia hills, and as *horu kapăh* (*lit.*, small cotton) in Nowgong possibly identical also with the *chotsá* of the Angámi Nágas. This species is sown annually, and can only be plucked once a year. The Lakhimpur variety has pale reddish flowers. That grown in the Jaintia hills is said to be the best cotton produced in the province. Its thread can be more closely woven than that of other kinds. The Nága hills variety is rated lowest of all, being very short in the staple and coming into the market in a very dirty condition.

"There is a pale *khaki* variety in Cachar and Manipur known as *khungájas* in the former and as *tissing anguangba* in the latter district. The pods are not a uniform *khaki*, but contain a few white threads here and there."

It is difficult to determine to what species all the above mentioned kindc belong. The large bolled, high grown cotton, however, is in most cases probably **G. neglectum**, *Tod.*; in the case of the Gáro hill cottton it certainly is so. Specimens which have been seen by the authors are perfect types of that species, and have the very large ovoid bolls referred to by **Mr. Darrah.** The statement that the cotton of Lakhimpur (*dhál*) has white flowers is curious, and would suggest its being a form of **G. neglectum** with which we are not familiar. It may be noticed, however, that **G. arboreum** is said by certain authors to occasionally bear white flowers. The large-bolled cottons of Majuli, Cachar, and Nowgong are probably all botanically identical with that of the Gáro hills.

Turning to the second form described by **Mr. Darrah** the difficulty is greater. The *chotsá* of the Angámi Nágas is undoubtedly **G. herbaceum,** *var.* **obtusifolium**; the small cottons of Cachar, the Jaintia hills, and Nowgong are probably also forms of **G. herbaceum** or **G. Wightianum,** *Tod.*; while the *shet* of Lakhimpur would appear, from its reddish flowers, to be a small round-bolled form of **G. arboreum.** The *khaki* variety of Cachar and Manipur is probably indigenous cotton, not Nankín.

The authors are indebted for the following descriptions of the nature of soil, methods of cultivation, cleaning, &c., to **Mr. Darrah's** valuable report dated September 1885 :—

Soils. (589) 209

SOILS AND MANURES.—"Cotton is most generally grown on forest clearings known as *jhúms.* The hill-men, as a rule, prefer bamboo and grass to tree jungle. The latter is more difficult to work, does not burn as thoroughly, and leaves obstructions in the shape of stumps and logs. The Kukis of Cachar, however, appear to prefer making their clearings in timber, and the Mikirs of Nowgong choose young forest, with saplings when they can. The soil should be calcareous, and the situation sunny. In the Gáro hills a species of small bamboo grows with great luxuriance, and the soil on which it is found is invariably selected, if the other conditions for cultivation are favourable. No manure is ever used, except the ashes of the burnt jungle. One reason why the bamboo is so appreciated by the Gáros is that it burns with much more completeness than tree jungle, and therefore affords better manure."

Methods. (590) 210

METHOD OF CULTIVATION.—HOEING.—"Land is never ploughed for cotton, except in the few places where it is grown in the plains. In these the ground is ploughed three or four times, and then hoed once, the latter process being considered indispensable in Mangaldai. A trench for drainage is usually dug round the plot. The hill-men always use the hoe, as the slopes on which cotton is grown are too steep for cattle to be employed. Moreover, between the stumps of trees and the half-burned logs which often litter the ground, a plough could not be worked. The Nágas generally give two hoeings, the Tipperahs and Gáros none at all. The other tribes generally hoe once. The jungle is usually cut in the cold weather and allowed to dry on the ground. It is burnt in March or April, and then, as a rule, hoed. As soon as possible afterwards the cotton is sown. It is scattered broadcast generally, not put down in drills. The Tipperahs and Gáros have a different custom. After the burning is finished, they go over the land with a pointed stick, and making small holes in the ashes drop a seed into each. In Cachar a similar process is employed, the seed being dibbled in with a pointed stick called *kuar*. In the first year only one kind of seed is placed in each hole. The soil is not further disturbed. It is fertilised by the ashes on the first shower of rain. If the rain is delayed, the value of the ashes, as a fertiliser, diminishes considerably." **CULTIVATION in ASSAM.** **Methods.**

Associated Crops. **211** (591)

Associated Crops.—"The associated crops are usually broadcast rice and *til*. But some forms, *e.g.*, *thumsá* of Cachar, are always sown alone. The other varieties are also sometimes sown alone; but, as a general rule, other crops are mixed with cotton,—mustard, Indian-corn, chillies, brinjals, linseed, jute, water-melons, are all used according to the wishes or convenience of the cultivator. But *áhu dhán* (unirrigated, broadcast rice) is the most commonly associated crop. In the Gáro hills, the usual practice is for *áhu dhán* to be sown broadcast the day after the fire. When the shoots show themselves above ground, the vacant spaces become apparent, and these are sown with cotton in the manner already described. The Rengma Nágas sow the cotton broadcast, with *áhu dhán*, the cotton seeds and the *dhán* being mixed up in the same basket. The reason for associating a second crop with cotton is said to be that the latter always grows best if shaded in the beginning."

Occupancy. **212** (592)

Occupancy.—"A jungle-clearing is rarely cropped for more than one season with cotton. In the second year upland rice is often sown alone, and when the crop has been gathered the *jhúm* is usually abandoned. If there is suitable land available in reasonable quantities, the clearing is not resumed for ten years. In no case is it re-occupied until at least five years have elapsed. No rotation of crops is ever observed. In Cachar in the second year the paddy straw of the previous year is burnt off, and the land having been cleaned and turned over with the *kuar*, rice and sesamum mixed are sown broadcast. A few days later Indian-corn and cotton mixed with earth are sown, and the ground is kept weeded till August. The Indian-corn ripens in July, the rice in August, the sesamum in November, but the cotton not till December. A similar course is followed in the third year, and in the fourth the land is abandoned. The Angámi Nágas, however, frequently crop a clearing with cotton for two or three years, according to the richness of the soil. The variety of *áhu dhán* known as *tidi* is generally sown by them the first year, while in the second year the kind called *teke* is usually planted. The Rengma and Lhota Nágas, who are the principal cultivators of cotton in the Nága hills, never crop a clearing with cotton for two successive years."

Irrigation. **213** (593)

Irrigation and Weeding.—"Irrigation is hardly ever practised, though occasionally required. The land is usually weeded once or twice, rarely oftener. When the crop germinates, sacrifices of eggs are offered by the Mikirs to the god Longlo Ahi. Similar ceremonies take place when the plucking begins."

Plucking. **214** (594)

Plucking.—"Plucking generally begins about November. It is usually over in January. *Kunghideva* is plucked from time to time as the pods open and burst. With *thumsa* the plucking is usually made once for all. The *borkapáh* of Nowgong is plucked twice—once in December and January, when the crop is called *bataria kapáh*, and once in May, when the

CULTIVATION in ASSAM. produce is known as *jetharia kapáh.* As a general rule, plucking lasts for about a month and a half, and usually an interval of six days is allowed to elapse between each picking."

Diseases. 215 DISEASES AND INJURIES.—"Cotton suffers considerably if rain holds off for long after it is put into the ground, as much of the seed does not then germinate. On the other hand, heavy rain, when the plant is well grown, rots the stems, and if the pods have formed, injures the cotton within. Cloudy and damp weather is always injurious, except at the very beginning, and for this reason the sunniest spots are invariably chosen. Insects also do a good deal of injury. Whole crops are sometimes destroyed by the *chilapok* of Lakhimpur, the *phalo jhinghi* of the Nowgong Mikirs, and the *mitchy* of the Nága tribes. White ants often cause serious loss in the Jaintia and Nága hills." **Mr. Darrah** unfortunately does not state what these pests are, of which he gives the vernacular names.

Yield. 216 YIELD PER ACRE.—**Mr. Darrah** writes, "The average produce of cleaned cotton per acre is a very difficult question to solve. From the nature of the crop, the experiment must extend over a considerable time, and it is impossible to have the field under trial watched night and day. Several experiments have, however, been made. Twelve of these in the Jaintia hills, on plots of a quarter of an acre each, yielded a minimum outturn of 80lb of cleaned cotton to the acre, and a maximum of 304lb, the average being 171lb. The Cachar estimate is 160lb of cleaned cotton per acre. In Nowgong experiments made by Inspectors of Police on areas of one *bigha*, gave an average of 150lb of cleaned cotton to the acre. In Sibsagar the estimate is 128lb; in Goálpara, 378lb to 450lb of uncleaned cotton, or about 150lb to 180lb of cleaned. The latest and most careful experiments in the Gáro hills, made upon four acres of land, gave 507lb of uncleaned cotton to the acre. At the Gáro hills proportion of 20 seers cleaned cotton to the maund of uncleaned, this would give nearly 260lb of the former to the acre. Even at 16 seers to the maund, the produce reaches the high figure of 202lb. Accordingly, for the whole province, 150lb of cleaned cotton to the acre may be assumed to be a fairly accurate estimate."

The figures of average outturn for the several districts in 1888-89 are returned as follows:—

DISTRICT.	Average outturn in lb.	DISTRICT.	Average outturn in lb.
Sylhet	150	Sibsagar	150
Cachar	160	Lakhimpur	150
Goálpára	262	Khásia Hills	104
Kamrup	50	Gáro Hills	180
Darrang	150	Nága Hills	Not reported.
Nowgong	50		

Regarding the proportion of cleaned to uncleaned cotton, **Mr. Darrah** writes:—"In working this out, the original figures, which of course shew uncleaned cotton, have been converted into cleaned on the assumption that five maunds of the former will yield two maunds of the latter. As a matter of fact, this estimate is rather under, than over, the mark. Using the ordinary native cotton gin, repeated experiments have given the outturn as 17, 17½, and 18 seers to the maund, whereas the proportion assumed above is only 16 seers to the maund. The Deputy Commissioner of the Gáro hills, experimenting with an English ginning machine, gives 20 seers to the maund as the correct proportion. On the other hand, experiments

CULTIVATION in ASSAM.

tried in Sylhet, with the view of making the export business pay, give an outturn of but 16 seers to the maund. It may, therefore, be assumed that a proportion of 2 to 5 is certainly not in excess of the fact." It may be remarked in passing, that for most other parts of India a proportion of 3 to 10 is accepted as the maximum. The yield of cleaned cotton from uncleaned, in the case of **G. arboreum**, however, is probably greater, and since that species produces a great part of the cotton of Assam, **Mr. Darrah's** estimate may possibly be, as he says, not over the mark. If this be the case, Assam is particularly suitable to the cultivation of cotton, for not only is the proportion of fibre obtainable from a given quantity of uncleaned cotton larger than in most other regions, but the yield of the latter per acre is also much greater.

Cleaning. 217 (597)

CLEANING.—The process employed in cleaning cotton is thus described by Mr. Darrah :—" The first step in the process of manufacture is cleaning, *i.e.*, separating the fibre from the seed. This is usually done in Assam by an instrument called *neothani* in Upper, and *neotha* in Lower, Assam. It is exactly the same in principle as the *charki* of Upper India, and consists of two horizontal rollers, one close above the other, generally both of wood, mounted on an upright stand. One end of one roller is formed into a screw, which catches a projection in the other and causes both to revolve in opposite directions when the handle at the other end is turned. Sometimes the ends of both rollers are alike and screw-shaped. The cotton passes through being caught by the rollers, but the seeds, being unable to get through the narrow slit, are left behind. Sometimes a comb, formed of the teeth of the *Bhorali* fish, is used to partially clean the cotton before it is passed through the *neothani.* Experiments have shown that this machine gives a result of from one seer to two and a half seers of cleaned cotton per diem. The cotton must be dried well in the sun before being ginned, or the outturn will not be good. In some places the seed is utilised as cattle-food. In the parts of Kámrúp near Gauháti it is a regular article of sale. Its price in the neighbourhood of Tura in the Gáro hills is 10 annas a maund. But this is not general. It is usually flung away as useless. If it were of more value in this country, it is probable that less uncleaned cotton and more cleaned would be exported. But partly owing to the scarcity and dearness of labour here, partly owing to the small value attached to the seed, together with the fact that labour is comparatively cheap in Calcutta and the seed there largely in demand, cotton is exported chiefly in the uncleaned state."

"After being cleaned, the cotton is subjected to a process of beating (*dhuna*) by means of a bow-string, in order that the fibres may separate and become loose enough to be spun into thread. The bow (*dhanu*) is usually a piece of bamboo, about three feet long; the string (*jor*) consisting of a twisted strip of the outside of the midrib of a plantain leaf. The bow is usually held so that the string may touch the cotton. The string is pulled and let go, and the vibration separates the fibres. The principle is precisely the same as that used in most parts of India."

"There is no special caste in the province, like the *Dhunia* of the North-Western Provinces and the *Pinja* of the Panjáb, whose occupation it is to prepare cotton for spinning in the manner described. Each household scutches, if the process may be so called, as much cotton as it requires for its own purposes."

Nearly all the cotton exported from Assam is uncleaned, probably owing to two reasons :—(1) that local labour is too dear; (2) that the seed is of much more value in Calcutta than it is in Assam. In any case the fact is certain that cleaning cotton for export with the native hand-gin does not pay. With the European machine gin, however, the case is differ-

CULTIVATION in ASSAM. Cleaning.

ent. Mr. Darrah states that in 1885-86 there were two working European gins in the province—one at Tura in the Gáro hills, the other at Doboka in Nowgong, each admirably placed for working the cotton of the Gáro and Mikir hills, respectively. The machines have been found capable of turning out one maund of cleaned cotton a day, each realising a daily profit of R5.

Reports from Calcutta, however, tend to shew that, though the Emery gin may be worked profitably, the fibre is considerably diminished in value by its action. Thus, in one case a bale of hand-ginned cotton was found to be worth R21 per maund, while the same fibre, saw-ginned, was valued at only R16.

Packing. (598) 218

PACKING.—Assam cotton is not, as a rule, made up into the heavy half-compressed bales commonly met with in other parts of India, but is packed either in conical baskets or in small bales for transportation by coolies. The former are packed as follows:—"The basket having been lined with leaves of the *tara* plant is placed upright in a hole in the ground. One man fills it with cotton, while another treads down the fibre. When filled, some leaves of the *tara* are placed on the top, and the whole firmly bound down with strips of bark or cane. A basket thus packed is called *dang*, by the Mikirs, and usually weighs half a maund." In the plains instead of these *dangs* the small bales above alluded to are generally made. They are suspended to the ends of a stout bamboo which is borne on the shoulder.

Trade. (599) 219

ANALYSIS OF THE TRADE.—I. INTER-DISTRICT TRADE.—"The trade in cotton is partly inter-district and partly provincial. The registration of the former is carried on at a few police-stations, but no attempts have as yet been made to summarise or elucidate the figures. The registration is really too partial to throw much light on the character of the trade. It is known, however, that the plains districts purchase largely for local consumption at the marts at the foot of the hills. The Gáro hills supply Goálpara and parts of Sylhet. The Jaintia and Lower Khásia hills meet the wants of South Kámrup and North Sylhet. Jaintia Hills cotton also goes down to Nowgong and parts of Sylhet. Darrang is supplied, like the greater part of Nowgong itself, by the Mikirs. North Cachar cotton goes into Nowgong by the Doyang and Kopili, and also into the plains of Cachar. The Nágá hills export to Sibsagar and parts of Lakhimpur. In all these transactions human beings are the beasts of burden, except when boats are used. Pack-bullocks are unknown, and ponies are not used for the purpose. The hill-men carry down the fibre, nearly always uncleaned, in conical bamboo baskets, which they support by a plaited cane strap across the forehead, partly on their heads and partly on their shoulders." "The plains people carry cotton in small bales suspended to the ends of a stout bamboo, which is borne on the shoulder. Boats are used when the trade route coincides with a river, and considerable quantities are floated down to the more important riverine marts. Wheeled carriage is absolutely unknown, and long strings of camels, laden with heavy, half-compressed bales, such as are so frequently met with in the cotton districts of Upper India, are never seen in Assam."

II. EXTERNAL TRADE.—"The export trade by boat is registered at Dhubri for the Brahmaputra, and at Bhairab Bazár for the Surma valley. The steamer traffic, which is inconsiderable, is known from the returns furnished by the steamer companies."

In the year under consideration the total exports amounted to 29,435 cwt., of which 21,964 cwt. went to Bengal, and 7,471 cwt. direct to Calcutta. The imports were unimportant, amounting to only 3,671 cwt., of which 49 cwt. came from Calcutta, 89 cwt. from Bengal, and 3,533 cwt. by trans-frontier routes from Bhutan, Hill Tipperah, and the Lushai Hills.

12. MYSORE.

CULTIVATION in MYSORE. 220 (600)

Reference.—*Gazetteer, I., 26, 57; II., 201, 293, 347, 456.*

Area. 221 (601)

REVIEW OF AREA, &c.—The average area for the five years ending 1886-87 was 29,814 acres; and the outturn during the same period 11,459 cwt.; the average yield would thus have been 43lb per acre.

The total outturn is insufficient for local consumption, consequently the imports exceed the exports. In 1888-89 the former amounted to 3,159 cwt. by rail, the latter to 1,120 cwt.; considerably greater quantities, however, are probably imported by road.

Races. 222 (602)

RACES OF COTTON.—The ordinary native cottons of Mysore are probably similar to those of Madras, *viz.*, forms of **G. Wightianum**, or *Uppam*, and of **G. herbaceum**, *var.* **obtusifolium**, and **G. neglectum**—the *Nadam* cottons.

Numerous experiments have been made at Bangalore to test the suitability of the soil and climate of that district for exotic cottons, with the result that Upland Georgian, Sea Island, New Orleans, and Egyptian were all found to do well. None of these have gained, however, any hold on the cultivation of the State.

Soils. 223 (603)

SOILS AND METHOD OF CULTIVATION.—The following is extracted from the *Gazetteer of Mysore and Coorg, Vol. I.*:—"The soil on which cotton is sown at Sira is a black clay containing nodules of limestone. In the two months following the vernal equinox, plough three times. At any convenient time, in the next two months, mix the seed with dung, and drop it in the furrows after the plough, forming lines about nine inches apart. A month afterwards plough again between the lines; and in order to destroy the superfluous plants and weeds, use the hoe drawn by oxen three times, crossing these furrows at right angles. The second and third times that this hoe is used it must follow the same track as at first; otherwise too many of the plants would be destroyed. Between each hoeing three or four days should intervene. In six months the cotton begins to produce ripe capsules, and continues in crop four more. The plants are then cut close to the ground, and after the next rainy season the field is ploughed twice in contrary directions. A month afterwards it is hoed once or twice with the same implement, and it produces a crop twice as great as it did the first year. In the third year a crop of *sáme* or *navane* (**Panicum frumentaceum** and **Setaria italica**) must be taken, and in the fourth year cotton is again sown as at first. The principal crop in the fine country towards Narsipur and Talkad is cotton, which in this locality is never raised in soil that contains calcareous nodules. The black soil that is free from lime is divided into three qualities. The first gives annually two crops, one of *jola*" (**Sorghum vulgare**, a name also used for **Zea Mays**) "and one of cotton; the two inferior qualities produce cotton only. Cotton is raised towards Harihar entirely on black soil, and is either sown as a crop by itself, or drilled in the rows of a *navane* field. In the former case two crops of cotton cannot follow each other, but one crop of *jola* at least must intervene. In the second month after the vernal equinox, the field is ploughed once, then manured, then hoed with the *heg kunte*, until the sowing season in the month preceding the autumnal equinox. The seed is sown by a drill having only two bills, behind each of which is fixed a sharp-pointed bamboo, through which a man drops the seed; so that each drill requires the attendance of three men and two oxen. The seed, in order to allow it to run through the bamboo, is first dipt in cow-dung and water, and then mixed with some earth. Twenty days after sowing, and also on the thirty-fifth and fiftieth days, the field is hoed with the *edde kunte*. The crop season is during the month before and that after the vernal equinox.

Cleaning. 224 (604)

CLEANING.—The floss is cleaned by means of a rude native gin or *charka*

CULTIVATION in MYSORE.

similar to that described elsewhere. After ginning, it is scutched by means of a bow, this operation being the special occupation of a class of Musalmans called *Pinjari*.

Trade. 225 ANALYSIS OF THE TRADE.—The railway-borne trade in cotton is very insignificant, and probably represents, but very imperfectly, the total traffic, most of which must be by road. In the year under consideration (1888-89) the exports by rail amounted to 1,120 cwt., of which 1,024 cwt. went to Madras seaports, 60 cwt. to the presidency, 35 cwt. to Bombay, and 1 cwt. to Bengal. The imports were larger, amounting to 3,159 cwt., of which 3,043 cwt. came from Madras, 104 cwt. from the sea-ports of that presidency, 9 cwt. from Bombay, and 3 cwt. from Bombay town.

CULTIVATION in BURMA.

13. BURMA.

References.—*Kurz, For. Fl. Burm., I., 129; Mason, Burma and Its People, 518, 756; Reports by Rev. & Agri. Dept. on Cult. of Cotton in British Burma; Rept. by Depy. Commr., Meiktila, Dec. 1888; Rept. by Burm. Govt., 1888; Agri-Hort. Soc. Ind. Transactions:—II., 124; V., 205; VI., 131; Journals, VII., 45-50; IX., 153; XI., 307-315, 584-586; XIV., 111-112; (New Series), I., Proc. xxxiv.*

LOWER BURMA. 226 I. LOWER BURMA.—It is necessary in this article to consider Lower and Upper Burma separately, since no accurate statistics are available for area, outturn, or trade in the case of the latter.

Area. 227 REVIEW OF AREA, &c.—Cotton cultivation has greatly fallen off in Lower Burma, notwithstanding the fact that the more inland districts possess soil and climate apparently well suited to the plant. In fact, in former times, cotton was largely cultivated in these districts. A considerable quantity was at one time exported from the Tharawaddy and Prome districts by Arakan to Dacca, where it was used in the manufacture of the fine muslins of that place. In the *Journal of the Agri-Horticultural Society of India, Vol. XI., 584*, it is stated that, "The species of cotton is still cultivated in those districts to a small extent. It is an annual, raised from the green seed variety and requiring from seven to nine months from the sowing of the seed until the crop arrives at maturity. The staple is of a peculiarly soft and rich appearance, and even now nearly the whole produce is exported to Tinian for the Chinese market." Now-a-days, however, the crop is an extremely unimportant one, and the produce is used almost exclusively for home consumption. During the five years ending 1885-86 the average area under the crop was 11,500 acres, while in 1888-89 it was only 10,191 acres. The total outturn is said to have been 9,494 cwt. The total imports were 55,295 cwt., the total exports 48,897 cwt., leaving a net import of 6,398 cwt. This, when added to the outturn, gives a total of 15,892 cwt. as available for local consumption. Dividing this amount by the population, 3,736,771, we obtain 47℔ per head of population as the average annual consumption. This figure, however, is probably an under-estimate, since it appears likely that considerably more cotton may be imported from Upper Burma than is shewn by the trade returns.

Methods. 228 METHODS OF CULTIVATION.—Cotton is not grown to any extent as a field crop except in parts of the Sandoway and Amherst districts; elsewhere, as in the hilly tracts of Thayetmyo and Northern Arakan, it is almost invariably grown mixed with other crops. Referring to the diminution in area, the Chief Commissioner in 1882-83 wrote as follows:—

"The shrinkage of the area is most conspicuous in the Thayetmyo district and appears to be due, in some measure at least, to the preference which the Burmese cultivator has for other crops, requiring less labour and trouble to produce, and to the cheapness of imported twist, which, in some parts of the country, is forcing the home produce out of the market altogether. In Kyaukpyu, for instance, no cotton whatever is now grown, and in Arakan generally the home product is rarely brought to market or even disposed of locally in a cleaned state."

CULTIVATION in LOWER BURMA.

Methods.

Further particulars regarding cotton in Burma are given in the subjoined extract from the *Lower Burma Gazetteer* published in 1880 (*Vol. I.*, 427):—

"Cotton is grown principally in hill-gardens, and in Thayet. In former years the Thayet produce was carried up-country and exported to Yunan, but the outbreak of the Panthay rebellion checked, if it did not altogether stop, trade by this route, and the price of rice in the husk rose so rapidly and has continued so high, and the importation of cotton goods and of twist has been so large that there was little or no demand for the home-grown article, and the cultivation of rice has become far more profitable. For many years persevering efforts have been made to introduce foreign cotton, especially Egyptian, Upland, Brazilian, and Carolina, which would give a longer staple, but these have been unsuccessful, not only for the reasons given above—of which undoubtedly the greater returns from rice, together with the large areas of rice land available on the easy terms on which land is granted are the most important—but also because the native plant is hardier and requires little care.

"The cotton grown in Upper Burma is still shorter-stapled than that grown here, and much of that imported uncleaned is taken to Prome, Allan-myo, and Kwa-toung, where are the principal cotton-ginning mills, and cleaned and mixed with Thayet and Prome cotton for export. The plants in British Burma are of two kinds—known to Burmans as the 'large' and the 'early,' respectively, 'The early kind is a plant which does not grow more than three or four feet high and its bolls are ripe in December and January. The large kind reaches a height of from six to ten feet and its seed does not ripen till a month or two later. The produce of the two kinds is hardly distinguishable.'"

Districts. 229 (609)

DISTRICTS.—The following special report, communicated by the Government of Burma, shows the extent of cultivation in certain districts:—

"In the *Prome District* it is grown to a small extent.

In the *Sandoway District* there were 282 acres of ground under cotton cultivation during the past year, and the cotton produced is believed to be of the indigenous Asiatic kind. The cultivating season commences about October. The quality of cotton hitherto grown has been very poor.

In the *Akyab District* the area of ground under cotton cultivation is about 940 acres. The seeds are planted with other miscellaneous growths on the hill-side, about May. The cotton can be used about February. The cost is about R15 per acre, and the profit about R30. The hill-sides are cleared of jungle and the soil turned up with rude ploughs. The floss is employed in the manufacture of cloths, and wearing apparel, thread, &c., and the production is worth about R30 per acre.

In the *Shwegyin District* it is grown in small quantities on the Karen hills, especially up Dondami Choung.

In the *Toungoo District* it is cultivated by Burmans and Shans to a moderate extent. There are two kinds, one with white and the other with yellow flowers. It is planted in May.

In the *Amherst District* it is grown on the banks of the Salween river from indigenous seed. The area under cultivation in 1887-88 was 556 acres. It is sown in September and October, and reaped in January and February. The seed is planted upon alluvial soil collected on the river banks during inundation. The produce is consumed locally in the native looms, and none is exported. The estimated outturn of clean cotton is about 1,430 cwt., and the retail price R1-8 to R2 per viss."

Races. 230 (610)

RACES.—The only specimens of cotton which the authors have seen

CULTIVATION in LOWER BURMA.

Races.

from Lower Burma correspond with **G. Wightianum**, *Tod.* In 1882-83 experiments were made in the Arakan Hill Tracts and Sandoway with seed of the Bamia and other Egyptian varieties of cotton (**G. hirsutum**). They were not, however, successful. In a report for 1868-69, published by the Cotton Commissioner for the Central Provinces and the Berars, it is stated that a variety of Persian cotton closely allied to the Egyptian grows exceedingly well about Rangoon at all seasons and yields a fine white cotton of strong staple. A more recent allusion to this exotic cotton cannot be discovered, though the subject is interesting. **Mason**, however, states in his "*Burma and Its People,*" that Peruvian and Sea Island cotton were at one time introduced experimentally, but that in Burma, as in most other parts of India, they appear to have taken little hold. But it may be added that in 1831, **Major Burney** contributed to the *Agri.-Horticultural Society of India,* samples of Pernambuco cotton, called in Burmese *Shembanwa*, or ship-cotton, which had been grown near Ava. The plant he described as growing into a lofty tree as thick as the thigh, yielding its cotton in February, March, and April (*Trans., II., 124*).

Mr. Edward Riley, in a *Paper on the Agricultural Capabilities of the Province of Amherst* (*1838*), refers to the same cotton as occurring in the lower jungles in a state of acclimatisation. Commenting on this plant, however, he admits that the attempt to introduce it as an agricultural crop had failed, though he expresses the opinion that the failure was due to a misjudgment of the kind of soil best suited to the crop. This is probably the cotton referred to, in the Note on Upper Burma, as having been introduced by King Min Doon Min. In 1841 **Dr. Wallich** brought to England samples of Martaban cotton collected from the Pernambuco stock. The floss was pronounced inferior to none, either in the quality of its staple or the ease with which it could be separated from the seed. (*Journ. Agri.-Hort. Soc. Ind.* (*Old Ser.*), *I., 195.*)

Trade. (611) 231

ANALYSIS OF THE TRADE.—In the *Provincial Cotton Report for 1882-83*, it is stated that none of the cotton grown in Lower Burma is exported from the Province, an assertion that would appear to be confirmed by the relation between the imports and exports. The land trade in cotton is wholly confined to imports by river and overland from Upper Burma. The greater part of the exports by sea is shipped to China, either direct or through the Straits Settlements.

UPPER BURMA. (612) 232

II. UPPER BURMA.—Trustworthy cotton statistics are not available regarding the area, outturn, and trade of Upper Burma, as a whole. The following valuable reports by the Deputy Commissioner of Meiktila, and from other districts, however, are sufficient to shew that cotton cultivation exists, to a considerable extent, at least in certain tracts, and is probably capable of extension:—

"Cotton is grown to a considerable extent in the Mahlaing sub-division—in fact it may be said to be the staple product. In other parts of the district it is grown to a small extent.

Races. (613) 233

RACES.—There are three forms of cotton grown here, *viz.* :—(1), *wa-gyi*, (2), *wa-galé*, and (3), *pinni*. The flowers of 1 and 2 are white, whilst that of 3 is red; 1 and 2 produce the pure white cotton, whilst 3 produces only cotton of a light brown tint, which eventually develops into a colour similar to the ordinary dark *khaki*. American cotton seed was introduced by King Min Doon Min, but in this part it has turned out a failure. The original seed did well, but it failed to acclimatise. Subsequent seed deteriorated and the cultivators ceased to plant it.

Area. (614) 234

AREA UNDER CROP.—It is a difficult matter to estimate the area at present under cultivation, but in Mahlaing it is stated at 30,000 *pès* (1 *pè*= 1·77 acres), and in the other parts of the district at 5,000. Taking a *pè* roughly to be an acre and a halt this will give 52,500 acres of land under cotton.

CULTIVATION in UPPER BURMA.

Sowing. 235 (615)

DATE OF SOWING AND REAPING—Cotton is sown in May and June and picked in November and December. To plant a *pè* of cotton from five to six baskets of seed are sown. The price of seed is ten baskets for R1. One yoke of bullocks can plough four *pès* of land, so in calculating the cost of cultivation it will be convenient to take the work of one man and a pair of bullocks as a datum. Therefore, to cultivate four *pès* of land the following may be estimated:—20 baskets of seed, value R2; 1 man's labour, R20; 2 bullocks, R30; picking, R10; total R62. This estimate is based on the presumption that labour and oxen have to be hired. As a rule, the work is done by the cultivator with his own cattle, and the cotton is picked by the family.

Yield. 236 (616)

YIELD PER ACRE.—The yield of uncleaned cotton varies from 250 to 500 *viss* (3·66lb) *per pè*; 250 is looked upon as a very poor crop, and 500 as a good one, so an average of 400 *viss* may be taken, for the purpose of arriving at some idea of the profits.

Profits. 237 (617)

PROFITS.—The cotton as picked from the plant, sells for R10 a 100 *viss* that is, R40 per *pè* or R160 for four *pès*. Therefore, the account stands as follows:—

	R
To cost of cultivating 4 pès of land	62
,, sale of 1,600 *viss* of cotton @ R10 per 100 *viss* . .	160
Balance in favour of Collector .	98

If the cultivator takes the trouble to clean his cotton he can obtain R60 per 100 *viss*, which would make his profits considerably greater.

Method. 238 (618)

METHOD OF CULTIVATION.—The land is ploughed and cleaned. The seed is sown broadcast, and beyond keeping the weeds down and shovelling the earth round the young plants, nothing more is done.

Soil. 239 (619)

NATURE OF SOIL.—It is difficult to say where cotton will not grow. It is found in the heavy black soils (what is commonly known as cotton soil), in the red marls, and in the alluvial deposits alongside rice lands. In fact, the people grow cotton and pulses wherever they cannot get water to grow paddy.

Outturn. 240 (620)

TOTAL OUTTURN.—Taking the acreage under cultivation at 35,000 *pès* and an average yield of 400 *viss* to the *pè*, the output for this district would be 14,000,000 *viss*, or in round numbers, say, 2,000 tons, equal to about 600 tons of cleaned cotton. This is principally exported, but a small amount is used for home consumption.

Cleaning. 241 (621)

CLEANING.—The floss is prepared by passing it through rollers worked by foot treadle, after which it is beaten and hand-picked. It is then in a marketable condition as raw cotton. For domestic purposes it is spun into twist on the old-fashioned wheel and then woven into cloth."

The specimens enclosed with the above report have been determined as follows:—

1. *Wa-gyi*, **Gossypium acuminatum.**
2. *Wa-galé*, **G. Wightianum,** *Tod.*
3. *Pinni*, **G. neglectum,** *Tod.* Kurz's 1232, collected in th Pegu Yomah and named **G. herbaceum,** *Linn., var.* **hirsutum,** is a hybrid between **G. Wightianum** and **G. neglectum.**

From a report recently received from the Government of Burma, it would appear that cotton is cultivated sparsely in the Mandalay District, extensively in the Lower Chindwin District, and in the Eastern Division of Upper Burma, and to a fair extent in the Bhamo District. It is also stated that in the Ruby Mines District two "varieties" are occasionally seen in gardens. No returns have been received of the area under the crop. In the Lower Chindwin District this crop is sown in March and

CULTIVATION in UPPER BURMA.

April, and collected from September to December, and the floss is said to sell from R13 to R14 per 100 *viss*.

"In the *Bhamo District* the plant grown has 3 to 5-lobed leaves (small), with a flower usually whitish, often yellow, and sometimes pale pink. It is grown pretty extensively."

Trade. (622) 242

ANALYSIS OF THE TRADE.—No statistics of the cotton trade of Upper Burma are available. A large proportion of the exported produce of the lower districts is imported into Lower Burma, either for use there, or for transmission to Rangoon. It is believed that a great part of the produce of the upper districts are exported trans-frontier to China. In the *Agri.-Horticultural Society of India Journal (Old Series), XI., 584*, it is stated that, "according to the records of the old Burmese customs, the exports at one time reached the large amount of 30,000,000℔ annually; and, in after years, the Resident at the Court of Ava, **Colonel Burney**, reported that Burma exported to China fully 4,000,000℔, and a like quantity *viâ* the Arakan mountains, Chittagong, and Bengal."

CULTIVATION in ANDAMANS. (623) 243

14. ANDAMAN AND NICOBAR ISLANDS.

References.—*Proceedings of Dept. of Rev., Agri. and Commerce, Govt. of India, Nos 6 and 7, July 22nd, 1871; Nos. 1, 6, and 7, July 1872.*

In 1869, **Colonel Man**, in his report on the Nicobar Islands, remarked that the soil and climate appeared to be well adapted for the growth of Sea Island cotton, and that it would be well to try the experiment. Seed was accordingly obtained and sown, together with that of what **Colonel Man** calls the indigenous cotton. Both did well, but the American plants were largely attacked by beetles and other insects. Samples of the "indigenous cotton" were subsequently forwarded to the Bombay Chamber of Commerce and pronounced much superior to any grown in India, superior, indeed, to any American race except the finest Sea Island. Its cotton was found to greatly resemble Brazilian, having adherent seeds, and an excellent long, strong staple. Specimens of the plant were examined by **Mr. C. B Clarke**, who pronounced them to be **Gossypium acuminatum**. *Conf. with p. 23*).

Samples of both were afterwards sent to England, where they were submitted for valuation to the Liverpool Cotton Brokers' Association, with the result that the "wild" cotton was valued at 11*d.* to 12*d.* per ℔, the Sea Island at 20*d.* to 24*d.* per ℔. The latter was said to be very irregular in staple; part of the sample was long and fine, worth from 33*d.* to 36*d.* per ℔, but other portions so much inferior that the whole would not fetch more than 16*d.* to 19*d.* per ℔.

In 1872 experiments were made with Hinganghát, Sea Island, Egyptian, *Jari*, New Orleans, and Dharwar, but none of these, except perhaps the form of **G barbadense** found by **Dr. Prain** in the Andaman Islands (*Conf. with p. 29*) proved equal to the reputed native cotton above alluded to.

GINS & GINNING. (624) 244

GINS AND GINNING.

References.—*Watson, Report on Cotton Gins, 1879; Agri-Hort. Soc. of India, Transactions, V., 187; VIII., 301; Journals (Old Series): II., Sel., 203; V., 83-100; VI., 222, proc., xx., xc; VII., 35, 189; VIII., 161-174, 175; XIII., proc., xlvi.; XIV., proc., xv., lvii.*

The methods of cleaning cotton in India, as will have been seen from the above provincial reports, are primitive in the extreme. Indeed up to twenty years ago mention is frequently made of cotton being picked by hand, in which manner one woman could clean one to one-and-a-half pounds of fibre per day. In all the chief cotton-growing districts, however, there have been in existence from time immemorial two simple forms of machine for the purpose, *viz.*, the foot-roller and the *churka*. Of these,

GINS.

the foot-roller is used only locally in a few places, and is alone adapted to the hard-seeded Indian cottons. By it the fibre is cleaned as follows:—the cotton is spread over a smooth flat stone of from one to two feet square, sometimes square, sometimes round; an iron rod 18 inches long, thick in the centre, tapering towards the extremities, is placed on the stone, and rolled forwards and backwards by the feet of the worker. Sometimes the rod is shorter and slightly conical, in which case the motion is circular, round and round the stone. Under the combined influence of the rolling motion and the pressure of the feet, the seed is squeezed out and pushed away in front of the roller, leaving the cleaned fibre behind. One worker can in this way turn out from four to six pounds of cleaned cotton in a day.

The usual implement employed, however, is a simple double roller machine called the *churka*. This has been already described (see pp. 94, 106, 123) and need not be further discussed. Suffice it to say, that the essential parts of the machine are two rollers, either both of wood, or one of wood and one of iron. To these a revolving motion towards each other is communicated by means of a crank or wheel at one or both ends. The effect is, that the cotton is presented at one side of the revolving cylinders, and the fibre is drawn through and pulled off the seed, which is too large and hard to pass through. On an average the *churka* produces from six to eight pounds of clean cotton per day for each man or woman engaged in the work.

The common *churka* is evidently liable to many imperfections. The feeding being done by hand it is necessarily impossible to keep the whole length of the rollers constantly supplied, so that the machine is never worked up to its actual ginning power, though, at the same time, there is now and then in places such an excess of seed-cotton that the rollers become clogged. As a consequence of this clogging, the surface of the wooden roller or rollers is liable to become worn in lines or narrow grooves, which tend to diminish the grip of the rollers on the cotton and eventually allow seed to pass through into the heap of cleaned fibre. In the case of the Guzerat *churka*, which has one small iron and one large wooden roller, the surface velocities of the two are unequal, consequently the tuft of cotton is exposed to an injurious grinding action.

The smooth and gentle working of the *churka* has suggested many mechanical adaptations of its principle, and numbers of roller gins have been constructed with the idea of preserving this gentle action, whilst increasing the outturn and diminishing the rapid wear of the rollers. "Improved mechanical feeding, arrangements for preventing the coiling of clean cotton round the roller, and contrivances for regulating the grip and the motion of the two rollers have been added to the simple native machine without, however, achieving any definite practical success." (*Watson.*)

This being the case, attention was directed to American gins, which have now been largely introduced into the country. In the year 1792, **Eli Whitney**, an American, invented the saw-gin, a machine which, under various modifications, is still employed for cleaning the greater proportion of the cotton grown in the Southern States. It consists of a series of "saws" (in reality discs with hooked blunt teeth), revolving between the interstices of an iron bed which forms the base of a large box or hopper. The fibre is caught up by the hooks and pulled through the interstices, leaving the seeds behind. Since the fibre of certain of the long-stapled cottons was found to be more or less cut, crushed, or broken by the action of the saw, a more recent American invention, the Macarthy gin, came largely into use for cleaning Sea Island, Egyptian, and Brazilian cotton. In this gin the fibre is drawn by a leather roller between a metal plate called the "doctor," fixed tangentially to the roller, and a blade

GINS.

called the "beater," which moves up and down in a plane immediately behind and parallel to the fixed plate. As the cotton is drawn through by the roller the seeds are forced out by the action of the moveable blade, which, in some machines, is made to work horizontally instead of vertically. Many improvements have been made on the original and simple forms of both machines, for a description of which the reader is referred to **Dr. Forbes Watson's** valuable Report on Cotton Gins.

The first notice of the introduction of foreign gins into India occurs in 1829, when the Court of Directors stated that they had at several times sent different machines to India, with the hope of bettering the condition in which their cotton reached England, and amongst these "two of a new instrument called **Whitney's** saw-gin." Other American gins had formerly been sent to India, and on one occasion under the charge of an American mechanic; two gins had also been made by **Mr. Maudsley** for the Company, but all alike had failed. Writing in 1839, the Court of Directors again state that the American machinery, which up to that time had been sent out to India, had not succeeded.

Attention was consequently again directed towards improvements in the native *churka*, and in 1848 a **Mr. Mather** was employed by the Government of the North-West Provinces to improve that machine. He shortly afterwards exhibited a working model of an improved *churka* which obtained the approval of a committee. This body recommended that twelve similar instruments should be constructed. These were, accordingly, made in Calcutta, and in 1849 eleven were sent to Agra for trial. In 1848, the *Agri.-Horticultural Society* awarded a prize of R500 to **Mr. Mather** for this machine, and in 1849 orders were issued by Government to construct ninety-six such *churkas*, being eight sets of twelve each. In 1850, these were procured at a total cost of over R20,000, and twenty-nine were distributed. After this large expense had been incurred the machines were found to be inferior in every way to the native *churka* and turned out less cotton for the power employed, and that too in a dirtier and less satisfactory state. Finally, all were advertised for sale in Calcutta, and the 65, which had not been distributed, were sold by auction to the highest bidder.

After this disheartening result it is not surprising to find that a reaction again set in in favour of saw-gins. **Dr. Royle** examined the then newly-invented "cottage saw-gins" worked by manual labour in Manchester and reported very highly on them, with the consequence that 48 were ordered to be sent to India in 1849. These were found to answer well when fitted with small gratings suited to the smaller size of the Indian cotton seed. In 1851, the Government of India, on the recommendation of the *Agri.-Horticultural Society of India*, offered a prize of R5,000 for the gin best suited to clean Indian cotton. This resulted in four gins being offered for competition, of which two American saw-gins gave very satisfactory results and were adjudged equal, each receiving half the award.

From that time up to the present day American saw-gins have been largely introduced, chiefly for cleaning Dhárwár and other acclimatised American cottons.

In 1871-72 and 1874-75, extensive trials were conducted by Government, under the supervision of **Dr. Forbes Watson**, in Manchester, Dhárwár, and Broach. That gentleman drew up an exhaustive report to which the reader is referred for full information on the subject. The following remarks on the relative advantages of the two classes of gin and their suitability for Indian work may, however, be quoted:—

"The relative merits of saw and Macarthy gins may be discussed irrespectively of the particular varieties of cotton for which they have been up to the present considered specially adapted. Some of the ideas com-

GINS.

monly entertained upon the subject are far from being correct. The great advantage of saw-gins has been supposed to consist in the rapidity of their work and their economy in the consumption of power. This opinion was well founded in the days when a 40-inch Macarthy gin was not expected to turn out more than from 8 to 12lb of clean cotton per hour. At the present time the construction of Macarthy gins of all kinds is greatly improved; in fact, the improvements effected in the knife-roller gin during the progress of the trials, and which may fairly be regarded as having resulted from the experience gained at the trials, have rendered it possible to obtain a rate of outturn of as much as 120 or even 200lb of clean cotton per hour, and this with a smaller consumption of power than that required for saw-gins, thus surpassing the latter both in outturn and in economy of power. Even the common double action Macarthy gin, when compared with saw-gins of a similar price, and worked at the moderate rate of speed at which alone it is safe to employ them, will probably have the advantage as regards rate of outturn though not as regards consumption of power. On the other hand, it was usually supposed that cotton was more liable to be cut by the saw-gins than by the roller gins, and a good deal of ingenuity has been devoted to the discovery of a form of tooth which would reduce the cutting action of the saw to a minimum. Without denying that the shape and size of the teeth may have some influence, and are, therefore, worthy of attention, it has been proved that, errors of adjustment excepted, the cutting and tearing action in all systems of gins is mainly determined by the striking speed, and that with a similar striking speed no great difference in this respect has been observed between the different systems of gins. In fact, the greatest amount of cutting action observed at the trials was exhibited, not by saw-gins, but by roller-gins running at an excessive speed. The great diameter of the striking part in the case of saw-gins, however, renders it particularly necessary to avoid any high rates of speed, as otherwise the permissible limits of striking speed may soon be exceeded. It may be observed that, in the United States, where the saw-gins are usually worked at a moderate speed and where the fibre of the cotton does not adhere very strongly to the seed, complaints as to the cutting action of the seed are always less frequent than in India, where the gins are frequently drawn at an excessive rate, and where the strong attachment of the fibre to the seed renders the staple specially liable to injury.

"But the one real drawback to the use of saw-gins is their tendency to 'nep' the cotton, although even this defect is greatly diminished by employing a moderate speed. A certain amount of nepping is produced even in the case of Macarthy gins, but the defects to which the Macarthy-ginned cotton is peculiarly exposed are curling and twisting, faults which may give rise to 'neps' during the subsequent carding of the cotton. It is thus only at the carding engine that the different samples of cotton can be compared in this respect" . . . . . "The observations gave a clear result in favour of the Macarthy gins, their average 'neppiness' amounting to 0·9 (out of a possible 3) against 3·4 degrees in the case of saw-gins. The increased 'neppiness,' moreover, appears to be usually accompanied by a certain diminution in the strength of the yarn.

"But, on the other hand, the use of saw-gins offers one undoubted advantage over the Macarthy gins in the greater cleanness of the cotton which, under certain conditions, it is possible to attain by their use. Most saw-gins possess mote-boards, fixed in such a position under the saws and brushes that a certain proportion of the leaf, crushed seed, and other coarse impurities, thrown off by the centrifugal force imparted by the rapidly revolving brushes is retained by them. . . . . As regards the quality

GINS.

of the cotton, if a saw-gin be driven at such a moderate speed as not to cut the cotton, the opinions for or against its use will depend in each particular instance on whether more importance is attached by the valuer or spinner to the cleanness of the cotton, or to its 'neppiness.'" . . . "The general conclusions just arrived at are of some importance in considering the future of cotton cultivation in the Southern Mahratta country. At the present time the whole of the acclimatised American cotton, produced in the Dhárwár and adjoining districts, is cleaned by the saw-gins which were originally supplied by the Dhárwár Government Factory, and it may be estimated that there are at the present time (1879) about 3,000 of these in the hands of the rayats.

"The cotton cleaned by these gins, however, even when they are in perfect condition, is proved by the results both of the valuations and of the spinning experiments to be inferior in quality to that prepared by the best American gins." . . . "Not only does the cotton prepared by the Dhárwár saw-gin present the unavoidable defects of saw-ginning (that is, nepping and cutting if the gin has been driven at too high a speed), but it does not possess, to any extent, the compensating advantage of cleanness. In fact it is one of the gins which produce the largest quantity of broken seed, considerably more indeed than a well-adjusted Macarthy gin. This is particularly the case with the Dhárwár-American cotton; in fact, the unexpected result was brought to light that the native varieties of cotton appeared to be, on the whole, less injured by the Dhárwár saw-gin than the American cotton for which it was specially intended." *(cff. p. 130.)* . . . "As the Dhárwár gin, even in the perfect condition in which it was tested at the trials, has been shown to be inferior to the American gins now in use, it follows that, in the neglected condition in which these gins are frequently found in the Dhárwár district, many of them not having been set or repaired for years, this inferiority must be immensely increased."

Dr. Forbes Watson then goes on to discuss the best means of remedying the then existing state of matters, which was year by year contributing so largely to lower the position of the Dhárwár-American cotton. After comparing the relative merits of the Macarthy and saw-gins, he arrives at the conclusion that "until the introduction of steam power on a large scale, gins of the type of saw-gins must be mainly relied on for cleaning the Dhárwár crop." He recommends a modification of the Emery gin as likely to prove most useful, and continues, "But in the meantime it appears possible to utilise, to some extent, the existing saw-gins. A certain advantage would be obtained by simply reducing the speed, which could be easily done by merely increasing the diameter of their driving pulleys. A speed of 200 revolutions per minute would very nearly correspond to the striking speed of 100 inches per second, which has been fixed as the approximate limit, every increase above which is injurious to the staple of the cotton." . . . "A more material improvement would be obtained by the substitution of the Emery saws and grids, or even by the replacement of the saws above, which could be effected at a moderate cost."

Up to the present day nearly all the indigenous cottons of India are hand-ginned by the ordinary *churka*; indeed, it appears doubtful if the American saw-gin will ever obtain a hold for this purpose. Notwithstanding the proof to the contrary, brought forward by Dr. Forbes Watson, reports are still frequent that the native cotton is much damaged by the saw-gin. It is possible that this may be due to imperfections in adjustment, excess of speed, and general carelessness, but it appears doubtful if it is entirely so, since by the same machines American cotton is improved in ginning. Moreover, the *churka* being, as already stated, almost entirely worked by women of the cultivator's household, whose labour

G. 244

(624)

GINS.

and time cost practically nothing, can never be replaced by an expensive machine, unless the increase of value of the product should prove sufficient to more than cover the outturn involved by its use. The only case in which the saw-gin is said to produce a higher quality fibre from indigenous "*kapás*" than the *churka*, is that of Sind, where the machine is reported to be consequently "much appreciated by the people." This fact is confirmed by the experimental trials, and **Dr. Forbes Watson** reports that "the percentage of cleaned cotton yielded by this variety is higher than in the case of any other of the native cottons."

Dr. Forbes Watson divides the Indian cottons into two classes: (1) those easily ginned, and (2) those ginned with difficulty. To the first belong the introduced American and certain Indian races, *viz.*, Broach, Dholleras, the *jari* form of Oomras, Westerns, Coomptas, Khandesh (Vilayti ?), and Tinnevellies; to the second Madras *Uppam*, Hinganghát, Khandesh, and the *bani* form of Oomras. Of all these, the Broach and Dhárwár growths are the only forms entirely cleaned by steam or hand saw-gins at the present time, but the quantity of Umras (*jari*) and Dholleras thus treated is yearly increasing. With the exception of *jari* and Dhollera, these are soft-seeded American races, which cannot be satisfactorily cleaned by the native *churka*. All other descriptions of Indian cotton are still entirely hand-cleaned by the *churka*, or by the foot-roller. It appears possible, however, that the Sind cotton also may soon be extensively cleaned by the saw-gin, and that the Westerns and Coomptas, classed by **Dr. Forbes Watson** as "easily ginned," may also be found to be more satisfactorily cleaned by the same means.

COTTON PRESSING.

PRESSING. 245 (625)

A good deal of the cotton produced in India is still sent to the seaports in bundles or *dokrahs* of about 200lb each, but the greater portion of the crop is pressed up-country ready for shipment. The method of packing these rough bundles is generally similar to that already described (see p. 106) and need not be further discussed.

The number of steam presses in India is increasing year by year. **Mr. Beaufort**, in his Indian Cotton Statistics, gives a list of 190 Cotton Steam-pressing Companies as existing in all parts of the country in 1889. **Mr. O'Conor** in his statistical tables, however, returns 249, of which 116 are in Bombay, 5 in Sind, 41 in Madras, 17 in the Panjáb, 1 in Burma, 36 in Berar, 13 in the North-West Provinces, 4 in the Central Provinces, 3 in Central India, 11 in Rajputana, and 2 in the Nizam's territories. The total capital employed (so far as known) is R1,87,07,125, and the value of the outturn R11,10,26,240. The head offices of the larger companies are in Bombay, but they possess steam presses in various important centres of the cotton districts such as Bombay Sind, Berar, and the North-West Provinces. Besides these, numerous smaller companies have their presses and head offices in the inland towns of Madras, Bengal, the Panjáb, the North-West Provinces, Central India, Rajputana, the Central Provinces, Berar, the Nizam's territory, Sind, Bombay, and Burma.

According to **Mr. Beaufort**, cotton is generally exported from Bombay and Karáchi in bales of 3½ cwt. each, or 392lb net. "As the staple is bought and sold by the *candy* of 784lb, two bales equal one *candy*. Formerly, at all other ports cotton was shipped in bales of 300lb net, but within the past fifteen years the trade has gradually, in Calcutta and Madras, changed the weight to 400lb net. From Coconada shipments are made in bales of both 300lb and 400lb, while at Tuticorin, of the steam presses at work, several have been converted so as to admit of

PRESSING.

pressing 500lb bales, the others confining themselves to 400lb bales. Since nearly the whole of the Tinnevelly crop is sent to Bombay by coasting steamers for transhipment to European ports, and many bales burst *en route* to Bombay, thus requiring to be re-pressed before re-shipment, it may be as well to point out to those interested that there is not in Bombay a single press capable of pressing a 500lb bale. A Bombay bale is, therefore, 2·04 per cent. smaller than a Calcutta or Madras one and 27·55 per cent. smaller than a Tuticorin bale.

On an average, at Bombay four bales (1,568lb net) go to the shipping ton of 40 cubic feet, but some of the more modern presses turn out bales 100 of which equal 21 tons measurement. Many of the older presses have now finishers attached which, after the bales have been pressed to a certain point in the larger presses, extract the cotton and complete the process. At Karáchi a space of 40 cubic feet also constitutes a steamer ton of freight, while at both ports 50 cubic feet is the sailing ship scale. On the other hand, both at Calcutta and the ports in the Madras Presidency, Coconada, Masulipatam, Madras, Pondicherry, and Tuticorin, a ton of freight, both by sailers and steamers, is on the single basis of 50 cubic feet.

The bales are covered with jute, 'gunny,' cloth of Calcutta manufacture, and lashed with iron or steel bands, the latter of late years generally superseding the former owing to their lightness and greater strength. In the Government Railway returns, railway freight being payable on the gross weight of the bales, five per cent. is deducted for tare, a correct percentage fifteen or twenty years ago, but obsolete now that iron and steel bands have taken the place of jute ones. The Bombay Chamber of Commerce in their daily returns of receipts from up-country deduct but two per cent., which is approximately correct."

MANUFACTURES.

COTTON MANUFACTURES.

Hand-loom. 246

1. HAND-LOOM.—Important and interesting as are the native cotton manufactures of India, it is impossible to enter into a discussion on the subject in an article such as the present which is designed to deal more especially with the raw product. The total number of hand-looms in the country, and the amount of raw material which they consume, can only be guessed at. With the increase in the number and outturn of Indian mills, and the gradual growth of the Manchester trade the manufacture of the finer textiles by the hand-loom, such as the Dacca muslins, once so famous all over the world, has almost, if not entirely, ceased. The local manufacture of the coarser kinds, however, for consumption in rural localities, and of ornamental varieties, such as the *do-pattas* of Benares, is likely to continue to a much greater extent. **Sir George Birdwood** (*Indian Arts, pp. 244—260*) gives an interesting and detailed account of Indian cotton goods arranged provincially. The reader is referred to that article for information regarding the technicalities of these Indian manufactures.

Power-loom. 247

2. POWER-LOOM.—**Mr. Beaufort** has written an excellent *résumé* of the available information on this subject, in his Cotton Statistics, from which the following may be quoted:—

"The first mill built in India was the Bombay Spinning and Weaving Company's, which was formed about 1851, but does not appear to have been in working order till 1854. The earliest return we have met with of the Indian mills, is one dated 1875, which gives the later date against the name of this concern, followed by the 'Oriental,' September 19, 1855, and the 'Throstle,' January 10, 1857. By 1861 the number in the country had increased to a dozen, containing 338,000 spindles, with an estimated annual consumption of 65,000 bales (of $3\frac{1}{2}$ cwts. each) of cotton. By 1879 or just five and twenty years after the starting of the first mill, the number had increased to 56 and the spindles to nearly 1½ millions. The progress in the succeeeding eleven years has been very remarkable.

MANUFACTURES. MILLS.

30th June.	Mills.	Spindles.	Looms.	Hands.	Cotton consumed. Bales.
1879	56	1,453,000	13,000	43,000	267,600
1889	124	2,763,000	21,600	91,600	888,700
Increase	61	1,310,000	8,600	48,600	621,100

In these eleven years the number of mills increased 121 per cent., the spindles 90, the looms 66, the hands employed 113, and the cotton consumed 232 per cent."

Bombay Island. 248

Bombay Island Mills.—Since 1865, or just a quarter of a century ago, the number of mills has risen from 10 to 69, equal to 590 per cent., the spindles from 250,000 to 1,591,000, or 536⅓ per cent., the looms from 3,380 to 13,380 or 298⅓ per cent., the hands employed from 6,600 to 52,500, or 700⅓ per cent., and the consumption of cotton from 42,000 to 564,000 bales, equal to nearly 1,243 per cent. Included with the 69 mills (on 30th June 1889) are the names of 14 in course of erection, the spindle power of which, when in working order, will add fully 300,000 to the 1,591,000 existing spindles.

Bombay Presidency. 249

Bombay Presidency (*Up-country*) *Mills.*—Of these, there are 22, most of which are situated in Guzerat, eight being at Ahmedabad, three at Broach, one at Baroda, three at Surat, one at Nariad, one at Veeramgaum, and one at Bhownugger. Another exists at Jalgaon in Khándesh, one at Sholapore in the South Deccan, and two in the Southern Mahratta country, one being at Hubli and the other, a water power concern, at the Gokak Falls near Belgaum.

Central India. 250

Central India.—During the past four years no returns have been received, though annually applied for, by the Bombay Mill-owners' Association, from the Government of His Highness the Maharaja Holkar, regarding the solitary mill owned by the State at Indore, which was started in 1872.

Central Provinces. 251

Central Provinces.—The three existing mills are at Jubbulpore, Hinganghát, and Nagpur. In eleven years the spindles have risen from 30,000 to 68,000, the looms from 32 to 600, the hands employed from 1,000 to 3,800, and the cotton consumed from 5,600 to 26,000 bales.

Berar. 252

Berar.—There is one mill at Budnera and another at Aurungabad, but the latter has only recently commenced work and gives no return of consumption.

Hyderabad. 253

Nizam's Territory (*Hyderabad*).—One at Hyderabad and one at Kalburga are the two mills in this Native State. Eleven years ago the former consumed less than 1,500 bales of cotton annually; last year the two required over 10,700 bales.

N.-W. Provinces. 254

North-West Provinces.—Of the five mills, four are working at Cawnpore and one is being put up at Agra. The first one was erected in 1871. In 1879, over 38,000 spindles in two mills consumed 4,800 bales of cotton; last year the four working at Cawnpore disposed of nearly 25,700 bales.

Bengal. 255

Bengal Presidency.—All the seven in this presidency are situated in Calcutta or its neighbourhood (the oldest dating from 1864 and the second from 1872). Six have been in the list for the past eleven years, but a new one has recently been added. From 164,000 the spindles have risen to nearly 250,000, but there are no looms working. The hands employed have increased from 4,300 to 6,300, and the consumption of raw material from 32,000 to 85,000 bales, of which about one-third is imported from

MANUFACTURES. MILLS.

Bombay by sea, the balance being mainly received from the North-West Provinces and the Panjáb.

Madras. (636) 256

Madras Presidency.—The first two mills were started in 1874 and 1875, and in the past eleven years the number has been trebled, the total now being nine, of which four are in Madras city, and one each at Bellary, Calicut, Coimbatore, Tuticorin, and at Ambasamudram near Tinnevelly. Of the nine, seven have been working throughout the year, the one at Tuticorin for six months only, while the mill at Coimbatore is still being erected. In the eleven years, the spindles have increased from less than 48,000 to nearly 153,000, the looms from *nil* to 470, the hands employed from 1,000 to 4,700, and the consumption of cotton from 9,300 to nearly 44,000 bales.

Travancore. (637) 257

Travancore.—The solitary one in this Native State is at Quilon. Its consumption has risen to 4,700 bales, while the number of spindles has recently been doubled.

Mysore. (638) 258

Mysore.—The two—one owned at Madras and the other at Bombay—are at Bangalore, and between them consume nearly 5,800 bales of cotton, the spindle power being over 23,000.

Pondicherry. (639) 259

Pondicherry.—If the returns are to be relied upon, the solitary mill in this little French Settlement consumes annually less and less cotton. Four years ago it took 5,400 bales, last year but 2,400 bales.

Mill-Consumption. (640) 260

Consumption of Cotton by the Mills.—In the first six of the last eleven years, the mill consumption of the total supply averaged 22·02 per cent; in the last five years it has risen to an annual average of 33·37 per cent. While the exports from India in the past three official years have ranged between 1,521,000 and 1,552,000 bales, the consumption by the Indian mills has risen from 726,000 to 889,000 bales, or from 31·87 to 36·87 per cent. of the available supply. The share of each province in the total consumption, during the year ending 30th June 1889, was as follows:—

Mills situated in	Bales of 3½ cwts.	Per cent.
Bombay Island	563,730	63·43
" Up-country Mills	103,968	11·69
Bombay Presidency total	667,698	75·12
Berars	5,536	·62
Central Provinces	26,006	2·93
Nizam's Dominions	10,732	1·21
Central India	9,864	1·11
Bengal Presidency	85,096	9·58
North-West Provinces	27,075	3·05
Madras Presidency	43,750	4·93
Travancore	4,718	·53
Mysore	5,778	·65
Pondicherry	2,400	·27
Total all India	888,654	100·00

(641) 261

Concluding Note on Indian Cotton Mills.—In the above report Mr. Beaufort includes several mills in course of erection in Bombay and elsewhere, and carries his statistics to the end of the commercial year June 1889, hence his figures differ somewhat from those given by Mr. O'Conor in his Statistical Tables for the year ending March 31st, 1889.* These shew 108 cotton mills as having been at work during the year, of which

* Conf. with foot-note, p. 52.

MILLS.

there were 75 in the Bombay Presidency, including 52 in the town and island of Bombay. There were 7 in Bengal, all in the vicinity of Calcutta; 8 in Madras, of which 4 were in the town; 6 in the North-West Provinces, all but one at Cawnpore; one at Indore; 3 in the Central Provinces; 3 in the Nizam's Territory; 2 in Mysore; 2 in Pondicherry, and one in Travancore. Of the 108, 50 are returned as both spinning and weaving; 53 as purely spinning mills; and 5 as entirely weaving mills. Two of the Bombay mills are stated to have manufactured hosiery in addition to yarn, piece-goods, &c. According to **Mr. O'Conor's** tables, the mills contained 22,156 looms and 2,669,922 spindles. "They consumed about 347 million pounds of cotton in the year, and employed a daily average number of 92,126 persons, of whom, as far as details have been obtained, there were 53,317 men, 18,031 women, 15,309 young persons, and 3,469 children. The nominal capital of the mills worked by joint stock proprietary is returned at 95⅓ million rupees; but there is no return of the capital of eight mills worked by private proprietary, and probably the whole capital invested in this industry is quite one hundred million rupees, or (in conventional exchange of ten rupees to the pound) 10 millions sterling.

REVIEW OF THE INDIAN TRADE IN COTTON AND COTTON MANUFACTURES.

TRADE. 262 (642)

In the table (p. 56) summarising the statistics regarding the production of, and trade in, raw cotton (for the year ending March 1889), India is shewn to have produced 9,219,493 cwt. of cotton. Of that amount 7,755,741 cwt. appear to have been exported from the provinces, the bulk of which naturally went to the ports, and constituted the supply from which the exports to foreign countries were derived, *viz.*, 5,331,904 cwt., which, deducted from the above outturn, would leave 3,887,589 cwt. as the amount available for local consumption in all India, *plus* 64,627 cwt. obtained from foreign countries. A critical examination, however, of the returns on every aspect of the cotton industry confirms the suspicion that either the area, 13,998,639 acres, is under the mark, or that the yield per acre is under-estimated. **Mr. Beaufort** gives the outturn for the year in question as having been 9,625,000 cwt., but it is probable that even that amount would not meet the total transactions shown in the trade statistics. It has already been pointed out, that there are at least two errors in all the returns of the cotton trade, *viz*,. the overlapping of the official year with the year of outturn, and the deficiency of statistics regarding road traffic and Native States. To these errors is doubtless largely due the fact that the net exports from the provinces are shown as having been less than the imports into the port towns by 700,000 cwt. The figures in the table alluded to, however, are relatively trustworthy. It will, therefore, be sufficient, in considering the trade of India, to accept the total outturn, the imports, exports, &c., there given as correct.

Conf. with pp. 54-55, 61.

In the annual statement of the trade and navigation of British India, for 1888-89, **Mr. J. E. O'Conor** gives the value of the exports to foreign countries as R15,04,56,476. Assuming that the amount retained in India was of equal proportionate value to the people as that which they sold for the foreign markets, the outturn may be valued at R26,01,75,009. It is, perhaps, not, strictly speaking, correct to add together the value of a raw with that of a manufactured material in estimating trade, but this may be allowed, since these items represent actual financial transactions. The total valuation of Indian-grown cotton, *plus* the value of the imports of raw and manufactured cotton, last year amounted to R57,66,93,851. The more important branches of the cotton trade were as follows:—

Value of the raw cotton exported, R15,04,56,476; estimated value of

TRADE.

Indian raw cotton used up in the country, R10,97,19,533; and the total value of imports of raw and manufactured cotton, R31,66,18,842. A considerable proportion of the cotton manufactured in India is, however, exported and has its value thus increased. The value of these exports amounted, in the year under consideration, to R6,37,45,636. But that amount is, after all, but the value of the exports (mostly mill manufactures), and takes no account of the cotton used up by hand-looms. From a consideration of all aspects of the question, it would seem safe to estimate the total value of Indian power and hand-loom manufactures at close on R18,00,00,000. But the raw cotton from which these manufactures are made has already been included in the figures given. While that is so the transactions connected with cultivation and manufacture are independent, and hence the value of the raw cotton employed, and the ultimate cost of the manufactures produced, may not incorrectly be added together; if so, the annual wholesale cotton transactions of India are probably close on £60,000,000. Carrying this analysis still further, by the acceptance of the above data, it may be said that India consumed last year approximately R10,00,00,000 of home manufactures and R31,00,00,000 of foreign goods.

Conf. with p 55.

The subject of the Indian cotton trade may, however, be dealt with in greater detail by reviewing the available information regarding A. INTERNAL TRANSACTIONS, B. EXTERNAL.

INTERNAL TRADE.

A. INTERNAL TRADE

Raw Cotton.

I. RAW COTTON.—The internal trade in raw cotton has been sufficiently reviewed in the provincial chapters and need not be further discussed.

Piece-goods & Yarns.

II. PIECE-GOODS AND YARNS.—European and Indian manufactures are treated separately in the "Report on the Inland Trade of India by rail and river," under the two heads of Twist and Yarn,' and 'Piece-goods, &c." As in the sea-borne trade so in the inland trade, English cotton manufactures exceed all other articles in commercial importance. The total value of the inland traffic in English goods, during 1888-89, was 2,547 lakhs of rupees, in Indian, 474 lakhs of rupees.

Twist & Yarn. (643) 263

TWIST AND YARN.—The total amount of European twist and yarn conveyed by rail and river during the year was 6,20,404 maunds, valued at R3,60,68,585, of Indian, 2,91,359 maunds, valued at R94,37,843.

A clear view of the importance of each block as an importer of twist and yarn may be briefly conveyed by the following figures extracted from Mr. Tucker's "*Review of Inland Trade*" for 1888-89:—

Net Imports of Twist and Yarn.

	European.	Indian.	Total.	Supplied from
	Rx.	Rx.	Rx.	
Madras	90	27	117	Madras and Bombay.
Bengal	85	31	116	Bombay.
Bombay	83	8	91	Bombay.
North-West Provinces and Oudh	32	12	44	Bengal.
Panjáb	28	3	31	N.-W. P. and Bengal.
Central Provinces	19	11	30	Bombay.
Nizam's Territory	11	4	15	Bombay.
Mysore	6	5	11	Bombay.
Sind	8	1	9	Bombay.
Assam	9	...	9	Bengal.
Berar	16	1	7	Bombay.
Rajputana and Central India	4	2	6	Bombay.

Rx.=lakhs of rupees (1,00,000.)

TRADE—INTERNAL.

Twist and Yarn.

"It must not be supposed," continues the author of the *Review*, "that the figures in the third column represent the whole consumption of twist and yarn of Indian manufacture. No estimate can be given of the produce of native spindles, which is still considerable; but putting the value of the annual outturn of Indian mills at 515 lakhs of rupees, and deducting 356 lakhs as the export to foreign countries, the home consumption of Indian mill produce alone comes to 159 lakhs against a foreign import valued at 343 lakhs."

The coastwise trade in Indian twists and yarns has been remarkably steady during the past five years. In 1884-85 the imports amounted to 20,231,819lb, valued at R93,67,197, in 1888-89 to 20,505,130lb, value R85,20,195. The largest importer in the year under consideration was Bengal, with 9,691,143lb, followed by Madras with 4,308,051, Burma with 3,860,193, Bombay with 2,153,209, and Sind with 492,534lb. The largest exporter was Bombay, with 18,897,068lb, followed by Bengal with 2,428,851, Madras with 1,030,575lb, Burma and Sind with very unimportant amounts.

The coastwise imports of foreign twist and yarn have also remained steady, but are much less important than those of Indian. In the year under consideration they amounted to 10,637,127lb, valued at R63,13,272, of which Bengal imported 4,035,666lb, Madras, 3,972,109, Sind, 934,718, Burma, 891,520, and Bombay, 803,144lb. The largest exporter was Bengal with 6,723,474lb, followed by Bombay with 3,044,486, and Madras, Burma, and Sind with smaller amounts.

Owing to its short staple the bulk of Indian cotton is unfit for the production of the finer descriptions of yarn, and local manufactures are therefore, as a whole, considerably inferior in value to that of the European article. For the same reason European yarns are largely used by native looms throughout the country in the manufacture of the finer classes of cloth.

Piece-Good. 264 (644)

PIECE-GOODS.—Mr. Tucker (*Review of the Inland Trade*) writes: "A correct representation of this important item in the inland trade returns is attended with special difficulties. In the case of goods passing through the Custom House the value is declared, and the nature and volume of consignments in pieces, yards, bales or other approximate ascertained. Railway invoices, however, shew only weights in the gross without distinction of kinds, and the fixation depends upon local enquiries of a tedious and difficult nature, and upon estimates which must be largely conjectural." "To study, however, the correlation between imports by sea and the distribution of English piece-goods over the interior of the country, the inland values with all their imperfections must be accepted, for no comparison is possible between a trade represented on the one hand by pieces, yards, &c., and on the other, by maunds."

The total value of English piece-goods, &c, conveyed by rail and river in India in 1888-89 was R20,26,99,428, of Indian piece-goods, R1,88,64,912. The proportion received by each province may be shortly represented by the following table:—

Net Imports of Cotton piece-goods

	European.	Indian.	TOTAL.
	Rx.	Rx.	Rx.
North-West Provinces and Oudh	399	—17	382
Bengal	836	—2	834

Rx. = lakhs of rupees (1,00,000).
— indicates net exports.

GOSSYPIUM. **Review of the Indian Trade in**

TRADE—INTERNAL.

Piece-goods.

Net Imports of Cotton piece-goods—contd.

	European.	Indian.	TOTAL.
	Rx.	Rx.	Rx.
Madras	138	—30	108
Panjáb	281	11	292
Bombay	91	—37	54
Central Provinces	50½	—10	40½
Nizam's Territory	18	4	22
Rajputana and Central India	84	31	115
Berar	27½	12	39½
Mysore	26	10	36
Assam	64¼	2¼	66½
Sind	43	34	77

Rx. = lakhs of rupees (1,00,000);
— indicates net exports.

By far the largest exporting centre of European goods was Calcutta, followed by Bombay, Karáchi, and the Madras seaports, named in order of importance. Of Indian goods Bombay, Karáchi, the North-West Provinces and Oudh, and Madras exported the largest quantities.

The coastwise transactions in piece-goods are usually large. The imports of Indian goods last year, for example, amounted in value to R71,03,138, of European to R2,16,63,818. The largest exporting town of the former was Bombay, which shipped goods to the value of R57,59,553, and was followed by the Madras seaports, the exports of which valued R15,81,717. Of European goods Bombay stood first, with a trade amounting to R1,66,29,594 in value, followed by Bengal with exports equal to R71,35,319.

Though, as already stated, the figures in the above statement of rail returns may not be absolutely correct, for purposes of comparison with the values of imports from foreign countries, the broad results indicated by the statistics of internal trade are substantially accurate. Thus they indicate very forcibly what is evident from the returns of foreign trade, *viz.*, the pre-eminence of Bengal as a consumer and distributor of English piece-goods. **Mr. Tucker** writes of the rail-borne trade: "Calcutta, which is one of the largest emporiums for piece-goods in the world, supplies the requirements of the vast population of Bengal, the North-West Provinces and Oudh, the Panjáb, and Assam. It exports 13,78 lakhs of rupees against 6,87 lakhs from the other provinces combined; but the disproportion between these figures is, in some measure, caused by the higher value rates applied in Bengal to the traffic in European piece-goods. Bombay town supplies its presidency, the Central Provinces, Berar, Rajputana and Central India, and Hyderabad, and to a less extent, the North-West Provinces and Oudh, and the Panjáb; while the Madras Presidency and Mysore have their demands met through the Madras seaports."

As far as can be gathered, the demand for European grey goods is much keener on the Calcutta than on the Bombay side; in bleached cloths and coloured prints it is about equal; while in handkerchiefs, shawls, &c., Bombay does the largest trade. Judging by the returns, the competition of Indian with European manufacture is much less pronounced in the case of piece-goods than in that of twist and yarn. The inland trade in country piece-goods is but one-tenth that of imported stuffs. But it must be remembered that, while the latter must necessarily all enter into the trade returns, a large local cloth manufacture exists in many parts of the country,

Cotton and Cotton Manufactures. (*Watt & Murray.*) **GOSSYPIUM.**

TRADE—INTERNAL. **Piece-Goods.**

the produce of which is consnmed locally or, at best, enters to a limited extent only into the statistics of inland trade. It may be safely stated that nearly all, if not the whole, of the Indian piece-goods, distributed by rail, is the produce of power-looms, and that the large outturn of hand-looms is nearly altogether unrepresented.

Trans-frontier Trade. 265 (645)

A considerable trans-frontier trade exists, the greater part of which is represented in the rail trade of frontier provinces. In 1888-89, the total value of imports of piece goods and yarns from trans-frontier countries amounted to R1,95,306; while the exports were valued R2,19,99,332. The largest receiving countries were Upper Burma, Kábul, Kashmír, Nepál, and the regions tapped by the Sind-Pishin Railway.

B. Foreign Trade.

FOREIGN TRADE. Raw Cotton. 266 (646)

I. Raw Cotton.—For an account of the early history of the Indian cotton trade the reader is referred to the chapter "History." (*pp.* 42—52.)

Imports. 267 (647)

Imports.—These amounted last year to 64,627 cwt., valued at R15,05,791, of which Persia supplied 54,013 cwt., Mekran and Sonmiáni, 6,000, Arabia, 1,477, and Turkey in Asia, 1,273; the remainder came from Egypt, the United Kingdom, and Japan in the order mentioned. The bulk was received by Bombay, which re-exported 45 cwt. to Persia.

In 1873, the import trade was even smaller and less important than it is now, the imports in that year amounting to only 14,500 cwt. In 1877-78, however, they had increased to 54,200 cwt., since which date they have preserved an average of about 50,200 cwt.

Exports. 268 (648)

Exports.—As already stated, the cotton export trade of India has remained fairly steady since the date of the American-war, before which time it presented fluctuations of the most extraordinary character.

Notwithstanding the general steadiness of the trade, however, progress has not been marked during the past fifteen or twenty years, a fact which is probably due, at least in part, to the largely increased demand of the Indian power mills, which, in 1888-89, are estimated by **Mr. Beaufort** to have consumed 3,111,024 cwt., as against an average of 1,089,735 cwt. in the five years ending 1881-82.

The increase in the trade during the past fifty years may be shortly shown by the following tabular statement:—

Quinquennial Averages of Exports of Cotton.

Quinquennial periods.	Weight in cwts.	Value in lakhs of Rupees.
(1) 1834-35 to 1838-39	1,191,345	205·8
(2) 1839-40 " 1843-44	1,574,751	234·6
(3) 1844-45 " *1855-56	?	?
(4) *1856-57 " †1861-62	2,858,427	631·5
(5) †1862-63 " 1866-67	4,959,959	2,885·2
(6) 1867-68 " 1871-72	5,809,586	2,001·0
(7) 1872-73 " ‡1876-77	4,816,023	1,350·3
(8) ‡1877-78 " 1881-82	4,108,521	1,132·3
(9) 1882-83 " 1886-87	5,367,237	1,352·2
1887-88 actual year	5,374,542	1,441·2
1888-89 ditto	5,331,536	1,504·6

The trade diminished considerably during periods (7) and (8) on the average of period (6), but recovered itself in period (9), and still more so during the last two years

* Indian Mutiny. † American War of Secession. ‡ Famine years.

TRADE—FOREIGN.

Raw Cotton —Exports.

The year of maximum exports was 1871-72, when 7,225,411 cwt., value 2,127·2 lakhs, was exported; but in 1864-65, though much less, *viz.*, 4,687,972 cwt., the highest value was attained, amounting to 3,757·3 lakhs. In the year under consideration the exports and mill consumption combined reached a maximum of 8,442,192 cwt.

The exports of the last two years were made chiefly to the countries noted below:—

COUNTRY.	Cwts. (000's omitted).	
	1887-88.	1888-89.
United Kingdom	2,140	1,776
Italy	774	781
Austria	688	755
Belgium	676	873
France	487	549
China	202	139
Germany	139	190
Russia	134	150

In consulting the above figures, however, it should be remembered that a large proportion (about three-fourths in 1888) of the cotton sent from India to England is reshipped thence to the Continent, which, in fact, consumes about fourth-fifths of our exports.

During the past five years, the supplies to the United Kingdom have been decreasing, while those to Austria and Belgium have shown a tendency to increase, owing, probably, to the fact that these countries are beginning to draw supplies more freely direct from India, than at second-hand from the United Kingdom. Since the opening of the Suez Canal, the opportunities thereby given for direct shipment to the continent, have caused England to steadily lose her position as the depôt from which the rest of Europe draws its requirements of Indian cotton. Part of the cotton sent to Italy is, it appears, destined ultimately for transport to South Germany, but the greater part is worked up in the cotton mills of Lombardy and elsewhere in Italy; a part of this cotton returns to India as Turkey red twist (see *Sea-borne Trade Review, 1877-78, 14*).

The reasons why Continental Europe takes so much larger a share of Indian cotton than England are explained, as follows, in Mr. O'Conor's *Seaborne Trade Review for 1874-75*:—

"In continental Europe Indian cotton is regarded with more favour than in the English manufacturing districts. English machinery was originally adapted to the treatment of cotton of longer staple than the Indian fibre, and as long as manufacturers can obtain a supply sufficient for their purposes of the longer-stapled American, Egyptian, or Brazilian cotton (of which not more than 9 or 10 per cent. of the imports into England are re-exported), the Indian fibre is regarded with disfavour. In some cases indeed, cotton operatives have refused to work with Indian cotton as being too troublesome to manage. This is not the case on the Continent, where mills exist possessing machinery specially adapted to the treatment of short, stapled cotton, and in these quarters there seems to be a growing demand for our staple."

The falling off in raw cotton exports to China within the past fifty years is very marked. In the thirteen years 1834 to 1846, the exports from Bombay alone averaged 471,590 cwt., in comparison with 533,421 cwt. to the United Kingdom; and, as late as 1860, over 700,000 cwt. went from

TRADE—FOREIGN.

Raw Cotton—Exports.

the same port to China. During the American War of 1863-64, however, not only was Surat cotton exported to America, but Bombay imported cotton from China. The erection of spinning and weaving mills after this date has gradually lessened the exports to China, and in the year under consideration they have fallen to only 139,000 cwts.

The shares taken by the seaports of India in the exports of the last two years have been as follows:—

	Cwt. (000's omitted).	
	1887-88.	1888-89.
Bombay	3,929	4,198
Madras	658	595
Bengal	618	375
Sind	131	130
Burma	37	32

The principal ports to which the exports from Bombay were consigned in 1888-89, named in order of importance, were:—Liverpool, Antwerp, Trieste, Genoa, Havre, London, Venice, Hamburg, and Odessa. From Coconada and Masulipatam nearly all went to Havre and London; from Madras town nearly the whole used to be taken by London which still secures about four-fifths, but the trade with Italy and Austria is steadily growing; from Tuticorin about four-eighths of the whole has recently annually gone to Liverpool, while smaller quantities are exported to London, Venice, Barcelona, and Antwerp, named in order of importance. The most important receiving port of cotton from Calcutta is London, though, in the past two years, Hamburg has been rapidly pushing to the front. Genoa, Venice, Antwerp, and Trieste follow in the order named. A little of the cotton exported from Calcutta still goes *viâ* the Cape, while from all other ports it goes entirely by the Suez Canal. About equal quantities of the exports from Karáchi go to Great Britain and the Continent. London is the principal importing port, Antwerp being next, followed by Marseilles. A small and, apparently, decreasing quantity is imported by China. Most of the exports from Burma are shipped to China either direct or through the Straits Settlements.

Twist and Yarn—Imports. 269 (649)

II. TWIST AND YARN.—IMPORTS.—The total imports of twist and yarn, in the year under consideration, was 52,587,181 ℔, valued at 374·68 lakhs of rupees. The trade is a steadily increasing one, notwithstanding the competition of Indian yarns, as will be seen from the table below:—

Quinquennial averages of Imports of Twist and Yarn.

	℔	R
1871-72 to 1875-76	31,934,495	2,72,68,650
1876-77 to 1880-81	36,339,902	2,96,16,344
1881-82 to 1885-86	44,542,928	3,31,97,401
1886-87	49,013,979	3,31,83,769
1887-88	51,542,549	3,58,19,061
1888-89	52,587,181	3,74,67,969

The above figures shew that, while the quantity imported has largely increased during the past twenty years, the value has been augmented in a smaller proportion, indicating a considerable reduction in prices.

TRADE—FOREIGN.

Twist and Yarn—Imports.

Up to 1879 all cotton twist and yarn was subject to a duty, but, in March of that year all twist of No. 32 mule, and No. 20 water, and lower numbers was declared free of duty. This was followed in 1880-81 by a large increase in the quantity imported, regarding which the Commissioner of Customs in Bombay wrote:—"The largest proportionate increase is in mule 32, importations of which have more than trebled, the reason, apparently, being now that this number can be imported free, the local mills, which formerly turned out considerable quantities, find it more profitable to confine themselves to 20, in which number they can readily sell as much as they can make, and have, accordingly, abandoned the market in 30 to importers."

The Tariff Act of March 10th, 1882, totally abolished the customs duty on all cotton goods with that on others, but the description of yarn and twist continued of qualities or counts superior to those made in the Indian mills. Thus, in the report for the year ending March 1884, **Mr. O'Conor** wrote: "Very little twist below No. 33 was imported, the lower qualities having been, to a great extent, displaced by Indian-made twist." In 1885 he remarks on the same subject, "Even the medium kinds are now diminishing, an indication that the Indian mills are beginning to make them too, and that hand loom weaving in India continues to decline as the spread of communication cheapens the cost of cloth in districts formerly unable to buy cloth, by reason of cost and difficulty of transport." In 1888 it is reported that the supply of even No. 40 was diminishing, but that the defect in that number was made up by the increase in that of other 'counts,' more especially in those higher than 40.

Nearly the whole of the imports come from the United Kingdom which, in the year under consideration, supplied 51,891,975℔ out of the whole. The next in importance was Austria with 299,148℔, followed by Italy with 146,500, Belgium with 103,085, and other countries with unimportant quantities.

The shares received by each of the importing provinces and presidencies in the last two years were as follows:—

	Thousands of ℔	
	1887-88.	1888-89.
Bengal	15,453	15,056
Bombay	15,374	15,005
Sind	1,676	2,683
Madras	13,758	15,813
Burma	5,281	4,029

The re-exports are unimportant. During the past five years those by sea have been as follows:—*

	℔	R
1884-85	958,419	6,55,160
1885-86	1,082,610	6,63,032
1886-87	1,319,248	8,11,466
1887-88	1,058,963	6,93,448
1888-89	1,601,596	11,15,140

* Imports of Foreign goods, Trans-frontier, have been included in Trans-frontier trade, see p. 165.

G. 269

(649)

Of the total amount in the year under consideration 946,394℔ went to Persia, 285,995℔ to Aden, 165,275℔ to Arabia, and small quantities to several other countries. The shares taken by each of the principal ports in the re-export trade were as follows:—Bengal, 29,632℔; Bombay, 1,523,301℔; Sind, 1,181℔; Madras, 34,627℔; Burma, 12,855℔.

TRADE—FOREIGN.

Twist and Yarn—

Exports.
270 (650)

EXPORTS.—The exports of cotton yarn and twist in 1888-89 amounted to 128,906,764℔, valued at R5,20,70,996. They have increased 158 per cent. in the last six years, that is, at a rate of more than 26 per cent. annually. The increase during the past twenty years may be shown by the following table:—

Quinquennial averages of Exports of Twist and Yarn.

	℔	R
1871-72 to 1875-76	4,370,329	20,96,007
1876-77 to 1880-81	19,524,665	86,55,305
1881-82 to 1885-86	54,035,981	2,06,16,340
1886-87	91,804,244	3,33,68,608
1887-88	113,451,375	4,07,73,865
1888-89	128,906,764	5,20,70,996

By far the largest proportion of the exports goes to China, as may be seen by the following table showing the distribution in the last two years:—

	℔ (000's omitted).	
	1887-88.	1888-89.
China	92,571	101,248
Japan	17,391	23,143
Aden	1,354	1,327
Straits	981	1,844
Asiatic Turkey	365	503
Java	283	327
Arabia	279	261
Persia	79	139
East Coast of Africa	66	64
Ceylon	32	26
Other countries	50	24

The export trade is virtually a monopoly of Bombay, but the other seaboard presidencies and provinces also contribute small amounts, as is shewn by the following statement of the shares taken by each during the last two years:—

	℔ (000's omitted).	
	1887-88.	1888-89.
Bombay	108,516	123,023
Bengal	3,780	4,351
Madras	1,132	1,493
Burma	20	38
Sind	3	1

The bulk of the yarn (about 82 per cent.) consists of No. 20, the rest is of lower count.

TRADE—FOREIGN.

Other Cotton Manufactures—Imports.

271

III. COTTON MANUFACTURES (EXCLUDING TWIST AND YARN.)—IMPORTS.—The imports of cotton manufactures show on the whole a steady though not very rapid increase, notwithstanding the yearly augmentation of Indian power-loom manufactured cloths. This is, probably, accounted for, partly by the increase of population, partly by the gradual ousting of the old hand-loom manufactures owing to the increase in popularity of the attractive, though less durable, goods of Manchester.

The average imports during the past twenty-five years have been as follows:—

Quinquennial averages of imports of Cotton Manufactures (excluding Twist and Yarn).

	Value (Rupees).
1866-67 to 1870-71	14,78,63,641
1871-72 to 1875-76	15,53,54,760
1876-77 to 1880-81	17,45,54,086
1881-82 to 1885-86	21,23,08,636
1886-87	25,84,65,080
1887-88	23,92,44,676
1888-89	27,76,45,082

In 1879 all grey piece-goods made of 30 s. yarn and lower counts were freed from import duty. As a consequence it was estimated that in the first year loss was sustained to revenue of 21 lakhs of rupees, and the import of medium and finer cloths almost entirely ceased. The Tariff Act of March 10th, 1882, entirely abolished all duty on cotton manufactures, and in that year we find the following remarks made by the Collector of Customs, Calcutta:—" It is very probable that the total abolition of the duty on cotton goods, along with the other duties on general imports, will have the effect of causing the trade to revert to the old descriptions of cloths made from finer yarns, though it is possible that the demand having been once created for the better class of cloths made of 30s. yarn, it may be found profitable, in certain cases, to continue to supply them. Generally speaking, however, it may be anticipated that shirtings and cloths of a similar description, which are the principal kinds that the partial remission of duty affected, will be manufactured as of old of 32s. and 50s. Whether this will be an advantage to the consumer, notwithstanding the greater cheapness, is a matter of opinion. Certain it is that, with the cloth made of the coarser yarn, he obtained a more durable article, and though this is a great advantage, it is doubtful if he will pay the higher price in comparison, when he can obtain a cloth of nice appearance and which, in many cases, may serve his purpose quite as well, at a lower price." In the Review of Trade for the year ending March 1883, **Mr. O'Conor** remarked that a reversion to the higher count classes had occurred, but not so generally as had been anticipated.

ANALYSIS OF IMPORTS.—The increase in the several items included under the heading of Cotton Manufactures, during the period for which trade returns are available, may be shown in tabular form:—

TRADE—FOREIGN.

Other Cotton Manufactures —Imports.

Table shewing increase in the various articles comprised under Imports of Cotton Manufactures.

YEAR.	Piece-goods, grey (unbleached).	Piece-goods, white (bleached).	Piece-goods, coloured, printed, and dyed.	Handkerchiefs and shawls in the piece.	Piece-goods, (other sorts).	Lace and Patent net.	Hosiery, pure and mixed.	Thread (sewing).	Canvas.	Other sorts.	TOTAL.
	R	R	R	R	R	R	R	R	R	R	R
1866-67* .	7,85,78,567	2,74,99,430	2,57,69,940	...	45,211	...	...	4,39,449	3,94,923	...	13,27,27,410
1875-76† .	10,63,97,122	2,68,77,064	2,83,72,506	9,16,754	31,169	4,18,962	4,57,176	9,99,536	4,405	27,427	16,45,02,121
1888-89‡ .	15,96,24,764	5,57,44,508	5,62,31,817	21,46,801	1,51,044	5,86,869	8,71,713	13,96,996	27,237	8,63,333	27,76,45,082

* First year of statistics. † First year of complete returns. ‡ Last year of available statistics.

TRADE—FOREIGN.

Other Cotton Manufactures—Imports.

It will be seen that nearly every article shows a steady and fairly proportional increase, with the exception of canvas, the imports of which have markedly diminished in value.

The United Kingdom virtually possesses the monopoly of this extensive trade. Thus, in the year under consideration, imports to the value of R27,34,92,942 came from that source, leaving only R41,52,140 to be accounted for by other countries. Of this small amount, Italy supplied goods to the value of R9,44,910; Austria, R8,37,331; the Straits Settlements, R7,25,928; the United States, R5,87,745; France, R5,33,167; and Germany, R1,83,561. Other countries supplied small amounts valuing less than one lakh.

In 1887-88, the amount derived from other countries than the United Kingdom, showed a marked correspondence in value with that of 1888-89, amounting to R40,05,262. In that year, however, Austria supplied much the largest quantity, while the value of the imports from the United States valued R1,41,134 only.

The shares taken in the imports, by each of the seaboard provinces, during the past two years were as follows:—

	1887-88.	1888-89.
	R	R
Bengal	12,62,81,551	14,54,77,039
Bombay	7,25,74,360	8,46,25,998
Madras	1,80,58,165	2,17,01,711
Burma	1,30,17,641	86,31,259
Sind	95,12,959	1,72,09,075

On analysing the above figures it will be found that, while Bengal receives nearly two-thirds of the imported grey goods, it imports less than half of the white goods, and a still smaller proportion of the coloured goods. The people of Bengal proper and of Assam and the bulk of the people of the North-West Provinces do not wear coloured clothing, while, on the other hand, the people of Western and Southern India are very partial to coloured cloths. The Burmese take a comparatively large quantity of white and coloured cloth to the proportion of grey, though the imports of the latter in these provinces have lately been increasing.

Re-exports.—The re-exports by sea are comparatively unimportant.* In 1888-89 they amounted to only R1,70,51,674, of which goods to the value of R68,54,523 went to Persia; R19,46,861 to the Straits Settlements; R17,10,702 to Turkey in Asia; R15,95,418 to Arabia; R12,70,548 to Zanzibar; R9,35,029 to Mauritius; R9,82,527 to Aden; R6,51,490 to Mozambique; R5,36,588 to Ceylon; R1,72,864 to Makran and Sonmiáni; and R1,43,683 to Natal. Smaller quantities, valuing less than one lakh, were despatched to other countries.

Nearly the entire trade is confined to Bombay, which, in the year under consideration, exported goods to the value of R1,63,31,313 out of the whole. Bengal and Sind exported to the amount of 2½ lakhs each, Madras to the value of 1½ lakh, and Burma an insignificant amount.

* Exports of Foreign Goods, by Trans-frontier, are included in Trans-frontier trade, see p. 165.

TRADE—FOREIGN.

Other Cotton Manufactures—Exports.

Exports.—The exports of all Indian cotton manufactures, exclusive of twist and yarns, in the year under consideration, valued R1,16,74,640. This amount was made up as follows:—

Nature of Goods.	lb	Yards.	R
I.—Piece-goods—			
1. Grey (unbleached) . .	...	54,938,813	64,66,979
2. White (bleached) . .	...	116,355	46,504
3. Coloured, printed, or dyed	...	15,189,059	43,18,741
4. Handkerchiefs and shawls in the piece.	...	1,073,077	5,23,515
5. Other sorts of piece-goods	...	21,122	15,888
II.—Hosiery, pure and mixed . .	...	...	19,036
III.—Thread, sewing . .	50,739	...	39,732
IV.—Canvas . . .	...	1,856	833
V.—Other sorts of manufacture . .	1,851,855	...	2,43,412
Total	...	...	1,16,74,640

The trade has been rapidly progressive within the last twenty years, as might naturally be expected from the increase in the number of mills. Thus, in the quinquennial period ending 1875-76, the value amounted to R34,99,927; in that ending 1880-81 to R51,12,644; in the five years ending 1885-86 to R82,14,879; in 1886-87 to R94,56,437; in 1887-88 to R1,15,05,419; and in 1888-89 to R1,16,74,640. Substantial as the progress has been, it is far short of the advance made in the export of yarns. Mr. O'Conor remarks on this point: "As long as the Chinese and Japanese maintain their preference for weaving their cloths themselves from imported yarn, instead of importing it ready-made, the export of our yarns must necessarily augment much faster than that of piece-goods."

The chief importers in the past year were:—Aden, with a total amounting in value to R20,61,942; China with R19,93,595; the Straits Settlements with R17,37,944; Ceylon with R16,99,511; Zanzibar with R11,79,446; Mozambique with R8,37,220; the United Kingdom with R6,19,520; Arabia with R5,27,202; Persia with R3,19,792; Turkey in Asia with R2,49,714; Mekran and Sonmiáni with R1,35,971; and Abyssinia with R1,13,104. Several other countries imported smaller quantities valuing less than one lakh of rupees.

The exports were distributed as follows, among the seaboard presidencies and provinces, during the last two years:—

	1887-88.	1888-89.
	R	R
Bombay	69,68,896	75,47,780
Madras	42,22,203	38,54,491
Bengal	1,69,443	1,41,403
Sind	1,44,334	1,30,562
Burma	543	404

Bombay thus possesses more than 64 per cent. of the trade in woven goods. It should be mentioned, however, that a considerable proportion of the exports from Madras consists of cloths dyed in that presidency, the pieces having been woven in Madras, in Bombay, or in England—that, in fact, Bombay is the centre of cotton-spinning and weaving industry in India.

TRADE—FOREIGN. Other Cotton Manufactures—Exports.

(See *Trade Review for 1888-89.*) The following remarks from **Mr. Beaufort's** *Indian Cotton Statistics* regarding the progress made by, and the distribution of, the chief cotton manufactures of Bombay may, therefore, be of interest:—"The demand for T. cloths and Sheetings, especially in the past three years, has increased considerably. Longcloths, too, have been in more favour in the past two years, but Domestics show no expansion. 'Drills' and 'Jeans' have fallen off largely. 'Dhoties' and 'Chudders' find but few buyers out of the country, but the total trade in the last three years has been growing. The increase in T. cloth is purely to China, the largest consumer of this description of cloth, Mozambique and Zanzibar being the next largest buyers. Of 'Sheetings' Zanzibar takes nearly half the exports, Mozambique being the next important outlet. 'Longcloths' go, as a rule, principally to Aden, for re-export; but last year China took the lead, whereas but two years ago her takings were so small as to be included under 'other ports.' Aden takes nearly all the 'Domestics.'"

GRACILARIA, *Grev.; Agardh., Sp. Alg., II., 584.*

A genus which comprises from 20 to 30 species, natives of the warmer seas of various parts of the world, and characterised, as its name implies, by the slenderness of its filaments.

(652) 272 **Gracilaria confervoides,** *Greville, Algæ Britann., 54;* ALGÆ.

Syn.—FUCUS ECHELIS, *Gmel.*

(653) 273 **G. lichenoides,** *Greville, Algæ, Britann., 54; Turner, Fuci, t. 118.*

Syn.—FUCUS AMYLACEUS, *O'Sh.;* F. LICHENOIDES, *Linn.;* F. GELATINOSUS, *Koeing;* SPHÆROCOCCUS LICHENOIDES, *Agardh* (not of *Grev.*); PLOCARIA CANDIDA, *Nees.*

These two species constitute the commercial product known as CEYLON MOSS, JAFFNA MOSS or AGAR AGAR, and may, for the purposes of this work, be considered conjointly.

Vern.—*Chinai-ghás,* BOMB.; *Dárya-ki-páchi,* DEC.; *Kadal-pách-chi,* TAM.; *Samudrapu-páchi,* TEL.; *Kujáv-pœn,* BURM.

References.—*Pereira's Mat. Med., II., Pt. I., 14; Pharm. Ind., 260, 465; O'Shaughnessy, Beng. Dispens., 668; Moodeen Sheriff, Supp. Pharm. Ind., 150; Dymock, Mat. Med. W. Ind., 2nd ed., 872; Fluck. & Hanb., Pharmacog., 749; Bent, & Trim., Med. Pl., 306; Mason, Burma and Its People, 507, 832; Balfour, Cyclop., I., 1154, 1242.*

Habitat.—Found on the coasts of the Indian Ocean, especially in Ceylon; extending to Burma, and the Malay Archipelago.

MEDICINE. Sea-weeds. (654) 274

Medicine.—The dried SEA-WEEDS have long been employed as a medicine in the localities in which they occur, and form an article of some considerable trade with China. The plants are chiefly collected during the south-west monsoon, when they become separated by the agitation of the water, and thrown on shore. They are dried in the sun, on mats, for two or three days, then washed several times in fresh water, and again exposed to the sun. By this means they are bleached, and become fit for use, or for exportation.

Ceylon moss is considered by the natives of Southern India and Ceylon to be nutritive, emollient, demulcent, and alterative, and especially valuable in pectoral affections. It has been described by many learned writers, and was for the first time submitted to a careful chemical examination by **O'Shaughnessy.** It has obtained a place amongst the officinal drugs of the Indian Pharmacopœia, but though occasionally employed in Europe and America, it is neither officinal in the Pharmacopœia of Great Britain, nor in the U. S. Dispensatory.

Composition (655) 275

GENERAL CHARACTERS AND COMPOSITION.—Officinal Gracilaria consists of light yellow or faintly purple filaments, from one to several inches in

MEDICINE.

length and varying in thickness from the size of a crowquill to that of ordinary thread. It has a feebly saline taste, a flavour of sea-weed, and a cartilagenous consistence.

Chemistry. 276 (656)

CHEMICAL CHARACTER.—Ceylon moss as examined by **O'Shaughnessy** yielded :—vegetable jelly 54·5 per cent., starch 15 per cent., cellulose 18 per cent., gum 4 per cent., and inorganic salts 7·5 per cent. The authors of the *Pharmacographia* write: "Some chemists have regarded the jelly extracted by boiling water as identical with pectin, but this statement requires proof. **Payne** called it *gelose*, and found it composed of carbon 42·77, hydrozen 5·77, and oxygen 51·45 per cent." The salts according to **O'Shaughnessy** consist of sulphates, phosphates, and chlorides, with neither bromide nor iodide.

Properties and Uses. 277 (657)

MEDICAL PROPERTIES AND USES.—In its properties it resembles Carragern moss, being emollient, demulcent, and nutritive. It may be used in the form of a decoction or jelly, as a light and readily digestible food for invalids and children, as an emollient in intestinal catarrh and irritation, and as a demulcent in pulmonary complaints and irritation of the throat. The jelly or decoction may be made more palatable by the addition of milk, wine or aromatics.

FOOD. 278 (658)

Food.—In the Indian Archipelago and China large quantities are employed in place of isinglass for the preparation of jellies, and, as already stated, considerable quantities are exported from India to China.

INDUSTRIAL USES. 279 (659)

Industrial Uses.—Occasionally employed in England as a jelly for dressing silk.

Grains, see **Cereals**, Vol. II., 257.

Grains of Paradise, see Greater Cardamom (**Amomum subulatum**, *Roxb.*); Vol. I., 222; and Lesser Cardamom (**Elettaria Cardamomum**, *Maton*): Vol. III., 227.

Gram, see **Cicer arietinum**, *Linn.*, Vol. II., 274.

Gram, Horse, see **Dolichos biflorus**, *Linn.*, Vol. III., 175.

GRANGEA, *Adans., Gen. Pl., II., 261.*

280 (660)

Grangea maderaspatana, *Poir., Fl. Br. Ind., III., 247; Wight, Ic. t., 1097;* COMPOSITÆ.

Syn.—ARTEMISIA MADERASPATANA, *Roxb.*; COTULA MADERASPATANA, *Willd.*; C. SPHÆRANTHUS, *Link.*; GRANGEA SPHÆRANTHUS, *Koch.*

Vern.—*Mustarú*, HIND.; *Namuti*, BENG.; *Máshipatri*, TAM.; *Savé*, TEL.; *Dovana*, KAN.; *Nelam-pata, nelampala*, MALAY.; *Wæl kolondu*, SING.; *Afsantín*, ARAB.; *Baranjásif kowhí*, PERS.

References.—*Roxb., Fl. Ind., Ed. C.B.C., 600; Dalz. & Gibs., Bomb. Fl., 124; Ainslie, Mat. Ind., I., 481; Dymock, Mat. Med. W. Ind., 2nd Ed., 434; S. Arjun, Bomb. Drugs, 78; Murray, Pl. and Drugs, Sind, 182; Atkinson, Him. Dist., 311; Gazetteers:—N.-W. P., I., 81; IV., lxxiii; Bomb., XV., 436.*

Habitat.—A common weed found throughout India, from the Panjáb eastwards and southwards. Distributed through Tropical and Sub-tropical Asia and Africa.

MEDICINE. Leaves. 281 (661)

Medicine.—The medicinal properties of this plant were first brought to the notice of Europeans by Ainslie, who wrote: "The Tamool doctors consider its LEAVES to be a valuable stomachic medicine; they also suppose them to have deobstruent and antispasmodic properties, and prescribe them in infusion and electuary in cases of obstructed menses and hysteria; they sometimes, too, use them in preparing antiseptic and anodyne fomentations." Apparently they are similarly employed to the present day, and are especially valued for their anodyne properties.

MEDICINE.

Special Opinion.—§ The juice of the leaves is employed as an instillation for earache. (*Honorary Surgeon P. Kinsley, Chicacole, Ganjam, Madras.*)

(662) 282 **GRANITE,** *Ball, in Man. Geology of India, III., 534.*

The following note has been kindly furnished by **Mr. H. B. Medlicott,** late Director of the Geological Survey.

Granite, *Eng.;* GRANITE, *Fr.;* GRANIT, *Ger.;* GRANITO, *Ital.*

Granite has as yet, economically considered, been only specialised as such in European countries and their off-shoots, and even there in many cases, the rock is not in reality a granite but some form of massive crystalline igneous or metamorphic rock. In India the term is even more general, and all crystalline rocks, as long as they are tolerably compact, granular, crystalline, and speckled in colour, are commonly called granites. Under such a definition, this rock might be set down as of universal occurrence in India; except in the Bombay Presidency and the Deccan, or in the flat alluvial districts. Keeping, however, to the proper idea of granite, that is, a crystalline igneous rock principally composed of quartz, felspar, and mica, either as a ternary compound of these minerals or as a binary compound of two of them, the occurrence of the rock becomes, for industrial purposes, very restricted. Without running the distinction too close, it may be well, on the economic view, to notice localities for stone which, for all practical purposes, and without infringing very much on the geological definition of the term, may be considered granite. With this proviso, the rock can be expected to occur only in the areas of crystalline rocks, which are, *par excellence*, the Himálayan region, in and the southern and eastern tracts of Peninsular India. In the Himálayan district the distinction of granite from gneiss is still a vexed question, but there seems to be no doubt that the rock occurs in the Central Himálaya in what is called the granitic axis of these mountains. To the east in Sikkim, and in the north-west, from the frontier of Nepal to Kulu, wherever examined, coarse white granite has been found in profusion along the line of peaks, near the present edge of the sedimentary basin of Tibet. It occurs in veins and dykes of every size, sometimes forming the massive core, up to the summit, of the highest mountains. The rock dies out completely to the west, being only very feebly represented at the Báraláchai pass. In the Simla regions many granitic intrusions are known, the principal being that of the Chor Mountain. Albite granite occurs in the gneiss of the Hamta pass, in the Pir Panjál chain. In the Panjáb hills west of the Jhelum, a prophyritic granitoid rock occurs; also in the Pakli Valley, Súsúlgali, Agror, &c., in northern and north-eastern Hazara.

In the lower Himálayas, granites occur here and there.

In the peninsular part of India, the range of granites may be considered as conterminous with the crystalline or gneissic area, though there are only comparatively few tracts in this great area which can be specialized as industrially granitiferous.

There is little or no information regarding the gneiss in Hyderabad, except that in the eastern half of the state a very massive form of granitoid gneiss is common and well seen in the large isolated humpy hills occasionally rising out of that part of the country. Granite veins are rare. Similar forms of granitoid gneiss are found on the south-western side towards the South Mahratta country. In Wainaad, there are some large exposures of granite in the Munny and Yedokal hills. Granite veins, but of an excessively coarse texture, are common in the Coimbatore

GRANITE.

and Salem districts, in the low country, especially in the Sunkerry Drug; but there is no intruded granite mass of large dimensions. A strong band of very coarse granite veins occurs in the Trichinopoly district on the north or left bank of the Cauvery river. In South Arcot, to the east of the Salem hill-groups, a considerable area is occupied by rocks having a very granitic aspect. In the eastern portions of the Bellary district, about Guti; and again in Dhone of the Karnúl district, an exceedingly compact close-grained form of pale, pink, and whitish rock occurs, which may be considered granite. In the Nellore district, about Gudur, a very coarse and typical granite occurs in force.

Among the gneissic rocks of Burma veins of granite are common, but little is known of them in detail.

All these granites with, perhaps, the exception of the Himálayan forms, appear to be worked only to a small extent, and to be exclusively used for local purposes; but there is no doubt that in certain places they were the favourite stone for the more ancient temple buildings, especially in the way of monolithic structures. Even in the future it is questionable whether they will come into greater demand, as long as the easier-worked gneisses with which they are associated, can be obtained close at hand. For further economic information see STONES, BUILDING, Vol. VI.

Grapes, see **Vitis vinifera,** *Linn.*; AMPELIDEÆ; also **List of Indian Fruits,** [Vol. III., 450.

Graphite, see **Plumbago,** Vol. VI.

Grass, see **Lists of Fodder,** Vol. III. 420—427 and 432—37.

Grass-cloth—A name given to the woven fibre of **Bœhmeria nivea,** *Hook. & Arn.*, Vol. I., 472; see also **Rhea,** Vol. VI.

Grass mats, see **Mats and Matting,** Vol. V.

Grass oil—Obtained from species of **Andropogon** (see Vol. I., 241—252).

Green or **American Hellebore,** see **Veratrum viride,** *Aiton*; LILIACEÆ.

Green-stone, see **Trap,** Vol. VI.

GREWIA, *Linn.*; *Gen. Pl., I., 233.*

A genus of shrubs or small trees belonging to the Natural Order TILIACEÆ, and comprising about 60 species, for the most part confined to the hotter regions of the Old World. Of these 36 are natives of India. Most of the species derive economic importance from the fact that their inner bark contains a valuable fibre.

Grewia asiatica, *Linn.*; *Fl Br. Ind., I., 386*; TILIACEÆ. 283 (663)

Syn.—GREWIA SUBINÆQUALIS, *DC.*

It is convenient, for economic purposes, to consider *var.* **vestita,** *Wall.*, separately.

Vern.—*Pharsá, phalsá, phulsá, pharoah, shukri,* HIND.; *Phalsá, shukri,* BENG.; *Singhindamin,* KOL.; *Jang olat,* SANTAL; *Fussi,* NEWARI; *Fursu, fulsa,* PARBUT; *Phalsa, pharsiya, dháman,* N.-W.P.; *Phálsa,* PB.; *Pastaoni, shikarim-ai-wah,* PUSHTU; *Dhamni,* AJMIR; *Pháraho, phalsa,* SIND; *Dhamru, dhamun,* C. P.; *Phálasi,* BOMB.; *Phálsi,* MAR.; *Phalsá,* GUZ.; *Phulsha, pulsha,* DEC.; *Tadáchi,* TAM.; *Putiki, phutiki,* TEL.; *Dowaniya,* SING.; *Porusha,* SANS.; *Fálseh,* PERS.

References.—*Roxb., Fl. Ind., Ed. C.B.C., 431; Brandis, For. Fl., 40, Kurz, For. Fl. Burm., 161; Gamble, Man. Timb., 55; Dalz. & Gibs., Bomb. Fl., 26; Rev. A. Campbell, Rep. on Ec. Prod., Chutia Nagpur; No. 8459; Elliot, Fl. Andhr., 161; Pharmacographia Indica, I., 238; U. C. Dutt, Mat. Med. Hind., 313; Murray, Pl. and Drugs, Sind,*

GREWIA excelsa.

Grewia Spirits.

65; *Irvine, Mat Med. of Patna*, 88; *Atkinson, Him. Dist.*, 738, 792; *Ec. Prod., N.-W. P., Pt. V.*, 48; *Lisboa, U. Pl, Bomb.*, 27 147, 230; *Birdwood, Bomb. Pr.*, 140, 191, 200; *Hove, Tour in Bombay*, 171; *Agri-Hort. Soc. Ind., Transactions, IV.*, 103; *Journal (Old Series), II., Sel.* 216; *IX., Sel.*, 40; *Cross. Bevan & King, Indian Fibres*, 89; *Balfour, Cyclop., I.*, 1253; *Kew Off. Guide to the Mus. of Ec. Bot.*, 21; *Indian Forester, III.*, 200; *VIII*, 417, 418; *XII., App.* 8; *Madras Man. of Administration, I.*, 313; *District Manuals:—Coimbatore*, 40; *Settlement Report, Lahore*, 12; *Gazetteers:—Panjáb, Peshawar*, 27; *N.-W. P., I.*, 79; *IV.*, lxix; *X.*, 306; *Mysore and Coorg, I.*, 58; *Bombay, V.*, 285.

Habitat.—A small hazel-like tree, cultivated throughout India; said to be indigenous in the Salt-Range, Poona, Oudh, and Ceylon. It flowers about the end of the cold season, and its fruit ripens in April and May.

FIBRE. Bark. (664) 284

Fibre.—From the BARK a fibre is extracted, which resembles European bast-fibre; it is much employed in rope-making.

MEDICINE. Fruit. (665) 285 Leaf. (666) 286 Bud. (667) 287 Bark. (668) 288 Root-bark. (669) 289

Medicine.—The FRUIT, which is one of the *phala-traya*, or fruit-triad of Sanskrit writers, is supposed to possess astringent, cooling, and stomachic properties. A spirit is distilled from it, and a pleasant *sherbet* prepared by mixing it with sugar. The LEAF is employed as an application to pustular eruptions, and the BUD is also prescribed by native practitioners. An infusion of the BARK is used as a demulcent. The **Rev. A. Campbell** states that the ROOT-BARK is employed by the Santals as a remedy for rheumatism.

FOOD. Berry. (670) 290

Food.—The small dark purple BERRY is much esteemed by Natives as a pleasant, acid fruit. As already stated, a *sherbet* and spirit are prepared from it in many parts of the country.

TIMBER. (671) 291

Structure of the Wood.—Yellowish-white, close-grained, strong and elastic, similar to that of *var.* **vestita**, and according to the **Rev. A. Campbell**, much prized in Chutia Nagpur, for making banghy-poles, and for other purposes for which combined lightness and strength are desired.

INDUSTRIAL USE. Juice. (672) 292

Industrial Use.—The mucilaginous JUICE of the bark is used in Saharanpur for clarifying sugar. (*Liotard*; *Atkinson.*)

(673) 293

Grewia asiatica, *var.* vestita, *Wall.*; *Fl. Br. Ind., I.*, 387.

Syn.—GREWIA ELASTICA, *Royle*; G. ASIATICA (in part) and G. OBTECTA, *Wall.*

Vern.—*Pharsia, dhamún, bimla, dhamáni*, HIND.; *Dhamáni*, BENG; *Olat*, SANTAL; *Sealposra*, NEPAL; *Kúnsúng*, LEPCHA; *Pershuajelah*, MICHI; *Dháman*, N.-W. P.; *Farri, phalwá, dhamán*, PB.; *Dhamun*, C. P.; *Potodhamun*, PALAMOW; *Pintayan, pintayo, pengtarow*, BURM.; *Dharmana, dhanvana*, SANS.

References.—*Brandis, For. Fl.*, 40; *Kurz, For. Fl. Burm., I.*, 160; *Gamble, Man. Timb.*, 55; *Stewart, Pb. Pl.*, 27; *Rev. A. Campbell, Ec. Prod., Chutia Nagpur*, No. 8790; *U. C. Dutt, Mat. Med. Hind.*, 296; *Baden Powell, Pb. Pr.*, 581; *Ind. Forester, IV.*, 321; *IX.*, 254, 274; *XII.*, 64; *XIV.*, 302; *Gazetteers:—Panjáb, Simla*, 11; *Gurdaspur*, 70; *N.-W. P., X.*, 306; *Settlement Reports:—Central Provs., Bhundara Dist.*, 19; *Seonee Dist.*, 10.

Habitat.—Met with in the Sub-Himálayan tract, Bengal, Central India, and Burma.

FOOD. Fruit. (674) 294 FODDER. Branches. (675) 295

Food and Fodder.—The BRANCHES are lopped for fodder. The FRUIT is eaten.

TIMBER. (676) 296

Structure of the Wood.—Tough and elastic, weight from 43 to 51lb per cubic foot It is used for shoulder poles, bows, spear handles, &c., and as it splits well, is also employed to make shingles.

(677) 297

G. excelsa, *Vahl.*; *Fl. Br. Ind., I.*, 385.

Syn.—GREWIA ROTHII, *DC.*; G. SALVIFOLIA, *Roxb.*; G. BICOLOR, *Roth.*

Vern.—*Kulo,* URIYA; *Bathar, nikki bekkar, garges,* PB.; *Pútiki, kolupu, Siri jana,* TEL.

References.—*Roxb., Fl. Ind., Ed. C.B.C., 431; Brandis, For. Fl., 43; Kurz, For. Fl. Burm., I., 160; Stewart, Pb. Pl., 27; Baden Powell, Pb. Pr., 582; Balfour, Cyclop., I., 1253; N.-W. P. Gaz., I., 79; Ind. Forester, III., 200.*

Habitat.—A shrub of East Bengal, Assam, and the Coromandel; found also in Sikkim and Bandelkhand. Distributed to Tropical Africa.

Food.—This shrub flowers in the hot season, and its fruit ripens a few months later. The DRUPE is small, agreeable to the taste, and is eaten by the natives. FOOD. Drupe. 298 (678)

Grewia lævigata, *Vahl.; Fl. Br. Ind., I., 389.* 299 (679)

Syn.—GREWIA DIDYMA, *Roxb.*; G. DISPERMA, *Rottl.*; G. MOLLOCOCCA, *Ham.*; G. LALPETA, *Ham.*; G. OVALIFOLIA, JUSS.

Vern.—*Kat bhewal, bhimúl,* N.-W. P.; *Kaki,* OUDH; *Alli payaru,* TEL.; *Kaori, karkisellimara,* KAN.; *Gulgollop,* KONKANI.

References.—*Roxb., Fl. Ind., Ed. C.B.C., 432; Brandis, For. Fl., 42; Kurz, For. Fl. Burm., I., 159; Elliot, Flora Andhrica, 13; Gazetteers:—Bomb., XV., 70, 428; N.-W. P., IV., 69; X., 306; Indian Forester, III., 200.*

Habitat.—A small evergreen tree, native of the Eastern Himálaya, the Khásia Hills, Southern India, and Burma. Distributed to Tropical Africa, the Malay Islands, and Australia.

Fibre.—In the *Bombay Gazetteer, Vol. XV.* (*Kanára*), it is stated that this tree yields a favourite fibre, a remark confirmed by a private communication from **W. A. Talbot, Esq.**, Conservator of Forests in the same district, who writes, "a fibre often used for cordage in Kanára." FIBRE. 300 (680)

Structure of the Wood.—Rather heavy, fibrous, but close grained, soft, white, turning yellowish, then brownish. (*Kurz*) TIMBER. 301 (681)

G. Microcos, *Linn.; Fl. Br. Ind., I., 392; Wight, Ic., t. 84; Ill., t. 33.* 302 (682)

Syn.—GREWIA ULMIFOLIA, *Roxb.*; G. AFFINIS, *Lindl*; G. BEGONIFOLIA, *Wall.*; G. GLABRA, *Jack*; MICROCOS PANICULATA, *L.*; M. MALA, *Ham.*; M. STAUNTONIANA, *G. Don.*

Var. rugosa (distinguished by its bullate leaves), G. RUGOSA, *Wall.*

Vern.—*Shiral, ansale,* BOMB.; *Myá za,* BURM.

References.—*Roxb., Fl. Ind., Ed., C.B.C., 432; Kurz, For. Fl. Burm., I., 157; Thwaites, En. Ceylon Pl., 32; Dalz. & Gibs., Bomb. Fl., 26; Rheede, Hort. Malab., I., t. 56; Lisboa, U. Pl. Bomb., 147, 230; Bomb. Gaz., XV., 428.*

Habitat.—A very variable tree, generally from 40 to 50 feet in height, native of Eastern Bengal, the Khásia Mountains, Chittagong, Burma, and Western India. In the Khásia Mountains it ascends to 4,000 feet. Flowers in the rains, and the fruit ripens during the cold season.

Fibre.—Beyond the fact that **Lisboa** includes this species in his list of fibrous plants, no record of its fibre exists. FIBRE. 303 (683)

Food.—The DRUPE, of the size of a gooseberry, is eaten by Natives, especially during times of scarcity. FOOD. Drupe. 304 (684)

G. multiflora, *Juss.; Fl. Br. Ind., I., 388.* 305 (685)

Syn.—GREWIA SEPIARIA, *Roxb.*; G. SERRULATA, *DC.*; G. CORIACEA, *Garcke.*

Vern.—*Pansaura,* HIND., BENG.; *Nilay,* NEPAL.

References.—*Roxb., Fl. Ind., Ed. C.B.C., 432; Brandis, For. Fl., 42; Gamble, Man. Timb., 55.*

Habitat.—A shrub or small tree of Eastern Bengal, Assam, the Sikkim Himálaya, the Khásia Mountains, and the Nilghiri Hills, ascending to 4,000 feet. Distributed through the Malay Archipelago and East Tropical Africa.

TIMBER. (686) 306 **Structure of the Wood.**—White, soft, similar in structure to that of **Grewia oppositifolia**, weight 42 lb per cubic foot. (*Gamble*).

AGRICULTURAL USE. Plant. (687) 307 **Agricultural Use.**—This PLANT is extensively employed in Bengal for making hedges, for which its close growth and evergreen leaves render it specially suitable.

(688) 308 **Grewia oppositifolia,** *Roxb.; Fl. Br. Ind., I., 384; Wight, Ic., t. 82.*

Vern.—*Biúl, biúng, bahúl, bhimal, bhengal,* HIND.; *Biúl, biúng, bewal, báhúl, bemal, bhímal, bhengúl,* N.-W. P.; *Dhámnú, dhámán, bímal, bhimal,* KUMAON; *Dháman, dhámun, tháman, pharwa, bhewal, bhenwal, bhímal, bahúl, biúl,* PB.; *Pastuwanne, pustawanai, pastawana,* PUSHTU; *Thidsal,* KAN.

References.—*Roxb., Fl. Ind., Ed. C.B.C., 430; Brandis, For. Fl., 37; Gamble, Man. Timb., 54; Stewart, Pb. Pl., 27; Baden Powell, Pb. Pr., 510, 582; Atkinson, Him. Dist., 306, 792; Drury, U. Pl., 235; Wardle, Dye Rep., Sec. 32; Note by W. Coldstream, Esq., C.S., Dep. Com., Simla; Cross, Bevan & King, Rep. on Indian Fibres, 9, 39; Balfour, Cyclop., I., 1253; Indian Forester, I., 384; IV., 389; XIII., 58; Agri.-Hort. Soc. of India, Transactions, VIII., 260, 274, 277; Journals (Old Series); X., Pro., 91, 108; Settlement Reports:—Hazara, 12; Kohat, 30 Simla, App. II., xliii; Kangra, 22; Gazetteers:—Mysore and Coorg, I., 58; Panjáb, Bannu, 23; Dera Ismail Khan, 18, 19; Hazara, 14; Rawal Pindi, 15; Shahpur, 69; Hoshiarpur, 11; Gurdáspur, 55; Simla, 11.*

Habitat.—A moderate-sized tree of the North-West Himálaya, from the Indus to Nepal, ascending to 6,000 feet; frequently cultivated as a hedge along roadsides and near Himálayan villages.

DYE. Leaves. (689) 309 **Dye.**—**Wardle** writes as follows in his Dye Report: "The LEAVES do not yield much tinctorial matter by themselves, but if mordanted with G. X. mordant they give a fine bright clean shade of dark fawn, approaching to yellow. The leaves possess tannin, shown by the infusion blackening with a salt of iron." The sample examined by **Mr. Wardle** having been sent from the Panjáb, the inference may be accepted that it is employed as a dye by the natives of that province, but the writer can find no mention of the fact in available literature.

FIBRE. Bark. (690) 310 **Fibre.**—A strong coarse fibre is obtained from the BARK by retting, and is much employed by the Natives of the North-West Himálaya for making ropes and nets. **Captain Huddleston** appears to have been the first to bring this fibre to the notice of Europeans. In his Report on Hemp cultivation, &c., in British Garhwal, published in the *Transactions of the Agri.-Hort. Soc. Ind., vol. viii., 260,* he wrote: "The branches of the tree are cut from July till March, or at all seasons save the spring. The sticks are soaked for a month or forty days in water, and when dry are beaten on stones, and the bark stripped off. One tree will give about five seers of the inner fibre fit for making ropes and string, which is used for tying up cattle and stringing cots. It is neither very strong nor durable, and is not to be had in any quantity." In another passage he remarked: "The *Bhímal* ropes get stronger from wet for the time, but do not last above eighteen months if kept constantly exposed to the weather." Eighteen years later (in 1858) **Lieutenant Pogson** submitted samples of the same fibre and rope from Simla, to the Agri.-Horticultural Society of India. These were examined by the Fibre Committee, one of the members of which reported as follows: "Weak and brittle; like all fibres from the bark of trees it will not bear dressing. Very probably, if worked by hand, a common sort of rope might be made of it for local use, but I do not think it would ever become an article of commerce." Later the fibre was employed for paper-making in the Kángra valley under European supervision, with a fair amount of success, but, as in the case of the fibre of **G. tiliæfolia** there would be little chance of its exportation for that purpose paying. In the report by **Messrs. Cross, Bevan, and King** it is stated that "owing to the very short length of the ultimate fibre, 1—1·5 mm., and the inferior character of the specimen it was not deemed worthy of further investigation." An analysis by the same investigators showed the fibre to contain 72 per cent. of cellulose to 10·6 of moisture.

Food and Fodder.—The LEAVES and YOUNG TWIGS are largely employed, wherever the plant is common, as fodder for cattle, sheep, and goats and are said to increase the milk-giving powers. It is also stored for winter fodder. **FOOD & FODDER. Leaves. 311**

Structure of the Wood.—White, with a small mass of irregular heart-wood, hard, giving out an exceedingly unpleasant odour, especially when **Young Twigs. 312** fresh cut, or burned, weight from 45 to 50lb per square foot. **TIMBER. 313** It is extensively used for making oar-shafts, axe-handles, banghy poles, cot-frames, and other articles in which elasticity and toughness are required.

Domestic Uses.—In the North-West Himálaya the dried BARK is, from its fibrous nature, valued for sandal-making, and the green bark is employed by women for cleaning the hair. **DOMESTIC. Bark. 314** The DRY TWIGS, after the bark has been separated by retting, make excellent torches, and are very largely used in the lower Himálayan ranges for that purpose. **Dry Twigs. 315**

Grewia polygama, *Roxb.; Fl. Br. Ind., I., 391.* 316

Syn.—GREWIA HELICTERIFOLIA, *Wall.;* G. VIMINEA, *Wall.;* G. LANCEOLATA, *Heyne;* G. LANCIFOLIA, *Grah.;* G. HIRSUTA, *Wall.;* G. ANGUSTIFOLIA, *Wall.;* G. BILOBA, *Wall.;* G. LANCEÆFOLIA, *Roxb.*

Vern.—*Kukur bicha,* HIND.; *Seta kata, seta andir,* SANTAL; *Gowli* or *gowali,* BOMB.

References.—*Roxb., Fl. Ind., Ed. C.B.C., 431; Brandis, For. Fl., 42; Thwaites, En. Ceylon Pl., 31; Dalz. & Gibs., Bomb. Fl., 26; Rev. A. Campbell, Ec. Prod., Chutia Nagpur, No. 7849; Christy, New Commercial Plants & Drugs, VII., 49; Lisboa, U. Pl. Bomb., 147, 230; For. Adm. Rep., Chutia Nagpur, 1885, 28; N.-W. P. Gaz.:—I., 79; IV., 69; X., 306; Ind. Forester, VIII., 417.*

Habitat.—A shrub or small tree of North-West India, extending along the Himálaya from the Salt-Range to Nepál, ascending to 4,000 feet; also met with in Chutia Nagpur, the Central Provinces, Konkan, Pegu, and Ceylon.

Medicine.—The **Rev. A. Campbell** reports that the FRUIT is employed as a medicine by the Santals, in cases of diarrhœa and dysentery. **MEDICINE. Fruit. 317** The ROOT pounded is also prescribed for the same diseases, and powdered in water is applied externally to hasten suppuration, and as a dressing for wounds. **Root. 318** The paste dries and forms a hard coating, thus effectually excluding air from the raw surface. The knowledge of the medicinal virtues of this plant in Chutia Nagpur is interesting, since an antidysenteric property is also attributed to it by the aborigines of North-West Australia. **Mr. Christy** commenting on its virtues writes: "**Dr. W. E. Armit** states that on one occasion having had to treat dysentery following on fever and ague, this plant was pointed out to him by a native as a sure remedy. He collected a quantity of LEAVES, and having made a pale sherry-coloured decoction, administered about two table-spoonfuls for a dose. **Leaves. 319** Repeating this every four hours throughout the night, the sixth dose made a complete cure. 'Since then,' says **Dr. Armit,** 'I have tried the remedy in scores of cases, and I have never known it to fail, however serious.'" As already stated, the leaves contain tannin, which, together with the mucilaginous properties peculiar to the whole genus, probably explains its special action.

Food.—The FRUIT is eaten in Chutia Nagpur. **FOOD Fruit. 320**

Edible Fruits.

TIMBER. 321 **Structure of the Wood.**—Similar to that of **G. oppositifolia,** but free from unpleasant odour.

322 **Grewia populifolia,** *Vahl.; Fl. Br. Ind., I., 385.*

Syn.—GREWIA BETULÆFOLIA, *Juss.;* G. RIGIDA, *Ham.*

Vern.—*Bursa,* KOL; *Ganger, kanger, gangi, inzarre, khircha,* PB.; *Shikári mewa,* KOHAT; *Khircha, indzar,* PUSHTU; *Gangerun, gangan, ganegan,* RAJ.; *Gango, gangi,* SIND.

References.—*Brandis, For. Fl., 38; Gamble, Man. Timb., 54; Stewart, Pb. Pl., 27; Bot Tour in Hazara, in Jour. Agri.-Hort. Soc., XIV. (Old Series), 33; Stocks, Report on Sind; Murray, Pl. and Drugs, Sind. 66, Baden Powell, Pb. Pr., 581; Balfour, Cyclop., I., 1253; Ind. Forester; IV., 228, 233; VIII., 418; XII. (App.) 2, 7; Gazetteers:—Panjáb, Bannu, 23; Dehra Ismail Khan, 19.*

Habitat.—A small shrub extending from the arid tracts of the Panjáb, Sind, Rajputána, and Western India, down to the Nilghiri Hills.

FOOD. Fruit. 323 **Food.**—The small, orange-red acid FRUIT is eaten by the Natives in Sind and the Panjáb. **Stewart** describes it as having neither substance, nor flavour, but **Stocks** says it is very palatable and might be improved by cultivation.

TIMBER. 324 **Structure of the Wood.**—Yellow, hard, close-grained; used for walking-sticks (*Gamble.*)

325 **G. salvifolia,** *Heyne; Fl. Br. Ind., I, 386.*

Syn.—GREWIA BICOLOR, *Juss.;* G. ARARIA, *Wall.*

Vern.—*Dhattiki,* URIYA; *Bursu, sita pelu, cheli,* KOL.; *Sitanga,* SANTAL; *Khorkhorendna,* MAL. (S.P.); *Bather, nikki-bekkar, gargas,* PB.; *Saras,* AJMERE; *Heriss, seriss, serissa, katang,* MERWARA; *Bihul,* SIND; *Jára,* CIRCARS; *Budamara, chipudi, putiki,* TEL.

References.—*Brandis, For. Fl., 43; Gamble, Man. Timb., 55; Elliot, Fl. Andhr., 31, 41, 160; Murray, Pl. and Drugs, Sind, 66; For. Adm. Rep., Chutia Nagpur, 1885, 28; N.-W. P. Gaz., IV., 69; Ind. Forester, IV., 228, 233; VIII., 418; XII. (App.) 7.*

Habitat.—A small tree met with on the North-West Himálaya from the Jhelum to Nepál, also in the Panjáb, Sind, the Central Provinces, and Southern India.

FOOD. 326 **Food.**—Fruit small, not succulent, sub-acid, eaten by the Natives.

TIMBER. 327 **Structure of the Wood.**—Yellow, heartwood brown, hard, close-grained, and similar to that of **G. tiliæfolia.**

328 **G. scabrophylla,** *Roxb.; Fl. Br. Ind., I., 387; Wight, Ic., t. 89.*

Syn.—GREWIA SCLEROPHYLLA, *Wall.;* G. CARREA, *Ham.;* G. SULCATA, *Wall.;* G. PILOSA, *Wall. (in part);* G. OBLIQUA, *Roxb., not of Juss.*

Vern.—*Pharsia,* fruit=*gúrbheli,* KUMAON; *Pándhari dhá man, khatkhati,* MAR.; *Darsuk,* KAN.

References.—*Roxb., Fl. Ind., Ed. C.B.C., 430, 432; Brandis, For. Fl., 39; Kurz, For. Fl., Burm., I., 162; Pharmacographia Indica, I., 238; Gazetteers:—N.-W. P., IV., 64; X., 306; Bomb., XV., 70.*

Habitat.—A small shrub of the Tropical Himálaya (in Sikkim and Garhwál), Assam, Chittagong, and Ava. It is also mentioned in the list of forest trees of the Kanára District in Bombay. It flowers in April, and ripens its fruit in October.

FIBRE. 329 **Fibre.**—**G. obliqua,** *Roxb.,* is said in the *Kanára Gazetteer* to yield a fibre from which ropes are made.

MEDICINE. Roots. 330 **Medicine.**—The authors of the *Pharmacographia Indica* write: "It is given in accordance with the theory of signatures as a remedy for leprosy in the Konkan; it appears to be simply mucilaginous like most of the genus. Its ROOTS are the 'althea' of the Portuguese in Goa, and are used as a substitute for althea." It is doubtful to what part of the plant the first sentence refers, but it may probably be to the scabrid LEAVES. MEDICINE. Leaves. 331 (711)

Food.—The FRUIT is of the size of a large cherry, nearly round, brownish-grey when ripe, with a glutinous, pale yellow, pulp. It is eaten by Natives. FOOD. Fruit. 332 (712)

Structure of the Wood.—In the *Kanára Gazetteer*, it is stated that **G. obliqua**, *Roxb.*, referred by the *Flora, British India*, to this species, is a small tree, the wood of which is used for field tools and posts. Some other species, however, is probably meant, since **G. scabrophylla**, *Roxb.*, is elsewhere a small shrub. TIMBER. 333 (713)

Grewia tiliæfolia, *Vahl.; Fl. Br. Ind., I., 386.* 334 (714)

Syn.—GREWIA ARBOREA, *Roth.*; G. VARIABILIS, *Wall. in part*; G. SUBINÆQUALIS, *Wall.*

Vern.—*Pharsa, dhamin, dhámani,* HIND.; *Dhamono, dhaman,* URIYA; *Dhamin,* KOL.; *Olat,* SANTAL; *Khesla, kasúl,* GOND; *Dhamni,* KURKU; *Dhamni, damún,* C. P.; *Damana, karakana,* BOMB.; *Dáman, dháman,* MAR.; *Dhamana,* GUJ.; *Thada, tharra,* TAM.; *Dhamnak,* BHIL.; *Charachi, tharrá, údúpai, tada,* TEL.; *Thadsal, dadsal, batala, bútále,* KAN.; *Sadachu,* MALAY.; *Daminne,* SING.; *Dhamni, dharmana,* SANS.

References.—*Brandis, For. Fl., 41; Kurz, For. Fl., Burm., I., 161; Beddome, Fl. Sylv., t. 108; Gamble, Man. Timb., 54; Thwaites, En. Ceylon Pl., 32; Dalz. & Gibs., Bomb. Fl., 26; Elliot, Fl. Andhr., 172; Rev. A. Campbell, Ec. Prod., Chutia Nagpur, No. 9411; Dymock, Mat. Med. W. Ind., 2nd Ed., 116; Pharmacographia Indica, Pt. I., 237; Atkinson, Him. Dist., 306; Lisboa, U. Pl. Bomb., 26, 147, 230, 277; Balfour, Cyclop., I., 1253; Kew Report, 1879, 34; Kew Off. Guide to the Mus. of Ec. Bot., 21; Indian Forester, III., 23, 200; X., 222; XIII., 119; For. Adm. Rep., Chutia Nagpur, 1885, 6, 28; Jour. Agri.-Hort. Soc. of India, V. (Old Series), Sel. 12; Madras, Man. of Admin., II., 82; Settlement Reports:—Chandwara (C. P.), 110; Gazetteers:—Mysore & Coorg, I., 52; Bombay, VII., 36; XIII., 24; XV., 40; XVII., 24; XVIII., 45; N.-W. P., I., 79; IV., lxix.*

Habitat.—A moderate-sized tree met with in the hot dry forests throughout the Sub-Himálayan tract, from the Jumna to Nepál, ascending to 4,000 feet; also in Central and South India, Burma, and Ceylon. It flowers in April and May, and its fruit ripens from June to October.

Fibre.—The BARK yields a fibre employed in many parts of India for cordage. It, like the fibres of other Grewias, has been recommended for paper-making, but is apparently little worthy of consideration. Specimens were sent to Kew in 1878, and were, amongst other fibre-yielding barks examined by the late Mr. Routledge of the Ford Works, Sunderland. He described the fibre as "strong, harsh, wiry, hard," and stated that the green-yield of the bark was 50 per cent., the bleached-yield 43·7. As in the case of other Grewia fibres, it would not pay to export it to Europe for paper-making purposes. FIBRE. Bark. 335 (715)

Medicine.—Dr. Dymock writes that, "In the Konkan the BARK, after removal of the suber, is rubbed down with water, and the thick mucilage strained from it and given in five tolá doses with two tolás of the flour of **Panicum miliaceum,** as a remedy for dysentery." Colonel Cox in his "Remarks on Certain Specimens of Wood from Central India" says that the WOOD reduced to a powder acts as an emetic, and is employed by the Natives as an antidote to opium poisoning. (*Journ. Agri.-Hort. Soc. Ind. loc. cit*). The BARK is also employed externally to remove the irritation in cow-itch. MEDICINE. Bark. 336 (716) Wood. 337 (717) Bark. 338 (718)

Food and Fodder.—The small DRUPES are of an agreeable acid flavour, and are eaten by Natives. The LEAVES and TWIGS are lopped for cattle fodder. FOOD. Drupes. 339 (719) FODDER. Leaves. 340 (720) Twigs. 341 (721)

TIMBER.
(722) 342 **Structure of the Wood.**—White, with a small brown heartwood; close-grained, hard, weight from 30 to 40℔ per cubic foot according to **Brandis**, 48℔ according to **Gamble**. It is easily worked, elastic, and durable, and is much used for shafts, shoulder poles, pellet-bows, axe-handles, masts, oars, fishing rods and other purposes for which elasticity, strength, and toughness are required.

Grewia vestita, *Wall.*, see p. 178.

(723) 343 **G. villosa**, *Willd.; Fl. Br. Ind., I., 388.*

Syn.—GREWIA ORBICULATA, *G. Don.*

Vern.—*Gaphni*, KOL.; *Tarse kotop*, SANTAL; *Jalidar, kaskúsri, thamther*, PB.; *Inzarra, pastuwanne*, PUSHTU; *Dhohan*, AJMERE.

References.—*Brandis, For. Fl., 39; Dalz. & Gibs., Bomb. Fl., 25; Stewart, Pb. Pl., 27; Rev. A. Campbell, Ec. Prod., Chutia Nagpur, No. 8712; Murray, Pl. and Drugs, Sind, 66; Lisboa, U. Pl. Bomb., 147; Ind. Forester, IV., 228, 233; VIII., 418; XII. (App.), 8; Gazetteer:—Bombay, V., 24.*

Habitat.—A small shrub of Western and Southern India, extending from the Panjáb and Sind to Travancore.

MEDICINE. Root.
(724) 344 **Medicine.**—The **Rev. A. Campbell** states that the ROOT is employed as a remedy for diarrhœa in Chutia Nagpur.

FOOD. Fruit.
(725) 345 **Food.**—**Stewart** mentions that the FRUIT, though very inferior, is eaten in the Panjáb.

Grislea tomentosa, *Roxb.*, see **Woodfordia floribunda**, *Salisb.*; LYTHRACEÆ.

Ground-nut, see **Arachis hypogæa**, *Linn.*, Vol. I., 282.

Guano, see **Collocalia nidifica**, *Gray*; Vol. II., 503; also **Manures**, Vol. V.

Guatteria longifolia, *Wall*, see **Polyalthia longifolia**, *Benth.*; ANONACEÆ.

Guava, see **Psidium Guyava**, *Linn.*; MYRTACEÆ.

GUAZUMA, *Plum.; Gen. Pl., I., 225.*

(726) 346 **Guazuma tomentosa**, *Kunth; Fl. Br. Ind., I., 375; Wight, Ill., t. 31;* STERCULIACEÆ,
THE BASTARD CEDAR.

Syn.—BUBOMA TOMENTOSA, *Spreng.*; GUAZUMA ULMIFOLIA, *Wall.*; DIUROGLOSSUM RUFESCENS, *Turcz.*

Vern.—*Bandóq-ké-jhár-ki-chhál*, DEC.; *Thain-púchie, thainpuche, rudrasum*, TAM.; *Rudraksha, údrik-patta*, TEL.; *Bucha-pattai, rudrakshe*, KAN.

References.—*Kurz, For. Fl. Burm., I., 149; Beddome, Fl. Sylv., t., 107; Gamble, Man. Timb., 45; Thwaites, En. Ceylon Pl., 29; Elliot, Fl. Andhr., 165; O'Shaughnessy, Beng. Dispens., 226; Moodeen Sheriff, Mat. Med. of Madras, 68; Drury, U. Pl., 236; Lisboa, U. Pl. Bomb., 26, 195, 229; Royle, Fib. Pl., 265, 266, 267, 268; Cross, Bevan, and King, Rep. on Indian Fibres, 53; Balfour, Cyclop., I., 1258; Kew Off. Guide to the Mus. of Ec. Bot., 21; Agri-Hort. Soc. of Ind., Trans., VII., 81, 83; Journal, I. (New Series), Sel., 52; Indian Forester, III., 233, 236; IX., 626; XIV., 29; Gazetteers:—Mysore and Coorg, I., 52, 68.*

Habitat.—A tree, stellately hairy on the young twigs, generally distributed and frequently cultivated in the warmer parts of India and Ceylon, but perhaps only introduced. It is probably a native of the West Indies.

FIBRE. Branches.
(727) 347 **Fibre.**—The straight, luxuriant, young BRANCHES yield a fibre which was submitted to experiments by **Dr. Roxburgh**, and found to be of con-

siderable strength, breaking at 100lb. when dry and 140lb when wet. **Messrs. Cross, Bevan, and King** have recently examined it, and state as the result of their observations that the ultimate fibre is 1 to 2 mm. in length, and that if the plant were cultivated for the purpose, it would probably yield a fibre which would prove useful for rope-making. FIBRE.

Medicine.—The medicinal properties of this tree appear to have been first brought to notice by **Lindley** who wrote: "In Martinique the infusion of the old BARK is esteemed as a sudorific, and as useful in cutaneous diseases, and diseases of the chest." **Moodeen Sheriff** in his forthcoming *Materia Medica of Madras*, proofs of which have been obligingly furnished to the editor, describes the bark as tonic and useful in some of those cases for which calumba and gentian are generally prescribed. The preparation recommended by him is a decoction prepared as follows: "Take of the inner bark, cut into small pieces or in coarse powder, four ounces; water one pint and a half; boil till the liquid is reduced to one pint, and strain when cool; dose from two to three fluid ounces." The inner bark is esteemed as a remedy for Elephantiasis in the West Indies. MEDICINE. Bark. 348 (728)

Food and Fodder.—The FRUIT, which contains a sweet and agreeable mucilage, is edible. The LEAVES are valued as fodder in the Bombay Presidency. FOOD. Fruit. 349 (729) FODDER. Leaves. 350 (730)

Structure of the Wood.—Loose-grained, light-brown or brown, streaked, coarsely fibrous, takes a good polish; weight, according to **Skinner**, 32lb per cubic foot. It is employed in Southern India for furniture, the panels of carriages, and packing-cases. TIMBER. 351 (731)

Industrial Uses, &c.—The glutinous decoction of the inner bark is employed in the West Indies for clarifying sugar. The tree is admirably adapted for avenues, and has been largely planted in Southern India for this purpose, as well as for fodder. INDUSTRIAL. 352 (732)

GUETTARDA, *Linn.; Gen. Pl., I., 99.*

Guettarda speciosa, *Linn.; Fl. Br. Ind., III., 126; Wight, Ic., t. 40;* RUBIACEÆ. 353 (733)

Syn.—CADAMBA JASMINIFLORA, *Sonner.*; NICTANTHES HIRSUTA, *Linn.*; JASMINUM HIRSUTUM, *Willd.*

Vern.—*Pannir*, DEC.; *Panir*, TAM.; *Panniru puvvu*, TEL.; *Rava-pu*, MALAY.; *Domdomah*, AND.; *Wal-pichcha, nil pitcha*, SING.; *Himma*, SANS.

References.—*Roxb., Fl. Ind., Ed. C.B.C., 230; Kurz, For. Fl. Burm., II., 37; Beddome, Fl. Sylv., Anal. Gen., t. 17, f 2; Gamble, Man. Timb., 229; Elliot, Fl. Andhr., 144; Rheede, Hort. Mal., t. 47, 48; Drury, U. Pl., 237; Trimen, Cat. Ceyl. Pl., 43; Balfour, Cyclop., I., 1259.*

Habitat.—A moderate-sized evergreen tree found in tidal forests along the shores of Southern India, the Andaman Islands, and Ceylon. Distributed along the tropical shores of the Old and New Worlds.

Domestic and Sacred Uses.—**Drury** quoting **Persoon** writes: "The flowers of this tree are exquisitely fragrant. They come out in the evening, and have all dropped on the ground by the morning. The natives in Travancore distil an odoriferous water from the corollas, which is very like rose-water. In order to procure it they spread a very thin muslin cloth over the tree in the evening, taking care that it comes well in contact with the flowers as much as possible. During the heavy dew at night the cloth becomes saturated, and imbibes the extract of the flowers. It is then wrung out in the morning. This extract is sold in the bazárs." The writer can find no other reference to this interesting product in available literature. The tree is considered by the Hindús to be sacred to Siva and Vishnú. DOMESTIC. Perfumery. 354 (734)

[Vol. II., 3.

Guilandina Bonduc, *W. & A.*, see **Cæsalpinia Bonducella**, *Fleming;*

Guinea-corn, see **Sorghum vulgare**, *Pers.;* GRAMINEÆ.

Guinea-grass, see **Panicum jumentorum**, *Pers.;* GRAMINEÆ.

GUIZOTIA, *Cass.; Gen. Pl., II., 382.*

(735) 355 **Guizotia abyssynica**, *Cass.; Fl. Br. Ind., III., 308; Wight, Ill., t.* [*132;* COMPOSITÆ.

NIGER SEED AND OIL.

Syn.—G. OLEIFERA, *DC.;* HELIANTHUS OLEIFERA, *Wall.;* BIDENS RAMTILLA, *Wall.;* RAMTILLA OLEIFERA, *DC.;* HELIOPSIS PLATYGLOSSA, *Cass.;* VERBESINA SATIVA, *Bot. Mag., t. 1017, Roxb.;* POLYMNIA ABYSSYNICA, *Linn. f.*

Vern.—*Kálá-tíl, surgúja,* HIND.; *Rám-til, sirgúja, sirgúza,* BENG.; *Surgúja,* KOL; *Surgúja,* SANTAL; *Rámatila, kerani,* BOMB.; *Kali-til,* MAR.; *Valesulú, vulisi,* TEL.; *Huchchellu, ram-til,* KAN.

References.—*Roxb., Fl. Ind., Ed. C.B.C., 606; Dalz. & Gibs., Bomb. Fl., 128; Elliot, Fl. Andhr., 193; Rev. A. Campbell, Ec. Prod. Chutia Nagpur, No. 7865; Ainslie, Mat. Ind., II., 256; Dymock, Mat. Med. W. Ind., 2nd Ed., 432; Drury, U. Pl., 238; Lisboa, U. Pl. Bomb., 163, 219; Birdwood, Bomb. Pr., 303; McCann, Dyes and Tans, Beng., 31, 34, 155; Cooke, Oils and Oilseeds, 46; Simmonds, Commercial Products, 535; Special Reports from Collector of Cuddapah; J. B Fuller, Esq., C.S., Central Provinces; Dir., Land Rec. and Agric., Bengal; Madras Board of Revenue Pros., June 1st, 1889, No. 2?6, p. 6; Watts, Dic. of Chem., II., 953; Balfour, Cyclop., I., 1260; Smith, Dic., 346; Kew Off. Guide to the Mus. of Ec. Bot., 85; Simmonds, Trop. Agri., 415; Agri-Hort. Soc. Ind., Journals (Old Series), VII., 40, 41, (Pro.), 21; VIII., 61, (Pro.), 8; IX., 47; Gazetteers:—Mysore and Coorg, I., 56; II., 11.*

Habitat.—A stout, erect leafy herb, native of Tropical Africa, but extensively cultivated as an oil-seed in various parts of India.

CULTIVATION. (736) 356 **Cultivation.**—Owing to the agricultural returns embracing all oil-seed crops under one general heading, it is difficult to obtain figures of the exact extent of the cultivation of NIGER OIL-SEED in India. In answer to enquiries, however, the following reports have been kindly furnished, and from these a fair idea of the amount grown in the more important regions may be obtained.

Bengal. (737) 357 BENGAL.—The Director of Land Records and Agriculture writes: "Is largely grown in Chutia Nagpur on *bari* land, *i.e.*, land near homesteads. It generally follows *gora dhan*, or upland rice. It is sown in August either alone or with some pulse like *kúlti* (**Dolichos biflorus**). Beyond a few ploughings the fields do not receive any other treatment. The amount of seed sown per acre varies from 20 to 30 seers. The crop is harvested in November-December. The produce on an average is four maunds of oil-seed valued at R5-8 per acre. The area under *Sirgúza* was found by **Mr. Slack**, Settlement Officer, to be 9½ per cent. of the total cultivated area. In Palamow it is not grown to this extent, its place being taken by *tíl* (**Sesamum indicum**). The oil is used extensively by the natives for cooking and anointing purposes. It is known to be extensively exported to Calcutta through Purulia."

Central Provinces. (738) 358 CENTRAL PROVINCES.—**J. B. Fuller, Esq., C.S.**, Officiating Junior Secretary to the Chief Commissioner, furnishes the following: "Niger seed is a very important product of the hill-tracts in these Provinces, and furnishes the hill tribes with the principal means of earning money. It is grown during the rains, generally on the flat stony land which lies along the

CULTIVATION.

summits of trap hills, and which at first sight would seem to be entirely unproductive. The surface of this land is covered with nodules of red laterite and what soil it carries is of the thinnest description. Yet with resting fallows after every three or four years it can yield very fair crops of niger or *kadon*.

"The districts in which the growth of niger seed is of the greatest importance are the following:—

DISTRICTS.	Area under Niger in	
	1885-86.	1886-87.
	Acres.	Acres.
Saugor	30,442	Not stated.
Seoni	20,485	18,701
Betul	64,356	54,540
Chhindwara	55,348	48,575
Chanda	74,795	74,148

Madras. 359 (739)

MADRAS.—The Collector of Cuddapah writes: "A three months' crop sown both in January and July, in dry lands (generally red loam), in Cuddapah (200 acres) and in Pulivendula. The seed is sold at from 14 to 15 seers a rupee; the cost of cultivation is R7 and the profit R15 an acre." The following note has also been furnished by the Madras Government: "Grown in many parts of the Presidency to a small extent. In South Arcot it is sown on red loam in both January and July and is a three months' crop. Cost of growth R7; profit R15. Seed sells at 18 measures per rupee; oil is extracted and mixed with gingelly oil." No further particulars regarding the total area under the crop in Madras are available, but the following account by Drury of the method of cultivation pursued in Mysore may be quoted: "The seed is sown in July or August after the first heavy rains, the fields being simply ploughed, neither weeding nor manure being required. In three months from the sowing, the crop is cut, and after being placed in the sun for a few days, the seeds are thrashed out with a stick. The produce is about two bushels an acre. In Mysore the price is about R3-8 a maund."

OIL. Achene. 360 (740)

Oil.—The shining, black, ACHENE yields a clear, limpid, pale, sweet oil largely employed for culinary and lighting purposes throughout India. From a chemical analysis of the achene, given in **Watts, *Dictionary of Chemistry***, it would appear that the actual percentage of oil which it contains is 43·22, but 35 per cent. is the maximum yield recorded, or 10 per cent. less than the yield from the *til* seed (**Sesamum indicum**). In the Deccan it is the chief substitute for *ghí* amongst the poorer classes, and the achenes after extraction of the oil are used to make oil-cake which is considered a valuable food for milch-cows. From its inferior quality and low price it is also extensively used to adulterate gingelly and castor oils. In 1851 the seed was reported on by brokers in London as follows: "It has been imported from Africa as Niger seed, and from Bombay as *kersaní* seed, and the present value (1851) of the quality represented by the sample is 37*s.* per quarter, taking the value of rape seed at 48*s.* The oil is good, but the yield is only about 16 gallons against an average of about 20 from rape seed per quarter. The cake is unfit for feeding, and could only be used for manure, owing to the horny excrescence (*sic*) and dry, heating property of it. The average weight of this seed is also deficient of that of rape seed by 8lb per bushel, which is of great moment in making calculations for shipment. There is no doubt it would meet with a ready sale at its relative

MEDICINE. Oil. value to rape-seed." (*Jour. Agri.-Hort. Soc. Ind.*) The present price of the "seed" in Bombay is about R6½ per *pharrah* of 35lb (*Dymock*).

361 **Medicine.**—The OIL may be used for the same pharmaceutical purposes as sesamum oil.

FOOD. Oil. 362 **Food.**—The OIL is much used for culinary purposes and as a substitute for *ghí* by the poorer classes in the regions where it is cultivated. In

Seeds. 363 Madras the SEEDS are sometimes fried with oil or *ghí* and eaten.

DOMESTIC. Oil. 364 **Domestic and Industrial Uses.**—The OIL is, when refined, fairly good for lighting purposes, and has been reported as "capital for painting and cleaning machinery." Like other sweet oils it is largely used in this country for anointing the body.

365 **Gulal,** a coloured powder used along with *Abír* at the *Holi* festival. It is generally prepared from sappan wood, alum, and flour, but has of recent years been largely supplanted by a cheap powder coloured by means of an aniline dye. See **Abír,** Vol. I., 6; also **Curcuma Zedoaria,** *Roscoe;* Vol. II., 669.

Gulancha, see **Tinospora cordifolia,** *Miers.;* MENISPERMACEÆ.

366 GUMS AND RESINS.

Under this heading the gums, resins, and inspissated saps yielded by plants are described in this work. These products will also (as mentioned in the short note under the heading Domestic and Sacred, III., 191) form the subject of a collective article in the Appendix, in which they will be classified, and an enumeration of the plants yielding them will be given. It may, however, be useful in this place to mention those of chief commercial value, and to refer to the views of experts in England as to the prospects of the Indian gum trade. For this purpose the following list may be given, which was selected by **Mr. C. Christy** (of **Messrs. Hall and Sons**) at the Colonial and Indian Exhibition, as most likely to meet the demands of the European market:—

Acacia arabica.
A. Catechu.
A. leucophlœa.
A. Sundra.
Anogeissus latifolia.
Balsamodendron Mukul.
Butea frondosa.
Canarium strictum.
Cochlospermum Gossypium.
Morus indica.
Odina Wodier.
Shorea robusta.
Sterculia urens.
Vateria indica.

The same expert remarked at a conference on the gums shown at the Exhibition that, owing to the troubles in the Soudan, and the consequent impairment of Egyptian trade, the supply of gums from India had greatly increased, and that if it could be possible to induce the Natives to collect any one of the gums above referred to, pure, and free from dirt, a large and important trade would immediately open out. The great disadvantage of Indian gums imported at that time consisted in the mixture of different kinds, resulting from the consignments having to be opened out, reassorted, and sold as broken packages which never realised anything like their proper value. **Mr. Christy** strongly advocated attention being paid to this fact, and urged that, if possible, the natives should, through the agency of Government, be taught to prepare honest consignments. It may be of interest to note that the subsequent trade statistics of Indian Gums and Resins fully substantiate **Mr. Christy's** statements. The exports from India excluding Cutch in 1883-84 immediately before the Exhibition were only 2,180 cwt., valued at R24,538, in 1885-86 they had increased to 55,407 cwt., valued at R15,97,997, while in 1887-88 they had again decreased to 37,132 cwt. of a greater proportionate value, however, amounting to R17,54,963. It is encouraging to notice that in the last year for which statistics are available, 1888-89, the amount had again increased to 62,136 cwt., valued at the considerable sum of R28,27,999. It may thus be justly inferred that the remarks of Mr. Christy at the conference above alluded to, had the desired result of attracting the attention of Indian merchants to the subject at the most favourable moment.

For further information see the description of the above enumerated plants in their respective alphabetical positions, also the account of **Bassora**, or **Hog-gums**, Vol. I., 417.

Gunny-bags, see **Corchorus**, Vol. II., 561; also **Jute**, Vol. IV., p. 558.

Gun-powder-charcoal. For a list of the timbers used in making this ingredient of gun powder, see **Charcoal**, Vol. II., 264.

Gun-stocks and **Gun-carriages.** The following are the principal timbers employed (or which might be so) in India for the construction of gunstocks and gun-carriages; for a complete description of which the reader is referred to the article on each in its respective alphabetical position:— 367 (747)

Adina cordifolia, *Hook. f. & Bth.*	**Juglans regia,** *Linn.*
Careya arborea, *Roxb.*	**Lagerstrœmia Flos-reginæ,** *Retz.*
Cedrela Toona, *Roxb.*	**Melanorrhœa usitata,** *Wall.*
Celtis australis, *Linn.*	**Mesua ferrea,** *Linn.*
Chloroxylon Swietenia, *DC.*	**Morus serrata,** *Roxb.*
Cordia Myxa, *Linn.*	**Pistacia integerrima,** *Stewart.*
Dalbergia latifolia, *Roxb.*	**Pterocarpus indicus,** *Willd.*
D. Sissoo, *Roxb.*	**Rhododendron arboreum,** *Sm.*
Dillenia indica, *Linn*	**Shorea robusta,** *Gærtn.*
Eriolæna Candollei, *Wall.*	**Taxus baccata,** *Linn.*
Ehretia acuminata, *Br.*	**Tectona grandis,** *Linn. f.*
Gmelina arborea, *Roxb.*	**Thespesia populnea,** *Corr.*
Hopea odorata, *Roxb.*	

Gurjun or Garjan, see **Dipterocarpus turbinatus,** *Gaertn. f.*, Vol. III., 161.

Guttapercha, see **Dichopsis Gutta,** *Bth. and Hook. f.*, Vol. III., p. 103.

GYMNEMA, *Br.; Gen. Pl., II, 769.*

Gymnema sylvestre, *Br.; Fl. Br. Ind., IV., 29; Wight, Ic., t. 349;* 368 (748)
[ASCLEPIADEÆ

Syn. **ecylanica;** MELICIDA, *Edgew.*; PERIPLOCA SYLVESTRIS, *Willd.*; ASCLEPIAS GEMINATA, *Roxb.*

Var. ceylanica; distinguished by its ovate cordate leaves, softly pubescent on both surfaces. **G. sylvestre** *var.* **Decaisneana,** *Thw.* (*excl. syn.*)

Vern.—*Méra-singi,* HIND.; *Méra-singi, chhóta-dúdhi-lata,* BENG.; *Parpatrah,* DEC.; *Kavali, wakandi,* BOMB.; *Shiru-kurunjá,* TAM.; *Podapatra, putla-podara,* TEL.; *Binnúg,* SING.; *Meshuringi,* SANS.

References.—*Roxb., Fl. Ind., Ed. C.B.C., 256; Thwaites, En. Ceylon Pl., 197; Dalz. & Gibs., Bomb. Fl., 151; Grah., Cat. Bom. Pl., 120; Elliot, Fl. Andhr., 154; Pharm. Ind., 143; Ainslie, Mat. Ind., II., 390; Moodeen Sheriff, Supp. Pharm. Ind., 151; U. C. Dutt, Mat. Med. Hind., 309; Dymock, Mat. Med. W. Ind., 2nd Ed., 521; Hooper, an Examination of the Leaves of Gymnema sylvestre, Ootacamund, 1887; Edgeworth, Pharm. Jour., VII., 551; Balfour, Cyclop., I., 1277; Gazetteers:—N.-W. P., I., 82; IV., lxxvi., Bombay, XV., 438.*

Habitat.—A stout, large woody climber, native of Central and Southern India, from the Konkan to Travancore; distributed to Tropical Africa.

MEDICINE. Root. 369

Medicine.—The ROOT has long been esteemed amongst the Hindús as a remedy for snake bite, the powdered root being applied to the part bitten, and a decoction administered internally. But the most curious property of the plant was first noticed and described by **Mr. Edgeworth,** who discovered that on chewing some of the LEAVES the power of taste for saccharine materials was completely destroyed for 24 hours; powdered sugar tasting like so much sand. **Dr. Dymock,** criticising this statement says, that in his experience sugar taken into the mouth after chewing the fresh plant, had a saltish taste, but was still easily recognisable. **Mr. Hooper,** however, repeated the experiment and in his paper above cited, states that "after chewing one or two leaves it was proved undoubtedly that sugar had no taste immediately afterwards, the saltish taste experienced by others was due to an insufficiency of the leaf being used." He also further discovered that the leaf had the valuable property of completely removing the taste of bitters, sulphate of quinine after a good dose of the leaf tasting like "so much chalk." He, and several friends who also tried the experiment, found however that the effect did not last for 24 hours as stated by **Edgeworth,** but for only one or two. **Mr. Hooper** also found that **Gymnema** had no effect on the appreciation of the taste of pungent or saline articles, nor of astringents and acids. Though of opinion that this property might prove of value in pharmacy for the purpose of destroying the taste of quinine, he writes, "I am not going to propose its use in the administration of nauseous drugs, until the properties of the **Gymnema** have been more studied, otherwise the quantity of the vehicle taken may be proved to counteract the effect of the medicines." The leaves have been subjected to a careful and thorough analysis by **Mr. Hooper,** the results of which have been published in the above cited paper, to which the reader is referred for further information regarding their chemical composition.

Leaves. 370

371 **Gymnema tingens,** *W. & A.; Fl. Br. Ind., IV., 31.*

Syn.—GYMNEMA TINGENS, *var.* CORDIFOLIA, *Wight, Ic., t. 593;* BIDARIA TINGENS, *Dcne.;* ASCLEPIAS TINGENS, *Roxb.*

References.—*Roxb., Fl. Ind., Ed. C.B.C., 258; Balfour, Cyclop., I., 1277; Agri.-Hort. Soc. Ind., Journals (Old Series), VI., 143, 144; X., 294, 303.*

Habitat.—A climbing shrub, native of the Tropical Himálaya (from Kumáon to Sikkim), As-am, Sylhet, Lower Bengal at Monghyr, Pegu, and Travancore.

DYE. 372

Dye.—In 1795 **Dr. Buchanan** brought specimens of this plant from Pegu to India, and informed **Dr. Roxburgh** that the Burmans extracted a green colour from the leaves. The latter examined them and wrote: "It is probable that those people forgot to inform the Doctor that it was necessary to dye the cloth yellow, either before or after the application of the colour prepared from the leaves of this plant; in which case it will be the second species of **Asclepias** described and figured by me, which yields Indigo; though for my part, I have not succeeded in procuring that material from the leaves."

Gymnosporia, *W. & A.,* see **Celastrus,** *Linn.;* Vol. II., 237.

GYNANDROPSIS, *DC., Gen. Pl., I., 106.*

373 **Gynandropsis pentaphylla,** *DC., Fl. Br. Ind., I., 171;* CAPPARIDEÆ.

Syn.—CLEOME PENTAPHYLLA, *Linn.*

Vern.—*Húrhúr, húlúl, karaila, churota,* HIND.; *Hurhuriá, kánálá, ansarishá, arkahuli, sádáhurhuriá,* BENG.; *Seta kata arak,* SANTAL.; *Kathal parhar,* N.-W. P.; *Halhal,* DEC; *Kinro,* SIND.; *Tilávana, mábli,* MAR.; *Velai, neivaylla, kadughú,* TAM.; *Váminta, vela kura,* TEL.; *Tai-vélá, kara-vélá, vélá,* MALAY.; *Wéla,* SING.; *Surjávarta, arkapushpiká,* SANS.

(For vernacular names Conf. with **Cleome viscosa,** *Linn.*; Vol. II., 370. The two plants are much confused by Indian writers —*Ed.*)

References.—*Roxb., Fl. Ind., Ed. C.B.C., 500; Wight, Ill., I., 34; Stewart, Pb. Pl., 18; Elliot, Fl. Andhr., 189; Rev. A. Campbell, Ec. Prod., Chutia Nagpur, Nos. 8102, 8783; Pharm. Ind., 25; Pharmacographia Indica, 132; Ainslie, Mat. Ind., II., 451; O'Shaughnessy, Beng. Dispens., 206; Moodeen Sheriff, Supp. Pharm. Ind., 151; Mat. Med. of Madras, 26; U. C. Dutt, Mat. Med. Hind., 292, 319,; Dymock, Mat. Med. W. Ind., 2nd Ed., 62; S. Arjun, Bomb. Drugs, 13; Murray, Pl. and Drugs, Sind, 52; Baden Powell, Pb. Pr., 330; Atkinson, Him. Dist., 738; Drury, U. Pl., 238; Lisboa, U. Pl. Bomb., 145; Stocks, Rep. on Sind; Darrah, Note on the condition of the People of Assam; Home Dept. Cor. regarding a revised Pharmacopœia, 239; Indian Forester, III., 237; XII. (App.), 5; Gazetteers:—Orissa, II., 79; Mysore and Coorg, I., 57; N.-W. P., II., 79; V., lxvii.; X., 305; Trimen, Cat. Ceylon Pl, 4; Rheede, Hort. Mal., IX., t. 34.*

Habitat.—A small, annual herb, abundant throughout the warmer parts of India, and all tropical countries.

Oil.—When crushed in the fresh state, the HERB yields an acrid volatile oil having the properties of garlic or mustard oil. (*Pharmacog. Indica.*) **OIL. Herb. 374**

Medicine.—**Ainslie** noticed the medicinal properties of the plant, thus:—"The small numerous, warmish, kidney-formed black SEEDS, as well as the LEAVES of the plant, are administered in decoction in convulsive affections and typhus (?) fever, to the quantity of half a tea-cupful twice daily." "The Cyngalese physicians use our article for nearly the same purposes that the Vytians do." **Sir W. Jones** (*Vol. V., 138*) observed that the sensible qualities of Cáravélla promise great antispasmodic virtues, since it had a scent much resembling asafœtida, but comparatively delicate. Later **Dr. Wight** and several other English writers, following his opinion, have written of the bruised leaves as rubefacient and vesicant, producing a very copious exudation, and affording in many cases the relief obtained from a blister, without the inconveniences of the latter. **Moodeen Sheriff**, however, in his forthcoming *Materia Medica of Madras*, states that the leaves of **G. pentaphylla** are neither distinctly rubefacient, nor do they possess a scent resembling asafœtida; he maintains that these properties have been assigned to them owing to confusion of the plant under consideration with **Cleome viscosa**. His observations seem, however, at variance with the chemical nature of the fresh leaves above described under OIL. The following properties and uses appear to be unquestioned. The seeds are anthelmintic and rubefacient and are employed internally for the expulsion of round-worms, and externally as a counter-irritant. The JUICE OF THE LEAVES is used in many parts of the country as an anodyne instillation for the relief of otalgia and catarrhal inflammation of the middle ear. In certain localities the leaves are applied externally to boils to prevent the formation of pus. A decoction of the ROOT is said to be mildly febrifuge. The following note kindly furnished by **Moodeen Sheriff** indicates the best methods of exhibition of the various medicinal parts of the plant.

MEDICINE. Seeds. 375
Leaves. 376
Juice. 377
Root. 378

SPECIAL OPINIONS.—§ "The seeds are anthelmintic and the leaves a remedy for a few diseases of the ear. They are employed internally for the expulsion of round-worms; and externally their application to the skin is attended with relief in all cases in which mustard is indicated. The juice of the leaves is often used by natives for the relief of otalgia and otorrhœa and occasionally with success; but the burning sensation it produces in the ear, particularly in the cases of the latter disease, is a drawback to its employment. The seeds are prescribed internally in powder, which should always be used with sugar; and externally in the form of a poultice or paste, by bruising them with vinegar, lime-juice, or hot water.

MEDICINE. For the use of the ear, the juice of the leaves is to be pressd out by bruising them without water. The dose of the powder is from 30 grains to one drachm, with sugar, morning and evening for two days, and followed on the third morning by a dose of castor oil. For children the dose is from five to twenty grains according to their age." (*Honorary Surgeon Moodeen Sheriff, Khan Bahadur, G.M.M.C., Trlipicane, Madras*). "The expressed juice of this plant, a lttile warmed, is a useful remedy for earache. It is also used internally in cases of bronchitis. It is a purgative, but rarely used for that purpose." (*Civil Surgeon J. H. Thornton, B.A., M.B., Monghyr*).

FOOD. Leaves. (759) 379 **Food.**—The LEAVES are eaten by natives in curries, and as a pot-herb. (*Roxb.*)

DOMESTIC. Seeds. (760) 380 **Domestic Uses.**—The SEEDS are used, rubbed up with oil, as a vermicide in dressing the hair. They are also said to be employed for poisoning fish.

GYNOCARDIA, *R. Br.; Gen. Pl., I., 129.*

(761) 381 **Gynocardia odorata,** *R. Br.; Fl. Br. Ind., I., 195;* BIXINIÆ.

THE CHAULMUGRA OIL.; LUCRUBAU OR LUKRABO SEEDS.

Syn. – CHAULMOOGRA ODORATA, *Roxb.;* CHILMORIADODECANDRA, *Ham.*

Vern.—*Chálmúgra, chhalmúgra, chavulmungri,* HIND.; *Chaulmúgri, petarkura,* BENG.; *Kadu,* NEPAL; *Túk-kung,* LEPCHA; *Chaulmugra,* BOMB.; *Túngpung,* MAGH.; *Taliennoe,* SING.; *Brinj-mógrá,* PERS.; *Ta-fung-tsze,* CHINESE.

References.—*Roxb., Fl. Ind., Ed. C.B.C., 740; Kurz, For. Fl. Burm., I., 76; Gamble, Man. Timb., 18; Pharm. Ind., 26; Pharmacog. Indica, I., 142; O'Shaughnessy, Beng. Dispens., 206; Moodeen Sheriff, Supp. Pharm. Ind., 151; Mat Med. of Madras, 33; Dymock, Mat. Med. W. Ind., 2nd Ed., 67; Fluck. & Hanb., Pharmacog., 75; U. S. Dispens., 15th Ed., 1658; Bent. & Trim., Med. Pl., 28; Waring, Bazar Med., 41; K. L. Dey, Indig. Drugs of India, 58; Irvine, Mat. Med., Patna, 72; Martindale and Westcott, Extra Pharmacop., 5th Ed., 270; British Medl. Journ., 1879, I., 431, 968; 1880, II., 844; 1881, I., 475, 559; Lancet, 1887, II, 604; Practitioner, XXI., 321; XXII., 241; Year Book, Pharm., 1875, 215; 1879, 523; Christy, Com. Pl. and Drugs, II., 3; III., 37; IV., 20; V., 72; VI., 6; VII., 95; VIII., 85; XI., 76; Balfour, Cyclop., 1278; Smith Dic., 106; Kew Report, 1878, 30; Home Dept. Cor. regarding revised Pharmacopœia, 222, 233, 281, 305; Agri-Hort. Soc. of India, Trans., VII., 73, 74; Journals (Old Series), IV., 204; VI., 38; IX. (Sel.), 41; Contrib. Mat. Med. and Nat. Hist., China (1871), F. P. Smith, 140.*

Habitat.—A moderate-sized evergreen glabrous tree, readily known by the hard, round, fruits, which grow on the stem and main branches. It is found from Sikkim and the Khásia mountains eastward to Chittagong, Rangoon, and Tenasserim.

OIL. Seeds. (762) 382 **Oil.**—The SEEDS yield under the hydraulic press from 25 to 30 per cent. of OIL, to ether 51·5 per cent.; the latter oil turns green with the sulphuric acid test. (*Dymock.*) Chaulmugra oil has long been known and used in India as a remedy for cutaneous diseases, and has lately become a drug of some importance in European practice. As obtained in the bazars, the oil is commonly very impure, and the means of detecting these impurities, has formed the subject of careful investigation by **Dr. Dymock.** When pure, the expressed oil is clear, of a pale sherry colour, with the odour of *chaulmúgra,* and a sp. gr. of 0·9450 at 85° Fh. On being kept for some time it throws down a granular white fatty deposit. The oil obtained by **Dymock** on boiling the powdered seeds in water, however, was of a golden sherry colour and formed no deposit. From **Dymock's** experiments it would appear that genuine chaulmugra oil shows two marked re-actions with sulphuric acid,

OIL.

twenty minims of the oil (either cold-drawn or obtained by means of heat) with one minim of strong sulphuric acid giving first a brown, afterwards a rich olive-green, colour. If cold-drawn a tenacious reddish-brown resinous mass, which will not mix with the rest of the oil, forms round the drop of acid, while, if obtained by heat, no such re-action takes place. As it is of great importance that the oil for medicinal purposes should be pure, the following test described by **E. Heckel** and **F. Schlagdenhauffen** may also be shortly detailed. "They direct the oil to be mixed with an ethereal solution of ferric chloride, and the mixture to be evaporated until the oil becomes of a dirty green colour; it is then allowed to cool, and a few drops of sulphuric acid are added, which produce a fine greenish-blue colour. The colouring matter may be dissolved out by chloroform, with which it forms a dichroic solution like that of chlorophyll. This solution gives a deep absorption band extending from 40° to 70° of the scale (the sodium-ray coinciding with 50°). In proportion as the solution is diluted the black band becomes fainter, until, with a very weak solution, only a narrow, very pale band can be seen, extending between 40° and 48° of the scale." (*Dymock, quoting from Journ. de Phar. et de Chim., April 1st, 1885.*)

CHEMISTRY. 383 (763)

Chemistry.—An exhaustive analysis of the oil by **Mr. J. Moss, F. I. C., F.C.S.**, an account of which was read at the British Pharmaceutical Conference in 1879, shows that it has a decidedly acid re-action, that the melting point is 42°C., that it contains no alkaloid, and that its chief constituents are, Gynocardic acid, 11·7 per cent., Palmitic acid, 63 per cent., Hypogœic acid, 4 per cent., and Cocinic acid, 2·3 per cent. These acids exist in combination with glyceryl as fats, and the two former in the free state as well. The acid burning taste of the oil is stated to be due to the first mentioned acid. For a full account of the chemistry, and the processes employed in analysis, the reader is referred to the paper by **Mr. Moss** in the *Year Book of Pharmacy, 1879, 523—533.*

MEDICINE Seeds. 384 (764)

Medicine.—The SEEDS of this plant and those of the nearly allied **Hydnocarpus Wightiana** (*see p. 309*) have long been used in the East as a remedy for leprosy. Dymock states that in the *Makhzan-el-Adwiya* they are shortly noticed under the name of *chawul mungri*, and that their use in leprosy and other skin diseases is mentioned both as an internal and external remedy. In native practice of the present day the seeds and OIL are both largely employed, and are as a rule administered mixed with *ghí*. They are supposed to be alterative and tonic internally, stimulant when applied externally. They are principally employed in the treatment of leprosy and other skin diseases.

Oil. 385 (765)

In 1856 *chaulmugra* was brought to the notice of Europeans in India as a remedy for secondary syphilis; later it was described by **Dr. R. Jones** of Calcutta as an efficient alterative tonic in cases of phthisis and scrofula; and in 1868 it obtained a place amongst the officinal drugs of the *Pharmacopœia of India*. In that work it is recommended as an alterative tonic in cases of leprosy, scrofula, other skin diseases, and rheumatism, in doses of six grains of the powered seed in pill three times daily, to be gradually increased till nausea is produced, or five to six drops of the oil similarly increasing the quantity. Of late years the knowledge and use of the drug has spread to Europe, where it appears to be increasing in favour and reputation.

During the past ten or fifteen years it has been very favourably reported on in many medical publications, especially as a remedy for leprosy, psoriasis, eczema, scrofula, phthisis, lupus, marasmus, chronic rheumatism, and gout. The preparations most in repute in Europe are the pure oil, gynocardic acid (which is supposed to be the active constituent), and

MEDICINE. an ointment prepared from the oil. The remedy is generally employed both externally and internally, and it has been recommended that patients taking the medicine should live generously, as a weakly and under-nourished person is said to be specially liable to the bad effects of the drug, *viz.*, loss of appetite, nausea, and intestinal irritation. The oil is best administered in capsules and should be taken immediately after meals, gradually increasing the dose from 4 grains to 30, 40, or even 1 drachm, as recommended in the *Pharmacopœia of India*; the best form of ointment is one prepared with vaseline. Perhaps the most satisfactory and trustworthy results have been those obtained in the treatment of chronic and acute eczema, and other forms of skin disease.

A selection of the more interesting reports from the experience of European practitioners in India may be here quoted:—

SPECIAL OPINIONS.—§ "The oil is alterative and tonic internally and stimulant externally. It is the best of all the remedies in use for leprosy. If the disease be slight and of short duration, the beneficial influence of the oil is remarkable and rapid; but, if severe and long-standing, no improvement takes place under its use until it is persisted in for months or years. It is useful in all the forms of leprosy (tubercular, anæsthetic and mixed), but the first-named variety is the one which is most benefited by it. The internal use of the oil should always be assisted by its external application to the affected parts. If used alone externally, it proves itself too strong in some cases and excoriates the tender and diseased parts, or renders the ulcers irritable and painful, and I have, therefore, generally employed it in combination with *Margosa oil*, in the proportion of one of the former to two or three of the latter. It is also very useful in lepra vulgaris, psoriasis, and tertiary syphilis. During the use of this oil the patient may be kept on a good and nourishing diet, but prohibited from taking the following articles—fish, prawn, beef, brinjal, greens, curdled milk, lime-juice, and ardent spirits.

The oil is best taken floating on water or milk. Its dose is from 5 to 15 drops or more (gradually increased) three or four times in the 24 hours. If it produces nausea, vomiting or looseness of bowels, it should be stopped for one or two days and commenced again in somewhat smaller doses." (*Honorary Surgeon Moodeen Sheriff, Khan Bahadur, G.M.M.C., Triplicane, Madras.*) "Used both internally and externally in leprosy, scrofula, skin-diseases, and chronic rheumatism." (*Civil Surgeon J. H. Thornton, B.A., M.B., Monghyr.*) "The oil is emollient, used in skin affections, leprosy, syphilis, &c." (*Surgeon G. Price, Shahabad.*) "I have used the oil with great advantage in the different forms of chronic skin disease, but especially those of a specific nature." (*Surgeon R. D. Murray, M.B., Burdwan.*) "I have frequently used chalmugra seed-pulp, as well as the oil, in leprosy and obstinate skin diseases internally and externally, with considerable success." (*Assistant Surgeon Nanda Lall Ghose, Bankipore.*) "I have used the oil externally and internally in leprosy with the improvement that all oily substances produce." (*Civil Surgeon D. Picachy, Purneah.*) "Chaulmugra oil has been highly recommended as a cure of leprosy. I persistently used it for more than a year in the treatment of the various forms of this disease, but found it destitute of any therapeutic property." (*Brigade Surgeon C. Joynt, M.D., Poona.*) "Useful in skin diseases. It proved efficient in a case of psoriasis where other applications failed" (*Assistant Surgeon Shib Chunder Bhattacharji, Chanda, Central Provinces.*) "I have used the oil both externally and internally in the treatment of leprosy with very great success, the eruptions fading away and the skin assuming an almost natural appearance in about three weeks." (*Civil Surgeon D. Basu, Faridpore, Bengal.*) "Very use-

(765)

ful in leprosy and skin diseases." (*Civil Surgeon S. M. Shircore, Moorshedabad.*) "Is beneficial when used (both internally and externally) in the incipient stage of anæsthetic leprosy." (*Surgeon Anund Chunder Mukerji, Noakhally.*) MEDICINE.

Structure of the Wood.—Hard, close-grained, yellow or light brown; weight 47℔ per cubic foot. Used in Chittagong for making planks and posts. TIMBER. 386

Domestic Use.—The SEED-PULP is employed in Sikkim to poison fish. DOMESTIC. Seed-pulp. 387

Trade.—Dymock writes "the seeds come from Calcutta, and cost in Bombay about R15 per Bengal maund of 80℔. They are collected in the Lower Himálaya in December." The following report has been furnished from Chittagong, from which the great proportion of seed is obtained in Calcutta. "From 1,200 to 1,500 maunds of *Chaulmúgra* seed are annually exported from the Hill Tracts to the collectorate, *viâ* the Karnafuli river alone. For the export *viâ* the Smigu, Matamori, and Fenny there are no records. The Chittagong price is R3 and the Calcutta price R5 per maund." **Moodeen Sheriff** gives R5 per pound as the wholesale, annas 6 per ounce as the bazár price of the oil. TRADE. 388

Gypsophila Vaccaria, *W. & A.*, see **Saponaria Vaccaria,** *Linn.*; CARYOPHYLLEÆ.

GYPSUM, *Ball, Man. Geol. of India, III., 450.*

The hydrous calcium sulphate has numerous varieties of form; when occurring in an amorphous mass it is known as Gypsum, when in a more massive form it constitutes alabaster, and when in transparent crystals it is called selenite. Gypsum on being calcined loses its water of combination and is easily powdered, forming the important industrial substance, Plaster of Paris. As is well known, the addition of water to this powder causes it to become hard and compact, and at the same time slightly to expand in volume, a property which renders it invaluable for making casts or moulds.

Gypsum.—SELENITE, ALABASTER; GYPSE, *Fr.*; GYPS, *Ger.*; GESAS, *Ital.* 389

Vern.—*Kulnar, kurpúra-silasit,* HIND.; *Sach,* KASHMIR; *Karsi,* SPITI; *Sang-i-jeráhat,* SALT RANGE; *surma safed,* JHELUM; *Gach*=Plaster of Paris, PB.

References.—*Mallet, Geology of India, IV. (Mineralogy), 143; Baden Powell, Pb. Pr., 41; Atkinson, Ec. Mineralogy, Hill Dists., N.-W. P., 1877, 34; Mason, Nat. Prod. Burma, 31; Burma & its People, 588, 735; Balfour, Cyclop., I., 1279; Quarterly Journal of Agric., I., 119; Gazetteers—Bombay, V., 6; Panjáb, Shahpur, 13; Rawal Pindi, 11; District Man., Trichinopoly, 67, 68.*

Occurrence.—The following short account of the chief localities in India in which Gypsum is to be found, has been compiled for the most part from the article on the subject in *Ball's Economic Geology, l.c.* LOCALITIES.

MADRAS.—*Trichinopoly District.*—The mineral is abundant in many parts of the cretaceous rocks of this district, but is generally somewhat impure, occurring in concretionary masses and plates. It would answer when made into plaster of Paris for taking moulds, but would not do for casts where whiteness is required; and seldom occurs in sufficient quantity to be worth collecting, though plates of pure selenite are to be found. Madras. 390

Chingleput District.—Masses of gypsum and crystals of selenite occur in the clayey estuarine beds to the north of Madras, but not in any great abundance. According to **Mr. Foote**, however, supplies for use in the School of Art at Madras have been obtained from this source.

Nellore District.—Crystals of greater purity than those found near Madras are said to occur in the Eastern coast districts, of which Nellore is one. **Mr. Foote** considered that they might, with profit, be collected in the neighbourhood of the canal and forwarded to Madras.

GYPSUM.

Localities where Gypsum occurs.

LOCALITIES.

Bombay. 391 Bombay.—Gypsum in the form of selenite is found in small quantities in the marine deposits about Bombay, and in Kathiawar, and is also said to occur in parts of the Deccan in connection with deposits of salt. But the principal source of Gypsum in the presidency is Cutch, where large deposits occur especially in subnummulitic shales. **Ball** writes: "Although much of it might be obtained without greater trouble than picking up the pieces, it does not appear to be utilised except to a slight extent by goldsmiths, who are said to use it in a powdered state for polishing their wares."

Sind. 392 Sind.—**Mr. Blanford** states that it is found in deposits, frequently 3 to 4 feet thick, near the top of the Gaj beds of the Kirthar range. **Dr. Buist** in 1852 drew attention to the interesting fact that the art of making plaster of Paris was then known to the natives of Sind, who employed it in casting lattices and open-work screens for places in which a free circulation of air was desired.

Rajputana. 393 Rajputana and Central India.—**Mallet** states that about seven miles north-north-west of Nagor, in Jodhpur, a bed of gypsum, probably not less than 5 feet thick, exists in the alluvium, which **Mr. Oldham** believes to have been formed in a salt lake. It is also to be found at Dakoria and Bhaddana. The Nagor gypsum is dug to some extent, and is used in the Jaipur School of Art, for coarser purposes.

Panjab. 394 The Panjáb.—*Bannu District.*—The mineral is found in Kalabagh and in the Kahsor range but is not at present utilised.

Kohát District—Contains gypsum in great abundance; **Mr. Wynne** in fact states that there is probably more of this mineral than of the rock salt with which it is associated. **Ball** writes: "It might be obtained by open quarrying in any quantity, but it is not worked." Though the crops raised on the soil resting on an expanse of the mineral at Spina are believed by the Natives to be finer than those in any other part of the country, they do not appear to draw the natural inference, and utilise the abundant gypsum as manure.

Salt-Range.—In the districts including these hills gypsum occurs in enormous quantities, associated with the salt marls of silurian or pre-silurian age. When powdered it is used to mix with mortar, and some of the more compact varieties near Sardi are manufactured into plates and small ornamental articles. Selenite is said to sell at Lahore for R3-14 per maund, the purpose for which it is used being probably medicinal. (*Ball.*) The writer can find no record of a trade in gypsum from this extensive source of the mineral, but it may be mentioned as a proof of its quality, that all the plaster of Paris models sent from India to the Colonial and Indian Exhibition by **Dr. Watt** were made from Salt-Range gypsum. It was found to be very pure, of a perfectly white colour, of admirable quality, and in every way suited for casting purposes.

Spiti.—Considerable deposits of gypsum are found in this valley. **Mr. Mallet** states that he believes these deposits to be derived from thermal springs. Some of the mineral obtained from this locality is of snowy whiteness, and apparently in every way suited for the most delicate ornamental work, or for the manufacture of the purest plaster of Paris.

N.-W. Provinces. 395 North-West Provinces.—Deposits occur in the Dehra Dún range, from which plaster of Paris has been made and employed with success in the internal decoration of houses at Dehra and Mussúrí. The mineral has also been found in the Kumáon and Garhwál Districts from which, according to **Atkinson**, good plaster of Paris has been derived. In the latter district a dark green variety occurs which is occasionally made into saucers and bowls by the Natives.

Burma. 396 Burma.—Available literature points to the supply in Burma being so small that as a commercial source of gypsum it is practically useless. At Kyauk Tyan and on Amherst island sparsely dessiminated crystals of selenite occur, and a granular variety is said to be found on the banks of the Tenasserim river. **LOCALITIES.**

MEDICINE. 397 (777) **Medicine.**—Gypsum is considered by the Hindús to have cooling properties, a gruel made from it being frequently administered in cases of fever. It is kept in small quantities as a drug in most Indian bazárs.

DOMESTIC. 398 (778) **Domestic and Industrial Uses.**—It is occasionally burnt and used for chewing with betel instead of Carbonate of Lime. It is also employed for whitewash and as a plaster; but only in one locality, *viz.*, Sind, do the natives appear to understand its value for making mouldings, hard cements, &c., when in the condition of Plaster of Paris. Of late years, however, the knowledge of its properties and decorative uses has spread to some extent; it is employed in all art schools, and there is every probability that its utilisation will increase. Should this be the case, it would seem that the gypsum obtained from the Salt Range would be most likely to supply the demand. As already stated, small ornamental articles are carved in a few localities from certain of the harder forms of the mineral.

Manure. 399 (779) Gypsum is a valuable manure especially for leguminous crops, in the ash of which it is an important constituent, *see* **Manures,** Vol. V.

GYROCARPUS, *Jacq.; Gen. Pl., I., 689.*

400 (780) **Gyrocarpus Jacquini,** *Roxb.; Fl. Br. Ind., II., 461; Bedd., Fl. Sylv., t. 196;* COMBRETACEÆ.

Syn.—GYROCARPUS ASIATICUS, *Willd.;* G. AMERICANUS, *Jacq.;* G. ACUMINATUS, *Meissn.;* G. SPHENOPTERUS and G. RUGOSUS, *R. Br.*

Vern.—*Zaitun,* HIND.; *Tanuku mánu, ponuku, kummara ponuku, kumar pulki,* TEL.; *Pinlethitkauk,* BURM.

References.—*Roxb., Fl. Ind., Ed. C.B.C., 149; Kurz, For Fl. Burm., I., 470; Gamble, Man. Timb., 187; Elliot, Fl. Andhr., 103, 155, 173; Drury, U. Pl., 239; Lisboa, U. Pl. Bomb., 77; Balfour, Cyclop., I., 1280; Mysore and Coorg Gaz., I., 52, 71; Ind. Forester, III., 202; Madras Man. of Administr., I., 363.*

Habitat.—A tall deciduous tree common in South India up to an altitude of 1,000 feet, also in the Malay Peninsula, the Andaman Islands, and Tenasserim; rare in Bengal; distributed through the tropics of the whole world.

TIMBER. 401 (781) **Structure of the Wood.**—Grey, soft, very light, weight 23lb per cubic foot. It is used in South India to make boxes and toys, and is preferred to all others for *catamarans.* (*Beddome.*)

DOMESTIC. Seeds. 402 (782) **Domestic Use.**—The SEEDS are made into rosaries and necklaces. (*Beddome.*)

HÆMATOXYLON, *Linn.; Gen. Pl. I., 567.*

1 **Hæmatoxylon campechianum,** *Linn.; Bentl. & Trim., II., 86;*
LOGWOOD. [LEGUMINOSÆ.

Vern.—*Bokkan,* BENG.; *Partanga,* KAN.

References.—*Gamble, Man. Timb., 135; Mason, Burma and Its People, 511, 770; Pharm. Ind., 67; O'Shaughnessy, Beng. Dispens., 310; Fluck. & Hanb., Pharmacog., 213; Indian Forester, VI., 240; XIV., 29.*

Habitat.—A small spreading tree, native of Central America and the West Indies; wood imported into India. The tree is mentioned by **Mason** as cultivated in a few gardens in Burma, where it is said to flourish as if an indigenous plant.

DYE. Heart-Wood. 2 **Dye.**—A decoction of the chips of the HEART-WOOD is used for dyeing violet and blue, certain shades of grey, and more especially blacks, to which latter it gives a lustre and velvety cast.

MEDICINE. Heart-wood. 3 **Medicine.**—For use in medicine the hard HEART-WOOD is cut up into small chips by the aid of powerful machinery. These have a feeble, sea-weed-like odour, and a slightly sweet, astringent, taste. Prescribed as a decoction and extract, it is a mild astringent and tonic, and is useful in chronic diarrhœa, atonic dyspepsia, and infantile diarrhœa. The decoction has been found a valuable injection in leucorrhœa. It is also said to be useful as an ointment in cancer and hospital gangrene, in which cases it probably acts as a feeble antiseptic, diminishing fœtor and purulent discharges. Long-continued use has sometimes been followed by phlebitis, hence caution is necessary in its employment.

SPECIAL OPINIONS.—§"In cases of menorrhagia, women are often treated with a decoction of log-wood; which is also used as an injection" (*Surgeon-Major C. W. Calthrop, M.D., Morar*). "In cases of chronic atonic diarrhœa, I have found the decoction of log-wood a valuable addition to astringent mixtures. Its powers as a tonic are undoubted." (*Surgeon S H Browne, M.D., Hoshangabad, Central Provinces*).

CHEMISTRY. 4 **Chemical Composition.**—The important part of the plant is the red HEART-WOOD, which contains *Hæmatoxylin*, also a volatile oil and tannic acid.

Hæmatoxylin, when pure, is nearly colourless, very soluble in hot water and alcohol, but only sparingly so in cold water and ether. It is sometimes found crystallised in clefts of the wood. When exposed to the air under the influence of alkalies, it becomes red, *hæmatein* being formed.

5 **Hair dyeing** or staining as practised in India.

1st Process.—Mix equal parts of chalk and soap and half the quantity of lime; rub in a leaden pestle and mortar until the mixture acquires a bluish colour; apply this to the hair, rubbing in; tie up the hair within a cloth for about an hour; wash; thereafter apply a paste, which has been allowed to ferment to some extent, made of wheat flour, pulverised iron filings, and yeast; tie again for another hour; wash in a strong infusion of galls or of *ámlá* (**Phyllanthus Emblica**), the latter being cheaper; thereafter apply an oil to give a gloss. The colour thus obtained is very black and perfectly fixed, being rendered useless only by the growth of the hair below which reveals the original colour.

2nd Process.—Rub *henna* leaves on the hair and tie for an hour; wash; apply thereafter a paste of indigo or indigo leaves; wash and fix with galls or *ámlá*. This gives a bluish black, but as the indigo becomes rubbed off, the *henna* gives the hair tips a red tinge.

Halicore dugong, *Cuv.;* THE INDIAN DUGONG, see **Whales.**; Vol. VI.

HALOXYLON, *Bunge; Gen. Pl., III., 70.*

A genus of shrubs or small trees, having, according to the *Flora of British India*, from 8 to 10 species, of which 4 are Indian; namely, **H. recurvum, H. Thomsoni, H. salicornicum, and H. multiflorum.** **Mr. Lace** has, however, in his collection of the CHENOPODIACEÆ of Baluchistan (Quetta, Sibi, &c.) added **H. Griffithii**, *Bunge*, and also what appears to be a new species to which the name of **H. Laceii** may be assigned. This somewhat resembles **H. multiflorum**, but is more slender, more profusely branched, and has large pink or yellow flowers. It is known in Sibi as *Shorái* and is recognised by the natives as quite distinct from **H. multiflorum** (the *lána*). Like the other species it is eaten by camels, but in this respect **H. Griffithii**, *Bunge*, is much preferred, even to *pilú* **(Salvadora olioides)**. (Compare with correction under **Caroxylon Griffithii**, Vol. II., 176.)

Haloxylon multiflorum, *Bunge; Fl. Br. Ind., V., 16;* CHENOPODIACEÆ. 6

Syn.—ANABASIS MULTIFLORA, *Moq.*

Vern.—*Shalme, lana,* TRANS-INDUS; *Lana,* SALT RANGE and RECHNAB DOAB; *Metra lana, gora lana,* BARI DOAB; *Dána, shori lána,* FEROZEPUR.

References.—*Stewart, Pb. Pl., 176; Baden Powell, Pb. Pr., 372; Settlement Reports:—Panjab, Jhang, 23; Montgomery, 18, 19.*

Habitat.—A low shrub, from 4 to 6 inches high, of the north-western Panjáb plains; distributed to Afghánistan and Balúchistan.

Medicine.—This species is mentioned in *Baden Powell's List of Drug-yielding Plants*, but no account is given of its properties or uses. MEDICINE. 7

Fodder.—Camels are fond of the PLANT. FODDER. Plant. 8

Domestic.—In the settlement report of the Montgomery district *gora lana*, or *metra lana* (both, according to **Stewart**, vernacular names for **H. multiflorum**) are mentioned as being used in the manufacture of BARILLA or *Sajji-khar*; but the plant referred to is probably **Salsola Kali**, *Willd.* (See article on BARILLA, *Vol. I., 396.*) **Stewart** writes, "It is used in some parts for washing clothes, and it is probably this which **Bellew** states to be used by women in the Peshawar valley for washing the head." DOMESTIC. 9

H. recurvum, *Bunge; Fl. Br. Ind., V., 15; Wight, Ic., t. 1794.* 10

Syn.—H. RECURVUM and H. STOCKSII, *Hook. f.;* CAROXYLON RECURVUM, *Moq.;* C. INDICUM, *Wight;* SALSOLA STOCKSII, *Boiss.;* S. LANA, *Stocks;* S. RECURVA, *Wall.*

Vern.—*Laghme,* TRANS-INDUS; *Khar,* CIS-INDUS, PB.; *Khári-láni* or *kári-láni,* SIND.

References.—*Stewart, Pb. Pl., 178; Aitchison, Cat. Pb. and Sind Pl., 127; Murray, Pl. and Drugs, Sind, 104; Settlement Reports:—Panjáb, Jhang, 23; Lahore, 13; Montgomery, 18, 19; Dera Ismail Khan, 8.*

Habitat.—A straggling bush, several feet in height, with long-spreading recurved branches; found in the Western Panjáb plains and the Salt Range, ascending to 2,500 feet; also in Sind, South-Western India, and Burma; distributed to Afghanistan.

Medicine.—The PLANT is used in the manufacture of *Sajji-khar* (see article on BARILLA, *Vol. I., 396*). MEDICINE. Plant. 11

Fodder.—It is a favourite food of camels, forming indeed their principal fodder in many parts of the Panjáb and Sind. The farmers of *sajji*, however, do not allow the camel-owners to take the plants *gratis* as a jungle product. Large quantities, notwithstanding this, are said by **Edgeworth** to be taken into Múltan for camel food (see Article on *Camel Fodder, Vol. II., 58*). FODDER. Plant. 12

13 HAMILTONIA, *Roxb.; Gen. Pl., II., 135.*

Hamiltonia suaveolens, *Roxb.; Fl. Br. Ind., III., 197; Bedd., [Fl. Sylv. t. 17 f. 3*; RUBIACEÆ.

Syn.—H. SCABRA, *Don;* H. MYSORENSIS, *W. & A.;* H. PROPINQUA, *Dcne.;* LASIANTHUS TUBIFLORUS, *Blume;* SPERMADICTYON SUAVEOLENS, *Roxb., Cor. Pl.;* S. AZUREUM, *Wall.*

Vern.—*Kudia,* KOL; *Bainchampa,* NEPAL; *Padera,* KUMAON; *Kanéra, pudári, phillu, muskei, kantálú, fisáúni, niggi, tulenni, phúl, gohinla,* PB.; *Mahabal,* C. P.

References.—*Roxb, Fl. Ind., Ed. C. B. C., 186, Hort. Beng., 15; Brandis, For. Fl., 278; Beddome, Fl. Sylv., 134; Gamble, Man. Timb., 119; Stewart, Pb. Pl., 115; Rev. A. Campbell, Cat. Ec. Prod., Chutia Nagpore, No. 8109; Atkinson, Him. Dist., 311; Indian Forester, II., 26; III., 203; XII., App. 15; Gazetteer of Mysore and Coorg, I., 56.*

Habitat.—An undershrub from 4 to 12 feet in height, with spreading branches; it occurs at altitudes from 1,000 to 5,500 feet in the Tropical and Subtropical Himálaya, from Kashmír to Bhotan, and the Salt Range; also found in Central India and the Western Peninsula, ascending to 4,000 feet on dry rocky hills from Marwar and Behar southwards to Mysore.

FODDER. Leaves. 14 **Fodder.**—Buffaloes eat the LEAVES. (*R. Thompson*).

TIMBER. **Structure of the Wood.**—Very small, but said by **Brandis** to be used in Chamba to make charcoal for gunpowder.

15 HARDWICKIA, *Roxb.; Gen. Pl., I., 586.*

16 Hardwickia binata, *Roxb.; Fl. Br. Ind., II., 270; Bedd. Fl. Sylv., [t. 26*; LEGUMINOSÆ.

Vern.—*Anjan, aojan,* HIND.; *Anjan,* KOL; *Chhota dundhera,* GOND; *Anjan,* C. P.; *Parsid,* BOMB.; *Anjan,* MAR.; *Bone,* KURKU; *Achá, alti, kat-udugu, Chita cha* (the fibre), TAM.; *Nar yepi, épe, nára épe, yapa,* TEL.; *Kamrá, karacho, karachi, asanagurgi,* KAN.

References.—*Roxb., Fl. Ind., Ed. C.B.C., 378; Gamble, Man. Timb., 143; Dalz. & Gibs., Bomb. Fl., 83; Elliot, Flor. Andhr., 51, 129; O'Shaughnessy, Beng. Dispens., 312; Drury, U. Pl., 240; Lisboa, U. Pl., Bomb., 64; Birdwood, Bomb. Pr., 329; Cross, Bevan, King, Rept. on Ind. Fibres, 9, 53; Indian Forester, II., 18, 19; IV., 292, 322, 366; VI., 105, 332; VIII., 103, 106, 112, 118, 119, 122, 123, 125, 126, 130, 369, 376, 400, 411, 412, 414, 417, 439; IX., 274, 283, 286, 451, 467; X., 546, 547, 550, 551; XIII., 120, 126; XIV., 269, 271; Spons, Encyclop, 961; Balfour, Cyclop, II., 17; Treasury of Bot., I., 569; Settlement Reports: Central Provinces; Nimar, 307; Upper Godavery Dist., 37; Chanda, App., VI; Bilaspore, 77; Bhandara, 19; Manual:—Madras Administration, I., 313; Trichinopoly Dist., 78; Coimbatore Dist., 401; Cuddapah Dist., 262; Gazetteers:—Mysore and Coorg, I., 50; II., 8; Bombay, XV., 70; XVI., 17.*

Habitat.—A tall deciduous tree, reaching a height of 100 feet, found in Behar and the Western Peninsula, ascending to 3,500 feet (*Fl. Br. Ind.*).

Found in the dry forests of South and Central India, but not everywhere; mostly gregarious in isolated belts or patches of greater or less extent. Very commonly found on sandstone, but also met with in trap and granite. Wanting in the Western moist zone, and not found in Northern India, though present as far north as the Banda district of the North-Western Provinces. (*Gamble.*)

GUM. 17 **Gum.**—It yields a gum. (*Gamble.*) Though no definite information exists regarding this fact, it appears probable that the so-called gum may really be a form of balsamic oleo-resin resembling that abundantly obtained from the other species.

TAN. Bark. 18 **Tan.**—In the Settlement Report of the Chanda district the BARK is said to be used for tanning.

Fibre.—The fibre obtained from the inner layers of the BARK is slightly woolly and of a brownish colour. Chemically it ought to be a valuable fibre, as it contains 63·4 per cent. of cellulose to 13·3 of water. It is strong, and used in the manufacture of ropes by the natives of the forests in which it abounds. Elliot, in his *Flora Andhrica*, writes, "I saw the people extracting the fibre in considerable quantities and employing it as cordage, without further preparation, in the island of Sivasamudram." It is also a very useful paper-making material, but it would not, at the present rates of freight, pay to send it to England for that purpose. FIBRE. Bark. 19

Fodder.—The LEAVES are given as fodder to cattle, and, owing to this, fine trees are often so much lopped as to be destroyed. Beddome referring to this subject, in his *Flora Sylvatica of Southern India*, writes, "It is naturally of straight growth, but, cattle being very fond of its leaves, it is pollarded to a frightful extent wherever it grows. It is heart-rending to see the damage done in the Cauvery Forests." FODDER. Leaves. 20

Structure of the Wood.—Sapwood small, white; heartwood extremely hard, dark red, often with a purplish tinge, and very close-grained; weight about 82lb per cubic foot. This wood, perhaps the hardest and heaviest in India, is extremely durable and does not warp, though it has a tendency to split. It is used for bridge and house posts and for ornamental work. It has been recommended for sleepers but is probably too hard, heavy, and difficult to work to be much in favour. (*Gamble.*) At the Dehri workshops it has been used instead of brass for bearings of machinery, and has been found to wear well. In the Sone River piles of *anjan* have been found after twenty years as sound as when first put in. It is much used for building purposes, also for making carts and agricultural implements in the districts where it occurs; but, owing to its hardness and weight, it is difficult to bring to market. The shoots grow very straight, and are hence valuable as rafters. [*t.* 255. TIMBER. 21

Hardwickia pinnata, *Roxb.; Fl. Br. Ind., II., 270; Bedd., Fl. Sylv.,* 22

Vern.—*Matúyen samprâni*, TRAVANCORE; *Kolávu*, TINNEVELLY; *Yenne*, MANJARABAD.

References.—*Roxb., Fl. Ind., Ed. C. B. C., 378; Gamble, Man. Timb., 143; Dymock, Mat. Med. W. Ind., 287* (under heading of **H. binata**); *Fluck. & Hanb., Pharmacog., 232; Cooke, Gums and Gum-resins, 116; Atkinson, Gums and Gum-resins, 34; Indian Forester, III., 22, 23, 24; VIII. 415; Spons, Encyclop., 1640, 1654; Balfour, Cyclop., II., 17; Treasury of Bot., I., 569; Gazetteer, Mysore and Coorg, I., 52.*

Habitat.—A very large tree of the Ghâts of Kanára, Travancore, and the Karnatic.

Balsamic Oleo-Resin.—An oleo-resin of a dark red colour is obtained by deep incision into the heart of the tree. In properties and chemical composition it much resembles the wood oil obtained from species of **Dipterocarpus**, and the balsam of Copaiva (see articleson **D. turbinatus**, Vol. III., 161). *In Beddome's Flora Sylvatica*, the following report on the analysis of the balsam by Mr. Broughton is given:—"The substance appears on examination to consist of a solution of certain chemically different resins in an essential oil, and is in fact an oleo-resin. Like the wood-oils from the different species of **Dipterocarpus**, it greatly resembles both in composition and properties the Copaiva balsam, though it lacks the transparency and light yellow colour of the latter. It is nearly entirely soluble in ammonia, but does not produce a clear solution. The essential oil has the same composition as that from Copaiva balsam. It boils (on the Nilghiris) at a temperature of 225°C. It rotates the plane of polarization to the left, but to a different degree from that found with the oil of Copaiva. This essential oil occurs in different amounts in the balsam, BALSAMIC OLEO-RESIN. 23

BALSAM.

and more abundantly in the fresher collected specimens; these are quite fluid, but other specimens are almost semi-solid, doubtless owing to the evaporation and oxidation of the oil. The oil is best obtained by prolonged distillation of the balsam with water. By this means I have obtained, from an apparently old specimen of balsam, 25 per cent. of oil, and in the most recently collected specimens, I have found over 40 per cent. I have made many attempts but have not obtained any crystals of copaivic acid from the balsam. The solid resins are of an acid character, but the balsam does not solidify so strongly as that of copaiva after being heated with magnesia. The oil can be separated from the balsam by **Ader's** process, but it is obtained in a very impure and coloured state." The authors of the *Pharmacographia* describe this oleoresin as a thick viscid fluid, which, owing to its intense tint, looks black when seen in bulk by reflected light, though it is perfectly transparent. It is dichromic by transmitted light, in a thin layer appearing yellowish green, in a thick layer vinous red.

MEDICINE. Oleo-resin. 24

Medicine.—The OLEO-RESIN forms an efficient substitute for copaiva, being useful in gonorrhœa and catarrhal conditions of the genito-urinary and respiratory tracts.

Dr. Dymock, in his *Materia Medica of Western India*, probably inadvertently quotes the *Pharmacographia* account of the oleo-resin of **H. pinnata** under the name of **H. binata**, but mentions that it is neither known nor used in Bombay. **Mr. Broughton**, in his account of the oleo-resin in *Beddome's Flora Sylvatica*, writes, "There appears little doubt that this balsam could effectually substitute copaiva balsam in medicine. But the appearance of the specimens that I have received is greatly inferior to the latter, and they could not, certainly under present circumstances, compete with the Brazilian balsam in the European market." It appears, however, that this balsam has by no means attracted the attention that it deserves; indeed, mention of it seems to have been made only in two out of the numerous list of works which treat of Indian Materia Medica. It is doubtful, however, whether the oleo-resin or the essential oil obtained from it could be manufactured and exported with profit; still there is no doubt of its valuable properties, and it is probable that, in India at least, it might become an efficient and economical substitute for copaiva balsam.

TIMBER. 25

Structure of the Wood.—Sapwood large; heartwood brown; weight 47℔ per cubic foot; used, in the neighbourhoods in which it abounds, for building purposes.

26

HARES.

The animals so known constitute the family **Leporidæ** of the Natural Order RODENTIA.

The family is divided into two genera—

1. **Lepus**, or true hares.
2. **Lagomys**, or mouse hares.

The family is generally distributed over the world except in Australasia, and many species are met with in India, more especially in the hills on the borders of Tibet, Afghánistan, and Balúchistan.

References.—*Jerdon, Mammals of India, 223; Blanford, Scientific Results of the Second Yárkand Mission, Mammalia, 60; Blanford, Eastern Persia, II., 81; Blyth, Cat. Mam. and Birds of Burma, 43; Murray, Vertebrate Zoology of Sind, 49; Sterndale, Mammals of Ind., 368; Forbes Watson, Indust. Surv. of Ind., 327, 382; Balfour, Cyclop., II., 704; Spons, Cyclop., 1031.*

Indian Hares. (*J. Murray.*) HARES.

Lepus, *Linn.* 27

The species of this genus found in India and the adjoining countries (the furs of which are utilised or brought into this country) are:—

1. **Lepus craspedotis,** *W. Blanf.*

THE LARGE-EARED HARE. 28

A native of Balúchistan, described and named by **Mr. Blanford** in *Eastern Persia, vol. II., p. 81.*

2. **L. hypsibius,** *W. Blanf.*

THE MOUNTAIN HARE. 29

Habitat.—Found during the Second Yárkand Mission in Northern Ladakh.

3. **L. hispidus,** *Pearson.* 30

THE HISPID HARE; THE BLACK RABBIT.

Habitat.—A rabbit-like hare, of the Himálayan Terai from Gorakhpur to Assam, extending south to Dacca, probably still further, even it is said to the Rajmahal hills. (*Jerdon*).

4. **L. nigricollis,** *F. Cuv.*

THE BLACK-NAPED HARE. 31

Vern.—*Khargosh,* HIND.; *Sassa,* MAR.; *Músal,* TAM.; *Kúndéli,* TEL.; *Malla,* KAN.; *Haba,* SING.

Habitat.—Southern India and Ceylon; said also to occur in Sind and the Panjáb, but according to **Jerdon** this statement requires confirmation.

5. **L. pallipes,** *Hodgson.*

THE PALE-FOOTED HARE.

Vern.—*Togh, toshkhen,* YARKANDI. 32

Habitat.—Yarkand, Tibet.

6. **L. pamirensis,** *W. Blanf.* 33

THE PAMIR HARE.

Habitat.— Described by **Blanford** as found near Lake Sirikal in Pamir.

7. **L. peguensis,** *Blyth.* 34

THE PEGU HARE.

Vern.—*Yung,* BURM.

Habitat.—A hare much resembling **H. ruficaudatus,** but with the tail black above; found in Pegu, inhabiting the open country within or beyond the range of forests. (*Blyth*).

8. **L. ruficaudatus,** *Geoff.*

THE COMMON INDIAN HARE. 35

Vern. —*Khargosh, lamma,* HIND.; *Kharra, sasrú,* BENG.; *Molol,* GOND.; *Kharra,* C. P.

Habitat.—From the foot of the Himálaya, southwards to the Godavery on the east, and on the west as far as the Taptí at least, perhaps further, extending from the Panjáb to Assam. It is also supposed to be found in Afghánistan.

9. **L. stoliczkanus,** *W. Blanf.* 36

Habitat.—A species closely allied to **L. pamirensis,** found in the Thian Shan range, north and north-west of Káshgar, by **Dr. Stoliczka.**

10. **L. tibetanus,** *Vigne, Waterhouse.*

THE TIBET HARE. 37

Habitat.—Little Tibet and Ladakh.

11. **L. yarkandensis,** *Günther.*

THE YARKAND HARE.

Vern.—*Toshkhan,* YARKANDI 38

Habitat.—The plains of Yárkand and Káshgar.

FOOD. 39 **Food.**—All the species of **Lepus** are probably used as articles of food, but they are as a rule dry and tasteless, and by no means so good for the table as the European hare. In his list of "Mammals used for Food" in the *Industrial Survey of India*, **Dr. Forbes Watson** mentions **Lepus nigricollis, L. pallipes, L. peguensis,** and **L. ruficaudatus.** Of these **L. nigricollis** is generally considered the best, one writer describing it as "nearly, if not quite, equal to the English hare in flavour."

DOMESTIC. Fur. 40 **Domestic.**—The FUR is valuable in trade, being used largely for linings, cloaks, &c., also in the manufacture of felt for hats. Referring to the latter use, the author of the article "Hare" in *Spons' Encyclopædia* writes: "It is about the best hair for hat-maker's purposes." In his list of "Furs and Fur-bearing Mammals" **Dr. Forbes Watson** gives only **Lepus nigricollis, L. pallipes,** and **L. ruficaudatus,** but the other species found in Thibet and Yárkand are doubtless equally valuable.

41 **Lagomys,** *Cuv.*

THE PIKAS, OR MOUSE-HARES.

A genus of small and robust animals, short-eared and tailless, referred by **Jerdon** to the family of **Leporidæ** or Hares, but separated by **Blanford,** in his account of the Mammalia of the second Yarkand Mission, as a distinct family, **Lagomyidæ.** It is convenient for the purposes of this work to consider them under **Jerdon's** classification as Hares. The species are:—

42 1. **Lagomys auritus,** *W. Blanf.*

Habitat.—A species differing from **L. roylei** in having much longer ears and a lighter colour. Found on the Panong lake, but probably also a native of other parts of Ladákh.

43 2. **L. curzoniæ,** *Hodgson.*

Vern.—*Phise-karin,* KORZAK.

Habitat.—The higher ranges of the Himálaya from 14,000 to 19,000 feet. It has been found in Ladákh and Sikkim.

44 3. **L. griseus,** *W. Blanf.*

Habitat.—Resembles **L. auritus,** but amongst other differences has longer and softer hair, of a grey colour. Found in the Kuenlun Range in Yárkand, south of the Sunju pass.

45 4. **L. ladacensis,** *Günther.*

THE LADAKH PIKA.

Vern.—*Zabra, karin, phise karin,* LADAKH.

Habitat.—A species found in the high plateaux of Ládakh, with a full and very soft fur, and separated by **Blanford** from **L. curzoniæ,** with which it was formerly supposed to be identical.

46 5. **L. macrotis,** *Günther.*

Retained by **Blanford** as a distinct species, though he is strongly disposed to suspect that **L. auritus** is the summer, and **L. macrotis,** the winter, form of the same species. First described by **Günther,** found by the Second Yárkand Mission, at a place called Dúba on the north side of the Kuenlun.

47 6. **L. roylei,** *Ogilby.*

THE HIMÁLAYAN MOUSE-HARE.

Vern.—*Rang-rúnt,* or *rang-dúni,* KUNAWAR.

Habitat.—The Himálaya from Kashmír to Sikkim. The species was first made known through a specimen obtained by **Royle** from the Chór Mountain, near Simla.

48 7. **L. rufescens,** *Gray.*

Habitat.—Afghanistan and Persia.

Several other species have been described by various writers, but all have been referred by **Blanford** to one or other of the above species.

Domestic Use.—Dr. **Forbes Watson** mentions **Lagomys roylei,** probably the only species known to him, as yielding useful FUR. It is probable, however, from the descriptions of the furs of the other species, that they also may be of value. DOMESTIC. Fur. 49

Hariali Grass, see **Cynodon Dactylon,** *Pers.*, Vol. II., 678.

Haricot Bean, see **Dolichos Lablab,** Vol. III., 184, also **Phaseolus,** Vol. V.

Haritaki Nut, see **Terminalia Chebula,** Vol. VI.

Hartshorn. 50

Vern.—*Sabarsinghadú,* GUJ.; *Swarsingh,* MAR.; *Mrigasringa,* SANS.

Medicine.—The ANTLER of the deer, incinerated in closed vessels, is used in painful affections of the heart, pleurodynia, sciatica, and lumbago. It is given in doses of about twenty-two grains with clarified butter. (*U. C. Dutt.*) In former times hartshorn was used for the manufacture of ammonia (see *Ammonic chloride,* Vol. I., 219). (For further information consult **Horns**). MEDICINE. Antler. 51

Hawthorn, see **Cratægus,** Vol. II., 583.

Hay and Hay-making, see **Fodder,** Vol. III., 407; also in Appendix.

Hazel Nut, see **Corylus,** Vol. II., 574.

HEDERA, *Linn.; Gen. Pl., I., 946.*

A genus of shrubs, which climb extensively on small trees, comprising two species, of which one is found in all the temperate regions of the Old World. The generic name HEDERA is probably derived from a Celtic word *hedra,* signifying a cord, while the English, Ivy, is from *iw,* a word of the same language, meaning green.

Hedera Helix, *Linn.; Fl. Br. Ind., II., 739;* ARALIACEÆ. 52

THE IVY.

Vern.—*Dúdela,* NEPAL; *Lablab,* BEHAR; *Halbambar, arbambal,* (JHELUM), *Kurol* (CHENAB), *Kuri, karúr* (RAVI), *Brúmbrúm dakári* (BEAS), *Karbaru, kaniúri, kadloli* (SUTLEJ), *Banda,* (KUMAON), PB.; *Parwata,* PUSHTU; *Karmora, mandia,* KASHMIR.

References.—*Brandis, For. Fl., 248; Gamble, Man. Timb., 210; Stewart, Pb. Pl., 110; also, Tour in Hazara and Khágán, Agri. Hort. Soc. Journ. (Old Series), XIV., 31; Irvine, Mat. Med Patna, 63; Atkinson, Him. Dist., 311; Birdwood, Bomb. Pr., 44; Mason, Burma & Its People, 434, 742; Indian Forester, XIII, 68; Balfour, Cyclop., II., 30; Smith, Dic., 224; Gazetteer:—Mysore and Coorg, I., 61; Shahpur Dist., 69; Bannu, 23; Dera Ismail Khan, 19.*

Habitat.—A large woody climber of the Himálaya, at altitudes from 6,000 to 10,000 feet, and of the Khásia Hills, from 4,000 to 6,000 feet.

Medicine.—The Ivy was at one time highly valued in medicine, but is now almost completely discarded. Dr. **Irvine,** however, in his *Materia Medica of Patna,* remarks, "Ivy LEAVES are used to stimulate sores, and the BERRIES to purge." MEDICINE. Leaves. 53 Berries. 54

Food and Fodder.—Dr. **Stewart** writes: "It is stated to be a favourite food of goats, and in Kulu the LEAVES are said to be added to the beer of the country to make it strong." Its BERRIES afford abundance of food for birds. FOOD & FODDER. Leaves. 55 Berries. 56

Structure of the Wood.—White, soft, porous. Weight 34lb per cubic foot. TIMBER. 57

58 **Hedges & Fences**—List of woody plants used for—

For fuller particulars on this subject consult **Cleghorn's** *Forests and Gardens of India, pp. 201—211.*

Agave americana and vivipara (extensively employed for railway hedges throughout India).

Acacia—various species, especially **Farnesiana** (cut down and used as dry fences).

Bambusa—various species of Bamboo are grown as hedges.

Bauhinia racemosa.

Cactus and **Opuntia** species (used in Madras and the hotter parts of India).

Cæsalpinia sepiaria, Sappan and **Bonducella.**

Carissa Carandas and **spinarium.**

Caparis (various species used in Upper India).

Dodonea viscosa.

Erythrina indica (grown in hedges).

Euphorbia Royleana (frequent on the lower hills).

E. Tirucalli in the plains).

Grewia multiflora and **G. oppositifolia.**

Helicteres Isora.

Hibiscus (various species).

Hippophæ rhamnoides (on the higher Himálaya).

Hypericum Hookerianum.

Jatropha Curcas.

J. glandulifera (extremely abundant in Bengal).

Lawsonia alba (very useful and ornamental).

Macaranga denticulata
M. gummiflua } (by the hill tribes of the Eastern Himálaya).

Morus (several forms).

Myrsine africana (occasional on the Himálaya).

Pandanus odoratissima.

Parkinsonia aculeata (largely used in Madras).

Pedilanthus tithymaloides (an introduced American plant, extensively used in Bengal).

Pithecolobium dulce.

Polyalthia longifolia (sown thick, is used as a hedge in Calcutta, and if well trimmed and kept down, it closely resembles the beech hedges of England).

Quercus pachyphylla.

Randia dumetorum.

Rosa macrophylla.

Sesbania ægyptiaca and **grandiflora.**

Spiræa sorbifolia and other species (in the Himálaya).

Streblus asper.

Zizyphus—various species both as hedges and dry fences.

HEDYCHIUM, *Kœn.; Gen. Pl., III., 642.*

A genus of plants belonging to the Natural Order Scitamineæ, in the Tribe Zingibereæ, which comprises about 28 Indian species, of which one is important economically, while some of the others are much cultivated as garden plants, and known as Garland Flowers. The finest of these is Hedychium coronarium, of which **Roxburgh** wrote: "This to me is the most charming of all the plants of this Natural Order that I have met with; the great length of time it continues to throw out a profusion of large, beautiful, fragrant blossoms, makes it particularly desirable."

Hedychium spicatum, *Ham.*; SCITAMINEÆ. 59

THE LESSER GALANGAL *of Ainslie, according to Royle but not of other Authors; see* **Alpinia officinarum,** *Hance, Vol. I., 195.*

Vern.—*Sit-ruti, kapúr kachri,* HIND.; *Kachúr-kacha, kapúr kachri, banhaldi,* N.-W.P.; *Kachur-kachu, ban kela, sáki, banhaldi, khor, shalwi, shedúri,* (*Bazár root*)=*kapúr kachri, kachúr,* PB.; *Sir, sutti,* BOMB.; *Kapur krachari,* MAR., GUZ.; *Shimai-kich-chilik kishangu,* TAM.; *Kapurakáchali,* SANS.

References.—*Stewart, Pb. Pl., 239; Pharm. Ind., 232; Ainslie, Mat. Ind., I., 140; O'Shaughnessy, Beng. Dispens., 652; Dymock, Mat. Med. W. Ind., 2nd Ed., 780; S. Arjun, Bomb. Drugs, 141; Med. Top. Ajmir, 142; Irvine, Mat. Med, Patna, 48; Atkinson, Him. Dist., 738, 774; Birdwood, Bomb. Pr., 87; Royle, Ill. Him. Bot., I., 358; Buck, Dyes and Tans, N.-W. P., 38; Liotard, Dyes, App., III; Balfour, Cyclop., II., 30.*

Habitat.—Common in parts of the Panjáb Himálaya, and Nepál, at an altitude of from 3,500 to 7,500 feet.

Dye.—The aromatic ROOT-STOCKS are often used as an auxiliary in dyeing, to impart a pleasant smell to the fabric, and are chiefly employed along with *henna* dye (**Lawsonia alba**) in preparing the cloth known in the North-West Provinces as *Malagiri.* They are sometimes confused with the yellow root-stocks of **Curcuma aromatica,** *Salisb.* (*Conf.* Vol. II., 655.) In the bazárs these root-stocks are known as *Kapúr-kachri,* or *kachúr.* DYE. Root-stocks. 60

Medicine.—*Kapúr-kachri* is found in the drug-shops of India in slices, varying in thickness, and from ½ inch in diameter to much smaller sizes. These slices are starchy, white internally, reddish brown externally, have a strongly aromatic odour, and a pungent, bitter taste. There are two varieties, the Indian and Chinese, the latter of which is larger, whiter, and less pungent. By the Natives they are much used as stomachic, carminative, tonic, and stimulant medicines. *Kapur-kachri* is non-officinal, but it is mentioned in the Indian Pharmacopœia. According to **Baden Powell,** it is principally employed in veterinary medicine. MEDICINE. 61

Royle considered the lesser Galangal of **Ainslie** to be this plant, but apparently incorrectly. The description, properties, and vernacular names in the *Materia Indica* seem to point to **Ainslie's** plant being the true lesser Galangal—**Alpinia officinarum.** The medicinal uses of the two are, however, very similar, and indeed considerable confusion seems to have existed in the species of **Curcuma, Alpinia, Amomum,** and **Hedychium.** The reader is referred for further information to the articles on those genera in Vols. I. and II.

Domestic and Sacred Uses.—*Kapúr-kachri* is the principal ingredient in the three kinds of ABIR or scented powder used as a perfume, and by the Hindus in their religious ceremonies. It is said also to protect cloth from the attacks of insects. In Garhwál it is employed to wash the newly married. **Madden** affirms that the pounded ROOT is smoked in the hookah with tobacco. The LEAVES (*sheduri*—Simla) are twisted and made into mats which constitute the chief sleeping mats used by the hill people. DOMESTIC AND SACRED USES. Abir. 62 Root. 63 Leaves. 64

SPECIAL OPINIONS.—§"Is used in cooking: when meat is tough, or when *dál* does not easily break up, a piece of the dried fruit is put in the *degchi.* Is known as *kachri* in the Lower Panjáb" (*Surgeon-Major C. W. Calthrop, M.D., Morar*). A large quantity is imported into Bombay from China; it is chiefly used as a perfume and in making *abir*" (*Dr. Dymock, Bombay*).

Trade.—As mentioned above, two forms are sold in India, one of home production, the other imported from China. According to **Dymock** the latter reaches India *viá* Singapore, while the former is received in Bombay TRADE. 65

TRADE. from Kumáon. The Indian variety is the more valuable, fetching R5 a maund of 37½lb, while the Chinese only sells for R4½. Since the plant is immensely abundant on the Western Himálaya, as, for example, in the immediate vicinity of Simla, it is somewhat surprising that the admittedly inferior Chinese root should be imported at all. **Sir Edward Buck**, in his *Dyes and Tans of the North-West Provinces*, gives the export from Kumáon in 1875-76 as 95⅛ cwt., and also states that in the same year 95⅛ cwt. was exported from Garhwál and 40½ from the Bijnor District. In **Davies'** Trade Report 25 maunds are given as the annual export *viá* Pesháwar to Afghánistan.

HEDYOTIS, *Linn.; Gen. Pl., II., 56, 1228.*

A genus of herbs, undershrubs, or shrubs, of the Natural Order RUBIACEÆ, comprising about 80 species, chiefly natives of Tropical Asia. Of these some 57 are Indian, but few are of known economic value. The famous dye-product, the *Chay-root* of Madras, was until recently supposed to be derived from **Hedyotis umbellata**, but this has now been reduced to **Oldenlandia umbellata.**

66 **Hedyotis Auricularia,** *Linn.; Fl. Br. Ind., III., 58; Bedd., Ic. Pl. Ind. Or., t. 27;* RUBIACEÆ.

Syn.—H. NERVOSA, *Wall.;* H. PROCUMBENS, *Wall.;* H. LINEATA, *Wall. not of Roxb.;* H. COSTATA, *Br.;* H. MULTICAULIS, *Schldl.;* H. VENOSA, *Korth.;* METABOLUS VENOSUS, *Bl.;* SPERMACOCE HISPIDA, *Miq.;* ? S. LINEATA, *Roxb.*

Vern.—*Gatta-colla*, SING.

References.—*Roxb., Fl. Ind., Ed. C.B.C., 122, 124; Rheede, Hort. Malab., X., t., 32; En. Pl. Zeyl., 142; Bombay Gazetteer (Kánara), XV., 435.*

Habitat.—Eastern Bengal from Nepál, Sikkim, and the Khásia Mountains to Assam, Chittagong, Manipur, Burma, and southward to Malacca; also a native of the Western Peninsula, from Kanára southwards; and abundant in Ceylon.

FOOD. Leaves. 67 **Food.**—According to the account in the *Enumeratio Plantarum Zeylaniæ*, "the LEAVES are boiled, after being cut very small, and eaten by the Cinghalese with their rice."

68 **H. capitellata,** *Wall.; Fl. Br. Ind., III., 56.*

Syn.—HEDYOTIS FINLAYSONIA, *Wall.;* OLDENLANDIA RUBIOIDES, *Miq.*

Vern.—*Bakrelara*, PAHARIA; *Kalhenyok*, LEPCHA.

References.—*Kurz in Journ. As. Soc., 1877, II., 135 (excl. var. γ.); Gamble, Trees, Shrubs, &c., of Darjiling, 48; McCann, Dyes and Tans, Beng., 139.*

Habitat.—In the *Flora of British India* it is stated that this climber occurs only in the Malay Peninsula from Tenasserim to Malacca. **Gamble** includes it in his *List of Trees, &c., of the Darjeeling District.* This seems to require confirmation, however, especially as he describes the plant as "a soft-wooded climber of the Terai." According to **Watt** (*Cal. Exhib. Cat.*) it is plentiful upon the Burma-Manipur frontier, which may be its most westerly habitat, but it is quite herbaceous, with hollow stems, and except in the root or the portion of the stem immediately above ground, does not possess anything that could be called *wood*; the stems are in fact hollow. It is probable that **Gamble** refers to **H. scandens,** *Roxb*, a climber of the tropical and sub-tropical Himálaya to the Khásia Hills, Chittagong, and Burma.

DYE. Leaves. 69 **Dye.**—**Gamble**, speaking of the plant referred by him to **H. capitellata,** *Wall.*, says that "it is used by the Lepchas as a green dye," and that "the green LEAVES are put into water and infused, and the cloth to be dyed steeped in the infusion." **Mr. Gamble** also states that the Paharias employ it mixed with the leaves of **Luculia** as a blue dye, but he expresses his opinion that it is of value more as a mordant than a real dye. **Dr. G. Watt**, however, in the Calcutta Exhibition Catalogue, remarks: "I found no trace of the use of either species (**H. capitellata** and **H. scandens**) as a blue dye among the Nagas; although both plants are very plentiful, the Nagas regularly import from Manipur and Assam the *room* dye (**Strobilanthes flaccidifolius**)." **Dr. Schlich** says of **H. capitellata**, *Wall.*: "The Lepchas grind up the green leaves and steep the article to be dyed in the infusion." "It yields a *green* dye." DYE.

From the preceding remarks, as also those under **Baccaurea sapida**, (Vol. I., 362) it is clear that there must be some mistake regarding this dye-stuff, and it therefore seems highly desirable that the whole subject should be re-investigated. Definite information is wanting as to the actual species of **Hedyotis** used by the Lepchas, as to whether it be used as a true dye or as a mordant, and as to the plant mixed with it (if any) to produce the green colour.

Food.—Gamble, in his account of the plant referred by him to **H. capitellata**, *Wall.*, writes: "The LEAVES are eaten by Lepchas." FOOD. Leaves. 70

Hedyotis nitida, *Wight. & Arn.; Fl. Br. Ind., III., 61; Bedd., Ic. Pl. Ind. Or., t. 36.* 71

Syn.—H. NEESIANA, *Arn.*; H. GLABELLA, *Br.*

Vern.—*Pittasúdúpala*, SING.

References.—*Thwaites, En. Pl. Zeyl., 142; Bombay Gazetteer (Kanára), XV., 435.*

Habitat.—Western Peninsula from Dharwar southwards to Travancore and Ceylon.

Food.—"The LEAVES of this plant are finely chopped up, boiled, and eaten with rice, by the Cinghalese." (*Thwaites*). FOOD. Leaves. 72

H. umbellata, *Wall.*, see **Oldenlandia umbellata**, *Linn.*

HEDYSARUM, *Linn.; Gen. Pl., I., 510.* 73

A genus of herbs, belonging to the Natural Order LEGUMINOSÆ, which numbers amongst its species the valuable European fodder plant, the *Sainfoin*, **H. onobrychis**, *Linn.* The eight Indian species, however, seem to be of no value as fodder; though a new species, **H. Wrightianum**, *Aitch. et Baker*, discovered by **Aitchison** while with the Afghan Delimitation Commission, is described by him as being greedily eaten by goats and sheep. No attempt seems up to this time to have been made to introduce the cultivation either of *Sainfoin* or of the Spanish *Sainfoin* (**H. coronarium**) into India, a fact probably due to the large number of indigenous fodder plants already available.

HELIANTHUS, *Linn.; Gen. Pl., II., 376.*

A genus of coarse tall herbs (of the Natural Order COMPOSITÆ,) which has no species indigenous to India. Two, however, are much cultivated, and are very important economically.

Helianthus annuus, *Linn.; Roxb., Fl. Ind., Ed. C.B.C., 607;* 74

THE SUNFLOWER. [COMPOSITÆ.

Vern.—*Surajmúkhi*, HIND.; *Surju múkl, shúria múkti*, BENG.; *súrij-ká-jhar*, DEC.; *Surajmaki, súryakánta*, BOMB.; *Brahmoka, surajmaka*, MAR.; *Aditya bhakti-chettu, podda-trin-gudda-chettu*, or *proddu-tiru-gudu chettu*, TEL.; *Suria-mukhi*, SANS.; *Gúli-aftab, áftábi*, PERS.

References.—*Elliot, Fl. Andhr., 12, 154; Moodeen Sheriff, Supp. Pharm. Ind., 22; Year Book Pharm., 1877, 169; Lisboa, U. Pl. Bomb., 163, 219; Crooke, Dyeing, 424, 513; Ain-i-Akbari, Blochmann's Trans., I., 85; Spons, Encyclop., 961, 1411; Balfour, Cyclop., II., 32; Smith, Dic., 398; Kew Off. Guide to the Mus. of Ec. Bot., 85; W. W. Hunter, Orissa, II., 179; Gazetteer, Mysore and Coorg, I., 62; Official Correspondence, Agri. Rev. and Com. Dept., 1873, 1877.*

HELIANTHUS annuus.

Sunflower Oil.

Habitat.—An annual herb, bearing large, flat, circular flower-heads; said to be a native of Mexico and Peru, and to have been introduced into Europe about the end of the sixteenth century. It may be of interest to note that the *Aftábí*, or sun-flower, is mentioned in the *Aín-i-Akbarí* as a flower cultivated for ornamental purposes during the reign of Akbar. It is largely cultivated in China and Tartary, also in Russia, Germany, Italy, and France, and to a small extent in India, where, however, it is chiefly grown in gardens. It succeeds well in gardens in the plains, flowering during the cold season, and in the hills during the summer; in many hill stations indeed, and markedly so in Simla, it has spread from the gardens to the hill-sides, and, with the *dahlia*, is rapidly becoming naturalised.

CULTIVATION. 75

Cultivation.—The experimental cultivation of the sun-flower in India was first undertaken at Bangalore in 1873, for the purpose of testing its value in the reclamation of marshy land, and its efficiency in removing malaria from swampy districts. This latter virtue had been drawn attention to in a paper presented to the Société de Thérapeutique, which described the wonderful effects obtained by its cultivation in Holland. These experiments, however, proved that the malaria-dispelling virtues, attributed to the plant, did not exist, though the drainage which had to be employed before it would thrive, naturally improved the climate of the district. Having thus begun, the cultivation of the sun-flower was continued, up to the end of 1877, with the view of utilising it profitably in the extraction of the oil, and other economic products to be obtained from its seeds, stem, &c. The last official report regarding the results of this continued cultivation, however, dated November 1877, contains a resolution by the Government of India, to the effect that pending the receipt of reports on the chemical analysis of the oil, it did not seem necessary to continue the experimental cultivation further. This decision was based on the fact, that the results, up to that time, proved that the cost of production of the oil did not permit of profitable cultivation. Since the date of this decision, no further efforts seem to have been made.

DYE. Blossoms. 76

Dye.—According to **Balfour** "the BLOSSOMS yield a brilliant, lasting, useful dye." **Crookes** states that the petals are peculiarly rich in the amorphous resinous substance—*Xanthin*—the base of the yellow pigment from which they derive their colour. The SEEDS contain *helianthic acid*,

Seeds. 77

which, when treated with hydrochloric acid in a current of hydrogen, is resolved into glucose and a violet dye.

FIBRE. Stalks. 78

Fibre.—A useful textile fibre is obtained from the STALKS. (*Spons' Encyclopædia; Balfour's Cyclopædia, &c.*).

OIL. Seeds. 79

Oil.—The OIL obtained by expression from the SEEDS is the most important product of the sun-flower, and is valuable for many purposes. When pure it is said to be equal to olive or almond oil for culinary and table purposes; indeed, it is much used for adulterating these more valuable oils, especially in Russia. The oil obtained from the plants grown at Bangalore, however, was found to be greatly inferior for table purposes. Further, in a report on the subject by the Superintendent of the Experimental Farm, it is stated that the Madras Railway Company condemned it as an illuminating oil, after a series of trials, since they found that its thinness unfitted it for fast running trains; the Superintendent of the Central Jail found that it dried too slowly to be useful in paint; the Ordnance Department found that it answered all the purposes required in the arsenal, but the price at which it was offered was greater than that of the equally good Rangoon oil. It was found also that four times the quantity as compared with castor oil was required for lubricating machinery, thus necessitating extra attention, and consequently increased expenditure. The author of the article on this subject in *Spons' Encyclopœdia* writes: "The chief industrial applications of the oil are for woollen dressing, lighting, and candle and soap-making; for the last mentioned purposes it is superior to most oils." In the form of OIL-CAKE it is a valuable food for cattle and poultry. OIL. Oil-cake. 80

A consideration of the uses of this product, and the attempts which have been made to employ it in India, seems to indicate that it is certainly of value, but that the cost of production by present methods is too great to encourage any hope of its competing as a lubricant, table oil, or illuminating material with other oils of perhaps greater utility, and certainly of lower price It is also doubtful whether it could be exported profitably for other industrial purposes.

Food and Fodder.—"In Russia the SEEDS are sold in the streets and eaten as nuts" (*Smith*). They are also roasted and used as a substitute for coffee. FOOD. Seeds. 81

They are highly valued for feeding sheep, pigs, poultry, pigeons, rabbits, &c., and are considered superior to linseed for cattle. In many parts of Europe, indeed, the plant is cultivated extensively owing to its SEEDS forming a useful food for domestic animals. The OIL-CAKE also, as already mentioned, is used to feed cattle. The LEAVES are said to be valuable as fodder for cattle and horses, but **Colonel Boddam** records the fact that "in Mysore the cattle do not take to them so readily as in Europe." FODDER. Seeds. 82 Oil-cake. 83 Leaves. 84

Domestic Uses.—The leaves and stalks make a valuable MANURE, either directly ploughed in or as a box manure after being used as cattle litter. The stalks are said to be useful as FUEL, and to yield potash from their ashes. DOMESTIC. Manure. 85 Fuel. 86

Smith observes that "it is an excellent plant for bees, large quantities of HONEY and WAX being obtained from the flowers. Honey & Wax. 87

Helianthus tuberosus, *Linn. DC. Prodr., V., 590.* 88

THE JERUSALEM ARTICHOKE.

Vern.—*Hattichók,* HIND.; *Brahmoka,* BENG.; *Atipich,* N.-W. P. & PB.; *Hattichók,* MAR.; *Hattichók,* GUZ.

References.—*DC. Origin Cult. Pl., 42; Ferminger, Man. of Gardening for India, 160; Atkinson, Ec. Prod. N.-W. P., pt. V., 18; Lisboa, U. Pl. Bomb., 162; Birdwood, Bomb. Pr., 164; Reports, Dir. Land Rec. and Agri., Bengal; Dir. Land Rec. and Agri., Poona; Collector of Cuddapah Dist., Madras; Smith, Dic., 228; Journ. Agri. Hort. Soc., 1873-74, Vol. IV., Secs. 1, 4; Gazetteer, Mysore and Coorg, I., 62.*

Habitat.—A perennial herb, attaining a height of 6 to 10 feet, with rodlike stems, and solitary terminal yellow flowers. It is, according to **DeCandolle**, a native of North-East America. Cultivated generally in the gardens of Europeans in India, and, according to **Atkinson**, "is slowly making its way amongst the Natives." The English name Jerusalem Artichoke is a corruption of the Italian *girasole,* meaning sunflower, combined with an allusion to the artichoke-like flavour of the root.

Cultivation.—The Jerusalem artichoke was first introduced into Europe at the Farnese Garden at Rome about 300 years ago, and rapidly spread over Europe. It was cultivated for its roots, which, before potatoes were known, were more used as an article of food than they are now. The plant is, however, still much cultivated both in Europe and in this country. Regarding the method of cultivation **Ferminger** writes: "The ordinary soil of the garden generally suits it without the addition of much manure. The tubers are put into the ground in May, in rows about a foot and a half apart, and with the same distance between each plant, and three inches deep. The plants grow to three or four feet high, and pro- CULTIVATION. 89

CULTIVATION. duce their sunflower-like blossoms in abundance; these possibly it would be of considerable advantage to remove before they open. When the tubers are taken up they should be stored away in large flower-pots, well covered with earth, or they will be liable to shrink and shrivel from exposure to the air." The Director of Land Records and Agriculture, Bengal, however, writing of the propagation of this plant by seed, directs that they should be sown in August so as to allow of the removal of the plants in November without disturbing the roots. The following report regarding its cultivation in Bombay has also been received from the Director of Land Records and Agriculture, Poona: "This vegetable is sparingly grown; it is propagated by tubers and requires rich loam. In the latter part of May the eyes are planted $1\frac{1}{2}$ feet apart in furrows at a depth of about three inches, and are watered every tenth day. As the plants grow they should be earthed up like potatoes. The plants flower by the end of August, when watering is stopped, and after the plants have dried up the crop is dug out. The tubers for the next year's propagation are kept buried in sand in the shade. The acre yield is at about 3 tons."

FIBRE. Stems. 90 **Fibre.**—The STEMS are said by **Balfour** to yield a rich textile fibre.

FOOD. Roots. 91 **Food.**—The TUBEROUS ROOTS are used as a vegetable. **Champlain**, in 1603, found the root thus employed by the natives of North America; he wrote, that they ate "roots which they cultivate and which taste like an artichoke."

The plant was brought to Europe as already described, thence probably it was introduced into India, where it is much eaten by Europeans and to a small extent by Natives. In Káthiawar it is boiled in milk, and is considered, by the Natives, a strengthening vegetable. It is generally said to be more wholesome and nutritious than the potato, and may be given to invalids when abstinence from other vegetable food is necessary.

HELICTERES, *Linn.*; *Gen. Pl., I., 220, 983.*

A genus of the Natural Order STERCULIACEÆ, comprising 10 Indian species, of which only one is of known economic value.

92 **Helicteres Isora,** *Linn.*; *Wight, Ic., t. 180*; *Fl. Br. Ind., I., 365*; STERCULIACEÆ.

THE EAST INDIAN SCREW TREE.

Syn.—HELICTERES CHRYSOCALYX, *Miq*; H. ROXBHURGHII, *G. Don.*

Vern.—*Maraphali, maror-phulli, jonkaphal, bhendu,* HIND.; *Atmorá,* BENG.; *Renta, sakomsang,* KOL.; *Aiteni,* KHARWAR; *Petchamra,* SANTAL; *Mori,* MAL. (S.P.); *Muri-muri,* URIYA; *Aita,* GOND; *Ainthia dhamin,* MONGHYR; *Marorphal, bhendu,* N.-W. P.; *Marorphali, kupási,* (BAZAR FRUIT=*marorphalí*), PB.; *Bottu ka,* C. P.; *Marorí-képhálli, maradsing, kevan, kewanné, dhamni,* DEC.; *Itah,* GODAVERI; *Khiran, kewan, kevana,* (the fruit=*murudásenga*), BOMB.; *Kewan,* MAR.; *Murdásing,* GUZ.; *Valumberi, valampúri, vadampiri,* TAM.; *Nuliti, kavanchi, syamali, ádasyamáli, gubadarra,* TEL.; *Kori-buta,* KURKU; *Anteri,* BANWARA; *Kavargi,* KAN.; *Khungiche, thu-guay-khyæ,* BURM.; *Zinia gaha, lineya-gass,* SING.; *Awartani, mriga-shinga,* SANS; *Kisht-burkisht, pechak,* PERS.

References.—*Roxb., Fl. Ind., Ed. C.B.C., 506; Kurz, For. Fl. Burm., I., 142; Beddome, Fl. Sylv. Anal. Gen., t. 5; Gamble, Man. Timb., 49; Dalz. & Gibs., Bomb. Fl., 22; Stewart, Pb. Pl., 25; Elliot, Fl. Andhr., 10, 89, 171; Ainslie, Mat. Ind., II., 447; O'Shaughnessy, Beng. Dispens., 228; U. C. Dutt, Mat. Med. Hind., 293; Dymock, Mat. Med. W. Ind., 2nd Ed., 113; Pharmacog. Ind., I., 231; S. Arjun, Bom. Drugs, 19; Murray, Pl. and Drugs, Sind, 56; Med. Top., Ajmir, 146; Year book Pharm., 1878, 290; Irvine, Mat. Med. Patna, 66; Moodeen Sheriff, Mat. Med., Madras, 67; Atkinson, Him. Dist., 306, 739; Lisboa, U. Pl.*

Bom., 229; Birdwood, Bomb. Pr., 10; Liotard, Paper-making Mat., 14; Cross, Bevan, King, Rep. on Indian fibres., 10, 53; Rev. A. Campbell, Ec. Pr., Chutia Nagpur, 8445; Mason, Burma and Its People, 502, 753; Indian Forester, IV., 228, 233, 323; VIII., 417; IX., 274, 325, 326; XII., App., 7; XIII., 119; XIV., 296; Balfour, Cyclop., II., 32; Treasury of Bot., I., 575; Kew Report, 1879, 34; For. Admn. Rep., Ch. Nagpur, 1885, 6, 28; Rep. Exp. Farm, Madras, 1884, 9; Madras Man. of Admn., I., 313; Gazetteers:—Mysore and Coorg, I., 58; Rajputana, 26; N.-W. P., I., 79; IV., lxix.; Bombay, XIII., 25; XV., 71.

Habitat.—A large arborescent shrub of the dry forests throughout Central and Western India, as far west as Jamu, the Central Peninsula, and Ceylon. (*Fl. Br. Ind.*). According to **Stewart** and other writers it is found wild in the Siwalik trac for some distance into the Panjáb.

FIBRE. 93

Fibre.—The FIBRE extracted from the bast is strong, thin, silvery, and reticulated, and when combed resembles jute in appearance. In the report on Indian Fibres by **Cross, Bevan, King and Watt**, the following analysis is given: moisture, 10·8; ash 3·1; cellulose, 56·7; and as the result of hydrolosis, *a* (by boiling for 5 minutes in a 1 per cent. solution of caustic soda) 12·0, *b* (by boiling for an hour) 20·2 per cent. of weight were lost. The length of the ultimate fibre is 1—2 mm. The fibre is very useful for cordage, rough sacking, and canvas, and would appear also to be suitable for paper-making. The following extract from the report of the Madras Experimental Farms for 1884 is of interest: "This fibre, a sample of which was exhibited at the Madras Agricultural Exhibition of 1883, and attracted the attention of the judges, was identified by **Professor Lawson** as **H. Isora**, and enquiries were instituted to ascertain its local distribution and the quantity that would be likely to be available for commerce. It is found chiefly in South Kanára, Malabar, Kurnúl, Ganjam, and Coimbatore, but there is nothing accurately known as to the quantity that could be made available for trade. A sample was procured from Kurnúl and forwarded to **Messrs. Arbuthnot & Company** for trial, but from a report that has been recently received from them, it would not appear that this fibre is suitable for weaving. It is probably adapted only for cordage." The report makes no reference to any trial having been made of it as a paper-making material.

MEDICINE. Capsule. 94

Medicine.—The CAPSULE, which consists of 5 linear, many-seeded carpels, spirally twisted, and terminating in a thick point to form a cone from 1½ to 2 inches long, has long been employed medicinally in India, and is still one of the commonest bazár drugs in most parts of the country. Its use, however, seems to depend almost entirely on the ancient "theory of signatures," the peculiarly twisted carpels being supposed to resemble the folds of the intestine. It is accordingly chiefly employed in intestinal complaints, entering into most native prescriptions for colic, flatulence, diarrhœa, and intestinal strangulation. In the Konkan it is said to be also used as a remedy for snake-bite, and in diabetes. **Ainslie** describes its employment also, in powder, mixed with a quantity of castor oil, as an application for discharge from the ear. The authors of the *Pharmacographia Indica* consider that its chief virtue lies in its harmlessness, and they have been unable to discover any therapeutic properties in the plant, beyond those of a demulcent and mild astringent. **Moodeen Sheriff**, on the other hand, describes it as "tonic and stomachic, and useful in all the diseases for which gentian and chiretta are ordinarily employed," and recommends it for these purposes in doses from one to two drachms twice or thrice a day. The **Revd. A. Campbell** describes the ROOT and BARK as also employed medicinally by the Santals for the same purposes as the fruit.

Root. 95
Bark. 96

HELIOTROPIUM indicum. **Heliotrope.**

MEDICINE. SPECIAL OPINIONS.—§"Used for worms and colic in children." (*Surgeon-Major Robb, Civil Surgeon, Ahmedabad*). "The fruit is used in diarrhœa as well as in constipation of newly-born children. A piece of the fruit tied on the body of a newly-born infant is believed to protect it from all kinds of bowel complaints." (*Civil Surgeon J. H. Thornton, B.A., M.B., Monghyr*).

FODDER. Leaves. 97 **Fodder.**—Buffaloes eat the LEAVES. (*R. Thompson*).

TIMBER. **Structure of the Wood.**—White, soft. Weight 35℔ per cubic foot. The branches are used for fuel, fencing, and thatching.

DOMESTIC and SACRED. Fruit. 98 **Domestic and Sacred Uses.**—The FRUIT is said to be used in the marriage ceremonies of the Vaisya caste, when it is tied upon the wrists of bride and bridegroom along with that of **Randia dumetorum.**

TRADE. 99 **Trade.**—**Atkinson**, in his account of the Himálayan District, mentions that about one ton of the fruit is exported from the Kumáon forests annually. **Dr. Dymock** gives the trade price in Bombay as R$3\frac{1}{2}$ per Surat maund of $37\frac{1}{2}$℔.

HELIOTROPIUM, *Linn.; Gen. Pl., II., 843.*

A genus of annual or perennial herbs belonging to the Natural Order BORAGINEÆ, and comprising 100 species, distributed over the tropical and temperate zones of both hemispheres. Of these 16 are known in India, and 4 are reputed to be of medicinal value.

100 **Heliotropium Eichwaldi,** *Steud.; Fl. Br. Ind., IV., 149;* BORAGINEÆ.

Syn.—HELIOTROPIUM ELLIPTICUM, *Ledeb.;* H. STRICTUM, *Ledeb.;* H. MACROCARPUM, *Guss.;* H. EUROPÆUM, *Aitch., Stewart, ? Linn.*

Vern.—*Nil kattei, bíthúa, atwin, popat búti, gidartamákú,* HIND.; *Nil kattei, bíthúa, atnún, popat búti, gidar tamákú,* PB.

References.—*Boiss., Fl. Orient., IV., 131; Benth., Fl. Austral., IV., 394; Stewart, Pb. Pl., 154; Aitchison, Cat. Pb. and Sind Pl., 94; Aitch., Afgh. Del. Com., 88; O'Shaughnessy, Beng. Dispens., 497; Dymock, Mat. Med. W. Ind., 2nd Ed., 575; Murray, Pl. and Drugs, Sind, 171; Indian Forester, XII., App., 17.*

Habitat.—Common throughout the Panjáb and Sind, found also in Kashmír ascending to an altitude of 5,200 feet.

MEDICINE. Leaves. 101 **Medicine.**—The LEAVES applied locally are used in cases of snake-bite, bite from a mad dog, and scorpion or wasp sting. They are also said to be of service as an application for cleansing and healing ulcers and for destroying warts; given internally they are emetic. The properties of the plant, therefore, are very similar to those of **H. europæum** with which the plant has been confused, and which was medicinally employed by the Greeks under the name of ηλιοτρόπιον τὸ μέγα, while by the Romans it was known as *Herba Solaris.*

102 **H. indicum,** *Linn.; Fl. Br. Ind., IV., 152; Wight, Ic., t. 171.*

Syn.—HELIOTROPIUM ANISOPHYLLUM, *Beauv.;* TIARIDIUM INDICUM, *Lehm.;* T. ANISOPHYLLUM, *G. Don.;* HELIOPHYTUM INDICUM, *DC.;* H. VELUTINUM, *DC.*

Vern.—*Hatta-júrie, hatta-súra, siriari,* HIND.; *Hátisurá,* URIYA; *Hátisurá,* BENG.; *Chappu tattu,* C.P.; *Búrúndi,* BOMB.; *Bhurundi,* MAR.; *Tel-kodduki, telmunnie,* TAM.; *Ek-sœtiya,* SING.; *Hatisunadá, srihastiní, bhúrúndí,* SANS.

References.—*Roxb., Fl. Ind., Ed. C.B.C., 152; Dalz. & Gibs., Bomb. Fl., 172; Ainslie, Mat. Ind., II., 414; O'Shaughnessy, Beng. Dispens., 497; U.C. Dutt, Mat. Med. Hind., 299; Dymock, Mat. Med. W. Ind., 2nd Ed., 574; Irvine, Mat. Med. Patna, 129; Indian Forester, VI., 238; W. W. Hunter, Orissa, II., 178, app. VI.; Gazetteers:—Mysore and Coorg, I., 89; Bombay, XV., 439.*

Habitat.—A small annual herb, common throughout India, especially in the moister parts. According to **Roxburgh** it is most often found in out-of-the-way corners, and amongst rubbish where the soil is rich.

Medicine.—**Irvine** describes the LEAVES as employed in Patna in cases of fever, the dose given being from half a drachm to three drachms. **Ainslie**, recounting its medicinal properties, says: "The JUICE of the leaves of the plant, which is a little bitter, the native practitioners apply to painful gum-boils, and to repel pimples on the face; it is also prescribed as an external application to that species of ophthalmia in which the tarsus is inflamed or excoriated." He further adds, "I find **Braham** tells us that it cleans and consolidates wounds and ulcers, and that, boiled with castor oil, it relieves the pain from the sting of a scorpion, and cures the bite of a mad dog." It thus, in its last given properties, resembles the immediately preceding species. **Dr. Dymock** writes that "in Bombay the PLANT is used as a local application to boils, sores, and the stings of insects and reptiles."

MEDICINE. Leaves. 103 Juice. 104 Plant. 105

Heliotropium strigosum, *Willd.; Fl. Br. Ind., IV., 151.* 106

In the *Flora of British India* **H. brevifolium,** *Wall.*, is reduced to be a variety of this species (**var. brevifolia**), and **H. compactum,** *Don*, is given as a synonym for that variety.

Vern.—*Safed-bhangra, chiti phúl,* HIND.; *Safed bhangra, kharai, tindu, gorakh pámo* (*chiti phul, gorakh pánu*=bazár plant), PB.

References.—*Stewart, Pb. Pl., 154; Aitchison, Cat. Pb. and Sind Pl., 94; Baden Powell, Pb. Pr., 360; Atkinson, Him. Dist., 314, 739; Indian Forester, XII., app., 17; Gazetteers, N.-W. P., IV., lxxv.; Bombay, XV., 439.*

Habitat.—A small plant, according to the *Flora of British India*, common throughout India, especially the variety **brevifolia.** It appears to be only described as of economic value, however, by writers on the Panjáb, North West Provinces, and Sind.

Medicine.—The whole PLANT has laxative and diuretic properties. According to **Atkinson** and **Baden Powell**, the JUICE is employed as an application in ophthalmia, presumably granular, and gum-boils, for the purpose of promoting suppuration, and to wounds and ulcers to hasten healing. Like that of the other species already described its juice is supposed to have curative powers in cases of poisonous bites and venomous stings.

MEDICINE. Plant. 107 Juice. 108

H. undulatum, *Vahl.; Fl. Br. Ind., IV., 150.* 109

Syn.—HELIOTROPIUM PERSICUM, *Lamk.*; H. CRISPUM, *Desf.*; H. ERIOCARPUM, *Delile*; H. RAMOSISSIMUM, *Sieb.*; H. AFFGHANUM, *Boiss.*; LITHOSPERMUM HISPIDUM, *Forsk.*

This species has, according to the *Flora of British India*, two varieties, one, the typical **F. undulatum,** the other, *var.* **suberosa,** to which **H. nubicum,** *Bunge*, has been reduced.

Vern.—*Pípat-buti, jati misák,* PB.

References.—*Boiss., Fl. Orient., IV., 143, 147; Stewart, Pb. Pl., 154; Aitchison, Cat. Pb. and Sind Pl., 95; Afghan Del. Com., 88.*

Habitat.—Found in the Panjab, Sind, and the Upper Gangetic plain, reaching a height of 1,000 feet; distributed to Western Asia and North Africa.

Medicine.—**Stewart** describes this PLANT as being, like **H. strigosum,** of value in snake-bite, in which cases it is both administered internally and applied externally to the wound. As an external application it is sometimes employed mixed with tobacco oil.

MEDICINE. Plant. 110

Hellebore, Green; see **Veratrum viride,** *Aiton*; LILIACEÆ.

HELLEBORUS, *Linn.; Gen. Pl., I., 7.*

111 **Helleborus niger**, *Linn.; DC., Prodr. I., 46;* RANUNCULACEÆ.

BLACK HELLEBORE or CHRISTMAS ROSE.

Vern.—*Kalá-kútki*, BENG.; *Khorasani kútki*, BEHAR; *Kala-kútki*, DEC.; *Kádágaruganie*, TAM.; *Katúkaró-gani*, TEL.; *Katuroum, katuróhini*, SANS.; *Khertic* or *kartick, kherbek aswed*, ARAB.; *Kherbeck-seeah*, PERS. It is doubtful whether the above vernacular names are really applied to HELLEBORE at all, though by many writers they have been given to it.

References.—*O'Shaughnessy, Beng. Dispens., 4; Bent. & Trim., Med. Pl., I., No. 2; S. Arjun, Bomb. Drugs, 3; Murray, Pl. and Drugs, Sind, 72; Irvine, Mat. Med. Patna, 53; K. L. De, Indig. Drugs Ind., 58; Birdwood, Bomb. Pr., 3; Kew Off. Guide Mus. of Ec. Bot., 7.*

Habitat.—A small, perennial herb, with black, jointed, definite rhizomes, having numerous interlacing rootlets. It is a native of Central and South Europe, extending eastwards to South Poland and westwards to Dauphiny and Provence. According to **Boissier** it is also indigenous to Greece.

MEDICINE. Root. 112

Medicine.—**Dr. G. Watt**, in his Calcutta Exhibition Catalogue, writes: "**Ainslie** says, with a degree of hesitation and doubt, that the plant which yields this drug is met with in Nepál, and he gives the above vernacular names for it. **Dr. W. O'Shaughnessy**, in his *Bengal Dispensatory*, gives an abstract of **Ainslie's** information, and is followed by **Birdwood** in his *Bombay Products*, **K. L. De** in his *Indigenous Drugs of India*, and later still, **Murray** in his *Drugs of Sind*. All these authors have simply compiled from **Ainslie**, altering the original information only so far as to remove the hesitation with which **Ainslie** published the statement that Hellebore was produced in Nepál. Bnt even **Ainslie** derived his information from **Dr. Kirkpatric**, who writes that it is known in Nepál as *kútka*, and that it also reaches India by way of the Red Sea. As a matter of fact, however, the plant has never been found in Asia in a wild state, and if it exists in India at all it could only be in flower gardens at our hill stations, but it is doubtful if it could grow even there much below an altitude of 10,000 feet. Muhammadan physicians do, however, prescribe a drug which is known as Hellebore, as also many others which neither grow nor are imported into India. **Dr. Royle** explains this by stating that the Muhammadans obtained their medical science from the Arabian physicians, who, in their turn, learned it from the Greek authors. European plants have thus come to be known by name in India, and **Royle** includes in his list of such, the roots of the black Hellebore, remarking that the druggist, trusting to the indifference of the physician and the ignorance of the patient, substitutes for such drugs others with supposed similar properties. This is remarkable and curious: the symbolic "Recipe" being apparently in the eyes of the physician of more importance than the ingredients of the mixture, for one cannot quite believe that he is wholly ignorant of the non-existence of the medicines he has prescribed. It is amusing how persistently even the commonest *hakim* will mysteriously mention the almost sacred Greek synonyms for his drugs. A knowledge of such matters seems to constitute his highest credentials.

"I have already drawn attention to the fact that **Actæa spicata** and **Cimicifuga fœtida** (two common Himálayan plants related to Hellebore) are apparently not known by the Indian doctors to possess medicinal properties. They are largely used as drastic purgatives in adulteration with Hellebore, both in Europe and America, besides possessing certain distinctive medicinal properties." **Dr. Watt**, revising the above paragraph, has kindly added: "According to **Drs. Moodeen Sheriff** and **Dymock**, however, Muhammadan physicians substitute the root of **Picrorhiza Kurooa**

MEDICINE.

for black Hellebore. There is no doubt of an external resemblance in these two roots, but **Picrorhiza** is a bitter tonic which is much more likely to be used as a substitute for **Coptis Teeta** (*see Vol. II., 521—524*). While **Dymock** attributes to **Picrorhiza** an aperient property (an opinion which **Moodeen Sheriff** does not concur in) it could never be substituted for the hydragogue purgative Hellebore. I am, therefore, strongly disposed to view the *Kádágar-oganie* of **Ainslie**—'the drastic cathartic' obtained from Nepál—to be a distinct plant from **Picrorhiza** and most probably **Actæa.** But as **Ainslie** speaks of his drug as also obtained from Syria, it seems probable that he may have confused two things, his Syrian root being **Helleborus orientale** (the habitat of which would agree with the source spoken of by him). It is, therefore, just possible that he may have incorrectly assigned the drastic property of the Syrian root to the Nepál substitute, and in that case his '*Kootka*' may have been **Picrorhiza.**" Through the kindness of **Dr. Gimlette**, Residency Surgeon, specimens of *Kútki* (both plant and root) were obtained from Nepál, and were identified as **Picrorhiza Kurooa.** The following remarks were made by **Dr. Gimlette**:—"Brought from the Thibetan border; root used in ague and enlargement of the spleen; dose four to eight *mashas.*"

Hellebore is a drastic hydragogue, cathartic, and also emmenagogue and anthelmintic. It was formerly held in high esteem and employed in the cure of mania, melancholia, epilepsy, dropsy, amenorrhœa, chronic skin affections, and worms. In large doses it is an acro-narcotic poison. It is to a certain extent used both internally and externally in veterinary pharmacy. As referred to above, the European root is often adulterated with **Actæa** and **Cimicifuga**, but from these it may be at once distinguished by the absence of the medullary rays so characteristic of these roots, and by an infusion not becoming black on the addition of a solution of persulphate of iron. *See* **Actæa** (*Vol. I., 103*), **Cimicifuga** (*Vol. II., 288*), **Coptis Teeta** (*Vol. II., 522*).

Special Opinions.—§ **Dr. Dymock** writes: "Is only obtainable in India from a few European druggists, and, as far as my experience goes, is unknown in the bazárs. The root of **Picrorhiza Kurooa** has been adopted as a substitute for **H. officinalis**, the Hellebore of the Greeks, by Muhammadan physicians. It is a harmless substitute, being febrifuge and aperient." "This drug, as far as my experience goes, is unknown in the bazárs. **Picrorhiza Kurooa** has been adopted as a substitute for **H. officinalis** (the Hellebore of the Greeks) by the Muhammadans. It is febrifuge and aperient, and has not the properties of Hellebore" (*Surgeon-General with the Government of Bombay*). "The plant is cultivated in India as a garden plant only. The gardeners are not aware of the medicinal properties of the rhizome. As far as I can learn the drug is a Continental import. Though it has been stated in my *Plants and Drugs of Sind* that it is a native of Nepál, I have reason to believe that it does not occur there" (*Mr. J. A. Murray, Author of "Plants and Drugs of Sind"*). "The powder used in eczema and dysentery" (*Surgeon-Major J. J. L. Ratton, M.D., Salem*). "This plant is not cultivated, as far as I am aware, in any part of Southern India, nor have I any information about it being found in Nepál" (*Mr. A. Lawson, Government Botanist and Director of Cinchona Plantations, Nilgiris*). "*Kalikutki.*—It must be noticed that two sources of this drug are given:—one under **Helleborus**, the other under **Picrorhiza.** The drug sold in Bombay on examination was found to resemble in structure a Hellebore. (See *Bombay Drugs, page 3*)" (*Sakharam Arjun Ravat, L.M., Assistant Surgeon, Girgaum, Bombay*). "I cannot say how far the cultivated plant is checking the importation of the drug. Certainly no specimen of wild Hellebore has been received at the Calcutta Herbarium, from Nepál." (*Surgeon D. Prain, M.B., I.M.S., Acting Superintendent, Royal Botanic Gardens, Sibpur*).

Helmia, see **Dioscorea bulbifera, D. dæmona, D. tomentosa**, and **D. triphylla**, Vol. III., 128—134.

HEMARTHRIA, *Br.; Gen. Pl. III., 1131.*

A genus of the Natural Order GRAMINEÆ, and belonging to the Tribe ANDROPOGONÆ. It comprises two species, of which the second, **H. fasciculata**, is probably only a variety of the first. They may, therefore, for the purposes of this work, be considered together under **H. compressa**, since their properties, habitat, and uses are identical. [*30*; GRAMINEÆ.

113 **Hemarthria compressa**, *R. Br.; Duthie, Fodder Grasses of N. Ind.*,

Syn.—ROTTBŒLLIA COMPRESSA, *Linn.*; R. GLABRA, *Roxb.*; *Duthie, Grasses of Plains of North-West India, pl. XVII.*

H. fasciculata, according to **Duthie** (*Fodder Grasses of Northern India*), is probably only a variety of this species, differing in having shorter leaves and smaller and more crowded spikes. The synonyms of the variety are **Rottbœllia fasciculata**, *Desf.*; **R. altissima**, *Poir.*; **Lodicularia fasciculata**, *Beauv.*

Vern.—*Ransherú, buksha, pansherú*, BENG.; *Biksa*, N.-W. P.; *shervú*, TEL.

References.—*Roxb., Fl. Ind., Ed. C.B.C., 118; Royle, Him. Bot., 422; Taylor, Top. of Dacca, 60; Duthie, Ind. Fodder Grasses of Plains of N.-W. Ind., 18; Mueller, Select Extra-Trop. Plants, 198; Balfour, Cyclop., III., 444; Treasury of Bot., II., 992; Gaz., N.-W. P., I., 86; IV., lxxx.*

Habitat.—A perennial grass of Bengal, inhabiting also the plains of the Panjáb and North-West Provinces, and the hills at low elevations. Roxburgh says that it is found on the borders of lakes, amongst other roots of long grass and brushwood; and he mentions the variety **R. glabra** as growing on pasture lands, the borders of rice-fields, and other moist places.

FOOD and FODDER. 114

Food and Fodder.—This grass forms a good food for cattle, and, according to **Taylor**, constitutes the principal article of cattle fodder in Dacca. **Baron Von Mueller**, in his *Select Extra-Tropical Plants*, writes: "This perennial grass, though somewhat harsh, is recommendable for moist pastures, and will retain a beautiful greenness throughout the year in dry climes; highly esteemed by graziers in Gippsland (Victoria); it is not injured by moderate frost;" and in a further passage he says: "Remarkably resistant to drought."

HEMICYCLIA, *Hook. et Arn.; Gen. Pl., III., 278.*

A genus of trees or shrubs belonging to the Natural Order EUPHORBIACEÆ, comprising nine Indian species, only two of which, however, are of known economic value. [*Ic., t. 1872*; EUPHORBIACEÆ.

115 **Hemicyclia sepiara**, *Wight & Arn.; Fl. Br. Ind., V., 337; Wight,*

Syn.—PERIPLEXIS, *Wall.*; P. RIGIDA, *Wall.*

Vern.—*Véré*, TAM.; *Wira-grass*, SING.

References.—*Beddome, Forest Man., 198; Gamble, Man. Timb., 347; Thwaites, En. Ceylon Pl., 287; Dalz. & Gibs., Bomb. Fl., 229; Trimen, Flowering Plants, &c., of Ceylon, 80; Lisboa, U. Pl. Bomb., 118; Indian Forester, X., 22, 23, 30; Balfour, Cyclop., II., 34.*

Habitat.—A rigid much-branched shrub, or low tree, with a curiously fluted stem. It belongs to the Deccan Peninsula from the Konkan southwards, ascending to an altitude of 3,000 feet. It is also one of the most characteristic trees of Ceylon forests, in which it is very common between the sea level and an elevation of 1,500 feet.

Structure of the Wood.—Hard, close-grained, resembling boxwood. The longitudinal flutings characteristic of the stem are, in Ceylon, split off and employed to make axe handles. There is no account of this close-grained hard wood being employed for the purposes of turning, &c., for which it would seem to be well suited. TIMBER. 116

Hemicyclia sumatrana, *Muell.; Fl. Br. Ind., V., 338.* 117

Reference.—*Kurz, For. Fl. Burma, II., 365.*

Habitat.—An evergreen tree, from 30 to 50 feet in height, inhabiting the tropical swamp forests of Martaban and the Irawaddi in Burma, and the Andaman Islands.

Structure of the Wood.—Heavy, pale greyish brown, coarsely fibrous but close-grained though soft. A fine wood. (*Kurz*). TIMBER. 118

HEMIDESMUS, *Br.; Gen. Pl., II., 740.*

A genus of the Natural Order ASCLEPIADEÆ, having only one Indian species.

Hemidesmus indicus, *R. Br.; Fl. Br. Ind., IV., 5; Wight, Ic., t. 594;* ASCLEPIADÆ. 119

INDIAN SARSAPARILLA.

Syn.—HEMIDESMUS WALLICHII, *Miq.*; PERIPLOCA INDICA, *Willd*; ASCLEPIAS PSEUDOSARSA, *Roxb., excl. syn.—Burm., Thes. Zeyl.; Rheede, Hort. Mal.*

Var. pubescens. SYN.—H. PUBESCENS *W. & A.*

Vern.—*Magrabu, jangli-chanbélli, hindi-sálsá,* HIND.; *Anantamul, anatomúl,* BENG.; *Ununta-múl,* BEHARI; *Sugandi-pálá, nannári, nát-ká-aushbah,* DEC.; *Uparsára,* BOMB.; *Anantamul, upalasari,* MAR.; *Nannári,* TAM.; *Gadisugandhi, pála-chukkani-déru, sugandhi-pála, tella sugan-dhipála, palasugandhi, muttapulgam,* TEL.; *Sogadaheru, sugandha-pálada-gidá,* KAN.; *Nannári-kizhanna, naru-ninti,* MALAY.; *Irimusu,* SING.; *Anantá, sugandhi, gópi-múlam, sárivá,* SANS.; *Zaiyán, aushbatunnár,* ARAB.; *Aushbahe-hindi, yása mine-barri,* PERS.

References.—*Roxb., Fl. Ind., Ed. C.B.C., 254; Gamble, Man. Timb., 266; Thwaites, En. Ceylon Pl., 195; Dalz. & Gibs., Bomb. Fl., 147; Aitchison, Cat. Pb. and Sind Pl., 90; Voigt, Hort. Sub. Cal., 544; Rheede, Hort. Mal., X., t., 34; Trimen, Cat. Flowering Plants, &c., of Ceylon, 55; Elliot, Flora Andhr., 56, 127, 142, 143, 170, 179; Pharm. Ind., 140; Ainslie, Mat. Ind., I., 381, 630; O'Shaughnessy, Beng. Dispens., 456; Moodeen Sheriff, Supp. Pharm. Ind., 152, 283; U. C. Dutt, Mat. Med. Hind., 195, 291; Dymock, Mat. Med. W. Ind., 2nd Ed., 509; Fluck. & Hanb., Pharmacog., 423; Bent. & Trim., Med. Pl., t. 174; S. Arjun, Bomb. Drugs, 86; Murray, Pl. and Drugs, Sind, 159; Waring, Bazar Med., 67; K. L. Dey, Indigenous Drugs of India, 59; Irvine, Mat. Med. Patna, 118; Baden Powell, Pb. Pr., 361; Drury, U. Pl., 241; Lisboa, U. Pl. Bomb., 260; Birdwood, Bomb. Pr., 53; W. W. Hunter, Orissa, II., app. 27, 159, 181; Spons, Encyclop., 823; Balfour, Cycop., II., 34; Smith, Dic., 369; Treasury of Botany, I., 580; Kew Offl. Guide to the Mus. of Ec. Bot., 97; Manual, Cuddapah District, 200; Settlement Report, Upper Godavery, 39; Gazetteers:—Mysore & Coorg, I., 62; Bombay, XV., 438; N.-W. P., IV., lxxiv.*

Habitat.—A climber of North India from Banda to Oudh and Sikkim, and southward to Travancore and Ceylon.

Fibre.—The author of the Settlement Report of the Upper Godavery District calls this plant the "Yellow Silk Cotton" and describes its fibre as "fine and strong." FIBRE. 120

Medicine.—The ROOT has long been employed in Southern India as an alterative and tonic. It is found in Indian commerce in the form of little bundles, which consist of the entire roots of one or more plants, often several feet long, tied up with a portion of the stem. The root itself is cylindrical, tortuous, from $\frac{2}{10}$ to $\frac{7}{10}$ of an inch in diameter, and seldom MEDICINE. Root. 121

MEDICINE.

branched. The fresh or freshly dried root has colour of tonka beans or melilot, and a sweet but slightly acrid taste. (*Dymock.*) It was first introduced to the medical profession in Europe by **Ashburner** in 1831, and in 1864 it was made officinal in the British Pharmacopœia, while in 1868 it was included amongst the drugs in the Pharmacopœia of India. **Dutt** informs us (*Hindu Mat. Med.*), that the roots of **Hemidesmus** and **Ichnocarpus frutescens** are both called *Sárivá* in Sanskrit medicine, and are described under the name of *Sárivádvaya* or the two *Sárivás*. They are often used together, since the Hindus consider both as tonic, sweet, demulcent, and alterative, and they prescribe them in dyspepsia, loss of appetite, skin diseases, syphilis, and leucorrhœa. By native practitioners the *Sárivá* is generally administered with other drugs, especially laxatives, tonics, and aromatics. **Ainslie** writes: "The *nunnarivayr* is recommended by the Tamool doctors in cases of gravel and strangury, given in powder, mixed with cow's milk; they also give it in decoction, in conjunction with cummin seeds, to purify the blood and correct the acrimony of the bile." He further mentions that in his time the decoction was employed by European practitioners in India as a substitute for true sarsaparilla, "in cutaneous diseases, scrofula, and venereal affections to the quantity of ℥iii or ℥iv three times a day."

Roxburgh wrote: "The natives employ them" (the roots) "in medicine more than we do, particularly for the thrush in children. For this disorder the dried bark is powdered and fried in butter, the proportion uncertain as is often the case with Hindu prescriptions, the quantities being in general guessed: about a drachm of this is given, night and morning."

The medicinal properties of **Hemidesmus** seem to be either unknown or unappreciated in the more northern parts of India, where it is very seldom employed. Thus **Dr. Dymock** writes: "In the northern part of the Bombay Presidency its medicinal properties appear to be ignored, as, though common everywhere in the neighbourhood of Bombay, it is never collected by the natives." In the more southern parts of the Konkan, however, he describes it as extensively employed. The MILKY JUICE is used in that region as an application to the conjunctiva in ophthalmia; the root, roasted in plantain leaves then beaten into a mass with cummin and sugar, and mixed with *ghí*, as a remedy in genito-urinary diseases, and also as an external application or *lép* for swellings. Great difference of opinion exists amongst European practitioners regarding the efficacy of this plant as an alterative, but it is generally regarded as at least an efficient substitute for Sarsaparilla. In England it is very little employed except at St. Bartholomew's Hospital, where it has for many years taken the place of Sarsaparilla. There seems to be no doubt that the roots ought to be employed when fairly fresh, and that their medicinal virtues decrease greatly as they become old and dried up. For this reason it appears probable that the specimens sent to Europe may have been inferior to those obtainable in this country: the authors of the *Pharmacographia* indeed observe, "the root found in the English market is often of very bad quality."

Juice. 122

SPECIAL OPINIONS.—§ "Used commonly as a substitute for Sarsaparilla in syphilitic and rheumatic affections" (*Assistant Surgeon Nehal Sing, Saharunpore*). "A cheap and very good substitute for Sarsaparilla." (*Surgeon G. Price, Shahabad*). "Is a useful alterative and tonic in syphilitic diseases, and a good substitute for Sarsaparilla" (*Surgeon R D. Murray, M.B., Burdwan*). "Useful in debility, chronic rheumatism, constitutional syphilis, skin-diseases, &c." (*Brigade Surgeon J. H. Thornton, B.A., M.B., Monghyr*). "In chronic cough and diarrhœa, the hot infusion of *Anantamúl* with milk and sugar acts as an alterative and tonic, specially

MEDICINE.

in children" (*Surgeon R. L. Dutt, M.D., Pubna*). "Probably as inert as Sarsaparilla itself" (*Civil Surgeon C. M. Russell, Sarun*). "I do not think this medicine has any alterative properties" (*Surgeon-Major H. J. Haslett, Ootacamund*). "A very good substitute for Sarsaparilla but not equal to it" (*Apothecary T. Ward, Madanapalle, Cuddapah*). "A decoction of this drug alone has been found to be ineffectual as an alterative in syphilis, unless combined with large doses of 10 grains of the Iodide of Potassium thrice daily" (*Honorary Surgeon P. Anderson, Guntur, Madras*). "Country Sarsaparilla is greatly used by natives" (*Surgeon-Major L. Beech, Cocanada*). "I have seen the fresh inner and white root-bark used with excellent results in syphilis" (*Honorary Surgeon E. A. Alfred Morris, Tranquebar*). "An excellent substitute for Sarsaparilla. Very useful in secondary syphilis in combination with Iodidé of Potassium" (*Assistant Surgeon Shib Chunder Bhuttacharji, Chanda, Central Provinces*).

Trade.—The root in Bombay costs 4 annas per ℔, the high price being due to the difficulty of digging out the long roots from the stony ground in which the plant grows. (*Dymock*). TRADE. 123

HEMIGYROSA, *Blume; Gen. Pl., I., 395.*

A genus of trees with pinnate leaves, bearing comparatively large flowers, in axillary branched racemes or panicles, belonging to the Natural Order SAPINDACEÆ and having three Indian species, of which one is of economic interest.

Hemigyrosa canescens, *Thw.; Fl. Br. Ind., I., 671;* SAPINDACEÆ. 124

Syn.—HEMIGYROSA TRICHOCARPA, *Thw.;* MOLINÆA CANESCENS, *Roxb.;* SAPINDUS TETRAPHYLLA, *Vahl;* S. BIJUGUS, *Wall.;* CUPANIA CANESCENS, *Pers.;* CUPANIA SP. *Wall.;* CUPANIA SP. 6, *H. f. and T.*

Vern.—*Lakhandi, lokaneli, kurpa,* MAR.; *Nekota, karadipongan,* TAM.; *Korivi, bodda mámidi, muru mámidi,* TEL.; *Kála-yatte, kurpah, karpa,* KAN.; (the two latter Telegu names are given by **Elliot**, with the accompanying caution that they are rather doubtful.

References.—*Roxb., Fl Ind., Ed. C.B.C., 320; Kurz, For. Fl. Burm., I., 290; Beddome, Fl. Sylv., t., 151; Gamble, Man. Timb., 93; Thwaites, En. Ceylon Pl., 56, 408; Dalz. & Gibs., Bomb. Fl 35; Elliot, Fl. Andhr., 29, 120; Lisboa, U. Pl. Bomb., 50; Indian Forester, III., 201; Balfour, Cyclop., II., 34; Gazetteers:—Mysore and Coorg, I., 52; Bombay, XV, 71.*

Habitat.—A shrub or good sized tree, with a trunk which may be of considerable girth, but is not straight. It is found in Tenasserim and Burma, the Western Peninsula from the Konkan southwards, also common on the east side of the Madras Presidency, and in Ceylon near Kandy.

Structure of the Wood.—White, moderately soft, even grained, containing no distinct heartwood. According to the writer of the *Kanára Gazetteer*, it is used in house-building. **Roxburgh**, however, writes: "the wood is white and not so serviceable as that of **Sapindus rubiginosus.**" TIMBER. 125

Hemlock, see **Conium maculatum** (Vol. II., 517).

Hemp, Ambari, see **Hibiscus canabinus,** p. 34.

Hemp, Bombay, see **Hibiscus canabinus,** p. 34.

Hemp, Deccani, see **Hibiscus canabinus,** p. 34.

Hemp, Indian, see **Cannabis sativa,** Vol. II., 103.

Hemp, Manilla, see **Musa textilis,** *Louis,* Vol. V.

Hemp, Sunn, see **Crotalaria juncea,** Vol. II., 595.

Henbane, see **Hyoscyamus niger,** *Linn.*, p. 319.
Henna, see **Lawsonia alba,** *Lamk.*; LYTHRACEÆ.

HEPTAPLEURUM, *Gærtn.*; *Gen. Pl.*, *I*, *942*.

Trees or tall shrubs often climbing, with digitate, or digitately compound, leaves, belonging to the ARALIACEÆ. The genus comprises 14 or 15 Indian species, of which some are of value as timber trees. [ARALIACEÆ.

126 **Heptapleurum elatum,** *C. B. Clarke*; *Fl. Br. Ind., II., 728*;

Syn.—HEDERA ELATA, *Ham.*; AGALMA ELATUM, *Seem.*; ARALIAD, SP. 6, *Herb. Ind. Or., H. f. & T.*

Var. Griffithii=HEPTAPLEURUM GLAUCUM, *Kurz*; AGALMA GRIFFITHII, *Seem.*

Vern.—*Chinia*, NEPAL; *Prongzam*, LEPCHA.

References.—*Kurz, For. Fl. Burm., I., 538*; *Gamble, Man. Timb., 209.*

Habitat.—A tree, from 30 to 40 feet in height, occurring on the Himálaya, from Kumáon to Bhután, between the altitudes of 5,000 and 7,000 feet. The variety **Griffithii** is a native of Martaban and is found at similar altitudes.

TIMBER. 127 **Structure of the Wood.**—White, soft, even-grained.

128 **H. impressum,** *C. B. Clarke*; *Fl. Br. Ind., II., 728.*

Syn.—AGALMA TOMENTOSUM, *Seem.*; HEDERA TOMENTOSA, *Ham.*; PANAX TOMENTOSUM, *DC.*; ARALIAD, SP. *13, Herb. Ind. Or., H. f. & T.*

Vern—*Balú chinia*, NEPAL; *Suntong*, LEPCHA.

Reference.—*Gamble, Man. Timb., 209.*

Habitat.—A handsome tree of the North-East Himálaya, from Kumáon to Bhután, between the altitudes of 6,000 and 11,000 feet. It is common in the forests of these regions, grows to the height of 60 feet, and may be recognised by its woolly leaves.

GUM. Bark. 129 **Gum.**—The thick, brown BARK yields a copious gum (*Gamble*).

TIMBER. 130 **Structure of the Wood.**—White or grey, and soft.

131 **H. venulosum,** *Seem.*; *Fl. Br. Ind., II., 729*; *Wight, Ic., t. 118.*

Syn.—HEPTAPLEURUM ELLIPTICUM, *Seem.*; PARATROPIA VENULOSA, *Wight & Arn.*; P. ELLIPTICA and MACRANTHA, *Miq.*; HEDERA VENOSA, *Wall.*; H. TEREBINTHACEA, *Wall.*; ARALIA DIGITATA, *Roxb.*; SCIADOPHYLLUM ELLIPTICUM, *Blume.*; ARALIAD, SP. *18, Herb. Ind. Or., H. f. & T.*

Var. macrophylla, *Wall., sp.*

Vern.—*Dain*, HIND.; *Sukriruin*, KOL.; *Singhata*, NEPAL; *Biluletwa, takya-lai-wa*, BURM.

References.—*Roxb., Fl. Ind., Ed. C.B.C., 276*; *Brandis, For. Fl., 249*; *Kurz, For. Fl. Burm., I., 538*; *Aitchison, Cat. Pb. and Sind Pl., 69*; *Rheede, Hort. Mal., VII., t., 28*; *Atkinson, Him. Dist., 311*; *Indian Forester, II., 23*; *XI., 370*; *For. Adm. Report, Ch. Nagpore, 1888, 31*; *Bombay Gazetteer, XV., 435.*

Habitat.—A small, glabrous tree, or evergreen climbing shrub, frequent in the mixed forests throughout tropical and sub-tropical India, from the North-West Himálaya to the South Deccan and Singapore; common all over Burma.

TIMBER. **Structure of the Wood.**—Light-brown, soft.

132
HERACLEUM, *Linn.*; *Gen. Pl., I., 921.*

A genus of the UMBELLIFERÆ, of which several species grow in the Himálaya. One of these (undetermined), which is common in parts of the Panjáb Himálaya from 8,500 to 11,000 feet in altitude, and is called by the Panjábis "*padáli*" or "*porál,*" is described by **Stewart** as collected by the natives of Bashahr and Chamba for winter fodder; he further quotes **Cleghorn** to the effect that it is believed to increase the milk of goats fed on it. (*Panjáb Plants, 107.*) **Mr. J. F. Duthie** also gives **Heracleum sp.** as an article of food for goats; see *Rough List of Indian Fodder-yielding Trees, &c.*

The following are the determined species of the locality in which this fodder plant is said to be collected :—

H. Jacquemontii, *C. B. Clarke*; **H. Brunonis**, *Benth.*; **H. canescens**, *Lindl.*; **H. candicans**, *Wall.*; and **H. nepalense**, *Don*. (*Fl. Br. Ind.*, *II.*, *711—717.*) Of these the two last are the most frequent, and it seems probable that **Stewart's H. sp.** may be either the one or the other.

Herba Bengalo. 133

Several old writers, amongst others **Linschoten, Mandelso**, the **Abbé Guyon, Fitch**, and **Hamilton**, give an account of a fibre obtained from a plant under the above name, from which hangings, coverlets, &c., were made. It is impossible, from the confused and contradictory descriptions of these early travellers, to identify the plant referred to; indeed, it seems probable that several fibre-yielding plants were indiscriminately called "**Herba Bengalo.**"

Herb Christopher, see **Actæa spicata**, Vol. I., 103.

Herb of Grace, see **Ruta graveolens**, *Linn.*; RUTACEÆ.

HERITIERA, *Ait.*; *Gen. Pl.*, *I.*, *219*.

A genus of trees belonging to the STERCULIACEÆ, comprising four or five species, natives of tropical Asia, Africa, and Australia. Of these, three are allowed to be Indian by the *Flora of British India*, but since the publication of the first volume of that work **Kurz** has restored **Heritiera macrophylla**, *Wall.*, to the position of a distinct species (*New Burmese Plants*, *Journ. Asiat. Soc. Beng.*, *XLI.*, *pt. II.*, *291*).

Heritiera Fomes, *Buch.*; *Fl. Br.*, *Ind. I.*, *363*; STERCULIACEÆ. 134

Syn.—BALANOPTERIS MINOR, *Gærtn*; HERITIERA MINOR, *Roxb.*

Vern.—*Súndri*, *súndari*, HIND.; *Shundri*, *súndra*, BENG.; *Pinlekanazo*, BURM.

References.—*Roxb.*, *Fl. Ind.*, *Ed. C.B.C.*, *506*; *Kurz*, *For. Fl. Burm.*, *I.*, *141*; *Gamble*, *Man. Timb.*, *47*; *McCann*, *Dyes and Tans*, *Beng.*, *156*; *Balfour*, *Cyclop.*, *II. 39*; *Mason*, *Burma and Its People*, *536*, *753*.

Habitat.—A gloomy-looking tree, abundant in the Gangetic plain, extending inland to Sylhet, and along the shores of the Eastern Peninsula (*Fl. Br. Ind.*). Frequent in the tidal forests all along the Burmese shores from Chittagong down to Tenasserim, ascending the rivers as far as the tidal waves (*Kurz*). According to **Captain Munro** it has given its name to the Sundarbans—*Sundarí-vana* or *ban* (Forest of the Sundarí tree).

Dye.—**McCann** writes, in his *Dyes and Tans of Bengal*, "a specimen of the BARK has been forwarded from Nadiya as a dye-stuff, but without any particulars." DYE. Bark. 135

Structure of the Wood.—Brown, strong, tough, and durable. It weighs 66℔ per cubic foot, and has a breaking weight of 1,132℔ (*Kurz*). **Mason**, in his *Burma and Its People*, writes: "It is the toughest wood that has been tested in India; when Rangoon teak broke with a weight of 870℔ Sundri sustained 1,312℔." Possessing these excellent qualities, it is naturally much employed economically, especially in the manufacture of boats, piles of bridges, house posts, rafters, and other articles which have to bear great weight. In Calcutta it is much used as firewood, and according to **Dr. Wallich**, it forms, when burnt, the best charcoal for the manufacture of gunpowder. TIMBER. 136

H. littoralis, *Dryand.*; *Fl. Br. Ind.*, *I.*, *363*. 137

THE LOOKING-GLASS PLANT.

Syn.—HERITIERA FOMES, *Wall.*, *in part*; H. TOTHILA, *Kurz* (Contributions towards a knowledge of the Burmese Flora); BALANOPTERIS TOTHILA, *Gærtn.*

Vern.—*Súnder, súndri*, BENG.; *Súndri*, BOMB.; *Penglai-kanazo, pinle kanazo*, BURM.; *Etúna*, SING.; *Mawtda*, AND.

References.—*Roxb., Fl. Ind., Ed. C.B.C., 506; Kurz, For. Fl Burm., I., 141; Gamble, Man. Timb., 47; Thwaites, En. Ceylon Pl., 28; Dalz. & Gibs., Bomb. Fl., 22; Rheede, Hort. Mal., VI., t., 21; Trimen, Cat. of Flowering Plants, &c., of Ceylon, 12; Mason, Burma and Its People, 536, 753; Indian Forester, II., 174; VI., 292, 366; VIII., 401; IX., 324; XI., 490; XII., 64, 329; XIII., 126; Balfour, Cyclop., II. 39; Gazetteers:—Mysore and Coorg, I., 58; Burma, I., 127.*

Habitat.—A small, gregarious tree found on the coasts of Bengal, the Eastern and Western Peninsulas, and Ceylon, extending inland as far as the Khásia Hills and Cachar. It is known to Europeans in some parts of India as the Looking-glass tree, owing to the dense silvery hairs with which the under-surface of its leaf is covered.

OIL. 138 **Oil.**—This TREE is reported to be the source of an oil in the Antilles, the use of which is not known.

TIMBER. 139 **Structure of the Wood.**—"Brown, durable, tough, and heavy, weighing about 65 ℔ per cubic foot. It is used for a great variety of purposes, such as beams, buggy shafts, planking, posts, furniture, firewood; but chiefly in boat-building, for which purpose it is very extensively used in Calcutta, and particularly in the Government Dockyard at Kidderpore. It is the chief timber of the Sundarban forests. Its reproduction is most favourable. On all lands flooded by ordinary flood tides, a new growth of jungle springs up immediately; but on land ordinarily above high-water mark, it only establishes itself by slow degrees. It soon spreads on newly formed islands on the sea edge of the forests. The *Súndri* forests are generally very well stocked; thus valuations made by Home in 1873-74 gave the average amount of material per acre of forest, as follows:—seedlings and saplings under 3 feet girth, 2,487; trees above 3 feet girth 182" (*Gamble*). The above description resembles that given by **Kurz** for the wood of **H. Fomes**, while his account of that of **H. littoralis** differs considerably. He writes: "Wood brown, rather light and close-grained." This account is confirmed by the *Burma Gazetteer, I., 127*, where it is reported "*Peng-le-ka-na-tso* (**H. littoralis**) is found in the tidal forests, and yields a brown, light, and rather loose-grained wood, used locally for house posts and rafters and for firewood by salt boilers." It seems probable that some mistake exists as to the exact botanical source of the *Súndri* wood of the Sundarbans. Both **H. Fomes** and **H. littoralis** seem equally frequent; they appear to have the same vernacular names, and have almost exactly the same habitat, all of which circumstances would render confusion very possible.

It is worthy of notice that **Mason**, in his *Burma and Its People*, fully describes **H. Fomes** under the vernacular name of *Súndri*, and, on the other hand, leaves the timber of **H. littoralis**, with which he seems unacquainted, undescribed. It is impossible with the present literature on the subject to determine where the error lies, but it is to be hoped that further information may be forthcoming.

TRADE. 140 **Trade.**—In the Forest Progress Report for Bengal, 1881-82, it is mentioned that "more than a lakh of rupees was received on *Súndri* wood (**H. littoralis**)." This indicates a very large consumption, which its highly useful qualities would seem to warrant.

141 **Heritiera Papilio,** *Beddome, Fl. Sylv., t. 218; Fl. Br. Ind., I., 363.*

Vern.—*Soundalay anna*, TINNEVELLY.

References.—*Gamble, Man. Timb., 48; Indian Forester, III., 23; Balfour, Cyclop., II., 39.*

Habitat.—A very lofty tree in the evergreen forests of Travancore and the Southern Carnatic, at Courtallam and Tinnevelly.

Structure of the Wood.—Red, very hard, similar to that of **H. littoralis,** but pores less numerous and smaller. It weighs 63℔ per cubic foot, and is used for building, cart poles, and agricultural implements. In Vol. III. of the *Indian Forester* it is given in a list of the most valuable timbers of the Tinnevelly District. TIMBER. 142

[*1873, 61; Trim. Journ. Bot., 1874, 66, f. 7.*

Heritiera macrophylla, *Wall.; Kurz in Journ. As. Soc., Beng.,* 143

Habitat.—Upper Tenasserim and Martaban. No information exists regarding the timber of this tree, nor of its economic uses, but as it has up to a recent date been confused with **H. littoralis,** *Dryand.*, it seems probable that its wood may resemble the timber of that tree in structure and qualities.

Hermodactylus, see **Colchicum sp.,** Vol. II., 501.

HERNANDIA, *Linn.; Gen. Pl., III., 164.*

A genus of evergreen trees belonging to the Natural Order LAURINEÆ, and comprising eight species, all tropical, of which one is Indian.

[*t. 1855;* LAURINEÆ.

Hernandia peltata, *Meissn.; Fl. Br. Ind., V., 188; Wight, Ic.,* 144

Syn.—H. SONORA, *Linn. in part.*

Vern.—*Uparanthi,* MYSORE; *Palati, palutu,* SING.

References.—*Kurz, For. Fl. Burm., II., 309; Beddome, Fl. Sylv., t. 300; Gamble, Man. Timb., 304; Thwaites, En Ceylon Pl., 258; Trimen, Syst. Cat. Flowering Pl., &c., Ceylon, 76; Indian Forester, VI., 238; Balfour, Cyclop., II., 40; Treasury of Bot., I., 585.*

Habitat.—An evergreen tree, with peltate leaves, found in the coast forests of Singapore, the Andaman Islands, and Ceylon.

Structure of the Wood.—Is very light and takes fire so readily from a flint and steel that it might be used as tinder (*Beddome*). TIMBER. 145

Medicine.—According to **Beddome** the JUICE is a powerful depilatory, removing the hair without pain, while the BARK and YOUNG LEAVES are cathartic. MEDICINE. Juice. 146 Bark. 147 Leaves. 148

HERPESTIS, *Gærtn. f.; Gen. Pl., II., 951.*

A genus of glabrous, often punctate, herbs, with yellow, blue, or white flowers, belonging to the Natural Order SCROPHULARINEÆ. It comprises about 50 species, all natives of warm countries, but of these only three are known to be Indian.

[PHULARINEÆ.

Herpestis Monniera, *H. B. & Kth.; Fl. Br. Ind., IV., 272;* SCRO- 149

Syn.—HERPESTIS SPATHULATA, *Blume;* GRATIOLA MONNIERA, *Linn.*

Vern.—*Barambhi, brahmi, jal-nim, shwet chamni, safed chamni,* HIND.; *Adha-birni, dhop-chamni, brihmi-sák,* BENG.; *Urishnapárni,* ORISSA; *Báma,* BOMB.; *Beami, nirpirimie, nir-brami,* TAM.; *Sambráni chettu,* TEL.; *Beami,* MALAY.; *Brahmi, mandúki,* SANS.

References.—*Roxb., Fl. Ind., Ed. C.B.C., 47; Voigt, Hort. Sub. Cal., 502; Thwaites, En. Ceylon Pl., 218, 426; Dalz. & Gibs., Bomb. Fl., 178; Aitchison, Cat. Pb. and Sind Pl., 107; Rheede, Hort. Mal., X., t. 14; Trimen, Syst. Cat. Cey. Pl., 62; Elliot, Fl. Andhr., 166; Pharm. Ind., 161; Ainslie, Mat. Ind., II., 239; O'Shaughnessy, Beng. Dispens., 476; U. C. Dutt, Mat Med. Hind., 213, 294; Dymock, Mat. Med. W. Ind., 2nd Ed., 579; S. Arjun, Bomb. Drugs, 99; Bidie, Cat. Raw Pr., Paris Exh., No. 620; Year Book, Pharm., 1874, 135; Atkinson, Him. Dist., 314, 739; Drury, U. Pl., 242; Balfour, Cyclop., II., 42; Treasury of Bot., I., 585; Gazetteers:—Bombay, III., 204; Mysore and Coorg, I., 56; Orissa, II., 159; N.-W. P., I., 83; IV., lxxv.*

Habitat.—An annual creeping plant, found in moist places near streams or on borders of tanks, throughout India, from the Panjáb to Ceylon and Singapore, ascending to an altitude of 4,000 feet.

MEDICINE. Root. 150 Stalks. 151 Leaves. 152 Juice. 153

Medicine.—**Ainslie** informs us that the JOINTED ROOT, as well as the STALKS and LEAVES, are used medicinally by the Hindus, who consider them to be diuretic and aperient, and particularly useful in "that sort of stoppage of urine which is accompanied by obstinate costiveness." **Roxburgh** wrote: "The natives used the expressed JUICE mixed with petroleum to rub on parts affected with rheumatic pains." **U. C. Dutt**, in his *Hindu Materia Medica*, affirms that it constitutes the *brahmi* of the native physicians, and assumes that it is also the *brahmi* of Sanskrit writers. He writes: "This plant is considered a nervine tonic, useful in insanity, epilepsy, and hoarseness. Half a tolá of the fresh juice of the leaves, with two scruples of *páchak* root and honey, is recommended to be given in insanity. The leaves fried in clarified butter are taken to relieve hoarseness." **Dr. Dymock**, however, thinks that **Dutt** may be mistaken in thus taking **Herpestis** to be the Sanskrit *brahmi*, and affirms that in Bombay **Hydrocotyle asiatica** is known by that name, while the plant under consideration is not used medicinally. The *Pharmacopœia of India* gives it a place amongst non-officinal drugs, but the writer evidently regards it as inert. He remarks that there is no trustworthy evidence of its value as an internal remedy, and that whatever value may be derived from the anti-rheumatic formula mentioned by **Roxburgh** is probably due to the petroleum employed.

SPECIAL OPINIONS.—§ "The root is an antidote for mineral poison" (*V. Ummegudien, Mettapollian, Madras*). "A tea spoonful of the juice of the leaves given to infants suffering from catarrh or severe bronchitis gives relief by causing vomiting and purging" (*U. C. Dutt, Civil Medical Officer, Serampore*).

HETEROPANAX, *Seem.; Gen. Pl., I., 945.*

154 **Heteropanax fragrans,** *Seem.; Fl. Br. Ind., II., 734;* ARALIACEÆ.

Syn.—PANAX FRAGRANS, *Roxb.;* ARALIAD SP. 47, *Herb. Ind. Or. H. f. & T.*

Vern.—*Kesseru*, ASSAM; *Hona*, CACHAR; *Lal totilla*, NEPAL; *Siriokhtem*, LEPCHA; *Tachansa*, BURM.

References.—*Roxb., Fl. Ind., Ed. C.B.C., 266; Brandis, For. Fl., 249; Kurz, For. Fl. Burm., I., 541; Gamble, Man. Timb., 208; List of Trees, Shrubs and Climbers, Darjeeling Dist., 44; Atkinson, Him. Dist., 311; Indian Forester, V., 212; Balfour, Cyclop., II., 42; Aplin, Report on the Shan States; Special report on Econ. Pl. of Assam.*

Habitat.—A small unarmed tree, found from the Sewalik Hills to Burma, and common in Bengal. It grows to an altitude of 4,000 feet.

TIMBER. 155

Structure of the Wood.—Light brown, or grey, rather heavy, fibrous, but close-grained, though very perishable (*Kurz*). There is no record of its being employed economically.

DOMESTIC. 156

Domestic Uses.—This tree is important as being that on the leaves of which the "Eri" silkworm is fed in Upper Assam.

HETEROPHRAGMA, *DC.; Gen. Pl., II., 1046.*

157 **Heterophragma adenophyllum,** *Seem.; Fl. Br., Ind., IV., 381;* [*Wight, Ill., t. 160;* BIGNONIACEÆ.

Syn.—BIGNONIA ADENOPHYLLA, *Wall.;* SPATHODEA ADENOPHYLLA, *A. DC.*

Vern.—*Petthan*, BURM.

References.—*Kurz, For. Fl. Burm., II., 236; Gamble, Man. Timb., 277; Aplin Report on the Shan States.*

Habitat.—A moderate-sized deciduous tree met with in Burma, Assam, Eastern Bengal, and the Andaman Islands.

Structure of the Wood.—Yellowish-white, moderately hard. Weight 41℔ per cubic foot. TIMBER. 158

Heterophragma Roxburghii, *DC.; Fl. Br. Ind., IV., 381.* 159

Syn.—SPATHODEA ROXBURGHII, *Spreng.*; BIGNONIA QUADRILOCULARIS, *Roxb.*

Vern.—*Pullung, warras,* BOMB., MAR.; *Baro-kala-goru,* TAM.; *Bondgu,* TEL.

References.—*Roxb., Fl. Ind., Ed. C.B.C., 494; Brandis, For. Fl., 350; Beddome, For. Fl., t. 169; Gamble, Man. Timb., 277; Dalz. & Gibs., Bomb. Fl., 160; Dymock, Mat. Med. W. Ind., 2nd Ed., 544; Lisboa, U. Pl. Bomb., 105; Birdwood, Bomb. Pr., 333; Indian Forester, XII., 312; Bomb. Gazetteer, XIII., pt., I., 27.*

Habitat.—A large tree met with in the Chanda district, Godaveri forests, and the West Deccan Peninsula generally, from Bombay southwards.

Fodder.—It is stated in the *Thana Gazetteer* that the LEAVES are much eaten by cattle. FODDER. Leaves. 160

Structure of the Wood.—Grey, rough, moderately hard. Weight 40℔ per cubic foot. It is employed in Bombay and Madras for the manufacture of planks, and for building purposes generally. TIMBER. 161

H. sulfureum, *Kurz; Fl. Br. Ind., IV., 381.* 162

Vern.—*Thitlinda,* BURM.

References.—*Kurz, For. Fl. Burm., II., 234; in Jour. As. Soc., 1873, Pt. II., 90; Gamble, Man. Timb., 277.*

Habitat.—A deciduous tree, met with in Burma, chiefly in Prome and Pegu Yomah.

Structure of the Wood.—Dark-grey, soft, even-grained, resembling in structure that of **H. Roxburghii.** Weight from 40 to 60℔ per cubic foot. Though no account is obtainable of this wood being employed for economic purposes, it seems probable from Gamble's description, that, like that of the former species, it might be used for planks and building. TIMBER. 163

HETEROPOGON, *Pers.; Gen. Pl., III., 1033.*

A genus belonging to the tribe ANDROPOGONEÆ, of the Natural Order GRAMINEÆ. It comprises five or six species, all inhabitants of warm countries, one of which, the well known Spear Grass, is common all over India.

[*Northern India, 32;* GRAMINEÆ

Heteropogon contortus, *R. & S.; Duthie, Fodder Grasses of* 164

THE SPEAR GRASS.

Syn.—HETEROPOGON HIRTUS, *Pers.*; H. ROYLEI, *Nees.*; ANDROPOGON CONTORTUS, *Linn.*

Vern.—*Kher,* BENG.; *Sauri ghás,* SANTAL; *Badopuncha, lamp, lampa, lampar, parba, parbi,* BUNDEL.; *Kanura, sarwála, paraura, riskawa, parba, banda, sarwar, musel, lap, banda,* N.-W. P.; *Barweza, sarmal,* TRANS.-IND.; *Suriala, snrari, sarudla, suriala, lamb,* PB.; *Sarwála,* RAJ.; *Pochati, saga,* BERAR; *Hukara gadi, kusal, kusáli, khar, lumpeil,* C. P.; *Yeddi, ... sheruu,* TEL.

References.—*Voigt, Hort. Sub. Cal., 706; Stewart, Pb. Pl., 255; Aitchison, Cat. Pb. and Sind Pl., 176; Duthie, Indig. Fodder Grasses of the plains of N. W. Ind., 19, pl. xix.; Symonds, Grasses of the Ind. Peninsula, 44, pl. 29; Elliot, Fl. Andhr., 49; Atkinson, Him. Dist., 321; Indian Forester, XII, App. 2; Maiden, Useful Native Plants of Australia, 90; Coimbatore District Manual, 198; Nursingpore Settlement Report, 57; Gazetteers:—N.-W. P., I., 85; IV., lxxx; Panjáb, Karnal Dist., 19.*

Habitat.—An erect, glabrous grass, with hygrometric awns, found growing in tufts on rich pasture ground. **Duthie** writes: "Common both in the plains and on the hills of the Panjáb and North-West Provinces. It grows in light soil about Banda, attaining a height of about 3 feet; in soil mixed with kunkur (*rakar*) it reaches 5 feet in height (*Miller*) It is common on the rock tablelands of the hilly country south of Allahabad (*Benson*). Abundant also on the warm lower slopes of the Himálayas, and up to an elevation of 7,000 feet in some parts."

FIBRE. 165 **Fibre.**—The grass is used as a fibre in the manufacture of coarse mats, &c.

FODDER. 166 **Fodder.**—It is largely used as fodder both before and after it has flowered, but chiefly when it is young and tender. In Rájputana and Bundelkhand, where it constitutes the principal fodder grass, it is cut and stacked after the rains, and will, it is said, keep good in stack for twelve years. On Mount Abu the people consider it the best fodder grass they have (*Duthie*). **Symonds** mentions that it is good fodder, when green, for cattle and horses, and **Maiden**, in his *Useful Native Plants of Australia*, writes: "A splendid grass for a cattle run, as it produces a great amount of feed." In the Settlement Report of Nursingpore, on the other hand, the grass is said to be "almost useless, as even horned cattle object to it when they can get anything else." This observation probably refers, however, to the plant which has been allowed to seed before it has been cut, in which condition its hard spear-like awns are naturally objectionable.

DOMESTIC. **Domestic Uses.**—It is much employed in many parts for thatching.

167 **Hevea braziliensis,** *Müll. Arg.*, see **India-rubber**, p. 365.

HIBISCUS, *Linn.; Gen. Pl., I., 207, 982.*

A genus of MALVACEÆ, including some 33 Indian indigenous, and several introduced, cultivated species, many of which are of considerable economic value, while nearly all might be utilised for the fibres they contain. Regarding the confusion which at present exists owing to the indiscriminate application of the name HEMP to plants of this genus, **Dr. Watt**, in the "Selections from the Records of the Government of India, Revenue and Agricultural Department, Volume I.," states: "So much confusion exists in the writings of Indian authors regarding the word 'Hemp' that it has been deemed desirable to make an effort to establish a more trustworthy mode of expression. It would, indeed, have been to the advantage of Indian commerce at the present day had **Dr. Royle's** remark on this subject, made nearly half a century ago, been carefully observed. He wrote: 'We should avoid calling every new fibre *Hemp*, when one only can properly be so named. In the same way, the South-sea Islanders called every new animal they saw a pig, because that was the only one with which they were acquainted.' The true hemp of European commerce is of course **Cannabis sativa,** a plant allied to the nettle, but having a leaf deeply cut into the shape of a hand (palmi-partite). The flowers of the hemp plant are small, green, and insignificant-looking—the males appearing on one individual, and the females on another. The Sunn-hemp of Indian commerce, sometimes called Brown-hemp, is **Crotalaria juncea**—a plant that belongs to the pea-family, and has large showy yellow flowers. Bombay-hemp and Deccani-hemp (and also, as sometimes, though incorrectly, called Roselle-hemp—the true Roselle being **Hibiscus Sabdariffa),** are terms often used loosely, but which should, in all probability, be restricted to **Hibiscus cannabinus,** the hemp-leaved **Hibiscus** (*ambári*). In point of foliage this plant closely resembles the true hemp (hence its botanical specific name), but in no other respect. The stems are slightly prickly, as also the sepals, and the flowers are large and yellow, with a deep purple centre. Any one who has seen a cotton plant in flower, or the common garden shoe-flower, could never for a moment mistake this plant for either *hemp* or *sunn*, and yet the literature of the fibres obtained from these plants is so disfigured through the application to all three of the popular word 'hemp,' that it is often quite impossible to discover to which an author is alluding. These plants not only differ materially from each other but their fibres are entirely different, and any attempt to sell the one in place of the other would have serious effects on the development of trade.

The Ambári is of course a bast fibre, and is thus more nearly allied to jute than to sunn-hemp or to true hemp. The latter is more nearly allied to rhea than to either of the others. Indeed, these three fibre-yielding plants may be accepted as the types of three of the most important classes of fibres,—the bast fibres, the nettle fibres, and the pea-fibres. Commercially, sunn-hemp is more nearly allied to flax than to hemp, and as stated, Bombay-hemp might with more reason be returned as a form of jute than of hemp."

It is to be hoped that this distinction will be observed by future writers, and that the word "hemp" may be restricted to **Cannabis sativa.** The reader is referred to the article on **Cannabis sativa**—hemp—and to that on **Crotalaria juncea**—sunn-hemp—for further particulars regarding these two fibres. It may be here added that the term Manilla-hemp is given in modern commerce to the fibre of **Musa textilis.**

Hibiscus Abelmoschus, *Linn.; Fl. Br. Ind., I., 342; Wight, Ic., t. 399;* MALVACEÆ. 168

THE MUSK MALLOW.

Syn.—ABELMOSCHUS MOSCHATUS, *Mœnch;* HIBISCUS FLAVESCENS, *Cav.;* H. SPATHACEUS, *Wall.;* H. RICINIFOLIUS, *Wall.;* H. CHINENSIS, *Wall.;* H. SAGITTIFOLIUS, *Kurz.*

The *Flora of British India* distinguishes two forms:—

Var. 1, multiformis;—BAMIA SP.; *Wall., Cat., 1917.*

Var. 2, betulifolius;—BAMIA SP.; *Wall., Cat., 1918.*

Vern.—*Mushk-dana,* HIND.; *Mushak-dana,* BENG.; *Mushk-bhéndi-ké-bij,* DEK.; *Mishk-dána, mushk-bhendi-ke-bij,* BOMB.; *Mushak-dana,* GUZ.; *Kasturi-vendaik-kay-virai, kattuk-kastúri,* TAM.; *Kastúri-benda-vittulu (néla benda,* Elliot), TEL.; *Káttu-kastúri, kastúri-venta-vitta,* MALAY.; *Ba-lu-wa* (Mason, *contradicted by* Moodeen Sheriff), BURM.; *Zatákasturiká (lata-kasturikam,* Moodeen Sheriff), SANS.; *Habbul-mishk, habbul-mushk,* ARAB.; *Mushk-dana,* PERS.

References.—*Roxb., Fl. Ind., Ed. C.B.C., 526; Thwaites, En. Ceylon Pl., 27; Aitchison, Cat. Pb. and Sind Pl., 20; Trimen, Cat. Fl. Pl. & Ferns, &c., Ceylon, 11; Rheede, Hort. Mal., II., t. 38; W. & A., Prodr., I., 53; Elliot, Fl. Andhr., 84, 131; Pharmacographia Indica, I., 209; Ainslie, Mat. Ind., II., 72, 335; Moodeen Sheriff, Supp. Pharm. Ind., 13; Mat. Med. of Madras, 52; U. C. Dutt, Mat. Med. Hind., 123; Dymock, Mat. Med. W. Ind., 2nd Ed., 102; U. S. Dispens., 15th Ed., 1664; Murray, Pl. and Drugs, Sind, 63; K. L. Dey, India Drugs, Ind., 60; Royle, Prod. Res., 230; Baden Powell, Pb, Pr., 332; Atkinson, Him. Dist., 306; Drury, U. Pl., 2; Royle, Fib. Pl., 259, 268; Liotard, Paper-making Mat., 17; Piesse, Perfumery, 156; Mason, Burma and Its People, 418; Rep. Agri. Dept., Bengal, 1886-87, 21; Smith, Dic., 280; Gazetteer, N.-W. P., IV., lxviii.*

Habitat.—A herbaceous bush, springing up with the rains and flowering in the cold season. Leaves of various shapes; the lower broad, ovate, cordate; the upper narrow, hastate, very hairy. Common throughout the hotter parts of India, now met with in most other tropical countries.

FIBRE. 169

Fibre.—Like other species of this genus this plant yields a fibre from its stem. **Dr. Roxburgh**, experimenting with cords made from it, found that wetting made no difference in its breaking point, the average weight required to break the rope whether wet or dry being 107lb. In the Annual Report of the Agricultural Department, Bengal, for 1886-87, an account is given of a series of interesting experiments with this fibre, conducted by the Agri.-Horticultural Society of India at the request of the Government of India. **H. Abelmoschus** yielded the best crop of all the fibre-yielding plants experimented with, and, with a Death and Ellwood's machine, 800lb of fibre was obtained as the yield per acre. By the ordinary system of retting, on the other hand, a yield from the non-seeded plant corresponding to 12 maunds 17 seers per acre was obtained, thus proving that the latter process gave a better outturn. The Society, however, arrived at the conclusion that the cultivation of **Hibiscus** had no advantage over that of Jute.

PERFUME. Seeds. 170

Odorous Principle.—The SEEDS yield about 6½ per cent. of an odorous principle and resin. The former is a light green non-volatile fluid having a strong odour, resembling that of a mixture of musk and amber, hence the Arabic name *hab-ul-mushk.* Owing to their possessing this principle, the musk mallow seeds are used in perfumery, and are known to the trade in Europe as "*Grains d'ambrette.*" **Piesse**, in his *Art of Perfumery*, writes: "Musk seed, when ground, certainly reminds our smelling sense of the odour of musk, but it is poor stuff at best," and he recommends it only for making cheap sachet powder. According to him the most valuable seeds are imported from Martinique.—(*See Article "Musk," Vol. III., 58 to 62.*)

MEDICINE. Seeds. 171

Medicine.—The aromatic brown SEEDS are employed medicinally by the natives of Northern India. They are kidney-shaped, with a hilum in the concave border, slightly compressed, striated, about two lines in length, and have a feeble odour of pure musk, which becomes more powerful on rubbing them together or between the fingers. By attention to these characters they may be distinguished from the seeds of **Psoralia corylifolia**, which are used as an adulterant or substitute in Madras. Arabic and Persian writers regard their therapeutical properties as cold, dry, tonic, and stomachic, an opinion which is still held by the Hindus. The author of the *Makhzan-el-Adwiya* recommends a mucilage made from

Root. 172 Leaves. 173

the ROOT and LEAVES in gonorrhœa. Though no part of the plant is recognised as of value by the Pharmacopœia of India, **Moodeen Sheriff**, in his *Materia Medica of Madras*, regards the seeds as stimulant, stomachic, and antispasmodic, and recommends their exhibition in cases of nervous debility, hysteria, atonic dyspepsia, and a few other conditions in which musk is indicated. He writes that the best and most efficacious method of using the drug pharmaceutically is as a tincture, an account of the preparation and usage of which will be found in his short note below.

SPECIAL OPINIONS.—§"The seeds of **Hibiscus Abelmoschus** or *Mushk-dánah* contain albumen, oil, and a peculiar principle on which their odour depends. They are stimulant, stomachic, and antispasmodic, and prove useful in some nervous affections in which musk is indicated. The seeds are best administered in tincture, which is prepared as follows:—Take of the *Mushk-dánah* in powder, two ounces and a half; rectified spirit, one pint. Macerate the seeds for forty-eight hours in fifteen fluid ounces of the spirit, in a bottle, agitating occasionally; then transfer to a percolator, and, when the fluid ceases to pass, continue the percolation with the remaining five ounces of spirit. Afterwards subject the contents of the percolator to pressure, filter the product, mix the liquids, and add sufficient rectified spirit, to make one pint. Dose, from one drachm to two drachms. In more than three drachm doses it produces headache or giddiness" (*Honorary Surgeon Moodeen Sheriff, Khan Bahadur, G.M.M.C., Triplicane, Madras*). "The juice of the fresh plant is used as a febrifuge and expectorant in domestic medicine, and a poultice made of the whole plant reduced to pulp is applied to the chest in bronchitis" (*Assistant Surgeon Sakharam Arjun Ravat, L. M., Girgaum, Bombay*). "Emollient and demulcent. Used as a cooling drink in fevers, gonorrhœa, &c., and as an inhalation in hoarseness, dryness of the throat, &c." (*Civil Surgeon J. H. Thornton, B.A., M.B., Monghyr*). "Useful in acute gonorrhœa" (*Surgeon-Major G. Y. Hunter, Karachi*).

FOOD. Seeds. 174

Food.—Ainslie mentions that the SEEDS are mixed with coffee in Arabia.

Domestic Uses.—The LEAVES are said by Royle to be employed in the North-Western Provinces to clarify sugar. DOMESTIC. Leaves. 175

Trade.—The seeds do not now form an article of Indian export. Those imported from the West Indies fetch about 6*d*. per ℔ in London (*Dymock*). TRADE. 176

Hibiscus cannabinus, *Linn.; Fl. Br. Ind., I., 339.* 177

DECCAN HEMP, AMBARI HEMP.

Syn.—H. WIGHTIANUS, *Wall.*

Vern.—*Ambári, pátsan, pulu, nálitá* (Conf. with C. 1861 in the Dictionary), HIND.; *Mesta pát* (*ambya pát*, the leaves, and *chandana*, the plant, according to Dr. B. Hamilton), BENG.; *Dare kudrum* (*kudrum* by Hindus of Chutia-Nagpur), SANTAL; *Kanuriya*, ORISSA; *Kudrum*, BEHAR; *Patsan, pitwa, rattia san* (*tukhm-i-bhang* at DELHI), N.-W. P.; *Shan* (in Jhelam basin), *sankokla, patsan, san kokra, sinjubárá*, PB.; *Sujjádo* (according to Stocks), SIND.; *Ambári* (in CHANDA), C.P.; *Ambári*, BOMB.; *Ambada*, MAR.; *Pulimanji, pulichi, pulicharkirai, palungu*, TAM.; *Góngúra* (according to Elliot), *gaynara* (GODAVERI), *gonkura*, TEL.; *Pundrike gida, holada*, KAN.; *Náli, garnikura*, SANS.; *Sujjádo* (according to Stocks), PERS.

References.—*Roxb., Fl. Ind., Ed. C.B.C., 528; Voigt, Hort. Sub. Cal., 117; Thwaites, En. Ceylon Pl., 401; Stewart, Pb. Pl., 22; Aitchison, Cat. Pb. and Sind Pl., 20; Drury, Handbook Fl. Ind., I., 71; Campbell, Report on Econ. Prod. Chutia Nagpúr, 17; Elliot, Fl. Andhr., 61; Rept. Bot. Gardens, Bangalore, by J. Cameron, 24; Roxb., Coromdl. Pl., II., p. 48, t. 190; U. C. Dutt, Mat. Med. Hind., 311; Dymock, Mat. Med. W. Ind., 2nd Ed., 104; Pharmacog. Ind., I., 213; Watt, in Selections from the Rec. of Govt. of Ind., Rev. & Agri. Dept., I., 266; H. C. Kerr, Cultvn. and trade in Jute, 3, 87, 91; Baden Powell, Pb. Pr., 332; Atkinson, Him. Dist., 709; Drury, U. Pl., 243; Duthie & Fuller, Field and Garden Crops, 86; Lisboa, U. Pl. Bom., 226; Birdwood, Bom. Pr., 316; Royle, Fib. Pl., 254; Liotard, Paper-making Mat., 5, 16, 17, 28, 34; Cross, Bevan, King, & Watt, Report on Indian Fibres, 39; Mueller, Extra-trop. Pl., 158; Manual, Coimbatore Dist., Madras, 238; Manual of Trichinopoly by Moore, 72; Madras Manual, 313, 360; Settlement Reports:—C. P., Godavery, 36; Chanda, 83; Panjáb, Kangra, 25; Montgomery, 107; N.-W. P., Azamghar 118; Gazetteers:—Orissa, II., 176 and App. VI.; Bom., XII., 153; Official Rept. of Kumáon by Madden, 280; Spons, Encyclop., 961; Special Volume of Madras Selections on Fibres; Report of Director of Land Rec. and Agricul., Bombay, 1885-86, App. XII.; 1886-87, App. XII.; 1887 88, App. XII.*

Habitat.—A small herbaceous shrub, apparently wild east of the Northern Gháts, but largely cultivated for its fibre throughout India, the produce being used by the agricultural classes locally. It is more specially cultivated in Chutia Nagpúr, the Central Provinces, Madras, and Bombay, and less abundantly in the North-West Provinces, Oudh, and the Panjáb. Stewart says of the last-mentioned province that the plant is not uncommon on the lower Himálaya, ascending to 3,000 feet in altitude, and Bellew reports its existence near Ghazni at about 7,000 feet above the sea.

It will be observed that the vernacular names given to this plant in the western, southern, and central parts of India have a more original character than those in use in Bengal, the North-Western Provinces, and the Panjáb. In the latter areas it is generally compared to Sunn-hemp or to Jute. This fact might be accepted as pointing to its being a native of the former rather than of the latter division of India. The plant is chiefly grown for its fibre, but occasionally, as in some parts of Bengal, as a vegetable; and an oil is also said to be expressed from the seeds in Bombay. (*Watt*)

Fibre.—The most recent investigations on this subject are those of Dr. G. Watt, which, as they afford a complete résumé of the subject, are here quoted in entirety:— FIBRE. 178

HIBISCUS cannabinus.

The Deccan Hemp.

FIBRE.

"This plant yields a soft, white, silky fibre, eminently suitable for the coarser textile purposes to which jute is applied. It is largely grown by the natives of India and employed for agricultural purposes—ropes, strings, and sacks being made of it. Were, however, a demand to be created for this fibre as distinct from that of Sunn-hemp or other fibres, the cultivation of the plant might be indefinitely extended, and with profit to many needy cultivators who are unable to produce either jute or cotton. But so long as there exists the risk of a supply of the fibre being ejected through a purchaser thinking he was ordering Sunn-hemp and rhad this fibre thrust on him, a healthy trade in the fibre as an article of export is never likely to be developed. It is, therefore, of the first importance that *Ambári* fibre should have its properties independently investigated; and that a place be established for it separate and distinct from all the other fibres with which it has been confused in modern commerce, or for which it has been employed only as an adulterant. **Messrs. Cross, Bevan, and King**, in their table of the Chemical Analyses of Indian Fibres, place **Hibiscus cannabinus** in their second list, which comprises such as are of less textile value, because of lignification, as for example, **Jute, Sida, Abroma, Pavonia**, &c. Under hydrolysis (*i.e.*, boiling in one per cent. caustic soda, Na_2O) for five minutes, they show it to have lost 14·0, and after continued boiling for one hour, 19·5 of its weight, and that under the same treatment, jute lost considerably less, *viz*. (*a*) 13·3, and (*b*) 18·6. Further, that it contains 73·0 of cellulose, in other words, only 3 per cent. less than jute, and in this respect it is a superior fibre to **Kydia calycina** (70·0 per cent.), **Anona squamosa** (62·3), **Hardwickia binata** (63·4), **Sterculia villosa** (36—40), **Bauhinia Vahlii** (61·6), and **Musa paradisiaca** (plantain) 64·6. These chemists contend that the percentage of cellulose is a more trustworthy criterion of fibre merit than any other single feature that can be cited. The facts of hydrolysis and cellulose give, therefore, to this fibre a considerably superior claim to recognition than most of the fibres enumerated above. It falls far short, however, of Sunn-hemp, a fibre that possesses 83·0 per cent. of cellulose, and of **Sida** 83·8 per cent. On being mercerized (*i e.*, thoroughly wetted in a solution of 33 per cent. Na_2O for one hour), it loses 16·0 per cent. of its weight, and undergoes important chemical and physical changes. The ultimate fibres (the filaments of **Cross, Bevan, and King's** Report) are from 1·5 to 4 mm. in length, that is to say, they are the same length as those of jute (1·5 to 3), from which, indeed, they can with difficulty be distinguished under the microscope, either in size or shape.

"To compare with **Messrs. Cross, Bevan, and King's** results of chemico-microscopic examination, it may serve a useful purpose to quote here the description of the fibre published in *Spons' Encyclopædia*. 'The dimensions of the filaments are:—Length: max., 0·236 in.; min., 0·078 in.; mean, 0·196 in; diameter: max., 0·00132 in.; min., 0·00056 in.; mean, 0·00084 in. The length of the extracted fibre varies between 5 feet and 10 feet. The fibre is somewhat stiff and brittle, and though used as a substitute for hemp and jute, it is inferior to both. The breaking strain has been variously stated at 115 to 190℔. It is bright and glossy, but coarse and harsh. It is sold with and as jute, and is employed in Bengal for the purposes of jute, including fishing-nets and paper. Samples of the fibre exposed for two hours to steam at two atmospheres, followed by boiling in water for three hours, and again steamed for four hours, lost only 3·63 per cent. by weight, as against flax, 3·50, Manilla hemp, 6·07, hemp, 6·18 to 8·44; jute 21·39. The average weight sustained by slips of sized paper weighing 39 gr. made from this fibre, was 71℔, as compared with Bank of England note pulp 47℔. It worked satisfactorily, and took ink well.'

FIBRE

Under jute in the same work it is stated that—'Slips of sized paper, weighing 39 gr., made from this fibre, bore 60℔, as against Bank of England note pulp, 47℔.' And in another passage the measurements of jute filaments are given as max., 0·196 in.; min., 0·057 in.; mean 0·078 in.; diameter: mean, 0·0008 in. 'The shortness of the filaments explains their inability to withstand long exposure.' According to the method of hydrolysis described in *Spons' Encyclopædia*, the fibre of **Hibiscus cannabinus** is fully six times as durable as jute, whereas by **Cross'** method of hydrolysis it is slightly less durable than jute. Reducing the measurements given by **Cross** to inches so as to admit of comparison with those published in *Spons' Encyclopædia*, the writer in the latter work gives the maximum as 0·08 in. greater, and the minimum 0·02 in. also greater than the measurements obtained by **Cross.** This may be due to some error in the instruments used by the two investigators. But should it be found that **Hibiscus** does actually possess longer filaments than jute, this fact should give to it a higher commercial position than it has as yet attained. It could be produced at the same price and be grown over a much wider area, hence some hopes of a greatly extended trade in *Ambári* fibre might be anticipated with the possession of a more thorough knowledge of its properties. In a Report recently issued by the Agri-Horticultural Society of India, on some experiments performed with '**Hibiscus, Abutilon, Sansevieria,** and **Sida,**' the Committee came to the conclusion that these fibres possess 'no advantages over that of jute.' But may it not be said that this conclusion was influenced by local necessities? It might not be of any special advantage to the jute-growers of Bengal to substitute for jute any of the fibres named; but from an imperial point of interest the question suggests itself whether it would not be advantageous for cultivators, in parts of India where jute cannot be grown, to adopt as a substitute or competitor either **Hibiscus, Malachra,** or **Sida.** It might, for example, be suggested that the efforts to cultivate jute in Madras could with advantage be abandoned in favour of **Hibiscus cannabinus,** and that the sacking trade of Bombay might, as once before suggested, be met by the organisation of a new industry in the fibre of **Malachra capitata. Sida** seems destined for a higher position than jute has as yet attained.

"To turn now from chemico-microscopic examinations to the results of the practical tests performed by **Dr. Roxburgh.** Two specimens of this fibre were experimented on by the Father of Indian Botany, *viz.*, (*a*) prepared from plants 'cut when in blossom and steeped immediately,' (*b*) 'the same cut when the seed was ripe.' Of (*a*) a line when dry broke with a weight of 115℔, when wet, of 133℔ the wet line gaining a tension represented by 18℔; of (*b*) a dry line broke with 110℔, a wet one with 118℔, gaining a tension of only 7℔ by the process. According to these experiments the *Ambári* fibre deteriorates by the plants being allowed to grow beyond the flowering season. With Sunn-hemp (**Crotalaria juncea**), **Abroma augusta,** and several other fibres, the mature or stronger fibres are obtained when the plants are in fruit. Thus comparing with the above figures, a line of Sunn-hemp prepared from plants in flower broke when dry with 130℔, when wet with 185℔, cut while in fruit the dry line broke with 160℔ and the wet line with 209℔, the latter gaining 35℔ in weight. **Royle** commenting on **Roxburgh's** experiments remarks:—'So in **Dr. Wight's** experiments, the fibre of **Hibiscus cannabinus,** which is sometimes called the jute of Madras, broke with 290℔, when Sunn (**Crotalaria juncea**) broke with 404℔. Both these, like **Dr. Roxburgh's** specimens, were probably grown in the same climate. But in the author's experiments, Sunn broke with 150℔, when Brown-hemp broke with 190℔;

FIBRE.

but the Sunn was from Bengal, and the Brown-hemp from Bombay. There is, however, some uncertainty about this, because, though no fibres can well appear more distinct than these two, yet the author has in his possession fibres of a **Crotalaria** which are hardly to be distinguished from **Hibiscus**—Brown-hemp. The Sunn fibre of Bombay was, even in **Dr. Roxburgh's** time, remarkable for its dark colour. But some specimens of *Ambári* fibre, sent by **Dr. Gibson** as those of **Hibiscus cannabinus**, and which closely resemble those of some kinds of ordinary Brown-hemp, were favourably reported on by a **Mr. Enderby**, then of the rope-manufactory at Greenwich. Brown-hemp is and has always been the peculiarly prepared Sunn-hemp of Bombay, otherwise known as Salsette Hemp. **Roxburgh** gave his own results of experiments with this, and admits that it is distinctly superior to Bengal Sunn-hemp, is of a dark colour, and the fibres more easily heckled because of being peeled off. Far from **Royle's** experiments upsetting therefore the experiments of **Roxburgh** and of **Wight**, they simply prove that **Royle** was comparing Bengal with Bombay Sunn-hemp, and not, as he affirms, Bengal Sunn-hemp with Bombay *Ambári*. **Royle** may thus be viewed as having seriously laid the foundation of much of the confusion in modern literature which has so materially retarded any possible formation and growth both of a Sunn-hemp and an *Ambári*-fibre industry." For Salsette or Brown Hemp the reader is referred to the article on **Crotalaria juncea**, Vol. II., 595.

"**Dr. Buchanan-Hamilton** wrote that the natives of Bhagulpur in his day considered this fibre stronger and more durable than jute. **Mr. Hem Chunder Kerr** says that, 'It is employed in Bengal for all the purposes of jute and also for making fishing-nets and paper.' In a further paragraph he gives an account of the process adopted in making paper from this fibre with lime—20 seers of fibre to 3 of lime and 20 of water being the proportions used."

CULTIVATION.

METHODS OF CULTIVATION AND SEASONS OF SOWING AND REAPING THE AMBÁRI-FIBRE PLANT.

Bengal. 179

"*In Bengal.*—**Roxburgh**, in his *Flora Indica*, says this plant is cultivated in the rainy season. Most writers affirm that it is grown as a mixed crop, only forming belts through or along the borders of *dál* fields and other such crops. **Dr. Buchanan-Hamilton** alludes to having observed it cultivated in Bhagulpur nearly as extensively as jute. In Purneah he found it grown by itself and not as a mixed crop.

N.-W. Provinces. 180

In the North-Western Provinces—**Messrs. Duthie and Fuller** admit that it is impossible to arrive at a definite knowledge regarding the extent of cultivation, 'but it is believed to be grown on a much smaller scale than Sanai. It is very rarely cultivated as a sole crop and most commonly occurs as a border to fields of sugar-cane, cotton, and indigo.' 'In ordinary Doab Districts it is only met with as a sparse bordering to some *kharif* fields, and is merely grown for the domestic uses of the cultivator.' 'Its cultivation will necessarily be similar to that of the crop with which it is associated, and thus it will be sown in February if as a border to sugar-cane, in May if a border to *jamowa* indigo, and in July if a border to cotton. When ripe the plants are cut down close to the ground or are pulled up by the roots. It is important that none of the lower part of the stem be lost, since this contains the best fibre. The stalks are then kept submerged in water for a period varying from six to ten days according to the weather, when the bark can be easily pulled off by hand in long continuous strips. The method of extraction is, therefore, much simpler than that of Sanai. If the stalks are kept too long in water, the fibre loses very greatly in strength, although gaining in colour.' The out-turn varies

CULTIVATION.

between 125 seers and 20 seers an acre according to the thickness of the sowings and the degree to which the belts of fibre-plants are overshadowed by the other crop. In the Azamgarh Settlement Report (*p. 118*), it is stated of this plant that 'it is grown throughout the district by cultivators of all castes round the edges of their sugar-cane fields. The seed is put into the ground at the beginning of the rains, and the plant is cut in October.'

Punjab. 181

In the Panjáb—**Stewart** writes: 'It is generally raised in narrow strips along the deegs of the fields of cotton or pulse, being sown either about April and irrigated up to the rains, or sown during the latter, the former giving much the best result (*Edgeworth*). Its fibre is used for the manufacture of ropes, twine, and sacking. In Sind its fibre is considered the best for nets and ropes, but it is rarely used for cloth.'

In the Kangra Settlement Report and also in the Montgomery Report the plant is described as an autumn crop.

Bombay. 182

In Bombay—**Lisboa** says: 'The fibres of this plant—which are prepared by steeping the stems in water, are hard, and more remarkable for strength than for fineness—might be considerably improved by care. A line made of them, 4 feet long, sustained, when dry, a weight of 115lb; in the wet state its tenacity was greatly increased, and it bore a strain of 133lb. Is only adapted as a mixture for the commoner description of paper.' The *Gazetteer of Bombay* (*Vol. XII., 153*) says the plant is sown in June and reaped in October."

In the Reports of the Director of Land Records and Agriculture, Bombay, it is stated that the area under cultivation in that Presidency was, in 1885-86, 53,488 acres, in 1886-87, 87,957 acres, and in 1887-89, 71,588 acres.

Madras. 183

In Madras—**Roxburgh** gives in his *Coromandel Plants* a full account of this plant. He says: 'The usual time of cultivation is the cold season, though it will thrive pretty well at all times of the year, if it has sufficient moisture. A rich loose soil suits it best. The seeds are sown about as thick as hemp, but generally mixed with some sort of small or dry grain, rendering it necessary to be sown very thin, that the other crop (which is one of those grains that does not grow nearly so high) may not be too much shaded. It requires about three months from the time it is sown, before it is fit to be pulled up for watering, which operation, with the subsequent dressing, is similar to that hereafter described for **Crotalaria juncea**.'

The most complete account of the cultivation of this plant which the writer has had the pleasure of reading is that given by **Mr. F. A. Nicholson** in his Manual of the Coimbatore District (*p. 238*). He says:—'*Pulimanji* is grown everywhere in gardens and on dry lands, but especially in red loams and gravels; in the former with various crops, such as turmeric, *ragi*, *kambu*, &c., in the latter with *cholam* and *kambu*. In gardens, as a mixed crop, it is planted on the ridges of the beds and watercourses, not indiscriminately; on dry land the seeds are mixed and sown broadcast. Occasionally it is grown as a separate crop in gardens, and is then sown thickly about two or three inches apart in beds, and watered twice a week.

'The fibre is prepared by bundling the stalks, which, after a few days, are steeped for nearly a week in water under stones; when sufficiently retted they are cleaned by beating them on the ground, the fibre stripped off, washed, and dried. Five hundred stems, about eight feet high, as grown *en masse* in gardens, were recently taken at random and the fibre removed and cleaned in the usual way; the result was 5½lb clean and good fibre. The stems when carefully dried weighed nearly 20lb.

CULTIVATION.

Assuming the acre to be 40,000 square feet, after allowing for waste patches, the number of stems at three inches apart would be 640,000; hence the yield in clean fibre at 1lb per 100 would be 6,400lb=2⁶⁄₇ tons; the stems would yield also 11 tons of poor fuel. The yield of three fine stems grown along the ridges in turmeric plantations, and measuring 16 to 17 feet high, was 3¼ ounces of clean fibre, or somewhat over one ounce each, instead of ⅛th of an ounce. The dried stems each weighed 5 ounces instead of less than ¾ths of an ounce. The fibre is very well spoken of; in some experiments it was stronger, in some weaker, than sunn-hemp. This fibre appears to offer a splendid field for agencies either for cordage, or more especially for paper stuff; it is grown with the greatest ease and abundance and on dry land it never fails, though less productive than when grown in gardens. At present, when produced only in small quantities and rudely extracted by hand, it is sold retail, including the rayat's profit, at R1 per 25lb, and could certainly be produced far cheaper by the new machines. Any quantity could be grown to order.'"

Area. 184 "AREA OF CULTIVATION AND TRADE IN AMBARI HEMP.—No information other than is conveyed by the above passages can be procured. It, therefore, seems desirable that all future reports on the so-called hemp fibres may deal with that of **Hibiscus cannabinus** separately from the sunn-hemp and the true hemp."

OIL. Seeds. 185 **Oil.**—The SEEDS of this plant have been frequently sent from India to England as an oil-seed. They are said, however, to yield only from 15 to 20 per cent., so in spite of their cheapness they are seldom crushed in this country. The oil is clear and limpid and, though coarse, forms a good lubricating and illuminating material.

MEDICINE. Juice. 186 Seeds. 187 Leaves. 188 **Medicine.**—The JUICE of the flowers mixed with sugar and black pepper is a popular remedy for biliousness (*Pharmacog. Ind.*). The SEEDS are said by Persian writers to be aphrodisiac and fattening, the latter quality being probably due to the edible oil they contain. They are also employed externally as a poultice for pains and bruises. The LEAVES are aperient.

SPECIAL OPINION.—§ "The leaves are used as a pot-herb by those suffering from constipation, and I am fully aware of its laxative properties." (*Honorary Surgeon Easton Alfred Morris, Tranquebar.*)

FOOD. Leaves. 189 Seeds. 190 Oil. 191 **Food and Fodder.**—The YOUNG LEAVES are employed as a pot-herb, and the SEEDS, which yield an edible OIL, are eaten when roasted, and in the Poona District are also used as cattle food.

DOMESTIC. Stalks. 192 **Domestic.**—It is reported that, in the Meerut District, the dried STALKS are used in making matches, being split into small pieces and tipped with some preparation of sulphur.

TRADE. 193 **Trade.**—See final paragraph of description of the fibre.

194

Hibiscus collinus, *Roxb.; Fl. Br. Ind., I., 338.*

Syn.—HIBISCUS ÆSTUANS, *Wall.*; H. ASPERATUS, *Wall.*; H. ACERIFOLIUS, H. ERIOCARPUS, *DC.*; H. SIMPLEX, *Roxb.*; PAVONIA ACERIFOLIA, *Lk. & Otto.*

Vern.—*Konda-gang* (Roxb.), ? *konda benda* (Elliot), TEL.

References.—*Roxb., Fl. Ind., Ed. C.B.C., 525; Thwaites, En. Ceylon Pl., 26; W. & A., Pl. Prodr., I., 51; Trimen, Syst. Cat Cey. Pl., 11; Royle, Fib. Pl., 261.*

Habitat.—An arboreous plant of the Western Peninsula, found in the Konkan, Circars, and Coromandel, also in Ceylon.

FIBRE Bark. 195 **Fibre.**—Roxburgh writes: "The natives of the Circars use the BARK as a substitute for hemp;" and in another passage, "The mountaineers use the bark of this species for cordage. I have often observed that the bark of most of the Indian plants of this class, particularly of this family, might

FIBRE.

be employed for the same purposes as hemp. It almost always peels off in very long strips, and is very tough."

Royle quotes this information, but gives no further particulars regarding the nature and value of the fibre.

Hibiscus esculentus, *Linn.; Fl. Br. Ind., I., 343.* 196

THE EDIBLE HIBISCUS; OCHRO OF WEST INDIES; GOMBO, *Fr.*

Syn.—HIBISCUS LONGIFOLIUS, *Roxb.;* ABELMOSCHUS ESCULENTUS *W. & A.*

Vern.—*Bhindi, katavandai, ranturi, rám-turái,* HIND.; *Dhenras, dhéras, rám-torái,* BENG.; *Bhindi, bhinda tori, rámturái,* PB.; *Bhendi,* DEC.; *Bhendi,* C. P.; *Bhénda* or *chendi,* BOMB.; *Bhenda,* MAR.; *Bhíndu,* GUZ.; *Vendi* (or *bhendi*), *vendaik-kay,* TAM.; *Venda-kaya, benda-káya,* TEL.; *Bendakainaru, bendé-káyi, bendekdi,* KAN. *Ventak-káya,* MALAY.; *Youn-padi-sí,* BURM.; *Bandaká,* SING.; *Gandhamula, dárivká* (**Moodeen Sheriff**), *tindisa* (**U. C. Dutt**), SANS.; *Bámiyá,* ARAB.; *Bámiyah,* PERS.

References.—*Roxb., Fl. Ind., Ed. C.B.C., 528; Stewart, Pb Pl., 21; DC., Origin Cult. Pl., 189; Mason, Burma and Its People, 418, 472, 505, 756; Pharm. Ind., 35; O'Shaughnessy, Beng. Dispens., 215; Moodeen Sheriff, Supp. Pharm. Ind., 13; Mat. Med., Madras, 52; U. C. Dutt, Mat. Med. Hind., 321; Dymock, Mat. Med. W. Ind., 2nd Ed., 103; Fluck. & Hanb., Pharmacog., 2nd Ed., 94; U. S. Dispens., 15th Ed., 1664; S. Arjun, Bomb. Drugs, 183; Murray, Pl. and Drugs, Sind, 63; Waring, Bazar Med., 9; Year Book Pharm., 1876, 232; Atkinson, Him. Dist., 306, 702; Drury, U. Pl., 1; Lisboa, U. Pl. Bomb., 147, 227; Birdwood, Bomb. Pr., 140; Royle, Fib. Pl., 259, 268; Liotard, Paper-making Mat., 14, 22, 24, 26, 28, 30; Atkinson, Ec. Prod. N.-W. P., Pt. V., 14; Indian Forester, IX, 203; XIV., 269, 271, 273; Spons, Encyclop., 961; Kew Off. Guide to the Mus. of Ec. Bot., 19; Jour. Agri-Hort. Soc., VII. (New Series), 224, 226, 227; Settlement Report, Chanda, 82; Gazetteers:—Sialkot, 68; N.-W. P., IV., lxviii.; Orissa, II., 180, App. VI.; Mysore and Coorg, I., 55; II., 11; District Manuals—Cuddapah, 199; Coimbatore, 40; Special Reports from Collectors of Cuddapah, Bellary, and Madura, in Madras, from the Directors, Land Rec. and Agric., Poona and Bengal, and from Chief Commr., Burma; Madras Board of Rev. Progs., 1889, No. 266.*

Habitat.—A tall herb, cultivated throughout India, and naturalised in all tropical countries. According to the *Flora of British India* it was probably originally wild in this country, but this opinion seems erroneous. **Flückiger** and **Hanbury** regard the plant as of African origin, bringing forward in support of their theory its description by **Abul-Abbas el Nabati**, as met with by him in Egypt in 1216 A.D., and the figure by **Prosper Alpinus**, published amongst his Egyptian plants in 1592. **DeCandolle** accepts this reduction to an African origin as undoubted, but adds that it does not appear to have been cultivated in Lower Egypt before the Arab rule.

METHODS OF CULTIVATION AND SEASONS OF SOWING AND REAPING THE EDIBLE HIBISCUS.

CULTIVATION.

In Bengal.—The Director of Land Records and Agriculture writes: "Grown all over Bengal for its edible capsules, chiefly in the districts of Eastern Bengal. It is usually confined to gardens." Bengal. 197

In Assam.—The Director of Land Records and Agriculture reports: "*Bhendi* is largely grown in gardens and around homesteads, but not as a field crop." Assam. 198

In Burma.—The Chief Commissioner has supplied the Government with the following information "In the UPPER CHINDWIN DISTRICT it is cultivated to some extent during the rains. The ground is broken and the seeds covered lightly with earth. It grows anywhere, but is not cultivated Burma. 199

CULTIVATION.

for exportation, the fruit being eaten locally. In the PROME DISTRICT the area under cultivation is 1·25 acres. The seed is sown in the month of June and the fruit gathered in October. The cost of cultivation for ¼ acre is R2 and the produce sells for R5. The seed is sown in soft soil by digging a hole of 4 inches depth. The annual production is calculated at 20,000 fruits, and the cost per 100, if purchased locally, is R1-4. IN THE HANTHAWADDY DISTRICT the area of cultivation is about 5 acres. It is generally sown in the months of December and January, and the fruit collected a month or so after. In the rains it is also sown about the month of June, and the fruit collected about the same time afterwards. The cost of its cultivation per acre is about R5-2 and the profit about R7. The soil best adapted for its cultivation is one of a light nature. Furrows are made in it and the seeds are sown in rows. The fruit is generally used for culinary purposes."

It is reported also to be more or less extensively cultivated under similar conditions in the districts of Sandoway, Kyaukpyu, Mandalay, Tharawaddy, Thongwa, Bassein, Shwegyin, Amherst, Ruby Mines, Toungu, Bhamu, and in the Eastern Division of Upper Burma. In all these districts it seems to be cultivated entirely for local demand, as the fruit is an important article of diet with all classes of Burmans.

N.-W. Provinces. 200 *In the North-West Provinces,* **Atkinson** writes: "Cultivated in gardens below the hills and consumed by all classes."

Punjab. 201 *In the Panjáb.*—**Stewart** mentions it as grown by natives and Europeans in the plains throughout the Panjáb, and the *Sialkot Gazetteer* gives it as one of the principal vegetables, grown all over the district.

Bombay. 202 *In Bombay.*—The Acting Director of Land Records and Agriculture, Poona, writes:—"Though it is grown in garden lands at other times of the year, the habit of the *bhendi* plant is adapted to the rains. From the form of its fruit the vegetable is divided into two kinds, one bearing six-ribbed fruits, and the other eight-ribbed. Besides this main division, it is sub-divided into "early" and "late" *bhendi*. The early variety is sown in June and bears from early August to December. The plant grows about 2 feet high, the leaves are large, and the pods short and thick. The late variety is sown in June or July along the edges of, or among, *bajri* crops. It begins to bear late in September and continues till the end of November. The plant reaches 6 to 7 feet in height. The leaf is small and the pod thinner than in the other variety, and prickly. The acre rate of seed varies from 5 to 10lb. It is sown on ridges 3 feet apart. Two or three seeds are put in holes one foot apart. The seed sprouts within a week. When the seedlings are a span high, they are manured with fish manure, farm yard manure or ashes, with a view to favour side growth; the plants are topped when a cubit high. Insufficiency of water as the plants are flowering causes an insect pest which is destroyed by sprinkling ashes over the plants. The acre yield varies from 6,000 to 6,500lb."

Madras. 203 *In Madras.*—The Collector of Cuddapah writes:—"The area under the crop is not available. It is sown in July, and the plants survive for a year. The soil requires to be richly manured. Cost of cultivation is R10 per *cawny*, and profit R40. No fibre is taken from the stem. The unripe fruit is sold at one anna a túk." The Collector of Bellary gives the following information: "The area of cultivation is given only in two taluks where it is estimated at 73 acres. There are two crops in the year, one sown from 3rd to 15th March and gathered from 3rd to 15th July, and the other sown from 15th to 30th July, and gathered in the second half of December. The cost of cultivation is about R5 per acre and the profit about R9. This is generally a mixed crop grown along with cereals,

CULTIVATION.

but in some places as Hadagali and Harpanapally it is grown as a garden crop. Red and black soils are both adapted for its growth. After ploughing, the land should be divided into beds wherein the seeds are sown and watered twice a week. It is eaten as a vegetable." In still another report, from the Madura District, it is stated that "The *vendai* is grown in backyards and gardens. The extent of cultivation is about 150 acres. Sowing begins in July and continues almost throughout the year, and the plant begins to yield in October. There are two varieties of this plant known in this district, one of which is taller than the other. The former is said to live for about six or seven months, the latter a month or two less. One acre is said to produce about 60 basketfuls of fruits, costing about R15, while the expense of cultivation is estimated at R10. Seeds of 100 fruits will sow one acre. Three ploughings are made to prepare the land; then the soil is manured and ploughed again. Plots are formed and the seeds buried one by one at a distance of 3 or 4 feet. Weeding is required after one month, and the plots are watered once every two or three days."

DYE. Stigmas. 204

Dye.—**Roxburgh** mentions that the STIGMAS are replete with a very beautiful deep purple juice which they communicate to paper, and which is tolerably durable. **Murray**, in his *Plants and Drugs of Sind*, confirms this, and adds that the stained paper forms a substitute for litmus paper. The writer can find no further record of the employment of this colouring matter as a dye.

FIBRE. 205

Fibre.—The bast yields a strong useful fibre of a white colour, which is long and silky, generally strong and pliant, and composed of very fine individual fibres. It is employed economically in some parts of India, but in many districts where the plant is much grown as a vegetable the excellence of the fibre seems to be unrecognised. It is undoubtedly valuable and seems to possess qualities specially fitting it for the purposes of paper-making. According to **Roxburgh** its breaking strain is 79℔ when dry, 95℔ when wet. It contains 74 per cent. of cellulose, and in **Messrs. Cross, Bevan, and King's** experiments it was found to lose 9·8 and 14·2 per cent. of its weight, when boiled in 1 per cent. solution of caustic soda for five minutes and one hour respectively. The Agri.-Horticultural Society of India, in the experiments referred to under **H. Abelmoschus**, found **H. esculentus** to yield a poor crop, sparse in low lying places. The average acre yield of fibre by Death and Ellwood's process was only 84½℔, while by retting, even from this poor crop, it amounted to 6 maunds and 17 seers. **Liotard**, in his *Paper-making Materials of India*, notices the fibre, mentioning that it is very fine and well suited for paper-making, and in another passage says that paper has been made with it, though only on a small scale, in the Lucknow Central Jail. In France, the manufacture of paper from this fibre is the subject of a patent; it receives only mechanical treatment, and affords a paper called *banda*, equal to that obtained from pure rags. In Burma, Madras, and other parts of India, the stem is allowed to rot unused. This valuable fibre, which could thus be obtained very cheaply, does not appear to have attracted the attention that it merits.

MEDICINE. Fruit. 206 Seeds. 207 Capsules. 208

Medicine.—The mucilage from the FRUIT and SEEDS is used medicinally as a demulcent in gonorrhœa and irritation of the genito-urinary system. Muhammadan writers describe it as cold and moist and beneficial to people of a hot temperament. **Roxburgh** considers the fruit nourishing as well as mucilaginous, and views it as one of the best of esculent herbs in India. The WHOLE PLANT, but particularly the CAPSULES, is replete with much mild mucilage, and might with advantage be applied to all diseases requiring emollients and demulcents **Roxburgh** strongly recommends its use in irritating cough. **Waring**, in his *Bazar Medicines*, calls this plant Abel-

Another Hibiscus Fibre.

MEDICINE.

moschus, and recommends a syrup composed of Abrus root, **Abelmoschus**, and sugar.

The fresh immature capsules have a place amongst the officinal drugs of the Indian Pharmacopœia, in which the fruit is described as a valuable emollient and demulcent, and as diuretic. It is there recommended as a drug which "may be resorted to with confidence in catarrhal affections, ardor urinae, dysuria, gonorrhœa, and other cases requiring demulcent and emollient remedies." **O'Shaugnessy's** *Bengal Dispensatory* contains a prescription for a Hibiscus lozenge which is recommended as highly soothing in irritable conditions of the pharynx. Both the fruits and leaves are also employed externally as an emollient poultice. The *U. S. Dispensatory* contains the following information regarding the ROOTS: "They are said also to abound in mucilage, of which they yield twice as much as the **Althæa** root from the same weight, free from any unpleasant odour. Their powder is perfectly white, and superior also to that of marsh mallow."

Root. 209

SPECIAL OPINIONS.—§ "Is sometimes used for dysentery" (*Surgeon-Major C. J. McKenna, I.M.D., Cawnpore*). "Used as a laxative and diuretic in dropsical affections" (*Surgeon-Major J. J. L. Ratton, M.D., Salem*). "The mucilage is used in Mysore for gonorrhœa, and it is also said to be aphrodisiac" (*Surgeon-Major J. North, Bangalore*). "This vegetable may be allowed where others are inadmissible in bowel complaints, owing to its demulcent properties" (*Assistant Surgeon Shib Chundra Bhattacharji, Chanda, Central Provinces*). "Demulcent, diuretic, doses ʒii to ʒiv, used in catarrhal attacks, irritability of the bladder, kidney, and gonorrhœa" (*Chuna Lall, 1st Class Hospital Assistant, Jubbulpore*). "The fruit is useful in gonorrhœa and dysentery. The bark of the stem is made into ropes, the stems being treated in the same manner as hemp. The young pods are eaten by the natives when suffering from spermatorrhœa" (*Narain Misser, Hospital Assistant, Hoshangabad, Central Provinces*).

FOOD. Fruit. 210

Food.—The unripe FRUIT is a favourite vegetable both with Natives and Europeans in India, and is also eaten in the south of France and the Levant. In European cookery its mucilaginous property is taken advantage of in the thickening of soups, &c. The more mature fruit is employed by the natives as a constituent of curry, and when quite young forms a good pickle. The SEEDS are said to be used as a substitute for pearl barley, and to flavour coffee, and are also eaten on toasted bread.

Seeds. 211

FODDER. Leaves. 212

Fodder.—The LEAVES are used in parts of Madras as cattle fodder.

DOMESTIC. 213

Domestic Uses.—The dried stalks are reported to be used by the poor as fuel in the Bellary District of Madras.

TRADE. 214

Trade.—As already stated, there is no direct export of the FIBRE from India, nor is it sold at all, except either as an adulterant of jute, or under the name of hemp. In Dacca and Mymensingh it is said to be exported under the latter appellation to the extent of a few thousand cwt. yearly. The price of the fruit varies in different parts of the country, and according to the demand for it. Thus in Bellary it is reported to vary from 2 to 3 annas a maund, in Cuddapah from 2 to 3 seers for quarter of an anna, in Pegu the price of from 8 to 10 fruits is 1 anna, and in the Ruby Mines it is sold at 4 annas a viss.

215 **Hibiscus ficulneus,** *Linn.; Fl. Br. Ind., I., 340; Wight, Ic., t. 154.*

Syn.—H. STRICTUS, *Roxb., Hort. Beng., 52;* H. PROSTRATUS, *Roxb., Fl. Ind.;* ABELMOSCHUS FICULNEUS, *W. & A., Prod., I., 53.*

Vern.—*Ban-dhenras* or *dheras*, BENG.; *Kapasiya*, N.-W. P.; *Dula*, PB.; *Parupu-benda, nella-benda*, TAM.

References.—*Roxb., Fl. Ind., Ed. C. B. C., 527 and 528; Voigt, Hort. Sub. Cal., 119; Thwaites, En. Ceylon Pl., 27; Stewart, Punjab Plants, 21*

Atkinson, Him. Dist., 791; Royle, Fib. Pl., 260; Cross, Bevan, and King, Report on Indian Fibres, 40; Watt, Hibiscus in Selections from the Rec. of Govt. of Ind., Rev. and Agri. Dept., I., 272; Balfour, Cyclop. Ind., II., 44; Journ. Agri.-Hort. Soc., New Series, VII., 224, 229; Indian Forester, XII., app. 7; Gazetteer, N.-W. P., IV., lxviii.

Habitat.—Indigenous in the hotter parts of India from the Panjáb, North-West Provinces, and Bengal to Madras.

A small, herbaceous, annual bush, which should be sown at the beginning of the rains. Roxburgh recommends a little earlier date for sowing, namely, in May, the plants being transplanted (from the seed-bed) in rows nine inches apart when about six inches high.

Fibre.—Dr. G. Watt, in his exhaustive account of **Hibiscus**, writes: "Like most other MALVACEÆ this plant yields a valuable fibre. Roxburgh says: 'In none have I found so large a quantity, equally beautiful, long, glossy, white, fine and strong as in this. To these properties may be added the luxuriant growth and habit of the plant, rendering it an object of every care and attention, at least until the real worth of the material is fairly ascertained.' Like many of Roxburgh's valuable economic discoveries, this has remained for nearly a century without a single fact having been added, or any progress made towards utilising the tons upon tons of valuable fibre lying useless on our waysides. Roxburgh planted 40 square yards in the manner above described, and obtained 'thirty-three pounds weight of the naturally very clean fibres.' In his experiments with this fibre he found that a dry line of it broke with a weight of 104lb, and a wet line with 115lb, the fibre having gained 12lb in weight by being wetted. FIBRE. 216

"This was one of the fibres recently experimented with by the Agri-Horticultural Society; and by the Death and Ellwood Machine, they obtained from 20 seers of green stems 2·81 per cent. of clean fibre and 1·41 waste fibre. Commenting on this experiment they say: 'Even the *bandhenras*, with its large amount of mucilage, was worked out with the greatest ease' Mr. Atkinson, in his *Himálayan Districts*, says of this plant that it 'affords a very large proportion of strong fibre of a white colour, useful for twine and light cordage.'"

Oil.—Lieutenant Hawkes mentions this plant as yielding one of the oils of South India. OIL. 217

Food.—The SEEDS are often put in sweetmeats, and are employed in Arabia for giving perfume to coffee. FOOD. Seeds. 218

Hibiscus furcatus, *Roxb.; Fl. Br. Ind., I., 335.* 219

Syn.—? HIBISCUS HISPIDISSIMUS, *Griff.*; H. ACULEATUS, *Roxb.*; H. BIFURCATUS, *Roxb. ic. ined., not of Willd. nor of Hort. Beng.*

Vern.—*Konda góngúra*, TEL.; *Hin nápiritta*, SING.

References.—*Roxb., Fl. Ind., Ed. C.B.C., 527; Thwaites, En. Ceylon Pl., 26; Dalz. & Gibs., Bomb. Fl., 19; Elliot, Fl. Andhr., 95; Murray, Pl. and Drugs, Sind, 62; Atkinson, Him. Dist., 306; Lisboa, U. Pl. Bomb., 335; Royle, Fib. Pl., 261, 268; Spons, Encyclop., 962; Bomb. Gazetteer, XV., 71.*

Habitat.—A prickly-stemmed shrubby plant, found in the hotter parts of India from Bengal to Ceylon.

Fibre.—Like most other species of this genus, this plant yields, from its bast, abundance of strong, white, flaxen fibres, which Roxburgh found to have a breaking strain of 89lb when dry and 92lb when wet. Owing, however, to the strong sharp spines with which the stem is covered, it is not so easily manipulated as other species of **Hibiscus**. FIBRE. 220

H. macrophyllus, *Roxb.; Fl. Br. Ind., I., 337.* 221

Syn.—H. SETOSUS, *Roxb.*; H. VESTITUS, *Griff.*

Vern.—*Kachia udal, kasaya palla,* BENG.; *Yet woon* (**Kurz**), *Bet-mwæ-shau* (**Mason**), BURM.

References.—*Roxb., Fl. Ind., Ed. C.B.C., 523; Kurz, For. Fl. Burm., I., 127; Mason, Burma and Its People, 520, 756; Indian Forester, XIV., 269.*

Habitat.—A tree or shrub of Eastern Bengal and the Eastern Peninsula, from Sylhet and Chittagong to Mergui and Penang.

FIBRE. Bark. 222 **Fibre.**—The BARK yields a strong cordage fibre, valued by the Burmans.

TIMBER. 223 **Structure of the Wood.**—Rather heavy, fibrous but close-grained, soft, white, turning pale-brown on exposure. **Kurz** expresses the opinion that it might be used for house-posts, and for other in-door house-building purposes.

224 **Hibiscus mutabilis,** *L.; Fl. Br. Ind., I., 344.*

THE CHANGEABLE HIBISCUS.

Syn.—H. ÆSTUANS, *Rottler in Herb.*

Vern.—*Shalapara, sthalkamal,* HIND.; *Thulpadma, sthalpadma,* BENG.; *Gul-i-ajaib,* PB.; *Padma-chárini, sthalapadma,* SANS.

References.—*Roxb., Fl. Ind., Ed. C.B.C., 525; Stewart, Pb. Pl., 22; Aitchison, Cat. Pb. and Sind Pl., 21; Rheede, Hort. Mal., VI., 38—42; U. C. Dutt, Mat. Med. Hind., 312, 319; Lisboa, U. Pl. Bomb., 227; Atkinson, Ec. Prod., N.-W. P., Pt. V., 14; Royle, Fib. Pl., 261, 269; Journ. Agri.-Hort. Soc., VII.* (*New series*), *224, 226, 227.*

Habitat.—A small tree without prickles, which has flowers that change in colour, almost white in the morning and red at night. It is a native of China, but is now largely cultivated in gardens all over India from the Panjáb to Burma and Madras.

FIBRE. Bark. 225 **Fibre.**—As with most of the members of the genus, the BARK yields a strong fibre, of which that from the inner layer is soft and silky, that from the outer layer hard and of a lead colour. **Roxburgh** found it inferior for cordage purposes.

FOOD. 226 **Food.**—**Atkinson** (*Economic Products of the North-West Provinces*) mentions **H. mutabilis** as being "cultivated for pickling."

227 **H. rosa-sinensis,** *Linn.; Fl. Br. Ind., I., 344.*

THE SHOE FLOWER, *Eng*; KETMI DE COCHIN CHINE, *Fr.*

Vern.—*Jásút, jásúm,* HIND.; *Juwa, joba, jiwa, oru,* BENG.; *Gudél, kudhal, jásút, jásúm,* DEC.; *Jásavanda,* BOMB.; *Jásavanda, dasinda-cha-phula,* MAR.; *Jasúva,* GUZ.; *Shappat-tup-pu,* TAM.; *Java push-pamu, japápushpamu, dásána,* TEL.; *Dásavala,* KAN.; *Chempa-rattip-púva, ayamparatti,* MALAY.; *Kaung-yan, koung-yan,* BURM.; *Joba, japa-pushpam,* SANS.; *Angharæ-hindi,* ARAB.; *Angaræ-hindi,* PERS.

References.—*Roxb., Fl. Ind., Ed. C.B.C., 523; Kurz, For. Fl. Burm., I., 126; Elliot, Fl. Andhr., 46, 73; Ainslie, Mat. Ind., II., 359; O'Shaughnessy, Beng. Dispens., 218; U. C. Dutt, Mat. Med. Hind., 300; S. Arjun, Bomb. Drugs, 200; Murray, Pl. and Drugs, Sind, 63; Bidie, Cat. Raw Pr., Paris Exh., No. 135; Moodeen Sheriff, Mat. Med. of Madras, 51; Taylor, Topography of Dacca, 57; Atkinson, Ec. Prod., of N.-W. P., Pt. V., 14; McCann, Dyes and Tans, Beng., 66; Piesse, Perfumery, 156; Balfour, Cyclop., II., 44; Smith, Dic., 53; Treasury of Bot., I., 589; Journ. Agri.-Hort. Soc., VII.,* (*New series*), *224, 226, 227; Gazetteers:—Mysore and Coorg, I., 57; Orissa, II., 179; N.-W. P., I., 79.*

Habitat.—A favourite ornamental bush, a native of China, but found in most flower gardens in the plains of India. There are numerous varieties, single and double, red, yellow, and white. The plant never seeds in India.

Dye.—Dr. Bidie reports that an infusion of the FLOWERS produces a purplish hue. Dr. McCann, in his *Report on the Dye-stuffs of Bengal*, writes that, "It is mentioned from Húglí that a red dye is obtained by children from the flower which is used in colouring paper. If the flowers are rubbed on paper they give at first a reddish colour, turning immediately into a pretty purple or lilac; the addition of a little acid turns this at once into a bright red." Roxburgh states that the petals are used to blacken shoes, hence the English name of the plant. The Chinese are said to utilise them in the same way, and also to make a black dye for their hair and eyebrows from the petals. DYE. Flowers. 228

Fibre.—The BARK yields a good fibre. FIBRE. Bark. 229

Medicine.—The FLOWERS are considered emollient, and an infusion of the petals is given as a demulcent. Moodeen Sheriff reports favourably of an infusion or syrup of the petals as a demulcent and refrigerant drink in fevers, and as a demulcent in cases of ardor urinae, strangury, and irritable conditions of the genito-urinary tract. He also recommends an oil, made by mixing the juice of the fresh petals and olive oil in equal portions, and boiling till all water has evaporated, as a stimulating application for the hair. O'Shaughnessy considered the LEAVES to be emollient, anodyne, and laxative. Murray (*Plants and Drugs of Sind*), and Taylor (*Medical Topography of Dacca*), both mention the employment of the flower by the Natives in the treatment of menorrhagia, the former describing it as administered fried in *ghí*. MEDICINE. Flowers. 230 Leaves. 231

SPECIAL OPINIONS.—§"The red variety is useful in colouring syrups" (*Honorary Surgeon P. Kinsley, Chicacole, Ganjam, Madras*). "SEEDS pounded into a pulp, and mixed with water, are given in gonorrhœa" (*Surgeon Anund Chunder Mukherji, Noakhally, Bengal*). Seeds. 232

Hibiscus Sabdariffa, *Linn.; Fl. Br. Ind., I., 340.* 233

THE ROZELLE OR RED SORREL of the West Indies.

Syn.—? H. SANGUINEUS, *Griff.*

Vern.—*Lál-ambári, patwa*, HIND.; *Mesta, patwa, lál-mista-bij*, BENG.; *Arak' kudrum, togot arak'* (Revd. A. Campbell), SANTAL; *Polechi*, MAL. (S.P.); *Lála ambádi*, SIND.; *Lal-ambari, patwa*, BOMB.; *Lál ambári, patwa*, DEC.; *Shvappu-káshuruk-virai, shimai-káshuruk-virai, Shivappu-káshuruk-kai*, TAM.; *Erra-gom-kaya, erra-gomgura, shima-gómgura, erra-góngúru, erra-góng-kúra, erra-góng-áka*, TEL.; *Pula-chakiri, pundibija*, KAN.; *Polechi, chivanna-pulachi-chira-vitta*, MALAY.; *Chin-paung-ni, chinbaung*, BURM.

References.—*Dals. & Gibs., Bomb. Fl., Supp., 7; Stewart, Pb. Pl., 22; Trimen, Syst. Cat. Ceylon Plants, 11; Pharmacog. Indica, I., 212; Ainslie, Mat., Ind., II., 335; O'Shaughnessy, Beng. Dispens., 12; Moodeen Sheriff, Mat. Med. of Madras, 50; Dymock, Mat. Med. W. Ind., 2nd Ed., 104; Murray, Pl. and Drugs, Sind, 63; Bidie, Cat. Raw Pr., Paris Exh., No. 138; Mason, Burma and Its People, 756; Lisboa, U. Pl. Bomb., 147, 227; Birdwood, Bomb. Pr., 140; Royle, Fxb. Pl., 260; Liotard, Paper-making Mat., 5, 15, 16, 30; Atkinson, Ec. Prod. N.-W. P., Pt., V., 14; Saidapet Experimental Farm Manual; Revd. A. Campbell, Ec. Prod. Chutia Nagpore, No. 7804, 7800; Rep. Dir. Land. Rec. & Agric., Bengal, No. 78; Spons, Encyclop., 962; Balfour, Cyclop., II., 44; Smith, Dic., 356; Kew Off. Guide to the Mus. of Ec. Bot, 18; Jour. Agri.-Hort. Soc., VII., (New Series,) 224; Gazetteers:—Mysore and Coorg, I., 55, 57; Bombay, VIII., 184; XV., 428; Orissa, II., 180, App. VI.*

Habitat.—A small, elegant shrub, of which there are two varieties, one with red stems and a red succulent, edible calyx, the other with green stems and a green calyx. Widely cultivated throughout the hotter parts of India and Ceylon.

CULTIVATION. METHODS OF CULTIVATION AND SEASONS OF SOWING AND REAPING.

Bengal. 234 *In Bengal.*—The **Revd. A. Campbell**, in his *Catalogue of the Economic Products of Chutia Nagpore*, writes :—"Largely cultivated; generally sown with some other crops." **Firminger** says it is sown in May and put out in the ground at a distance of four feet from each other. The fruits are ripe in November or December.

Burma. 235 *In Burma.*—The Government has been supplied with the following information by the Chief Commissioner :—"In the *Prome District* the area under cultivation is 1·75 acres. It is sown in June and the fruit as well as the leaves are gathered in August and September. The cost of cultivation for one-fourth of an acre is R1. The probable amount realised on account of sale proceeds of the product of one-fourth of an acre is R3, and the profit is therefore R2. The seed is sown in soft soil by digging a hole 4 inches in depth. The annual production is 10,000 plants sold at the place of growth for R2 per 1,000 plants including fruit. In the *Pegu District* it is so plentiful as to be found growing in or near every Burman's compound. It is also to be seen growing wild in the jungles. There is a great demand for the plant among the natives. The cost of cultivation of an acre would be between R2 and R3. The outturn yields a profit of R17 or R18. In the *Hanthawaddy District*, it is sown in or about the month of November in the dry season, and the tender leaves are cut in December. In the wet weather it is sown about June, and the leaves are cut about July. The cost of its cultivation is about R6, and the profit about R14. After the soil, which is best when of a light nature, has been ploughed, the seeds are scattered and a little watering is done in the cold weather. In the *Sandoway District* it is cultivated to a small extent in taungyas and private vegetable gardens. In the *Akyab District* it is sown early in June and ripens in about three months, but the leaves are used after one month. The cost of cultivation is about R20 per acre, the profit about R30. The soil used is sandy and has to be ploughed. In the *Kyaukpyu District* it is sown in manured soil in April and May, and is fit for consumption from June to December. In the *Mandalay District* it is extensively cultivated with other products in *taungyas.* In the Shewgyn Division of the *Bhamo District* there are two kinds, the white and the red. The red kind is grown in all the *taungyas* and in most house-gardens." Reports of its cultivation to a greater or less extent on the same system in Bassein, Amherst, the Ruby Mines District, and the Eastern Division of Upper Burma, are also given.

N.-W. Provinces. 236 *In the North-Western Provinces.*—It is grown as a food plant. Mention is specially made of it in the *Agra Gazetteer.* **Firminger** says it succeeds better in the damp climate of lower India than in the upper provinces, where the winter cold proves injurious.

Panjab. 237 *In the Panjáb.*—**Stewart** writes: "Is cultivated for its succulent acid calyces, which make excellent jelly. These come to maturity in the Panjáb, but even at Lahore the seeds hardly ever ripen."

Bombay. 238 *In Bombay.*—Regarding this Presidency the Acting Director of Land Records and Agriculture has furnished the following :—"This is found in gardens near large towns, but is common in Gujarat, where it grows self-sown in the borders of garden land. It grows 3 to 4 feet high, and fruits in November to December, occupying the land for about seven months. It is also grown in kitchen gardens in the cold season. It is sown thickly broadcast, and watered sparingly. This mode of culture is most productive of fruit." **Lisboa** incidentally mentions the plant in his *Lists of Fibre-yielding and Food Plants.*

Madras. 239 *In Madras.*—In the Selections from the Records of the Madras Government published in 1856, information regarding the cultivation and method of preparation of the fibre is furnished, and it appears that even at that time the commercial value of Rozelle was recognised. The reports then called for show, however, that the plant was grown only to meet the local demands for the fibre and fruit; but of several districts the remark was made that should a demand arise, any quantity could be produced on waste lands, as the crop grew with little care or attention. CULTIVATION.

The following recent reports may be quoted:—In the *Madura District:* "The extent of cultivation is probably 50 acres. It is not cultivated separately but mixed with *cholam, ragi,* onions, &c., in gardens. No expenditure is separately incurred on account of this cultivation. It is sown in August and March, and reaped in November and December. One acre is said to yield one maund of fibre costing R1-4. *In Cuddapah:* "It is grown in the western parts of the district, is a nine months crop, and is sown in January. The cost of cultivation is R5 per *cawny,* and the profit is R10. The extent of cultivation cannot be given."

The following account is quoted from the Report of the Madras Experimental Farms for 1875:—"A plot of land, 1,210 square yards, was sown with rozelle seeds. The soil was a free sandy loam. It was prepared for the crop in the following way: twice ploughed, once harrowed and ridged, then manured with 3½ cart-loads of farm manure, and again ridged over the manure; on the 1st February 6lb of seed was sown on the ridge. During the growth of the crop the land was once hand-hoed, once bullock-hoed, and eighteen times irrigated. The seed germinated well and the plants grew with great vigour; they were straight and almost without side branches, and averaged between 5 and 6 feet high when ready for cutting, about three months after being sown."

Fibre.—The stems yield a good, strong, silky FIBRE, the Rozelle Hemp of Commerce, obtained by retting the twigs when in flower. The process is described as follows in a statement from the Nellore District of Madras: "After the plants are supposed to be properly dried, they are made into bundles and soaked in water, in which state they are allowed to remain for a period varying from 15 to 20 days. After that time the bark is separated by the hand, and well washed to free it from any impurities; it is then allowed to dry, and becomes available for use." It is employed by the natives for the purposes of cordage, being twisted into a rope of varying thickness called *núlaka.* The thinner varieties of rope are substituted for tape and rattaning for cots, the thicker is generally employed as a strong rope, for agricultural purposes, tying up cattle, &c. The fibre is also said to be employed in the manufacture of gunnies in certain districts of Madras. FIBRE. 240

Liotard (*Paper-making Materials of India*) mentions the Rozelle as yielding a fibre likely to be valuable to paper-makers, but no account exists of its actually having been so employed. On the whole, considering the easy growth of the plant and the cheap rate at which it could be procured, the fibre seems worthy of more attention than it has hitherto received.

Oil.—An OIL is prepared from this plant at the Allahabad Jail, of the preparation and economic uses of which, however, the writer has no information. OIL. 241

Medicine.—The SEEDS of the Rozelle are used medicinally, and have demulcent, diuretic, and tonic properties. A decoction of them is recommended by **Moodeen Sheriff** as a draught, in doses of from one to two drachms, three or four times a day, in cases of dysuria and strangury, and also in some mild forms of dyspepsia and debility. The authors of the *Pharmacographia Indica* observe: "In this plant we have the emollient and demulcent properties of the Malvaceæ, combined with a large amount MEDICINE Seeds. 242

MEDICINE. of acidity, which stimulates, and at the same time neutralizes, the bilious secretion."

Fruit. 243 From the FRUIT (or rather succulent calyx) a drink may be made, which is useful in biliousness, by boiling it with water and adding a little salt, pepper, assafœtida, and molasses, and for convalescence, or in mild cases of fever, an acid refreshing drink may be prepared from it. The French use it in the manufacture of an astringent syrup.

A general consensus of opinion also appears to exist regarding the valuable antiscorbutic properties of the fruit, either fresh or dried. With reference to this, it may be of interest to give here the following analysis of its chemical composition originally published by **Dr. Lyon**: Water 8·29, hot water extract 65·96, cellulose 7·68, ash 6·32, the alkalinity of which, calculated as potash, was ·75, tartaric acid 9·9, and malic acid 15·54 per cent. No citric acid was detected.

Leaves. 244 The LEAVES are regarded as emollient.

SPECIAL OPINIONS.—§"It is also considered to be antiscorbutic taken in the form of *Chatrice*." (*Surgeon Anund Chunder Mukerji, Noakhally, Bengal.*) "It is eaten with curry as an acid like tamarinds, &c. The dry calyx is an article of commerce." (*Dr. Dymock, Bombay.*) "*Mesta* is a good antiscorbutic. Its different preparations are now much used in Bengal Jails." (*Surgeon R. L. Dutt, M. D., Pubna.*)

FOOD & FODDER. Calyx. 245 Capsule. 246 Leaves. 247 **Food and Fodder.**—The fleshy CALYX and CAPSULE are largely made into jam and other preserves, and in the fresh state are very acid but refreshing. A decoction of them, sweetened and fermented, is commonly called in the West Indies "Sorrel drink." The LEAVES are used in salads, and by natives in their curries, a fact alluded to by **Strattel** in his *Narrative of Travels in Burma.*

Seeds. 248 The SEEDS are said to be an excellent food for cattle.

TRADE. 249 **Trade.**—The FIBRE is as a rule extracted only for local consumption, and the amount of profit derived may be found under the heading "Cultivation." The FRUIT is sold in most bazárs either in the fresh or dry state. The wholesale price of the fruit is given by **Moodeen Sheriff** as R8 per maund, and the retail or bazár price as annas 5 per pound.

250 **Hibiscus surattensis,** *Linn.; Fl. Br. Ind., I., 334; Wight, Ic., t. 197.*

Syn.—H. FURCATUS, *Wall.*, not of *Roxb.*

Vern.—*Rhan-bhendy*, BOMB.; *Kashlikire*, TAM.; *Mulu gogu*, TEL.; *Welmá chinpoung*, BURM.; *Naapiritta, hin-nápiritta*, SING.

References.—*Roxb., Fl. Ind., Ed. C.B.C., 527; Thwaites, En. Ceylon Pl., 26; Dalz. & Gibs., Bomb. Fl., 20; Trimen, Cat. Ceylon Plants, 11; Elliot, Fl. Andhr., 119; Lisboa, U. Pl. Bomb., 227; Balfour, Cyclop., II., 44.*

Habitat.—A weak-stemmed, trailing plant, covered with soft hairs and scattered prickles, found in the hotter parts of India from Bengal to Penang, also in Ceylon.

FIBRE. Bark. 251 **Fibre.**—**Lisboa** mentions this plant in his list of those yielding fibre, but beyond observing that the fibre from the BARK is strong, he gives us no information on the subject.

FOOD. Leaves. 252 **Food.**—According to **Balfour** the LEAVES are used as greens.

253 **H. tetraphyllus,** *Roxb.; Fl. Br. Ind., I., 341.*

Syn.—H. CANARANUS, *Miq.*; BAMIA TETRAPHYLLA, *Wall.*; ABELMOSCHUS TETRAPHYLLUS, *Grah.*; A. WARREENSIS, *Dalz.*; EREBENNUS CANARANUS, *Alefeld.*

Vern.—*Rán-bhenda*, MAR.; *Ong-ká-to*, BURM.

References.—*Roxb., Fl. Ind., Ed. C.B.C., 529; Dalz. & Gibs., Bomb. Fl., 19.*

A useful Cordage Fibre. (*J. Murray*)

Habitat.—An erect, large, racemose, annual species, from 4 to 6 feet high, found in Bengal near Calcutta, and in Concan and Kánara.

Fibre.—Is said to yield a tough fibre, regarding which, however, definite information is wanting. FIBRE. 254

Hibiscus tiliaceus, *Linn.; Fl. Br. Ind., 343; Wight, Ic., t. 7.* 255

Syn.—PARTIUM TILIACEUM, *W. & A.*

Vern.—*Bola, chelwa,* BENG.; *Baniá,* or *bariá,* ORISSA; *Belli-pata,* BOMB.; *Thinban, thengben,* BURM.; *Beligobel, bellipattá,* SING.

References.—*Roxb., Fl. Ind., Ed. C.B.C., 522; Kurz, For. Fl. Burm., I., 126; Beddome, Fl. Sylv. Anal. Gen., 14; Gamble, Man. Timb., 42; Thwaites, En. Ceylon Pl., 26; Dalz. & Gibs., Bomb. Fl., 17; Irvine, Mat. Med. Patna, 123; Hunter, Orissa, II., 175, App. VI; Lisboa, U. Pl. Bomb., 15, 194, 227; Christy, Com. Pl. and Drugs, VI., 33; Royle, Fib. Pl., 261; Watt, Hibiscus in Sel. from Rec. of Govt. of Ind., Rev. & Agri. Dept., 1889, 273; Indian Forester, XII., 329; XIV., 269.*

Habitat.—A much branched bush of the Coasts of India, Burma, and Ceylon. Particularly plentiful in the Sundarbans, and on the river banks of Burma. The ease with which this plant might be propagated over wide tracts of swampy country, where little else will grow, should afford a powerful incentive to an effort being made to utilise the fibre it affords.

Fibre.—Dr. Watt, in his paper on **Hibiscus** already referred to, writes of this species that it "yields a useful fibre, extensively employed locally for cordage. It is said to gain in strength when tarred. Roxburgh's experiments with this fibre gave the following results:—A line broke when white with a weight of 41℔, after being tanned with 62℔, and after having been tarred with 61℔. A similar line macerated in water for 116 days broke when white with 40℔, tanned 55℔, and tarred 70℔. These observations are of great interest, for, of the other fibres experimented with by Roxburgh, the majority were rotten after maceration, and no other fibre showed so marked an improvement for cordage purposes when tarred. English hemp and Indian grown hemp, treated in the same manner, were found to be rotten, and Sunn-hemp broke with 65℔, and Jute with 60℔. The power of endurance under water is, therefore, a point of great importance. FIBRE. 256

The fibre is readily separated from the green or unsteeped branches, the work of preparation being less tedious than applies to the other fibre-yielding plants of this genus. It appears to be well adapted for making ropes, mats, and possibly paper (*Cameron*). The fibre seems highly suitable for the paper trade, and immense quantities of it might be conveyed in boats to the paper mills of Bengal."

Medicine.—The BARK is used in medicine. The ROOT is also said by Irvine in his *Materia Medica of Patna* to be febrifuge, and employed in the preparation of embrocations. MEDICINE. Bark. 257 Root. 258

Food.—Forster states that the BARK is sucked in times of scarcity when the bread fruit fails. FOOD. Bark. 259

Structure of the Wood.—The wood is only employed in India as fuel. It is said to be used in Tahiti, however, for planking and building light boats. TIMBER. 260

H. tricuspis, *Banks; Fl. Br. Ind., I., 344.* 261

Syn.—PARITIUM TRICUSPE, *G. Don.*

Vern.—*Gurhul,* HIND

References.—*Roxb., Fl. Ind, Ed. C.B.C., 526; Kurz, For. Fl. Burm., I., 126; Gazetteer, Mysore and Coorg, I., 57.*

Habitat.—A tree, introduced from the Society Isles, cultivated in gardens in Bengal and the North-Western Provinces.

FIBRE. 262 **Fibre.**—A strong bast-like FIBRE is obtained from the inner bark of the trunk, and branches. A sample produced at Bangalore was prepared by steeping in water for 13 days (*Cameron*).

263 **Hibiscus vitifolius,** *Linn.; Fl. Br. Ind., I., 338.*

Syn.—HIBISCUS OBTUSIFOLIUS, *Willd.*; H. TRUNCATUS, *Roxb.*; H. SERRATUS, *Wall*; H. CUSPIDATUS, *Edgew.*

Vern.—*Ban-kápás*, BENG.; *Karu-patti*, TEL.; *Bháradváji, vanakárpása*, SANS.

References.—*Roxb., Fl. Ind., Ed. C.B.C., 525; Kurz, Paper on Burmese Flora, in Journ. As. Soc. of Bengal, 1874, XLIII, Pt. II., 108; Thwaites, En. Ceylon Pl., 26; Dals. & Gibs., Bomb. Fl., 20; Elliot, Fl. Andhr., 87; U. C. Dutt, Mat. Med. Hind., 294, 322; Cross, Bevan and King, Rep. on Indian Fibres, 41; Indian Forester, XIV., 269; Gazetteers:—Mysore and Coorg, I., 57; Bomb. XV., 428; N.-W. P., I., 79; IV., lxviii.*

Habitat.—A common, herbaceous bush, in the jungles and brushwoods of the hotter parts of India, from the North-West Provinces to Ceylon. It bears large, yellow flowers having a deep rose purple eye at the base of the corolla. The leaves are often much perforated by insects. **Kurz,** in the paper quoted in the references, describes it as common along borders of fields, in shrubberies, rubbishy places round villages, &c., also in dry forests, all over Burma from Chittagong and Ava down to Pegu.

FIBRE. Bark. 264 **Fibre.**—The BARK yields a strong silvery fibre.

HIDES & SKINS.

265 The name Hides commercially comprises the raw, dressed or tanned, skins of the cow, bullock, or buffalo, while the term Skins is applied to those of goats, deer, and animals other than horned cattle. For a full list of these the reader is referred to the articles on **Furs** and **Skins.**

Vern.—*Chamrá*, HIND.; *Toll, tolu*, TAM.; *Tol, tolu*, TEL.; *Balullang, kulit*, MALAY.; *Charma*, SANS.

References.—*Royle, Prod. Res., 3; Spons, Encyclop., 1757; Balfour, Cyclop., II., 44; Madras Manual of Administration, I., 360; Settlement Reports:—Azamgarh, N.-W. P., 171; Jubbulpore, Central Prov., 87; Bombay Administration Rept., 1871-72, 372—374, 378, 380, 387, 394, 398; Gazetteer, Shahpur, 74; Statistics on Hides and Skins, Rev. and Agri. Dept., 1886.*

TRADE. 266 **Trade.**—The whole question of the production and preparation of hides and skins may be most conveniently discussed under this heading.

The trade in these articles is now a very large and important one, and has expanded considerably during the past few years. It is questionable, however, how far this increase in the exportation of hides indicates an increase in prosperity of the *rayats* or of the country at large. **Mr. O'Conor,** commenting on this in his Review of the Sea-borne Trade Returns for 1874-75, writes: "The trade is in the hands of a few Muhammadans (Hindus will not deal in hides by reason of religious prejudices), and some of these men are very wealthy, for the trade is a highly profitable one. They purchase hides from *chamars*, one of the lowest classes of the people, who go about the country skinning dead cattle, and reap a rich harvest in bad seasons. It is well known that this is an unscrupulous class of men who do not hesitate systematically to poison cattle, while feeding in the pastures, for the sake of the hide, which is abandoned to them. Cattle poisoning by this class is in fact a profession, to which the attention of the authorities has long been directed with a view to its suppression, but with little effect as yet. A considerable proportion of the cattle whose hides are exported from India every year, die from preventible causes, such as neglect (for the rayat is not careful of his cattle), drought, murrain, and

Traders in Hides. 267

Cattle-poisoning. 268

TRADE.

poison, and these beasts represent an actual loss to the country which outweighs the value of the whole number of hides put together." There is little doubt that the same state of things still prevails, and as the trade has much increased, it is to be feared that it must exist to a still greater extent.

Inter-Provincial Trade. 269

On referring to the Inter-Provincial trade returns it is found that much the largest exporters, as far as hides are concerned, are the North-West Provinces and Oudh, and that next to these come the Panjáb and Central Provinces. From the two former numerous dressed hides are exported, owing to a considerable and increasing tanning industry, especially at Cawnpcre. The Panjáb and Central Provinces, on the other hand, supply principally the undressed material. The hides (raw and dressed) exported from the North-West Provinces are nearly all sent to Bengal; the skins which are for the most part undressed are sent to Madras. Bombay imports the greater part of its hides, both raw and dressed, from the Central Provinces, its raw skins from Rájputana and Central India, and its dressed skins from Madras, the Panjáb, Rájputana, and Central India.

By road and rail, &c. 270

The following quantities of hides and skins, dressed and undressed, conveyed by road, rail, &c., from the provinces named, in 1886-87, may be given for the sake of comparison:—North-West Provinces and Oudh, 2,79,713 maunds, Panjáb 1,69,086 maunds, Rájputana and Central India, 67,705 maunds (principally skins), Mysore 62,511 maunds, Nizam's territory 51,036 maunds, Central Provinces 42,351 maunds (principally undressed hides), Sind 21,148 maunds, Berar 17,738 maunds. Besides these sources of supply, Bengal, Madras, and Bombay yearly send large consignments to their respective capitals and ports, Bengal alone supplying 3,33,446 maunds of dressed hides to Calcutta. During the same year the following were the imports by road, rail, and river, of hides and skins into the principal shipping towns: Calcutta 8,24,459 maunds, Madras 4,19,094 maunds, Bombay 74,423 maunds, Karachi 51,261 maunds. Of its hides Calcutta obtained 25,32,412 in number from Behar, 23,97,199 from Bengal, 11,48,864 from the North-West Provinces, and smaller numbers from the other districts above named.

Coasting trade. 271

A review of the coasting trade for the same year shows that Calcutta imported 40,637 cwt., and exported 47,671 cwt. Madras imported 55,598 cwt., and exported 48,450 cwt. Bombay imported 33,274 cwt., and exported 4,441 cwt. Sind imported 1,032 cwt., and exported 25,089. Burma imported 3,765 cwt., and exported 7,525. On analysing these figures it will be found that the principal part of Calcutta's imports consists of raw hides from other ports within the Presidency, the greater part of those of Madras also consisted of raw hides from British ports within the Presidency, but it also received 17,353 cwt. of raw skins from Bengal, presumably to be tanned and re-exported as the more valuable Madras article. Of Bombay's total import of 33,274 cwt., 26,422 cwt. consisted of raw skins from Sindh. The greater part of the coasting exports of Calcutta were 27,337 cwt., to other British ports within the Presidency, and the raw skins to Madras already referred to. Madras also sent 22,910 raw hides to other ports in the Presidency, while Sindh exported the large number of raw skins above mentioned to Bombay. The coasting exports of the latter port were unimportant.

Trans-Frontier. 272

The Trans-Frontier trade has considerably increased during the past few years, the largest source of hides being Upper Burma, of skins Kashmír and Nepal. The total Import trade of hides for 1886-87 was 95,186 pieces, 59,866 cwt., valued at R17,73,241, a very great increase on the imports of 1881-82, which only amounted to $5\frac{3}{4}$ lakhs of rupees. That of skins for the same year (1886-87) was 71,103 pieces,

HIDES. **Trade in Hides and Skins.**

TRADE. Imports. 273 5,040 cwt., valued at R2,11,796, while the value of this import in 1884-85 was only R1,27,541. The very great increase in trans-frontier trade is mainly due to the large imports from Upper Burma, the Burmans having evidently discovered that, with increased facilities for intercourse and trade, their formerly valueless cattle hides might form an article of large and profitable commerce. In 1881-82, the imports from Burma valued only $3\frac{1}{2}$ lakhs of rupees; last year they amounted to $13\frac{1}{2}$ lakhs, or more than $\frac{2}{3}$rds of the total trans-frontier hide imports.

Exports. 274 The Trans-Frontier export trade in hides appears to be decreasing; thus in 1884-85 it amounted to R24,093, in 1887-88 it was only R14,033, or little more than half. The export trade in skins is a small and fluctuating one, and amounted to only R19,599 in 1886-87.

Exports from India. 275 A review of the statistics, relating to the exports of hides and skins to Foreign countries, shows that a marked increase has taken place during the past ten or twelve years, due principally, during the last few years, to greatly increased activity in the skin trade, while that in hides appears to be falling off. Thus in 1887 the total value of exports amounted to R4,85,23,818, an increase of more than 100 lakhs of rupees on that of 1877-78, which valued R3,75,68,878. The greatest increase is seen to be in the export of dressed skins, which has risen in value from 92 lakhs in 1877-78, to 192 lakhs ten years later. The value of undressed hide exports increased 30 lakhs, and that of raw skins 10 lakhs, while the exports of dressed hides decreased 30 lakhs.

Analysing these statements it is found that during the same ten years the value of exports from Bengal, increased from R1,72,13,965 to R2,07,68,574; from Bombay it decreased from R56,75,427 to R46,17,119; from Madras it became during the same period very greatly augmented, namely, from R1,46,59,521 to R2,07,26,638, the total value having thus become very little short of that of Bengal; from Sind it increased from $4\frac{1}{3}$ to $9\frac{2}{3}$ lakhs, and from Burma it rose from $\frac{1}{2}$ lakh to $14\frac{1}{3}$ lakhs of rupees. Bengal appears to have almost a monopoly of the export trade in raw hides and skins, while Madras, as might be expected from the excellence of its tanning, exports nearly all the dressed hides and skins. With reference to this, the most important export trade of Madras, the following passage from the Madras Manual of Administration, Vol. I., may be quoted: "The tanning and export of skins have become of late years very large industries in South India, and Madras dressed skins are well known and highly appreciated both in European and American markets. In fact, South Indian skins fetch so much higher a price than those shipped at other ports in India, that salted hides are sent from Bengal to be tanned and exported from Madras. The superiority of the Madras article is partly due to the fine quality of the bark used for tanning it, and partly to the superior methods of manipulation. The bark is obtained from a small shrub, **Cassia auriculata** (*see Vol. II., 215*), which grows wild all over the country and is in great demand."

Foreign markets. 276 Turning to the markets to which these exports are shipped, it is found that by far the largest importer of Indian raw hides is the United Kingdom, which, in 1887-88, imported to the value of R1,21,57,552. Next in importance is the United States with an import valued at 24 lakhs; then Italy with 19 lakhs, followed by Austria with $16\frac{1}{3}$ lakhs, and the Straits Settlements with 11 lakhs. Regarding the Italian export trade, **Mr. O'Conor,** in his Report on the Trade of India for 1888-89, writes: "The British Consul at Venice reports that there was a decline there in the price of Calcutta hides, 'which are alleged to be of inferior quality and are gradually losing favour with merchants.' In this case the matter is of some importance, for the exports of hides to Italy has been a large trade, and it seems to be

TRADE IN HIDES AND SKINS.

declining." The statement is confirmed by statistics, for, since 1885-86, the exports to that country have decreased from 36·96 lakhs of rupees, to 19·27 lakhs in 1887-88, and 19·14 in 1888-89. Smaller quantities are shipped to many other countries, including France, Germany, Belgium, Greece, and Egypt, but doubtless a great proportion of the immense consignments to the United Kingdom are reshipped to continental ports. In the case of dressed hides the only direct importer of any consequence is also the United Kingdom, which receives an amount valued at R60,40,222 out of a total export valued at R60,58,092

On the other hand, Indian raw skins are imported in much the greatest quantities by the United States of America, which, in the year under consideration, imported a quantity valued at R27,22,314 out of a total export of 30½ lakhs.

Of dressed skins the United Kingdom again receives much the largest proportion, valued R1,52,07,141, out of a total of R1,92,08,570, and the United States of America comes next with an import valued at 19 lakhs. Thus in the year under consideration the United Kingdom alone imported Indian hides and skins, dressed and undressed, to the value of R3,36,50,155, out of the total export of R4,85,23,818.

The foregoing short account of the present state of the hide and skin trade in India shews that a steady increase is taking place, especially in the industry of dressing skins. There seems certainly to be a large and profitable field for this branch of Indian trade, and it is to be expected that the acknowledged expertnees in manipulation of the tanners of Southern India may enable them, with improvements in the methods of preparation, to supply skins of such a quality as to still further increase the present great demand.

For a description of the methods of preparation and respective economic values of various SKINS, the reader is referred to the article on that subject, and for similar information regarding hides to that on LEATHER.

Hing, see **Ferula alliacea,** *Boiss.*, Vol. III., 333.

Hippion orientale, see **Enicostema littorale,** *Blume*, Vol. III., 245.

HIPPOPHAE, *Linn.; Gen. Pl., III.*, 204.

A genus of diœcious shrubs or small trees belonging to the Natural Order ELÆAGNACEÆ, and comprising two species found in Europe and North Asia, both of which are natives of India.

Hippophae rhamnoides, *Linn.; Fl. Br. Ind., V., 203;* 277
THE SEA BUCKTHORN [ELÆAGNACEÆ.

Syn.—HIPPOPHAE TIBETANA, *Schlecht.*

Vern.—*Dhúrchúk, tárwá, chúk, chuma,* N.-W. P.; *Kála bísa, bánt phút, amb, tswak, kando, tserkar, starbú, milech, miles, suts, rúl, tsarmang, tsuk, tsarap, sirma, tarrú, tsarma níechak* (fruit, *tirkú*), PB; *Star-bu,* TIBETAN.

References.—*Brandis, For. Fl., 388; Gamble, Man. Timb., 317; Stewart, Pb. Pl., 190; Cleghorn, Panjáb Forests, 151; Aitchison, Fl. Lahoul (Lin. Soc. Jour., X., 75), also Fl. Kuram Valley, 92; Ind. For., XI., 4; Balfour, Cyclop., II., 85; Smith, Dic., 67; Treasury of Bot., I., 592.*

Habitat.—A thorny shrub, or small tree, met with in the inner tract of the North-West Himálaya, and Western Tibet, chiefly in moist, gravelly stream-beds, from 7,000 to 15,000 feet; distributed to Afghánistan and westwards to North and Central Asia.

MEDICINE. Fruit. 278

Medicine.—The small sour FRUIT is prepared as a syrup, and is believed by the natives of many parts to be a valuable remedy for lung complaints. Stewart mentions it as thus employed by the Tibetans.

FOOD. Fruit. 279 **Food.**—The small intensely acid FRUIT may be made into a good jelly, which **Aitchison** describes as really excellent when prepared with half its weight of sugar. It is, however, only eaten by the natives in certain localities; thus **Longden** mentions it as eaten by the natives of Kanáwar as a sort of *chatní*, while **Aitchison** remarks that in Lahoul it is not employed at all as an article of food. **Smith**, in his *Dictionary of Economic Plants*, erroneously describes it as acrid and poisonous.

TIMBER. Fuel. 280 **Structure of the Wood.**—Heartwood yellowish brown, mottled, moderately hard, close-grained. Weight 38 to 54lb per cubic foot (*Gamble*). In the dry almost treeless tracts of the Inner Himálaya it is very valuable as a fuel and charcoal-supplying plant; indeed, so much is it esteemed in these districts, that, as **Cleghorn** mentions of Lahoul, it is considered village property. The thorny branches are also employed, piled up, to form hedges or fences.

281 **Hippophae salicifolia,** *Don.; Fl. Br. Ind., V., 203.*

Syn.—H. CONFERTA, *Wall.*

A considerable amount of doubt may be entertained as to whether this species is really distinct from the former. The vernacular names of the two are identical, and the parts of the plant are similarly employed for economic purposes. Thus **Stewart** only recognises **H. rhamnoides,** while **Atkinson** and **Baden Powell** describe only **H. salicifolia,** giving the same vernacular names, habitat, and uses. **Hooker,** though acknowledging the two species, remarks, "different as this plant looks in its ordinary condition from **H. rhamnoides,** I expect that it will prove a form of that plant, due to the moister climate which it affects," an opinion which was also promulgated by **Brandis** in his *Forest Flora*. As most modern writers, however, have agreed in separating the two species, the following facts may be given as most probably referable to this form:—

Vern.—*Ashúk,* NEPAL; *Lhála,* BHOTAN, LEPCHA; *Súrch,* BASSAHIR; *Súrch, súts, kálá bís, tserdkar* (**Baden Powell**), *dhúr-chuk, tarwa-chuk, chuma,* PB.

References.—*Brandis, For. Fl., 387; Gamble, Man. Timb., 317; Royle, Ill. Him. Bot., 323; Baden Powell, Pb. Pr., 272, 582; Atkinson, Him. Dist., 316; Indian Forester, V., 186; XI., 4; Balfour, Cyclop., II., 85; Settlement Report, Kangra, 163; Gazetteer, Simla, 12.*

Habitat.—A willow-like shrub, from 10 to 20 feet in height, found in the Temperate Himálaya, from Jamu to Sikkim, between the altitudes of 5,000 and 10,000 feet.

MEDICINE. Fruit. 282 **Medicine.**—The FRUIT is mentioned by **Baden Powell** as being employed, like that of the former species, in cases of lung disease.

FOOD. Fruit. 283 **Food.**—The ripe FRUIT forms a good jelly or syrup, when prepared with sugar. When unripe it is very acid and unpalatable.

TIMBER. **Structure of the Wood.**—The thorny branches, cut and dried, are employed to make piled-up hedges, and also as fuel.

284 **Hips,** see **Rosa canina,** *Linn.* ROSACEÆ.; Vol. VI.

HIPTAGE, *Gærtn.; Gen. Pl., I., 258.*

A genus of climbing or sub-erect shrubs, or trees, belonging to the MALPIGHIACEÆ, and embracing five tropical Asiatic species. [*50;* MALPIGHIACEÆ.

285 **Hiptage Madablota,** *Gærtn.; Fl. Br. Ind., I., 418; Wight, Ill., t.*

Syn.—MOLINA RACEMOSA, *Lamk.*; GÆRTNERA RACEMOSA, *Roxb.*; BANISTERIA BENGALENSIS, *Linn.*; B. UNICAPSULARIS, *Lamk.*; B. TETRAPTERA, *Sonnerat.*

Vern.—*Kampti, madmalti, mádhavilatá,* HIND.; *Madúbhi, madúbhlúta, mádhavilatá, bos-anti,* BENG.; *Boromali,* URIYA; *Shempati,* NEPAL; *Aita-lugala,* N.-W. P.; *Endra, chopar, benkar, khúmb, chábuk, churi,* PB.; *Kampti,* C. P.; *Haladwail,* MAR.; *Mádhavi tige* (**Elliot**), *vadlayárála, pótuvadla,* TEL.; *Ati muktamu,* SING.; *Mádhavi,* SANS.

References.—*Roxb., Fl. Ind., Ed. C.B.C., 360; Brandis, For. Fl., 44; Kurz, For. Fl. Burm., I., 173; Gamble, Man. Timb., 58; Stewart, Pb. Pl., 30; Rheede, Hort. Mal., VI., t. 59; Elliot, Fl. Andhr., 17, 109, 157, 188; Moodeen Sheriff, Supp. Pharm. Ind., 154; Mat. Med. of Madras, 70; U. C. Dutt, Mat. Med Hind., 308; Murray, Pl. and Drugs, Sind, 68; Atkinson, Him. Dist., 306, 739; Indian Forester, VI., 240; XIV., 391; Balfour, Cyclop., II., 86; Works of Sir W. Jones, V., 123; Journal A. H. Soc. of Ind. XIV., 33; Gazetteers:—N.-W. P., I., 79; IV., lxix.; Bombay, XV., 428; Mysore and Coorg, I., 58; II., 7.*

Habitat.—A common climbing shrub with beautiful and fragrant flowers, found in the hotter parts of India, Burma, and Ceylon.

Medicine.—The LEAVES are esteemed useful in cutaneous diseases, and the JUICE is specially mentioned by **Moodeen Sheriff** as an effectual insecticide, and a valuable application in scabies. **MEDICINE. Leaves Juice. 286**

SPECIAL OPINIONS.—§ "Useful in chronic rheumatism and asthma." (*Surgeon-Major J. M. Houston, Travancore.*)

Fodder.—The LEAVES are said to be eaten by cattle (*R. Thompson*). **FODDER. Leaves. 287**

Structure of the Wood.—Red in the centre, yellowish white in the outer portion; moderately hard. **TIMBER. 288**

Hirudo, see **Leech, Vol. IV.**

HOG. **289**

Is the name applied to different animals of the family of pigs or **Suidæ.** Besides the numerous domesticated varieties, India has, according to **Jerdon,** two species of wild pig, each representing a different genus. The first of these is the Indian Wild Hog or Boar, **Sus indicus,** belonging to the genus SUS, *Linn.*; the second the Pigmy Hog, **Porculia salvania,** of the genus PORCULIA, *Hodgson.* **Blyth** has described what he believed to be three new species of the former, which he found in Tenasserim, Ceylon, and the Andamans, and has designated the two last **Sus zeylanensis** and **Sus andamanensis.** As the only species of importance or interest economically is the **Sus indicus,** it alone, together with the domesticated pig, **Sus scrofa,** *Linn.,* will be treated of in this article.

Vern.—*Súr, jangli súr, súwar, bad* or *búra janwar* (lard=*súr-ki-charbi*), HIND.; *Varaha,* BENG.; *Paddi,* GOND; *Dukkar,* BOMB.; *Dúkar,* MAR.; *Pandi,* TEL.; *Handi, mikka, jewadi,* KAN.; *Tan-wet,* BURM; *Walúra,* SING.

References.—*Jerdon, Mammals of India, 241; Sterndale, Mammalia of India, 416; Pharm. Ind., 284; O'Shaugnessy, Beng. Dispens., 690; Spons, Encyclop., 1366; Balfour, Cylop., II., 92, III., 211; Collin, Hides and Leather in Bevan's Brit. Manuf. Ind. Series, 41; Bomb. Gazetteer, XII., 31.*

Habitat.—The wild hog is found throughout India, from the level of the sea to an altitude 12,000 feet, wherever there is sufficient shelter, either of long grass, low jungle, or forest. It varies in abundance in different parts, but wherever it exists it does much damage to the crops. The following passage from the *Bombay Gazetteer, Vol. XII.,* may be quoted, as bearing on this: "The Hog, of all wild animals, causes most loss to the cultivator. Though much like the domestic village pig, he differs from him widely in his habits. A pure vegetable eater, he is most dainty in his tastes. He must have the very best the land affords, and while choosing the daintiest morsels, destroys much more than he eats. Sugarcane, sweet potato, and other roots, juicy millet, and Indian corn stalks are his favourite food. A few years ago herds of wild pig were found everywhere, but their numbers are now much smaller. From the border hills they still sally at night to ravage the crops in the neighbourhood, but are no longer so destructive as they were . . . Though comparatively few are left, herds of fifty and upwards are still occasionally seen."

Several varieties of domesticated pig are kept by villagers, when their religious and caste beliefs permit, in many parts of India.

OIL. Lard. 290 **Oil.**—The fat of the pig, freed from its containing cellular tissue, forms the valuable commercial product LARD. The best, or Leaf Lard, is yielded by the fat immediately surrounding the kidneys, and by the flaky layers below the skin. In European trade two other qualities are prepared from the softer and more fusible parts of the fat, and are known as second and third quality lard. Leaf Lard is much used for cooking and for preparing ointments, &c., the second quality is used in the manufacture of Lard oil, the third is employed as a low grade oil in soap-making. This lard oil is manufactured by exposing the lard, in woollen bags between wickerwork, to a pressure of about 10 cwt. a square inch in the cold for about 18 hours. The oleine thus obtained is pure, colourless, and limpid, and averages about 62 per cent. of the original lard. It is employed as an adulterant for olive oil in France, and for sperm oil in the Eastern States of America; is esteemed as a lubricant, and is said to be used for illuminating. America is the great seat of lard and lard-oil manufacture (*Spons*).

MEDICINE. Lard. 291 **Medicine.**—Lard, under its classical name of *Adeps*, or *Adeps Suillus*, has long held the principal place in medicine as a medium for the exhibition of other substances as ointments, &c. As an external application it possesses of itself virtues as an emollient, and is extensively employed as an inunction in external inflammation, bruises, sprains, in various skin eruptions, as eczema, erysipelas, and in several exanthemata, such as scarlatina. Recently it has been considerably superseded by other preparations which are not susceptible of decomposition, and do not turn rancid, such as vasseline. Owing to the religious and caste prejudices of sections of the people, it has never been employed extensively by native practitioners in India, and it is advisable in general work amongst a mixed community to avoid prescribing it.

FOOD. Pork & Lard. 292 **Food.**—The flesh is eaten by natives of many parts of India, that of the Wild Hog being generally preferred, and naturally so, as the animal is a much cleaner and more careful feeder. Little or no endeavour seems to be made by the natives to improve the condition, as a food producing animal, of the domesticated animal, which is generally allowed to roam about and feed wherever it can get anything to eat. Lard is one of the most extensively used articles in European cookery, and forms a cheap substitute for butter.

DOMESTIC. 293 **Domestic Uses.**—The skin forms when tanned a leather which is principally valued by saddlers. That of the wild boar is much thicker than that of the domesticated animal, and consequently offers more difficulty in the processes of tanning and preparation. Much of the so-called pig skin of commerce is merely imitation, the skins of other animals being passed between plates or rollers with a number of projecting points, by which the peculiar punctated or dotted appearance of the real skin is imitated. Pigskin is chiefly employed by saddlers in covering the seats, and making the flaps of saddles, for which, owing to its pliancy and durability, it is eminently suited. It is also, however, used in the manufacture of many small articles of leatherware, such as purses, coverings of pocket books, &c., &c.

Hog-Deer or **Axis porcinus**, see article **Deer**, Vol. III., 57.

Hog-gum, see **Bassora**, Vol. I., 417; also **Cochlospermum Gossypium**, [Vol. II., 413.

Hog Tragacanth, *Pers.*, see **Prunus Amygdalus**, *Baill.*; Vol. V.

HOLARRHENA, *Br.; Gen. Pl., II., 708.*

A genus of trees or shrubs with white flowers, belonging to the Natural Order APOCYNACEÆ, comprising seven or eight tropical Asiatic and African species, of which two are allowed, by the *Flora of British India*, to belong to India and Ceylon.

[*Wight, Ic., tt. 1297, 1298*; APOCYNACEÆ.

Holarrhena antidysenterica, *Wall.; Fl. Br. Ind., III., 644;* 294

KURCHI OR CONESSI BARK.

Syn.—HOLARRHENA CODAGA, *G. Don.*; H. PUBESCENS, *Wall.*; H. MALACCENSIS, *Wight*; ECHITES ANTIDYSENTERICA, *Roxb.*; WRIGHTIA ANTIDYSENTERICA, *Grah.*; (HONEMORPHA? ANTIDYSENTERICA, *G. Don.*

Vern.—*Kureyá, kaureyá, karvá-indarjou, hat, karra, kaura, kora, kúra, kúar, kari, karchi, dhúdi,* (*indrajab*=seeds), HIND.; *Títá-indarjou, kurchi,* (*indrajab*=seeds), BENG.; *Hat,* KOL.; *Hat,* SANTAL; *Dudcory,* ASSAM; *Madmandi,* GARO, CACHAR; *Indrajao* (seeds), *kat-bij* (bark), *kirra, karingi,* NEPAL; *Dudhali, dudhkuri,* MICHI; *Samoka, gırchi,* GOND.; *Lisan-ul, as-ufir, inderjan tulkh, indrajab,* BEHARI; *Kuár, kuer, kura, moriya,* N.-W. P.; *Kachri,* OUDH, KUMAON; *Kewar, koeva, kogar, kawar, kúra* (Bazar seed=*indarjau*), PB.; *Karvá-indarjou, pandhra kúra, dola kúra,* DEC., BOMB.; *Kadú-indra-jou, kudá,* MAR.; *Kadvoindarjou, kuda, dhowda, hatbaha, hath,* GUZ.; *Kashappu-vetpálarishi, kulap-pálai-virai* (seeds), TAM.; *Chédu-kodisha-vittulu, amkuduvittulu* (**Moodeen Sheriff**), *istaráku pála, vistaráku pála* (**Elliot**), *amkudu-vittum, palakodsa,* TEL.; *Beppale,* KAN.; *Kaipa-kotakap-pálavitta,* MALAY.; *Letonkgyi, letou-kgyi,* BURM.; *Kutaja, kálinga, indrayava* (=seeds), SANS.; *Lasánul-aasáfirul-murr,* ARAB.; *Zabáne-kunjashke-talkh, indar-jave-talkh, tukhme-ahare-talkh,* PERS.

References.—*Brandis, For. Fl., 326, t. 49; Kurz, For. Fl. Burm., II, 182, 183; Beddome, Fl. Sylv., Anal. Gen., XX., fig. 6; Gamble, Man. Timb., 263; Dalz. & Gibs., Bomb. Fl., 145; Stewart, Pb Pl., 142; Aitchison, Cat. Pb. and Sind Pl., 88; Mason, Burm. and Its People, 479, 799; Elliot, Fl. Andhr., 71, 193; Pharm. Ind., 137, 455; Ainslie, Mat. Ind., I., 88; O'Shaughnessy, Beng. Dispens., 53, 54; Moodeen Sheriff, Supp. Pharm. Ind., 155; U. C. Dutt, Mat. Med. Hind., 192, 300, 307; Dymock, Mat. Med. W. Ind., 2nd Ed., 497; S. Arjun, Bomb. Drugs, 87; Murray, Pl. and Drugs, Sind, 149; Irvine, Mat. Med., Patna, 40, 50, 62; Med. Top., Ajmir, 138; Flemings, Med. Pl. & Drugs. in As. Res., XI., 166; K. L. Dey, Indig. Drugs., Bengal, 123; Atkinson, Him Dist., 313, 739; Drury, U. Pl, 245; Lisboa, U. Pl. Bomb., 101; Birdwood, Bomb. Pr., 54; Christy, Com. Pl. and Drugs, IV., 27; V, 72; VI, 96; Indian Forester, III., 203; IV., 228; X., 325, XIII., 121; XIV., 112; Ind. Agric., Aug. 24th, 1886; Kew Off. Guide to the Mus. of Ec. Bot., 96; Forest Ad. Rep. Ch. Nagpore, 1885, 32; Rep. by Conservator of Forests, S. Circle, Madras; Gazetteers:—Bombay, XV., 438; Orissa, II., 159, 181, App. IV.; N.-W. P., I., 82; IV., lxxiv; District Manual, Trichinopoly, 80; Settlement Reports, Kángra, 22; Cent. Prov., 39.*

Habitat.—A small deciduous tree, with a pale bark, native of the Tropical Himálaya, from the Chenab westwards, ascending to 3,500 feet, extending throughout the dry forests of India to Travancore and Malacca. Great confusion existed for years as to the identity of this plant, and in India the drug was adulterated with, or mistaken for that obtained from a species of **Wrightia**, a plant medicinally inert. For this reason the very valuable medicinal bark and seeds have fallen into undeserved disrepute. In the *Pharmacopœia of India, p. 455*, an interesting account by **Wight** of the origin of this confusion between **Wrightia** and **Holarrhena**, will be found. From this it appears that the mistake was originally due to **Linnæus** who, in preparing his *Flora Zeylanica*, found a specimen of what was then known as **Nerium indicum siliquis angustis**, in **Herman's** Herbarium. To that plant he subjoined as a synonym **Rheede's Codagapala**, a continental and widely different plant, reported even at that early time to be possessed of great medicinal virtues as a remedy in dysentery. On the strength of this report he named the double, or spurious, species, **Nerium antidysentericum.**

This formation of one species from types of two different genera was not corrected till **Brown** in 1809 published his revision of the APOCYNACEÆ. **Herman's** Ceylon species was then assigned to the new genus **Wrightia**, under the name of **W. zeylanica**, *Br.*, and **Rheede's** Malabar plant fell into another new genus, **Holarrhena** becoming **H. antidysenterica.** About the same time **Roxburgh**, while writing his *Flora Indica*, prepared drawings of three Apocynous trees, one **Nerium tinctorium**, *Roxb.*, which **Brown** referred to **Wrightia**, another of **Echites antidysenterica**, the **Codagapala** of **Rheede**, and the **Holarrhena** of **Brown** and the third, **Alstonia**, which **Roxburgh** also placed in **Echites**. In his description of these species he commented on the confusion between these plants, and expressed it as his opinion that the bark of the **Nerium (Wrightia) tinctorium** was gathered and sold as Conessi bark, and thus had given rise to the disrepute into which that medicine had even then fallen. The distinction between **Holarrhena antidysenterica** and **Wrightia tiuctoria, tomentosa** and **zeylanica** has thus been made for nearly 100 years, yet notwithstanding this fact confusion has always existed between them, and still exists in the works of certain Indian writers, the species of **Wrightria** being accepted as identical with **Holarrhena**. To the native drug-collectors these plants are, however, quite distinct, the bark of **W. tinctoria** being professedly the chief adulterant of Conessi. As the error here indicated is one seriously affecting the reputation of the true drug, and contracting the extent of its employment, the following short account of the distinctive characters of the two genera, may be given :—

WRIGHTIA.	HOLARRHENA.
(1) *Corolla* not more than twice the length of the calyx, mouth surrounded by a corona or teeth.	(1) *Corolla* three or four times the length of the calyx; mouth naked.
(2) *Stamens* inserted within the mouth of the corolla, anthers protruding, twisted and surrounded by the corona.	(2) *Stamens* inserted at the bottom of the tube and therefore not protruding.
(3) *Seeds* straight, oblong, compressed with a coma of hairs at the base, the apex being pointed and naked.	(3) *Seeds* linear, oblong, compressed, concave, with a coma of hairs on the apex.

Dr. Dymock, alluding to the confusion of **Wrightia** with **Holarrhena**, remarks that the barks may be distinguished pharmaceutically; the stem-bark of **Holarrhena**, he says, is thick and dirty white or buff coloured, its root bark reddish-brown and nearly smooth; the stem-bark of **Wrightia** is reddish-brown, and nearly smooth, the root dark brown or nearly black, and both are less bitter than those of **Holarrhena**.

In **Alstonia**, a genus which has also been confused with the preceding, the seeds are attached to the fruit in the middle, and have a coma of hairs at both extremities.

OIL. Seeds. 295 **Oil.**—The SEEDS yield about 30 per cent. of a fixed oil, of a greenish yellow colour and having a peculiar odour, but bland to the taste. The **Revd. A. Campbell** mentions this seed oil as being used medicinally by the Santals

DYE. 296 **Dye.**—"The wood ash is used in dyeing in Chutia Nagpore" (*Revd. A. Campbell*).

MEDICINE. Bark. 297 Seeds. 298

Medicine.—The BARK both of the stem and root and the SEEDS are amongst the most important medicines of the Hindu Materia Medica, the former being considered a powerful antidysenteric, while the latter have ascribed to them astringent, febrifuge, antidysenteric, and anthelmintic properties. By Arabian and Persian writers the seeds are considered carminative and astringent, valuable in pulmonary affections, tonic, lithontriptic, and aphrodisiac. Both seeds and bark enjoyed at one time a European reputation, but probably owing to confusion with, or adulteration by, the cheaper and comparatively inert species of **Wrightia**, they are now almost entirely neglected. The *Pharmacopœia of India* classes **Holarrhena** amongst the non-officinal drugs, but reports favourably on its therapeutic qualities; indeed, Indian writers on therapeutics and materia medica, generally, are unanimous in recommending the drug. **U. C. Dutt**, in his *Hindu Materia Medica*, gives the principal methods of prescription followed by the Hindus and Sanskrit writers, from which it appears that a fluid extract, an expressed juice, a compound decoction and confection are prepared from the bark; from the seeds a compound powder and a fermented liquor. An oil for external application is also made by mixing a decoction of the bark, a number of astringent and aromatic substances in small quantities, and sesamum oil. The **Revd. A. Campbell**, in his *Economic Products of Chutia Nagpore*, writes: "The bark, dried and ground, is rubbed over the body in dropsy. The fruit is applied in snake-bite to allay swelling and irritation, and the seeds yield a medicinal oil." **Mr. U. C. Dutt** prefers for administration a watery extract of the root-bark, in doses of about three grains in combination with half a grain of opium. **Dr. Dymock** mentions that the root-bark alone is employed by the natives of the Goa territory and the Konkan. This selection of the root-bark as the most powerful antidysenteric part of the plant appears to be confirmed by several of the special opinions quoted hereafter. The *Indian Medical Gazette, 1886, Vol. I., 352*, states that during the cattle plague epidemic at Backergunge the seeds were extensively employed, being supposed to possess certain specific virtues, but the results are not given. As both the bark and seeds contain a basic substance (*Wrightine* or *Conessine*), which seems to be the important therapeutic constituent, it appears desirable that a careful quantitative analysis of the different medicinal parts of the plant should be made, so that it might be finally decided which is the most valuable. Such an analysis might result in the discovery of a cheap method of separation of an antidysenteric, and possibly antipyretic and antiperiodic alkaloid of very considerable value. **Dr. Warden**, in a short account of the qualitative analysis of the barks and seeds, writes that a solution containing the partly purified alkaloid has been used with success in the treatment of fevers and dysentery. Should it, even in a smaller degree, possess the specific properties of Quinine and Ipecacuanha a most valuable drug would be added to our remedies for tropical diseases.

SPECIAL OPINIONS.—§ "Is a very useful medicine in chronic dysentery. In acute cases is not to be compared with ipecacuanha. Assistant Surgeon Moti Lall Mukerjee (of the hospital here), who uses it largely, tells me that the powder is a more useful preparation than any other. Dose 10 to 20 grains three times daily in a little water for an adult" (*Surgeon-Major E. Sanders, Chittagong*). "I use the decoction in chronic dysentery with excellent results" (*Surgeon R. L. Dutt, M.D., Pubna*). "Useful in chronic dysentery and diarrhœa" (*Surgeon-Major E. E. Bensley, Rajshahye*). "In fourteen consecutive cases of dysentery in the jail hospital I have used the decoction of the bark with success. The average period of treatment was five days. Ipecacuanha was given in no case" (*Civil Surgeon D. Basu, Faridpore, Bengal*). "I have used this drug with great

MEDICINE.

advantage in dysentery among prisoners" (*Surgeon R. D. Murray, M.D., Burdwan*). "Decoction of bark is very valuable in dysentery" (*Bolly Chunder Sen, Teacher of Medicine*). "The bark of **H. antidysenterica** called *kurchi* is useful in dysentery, given as an infusion in doses of from 1 to 2 ounces" (*Surgeon E. S. Brander, M.B., F.R.C.S.Edin., Rungpore*). "*Koorchee bark* is used in cases of dysentery; it may be administered together with opium, ipecac., tannic acid, &c." (*Surgeon J. Ffrench Mullen, M.D., Saidpore*). "Almost a specific in chronic dysentery even if there be sloughing of the mucous membrane. The decoction is useful in cases of piles, bleeding, or otherwise. The seeds are given in the acute stage of dysentery whether accompanied or not by fever" (*W. Forsyth, F.R.C.S., Edin., Civil Medical Officer, Dinajpore*). "Have used the decoction of the bark largely in dysentery after the acute stage with satisfactory results; also in chronic diarrhœa" (*Surgeon C. H. Joubert, Darjeeling*). "I seldom use ipecacuanha in dysentery in my practice now. If the decoction of the bark be used in acute dysentery it should be given in 2 oz. doses with a little opium every two hours or so. But the alkaloid *Kurchiene* is very efficacious in the acute stage of dysentery" (*Surgeon K. D. Ghose, Bankura, Bengal*). "There is nothing to remark, in addition to what has already been noted, regarding this useful remedy except to point out that in a hœmorrhagic, sub-acute, and chronic dysentery, it should be prescribed. In ordinary acute inflammatory dysentery this remedy should not be employed; *kurchi*, in common with every other active astringent, should be avoided in the early stages of the disease. The indiscriminate use of this drug in every stage of dysentery and diarrhœa has, in my opinion, led many practitioners, both European and Native, to doubt the efficacy of this medicine. It is a most useful remedy when prescribed with judgment" (*Civil Surgeon R. G. Mathew, Mozufferpore*). "Used in chronic rheumatism" (*Surgeon-Major J. J. L. Ratton, M.D., Salem*). "The seeds in doses of 1 to 2 grains are useful in the diarrhœa of young infants, and as a tonic in cases of emaciation and general debility in doses of 4 to 5 grains. A combination of the seeds, old cotton, and honey is used by the natives as a local application to the os uteri in cases of inflammation of the uterus" (*First Class Hospital Assistant Lal Mahomed, Hoshangabad, Central Provinces*). "I can also bear testimony to the effects of this remedy in dysentery, especially in children" (*Surgeon-Major P. N. Mookerjee, Cuttack, Orissa*). "The decoction of the bark is very useful in acute and sub-acute dysentery. In chronic cases, however, I found it fail, in two instances, like other medicines. The bark must be used fresh from the tree" (*Assistant Surgeon Shib Chunder Bhattacharji, Chanda, Central Provinces*). "I can with great confidence add my testimony to the value of this remedy in dysentery. I use a decoction of the bark with or without Ipecac.: and I have always found the result most satisfactory. It is specially useful in acute cases after large doses of ipecac. given in the early stage of the disease and is largely used in charitable dispensaries and hospitals" (*Civil Surgeon S. M. Shircore, Moorshedabad*). "The decoction in 1 ounce doses, as well as the fluid extract of the bark of the root, in doses of 10 to 20 minims, has been frequently used in dysentery, hæmorrhagic as well as chronic" (*Assistant Surgeon Nundo Lal Ghose, Bankipore.*) "Have employed the decoction of the bark in ounce doses in dysentery with very good results" (*Surgeon D Picachy, Purneah*).

TRADE. 299

Trade.—Dr. **Dymock** states of Bombay: "The bark and **seeds** are both articles of local commerce. Value, bark R1½ per maund of 37½℔; seeds R25 per maund of 37½℔."

FODDER. Leaves. 300

Fodder.—In the Panjáb the LEAVES appear to be used as fodder (or litter)? (*Stewart*). The Conservator of Forests, Panjáb, confirms this fact, writing that it is a fodder tree in the Kángra District.

Structure of the Wood.—White, soft, even-grained, weight about 40lb per cubic foot. It is largely used for carvings, especially at Saharanpur and Dehra Dún; in Assam for furniture; in South India for turning. Hamilton mentions that beads made of the wood are worn round the neck in Assam as a charm. TIMBER. 301

Holarrhena mitis, *Br.; Fl. Br. Ind., III., 645.* 302

Vern.—*Kiriwalla, kirri-walla-gass,* SING.

References.—*Beddome, Fl. Sylv., Anal. Gen., 161; Gamble, Man. Timb., 263; Thwaites, En. Ceylon Pl., 194; Balfour, Cyclop., II., 93.*

Habitat.—A medium-sized tree, found not uncommonly in Ceylon up to 1,500 feet.

Structure of the Wood.—White, close-grained, soft, in structure resembling that of **H. antidysenterica.** Weight about 35lb per cubic foot. It is employed in Ceylon for fine wood-work, especially for inlaying. TIMBER. 303

HOLBŒLLIA, *Wall.; Gen. Pl., I., 42.*

Holbœllia latifolia, *Wall.; Fl. Br. Ind., I., 108;* BERBERIDEÆ. 304

Var. 1, latifolia.

Syn.—H. ACUMINATA, *Lindl.;* STAUNTONIA LATIFOLIA, *Wall.*

Var. 2, angustifolia.

Syn.—H. ANGUSTIFOLIA, *Wall.;* STAUNTONIA ANGUSTIFOLIA, *Wall.*

Vern.—*Bagul,* NEPAL; *Pronchadik,* LEPCHA; *Domhyem,* BHUTIA; *Gophla,* KUMAON.

References.—*Atkinson, Him. Dist., 304; Treasury of Bot., I., 594.*

Habitat.—A climber found in the Himálaya, altitude 4,000 to 9,000 feet, from Kumáon eastward to Bhután, also in the Khásia Hills; and in Upper Assam at low elevations.

Food.—Produces a large, edible FRUIT. FOOD. Fruit. 305

Holcus Sorghum, *Linn.*

THE GREATER MILLET, or *Juar,* see **Sorghum vulgare,** *Pers.*

HOLIGARNA, *Ham.; Gen. Pl., I., 425.*

A genus of lofty trees belonging to the ANACARDIACEÆ, and comprising seven species, all Indian.

Holigarna Arnottiana, *Hook. f.; Fl. Br. Ind., II., 36;* ANACARDIACEÆ. 306

Syn.—HOLIGARNA LONGIFOLIA, *Wt. & Arn.*

Vern.—*Bibu, hulgeri,* BOMB.; *Holgeri,* MAR.; *Kagira, kutugeri,* KAN.

References.—*Beddome, Fl. Sylv., t. 167; Gamble, Man. Timb., 112; Dalz. & Gibs., Bomb. Fl., 51; Graham, Cat. Bomb. Pl., 41; Lisboa, U. Pl. Bomb., 55, 250.*

Habitat.—A tree with stout branches, of the Western Gháts, from the Konkan southwards.

Oleo-resin.—Beddome states that "a very acrid black juice is obtained from the trunk and fruit rind, which is used by painters, and as a black varnish." OLEO-RESIN. 307

Medicine.—The FRUIT and BARK are employed medicinally (*Beddome; Lisboa*). No information regarding their actual therapeutic properties is, however, obtainable. MEDICINE. Fruit. 308 Bark. 309

Structure of the Wood.—The tree is given by Lisboa in his *List of Timber Trees.* He writes: "Wood grey or yellowish-brown, close-grained, but soft. It is stated that in some parts it is used for house and boat-building." Beddome writes that he has never seen it used, but has heard that it is employed for similar purposes. TIMBER. 310

311 **Holigarna Grahamii,** *Hook. f. (not of Kurz); Fl. Br. Ind., II., 37; [Wight, Ic., t. 235; Ill., t. 185.*

Syn.—SEMECARPUS GRAHAMII, *Wt.*

References.—*Gamble, Man. Timb., 112; Dalz. & Gibs., Bomb. Fl., 52 Lisboa, U. Pl. Bomb., 55; Bombay Gazetteer, XV. (Kanára), 431.*

Habitat.—A tree 20 to 30 feet in height (80 to 100, *Kanára Gazetteer, l. cit.*), found in parts of the Western Peninsula.

OLEO-RESIN. 312 **Oleo-resin.**—The trunk and fruit-rinds yield a black viscid juice similar in properties to that obtained from **H. Arnottiana.**

TIMBER. 313 **Structure of the Wood.**—Although Lisboa mentions this in his list of timber trees, he says that its wood is not known to be of any value.

314 **H. Helferi,** *Hook. f.; Fl. Br. Ind., II., 37.*

Reference.—*Kurz, For. Fl. Burm., 315.*

Habitat.—A tree with robust branchlets, frequent in the tropical forests of Martaban, Tenasserim, and the Pegu Yomah.

OLEO-RESIN. 315 **Oleo-resin.**—" It yields a black varnish " (*Kurz*).

TIMBER. 316 **Structure of the Wood**—" Heavy, brown, soft, close-grained, perishable, and soon attacked by xylophages." (*Kurz*).

317 **H. longifolia,** *Roxb.; Fl. Br. Ind., II., 37.*

Vern.—*Barola,* BENG.; *Húlugiri, húlagiri,* BOMB.; *Sudra bibo,* MAR.; *Hole ger,* KAN.; *Khreik,* MAGH.; *She-che,* BURM.

References.—*Roxb., Fl. Ind., Ed. C.B.C., 267; Kurz, For. Fl. Burm., 315; Gamble, Man. Timb., 112; Mason, Burma and Its People, 514; Drury, U. Pl., 246; Liotard, Dyes, app. ix.; Watson, Report on Gums, 22, 32; Balfour, Cyclop., II., 93; Treasury of Bot., I., 594; Burm. Gaz., I., 134; Bombay Gaz., XV., 71.*

Habitat.—A tall tree, native of Eastern Bengal, Chittagong, and Pegu. The Bombay form seems most probably to be **H. Arnottiana,** *Hook. f.*

GUM-RESIN. 318 **Gum-resin.**—A resinous highly acrid and poisonous juice is obtained from the trunk and fruit-rind which on drying forms the well-known black lacquer varnish. **Mason** states that it is used by the natives of Burma to varnish shields, and for other similar purposes. The prepared juice is imported from Manipur to be employed for lacquer-work.

MEDICINE. Fruit. 319 Bark. 320 **Medicine.**—**Morton** states that the FRUIT and BARK are employed medicinally, but require to be prescribed with caution, as they are apt to give rise to dangerous symptoms. He makes no mention, however, of their therapeutic properties, nor does further information regarding this appear to be available.

TIMBER. 321 **Structure of the Wood.**—Grey with yellowish streaks, soft. **Balfour** mentions it as being employed for house-building and making beams, while **Drury** states that it is used for making small boats. Other authors regard it as useless.

Holly, see **Ilex,** p. 327.

Hollyhock, see **Althæa rosea,** Vol. I., p. 200.

HOLMSKIOLDIA, *Retz.; Gen. Pl., II., 1156.* [VERBENACEÆ.

322 **Holmskioldia sanguinea,** *Retz.; Fl. Br. Ind, IV., 596;*

Syn.—HOLMSKIOLDIA RUBRA, *Pers.*; HASTINGIA COCCINEA, *Smith;* H. SCANDENS, *Roxb.*; PLATUNIUM RUBRUM, *Juss.*

Vern.—*Kul tolia,* KUMAON.

References.—*Roxb., Fl. Ind., Ed. C.B.C., 480; Brand., For. Fl., 370; Kurz, For. Fl. Burm., II., 257.*

Habitat.—A straggling shrub of the Sub-Tropical Himálaya from Kumaon to Bhotan, ascending to an altitude of 4,000 feet; found also in the Prome Hills.

Fodder.—See Vol. III., 429. 323

HOLOPTELEA, *Planch.; Gen. Pl., III., 352.*

Holoptelea integrifolia, *Planch.; Fl. Br. Ind., V., 481; Wight, Ic., [t. 1968;* URTICACEÆ. 324

THE ENTIRE-LEAVED ELM.

Syn.—ULMUS INTEGRIFOLIA, *Roxb.*

Vern.—*Papri, dhamna, kúnj, karanji, chilbil, chilmil, kúmba, kúnja náli, kandru, begana,* HIND.; *Karinji,* GOND; *Karanjel,* KURKU; *Papar, kanju,* KUMAON; *Pápri, khulen, arján, rajáin, kachám,* PB.; *Chilla,* BANDA; *Wawali,* MAR.; *Aya, tam bachi-marum,* TAM.; *Namli, navili, nali, pedda-nowlieragu,* TEL.; *Ras bija,* KAN.; *Thapsi,* MYSORE, COORG; *Kaládri,* HASSAN; *Myaukseit,* BURM.; *Dadahirilla,* SING.

References.—*Roxb., Fl. Ind., Ed. C.B.C., 263; Brandis, For. Fl., 431; Kurz, For. Fl., Burm., II., 473; Beddome, Fl. Sylv., t. 310; Gamble, Man. Timb., 342; Dalz. & Gibs., Bomb. Fl., 238; Grah. Cat. Bomb., Pl., 644.*

Habitat.—A large, spreading, deciduous tree, met with on the outer lower ranges of the Himálaya from Jamu to Oudh, ascending to 2,000 feet; also from Banda and Behar to Travancore, and from Pegu to Martaban. It is often planted in North and Central India, and prefers a dry, sandy or shingly soil.

Oil.—"An oil is expressed from the SEED in the Melghát" (*Gamble*). OIL. Seed. 325

Fodder.—See Vol. III., 414. FODDER. 326

Structure of the Wood.—Light, yellowish grey, moderately hard, no heartwood. Employed for building purposes, cart making, &c.; also for carving, but its durability is uncertain. It is much utilised for fuel and for making charcoal. TIMBER. 327

HOLOSTEMMA, *Br.; Gen. Pl., II., 760.*

A small genus of the ASCLEPIADEÆ, consisting of twining glabrous shrubs, with large purple flowers. There is only one Indian species.

Holostemma Rheedei, *Wall.; Fl. Br. Ind., IV., 21; Wight, Ic., t. [597;* ASCLEPIADEÆ. 328

Syn.—HOLOSTEMMA FRAGRANS, *Wall.*; H. BRUNONIANUM, *Royle*; H. ADAKODIEN, *Roem & Sch.*; ASCLEPIAS ANNULARIS, *Roxb.*; SACROSTEMMA ANNULARE, *Roth.*; GOMPHOCARPUS VOLUBILIS, *Herb. Ham.*

Vern.—*Apúng,* KOL.; *Apúng, morou raak* (Rev. A. Campbell), SANTAL; *Tultuli, sidori, dudurli,* BOMB.; *Tulatule, sidodi,* MAR.; *Palay kirai,* TAM.; *Pala kura pála gurugu, istara kula,* TEL.; *Ada modien, ada kodien,* MALAY.

References.—*Roxb., Fl. Ind., Ed. C.B.C., 253; Dalz. & Gibs., Bomb. Fl., 148; Graham, Cat. Bom. Pl, 121; Rheede, Hort. Mal., IX., t. 7; Campbell, Cat. Ec. Prod., Chutia Nagpur, No. 7573; Elliot, Fl. Andhr., 142; Dymock, Mat. Med. W. Ind., 2nd Ed., 525; S. Arjun, Bomb. Drugs, 200; Drury, U. Pl., 246; Lisboa, U. Pl. Bomb., 165, 233; Royle, Ill. Him. Bot., 276, t. 66; Royle, Fib. Pl., 306; Indian Forester, III., 237; Balfour, Cyclop., II., 96; Treasury of Bot., I., 594; Gazetteers:—Mysore and Coorg, I., 56; Bombay, XV., 438; N.-W. P., IV., lxxiv.*

Habitat.—An extensive climber, met with in the forests of the Tropical Himálaya from Sirmore to Sikkim at altitudes of from 3,000 to 5,000 feet; also in the Deccan Peninsula from the Circars and Kanára southwards, and in Pegu and Burma.

Fibre.—The PLANT yields a fibre about which very little information is available. Lisboa, however, describes it as fine and silky and adapted FIBRE. Plant. 329

for cordage and paper-making. **Royle** states that it attains its best condition after the rains.

MEDICINE. Root. 330 **Medicine.**—**Rheede** first drew attention to the medicinal virtues of the ROOT, mentioning its value as an application for ophthalmia and "dimness of vision" (presumably that form due to keratitis). **Dymock** writes regarding its utilisation in Western India: "The roots are used as a remedy for scalding in gonorrhœa, and, beaten into a paste, are applied to the eyes in ophthalmia. In diabetes, the root rubbed to a paste, is given in cold milk. In spermatorrhœa the dried root, with an equal quantity of the root of **Eriodendron anfractuosum** powdered, is given in 6-massa doses, with milk and sugar, twice daily." The **Revd. A. Campbell** describes it as employed in decoction by the Santals, as a remedy for cough, and also for orchitis.

FOOD AND FODDER. Leaves. 331 Flowers. 332 **Food and Fodder.**—The LEAVES and FLOWERS are eaten as a vegetable in parts of Bombay. **Campbell** mentions that the leaves are similarly employed in Chutia Nagpur, and that cattle eat the plant.

HOMALIUM, *Jacq.; Gen. Pl., I., 800.*

A genus of the SAMYDACEÆ, of shrubs or trees, which comprises 30 species, scattered over the tropical regions of nearly the whole globe. Of these, 11 or 12 are Indian.

333 **Homalium tomentosum,** *Bth.; Fl. Br. Ind., II., 596;* SAMYDACEÆ.

Syn.—BLACKWELLIA TOMENTOSA, *Vent.; Brandis, Burma Catalogue, 1862, No. 58;* B. SPIRALIS, *Wall., Cat., 4897, partly.*

Vern.—*Myaukchyaw,* BURM.

References.—*Brandis, For. Fl., 243; Kurz, For. Fl. Burm., I., 531; Gamble, Man. Timb., 207; Indian Forester, XII., 73; Balfour, Cyclop., II., 96; Burm. Gaz., I., 130.*

Habitat.—A large deciduous tree of Chittagong and Burma, growing to a height of 80 to 90 feet, with a clean stem of from 40 to 50, and a girth of from 8 to 10 feet. The bark is fine-grained, of a pale greyish-green colour, and marks the tree very distinctly and unmistakably.

TIMBER. 334 **Structure of the Wood.**—"Brown, with dark-coloured heartwood, very hard, heavy, and close-grained; splits in seasoning" (*Gamble*). "Light yellow, turning pale to greyish brown; very heavy (average 58lb per cubic foot), very close grained, but of unequal fibre; rather soft, takes a very fine polish" (*Kurz*). It appears to be durable and susceptible of a high polish, and is said to be employed for making the teeth of harrows, and to be particularly adapted for furniture-making. **Dr. Romanis,** the Chemical Examiner for Lower Burma, gives an analysis in the *Indian Forester, Vol. XII,* from which it appears that the wood is peculiarly rich in mineral matter. The ash of the sapwood contained 39 per cent. of potash, 38 per cent. of lime, and 13 per cent. of phosphoric acid; while that of the heartwood contained 12 per cent. of potash and 72·88 per cent. of lime.

335 **H. zeylanicum,** *Benth.; Fl. Br. Ind., II., 596.*

Syn.—BLACKWELLIA ZEYLANICA, *Gardn.;* B. TETRANDRA, *Wt., Ic., t. 1851.*

Vern.—*Liang,* SING.

References.—*Beddome, Fl. Sylv., 210; Gamble, Man. Timb., 207; Thwaites, En. Ceylon Pl., 79, 410; Trimen, Cat. Cey. Pl., 37; Lisboa, U. Pl. Bomb., 81; Indian Forester, III., 23; VIII., 415.*

Habitat.—A tree of from 40 to 50 feet in height, not uncommon in the forests of Malabar up to an altitude of 4,000 feet, and in the moister parts of Ceylon up to 3,000 feet.

TIMBER 336 **Structure of the Wood.**—Both **Thwaites** and **Beddome** describe this as very strong and valuable for building purposes.

Hominy meal, see **Zea Mays,** *Linn.;* GRAMINEÆ.

HOMALONEMA, *Schott; Gen. Pl., III., 983.*

A genus of robust herbs belonging to the AROIDEÆ, which comprises about 20 species, natives of Tropical Asia and America. Of those indigenous to India, only one is of economic value. **Schott** and also **Engler** give the genus as HOMALOMENA, but the *Genera Plantarum* adopts the form here given, *viz.*, HOMALONEMA.

Homalonema aromaticum, *Schott; Engler, in DC. Monog. Phanero. II., 335; Wight., Ic., t. 805;* AROIDEÆ. 337

Syn.—CALLA AROMATICA, *Roxb., Fl. Ind., Ed. C.B.C., 630.*

Vern.—*Kuchu gundubi,* BENG.

References.—*O'Shaughnessy, Beng. Dispens., 625; Drury, U. Pl., 247; Balfour, Cyclop., II., 97; Treasury of Bot., I., 595.*

Habitat.—A herb which when cut diffuses a pleasant aromatic scent; indigenous to Chittagong.

Medicine.—The large RHIZOME, which is invested with the old withered leaf-scales, bears numerous white, long rootlets issuing from its surface, and is said to be held in high estimation by the natives as an aromatic stimulant. MEDICINE. Rhizomes. 338

HOMONOIA, *Lour.; Gen. Pl., III., 322.*

A genus of rigid shrubs belonging to the EUPHORBIACEÆ, comprising two Indian species.

Homonoia riparia, *Lour.; Fl. Br. Ind., V., 455; Wight., Ic., t. 1808;* EUPHORBIACEÆ. 339

Syn.—ADELIA NERIIFOLIA, *Roth.;* RICINUS SALICINUS, *Hassk.;* SPATHIOSTEMON SALICINUS, *Hassk.;* S. JAVENSE, *Blume.;* HEMATOSPERMUM SALICINUM, *Baill.;* H. NERIIFOLIUM and RIPARIUM, *Wall.;* CROTON SALICILIFOLIUS, *Geisel.*

Vern.—*Khola ruis,* NEPAL; *Mongthel,* LEPCHA; *Kat-alluri,* MAL. (S.P.), *Sundeh,* GOND.; *Kandagar,* KUMAON; *Jeljambu,* KURKU; *Taniki,* TEL.; *Momaka, yae-ta-kyee, yae-ta-gyiben,* BURM.

References.—*Roxb., Fl. Ind., Ed. C.B.C., 744; Brandis, For. Fl., 445; Kurz, For. Fl. Burm., II., 401; Beddome, Fl. Sylv., t. 212; Gamble, Man. Timb., 364; Thwaites, En. Ceylon Pl., 273; Dalz. & Gibs., Bomb. Fl., 231; Grah., Cat. Bomb. Pl., 185; Trimen, Cat. Cey. Pl., 82: Atkinson, Him. Dist., 317; Indian Forester, III., 204; Gazetteers:—Bombay, XV., 443; N.-W. P., IV., lxxvii.*

Habitat.—A small rigid evergreen shrub, found on the rocky and stony river-beds of the Sikkim Himálaya at an altitude of 1,000 to 2,000 feet; also in Assam, the Khasia Hills, and southward to Burma, Tenasserim, the Andaman Islands, and Bundelkhund, and in the Deccan Peninsula from the Konkan southward. It is also common in Ceylon up to an elevation of 2,000 feet.

Structure of the Wood.—Grey or greyish brown, moderately hard, close-grained. Weight 40lb per cubic foot. TIMBER. 340

HONEY AND BEES'-WAX. 341

Honey is the saccharine substance obtained from the honeycomb of **Apis mellifica** and other species of bees, for an enumeration of which the reader is referred to the article "Bees," Vol. I., p. 434. The saccharine matter is gathered by the bees from the nectaries of many species of flowers, and is sucked into the crop, where it undergoes certain chemical changes, the cane sugar or saccharose of the nectar being converted into "invert" sugar, a compound of dextrose and lævulose. According to certain chemists, this conversion of cane sugar is complete, leaving none in the prepared honey, while others contend that cane sugar is always present. This difference of opinion probably arises from the samples examined having

BEES-WAX. 342

been of different ages, as it seems likely that a certain amount of change may go on after deposition in the comb.

Bees'-wax is a secretion elaborated within the insects' body from the saccharine elements of the honey, which, extended in plates from beneath the rings of the abdomen, and mixed with certain impurities, pollen, &c., constitutes the comb or framework of the nest.

Vern.—*Shahad, madh,* wax=*móm,* HIND.; *Madhu, modhu,* wax=*móm,* BENG.; *Sahut,* BEHARI; *Saht, shahd,* PB.; *Polie,* C. P.; *Shahad,* wax=*móm,* DEC.; *Mada,* wax=*ména,* MAR.; *Madh,* wax=*mín,* GUZ.; *Tén,* wax=*mozhukku,* TAM.; *Téne,* wax=*mainam,* TEL.; *Jénu,* wax=*ména,* KAN.; *Tén,* wax=*mezhuka,* MALAY.; *Piyá-ye,* wax=*phayonii,* BURM.; *Páni,* wax=*itti,* SING.; *Madhu, mákshika,* wax=*madhujam,* SANS.; *Aasl, aaslun-nahal,* wax=*shama,* ARAB.; *Shahad, angabín,* wax=*móm,* PERS.

References.—*Hooker's Himálayan Journal, I., 201; II., 16, 276; Pharm. Ind., 277; Ainslie, Mat. Ind., I, 172; O'Shaughnessy, Beng. Dispens., 684; Moodeen Sheriff, Supp. Pharm. Ind., 172; U. C. Dutt, Mat. Med. Hind., 277; S. Arjun, Bomb. Drugs, 177; K. L. Dey, Bengal Drugs, 71, Irvine, Mat. Med., Patna, 102; Johnston, Chemistry of Common Life, Ed. Church, 179; Spons, Encyclop., 1127, 2042; Balfour, Cyclop., II., 98; III., 1050; Davis, Rept. on the Trade and Resources of the Countries on the N.-W. Boundary of British India, App. ccxiii; Special Reports from:—Conservator of Forests, S. Circle, Madras; Com. of Settlements and Agric., Panjáb, 1883; Dir. Land Rec. and Agric., Bengal; Collector of Cuddapah Dist., Madras; Collector of Trichinopoly Dist., Madras; Chief Commissioner, Burma; Govt. of Madras, Madura Dist.; Dir. of Land Rec. and Agric., Assam; Settlement Reports:—Jubbulpore, 87; Hoshungabad, 285; Bhundara; 20; Chanda, 109; Nimar, 1225; Upper Godavery, 40; Montgomery, 24; Gazetteers:—Hazara, 13; Montgomery, 23; Bombay, VI., 53; XV (Kanara), Pt. I., 104-106; Manuals of Administration, Madras, I., 314, 361; II., 24; District Manuals:—Coimbatore, 5; Trichinopoly, 16.*

SOURCE.

Source, Collection, &c.—Honey is a very plentiful wild product over the greater portion of India, but is obtained artificially from domesticated bees, only in the hills north of the Panjáb, and in certain districts of Burma. The following passages from reports published by the Governments of the several Provinces and Presidencies, also from Gazetteers and other sources, may be given, as indicating the annual amount procured, the methods of collection, and trade in honey and wax in the different parts of the Empire.

MADRAS. 343

Madras.—"The hive bee of Europe (**Apis mellifica**) is unknown in South India. There are four species of wild bee which attach their combs to branches, or place them in crevices of rocks or ruins. **A. indica** and **A. floralis** are migratory, moving especially to localities where any plants of the **Strobilanthes** genus are in flower. **A. nigrocincta** remains in the same spot throughout the year. A small variety of this species is known as the 'Mosquito Bee' frequenting the plains and rocky hills throughout the Presidency The quantity of honey in its nest is small and not of good quality. The fourth species, **A. dorsata,** or the 'Rock Bee,' is twice as large as **A. indica.** The honey is collected by jungle tribes who are familiar with the localities which the swarms frequent and refuse to leave." Regarding the Cuddapah district, the Collector writes: "Honey is collected in the forests in almost all the taluks and is sold at R2½ to R4 a maund. The wax is here sold at R15 a maund. The right to collect honey is either farmed out or managed under *amáne* by the Forest Department. In 1887 and 1888, 243lb valued at R35 were collected by the Department, and 840lb valued at R29 by license." The Collector of Trichinopoly reports that honey is generally collected in the mountainous and thick jungles, such as those of the Pachamalais, the Marnagapuri, and the Korlavúr hills. In the District Manual it is mentioned as one of the principal products of the Pachamalai hills.

MADRAS. 344

In a memoir by the Revenue Department on the Madura District the following occurs: "Honey is collected to a large extent in the Saduragi and Varshanad hills, and to a small extent from the trees on the plains. A measure of honey costs between 12 annas and R1 The Conservator of Forests, in an interesting report furnished in July 1889, writes: 'Honey and wax are forest products, and are either farmed out or collected departmentally, but in reality only a small portion of the quantity actually produced ever reaches the market' The following return has been furnished for South Arcot: I. *Collected by Department.*—In 1884-85, 1,484℔ of honey valued at R76-12-7, 840℔ of wax at R360, or a total of R436-12-7. In 1885-86, 171℔ of honey, R21-4-5; 29℔ of wax, R10-3-0,—total R31-7-5. In 1886-87, 793℔ of honey, R82-3-8; 209℔ of wax, R80-1-11,—total R163-1-7. In 1887-88, 244℔ of honey, R35-1-8; 61℔ of wax, R26-11-0,—total R61-12-8. In 1888-89, 409℔ of honey, R61-2-6; 57℔ of wax, R24-10-11,—total R85-13-5. II. *Collected by private consumers and purchasers.*—In 1885-86 and 1886-87 the sum annually realised by leasing out the right of collecting honey and wax was R80; in 1887-88, R47 was obtained from this source for 1,120℔ honey and 560℔ of wax. The total amount realised by the Department during five years for honey and wax was thus R985-15-8."

It appears from the foregoing that the amount obtained for the produce collected by the Department itself very greatly exceeded proportionately that obtained by the system of leasing out the trees.

Coimbatore. 345

The quantity collected in South Coimbatore in 1888-89 was 1,192 small measures, and the revenue realised R264. The District Forest Officer of Tinnevelly reports: 'Three descriptions of honey are found in the forests, of which one is said to be used only for medicinal purposes. The cost of collection of the other two kinds is eight annas per Madras measure, and the selling price R1·4, for the best quality. The total yield varies considerably, but including that from Zamin forests, it might amount to about 1,800 measures in a good year.' In South Kanára, 'bee-keeping is unknown; wild honey and wax are collected from the jungles, the combs being generally on large trees. The honey of the small bees is preferred for medicine. Honey is collected by Government agency at a cost of R1-6 to R1-14 a maund of 28℔. The quantity annually collected is 29 maunds, and is generally sold at R2-10 a maund.'"

CENTRAL PROVINCES. 346

In the Central Provinces.—The Report on the Settlement of the Upper Godaveri District contains the following: "There are four different sorts of honey produced in the jungles of this district: (1) *Kurra teneh,* (2) *Moosur teneh,* (3) *Tonudee teneh,* (4) *Pitwar teneh,* (5) *Kunagole teneh,* of which 1 and 5 are the most delicate, and the bees' nests of the same (the former being found on branches, the latter in holes in the trunks of trees) yield good wax 2 is also found in holes in trees, 3 in holes in the ground and ant-hills, while 4 is the honey of the large bee the nests of which are found suspended as large combs from lofty trees and rocks. The wax of all is good and is collected by *Gottohs* and *Kays,* and sold or bartered to traders; and turmeric is sometimes employed to dye it yellow. The honey is not exported." A similar account is given in the Settlement Report of the Chanda District, in which it is, however, stated that the honeys above described as Nos. 3 and 4 are alone of importance. A comb of the former is said to yield on an average about one seer of honey and two chittacks of wax; that of the latter, 10 seers of honey and 4 seers of wax. The Report on the Nimar District describes the Ratgurh hill as a "perfect hive of bees' nests," and further states that the honey and wax are collected by a class of people called *Dhanook* and *Nahal,* who perform their dangerous work with great dexterity. Honey is also reported to be plentiful and good in the Jubbulpore, Bhundara, and Hoshungabad Districts. No information is available of the total yield of honey and wax in the Central Provinces, but from the foregoing accounts the production would appear to be very great.

BENGAL. 347 *In Bengal.*—The Director of Land Records and Agriculture, in an interesting report on honey, states that the bee is never domesticated, but that a large amount of honey is gathered in the jungle and fields, especially of Chutia Nagpur and the Sanderbans. The men who collect it belong to the Bagdee caste, and are called *moulays* or "honeymen." The principal honey season commences in January and lasts till the end of March, but there is another less important season in April and June. The honey obtained in February is considered the finest, is known as *dinhar modhu*, and fetches a high price, while that collected in June is the worst.

The average amount collected and sold by the hill tribes of Chutia Nagpur is about 150 maunds at R4 a maund; while about 5,000 maunds are sold annually at Naluah, Khorumpara Ghât, Khari, and Harwa, at prices ranging from R8 to R10 per maund. The preparation and sale of wax are incidentally referred to in the report, but no information is given as to the extent of trade in that article.

BURMA. 348 *In Burma*—Honey is found in large quantities in the jungles, and is collected and much used by the natives. No reports of the annual amount gathered are available. In the Amherst District, as in Madras, licenses are granted for the right to collect honey and wax. In the Tongu District wax is mentioned as being used in the manufacture of candles, and probably it is similarly employed in other parts. The report from the Chief Commissioner shows that in the greater number of the Burman districts bee-keeping is not known, but it appears that in the Thongwa District bees are domesticated and honey collected from their hives. **Strettell** in his *Narrative of Travels in Upper Burma* describes a wild bee, which builds large nests suspended from ledges of rocks. The nest is conical in shape, 3 feet long, and as broad at the top. From these hives, he says, large quantities of honey are collected. For the curious wax-like POON-YET of Burma,—see the articles on "**Dammar**," Vol. III., p. 17; also "**Poonyet**, Vol. VI."

BOMBAY. 349 *In Bombay.*—The Gazetteers contain very few references to this product, from which it may be presumed that it is not collected by any means

Kanara. 350 to the same extent as in Madras and the Central Provinces. *Kanára,* however, appears to differ markedly in this respect from the rest of the Presidency, for not only is honey collected to a large extent in that district, but a rudimentary form of bee-farming is carried on. The Gazetteer contains an interesting description of honey and wax gathering, from which the following has been compiled: Kanára bees are said to be of four kinds, named in the vernacular *togar-jeinu*, *tudabi-jeinu*, *kol-jeinnu* or *kotti-hulla*, and *nusarri-jeinu* or *misri*, of which the first is the largest and is described as particularly fierce. It builds its comb either on the upper limbs of the tallest trees (as many as twenty to thirty sometimes occurring on a single tree), or attached to precipitous cliffs, the sides of high bridges or houses, and similar inaccessible positions. This bee is specially fond of the nectar secreted by the flowers of the *kárvi*, a species of **Strobilanthes**, and always abounds in the neighbourhood of these plants. Owing to the inaccessible position of the combs, collection is an extremely difficult and dangerous proceeding. It is effected during dark nights twice a year,—once just before the setting in of the rains, in April-May, and again in October-November, the honey of the latter period being known as the "grass harvest," and that of the former as the "main harvest." The combs are obtained from high trees by

BOMBAY.

means of bamboos, the branches of which are lopped off to serve as steps; while from cliffs, bridges, &c., the honey is generally collected by a gatherer let down from the top in a basket or net suspended from a long rope. Though the method of collection is doubly dangerous from the inaccessible positions of the combs and the fierce nature of the bees, the harvest is a plentiful one, and well repays collection. One comb is said to yield on an average eight to fifteen "beer bottles" of reddish-brown honey, and from one to two and a half pounds of wax.

The second or *tudabi* bee is much smaller, and builds a small comb, only yielding from one to three "beer bottles" of honey, which, however, is of superior quality and flavour. The combs of this variety are found in the hollows of trees or old walls. The third or *kol* bee is still smallar, and its comb, which is built on thorn bushes or small plants, yields only about a tea-cupful of honey The fourth or *misri* bee is said to be about the size of an ant, and to collect a honey much prized in medicine. It seems doubtful if this is in reality a true bee.

The right to collect honey and wax is farmed out as in Madras, and is productive of a fairly large but varying revenue, since higher bids are made in years when the **Strobilanthes** plants are in flower. In the four years ending 1880-81, R19,140 were thus obtained, or an annual average of R4,780. The estimated yearly outturn is about 568 cwt of honey and 290 cwt. of wax: the latter is nearly all sent to Goa, where it meets a ready sale for the manufacture of wax candles for the Roman Catholic churches; while the former is nearly all sent to Bombay, selling at 1 to 4 annas the ordinary quart bottle according to quality.

The system of bee-keeping pursued in Kanára, though rudimentary and not extensive, is interesting as forming the only instance of its being even attempted south of the Panjáb Himaláya. It is carried on only in some of the small hill-villages, and the method is very crude. An empty earthen pot is placed mouth downwards in a white-ants' nest, or, more rarely, in a hole made for the purpose, a small opening is made at one side, and the hive, thus completed, is left to take its chance of becoming the selected quarters of the bees. If bees do come and build, they are left unmolested for some time, after which the comb is extracted as carefully as possible, every endeavour being made not to disturb the parts containing the young. By these means the bees are not frightened, and generally stay on for a year or two.

The honey thus obtained is principally employed for home medicine.

A report recently received from the Acting Director of Land Records and Agriculture contains the information that the honey collected in Kanára and sent to Bombay is purchased by grocers, who adulterate it with treacle, or syrup, and that Mahableshwar honey is much prized on account of its purity.

PANJAB. 351

In the Panjáb.—Bee-culture is extensively carried on in the Murree hills, in the tahsil of Rawalpindi, in Házara, in the Simla hills and in Kulu. In Házara and the Pindi District the hive is constructed of mud, moulded in a cylindrical shape, about 8 inches in diameter at one end and 16 to 20 inches at the other, and some 15 to 20 inches in length. This is inserted by the bee-keeper into a hole in the wall of his hut, made to fit the larger end. He then closes it up from the inside of the house, with a sort of door made of grass and mud, and stops the smaller end also with mud, leaving only a small entrance hole, about 1 inch in diameter. Having thus prepared a hive, he, during the month of April, smears honey, green *bangh*, or a mixture of ghur and milk, over its mouth. This generally attracts occupants, which flying away return with a swarm, or should this means fail, a captured swarm is shaken into the previously smeared hive.

HONEY. **Bee-keeping in the Simla Hills.**

PANJAB.

During winter the entrance hole is closed at night, and the bees occasionally, but rarely, require to be artificially fed. The comb is generally ready for removal in October or November, a process which is very simply accomplished by making a small hole in the back of the hive, and burning some old cloth, tobacco, or cowdung below it. The bees fly out by the external orifice and cluster round the hive, after which the whole back of the hive is removed and the comb extracted. A hive generally contains from 2 to 4 seers of honey and from ½ to ¾ seers of wax. These are separated from each other. The "run" honey is sold at from 2 to 3 seers a rupee, the uncleaned wax at 2 to 2½ seers a rupee, or when cleaned at 1lb a rupee.

Bashahr. 352

Bee-keeping is carried on on a still larger scale in the Simla hills. **G. G. Minniken, Esq.**, Deputy Conservator of Forests, Bashahr Division, in an interesting communication on this subject, writes: "The management of the bee is carried on in most villages as far up the valley" (of the Sutlej) "as Zipe and the Runang pass. The largest returns of honey and wax are obtained from the side valleys on the north and south of the river below Wangtu, where the rainfall is about 66 inches. In these valleys, houses—one, two, or three stories high—are especially kept for rearing bees,—small recesses, 1 foot by 1 foot by 9 inches, being let in along the walls at 2 feet apart and closed on the outside by a wooden panel in which an entrance hole is made. A man is usually in charge of each house, whose duty it is, *first*, to prevent excessive swarming, which is done by giving each colony ample room, and sometimes by clipping the wings of the queen bee; *second*, to keep the apiary well stocked with early swarms, and to guard it against the attacks of bears, martins, hornets, wasps, and caterpillars. Stocking is effected by rubbing the inside of the recesses with a paste prepared from honey, white wild rose, roots of the **Jurinea macrocephala**" (*dhup*), "and the petals and seed of the **Pleurospermum Govanianum** (*espouse*), which is said to be most attractive to the queen bee; but as sufficient swarms are not caught in this way, cylindrical boxes formed of two or three lengths of hollow trunks, 2 to 3 feet in circumference, covered at the top, and with an opening on one side, are rubbed with the paste, and set out in different places, 2 to 3 miles away, in order to catch new swarms, which, when established, are taken to the apiary. In Bashahr proper, *gurrahs* are used instead of boxes. Besides these special bee establishments, the zamindars have places in the lower parts of their houses reserved for the bees, and the boxes are sometimes used as permanent hives. It is believed that the *Yung*" (the bee domesticated in the district) "extracts honey and pollen from almost every flower, except the Jessamine (*chumbélí*); that the honey from **Plectranthus rugosus** (*pekh*) is the best and purest, and that after a rainy night, if the bees return laden with honey from the flowers of **Pyrus Pashia** (*shegul*), numbers of them eat it and die; but no precaution appears to be taken by the villagers against this happening. It is also said that, when the bees have collected honey and pollen from the male catkins of the Deodar, two-thirds of the honey in the comb is quite bitter and useless, and that the honey made from April to July is watery and of a yellow colour, but that the autumn store is of good quality. The honey is taken either at night or in the day-time by smoking out the bees, and to induce them to return water is sprinkled about, and a noise made by whistling and rapping on brass plates; but if this plan does not answer, the queen, with the bees which cluster around her, is swept into a cloth or basket and returned to the hive, from which, however, all the combs on the outside of the recesses have been removed, leaving those in the recesses for the support of the bees, and more is left in autumn than in summer. It frequently happens that the bees

PANJAB.

whose store has failed have to be fed during winter on a mixture of honey and fine buckwheat flour.

"The honey is extracted by squeezing the combs over a fine bamboo sieve, through which the honey drips into a vessel placed below, and the wax is obtained by boiling the comb, and whilst boiling skimming off the wax, and further refining it by straining, while it is liquid, through a thin cloth. The refuse is about one quarter of the whole and is of no value. The yield of honey from an apiary is from 30 to 50 pucka maunds, and from a single hive 5 seers to 1½ maunds, and from ½ seer to 5 seers of wax. About one-fourth of the annual produce is exported, and three-fourths consumed by the people."

Kulu. 353

In Kulu, bees are kept in the upper verandahs of houses in bee-hives formed of short lengths of the hollow trunks of trees, covered at the top, and with an entrance hole at the side. The insects thrive best in the highest villages, from which the flowery slopes above the forests are accessible. Honey is also obtained in many parts of the Panjáb from the different varieties of wild bee, but no returns of the total amount annually collected are available. In Davies' *Trade Report of the Countries on the North-Western Border of India, 1862,* it is stated that more than 100 maunds were then exported from Kangra and Kulu to Ladakh and Yarkhand, and that a large quantity was also annually exported from Afghánistan to Turkish China where it was used in making "*Goolkund*" or Preserve of Rose flowers.

It seems possible that should such a system of bee-keeping as that just described be introduced into Bengal, Madras, and the Central Provinces, the amount of wax and honey collected in India might be very greatly increased, and bee-culture become an important industry. The demand for wax for several manufacturing purposes is very large, and notwithstanding the large amount produced in this country, the imports are considerable. It may be interesting in connection with this to give a short review of the trade during the past few years.

TRADE. Wax. 354

Trade.—In 1875-76 the exports of wax were valued at R6,21,890, in 1880-81 at R5,45,110, in 1885-86 at R4,77,230, and in 1888-89 at R4,20,959. The trade would thus seem to be falling off, but, perhaps, a larger amount of wax produced in India is now being used up within the country. The imports of foreign wax (excluding wax candles) were valued in 1875-76 at R20,010, in 1880-81 at R1,45,467, in 1885-86 at R34,244, and in 1888-89 at R99,272, showing a marked increase. The exports of foreign wax are unimportant.

Honey. 355

The trade in honey is small and local. For the prices, &c., in different parts of the country, the reader is referred to the part of this article dealing with SOURCE, COLLECTION, &c.

PREPARATION. 356

Preparation.—Honey is sold either in the comb, as "comb honey" or as "run honey" after extraction This is ordinarily accomplished by cutting the covering off the cells, and allowing the honey to flow from the comb into a receptacle beneath, or by some method of compression and straining. With improved methods of bee-keeping, which have lately greatly developed in Europe and America, it is possible, by a simple adaptation of the principle of centrifugal force, to completely empty a comb n a few minutes, without destroying the fine structure of the cells. The empty comb is then returned to the hive to be refilled.

Commercial bees'-wax is of two principal kinds, yellow and white. Yellow commercial wax is obtained by melting the combs after expression of the honey in boiling water. It rises to the surface, leaving a great part of the impurities behind, and is either skimmed off or allowed to cool in the form of a cake. This process is repeated till the desired degree of purity

PREPARATION.

is obtained. The addition of nitric acid to the boiling water accelerates the purification. White wax is manufactured from the yellow by rolling it into sheets or ribbons and exposing these to the bleaching action of the sun. By this means the surfaces of the ribbons are whitened, after which they are again gathered together, remelted and again made into ribbons, in order to expose a fresh surface, and this is repeated till the bleaching is completed. The same result may also be more rapidly effected by treating the wax with sulphuric acid and bichromate of potash, when the liberated chromic acid bleaches the whole mass in a few hours.

FOOD. 357

Food.—Honey is highly appreciated by the natives of many parts of India as an article of food. There seems to be little doubt, however, that certain kinds of honey, or honey collected from certain flowers, is more or less poisonous. This fact has long been recognised in India; thus **Ainslie** writes of a peculiar dark greenish coloured kind, "which, according to the Vytians, cannot be eaten with impunity." In the *Madras Quart. Journ. Med. Sc., Oct. 1861, 399,* **Dr. Bidie** also gives an account of a small quantity of honey, obtained from the jungle of Coorg, having produced symptoms of urtication, headache, extreme nausea, prostration, and intense thirst. Again, **Hooker**, in his *Himálayan Journal*, alludes to poisonous honey, mentioning that honey is much sought for in Eastern Nepál, except in spring, when it is said to be poisoned by rhododendron flowers, just as that eaten by the soldiers in the retreat of the Ten Thousand was by the flowers of **R. ponticum. Strettell** also mentions that at certain seasons the honey obtained in Burma is poisonous. In the Panjáb Himálaya honey is principally used in the manufacture of an alcoholic liquor, which the natives prepare by mixing one part with an equal quantity of water, and leaving it to stand for a year in a closed earthen vessel. The resulting liquor is very potent, a cupful is said to be sufficient to produce intoxication.

MEDICINE. 358

Medicine.—Honey forms the basis of several very popular preparations and has long been an important vehicle for other medicines in the Hindu Materia Medica. By Sanskrit writers new honey is considered to be demulcent and laxative, while, when more than one year old, it is said to be astringent and demulcent. Applied externally it is supposed to be a useful detergent. The Koran also, in the chapter on The Bee, contains the following; "There proceedeth from their bellies a liquor of various colour, wherein is a medicine for man." In European medicine, however, it is employed only as a flavouring agent in cough mixtures, gargles, confections, and in the preparation of oxymels.

Wax, like honey, is supposed to be emollient and demulcent, and the white variety is sometimes prescribed in doses of 10 to 20 grs. Its principal value, however, is in the preparation of ointments. A useful substitute for lard or simple ointment in India (where the latter is not only objected to from caste and religious principles, but also rapidly becomes rancid) is Ceromel, a mixture of one part of yellow wax and four parts of honey.

DOMESTIC, INDUSTRIAL, & SACRED USES. 359

Domestic, Industrial, and Sacred Uses.—Honey is valued in India as an antiseptic, for the preservation of fruits, and for making cakes, sweetmeats, &c., which are required to keep for some length of time. A curious adaptation of this property is mentioned by **Hooker** as employed by the Khásias. He writes, in describing their method of disposing of the dead: "The body is burned, though seldom during the rains, from the difficulty of obtaining a fire; it is therefore preserved in honey (which is abundant and good) till the dry season, a practice I have read of as prevailing among some tribes in the Malay Peninsula." The reader will remember its extensive employment in ancient Egypt for the same purpose. Honey plays an important part in many of the ceremonial customs of the Hindus. Thus mixed with milk, curds, or clarified butter it is ordered to be given as a respectful offering to a guest, or to a bridegroom on his arrival at the door of his bride's father. Honey sipping forms part of the marriage ceremony of certain castes, and one of the purificatory ceremonies is performed by placing a little honey in the mouth of a newly-born male infant. DOMESTIC, &c.

Yellow wax is employed for polishing floors, and in the manufacture of sealing wax, lithographic crayons, and mastics. When bleached it is used to make candles, and for modelling figures, flowers, and other objects (*Spons' Encyclop.*). It is also employed, by calico-manufacturers and dressers, to impart a fine gloss to fabrics.

For descriptions of Indian White Wax, Chinese Wax, and all others not the product of Bees, the reader is referred to the articles on **Oils**, Vol. V; and **Wax**, Vol. VI.

Hop, see **Humulus Lupulus**, *Linn.*; URTICACEÆ; p. 302.

HOPEA, *Roxb.*; *Gen. Pl.*, *I.*, *193*.

A genus of trees mostly resin-yielding, which comprises about 11 Indian species. It belongs to the Natural Order DIPTEROCARPEÆ, and is very nearly allied to SHOREA.

Hopea glabra, *W. & A.*; *Fl. Br. Ind.*, *I.*, *309*; DIPTEROCARPEÆ. 360

Syn.—HOPEA WIGHTIANA, *Var. β.* GLABRA, *Wight.*

Vern.—*Kong, kongu*, TAM.

References.—*Beddome, Fl. Sylv., 96 (described under* **H. Wightiana**); *Gamble, Man. Timb, 40; W. & A., Prodr., 85; Drury, U. Pl., 248; Balfour, Cyclop., II., 101.*

Habitat.—A large tree of the Western Peninsula (*Wight*), of the Southern Carnatic at Tinnevelly (*Beddome*).

Structure of the Wood.—Beddome describes it as, *par excellence*, the timber of the Tinnevelly district, a statement repeated and confirmed by Drury. TIMBER. 361

H. micrantha, *Hook. f.*; *Fl. Br. Ind.*, *I.*, *310*. 362

Vern.—*Dammar-mata-kooching, dammer batu*, MALAY.

References.—*Trans. Linn. Soc., xxiii., 160; A.DC., Prodr., XVI., 2, 634; Cooke, Gums and Gum-resins, 92; Smith, Dic., 150.*

Habitat.—A very lofty tree of Malacca, distributed to Borneo.

Resin.—This tree, along with **H. odorata**, yields the Rock Dammar, for a description of which the reader is referred to the article on the latter species. Cooke states, however, that the resin of **H. micrantha** is darker and not so friable as that obtained from **H. odorata**. RESIN. 363

H. odorata, *Roxb.*; *Fl. Br. Ind.*, *I.*, *308*; *Wight, Ill.*, *88*. 364

THE ROCK DAMMAR OF COMMERCE.

Syn.—HOPEA FAGINEA, *Hort. Calc.*; H. EGLANDULOSA, *Roxb.*; H. DECANDRA, *Buch.*; H. WIGHTIANA, *Miq., not of Wall.*

Vern.—*Théngan*, BURM.; *Rimda*, AND.

References.—*Roxb., Fl. Ind., Ed. C.B.C., 438; Kurz, For. Fl. Burm., I., 120; Gamble, Man. Timb., 40; Mason, Burma and Its People, 527, 757; Pharm. Ind., 33; Cooke, Gums and Gum-resins, 91; Watson, Report on Gums, 34, 37; Indian Forester, I., 109, 113, 363; III., 22; VI., 125; VII., 250; XI., 321; XII., 73; Spons, Encyclop, 1645; Balfour, Cyclop., II., 100; Burm. Gaz., I., 127, 133; Journ. Agri.-Hort Soc., X., Proceed., cxxvii.*

Habitat.—A large evergreen tree, common in the forests all over Burma, from Chittagong and Martaban down to Tenasserim, and found also in the Andaman Islands. In the *Burma Gazetteer* it is described as "found principally near mountain streams and in the evergreen forests, compara-

HOPEA odorata.

Rock Dammar.

tively scarce in Pegu, but plentiful in Tenasserim, and large specimens are common east of the *Trit-toung*.

RESIN. 365

Resin.—This species is the principal source of the copalline resin known as Rock Dammar (see III., 17). The first mention of this resin, the writer can discover, is that in the *Proceedings of the Agri.-Horticultural Society, August 1858* (*Jour., Vol. X.*), in which it is recorded that a sample of the resin of the *Thengan* of Burma (**Hopea odorata**) was presented by the **Rev. C. S. Parish** In an accompanying letter that gentleman commented on its resemblance to Dammar, and remarked, "this seems such a nice, pure, white resin that I cannot but think that it might be turned to many useful purposes," and further, "as every third or fourth large tree is a *Thengan*, the resin might be got, I believe, in any quantities." The resin was examined and reported on very favourably by **Dr. Barry**, who wrote, "although not soluble in spirits of wine, it is perfectly so in benzole or turpentine. The varnish afforded by both reagents is clear and limpid, and dries almost instantaneously. It should take very well at home." Very little has been done during the last thirty years in investigating the methods of collection and preparation, or in developing the trade in this resin. **Dr. Cooke**, however, in his report published in 1874, described its properties as follows: "The resin derived from this source occurs in nodules about as large as a walnut, rounded, of a pale straw colour, sometimes almost colourless, brittle, with a shining resinoid fracture, scarcely distinguishable in appearance from the East-Indian dammar of the London markets. We have found it to dissolve with equal freedom in spirits of turpentine or benzole, producing a clear bright solution, which, when used as a varnish, dries rapidly and smoothly. In all essential qualities it seems to equal East-Indian dammar, even for microscopical purposes, and hence gives good promise of taking a similar place and rate. Externally it has also sufficient resemblance to make it difficult to discriminate the one from the other. The advantage in hardness is rather in favour of the *Thengan*. The London brokers class it as a copal, and value clear pale samples at a rate equal to that of East-Indian dammar, *viz.*, about 40 per cwt." There is no record of the employment of this *Thengan* resin for economic purposes by the natives of Burma, but it is probable that, like others of the same class, it is used for caulking boats, and as a rough varnish to woodwork, &c. **Major Protheroe,** however, states that it is employed by the Andamanese, mixed with bees wax and red ochre, to make a wax used to fasten their spear and arrowheads. East-Indian Dammar, which apparently this product ought to rival in commerce, is extensively employed in the manufacture of varnishes for coach-builders and painters, in mounting microscopic objects, and for other similar purposes, while inferior qualities are employed in the locality of production for caulking boats and making torches. For an account of the interesting honey-combed black resin of Burma, called locally *poon-yet* or *pwai-nget*, which is supposed to be mainly derived by a peculiar bee from this, and other allied species, the reader is referred to the article on **Dammar** (Vol. III., 17), and to **Poon-yet** in Vol. VI.

MEDICINE. Resin. 366

Medicine.—The *Indian Pharmacopœia* places the RESIN amongst its non-officinal drugs, and describes it as a clear and fragrant substance, which, reduced to powder, forms, amongst the Burmese, a popular styptic, but its action is supposed to be probably purely mechanical.

TIMBER. 367

Structure of the Wood.—Yellow or yellowish brown, hard, close, and even grained, weight about 50℔ per cubic foot; transverse strength 7′ × 2″ × 2″ about 800℔ **Gamble** writes: "It is very durable, *e.g.*, the specimens brought by **Wallich** from Tavoy in 1828, which, though now 20 years old, are perfectly sound and good. Boats made of it are said to last 20 years, and it is principally used for making canoes and houses." The *Burma Gazetteer* contains an account of the tree, in which it is said to be one of the finest timber trees of the country, and to be employed for making cart wheels, canoes, and boats, and in house-building. In the *Burma Forest Report, 1880-81, 13,* a tree met with in narrow belts along streams, the wood of which is in great demand for boat hulls, is referred to. It is said to be **Hopea, sp.**, and has the vernacular name of *Thinganshway*, the similarity of which to *Thingan*, and the resemblance of the habitat and timbers of the two trees seem to indicate that the tree meant is really **H. odorata.** TIMBER.

Hopea parviflora, *Beddome ; Fl. Br. Ind., I., 308.* 368

Vern.—*Kiral boghi, tirpu,* KAN. ; *Irubogam,* MALABAR.

References.—*Beddome, Fl. Sylv., t. 7 ; Anal. Gen., xxviii., in part ; Gamble, Man. Timb., 40 ; Drury, U. Pl., 247 ; Indian Forester, II., 21 ; Balfour, Cyclop., II., 100 ; Indian For. Rep., 1863-64 ; Madras Man. of Administr., II., 64.*

Habitat.—A large handsome tree, common in both the moist and dry forests of Malabar and South Kanára, up to 3,500 feet.

Resin.—The Conservator of Forests, Madras, mentions, in **Dr. Watson's** report, the *Kiral boghi* as "yielding a gum resin" (probably a true copalline resin), "of which the uses are not known, but of which a considerable amount would be available annually, at a cost of about R 10 a maund at the coast." With the exception of the republication of the above remark, by **Colonel Drury**, the writer cannot find any other reference to this resin. From the abundance of the tree, and the value of the resin yielded by allied species, it seems desirable that its properties, chemical and economic, should be accurately determined. RESIN. 369

Structure of the Wood.—Hard, brown, and close-grained, weight from 62 to 63lb per cubic foot (*Gamble*). **Beddome** remarks, "The wood is hardly known commercially, but is much valued by the natives of South Kanára, and I believe it will be of great value for gun carriage purposes; it will also answer well for sleepers. In the district above mentioned it is much valued for temple building purposes." In the *Indian Forester, Vol. II. (1876-77)*, it is mentioned as one of the most valuable timber trees in the moist evergreen forests of the Nilghiris. TIMBER. 370

H. Wightiana, *Wall. ; Fl. Br. Ind., I., 309; Wight, Ill., t. 37.* 371

Syn.—The inflorescence is often diseased and condensed into a globular mass; the plant in that state was described by **Roxburgh** under the name of ARTOCARPUS LANCEÆFOLIA.

Vern.—*Kavsi,* MAR. ; *Kalbon, kiralbogi, haiga,* KAN.

References.—*Roxb., Fl. Ind., Ed. C.B.C., 635 ; Beddome, Fl. Sylv., t. 96 ; W. & A., Prodr., 85 ; Drury, U. Pl., 248 ; Lisboa, U. Pl. Bomb., 14 ; Balfour, Cyclop., II., 101 ; Bomb. Gaz., XV., Pt. I., 71.*

Habitat.—A large tree of the Western Peninsula from the Konkan southwards. It is most abundant in South Kanára.

Structure of the Wood.—"Very valuable, and similar to that of **H. parviflora**, a first-rate coppice firewood" (*Beddome*). The *Kanára Gazetteer* describes it as yielding "a good wood, very hard, and lasting and much used." TIMBER. 372

HORDEUM, *Linn. ; Gen. Pl., III., 1206.*

A genus of the Natural Order GRAMINEÆ, distributed over Europe, Northern Africa, Temperate Asia, and Extra-tropical America. There are many strongly marked forms, which, however, are now generally regarded by botanists as referable to one species. The subject of Barley will accordingly be treated in this work under the heading of **H. vulgare,** *Linn.*, subordinating the non-typical forms as varieties.

373 **Hordeum vulgare,** *Linn.; Duthie, Fodder Grasses of N. India, 69, pl. F., fig. 32; Duthie & Fuller, Field and Garden Crops of N.-W. P. and Oudh, 9, pl. II.*

BARLEY, *Eng.;* ORGE, *Fr.;* GERSTE, *Ger.*

VARIETIES.

374 **H. ægiceras.**—A peculiar form with cylindrical ears, arranged in a confused manner, not in rows; found in Tibet (*Royle; Thomson*), and in some parts of the inner Himálaya (*Stewart*).

375 **H. cœleste.**

Vern.—*Uá, úá jab, újan, grim,* PB.; *Grim, nas,* LAD.; *Chám̀a,* BHOTIYA.

A variety grown in villages bordering on the snowy ranges of the North-Western Himálaya, at high elevations from 7,000 to 12,000 feet. **Atkinson** remarks: "The seed is sown in first class unirrigated land in October and ripens in May. The average yield per acre is about fifteen loads, worth R1 per load, and raised at a cost of about R8 an acre. The produce is consumed locally by the Bhotiyas, being esteemed much too poor a food for the lowland folk" (*Atkinson, Him. Dist., 684; Stewart, Pb. Pl., 255*).

376 **H. distichum.**—The variety most cultivated in Europe having only two developed rows of spikelets, the lateral ones being rudimentary or barren. Its growth and cultivation have been experimentally tried in several parts of India, for an account of which the reader is referred to the paragraph CULTIVATION.

377 **H. gymnodistichum.**

Vern.—*Paighambari, rasuli,* TIBET.

A beardless variety, having, like the preceding, only two rows of spikelets, and presenting further the curious character of having the flower-scales non-adherent to the grains. These scales drop off in threshing, leaving the grains naked like those of the wheat. Three sub-varieties are said to be largely culivated in Tibet, a dull green, a white and a dark or chocolate brown. It is said to be largely grown in the hills near Kotgarh, but is rare in the plains. With regard to the origin of this variety, **Captain Pogson** communicated a report, at the meeting of the Agri.-Horticultural Society in January 1886, in which he stated that Indian wheat-barley, as he termed it, was introduced into the Panjáb and North-West Provinces during 1881, by seed obtained from Poo in Tibet. In **Baden Powell's** *Panjáb Products,* however, written in 1868, *paighambri jau,* both white and black, are mentioned as having been obtained from the inundated lands west of the Sutlej, from the Lahore District, from Gujranwala, Gugaira, Dera Ghazi Khan, and Kashmir. The description of the grain there given appears to coincide with that of this variety. In 1884 **Mr. J. F. Duthie** presented specimens of this form to the Kew Museum, reared from a sample exhibited at Sahâranpur by a zamindar of the Muzaffarnagar District. This was reported by him to have yielded 15 maunds of grain, and 12½ maunds of straw per acre. The samples sent were of two kinds (both referred to the same variety), one brown or chocolate coloured, the other white. Both were examined by experts and were submitted for inspection in the London market, where they were received with great interest, and the opinion was expressed that they might prove of very considerable value. In the *Kew Bulletin* for 1888 a recent examination of these grains by **Mr. H. T. Brown** of Burton-on-Trent is recorded, in which it is stated that both the black grain and the white from Kotgarh germinated well, but that the value of the former as a malting grain was considerably impaired owing to the spontaneous rupturing of the pericarp during ripening, which would render the embryo very liable to be separated during the process (*Kew Bulletin, 1888, 271*). Further, it would appear from the report that this curious wheat-like variety, especially the white form, is likely to prove of commercial value in Europe as a malting grain. It is stated that Messrs. MacDougall considered that if it could be possible to send this variety from India in a sound condition, it would command a ready sale at a high price.

H. hexastichum, *Linn.* 378

The barley, *par excellence*, of India. As this is almost the only cultivated form in India, the enumeration of its vernacular names, description of its cultivation, &c., will be given under the general account of Barley. Botanically, this variety is characterised by having six rows of developed spikelets arranged in two rows of groups of three, on each side of a flattened rachis.

Besides the above purely Indian varieties it may be of interest to mention three, collected by **Dr. Aitchison** during the **Afghán Delimitation Commission**.

H. Caput-medusæ, *Benth. & Hook.f.* 379

Found in great abundance on the downs of Badghis, growing over a foot in height.

H. murinum, *Linn.* 380

Common in cultivated land and along the banks of irrigation canals in the Hari-rud valley.

H. ithaburense, *Boiss.* 381

Very characteristic of the rolling downs of the Badghis, growing in great clumps up to 3 feet in height, and resembling cultivated barley in habit.

BARLEY. 382

Vern.—*Jav, jao, jawa, súj*, HIND.; *Jab, jau, jóo*, BENG.; *Jowa khar*, BEHARI; *Nas*, BHOT.; *Soah*, LASSA; *Tosa*, NEPAL; *Jau, indarjau, yurk*, N.-W. P.; *Thansatt, nái, jawa, chak, jau* (cut as fodder, *kawid, kasíl pathá, soá, jhotak, shiroka, tro, ne, chung, lúgar, búza, chang*), (spirits=*arrak*), (ashes=*iáwa khar*), PB.; *Jao-tursh, jao* (**H. hexastichum**=*jao-shirin*), AFG.; *Satú*, DEC.; *Sátu, jav*, BOMB.; *Java, sátu, jav*, MAR.; *Jau, jav, ymwah*, GUZ.; *Barli-arisi, barli-arishi*, TAM.; *Pachcha yava, yava, dhánya bhédam, yavaka, yavala, barli-biyam*, TEL.; *Javegodhi*, KAN.; *Mu-yau*, BURM.; *Yava, yavaka, situshúka*, SANS.; *Shaair*, ARAB.; *Jaó*, PERS.; *Arpa*, TURKI.

References.—*Roxb., Fl. Ind., Ed. C.B.C., 120; Stewart, Pb. Pl., 256; Aitchison, Cat. Pb. and Sind Pl., 171; DC., Origin Cult. Pl., 367, 368, 369; Aitch. Afgh. Del. Com. Rep. in Trans. of Linn. Soc., 2nd Ser., Vol. III., Pt. I., 127; Elliot, Fl. Andhr., 46, 73, 140, 194; Mason, Burma and its People, 477, 818; Pharm. Ind., 253; O'Shaughnessy, Beng. Dispens., 632; Moodeen Sheriff, Supp. Pharm. Ind., 155; U. C. Dutt, Mat. Med. Hind., 270, 324; Fluck. & Hanb., Pharmacog., 722; U. S. Dispens., 15th Ed., 742; Bent. & Trim., Med. Pl., 293; S. Arjun, Bomb. Drugs, 153; Murray, Pl. and Drugs, Sind, 14; Irvine, Mat. Med., Patna, 42; Med. Top. Ajmir, 139; Baden Powell, Pb., Pr., 228, 383; Atkinson, Him. Dist., 320, 684, 739; Lisboa, U. Pl. Bomb., 189; Birdwood, Bomb. Pr., 110; Church, Food Grains of India, 99; Royle, Prod. Res., 214, 220, 230; Ayeen Akbary, Gladwin's Trans., II., 135, 169, 174; Balfour, Cyclop., II., 101; Smith, Dic., 40; Treasury of Bot., I., 596; Morton, Cyclop., Agri., I., 176; Gazetteers:—Bombay, IV., 53; VIII., 182; 187; XVII., 267; Orissa, II., 133; N.-W. P., I. (Bundelkhand), 86, 198; IV. (Agra), lxxx; Rajputana, I., 109, 150, 227, 254, 256, 278; Adm Rep. Bengal, 1882-83, 12; Yarkand Mission, Rep., 1873, 77; Settlement Rep.:—Panjáb, Jhang, 84; Kohát, 122; Kangra, 24; Simla, App. II., xxxix; Rohtak, 92; Kumáon, App., 32d; N.-W. P., Bareilly, 82; Allahabad, 32; Central Prov., Nursingpur, 52; Chanda, 81; Records of Finance Commissioner, No. 22, 1878; Reports:—Govt. of Burma; Dir.*

Habitat of Barley.

Land Rec. and Agri., Bengal; Reports of the Experimental Farms:—N.-W. P. and Oudh, 1872, 9, 23; 1878, 3; Khandesh and Hyderabad, 1887, 3; Hyderabad in Sind, 1885-86, 35; 1886-87, 3, 14; Reports of the Agric. Dept.:—Bombay, 1886-87, App. VI.; N.-W. P., 1877-78, 66, 74; 1882, 47, 54; 1883, 21; 1885, 26; 1886, 17, 20; Bengal, 1886, App. II., xxvii, xlviii, xcix; Madras, 1885-86, 8, 34; 1886-87, 7; 1887-88, 4; District Manuals; Madras, Cuddapah, 53; Madras Man. of Administr., II., 109, 119; Bomb. Administr. Rep., 1872, 73, 352; Jour. Agri.-Hort. Soc., 1843, 279; 1844, 537.

HABITAT & HISTORY. 383

Habitat & History.—An annual grass, producing many stems, from 2 to 3 feet long, from a single grain. The typical form has the spikelets arranged along the sides of a flattened rachis in groups of three; but, as already described, in several of the varieties the lateral spikelets are more or less abortive. In the typical form they are developed, but are more or less irregularly arranged, while in the Indian cultivated barley they are grouped in six symmetrical rows. The only variety hitherto found wild, in any part of the globe, is **H. distichum**, *Linn.*, which is probably indigenous to Western Temperate Asia. According to **DeCandolle**, "It has been found wild in Western Asia, in Arabia Petrea, near Mount Sinai, in the ruins of Persepolis, near the Caspian Sea, between Leukoran and Bakú, in the desert of Chirvan and Awhasia, to the south of the Caucasus, and in Turcomania." Its modern indigenous area is, therefore, from the Red Sea to the Caucasus and the Caspian Sea. As no other cereal can be cultivated under so great a variety of climate, its distribution as an agricultural plant is very extended. It occurs over all the temperate and extra-tropical regions of the world, growing at altitudes of 11,000 (and according to **Baden Powell** 15,000) feet on the Himálaya, and as far north as 68·38° latitude in Lapland. Barley is amongst the most ancient of cultivated plants; but as its forms resemble each other in their economic properties, and seem to have had, in all languages, a common name, it is very difficult to ascertain which variety is referred to by ancient writers. Proofs exist in abundance, however, that barley in one or more of its forms was cultivated in the remotest times. According to **Bretschneider** it is included in the list of five cereals sown by the Emperor Shen-nung of China who reigned about 2700 B.C. **Theophrastus** was acquainted with several sorts of barley (Κριθή). It is frequently alluded to in the Bible, and must have been an important article of food in the time of Solomon (B.C. 1015). **DeCandolle**, quoting **Unger** and **Heer**, tells us that the variety **hexastichum** which was known to **Theophrastus**, "has also been found in the earliest Egyptian monuments; in the remains of the lake dwellings of Switzerland (age of stone), of Italy and of Savoy (age of bronze)." A six-rowed barley is also represented on the medals of Metapontes, a town in the south of Italy, six centuries before Christ. There appears to be little doubt that the variety most cultivated in antiquity, as it is to this day in India, was **H. hexastichum**, *Linn.* **H. distichum**, *Linn.*, was also grown and employed as a food-grain in prehistoric times, while the cultivation of **H. vulgare** proper seems to date from more modern times.

Since **H. distichum** is the only one of these which appears, at present, to exist in a wild condition, it would seem that the four and six-rowed barleys are either cultivated varieties of the two-rowed form, or that they owe their origin to a wild ancestor of their own type which has since become extinct.

The only variety cultivated to any extent in India is, as has already been stated, **H. hexastichum**. Little is known regarding its introduction, or the origin of its cultivation in this country, but from its having evidently been well known and valued by other Eastern peoples, it is probable

HISTORY.

that it was also grown at least in the north of India in very remote times. This supposition is confirmed by the intimate connection of the grain with several of the rites and beliefs of the Hindú religion. Thus the God Indra is called, "He who ripens Barley," and the grain is employed in the ceremonies attending the birth of a child, weddings, funerals, and in certain sacrificial rites. It is further supported by the antiquity of the Sanskrit name, *Yava*, which in the earliest times was probably applied as a general term to any grain or corn-yielding plant, and was only at a later date restricted to what, at that time, must have been probably the most important cereal.

In Gladwin's translation of the *Aín-i-Akbarí* the crop is said to have been one of the most important in Afghánistan and Kashmír, and a large part of the revenue from these countries was obtained from barley, by exacting the usual two out of every ten *kherwars* produced.

CULTIVATION.

CULTIVATION. Area. 384

Area.—It appears from the Agricultural Statistics of British India, that by far the largest quantity of this cereal is grown in the North-West Provinces and the Panjáb; Ajmere-Merwara, Bombay, the Central Provinces, and Madras following on the list very far behind the two first, but in the order given. For the sake of comparison the following statements of the acreage under cultivation in these Provinces, during the last three years, may be here given:—

	1885-86.	1886-87.	1887-88.
	Acres.	Acres.	Acres.
North-Western Provinces and Oudh.— Unmixed	1,349,000	1,716,400	1,670,800
Wheat and Barley	1,145,300	1,233,100	1,210,800
Barley and Gram	2,347,400	2,463,600	2,642,000
North-Western Provinces and Oudh	4,841,700	5,413,100	5,523,600
Panjáb	1,852,500	1,163,400	1,410,600
Ajmere and Merwara	63,300	50,600	36,200
Bombay	41,200	38,000	41,200
Central Provinces	8,600	Not given	Not given
Madras	3,900	...	3,700
TOTAL	6,811,200	6,665,100	7,015,300

Thus, excluding Bengal, and the Native States—tracts from which reliable statistics are not forthcoming—the total area averages about 7 million acres under barley, mixed or unmixed, of which more than three-fourths belong to the North-West Provinces. In these Provinces, Benares, Allahabad, Agra, and Rohilkhand are the most important barley-growing districts; in the last the grain is grown mixed with wheat, in Allahabad and Agra it is generally mixed with gram, while in the first it is most commonly grown alone. Barley, mixed and unmixed, occupies about 20 per cent. of the total calculated area in the thirty temporarily-settled districts of these Provinces, or 42 per cent. of the total area under spring crops.

METHODS. N.-W. Provinces. 385

Methods.—*In the North-West Provinces.*—The method of cultivation in this, the most important barley-growing region may be taken as typical and will accordingly be fully described. Barley is a *rabi* or spring crop, being sown in October and reaped in March or April. As already mentioned it is not only grown alone, but also mixed with wheat, when the crop is called *gojai*, or with gram, peas, or lentils, when it is known as *bijra*, or (in the tracts below the hills) *gochin*. Messrs. Duthie and Fuller state that the areas under barley alone, barley-gram, and barley-

CULTIVATION.

wheat stand in the relative proportions of 15, 22, and 10. One of these crops is usually the *rabi* accompaniment of indigo or rice, since it is considered much better adapted than wheat alone, for growth on a soil which has not been allowed to recuperate itself by even six months' fallow. Rape (**Brassica campestris**), Indian mustard (**Brassica juncea**), and *duán* or *tara* (**Eruca sativa**) are also commonly sown in barley fields either as a border, or in rows some 15 feet apart; linseed is also occasionally grown as a border. The soil (which is generally light and sandy, and but sparsely manured, except when the wheat-barley mixed crop is to be grown) is prepared in *Assauj* (September-October) by ploughing and cleaning, and when practicable the fields are irrigated by turning a stream into them from some neighbouring river. The amount of ploughing necessary seems to vary much in different districts; thus in Rohilkhand, the operation is frequently performed as often as twelve times, while in Bundelkhand two or three are considered sufficient. Probably four ploughings is the average throughout the Provinces. Sowing in the irrigated fields takes place in October-November, and in the uplands in November-December, a little later than the sowing time of gram, and earlier than that of wheat. The seeds are sown in plough furrows either by hand direct, or down a hollow bamboo fastened to the plough stilt, to the average amount of about 100℔ an acre. After the sowing is completed the seeds are covered in by the plough, the coarse clods of earth are broken by the *dálaya*, and again smoothed by a heavy flat wooden log, drawn by oxen, and kept steady by a man standing on it. Irrigation may or may not be employed, according to the amount of winter rainfall, and in districts which enjoy a tolerable certainty of these rains, it is but rarely carried out. A comparison of the total irrigated and unirrigated crops of the several divisions of the province show that irrigation is carried on to the greatest extent in Agra and Benares, to the smallest in Meerut and Rohilkhand, while the total irrigated crop is less than two-fifths of the whole.

Barley requires very little weeding, or at least receives little, so with the exception of possible irrigation, the crop is left very much to itself, till the season of ripening in March-April when it is reaped like wheat, by being cut in the middle of the stalk with a sickle, tied in sheaves, and stacked near the homestead to dry. When quite dry these sheaves are unbound and threshed out by a flat board with a short handle, termed a *mungra*, or in some of the north-eastern parts of Kumáon by a primitive form of flail, consisting of a long pliant stick. The grain thus separated is said to be mixed with the ashes of cowdung before being stored, to prevent the attacks of insects. In the case of mixed crops the barley and gram, or barley and wheat, are reaped, stored, and eaten as one.

COST. 386

Cost and average outturn.—Messrs. **Duthie** and **Fuller** give the following estimate of the total cost of growing an acre of barley :—

	R	a.	p.	R	a.	p.
Ploughing (four times)				3	0	0
Clod crushing (four times)				0	8	0
Seed (120 ℔)				2	8	0
Sowing				0	14	0
Reaping				1	8	0
Threshing				3	0	0
Cleaning				0	6	0
Total				11	12	0
Irrigation (twice)—						
Canal dues	1	8	0	4	0	0
Labour	2	8	0			
Rent				5	0	0
Grand Total				20	12	0

OUTTURN. 387

The outturn of barley is, under similar conditions, from one-fourth to half again as much as that of wheat. When irrigated the average may be taken as 16 maunds to the acre of unmixed barley, 15 of wheat-barley, and 14 of barley-gram, while on unirrigated soil the average is from 8 to 11 maunds for unmixed barley, from 7 to 10 for barley-wheat, and from 6 to 9 for barley-gram, the amount varying with the winter rainfall of the district. The average outturn of straw is about one and a half times that of the grain. In mixed crops about three-fifths of the total produce is barley. (*Duthie and Fuller, Field and Garden Crops.*)

Experiments were conducted in the Government Experimental Farm of the North-Western Provinces and Oudh at Allahabad in 1872 for the purpose of introducing the English two-rowed barley into that province. In the report on these experiments it is stated that the country barley, **H. hexasticum**, gave a yield of nearly double the English grain. These results discouraged further attempts, but they appear strangely at variance with the outcome of later experiment carried out in the Nilghiris, and described below in treating of Madras.

The average price realised for Barley during the past twenty years in this province has been 1 rupee for from 27 to 28 seers.

Panjab, 388

In the Panjáb.—The largest areas are in the districts of Peshawár, Sirsa, Ferozepore, Hissar, and Gurgaon. The system of cultivation is very similar to that already described as pursued in the North-West Provinces; but the practice of topping an over-leafy crop, which is never followed in the latter region, is said to be common, and the crop is generally grown unmixed, *gojai* and *bijra* being almost unknown. As in the North-Western Provinces the crop in this province receives very little special attention. Thus in the *Karnal Gazetteer* the system is described as follows: "Men may be seen sowing barley at the very end of the season on the edges of a swamp, still too wet to plough, with the intention of ploughing it as the soil dries. The limit to sowing is expressed by the proverb '*boya Po, diya kho*,' 'Sow in Po, and you lose your seed.' The field is ploughed two to four times, the *sohagga* is passed over it, and the seed sown broadcast. Manure is given if there is any to spare, which there seldom is, and water is given if the needs of the other crops allow of it. It is seldom weeded unless the weeds are very bad."

In 1876 an important and interesting official correspondence was carried on regarding the quality of Panjáb barley, which arose from complaints made regarding it by the Veterinary Department. From this correspondence it appears that the quality of the grain, at least at that time, suffered severely from its being cut before it was quite ripe, and from its being carelessly stored while damp, or in a damp place. One firm of brewers who were applied to for an opinion reported that, by buying the barley immediately after it was reaped and threshed, and storing it themselves, they obtained a most excellent grain; while two other firms who seemingly bought the barley when required for use in October, had great difficulty in obtaining even fairly good barley, a large percentage being dead, mouldy, or weevil-eaten. These facts point to careless and ignorant methods of storing as the principal cause of the frequently inferior quality of the grain, and to this being the point requiring attention in endeavouring to improve it.

The outturn per acre varies greatly in different districts, but the average agrees with that in the North-West Provinces. The average price of the grain during the past twenty years has been between 31 and 33 seers per rupee.

Central India and Rajputana, 389

In Central India and Rájputana.—Barley is largely cultivated both in Ajmere-Merwara and in Rájputana proper, especially in its more northerly

HORDEUM vulgare.

Cultivation of Barley in India.

CULTIVATION.

Rajputana. portion. The system of cultivation is, on the whole, similar to that previously detailed, but the crop appears to be generally ploughed about six times, manured, weeded at least once, and irrigated; in fact, to receive altogether more attention than in many other parts. The cost of production is about R28, the value of the crop about R34, and the consequent profit about R6 per acre.

Bombay. 390 *In Bombay.*—It is principally grown in Gujarat. In an interesting report on the subject the Acting Director of Land Records and Agriculture writes: "In 1887-88 Gujarat alone had more than three-fourths of the total area of barley-cultivation in Bombay. It is less important in the Deccan. In Gujarat it is generally an after-crop in garden rice lands or in soils too sandy and open for wheat. It is irrigated and manured. The garden rice land is, previous to cultivation, manured to the extent of 12 cart-loads, watered on the stubble in December, then ploughed twice and levelled. The seed to the extent of 80 to 100℔ is immediately thrown by hand into the furrows, the sower following the plough. It is then rolled and watered six or eight times, and harvested in April. The estimated acre yield is from 1,020 to 2,270℔. According to experiments conducted on this crop, the cost per acre comes to about R20, and the value of produce to about R21. The fact that it is not subject to the wheat blight renders it a favourite rain crop in the light lands of Ahmadnagar and Kaira. In the Deccan it is grown as a garden crop, and rarely as an after-crop in rice fields."

Central Provinces. 391 *In the Central Provinces.*—The crop of barley in these provinces is so insignificant that during the past two years no returns of the area under cultivation have been made.

Madras. 392 *In Madras.*—As might be expected from the climate of this province, barley is a very unimportant crop, only from 3,000 to 4,000 acres being cultivated on the spurs and slopes of the Nilghiris and Pulneys. Endeavours were, however, made by the Agricultural Department in 1885-86, to improve the quality of Nilghiri barley, and to increase the extent of its cultivation, by introducing good seed both from the Panjáb and England, and distributing it to the *rayats*. To ensure success and to help the people with advice and instruction, these experimental trials of foreign seed, though conducted by the *rayats* themselves, were placed under the supervision of an agricultural inspector. In the report of the Department for 1886-87 it is stated that the results with English barley (**H. distichum**) were very satisfactory, a good crop having been obtained which was purchased by the Nilghiri Brewery Company. Panjáb barley on the other hand, was found to be inferior to that already grown in the district, and accordingly attempts to introduce the latter have been stopped. English barley is further favourably commented on in the last report (1887-88), in which it is mentioned that during 1887 nearly 5,000 bushels of seed were sold at an average price of R1-12 per bushel. Every endeavour is being made to increase the cultivation of this important malting grain, for which there is great demand, and, as a consequence, a considerable augmentation of the area under cultivation and yearly outturn in the Presidency may be expected.

Bengal. 393 *In Bengal.*—No returns are available of the area under cultivation, and the crop as a whole is unimportant. The Director of Land Records and Agriculture mentions that it is cultivated in the Bhagulpore Division and to a limited extent in Chutia Nagpur. As in other districts, it is a *rabi* crop, being sown in October-November and reaped in March-April. The ground is prepared by three or four ploughings, at intervals of four or five days. The seed is then sown, and the soil harrowed, and ploughing and harrowing are repeated after three days. No mention is made of any further attention, in the form of irrigation or weeding, being paid to the crop.

CULTIVATION. Burma. 394

In Burma.—The Government of Burma reports: "In Mandalay it is the margin paddy of the district. It is sown in the month of January, and reaped in the month of May, averaging 120 days; the cost of cultivation being about R12 per acre, and the proceeds about R25. It is an inferior kind of grain and only eaten by the poorer classes. In the Thongwa District it is grown in a very small area, there being not even 50 acres in all. It is cultivated indiscriminately, and at little cost. In the Yamethin and Meiktila Districts margin paddy is grown extensively. No returns are available regarding the total area under cultivation."

PRODUCTION and CONSUMPTION. 395

Total Production and Consumption.—By far the greater proportion of barley grown in the country is consumed locally, a very small quantity being exported. An attempt has been made by the calculations shown in the adjoining table to arrive at a fairly accurate estimate of the actual total outturn of the grain in India during the year 1887-88 excluding Bengal, Native States, and tracts from which no reports are available. The result is necessarily (from the number of averages which had to be employed in compiling it), only at best an approximation, but it errs, if to any noteworthy extent, on the side of under-estimation, and may be interesting for purposes of comparison with the trade statistics.

Such a comparison shows that of the total production of 46,604,007 cwt., 42,179,869 cwt. were consumed in the country during the year under consideration.

Table showing calculation of total production of Barley in India, excluding Bengal, Native States, and Tracts from which Statistics are not forthcoming in 1887-88 (Maund=82lb).

Crop.	Approximate Acreage. (Returns of areas outside the North-West Provinces taken as Unmixed Barley.)	Approximate Yield per Acre.	Total Yield of Crop.	Total Yield of Barley. (In mixed crops barley = ⅔ of the whole.)	
		Maunds.	Maunds.	Maunds.	Cwt.
Unmixed—					
Irrigated	1,581,250	16	2,53,00,000	2,53,00,000	18,523,214
Unirrigated .	1,581,250	9¼	1,46,26,562	1,46,26,562	10,708,732
Barley-wheat—					
Irrigated .	484,320	15	72,64,800	43,58,880	3,191,322
Unirrigated .	726,480	8¼	59,93,460	35,96,076	2,632,841
Barley-Gram—					
Irrigated . .	1,056,800	14	1,47,95,200	88,77,120	6,499,320
Unirrigated .	1,585,200	7¼	1,14,92,700	68,95,620	5,048,578
Total .	7,015,300		7,94,72,722	6,36,54,258	46,604,007

TRADE. Imports. 396

Trade.—Imports.—Separate returns of the imports of this cereal are not made either by sea or trans-frontier; it is therefore to be presumed that the quantity is so small as to be unworthy of special notice.

Exports. 397

Exports—Are also very small in comparison with the approximate annual production.

The grain prior to 1887-88 was considered, in the Annual Reports of Trade and Navigation, under the heading of "Other sorts of grain," hence it is only possible to give the figures for last year. From these it appears that the total exports by sea to foreign countries during that period were 29,575 cwt., valued at R89,776, of which Bombay shipped 18,688 cwt.

HORDEUM vulgare. **Chemistry and Medicinal Properties.**

Bengal 6,873 cwt., and Sind, 4,014 cwt., valued at R58,632, R20,556, and R10,588, respectively. The country which imported most largely was Persia, with 10,358 cwt., following on which were Arabia with 7,675 cwt., Ceylon with 7,539 cwt., and Aden, the United Kingdom, Zanzibar, and "other countries," with insignificant quantities.

MEDICINE. Seeds. 398 **Medicine.**—The husked SEED of **H. distichum** occupies a place in the Pharmacopœias of England, the United States and India, under the name of HORDEUM DECORTICATUM or Pearl Barley. The grains of this variety of prepared barley are sub-spherical or ovoid, about two lines in diameter, of a white farinaceous appearance, and have the peculiar taste and odour common to most of the cereal grains. An account of the preparation of "pearl barley" will be found under the heading FOOD.

Chemistry. 399 CHEMISTRY.—The grain of **H. hexastichum** is frequently used in this country instead of that of the officinal plant, and is exactly similar in medicinal properties. Chemically the grains contain (along with the food-forming constituents, which will be discussed hereafter) from 13 to 15 per cent of water, and after drying they yield about 3 per cent. of fat oil, with insignificant proportions of tannin and bitter principles, residing chiefly in the husks, and 2·4 of mineral ash. The oil, according to **Hanamann**, is a compound of glycerin with palmitic and lauric acids, or less probably with a peculiar fatty acid. The ash was found by **Lermer** to contain 29 per cent. of silicic acid, 32·6 of phosphoric acid, 22·7 of potash, and 3·7 of lime. Besides these constituents **Lintner** in 1868 demonstrated the presence of a little cholesterin in the grain, and still later **Kühnemann** extracted from it a crystallized dextrogyrate sugar, and an amorphous lævogyrate mucilaginous substance, sinistrin, and demonstrated the absence of dextrin (*Flückiger and Hanbury*).

MEDICINAL PROPERTIES.—Barley is demulcent, and easy of digestion, and is for these reasons much used in the dietary of the sick. In India

Grains. 400 *saktu*, or powder of the parched GRAINS, is much employed in the form of a gruel in cases of painful and atonic dyspepsia. In European practice Barley water, a decoction of the grain, is principally prescribed, and is valuable in cases requiring demulcent treatment. **Dr. Irvine** states that

Leaf. 401 in Patna the ASHES OF THE LEAF are employed in the formation of cooling sherbets; and **Stewart** writes that the ASHES OF THE STALKS are pre-

Stalks. 402 scribed for indigestion in the plains of the Panjáb. Preparations of MALT

Malt. 403 have acquired some reputation of late years in Europe and America, since they are more demulcent and nutritious than those of the unmalted barley. Malt extract may be prepared by boiling two to four ounces of the germinated and dried grain in a quart of water and straining. When

Wort. hops are added, the decoction becomes WORT, and acquires tonic properties, which have been found especially valuable in cases of debility following on long continued chronic suppuration.

FOOD. 404 **Food.**—The food-forming constituents of average husked Indian barley are, starch 63 per cent., cellulose 7 per cent., albumenoids 11·5 per cent., with small quantities of oil, ash, fibre, and 12·5 per cent. of water. The nutrient-ratio is given by **Church** as 1 : 6·3 and the nutrient value as 84·5. On comparison of the above with the results of analyses of English barley it will be found that the latter show a smaller percentage of albumenoids. The process of cleaning barley for food purposes is, in this country, carried out, as a rule, by pounding in wooden mortars and winnowing, or, in the North-West Provinces as already described, by beating with a flat board. It is then ground into coarse meal, from which either alone, or mixed with the meal of wheat or gram, *chapattis* are made and baked; or a thick gruel or pasty mass is made, to which a little salt is added, and the preparation is eaten with garlic, onions, or chillies. In either of these

FOOD.

forms it is a staple article of food of the poorer classes in many parts of the country, especially in the North-West Provinces, the Panjáb, and Oudh.

The grain thus roughly cleaned and ground is much more rich in albumenoids than the more carefully-prepared culinary barley of Europe, partly owing to the higher percentage of nitrogen naturally existing in the former, partly to the amount of the richly albumenoid husk left by the imperfect methods of cleaning it. Though this fact renders Indian barley meal of higher nutritive value than English, it at the same time makes it more difficult of digestion, and hence partly unsuits it for the dietary of dyspeptics and invalids generally.

The grain in England undergoes an elaborate process of cleaning, by which several different sorts of barley are produced, to which various names are given in commerce. This process consists essentially in passing the hard dried barley between horizontal millstones, placed so far apart as to rub off its integuments without crushing it. The millstones may be a little approximated, and by this means results are obtained varying in cleanness. **Church**, in describing the process, writes: "100lb of barley yield 12½lb of COARSE DUST and become "BLOCKED BARLEY." By closer and longer grinding Blocked Barley yields 14¾ lb of FINE DUST and becomes "POT" or "SCOTCH BARLEY," which again on being further ground yields 25¼lb of "PEARL DUST" and becomes "PEARL BARLEY." Thus from 100lb only 37⅝lb of pearl barley is obtained, 10 per cent. of actual loss unaccounted for occurs, and the remainder consists of the "dusts" above mentioned. These dusts, though generally considered waste products, are of considerable nutritive value, and might be utilised, as the following figures given by **Church** will show:—

	Coarse Dust.	Fine Dust.	Pearl Dust.
Water	14·2	13·1	13·3
Albumenoids	7·0	17·6	12·1
Oil	1·7	6·	3·4
Starch	46·9	50·5	67·2
Fibre	24·5	8·5	1·8
Ash	5·7	4·3	2·2

Church, however, takes care to mention that in point of fact the figures representing percentage of albumenoids are in all these cases too high, because the nitrogen from which they are calculated does not exist altogether in the albumenoidal constituents, aud suggests the deduction of ⅓ from the totals given as probably likely to yield a more accurate result.

Barley in one or other of these forms is much employed in European cookery as a bland demulcent grain in the preparation of soups, &c. It is also employed by the poorer classes in certain localities to make bread.

FODDER, 405

Fodder.—Barley has of late years attracted considerable attention owing to its cheapness and value as a fodder. It has long been employed in the Panjáb for this purpose, the crop being cut two or even three times when quite young without marked injury to the final yield of grain (*Stewart*).

In the report on the Experimental Farm of Hyderabad in Sind for 1886-87, it is recorded that experiments were tried, during the previous year, to determine the pecuniary benefit of growing it purely as a fodder, with the result that the balance was very much in favour of this method, even in a year when fodder was very cheap. The exact outcome of

Alcoholic Liquors from Barley.

FODDER.

this interesting trial may be shortly given in tabular form :—

No.	Area.	Description.	Yield per Acre.	
			Grain.	Straw.
	Gunthás.		lb	lb
1	8 . .	Cut before ripening . .	...	21,740
2	8 . .	Cut in grain, green . .	...	14,925
1,2	(Same plots) .	Second cutting .	56 62	280 310
3	8 . .	Cut ripe .	2,145	4,270
4	8 . .	Cut ripe .	2,931	4,428

The total value of the yield of plots 1 and 2 was R94-5-2, of 3 and 4 R77-0-8, or a balance in favour of the fodder crop of R17-4-6.

The straw of even ripe barley is a fairly good fodder when cut up as "*bhúsa,*" but is inferior to that of wheat. The grain is a good feed for both horses and cattle, either given alone, or mixed with gram, when it is known as "*adaur.*"

DOMESTIC, INDUSTRIAL. and SCARED USES. 406

Domestic, Industrial, and Sacred Uses.—The grain is much employed in parts of India in the preparation of a kind of beer, or spirituous liquor, and its value for this purpose has been long known. The Sanskrit *Yava-sura* is an intoxicating drink prepared from barley. **Stewart** writes regarding the employment of barley for this purpose in the Panjáb : "In Lahoul and on the Sutlej a kind of beer is made from its grain, the ferment in the former case, being brought from Tibet as little farinaceous-looking cakes, the size of a fig, called *pab* or *phap*. In Ladák also, a similar beverage is made bv the aid of the same substance which is said to be made in Drás, to the West, from barley flower, mixed with cloves, cardamoms, ginger, and an herb which is probably an Umbellifer (and then fermented ?). On the Sutlej, **Moorcroft** states that in the preparation of the beer, rice is mixed with it, and the root of a bitter aromatic obtained from higher altitudes is added to prevent indigestion. In Lahoul, **Aitchison** mentions that spirits made from barley are used by some of the richer inhabitants, and spirits are also made from it in Ladak"

The employment of barley for similar purposes in the manufacture of European Beer is well known, and will be fully described in the article on MALT LIQUORS. The straw is used in many parts of the Panjáb hills in the manufacture of sandals.

Malt Liquors. 407

Barley is required for many of the ceremonials of the Hindu religion. It is considered a symbol of wealth and abundance, is claimed by astrologers as a "notable plant of Saturn," and is in India particularly associated with the God Indra. It is specially introduced in the ceremonials attending the birth of an infant, weddings, funerals, and at certain sacrifices; on the fourth day of the light half of the month "Vaisákha" a sort of game is played in which people throw barley-meal over each other, known in the Sanskrit as *Yava-cáturthí*.

Hornbeam, Indian, see **Carpinus viminea**, *Wall.*; Vol. II., 182.

408

HORNS, ANTLERS, AND HORN-WORK.

BOIS, CORNE, *Fr.*; HORN, *Ger.*

Horns are largely utilised in the manufactures of Southern and Eastern Asia, and form an important article of export to Foreign Countries from India. The animals of India, which yield horns of commercial and industrial importance, may be here enumerated, leaving the reader for information regarding their habitats, vernacular names, &c., to refer to articles on the animals as they occur in this work, grouped under their popular names. For information regarding ANTLERS, see DEER, Vol. III., 55; and for "IVORY," see ELEPHANT, Vol. III, 208.

Axis maculatus, *Gray*. The Spotted Deer.
A. porcinus, *Zimm*. The Hog Deer.
Bubalus arni, *Shaw*. The Wild Buffalo.
Capra hircus, *Linn*. The Domestic Goat.
C. megaceros, *Hutton*. The Markhor.
C. sibirica, *Meyer*. The Himálayan Ibex.
Cervulus aureus, *Ham. Smith*. The Barking Deer.
Cervus affinis, *Hodgson*. The Sikkim Stag.
C. wallichii, *Cuv*. The Kashmír Stag, or Barasingh.
Gavæus frontalis, *Lambert*... The Gayal.
G. gaurus, *Ham. Smith*. The Gaur or Bison of Anglo-Indian Sportsmen.
G. sondaicus, *Muller*. The Burmese Wild Bull.
Gazella benettii, *Sykes*. The Ravine Deer, Goat Antelope, or Indian Gazelle Antelope.
Hemitragus hylocrius, *Ogilby*. The Nilghiri Wild Goat, or "Ibex."
H. jemlaicus, *Smith*. The Ther, or Himálayan Wild Goat.
Nemorhœdus bubalina, *Hodgson*. The Serow or Forest Goat.
N. goral, *Hardu*. The Gúral, or Himálayan Chamois.
Pœphagus grunniens, *Linn*. The Yak.
Rhinoceros indicus, *Cuv*. The Great Indian Rhinoceros.
R. sondaicus, *Sol. Muller*. The Lesser Indian Rhinoceros.
R. sumatranus, *Cuv*. The Indian two-horned Rhinoceros.
Rucervus duvancelli, *Cuv*. The Swamp Deer.
R. eldii, *Guthrie*. Eld's Deer, the Manipúr or Burma Stag.
Rusa aristotelis, *Cuv*. The Samber Deer, or Elk of Indian Sportsmen.

Besides these wild animals, the different varieties of domesticated ox, buffalo, and sheep, yield horns which are collected with their hides or skins, and, indeed, these form the major part of the commercial exports of horn.

Vern.—*Sing*, HIND.; *Sing*, GUZ.; *Kombú*, TAM.; *Kommú*, TEL.; *Tanduk sangú*, MALAY.

References.—*Baden Powell, Pb. Pr., I., 159; Royle, Prod. Resour., 4; Forbes Watson, Ind. Surv., I., 386; Spons, Encyclop., 1132; Balfour, Cyclop., II, 102; Ure, Dic. Indus. Arts and Manu., II., 597; Hoey, Trade & Manufactures of N. India, 130; Madras Man. of Administr., 361; Bombay Administr. Rep., 1871-72, 378; Settlement Reports, Upper Godavery Dist., 40; Gazetteer, Bom., X., 188; XV., pt. ii., 70.*

CHARACTERS, COMPOSITION, &c.

CHARACTERS. 409

The name horn is applied to the organs of attack and defence, which project from the heads of various species of animal. Anatomically they are of very various structure, and, according to Professor Owen, "belong to two organic systems as distinct from each other as both are from teeth. Thus the horns of deer consist of bone, and are processes of the frontal bone; those of the giraffe are independent bones, or 'epiphyses,' covered by hairy skin; those of oxen, sheep, and antelopes are 'apophyses" (or direct processes) "of the frontal bone, covered by the corium" (or skin) "and by a sheath of true horny material;" those of the prong-horned antelope, consist at their bases of bony processes covered by hairy skin, and are covered by horny sheaths in the rest of their extent. They thus combine the character of those of the giraffe and ordinary antelope, together with the expanded and branched form of the antlers of the deer. Only the horn of the rhinoceros is composed wholly of horny matter, and this is disposed in longitudinal fibres, so that the horns seem rather to consist of coarse bristles, coarsely matted together in the form of a more or less elongated sub-compressed cone." So-called horns may thus be roughly divided into two large classes, comprising five sub-divisions :—

I.—Horns consisting of bone, and having no true horny matter in their structure.

(1) True processes of bone, *e.g.*, antlers of the deer.

(2) " Epiphyses " or separate pieces of bone, covered by skin, *e. g.*, horns of the giraffe.

II.—Horns more or less consisting of true horny matter.

(1) Processes of bone covered at the base by hair, and tipped by horn, *e g.*, horns of the prong-horned antelope.

(2) Processes of bone covered by horn, *e.g.*, horns of the ox.

(3) True horn, *e.g.*, nasal horn of the rhinoceros.

True horny matter, found in class II, is formed by a modification of epidermal tissue (the superficial layers of the skin), and consists of an albumenoid material called ***keratin.***

USES. 410

INDUSTRIAL USES.

Horns of the 1st class—Are largely exported from India, which is the principal source of the supply of Great Britain. The antlers principally exported are those of **Axis maculatus and Rusa aristotelis.** Those of the latter rarely exceed 40 inches in length, and are generally under three feet, but four feet along the curvature has been recorded. They consist of a basal antler springing directly from the burr or base of the horn, and pointing forwards, upwards and outwards, the beam bifurcating at the extremity, and a snag separating posteriorly and pointing obliquely to the rear. The horns of **Axis maculatus** are about 30 inches in length, have a long antler sweeping upwards and backwards, with one basal and one subterminal snag, the former projecting forwards and upwards from the beam. The whole horn is pale and somewhat smooth.

These antlers and other bony horns are extensively employed in Europe for the manufacture of handles for cutlery, umbrellas, and sticks, &c. The Gazetteers of Ratnagiri and Kánara in Bombay, and the Madras Manual of Administration treating of Vizagapatam, contain accounts of the antlers of Samber and other deer, being employed locally in the manufacture of small fancy articles, such as those mentioned above, but the industry seems to be a very poor and unprofitable one.

Horns of the 2nd class—Are employed principally in the manufacture of " horn " for the purposes of comb-making. The keratin, of which the commercially valuable part is composed, is eminently suited to a number of purposes, owing to its elasticity, flexibility, and toughness, together with its physical property of softening under the application of heat, and its capability while in this condition of being moulded and welded into various forms under pressure. The horns of sheep and goats are whiter and more transparent than those of other animals, and are, therefore, most valued for comb-making purposes, while of the Indian horns, those of the arna buffalo (**Bubalus arni**) are the handsomest and best for more ornamental work. According to the authors of ***Spons' Encyclopœdia*** about one-fifth of the total imports of horn into England are employed in comb-making, while a small portion is converted into shoe-horns, scoops, cattle-drenches, drinking cups, &c. The solid tips and the hoofs of cattle (which like horn consist largely of keratin) are made into buttons.

Combs. 411

In the preparation of horns for comb-making, &c., the following is the process usually adopted :

" Horns which are to be manufactured are first thrown into water by which means slight putrefaction is caused, ammonia is liberated, and the

USES. **Combs.**

horn begins to soften; the softening is then continued by immersion in an acid bath, for a period of about two weeks. When sufficiently soft, they are cleaned and split into two parts by a circular saw. These slices are introduced between heated plates, and the whole is subjected to a pressure of several tons a square inch. The plate may bear devices, or be of varying form, thus producing at once any desired effect. The horn may then be dyed black or brown by dipping it into a bath containing a weak solution of mercury or lead salts, and rubbing on hydrosulphite of ammonia; or it may be mordanted in an iron bath and dyed by logwood. Fancy markings are produced by immersing the horn in a bath of lead salt, and then in hydrochloric acid, thus forming white lines in the interstices of the horn.

"The manufacture of combs is by far the most important application of horn. The laminatory character of the horn, its very diversely running grain, and the raising up of the fibres by the use of the various tools, render it very difficult to apply machinery in its conversion, and the large amount of hand labour required helps to cause the proportionately high price of the manufactured article. The softened horn is first split lengthwise, in the direction of the grain; it is then warmed in hot water, laid out flat, between cold iron plates, and pressed level. If the goods are to be subsequently stained, the slices are further placed between hot steel plates, and very strongly pressed, to reduce the thickness and destroy the superficial grain. The prepared slices are next stamped out by cutters, arranged to form as many combs as possible, of various sizes and shapes so as to fully economize the material. The slices are again pressed, straightened and ground, ready for cutting the teeth, which operation is performed by a 'parting-engine,' or dye-stamping machine, in the case of coarse combs, and by circular saws in that of fine-toothed combs." (*Spons' Encyclopœdia.*)

The employment of this class of horn in India, like that of the first, is small and local, records only being made of special horn industries in Ratnagiri and Kánara in Bombay, and Vizagapatam in Madras. Of the latter place it is stated, "ornamental articles are neatly executed in horn, but they have no sale in England, being surpassed in cheapness and workmanship by articles of German manufacture" In the *Bombay Gazetteer, XV.*, (Kánara) the following passage occurs: "Fancy articles of cattle, deer, and bison horn, are made by some carpenters and *gudigars* with considerable skill at Kumta, Honawar, Siddapur, Bilgi, Sirsi, and Sonda. The demand for the work is small, and in no place employs more than a few families. The horn is collected in the district, the price of a horn varying from 4 annas to R1. The articles made are small jewel-boxes, combs, snuff-boxes, cups, handles for sticks and knives, buttons, rings, and toys. A jewellery box costs about R5, and a comb or small snuff-box 2 to 4 annas." Similarly, it is reported from Ratnagiri: "Fancy articles of Bison's horn are made by a few carpenters' families with considerable skill at Vijaydurg, Maban, and Rajapur. The industry is said to have been started some 200 years ago at Vijaydurg. The horn is imported in small quantities from Malabar and Cochin, the price varying from R1 to R2 according to size. The horn is heated over a moderate fire, and to make it malleable is softened with cocoanut oil and wax. The articles made, varying in price from R5 to R8, are card trays; inkstands; snuff-boxes; cups for idols, decorated with bulls, cobras, and deer; combs; chains; handles for sticks; and different kinds of birds and animals. The demand for the work, perhaps the only specialty in the district, is very limited and the workers few and much indebted."

From the above extracts it will be seen that, with their present deficient

USES.
Combs.

methods and want of machinery, the Natives of districts even in which the industry is an old and well known one, cannot compete with the cheap horn articles of European manufacture, now so easily obtained in every bazár, at least of the larger towns. A small quantity of horn is also utilised in most parts of India by the *kangi-saz* or comb-maker, who employs, amongst other materials, the horn of the buffalo in the manufacture of his wares. Mr. Hoey, in his description of this industry, as carried on in Northern India, writes: "Buffalo horns of the description used for combs are usually sold at R25 per 100. If a horn be a seer in weight it will make 20 combs, *viz*, ten 1st class combs, which sell wholesale at R5 per hundred, five 2nd class combs, which sell wholesale at R3-8 per hundred, and five 3rd class combs, which sell wholesale at R2 per hundred. A single horn may not be one seer in weight, but a seer of horn will generally yield as above detailed. Retail vendors sell these combs at 1 anna each, first quality, 9 pies each, second quality, 6 pies each, third quality." In a further passage he writes that ebony combs are preferred by all classes, but that they are too expensive for general use; and that, of other kinds, women prefer horn combs, and men those made of wood. "This is so to such an extent," he remarks, "that *zanana kanghis* (which are made with teeth on both sides) are made of horn exclusively, and *mardana kanghis* (made with teeth on one side only) are constructed of wood." Near Hugli, in Bengal, a small trade also exists in locally-manufactured combs, some half a dozen families being so employed. It is probable that throughout India, here and there, a few comb-makers still practise their art, though it would appear that the European-made combs are rapidly displacing those of home manufacture.

TRADE.
Exports.
412

Trade.—The export trade in horns is large and important and appears to remain fairly stationary. During the past five years the total amount has varied from 48,000 to 68,000 cwt., and the total value from 12 to 16 lakhs of rupees. The average during that period was 58,568 cwt., value R15,00,933; the minimum export was in 1886-87, 48,435 cwt., value R12,27,082, and the maximum was in 1887-88, 68,018 cwt., value R,16,43,937. The countries which form the chief markets for Indian horns, are France, the United Kingdom, and the Straits Settlements, but returns shew a small though fairly steady trade with Belgium, Italy, Egypt, Germany, Austria, Ceylon, and China, named in order of importance. The average amounts imported by these during the period under consideration were, France, 26,253 cwt, value R7,33,528; United Kingdom, 25,205 cwt., value R5,72,624; Straits Settlements, 4,899 cwt., value, R1,26,107-8; the others are unimportant, and need not be enumerated.

The provinces from which the horns thus exported are principally obtained are Bombay, the North-West Provinces and Oudh, the Central Provinces, the Panjáb, Bengal, and Madras, of which the first appears to be the most important. Thus in the returns of the total quantities carried by rail and river in 1886-87, it will be found that Bombay exported 10,531 maunds, the greater part of which went to its own seaport town; the North-West Provinces and Oudh, 7,859 cwt.; the Central Provinces, 7,374 cwt.; the Panjáb 5,223 cwt.; Bengal (excluding Calcutta) 3,570 cwt., and Madras (excluding Madras Ports) 3,387 cwt. By far the largest quantity thus collected goes for exportation to Bombay; thus the average total exports from Indian ports during the four years under consideration were: from Bombay, 31,766 cwt, value R8,29,289; from Bengal, 14,328 cwt., value R3,75,374; from Madras, 9,195 cwt., value R2,14,397; from Burma, 4,939 cwt., value R1,34,070; from Sind, 466 cwt., value R3,702; and in every year the exports from these ports retained much the same relative positions. The provincial value rates of horns per maund is given by the returns of the

TRADE.

Agriculture and Revenue Department for 1886-87 as follows: Madras R17-12; Bombay R20-7; Sind R10; Bengal R10-12; North-West Provinces and Oudh R8; the Panjáb R4; the Central Provinces R19-4; Berar R15, and Assam R11-5.

The writer of the article on horns in *Spons' Encyclopædia* records that the approximate value in England of the different classes of horn, were in 1879: deer, East Indian, 40-120 *s.* a cwt.; buffalo, East Indian, 20-60 *s.*; tips, East Indian (consisting of solid horn for the manufacture of buttons &c.,) 18-40 *s.*

Imports & Re-exports. 413

IMPORTS FROM FOREIGN COUNTRIES AND RE-EXPORTS.—On reviewing the trade returns of imports by sea it will be found that the quantity thus annually received in India is a very fluctuating one. Thus in 1883-84, 2,465 cwt. were imported; while in 1887-88, only 368 cwt. came into the country. The average for the past five years has been 1,116 cwt., valued at R62,014. The countries from which these have been principally obtained are Ceylon, from which the average imports for the same period was 725 cwt., valued at R28,652; Zanzibar 283 cwt., valued at R35,645; and Mozambique 37 cwt., valued at R4,507. The horns obtained from Eastern Africa are thus evidently much more valuable than those from Ceylon. The trade returns of re-exports shew that only a small proportion of this import has been re-exported, at least as foreign merchandise, since the average during the same period is found to be only 321 cwt., valued at R38,022. The whole of the re-exports are shipped from Bombay, and nearly all go to Hong-Kong, the average for the past five years being 296 cwt., value R37,683. In 1884-85 a small consignment was sent to Great Britain, while in 1883-84, Aden, and in 1886-87 the Straits Settlements, received 6 and 2 cwt., respectively. The greater part of the horns imported by trans-frontier trade comes from Kashmír, but the quantity thus recorded is small. In 1887-88 it amounted to 147 cwt., valued at R1,973. There is no record in the trade returns of any exports to trans-frontier countries.

HORSES, ASSES, & MULES. 414

The Horse and its near allies, the several species of Ass and Zebra, form the genus **Equus** of **Linnæus.** They are all indigenous to various parts of the Old World, though fossilised remains of an animal about the size of a fox or sheep, but unmistakeably belonging to the genus **Equus,** have been found in the Eocenes of North America. They are very distinct in structure from other genera of Mammalia, and in the old systems of classification constituted a separate Natural Order under the name of MONODACTYLA or SOLIDUNGULA. Researches in comparative anatomy have, however, disclosed the fact that their structure, seemingly so singular and exceptional, is in reality nearly allied to that of the rhinoceros and tapir, a result which has been confirmed by numerous palæontological discoveries, and restorations of extinct species forming intermediate stages between a remote many-toed ancestor, and the present one-toed type. It is not proposed in this article to enter into the Anatomy, Physiology, and Pathology of the Horse and Ass, for accounts of which the reader is referred to the many works of authors who deal with veterinary science. Nor is it necessary to discuss the extensive and complicated subject of Government Horse Breeding in India, since the numerous and valuable reports issued on the subject are accessible to the public. The species of the genus which present features of economic interest are only two—**Equus caballus,** *Linn.*, The Horse; and **E. asinus,** *Linn.*, the domestic Ass. Four species of ass and three of zebra have, however, been described by naturalists, *viz.*, **E. tœniopus,** *Heuglin.*, the Wild Ass of Abyssinia; **E. hemionus,** *Pallas.*, the Kiang or Dzeggetai of the high tablelands of Tibet; **E hemippus,** *Geoff.*, the Syrian wild ass, and **E. onager,** *Pall.*, the Onager of Persia, the Panjáb, Sind, and the desert of Cutch. The first of these is undoubtedly the ancestor of our domesticated animal, while the other three have been by several Zoologists reduced to one species. The zebras have by **Hamilton Smith** been separated as a genus

Hyppotigris, but they only differ from the asses in their colouration, and one of them **E. quagga,** *Gmel.*, the Quagga, is intermediate between the zebras and true asses. The other two species are **E. burchelli,** *Gray;* the Dauw, and **E zebra,** *Linn.*, the Mountain Zebra. All three species are natives of South Africa. Though there are thus seven distinct species, or species considered distinct by the majority of naturalists, they will all, at least in a state of captivity, breed with perfect freedom with any of the others. Thus the writer of the article "Horse" in the *Encyclopædia Britannica* states that cases of fertile union are recorded between the horse and the quagga, the horse and the dauw, the horse and the hemionus or Asiatic wild ass, the domestic ass and the dauw, the domestic ass and the zebra, the domestic ass and the hemionus, the hemionus and the zebra, and the hemionus and the dauw. As is well known, also, the horse and the ass, perhaps the two most widely distinct of the species, produce the mule, which in many useful qualities excels both its progenitors. These hybrids or mules do not breed *inter se*, and only very rarely have instances been recorded of a fertile union between a female mule and a male horse or ass.

Vern.—*Ghora* (*gadda* = ass), (*ghur-khur* = **Equus onager**), HIND.; *At*, KIRGHIZ; *Kudri* (*kalda*=ass), TAM.; *Guramu* (*gardhi*=ass), TEL.; *Son, hnyet, myen*, (*myai*=ass) BURM.; *Asu, hya, aswa, túrg, vag, vaji, ghotak, piti*, SANS.; *Hisán* (*khamar*=ass), ARAB.; *Asp*, (*ghour*=**Equus onager**), PERS.

References.—*Wallace, India in 1887, 127; Darwin, Animals and Pls. under domestication, I., 49; Jerdon, Mam. of India, 236; 1st Yarkhand Mission Rep., 70; Lahore to Yarkhand, 135; Adams, Naturalist in India, 269; Drew, Jammu and Kashmir, 386; Voyage of John Huyghen Van Linschoten, Ed. 1885, I., 54; Ain-i-Akbari, Blochmann's Trans., 132-139, 215, 233, 250, 255; Gladwin's Trans., I., 144, 196, 207; II., 78, 95, 118, 136, 177; Spons, Encyclop., 1098; Balfour, Cyclop., I., 188; II., 103; Morton, Cyclop. Agr., II., 71; Ind. Agri. Gazette, 1887, 531; Rep. of Agri. Dept.:—Madras, 1884-85, 10; 1885-86, 9, 46; Bombay, 1885-86, 44; 1886-87, App. xvi.; N.-W. P., 1882, 52; Burma, 1887-88, 29; Rep. Exper. Farms;—Madras, 1884-85, 54; Bombay Admin. Rep. 1872-73, 341; Gazetteers:—Panjáb;—Rawal-Pindi, 85; Jhelam, 113; Dera Ghazi Khan, 86; Bombay;—IV., 24; V., 28; VIII., 97, 311; XII., 28; XVI., 21; XVII., 35; XVII., Pt. I., 61; XXII., 37; Burma, I., 435.*

HISTORY. 415 **Habitat, History, &c.**—The horse has been a domesticated animal from pre-historic times, which renders it difficult to determine its original habitat. No aboriginal or truly wild horse is known positively to exist, though many authors assert that the wild horses of the East are not descendants of escaped domestic horses, but are in reality of aboriginal origin. Thus Prejevalski described in 1881 a wild horse found by him in the deserts of Central Asia. It, of all others, appears to have the strongest claim to being a genuine wild horse. In colour it is whitish grey, paler and whiter beneath, and reddish in the head, which is also large and heavy in proportion to the small body. It approaches the ass and zebra in the structure of the tail, the hairs of which begin half way down from the base. The ears, however, are horse-like, and it has warts on the hind, as well as on the fore-legs, which characteristics are sufficient to classify it as a true horse. Intense heat and extreme cold appear to be equally well borne by the horse, which readily becomes acclimatized and suited to the most diversified surrounding conditions. Thus wild horses (horses which have become wild from a state of domestication) are found in Siberia at lat. 56° N., while, on the other hand, in Arabia, with its intense heat, the horse abounds in the highest state of perfection and with great powers of endurance. There seems to be little doubt that all the varieties of horse, domesticated and feral, existing at the present day, are descended from one common stock, which Darwin, in his very interesting chapter on "Horses and Asses" (*Animals and Plants under Domestication*), has proved to have been probably dun-coloured and more or less striped. In the Neolithic or Polished-Stone Period, wild horses were certainly abundant in Europe, as is evi-

HISTORY.

denced by the discovery of their remains, associated with those of men; and the rude representations of the animal on reindeer horns, &c., found in lake dwellings, tend to prove that the chase of the horse must have been one of the chief occupations of pre-historic man. Bones, and the rough etchings above alluded to, show that the horse of that period must have been a small, heavy animal, with shaggy hair, mane, and tail, and a large head.

That these horses were domesticated or partly domesticated in pre-historic times is proved by their remains having been found in the lake dwellings of Switzerland. It appears probable, however, that the greater number of breeds of horses now existing in Europe are not direct descendants of these European pre-historic races, but are derived from horses imported through Greece and Italy from Asia. These Asiatic ancestors were, on their part, probably the outcome of a still earlier domestication, accompanied by long continued attention to breeding and selection, which resulted in a great improvement in the qualities which render the animal so valuable to man. As these qualities have always been the points towards the improvement of which the breeding of horses has been directed, all existing breeds present some peculiar character, specially fitting them for their surroundings and work. The process of artificial selection has thus resulted in the production of such various types as the English thorough-bred, which may be called a highly perfected galloping machine, the English draught horse, with its enormous power and working capacity, and the little Shetland, or to take an Indian example, the Manipur pony, so admirably fitted by its small size, strength, and powers of endurance, for work in mountainous countries. Horse-breeding, or the artificial selection of stallions and mares having certain qualities or peculiarities which it is desired to perpetuate, together with careful rearing of the foal, is historically of very ancient date. Thus Solomon imported horses for breeding purposes from Egypt at high prices, and the Arabs have for centuries been extremely careful in their breeding operations, and have taken every precaution to keep certain strains of blood pure and unmixed. There is abundant philological evidence that the horse was known to the Aryans before the period of their dispersion, and it appears probable that the Semetic peoples as a whole are indebted for the horse to the lands of Iran. Thus the earliest Egyptian monumental record of the animal dates from about the eighteenth century B.C. the earliest Hebrew reference is not older, and **Hehn** has pointed out that literature affords no trace of the horse in Arabia before the fifth century B.C. The present Indian breeds of horses have probably been derived, for the most part, from Arabian, Persian, and Turkoman ancestors, and present, as in other countries, many distinct races. They are, as a rule, small, lightly built, and possess many of the characters of the Arab. As the horse is never employed in India for heavy draught work, like the dray or farm horse of England, no similar strong, heavy breed exists. The question of the origin of existing races, owing to the want of records and literature on the subject, is one of extreme difficulty. There seems little doubt, however, from the earliest reports we possess that the Arabian horse in this country, as in Europe, has always been the mainstay of breeders. Thus **John van Linschoten**, in the account of his travels in India in the sixteenth century, mentions that, "out of Arabia diverse goodly horses are imported, that are excellent for breeding," and **Abul Fazl**, in a long and accurate description of Akbar's stables written about the end of the same century, writes: "His Majesty is very fond of horses, because he believes them to be of great importance in the three branches of the government, and for expeditions of conquest, and because he sees in them a means of avoiding much inconvenience. Merchants bring to court good horses from 'Iráq i 'Arab and 'Iráq i 'Ajam, from

HORSES.

Breeds of Indian Horses.

HISTORY.

Turkey, Turkestan, Badakhshán, Shirwán, Quirghiz, Tibet, Kashmír, and other countries. Droves after droves arrive from Túrán and Irán, and there are now-a-days twelve thousand in the stables of His Majesty. Skilful experienced men have paid much attention to the breeding of this sensible animal, many of whose habits resemble those of man; and after a short time Hindustan ranked higher in this respect than Arabia, whilst many Indian horses cannot be distinguished from Arabs or from 'Iráqí breed. There are many fine horses bred in every part of the country, but those from Cutch excel, being equal to Arabs. It is said that a long time ago an Arab ship was wrecked and driven to the shore of Cutch; and that it had some choice horses from which, according to the general belief, the breed of that country originated. In the Panjáb horses are bred resembling 'Iráqís, especially between the Indus and the Bahat (Jhelum); they go by the name of *Sanúji*" (or *Satúji*); "so also in the district of Patí Haibatpúr" (near Moonuk), "Bajwárah, Tahárah" (Patiala), "in the Súbah of Agra, Mewát, and in the Súbah of Ajmír where the horses have the name of *Pachwariya*. In the northern mountainous district of Hindustan, a kind of small but strong horse is bred, which is called *gút*; and in the confines of Bengal near Kúçh" (-Behar), "another kind of horses occur, which rank between the *gút* and Turkish horses, and are called *táng'han*; they are strong and powerful." The writer then goes on to eulogise the efficient means taken by the Emperor to secure good breeding, from which and many other passages it is evident that Akbár was not only a keen admirer of horses, but a careful and competent horse-breeder, and it seems certain that the general breed of horses in India must have considerably improved during his reign. In his further detailed description of the various provinces of the Emperor's territory, **Abul Fazl** again comments on the fine breeds produced in Cutch, Patí, Haibatpúr, Bajwarah, Taharah, Agra, Mewát, and Ajmir, and also mentions the horses of Delhi as being good, those of Lahore as resembling 'Iraqis and very fine, those of Kashmír as small, hardy, sure-footed, and cheap, and those of the more eastern Himálaya and Kúch (Behár) as having the same qualities. It would appear, therefore, that, in the sixteenth century except in Cutch and Kúch Behár, all the finest Indian breeds were produced in the Panjáb. **Linschoten**, commenting on the horses of Western India of the same period, remarks, "No man may send any horses in India, but only the Captaine (appointed by the King of Portugal), or such as have authorities from him, whereby he raiseth a great commodity, for that horses in India are worth much money, those that are good are sold in India for four or five hundred pardauwen, and some for seven, eight, yea one thousand pardauwen and more, each pardauwen being accounted as much as a Réekes Dollar, Flemish money." **Abul Fazl**, in a passage relating to the value and classification of horses, writes that they were divided in Akbar's stable into two great classes—*Kháçah* and those that were not *Kháçah*. The former consisted entirely of horses of fine Arab or Persian blood, imported or bred in the imperial stud, while the latter consisted, presumably, of country-breds, and were divided into separate stables according to their value, which seems to have ranged from about ten to thirty mohurs. In another place he states that the price of horses in the country generally varied from R200 to 500 mohurs.

BREED OF HORSES. 416

PRINCIPAL BREEDS AT PRESENT IN INDIA.—There seems to be little doubt that the native breeds of horses have diminished and deteriorated under British rule, a circumstance which the overthrow of the Mahrattas and Sikhs rendered almost unavoidable. Up to the beginning of the nineteenth century, several horse fairs were periodically held in Rájputana, especially those of Bhalotra and Poshkur, to which the horses of Cutch and Káthiáwár, the Lakhi jungle, and Múltán were brought in great numbers,

BREEDS.

Valuable horses were then bred also on the western frontier, on the Lúni, those of Rardurro being especially esteemed. But with the conquests of the British the steeds of Rardurro, Cutch, and Lakhi, became almost extinct, and the horses from the west of India were carried to the Sikhs. The suppression of the predatory system, which had been such a large source of demand, lessened the supply, and the superior Lakhi and Cutch breeds, which for centuries were famous, became almost if not altogether extinct. It is possible, also, that the demands of the British for a larger class of animal than the country naturally produced, led to a system, at least in some cases, of artificial breeding, in which size was the object mainly aimed at, an object which resulted in the deterioration of the original small, hardy breed, and the production of badly-formed and weedy animals. Within recent years, however, the constant endeavours of the Government to improve the breeds of Indian horses, by importing stallions which possess in a marked degree the qualities wanting in the country-bred mare, and by encouraging breeding on the part of the zamindars, have resulted in great improvements in the weight-carrying powers and endurance of the country-bred of the present day.

Punjab. 417

Panjáb.—Horse-breeding, in this, once upon a time the chief producing province of India, has greatly fallen off, owing to several reasons. Thus at the time of the annexation, many of the best brood-mares were withdrawn; extensive demands were made on the province both for horses and mares during the mutiny; and breeding from mares in the ranks, a system always carried on by the Sikhs, has, under British Government, been stopped owing to the necessity of every horse in a regiment being always ready for service. Owing to the last cause many well-bred and tractable mares, which would have been invaluable for brood purposes, have been lost to the province. The best brood-mares are probably those in the Rawal Pindi, Jhelam, Gujerat, Gugaira, and Lahore districts, and many very good animals are also to be found in the frontier districts, such as Bunnu, Kohat, Dera Ismail Khan, and Dera Ghazi Khan. The average Panjábi country-bred is small, but fairly well-bred and possesses great powers of endurance. Breeding operations under the superintendence of Government are now largely carried on in the several districts above mentioned, both for the production of horses and mules. In the Rawal Pindi District alone the *Gazette* (1883-84) reports that there are 3,228 branded brood-mares, of which 1,090 are for horse-breeding, and 2,138 for mule-breeding, also 25 horse stallions and 52 donkey stallions. The mules thus bred in Rawal Pindi are perhaps the best procurable for artillery purposes in India.

The horses of the Jhelam district have long been held in high esteem, especially those of the Dhan. The Gazetteer of the district contains the following interesting paragraph regarding them: "In former days the greater part of the Sikh cavalry was horsed from the Dhanni plains north of the Salt Range, and even now large numbers of remounts are drawn thence by the British Army; but the fall of the rich Sikh chiefs has removed the incentive for breeding large and powerful horses, such as the native gentleman delights in. Although the Dhan is best known for its horse-breeding, yet very good animals are to be found all over the district. Some of them are fast, and nearly all remarkably enduring and able to go over the stoniest ground without shoes. It may be doubted whether the Dhani and Talagong breeds are not deteriorating. Owing to the spread of cultivation and other causes the animals are allowed much less liberty than formerly, and the method of tethering them up is very bad." In 1884 there were eight Government horse stallions, and seven donkey stallions in the district, with 263 brood-mares branded for horse-breeding, and 109 for mule-breeding. The Biloch mares of the Dera Ghazi Khan district are famous for their endurance, some also being very handsome, and the class

BREEDS. Punjab.

of remount obtained out of them by Government stallions promises to be a very good one. There were twenty-two of the latter in the district in 1884, and nine stallion donkeys.

Bombay. 418

Bombay.—As above mentioned in recounting the history of the horse in India, the horses of CUTCH were at one time perhaps the best bred in the country. With the increased ease of importing horses from Arabia, Persia, and Australia, however, along with the reasons influencing horse-breeding incident on the British successes over the Mahrattas and Sikhs, the value of the breed has much declined. The Cutch horse is generally a little over 14 hands, well made, spirited, showy in action with clean limbs and good bone, thin long neck, large head, outstanding ram-like brow, and small ears (*Cutch Gazetteer*). His chief defects are the length of his cannon bone, his ugly heavy brow, and bad temper. As in other districts attempts have been made by the introduction of Government stallions to improve the breed, but up to date seemingly with indifferent success.

The Káthiawár breed was also a famous one, for many centuries, and is closely allied to the former; indeed, the Persian name applied to the race, *viz.*, *Kachhí* (Cutchí) would imply a common origin. In former days it was the custom of the *Subahdárs* of Ahmadabad and the *Faujdárs* of Janágad to send presents of horses of this breed to the Emperor. The best horses of the true Kathí or Káthiawár type are found even to the present day on the shores of the Ran, and since in former times, the Arabs imported for the imperial stables were landed at Verával, Somnáth-Patan, Diu or Surat, it appears probable that the superiority of the breed is due to an Arab strain then introduced, the Káthís having availed themselves of the services of these Arab stallions on their way through the province. The breed, however, like many others of the old Indian country-bred, has apparently fallen off greatly during the past century. The best horse-breeding region in the district is said, in the Gazetteer, to be Panchál, in the heart of the province, containing as it does every requirement essential to successful breeding, *viz.*, favourable soil for the formation of the feet, hilly ground for the development of muscle, plenty of good fodder and running water, and a hot, dry climate. The author of the Gazetteer further remarks: "In spite of neglect, confining the young stock, and want of care in the choice of stallions, handsome specimens of the Káthí breed may still be found. The peculiarities of the animal are that it is generally undersized and small-boned; its distinctive marks are a black cross down the back and black bars on the legs, the colour of the coat varying through every shade of dun. A well-bred Káthi is teachable and honest, free from vice, full of spirit, and wonderfully lasting and hardy." Mares are most esteemed by native breeders, as the horses are noted screamers, and are consequently not liked by the purchaser. The endeavours of Government to improve the breed have not as yet met with marked success, owing to the fact that the Káthís have shown themselves averse to showing their best mares, and will very rarely sell a first-rate animal. The animal varies greatly in price, as much as R3,000 being still paid, according to the Gazetteer, for a horse of the best blood, while a serviceable, sound, and good-looking nag of from 14 to 14½ hands may be secured for from R100 to R300. Considerable numbers of Káthí horses are kept in Ahmedabad, and as that district is one of the best for horse-breeding in the Presidency, it has attracted the special attention of Government, and appears likely to furnish good remounts.

The Baimthadi or Deccan horse is the remaining special breed of country-bred in Bombay, and probably, like the others, owes much to foreign blood, arising from Arabian, Persian, and Turki sires. The finest specimens of the breed are reared on the banks of the Bhíma and Uíra in Poona and Ahmadnagar, but, as in the case of other districts of Bombay,

BREEDS. Bombay.

well-bred animals are neither so common, nor is the general breed so good as in former times. The well-bred Deccan horse is of middle size (rarely over 14·2 hands high), strong, rather handsome, generally of a dark bay colour with black legs, and has the fine limbs, broad forehead, and much of the docility and endurance of the Arab. The great mart for this animal is Malligaum, about 25 miles from Ganga Kheir, on the Godavery. This breed is certainly one of the best in India, and is deservedly held in high esteem, a regard which is fully warranted, when the long marches of the Mahratta and Pindari horsemen, mounted on Deccan and Káthí horses, are remembered. Government, consequently, as in other places, is at present making great endeavours to encourage the development of the breed. The ponies of the same district are also much valued on account of their strength, endurance, and frequently surprising speed. The Deccan pony was much more bred, however, 30 years ago before the days of railways when he was of great value for carrying mails, than now. They are now principally in demand as hacks and *tongá* ponies, and a good one will fetch R150 to R200. The Deccani pony is thick-set, short-legged, and hardy, varies from 12 to 13½ hands or a little more in height, and is generally bay, brown, or chesnut, seldom grey, and still less often dun. Of all the Deccan breed the best pony is the Dhangar or Khilári. They were generally thought to be a special breed, but the general belief of authorities in the present day is that their superior excellence is due to the Dhangar's practice of castrating them. A small but hardy breed of ponies is also raised by the Thiláris, a tribe of wandering herdsmen chiefly inhabiting the west of Khándesh. Dhárwár in the Karnátak was also once famous for its breed of ponies running up to 14 hands high, but this has much deteriorated within recent times. The breed is now small, under 13 hands, often ill-shaped and vicious, but hardy. All these ponies of the Deccan and Karnátak were much employed in the Persian campaign of 1856-57, the Abyssinian campaign of 1867-68, and the Afghan campaign of 1879-80, and on each occasion their numbers suffered considerable diminution.

Himalayan Regions. 419

Ponies of the Himálayan Regions.—Several distinct breeds occur in the North of India and adjoining territories, all of which are valuable owing to their powers of endurance, their weight-carrying capabilities and their sure-footedness. The *Ghunt* or *Khund* is a breed of the districts adjoining Lahoul and Spiti, employed almost entirely for saddle purposes. It is of two varieties, one small, never above 12 hands high, the other 13 to 13½ hands, which is imported from China and is more highly esteemed than the small breed peculiar to the country. The *Ghunt* is strongly made, exceptionally sure-footed and hardy, but is said to be very hard-mouthed and frequently unmanageable. It is, however, specially valuable in hilly regions, owing to its extreme hardiness; indeed, it requires little if any care. The ponies, except the yearlings, which are housed, are kept out all winter, and live on the roots of stunted bushes or any other nourishment they can pick up after scraping away the overlying snow.

Ghunt. 420

Yarkand. 421

The YARKAND pony is a breed that is raised in great numbers all over its native country, and is distinct from, though it very much resembles, the Yábú of Afghánistan. Its chief points are short or medium height, round barrel and deep chest, with full quarters and thick limbs, and its general appearance is that of a miniature English cart-horse. Unlike the *ghunt* it is an admirable pack-horse, and is inured to carrying heavy loads at a peculiar jog-amble of about five miles an hour. When not urged beyond its natural pace it has great powers of endurance. In the vernacular of the country this pony is known as *topichác* or "*roadster*," in distinction to the *arghumác* or "thorough-bred," the latter of which are Turkoman, or Andijan horses, in the possession of the chiefs or more

BREEDS. Himalayan Regions.

wealthy inhabitants. The pony in Yarkand, unlike that in India, is not only employed under the saddle or pack, but is also used in the plough, and in harness. The natives have distinctive names for the several colours of pony, and have preferences for certain sorts, "*tunk*" or chesnut being considered the best and most hardy. At the time of the Mission in 1873, the price varied from R40 to R150.

Kashmir. 422

The KASHMIR pony has long been known and appreciated for pack and riding purposes in the hills. **Drew** mentions those of Baltistán as being particularly good, writing—"They stand about 12·3 or 13 hands high; for their size they are rather large boned; they are compact in make; they have a broad chest, a deep shoulder, a well-formed barrel well ribbed up, and good hind quarters, and a small, well-shaped head. They are good at hill climbing and at polo they are very active; they are of good heart, going long without giving in."

Bhutia. 423

The BHUTIA pony is a strong-boned, powerful, and very useful, hardy breed, largely employed as a hill pony in North-Eastern India. It is larger than those of the more Western Himálaya, reaching a height of 14 hands.

Burma. 424

BURMA.—The ponies of Burma, which are small, hardy, and exceedingly tractable, are said to be all importations from the Shan States, from whence large numbers are exported to Burma every year. They are, however, generally known as Pegu ponies, presumably because they are exported thence to India. In Burma they are generally employed for drawing carriages, which, however, are only used in towns with roads. Their characteristic pace is an unbroken run, in which the shoulders seem to roll from side to side. A few of the officials and richer people, ride, and the common people are all fond of pony-racing, not because they ride, or are fond of horses, but because a race meeting affords them an opportunity of indulging their inveterate gambling propensities. About 100 are exported to India and the Straits yearly, where they are valued for their wonderful weight-carrying and enduring powers, and make good harness ponies. Burman stallions have been introduced into Ahmadnagar and Poona with the view of improving the breed of Deccan ponies.

Manipur. 425

The MANIPUR pony is generally considered the best of all Indian breeds under 13 hands. It is a small, mostly dun-coloured, animal, rarely exceeding 12 hands in height, but possessed of wonderful powers of endurance and weight-carrying capabilities. It is probable, however, that the breed has deteriorated since the beginning of this century, at which time the Burmans, during their invasion of the country, made every endeavour to exterminate the ponies which gave such powerful assistance in the hill warfare of the Manipuris. The Manipuris who, at that time took refuge in Kachar (many of whom have since settled there), withdrew with them, in their flight, as many of the ponies as possible. From that stock has been derived the Manipuri pony as met with in India, but the Cachar Manipuri is a very much inferior animal to the more carefully bred pony in Manipur itself. In Manipur the strictest system of protection against inter-breeding is followed. The Maharaja allows no pony to leave the State, and appropriates the best colts and fillies for his own stables, hence the produce of his stud is extremely good. The ponies of the general inhabitants, however, are allowed to roam the country in flocks, are very much left to themselves, and uncared for, and **Dr. Brown**, in the annual report for 1872, lamented the fact that the general breed was deteriorating. In 1839 Government sent one Arab stallion and eight stud-bred mares to the then regent, Nur Sing, but they and their descendants have completely died out, and seem to have left no trace of their strain in the present breed. Before the Burman invasion at the beginning of this century, the

BREEDS.

traffic in ponies between Manipur and that country formed a large and important trade, so it is probable that the Burman and Manipur breeds are closely allied, though now in many points dissimilar.

Foreign. 426

FOREIGN BREEDS.—Besides these different breeds of Indian horses and ponies, many foreign breeds are yearly imported in large quantities, of which the most important are Australians, New Zealand horses, Arabs, Persians, Gulf Arabs (out of Persian mares by Arab horses), Turkomans, Cape horses, and Afghâns or Kábulis. A few English horses are also annually imported for stud purposes, principally thorough-breds, and Norfolk Trotters, the latter of which are extremely popular with native breeders. As an outcome of the large and general importation of these different breeds, the foal at present reared by the zamindar contains a mixture of many bloods, with the natural consequence that the old characteristic breeds of various parts of the country are dying out. For accounts of the results of this cross-breeding, and the relative values of the English thorough-bred, Norfolk Trotter, Arab, or Australian as a sire, the reader is referred to the numerous and valuable Government Reports on horse-breeding.

ASSES. 427

Domestic Asses exist in all parts of India, and are largely employed as beasts of burden, especially by the *Dhobis* or washermen, potters, and tinkers. As a rule the Indian donkey is badly cared for, and absolutely no attention is paid to its breeding, with the natural consequence that it is small and unfit for saddle work. It is, however, cheap (costing from R15 to R50), hardy, and readily picks up a sustenance from the most unlikely ground, and consequently forms a valuable pack animal for the poorer classes of itinerary merchants, such as those above enumerated. A specially fine breed occurs in Kathiáwár, of which the Hálár or Jhálávád white variety is one of the strongest and largest. The Bhutias, who come down from the higher Himálaya to seek employment during winter at Simla and other hill stations, bring with them a very small, dark or almost black donkey, with long shaggy hair. These appear to be useful animals for load-carrying, but as they are not much larger than a good sized goat, they can only carry small loads.

The donkey stallions employed by the Government in mule-breeding are not country-bred, but of the much more valuable Arab, Persian, Italian, and Spanish breeds, which may cost as much as R300.

MULE-BREEDING. 428

Mule-breeding has been very extensively carried out by the Government, since large numbers are required for use with transport and mountain artillery. Considerable difficulty is, however, experienced in many parts of the country in inducing the zamindar to employ his mares for mule-breeding purposes, owing to the fact that rearing mules is considered a degrading occupation. With good prices and every encouragement from those in authority, this objection is, however, beginning to lose its force, and large operations are now conducted, especially in the Rawalpindi and Jhelam districts of the Panjáb. An interesting paper on mules was recently read by Mr. J. A. Steel, A.V.D., before the Bombay Natural History Society, in which he contended that mules should be obtained for Government by importation from abroad rather than by mule-breeding in districts where valuable races of horses are bred, but that it might be encouraged in places such as Madras, Bengal, and Burma, which supply no horses and few ponies. Of late years a considerable number of mules have from time to time been imported by the Government from Persia, the supply from India not having proved sufficient for the demand. This demand has steadily increased, and will in all probability continue to do so, owing to the superiority of the mule as a strong, hardy, surefooted, and healthy baggage animal, suited for transport purposes in almost any description of country.

HORSE-FOOD. 429 **Food and Fodder of Horses, &c.**—See articles on **Fodder**, Vol. III., 407-487, **Gram**, II., 274, **Madras Horse Gram**, III., 175, **Barley**, p. 77; also **Oats**, **Maize**, and **Fodder** in Appendix.

HIDE. 430 **Hide.**—For horse hides and their special qualities see articles on **Hides**, also **Leather.**

HAIR. 431 **Hair**—Is apparently little employed in India, nor does the Indian article form part of the commercial supply in Europe. The hair employed in trade is derived from the mane and tail only. In Europe the long hair is used chiefly in making hair-cloth for purposes of upholstery, &c., also for the manufacture of straining-bags and cloths, seives, plumes, wigs, fishing-lines, and ropes. The short hair is curled for stuffing seats, mattresses, &c., while medium coarse hair is sometimes used in brushes (*Spons' Encyclopædia*). The market price varies from *4d.* to *4s.* a pound in England, depending on the cleanness, quality, and length of the hair. Horse-hair is employed by the natives of parts of India in ornamental embroidery of leather work, and for the manufacture of the commoner sorts of fly flapper or "*chauhrí*."

FOOD. Flesh. 432 **Food.**—The FLESH of the horse is largely utilised as food by many castes in various parts of the country, and is considered specially good by the hill tribes of Northern India.

MEDICINE. Milk. 433 **Medicine.**—The MILK both of mares and she-asses is recommended in Sanskrit literature, as slightly acidulous, easily digested, and specially suitable for the dietary of dyspeptics and convalescents. **U. C. Dutt**, in his *Hindu Materia Medica*, also mentions the URINE of the horse as being employed medicinally by Hindu practitioners, but does not describe its supposed virtues.

Urine. 434

DOMESTIC & SACRED USES. 435 **Domestic & Sacred Uses.**—Horse sacrifice is considered one of the most efficacious of propitiatory offerings in Hindu religion. Thus in the ordinances of Manu it is laid down as one of the methods of purification on the part of one of the Hindu castes, after unintentionally slaying a Brahman. Ancient Sanskrit literature relates that the gods killed a man for their victim, but from the man the part fit for sacrifice went out and entered a horse, after which the horse became an animal fit for sacrifice. In like manner the ox, sheep, and goat in turn became sacrificial animals, and according to Hindu belief the specially sacrificial part remained longest in the last. **Sir Monier Williams**, commenting on this subject, writes: "The *Asva medha* or 'horse sacrifice' was a very ancient ceremony, hymns 162 and 163 in Mandala I of the *Rigveda* being used at the rite. It was regarded as the chief of all animal sacrifices, and in later times its efficacy was so exaggerated that a hundred horse-sacrifices entitled the sacrificer to displace Indra from the dominion of heaven."

TRADE. 436 Imports. **Trade.**—A large and important import of horses takes place annually, principally of Australians (called "Walers"), Persians, and Gulf Arabs, and Cape Horses, from which the remounts for British Cavalry, and Horse and Field Artillery are derived. The number of animals thus brought into the country has steadily increased up to 1887-88, in which year there was a slight falling-off. Thus the average number imported during the five years ending 1882-83 was 4,224, value R14,90,458, while in the period ending 1887-88 the average was 6,202, valued at R23,14,148. The countries which form the source of supply are Australia, Persia, Turkey in Asia, Cape Colony, and the United Kingdom, and to a very small extent Ceylon, South America, Arabia, Austria, Italy, and Aden. As an example of the relative proportions contributed by each country, the figures for 1887-88 may be here quoted :— TRADE.

	Number.	Value.
		R
Australia	3,924	14,51,300
Persia	1,589	5,32,600
Turkey in Asia	946	3,35,400
Cape Colony	150	30,000
United Kingdom	33	1,27,520
Ceylon	10	3,135
South America	6	3,500
Arabia	5	1,250
Austria	5	2,000
Italy	2	5,000
Aden	1	700
TOTAL	6,671	24,92,405

Of the total number, Bengal imported 2,591, value R10,41,500; Bombay 2,705, value R10,31,270; Sind 28, value R13,000; Madras 1,346, value R4,06,235; Burma 1, value R400.

The re-exports are small and unimportant, the average during the same period having been 98 horses, value R36,801; a considerable falling-off from the average for the preceding five years, which was 158 horses, value R52,784. The countries receiving these re-exports were the United Kingdom, Ceylon, Straits Settlements, Zanzibar, and Australia, the former country importing those of greatest value. Thus in 1887-88, though the total export of foreign horses to the United Kingdom was only 10, the value was R17,000, or an average of R1,700 per horse. The four chief ports of India all take part in this re-export trade, Calcutta having the chief share. Re-exports.

While the exportation of foreign horses shows a falling off during the past five years, that of country-breds, though small, has markedly increased during the same period. Thus the average for the period under consideration was 670 horses, value R1,14,442 in comparison with 362 horses, value R51,858 for the five years ending 1882-83. Burma appears to have the largest share of this export trade, having in 1887-88 shipped 510 horses, value R88,131, out of a total of 770, value R1,43,924. In the same year Calcutta exported 207 horses, value R52,673; Madras 51, value R2,870; and Bombay 2, value R250. The chief importing countries in the same year were the Straits Settlements with 717, value R1,38,834; Ceylon with 43, value R2,540, and China with 1, value R1,000, while other countries imported an aggregate of 9, value R1,550. Exports.

It would seem, from these statistics, that the only demand for country-breds is for low priced, and probably small, ponies, a demand which is principally met by the supply from Burma.

Horse chesnut, see **Æsculus Hippocastanum,** *Linn.*; Vol. I., 127.

Horse gram, see **Dolichos biflorus,** *Linn.*; Vol. III., 175.

Horse-radish, or **Cochlearia Armoracia,** *Linn.*; CRUCIFERÆ. 437

A perennial plant, common in moist places in Europe, and occasionally grown as a vegetable in India. Its root is officinal in the Indian Pharmacopœia, and is imported for use. Internally it is stimulant, sudorific, and diuretic; externally applied, irritant and vesicant; when chewed it is sialagogue. It is prescribed for atonic, dropsical, and rheumatic affections, scurvy, &c., internally and locally, chewed in substance, for toothache, and as a stimulant embrocation or cataplasm in paralysis, &c. (*Indian Pharmacopœia, 23*).

Horse-radish Tree, see **Moringa pterygosperma,** *Gærtn.;* MORINGEÆ; [Vol. V.

Horse-tails, see **Equisetum,** VOL. III., 252.

438 **House-Building.**—The following list contains the chief timber trees employed in India for this purpose. For a description of their respective properties and uses, &c., the reader is referred to the separate articles on each in their alphabetical positions in this work :—

439 **House-Building, Timbers used for—**

Abies Webbiana.
Acacia Catechu.
A. dealbata.
A. ferruginea.
Acrocarpus fraxinifolius.
Adenanthera pavonina.
Adina sessilifolia.
Æsculus indica.
Afzelia bijuga.
Albizzia amara.
A. Lebbek.
A. procera.
A. stipulata.
Alseodaphne sp.
Altingia excelsa.
Amoora cucullata.
Anogeissus acuminata.
Anthocephalus Cadamba.
Artocarpus hirsuta.
Averrhoa Carambola.
Bambusa Balcooa.
Barringtonia racemosa.
Bassia latifolia.
Bauhinia purpurea.
Beilschmiedia Roxburghiana.
Betula Bhojpattra.
Bischofia javanica.
Briedelia montana.
B. retusa.
Bruguiera gymnorhiza.
Buchanania latifolia.
Bucklandia populnea.
Calophyllum spectabile.
C. tomentosum.
Capparis aphylla.
Carapa moluccensis.
Careya arborea.
Casearia glomerata.
Cassia timoriensis.
Castanopsis rufescens.
Cedrela Toona.
Cedrus Libani, var. **Deodara.**
Ceriops Candolleana.
Chætocarpus castaneæcarpus.
Chrysophyllum Roxburghii.
Cinnamomum sp.
Cinnamomum sp. (C. ? Parthenoxylon. *Meissn.*).
Cocos nucifera.
Cordia Rothii.
Cratoxylon neriifolium.
Cupressus torulosa.
Cynometra ramiflora.
Dalbergia lanceolaria.
D. paniculata.
D. Sissoo.
Daphnidium elongatum.
Dendrocalamus strictus.
Dillenia indica.
D. pentagyna.
D. retusa.
Diospyros ehretioides.
D. melanoxylon.
Dipterocarpus alatus.
D. tuberculatus.
D. turbinatus.
D. zeylanicus.
Dolichandrone falcata.
D. stipulata.
Doona zeylanica.
Ehretia lævis.
E. Wallichiana.
Elæocarpus lanceæfolius.
Engelhardtia spicata.
Eucalyptus Globulus.
Eugenia Jambolana.
E. operculata, *Roxb.;* var. operculata, proper.
E. tetragona.
Fagræa fragrans.
F. racemosa.
Feronia Elephantum.
Ficus bengalensis.
F. retusa.
Filicium decipiens.
Garcinia speciosa.
Garuga pinnata.
Gluta elegans.
Gmelina arborea.
Hardwickia binata.
H. pinnata.
Heritiera littoralis.

HOUSE-BUILDING TIMBERS.

Heritiera Papilio.
Hopea odorata.
H. parviflora.
Ixora parviflora.
Juglans regia.
Juniperus excelsa.
Kydia calycina.
Lagerstræmia Flos-Reginæ.
L. hypoleuca.
L. parviflora.
Litsœa zeylanica.
Lophopetalum Wightianum.
Machilus odoratissima.
Mæsa montana.
Mangifera indica.
Marlea begoniæfolia.
Melanarrhœa usitata.
Melia dubia.
Melocanna bambusoides.
Mesua ferræa.
Michelia Champaca.
M. excelsa.
Mimusops Elengi.
M. indica.
M. littoralis.
Myristica malabarica.
Nyssa sessiliflora.
Ougeinia dalbergioides.
Phœbe attenuata.
Phœnix sylvestris.
Phyllanthus Emblica.
Pinus excelsa.
P. Khasya.
P. longifolia.
Populus euphratica.
Premna longifolia.
Prosopis spicigera.
Pterocarpus Marsupium.
P. santalinus.
Pterospermum suberifolium.
Quercus annulata.
Quercus dilatata.
Q. fenestrata.
Q. Griffithii.
Q. incana.
Q. lamellosa.
Q. lanceæfolia.
Q. semecarpifolia.
Q serrata.
Q. spicata.
Rhododendron arboreum.
Salix daphnoides.
Salvadora oleoides.
Sapindus emarginatus.
Schima Wallichii.
Shorea obtusa.
S. robusta.
S. siamensis.
Shorea Talura.
S. Tumbuggaia.
Sonneratia apetala.
Soymida febrifuga.
Stephegyne parviflora.
S. sp.
Stereospermum chelonoides.
S. suaveolens.
Strychnos potatorum.
Symplocos lucida.
Tectona grandis.
Terminalia Arjuna.
T. belerica.
T. Chebula.
T. citrina.
T. myriocarpa.
T. tomentosa.
Ulmus integrifolia.
Viburnum erubescens.
Vitex alata.
V. altissima.
Wendlandia exserta.
Wormia triquetra.

HOVENIA, *Thunb.; Gen. Pl., I., 378.*

A genus of the RHAMNEÆ which has only one species, distributed over China, Japan, and the Himálaya.

440 **Hovenia dulcis,** *Thunb.; Fl. Br. Ind., I., 640; Lamk., Ill., t. 131;* THE CORAL TREE. [RHAMNEÆ.

Syn.—H. ACERBA, *Lindl.*; H. INÆQUALIS, *DC.*

Vern.—*Sicka*, HIND.; *Chamhún*, PB.

References.—*Roxb., Fl. Ind., Ed. C.B.C., 211; Brandis, For. Fl., 94; Gamble, Man. Timb., 88; Royle, Ill. Him. Bot., 123; Atkinson, Him. Dist., 307, 717; Atkinson, Ec. Prod. of N.-W. P., Pt. V., 55; Balfour, Cyclop., II., 116; Treasury of Bot., I., 599.*

Habitat.—A small unarmed tree, about 30 feet in height, with a straight trunk, and broad rounded head, found in the subtropical Himálaya from

Chamba and Hazára to Bhotán, at altitudes of from 3,000 to 6,500 feet. It is also frequently cultivated. The tree flowers in April-May, and produces a ripe fruit in July.

FOOD. **Food.**—The fruit consists of a capsule with three seeds, resting on an enlarged, arched pedicle, the size of a pea, which is soft, fleshy, and contain a sweet juice. This PEDICLE is edible, having a pleasant flavour like that of a Bergamot pear.

Pedicle. 441

HOYA, *Br.; Gen. Pl., II.*, 776.

A genus of twining, pendulous, or rambling and rooting, rarely erect, shrubs, comprising about 60 species, natives of tropical Asia, the Malay Peninsula, and Australia. Of these about 40 are indigenous to India, but few are of economic interest. Several of the species, however, are cultivated on account of the beauty of their flowers, under the popular name of WAX-PLANTS.

442 **Hoya pendula,** *Fl. Br. Ind., IV., 61; Wt., Ic., t. 474;* ASCLEPIADEÆ.

Syn.—ASCLEPIAS PENDULA, *Roxb.*

Vern.—*Dodi, hiran dori,* BOMB.; *Nasjera patsja,* MALAY.

References.—*Roxb., Fl. Ind., Ed. C B.C., 253; Dalz. & Gibs., Bomb. Fl., 152; Lisboa, U. Pl. Bomb., 233; Balfour, Cyclop., II., 116.*

FIBRE. 443 **Fibre.**—The plant is said by Lisboa to yield a useful fibre.

H. viridiflora, *R. Br.*, see **Dregea volubilis,** *Benth.*; Vol. III., 193.

HUGONIA, *Linn.; Gen. Pl., I.*, 243, 987.

A genus of climbing shrubs of the Natural Order LINEÆ, comprising about SIX species, natives of Tropical Asia, Africa, and Australia.

444 **Hugonia Mystax,** *Linn.; Fl. Br. Ind., I., 413; Wight, Ill., t. 32;* [LINEÆ.

Syn.—HUGONIA OBOVATA, *Ham.*

Vern.—*Agúre,* TAM.; *Gatrinta, tivoa potike, vendapa, káki bíra,* TEL.; *Modera canni,* MALAY.; *Búgatteya,* SING.

References.—*Dalz. & Gibs., Bomb. Fl., 17; Rheede, Hort. Mal., II., t. 19; Elliot, Fl. Andhr., 58, 182, 193; Drury, U. Pl., 249; Balfour, Cyclop., II., 118.*

Habitat.—A rambling leafy shrub, with handsome, golden yellow flowers, found in the Western Peninsula from the Konkan to Travancore, and in Ceylon.

MEDICINE. Roots. 445 **Medicine.**—The bruised ROOTS are employed externally in reducing inflammatory swellings, and as an antidote to snake-bites. In the form of a powder it is administered internally as an anthelmintic and febrifuge.

Bark. 446 The BARK OF THE ROOT is also employed as an antidote to poisons.

HUMBOLDTIA, *Vahl.; Gen. Pl., I.*, 579.

A genus of small unarmed erect trees, of the Natural Order LEGUMINOSÆ, comprising four Indian species, natives of Southern India and Ceylon.

Humboldtia unijuga, *Bedd.; Fl. Br. Ind., II., 274;* LEGUMINOSÆ.

References.—*Gamble, Man. Timb., 135; Beddome, Fl. Sylv., t. 183.*

Habitat.—A large tree of the Travancore Mountains, found at altitudes from 3,000 to 4,500 feet.

TIMBER. 447 **Structure of the Wood.**—Hard and durable (*Beddome*).

HUMULUS, *Linn.; Gen. Pl., III.*, 356.

A genus of perennial, twining, scabrid herbs, of the Natural Order URTICACEÆ, comprising two species, one Chinese and Japanese, the other the much cultivated Hop plant.

448 **Humulus Lupulus,** *Linn.; Fl. Br. Ind., V., 487; Bent. & Trim.,* THE HOP. [*Med. Pl.*, 230; URTICACEÆ.

References.—*Stewart, Pb. Pl., 216; DC., Origin Cult. Pl., 162; Aitchison, Botany of Afgh. Del. Comm., 109; Pharm. Ind., 214; Ainslie, Mat. Ind., I., xxii.; O'Shaughnessy, Beng. Dispens., 578; Flück. & Hanb., Pharmacog., 551; U. S. Dispens., 15th Ed., 629, 744, 905; Year-Book Pharm., 1874, 629; 1878, 223; 1879, 466; Baden Powell, Pb. Pr., 247; Atkinson, Him. Dist., 707; Royle, Prod. Res., 427, 428; Liotard, Note on Hop Culture in India, 1883; Balfour, Cyclop., II., 101; Smith, Dict., 215; Treasury of Bot., I., 601; Morton, Cyclop. Agri., II., 42; Kew Off. Guide to the Mus. of Ec. Bot., 120; Journ. Agri. Hort. Soc., I., 204; Transactions, V., 46; Quartly. Journ. of Agri., V. (1834-35), 519; XI. (1840), 41, 122; Madras Board of Rev. Procgs., June 1st, 1889, No. 266, 7; Nilgiri Bot. Gardens Rept., 1885-86; Chamba Administr. Rept., 1883-84; Special Reports:—Conservator of Forests, Panjáb, 1889; Dir. of Land. Rec. and Agri., N.-W. P. and Oudh; J. H. Lace, Esq., Quetta.*

Habitat.—A perennial, twining, scabrid herb, much cultivated in many parts of the world. A considerable amount of difference of opinion regarding the origin and present indigenous occurrence of the hop appears to exist amongst botanical writers. Thus in the *Flora of British India* it is described as native of North America, and perhaps of Northern Asia, while other authors agree in its being also indigenous to Europe, and certainly to Northern Asia. **DeCandolle** writes: "The hop is wild in Europe from England to Sweden as far south as the mountains of the Mediterranean basin, and in Asia as far as Damascus, the south of the Caspian Sea, and of Eastern Siberia, but it is not found in India, the North of China, or the basin of the river Amur." He also mentions in a footnote, and apparently concurs in, the statement of **Maximowicz** that the plant is indigenous also in the east of the United States, and the Island of Yeso. The learned French writer, in support of his theory against the plant being an introduction into Europe from Asia, points out that its names in the different languages of Europe cannot possibly have been derived from one root, a circumstance which tends to confirm the idea that the hop existed in Europe before the arrival of the Aryan nations. The general consensus of opinion appears to be in favour of **DeCandolle's** view. With reference to the statement of the *Flora of British India* that it is perhaps indigenous to Northern Asia, it may be remarked, that the plant was found by **Aitchison** while with the Afghán Delimitation Commission, forming, with other shrubs, impenetrable hedges over the whole country between the foot of the mountains at Asterabad and the Caspian Sea.

History.—It would seem, from the evidences afforded by philology, that the Hop plant was known and employed for some purpose long before it was used in brewing, but on this subject ancient authors throw very little light. **DeCandolle** remarks that, judging from the Italian name *lupulo*, however, it would appear probable that the plant was one of the vegetables mentioned by **Pliny** under the name *lapus salictarius.* Hop gardens are first mentioned in an act of donation made by Pepin, father of Charlemagne in 768. In the eleventh century Bohemian and Bavarian hops became famous, and in 1069 William the Conqueror is said to have made a grant of hops and hop lands in the county of Salop. **Flückiger** and **Hanbury**, in their interesting account of the history of the plant, write: "As to the use made of hops in these early times it would appear that they were regarded in somewhat of a medicinal aspect. In the *Herbarium of Apuleius*, an English manuscript written about A.D. 1050, it is said of the hop (*hymele*) that its good qualities are such that men put it in their usual drinks; and **St. Hildegard**, a century later, states that the hop (*hoppho*) is added to beverages, partly for its wholesomeness and bitterness, and partly because it makes them keep." As in the case of several other important plants, the use of the hop appears to have come under legal restric

HISTORY. 449

HISTORY.

tions as it began to be largely employed. Thus in the reign of Henry VI. (1425-26) an information was laid against a person for putting into beer "an unwholesome weed called an hopp," and in the same year Parliament was petitioned against "that wicked weed called hops" (*Flückiger and Hanbury*). Notwithstanding these restrictions and regulations, the degree of utilisation of the plant increased, till in the reign of Edward VI., its cultivation was directly sanctioned by an Act of Parliament. Since that time it has become more and more regarded as an essential constituent of beer, and the cultivation of the plant has steadily increased in extent and importance, till in 1873 the area under the crop in Kent alone was 39,040 acres.

CULTIVATION. 450

Cultivation.—As already mentioned, the hop is not indigenous in India, and its cultivation is still only experimental. The first trial appears to have been made at Dehra Dún, with plants obtained by Lord Auckland, presumably about the year 1840. Since then it has been attempted in the Upper Chenab region of the Panjáb, Kashmír, and many other places with more or less success. Owing very probably to ignorance in selecting the best localities for the growth of the plant, and partly also to want of knowledge of the best methods of cultivation, the attempts made in many parts have not been pecuniarily successful. Thus **Dr. Stewart** writes: "In 1851 **Lowther** proposed its introduction into Kashmír. It has been successfully cultivated in Dehra Dún for many years, so far as mere growth is concerned, but heavy rain at the flowering period prevents the flower from reaching perfection. As to the quantity and quality of the powder on which its value depends, the results have on the whole been unsatisfactory. Within the last few years the plant has been tried at Kyelang and Kilar in the arid tracts of the Upper Chenab, and it has flourished. But, unfortunately, it has been found out, after several years' care, that the sets, introduced at the latter place, were those of male plants, so that the experiment has still to get a fair trial there!" Up to the present time the efforts to cultivate hops appear to have been accompanied by the greatest degree of success in Kashmír (which in 1883 exported 15,000℔. to one of the Himálayan breweries), and in Chamba, the Administration Report of which for 1883-84 contains the information that the cultivation of hops, recently tried as an experiment, had proved successful, the crop raised having been of such excellent quality as to obtain a medal at the Amsterdam Exhibition; and in 1884-85 it is further reported that efforts continued to be rewarded by success, 63 maunds having been sold to the Murree Brewery Company for R2,770. **Mr. Liotard** in 1883 wrote a note on the subject, pointing out that owing to the rapidly increasing brewing industry in India and the high price of imported hops, the cultivation of the crop offered an admirable field for enterprise. The following extracts from his report show the points in which Indian-grown hops were found to be deficient, and indicate the localities in which the cultivation ought to prove most successful:—

"In October last, small samples of hops received from the Chamba State (Panjáb), were distributed to some of the Himálayan breweries for opinion.

"The opinions received vary somewhat as to the quality of the samples, but they lead to the following conclusion:—

"Some of the samples were over-ripe and discoloured; others were of good quality and aroma, worth about £15 per cwt., and quite suitable for brewing ordinary beer. The flower was apparently of healthy growth on the whole, but was poor in pollen, owing probably to its having either been gathered over-ripe or to its having suffered by damp or cold winds. The general conclusion is that, if the samples had been gathered properly, they would have answered very well. One brewer wishes to have a few cwts.

CULTIVATION.

at £15 per cwt, in order to make a separate brew with them; and others ask for quantities larger than those sent. The suggestion is that samples should be procured from the vines directly the burr turns a fair straw colour (if left until brown they are spoilt); and they should be lightly packed in cases or bags, the drying to be done within six hours of picking.

"As to future action it is said that if hops, of such quality as the samples must have been before they were spoilt, can be produced in any quantity in the Panjáb or in districts in the neighbourhood of breweries, there is no doubt that they would handsomely pay to cultivate.

"Experience has, however, proved the inutility of attempting to grow hops in any part of the Himálayas visited by heavy rainfall at a time of the year when the pollen is exposed to be washed off. But in Himálayan tracts which escape the violence of the summer monsoon there appears to be a very good chance of success. The question whether any tracts in the Nilghiris are suited to hop-growing is one which is under enquiry.

"There does not appear to be any reason why India should not produce hops for its own breweries in those Provinces in which suitable localities can be found, and European settlers in the hills might be induced to take up the subject. That growers would find a ready market for their hops is sufficiently shown by the large and increasing demand."

Since the date of **Mr. Liotard's** paper, enquiries have been made regarding the success or otherwise of hop cultivation in the various regions in which it has been attempted; in answer to which the following replies have been received.

1. *From the Conservator of Forests, Panjáb.*—"The cultivation of Hops was tried by **Colonel Rennick** in Bajaura, but was abandoned on account of the cost of hop poles. The Murree Brewery also made an attempt near Abbottabad, which was apparently successful; but was not continued. Hops were also introduced into Kilar in Chamba by **Mr. Elles**, where there are now some five acres under white line cultivation; the gardens chiefly belong to the Raja, though some *zamindars* at Re and Pontu have grown them remuneratively: the chief market is at Dalhousie."

2. *From the Director of Land Records and Agriculture, North-West Provinces and Oudh.*—"Hops have not been grown with any commercial success in any part of these provinces. There are certain tracts in the Raja of Tehri's territory and in the villages of Badrinath and Kedarnath in British Garhwal, where it is believed that hops might be cultivated with success, but until the roads are improved and rendered fit for mule carriage the cost of transport must absorb all profit."

With reference to **Mr. Liotard's** suggestion that the crop might possibly be profitably cultivated on the Nilghiris, it is disappointing to find in the report of Proceedings of the Madras Board of Revenue, dated 1st June 1889, that several attempts have been made to introduce the cultivation but without success.

It would thus appear that the only districts which have up to this time shown any promise of attaining success with hops are Chamba and Kashmír. The present rapid increase in the demand for hops for brewing purposes is likely, however, to induce further endeavours on the part of planters, European and Native, and it is to be hoped that with greater care in the selection of suitable localities, and in methods of cultivation, the crop may yet become a more generally remunerative one.

MEDICINE. Strobiles. 451

Medicine.—The officinal parts of the hop are, the dried STROBILES of the female plant called *Lupulus* or hop, in the British and Indian Pharmacopœia, and *Humulus* in the United States Dispensatory; and also in the latter, *Lupulina*, or the yellow powder separated from the strobiles.

MEDICINE.

The strobiles, as found in commerce, are more or less broken up and compressed When fresh they have a pale yellowish green colour, an agreeable, peculiar, somewhat aromatic and narcotic odour, and a bitter, aromatic, pungent, and feebly astringent taste. When handled, or more especially when rubbed between the fingers, they have a sticky feel, and their odour becomes more evident. But by keeping, their odour becomes less agreeable, or even unpleasant, owing to the formation of a little valerianic acid. At the same time they lose their greenish yellow colour and fresh appearance, acquire a brown tint, and frequently a spotted appearance, and finally become weaker and of inferior value (*Bentley and Trimen*),

Chemistry. 452

CHEMICAL COMPOSITION.—Over and above the constituents of the glands, or lupulina, which will be described below, the scales of the strobiles contain *lupulo-tannic acid* in the proportion of from 3 to 5 per cent., a minute quantity of *trimethylamine*, and a liquid volatile alkaloid, named by **Griessmayer** *lupuline*, and said to have the odour of the opium derivative *conia*. **Etti** further demonstrated the existence in hops of arabic acid, phosphates, nitrates, malates, citrates, and also sulphates, chiefly of potassium.

The yellow powder separated from the strobiles,—Lupulina – consists of minute, shining, translucent glands, in which the medicinal properties of the plant essentially reside. The powder is obtained by stripping off the bracts, shaking and rubbing them, and then separating it by means of a seive. The product thus obtained is yellowish brown and granular, with the agreeable odour and bitter taste of hops, easily wetted by alcohol or ether, but only gradually by water. Chemically, Lupulina consists principally of *wax* and two *resins*, one of which is crystalline and unites with bases, while another is amorphous and brown, and is closely associated with the bitter principle. The medicinal virtues, however, appear to reside chiefly in a *volatile oil* and a *bitter principle*, both of which may be extracted by alcohol. The latter has been variously named *lupulin*, *lupulite*, and *humuline*, but was first separated in a pure condition by **Lermer**, who called it the *bitter acid of hops*. It crystallizes in large brittle and rhombic prisms, possesses in a marked degree the peculiar bitter taste of beer, and is probably the tonic principle of the plant. The volatile oil may be readily obtained by distilling hops with water, the yield varying from 1 to 2 per cent. It contains a substance called *valerol*, which readily passes into *valerianic acid* (the cause of the disagreeable odour of old hops), and the entire oil rapidly resinifies on exposure to the air. For this reason Lupulina, which owes much of its medicinal properties and aroma to this oil, ought to be used fresh, and when kept for some time ought to be preserved in closely-stoppered bottles.

Properties. 453

MEDICINAL PROPERTIES.—Therapeutically hops possess tonic, sedative, and also to a small extent astringent, properties. They are administered chiefly in the form of tincture, infusion or extract in cases of atonic dyspepsia, nervous affections attended with sleeplessness, hysteria, intermittent fevers, and rheumatism; while lupulina exercises a markedly beneficial influence in cases of spermatorrhœa, chordee, and enuresis. Since the narcotic property appears to reside chiefly in the volatile oil, a pillow stuffed with the strobiles, but wetted with alcohol to prevent rustling, is sometimes employed to prevent restlessness and induce sleep. Advantage has also been taken of the anodyne property of the oil in the preparation of external applications, poultices, or fomentations for painful swellings and inflammation, and an ointment for cancerous tumours. Neither the strobiles, nor the granular powder, are, however, employed extensively alone in European medicine, though they are valued as adjuncts to other drugs.

FOOD. 454

Food.—The chief consumption of hops is in the manufacture of beer and ale, for an account of their employment in which the reader is referred to the article on "MALT LIQUORS." FOOD.

Trade—The steady increase of brewing in India has occasioned a continuous and rapid development of the amount and value of the imports of hops into the country. Thus the average import for the five years ending 1882-83 was 2,525 cwt., valued at R2,95,108, while that for the five years following was nearly double, *viz.*, 4,401 cwt., value R3,59,022. The highest import up to the year 1887-88 was in that year, and amounted to 5,838 cwt., value R4,74,345. The countries growing and exporting hops appear to be increasing in number, thus in 1882-83 our imports were obtained only from the United Kingdom, Italy, and China, while in 1887-88, Austria, Australia, and Ceylon each formed an additional source of supply, and the article obtained from Australia appears to have been much the most valuable; at all events it fetched much the highest proportionate price. Out of the 5,838 cwt. imported in that year, the United Kingdom furnished 3,968 cwt., value R3,10,349; Italy 1,163 cwt., value R79,642; China 357 cwt., value R50,264; Austria 203 cwt., value R11,750; Australia 125 cwt, value R20,720; and Ceylon 22 cwt., value R1,620. TRADE. 455

HUNTERIA, *Roxb.; Gen. Pl., II., 698.*

A genus of glabrous trees belonging to the Natural Order APOCYNACEÆ, and comprising three Indian species, of which only one is important.

Hunteria corymbosa, *Roxb.; Fl. Br. Ind., III., 637; Wight, Ic., [t. 428;* APOCYNACEÆ. 456

Syn.—HUNTERIA ZEYLANICA, *Gardn.;* H. LANCEOLATA, *Wall., Cat., 1611;* ALYXIA ROXBURGHIANA, *Wight, Ic., t. 1294;* GYNOPOGON LANCEOLATUM, *Kurz, For. Fl., II., 177;* TABERNÆMONTANA SALICIFOLIA, *Wall.;* T. PARVIFLORA, *Herb. Heyne.*

References.—*Roxb., Fl. Ind., Ed. C B. C., 233; Kurz, For. Fl. Burm., II., 177; Beddome, For. Fl., II., t. 265; Thwaites, En. Ceylon Pl., 191; Balfour, Cyclop., II., 121.*

Habitat.—A tree with slender smooth branches, inhabiting the Deccan Peninsula, Tavoy, Penang, Ceylon, and Tenasserim.

Structure of the Wood.—Fine, close-grained, and hard, resembling boxwood (*Thwaites*). It answers well for engraving (*Beddome*). TIMBER. 457

HURA, *Linn.; Gen. Pl., III., 339.*

A genus of the EUPHORBIACEÆ, having no Indian representative, but of which one species has been introduced from the West Indies.

Hura crepitans, *Linn.; DC., Prodr., 1229;* EUPHORBIACEÆ. 458

SANDBOX TREE.

References.—*Voigt, Hort. Sub. Cal., 161; Dalz. & Gibs., Bomb. Fl. Suppl., 76; Smith, Dic., 365; Treasury of Bot., I., 602.*

Habitat.—A small armed tree, from 30 to 40 feet high, having a few prickles on the stem, indigenous in Tropical America, and introduced into India from Jamaica.

Oil.—A clear, pale coloured fluid oil is obtained from the SEEDS, but its properties are as yet not well known. Since the whole tree abounds in an extremely poisonous milky juice, which when applied to the eye causes almost immediate blindness, it appears probable that the oil may partake of the same deleterious properties. OIL. Seeds. 459

Medicine.—The SEEDS are emetic, and in a fresh state violently purgative, a property which, however, they seem to lose when dried. The OIL is said to be a useful though very drastic purgative, 20 drops equalling in action half an ounce of castor oil. MEDICINE. Seeds. 460 Oil. 461

TIMBER. 462 **Structure of the Wood.**—Soft and friable; the hollowed trunks are said to be employed in the West Indies as vats for cane juice.

Husks of the Scripture, see **Ceratonia Siliqua,** Vol. II., 254.

HYDNOCARPUS, *Gærtn.; Gen. Pl., I., 129.*

A genus of trees belonging to the BIXINEÆ, and comprising about six species, inhabitants of Tiopical Asia. [*t. 942*; BIXINEÆ.

463 **Hydnocarpus alpina,** *Wight; Fl. Br. Ind., I., 197; Wight, Ic.,*

Vern.—*Kastel,* MAR.; *Maratatti,* NILGHIRIS; *Torathi, sannasolti,* KAN.

References.—*Bedd., Fl. Sylv., t. 77; Gamble, Man. Timb, 16; Thwaites, En. Ceylon Pl., 19; Trimen, Cat. Ceylon Pl., 6; Indian Forester, II., 22, 23; Balfour, Cyclop., II., 136; Bombay Gazetteer, XV. (Kanara), Pt. I., 71.*

Habitat.—A large tree, 70 to 100 feet in height, common on the Nilghiris up to 6,000 feet, found also on the Calcad hills, Tinnevelly, at an elevation of 1,500 feet, and probably throughout the western gháts of Madras; also in Ceylon at an elevation of 1,500 feet (*Beddome*). It is also mentioned in the *Kánara Gazetteer* as growing in the forests of that district.

OIL. Seeds. 464 **Oil.**—The SEEDS yield an oil, which is employed for burning in Kárwár (*Kánara Gazetteer*).

TIMBER. 465 **Structure of the Wood.**—Splits readily, much used on the Nilghiris and in Kánara, for beams and rafters in house-building, also employed in making packing cases and for firewood.

466 **H. heterophylla,** *Bl.; Kurz, Fr. Fl. Burm., I., 77.*

Vern. *Kal-law-hso,* BURM.

Habitat.—An evergreen tree from 40 to 50 feet in height, very frequent in the tropical forests of Martaban, less so along the eastern and southern slopes of the Pegu Yomah; found also in Tenasserim. It flowers in April and fruits in February and March (*Kurz*).

TIMBER. 467 **Structure of the Wood.**—Heavy, strong, close-grained, of short fibre, yellowish white, turning light brown (*Kurz*).

H. odorata, *Lind.*; see **Gynocardia odorata,** *Roxb.*; p. 192.

468 **H. venenata,** *Gærtn; Fl. Br. Ind., I., 196.*

Syn.—HYDNOCARPUS INEBRIANS, *Vahl.; non Wall.*

Vern.—*Jangli-badam?* (Bidie), HIND.; *Jangli-badam?* (Bidie), DEC.; *Kauti,* MAR.; *Niradi-mattu, marra vittai,* TAM.; *Niradi,* TEL.; *Moratti,* MALAY.; *Makúlú,* SING.

References.—*Thwaites, En. Ceylon Pl., 18; Rheede, Hort. Mal., I., t. 36; Trimen, Cat. Cey. Pl., 6; Bidie, Cat. Paris Exhib., No. 639; Fluck. & Hanb., Pharmacog., 76; Drury, U. Pl., 249* (partly refers to **H. inebrians,** *Wall.*); *Cooke, Oils and Oilseeds, 17; Balfour, Cyclop., II., 136; Treasury of Bot., I., 604.*

Habitat.—A large tree, found in Ceylon by the bank of rivers up to 2,000 feet, also in Malabar, Tinnevelly, and Travancore.

OIL. Seeds. 469 **Oil.**—The SEEDS yield an oil of the consistence of ordinary hard salt butter, called the *Thortay* oil in Kánara.

MEDICINE. Oil. 470 **Medicine.**—The OIL above mentioned is used as an external application in certain cutaneous diseases, and has a special reputation in leprosy. It has been recommended as a substitute for *chaulmugra* oil obtained from **Gynocardia odorata. Dr. Bidie,** however, thinks that its good effects are doubtful. The SEEDS, if eaten, produce giddiness, and are employed by the natives to poison fish. Their poisonous properties, however, are so strong that fish, thus killed, are unfit for food.

Seeds. 471

472 **H. Wightiana,** *Bl.; Fl. Br. Ind., I., 196; Wight, Ill., I., t. 16.*

Syn.—H. INEBRIANS, *Wall. non Vahl.*; MUNICKSIA, *Dennstd.*

Vern.—*Jangli-bádám* (=seeds), *jangli-bádám ká tel* (=oil), DEC.; *Kowti, kadu-kavatha, kova, kauti* (*kosto* at GOA), BOMB.; *Kowti, kadukavata* (*kavatela*=oil), MAR.; *Yetti, maravetti* (*niradimuttu*=seeds), (*niradimuttu-enney*=oil), TAM.; *Níradi-vittulú* (=seeds), *níradi-vittulú-núne* (=oil), TEL.; *Tamana, maravetti,* MALAY.; *Makúlá* (*ratakekuna* =seeds), SING.

References.—*Dalz. & Gibs., Bomb. Fl., 11; Rheede, Hort. Mal., I., 65, t 36; Pharm. Ind., 27; Ainslie, Mat. Ind., II., 235; Moodeen Sheriff, Mat. Med Madras, 34; Dymock, Mat. Med. W. Ind., 2nd Ed., 71; Pharmacog. Indica, Pt. I., 148; Fluck. & Hanb., Pharmacog., 76; S. Arjun, Bomb. Drugs, 14; Bidie, Cat. Paris Exhib., No. 633; Lisboa, U. Pl. Bomb., 260, 272; Gazetteer, Bombay, XVII.* (*Ahmednagar*), *25.*

Habitat.—A common tree of the Western Peninsula from the Konkan along the coast ranges.

Oil.—The SEEDS yield by expression, or on boiling them in water, about 44 per cent of oil, which has a sherry-yellow colour, is devoid of characteristic taste or odour, and has a sp. gr., at 85° Fh., of ·9482. Unlike the *chaulmugra* oil (obtained from the seeds of the allied **Gynocardia odorata**), which it otherwise closely resembles, it does not, at least at ordinary temperatures, deposit a crystalline fatty acid. Treated with sulphuric acid the oil affords the gynocardic acid reaction (see **Gynocardia odorata**, *p. 192*); but in a less degree than *chaulmugra* (*Pharmacographia Indica*). According to **Lisboa** it is chiefly employed as a lamp oil in Goa. **OIL. Seeds. 473**

Medicine.—The SEED has long been employed as a domestic remedy, by the natives of the Western coast ranges, in cases of skin disease and ophthalmia, and as a dressing for wounds and ulcers. At the present day the OIL is employed as an external application to scabby eruptions, after being mixed with an equal portion of **Jatophra Curcas** oil, sulphur, camphor, and lime-juice. For scald-head equal parts of the oil and lime-water are used as a liniment. In the Konkan also the oil has a reputation as a remedy for *Barsati* in horses. In Travancore half teaspoonful doses are given internally in leprous affections, and the oil, beaten up with the kernels and shells of castor-oil seeds, is used as a remedy for itch (*Pharmacographia Indica*). The medicinal properties of the seeds depend wholly on the oil they contain, which is much more useful and convenient as a drug than the seeds themselves. Though the seeds of this plant are evidently those described by **Ainslie** under the name of *Neeradimootoo*, the medicinal properties of the oil they yield were neglected, till recently, when the oil was again brought to the notice of Europeans as a substitute for the more expensive *chaulmugra* oil. Since then it has been tried and appears to have given satisfactory results. In physiological action it is alterative, tonic, and a local stimulant, and appears also to have a specific effect on certain skin diseases. It has been recommended for trial as a local application in rheumatism, leprosy, sprains and bruises, sciatica, chest affections and phthisis, ophthalmia, and in various forms of skin disease. Internally it may be prescribed in doses of from 15 minims to 2 drachms in cases of leprosy, various forms of cutaneous disease, secondary syphilis, and chronic rheumatism. It must, however, be employed with caution, as in certain cases it is said to act as a gastro-intestinal irritant, producing vomiting and purging. **MEDICINE. Seed. 474 Oil. 475**

Trade.—The price of the SEEDS, according to **Dr. Moodeen Sheriff**, is R2-8 per maund wholesale, and annas 2 per pound retail in the bazár. The authors of the *Pharmacographia* write: "The seeds are not an article of trade, but if ordered may be obtained at about half the price of those of *chaulmugra* (R12 per Bengal maund of 80lb). The OIL has been sold in Madras at 2½ annas per seer." **TRADE. Seeds. 476 Oil. 477**

Hydnum coralloides, see **Mushroom,** also **Agaricus campestris,** Vol. [I, 131.

HYDRANGEA, *Linn.; Gen. Pl., I., 640.*

A genus of large shrubs or trees, often subscandent when young, belonging to the Natural Order SAXIFRAGACEÆ, and comprising 33 species, distributed from Java to the Himálaya and Japan (the centre of the genus), Eastern North America, and Western South America. Of these species five or six are indigenous to India.

478 **Hydrangea altissima,** *Wall.; Fl. Br. Ind., II., 404;* SAXIFRAGACEÆ.

References.—*Gamble, Man. Timb., 172; Brandis, For. Fl., 211*

Habitat.—A tall spreading shrub, of the Himálaya, from the Sutlej to Bhotán, above 5,000 feet.

DOMESTIC USES. Bark. 479 **Domestic Uses.**—Its BARK is employed as a substitute for paper. This economic use of the bark was first pointed out by **Thomson,** and is quoted by **Stewart** under the heading of **Hydrangea sp.**

480 **H. robusta,** *H. f. & T.; Fl. Br. Ind., II., 404.*

Syn.—HYDRANGEA CYANEMA, *Rutt.*

Vern.—*Bogoti,* NEPAL.

References.—*Gamble, Man. Timb., 172; Thwaites, En. Ceylon Pl., 103; Gamble, List of Trees, Shrubs, &c., in Darjiling, 38.*

Habitat.—A small deciduous tree, found in the Eastern Himálaya and Sikkim, from 5,000 to 7,000 feet, generally as undergrowth in the oak forests.

TIMBER 481 **Structure of the Wood.**—White, moderately hard, close-grained. Weight 42℔ per cubic foot, easily worked.

H. sp.; *Stewart, Panjáb Plants, 103,* see **H. altissima.**

482 **H. vestita,** *Wall.; Fl. Br. Ind., II., 405.*

Syn.—HYDRANGEA HETEROMALLA, *Don;* H. KHASIANA, *H. f. & T.*

Vern.—*Pokuttia,* NEPAL; *Kulain,* BHUTIA.

References.—*Brandis, For. Fl., 211; Gamble, Man. Timb., 172; List of Trees, Shrubs, &c., of Darjiling, 38; Wall, Cat., 440 a; Indian Forester, XI., 3.*

Habitat.—A small deciduous tree, met with in the Himálaya from Kumáon to Bhotán between 8,000 and 11,000 feet; also in the Khásia Hills from 4,500 to 5,500 feet.

TIMBER. 483 **Structure of the Wood.**—Pinkish white, moderately hard, annual rings indistinct, weight 45℔ per cubic foot.

Hydrargyrum, see **Mercury;** Vol. V.

HYDRILLA, *L. C. Rich.; Gen. Pl., III., 450.*

A genus of submerged leafy diœcious water herbs, belonging to the Natural Order HYDROCHARIDEÆ, and having only one species, which is distributed in the still and slowly running waters of Tropical Asia and Australia.

484 **Hydrilla verticillata,** *Casp.; Fl. Br. Ind., V., 659;* HYDROCHARIDEÆ.

Syn.—HYDRILLA OVALIFOLIA, *Rich.;* H. DENTATA, *Casp.;* H. WIGHTII, *Planch.;* H. ANGUSTIFOLIA, *Blume;* LEPTANTHES VERTICILLATA, *Herb. Wight;* SERPICULA VERTICILLATA, *Linn. f.;* VALLISNERIA VERTICILLATA, *Roxb.;* HOTTONIA SERRATA, *Willd.*

Vern.—*Jhangi, kureli,* HIND.; *Jhanjh, jála,* PB.; *Púnachu, páchi, náchu* (these names are according to **Elliot** applied indifferently to all sorts of herbaceous, aquatic plants), TEL.

References.—*Roxb., Fl. Ind., Ed. C.B.C., 653, 711; Cor. Pl., II., t. 164; Thwaites, En. Ceylon Pl., 331; Dalz. & Gibs., Bomb. Fl., 277; Stewart, Pb. Pl., 241; Aitchison, Cat. Pb. and Sind Pl., 146; Boiss., Fl. Orient., V., 8; Wall., Cat., 5048; Elliot, Fl. Andhr., 160; Gazetteers:—Mysore and Coorg, I., 56; N.-W. P., I. (Bandelkhand), 84; IV. (Agra), lxxvii.*

Habitat.—A submerged leafy herb, forming large masses in the still and slowly running waters throughout India and Ceylon.

INDUSTRIAL USES. Sugar refining. 485

Industrial Uses.—Along with other water plants this herb is employed in many parts of India in the process of sugar refining. The surface of the sugar is covered with it, as it is with clay in the West Indies, to ensure slow percolation of the water afterwards applied (see **Saccharum officinarum**).

Hydrochlorate of Cocaine, see **Erythroxylon Coca**; Vol. III., p. 270.

HYDROCOTYLE, *Linn.; Gen. Pl., I., 872.*

A genus of prostrate herbs belonging to the UMBELLIFERÆ, and comprising about 70 species which are found in wet places in temperate and tropical regions, more especially of the Southern hemisphere. The generic name is derived from ὕδωρ = water and κοτύλη = a cavity or vase, in reference to the peculiar rounded concave leaves of the plants. [*565*; UMBELLIFERÆ.

Hydrocotyle asiatica, *Linn.; Fl. Br. Ind., II., 669; Wight, Ic., t.* 486

ASIATIC PENNY-WORT, *Eng.*; BEVILACQUA, *Fr.*

Syn.—HYDROCOTYLE WIGHTIANA, *Wall.*; H. LURIDA, *Hance.*

Vern.—*Bráhmamanduki, khulakhudi,* HIND.; *Thol-kuri, bráhmamanduki,* BENG.; *Bhika-purni,* DACCA; *Mani-muni,* ASSAM; *Vallári,* DEC.; *Karivana, karinga,* BOMB.; *Bráhmi,* MAR.; *Barmi,* GUZ.; *Vallárai, babassa,* TAM.; *Mandúka-bramha-kúráku, pinna-élaki-chettu, babassa, bokkudu-chettu, bokkudu* (Elliot), TEL.; *Von-de-lagá,* KAN.; *Kutakan, kodagam,* MALAY.; *Minkhuá-bin,* BURM.; *Hingotukola,* SING.; *Mandukaparn, bheka-parni,* SANS.; *Artániyde-hindi,* ARAB.

References—*Roxb, Fl. Ind., Ed. C.B.C., 270; Dalz. & Gibs., Bomb. Fl., 105; Wall, Cat., 560; Kurz, in Journ. As. Soc., 1877, Pt. II., 113; Rheede, Hort. Malab., X., t. 46; Elliot, Fl. Andhr., 29, 153; Mason, Burma and Its People, 502; Taylor, Topography of Dacca, 58; Pharm. Ind., 107, 448; Ainslie, Mat. Ind, II., 473; Moodeen Sheriff, Supp. Pharm. Ind., 158; U. C. Dutt, Mat. Med. Hind., 176, 309; Dymock, Mat. Med. W. Ind., 2nd Ed., 361; Fluck. & Hanb., Pharmacog., 297; U. S. Dispens., 15th Ed., 1666; Bent. & Trim., Med. Pl., 117; S. Arjun, Bomb. Drugs, 61; Waring, Bazar Med., 69; K. L. Dey, Indig. Drugs of India, 62; Journal de Pharm., July 1855, 49; Dr. Clément Daruty de Grandpré, in the Nouveaux Remedies, April 8th, 1889, 146; Hunter, Madras Med. Rep., 1855, 356; Waring in Pharm. Journ., XVII., Ser. I., 312; Atkinson Him. Dist., 316; Drury, U. Pl., 250; Lisboa, U. Pl. Bomb., 260; Birdwood, Bomb. Pr., 41; Christy, Com. Pl. and Drugs, VII., 52; VIII., 4, 58; Darrah, Rep. on Condition of the People of Assam; Atkinson, Ec. Prod., N.-W. P., Pt. V., 95; Balfour, Cyclop., II., 137; Kew Off. Guide to the Mus. of Ec. Bot., 73; Report by Mr. D. Hooper, Govt. Quinologist, Madras; Gazetteers:—Mysore and Coorg, I., 61; North-West Prov., IV., lxxii.; Bomb., VI., 14; XV., 435; District Manuals, Cuddapah, 200.*

Habitat.—A small herbaceous plant, found throughout India, from the Himálaya to Ceylon at altitudes up to 2,000 feet, particularly abundant in damp places in Bengal; a common weed in the vicinity of Calcutta. It is also distributed to Malacca.

HISTORY. 487

History.—The plant was known to Sanskrit writers of very remote times, its properties being supposed to resemble those of *Brahmi* (**Herpestis Monniera**), both being regarded as alterative, tonic, and useful in diseases of the skin, nervous system, and blood. **Dr. Dymock**, however, believes the word *Brahmi* to refer in Hindu and Sanskrit literature to the **Hydrocotyle**, owing to the fact that it is applied in Bombay to that plant. The earliest European writers on Indian Materia Medica, **Rheede, Rumphius**, and **Ainslie**, were all acquainted with the medicinal properties of the plant as an alterative, tonic, and astringent, but it was not till **Boileau** in 1852 made known its virtues in the treatment of leprosy, that the drug came under special attention. In 1885 **Dr. Hunter** experimented with it for the same disease in Madras, with sufficiently satisfactory results to

HISTORY

bring about its admission as an officinal drug into the Indian Pharmacopœia. Of late years attempts have been made to introduce the plant into European medicine, but up to this time it has not appeared in either the Pharmacopœia of the United States, or that of Great Britain.

MEDICINE. 488

Medicine.—The parts of the plant generally employed are the leaves, deprived of their petioles, dried by exposure to the air in the shade, and ground to a powder. It appears that if dried in the sun, or by any method of artificial heat, the leaves lose a great part of their medicinal properties, owing to the volatilisation of the oil which is their active principle. The powder thus carefully prepared ought to be kept in well-stoppered bottles to prevent the access of moisture. When fresh, the leaves have scarcely any smell, but emit a peculiar faint, aromatic odour when crushed between the fingers. They have a harsh, bitter, and disagreeable taste, which, however, becomes scarcely perceptible after the leaves have been well dried. The powder, $3\frac{1}{2}$ to 4℔ of which may be obtained from 30℔ of the fresh leaves, is of a pale green colour, and exhales a slight characteristic aroma. It appears possible that a mistake has been made in only employing the leaves in the Indian Pharmacopœia, since **Boileau** particularly mentions that the entire plant (roots, twigs, leaves, and seeds) ought to be used, and as will be seen below, the supposed active principle *Vellarin* is said to exist principally in the roots.

Chemistry. 489

CHEMICAL COMPOSITION.—A careful analysis of hydrocotyle was made, in 1855, by **Lêpine**, a pharmacien of Pondicherry, who found that it yielded a body which he called *vellarin*, from *valárai*, the Tamil name of the plant, and regarded as its active principle. This *vellarin* was said to exist in the dry plant to the extent of 0·8 to 1 per cent., to be an oily, non-volatile fluid, with the taste and smell of fresh hydrocotyle and to be soluble in spirits of wine, ether, caustic ammonia, and partly also in hydrochloric acid. The authors of the *Pharmacographia*, however, remark, "these singular properties do not enable us to rank *vellarin* in any well-characterized class of organic compounds," and, moreover, they did not succeed in obtaining anything like it from the dried plant (presumably the officinal parts only, *viz.*, the leaves, were employed in the analysis), but "simply a green extract almost entirely soluble in warm water, and containing chiefly tannic acid, which produced an abundant green precipitate with salts of iron." In a very recent note on Hydrocotyle, by **Dr. Clement Daruty**, however, *vellarin* is described, and appears to have been again obtained. He remarks: "*vellarin* is an inspissated oil of a pale yellowish colour, with a bitter, pungent, and persistent taste and a marked odour of Hydrocotyle; but which is subject to variations under the influence of heat, humidity, and even of the atmosphere, volatilising at 120°C. It is soluble in alcohol. Of this active principle the plant contains ·07 per mille. There are also found in it two resinous ingredients, one green, in the proportion of ·085 per mille, the other brown, in the proportion of ·3 per cent., and traces of tannic acid. *Vellarin* is obtained principally from the roots of the plant." It would thus appear that as the analysis of the plant at present is understood, it contains *Vellarin* in much smaller amount than that obtained by **Lepine**, chiefly in the roots, a part of the plant at present not officinal in India. An analysis in 1888 of the carefully dried leaves of the plant by **Mr. D. Hooper**, Government Quinologist, Madras, shewed them "to contain 8·9 per cent. of resinous and oily substances, the latter of a volatile and pungent nature, 24·5 per cent. of tannic acid and sugar, 11·5 per cent. of mucilage and extractive, 12·5 per cent. of pectin and albuminous matter, and 12 per cent. of ash, consisting largely of alkaline chlorides. The leaves on drying lost four-fifths of their weight."

MEDICINE. Physiological Action. 490

PHYSIOLOGICAL ACTION.—**M. Clement Daruty**, in the note above alluded to, gives an interesting account of the effects of the drug on lepers, and on his friend, **Dr. Boileau,** who experimented on himself. The following translation of his remarks may be here given, as his investigations afford the latest and most complete account of the physiological action of the plant:—"The first effect produced on lepers by the administration of **Hydrocotyle** is a sensation of warmth and tingling of the skin, especially that of the extremities, followed after some days by a general increase in the temperature of the body, amounting in some instances to an intolerable itchiness accompanied by cutaneous redness. The capillary circulation is accelerated and the pulse becomes stronger and fuller. After a week's treatment, the patient's appetite sensibly improves and the functions of the principal viscera are performed more easily. As the result of a more prolonged treatment, the skin becomes more supple and uniform, the epidermis gradually peels off in small scabs, or, in severe cases, in large scabs, perspiration is restored, the excretory functions resume their normal action, the digestion becomes improved, and the appetite increases. Administered experimentally in small doses to healthy persons, it produces, within a short space of time, diuretic effects, then a general stimulation of the circulation, and eventually intense itching. In doses of 1 to 2 grammes of the powder it produces considerable giddiness, accompanied by cephalalgia, which sometimes lasts for a whole month, even after the medicine has been discontinued. It has also been known to give rise to dangerous dysenteric symptoms. Thus **Dr. Boileau** who, in treating himself, progressively increased the dose, found that after two months the drug had produced all the effects of a violent, cumulative poison. He writes, 'Yesterday in the morning I was seized with a violent trembling, so severe that I was forced to lie down in bed, and in spite of the many blankets with which I covered myself, it was more than an hour before I could recover the warmth of my body. But this was unimportant in comparison with what followed. The other symptoms were violent, spasmodic contraction of the larynx, which made me believe suffocation to be imminent, palpitations of such violence and frequency, as to make me fear a rupture of the heart, and tetanic spasms of the trunk and limbs. Towards evening a fit of vomiting and hæmorrhagic diarrhœa supervened, which soon abated and finally disappeared. I awoke in the morning free of dangerous symptoms, and with no result of the attack remaining, save a state of intense lassitude and a slight pain in the neck.'" To sum up it would appear that **Hydrocotyle,** properly prepared and administered, is a powerful stimulant of the circulatory system, its action chiefly affecting the vessels of the skin and mucous membranes In larger doses it is a stupefying narcotic, and in some cases produces cephalalgia or vertigo with a tendency to coma.

Applications. 491

APPLICATIONS IN MEDICINE.—The principal value of the drug appears from its physiological action to be as a stimulant to the cutaneous circulation in skin diseases, and for this purpose it will be found to have been chiefly employed. Though it appears to have no specific effect in leprosy nor in syphilis, there can be little doubt that by its action in stimulating circulation it is of value in these affections. **Dr. Shortt** speaks of the drug in high terms, and considers it to have a powerful action in all leprous affections, but later writers agree that its effects are most marked in the preliminary anæsthetic stages of the disease. **Drs. Lolliot, Cazenove,** and **Bertin** find it of little value in advanced cases of tubercular leprosy, but extol its virtues in the treatment of chronic and obstinate eczema, the latter remarking: "The eczemas treated by me with the preparations of **Hydrocotyle** were of the most rebellious type, *viz.*, localised eczemas; nevertheless cures were effected in every case and that within a very brief space of time" It

MEDICINE.

has also been prescribed with excellent result in cases of secondary and tertiary syphilis, accompanied by gummatous infiltration and ulceration; in chronic and callous ulcers; as a stimulant to healthy mucous secretion in infantile diarrhœa and ozœna; in cases of scrofulous ulceration, enlargement of glands and abscess; and in chronic rheumatism. It has also been employed with success as a diuretic in several diseases, and as an emmenagogue in cases of amenorrhœa. In all cases in which a constitutional or general disease is accompanied by a local lesion, the drug ought to be not only administered internally, but also applied locally as a powder, poultice or ointment. Numerous descriptions of cases so treated by practitioners in many parts of the world have appeared of late years, with the result that the use of the plant is becoming daily more wide-spread, and the belief in its therapeutic value more universal. Under these circumstances it would certainly seem advisable to give the whole plant a careful trial, as recommended by **Boileau**, and more recently again strongly advised by **Daruty**, as it is quite possible that much of the disrepute into which it has fallen in India may be the result of the Pharmacopœia recognising only the least active part, *viz.*, the leaves.

Preparations. 492 PHARMACEUTICAL PREPARATIONS.—The *Indian Pharmacopœia* describes only two, *viz.*, POWDER OF HYDROCOTYLE to be dusted on ulcerated surfaces, or given internally in doses of from 5 to 8 grains thrice daily; and HYDROCOTYLE POULTICE, a stimulant application.

Powder. 493 Leaves 494 Roots. 495 Dr. **Daruty** recommends several preparations; (1) THE POWDER prepared from the LEAVES and ROOTS together, by washing the latter, and drying all in a well-ventilated room, completely shaded from the sun. He advises that a dose of 10 grains be given three times a day. **Dr. Boileau** recommends the following doses of the same powder in cases of leprosy, for an adult of 20 to 40 years of age and proportionately less for a child: "20 grains to be taken daily for two weeks, the dose to be then increased by weekly increments of 5 grains up to 60 grains, at which daily rate it is to be continued for one month. It should then be reduced by 5 grains a week till it is again brought down to 10 grains, after which the treatment should be suspended for an entire month" (to guard against the cumulative poisonous action of the drug). "It should then be resumed and progressively increased and diminished according to the sliding scale detailed above. The powder should be taken before going to bed, in hot wine sweetened, but on reaching 30 grains it is better to divide it into two equal portions, one to be taken in the morning, the other in the evening." (2) THE

Plaster. 496 Syrup. 497 Ointment. 498 Fluid Extract. 499 Decoction 500 Baths. 501 PLASTER is prepared from the leaves by trituration with *cold* water in a mortar, sufficient of the latter being added to form a thick paste. A stimulating application. (3) THE SYRUP is prepared from 90 grammes of the powder boiled in a quart of water, till the liquid is reduced to a pint, to which 2℔ of sugar is added and thoroughly mixed at 31°C. till a syrup is formed. Dose, one teaspoonful, increased in the same ratio as **Boileau** recommended for the powder. (4) THE OINTMENT.—Four grammes of the powder mixed with 30 grammes of vaseline. (5) THE FLUID EXTRACT.—Prepared so that one ounce represents one ounce of the powder. Dose, 10 to 15 drops thrice a day. (6) THE DECOCTION.—Prepared from 30 grammes of the dried plant to one pint of water. (7) BATHS are employed advantageously in skin diseases, 1,500 grammes of the fresh plant being added to the hot water.

SPECIAL OPINIONS.—§ "Is largely used in brain affections and insanity. A decoction of the leaves is evaporated with butter and the resulting preparation is then given internally" (*Surgeon-Major Robb, Civil Surgeon, Ahmedabad*). "This drug deserves attention. Its efficacy is well reported on. I have used it with great success in secondary and tertiary

MEDICINE.

syphilitic skin affections. Its efficacy in these cases is well marked" (*U. C. Dutt, Civil Medical Officer, Serampore*). "A very useful alterative in constitutional syphilitic ulcers and skin disease. Dose, 5 to 10 grains of powdered leaves. Externally stimulant" (*Thomas Ward, Apothecary, Madanapalle, Madras*). "Much used in the hospital, similar to **Calotropis**; and always combined with it. A valuable alterative tonic used in all cases of leprosy and obstinate forms of skin diseases. The following formula is employed here. Pulvis Calotropis, grains iii, Pulvis Hydrocotyle grains x., twice a day. At times combined with iron preparations" (*J. G. Ashworth, Apothecary, Kumbakonom, Madras*).

TRADE. 502

Trade.—Of recent years enquiry for the plant has been made in India which has led to its cultivation on a small scale in gardens near Bombay, in the neighbourhood of which town, **Dr. Dymock** remarks, the plant is rare in a wild state. Should any increased demand arise, Calcutta, near which the plant abounds as a common weed, would probably become the source of supply. The trade value of the dry herb in Bombay is given by **Dr. Dymock** as R7-9 per Surat maund of 37½lb.

FOOD. Leaves. 503

Food.—The LEAVES are sometimes made into a soup, which, however, probably serves more as a medicine than as a food.

HYDROLEA, *Linn.; Gen. Pl., II., 831.*

The single Indian genus of the HYDROPHYLLACEÆ, a Natural Order closely allied to the GENTIANACEÆ.

504

Hydrolea zeylanica, *Vahl.; Fl. Br. Ind., IV., 133; Wight, Ill., t. 167, & Ic., t. 601;* HYDROPHYLLACEÆ.

Syn.—HYDROLEA JAVANICA, *Bl.*; NAMA ZEYLANICA, *Linn.*; NAMA, *Linn.*

Vern.—*Kasschra, isha-langulya,* BENG.; *Tsjeru-vallel,* MALAY.; *Deyakirilla,* SING.; *Langali,* SANS.

References.—*Roxb., Fl. Ind., Ed. C.B.C., 265; Thwaites, En. Ceylon Pl., 209; Dalz. & Gibs., Bomb. Fl., 170; Wall. Cat., 4398; Rheede, Hort. Malab., X., 28; Works of Sir W. Jones, VI., 106; O'Shaughnessy, Beng. Dispens., 507; Drury, U. Pl., 251; Balfour, Cyclop., II., 137; Treasury of Bot., I., 606; Gazetteers:—N.-W. P., I., 83; IV., lxxiv; Bombay, XV., 438.*

Habitat.—A procumbent, branched herb, abundant in wet and marshy places throughout India, ascending to an altitude of 4,000 feet.

MEDICINE. Leaves. 505

Medicine.—The LEAVES, beaten into a pulp and applied as a poultice, are considered to have a cleaning and healing effect on neglected and callous ulcers. They apparently possess some antiseptic property.

HYGROPHILA, *Br.; Gen. Pl., II., 1075.*

A genus of herbs of the Natural Order ACANTHACEÆ, comprising eighteen tropical and sub-tropical species, of which eight are Indian.

506

Hygrophila salicifolia, *Nees; Fl. Br. Ind., IV., 407; Wight, Ic., t. 1490;* ACANTHACEÆ.

Syn.—RUELLIA SALICIFOLIA, *Vahl.*; R. LONGIFOLIA, *Roth.*

Var. assurgens. SYN. HYGROPHILA ASSURGENS, *Nees, excl. syn. and var. β*; H. RADICANS, *Nees*; RUELLIA RADICANS, *Wall.*

Var. dimidiata. SYN. H. DIMIDIATA, OBOVATA, and UNDULATA, *Nees.*

Vern.—*Mathom arak',* SANTAL; *Lai pin kha,* BURM.

References.—*Roxb., Fl. Ind., Ed. C.B.C., 475; Thwaites, En. Ceylon Pl., 225; Dalz. & Gibs., Bomb. Fl., 184; Wall. Cat., 2373, 2410, 7150; Rev. A. Campbell, Cat. Ec. Prod., Chutia Nagpur, No. 8226.*

Habitat.—Herbs from 1 to 3 feet in height, very common throughout India and Ceylon. Var. **assurgens** is a native of Eastern Bengal and Malacca; *var.* **dimidiata** of Prome and Mergui.

FOOD. Leaves. 507 **Food.**—The LEAVES are eaten as a pot-herb by the Santals.

508 **Hygrophila spinosa,** *T. And.; Fl. Br. Ind., IV., 408; Wight, Ic., t. 449.*

Syn.—H. LONGIFOLIA, *Kurz;* BARLERIA LONGIFOLIA, *Linn.;* B. HEXACANTHA, *Mori;* RUELLIA LONGIFOLIA, *Roxb.;* ASTERACANTHA LONGIFOLIA, *Nees.;* A. AURICULATA, *Nees.*

Vern.—*Tálmakhána, gokhula kanta, gókshura,* HIND.; *Kuliákhárá, kantakalika,* BENG.; *Tal-makhana, kanta kúlika,* BEHAR; *Gokhula janum,* SANTAL; *Tálimkhana, kolsunda,* BOMB.; *Tálimakhána,* MAR.; *Ekharo, gokhru,* GUZ.; *Nirmalli,* TAM.; *Nirguvi veru,* TEL.; *Kalavankabija,* KAN.; *Bahel-schulli,* MALAY; *Katre-iriki,* SING.; *Ikshugandhá, kokiláksha,* SANS.

References.—*Roxb., Fl. Ind., Ed. C.B.C., 475; Thwaites, En. Ceylon Pl., 225; Dalz. & Gibs., Bomb. Fl., 189; Aitchison, Cat. Pb. and Sind Pl., 111; Kurz in Journ. As. Soc., 1870, II., 78; Wall., Cat., 2505; Rheede, Hort. Malab., II., t. 45; Rev. A. Campbell, Ec. Prod. Chutia Nagpur, No. 8450; Pharm. Ind., 162; Ainslie, Mat. Ind., II., 236; U. C. Dutt, Mat. Med. Hind., 215, 305; Dymock, Mat. Med. W. Ind., 2nd Ed., 585; S. Arjun, Bomb. Drugs, 107; Irvine, Mat. Med., Patna, 110; Atkinson, Him. Dist., 315; Birdwood, Bomb. Pr., 67; Home Dept. Cor., 225, 238; Gazetteers:—Bombay, VI., 14; N.-W. P., IV., lxxvi.*

Habitat.—A small spiny bush, common in moist places everywhere throughout India from the Himálaya to Ceylon.

MEDICINE. Plant. 509 Root. 510 Seed. 511 **Medicine.** The WHOLE PLANT, ITS ROOT, and SEED, are the parts used medicinally. Their value has been long recognised both by Sanskrit and Muhammadan writers, the former classing the plant as cooling and diuretic, and employing the whole herb, or its ashes, and the roots, as a cooling medicine and diuretic in cases of hepatic obstruction, dropsy, rheumatism, and diseases of the genito-urinary tracts; while the latter employ it in the same way, but also consider the seeds aphrodisiac. In native practice to the present day they continue to be employed for the same purposes, and the seeds may be found in almost every bazár. The attention of Europeans was first drawn to **Hygrophila** by **Rheede,** who wrote that a decoction of the root, in doses of half a tea-cupful twice daily, was employed by the natives on the Malabar Coast as a diuretic in "dropsy and gravelish diseases." Since that date **Ainslie, Kirkpatric,** and **Gibson** have borne testimony to its diuretic properties. The different parts of the plant appear to have the following therapeutic actions: The seeds are demulcent, diuretic, and possibly tonic, and are therefore indicated in diseases of the genito-urinary tract. The roots are cooling, bitter, tonic, and diuretic, and the leaves, cooling and diuretic. The seeds may be prescribed in doses of 10 grains to 2 drachms in infusion; and the root or whole plant in doses of 1 to 2 ozs. daily, of a decoction prepared from one ounce of the drug and one pint of water boiled down to 14 ozs.

SPECIAL OPINIONS.—§ "The leaves are boiled overnight and taken next morning in cases of dropsy. It is a good diuretic" (*Surgeon W. F. Thomas, 33rd Madras Native Infantry, Mangalore*). "Found useful as a diuretic in dropsical affections. Preparations:—Take of **Hygrophila** fresh roots one ounce; water ten ounces. Boil for fifteen minutes in a covered vessel and strain. Dose from 1 to 2 ounces. Invariably used in combination with other diuretics" (*J. G. Ashworth, Apothecary, Kumbakonum, Madras*).

TRADE. 512 **Trade.**—**Dr. Dymock** states that the seeds alone form an article of commerce in Bombay, and fetch R6 per maund of 37½ lb.

HYGRORYZA, *Nees; Gen. Pl., III., 1116.*

A small genus belonging to the tribe ORYZEÆ, of the Natural Order GRAMINEÆ having only one species, and that confined to India.

Hygroryza aristata, *Nees; Duthie, Fodder Grasses of N. India, 20;* [GRAMINEÆ. 513

Syn.—LEERSIA ARISTATA, *Roxb.*

Vern.—*Jangli dal,* HIND.; *Passai, passari, passáhi, parsál, tinni,* N.-W. P.; *Pastál,* PB.; *Nir-valli-pullu,* MALAY.; *Gojabbá,* SING.

References —*Roxb., Fl. Ind., Ed. C.B.C., 308; Thwaites, En. Ceylon Pl., 356; Aitchison, Cat. Pb. and Sind Pl., 157; Rheede, Hort. Mal., 10, t. 12; Duthie, Grasses of N.-W. India, 12; Duthie & Fuller, Field and Garden Crops, 16.*

Habitat.—An aquatic grass, found either floating on the surface of water, or creeping on wet ground, frequent in most parts of the plains of India. The grain ripens in September.

FOOD & FODDER. 514

Food and Fodder.—The grain, which in Bengal is called "wild rice," is eaten by certain of the poorer classes, who collect it by sweeping the heads of the grass with baskets. **Roxburgh** says that cattle are fond of the plant.

HYMENOCARDIA, *Wall.; Gen. Pl, III., 285.*

A genus of shrubs or trees belonging to the EUPHORBIACEÆ, and comprising five species, natives of Tropical India, the Malay Archipelago, and Africa.

Hymenocardia punctata, *Wall.; Fl. Br. Ind., V., 377;* EUPHORBIACEÆ. 515

Syn.—HYMENOCARDIA WALLICHII, *Tulasne;* SAMAROPYXIS ELLIPTICA, *Miq.*

Vern.—*Ye-kin,* BURM.

References.—*Kurz, For. Fl. Burm., II., 394; Gamble, Man. Timb., 347.*

Habitat.—A deciduous shrub or small tree common in the swamp forests and along the marshy borders of choungs in the savannahs, from Pegu and Martaban down to Tenasserim; also found in Upper Assam by Griffith.

TIMBER. 516

Structure of the Wood.—Rather heavy, of unequal fibre, pale brown, becoming red brown on exposure; rather hard and brittle.

HYMENODICTYON, *Gen. Pl., II., 35.*

A genus of trees or shrubs belonging to the Tribe CINCHONEÆ, of the Natural Order RUBIACEÆ, and comprising four to five species, natives of Tropical Asia and Africa.

Hymenodictyon excelsum, *Wall.; Fl. Br. Ind., III., 35; Wight, Ic., t. 79, 80;* RUBIACEÆ. 517

Syn.—HYMENODICTYON THYRSIFLORUM, *Wall;* H. UTILE, *Wight;* H. OBOVATUM, *Wight;* H. HORSFIELDIANUM, *Miq.;* CINCHONA EXCELSA, *Roxb.;* C. THYRSIFLORA, *Roxb.*

Vern.—*Bhaulan, bhalena, bhamina, dhauli, kúkúrkat, bhúrkúr, phaldu, bhohár, potúr, bandárú, phargur,* HIND.; *Bodoka, konoo,* URIYA; *Sali,* KÓL.; *Bhorkhond,* SANTAL; *Kúkúrát, bhúrkúr,* N.-W. P.; *Bartu, barthoa, thab (?) manabina,* PB.; *Bohar pótúr, pútúr,* C. P.; *Dondru, dandelo,* PANCH MEHALS; *Kalá-kadu,* BOMB.; *Bhoursál,* MAR.; *Bandári, jangli-anár-ká-jhar,* DEC.; *Sagapu, vilári,* TAM.; *Dudiyetta, dudippa, chétippa, búrja, bandára,* TEL.; *Bandaray anni,* KAN.; *Vallári,* MALAY.; *Khoyari, kusan,* BURM.; *Burkunda,* BHUMIJ.

References.—*Roxb., Fl. Ind., Ed. C.B.C., 178; Brandis, For. Fl., 267; Kurz, For. Fl. Burm., II., 71; Gamble, Man. Timb., 224, 225; Dalz. & Gibs., Bomb. Fl., 117; Stewart, Pb. Pl., 115; Rev. A. Campbell, Ec. Prod. of Chutia Nagpur, No. 7585; Elliot, Fl. Andhr., 21, 32, 37; Pharm. Ind., 117; Ainslie, Mat. Ind., II., 341; O'Shaughnessy, Beng. Dispens., 394; Moodeen Sheriff, Supp. Pharm. Ind., 158; Dymock, Mat. Med. W. Ind., 2nd Ed., 404; U. S. Dispens., 15th Ed., 444; Atkinson, Him. Dist., 311, 739; Drury, U. Pl., 251; Lisboa, U. Pl. Bomb., 84, 246, 278; Indian Forester, III., 203; IX., 438; X., 325; XII., App., 4; Kew Off. Guide to the Mus. of Ec. Bot., 81; For. Adm. Rep. Ch. Nagpur, 1885, 31; Gazeteers:—Mysore & Coorg, III., 28; N.-W. P., IV., lxxiii.*

Habitat.—A large deciduous tree, 30 to 40 feet high, met with on the dry hills at the base of the Western Himálaya, from Garhwál to Nepál, ascending to 2,500 feet; and throughout Chutia Nagpur and the Central Provinces to the Deccan, Central India, and the Anamalays; also found in Tenasserim and Chittagong.

DYE & TAN. Leaves. 518 Bark. 519 **Dye and Tan.**—The **Rev. A. Campbell** writes that the LEAVES are employed as a dye by the Santals, but he does not state the colour produced. **Ainslie** mentions that the BARK, which has powerfully astringent properties, is used by tanners, a statement which is repeated by **Lisboa.**

MEDICINE. Bark. 520 **Medicine.**—The inner coat of the bark, like that of the true cinchonas, is bitter and astringent, and has been long employed by the Hindus as a febrifuge and antiperiodic, especially in cases of tertian ague. **Roxburgh** commenting on this writes: "The infusion of one fresh leaf in water all night had little taste, but struck quickly a deep purplish blue with a chalybeate. The two inner coats of the bark (the outer light spongy stratum is tasteless) possess both the bitterness and astringency of Peruvian bark, and I think, when fresh, in a stronger degree. The bitterness is not so quickly communicated to the taste on chewing the bark as that of the former, but is much more durable, and chiefly about the upper part of the fauces." Since the above was written the bark has attracted the attention of many European Chemists and Scientists, who have naturally been hopeful of discovering in it alkaloidal elements similar to those existing in the true cinchonas. The *Pharmacopœia of India*, published in 1868, contained the name of the plant amongst its non-officinal drugs, and the author of the article recommended that, "in all future inquiries into the subject, this bark should be one of the first to which attention should be directed." The first chemist who appears to have made an analysis was **O'Shaughnessy,** who found that, notwithstanding the excellent antiperiodic effects yielded by use of the drug in his hands, a specimen obtained from the Calcutta Botanic Gardens contained no alkaloid.

In 1870 an analysis was again made by **Broughton,** who found that the bitter taste was due to the existence of *æsculin*, a substance which, when the bark became dry, underwent transformation into *æsculetin*, an almost tasteless compound. **Dr. Dymock,** commenting on this analysis, suspects that the bark actually examined was that of **H. obovatum,** not that of **H. excelsum,** as the latter, when dry, is extremely bitter. The latest analysis, one which has apparently definitely settled the question of the chemical constitution of the bark, was performed by **Mr. W. A. H. Naylor** in 1883. Of this the following summary is given by **Dr. Dymock:** "It appears that the bitter principle is not the glucoside *æsculin* nor its decomposition product, *æsculetin*, but an alkaloidal substance. That as such it is allied to *quinoidin*, *berberin*, and *paracin*. From *quinoidin* it differs in being optically inactive, and from its double compound containing relatively less platinum; from *berberin* it differs in that it contains a higher percentage of carbon, while its double compound also yields a larger amount of platinum; and from *paricin* it differs only in the percentage of hydrogen it gives. **Mr. Naylor** considers it to be a new alkaloid having a composition corresponding to the empirical formula $C_{24} H_{40} N_3$, and therefore to be an addition to the small class of bases devoid of oxygen. Besides "*Hymenodictyonine*," which is the name given to the new alkaloid, **Mr. Naylor** has separated a bitter neutral principle, represented by the formula $C_{22} H_{43} O_{10}$, which he thinks may be a decomposition product of a glucoside. **Mr. Naylor** has since succeeded in obtaining *hymenodictyonine* in a crystalline form, by extremely slow evaporation of its ethereal solution. When placed in contact with cold sulphuric acid the solution by transmitted light assumed a lemon-yellow colour, passing to a wine red,

MEDICINE.

and ultimately to a deep claret; by reflected light it presented a bronze appearance. It was found difficult to burn off the carbon, but combustion with chromate of lead yielded figures which agreed with those deduced from analysis of the platinum salt, estimation of the chlorine in the double salt, in the hydrochlorate of the alkaloid, and in the composition of its diethyl-iodide. The alkaloid proves to be dibasic with the formula $C_{23}H_{40}N_2$, and it is apparently a tertiary diamine. Its chemical properties display a close analogy with those of nicotine."

No experiments appear to have been made to determine the therapeutical virtues of either the alkaloid or of the bitter neutral principle when separated from the bark, so no definite comparison with the cinchona alkaloids can be made.

FODDER. Leaves. 521

Fodder.—The LEAVES are used as cattle fodder.

TIMBER. 522

Structure of the Wood.—Fine, close-grained, of a pale mahogany colour, weight 31lb per cubic foot. It is employed for making agricultural implements, scabbards, grain measures, palanquins, toys, and similar articles.

523

Hymenodictyon obovatum, *Wall.; Fl. Br. Ind., III., 36; Wight, Ic., t. 1159.*

Vern.—*Karvi, kari,* CH. NAG.; *Karwai* (bitterness), *sirid,* BOMB.; *Yella malla kai,* TAM.; *Mallai tanák,* MADURA.

References.—*Brandis, For. Fl., 268; Beddome, Fl. Sylv., t. 219; Dalz. & Gibs., Bomb. Fl., 117; Lisboa, U. Pl. Bomb., 85; For. Admins. Rep. Chutia Nagpur, 1885, 32; Indian Forester, III., 23; Gazetteer, Bombay, XV., Pt. I., 71.*

Habitat.—A handsome tree of the Western Peninsula, from Bombay to Travancore.

TIMBER. 524

Structure of the Wood.—Pale, mahogany-coloured, and close-grained. Beddome remarks that its timber is used by the natives for many purposes, and the writer of the *Kánara Gazetteer* mentions it as worthy of attention.

HYOSCYAMUS, *Linn.; Gen. Pl., II., 903.*

A genus of erect, coarse herbs, belonging to the Natural Order SOLANACEÆ, and comprising nine species, natives of Europe and Central Asia.

525

Hyoscyamus niger, *Linn.; Fl. Br. Ind., IV., 244;* SOLANACEÆ.

HENBANE.

Vern.—*Khurásání-ajváyan, khurásání yamání, khurásání jamáni* (*bazrúl* =seeds), HIND.; *Khorasani ajowan* (*bazrúl*=seeds), BENG.; *Khorasáni ajwáin,* N.-W. P.; *Dandúra, bazrbang, dentúrá, datúra, súra, damtúra, bangidewána* (*bazrul banj*=seeds in bazar), *Telingchi,* SIMLA; PB.; *Damtura,* SIND.; *Khurásáni-ajvan,* DEC.; *Khorasá' i-owa,* BOMB.; *Khórásáni-vóvá,* MAR.; *Khorásáni-ajmo, khorásáni-ajván,* GUZ.; *Kúrá-sáni-yomam,* TAM.; *Kúráshání-vámam, kurinji vámam,* TEL.; *Khurá-sáni-vómá, khurásáni-vádakki,* KAN.; *Párasikaya, máni,* SANS.; *Bazrul banj,* ARAB.; *Bazrul-bang,* PERS.

References.—*Roxb., Fl. Ind., Ed. Carey and Wall., II., 1237; Dalz. & Gibs., Bomb. Fl. Suppl., 62; Stewart, Pb. Pl., 156; Boissier, Fl. Orient., IV., 294; Pharm. Ind., 177, 460; Ainslie, Mat. Ind., I., 167; O'Shaughnessy, Beng. Dispens., 470; Moodeen Sheriff, Supp. Pharm. Ind., 159; U. C. Dutt, Mat. Med. Hind., 211, 312; Dymock, Mat. Med. W. Ind., 2nd Ed., 629; Fluck. & Hanb., Pharmacog., 463; Bent. & Trim., Med. Pl., t. 196; S. Arjun, Bomb. Drugs, 97; Murray, Pl. and Drugs, Sind, 156; Irvine, Mat. Med., Patna, 6; Med. Top., Ajmere, 124; Med. Top., Oudh, 31; Year-book, Pharm., 1876, 306; 1878, 288; Atkinson, Him. Dist., 739; Drury, U. Pl., 252; Birdwood, Bom. Pr., 61; Royle, Ill. Him. Bot., 280; Royle, Prod. Res., 237; Balfour, Cyclop., II., 138; Smith, Dic., 211; Kew Off. Guide to the Mus. of Ec. Bot., 101; Report of Agric. Dept., N.-W. P., 1882, 20; Report on Nilgiri Bot. Gardens, 1884-85, para. 5; Jameson, Saharanpur Garden's Report, 1854; Special*

Reports from :—M. A. Lawson, Esq., Govt. Botanist, Madras ; D. Prain, M.B., Roy. Bot. Gardens, Calcutta ; Surgeon Genl. with Govt. of Bombay, 1888 ; Dr. Moodeen Sheriff, J. F. Duthie, Esq., Dir. Bot. Dept., Northern India ; D. Hooper, Esq., Govt. Quinologist, Madras ; Govt. of Madras, 1882 ; Gazetteers : - N.-W. P., 173 ; II., 173.

Habitat.—A herb of the Temperate Western Himálaya from Kashmír to Garhwál, from 8,000 to 11,000 feet, distributed over Western and Northern Asia, Europe, and Northern Africa.

CULTIVATION. 526

Cultivation.—The first record of the cultivation of **Hyoscyamus** in India is made by **Dr. Royle**, in his *Illustrations, l. c.*, in 1839. That author remarks that Henbane, **Datura Stramonium**, and **Nicandra indica** (? **Physalis minima**, *Linn.*, VAR. **indica**) were at that time successfully grown and converted into extract for the medical depôts, in several stations in the plains of Northern India and in the Himálaya. He concludes his remarks as follows: "The extract of Henbane particularly was highly approved of by several medical officers, and was pronounced by **Mr. Twining**, after trial in the General Hospital of Calcutta, to be of most excellent quality." Later records exist of its more or less successful cultivation in Calcutta, at the Royal Botanical Gardens, Saharanpur under **Dr. Jameson**, in the neighbourhood of Agra and Ajmere, in Bombay at Hewra and Dapúri, and in the Nilghiris. The crops thus produced seem to have been almost entirely used in supplying the Government Medical depôts with leaves and extract, and no market with European druggists appears to have been formed. Thus in a report furnished by the Superintendent of the Botanical Gardens, Saharanpur, on the cultivation of **Hyoscyamus** during the past ten years, extract to the value of R1,939 and leaves to the value of R2,186 are said to have been supplied to the Government Medical depôts, and an entry occurs to the effect that in 1882-83 leaves to the value of R40, and extract to the value of R30, were supplied to an European firm in Bombay, but since then no extension of trade with commercial firms has been made. The seeds of the bazár also appear to be principally imported from Persia and Afghánistan, so that at the present day all **Hyoscyamus** employed in European medicine, except by Government, comes from Europe, and even the seeds offered for sale in bazárs are not obtained in this country. It is difficult to understand why this should be the case. The plant grows wild abundantly on the Himálaya; it is fairly extensively cultivated, apparently with a considerable amount of success in various parts of the country, and the supply ought to more than equal the demand. It seems possible, however, that the plant artificially cultivated, as it is in the hot climates of Saharanpur and parts of Bombay, where the surrounding conditions are very different to those which the wild plant enjoys, may be deficient in some of the properties which render it medicinally valuable. It is, therefore, highly desirable that an accurate analysis of the Indian plant, in comparison with the European, and of the cultivated plant of the plains in comparison with the wild Himálayan stock, should be made. Such an analysis would definitely determine the relative values of each, and should it prove favourable, might afford a stimulus to the employment of the native drug, not only in India, but in other countries. The plant can be grown and the extract prepared so that the product may be sold much more cheaply than that imported from Europe. Thus the average rate at Saharanpur during the past ten years has been R1-4 per ℔ for the extract, and annas 4 per ℔ for the leaf, while the last quotations in England for the latter is from 10*d.* for foreign to 8*s.* per ℔ for English biennial, and for the former 6*s.* 3*d* per ℔.

OIL. Seeds. 527

Oil.—The SEEDS are mentioned by **Dr. Cooke** as yielding an oil. **Dr. Dymock** describes a specimen he obtained from the Poona Botanical Gardens as bright yellow and rather thick. It is apparently not employed for any purpose.

MEDICINE. History. 528

Medicine.—HISTORY.—Henbane has been employed and valued as a drug in Europe and Northern Asia from remote times. **Baron Hammer-Purgstall** believes that the "*bendj*" (or *banj*), the Coptic of which is "*nibendj*" is undoubtedly the "*nepenthe*" of **Homer.** It is the ὑοσκύαμος of the Greeks, and was particularly commended by **Dioscorides** as a medicine. By **Celsus** it was employed as an external application in ophthalmia, and as an anodyne, and **Pliny** mentions it in various passages. In oriental literature the earliest record of the plant is, as is natural from its habitat, to be found in Arabic and Persian writers, and the evidently foreign origin of the Sanskrit names indicate that it was unknown to ancient Hindu medicine. The Muhammadan writers of Arabia and Persia regarded the plant as intoxicating, narcotic, and anodyne, and employed it extensively as an external application in cases of inflammatory swellings, rheumatism, and gout. **Dr. Dymock,** in his interesting account of the drug, also mentions that a mixture of the powdered seeds and pitch was employed to stop painful hollow teeth, and as a pessary in painful affections of the uterus; that the juice or a strong infusion of the seeds was dropped into the eye to relieve pain; and that the seeds made into a paste with mare's milk and tied up in a piece of wild bull's skin was supposed to prevent conception when worn by a woman. In the early centuries of the middle ages henbane was also one of the principal medicines of the physicians of Europe, by whom it was known as Hyoscyamus, Symphoniaca, Jusquiamus, Henbane, Henbell, or Chenille; but it appears to have fallen into disuse till the experiments and recommendations of **Storck** in the beginning of the nineteenth century brought about its reintroduction into medicine. At the present day in India, the only part of the plant generally employed is the seed, which is imported from Persia. **Dymock** considers that the greater quantity, if not all of the seed thus obtained, is really the product of **H. albus,** *Linn.* They are believed by the natives to be stimulant or heating, narcotic, astringent, digestive, and anodyne.

Foliage. 529 **Green tops. 530** **Seeds. 531**

The parts of medicinal value are the FOLIAGE and GREEN TOPS, and the SEED, but the latter is not officinal in the Indian Pharmacopœia, in which the preparations laid down for general use are a tincture, dose 20 to 60 drops, and an extract, dose 5 to 10 grains, both prepared from the leaves. In Europe dried henbane is sold in three forms: *1st,* the foliage and green tops of the annual plant, *2nd,* the leaves of the first year of the biennial plant, and *3rd,* the foliage and green tops of the biennial plant; of these the last form is regarded as the best.

Chemistry. 532

CHEMICAL COMPOSITION.—The most important constituent of Henbane, *Hyoscyamin,* was first obtained by **Geiger** and **Hesse** in 1833. It exists most richly in the seeds, from which it was first separated by **Hohn** in 1871, and again by an improved method by **Thibaut** in 1875. The authors of the *Pharmacographia* describe the alkaloid as follows: "*Hyoscyamin* ($C_{15}H_{23}O_3$) is easily decomposed by caustic alkalis. By boiling with baryta in aqueous solution, it is split into *Hyoscin,* $C_6H_{13}N$, and *Hyoscinic acid,* $C_9H_{10}O_3$. The former is a volatile oily liquid of a narcotic odour and alkaline reaction; by keeping it over sulphuric acid it crystallizes and also yields crystalline salts. *Hyoscin* may be closely allied to *conin,* $C_8H_{15}N$. *Hyoscinic acid,* a crystallizable substance, having an odour resembling that of empyreumatic benzoic acid, melts according to **Hohn** at 105°; *tropic acid* (a product of a similar decomposition of *atropin*) melting at 118°, agrees so very nearly with hyoscinic acid that further researches will probably prove these acids to be identical." The seeds also contain 26 per cent. of fatty oil, and the extract employed in medicine is rich in potassic nitrate and other inorganic salts.

MEDICINE. Physiological action. 533 PHYSIOLOGICAL ACTION.—**Dr. John Harley**, who has made special investigations into the action of the drug under consideration, writes: "Henbane, like belladonna, produces dilatation of the pupil, somnolency, a parched condition of the tongue and mouth, and in sufficient dose delirium. The general action on the secretions and nervous system agree in all respects with that of belladonna, and the results of its action is the same; but the influence of henbane on the cerebrum and nerve centres is somewhat greater, while its stimulant action on the sympathetic is less. Both drugs directly stimulate the heart, but after moderate doses the action of henbane results in a sedative effect. Small doses are sedative and tonic to the heart; large doses excite its action, and excessive doses depress it almost as readily as those of belladonna. Both drugs produce relaxation of the voluntary muscles, and of the occluding fibres of the intestine and bladder."

Uses. 534 MEDICINAL USES.—**Hyoscyamus** is especially valuable as a nervous, sedative, and narcotic in cases of maniacal excitement, sleeplessness, and nervous depression. This action appears to be most largely due to the syrupy alkaloid above mentioned, *viz.*, Hyoscin, which may be administered alone, by hypodermic doses of $\frac{1}{100}$ gr. or more of the Hydriodate. The drug is also a valuable sedative in cases of cough, especially in the early stage of bronchitis and in asthma, in cases of which its relaxing effect on muscular tissue is highly beneficial. It is a decided laxative and carminative, and appears to possess a peculiarly sedative effect in irritable conditions of the genito-urinary system. As an external application it has been employed in neuralgia and rheumatism, painful glandular enlargements, irritable ulcers and hæmorrhoids. In diseases of the eye its sedative, anodynec and midriatic actions are most valuable. **Mr. Duthie** mentions the employment of henbane by natives near Saharanpur as an ingredient in a *massala* administered to trotting bullocks during the hot and rainy seasons, as a sort of "condition powder."

SPECIAL OPINIONS.—§ "I have found the tincture in drachm doses in combination with Bromide of Potassium useful in cases of maniacal excitement, sleep being produced" (*Surgeon S.H. Browne, M.D., Hoshangabad, Central Provinces*). "The tincture of this has proved successful in three cases of Beri Beri" (*Surgeon W. F. Thomas, Madras Army, 33rd M. N I., Mangalore*). "A very useful anodyne, sedative, and antispasmodic" (*Surgeon G. Price, Shahabad*). "I have found the extract very useful in irritating cough accompanied with tenacious mucus in the air passage" (*Civil Surgeon W. Moir, Meerut*). "Very useful in mania and in delirium tremens" (*Surgeon-Major E. G. Russell, Superintendent, Asylums, Calcutta*).

FODDER. Leaves. 535 **Fodder.**—The young LEAVES are said to be eaten by cattle.

TRADE. 536 **Trade.**—The value of the seed in Bombay is R7 per maund of 37½lb. As already mentioned the leaves may be obtained at Saharanpur at an average cost of 4 annas per lb, and the extract, carefully prepared, for R1-4.

HYPECOUM, *Linn.; Gen. Pl., I., 54, 965.*

A small genus of the Natural Order FUMARIACEÆ, comprising four or five species, natives of the Mediterranean region and Temperate Asia.

537 **Hypecoum procumbens,** *Linn.; Fl. Br. Ind., I., 120;* FUMARIACEÆ.

THE HORNED CUMIN.

References.—*Boiss., Fl. Or., I., 124; Aitchison, Bot. of Afgh. Del. Com., 32; Pharmacog. Indica, I., 117; Dymock, Mat. Med. W. Ind., 2nd Ed., 54; Murray, Pl. and Drugs, Sind, 77.*

Habitat.—A low, annual weed of cultivation of the drier parts of the Panjáb and the Salt Range, distributed to Western Asia and the Mediterranean Region.

Medicine.—Murray, in his *Plants and Drugs of Sind*, describes the horned-cumin as yielding a juice which has the same effect as opium, and its leaves as diaphoretic. The authors of the *Pharmacographia* are of opinion that it is the ὑπηκόον of Dioscorides and Hypecoum of Pliny. **MEDICINE. Juice. 538 Leaves. 539**

Hyperanthera Moringa, *Vahl.*, see **Moringa pterygosperma,** *Gærtn.*, [Vol. V.

HYPERICUM, *Linn.; Gen. Pl., I., 165.*

A genus of herbs, shrubs, or small trees, belonging to the Natural Order HYPERICINEÆ, and comprising 160 species distributed over the temperate regions of the globe. Of these about 20 are Indian, natives of the hills, especially the Himálaya, but very few are at present known to possess properties of economic importance. The writers of the *Pharmacographia Indica* state that a plant exists in Persia described by Muhammadan writers as a species of Hyefárikún (Hypericon) which is locally named *Dádi* and *Jao-i-jádú*, or "magic barley," to which is ascribed the medicinal properties formerly attributed to the St. John's worts of Europe, namely, the power of expelling the demon of hypochondriasis, and of acting as a charm against witch-craft. The red juice of the flowers of Hypericum was also considered by ancient classical writers to be a signature of human blood; it was called ἀνδρόςαιμον by the Greeks, and was employed as an application to wounds. Several species are cultivated as ornamental shrubs.

Hypericum Hookerianum, *W. & A.; Fl. Br. Ind., I., 254; Wight, [Ic., t. 949;* HYPERICINEÆ. **540**

Syn.—H. OBLONGIFOLIUM, *Hook.*

Var. **Leschenaultii,** *Choisy;* H. TRIFLORUM, *Blume;* H. OBLONGIFOLIUM, *Wall.;* H. CHOISIANUM, *Wall.*

Vern.—*Tumbomri*, LEPCHA.

References.—*Gamble, Man. Timb., 21; List of Trees, Shrubs, &c., of Darjiling, 6.*

Habitat.—A glabrous shrub from 6 to 8 feet in height, found in the Sikkim Himálaya from 8,000 to 12,000 feet; the Khásia mountains, from 4,000 to 6,000 feet; and the Nilghiris.

Structure of the Wood.—Close-grained, moderately hard, with marked annual rings. **TIMBER. 541**

Arboriculture.—Gamble mentions that this shrub is frequently employed for hedges in Sikkim. **ARBORICULTURE. 542**

H. patulum, *Thunb.; Fl. Br. Ind., I., 254.* **543**

Syn.—HYPERICUM URALUM, *Ham.;* H. OBLONGIFOLIUM, *Wall. non Choisy.*

Var. **attenuatum,** *Choisy.*

Vern.—*Túmbhúl*, BEHARI.

References.—*Irvine, Mat. Med. Patna, 113; Atkinson, Him. Dist., 306.*

Habitat.—A small glabrous shrub found throughout the Temperate Himálaya (Sikkim excepted) from Bhotán to Chamba and Simla, also in the Khásia mountains and Yunan.

Medicine.—The scented SEEDS are employed as an aromatic stimulant in Patna, to which they are imported from Nepál. Dose 2 to 10 grains; price 2 annas per ℔ (*Irvine*). **MEDICINE. Seeds. 544**

H. perforatum, *Linn.; Fl. Br. Ind., I., 255.* **54**

Var. **debile,** *Royle, mss.*

Vern.—*Bassant, dendlu*, HIND.; *Bassant, dendlu*, PB.

References.—*Stewart, Pb. Pl., 30; Journ. of a Bot. Tour in Hazara and Khágán, in Journ. Agri.-Hort. Soc. of India, XIV., 36—42; Yearbook Pharm., 1874, 623; Atkinson, Him. Dist., 306; Balfour, Cyclop., II., 138; Kew Off. Guide to the Mus. of Ec. Bot., 15.*

Habitat.—A perennial herb, found in the Temperate Western Himálaya from Kumaon, altitude 6,000 to 9,000 feet, to Kashmír, altitude 3,000 to 6,500 feet.

DYE. 546 **Dye.**—The dried herb boiled in alum communicates a yellow or yellowish-red colour to wool, silk, &c.

MEDICINE. 547 **Medicine.**—The herb is bitter and astringent, and was recommended by Arabic writers as a detersive, resolutive, anthelmintic, diuretic and emmenagogue and externally as excitant, but it does not appear to be used in modern medicine. The authors of the *Pharmacographia Indica* give the following account of the chemical composition of the flowers: "When the flowers of **H. perforatum** freed from their calices and dried are exhausted with absolute alcohol and the tincture is evaporated, a soft residue is left of a red colour (hypericum red) together with volatile oil. If the flowers are exhausted with water, then with dilute alcohol, well dried after exhaustion, and the colouring matter extracted from them by ether, it remains on evaporation as a blood red resin, having an odour of chamomile. It melts below 100° and does not yield ammonia by dry distillation. It is insoluble in water and in dilute acids. By aqueous ammonia, potash, and soda, it is coloured green and dissolved; the saturated solution is red by reflected light, but exhibits after dilution a green colour by transmitted light. The ammoniacal solution leaves on evaporation a neutral blood-red resin having the odour of hypericum, soluble with yellow colour in water, and giving off ammonia when treated with potash. The red combines also with the alkaline earths, earths proper, and heavy metallic oxides, its alcoholic solution precipitates the alcoholic solution of chloride of calcium, also neutral acetate of lead, and ferric chloride. It dissolves in alcohol, more readily in ether, with wine-red to blood-red colour, also in volatile oils, and in warm fixed oils (*Buchner*). According to **Marquart** the colouring matter of the fresh flowers is a mixture of anthocyan and anthoxanthin separable by exhausting with alcohol and treating the residue with water."

Hypoxis orchioides, *Kurz,* see **Curculigo orchioides,** *Gærtn.*; Vol. II., 650.

HYPTIANTHERA, *W. & A.; Gen. Pl., II., 94.*

548 **Hyptianthera stricta,** *W. & A.; Fl. Br. Ind., III., 121;* RUBIACEÆ.

Syn.—RANDIA STRICTA, *Roxb.*; MACROCNEMUM STRICTUM, *Willd.*; RONDELETIA STRICTA, *Roxb.*; HYPOBATHRUM STRICTUM, *Kurz*; RUBIACEA, *Wall., Cat., 8138 in part, 8313, 8307.*

References.—*Roxb., Fl. Ind., Ed. C.B.C., 177; Brandis, For. Fl., 274; Kurz, For. Fl. Burm., II., 50; Gamble, Man. Timb., 229; Indian Forester, XI., 370; XIV., 343.*

Habitat.—An evergreen branched shrub, 3 to 6 feet high, common in Northern India and Bengal, ascending the outer Himálaya and Khásia Hills to 4,000 feet; also a native of Chittagong and Burma.

TIMBER. 549 **Structure of the Wood.**—Brown, moderately hard, close grained, weight 56℔ per cubic foot.

HYSSOPUS, *Linn.; Gen. Pl., III., 1187.*

550 **Hyssopus officinalis,** *Linn.; Fl. Br. Ind., IV., 649;* LABIATÆ.

Vern. *Zúfah yabis,* HIND., ARAB., PERS.

References.—*Boiss., Fl. Orient., IV., 584; Stewart, Bot. Tour in Hazara in Journ. Agri.-Hort. Soc. of India, XIV., 26; Ainslie, Mat. Ind., I., 177; O'Shaughnessy, Beng. Dispens., 488; Dymock, Mat. Med. W. Ind., 1st Ed., 508; S. Arjun, Bomb. Drugs, 100; Year-book of Pharmacy, 1879, 466; Birdwood, Bomb. Pr., 62; Balfour, Cyclop., II., 139; Smith, Dic., 218; Treasury of Bot., I., 616.*

The Hyssop. (*J. Murray.*) **HYSSOPUS officinalis.**

Habitat.—An undershrub of the Western Himálaya, from Kashmír to Kumáon. Distributed to Eastern Europe and Western Asia.

Medicine.—It appears doubtful whether the plant imported into India from Syria and bearing the above vernacular name is really identical with the true **Hyssopus officinalis.** Alston was of the opinion that at any rate it did not correspond with the ὑςωπος of the Greeks, and the Esof of the Hebrews. Ainslie also thought the plants were probably different, and of later years the doubt has been again raised by **Dymock.** The true Hyssop also, though fairly plentiful in many parts of the Himálaya, does not appear to be employed locally by the natives, nor does it form an article of trade with the plains. Thus **Mr. Lace,** in a note on a specimen collected by him near Quetta, remarks that the natives do not appear even to give a name for the plant. Whatever the plant *Zúfah yabis* may be, it is classed by the Arabians amongst their anthelmintics, stimulants, and deobstruents. The officinal hyssop of Europe has long been known as a tonic and stimulant, and was at one time in great repute as a remedy for nervous diseases. **Pliny** also thought it useful in chest affections, and **Celsus** regarded it as anthelmintic. MEDICINE. 551

ICELAND MOSS.

Iceland Moss, see **Cetraria islandica,** *Achar.;* LICHENES; Vol. II., 262.

ICHNOCARPUS, *R. Br.; Gen. Pl., II., 717.*

Ichnocarpus fragrans, *Wall.*, see **Trachelospermum fragrans,** *Hook. f.:* [APOCYNACEÆ.

I **I. frutescens,** *R. Br.; Fl. Br. Ind., III., 669; Wight, Ic., t. 430.*

Syn.—ECHITES FRUTESCENS, *Roxb.;* APOCYNUM FRUTESCENS, *Linn.;* ICHNOCARPUS RADICANS, *Wall.;* I. DASYCALYX, *Miq. Syama, Roxb. in As. Res., I., 261.*

Vern.—*Dudhi* or *siama-latá,* HIND.; *Dudhi, shyámá* or *syama latá,* BENG.; *Dudhi,* KOL.; *Dudhi lota,* SANTAL; *Bhori* (Chanda), C. P.; *Nallatiga* (of **Roxb.**), *illu katte, nalla tige, munta gajjanamu* (of **Elliot**), TEL.; *Tansapai,* BURM.; *Kirri-wel,* SING.; *Sárivá,* SANS.

References.—*Roxb., Fl. Ind., Ed. C.B.C., 245; Brandis, For. Fl., 327; Kurz, For. Fl. Burm., II., 185; Dalz. & Gibs., Bomb. Fl., 147; Grah., Cat. Bomb. Pl, 113; Sir W. Jones, V., 95; Sir W. Elliot, Fl. Andhr., 70, 119, 127, 165; Royle, Ill. Him. Bot., I., 271; Wight, Illust., Vol. II., 162; Rev. A. Campbell, Econ. Prod., Chutia Nagpur, No. 9220; Pharm. Ind., 138; O'Shaughnessy, Beng. Dispens., 54; U. C. Dutt, Mat. Med. Hind, 195, 317, 320; Dymock, Mat. Med. W. Ind., 2nd. Ed., 509; Atkinson, Him Dist. 313, 739; Jour. Agri.-Hort. Soc. Ind.* (Old Series), *VI., 50; X., 16; Gazetteer, Bombay* (*Kanara*), *XV., 438; Gazetteers:—N.-W. P., Bundelkhand, 82; of Agra, lxxiv.; Mysore and Coorg, I., 56; Settlement Rept.:—Chanda, Central Provinces, App. VI.*

Habitat.—An extensive climber, met with on the Western Himálaya, from Sirmur to Nepál, altitude 1,000 to 2,000 feet; and from the Upper Gangetic plains to Bengal, Assam, Sylhet, Burma, South India, and Ceylon. In some of the hotter valleys of the North-West Himálaya, it ascends to 4,000 feet, being, for example, frequent below Simla.

FIBRE. Bark. 2 **Fibre.**—The BARK yields a good fibre.

Medicine. Root. 3 Stalks. 4 Leaves. 5 **Medicine.**—The ROOT possesses alterative and tonic properties, and is employed as a substitute for Sarsaparilla (*Conf.* with **Hemidesmus indicus,** p. 219). The STALKS and LEAVES are used in the form of decoction in fevers. An INFUSION is recommended to be prepared from 2 oz. of the root in a pint of boiling water: of this the dose should be 2 or 3 oz. twice daily. **Taylor** in his *Medical Topography of Dacca* (*p.* 55) says: "This plant, which derives its name from its creeping stems, abounds with an acrid milky juice. The stalks and leaves are used in the form of decoction in fever." **Dymock** (*Materia Medica of Western India*) discusses this drug conjointly with **Hemidesmus indicus.** He writes: "They are used together and are considered to have similar properties. When, however, *sárivá* is used in the singular number, it is the usual practice to interpret it as meaning **Ichnocarpus frutescens** (*Dutt, Hind. Mat. Med., p. 915*). **Ichnocarpus frutescens** is common in the southern part of the Presidency. The Hindus consider its roots and those of **Hemidesmus** to be demulcent, alterative and tonic, and prescribe them in dyspepsia, skin diseases, syphilis, &c. They are generally combined with bitters and aromatics." The wide area over which a knowledge of the medicinal properties of this plant occurs argues actual merit. Thus from the extreme southern parts of India to the north and east, the people use a decoction in the treatment of fever. In the *Journals of the Agri.-Horticultural Society,* for example, repeated mention is made of this property; even the wild Naga tribes, on the north-eastern frontier of Assam, are familiar with it. The **Rev. J. Long** (*Indigenous Plants of Bengal*) states that the roots are commonly employed in hospitals under the name of country Sarsaparilla.

Structure of the Wood.—White, soft, without heartwood. TIMBER. 6

Icthyocolla, see **Acipenser,** Vol. I., 83; also **Fish,** Vol. III., 369–397;
Ignatia amara, and [**Isinglass; & Sharks' Fins.**
Ignatius' Bean, see **Strychnos Ignatii,** *Bergius*; LOGANIACEÆ; Vol. VI.
Ilang-ilang, see **Cananga odorata,** *H. f. & T. T.*; Vol. II., 93.

ILEX, *Linn.; Gen. Pl., I., 356, 997.*

Ilex Aquifolium, *Linn.*; ILICINEÆ. 7

THE EUROPEAN HOLLY.

References.—*O'Shaughnessy, Beng. Dispens., 271; U. S. Dispens. (Ed. XV.), p. 1668; Indian Forester, V., 465; XIII., 60, 68; Clarke, Linn. Soc. Jour., XXV., 11.*

Habitat.—A small tree, native of Europe; introduced into India and commonly found in gardens within the temperate tracts. According to **Mr. C. B. Clarke,** it may also be a native of the Naga Hills on the frontier of Assam, for he collected a holly, not in flower or fruit, but which possessed leaves that matched accurately those of the common holly. The writer found in many parts of Manipur and the Naga Hills, bushy, sterile forms of what he took to be **Ilex dipyrena,** but possessing very thorny, broad ovate leaves which much resembled in form and texture those of the true holly. Manipur is peculiarly a country of hollies, no less than five species having been collected, all of which in their higher limits were seen to become flowerless and form stunted bushes with very spinose leaves. Of these five species three are undescribed. The Manipur hollies are: **I. dipyrena,** *Wall.*; **I. Thomsoni,** *Hook. f.*; **I. sikkimensis,** *King, mss.*; **I. Duniana,** *Watt, mss.*; and **I. monopyrena,** *Watt, mss.*

Dye.—The common holly contains the principle *ilixanthine*, a yellow colouring substance soluble in concentrated hydrochloric acid. With alum it dyes yellow and with ferric chloride green. DYE. 8

Medicine.—**O'Shaughnessy** (*Bengal Disp.*) remarks that according to **Dr. Rousseau** the LEAVES and BARK form a useful remedy in intermittent fever; they are also said to be emollient and diuretic. The *United States Dispensatory* gives an interesting account of this remedy which formerly enjoyed in France a high reputation. Its property was said to depend upon a bitter principle *ilicin*. A yellow colouring substance called *ilexanthin* and a peculiar acid known as *ilicic acid* have been obtained by **Dr. F. Moldenhauer.** The BERRIES are said to be purgative, emetic and diuretic, 10 or 12 being enough to act on the bowels. MEDICINE. Leaves. 9 Bark. 10 Berries. 11

Structure of the Wood.—White and hard, much used for ornamental purposes and for calico-printers' blocks. TIMBER. 12

Domestic.—The inner bark affords a viscid substance used as BIRD LIME. The juice is boiled and mixed with a third part of nut-oil. DOMESTIC. Bird-lime. 13

I. dipyrena, *Wall.; Fl. Br. Ind., I., 599.* 14

Vern.—*Kaula, karaput, munasi, gulsima,* NEPAL; *Shangula, kandlar, kaniru, karelu, dinsu, krúcho, kalúcho, diúsa, dodru, drúnda, kandara, kadera, kateru* (SIMLA), PB.

References.—*Brandis, For. Fl., 76; Gamble, Man. Timb., 82; Stewart, Pb. Pl., 40; Journal of Bot. Tour. in Hazara by Stewart (Agri.-Hort. Soc. Jour., XIV., p. 31); O'Shaughnessy, Beng. Dispens., 272; Indian Forester, IX., 198; XI., 284; XIII., 60; Atkinson, Him. Dist., 307; Kew Off. Guide to Bot. Gardens and Arboretum, 145.*

Habitat.—A small evergreen tree of the Himálaya, from the Indus to Bhután and Manipur; found in temperate tracts between 5,000 and 9,000 feet above the sea.

A remarkable **Ilex**, which the writer provisionally accepts as a form of this species, runs to a large wide spreading tree, the leaves being sparingly spinescent. A good example of this exists in Simla, which is peculiar in forming minute yellow seedless fruits in March (as the flowers of next year appear) instead of producing the large orange coloured fruits of **Ilex dipyrena** which in December constitute a striking feature of the winter vegetation of many parts of the Western Temperate Himálaya. This may be but the staminate condition of the species, a condition which **Darwin** (*Origin of Species, p. 73*) describes as observed in **I. Aquifolium**—the European Holly. But it is worth adding that, in the Simla neighbourhood, all the other examples of **I. dipyrena** (so far as the writer has observed), with the single exception of the tree here described, are hermaphrodite. (*Conf.* with *Darwin, Forms of Flowers, p. 297,* where the English holly is stated to be diœcious or as having only imperfect stamens in the pistilate flowers.)

FODDER. 15 **Fodder.**—The leaves are occasionally given as fodder for sheep.

TIMBER. 16 **Structure of the Wood.**—White, hard, close-grained. Weight 46℔ per cubic foot. Is not much esteemed.

17 **Ilex Godajam,** *Colebr.; Fl. Br. Ind., I., 604.*

TIMBER. 18 **Structure of the Wood.**—Kurz remarks that this species yields a whitish wood, which turns grey, is rather heavy, fibrous and tough, but close-grained.

19 **I. insignis,** *Hook. f.; Fl. Br. Ind., I., 599.*

Vern.—*Lasuni,* NEPAL.

Reference.—*Gamble, Man. Timb., 83.*

Habitat.—A small evergreen erect tree, met with in Nepal and Sikkim at an altitude of 7,000 feet.

TIMBER. 20 **Structure of the Wood.**—White, soft, close-grained. Weight 40℔.

21 **I. malabarica,** *Bedd., Fl. Sylv., t. 143; Fl. Br. Ind., I., 600.*

Syn.—ILEX WIGHTIANA, *Dalz. & Gibs.* (not of *Wall.*)

Habitat.—A large tree met with on the Western Gháts from the Konkan southwards, ascending to 3,000 feet in altitude.

TIMBER. 22 **Structure of the Wood.**—Hard, yellowish-white. **Mr. Lisboa** (*Useful Pl. Bombay, 48*) says it is much used for planks, platters, and building purposes.

23 **I. paraguayensis,** *St. Hilaire.*

PARAGUAY OR MATE TEA; JESUITS' TEA.

Habitat.—A native of Brazil and of Paraguay, now extensively cultivated experimentally in most countries.

FOOD. 24 **Food.**—From time immemorial, the inhabitants of Brazil and Paraguay have been in the habit of drinking a beverage prepared from the leaves of this holly much after the fashion of the use of tea in China. These leaves possess a pleasant aroma and a principle analogous to that in tea and coffee. They have the advantage of cheap and easy preparation, since it has been found unnecessary to cure and roll them by so expensive a process as that adopted with tea. They are scorched and dried while still attached to the branches; then removed by being threshed, ground to a powder and packed in skins or leathern bags. The leaves are collected from the bushes every two or three years from December to August. Although originally and even now to a large extent collected from the wild plant, this holly is also cultivated to afford a more accessible supply of leaf, but the trade though large cannot be said to have extended to Europe. In 1884, it is stated that 8,000,000℔ of leaves were used in South America (the bulk being exported from Paraguay) in place

FOOD.

of tea, which though gaining ground has by no means displaced the indigenous article. In the Argentine Republic it is stated the consumption amounts to 27,000,000℔ per annum, or about 13℔ a head, while of coffee and tea only 2℔ a head are consumed.

The use of *mate* leaves has been long known and various attempts have been made to introduce this cheap tea into Europe with little or no practical results. It has even been cultivated in France and other parts of the Continent, but interest is if anything lessening, a result doubtless largely due to the cheapening of tea. The first mention apparently (by an Indian writer) of *mate* is that given by Dr. Ainslie (*Mat. Med., I., 436-437*), but *he* only refers to the substance in an account of tea substitutes. **St. Hilaire,** Dr. Ainslie remarks, urges that there is a great difference between the Brazilian and the Paraguayan *mate,* and he further states that **Lambert** figures and describes the plant in an appendix to the second volume of his description of the GENUS **Pinus.** In the writer's copy of that work the appendix in question is not given, but there can be no doubt the Paraguay Holly was experimentally cultivated in England at the beginning of the century. The first plant grown in India was apparently that obtained by the Agri.-Horticultural Society of Madras from Kew in the year 1870. It may now be seen in most public parks and gardens, and in Lucknow it is said to flower and fruit. It is thus probable that the plant could be cultivated in any of the provinces of India, and afford a cheap and wholesome beverage to the millions of people who for generations will most probably be too poor to purchase tea. Various reports have appeared, and an extensive correspondence been conducted, but as yet the cultivation of *mate* in India has not passed beyond the experimental stage, and the plant is only grown for ornamental purposes. (*Conf.* with *Johnston's Chemistry of Common Life* (*Church's Ed.*), *pp. 135-140.*)

The decoction is prepared by pouring boiling water over a small quantity of the powder. The infusion is imbibed through a straw or tube with perforated end to prevent the powder being drawn up. The beverage is said to be pleasant and palatable, but if too largely consumed is purgative or even emetic. According to the published analysis of this substance it contains from 0·450 to 1·5 of thiene, and 20·880 of caffeotanic acid, also 2·830 of gum and 5·902 of resin with other less important ingredients. (For the analysis of *mate,* see *Year-book of Pharmacy, 1878, p. 122,* also *1884, pp. 218-220.*) It has a balsamic odour and bitter taste, but may be softened with sugar and milk. It is said to have an advantage over coffee and tea in its mild aperient property and its bitter flavour is regarded as giving it a tonic stimulating influence. **Mr. O'Conor** of the British Legation, Brazil, calls attention to *mate,* and says it is more fortifying and alimentary than either tea or coffee and much more wholesome. **Mr. Reid** remarks that it is more invigorating and sustaining than tea. **Christy** (*New Commercial Plants*), from whom the last two statements have been compiled, gives numerous opinions in favour of this substance, and advertises his willingness to supply it to the trade. One of the most instructive accounts of this substance appeared in the *Journal of Botany, January 1842,* and was reprinted in various publications including the *Journal of the Agri.-Horticultural Society of India* (*Vol. II., Selections, 5 to 16*).

Ilex theæfolia, *Wall.; Fl. Br. Ind., I., 601.* 25

Syn.—I. GAULTHERIÆFOLIA, *Kurz.*

References.—*Kurz, For. Fl. Burm., I., 245; Gamble, Man. Timb., 82.*

Habitat.—A moderate-sized evergreen tree, found near Darjīling and distributed thence to the Khásia Hills, Manipur, and Tenasserim.

TIMBER. 26 **Structure of the Wood.**—White, soft, close-grained, with white concentric lines, which seem to correspond to annual rings. Weight 39lb per cubic foot.

27

ILLICIUM, *Linn.; Gen. Pl., I., 18.*

A considerable Indian trade is done in importing into Bombay and Calcutta and distributing over the country, the Chinese fruit known as Star Anise. Until very recently, it has been customary to attribute that fruit to a plant named and described by **Linnæus** as **Illicium anisetum,** but **Sir J. D. Hooker** (*Bot. Mag., July 1888*) has shown that the true source of the spice is a plant not hitherto described and to which he has assigned the name of **I. verum.** He points out that **Linnæus** is not responsible for his species, having attributed to it the officinal fruit obtained from China, and that the Japanese species instead of now receiving the older name of **I. anisetum,** *Linn.*, might with propriety be called **I. religiosum,** *Sieb. et Zucc.* The South Chinese plant, referred by **Loureiro** to **I. anisetum,** *Linn.*, **Sir Joseph Hooker** regards as a doubtful species, but since it has yellow flowers, a six-leaved calyx and 30 stamens, it cannot, he thinks, be **I. verum.** The Japanese species (**I. anisetum vel religiosum**) has the peduncle arising single or in clusters of threes, and enclosed by bracts. It also possesses long spreading inner perianth segments, and hence the plant or plants referred to is precluded from belonging to the same section of the genus with **I. verum. Sir Joseph Hooker** figures his **I. verum** as having solitary stout axillary flowers borne on naked peduncles, the perianth segments being broad, rounded, and ciliate. He says that it is more nearly allied to the Indian species **I. Griffithii** and **I. majus** than to the Japanese, and that these species have globose flowers like **I. verum**; "but all differ from **verum** in the increased number of perianth-segments, stamens, and carpels." The **I. cambodgianum,** *Hance*, **Sir Joseph** remarks, is "a broad-leaved species with long-peduncled flowers," and is "a native of the Elephant Mountains in Cochin China."

Of **Illicium** there are known in all seven or eight species—two in China (or perhaps three including that of Cochin China); one in Japan; three in the Eastern Peninsula of India; and one or two in the lower basin of the Mississipi. The species discovered by the writer in Manipur, along with **I. cambodgianum,** connects the Chinese and remaining Indian forms with the Japanese, and it is somewhat remarkable that after passing round the globe, a form allied to the Japanese and like that species said to be poisonous, should reappear in almost the same latitude as its congeners. While it is probable that **I. verum** is the only species that possess the properties for which its fruits are so extensively traded in, the question as to the extent India could meet her own market cannot be definitely settled until the indigenous species have been more carefully studied. **Dr. Dymock** in his *Materia Medica* states that a form with 13 carpels and possessing none of the odour of the true Star Anise was offered for sale by a Calcutta merchant; commenting on that sample **Dr. Dymock** says there can "be little doubt that that article must be the produce of **I. Griffithii.**" Acting on that suggestion the writer instituted an enquiry on the subject, but the replies emphatically assert that in the Khásia Hills the fruits of **I. Griffithii** are never collected and consequently that there is no trade done in them. It appears therefore likely that India imports from Japan, the Straits or China, substitutes for, as well as the true anise from China, and that these substitutes may be the produce of **I. cambodgianum** and **I. religiosum,** or some undescribed species, but that they are in all probability not Indian fruits, since there is no evidence that any of the Indian trees are understood by the natives to be of economic value.

TRADE. 28

One of the earliest notices of the Indian trade in Star Anise occurs in *Milburn's Oriental Commerce (Ed. 1813, Vol II., p. 499).* The fruits are there said to be the produce of a tree growing in China and the Philippine Islands. The East India Company's sales of imports are given as follows:—

1805 . cwt. 266 valued at £2,947 or £11-1-7 a cwt.
1806 . " 101 " " £1,590 " £15-14-10 "
1807 . " 151 " " £1,943 " £12-17-4 "
1808 . " 711 " " £4,897 " £ 6-17-9 "

Milburn adds that 8 cwt. of aniseeds are allowed to the ton, and that the permanent duty thereon is £1-8-6, and the temporary or war duty 9*s.* 6*d.*, making £1-18 per cwt.

The earliest European mention of Star Anise (*Pharmacographia, p. 21*) is connected with the voyager **Candish** who brought it from the Philippines about A.D. 1588. **Clusius** purchased the fruit in London in 1601, and during the seventeenth century it reached England by way of Russia, hence known as *Cardamomum Siberiensis.* In 1690-92 **Kæmpfer** figured and described the Japanese plant *Somo* or *Skimmi*, which, as the learned authors of the *Pharmacographia* say, was, by subsequent writers, assumed to be the source of the Star Anise. **Thunberg** was the first to point out that the fruits of the Japanese plant were not so aromatic as those met with in trade. **Von Siebold** made the same observation and maintained that the Japanese plant was distinct from the Chinese species described by **Loureiro.** It was left, however, for **Dr. Bretschneider** to point out that the Japanese species was poisonous and to **Dr. Hance** in 1881 and about the same time to **Mr. Ford** (of the Hong Kong Botanic Gardens) to establish the independence of the true Star Anise of South China—to which **Sir J. D. Hooker** has given the name of **I. verum** – from **Loureiro's** Chinese plant and **Siebold's** Japanese species.

Illicium verum, *Hook. f.;* MAGNOLIACEÆ. 29

THE STAR ANISE OF CHINA; BADIANE, ANISETOILÉ, *Fr.;* STERNANIS, *Germ.*

I. (anisetum, *Linn.*) **religiosum,** *S. & L.* 30

THE JAPANESE SACRED ANISE TREE; THE SKIMMI OF JAPAN.

Except as a probable adulterant for the former, the latter (the Japanese) species is of little interest to India, and even the Japanese themselves use the Chinese fruit instead of that from their indigenous species.

The true Star Anise of Pakhoi in South China is by the Chinese, known as *Pakionui hiang* or eight-horned Fennel, a fact which, as **Sir J. D. Hooker** remarks, is due to the fruits, though commonly compared to anise, resembling more closely in point of test that of fennel.

Vern.—*Anásphal, sonf,* HIND.; *Bádián,* BOMB.; *Anásphal,* DECCAN; GUZ.; *Anáshuppu (anasí-pú,* according to **Ainslie**), TAM.; *Anásapuvu,* TEL.; *Nanat-poén,* BURM.; *Riziya naje (badiáne huttáic,* according to **Ainslie**), ARAB.; *Bádiyán, raziyanahe-khatái,* PERS.; *Bádiyáne-khatái-* TURKEY; *Skimoni, somo,* JAPANESE.

References.—*Gamble, Man. Timb., 4; Baillon, Nat. Hist. Pl., Vol. I., 146-151; Hooker, Bot. Mag., July 1888; Bretschneider, China Review, Vol. IX., 283; Pharm. Ind., 5; Ainslie, Mat. Ind., II., 19-20; O'Shaughnessy, Beng. Dispens., 7, 191; Moodeen Sheriff, Mat. Med. South India (a forthcoming work), 7; Dymock, Mat. Med. W. Ind., 2nd Ed., 22; Fluck. & Hanb., Pharmacog., 20-22; U. S. Dispens., 15th Ed., 195, 787, 1669; Bent. & Trim., Med. Pl., 10; S. Arjun, Bomb. Drugs, 5; Bidie, Cat. Raw Pr., Paris Exh., 1, 84, 103; Pharmacog. Ind., 39; Year Book, Pharm., 1873, 348, 349; 1879, 465; 1881, 163; 1885, 171; Irvine, Mat. Med. Patna, 14; Bidie, Report of Madras Drugs shown at the Calcutta International Exhibition; Baden Powell, Pb. Pr., 326; Birdwood, Bomb.*

Species of Star Anise.

Pr., 3; *Piesse, Perfumery*, 92, 93; *Agri.-Hort. Soc. Ind.* (*Trans.*), *V.*, 120; *Smith, Dic.*, 17, 18; *Treasury of Bot.*, *Vol. I.*, 618; *Kew Bulletin*, 1888, 173-175; *Kew Off. Guide to the Mus. of Ec. Bot.*, 8; *Kew Off. Guide to Bot. Gardens and Arboretum*, 78.

Habitat.—The Star Anise tree is a native of Pakhoi, South China. There are, however, three Indian species which may be briefly noticed in this place since they are not of such importance as to necessitate separate and independent accounts.

These Indian species are:—

31 1. **I. Griffithii**, *H. f. & T.; Fl. Br. Ind.*, *I.*, 40; *Griffith, Trans Agri.-Hort. Soc. Ind.*, *V.*, 120.

A shrubby species met with in copses in Bhután and the Khásia Hills.

Leaves elliptic-lanceolate, 2 to 4 by 1 to 2 inches in size, acute at both ends. *Flowers* 1½ inches in diameter, with a perianth of about 24 leaves, the outer whorl of 6 orbicular sepals, the inner of 18 oval petals, the innermost small and narrow: "Carpels 13, beak short, incurved; terminal depression well marked. Taste, at first none, but shortly bitter, with some acridity, and a flavour between that of cubebs and bay leaves" (*Holmes*).

32 2. **I. majus**, *H. f. & T.; Fl. Br. Ind.*, *I.*, 40.

A shrubby species (about 30 feet high) met with in Tenasserim at altitudes of about 5,500 feet.

Leaves obovate-lanceolate, 4 to 6 by 1½ to 2 inches in size, sharply acuminate and petiole 1 inch long. *Flowers* pink, pedicles 1 to 3 inches long, subterminal solitary or fascicled; sepals and petals about 16, orbicular, ciliate, inner broad-oval; filaments short, broader than the oblong anthers; ovaries spreading: "Carpels 11 to 13; terminal depression longer and shallower, and beak short and less incurved, than in the preceding. Taste strongly resembling mace, not bitter" (*Holmes*).

33 3. **I. manipurale**, *Watt mss.*

A large tree collected by the writer in the mixed Magnolia and Michelia forests of north and north-east Manipur bordering on the Naga Hills and Upper Burma.

Leaves elliptic, tapering at both ends (almost acuminate) and 3 to 4 by 1¼ to 2 inches in size; petiole ½ inch long. *Flowers* solitary or fascicled, axillary in the last two or three leaves of the twigs; flower buds bracteate scales five in number, 3 inner ovate obtuse, brown and caducous, and two longer, narrower, green and persistent ones at the base of the peduncle; peduncle 1 to 1½ inch long; outer perianth-segments 4 broad linear obtuse lemon-green with purple tips, inner 9 to 12 narrow linear (1 inch long and ¼ broad) pure white; stamens 20 short stout: carpels 8 to 12 tapering upwards and recurved outwards. Ripe fruits not seen.

The last mentioned species forms almost lofty trees and has a stem 30 to 40 feet high, before the branches occur, and these bear a dense dome of thick bright shining leaves. It is plentiful in the Kupra and Kunho forests near Mao—the frontier town between Manipur and the Naga Hills. It is not known to possess any useful properties, nor apparently has it assigned to it any distinctive name by the savage tribes in whose country it abounds. In most of its characteristics, especially the bracteated pedicles, it approaches most nearly to the Sacred Anise of Japan, but the writer while questioning the people around him found no one who attributed to it a poisonous quality such as the Japan and Florida species are said to possess. It was in flower in February at an altitude of 7,500 feet near Mao, and in April at 8,500 feet, far to the east, at Ching Sow, on the Burmese frontier.

Mr. E. M. Holmes, from whom the description of the fruit of **I. Griffithii** and **I. majus** have been derived, affirms that the fruit of the Japanese Anise when wetted and laid on a piece of blue paper reddens it immediately and strongly, while the Chinese Star Anise causes only a very faint red colouration. The fruits of **I. Griffithii** and **I. majus** produce no such reaction.

Oil.—The eight, carpelled fruit of this plant yields by distillation with water an essential OIL somewhat resembling that of aniseed for which it is used in England. Oil. 34

The authors of the *Pharmacographia* remark that "the volatile oil amounts to four or five per cent. Its composition is that of the oils of fennel or anise." It remained fluid at a temperature below 8°C., but soon solidified when a crystal of anethol was brought in contact with the oil. "The crystallized mass began to melt again at 16° C.; the oils of anise and Star Anise possess no striking optical differences, both deviating very little to the left." **Eykman** (*Phar. Jour. & Trans., XI., 1048*) gives a comparative analysis of the oils of Anise, Fennel, the true Star Anise of China and the Sacred Star Anise of Japan. The differences between the true Star Anise and that of Japan may be here briefly indicated. The former consists (according to **Eykman**) chiefly of solid and liquid anethol; its melting point is O°C.; specific gravity 0·978; molecular rotation O° to 0·4°; action of alcoholic hydrochloric acid—colourless; chloral reagent—colourless, then beautifully red; ammoniacal silver solution—after 24 hours no reduction. The latter consists of much of a turpentine, boiling at 173° to 176° C., liquid anethol boiling at 232° to 233° C.; melting point—not solid when cooled to 20° C.; specific gravity 1·006, molecular rotation—8·6°; action of alcoholic hydrochloric acid—colourless, afterwards blue; chloral reagent—colourless, afterwards dirty brown-yellow; ammoniacal silver solution—reduction in a few hours. In the volatile oil of the leaves **Eykman** discovered eugenol, shikimen and shikimol—the second a terpine and the last identical with safrol.

The seeds also contain a large quantity of a FIXED OIL. **I. religiosum** was found by **Eykman** to contain as much as 30 per cent. of fixed oil. By experiments on dogs he discovered that neither the essential nor the fatty oil represent or contain the poisonous principle. He succeeded, however, isolating a poisonous principle as a crystallizable substance. This he named *Sikimine*. This analysis has been carried still further by **Eykman** and the Japanese plant found to contain proto-catechuic acid, shikiminic acid, and shikimipicrin, the last mentioned being the actual poisonous principle. It forms large transparent crystals, freely soluble in water and may be represented by the formula $C_7H_{10}O_3$.

Medicine.—While the *Star Anise* has long been employed in China and Japan, but has in India only come into use in modern times. It is chiefly employed in Europe to flavour spirits, the greatest consumption being in Germany, France, and Italy. It is not used as a medicine in Europe, but in India is regarded as stomachic and carminative. It is viewed as of great service in flatulent and spasmodic affections of the intestinal canal. **Ainslie** remarks that in his time (1826) Europeans in India were not familiar with the drug, but the *Vytians* consider the FRUITS powerfully stomachic and carminative, and the Muhammadans occasionally prepare from them a very fragrant volatile oil. MEDICINE. 35 Flavouring Spirits. 36 Fruits. 37

Dr. Dymock mentions that a consignment of a kind of *Star Anise* reached Bombay which was found unsaleable, and chiefly because the fruits were almost devoid of scent. He formed the opinion that they were the produce of **I. Griffithii**, and apparently because of the fact of their consisting of 13 instead of 8 carpels. In a correspondence on this subject **Dr. Dymock** informed the writer that the consignment in question was first offered for sale at Calcutta, but not finding a purchaser there, was sent on to Bombay A subsequent enquiry, alluded to under the generic note above, throws considerable doubt, however, on the possibility of the 13-carpelled consignment in question having been the fruits of **I. Griffithii.** The *Flora of British India* does not mention the number of carpels

MEDICINE.

possessed by the two hitherto published Indian species, but, **I. imphalense** has 8 to 12, and according to **Holmes**, **I. Griffithii** has 13, and **I. majus** 11 to 13. It is, as suggested above, much more likely, therefore, that the inodorous form described by **Dymock** is a foreign imported fruit, most probably from the Straits or Cochin China. **I. verum** and **I. religiosum** have only 8 carpels and the two American species have also that number; **I. floridanum** like **I. religiosum** are poisonous, at least so it is stated of the plant found in Alabama. **I. parviflorum**, the Georgian species, has a flavour closely resembling that of Sassafras root.

SPECIAL OPINIONS.—§ "Extensively prescribed by Hakims as carminative and stomachic. Seldom, if ever, prescribed by Kobirajs" (*Civil Surgeon S. M. Shircore, Moorshedabad*). "After roasting and mixing with equal parts of sugar is thought to be useful in ordinary dysentery. My personal experience is limited" (*Surgeon-Major C. J. McKenna, I. M. D., Cawnpore*).

FOOD. 38

Food.—Being highly aromatic, it is in great repute in China and other Eastern countries in the manufacture of condiments and flavouring of spirits.

IMPATIENS, *Linn.; Gen. Pl., I., 277, 989.*

A genus of succulent herbs, rarely shrubs, comprising in all some 150 species, chiefly distributed throughout the mountains of Tropical Asia and Africa. They become rare in Temperate Europe, North America, and South Africa, though in India some of the most prevalent species are those met with on the Temperate Himálaya, the under vegetation of the mixed forests of the Western division often consisting almost exclusively of a rampant annual growth of one or perhaps two or three species of Balsam. Indeed, India alone possesses 123 of the known species, and of these the vast majority occur on the Western Ghâts the mountains extending from Bombay and Madras to Ceylon. Some of these species are distributed throughout the area indicated, but Madras possesses 40, Bombay 35, and Ceylon 21; while the lower mountains of Bengal, the Central Provinces, the North-West Provinces and the Panjáb each have only one or at most two species and these are common also to all the other provinces of India. In Assam (chiefly on the Khásia Hills) there are 23 species and 9 in Burma. The Burmese species may be described as the most tropical Indian members of the genus, the majority occurring at the sea's level near Moulmein. Of the Himálayan species the Western Division possesses 14, the Central 16, and the Eastern 26, some of the last being also found in the Khásia Hills. Many of the Western Himálayan species extend from Kumáon to Kashmír, a few only going eastward to Sikkim. These Western forms are also by far the most prevalent, and, as already remarked, miles of country during the monsoons (June to September) are literally covered with one or two species. This is peculiarly the case near Simla where **I. amphorata**—a rose-coloured species with yellow and red veins—and **I. amplexicaulis**—a magenta coloured one, dispute possession with the yellow-flowered **I. scabrida.** It is significant also that with the exception of **I. Balsamina** all the West Himálaya species belong to the section with linear capsules, and most of them possess in addition edible seeds.

39 **Impatiens Balsamina,** *Linn.; Fl. Br. Ind., I., 453;* GERANIACEÆ.

THE GARDEN BALSAM.

Vern.—*Gúl-mendí*, HIND.; *Dúpati*, BENG.; *Haragaura*, URIYA; *Mujethi*, (a name indicative of its yielding a red dye), N.-W. P.; *Bantil, trúal, hálú, tatúra, pallú, tílphár, júk*, PB.; *Teradá*, BOMB.; *Tiradá*, MAR.; *Panshít, dan-da-let*, BURM.

References.—*Roxb., Fl. Ind., Ed. C.B.C., 219; Voigt, Hort. Sub. Cal., 189; Thwaites, En. Ceylon Pl., 65; Dalz. & Gibs., Bomb. Fl., 44; Stewart, Pb. Pl., 36; Mason's Burma & Its People, 433, 765; U. S. Dispens., 15th Ed., 1669; Atkinson, Him. Dist., 774; Gazetteers:—Orissa, II., 178; App. VI.; Mysore & Coorg, I., 69; Bombay (Kanara), XV., 429; N.-W. Provinces, Bundelkhand, 79; Himalayan Districts, 307; Indian Forester, IX., 292; XII., App. 8.*

Habitat.—A native of India, introduced into England about 1596, since which date its cultivation has been prosecuted with much success as an ornamental exotic. In the North-West Himálaya it often occurs so plentifully in fields, at about 3,000 feet above the sea, as to give the fields a rose-purple tint. The cultivated plant may be said to have become naturalised in the vicinity of gardens, in the plains of India, but the wild form occurs only on the lower hills, and is not met with on the plains proper.

The *Flora of British India* describes six varieties, *viz.*, (1) **vulgaris**, (2) **coccinea**, (3) **arcuata**, (4) **macrantha**, (5) **micrantha**, and (6) **rosea**. The last mentioned is the form met with in the Western Himálaya, while Nos. 3, 4, and 5 are found in the Western Peninsula.

Firminger says the best time to sow Balsams, in the plains of Bengal, is September, and that the plants should be retained in pots until they are ready to be planted out in the border. If sown earlier the severe rains destroy them, and if later they do not obtain sufficient heat to allow of their flowering properly.

Dye.—**Madden** says that the flowers of this plant are in Garhwál used for a dye, whence it is called *Majiti* or *Mujethi*. The writer received from the Jhaintiya hills a few samples of plants used by the inhabitants for dyeing. Among these was found a specimen of this balsam, with the remark "Leaves when bruised together with *Metchta langa* give a red colour." The writer cannot discover a detailed account of the chemistry of this plant, and is therefore unable to offer any opinion as to the nature of the red dye it contains. That it does possess a colouring principle and one closely allied to madder in appearance, there seems to be no doubt. The subject is well worthy of careful examination, since it is probable the cultivated plant (to be found in Europe) is likely enough to preserve this property. A use for the wild plant would be a positive boon to the Himálaya cultivators whose crops are often seriously injured by the weed. **DYE. Flowers. 40**

The flowers of the British **I. Noli-me-tangere** were once upon a time employed as a yellow dye. **Baillon** (*Natural History of Plants, V., 33*) remarks of **Impatiens**: "Several Balsams are tinctorial plants, especially **I. fluva**, *Nutt*, and **I. tinctoria**, *A. Rich*. The Tartars are said to colour their eyes and nails with the juice of several balsams with alum."

Oil & Food.—The seeds are eaten in Chumba, and the oil expressed is eaten and also burned. **OIL & FOOD. 41**

Medicine.—It is not known whether any of the Indian species of **Impatiens** have attributed to them medicinal properties; **I. Noli-me-tangere** has an acrid burning taste, and when taken internally, acts as an emetic, cathartic, and diuretic. It is considered too dangerous, however, to be of much use. The *United States Dispensatory*, after having previously discussed the properties of **I. fulva**, **I. pallida**, and **I. Noli-me-tangere**, states that I **Balsamina** resembles the other species in its effects **Baillon** says of **I. Noli-me-tangere** that it was formerly valued as a diuretic and antihemorrhoidal. It was topically used for pains in the joints and was said to cure diabetes, but it is not much thought of at present. In Japan **I. cornuta** is said to make the hair grow. **MEDICINE. 42**

Impatiens racemosa, *DC.; Fl. Br. Ind., I., 479.* 43

Habitat.—A small, herbaceous plant (2 to 3 feet high) common on the temperate Himálaya, altitude 5,000 to 7,000 feet from Simla to Sikkim, at times ascending in Sikkim to 12,000 feet.

Dye.—Specimens in the herbarium stain the paper on which they are mounted as well as any plants they come in contact with of a bright rose pink resembling madder. (*Conf.* with **I. Balsamina.**) This property was first pointed out by **Edgeworth**, but the writer has found it to be the case **DYE. 44**

with his own specimens, the dye being specially washed out with the spirit and corrosive sublimate employed in poisoning.

OIL. Seeds. 45 **Oil.**—The SEEDS yield oil which is used for burning; it is also edible.

46 **Impatiens Roylei,** *Walp.; Fl. Br. Ind, I., 468.*

Habitat.—A handsome bush, often 10 feet in height, common on the temperate Western Himálaya from Nepál to Marri, altitude 6,000 to 8,000 feet.

OIL & FOOD. Seeds. 47 **Oil & Food.**—The raw SEEDS are edible, tasting like nuts; from them an oil is prepared which is both eaten and used for lamps.

48 **I. sulcata,** *Wall.; Fl. Br. Ind., I., 469, 475.*

Habitat.—A gigantic annual, often 15 feet in height, frequent on the temperate Himálaya, altitude 7,000 to 12,000 feet. The writer came into an expanse of grassy hills in Kullu where these plants formed an almost impenetrable mass. On endeavouring to push his way through it, the seeds shot off in every direction in such showers that it was almost impossible to open the eyes. He was guided to the spot by a native who assured him that pigeons in great numbers would be found. This proved the case, the birds rising in great flocks from the balsam upon which they were feeding. At another place he found the people collecting the seeds, and was told they were to extract oil from them.

OIL & FOOD. Seeds. 49 Husks. 50 **Oil & Food.**—The SEEDS are edible and yield an oil. **Dr. Aitchison** says that in Lahoul the HUSKS are also eaten raw.

IMPERATA, *Cyr.; Gen. Pl., III., 1125.*

[GRAMINEÆ.

51 **Imperata arundinacea,** *Cyrill.; Duthie, Fodder Grasses, N. Ind., 23;*

Syn.—SACCHARUM CYLINDRICUM, *Linn.;* LAGURUS CYLINDRICUS, *Linn.;* IMPERATA CYLINDRICA, *Beauv.;* I. KŒNIGII, *Beauv.*

Vern.—*Usirh, sir, sil, bharwai,* UPPER IND.; *Dábh,* HIND.; *Ulú,* BENG.; *Shiro* (Bhábar and Lower Hills), N.-W. P.; *Sirú, úlú,* PB.; *Dáb,* SIMLA; *Tharpai-pullu,* TAM.; *Barumbiss* (according to **Roxburgh**), *modava gaddi* (according to **Elliot**), TEL.; *Thek-kay-nyen,* BURM.; *Iluk,* SING; *Darbha* (according to **Dutt**) (*balbajamu, barhissu,* (according to **Elliot**), SANS.

References.—*Roxb., Fl. Ind., Ed. C.B.C., 78; Voigt, Hort. Sub. Cal., 704; Thwaites, En. Ceylon Pl., 369; Stewart, Pb. Pl., 257; Aitchison, Cat. Pb. and Sind Pl., 173; Mason, Burma and its People, 524, 817; Elliot, Flora Andhr., 20, 23, 116; Kurz, Prelim. Rept. Pegu; U. C. Dutt, Mat. Med. Hind., 296; Bidie, Cat. Raw Pr., Paris Exh., 76; Atkinson, Him. Dist., 321, 808; Jour. Agri.-Hort. Soc. Ind., X., 357; XIII., 337; (New Series), I. (Proc.), 62; II., 300; Indian Forester, I., 22; IV., 168; IX., 245; X., 26.*

Habitat.—A small perennial grass inhabiting the plains and hills of Bengal, the North-West Provinces, the Panjáb and Sind, in moist, stiff, pasture ground It is particularly common over Bengal, and ascends the Himálaya to altitudes of 7,500 feet. When in flower about April and May, the fields, roadsides, and railway embankments become white with its silky heads. It is, in fact, a very characteristic grass of the neighbourhood of Calcutta, the *maidan* being chiefly carpetted with it.

FIBRE. 52 **Fibre.**—Used for the same purpose as that obtained from the *Munja* (**Saccharum ciliare,** *Anders.*), namely, to prepare the sacrificial thread of the Hindus; and the leaves are employed for thatching (*Atkinson's Himalayan Districts*). **Roxburgh** says that the Telingas make use of it in their marriage ceremonies. He adds that it is much used in Bengal for thatch, and **Mason** remarks that this is the case in Burma also. **Thwaites** states that in Ceylon this grass is viewed as making excellent thatch.

Fodder.—Not of much use as a fodder, because cattle refuse it except when quite young, or when no other forage can be obtained. Mr. Duthie writes: "In Australia it is called 'blady-grass,' and the young succulent foliage which springs up after the occurrence of a fire is much relished by stock. I have observed the same effect resulting from periodical fires on certain parts of the Himálaya where this grass is plentiful." FODDER. 53

Implements and Machinery, &c., Woods used for. See **Agricultural Implements,** Vol. I., 145.

Incense, an odoriferous preparation burned at religious ceremonies. Frankincense (= pure incense) may be regarded as the chief and more expensive ingredient (see **Boswellia,** *Vol I., pp. 511 to 514*), the following being the materials of incense used by the poor or employed as adulterants with Olibanum:— 54

(1) **Boswellia serrata,** *Roxb.* Gugul or Salhi gum, the Indian Olibanum.

(2) **Cupressus torulosa,** *Don.*

(3) **Juniperus communis,** *Linn.* The smoke of the green roots was considered by the ancients to be most acceptable to the infernal gods.

(4) **J. macropoda,** *Boiss* (? **J. excelsa,** *Rieb.*).

(5) **J. recurva,** *Ham.*

(6) **Jurinea macrophylla,** *Benth.* The roots of this herbaceous plant are largely collected in Kumáon and Garhwál under the name of *Dhúp*, and either exported to the plains or made into a resinous preparation which in that condition is exported.

(7) **Morina Coulteriana,** *Royle.*

(8) **Pinus longifolia,** *Roxb.* Known in Oudh as *Dhúp.*

(9) **Saussurea Lappa,** *Clarke.* The sacred Costus root largely exported from Kashmír and used in the preparation of perfumery, medicine, and expensive forms of incense.

(10) **Taxus baccata,** *Linn.*

Though mentioned by writers on this subject as ingredients in incense, it is not known to what extent the coniferous plants in the above list are used in the incense burned at Hindu temples. After Olibanum, Nos. 1, 6, and 9 are the most important and in that order relatively.

Indian Fig or Prickly Pear, see **Opuntia,** Vol. V.

Indian Red, see **Pigments,** Vol. V.

Indian Shot, see **Canna indica,** Vol. II., p. 102.

INDIA-RUBBER, CAOUTCHOUC or GUM ELASTIC.

GOMME ELASTIQUE, CAOUTCHOUC, *Fr.*; KAUTSCHUK, *Germ.*; VERDERHARS, *Dutch*; GOMMA ELASTIC, *It. & Port.*; GOMA ELASTIC, *Sp.*; GUMMI, *Da. & Sw.*; CHIRIT MURAI, *Malay.*; SIANG-PI, *Chin.* DESCRIPTION. 55

Caoutchouc or **India-rubber** is the inspissated milk or sap obtained from at least six genera of plants which belong to three widely different Natural Orders, *viz.*, **Landolphia** and **Willoughbeia,** to APOCYNACEÆ; **Castilloa** and **Ficus,** to URTICACEÆ (sub-order ARTOCARPEÆ); and **Hevea** and **Maninot,** to EUPHORBIACEÆ. In the plant tissue Caoutchouc is found to circulate within certain anastomosing vessels which are distributed throughout the middle or more rarely the inner layer of bark. A far larger number of plants possess a milky sap (and even a caoutchouc-yielding sap) than those generally viewed as the sources of India-rubber. The term Caoutchouc is sometimes used synonymously with that of India-rubber, but it is more correctly the pure hydrocarbon isolated from the

DESCRIPTION.

other materials with which it forms the impure rubber of commerce. Under the microscope the milk appears as a kind of emulsion with minute globules, the caoutchouc *granules* suspended in it. These range from $\frac{1}{204000}$ to $\frac{1}{5000}$ of an inch in diameter. Caoutchouc is highly elastic, lighter than water, has neither taste nor smell, is fusible at about 120°C. (248° Fh.) and inflammable at higher temperatures. According to Faraday it consists of 87·2 parts of carbon and 12 8 of hydrogen. The sample he examined came from India, and was thus probably obtained from **Ficus elastica.** It is interesting to add, therefore, that Faraday makes no mention of having detected ammonia in the rubber he examined.

When the bark of plants containing this substance is cut, the milk exudes, and in time hardens on exposure to the air. This agglutination may be hastened by adding salt water or alum to the milk, but the former increases one of the defects of rubber, *viz.*, its hygroscopic property, and thus injures it commercially. It would seem that the chemist has still to perform a service to the planter and extractor of rubber that is likely to prove of the greatest importance, *viz.*, to analyse the fresh milk of each and all the rubber-yielding plants. The methods of coagulation in practice are very much the hereditary arts of savages, instead of being based, as they should be, on scientific principles. It does not follow that the acid process, though applicable to one milk, is at all suited to another; in fact, systems of extraction have been brought about by experience blindfoldly instead of based on demonstrated science, hence the full amount of rubber contained in the milk is not, as a rule, removed. It is commonly stated that the best rubber is obtained by boiling the milk, but delay in agglutination is reported to be injurious, since destructive oxidation takes place. Where the climate of the rubber forests is detrimental to human life, the collectors frequently add alum to the fresh milk, as a temporary measure, and effect the complete separation of the rubber with greater deliberation on reaching their homes at some distance from the forests. It is believed by most writers that the milk is kept in its fluid state within the plant tissue through the agency of ammonia—a theory in part supported by the facts that the fresh milk has often an ammoniacal odour, and that a little ammonia * added to the milk has been observed to facilitate its retention in the liquid state. On coagulation, the milk changes colour, assuming a more or less reddish tint, and forming a plastic leathery substance. A thin piece is seen under the microscope to contain irregular rounded pores which partially communicate with each other and dilate under the influence of fluids.

Chemically India-rubber may be said to approximately consist of two substances—an elastic material, on which its merit depends, and a viscid, resinous, readily oxidisable principle, to which it owes its depreciation. The greater the amount of resin present in a sample of rubber, the less is its value, but it is remarkable that the amount of this oxidisable substance seems to be more consequent on a tardy precipitation of the caoutchouc from the milk, than on individual or even specific differences in the plants from which it is obtained. The property of the elastic substance also varies, and in a marked degree, between that obtained from one genus of plants and another, for every gradation exists between the non-elastic hydrocarbon known as Gutta-percha and the finer qualities of Gum Elastic, such as the Para and Ceara rubbers which contain the smallest known amount of resinous matter. The African and Guatemala rubbers possess the highest percentage, and are consequently the least valuable.

Caoutchouc is quite insoluble in water, alcohol, and acids. It is, how-

* *Conf.* with remarks at page 360.

SOLVENTS for CAOUTCHOUC.

ever, oxidised by strong nitric and charred by strong sulphuric acid. Ether, benzone, rock-oil, and sulphide of carbon penetrate it rapidly, causing it to swell up and ultimately assume an apparent though not actual condition of solution. "The liquid thus formed is not, however, a complete solution, but a mixture formed by the interposition of the dissolved portion between the pores of the insoluble substance, which is considerably swelled up, and has thus become easy to disintegrate. By employing a sufficient quantity of these solvents, renewed from time to time, without agitation, so as not to break the tumefied portion, the caoutchouc may be completely separated into two parts, *viz.*, a substance perfectly soluble, ductile, and adhering strongly to the surface of bodies to which it is applied; and another substance, elastic, tenacious, and sparingly soluble. The proportions of these two principles vary with the quantity of the caoutchouc and the nature of the solvent employed. Anhydrous ether extracts from the amber-coloured caoutchouc 66 per cent. of white soluble matter; oil of turpentine separates from common caoutchouc 49 per cent. of soluble matter having a yellow colour. The best solvent for caoutchouc is a mixture of 6 to 8 pts. of absolute alcohol and 100 pts. of sulphide of carbon." "Caoutchouc yields by *dry distillation* an empyreumatic oil called OIL OF CAOUTCHOUC or CAOUTCHOUCIN, which forms an excellent solvent for caoutchouc and other resins. It is a mixture of a considerable number of hydrocarbons. Ordinary impure caoutchouc likewise yields small quantities of carbon anhydride, carbonic oxide, water and ammonia" (*Watts' Chemistry*).

Continued boiling in fixed oils also causes the separation of the caoutchouc particles thus seemingly to dissolve the rubber. Of this nature doubtless is the action of *garjan* oil, the property of which, as a solvent for caoutchouc, the writer has already alluded to under **Dipterocarpus turbinatus** (*Vol. III., 165*). **Mr. Laidlay** in 1839 reported that when fragments of rubber are dropped into *garjan* oil (especially if the oil be heated or boiled) the rubber swells and dissolves. Since the date on which this republication was made of **Mr. Laidlay's** discovery, the writer has come across another passage which would seem to refer to the solvent property of *garjan* oil, and he has also performed certain preliminary experiments to test the accuracy of the reports regarding *garjan* oil as a solvent. **Mr. Strettell**, in his *Narrative of my Journey in Search of Ficus elastica in Burma Proper*, states (p. 184) that a **Mr. Henri** who was employed by the King to work the serpentine mines bethought him of a method of repairing the gutta-percha tubing that had got out of order. Observing water-proof baskets used by the people for baling water he enquired as to the material used in rendering these waterproof. He found it to be the India-rubber of **Ficus elastica**. Further on, **Mr. Strettell**, commenting on **Mr. Henri's** discovery, remarks: "I had endeavoured to find out the process adopted by **Mr. Henri** to utilize caoutchouc in the form of a cement for repairing the gutta-percha tubing of his pumps, but no reliable information could be obtained. One account was that he boiled ten *tickals* of caoutchouc with two-and-a-half *tickals* of *kunnin* oil, but this is impossible." The writer's elementary experiments, suggested on reading **Mr. Laidlay's** report, have satisfied him that far from being impossible it is probable that the above report was correct, and he strongly suspects that the Burmans use more extensively than we have any positive knowledge of, a solution of caoutchouc dissolved in *In* or *Kanyin* oils. Procuring in the Simla bazár a bottle of the substance sold as *garjan* oil, the writer performed **Mr. Laidlay's** experiments, but was more successful in dissolving gutta-percha than India-rubber. It was thought, however, desirable to obtain from Burma well authenticated samples of the various forms of wood oil; accordingly **Mr. J. W.**

DESCRIPTION.

Oliver, Conservator of Forests, Upper Burma, was addressed on this subject, and that officer very obligingly furnished samples. **Messrs. Slade, & Toussaint** kindly collected these samples at **Mr. Oliver's** request, but unfortunately the jar containing *In* oil got broken on its way to Simla, so that the experiments could be tried only with the various forms of *Kanyin* oils. These oils after prolonged boiling all acted as solvents both of India-rubber and Gutta-percha, and it was observed that the pure oils decanted from the resin possessed this property to a less extent than when used in the impure form—resin and oil mixed.

Thinking that perhaps a solution of rubber in *Kanyin* oil might be used in the water-proofing of Burmese umbrellas, the writer also invited **Mr. Oliver** to investigate that subject, remarking that it was customary to read of umbrella water-proofing as accomplished by a decoction prepared from the fruits of a **Diospyros. Mr. H. Jackson**, Deputy Conservator of Forests, Prome, furnished the report and samples which this enquiry elicited. Two bottles were received, one containing a strongly adhesive balsamic substance, said to have been prepared from the fresh fruits of **Diospyros burmanica**, and the other a varnish made from earth-oil, rosin, wax, and vegetable oil. There is thus no India-rubber used in the Burmese umbrella water-proofing, and the process and manipulation pursued by the Burmese appears to differ from that followed by the Chinese mainly in that the latter people use the fruits of **Diospyros pyrrhocarpa** (*Conf.* with *Vol. III., 154*). Whether or not the above facts regarding the solvent property of the **Dipterocarpus** oils on caoutchouc may prove of industrial value, they are of interest in amplifying what has been said under **Dipterocarpus** and **Diospyros.** It may, therefore, be added that the writer observed that hot *garjan* oil painted on cloth previously dyed with cutch, and mordanted with alum, dried in a few hours. This latter fact may be of considerable utility, since (as remarked under **Dipterocarpus**) the chief drawback to the utilisation of *garjan* oil has hitherto consisted in the difficulty of drying the oil. A cloth painted with cold or hot *garjan* oil took some months to dry and when dry became stiff, so that on being folded it cracked and gave out the resinous principle of the oil. The cutch-dyed cloth, on the other hand, dried in a few hours and never became stiff nor parted with its resin. It was, however, not completely water-proof, and the *garjan* oil seemed to soften when water was left on the cloth for some hours.

The reader should consult the article **Gutta-percha** (*Vol. III., 101—109*). A comparison of the list of plants there given, as yielding that substance with those that afford rubber, will reveal the fact that the majority of the former belong to the Natural Order SAPOTACEÆ, while the latter are mainly obtained from members of the URTICACEÆ and EUPHORBIACEÆ. The APOCYNACEÆ and ASCLEPIADACEÆ appear sometimes to yield guttas, at other times rubbers. The exact nature of the hydrocarbons obtained from these families, however, must remain undetermined until they have been subjected to a careful chemical examination. The allotment given in this work of certain species as yielding gutta and of others rubber, may therefore have to be adjusted hereafter. **Mr. W. T. Thiselton Dyer** read a paper on this subject before the Linnean Society (*Proceedings, June 1882, p. 35*) in which he discussed the Apocynaceous India-rubber plants of the Malaya and of Tropical Africa. He pointed to the interesting fact that these all belong to the TRIBE CARISSEÆ. **Mr. J. G. Baker**, in an article which appeared in the *Gardener's Chronicle* (*1886*); dealt with the sources of the India-rubber of commerce. "Part of the supply," he wrote, "comes from South America (shipped principally from Para and Carthagena), part of it from Sierra Leone, Mozambique, and Madagascar, and the remainder from tropical Asia." Besides the two genera of APOCYNACEÆ (**Landol-**

DESCRIPTION.

phia and **Willoughbeia**) there are, he says, "at least six others which yield a similar milky juice not at present utilised to any considerable extent. In the United States in 1883 there were 120 India-rubber factories, employing 15,000 hands. The total importation of raw material into the States in that year was 30,000 tons, worth about £6,000,000 sterling. The value of the manufactured goods made in a single year is estimated at £50,000,000. The quantity of unwashed rubber imported into the United Kingdom in 1883 was more than 10,000 tons, worth about £3,000,000, but in 1885 it had sunk to less than £2,000,000. None of the trees which yield India-rubber have yet been brought into cultivation on a large scale, and the time will soon come when either this will have to be done or the supply will gradually lessen. There are about sixty distinct species of the rubber-yielding genera, and the botanists and foresters will have to settle between them which of these are best worth cultivating and where it will pay to grow them. Unfortunately, at the present time, the price of India-rubber of all kinds is exceptionally low, the best Para rubber being now worth only about 2*s*. 6*d*. per pound in London against 4*s*. in 1884, and the best African and Asiatic kinds about 2*s*. per pound." **Mr. Baker** then furnishes a most instructive table of the India-rubber-yielding plants, showing their native countries and the imports into Great Britain in 1880 of the rubber from each. The table may be here reproduced :—

Natural Order.	Genus.	Number of species.	Native country.	Imports in Tons.
Apocynaceæ .	Willoughbeia . .	9	Tropical Asia . . .	530
" .	Landolphia (including Vahea).	16	Africa and Madagascar .	2,200
" .	Hancornia . .	1	Brazil.	
" .	Urceola . . .	7	Malay Peninsula and Archipelago.	
" .	Dyera . . .	3	Malay Peninsula.	
" .	Couma (Collophora)	4	Guiana and Brazil..	
" .	Alstonia . .	3	Malay and Fiji.	
" .	Cameraria . .	2	West Indies.	
Urticaceæ (Artocarpeæ).	Castilloa . .	3	Central America and Cuba.	100
" .	Ficus . . .	2	Africa and Tropical Asia .	370
Euphorbiaceæ	Hevea . . .	9	Amazon region	5,768
"	Manihot . . .	1	Brazil . . .	35
TOTALS .	12 genera	60	......	9,003

The facts in the above tabulated statement, andthe arguments used by **Mr. Baker** regarding the present immense trade and future possible decline in supply, should cultivation be not resorted to, have been urged by many writers and have proved the keynote to the extensive experiments conducted in India and other tropical countries. An attempt will be made, in the subsequent pages, to discuss very briefly the nature of these experiments, but in concluding the present introductory remarks it may safely be said that the Assam plantations of **Ficus elastica** represent the most rational efforts that have been made, though high expectations are also entertained of the Tenasserim **Hevea** (Para) plantations. So long ago as 1880 **Dr. King**, while reporting the successful introduction of the climbing **Landolphia** rubber plants into the Botanic Gardens, Calcutta, remarked :

DESCRIPTION.

"But I fear, even if it were to turn out to be suited to the climate of Calcutta, **Landolphia** would prove rather an unmanageable crop; for it is described to be an enormous creeper, climbing to the tops of the highest trees. With regard to all these exotic rubbers, it must be remembered that 'with the exception of Ceara' they are either very large trees or climbers; and although it may pay well to collect rubber from them in their native forests, where they have grown to maturity without cost to the collector, it is quite a different matter when their planting and protection have to be paid for, and their coming to maturity has to be awaited for years." This opinion has since been only too truly confirmed, so far as the interest of European planters (who are at most but temporary residents in the tropics) are concerned, though perhaps the plantations undertaken by Government may ultimately prove of considerable value to the country.

CAOUTCHOUC PLANTS. 56

PLANTS WHICH YIELD CAOUTCHOUC.

It is perhaps unnecessary to deal exhaustively with this subject, since during what may be called the various spasmodic India-rubber fevers, that have periodically engrossed the attention of planters, the cry has been raised needlessly, as it would now seem, for more and still more Caoutchouc-yielding plants. The natural result has followed, namely, small fortunes made by adventurers and powerful advertisements obtained for would-be public benefactors, at the expense of the earnest workers in tropical fields, who now suffer in their present apathy the consequences of undue haste. It is probable that much time and money were spent on experiments with worthless milky shrubs, or climbers, that necessitated a more expensive system of cultivation than the value of the produce obtained. The public were thereby compelled to consider as caoutchouc-yielding a far larger number of plants than there was any occasion for, and the merits of the commercial-rubber trees were thereby largely neglected. Even with the true caoutchouc-yielding plants legitimate experiments in testing the suitability of soils, climates, and methods of treatment were subordinated to hallucinations of immediate fortunes which were to flow from saplings of a few years' growth. Indeed, few subjects of commercial importance have been dealt with in the same spirit of precipitation as the India-rubber enquiry. But this is perhaps accounted for by the practically unprecedented growth of the European manufactures to meet which an almost inexhaustible supply of raw material existed and could be remuneratively furnished by the indigenous forests of tropical America.

The caoutchouc-yielding plants that alone need be discussed here may be dealt with very briefly in alphabetical order, under two sections—(A) Indigenous, (B) Introduced.

(A). INDIGENOUS CAOUTCHOUC-YIELDING PLANTS. (For B. see page 364.)

INDIGENOUS CAOUTCHOUC. 57

1. **Artocarpus Chaplasha,** *Roxb.; Dic. Econ. Prod., I., 329;* URTICACEÆ.

Vern.—*Chaplash,* BENG.; *Sam,* ASS.; *Cham,* CACHAR; *Pani,* MAGH.; *Taungbeinné,* BURM.; *Kaita-da,* AND.

References.—*Fl. Br. Ind., V., 543; Roxb., Asiatick Researches, V.; also Fl. Ind., Ed. C.B.C., 634, 640; Kurz, For. Fl. Burm., II., 432.*

Habitat.—A lofty deciduous tree of the trans-Gangetic regions from the Nepál Tarai to Assam, Chittagong, Burma, and the Andaman Islands.

Caoutchouc.—Kurz remarks that it yields a tenacious milky caoutchouc. This same fact is alluded to by other writers, but the substance does not appear to have been commercially investigated.

58

2. **A. integrifolia,** *Linn.; Dic. Econ. Prod., I., 330.*

THE JACK-FRUIT TREE.

Vern.—*Kánthál, katol, panas,* HIND., BENG., and ASS.; *Poros,* KOL.; *Phanas,* MAR., BOMB.; *Pilá,* TAM.; *Peinné,* BURM.

INDIGENOUS CAOUT-CHOUC.

References.—*Fl. Br. Ind., V., 541; Roxb., Fl. Ind., Ed. C.B.C., 633; Brandis, For. Fl., 425; Kurz, For. Fl. Burm., II., 432; Dalz. and Gibs., Bomb. Fl., 244; Agri.-Hort. Soc. Trans. (1837), Vol. V., 38-42 (an interesting paper by Dr. Helfer on various Assam rubbers): report on the samples, pp. 74-78.*

Habitat.—A large evergreen tree, a native of the forests of the Western Ghâts: cultivated for its fruit throughout the hotter parts of India.

Caoutchouc.—A writer in the *Indian Agriculturist* discusses the properties of this rubber. He says it is elastic, leathery, water-resisting, &c. Each green fruit, he remarks, yields about two ounces of milk from which a drachm and a half of caoutchouc may be obtained. Many years ago (1837) **Dr. Helfer** experimented with this rubber in Assam. He wrote, "The milky juice emanating in moderate quantity from every part of the tree, but most copiously from the immature fruits, has till now not been transformed by me in a perfectly solid state. Pouring the milky juice fresh from the tree in earthen moulds as is done in America, did not answer the purpose; the moulds exposed to a slow fire, instead of drying the substance, rendered it more fluid, and so it continued for weeks." **Dr. Helfer's** process for obtaining the Caoutchouc, he thus describes:—"The milky juice brought in a flat basin is mixed with an equal quantity of water, and then agitated for two hours with sticks, by which a good deal of carbonic acid is evolved. In about an hour the caoutchouc substance begins to separate from the water and extractive matter and coagulates; in one hour more the separation is perfectly effected, the caoutchouc gets a milk-white appearance and a tolerable consistency, but it does not dry perfectly." **Dr. Helfer** also describes a further process: "The juice, after having been agitated about one hour as above mentioned, I added to the fluid, acetic acid, which had the property of separating the thinner parts of the juice at once by coagulating it. I put it then on a moderate fire in a tin vessel, till the aqueous parts were evaporated and obtained the caoutchouc substance more elastic, but perfect dryness remained a desideratum."

3. **Carissa** (several species); *Dict. Econ. Prod., II., 165;* APOCYNACEÆ. 59

These milk-giving shrubs are said to afford caoutchouc, and if of sufficiently good quality and in such abundance as to make its collection remunerative, these plants should prove of great value to India, since they exist in great abundance though hitherto viewed as worthless. **Mr. T. Christy** is reported to have sold the seeds of an African species at six pence each, but nothing more has been heard of these reputed caoutchouc plants. A climber spoken of as **Carissa macrophylla** (*see Report, Botanic Gardens, Nilghiri Hills, 1883-84, p. 9*) was experimentally cultivated as a possible source of rubber, but with no result. It is probable the plant in question was **Carissa suavissima**, a lofty climber of the Deccan Peninsula. The rubber of an African species was collected by **Sir J. Kirk** (Livingstone Expedition), and much correspondence ensued for some years after, the seeds of perhaps several species having been distributed to planters in the tropics. [APOCYNACEÆ.

4. **Chonemorpha macrophylla**, *G. Don.; Dict. Econ. Prod., II., 271;* 60

Vern.—*Gar badero*, HIND.; *Yokchounrik*, LEPCHA; *Harké*, SYLHET.

References.—*Echites macrophylla, Roxb., Fl. Ind., Ed. C.B.C., 246.*

Habitat.—A large climbing shrub met with in North and East Bengal and the moist sub-tropical forests of India, generally from Kumáon (4,500 feet) to Travancore, the Andaman Islands, Malacca, and Ceylon.

Caoutchouc.—Gamble (*List of Trees, Shrubs and Climbers of Darjeeling, 56*) says that this climber gives a good sort of caoutchouc. Other writers allude to this fact, but apparently derive their information from the

INDIGENOUS CAOUTCHOUC.

statements made by **Mason** and **Parish** regarding the plant they identified as **Echites macrophylla,** *Wight.*, which see under **Urceola esculenta.** [ASCLEPIADACEÆ.

61 5. **Cryptostegia grandiflora,** *R. Br.; Dict. Econ. Prod., II., 625;*

Vern.—*Vilarjati vakundi,* MAR.; *Palay,* MAL.

References.—*Trans. Agri.-Hort. Soc. Ind., IV.* (1837), 223; *Wight, Ic., t. 832; Dalz. and Gibs., Bomb. Fl. Supp., 54; Agri.-Hort. Soc. Jour. Mad., 1883-84; Rep. Bot. Gard. Hyderabad, Sind, 1882; Rep. Dir. Agri. Dept. Bomb., 1883-84; Tropical Agriculturist, II., 160; Indian Forester, VIII., 202.*

Habitat.—An extensive climber supposed to be a native of Africa, but very generally cultivated throughout the hotter parts of India.

Caoutchouc.—A full account of the recent experiments made in Bombay with this plant, as a source of rubber, will be found in this work, Vol. II., 625. The Caoutchouc is said to be good, but the opinion (a natural one) seems to have been arrived at, *viz.*, that it will not pay to cultivate a climber as the source of a substance that has to be produced in large quantities and at a low price. **Dr. Wallich,** writing of a specimen of this climber grown in the Botanic Gardens, Calcutta, in 1837, says it yielded a rubber superior in whiteness and elasticity to that of **Ficus elastica.** This opinion was given at the commencement of the enquiry into Indian sources of rubber and was probably very much overstated. The Conservator of Forests, Southern Circle, Madras, reports that this climber "grows all well over the plains and yields a fair amount of sap which on coagulation produces a very firm and superior rubber."

62 6. **Ficus elastica,** *Roxb.; Dict. Econ. Prod., III., 350;* URTICACEÆ.

THE CAOUTCHOUC TREE OF ASSAM.

Vern.—*Bor, attah bar,* BENG., ASS.; *Kagiri, kasmir,* KHASIA; *Ranket,* GARO; *Lesu,* NEPAL; *Yok,* LEPCHA; *Nyaung bawdi,* BURM.

References.—*Roxb., Fl. Ind., Ed. C.B.C., 640-641, also his earlier paper on Caoutchouc in the Asiatick Researches, Vol. V.; Royle, Prod. Res. India, 14, 68, 75-77; Collins, Report on Caoutchouc, 19-22, 32-39, 48-54 (the last passage containing Sir D. Brandis' Review of Mr. Collins' report); Griffith, report on Caoutchouc tree of Assam, Jour. Asiatic Soc. Beng., VII., 132; Strettell, Narrative of my Journey in search of Ficus elastica in Burma proper; Mason, Burma and Its People, 523; Kew Report, 1880; J. E. O'Conor's Trade Reviews: 1874-75, 50; 1880-81, 44; 1881-82, 55; 1885-86, 43; 1886-87, 57; Transactions, Agri.-Hort. Soc. Ind. (from 1832), Vols. II., 181; IV., 221-322 (Proc.), 53; V., 38-42, 73-80, 194-200, (Proc.), 39, 40, 42, 85, 86, 95-96, 123; V., 25-29, 103-106, 127, (Proc.), 4, 25, 34, 40, 41; VII. (Proc.), 23-24, 41, 137, 150, 163, 208; VIII., 345; Journal (Old Series), Vol. I., 389; II. (Proc.), 49, 314, 464; IV. (Selections), 128; VIII., 102-107; X., 41, (Proc.), 125-126; Journal (New Series), I., (Selections), 70-81, 311-328; Tropical Agriculturist (1881-82), I., 271, 272; II., 157; IV., 9, 88 (F. elastica in Java), 95 (Mann's report of 1884), VII., 110, 246, 747; VIII., 213-214, 293 (Burma early trade in rubber to China); Indian Forester, I., 86, 124-141 (gives Mann's report for 1875), 141 (Schlich's Appendix to the report), 188; IV., 40-41 (Arakan); VII., 241 (Burma); VIII., 203; IX., 225; X., 403; XI., 256-261, (contains Mr. Campbell's report on tapping Assam trees in 1883-84); XII., 563-564 (Nurseries for F. elastica); XIII., 550 (Chindwin, Burma); XIV., 297 (Manipur); Forest Department Annual Reports. Assam, 1873-74, para. 77-146; 1874-75, 250-256, 272-306; 1875-76, 65, 68-77; 1876-77, 45, 83, 96-110; 1877-78, 106, 122-131, 175-178; 1878-79, 111, 127-136, 199-201; 1879-80, 125-127, 146-151, 220-221; 1880-81, 113-118, 137 145, 214-216; 1881-82, 110-115, 136-142, 207-208; 1882-83, 77-78, 83-85, 113, 147, App. IX.; 1883-84, 102-105, 106-114, 166, 167, 184, App. VI., VII., and VIII.; 1884-85, 119-121, 124, 125-136, 198-230, App. III., IV, and V.; 1885-86, 139-142, 145-156, 211, 244, App. III., IV., and V.; 1886-87, 133-136, 139-152, 200, 201, 224, App. VII., VIII. and X.; 1887-88, 120, 125, 129, 131-139, 191, 192, 215, App. VII., VIII., and IX.,*

ASSAM RUBBER.

Gazetteers:—Mysore and Coorg, I., 70; III., 25; Madras Man. Admin I., 360; Official Correspondence in Revenue and Agricultural Department (Forests) on Ficus elastica and also introduced rubber plants from 1869 to present date.

Habitat.—A large evergreen tree, usually epiphytic in its young stage, but finally or originally rooting in the ground and sending down aerial roots from its branches. It is found in the damp forests at the base of the Sikkim Himálaya and eastward to Assam, Chittagong, and Burma.

Fuller particulars regarding the natural and acclimatised habitat of this tree will be found in the details given in a further paragraph regarding its CULTIVATION as a source of rubber; see also under **Ficus**, *Vol. III.*, 350.

Caoutchouc.—The rubber of Indian commerce, originally derived from the regions inhabited by savage tribes on the frontier of Assam and Chittagong, is produced from this tree. It is now extensively cultivated and is perhaps the only rubber plant in India which can be said to give indications of a future greatly extended commercial importance.

HISTORY OF THE CAOUTCHOUC OF FICUS ELASTICA.

HISTORY. 63

The earliest writer on this subject was the late **Dr. Roxburgh**, whose attention was directed to the rubber on a vessel containing honey sent to him from Sylhet. He described the plant that yielded the Sylhet elastic water-proofing gum in the 5th Volume of the *Asiatick Researches*, but his remarks in his subsequent *Flora Indica* contain the main facts, and as being of historic interest these may be here quoted: "Towards the close of 1810, **Mr. Matthew Richard Smith** of Sylhet sent me a vessel, there called a *Turong*, filled with honey in the very state in which it had been brought from the Pundua or Juntipoor Mountains north of Sylhet. The vessel was a common or rather coarse basket in the shape of a four-cornered, wide-mouthed bottle, made of split rattans, several species of which grow in abundance among the above-mentioned mountains, and contained about two gallons. **Mr. Smith** observed that the inside of the vessel was smeared over with the juice of a tree which grows on the mountains. I was therefore more anxious to examine the nature of this lining than the quality of the honey. The *Turong* was, therefore, emptied and washed out, when to my gratification I found it very perfectly lined with a thin coat of caoutchouc." Several more recent writers assert that the Natives of Assam have for ages been acquainted with the properties of gum-elastic, using it to water-proof their baskets and to burn as candles. **Mr. Smith** was thus the actual European discoverer of Assam caoutchouc, since he thought it necessary to draw **Dr. Roxburgh's** attention to the same, and he was also the agent through whom **Dr. Roxburgh** was enabled to prosecute the enquiry that finally led to its determination. Interest in India was thus awakened in caoutchouc, and the writings of subsequent investigators for several years were replete with reports on the subject. The resources of Assam were at once drawn upon, and a system of reckless extermination of all accessible trees took place without any real progress towards establishing a commercial industry. With the decline of the Assam supply interest died out as suddenly as it had been awakened, but it may serve a useful purpose to review the early literature of Indian caoutchouc prior to the revival which was consequent on the introduction of the American rubber plants into Asia.

In the *Transactions of the Agri.-Horticultural Society of India* (*Vol. VI.*, 29) mention is made of a sample of Sylhet rubber sent to **Sir D. Brewster** by **Mr. G. Swinton**, which, after being in Edinburgh for eleven years, was found to possess its elasticity and to be superior to the South

HISTORY.

American rubber in lightness of colour and freedom from smell. The Calcutta firm to whom **Mr. Swinton** submitted the sample regretted their inability to give an opinion as to its value. In that year, however, rubber was selling in London at 2 shillings a pound.

In 1832, **Lieutenant Charlton** wrote that **Ficus elastica** was abundant throughout Assam and yielded a copious supply of the gum-elastic.

In 1837, **Dr. N. Wallich** placed before the *Agri.-Horticultural Society of India* an extract from a letter received by him (dated 8th December 1836) from **Dr. Royle**, in which particulars were given of the formation in London of a Caoutchouc Company which was prepared to give a reward of £50 to any person who would send a cwt. of Assam rubber to England. **Dr. Royle** then urged the desirability of starting rubber cultivation in Assam, and stated that the facts made known by **Roxburgh** regarding **Ficus elastica** had astonished the home authorities who had formed the opinion that rubber was exclusively an American product. Shortly after the publication of the above facts, a sample of Assam caoutchouc was collected by **Lieutenant Vetch** and forwarded to the Society by **Captain Jenkins.** On this a highly complimentary report was given by the Caoutchouc Committee of the Agri.-Horticultural Society. The Committee wrote: "This proof of one of the many great commercial resources abounding in Assam is highly encouraging." Again, "Your Committee regret that they have it not in their power to afford **Captain Jenkins** any information as to the *value* of caoutchouc in the Calcutta market, since it is altogether a new article, and has as yet been exported only by parties interested in keeping their proceedings quiet; but without wishing to inspire extravagant hopes of success, they feel justified in offering it as their opinion, that there is a ready market in England to take off a very large quantity, and that there would be no great difficulty in finding parties in Calcutta willing to make advances to a reasonable extent. The Committee would be glad, however, to ascertain from **Captain Jenkins**, by actual test, at what rate it can be produced in Assam per bazár maund, without which data it would be vain to hold out any inducement. The quantity of caoutchouc exported from Calcutta up to the 30th April 1836, which may be taken as the total exportation of the article to that date, amounted to 514* bazár maunds, paying duty on a nominal value of R4,112, or R8 per maund, a valuation which your Committee believes to be far below the *market price*, if such market price were known, and which at present is known only to those who have advanced for, and procured it from, the forests." In a later report, on samples prepared in Assam by **Dr. Scott**, the Committee stated that "no doubt can be now entertained that Assam is quite capable of competing with any part of the world, and it only remains to be ascertained what quantity the country is equal to the production of."

In 1838, **Captain A. Bogle** furnished samples of Caoutchouc from Akyab; **Lieutenant Wemyss** from Assam; and so also **Captain M. Macfarquhar** from Tavoy. In the same year, the Agri.-Horticultural Society presented its gold medal to **Lieutenant Vetch,** for one maund of Assam rubber, and the London Caoutchouc Company, while not awarding their full premium of £50, gave **Lieutenant Vetch** 25 guineas, remarking, that the sample was not prepared in such a manner as to allow their giving him the prize they had offered. The Agri.-Horticultural Society also received from **Mr. Robert Smith** pieces of cloth which he had rendered water proof to be used in place of wax cloth. **Mr. Smith** then offered to purchase all the rubber the Society could spare at R18 per maund. **Dr. Royle** in letter dated 27th October wrote: "I may repeat that there is

* *Conf.* with the modern trade, page 356.

ASSAM RUBBER.

not the smallest doubt of this becoming an extensive article of Indian commerce, if managed with moderate prudence."

In their annual report for the year 1839, the Agri.-Horticultural Society stated that the caoutchouc from **Ficus elastica** is liable to decomposition. In the same year a Committee of the Society of Arts reported on **Captain Jenkins'** and **Lieutenant Vetch's** samples of Assam rubber, pronouncing the former superior to the latter. In the Proceedings of 1843 it is complained that few of the samples of Assam rubber examined by the Society have been carefully enough prepared to resist the damp and heat of the Calcutta climate.

In 1852, **Dr. Falconer**, commenting on a decline in the Assam traffic in Caoutchouc, and in answer to certain questions as to the best modes and position of tapping the trees, threw out the suggestion that the decomposition Assam Caoutchouc was liable to was probably due to the mixing of the milk obtained from the roots with that from the stem. He contended that the milk being obtained from the descending sap of the plant which had been elaborated in the leaves was probably in its more perfect state high up the stem than in the lower portions, and hence he urged that tapping should be made on the branches instead of on the roots and lower part of the stem. He recommended that this theory should be investigated by keeping the milk from the various parts of the plant distinct until it was seen which of these imparted to the mixture the soft and adhesive property complained of by the manufacturers. Careful preparation and above all the avoidance of mixing *other* milky saps with that of the true rubber plant would effect great improvements, but it must be here added that **Dr. Falconer's** suggested explanation is at variance with the accepted physiological conception of the character and movement of the laticiferous fluid. It is quite distinct from the elaborated descending sap of the plant, and its motion is *in* reality from the older to the newer structures or upwards instead of downwards. The caoutchouc granules contained within it are useless excretary materials which when once formed are probably never again chemically altered, so that but for the other ingredients of the milk it would be quite immaterial from what part of the stem or root it was drawn. The question, however, as to the part which could be tapped with least injury to the plant would have to be answered in favour of **Dr. Falconer's** recommendations.

In a most instructive report, from the pen of **Mr. G. Mann**, Conservator of Forests, Assam (*Progress Report of Forest Administration in Bengal for the year 1868-69; reprinted Jour. Agri.-Hort. Soc., India* (*New Series*), *Selections, 70-81*), many facts are given regarding the destructive methods of tapping pursued in Durrang and other districts of Assam. In 1865, the privilege of tapping in the Government Forests between the Bor Nuddy and the Moora Dhunseeree Nuddy was sold for R1,012 to Kyabs in the Mungledye Bazár. These contractors purchased from the people who worked the forests 2,500 maunds of rubber, "but had not," **Mr. Mann** remarks, "the slightest control over the tapping of the trees, and encouraged the latter as much as in their power to obtain the largest possible quantity during the short time they held the monopoly of buying it, consistent with their interests without any regard, however, for future supplies, which was of most disastrous consequence, in so far as it induced the men who tapped and collected the rubber to indulge in the most outrageous wholesale destruction of these valuable trees, by either felling them with axe, or, if this was too troublesome, to collect firewood and burn them down, so as to render the operation of tapping more convenient than it would have been had the trees been left standing. In consequence several hundred magnificent trees were counted, in all direc-

HISTORY.

tions, lying on the ground with cuts across their trunks and roots from 6 to 18 inches long, 3 inches broad, and a foot to 18 inches apart, with smaller cuts on the upper branches, by which all that they could yield was extracted immediately after they were felled, with utter disregard for future wants." A less immediately destructive system was pursued in other parts of the province, which, nevertheless, in time resulted in the trees losing their power of yielding the milk. **Messrs. Martin, Ritchie & Co.** of Tezpore, for example, got the exclusive right of tapping over a certain tract of forests for fifteen years from 1852 and free of payment, but on condition that tapping should only be made between the 1st of November and the 30th April, and that yearly they should plant 200 caoutchouc-yielding trees. The trees, however, embraced by this concession, got exhausted, and one year before the term fixed **Messrs. Martin, Ritchie & Co.** discontinued tapping them, because, as **Mr. Mann** remarks, rubber could not now be drawn from them at a remunerative price. Continuous tapping for six months year after year, **Mr. Mann** affirms, will kill the trees, and accordingly he urged either that tapping should be restricted to three months a year (January, February, and March), or that a regulation should be made prohibiting the tapping of forests more frequently than once every three years.

Mr. Mann concludes his remarks on the exhaustion of the Assam rubber supply by saying that "unless the time of collecting is much more limited and some plantations are started to provide for future supplies, there will be no more rubber procurable from these forests a few years hence, and the rubber tree will gradually disappear and a valuable commodity be lost to the public and the revenue derived therefrom to Government."

Mr. Mann further gives instructive figures as to the value of the rubber trees and their yield of caoutchouc. "Assuming that a tree reaches its full size at fifty years without tapping, and would after that yield every third year one maund of rubber, which would be collected, manufactured, and delivered in Calcutta at R15 per maund, and should realise the present price of good rubber, *viz.*, R35 per maund, it would have a net profit of R20 per tree every third year. Besides this one maund of lac may be reckoned on from every tree per year, which, if collected at its present rate, could be delivered in Calcutta at R10 per maund, whilst it fetches R15 to 20 per maund there now, which is a profit of R5 at least per tree yearly. All these figures are the very lowest, and the tapping the most cautious; still if the tree planted lives a second fifty years, which it is sure to exceed, it produces R320 for rubber and R250 for lac, which is more than any two timber trees of fifty years each, which might be grown in that time could equal."

Mr. Mann then deals with the two kinds of rubber manufactured by the people of Assam, *viz.*, one in irregular solid lumps or loaves about 16 to 20 oz. in weight and the other in balls of rubber threads each weighing 12 to 16 oz. The price paid (in 1869) for the two kinds varied, he says, from R8 to R12, but this was paid for by pieces of Eri silk cloth of that value in exchange for a maund of rubber. This fetched in Calcutta from R20 to R40 per maund, but **Mr. Mann** adds "if care were bestowed on the manufacture, it beyond doubt would fetch much higher prices." **Messrs. Martin, Ritchie & Co.**, however, purchased their rubber only in the fluid state from the people who tapped the trees. It was brought to them either in earthen pots or cane baskets made water-proof with a previous coating of rubber. This coating of rubber, **Mr. Mann** states, was held to retain the sap in its fluid state. He then goes on to say that "Rubber in this fluid state was first purchased at R1-8 per maund, but

ASSAM RUBBER.

soon rose to R5 for *the best or thickest procured from the roots, and R4 for the next best procured from the lower part of the stem, and R3 for the worst supposed to come from the upper branches of the tree and to have been mixed with the juice of other species of Figs and water.*" The italics have been given here to contrast this opinion with that given above in **Dr. Falconer's** suggested explanation of the inferiority of Assam rubber; **Dr. Falconer** held that it would be found that the most inferior rubber would likely be found to be that drawn from the roots and lower part of the stem.

Since the year 1869 (the date of **Mr. Mann's** report which has been so frequently placed under quotation above), the Government have started caoutchouc plantations of **Ficus elastica** in Assam. But it seems likely that, had the similar recommendation made by **Dr. Royle** in 1839 been acted upon, India would have now possessed a flourishing caoutchouc industry. **Mr. Mann** tells us that the total revenue derived by Government from Assam rubber between the years 1861 and 1870 amounted to R83,733, the total for the last year (1869-70) having been R23,940. That small sum, therefore, practically (or only a portion of it) was the purchase money for a right to exterminate or ruin the **Ficus elastica** trees within British Indian territory. And so completely had this been accomplished that we read of contractors abandoning rubber collection before the expiration of the terms of their concessions.

Mr. Charles Brownlow in 1869 (*Jour. Agri.-Hort. Soc., New Series, Vol. I., 311*) started on a journey up the valley of the Jatinga in Cachar, with the object of studying among other things the India-rubber of that neighbourhood. His account is most instructive, but want of space prohibits more than a mere mention of one or two of the points he brings out. The tree, germinating on other trees as an epiphyte, sends down immense aerial roots and from the top of these (often at a great height above the ground) produces its horizontal branches and dome of boughs and leaves. The roots that are tapped, and to which allusion has been made above, are these aerial roots which at first sight look like unbranched stems. Speaking of this great tree **Mr. Brownlow** says:—"Among forest trees and in regard to dimensions, this is 'facile princeps,' and there is no other, not even the Banyan, that approaches it in dimensions and grandeur." Again, "It will be observed that every portion below the head of the foster tree is strictly root and incapable of throwing out a branch, and as the head is rarely less than 60 to 100 feet high, it is no easy matter to procure a branch. . . . These cables and buttresses as they approach the ground throw out smaller and subsidiary rootlets of all thicknesses down to that of twine. If any of these be cut they die below, but from above grow again downwards. . . . It is only necessary to see the tree to appreciate the fearful risk encountered by the gum gatherers, who by no means confine their operations to the base, but climb up as high as the roots extend and higher along the horizontal branches, chopping with their *dhaus* at intervals of every few inches, the cuts answering as well for their foothold as for the sap to exude from. . . . Were the base of the tree alone tapped the yield would be very insignificant, especially in trees that have been frequently tapped before. And as the trees occur very sparsely, and long distances have to be gone over to meet them, it becomes an object to get as much off at each cutting as possible. The tree must be twice climbed, once to cut it and a second time after the gum has dried (which takes a day or two) to gather it. This is done by pulling off the tear which gathers below the wound, which brings away with it all the gum that has exuded, and these tears have only to be moulded together to agglutinate into a ball. The quantity that can thus be collected at one cutting does not exceed 4 to 5

HISTORY.

seers (8 to 10lb). Of course, no mercy is shown to the trees, all of which (at least those that I saw) suffer severely; many, I was assured, are killed outright. The damage they sustain is apparent in the large cankers, and buttresses rotted off, owing to the bark being unable to heal over the frequent wounds they have received all round. The foliage is wanting in luxuriance, and dried branches and roots lying about testify to the injury in health that the tree has sustained. . . At present the trees are having a long respite which will no doubt do them good, but it is only owing to a clause in the conditions of Government which provides that the monopolist shall, at his own cost, plant a certain number of trees yearly . . Practice will soon show how the gum should be gathered; probably the best plan would be by scoring the bark on one side (not chopping it and leaving the other half of the bark to carry on the circulation . Of course, the only season in which the gathering is practicable is the dry months; during the rains the tears would be washed away before they had time to solidify. The rubber trade of Cachar amounts to about 2,000 maunds per month during the dry months (November to April). The price at which the gatherers sell averages from R20 to R25 per maund." **Mr. Mann** states that in Assam in 1868 there were sold in Mungledye Bazár 2,500 maunds, in Tezpore about the same quantity, and in Chydooar 1,500 maunds. For the year in question there was sold in Assam and Cachar close on 9,000 maunds, but of course a large proportion of that amount was drawn from trees beyond the Assam frontier.

It will thus be observed that from 1810 to 1839 the keenest possible interest was taken in the subject of Assam rubber. From the latter date the defects of the substance were spoken of as damaging its further development as an article of European consumption, and various theories were advanced as to the cause of this. It seems only natural since the earlier reports speak highly of Assam rubber and recommend as necessary a peculiar method of collection, a method adapted to the then known way of utilising rubber, that the defects subsequently complained of were due to adulteration or selfish and exhaustive extraction that sacrificed quality to bulk. About this time also the alarm of extermination was established and down to 1869 various reports and regulations of Government appeared which aimed at checking the danger anticipated. The *Journals of the Agri.-Horticultural Society of India* which, for twenty years became the enlightened medium of disseminating information regarding Assam rubber, were from 1839 practically silent on the subject, until a new interest, and perhaps, though this could not have been foreseen, an unfortunate one, was awakened in 1872-73 by the expectations of immediate fortunes to be made from the superior rubber-yielding trees of tropical America and the gigantic climbers of Africa. Some of these introduced plants have taken kindly, however, to certain districts of India, and may yet come to be a source of wealth in their new homes; but they have by no means fulfilled the high expectations once entertained. Indeed, it may confidently be stated that even from the present date it will take some time before the money spent on their introduction will be recouped to the planter or to Government. Greater success may, in fact, be looked for from the efforts put forth by the Assam Government to regulate and resuscitate the forests of **Ficus elastica.**

The following extract from **Sir D. Brandis'** review of **Mr. J. Collins'** report on Rubber may be here given, since it deals with the question of the adulteration of Assam rubber, and was the chief cause of the establishment of plantations in Assam:—"Assam rubber stands low in the list at present, but there seems good ground to believe that this is mainly due to the large proportion of impurities (bark, sand, stones) with which it is

ASSAM RUBBER.

commonly mixed. The caoutchouc which was collected and prepared by **Messrs. Martin, Ritchie & Co.** of Tezpore, while they had the lease of the caoutchouc forests previous to 1865, and which was known in the London market under the name of 'Fine slab Assam,' was a very superior article, and quite lately an improvement has again taken place in the quality of the Assam product.

"**Mr. Mann** thinks that apart from the accidental (and sometimes intentional) impurities, such as pieces of bark, wood, sand, stones, the Assam article is often adulterated with the milk of other species of **Ficus**, which is of a quality much inferior to the milk of **F. elastica.** **Messrs. Martin & Ritchie** are said to have given up their lease before it had expired, because the supply had diminished so far that their business was no longer remunerative, and from late reports which I have seen on the subject I gather that the number of caoutchouc trees remaining in British territory is believed to be limited. Under these circumstances, it does not appear likely that any considerable improvement of the article can now be expected to be effected, through the agency of private enterprise, except at the risk of exhausting the remaining sources of supply. The question then arises, whether it is possible in some way or other to place the collection and preparation for the market of this valuable article under the control of public officers who will devote their whole time and attention to this subject. No great skill is required for the collection and preparation of a pure and valuable article, the facts stated in **Mr. Collins'** report and in **Mr. Mann's** previous papers on the subject will enable any intelligent and careful person to arrange and superintend the collection and preparation.

"**Mr. Mann** specially insists on the following points being observed :—

"(1) Fresh cuts to be made only in February, March, and April, and the trees to have rest for two years between each tapping.

"(2) The cuts to be at least 18 inches apart, to penetrate into the bark only, not into the wood, and to be made with an instrument more suitable than the ones at present used. **Mr. Mann** prefers the German timber scoring knife.

"(3) As far as possible, the milk to be collected in a fluid state in narrow-mouthed rattan baskets, and to be brought to central manufactories.

"(4) Endeavours to be made to convert the milk into a solid state by a process of slow drying similar to that practised in Pará.

"(5) In case this method should not succeed, then the process employed by **Messrs. Martin & Ritchie** to be followed.

"(6) Those varieties of caoutchouc which dry naturally on the tree to be collected with care, and to be picked so as to get rid of all impurities.

"So much, however, is evident that unless the collection and preparation of the caoutchouc produced in British territory is placed under the control of public officers who have an interest in the protection and improvement of the forests, no satisfactory result can be expected. If this is not practicable, then we must for future improvements entirely rely on the caoutchouc plantations to be established.

"Under all circumstances plantations of the **Ficus elastica** should be commenced at once in Assam on a large scale. The tree strikes readily from cuttings; its cultivation therefore is easier than that of most other trees. In one of his first reports on the subject **Mr. Mann** suggests that lines be cleared through the forest, and that cuttings, as large as possible, be planted at convenient distances on either side of these lines. Very likely this would be a good plan to begin with, and as the carriage of big cuttings over long distances would be expensive, nursery beds should be prepared and enclosed for the growth of such cuttings from small slips. If suitable soil and localities are selected, and if these plantations are at

HISTORY.

once placed under efficient supervision, there ought to be no difficulty in this undertaking, and operations should in my opinion be commenced as soon as possible on a large scale, and in accordance with a well-considered plan."

Mr. **Mann's** report for 1884 gives the result of the experiments in the cultivation of **Ficus elastica** in Assam; the following passages from that report may be given here:—

"The past history of rubber plantations in Assam, and, for the matter of that, in India, dates from the year 1872, when **Mr. James Collins** was charged by Her Majesty's Secretary of State for India to prepare a report on the caoutchouc or India-rubber of Commerce, the plants yielding it, their geographical distribution, and the possibility of their cultivation and acclimatisation in India. The only rubber-tree indigenous in India (**Ficus elastica**) is noticed on pages 19 to 21, 32 to 39, and 48 to 54 of that report.

"This report was largely circulated by Government in this country, and the attention of Local Governments was directed to the necessity of protecting the trees which yielded this valuable commodity, because it had become quite evident that the caoutchouc-trees were being recklessly destroyed in all parts of the world, and particularly so in Assam, which is, so to speak, the only province in India where caoutchouc-trees grow, and the experimental cultivation of the indigenous rubber-tree (**Ficus elastica**) was accordingly ordered in May 1873 by His Honour the Lieutenant-Governor of Bengal. But by the time these orders reached the Commissioner of Assam the season was so much advanced that but little could be done that summer, because there was only one small forest plantation with a resident forest officer in existence at that time, and this was at the Kulsi river, which is not as favourable a locality as the Charduar in the Darrang district. In the latter district forest work had not been started, and, consequently, the first commencement in the present Charduar rubber plantation was not made until the next cold season. A detailed account of these first attempts at planting rubber will be found in paragraphs 80 to 114 of the Assam Forest Report for 1873-74.

"A particular impetus was given to this work by the complications and difficulties that had arisen at about the same time in the proper management of and control over the India-rubber trade in this province, brought on by competing speculators, which had necessitated an order from the Supreme Government that the operations of the Forest Department should be limited to conservancy and reproduction of the rubber-trees in certain well-defined areas, and to the collection and manipulation of the produce in such limited areas through their own agency.

"This order of the Government of India was repeated in 1876, and has been acted up to until now: all work in the way of rubber plantations is based on it, and, what is more, the experience gained in the twelve years that have elapsed since the issue of that order has made it clear that the effectual protection of selected areas, *with naturally grown rubber-trees on them*, is next to impossible, on account of the localities where these trees grow being, generally speaking, very inaccessible, and the unequal way in which these trees are scattered about in the forests, as it would mean the protection of enormous areas to ensure anything like the present export of rubber from Assam, and this in turn would mean the employment of very large establishments to watch over the forests, because rubber is so very portable, and its removal not necessarily confined to roads or tracks, rivers, and so forth, as is the case with timber, and the cost of such establishments would altogether exceed the advantages arising from the rubber trade. This simply reduces the whole question of permanently keeping up the export of this valuable product from India to making plantations of the tree that yields it.

"In April 1874 the Government of India called for a special progress report on the caoutchouc plantations in Assam, which was furnished . . . . , and subsequently printed and circulated with the Government of India, Department of Revenue, Agriculture, and Commerce (Forests) letter No. 22, dated the 31st August 1875. The efforts made up to that time in the way of planting caoutchouc-trees in Assam, and all information regarding the yield of caoutchouc-trees then available, have been fully stated in that report. . . . .

"The Charduar plantation has, as was maintained from the commencement, proved in every respect the best locality in Assam where the rubber-tree has been planted; the land, it is true, is not high, and so we must, no doubt, have some area planted on higher ground, if for no other reason than to enable us to make comparisons; this is to be done at once on the high land immediately to the west of the present plantation,

ASSAM RUBBER.

as the Chief Commissioner has sanctioned an extension of 200 acres. The present area under cultivation is fully stocked, containing 12,511 trees; they have been planted at 25 feet apart in the lines, which latter are 100 feet apart; this is double the number of trees that was planted on an acre at the commencement. The oldest trees are about 30 to 40 feet in height, and a few from 45 to 50 feet, but this cannot be put down as the average growth of **Ficus elastica** in ten years, since half this time, and longer, these plantations were entirely experimental and everything had to be learned, as, for instance, the first trees were all raised from cuttings, which mode of propagation has entirely been given up, since the trees raised from seed have proved much hardier and faster growing, and as to the planting of rubber seedlings high up in the forks of other trees, this also has almost entirely been given up, because such trees in most instances did not make more than a few leaves in the year, and it would, as a matter of course, be out of the question to plant rubber-trees where they would take a century to become large enough for tapping, when such trees can be grown in a different way in one-fourth this time.

"On the other hand, it has been found that trees planted on small mounds of earth, 3 to 4 feet in height, grow very much better than if they are planted on ordinary level ground, and this plan has therefore also been adopted, although it adds considerably to the cost of making these plantations, but the faster growth of the trees amply compensates for the higher expenditure.

"The method of planting adopted from the beginning has been to clear lines from east to west through the forest for the young trees a hundred feet apart; the width of the lines is 40 feet, so that a broad strip of forest 60 feet wide is left standing between these lines, to ensure the utmost amount of moisture in the atmosphere for the young rubber-trees. At first, the lines were only cleared 20 feet broad, but it was found after a few years that these closed up very soon, and thus retarded the growth of the young trees by shutting out the requisite amount of light. However, the widening of the lines also brought about the faster growth of the scrub in them, besides that of the rubber-trees, and more money, time, and attention has in consequence to be spent, especially in the rainy season, on those plantations, than had at first been anticipated; but the greatest and most costly difficulty that had to be overcome was the effectual protection of the rubber-trees against deer, which during the first few years constantly bit off the young plants, and, where they were not entirely ruined by this, they were so much injured and retarded in growth that a considerable increase in the expenditure on these plantations had to be incurred on fencing to prevent it. But for the future this expenditure will not be necessary, since it has been found that saplings 10 feet and more in height can be transplanted without difficulty and with perfect success, and if such saplings are tied firmly to stakes the deer can do little or no damage to them.

"The efforts made to interplant with timber trees besides the rubber, so as to obtain a yield of timber in addition to that of caoutchouc, have up to the present met with but partial success in the Charduar plantation, but there is no reason to doubt that this will soon improve, as the officer in charge gains more experience; in the rubber plantation at the Kulsi in the Kamrup district, this work has been most successful.

"The total area of the Charduar caoutchouc plantation is now 892 acres, and has cost R64,351, or R72 per acre; this is abnormally high, since much of the work during the first five years had to be done twice over, and sometimes oftener, because the planting of caoutchouc trees was new, and everything had to be learned and found out by experiments, which naturally took some time. But matters have changed in this respect; we know now what we are about, and the officer in charge of this work, **Mr. T. J. Campbell**, has estimated the cost of the extension which is at present being carried out at R29 per acre, to which another R6 for subsequent cultivation and clearing should be added, bringing the cost, including everything, up to R35 per acre.

"Besides the experimental nature of the work, to which the cost of R72 per acre of this plantation must to a great extent be attributed, we have also prepared extensive nurseries, covering an area of about 23 acres, and containing some 184,000 plants of different sizes, which is sufficient to extend the plantation by 200 acres per annum for the next fifteen years, or a square mile per annum for the next five years, if desired and these nurseries have been so planted that, if for special reasons it is considered advisable not to extend the plantation at any particular time, the trees can be kept almost stationary for fifteen to twenty years, without becoming less suited for transplanting, a particular advantage enjoyed by **Ficus elastica** in common with other semi-epiphytes, as compared with ordinary trees. . . . .

HISTORY.

"Thus far I have given an account of the Charduar rubber plantation as an experimental undertaking only, and shown that it has been a perfect success as far as the growing of the trees is concerned; but it remains to be considered what the financial results of the undertaking are likely to be, since, as I have always held, and do now hold, the financial success of forest management is the only sound basis on which it can be permanently established and maintained. To make an even approximately correct estimate of the probable revenue that may be expected from these plantations, it is first and foremost necessary to know what a rubber-tree will yield, and on this point our information is most imperfect. The statements made by rubber-collectors are quite unreliable, and the exhausted state of the naturally grown rubber-trees has prevented us until last year from making experiments; the result of last year's experimental tapping, as recorded in Appendix IX, of last Annual Forest Report (1882-83), interesting as it is, and much as it has increased our knowledge of the yield of caoutchouc from **Ficus elastica,** still leaves us in considerable doubt on the subject, as has been stated in paragraph 118 of that report. However, so much is certain, that a full-grown rubber-tree of about fifty years old will yield at the very lowest 5 seers of rubber, if very carefully tapped, and this quantity may be expected about sixteen times, which will be an equally safe estimate for calculating the yield of a rubber-tree. To be quite on the safe side, I will only calculate ten trees per acre, which would give us about 20 maunds of rubber from every acre. This, at the price at which rubber was collected last year in the Darrang district and sold, and deducting the expenditure incurred on collecting it, would give us a net profit of R54 per maund, or R1,080 per acre in fifty years, and if the rubber-trees have a longer life, the yield may be reckoned for their remaining years of life at the same if not at a higher rate.

"An acre of first-class timber trees would cost about double as much to plant and maintain; at the rate of sixty trees per acre and taking the value of the trees at R10 each (the present royalty charged), this would give us R600 only, as compared with R1,080 from rubber, and most of the first-class timber trees will require one hundred years to reach maturity, or double the time of a rubber-tree; this means, in other words, especially if the compound interest on the capital used is taken into consideration, that an acre planted with rubber-trees will give about four times as much revenue as an acre planted with first-class timber trees.

"It may be and in fact has been argued that rubber might be produced artificially, and that thus a fall in the price might be brought about. I think there is little to be feared in this respect, not more so than timber has to fear from the extended use of iron, and rubber, being a raw product, has a great advantage, inasmuch as the artificially produced article would have all the cost of manufacture added to the cost of the raw materials, and I myself have not the slightest fear in this respect. The price of rubber has been very high for many years now, and during this time it is known that efforts have been made to produce artificial rubber, but that they have failed.

"It now only remains for me to consider the value of **Ficus elastica,** as compared with other trees yielding rubber, both as regards quality and quantity, and although it must be admitted that the rubber yielded by our indigenous tree is slightly inferior to that from some other rubber-trees, the difference is so little that, in my opinion, it has nothing to fear in this respect, and, as to the quantity yielded by other species we have positively no authentic information to make comparison; but I am very doubtful whether any of them will yield more than **Ficus elastica,** and, certainly, the difference, if any, could not be so much as to make the cultivation of the latter unadvisable.

"Of the two exotic rubber-trees which have been tried in Assam, *viz.*, **Hevea brasiliensis,** the Para rubber, and **Manihot Glaziovii,** the Ceara rubber, the former has failed completely, as the climate of Assam is altogether too cold for it, and although the latter tree grows remarkably fast during the first year or two, and seems to thrive very well, its appearance is not at all such as to make me hope that it will do as well as our indigenous trees, much less that it will do better. Nothing positive can be said on this score until experiments with both have been made under careful supervision by a competent and responsible officer. . . . . .

"Under the circumstances as sketched above, and considering that Assam is the only province in India in which rubber grows to a considerable extent naturally, I am of opinion that it is a duty of the State to have rubber plantations in this province."

The Government of India supported Mr. Mann's opinion as to its being the duty of the State to organise rubber plantations. Accordingly directions were given in May 1884, that for five years from that date the Assam plantations should be increased by 200 acres a year. Part of that extension, it was recommended, should be situated on higher ground than

ASSAM RUBBER.

hitherto planted, at Charduar. At the same time, it was added, endeavours should be made to induce private persons to plant India-rubber trees on their estates, seedlings being supplied to them for this purpose at cost price. It was also suggested "that the experimental plantation of **Ficus elastica** as an epiphyte might with advantage be undertaken by the Forest Department on a somewhat larger scale than has been done up to date. Naturally, rubber generally reproduces itself in this way; and although the growth of the seedlings thus raised is slow at first, the trees are likely to grow to much larger dimensions ultimately. This method of reproduction is, moreover, inexpensive, as the seedlings do not require any attention after they have once been deposited in the upper forks of trees." The Government of India also desired that in order to test the financial results of the cultivation of this rubber fifty mature trees should be experimentally tapped, year by year. Acting up to these suggestions, the subsequent reports give the results. The area under **Ficus elastica** was steadily increased and the tapping of the trees proceeded with, showing an inexplicable irregularity year by year, in the amount obtained from each tree. The trees experimentally tapped were numbered from 1 to 50, and the record of each tree kept, as also the average of all the fifty trees year by year. In 1882-83, the average yield was 4 seers 4 chataks; in 1883-84, 1 seer 1 chatak; in 1884-85, 2 seers 5 chataks; but in this year six new trees (51 to 56) were tapped and gave 12 seers 10 chataks; in 1885-86, 3 seers 4 chataks; but Nos. 51 and 52 were also tapped giving 5 seers 13 chataks; and in 1886-87, 1 seer; but Nos. 55 and 56 were also tapped giving 3 seers 3 chataks. Commenting on these results the Conservator in his annual report says: "The above figures show that the average quantity of rubber yielded by one tree has been less than in any previous year, it being 1 chatak less than in 1883-84. The value in money depends of course on the market, and on this account is higher than it was in 1883-84. A close examination and comparison of the detailed statement (appendix X.) with those of former years shows that the fluctuations in the yield of one and the same tree, in different years, is often very considerable, and remains up to the present inexplicable, since the officers, under whose personal supervision these experiments were made, have up to date not been able to find out any reasons for, or causes of, these very material fluctuations. It is quite evident that we have as yet much to discover to be able to make an even approximately correct estimate of the yield of rubber trees. As pointed out by **Mr. Jellicoe**, the difference in yield has been sometimes very remarkable, as, for instance, tree No. 50 gave in 1884-85, 4 seers; in 1885-86, 5 seers 4 chataks, when suddenly in 1886-87, its yield came down to 6 chataks only, and there are other trees, as, for example, Nos. 12, 17, 18, 20, 21, 42, and 49 that show equally remarkable differences in yield between last year and this, which **Mr. Jellicoe** has not been able to account for in any way. The Conservator feels sure these are not ordinary fluctuations in the yield of rubber trees, and that a much closer watching of this tapping will be necessary in future. The slight but general falling off in the yield of these trees is probably natural, and the figures seem to show that this difference is not due so much to the differences in the seasons as to that in the individual trees. The falling off in the average yield per tree of the six, Nos. 51 to 56, has even been more marked than of the other fifty trees." The Conservator is careful to explain that all the tappings were made on the stem, and that no roots or branches were touched. The tappings commenced about four feet from the ground and were never less than two feet apart. This careful and considerate system is pointed to as the reason why much less was obtained than is currently reported by native dealers as

HISTORY.

the yield. In the report for 1883-84, for example, it is stated, "Regarding the possible yield of a rubber-tree, instances are well known in this district where a large tree has been tapped for the third time, and yielded between 2 and 2½ maunds. Subsequent tappings gave from 4 to 10 seers, but these have not been carried over a sufficiently large number of years to lead to exact conclusions. Of course, this yield was obtained from a mode of tapping very different to what is departmentally adopted : roots, branches, and trunk, receiving as many incisions as they were capable of bearing."

It is difficult, and perhaps unwise, while compiling from books and reports, to attempt to offer practical suggestions, but it would seem, since the yield of rubber fluctuated and even declined, on the trees which were subjected to the moderate tapping pursued in the Government reserves and plantations, that we have by no means learned the maximum yield commensurate with financial success. The formation of caoutchouc is what may be called a by-product in vegetable economy, though, of course, the milk contains other substances of importance to life. Severe tapping may set up what for the moment could be called an abnormal or diseased condition. But it would seem desirable to ascertain the effect of a much more severe system than that hitherto followed and to tap year by year the maximum amount short of reckless injury to the growing woody structure or complete destruction of the tree. It would, from past experience, be scarcely safe to affirm that the fluctuations or even the decline established the conviction of an injury done to the trees, but rather that whether tapped or not the formation of milk is not constant. A large amount of milk remained in the structures not tapped which might have been removed without doing any greater harm than was done by the tapping of the lower part of the stem. The current of the lacticiferous fluid being upwards rather than downwards it cannot be held that the few cuts made on the stem drained the milk from the whole tree. If these ideas be correct in a year of less formation a smaller quantity would be obtained from a uniform degree of tapping, whether that degree was injuring the tree or not. If we possessed data by which an opinion could be formed as to the maximum yield, until death supervened, it would be possible to fix the degree of tapping that was desirable and the rate of reproduction necessary to establish a profitable enterprise. It may be the case that a higher yield in a fewer number of years' existence would prove more profitable than a smaller yield distributed over a longer period. A system of severe tapping alternating with a period of rest might also prove more remunerative than an annual restricted tapping. It would thus seem desirable to know the maximum yield and the number of years during which obtainable.

TRADE.

Trade in India-Rubber from Ficus elastica.

Assam. 64

I. Assam.—The following table gives the imports from across the frontier into Assam and the total exports from Assam during the years 1882—87. In the Forest Department reports the difference between these figures is taken as the probable yield from forests within the British territory :—

	1882-83.	1883-84.	1884-85.	1885-86.	1886-87.
	Mds.	Mds.	Mds.	Mds.	Mds.
Exports	9,329	9,792	7,529	6,658	4,039
Imports	4,586	1,642	802	983	1,597
Assam Production	4,743	8,150	6,727	5,675	2,442

II. BENGAL.—The following note on the caoutchouc of the Chittagong Hill Tracts has been contributed by the Deputy Conservator of Forests: "The table herewith furnished gives the quantity of rubber brought by the hill tribes to Demagiri, the frontier police post, and sold there to traders:— **BURMA RUBBER. 65**

Year.	Maunds.	Seers.	Chataks.
1879-80 .	964	7	...
1880-81 .	906	19	8
1881-82 .	222	4	14
1882-83 .	183	4	14
1883-84 .	125	17	...
1884-85 .	39	29	8
1885-86 .	14	20	...
1886-87 .	66	22	8
1887-88 .	174	21	...

"The above figures show a rapid decline from 1880-81 to 1885-86 of the amount brought in, due probably to excessive and wasteful tapping in previous years. The rubber comes from the Lushai hills north and north-east of Demagiri. This is a thickly wooded and thinly peopled tract, so that it is difficult to conceive the complete destruction of all the rubber trees, but the tendency is certainly in that direction. The **Ficus elastica** has been introdued into the Sitaphahar reserves and promises well. Price per maund at Demagiri R40. It is thence exported to Calcutta."

III. MADRAS.—The Conservator of Forests, Southern Circle, reports that **Ficus elastica** has been introduced into Malabar from Assam, and several thousand trees have been raised. "There is also a good stock of healthy young plants coming on and a well supplied nursery besides. The experiment has been successful, proving that the tree can be grown in Malabar." **Madras. 65**

IV. BOMBAY.—The Conservator of Forests, Southern Circle, thinks this fig would grow well in some parts of Bombay: it thrives in gardens at Belgaum. **Bombay. 66**

V. BURMA.—**Mr. Strettell** (*Narrative of my Journey in Search of Ficus elastica in Burma Proper*, pages 9 to 12) quotes a report by **Mr. Hough** regarding the Prome plantations of the India-rubber tree. The plants came from Bhamo in 1872; twenty-three plants in all, of which six were sent to Rangoon for the Division. No further information is available as to the success or otherwise of this experimental cultivation in Lower Burma of **Ficus elastica.** The following communication has been obtained, however, from the Government:—"There is a very large trade carried on in India-rubber in the Mogaung district of Upper Burma. The collection of the rubber was till lately in the hands of a monopolist who paid Government R1,00,000 a year, but the collection is now open to all. About 100,716 viss of rubber appear to have been collected last year. A very fair quantity is also collected in the Upper Chindwin district. There is another **Ficus** in the Mogaung district which the natives say gives just as good rubber as the **Ficus elastica.** This appears to be the **Ficus laccifera.** There are no India-rubber plantations of any importance in Burma. Foreign rubber-yielding trees have been grown in gardens in the Mergui district with fair success—the **Hevea,** *Para,* and **Castilloa** have succeeded best. **Parameria glandulifera** grows in the Mergui and Tovoy districts, but the cost of collecting the rubber is too great ever to make the growing of it a paying speculation. There is another rubber creeper in these two districts called the *Talaing Sok,* but this also is too costly to collect." **Burma. 67**

BURMA RUBBER.

The following passages appeared in *The Times* (1888) regarding Burmese Rubber:—" Mr. Warry, of the British Consular service in China, who is at present stationed as political officer at Bhamo, has made a report to the Chief Commissioner on the India-rubber trade of the Mogaung District. Rubber was first exported from Upper Burma in 1870; and until 1873 the trade was free to all. Since the latter year, however, the forests have been worked under the monopoly system, five Chinese firms being the joint concessionaires, two supplying the money, and three superintending the work. The price ranged from R60,000 to R90,000 per annum, but in the present year the sale of the right produced a lakh of rupees. The forests occupy an extensive Kachin district north of Mogaung and stretching east across the Chinese border. The Kachins are exceedingly jealous of interference with their trees, and although at first they made the mistake of overbleeding them, they are more careful now, and though the trees seen by Mr. Warry were covered with innumerable small incisions even up to the tiny topmost branches, they were obviously not drained to the extent of half their power.* Mogaung is the head-quarters of the trade; four-fifths of the yearly supply is brought in there by Kachins in the employ of Chinese, the remaining fifth is purchased in the districts by Chinese agents of the lessees. The practice is for the Chinese manager in Mogaung to make liberal advances to the Kachins to defray expenses during the collecting season; when the rubber is brought in, the refund is made by selling the rubber to the manager at half the market price. Formerly the Kachins were cheated in the weighing, and they retaliated by passing off India-rubber balls the insides of which were mostly stones and dirt. The travelling Chinese agents who also collect rubber merely travel from place to place buying such quantities as the Kachins offer, but as the latter have no standard weights they are usually cheated to the extent of about 70 per cent. The profit on this difference of weight more than pays the expenses of the agents. In most cases rubber is the subject of certain transit charges through the Kachin districts, the Tsawbas, or local chieftains, levying a certain tax—perhaps two or three balls out of a hundred. So long as these charges do not amount to 10 per cent. there is no complaint; they sometimes exceed this and then the Chinese protest usually with success. An ex-chief of Mogaung who is now a fugitive was of great service in arranging disputes of this nature, and now a regular expenditure in presents to the chiefs is necessary in order to keep the amount of the transit dues at a reasonable level. Whatever the toll, the Chinese manager and Kachin owner bear the loss in equal shares, but the latter is amply compensated by being housed and fed at the expense of the Chinese during his stay in Mogaung. Last year a new district was opened, and a Chinese capitalist employed 400 Chinese and Shan labourers to work the forests in the neighbourhood of the amber mines. The local Kachins objected to the inroad and insisted on their right to the forest. A compromise was reached, 200 of the labourers being sent back at once, the remainder collecting rubber under Kachin supervision, to whom 10 per cent. was to be paid, and 200 Kachins paid at the current rate took the places of the 200 dismissed coolies."

YIELD. 68

EFFECTS OF CULTIVATION WITH REFERENCE TO YIELD.

In the remarks below regarding the yield of rubber from the introduced trees, Mr. Lawson, Government Botanist, Madras (supported by other writers), says that he cannot believe the yielding powers of the Cearas could have decreased on account of their having been transplanted from Brazil to Southern India. " At any rate," he remarks, " I can give no scientific

* *Conf.* with the remarks at page 356.

YIELD.

reason for believing that such is the case." A few jottings may be here given which have a bearing on this subject. **Dr. Helfer** in experimenting with the wild caoutchouc-yielding plants of Assam found that **Ficus religiosa** produced a different kind of rubber, and in varying proportions according to position and nature of soil. On elevated ground it yielded a very coarse substance without value: growing close to water it afforded a reddish fluid which remained unaltered for several days (*Trans. Agri.-Hort. Soc. India, Vol. V., 40*). So also **Mr. G. Mann** (Report on the Caoutchouc Tree in the Darrang District, Assam) says: "As the distance from the hills increases, and the atmosphere in which the tree grows, gets drier, the quantity of rubber to be obtained from a tree decreases; and whilst it is stated by the men who fetch it from the hills, that one tree is able to produce from 2 to 3 maunds, the men who gather it from the forests at the foot of the hills only get from 20 to 30 seers per tree, and if far away from the hills, only half that quantity is obtained, especially if the ground is gravelly or otherwise severely drained" (*Journ. Agri.-Hort. Soc.* (*New Series*), *I., 73*). Again in the Proceedings of the Agri.-Horticultural Society of India (*Vol. V., New Series, lii*) information is given regarding the yield of rubber from **Ficus elastica** in Furreedpore, the Collector being so encouraged by the result as to have commenced planting the tree on road-sides. In the Kew Report for 1875 this action, in the light thrown on it by the passage given above from **Mr. Mann's** report, is thus commented on:—"It is found that although the **Ficus elastica** will grow with undiminished rapidity and luxuriance in situations remote from the hills, it fails to yield caoutchouc. **Mr. Mann** concludes that no greater mistake could be made than to start plantations of **Ficus elastica** in any part of Bengal. It appears, therefore, judging from this case, that conditions which may ensure the successful growth of caoutchouc-yielding trees may not be sufficient to determine their producing caoutchouc." Although this may be discouraging to planters who, daily looking upon their healthy plantations of Para and Ceara rubber trees, await the discovery of some successful mode of tapping them, it is feared the doubt implied by the above facts has to be reiterated. The defect complained of may be a permanent one, namely, that while the trees will grow freely enough, they may not form their milky sap in sufficient abundance or they may produce a sap containing ingredients that prove detrimental to the rubber extraction, or a sap so deficient in caoutchouc grains as to make cultivation in India unprofitable. **Mr. Thomas J. P. Bruce Warren** (*Planters' Gazette*) has asked the question whether Para rubber trees, acclimatised in tropical countries, produce free ammonia in their milk. The only rubber milk, he says, which he has examined that contains a large amount of ammonia, is the **Syphonia elastica** grown in Para. As the soil contains comparatively little nitrogenous matter, he suggests an explanation of this nitrogenous sap in the diffused lightning that occurs within the basin of the Para river. If tropical countries, where the tree is acclimatised, do not have such a meteorological condition, he further suggests that nitrogenous manuring should be tried. In this connection also a remark in the *Kew Bulletin* (1888, p. 259) regarding the West Coast of Africa as a future field of rubber cultivation may be quoted:—"In a locality so favourable for the growth of India-rubber producing plants, it would be interesting to know whether any of the plants yielding good descriptions of rubber could be acclimatised successfully *without invalidating the product.*" In the French Bulletin of Agriculture (Dec. 1888) **M. Naudin** gives much useful information, but regarding **Ficus elastica** he remarks that although the tree thrives admirably for landscape and ornamental purposes at Algiers, it does not form its milk in such abundance as to make it a profitable source of rubber.

CAOUTCHOUC. In the account given below of the rubber of **Ficus Vogelii** a powerful argument is advanced regarding the composition of the milk of caoutchouc-yielding plants. The necessity for more chemical analyses has been urged (at page 340) and it need only be here added that an important service would be rendered by the chemist furnishing an analysis of the fresh milk of each and every rubber-yielding tree. A comparison of these results with those obtained by the analysis of the same milks from the acclimatised stock would save needless expenditure in forcing cultivation, if adverse departures were found, and the chemist would at the same time afford the data on which a rational system of coagulation might be framed. (*Conf.* with the remarks regarding the variability of both **Hemp,** Vol. II., 110, 118, and **Flax,** Vol. V, in their production of fibre, &c., under different climatic conditions.)

69 **7. Ficus infectoria,** *Roxb.; Dict. Econ. Prod., III., 355.*

Little more can be said regarding this species than will be found in *Vol. III., 355,* of this work. A writer in the *Tropical Agriculturist, Vol. I., 121,* mentions it as affording caoutchouc.

70 **8. F. laccifera,** *Roxb.*; a synonym for **F. altissima,** *Bl.; Dict. Econ. Prod., III., 342.*

Nothing further need be said of this plant than will be found in the third volume of this work, where the opinions of several writers are given regarding its rubber. (*Conf.* also with *Kurz, For. Fl. Burm., II., 442.*)

71 **9. F. obtusifolia,** *Roxb.; Dict. Econ. Prod., III., 356.*

Vern.—*Krapchi,* MICHI; *Date,* MAGH.; *Ngaungyat,* SHAN; *Nyoungkyap,* BURM.

Habitat.—A small-leaved, epiphytic large tree of the tropical forests at the base of the Eastern Himálaya from Sikkim to Manipur, Assam, Chittagong, and Burma.

Caoutchouc.—This tree (and less frequently species No. 8 above) is alluded to by **Strettell,** as affording part of the caoutchouc of Burma. (*Conf.* also with *Kurz, For. Fl. Burm., II., 443.*)

72 **9*a*. F. Vogelii,** *Miq.*

A West African tree, supposed to be the source of the so-called Abba rubber that is being experimentally introduced into commerce from Badagry. A sample of this rubber was recently examined by the India-rubber, Gutta-percha and Telegraph Company at Silverton. The report will be found in the *Kew Bulletin for 1888, pages 253—261.* That report is of special interest to India, since the properties of the rubbers derived from the genus **Ficus** will most probably be found to be more nearly allied to each other than to the rubbers of the other plants resorted to as sources of the commercial article. Interest may also be viewed as existing in the fact that it seems probable plantations of this Fig will soon be started. In the report on the samples of Abba rubber it is stated that "The sap of a tree may contain a large quantity of caoutchouc, but the same may be associated with other principles contained in the same or other plant tissues, which completely modify its character." Again the author of the report remarks:—"I am not aware of any native India-rubber with an acid re-action; even the juice or the Para-rubber tree, **Hevea brasiliensis,** is distinctly alkaline when drawn, and exudes a strong smell of ammonia. The rubber from this source is strongly acid. In roasting the nuts of the Urucari palms, a large quantity of acetic acid is given off, which probably, by neutralising the ammonia, brings about the coagulation of the caoutchouc; the excess of acid from the roasting of the nuts may help to explain the acid re-action of the Para rubber, but as the negrohead variety is obtained from the same source, and is *not* smoked although it is strongly acid, we must consider the generation of acid as due to fermentation, at least in a very great measure. The samples obtained from the 'Abba' tree are not acid, but whether the product could be improved by precipitation with ordinary crude acetic acid which at the same time would arrest those changes that are liable to go on afterwards, to the detriment, probably, of the rubber, is worth finding out." **SECOND BEST RUBBER of BURMA.**

The suggestion here offered would seem of much value to India. Assam rubber is often complained of as undergoing injurious changes even in the prepared state. It is difficult to dry, and remains adhesive for a long time instead of becoming perfectly dry, like Para rubber. It would seem therefore that before the home markets are prejudiced against Assam rubber and a price comes to be affixed to it which may be below its real value, the chemistry of the milk and of the rubber should be worked out with the view to discovering whether the defects complained of are due to other than caoutchouc ingredients in the milk which some process of coagulation and washing might overcome.

10. **Parameria glandulifera,** *Bth.*; Apocynaceæ. 73

References.—*Fl. Br. Ind., III., 660; Kurz, For. Fl., II., 189; Wight, Ic., t. 1307; Kew Report for 1881, 47—48; Official Correspondence between Govt. of India and the Govts. of Assam, Burma, Madras, Coorg, &c., from 1881 to 1884; Mr. Pierre's account of this creeper in Cochin-China and Correspondence thereon, see Tropical Agriculturist, II., 79, 158; III., 115, 687; IV., 86, 159.*

Habitat.—An extensive climber found on the trees of the tidal forests of Martaban, Malacca, Singapore, and the Andaman Islands, but by mistake spoken of (in the *Kew Report*) as abundant in Southern India. The above quoted official correspondence, from this mistake of habitat, resulted in many futile attempts to find the plant in India proper. It has been experimentally cultivated in Burma, but interest in the climber appears to have died out.

Caoutchouc.—Rubber of excellent quality was said to have been obtained from this creeper.

(*Conf.* with the concluding remarks under **Urceola.**)

11. **Urceola elastica,** *Roxb.*; Apocynaceæ. 74

References.—*Fl. Br. Ind., III., 657; Sir D. Brandis, Review of Mr. Collins' Report; Kurz, For. Fl. Burm., II., 162; Indian Forester, VII., 241; Agri.-Hort. Soc. Journ. (Old Series), VIII., 104; Kew Report, 1880, 45.*

Habitat.—A climbing shrub met with in Malacca and Penang; distributed to Sumatra.

This is reported by some writers to yield part of the Borneo rubber, but according to the *Kew Report* (*1880, p. 45*) the North-West Borneo rubber is the produce of **Willughbeia Treacheri** and probably therefore **Howison's** discovery of rubber in Asia in 1799 was the rubber of **Willughbeia**, not of **Urceola.**

12. **U. esculenta,** *Benth.* 75

Syn.—Chavannesia esculenta, *A.DC.*

Vern.—*Kyet-poung-hpo*, Burm.

References.—*Fl. Br. Ind., III., 658; Strettell, Note on Caoutchouc (Rangoon, 1874); Tropical Agriculturist, II., 79; Kurz, For. Fl. Burm., II., 162, 184; Kew Report, 1877, p. 31; Indian Forester, I., 186—190 (republishes Strettell's Paper), VII., 241; Agri.-Hort. Soc. Journ. (Old Series), Vol. X.,* (Proc.) *125—126.*

Habitat.—A troublesome climber, which, after **Ficus elastia,** is said to yield the best India-rubber. It is found in Martaban and Tavoy; and is common all over Pegu.

CAOUTCHOUC Burma. 76

Caoutchouc. I. BURMA.—A sample of very superior rubber obtained from a plant which **Dr. Thomson** thought might be the above was procured from Rangoon by the *Agri.-Horticultural Society of India* in 1858. It was furnished by **Captain E. H. Power**, and the report on it affirms that it was free from that stickiness which characterises and deteriorates the India-rubber obtained from Assam.

Dr. Mason (*Burma and Its People, 523*) describes a superior rubber derived from a creeper which the Rev. **Mr. Parish** thought might be **Echites macrophylla** of *Wight's Ic., t. 432:* but that determination appears to have been incorrect, and hence perhaps the confusion with **Chonemorpha macrophylla**, which exists in the literature of this subject. **Major Macfarquhar** sent a sample of what appears to have been this rubber from Tavoy to the *Agri.-Horticultural Society of India.* It was reported on as follows:—"With care in preparation, it would be equal to the best South American."

In 1874 fresh and more pointed attention was directed to this subject by the publication of **Mr. G. W. Strettell's** pamphlet on the creeper. He tells us that his attention was first directed to the subject by observing a Kachyen girl dyeing thread an indigo blue by means of the leaves of **Strobilanthes flaccidifolius** (the *rúm*, indigo of Assam and Burma) with an equal quantity of the leaves of **Urceola esculenta.** This fact has a curious interest since the allied plant **Wrightia tinctoria** and several other milk-giving plants, such as **Marsdenia tinctoria** and **Gymnema tingens,** are known to afford indigo. It is thus highly probable that the leaves of **Urceola esculenta** contain indigo, and that the combination with **Strobilanthes** serves a direct tinctorial purpose and is not a mere adjunct. If this be so, it is probable that **Urceola** might afford indigo as well as rubber. But **Mr. Strettell** being then engaged on an enquiry into the source of Burmese rubber, was naturally more interested in that property of the plant. He collected a sample of the rubber, submitted it to **Messrs. Galbraith, Dalziel & Co.**, and obtained a report of a highly encouraging nature. "We consider the quality to be very fair, and at the present market value of about R200 per 100 viss." Stimulated by the prospect held out **Mr. Strettell** went into the question of cultivation of the creeper and drew up the following estimates of expenditure and problematic profit: "Now allowing," he writes, "the foregoing data to be approximately correct, and assuming the trees to be 30 feet apart, the following details will enable us to form a fair idea of the probable financial results. Area to be cultivated 400 acres. Trees at 30 feet by 30 feet, equal per acre to 48, or 19,200 creepers in 400 acres. Minimum yield of caoutchouc per annum, estimated at one viss per creeper, equal 19,200 viss, or at R200 per 100 viss, R38,400 per annum. The cost of starting this project will be trifling in the extreme. All that will be necessary ought not to exceed, on an average of seven years, R4 per acre per annum. After the first year the creepers will have attained a sufficient height to require little or no further attention, beyond, of course, protection from fire. Thus, at the end of seven years, the cost on 400 acres would represent R11,200; and even this expenditure might be reduced if Shans or others were induced to sell their labour for the privilege of cultivating within the area free of taxes; while a still further reduction might be brought about by intermediate sowing, tapping each alternate creeper to death immediately it commenced to interfere with its neighbour. At the expiration of seven years the expenses will embrace tapping, pressing, and preparing the caoutchouc, which I estimate at 12½ per cent. of the profits. According to these figures and the present market value of the India-rubber of this creeper, the net assets of this scheme may be approximated at R33,600 per annum."

Mr. Strettell affirms that the milk of the creeper more readily coagu-

WILLUGHBEIA RUBBER.

lates than that of **Ficus elastica,** "for I have known it," he says, "resolved into a coagulum floating in an aqueous solution within a few hours after collecting, and without exposure to the direct rays of the sun, or artificial heat of any kind. This consolidated mass should be collected at once, and all moisture expelled by means of graduated pressure." The season for tapping, **Mr. Strettell** recommends, is from the end of April: from October to March, he says, the circulation is slow and the milk scarce. **Mr. Strettell's** pamphlet concludes with the statement that **Messrs. Galbraith, Dalziel & Co.** report that the rubber he has furnished them with is superior to that from the India-rubber tree, being purer and better suited to their purposes.

This species of rubber yielding plant is cultivated to some extent in Burma on account of its fruit which finds a market as a substitute for tamarinds: if to this fact it were found that the leaves also had a value, additional arguments might be advanced in favour of an experimental cultivation of this creeper to those used by **Mr. Strettell.** At all events no action seems to have taken place and the subject has once more lapsed into the oblivion into which it fell, after the appearance of the reports (quoted above) which were published nearly a quarter of a century before the date of **Mr. Strettell's** paper. This is the more to be regretted since **Landolphia** and other foreign climbers and trees have received every attention, and money has been spent in their acclimatisation which might more profitably have been used in dispelling the uncertainty that still exists regarding **Urceola elastica.** The force of this recommendation will be apparent when it is added that **Mr. Strettell** affirms that an annual budget provision is made to exterminate this most destructive and persistent climber from the teak forests. It is thus probable that a considerable amount of rubber, which might even be collected from the wild stock, is being wasted. Without, therefore, corroborating, in any way, **Mr Strettell's** statement of yield or his estimates of profit (which of course the writer has no means of doing), it would seem highly desirable that this subject should once more receive the consideration it seems to deserve, so that the public may be saved from other spasmodic reports in the future by the publishing of results either in support or in refutation of **Mr. Strettell's** arguments. If this be the *sok* of the Talaing-speaking people, a recent report states that like the rubber from **Parameria** it will not pay its extraction and preparation.

II. MADRAS.—The Conservator of Forests, Southern Circle, reports that this rubber plant "has been introduced by **Mr. Morgan** into Malabar from Burma. The growth is very slow, the largest plant, now four years old, is only 15 feet high and 1½ inches in diameter at the base. It does not promise to be of much value as a rubber-producer." Madras. 77

13. **Willughbeia edulis,** *Roxb.; Dic. Econ. Prod., I., 273—277;* APOCYNACEÆ. 78

Syn.—W. MARTABANICA, *Wall.*

The rubber from this and one or two allied species is commercially designated MALAYA or BORNEO RUBBER.

Vern.—*Lati-am,* BENG.; *Thit kyouk nway,* BURM.

References.—*Fl. Br. Ind., III., 623; Roxb., Fl. Ind., Ed. C.B.C., 260; Kurz, For. Fl. Burma, II., 165; Report on Pegu, App. XII.; Report, Botanic Gardens, Ceylon, 1881 (see Tropical Agriculturist, Vol. I., 1047); Tropical Agriculturist, II., 79; III., 81; V., 767; Indian Forester, III., 240, 242; VIII., 202; Drury, Useful Plants of India, 445; Kew Report, 1879, 18; 1880, 38, 43, 44, 45.*

Habitat.—An immense climber met with in Assam, Sylhet, Cachar, Chittagong, Pegu, Martaban, Malacca, and Borneo. Yields a large edible fruit which, apparently from an external resemblance to the mango, has received its Bengali name of climbing-mango.

CAOUTCHOUC. **Caoutchouc.**—**Roxburgh** was seemingly the first writer who described the rubber of this climber. Every part of the plant, he writes, on being wounded, discharges copiously a very pure white viscid juice which is soon, by exposure to the open air, changed into an indifferent kind of elastic rubber. **Trimen** in his report of the Ceylon Botanic Gardens for 1881 speaks of **W. zeylanica,** *Thw.*, as yielding a caoutchouc which after passing through a viscous sticky condition dries into a putty-like substance of no great tenacity and scarcely any elasticity. **W. Burbidgei** of Singapore yields the rubber known in commerce as **Gutta-singgarip.**

INTRODUCED CAOUTCHOUC. (B). INTRODUCED CAOUTCHOUC-YIELDING PLANTS. (For A. see page 342)

79 80 14. **Castilloa elastica,** *Cerv.*; *Dic. Econ. Prod*, *II.*, *229*; URTICACEÆ.

THE CENTRAL AMERICAN RUBBER TREE, known also as NICARAGUA RUBBER.

References.—*Tropical Agriculturist, Vols. I., 79, 1048; II., 14, 927; III., 850; IV., 696; V., 50; VI., 814; VII., 783; VIII., 135; Hooker in Trans. Linnean Soc., 1886, 209—215; Trimen, Notes on India-rubber Trees (1880), 7; Colony of British Honduras by D. Morris, 76; Indian Forester, IV., 44; Reports, Royal Botanic Gardens, Calcutta, 1881-82, 2; Kew Report, 1879, 19; 1881, 13; 1882, 40-41; Kew Bulletin, 1887, Dec., 14; 1888, 255. For references to other works and a general account of the plant consult the article* **Castilloa,** *Vol. II., 229, of this work.*

Habitat.—A tree, found in the Central American forests, from the level of the sea to altitudes of 1,500 feet (on the Pacific coast). It has been remarked that a dry or a rainy climate seems equally well suited to it, provided the temperature does not sink below 60° Fh. It is a rapidly growing tree, and attains a height of 160 to 200 feet with a girth of 12 to 15 feet. Trees two years old were observed to have obtained the height of 23 feet. It may be propagated by seeds or cuttings, and **Mr. Cross** recommends that on transplantation the lowest leaf be buried in the soil. In this way the young plants make better progress.

Collection of Milk. 81 COLLECTION OF MILK.—**Dr. Trimen** writes on this subject:—"Milk is abundant and flows readily, but it is of a somewhat more watery consistence than that of the Para rubber. In consequence of the large size of the trees, it is the practice of the collectors in Panama and other parts to cut them down. A groove or ring is first cut round the base of the trunk and the milk received into large leaves. The tree is then felled, rings or channels are cut around the prostrate trunk at about 12 or 15 inches apart and the rubber allowed to run into leaves or vessels. In Nicaragua the trees are tapped with sharp axes in various ways, so much injury resulting that the process is repeated only at intervals of three years. The milk is received into iron pails. It does not appear that this species is tapped until it has a diameter of 16 or 18 inches." A writer in the *Tropical Agriculturist* discusses the merits of tapping as compared to cutting down the trees. By tapping, the trees are in any case killed in five or six years; by cutting down the same amount is obtained as would be by tapping for that period (*Vol. IV., 696*).

Mr. D. Morris, in his *Account of the Colony of British Honduras,* describes the method there pursued in preparing the rubber from the milk: "At the close of the day the rubber-gatherer collects all the milk, washes it by means of water, and leaves it standing till the next morning. He now procures a quantity of the stem of the moon-plant (**Calonictyon speciosum**), pounds it into a mass, and throws it into a bucket of water. After this decoction has been strained, it is added to the rubber milk, in the proportion of one pint to a gallon, or until, after brisk stirring, the whole of the milk is coagulated. The masses of rubber floating on the surface are now strained from the liquid, kneaded into cake

CASTILLOA RUBBER.

and placed under heavy weights to get rid of all watery particles. When perfectly drained and dry the rubber cakes are fit for the market, and exported generally in casks." Alum is sometimes used in place of the moon-plant, or the milk is freely washed in water and slowly pressed. This process of preparing by means of water was described by **Mr. Cross**, the milk after washing being left to dry in vats. **Mr. Niellson** in a report on Abba rubber of the West Coast of Africa, states that in his experience of Castilloa rubber in British Honduras, one gallon of milk yielded 3℔ of rubber.

CULTIVATION OF CASTILLOA RUBBER IN INDIA.

Bengal. 82

I. BENGAL.—In his annual report of the Botanic Gardens for 1881-82 **Dr. King** wrote that of this rubber tree he had up to that date only eight plants, but added it was being propagated as fast as possible. Other reports followed, but apparently no material progress was made in the acclimatization of this rubber.

Madras. 83

II. MADRAS.—In the report of the Botanic Gardens, Nilghiri hills, for 1881-82, **Mr. Jamieson** reported receipt of the first plants. Two healthy young **Castilloas** were then obtained from Ceylon, **Mr. Jamieson** remarking regarding them :—"The reason I value these plants so much is that I have had to wait for them so long, having been led to expect a supply of **Castilloa** plants so far back as 1878. Shortly after receipt of the plants I took off a few cuttings, which rooted in less than a month. Since then every available cutting has been successfully propagated, and our stock now consists of fourteen healthy plants, including the two from Ceylon. The cuttings require constant care and close personal supervision for the first few days, being liable to damp off by the slightest excess of moisture. It, however, seems a strong growing plant when once rooted. The only drawback we have had to contend with in growing the **Castilloa** under glass is to keep it free from the attacks of the red spider, which, if allowed to get hold of the plants, would considerably retard their growth. One of the Ceylon plants has been put out in the Barliyár gardens, where I hope it will thrive, as in these days of uncertain coffee crops, and low prices, planters are anxious to cultivate any plant that will return a small interest on their outlay. It is therefore of the utmost importance that every effort should be made to propagate this plant for distribution. I have no doubt many localities in Wynaad and on the slopes of the hills will be found to suit the **Castilloa** and where it will yield a profitable return to the cultivator."

The Conservator of Forests, Southern Circle, reports that the trees in Nilambúr "are far too young to produce any quantity of rubber." In a recent report (1888) by **Mr. M. A. Lawson**, on the present condition of the introduced rubber trees of South India, the following passage occurs regarding this species :—"With respect to the **Castilloa**, there are in the Barliyár Gardens some five or six trees of about twenty feet in height, and with stems of about two feet in circumference. These plants flowered and seeded for the first time last year, and the seed germinated freely. They look as healthy as possible, and grow with great rapidity. Besides these trees at Barliyár, there are a few in the Teak Plantations at Nilambúr, and **Mr. Ferguson** has several in his garden near Calicut. They are easily propagated from cuttings, and could be, by this means, increased indefinitely, but the same difficulty has been experienced in extracting rubber from them as has occurred in the case of the **Cearas**."

PARA RUBBER. 84

15. **Hevea brasiliensis,** *Müll. Arg.*

THE PARA RUBBER TREE.

References.—*Tropical Agriculturist, Vols. I., 79, 665, 1048; II., 15, 682, 927; III., 21,109, 326, 851; IV., 528, 689, 690; V., 660, 824; VIII., 135,*

PARA RUBBER. *692; Indian Forester, IV., 42, 46; VI., 150, 187, 232; VII., 244; VIII., 58, 62; XIII., 285; Kew Report, 1879, 19; 1881, 15; 1882, 40-41; Report on Govt. Botanic Gardens & Park, Nilghiri Hills, 1881-82; Report, Botanic Gardens, Calcutta, 1875-76, 2; 1877-78, 2; 1878-79, 1; Report, Andaman Islands, 1885-86, 53.*

Habitat.—This and one or two allied trees occur in the forest of Central and Northern Brazil and in the forest of Para from the mouths of the Amazons, south and west. The region thus indicated has a climate remarkable for its uniformity of temperature, the mean being 81° F. and the greatest heat 95° F. (Plantations should not therefore be made in a region where the temperature at any time falls below 60° F.) The rainfall occurs principally in January to June, the maximum being in April, when it reaches 15 inches. **Mr. Cross** describes the region as having a deep rich soil, large tracts being flooded in the wet season and in the dry season completely intersected with water-courses. The most productive tracts are at most only 80 to 100 feet high, the soil full of moisture, rich and fertile, but the atmosphere very unhealthy.

The tree grows rapidly, attaining some thirty feet in three years; it is readily propagated by cuttings. **Mr. Cross** recommends plantations to be made in seasons of inundation or flood, the cuttings being forced into the mud for half their length, leaving enough above ground to save the tops from being actually submerged. Propagation by seed is not so successful.

Suitable Regions. 85

REGIONS MOST LIKELY TO PROVE SUITABLE FOR THE PARA RUBBER TREE.

The following extract from **Sir D. Brandis'** review of **Mr. Collins'** report gives the tracts of India which **Sir D. Brandis** anticipated, and which all subsequent experience (since 1873) has confirmed, as being most suited to Para rubber. Having compared the extremes of climate and also the means, on the west coast of India and in Burma, with that of Brazil, he remarks: "We may, therefore, conclude that Kánara, Malabar, Travancore, and the Burma coast, from Moulmein southwards, offer the desired conditions as regards temperature, for the successful cultivation of the caoutchouc-yielding species of **Hevea.** I would specially draw the attention of forest officers in this respect to the moist, evergreen forests at the foot of the Coorg Gháts, and in Kánara, as well as to the Attaran valley, and similar localities in Tenasserim."

Collection. 86

Collection of Rubber.—**Mr. Cross** gives a detailed account of the method pursued at Para, of which the following are the main facts. The milk is drained into small earthen vessels attached to the trees by an adhesive clay. The contents of fifteen of these cups make one English imperial pint. "Arriving at a tree, the collector takes the axe in his right hand, and, striking in an upward direction as high as he can reach, makes a deep upward sloping cut across the trunk, which always goes through the bark and penetrates an inch or more into the wood. The cut is an inch in breadth. Frequently a small portion of bark breaks off from the upper side, and occasionally a thin splinter of wood is also raised. Quickly stooping down he takes a cup, and pasting on a small quantity of clay on the flat side, presses it to the trunk close beneath the cut. By this time the milk, which is of a dazzling whiteness, is beginning to exude, so that if requisite he so smooths the clay that it may trickle direct into the cup. At a distance of four or five inches, but at the same height, another cup is luted on, and so the process is continued until a row of cups encircle the tree at a height of about six feet from the ground. Tree after tree is treated in like manner, until the tapping required for the day is finished. This work should be concluded by nine or ten o'clock in the morning, because the milk continues to exude slowly from the cuts for three hours or perhaps

HEVEA BRASILIENSIS.

longer." . . The quantity of milk that flows from each cut varies, but if it is large and has not been much tapped, the majority of the cups will be more than half full, or occasionally a few may be filled to the brim. But if the tree is much gnarled from tapping, whether it grows in the rich sludge of the *gapó* or dry land, many of the cups will be found to contain only about a tablespoonful of milk, and sometimes hardly that. On the following morning the operation is performed in the same way, only that the cuts or gashes beneath which the cups are placed are made from six to eight inches lower down the trunks than those of the previous day. Thus each day brings the cups gradually lower until the ground is reached. The collector then begins as high as he can reach, and descends as before, taking care, however, to make his cuts in separate places from those previously made. If the yield of milk from a tree is great, two rows of cups are put on at once, the one as high as can be reached, and the other at the surface of the ground, and in the course of working, the upper row descending daily six or eight inches, while the lower one ascends the same distance, both rows in a few days come together. When the produce of milk diminishes in long wrought trees, two or three cups are put on various parts of the trunk, where the bark is thickest. Although many of the trees of this class are large, the quantity of milk obtained is surprisingly little. This state of things is not the result of overtapping, as some have suggested. . . The best milk-yielding tree I examined had the marks of twelve rows of cups which had already been put on this season. The rows were only six inches apart, and in each row there were six cups, so that the total number of wood cuts within the space of three months amounted to seventy-two." There are other methods occasionally pursued, but the cup process is the most general and gives the best result. **Dr. Trimen** adds to the above account that the trees are tapped if they have a circumference of 18 to 24 inches, and the rough process described is carried on for many years, until the constant and extensive injury to the young wood causes their death, for some years previous to which event they almost cease to yield milk and are practically abandoned. A writer in the *Tropical Agriculturist* fixes the age of 25 years as that at which tapping should commence. The actual yield of a tree per annum and for each year of its life, does not appear to have been worked out by **Mr. Cross**, but assuming 72 cuts to be a high average, and that each cut yields half a cup full (also from **Mr. Cross's** statements a very high average) about six imperial pints of milk would be obtained from a tree in good bearing condition. **Mr. Cross** is careful to point out, that although the trees are large, the yield is surprisingly little, and it would thus seem probable that the Para rubber may be remunerative at present prices, when dealing with wild plants in their indigenous habitat, but might not be so if expensive cultivation and supervision were necessary.

Cultivation of Para Rubber in India.

CULTIVATION. Introduction. 87

History of Introduction into Asia.—In a letter dated April 1878, **Mr. W. T. Thiselton Dyer** gives the history of the attempt to introduce **Hevea braziliensis** into India and Ceylon. On the 4th of June 1873, the Director of Kew Gardens received from **Mr. Markham** some hundreds of seeds which had been collected by **Mr. James Collins**. Of these less than a dozen germinated, and six were in that year taken out by **Dr. King** to Calcutta. These did not succeed well in Calcutta, and it was accordingly arranged that Ceylon should be established as the depôt for supplying young plants to the parts of India where **Hevea** cultivation was thought possible. On June 14th, 1876, 70,000 seeds were received at Kew from **Mr. Wickham** (who was paid for them at the rate of £10 per 1,000); 4 per cent. germinated. Of these, 1,919 plants were sent to Ceylon in 38

PARA RUBBER.

Wardian cases, in charge of a gardener, and 90 per cent. reached in excellent condition. On August 23rd, 50 plants were sent to **Colonel Seaton** in Burma, but arrived in bad condition. A further supply of 100 plants was taken out to Ceylon from Kew by **Mr. J. F. Duthie.** Similar consignments were also issued from Kew, amounting in all to upwards of 2,000 plants sent to Ceylon, and smaller numbers to Burma, Calcutta, Mauritius, and Singapore. **Mr. Dyer** adds in his review of these endeavours to secure the acclimatisation of Para rubber that the plant is now therefore to be regarded as definitely established in the East Indies. So again, in November 1876, **Mr. Cross** arrived at Kew with about 1,000 plants, brought direct from South America, but only about 3 per cent. of these plants survived, so that it would appear **Mr. Cross's** share in introducing Para rubber was a very small one.

On the subject of Para rubber introduction **Dr. Trimen** wrote:—"The cost of procuring the seeds of Para rubber, freight, and other expenses appears to have been no less than £1,505-4-2, the Wardian cases alone costing £120, and the gardener and his passage £163. The whole of this large expenditure was borne by the Indian Government. An undertaking involving such an outlay as this, it is obviously beyond the power of the executive of this Colony (Ceylon) to carry out; but in this case, it is Ceylon which 'from climatic causes chiefly' appears likely to benefit most largely from the successful action of the Government of India" (*Tropical Agriculturist, Vol. I., 399*). This view appears to have been so far correct, for **Dr. King**, in his Annual Report of the Botanic Gardens, Calcutta, for 1881-82, wrote that the cultivation of the Para rubber plant had been abandoned in Bengal.

But even in Ceylon, and after some eight or ten years of the most enthusiastic attempts at cultivation, the opinion was finally arrived at that Para rubber would not pay the planter to grow it. The pages of the ***Tropical Agriculturist*** from 1881—86 literally teem with contributions, enquiries, and reports on Para and Ceara rubber, with only one or two passing notices devoted to **Ficus elastica.** The planter's verdict ultimately was, as stated, that Para and Ceara rubbers would not pay, and in many parts of the country the trees have since been hewn down, being viewed as worthless. In 1888-89 fresh interest was, however, awakened, in consequence of favourable reports having been furnished on samples of Ceylon rubber sent to Europe. **Dr. Trimen** in his annual report for that year urges the Ceylon Government to take the matter in hand as a State enterprise. He had tapped a tree (eleven years old having a stem circumference of 4 feet $2\frac{1}{2}$ inches), during three periods of dry weather, and obtained 1lb $12\frac{3}{4}$ oz. of dry rubber at a cost of 62 cents. **Dr. Trimen** comments on this result:—"As the rubber obtained is probably worth at present prices about 4*s*. per lb, there is clearly a large profit to be made here out of a **Hevea** plantation. I urged the formation of such upon Government so far back as 1882, but in the absence at that time of any organised Forest Department nothing could be done. As a valuable forest product, Para rubber may be confidently reckoned upon as a steady source of future revenue, and I strongly recommend that large plantations of it be formed in suitable places."

It may thus be inferred that as a State undertaking Para rubber would likely prove as remunerative as most other trees, and it is perhaps the duty of all Governments in tropical countries to assume the responsibility of providing for the world's future wants in rubber by organising suitable forests, but it may be accepted as established that such forests are not likely to be of interest to the planter who requires to turn over his capital in as short a period as possible. In January 1888, the Government of India (in consequence of a despatch from the Secretary of State in which

PARA RUBBER.

attention was drawn to the fact that the Royal Botanic Gardens, Ceylon, possessed a plentiful supply of the seeds of **Hevea brasiliensis**), issued a circular letter to Local Governments calling for a short *résumé* of the success which had attended the cultivation of that tree. The following extracts from the replies bring our knowledge of the cultivation of the **Hevea** rubber tree down to the most recent data :—

Bengal. 88

I. BENGAL.—"It has been ascertained from the Superintendent of the Royal Botanic Garden, Calcutta, that nearly ten years ago a quantity of the seeds of this tree (one of the species from which South American India-rubber is obtained) was sent to the garden from Kew. Some of these seeds germinated, but it was found that the resulting seedlings were extremely sensitive to cold, and that the low night temperature of the cold weather of Bengal proved fatal to them, even when planted in the most sheltered situations. Some of the seedlings were supplied to Tea Planters likely to give attention to their culture, and some were sent to Mr. Mann, the Conservator of Forests in Assam. The results obtained by these gentlemen were, however, the same as those obtained by Dr. King, and it was on his recommendation that subsequent supplies of **Hevea** seeds were sent to Ceylon, instead of to India. The Superintendent thinks that **Hevea** will grow well in Provinces like Malabar or Lower Burma, where the climate is said to be moist and equable; but in Northern India, where there is a distinct cold season, Dr. King is of opinion that its cultivation is not likely to prove successful."

"The Conservator of Forests, Bengal, who has also been consulted on the subject, reports that it does not appear that any experiments in the culture of the tree in question have ever been undertaken by the Forest Department in this Province."

Burma. 89

II. BURMA.—Colonel W. J. Seaton wrote of Tenasserim :—

"*Early experiments.*—Experiments on a small scale were commenced at Mergui in 1877, with eight seedlings, the survivors of a small batch received from Dr. King, of the Botanical Gardens, Calcutta. They were successfully set out in the Forest Office compound at Mergui, and although on a low hill, a not very desirable site, yet their growth was for some time satisfactory. In 1879, a large number of **Hevea** plants, believed to be well-rooted cuttings, were forwarded by Dr. Thwaites, of the Royal Botanical Gardens, Ceylon, and although in the charge of a subordinate who had been sent to Ceylon for special instructions, only 178 survived the voyage. These were set out in the plantation area selected, about 1½ miles inland from Mergui, on somewhat low ground drained by the sources of the Boke Chaung, a small tidal creek. Only 64 of the healthiest plants survived the planting operation, and of these again casualties continued to take place yearly, owing chiefly to attacks of white-ants until the number was reduced to 50 in 1886, since when there have been no further casualties. The following were the sizes of ten of the largest trees of 1879 measured on 29th March 1888 :—

No.	Height in feet.	Girth in inches at 2 feet from ground.	REMARKS.
1	39	29½	Forked into two branches 4 feet from ground.
2	34½	37	Clean bole of 9 feet.
3	40	38	Ditto 8 ,,
4	43½	40½	Ditto 12 ,,
5	36½	39½	Forked at 3 feet from ground.
6	38½	27½	Clean bole of 8 feet.
7	36¾	31	Ditto 10 ,,
8	30	18	Ditto 6 ,,
9	31	27	Ditto 6 ,,
10	21½	18½	Ditto 8 ,,

Propagation with cuttings—In the rains of 1879, 24 cuttings from the young trees in the Forest Office compound were set out in the plantation, but the experiment proved unsuccessful. Subsequent attempts, made from time to time, met with no better success, the cuttings generally dying off during the second year.

HEVEA RUBBER.

Propagation with seed.—In 1884, a few of the older trees having commenced to seed, experiments were made, with the result that 51 seedlings were successfully raised. These, however, when transplanted into the main plantation, were speedily reduced in number to 28, by attacks of white-ants and the browsing off of the young shoots by deer. The following year a large quantity of seed was procured from the 50 older trees, but not being sown immediately after collection, a great portion of it failed to germinate, and only 121 seedlings were raised. In the rains of 1886, better results were obtained by the timely sowing of the seed obtained from the older trees and by the part removal of the husk enclosing the seed. As many as 7,030 seedlings were raised, germination occupying three to four days. Experiments were continued in 1887, and 8,430 additional seedlings obtained. From Ceylon 54 seeds were received in October 1887, of which only 31 were fit to sow, but all failed to germinate.

Stock on hand at end of March 1888.—The stock of, trees and plants in the plantation and nurseries was as follows:—

Trees set out in 1879	. . . .	50
Seedlings of 1884 to 1886, set out in the main plantation at 20′ × 10′	. . . .	2,752
In the nurseries, ready for transplanting and distribution.	of 1886 . .	3,609
	of 1887 . .	8,430
	GRAND TOTAL .	14,841

General remarks.—The 50 older trees appear to be in perfect health, with evidence of such vigour as to leave no doubt that they are fully established and have outgrown all danger from attacks of white-ants. They yield an abundant supply of seed, some of which, if allowed to fall, occasionally germinate under the trees. The flowering takes place generally in January in the cool season, the fruit forms in March and April, and ripens in July and August about the middle of the rainy season. It will be seen that the propagation of the **Hevea brasiliensis** in this part of Burma is now quite independent of external assistance, and that its acclimatisation has been successfully demonstrated. It now only remains to subject the larger trees to periodical tapping to ascertain the yield in caoutchouc, after which the question will have to be determined as to the precise area which it may be advisable to plant up at Mergui and other suitable localities with this valuable tree.

Yield. 99

In a communication, dated 28th January 1889, **Colonel W. J. Seaton** gave the following account of the method of tapping and the yield of rubber obtained by him from the Para-rubber trees of Tenasserim:—

"The tapping experiment was first undertaken in July, under the impression that the flow of milk would be more abundant during the rainy season. Small bamboo pots were in the first instance affixed to the trees by means of well-wrought potter's clay, and above them small pieces of tin were also placed in such a position as to protect them from the rain; but, as the clay yielded to the rain and fell to the ground, tapping had to be undertaken at intervals between the showers, the bamboo pots being affixed by sharpening the upper end and forcing them into the bark in the manner followed by the '*Thitsi*' collectors. In order to obtain the largest quantity of milk in the shortest time possible, numerous incisions were made on the trees. The incisions were made in an upward direction and converging as required. The quantity of milk collected was so small in the intervals between the showers that it was deemed necessary to limit the experiment finally to five of the larger trees on the west bank of the Bôkchaungale, which flows through the plantation. The milk was found to flow much more freely from these trees, although not much larger than the trees first experimented on. They have, however, a thicker bark, and it was observed that the exudation of milk was greatest near the ground, where the bark was thickest, while at a height of six or seven feet it was almost *nil*. Owing to continued wet weather, it was found necessary to dry the milk over a fire and keep it subsequently in a warm place near the fire for about three weeks. The experiment was renewed between 22nd and 26th November, when the rains had fully ceased, 42 trees being operated on, *viz.*, 5 to the west and 37 to the east of the Bôkchaungale. The method of tapping was the same as that followed previously; but the yield from each incision being small (less in fact than was the case in the rains), the several trees were tapped to their utmost extent, and, by constantly collecting the milk before it had time to dry, the quantity now forwarded was obtained, *viz.*, 3 oz. from the 5 trees to the west and 9 oz. from the 37 trees to the east of the Bôkchaungale."

It may here be remarked that while Para was early seen to be unsuited to Bengal, and accordingly the Botanic Gardens of Ceylon were

PARA RUBBER.

selected as the best locality for forming a nursery, the Ceylon planters took a far keener interest in **Ceara**. It has since transpired that Burma stands the best chance of becoming the successful Asiatic acclimatisation home of Para, not Ceylon, and, it may be pointed out that **Sir D. Brandis** at the very commencement of the interest in foreign rubbers suggested Tenasserim as a likely locality for the **Heveas** but made no mention of Ceylon.

Madras. 91

III. MADRAS.—In 1881-82 **Mr. Jamieson** reported of the Botanic Gardens on the Nilghiri hills that he had received from **Mr. J. Fergusson** a few cuttings of this rubber tree. From these he succeeded in raising one plant. It was sent to Barliyár, and **Mr. Jamieson** adds, "this garden now contains plants of the three most valuable species of rubbers." Later on **Mr. M. A. Lawson** reported :—"There are three young trees of this species in the Barliyár Gardens. They are about twenty feet in height, and have stems of about eighteen inches in diameter at the base. Like the **Ceara** and the **Castilloa,** they grow vigorously, and have flowered for the first time this spring, but so far I have been unable to extract rubber from them in any large quantity." In a special communication for this work the Conservator of Forests, Southern Circle, Madras, states : "There are only two young trees in the experimental garden at Manontoddy averaging 20 feet in height by 3½ inches in diameter at 4 feet from the ground. Their growth is decidedly good, and the climate seems suited to them. There are numerous trees in Nilambúr which have made excellent growth, but all are too young to be tapped."

It will thus be seen that in Tenasserim the Para-rubber tree would appear to have found a congenial home; according to **Colonel Seaton,** the tree is quite acclimatised. India is thus independent of external assistance, and from Tenasserim, seed or even seedlings might be furnished to all other parts of India, where its cultivation was thought likely to prove successful and desirable. It cannot be said, however, that all has been learned that is necessary, to place Para cultivation in India on a commercial basis. Cultivators constantly raise the questions put by **Mr. Lawson** in his remarks quoted below under **Manihot Glaziovii,** and it seems probable that acclimatisation in a new country may have lessened the volume of milk or even altered its property. The season and age at which the trees should be tapped and the method best suited to India, are points that future experience alone can solve.

INDIAN-GROWN PARA RUBBER.

Indian-grown. 92

Samples of **Hevea** rubber collected in Tenasserim were sent to the Secretary of State for favour of report, and the following replies were received from the Kew authorities in June 1889 :—

"I am desired by **Mr. Thiselton Dyer** to acknowledge the receipt of your letter of the 26th April last (R S. and C. 614), forwarding a copy of a letter received from the Government of India, with enclosures, reporting the results obtained from tapping trees of **Hevea brasiliensis** near Mergui in Tenasserim.

2. The specimens of caoutchouc referred to were duly received by Parcels Post and they were subsequently submitted for valuation and report, through **S. W. Silver, Esq., F.L.S.,** to the India-Rubber, Gutta-percha and Telegraph Works Company (Limited) at Silvertown.

3. I enclose herewith a description of the specimens, and a copy of the valuation and report received respecting them. On the whole this report is favourable. The small quantity of rubber available (in no case exceeding a few ounces in weight) rendered its manipulation somewhat difficult; but, bearing this fact in mind, the result, as shewn in the samples of prepared rubber sent in a separate cover, is very encouraging.

4. It will be noticed that the best quality, valued at 2s. 3*d*. per pound, is nearly

PARA RUBBER.

equal to the best South American rubber. This was labelled "Sernamby," and was formed by milk which coagulated immediately on the trees in the dry season.

5. The rubber (marked No. 3) obtained from trees during the rainy season was dried over a fire. The quality of this appears to be better than either No. 1 or 2, and it approaches very near to No. 4. Except as regards the difficulty of coagulating the rubber, there appears from these experiments to be little difference between the specimens collected during the rainy season and those collected when the rains had fully ceased.

6. All the trees tapped were young, and few were more than twelve inches in diameter. **Mr. Thiselton Dyer** is of opinion that it is very desirable these interesting experiments should be continued, if there are sufficient trees available. If during the dry season the milk is found to coagulate readily on the trees, this method might be provisionally adopted with the view of testing on a larger scale its suitability for general use in India. Where, however, the milk does not coagulate readily, it might be advisable to try the cautious application of dry heat in the most convenient manner locally available. Mere sun-heat, especially during the rainy season, does not appear to produce good rubber.

7. In South America the milk of **Hevea brasiliensis** is collected generally at the beginning of the dry season. Where the quantity collected is large, it is necessary, in order to prevent decomposition, to obtain the caoutchouc in a solid mass as soon as possible. The best Para rubber is prepared by dipping a wooden paddle in the milk and holding it in the thick hot smoke from burning wood and palm nuts. When the first layer is dry, the paddle is dipped again and the process repeated until a thick solid mass of caoutchouc is obtained. A slit is made down one side, the rubber is peeled off the paddle, and hung up to dry."

Report from the India-Rubber, Gutta-percha and Telegraph Works Company (Limited), dated Silvertown, 30th May 1889.

The four samples of **Hevea** rubber received from Kew have been treated with sulphur in the same way as that adopted in the case of the better kinds of Brazilian rubber.

Allowance must be made for the smallness of the quantity experimented upon.

Eight samples sent herewith, four each, "washed" and "cured."

No. 1—Has the appearance of that imported some twelve months since, and known as Rio rubber; is soft, and would decompose if exposed to the necessary heat after washing, losing 12 per cent. in that process: its commercial value would be, say, 1*s*. 11*d*. to 2*s*.

No. 2—Slightly firmer: in other respects the same as No. 1.

No. 3—Percentage of loss somewhat less, and therefore of a trifling increased value.

No. 4—Is found to be stronger and firmer; not so likely to decompose when drying: worth 2*s*. 3*d*. Owing to the scrappy nature, the loss is greater than it otherwise would be.

[Apocynaceæ.

ACCRA RUBBER.

93

16. **Landolphia florida, Kirkii, owariensis,** and other species;

These are known in Commerce as Accra Rubbers; whether obtained from East or West Africa.

References.—*Tropical Agriculturist, Vols. I., 40, 271, 291, 1047; II., 15, 158; III., 727, 851; V., 388, 526, 767, 864; VI., 814; VII., 854; Collins' Report on Caoutchouc of Commerce; Reports, Royal Botanic Gardens, Calcutta, 1880-81, 2; Botanic Gardens, Nilghiri hills, 1882-83, 13; Kew Reports, 1877, 32; 1879, 18; 1880, 38-42; 1881, 15; Kew Bulletin (1888), 260; Official Correspondence from* **Sir J. Kirk** *with the Government of India, &c., &c., &c.*

Habitat.—A genus of climbing apocynaceous plants found on both the East and West Coasts of Tropical Africa and the adjacent islands. The following classification of the species is from the Kew Report:—*West Coast*—**L. owariensis,** a species possessing a great latitudinal range, *viz*., from Oware to Angola: **L. Mannii,** found at Corisco Bay: **L. florida** originally found in the Niger basin and now distributed over the whole of Central Tropical Africa: *East Coast*—**L. Kirkii,** common along the maritime region, and abundant at the mouth of the Zambesi

ACCRA RUBBER.

(shipped from Mozambique): **L. florida** found on the coast near Dar-es-Salam and extending inland, and **L. Petersiana** growing near Tanga on the coast of the main land opposite Pemba.

94

Caoutchouc.--The species of these African climbers have been here collectively discussed, since, for the purpose of rubber-yielding, their differences are very small, the more so in the light of their possible acclimatisation in India It may be here remarked that the world is indebted to the exertions of **Sir J. Kirk** for the East African supply of rubber. But as the species of **Landolphia** cannot be said to have been established commercially in India, only a few jottings need be given, and chiefly the facts regarding the attempts at their acclimatisation in India. The methed of collection of West Coast rubber from **L. florida** is peculiar. The milk dries so quickly as to form a ridge on the wound, which stops the further flow, so that it cannot be collected into vessels placed on the stem. The natives collect it by making long cuts on the bark with a knife, and as the milky juice gushes out, it is wiped off continuously with the fingers, and smeared on the arms, shoulders, and breast till a thick covering is formed. This is peeled off their bodies and cut into small squares which are then said to be boiled in water (*Journ. Soc. Arts, Notes by C. G. Warnford Lock*). The West African rubber, **Mr. F. T. Valdez** states, is collected in June, July, and August, but he describes a process different from the above. An incision is made on the tree, and a vessel placed under it, which, by means of a conductor, is filled, in about 24 hours. From this the caoutchouc is poured into moulds of various forms, which have been well smoked with massic or palm tree, from which the gum that gives the black colour, which they consider indispensible, is procured. "The natives of West Africa are said to make playing balls from the juice of **L. florida.**" It is also reported that in tapping these **Landolphias** the rubber is spoiled if the cut be made too deep as the milk obtains an injurious gum.

Sir J. Kirk describes the process of collecting rubber as pursued on the East Coast of Africa from *matere*—**L. Kirkii**—thus:—"The operation of tapping is very simple: a slice of the rough bark is cut off the surface, also just enough of the tree bark, when the juice starts out in drops. These dry as they are taken from the wound with the finger, and applied to the surface of the ball of caoutchouc. There is no evaporation required by exposure in vessels." An improved process is thus described by **Mr. F. Holmwood**: "A quantity of milk is daubed upon the forearm and being peeled off forms a nucleus. This is applied to one after another of the fresh cuts, and being turned with a rotary motion, the exuding milk is wound off like silk from a cocoon. The affinity of this liquid for the coagulated rubber is so great that not only is every particle cleanly removed from the cuttings, but also a large quantity of semi-coagulated milk is drawn away from beneath the uncut bark, and during the process a break in the thread rarely occurs." **Lieutenant O'Neill** found **L. florida** not easily killed by the process of collection. He observed hundreds of plants so thickly scarred with cuts that nearly ⅔rds of the bark must have been stripped from the tree. **Mr. Holmwood** speaks of the *mbunga* (**L. florida**) as yielding a milk that does not, like **L. Kirkii**, coagulate quickly. Accordingly, the Maknas who were collecting it, used sand to assist this process, and not for the purpose of adulteration as was originally thought. The milk of **L. Petersiana** is also gathered in a fluid state by tapping and coagulated by heat. The produce is inferior to that of **L. Kirkii.**

In the annual report of the Royal Botanic Gardens, Ceylon, for 1880-81, it is stated that "**Sir J. Kirk** thinks the **Landolphias** (especially **L.**

ACCRA RUBBER.

Kirkii), by far the most promising of the rubber plants for cultivation, their stems can be cut down at frequent intervals for the rubber, and fresh shoots readily spring up from the stools." The suggestion is made that the rubber still in the stems might be removed by crushing them and removing the rubber by means of bisulphide of carbon. Owing to this favourable opinion a demand at once arose for the seeds, and **Sir J. Kirk** in a paper read before the Linnean Society on the **Landolphias** of the East Coast of Africa said, that the natives, in obedience to this demand, had collected the seeds promiscuously, so that he feared much disappointment would be felt by planters. **Sir John** values first and foremost **L. Kirkii,** because it hardens very rapidly, next **L. florida,** because it yields a very fine rubber. The other forms, he states, are of little value. In 1883 the **Landolphias** in the Ceylon Botanic Gardens flowered. It may be said in conclusion that **Sir J. Kirk** stated that in 1882 the value of the exports of African rubber amounted to £300,000.

CULTIVATION Bengal. 95

CULTIVATION IN INDIA.

I BENGAL.—**Dr. King,** in his report of the Royal Botanic Gardens, Calcutta, for 1880-81, says that, thanks to the kind exertions of **Dr.** (now **Sir**) **John Kirk,** seeds of this rubber climber have been received, and some of them have germinated. In 1882-83, **Dr. King** reported that the **Landolphias** appear to have found a congenial home in Lower Bengal.

N. W. Provinces. 96

II. NORTH-WEST PROVINCES.—The Horticultural Society of Lucknow in 1883-84 reported receipt of a dozen **plants** of the African **Landolphia** from the Botanic Gardens, Calcutta.

Madras. 97

III. MADRAS.—In the report of the Botanic Gardens, Nilghiri hills, for 1882-83, **Mr. Jamieson** writes:—"Four of the six **Landolphias** kindly presented to the gardens by **F. J. Fergusson, Esq.,** Calicut, have lately been planted at Barliyár. The plants are rather small and weak, but I trust with care and attention to establish them." In the following year he reported that all these plants were dead, but that this rubber had been successfully established at Nilambúr, and that a second species received from the Agri.-Horticultural Society, Madras, was doing well. The Madras Forest Department in 1883 reporting on the results of **Landolphia** cultivation, stated that their plants were sickly. Further **Colonel Campbell Walker** in 1886 informed the Government of India that of **Mr. Fergusson's** plants of **Landolphia** at Calicut, only two remained alive, and that they showed no growth or vigour. So far as is known, all the plants distributed by **Mr. Fergusson** in the Wynaad and elsewhere are dead. The twelve plants planted by **Mr. Morgan** in the experimental garden at Madantoddy all died within 18 months, but the plant received from Ceylon Botanical Gardens and planted at Nilambúr in August 1882 has grown two feet nine inches and is now six feet six inches in height and is healthy and robust." In a report to hand, however, the Conservator of Forests, Southern Circle, gives it as his opinion that the climate of Madras is too cold for the **Landolphia.**

CEARA RUBBER. 98

17. **Manihot Glaziovii,** *Müll. Arg.*; EUPHORBIACEÆ.

THE CEARA RUBBER-TREE: SCRAP-RUBBER.

References.—*Journal of Botany, Nov. 1880; Tropical Agriculturist, Vols. I, 79, 361, 401, 437, 567, 576, 665, 982, 1048; II., 14, 69, 273, 331, 369, 381, 405, 629, 651, 751, 908, 927; III., 7, 154, 179, 262, 272, 283, 298—299, 326, 354, 434, 449, 482, 620, 850, 926; IV., 72, 85, 417, 547; V., 50, 633; VIII., 135; Reports, Royal Botanic Gardens, Calcutta, 1880-81, 2; 1881-82, 2; Report on the Horticultural Gardens, Lucknow, 1882-83, 9; Kew Report for 1879, 19; 1881, 13-15; 1882, 23, 40; Indian Forester, IV., 44; VI., 150; XI., 258.*

Habitat.—**Mr. Cross** describes the Brazilian tracts on which he found this plant as possessing "a very dry arid climate for a considerable part of

CEARA RUBBER.

the year." The rainy season begins in November and continues to May and June. The temperature during the period of **Mr. Cross's** observations ranged from 82° to 85° F. The tract of country, he adds, nowhere seemed to be elevated more than 200 feet above the sea. The soil was in places a sort of soft sandstone or gravel. In another locality somewhat further from the coast large boulders of grey granite occurred between which many good-sized rubber trees were growing.

Mr. Cross recommends propagation by means of seed (sown in brown sand kept pretty moist); but he adds that cuttings will also take root as easily as willow. **Dr. H. Trimen** of Ceylon commenting on **Mr. Cross's** recommendations says: "Experience of the plant in the Botanic Garden here has proved the general accuracy of the above remarks. There can be no doubt of the hardiness of the species, its readiness of culture, and adaptability to circumstances. It grows equally readily from seed or from cuttings, and, though a native of a tropical sea-level, thrives well here in Ceylon up to at least an altitude of 3,000 feet, and on the most barren soils. It has succeeded equally in Calcutta and Madras, but the wet season appears to have killed it at Singapore." **Dr. Trimen** adds that it would not be wise to risk it in localities where the temperature is liable to fall below 60° F. In Ceylon, in 1884, there were stated to be 977 acres under this rubber, but about that time the opinion seems to have been formed by the planters that the undertaking was a financial failure. The plant either did not form enough caoutchouc, or the methods of tapping were so defective as to prove futile. The following passage expresses very nearly the universal opinion of planters: "As to rubber cultivation, my advice to those intending to plant Ceara rubber is 'don't.' To those who have a large area under Ceara rubber trees only, my advice is let them grow, but spend nothing on their cultivation such as on weeding. I have not yet found that it pays even the cost of tapping and curing of the rubber, and some of the trees in my charge are 5½ years old. We have the assurance that the trees give a plentiful supply of rubber when they are older, and I have no reason to doubt it. Meantime, I am not aware of its having proved a paying investment to anyone in Ceylon, by harvesting the rubber, therefore, I do not recommend its cultivation. It is my honest opinion, there are far too many acres under the product already, and as regards rubber, any one with land suitable for rubber, would do better to select some other product" (*Tropical Agriculturist*).

Collection. 99

COLLECTION OF RUBBER.—The process described by **Mr. Cross** differs materially from that pursued with the Para-rubber tree. After carefully sweeping the ground around the base of the tree a few large leaves are placed on the ground. The outer surface of the bark is pared or sliced off to a height of four or five feet. "The milk then exudes and runs down in many tortuous courses, some of it ultimately falling on the ground. After several days the juice becomes dry and solid, and is then pulled off in strings and rolled up in balls or put in bags in loose masses. Only a thin paring should be taken off, just deep enough to reach the milk vessels; but this is not always attended to. Nearly every tree has been cut through the bark, and a slice taken off the wood. Decay then proceeds rapidly, and many of the trunks are hollow." **Mr. Cross** incidentally mentions that ceara-rubber trees may be tapped on attaining "a diameter of four or five inches."

CULTIVATION Bengal. 100

CULTIVATION OF CEARA RUBBER IN INDIA.

I. BENGAL.—**Dr. King**, in his annual report of the Botanic Gardens, Calcutta, for 1880-81, wrote that this plant was growing vigorously. It is the only introduced species that can be said to have taken kindly to

MANIHOT RUBBER. Bengal, and its rapid and easy growth and adaptability to dry soils are points which strongly recommend it. In his previous report **Dr. King** wrote of his first attempts to cultivate this plant that when freely exposed to sun and rain, this **Manihoot** is a wonderfully hardy plant, capable of standing the roughest treatment, easily grown, and rapidly propagated. If its rubber turns out really good, the cultivation of this species will, no doubt, be taken up by tea-planters whose gardens are on the plains, and by indigo planters, as adjuucts to their othet cultivation. In expectation of large demands for young plants and seeds, I have made a Ceara plantation on a suitable piece of ground on the outskirts of the garden." In 1881-82, **Dr. King** reported that the trees had begun to seed and that in consequence he was able to distribute a good many seedlings to planters and other in Assam, Chittagong, and elsewhere. In 1882-83, he reports progress in the propagation of this tree, but remarks that it is rather easily blown down or broken by the wind, especially during wet weather. From that date the annual reports of the Botanic Gardens of Calcutta make only passing allusions to the exotic rubber trees; but in a report of Chittagong it is said there are a number of trees on the road to Nazirhat "which were planted out many years ago by the road cess officials, and these are now in a flourishing condition." As large numbers of these trees were distributed by **Dr. King**, they doubtless exist here and there all over Bengal.

N. W. Provinces. 101 II. North-West Provinces.—Two plants of this rubber tree were received in 1880 by the Horticultural Society and planted out; in 1882-83, it was stated that while they grew they did not appear to thrive well. Seeds received from Ceylon did not germinate. In 1886-87, the above plants appeared to be dying, but other plants put in the open and not watered were doing better and were 24 feet high, the superintendent of the gardens adding: "It may safely be assumed that the tree can be grown in this part of India, but whether profitably remains to be proved."

In 1881-82, the Superintendent, Government Botanical Gardens, Saharanpore, reported: "That about 100 seedlings of this valuable rubber plant have been raised from the seed sent last year by **Dr. Trimen** from Ceylon." In 1882-83, the Superintendent again wrote that eight of these plants kept under glass survived the winter, but all those planted in the open or kept in pots under the shade of trees were killed by frost. Some of those planted out were from 5 to 7 feet high at the close of the rains. They were carefully protected with straw, but they all died off during the cold season." He adds "No one, therefore, need attempt to cultivate this tree in the North-West India with any hope of success."

Sind. 102 III. Sind.—In the report of the Government Farm and Economic Gardens at Hyderabad for 1883, it is stated that up to date 5 out of the 100 seeds received had germinated. In 1885-86, it was further reported that "the largest plant we have is 7 feet high. It is subject to injury from frost. During the hot season it grows freely." In 1887, the report states that all the plants had been killed by frost.

Burma. 103 IV. Burma.—This rubber tree was introduced by **Colonel Seaton** who sent "an intelligent lad to Ceylon to receive instructions from **Dr. Thwaites**" on its cultivation.

Madras. 104 V. Madras.—**Mr. Cross** in 1881, after inspecting the Ceara trees of Nilambúr, wrote: "This region is without doubt admirably adapted for the growth of the tree." He recommended that cultivation should be confined "to rather dry arid situations and poor soils." He tapped a young tree by the usual process of slicing off the bark: "The milk exuded freely, but next day on examination it was found that the greater portion had evaporated, showing the watery and immature state in which milk exists

CEARA RUBBER.

in young growing plants." In a recent report (June 1888) Mr. M. A. Lawson furnishes the following information regarding this rubber plant in South India at the present date:—

"In the Government Gardens at Barliyár there are some fifty or sixty trees, the largest of them being about thirty feet in height, with stems having a circumference of three to four feet. They flower and fruit freely, and are spreading through the forest of the neighbouring ghát. Besides these, there are some thousands on an estate at Kullar, situated about four miles below Barliyár, belonging to Messrs. Gordon, Woodroffe & Co. of Madras. There are also some thousands of trees on th Government Teak Plantations at Nilambúr; and in addition to these, there are many spread about over the Malabar and Wynaad districts, notably some of the finest on Mr. Fergusson's grounds in the vicinity of Calicut. The tree grows very rapidly, and, to all appearance, thrives well; but I have been wholly unable to extract rubber from it in anything like a paying quantity, and every one else hitherto has also failed. The following is a letter which I wrote a short time ago to Messrs. Gordon, Woodroffe & Co. on the subject, and which gives in full the difficulties I have to contend with:—'In continuation of my former letters on the subject of extracting rubber from the *Ceara*, I am very sorry to state that I have little or no information of any value to give you.

I have written to Kew and Ceylon to know if they could help me; the authorities at the former place have promised to make inquiries on the subject, of their West Indian correspondents, but I have not as yet received news from them. Dr. Trimen of Ceylon is in the same straits as myself, and has failed in all his experiments to extract the rubber in paying quantities.

The following are the chief points which I have paid special attention to:—

(1) Do our trees in India really produce as much rubber as they do in their native country?
(2) Are they as yet too young to yield rubber in large quantities?
(3) What is the best time of the year for tapping the trees?
(4) What is the best mode of tapping?

With respect to the first consideration, I cannot believe that the yielding properties of the *Cearas* should have decreased on account of their having been transplanted from Brazil to Southern India; at any rate, I can give no scientific reason for believing that such is the case.

With respect to the second consideration, we have now trees that are over three feet in girth; and one would think that a tree of that size would be sufficiently old to show what its full yield would be.

The third point is the one to which I desire particularly to call your attention. I have made myself a large number of experiments with the view of finding out what the best time of year for tapping the trees is likely to be. The trees, as you know, lose their leaves at the beginning of the year; and do not regain them till after the early rains have set in. During this period of the year, the sap is quiescent, and there is very little exudation of the milky juice, which contains the rubber.

The least want of success which I have gained in my experiments has been when I have tapped the trees just at that time when the new leaves are beginning to show themselves, and so on till they have become matured; that is, during the months of May and June; and I would suggest that you should try the experiment upon a large scale in your *Ceara* forest at Kullar during the present season.

With reference to the fourth consideration, *viz.*, the best mode of tapping, I can only tell you how I have proceeded. I have cut out wedges with an axe, and made incisions of various sorts, with knives and chisels, and I have followed the process which was described by Mr. Cross as being the one adopted by the natives in Brazil. This latter process is conducted as follows:—

(1) Peel off the outer brown bark, when the smooth green inner bark will be exposed. In this latter bark make incisions with a very sharp knife, forming a series of V's. The latex, it is said, which springs from the several gashes, unites into one stream, and is deposited at the apex of the V, where it is collected in tins, stuck into the tree at that point; but in all my experiments, I have never found the latex to flow in such quantities as to form anything like a stream.

(2) The other method which I have adopted was one which was suggested to me by Dr. Trimen. He took an ordinary wood-marking implement, and cut out deep channels in the bark, instead of making gashes; the latex, when the tree is under this treatment, is certainly given up in larger quantities, and the rubber is more easy

CEARA RUBBER.

to collect, as the size of the groove is sufficient to allow the workman to scoop out the rubber with a stick. I would recommend your trying this form of extracting the rubber on a larger scale than I have been able to do; you might also attempt to facilitate a greater exudation of the latex by beating the tree with a mallet, though of course not too violently, otherwise you might destroy the tree itself.

I am very sorry not to be able to give you any more satisfactory information, but I hope that, in time, we may discover where it is that our want of success lies.

I do not know if you are aware that the roots of the *Ceara* often swell up into tuberous-looking masses, resembling a yam; they are full of starch very like that of tapioca, both in taste and other properties; the tubers if boiled, or roasted, are very palatable, and they have this advantage over the tapioca plant, that they contain no poison, something perhaps might be made from these roots."

The Conservator of Forests, Southern Circle, Madras, has furnished for this work, the following remarks regarding the South Indian Ceara rubber trees: "The Ceara rubber grows both in Wynaad and in Nilambúr like a weed. The largest tree in the experimental garden at Manontoddy out of several hundred is about 30 feet high with a crown of foliage 40 feet in diameter and a trunk 45 inches in circumference at 4 feet from the ground. This tree is only five years old. On being tapped, the trees yield very little rubber and the cost of collecting the milky sap is greater than the value of the rubber obtained from it. Commercially this tree is, therefore, a failure, at any rate for the present"

The inability to make Ceara-rubber tree a commercial success seems, therefore, to be the only difficulty that now exists. The trees have been quite acclimatised, but either they are too young or do not yield in India the same amount of milk as in Brazil. It is possible that the defect may lie in imperfect modes of tapping, or perhaps from the tree being grown in a region where a peculiar condition of vegetation takes the place of milk-forming. Cultivation of acclimatised stock in parts of India, where initial acclimatisation would have been impossible, might correct the defect that now retards the enterprise of Ceara and other foreign rubbers. Many examples might be cited in support of the idea of improved or defective yield accompanying successful cultivation in a new country, for example, the behaviour of **Cannabis sativa** (Hemp) in Europe, India, and Central Asia* or **Linum usitatissimum** (Flax),† in Europe and India. If it be the case that trees of equal age in Brazil and the localities of India where Ceara is being cultivated yield different proportions of milk, the cultivation of the tree should not be abandoned until it had been tried in every district of India, where its cultivation was at all possible. If it be found that in no available tract of country in India can Ceara be grown profitably, then, but not till then, should the hope of ultimate success be abandoned. It seems probable, however, that greater experience in methods of treatment and modes of tapping, together with more precise information as to the age and season at which the milk flows in greatest abundance, may result in the attainment of better results than have as yet been reported.

DISCOVERY & UTILISATION. 105

HISTORY OF THE DISCOVERY AND UTILISATION OF INDIA-RUBBER.

The earliest known mention of India-rubber is the remark by **Herrera** in connection with Columbus' second voyage (nearly 500 years ago) of the inhabitants of Hayti playing a game with balls made from the gum of a tree. These balls,‡ he remarks, though large, were lighter and bounded better than those of Castile. A more direct allusion to rubber occurs, however, in **Torquemada's** *De la Monarquia Indiana* (published at Mad-

* (*Conf.* with the remarks in Vol. II., pp. 104-105.)
† (*Conf.* with flax culture, Vol. V.)
‡ (*Conf.* with the remark regarding playing balls made of rubber under p. 373.)

DISCOVERY & UTILISATION.

rid in 1615) where the tree is said to be known to the Mexican Indians as *Ulequahuitl.* A white milky substance, **Torquemada** says, which was thick and gummy and found in great abundance is got from it. This is collected in calabashes and afterwards softened in hot water or the juice is smeared over the body* and rubbed off when sufficiently dry. He also remarks that an oil is extracted from the *ulli* or rubber by heat which possesses soft and lubricous properties, and is of special effect in removing tightness of the chest. It was also drunk by the Mexicans with cocoa to stop hæmorrhage.

Torquemada mentions that the Spaniards at the time of his visit used the juice of the *ulé* to waterproof their cloaks, and he adds that "it is of great effect in resisting water, but not so the sun, for the rays thereof melt it." Perhaps the earliest accurate description of the South American caoutchouc-yielding plants is that given by **Condamine** in 1735. He speaks of the natives moulding the fresh juice in various forms, producing bottles, boots, and bowls, which may be squeezed flat and yet recover their form when no longer under restraint. The Portuguese of Para learnt from the Omaquas to make syringes which had no need for a piston or sucker. They were hollow, pear-shaped, with a pipe at the mouth. When compressed and held under water they filled with the fluid on being freed from pressure, and squirted it through the pipe when again compressed. This use of the milk led to the Portuguese name *Pao di Xirringa, i.e.*, "Syringe Tree."

These facts did not, however, attract attention in the Old World till nearly a century later. **Priestley**, for example, mentions caoutchouc in the preface to his work on Perspective (1770) as a valuable material for erasing pencil marks, which might be purchased from **Mr. Hairne**, Mathematical Instrument Maker, opposite the Royal Exchange, at a cubic piece of about half an inch for three shillings. Hence the name India-rubber. The slow progress made by caoutchouc in the early stages of its European history is the more remarkable when it is added that the need had long been felt for some water-proofing material and a patent was taken out in 1627 for "a newe invencon for the making and pparing of ctaine stuffe and skynns to hould out wett and rayne." The first patent granted for the use of India-rubber water-proofing was held by **Samuel Peal** in 1791. In 1813, **Mr. John Clark** obtained a patent for making air-beds, the inner layer of which was made air-tight by being coated with a solution of caoutchouc dissolved in spirits of turpentine boiled in linseed oil. The dawn of the great India-rubber era may, however, be said to have broken with the first of **Mr. Thomas Hancock's** patents in 1820. **Hancock** has accordingly been spoken of as "the father of this important and wonderfully increasing branch of the arts." A useful service was also rendered by **Mr. Hancock** in the publication of his '*Personal Narrative of the Origin and Progress of the Caoutchouc or India-rubber Manufacture in England*, London, 1857.' In 1823 **Mr. Charles Mackintosh** of Glasgow obtained a patent for making two fabrics water-proof by uniting them with a solution of rubber called 'water-proof double textiles,' the material used in the preparation of the now famous Mackintosh over-coats. In India about this time also **Sir W. O'Shaughnessy** reported that he had made a water-proofing fabric of a double layer of cloth with rubber between which might be used for surgical purposes. One of the greatest advances, however, was the invention of **Hancock's** '*masticator*' which tears the rubber to pieces and afterwards consolidates it into a homogeneous mass from which all impurities may be washed out and the rubber pressed into blocks or

* (*Conf.* with method of collecting **Landolphia** rubber in Africa, p. 373.)

DISCOVERY & UTILIZATION.

sheets. Prior to this invention the crude rubber had to be prepared in a peculiar manner so as to allow of its being cut into sheets and strings. The passages quoted above, from the *Agri-Horticultural Society's Journals*, contain directions for collection, so as to adapt the rubber to the then only known way of utilising it. **Hancock's** masticator dispensed with that necessity. An important feature of **Mackintosh's** discoveries was the utilisation of naphtha as the solvent, and **Hancock's** masticator having purified the caoutchouc, thus facilitated its rapid solution in naphtha. The curious discovery that caoutchouc stretched and retained for some time while kept cold, loses its elasticity, led to the invention of the elastic web by which fine threads of rubber could be worked up in a braiding or other machine and have their resilience restored by a hot iron being passed over the fabric. But the full utilisation of India-rubber was only realised by **Mr. Walter Hancock's** discovery of what is now known as Vulcanizing. This was attained in connection with experiments conducted with the object of divesting caoutchouc of its adhesiveness, thus allowing of water-proofing garments of single textiles having a surface coating of rubber instead of **Mackintosh's** double textiles. A sample of rubber obtained from America smelt of sulphur, but though this fact was noted **Hancock** followed up his own researches and produced inadhesive rubber by combining it with silicate of magnesia and other substances such as fuller's earth, whiting, ochre, asphalte, &c. Having ascertained that rubber could be subjected to a temperature (240° Fh.) sufficient to melt sulphur without destroying the rubber he plunged pieces of rubber into a sulphur bath and obtained thereby a hard horny material penetrated but not chemically altered by sulphur. By raising the heat to 270° or 280° Fh., at the end of an hour, he found the caoutchouc had taken up all the sulphur that it could retain, *viz.*, about 2 to 3 per cent. by weight. On subjecting these pieces of caoutchouc in an inert medium to a temperature of 275° to 320° Fh., a chemical reaction took place, the rubber losing all its defective properties but retaining all the peculiarities to which it owes its high position in the arts. By causing the penetration with sulphur to be effected by means of his hot rollers and masticator, **Mr. Hancock** obtained a substance which could be dissolved and applied to textiles. Commenting on the importance of Vulcanization, **Mr. James Collins** writes:—"Many as were the applications of caoutchouc in its natural or unaltered forms, it seemed at one time as if it were likely to fall into disuse, or that its applications would have to be circumscribed from certain defects. Its manufacture had been carried on in two ways: first, by direct mechanical treatment of the gum after simple cleansing; secondly, by the action of some volatile solvent, and the application of the solution to some material, so that a film of caoutchouc was left on the surface. But it was found that such caoutchouc became rigid and inflexible under the influence of cold, while it softened and decomposed in the sun and hot weather, this defect alone rendering many manufactured goods a total loss, and causing, it is said, the failure of several American firms. Contact with grease or any kind of fatty or essential oil quickly dissolved it, and perspiration had the same effect though slower in its action. It was also very adhesive and sticky, so that any substance pressed against its surface was quickly joined and could not be separated. By continued use or tension it lost its elasticity, and, besides this, an unpleasant odour produced by its solution pervaded the goods. But at the most critical time the method of Vulcanization was discovered, which, while preserving and even heightening the properties for which it was valued, added to its stability and divested it of the defects which were so many barriers to its present success and future extensions. Caoutchouc so treated was found to have

I. 105

DISCOVERY & UTILISATION.

almost perfect elasticity at *all* temperatures below 239° Fh., and not to become adhesive below a heat of 212° Fh., while fresh cut edges did not unite as in the unchanged substance. Its resistance to solvents too was very marked; though not absolutely insoluble, and its capability of being inadhesive and unaltered by cold, heat, water, or solvents, opened at once a new field of industry."

The honour of this discovery is claimed in Germany by **Hindersdorf**, and in America by **Goodyear**, and it is probable many persons were working on the same lines about the period of **Hancock's** discoveries. Two years later (1846) **Mr. Alexander Parkes** took out a patent that gave an ulterior mode of Vulcanizing. By the employment of a solution of chloride or hypochloride of sulphur in bisulphuret or sulphuret of carbon, coal, naphtha, or turpentine, he obtained the same change as described above in **Hancock's** process of Vulcanization Speaking of this process **Mr. Collins** writes :—" Caoutchouc in the dry state, if kneaded with a certain amount of dry chloride of sulphur, slowly underwent the same change, and whilst hot could be pressed into any desired shape. Caoutchouc, previous to being submitted to the "changing," could be combined with wool, flax, cotton, wood or cork dust, earths, oxides of metals and with other gums. By this process, however, only thin sheets could be subjected to this method, as in the case of thick sheets, the action on the outside would have proceeded too far before it had penetrated to the centre. After the articles were taken out of this bath, they were placed in a hot chamber and then washed in water or in a solution of caustic potash, or soda.

Vulcanized rubber may be hardened by various processes until it can be cut like ivory and hence it became applicable to a numerous series of new purposes such as the manufacture of combs, backs of brushes, in fact it was rendered suited for all the industries in which bone, ivory, whalebone, tortoiseshell, &c., are required. It might even be coloured to any shade and built up in mosaics, made into railway buffers, used as pavements, or a thousand new applications for which the natural rubber was quite unsuited. It could be moulded or engraved, or have its surface only hardened by **Parkes'** process. Goloshes manufactured and then Vulcanized were found to be superior to those made by any other process. Various patents were also taken out for other methods of hardening caoutchouc, such as by means of iodine or of bromine or of both and with or without sulphur, the discoveries thus widening and extending the uses of India-rubber until it may be said that substance has become an indispensable necessity of modern commerce.

The above brief sketch, compiled from most modern writers but mainly from **Collins'** various papers (in the British Manufacturing Industries, in the *Encyclopœdia Britanica*, &c.) will suffice to give some idea of the properties of caoutchouc as now manipulated; and to show that its extensive modern application justifies the importance given to it in tropical agriculture and forestry. To complete this sketch, it need only be necessary to briefly review the Indian trade statistics of the substance.

INDIAN TRADE IN CAOUTCHOUC.

INDIAN TRADE. 106

Mr. J. E. O'Conor, in his Review of the Trade of British India for 1874, gives a retrospective account from 1868-69. "The trade, he says, for 1874-75, was somewhat less than in previous years and still less than in 1872-73, the supply of the market of shipment being somewhat uncertain under the conditions in which the rubber is collected by jungle tribes." "The exports of Bengal are much larger than those from Burma. All the caoutchouc brought down to Calcutta is produced in Assam." **Mr. O'Conor** then gives the figures of exports, the average for the five

INDIAN TRADE.

years ending 1872-73 being 12,003 cwt., valued at £63,217, but for 1873-74 they were 16,837 cwt., valued at £117,775 and for 1874-75, 15,893 cwt., valued at £108,645. Mr. O'Conor then remarks: "The trade is shared by Bengal and Burma, but by the former in much the larger proportion. The exports from Bengal fell off during the year from 16,255 cwt. worth £115,754 to 13,938 cwt. worth £96,492. The exports from Burma, on the other hand, increased from 582 cwt., worth £2,020 to 1,954 cwt. worth £12,104. The export trade from Burma only commenced in appreciable quantity, in 1873-74. All or nearly all the caoutchouc exported from that province is produced in Upper Burma." In the correspondence quoted at page 346 the exports from Calcutta were 514 maunds in 1836. In 1880-81, Mr. O'Conor reported a decline in quantity, but increase in value, and in the following year he remarked that the exports were 10,699 cwt., valued at R10,88,426. "The shipments, he there states, are entirely made from Calcutta and ports in Burma, and the countries to which the article is exported are the United Kingdom and the United States of America, small quantities being sent also to the Straits Settlements and Egypt. The shipments to the United States show a considerable increase compared with the previous year." In 1885-86 the exports are pointed out as continuing to decline; in the following year the trade had slightly improved. In 1888-89 the IMPORTS FROM FOREIGN COUNTRIES were 53 cwt. valued at R6,147, and the EXPORTS of FOREIGN MERCHANDISE were 52 cwt., valued at R8,630. These imports came mainly from Madagascar to Bombay and were exported to the United Kingdom and France. These transactions show, therefore, a certain amount of the new African trade in rubber falling into the re-export route, that much of the African and Madagascar produce has followed from time immemorial. Of the Rubber of Indian produce there were EXPORTED to FOREIGN COUNTRIES, in 1888-89, 8,673 cwt., valued at R9,67,348: this went from Bengal, 5,609 cwt, and from Burma, 3,064 cwt. As in former years the Indian caoutchouc of 1888-89 went chiefly to the United Kingdom, United States, and Egypt.

The coastwise transactions last year were 118 cwt., valued at R11,010 IMPORETD and 2,842 cwt., valued at R3,58,113, EXPORTED. The imports were mainly from other Bengal ports (Chittagong) into Calcutta. And the exports were from Burma into Bengal (namely, 2,731 cwt.). These transactions, of course, overlap each other and to a certain extent reappear in the FOREIGN EXPORTS. Thus the Bengal exports to foreign countries would doubtless comprise Assam, Chittagong, and Burma rubber, but the exact amounts of each kind cannot be made out since the Bengal consumption is not known.

The trans-frontier trade is mainly into Assam and Chittagong, and the figures for this trade have already been given (see page 356). To what extent, if any, a transfrontier export trade may take place from Burma to China and other countries beyond that frontier, is not known; but we are told the contractors paid Government R1,00,000 for the right to collect and purchase rubber while the declared value of the foreign and coastwise exports from Burma were valued at R7,06,378.

It will thus be seen that the Indian trade in caoutchouc sprang suddenly into importance, and from 1872 fluctuated forward until 1882-83, when it reached its maximum in the value of R12,59,165, and from which date it may be said to have fluctuated downwards. The whole of this rubber since about 1875 has been drawn from the indigenous tree and mostly beyond the British frontiers of Assam, Burma, and Chittagong, the cultivated and acclimatised trees not having as yet contributed to the supply.

INDIGOFERA, *Linn.; Gen. Pl., I., 494.*

A genus of herbs or shrubs belonging to the LEGUMINOSÆ, which comprises some 250 or 300 species, distributed throughout the tropical regions of the globe, India having about 40. They may be readily recognised as being papilionaceous plants, with imparipinnate leaves which bear laterally attached hairs: stamens diadelphous; pods round, dehiscent, not jointed. Of the 40 indigenous species 10 are said to occur throughout the plains, ascending the Himálaya and hills of India to altitudes of 3,000 to 5,000 feet. About half that number are temperate and occur between 6,000 and 8,000 feet. Including the species that are spoken of as widely distributed in the tropical regions of India—Bombay and Sind possess in all 25 species, so that these provinces have about 10 or 12 peculiarly local forms, some of which are distributed to Madras and Ceylon; the Panjáb and the North-West Provinces each possesses two local species; and Madras three, while Bengal cannot be said to have any species that is confined or peculiar to it. The head-quarters of the Indian **Indigoferas** is thus in the Western Presidency, and from Bengal to Assam and Burma there is a marked decrease in the number of forms and abundance of individuals. This sketch of the distribution of the Indian species may be viewed as of considerable interest when it is added that Bombay is the only region where a botanist has spoken of **Indigofera tinctoria** (the true Indigo) as having been seen in what appeared a wild condition. The older records of commercial indigo also point to the Western Presidency (or the basin of the Indus) as the region of supply and earliest manufacture.

The following enumeration of the species of **Indigofera** comprises only those regarding which economic facts have been published. About as many more are known to the Natives and bear specific vernacular names in the regions where they occur.

Indigofera Anil, *Linn.; Fl. Br. Ind., II., 99;* LEGUMINOSÆ. 107

Vern.—*Vilaiti-nil* (European Indigo), HIND.; *Shimaiya-viri,* TAM.; *Shúnanili,* TEL.; *Shímenili,* KAN. (Foreign Indigo in TAM., TEL., and KAN.); *Vishashódhani,* SANS. The Sanskrit name is derived from *shód* to clear and *visha* poison.

Habitat.—This species is admitted by botanists to be purely American. It was cultivated by the Spaniards in Mexico and by the Portuguese in Brazil as a source of indigo. It nowhere exists in a wild state in India, and was probably introduced during the period of Portuguese ascendancy in the Western and Southern Presidencies. It is nearly allied to **I. argentea** *var.* **cœrulea** as may be seen from the fact that **Kurz** reduces that form to this species.

Dye.—The dry leaf indigo of Madras is to some extent obtained from this species. The reports furnished by the Madras Government, in connection with this work, do not afford sufficiently detailed information to allow of an opinion being formed as to the relative degree of cultivation of this species and of **I. tinctoria.** Nor do they supply the means of enabling us to judge as to their relative value as sources of the dye, but it may be presumed that **tinctoria** is superior, from the fact that mention is made of that species displacing **Anil.** The reader is referred for further information to the account of INDIGO in the pages below. DYE. 108

I. argentea, *Linn.; Fl. Br. Ind., II., 98;* also *var.* **cœrulea.** 109

Often called WILD INDIGO.

Syn.—INDIGOFERA CŒRULEA, *Roxb.;* I. RETUSA, *Grah.;* I. BRACHYCARPA, *Grah.;* I. TINCTORIA, *var.* BRACHYCARPA, *DC. Prod.;* I. TINCTORIA, *Forsk., Egypt.*

Vern.—*Surmainil,* HIND; *Nil,* MERWARA; *Káru-nili,* TEL.; *Kát-avéri* (according to **Ainslie**), TAM.; *Karu-níli,* KAN.; *Kálaklítaka,* SANS. The above names mean black indigo.

References.—*Roxb., Fl. Ind., Ed. C.B.C., 583, 584; Ainslie, Mat. Ind., II., 34; Dalz. & Gibs., Bomb. Fl., 59; DC., Orig. Cult. Pl., 137; Elliot, Fl. Andh., 87; Murray, Pl. and Drugs, Sind, 116; Indian Forester, XII., App. 11; Duthie, Report, Flora Merwara.*

Habitat.—A shrubby plant, according to the *Flora of British India*, found in the plains of Sind, the variety **cœrulea** occurring on the plains at Banda. It was, however, apparently found by **Roxburgh** and **Elliot** in the Telegu country, and **Mr. Duthie** has recently reported its discovery at Merwara. **Kurz's I. tinctoria** *var.* **Anil** (*Jour. Asiatic Soc., Bengal, XLV., 1876, 269*) is apparently *var.* **cœrulea**, and if so, it occurs all over Burma from Ava to Martaban and Tenasserim. It is distributed to Arabia, Egypt, and Abyssinia; and according to **DeCandolle**, is cultivated in Egypt and Arabia.

DYE. 110 **Dye.**—**Roxburgh** states that, although he was able to extract indigo of good quality from this species by the ordinary process, he could not discover that it was employed for that purpose by the natives of India. Subsequent writers (**Ainslie** for example), however, mention that it is so used, and **Dalzell** and **Gibson** were of opinion that the variety **cœrulea** may probably have been the source of the plant (**I. tinctoria**) now cultivated for its dye property.

111 **Indigofera aspalathoides,** *Vahl.; Fl. Br. Ind., II., 94; Wight, Ic., t. 332.*

Syn.—I. ASPALATHIFOLIA, *Roxb.*; ASPALATHUS INDICUS, *Linn.*; LESPEDEZA JUNCEA, *Wall.*

Vern.—*Níl*, PB.; *Shevenar-vaymbú*, or *shivánarvembu* (*shivanar*, honorific for *Shiva*, and *vembu*, the *Neem Tree*), TAM.; *Shiva-malli* (*Shiva's Jasmine*), KAN.; *Manneli*, MALAY.; *Shéváníarba* (according to **Ainslie**)—*Shiva's neem*, SANS.

References.—*Roxb., Fl. Ind., Ed. C.B.C., 582; Thwaites, En. Ceylon Pl., 83; Dalz & Gibs., Bomb. Fl., 58; Rheede, Hort. Mal., IX., t. 37; Ainslie, Mat. Ind., II., 385; Dymock, Mat. Med. W. Ind., 2nd Ed., 215; Bidie, Cat. Raw Pr., Paris Exh., 52; Drury, U. Pl., 277; McCann, Dyes and Tans, Beng., 93; Cooke, Oils and Oilseeds, 49; Balfour, Cyclop., 337.*

Habitat.—A low under-shrub of the plains of the Carnatic and Ceylon. **Roxburgh** speaks of it as found on dry lands near the sea.

OIL. Root. 112 **Oil**—**Rheede** says that an oil is obtained from the ROOT, which is used to anoint the head in erysipelas.

MEDICINE. Leaves. 113 **Medicine.**—**Dymock** compiling from **Ainslie** and **Rheede** writes:—

Flowers. 114 Shoots. 115 Root. 116 Ashes. 117 Oil. 118 The LEAVES, FLOWERS, and tender SHOOTS are said to be cooling and demulcent, and are employed in decoction in leprosy and cancerous affections. The ROOT is chewed as a remedy for toothache and apthæ. The whole plant, rubbed up with butter, is applied to reduce œdematous tumours. A preparation is made from the ASHES of the burnt plant to remove dandriff from the hair. The leaves are applied to abscesses, and the OIL obtained from the root, is used to anoint the head in erysipelas. **Dr. Dymock** adds that he has not seen the plant used medicinally in Bombay.

119 **I. atropurpurea,** *Ham.; Fl. Br. Ind., II., 101; Wight., Ic., t. 369.*

Syn.—I. HAMILTONI, *Grah.*

Vern.—*Bankati, kala, sakena, sakna*, HIND.; *Khenti, jund*, KAGHAN; *Kathi, gorkatri*, KASHMIR.

References.—*Roxb., Fl. Ind., Ed. C B.C., 586; Voigt, Hort. Sub. Cal., 212; Brandis, For. Fl., 136; Gamble, Man. Timb., 117; Stewart, Botanising Tour in Hazara, 67; Atkinson, Him. Dist., 308, 340, 472; Duthie and Fuller, Field and Garden Crops, 43; Gaz., Mysore and Coorg, I., 59.*

Habitat.—A small shrub of the Salt Range, from 2,500 to 5,000 feet, and outer Himálaya from Hazara to the Khasia Hills, ascending to 9,000 feet but found as low as 1,200 feet on the Siwalik Hills.

Fibre.—The TWIGS are used for basket-work and bark bridges. FIBRE. Twigs. 120

Indigofera cordifolia, *Heyne; Fl. Br. Ind., II., 93.* 121

Vern.—*Vekriavas* (Merwara), RAJ.; *Godadi, bodaga, botsaka*, BOMB.

References.—*Dalz & Gibs., Bomb. Fl., 58; Aitchison, Cat. Pb. and Sind Pl., 40; Murray, Pl. and Drugs, Sind, 114; Drury, U. Pl., 276; Lisboa, U. Pl. Bomb., 197; Gaz., Mysore and Coorg, I., 56; Gaz., N.-W. P., I., 80; IV., lxx.; Indian Forester, XII., App. 2, 11*

Habitat.—Plains throughout India proper, ascending to 4,000 feet in the Chenab Valley. Distributed to Afghánistán, Baluchistán, Nubia, the Malay Isles, and North Australia.

Food.—SEEDS eaten in times of scarcity and famine. FOOD. Seeds. 122

I. Dosua, *Ham.; Fl. Br. Ind, II., 102.* 123

Syn.—I. HETERANTHA, *Wall.*; I. VIRGATA, *Roxb.*; also var. TOMENTOSA, *Grah.*; I. STACHYODES, *Lindl.*

Vern.—*Khenti, shagali, mattu, kaskeri*, PB.; *Theot, kathewat, kati*, SIMLA.

References.—*Roxb., Fl. Ind., Ed. C.B.C., 586; Voigt, Hort. Sub. Cal., 212; Atkinson, Him. Dist., 308, 340, 472; Settle. Rept. Simla, App. 2, xliv.*

Habitat.—A shrub of the Temperate, Central and East Himálaya—Simla to Bhotan and Assam; 6,000 to 8,000 feet.

Food and Fodder.—The FLOWERS are said to be eaten in Kangra as a pot-herb. Prized as a fodder for sheep and goats; buffalos are also said to be fond of it. FOOD & FODDER. Flowers. 124

I. enneaphylla, *Linn.; Fl. Br. Ind., II., 94; Wight, Ic., t. 403.* 125

This is the RED NERINJY of Madras.

Vern.—*Bhingule*, MAR.; *Sheppunerunji*, (=Red Plant), TAM.; *Yerra palleru, chalapachchi, chera-gaddam*, TEL.; *Kenneggilu*, KAN.; *Cheru-pullate*, MALAY.; *Vasuka*, SANS.

References.—*Roxb., Fl. Ind., Ed. C.B.C., 584; Voigt, Hort. Sub. Cal., 211; Thwaites, En. Ceylon Pl., 411; Dalz. & Gibs., Bomb. Fl., 58; Aitchison, Cat. Pb. and Sind Pl., 40; Ainslie, Mat. Ind., II., 74; O'Shaughnessy, Beng. Dispens., 293; Dymock, Mat. Med. W. Ind., 2nd Ed., 215; S. Arjun, Bomb. Drugs, 201; Drury, U. Pl., 277; McCann, Dyes and Tans, Beng., 93; Balfour, Cyclop., 337.*

Habitat.—A small, trailing, much branched shrub. Met with in the plains of India and ascending the Himálaya to 4,000 feet; distributed to Ceylon and Burma, the Malay isles, and North Australia.

Medicine.—Ainslie remarks that the JUICE of this plant is used as an antiscorbutic and diuretic, and is considered alterative in old venereal affections. MEDICINE. Juice. 126

Fodder.—Roxburgh regards it as one of the most useful plants of pasture lands; it is employed as human food in times of famine. FODDER. 127

I. Gerardiana, *Wall.; Fl. Br. Ind., II., 100.* 128

Var. heterantha.

Syn.—I. DOSUA, *Wall.*; I. VIRGATA, *Roxb.*; I. QUADRANGULARIS, *Grah.*; I. POLYPHYLLA, *DC.*

Vern.—*Kati, khenti, mattu, kutz, shágali, khenti, kátsú*, PB.; *Kathi, theot, kathu*, SIMLA; *Kaskai* (PUSHTU), AFG.

References.—*Roxb., Fl. Ind., Ed. C.B.C., 586; Brandis, For. Fl., 135; Gamble, Man. Timb., 117; Stewart, Pb. Pl., 70; Stewart, Bot. Tour in Hazara, 15; Atkinson, Him. Dist., 308, 340, 472; Gazetteers:—Dera Ismail Khan, 19; Bunnu, 23; Rawalpindi, 15; Balfour, Cyclop, 337; Jour. Agri. Hort. Soc. Ind., VIII. (Sel.), 155.*

Habitat.—A low copiously branched shrub of the North-West Himálaya and the Eastern skirts of the Suliman Range, ascending to 8,000 feet. Distributed to Afghánistán. The variety **heterantha** is also found in the Khasia Hills and Bhotan.

FIBRE. Twigs. 129 **Fibre.**—TWIGS are employed for basket-work and in making rope bridges.

TIMBER. 130 **Structure of the Wood.**—Hard, white, with an irregular heartwood of dark colour. Weight 56lb per cubic foot.

131 **Indigofera glandulosa,** *Willd.; Fl. Br. Ind., II., 94; Wight, Ic., t. 330.*

Vern.—*Vékhariyo,* MAR.; *Barbed,* SHOLAPUR; *Gavacha malmandi,* KALADGI, BOMB.; *Vekhariyo, baragadam, barapatálu, boomidapu, barapatam* (Madras), TEL.

References.—*Roxb., Fl. Ind., Ed. C.B.C., 583; Voigt, Hort. Sub. Cal., 211; Dalz. & Gibs., Bomb. Fl., 58; Fl. Andh., Elliot, 23, 30; Dymock, Mat. Med. W. Ind., 2nd Ed., 216* (under **I. trifoliata**); *S. Arjun, Bomb. Drugs, 41; Drury, U. Pl., 276; Lisboa, U. Pl. Bomb., 197; Bomb. Gaz., XV., 431; N.-W. P. Gaz., I., 80; Balfour, Cyclop., 337.*

Habitat.—An annual with elongated slender branches, found on the plains of the Western Peninsula and Bundelkhand.

MEDICINE. Seeds. 132 **Medicine.**—The SEEDS of this species, as also of **I. trifoliata**, are said to be employed as a nutritive tonic.

FOOD. Seeds. 133 **Food.**—The natives make flour of the SEEDS, and when baked into bread use the flour as an article of diet in times of scarcity. The seeds were extensively employed in that manner during the Bombay famine of 1877-78. **Lisboa** regards them as highly nitrogenous.

134 **I. linifolia,** *Retz.; Fl. Br. Ind., II., 92; Wight, Ic., t. 313.*

Vern.—*Torki,* HIND.; *Bhangra,* BENG.; *Tandi khode baha,* SANTAL; *Torki,* PB.; *Burburra, pandhari pale, bhangra, torki,* BOMB.; *Pandhí,* NASIK; *Jawarich malmandi,* KALADGI, BOMB.

References.—*Roxb., Fl. Ind., Ed. C.B.C., 582; Voigt, Hort. Sub. Cal., 211; Brandis, For. Fl., 136; Thwaites, En. Ceylon Pl., 83; Dalz. & Gibs., Bomb. Fl., 58; Stewart, Pb. Pl., 70; Aitchison, Cat. Pb. and Sind Pl., 40; Dymock, Mat. Med. W. Ind., 2nd Ed., 887, 889; Atkinson, Him. Dist, 308, 340; Lisboa, U. Pl. Bomb., 197; Gaz. Bomb., V., 25; N.-W. P., I., 80; IV., lxx.; Balfour, Cyclop., 337.*

Habitat.—A small, prostrate plant, common in the plains of India and Ceylon; ascending the Himálaya and distributed to Abyssinia, Afghánistán, &c.

MEDICINE. 135 **Medicine.**—It is given medicinally in febrile eruptions. The **Rev. A. Campbell** says that the Santals use the plant in amenorrhœa along with **Euphorbia thymifolia.**

FOOD. Seeds. 136 **Food.**—The SEEDS were largely consumed during the Deccan famine of 1877-78, by the people of Kaladgi, Dharwar, Sholapur, Ahmednagar, &c., though they are unpleasant to the taste. After being ground to flour they were, either alone, or mixed with cereals, made into cakes. **Dr. Lyon's** analysis proves the uncultivated pulse to be rich in nitrogen.

137 **I. paucifolia,** *Delile; Fl. Br. Ind., II., 97; Wight, Ic., t. 331.*

Syn.—I. ARGENTEA, *Roxb.*; I. HETEROPHYLLA, *Roxb. mss.*

Vern.—*Kuttukkárchammatti,* TAM. (from *Kuttukkal,* an upright post, and *chammatti,* a smith's hammer).

References.—*Roxb., Fl. Ind., Ed. C.B.C., 383; Dalz. & Gibs., Bomb. Fl., 59; Murray, Pl. and Drugs of Sind, 116; Gaz., N.-W. P., I., 80; IV., lxx.; Bhandara Settle, Rept., 20; Madras, Man. Admn., II., 123; District Man., Coimbatore, 247; Sydapet Ex. Farm Man. and Guide; Duthie on Merwara, 11.*

Habitat.—A shrub 4 to 6 feet high, found in the plains of Sind and the Upper Gangetic basin. Distributed to Baluchistán, Arabia, Tropical Africa, and Java.

Medicine.—Mr. Maclean writes that it is considered an antidote to poisons of all kinds. The ROOT boiled in milk is used as a purgative and a decoction of the STEM as a gargle in mercurial salivation. MEDICINE. Root & Stem. 138

Fodder.—Sheep are fond of this plant, and it grows freely on the poorest soils. It has been experimentally cultivated at the Sydapet Farm, Madras, with satisfactory results. The opinion has been expressed in the Manual and Guide to the Farm that it may possibly become a valuable crop. FODDER. 139

Manure.—It is extensively employed as a manure for wet lands in South India. MANURE. 140

Indigofera pulchella, *Roxb.; Fl. Br. Ind., II., 101; Wight, Ic., tt. 367, 368.* 141

Syn.—I. PURPURESCENS, *Roxb.;* I. CASSIOIDES, *Rottl.;* I. ELLITTICA, *Roxb.;* I. VIOLACEA, *Roxb.;* I. ARBOREA, *Roxb.*

Vern.—*Sakena, hakna,* HIND.; *Uterr, jhurhur,* KOL.; *Dare-huter, lili bichi,* SANTAL; *Hikpi,* LEPCHA; *Jirhúl,* KHARWAR; *Baroli,* MAR.; *Togri,* BHIL; *Taw maiyain* or *tawmeyiang,* BURM.

References.—*Roxb., Fl. Ind., Ed. C.B.C., 585, 586; Voigt, Hort. Sub. Cal., 212; Brandis, For. Fl., 136; Kurz, Fl. Burm., I., 361; Beddome, For. Man., 85; Gamble, Man. Timb., 117; Dalz. & Gibs., Bomb. Fl., 60; Aitchison, Cat. Pb. and Sind Pl., 41; Kurz, Prelim. For. Report, Pegu, App. C., v.; Grah. Cat. Bomb. Pl., 46; Rev. A. Campbell, Rept., Econ. Prod. Chutia Nagpur, Nos. 8480, 9234; Atkinson, Him. Dist., 308, 340, 472; Drury, U. Pl., 280; Gaz. Bomb., XV., 431; Gaz., N.-W. P., I., 80; IV., lxx.; Gaz., Simla, 12; For. Admn. Report, Chutia Nagpur, 1885-86, 29; Jour. Agri.-Hort. Soc. (Sel.), VIII., 155.*

Habitat.—A large shrub, 4 to 6 feet high, found throughout the Himálayan tract and hills of India, ascending to 5,000 feet. Kurz mentions it as not unfrequent in the dry and open, especially in the *In,* forests from Ava and Prome down to Pegu and Martaban.

Medicine.—A decoction of the ROOT is given by the Santals for cough, and a powder of the same is applied externally for pains in the chest. MEDICINE. Root. 142

Food.—Its pink FLOWERS are sometimes eaten in Central India and also in Chutia Nagpur, as a vegetable. FOOD. Flowers. 143

Structure of the Wood.—Similar to that of **I. Gerardiana,** *var.* **heterantha.** TIMBER. 144

I. tinctoria, *Linn.; Fl. Br. Ind., II., 99; Wight, Ic., t. 365.* 145

THE INDIGO PLANT OF COMMERCE.

Vern.—*Nil* (*nil patie,* the plant, and *nil mal,* the dye stuff), HIND., BENG.; *Níl* (*basma, e.g.,* the leaves, as a medicine), PB.; *Wasma* (the leaves), BHOTE; *Osma* (the leaves), TURKI; *Jil, níl, nir,* SIND; *Nila, guli,* BOMB.; *Níli,* MAR.; *Gali, nil,* GUZ.; *Nilam,* TAM.; *Níli-mandu,* TEL.; *Níli,* KAN.; *Nílam,* MALAY.; *Mainai,* or *shán mai,* BURM.; *Níli, nilini, tuli, kálá, dolá, nílika, sriphali, tuttha, grámíná, ranjani* (*e.g.,* that which dyes), SANS.; *Nílaj,* ARAB.; *Nil, nílah,* PERS.

References.—*Roxb., Fl. Ind., Ed. C.B.C., 256, 585; Voigt., Hort. Sub. Cal., 212; Brandis, For. Fl., 135; Beddome, Fl. Sylv., Man., 85; Gamble, Man. Timb., 117; Thwaites, En. Ceylon Pl., 411; Dalz. & Gibs., Bomb. Fl., 59; Stewart, Pb. Pl., 70; Aitchison, Cat Pb. and Sind Pl, 41; DC., Origin Cult. Pl., 136; Linnæan Soc., Jour. VIII., 64; XX., 404; Mason, Burma and Its People, 510, 532, 776; Kurz, in Jour. Asiatic Soc., XLV., 1876, 269, also Kurz, Prelim. For. Rept., Pegu, App. C., v.; Pharm. Ind., 393, 396; Ainslie, Mat. Ind., I., 179; II., 33, 74; O'Shaughnessy, Beng. Dispens., 292; Moodeen Sheriff, Supp. Pharm. Ind., 94, 161; U. C. Dutt, Mat. Med. Hind., 311; Dymock, Mat. Med. W. Ind., 2nd Ed., 213; U. S. Dispens., 15th Ed., 1671; Bent. & Trim. Med. Pl., 72; Murray, Pl. & Drugs, Sind, 114; Bidie, Cat. Raw Pr., Paris Exh., 111; K. L. Dey, Indig. Drugs of India, 163; Ain-i-Akbari, Gladwin's Translation, II., 28, 41; Year-book of Pharmacy, 1878, 148; 1880, 148; Irvine, Mat. Med., Patna, 61; Voyage of John Hughen van Linschoten, I., 62; II., 91, 230, 280; Hove's tour in Bomb., 1787, 47, 48, 49, 58, 67, 107, 116, 162; Baden Powell, Pb. Pr., 339; Atkinson, Him. Dist., 308; Drury,*

INDIGOFERA tinctoria.

The Indigo Plant of Commerce.

U. Pl., 254, 473; Duthie & Fuller, Field and Garden Crops, 43; Lisboa, U. Pl., Bomb., 242; Birdwood, Bomb. Pr., 297; Royle, Ill. Him. Bot., 248; Atkinson, Dyes and Tans, 1; McCann, Dyes and Tans, Beng., 93, 126; Liotard, Dyes, 96 to 109; Crooke, Dyeing, 438 to 489; Wardle, Dye Report, 18; Hummel, Dyeing & Textile Fab., 295—318; British Manuf. Industries, Dyeing, 141; Needham, Kampti Dyes; Pandit Bhág Rám, Report on Indian Dyes, 1; Shortt's Prize Essay on Indigo; Royle, Prod. Resources, 37, 52, 94, 387, 392; Kew Rept., 1860, 49; Schunck on Indigo-blue in Philosophical Magazine, Aug. 1855, 73; January to June 1858, 29, 117, 183; Indigo, by W. M. Reid; Watts, Dict. of Chemistry, 246—264; Milburn's Oriental Commerce, Ed. 1825, pp. 289—291; Spons' Encyclop., 8, 58; Balfour, Cyclop., II., 338; Smith, Dic., 220; Treasury of Bot., 621; Ure, Dict. Indus. Arts and Manu., II., 656; Kew Repts., 1881; Kew Off. Guide to the Mus. of Ec. Bot., 40; Kew Off. Guide to Bot. Gardens and Arboretum, 69; Trop. Agri., 1st March 1882, 770; 1st April 1882; 23rd February 1889; Simmonds, Trop. Agri., 356; Kew Bulletin, March 1888, 74; Selections from Rec. of Bengal Govt., No. 33, Pts. I., II., III.; Indigo Planter's Gaz.; Wallace, India in 1887, 231—234; The Calcutta Review; Indian Field; Proc. Agr. and Hort. Soc., Burma, 1872; Manson, Report on Cultivation of Indigo in Rajmehal, 1877; Experimental Farm Report, Saidapet, 1878, 19; Trade and Navigation Reports; Rep. by Dir. of Land Records and Agri. of Bengal; Reports of Rev. and Agrl. Dept., 5th August 1889; Special Reports: H. Sewell, Esq., Collector of Cuddapah, 21st Oct. 1888; Burma Govt., 1888; H. Willock, Esq, Collector of Trichinopoly, 30th Oct. 1888; H. Goodrich, Esq., Collector of Bellary, 4th Oct. 1888; Memo. by Rev. Dept., on Madura District, 1888; No. 1431 of 1851, Bombay—Official Papers regarding the cultivation of Indigo in Sind; Rept. of Indigo Commissions, 1861 and 1877; Papers relating to Indigo cultivation in the Presidency of Bengal, 1860, printed Bengal Military Orphan Press; Selections from Papers on Indigo Cultivation in Bengal by a Ryot; Selections from Records of E. I. Company; Dictionnaire Universel, Théorique et Pratique, II., 165; Settlement Reports: Chanda, App., vi.; Allahabad, 26, 27; Bareilly, 82; Banda, 49; Azamgarh, 134; Dera Ghazi Khan, 9, 10; Lahore, 7; District Manuals: Trichinopoly, 74; Godavery, 69, 140; Chingleput, para. 60; South Arcot, 109; Cuddapah, 204; Madras Manual of Administration, 290, 337; Gazetteers: Mysore and Coorg, I., 59, 437; Bombay, IV., 58; XII., 164; XV., 431; XVI., 170; North-West Provinces, I., 80; II., 378; III., 463; IV., lxx; Punjab;—Lahore, 87; Ludhiana, 141; Muzaffargarh, 28, 90; Multan, 97; Dera Ghazi Khan, 82; Bengal, several volumes; Westland's Jessore, 135; Agri. Statistics of Jessore by Sen, 12—23; Grierson, Behar Peasant Life, 60—62, &c., &c.

In the Transactions, Journals, and Proceedings, Agri-Hort. Soc. of India, Indigo is dealt with in almost every volume. The following may be specially consulted:—Transactions, II., 31; III., 60—68; IV. 184, (Proc.), 6; VI., 126; VII. (Proc.) 26, 102, 112; Journals, I., 102, 347; II., (Sel.) 213; III., 231—232; IV., (Sel.) 86—88, 129, 130, 135, 153; V., 106—118, (Sel.) 28, 29, 77; VI., 51, 68, 142; VIII., 142; IX., 414, (Sel.) 41, 149; X., 111, 275—388, (Proc.) 25; XIX., 101—110, 140—143.

DYE. Leaves. 146

Dye.—The infusion of the LEAVES of this species, as also of **I. Anil**, yields the Indigo of commerce. See the special article on INDIGO below.

OIL. Seeds. 147

Oil.—The SEEDS are said to yield an oil which is used medicinally.

MEDICINE. Extract. 148

Medicine.—The EXTRACT is given in epilepsy and nervous disorders. It is also used in bronchitis and as an ointment in sores. It is largely employed as a test reagent.

SPECIAL OPINIONS.—§ The following communications, as a selection from many other similar medical opinions, may be here given:—"Used by native practitioners in chronic affections of the brain" (*Civil Surgeon J. Anderson, M.B., Bijnor, North-Western Provinces*).

Leaves. 149

"A poultice made of the LEAVES and applied to the perinæum is said to cause a flow of urine in spasmodic stricture" (*Civil Surgeon C. M. Russell, Sarun*). "A chief remedy for mineral poison" (*V. Unmegudien Mettapollian, Madras*). "I have used it frequently for sores on horses; it is sup-

MEDICINE.

posed to promote the growth of hair" (*Surgeon-Major C. W. Calthrop, M.D., Morar*). "It is used as an external application in the form of paint over the abdomen in cases of tympanites and retention of urine. In the form of paint or ointment it is largely used in sores and diseases of cattle" (*Civil Surgeon J. H. Thornton, B.A., M.B., Monghyr*). "Indigo is used by natives as a cooling application to burns and sores of horses" (*Assistant Surgeon Bhagwan Dass, Civil Hospital, Rawalpindi, Panjáb*). "An infusion of the ROOT is given as an antidote in cases of poisoning by arsenic" (*Surgeon W. F. Thomas, Madras Army, 33rd Regiment, M. N. I., Mangalore*). "Used as a medicine as well as a dye" (*Surgeon-Major J. E. T. Aitchison, C.I.E.*)

Root. 150

INDIGO.

HISTORY.

HISTORY. 151

The chief source of the Indigo of Commerce is the **Indigofera** distinguished as **tinctoria**. The origin of that species is involved in the greatest obscurity. Early botanists recorded having found it wild in India, Africa, and Arabia; but it is now admitted that the form they referred to was **Indigofera cœrulea**—a plant by some writers viewed as the original wild condition from which was derived through cultivation the modern indigo-yielding stock. By others these plants are regarded as distinct. There are in all some 300 species of **Indigofera**, distributed throughout the tropical regions of the globe, with Africa as their head-quarters. India possesses some 40 species, but with reference to **Indigofera tinctoria** there is no authentic record of its having been found far removed from human influence, still less of its having been observed in what could be viewed as an indigenous habitat. It may perhaps be desirable to review here the chief botanical opinions on this point. **Bentley and Trimen**, following the view once generally maintained, say, it is a native of the west coast of Africa, where it is also extensively cultivated. But **De Candolle's Indigofera tinctoria**, *var.* **macrocarpa** (upon which that opinion probably depended) is now regarded as **I. cœrulea. Boissier** (*Flora Orient.*) describes nine species of **Indigofera** as met with in Central Asia, Arabia, and Egypt; but he makes only an accidental allusion to **Indigofera tinctoria** as met with under cultivation. In the recent enumeration of the plants of China by **Forbes and Hemsley, tinctoria** is described as "generally cultivated and wild in the tropics; but it is uncertain where it is really wild." **Linnæus**, in his *Flora Zeylanica*, refers to it, but whether wild or only cultivated he does not state. **Thwaites** speaks of it as occurring "in the hotter parts of the island."

Wild Indigo. 152

Turning now to Indian authors, **Rumphius** appears to have regarded **I. tinctoria** as indigenous to Gujarat. It seems probable that the **Indigofera** first cultivated in South India, at least in the Carnatics, was some other species of **Indigofera**. The writer indeed suspects that in Western and Central India a **Tephrosia** may have been employed as a source of the dye before the true indigo plant was known. A striking fact may in this connection be mentioned, *viz.*, that even to this day **Indigofera Anil** (an admittedly American species) is largely grown in Southern India, while it is quite unknown in the great modern Indigo area of the Eastern and Upper tracts of India. **Roxburgh** has no hesitation in saying that the native place of **I. tinctoria** is unknown: "though now common in a wild state over most parts of India, yet it is generally not remote from places where it is, or has been, cultivated." **Dalzell and Gibson**, in their *Bombay Flora*, say, it is "found apparently wild in many parts of the Concan," but they add that **I. cœrulea** is probably the original wild condition. It is

HISTORY.

likely that their so-called "apparently wild" plant was one of the forms of **I. cœrulea**—these botanists thus making the same mistake that **De Candolle** did regarding Western Africa. **Kurz** speaks of **I. tinctoria** as frequently cultivated on the alluvium of the Irrawadi in Burma; but he incorrectly refers **I. cœrulea,** *Roxb.*, to be a synonym of **I. Anil.** The plant he deals with, he says, occurs all over Burma from Ava to Martaban and Tenasserim. And significantly he adds that he could not find any sufficient grounds for separating specifically the two forms. **Kurz** was far too accurate a botanist to make such a statement without sufficient reasons, and it may safely be concluded that **I. tinctoria** and **I. cœrulea** are very easily confused one with the other, and that the popular Indian writers, who speak of the Indigo plant being wild in India, allude to **I. cœrulea** and not to either **I. tinctoria** or **I. Anil.**

There is thus no botanical evidence that **Indigofera tinctoria** is indigenous to India. **DeCandolle,** however, lays stress on the fact, that it is the "*Nili*" of Sanskrit authors, from which circumstance he concludes that it must be of Asiatic origin. On the other hand, it may with equal force be urged that had it been a native plant of India, like the truly indigenous Indigoferas, it would have had its specific names in most of the languages and dialects of this country, more especially among the aboriginal races. The singular uniformity with which the word *Nílá* is associated both with the plant and the dye argues a common origin in a centre from which the knowledge of so valuable a product was diffused—its name passing from tongue to tongue with the gift of a few seeds. According to **Aitchison** and other travellers, **Isatis tinctoria**—the woad—is the source of the Indigo prepared in Upper Asia, a region where **Indigofera tinctoria** does not apparently exist. The Sanskrit people may have accordingly been first made acquainted with **Indigofera tinctoria** in India, and it is therefore just possible that their *Nílá*, of the earlier epochs, may have been the woad which with the ancient Britons was used, like the Indigo of the American Indians, to dye the skin and hair. Complex and difficult as the art of dyeing with Indigo is, it is more intimately associated with the early human race than any other dye or pigment. But in India one of the most striking features of the use of a blue dye is the large number of plants which are not only known to afford such a dye, but which to this day are regularly so employed in place of the modern commercial article. And it may be added that every source of blue dye (chemically indigo) has its special and peculiar names in the regions where it is found or utilised. These blue-dye plants are purely indigenous, and were probably the sources of the blue-dye of India long prior to a knowledge in *Nílá*. (*The reader is referred to the paragraph on Indigo Substitutes, for further information on this point, see page 449.*)

Application of Nili. 153

But there are other considerations of equal weight. The word *Níla* simply denotes dark blue colour, and is, with many of the early writers, practically synonymous with *kála* (black). Adjectively it is applicable to animals, plants, minerals, according to their colour. Thus the Sanskrit authors speak of *níla* flies, birds, cows (*e.g. nílgao,* the blue-bull or **Portax pictus** of Zoologists); of blue stones (*nílopala,* the Lapis lazuli, and *nílamani, nílaratna,* the Sapphire); of blue flowers, rivers, seas, hills, and clouds (*nílabha*). *Níla, níl, nel* is associated with a large number of plants besides **Indigofera.** Thus we have the blue water-lily (**Nymphæa Lotus**), in Arabic the *nílufar,* from which we derive the generic name for a small group of water-lilies, **Nuphar.** So, again, we have it in the botanical name for the sacred Lotus—**Nelumbium**; and in botanical nomenclature and oriental literature we have the **Pharbitis Nil** or *nabbun-níl* of Arabic, *tukhme-níl* of Persian, and *níl kolomi* of Bengal. A curious

HISTORY.

instance (having perhaps a special significance in connection with Indigo since one species is actually used along with indigo) exists in the fact that some of the species of **Senna** bear as a prefix to their specific names, the syllable *níl*. But *níla* carries with it also the abstract meaning of darkness. *Nílá* substantively is the plant which yields a blue or dark-coloured dye. Thus in its range of meanings it passes from the general to the specific, and probably obtained its restricted signification of *the blue dye from Indigofera* in India, and that too perhaps only during comparatively modern times. In addition to the fact that there are no specific primitive names in the languages of India for **Indigofera tinctoria**, such as exist for the other species of the genus, there are in some of the languages or dialects descriptive and comparative names for it that would seem to indicate a previous knowledge in other blue dyes. Thus in Kánarese it is *olléníli* (*olle*, good) : again, *hennuníli* (female indigo) : in Tamil, *aviri* (*avi*, boiled), and *karundóshi* (black indigo): so also even in Sanskrit out of a long list of descriptive names, besides *nílá*, it is *vajraníli* (hard blue). Even the Gujarati name *galí* may simply mean "the decoction," and thus indicate the method of preparation. In part support of this contention for a comparatively modern application of *Nílá* specifically, we have the further fact that Indigo did not, except in Portuguese and during the sixteenth century, carry with it to Europe the name *Nílá*. That an indigo or blue dye was used in remote times there can be no doubt since it has been detected in the coloured borders of some of the Egyptian mummy garments, but that the word *Nílá* is of a like antiquity, or that the mummy dye was prepared from **Indigofera tinctoria** are problems that perhaps can never be solved. **Dioscorides** (A.D. 60) speaks of Indigo as Ἰνδἰκον; **Pliny** calls it *Indicum*; and in the *Periplus* it is Ἰνδικον μελαν or the Indian Black, exported from Barbaricon on the Indus. The word "black" is instructive both from its association with *Nílá*, and from the circumstance that the dye was used as a black colour before its property of affording blue was discovered. In the thirteenth century **Marco Polo** recorded having seen it at Colium (an ancient port in the Native State of Travancore) where he says, "they also have abundance of a very fine indigo (*ynde*)." **John Huyghen van Linschoten**, whose journal of Indian travel was published in 1596, very carefully describes the manufacture of what he speaks of as "*Annil* or Indigo, by the Guzuratis called *Gali*, by others *Nil*." **Conte** in the fifteenth century, and **Tavernier** in the seventeenth, also minutely describe the manufacture of Indigo. Up to the time when the discoveries of the fifteenth century established the new commercial route to India, Indigo reached Europe by the Persian Gulf and Alexandria. In the commercial reports of Marseilles as early as A.D. 1228, it is described as the Indigo of Bagdad. But when first brought into Europe it was used only in small quantities to heighten and deepen the blue colour obtained from woad—a dye which, for several centuries, was the object of an important cultivation and manufacture in Germany, France, Prussia, Italy, and England. Towards the end of the sixteenth century, however, the European dyers had begun to recognise that Indigo afforded the means of a great economy, and that the dye was at the same time of a superior quality. At this period the dyers of Holland were the most famous and prosperous in all Europe. Even down to the beginning of the seventeenth century the English manufacturers sent their white cloths to be dyed in Holland : hence known as Hollands. From this trade the fortunes of the great Dutch capitalists were made; but difficulties with Portugal led to the Dutch making an effort to procure their supplies direct, instead of through the Portuguese merchants. For nearly a century Lisbon rivalled even Venice as a depôt for Eastern produce; but

Early mention of Indigo in India. 154

HISTORY.

the skill of the Portuguese stopped short of utilising in home industries the materials which their maritime enterprise brought to their shores: hence their Indian Indigo found its way first to Holland, thence over all Europe. But in 1631 the Dutch East India Company was formed, and enough Indigo was brought direct to Holland to supply the whole world, and an immense Dutch trade was created at the expense of the Portuguese. The envy of the merchants of all Europe was thereby provoked. The great woad-growers and traders were ruined. It was not unnatural, therefore, that the success of the Holland dyers, rivalling and replacing a branch of ancient agriculture and trade, brought upon themselves a vigorous persecution. In France the rich princes of Languedoc depended for their fortunes on the cultivation of the indigenous dye, and even the guarantee for a considerable portion of the ransom of King Francis I. was to be met from that industry. Indigo was accordingly interdicted in that country (1598), and Henry IV. issued an edict in 1609 sentencing to death any person who should be found using the dye. In Germany, in a like manner, stringent measures were taken to stifle the growth of a trade in Indigo, for the wealthy woad merchants there enjoyed the proud distinction of the "Waid Herrn"—the gentlemen of woad. The Emperor Rudolf in 1607, and the Elector of Saxe in 1650, prohibited its use. But notwithstanding all efforts to suppress indigo, woad plantations and factories rapidly disappeared, and the Indian dye took the place of the ancient European substance. In England during 1608, the art of indigo-dyeing had been learnt, and an exclusive right granted by the king to one merchant, led to the starvation of many of the weavers who were prohibited from having their goods dyed in Holland. The privilege was withdrawn, and British goods once more were sent to Holland to be dyed. Mention is first made of indigo-dyeing in England in 1581, during the reign of Queen Elizabeth, as permissible along with woad, or alone for the black ground colour in dyeing wool. Its employment by itself as a blue-dye was then unknown. Soon, however, the opposition raised in England against Indigo, procured a verdict that it was poisonous, and that its use should therefore be interdicted in the interests of the public. This was done and the Act remained in force until 1660, when Charles II. was compelled to procure dyers from Belgium, to teach the English the art of using indigo, as a measure to protect the greater interest of the British manufacturing industries. The East India Company, accordingly, soon after commenced to import Indigo, and from 1664 to 1694 they had brought from Surat and Bombay 1,241,967lb of the dye, which is said to have been obtained chiefly from Agra and Lahore, with 510,093lb from Ahmadabad. In these early records, however, there is no mention of Bengal Indigo, and according to Linschoten the bulk of the dye was produced in "Cambaia." It may have been observed that the references mentioned above to European classic writers and early European visitors to India regarding Indicum, Indico, and Indigo, all refer to the Western or South-Western side of India. In the Institutes of Manu "Indigo" is mentioned (in *Burnell's Translation*) as one of the articles which a Brahman should not sell when reduced to the position of having to earn a living. Even in so comparatively modern a work as the *Ain-i-Akbari* (the Persian account of Akbar's Administration in 1590), the Indigo of Agra is only accidentally spoken of as of a superior quality. Tirhút and Bengal are described, but no mention is made of Indigo in the Lower Provinces, and indeed from the very slight notice taken of the dye in the *Ain-i-Akbari* the inference is almost unavoidable that Indigo cultivation and manufacture were not viewed by Akbar as of great importance. A special chapter is devoted to dyes and colours without the mention even of

Woad cultivation of Europe. 155

Early trade in Indigo from India. 156

HISTORY.

the name Indigo. The writer is thus disposed to regard Indigo (so far as India is concerned) as having been first manufactured in Western India—Gujarat or Sind being its home—a region where it is now rarely, if at all prepared, but where it still bears its vernacular name, *gali*. The Persian and African influences (especially in the dyeing and weaving industries) are so strong on the Western side of India as to give countenance to the idea of Indigo-growing and manufacture having been introduced—an idea almost confirmed by the rapidity and completeness of its migration to other tracts of India subsequently found better suited. Indeed, it may be suggested that an indigenous industry would almost naturally be expected to have survived with greater pertinacity than indigo has manifested.

About the middle of the eighteenth century, the injunctions against the use of Indigo were in most European countries removed, though in Nüremburg the dyers were required, so late as the end of that century, to take an oath that they would not use it. Gradually the knowledge of the dye extended, and it may safely be said that, though attempts have been made to discover a substitute, all have failed, and at the present day Indigo is perhaps the world's most important tinctorial re-agent.

Two influences, however, early began to militate against the Indian market—the discovery of a source of the dye in America, and the adulteration practised by the native manufacturers. The high price paid for the article gave birth to a wholesale system of adulteration, while at the same time the dye had become indispensable. The colonists, French, Spanish, Portuguese, and English, accordingly took to Indigo cultivation. European skill and capital soon placed the enterprise on a footing which killed, or all but killed, the Indian trade. The British West Indian colonists, however, soon found that coffee, sugar, and other products were more remunerative; and at the same time the severance of America from Great Britain left the British dyers at the mercy of foreign countries for their supplies of Indigo. According to some authorities, the first efforts to revive an Indian Indigo industry proceeded from a small French factory established near the French settlement of Chandernagore on the Hugli, a few miles from Calcutta. But if the idea had been thus actually entertained, very little was done until the Directors of the East India Company, seeing the prospect of renewing their Indigo transactions and at the same time of saving the British manufacturers from dependence on French and Spanish colonists, resolved to take active steps towards starting Indigo cultivation in Bengal. European Indigo planters were brought from the West Indies and established in selected districts of Bengal. The Company's officers were permitted to trade in the article, and home remittances were made in Indigo for a series of years, even although this resulted in heavy losses. It was early seen to be unnecessary for the Company to directly own factories. Many gentlemen were found willing to undertake the responsibility upon obtaining grants of land on favourable leases, the Company purchasing the produce at contract rates. In the Proceedings of the Hon'ble Company for 1780 we read of a contract having been made with a **Mr. Prinsep** "who at this time conceived that India might supply Europe with sugar and cotton as well as Indigo;" and for a supply of the latter the Company continued to make similar other engagements until 1788. But in reviewing the issue of all the sales prior to the year 1786, it was ascertained that the several parcels yielded a remittance of only "*1s. 7d.* 67 *dec.* for the current rupee, which was a loss in the first instance of upwards of 17 per cent., independent of freight and charges which may be reckoned at full 10 per cent. more." A **Mr. J. P. Scott**, however, produced Indigo that gave a profit of *11d.* 01 *dec.* per pound. To overcome

Re-ntroduction of Indgio cultivation into India. 157

HISTORY.

these losses every encouragement was given to improve and cheapen the process of manufacture, and the Company continued to subsidise the industry until the quality of, first the French and then the Spanish Indigo, had been surpassed by the Indian. In 1787 **Dr. Hove** visited Western India, and in his Journal he describes an Indigo industry on a scale that there no longer exists. Thus the establishment of the *Bengal Indigo trade may be viewed as having given the final blow to the extinction of the Bombay cultivation. Even so late as 1820, Indigo was a fairly important crop in Gujarat, and the number of unused pits near old villages and among the buried cities of the Sátpuda mountains bear testimony to the abandoned industry. Early, however, we read that the new Bengal trade showed indications of spreading northwards, and that the natives of Upper India had even taken to producing large quantities. The good name earned through the European skill and care in manufacture thus began to be injured by the inferiority of the Native stuff that found its way to Europe along with the highly-prized East Indian European-made, article. The success of the new industry soon also created another danger. European factories multiplied until alarm was felt that the markets were being glutted. The East India Company then felt called upon to restrict the traffic, especially from territories outside their own possessions. Accordingly in a despatch dated August 1800 we read:—"European skill and enterprise have formed the present indigo manufacture and indigo trade of India. To these both Oudh and Bengal are indebted for the share they possess in the exports of that article, and on these, there seems reason to believe the trade in both countries will continue to depend. A trade thus raised and supported greatly to the benefit of those countries, by persons not natives of them who have a right to be there only by our permission—a trade which on its present great scale can be conducted only by shipping employed or licensed by us, a trade in short which the Company have fostered at a considerable expense—such a trade we regard as naturally more subject to our direction and modification than if it had been established by Natives themselves; and this observation, we think, applies still more strongly to the allied country of Oudh than to Bengal."

Indigo purely a European trade. 158

"In whatever degree, also, the indigo trade of Oudh is carried on by the capital of Bengal, that is, by advances furnished from thence, so far the Government of Bengal acquires an additional right of interference in this trade, especially if Bengal itself is capable of employing that capital in the indigo manufacture."

"If these observations are just, with respect to Oudh, they will apply with still greater force to countries beyond it, not at all connected with us, whence, however, we are told not only that much of the indigo exported by Oudh comes, but that the profits on indigo raised in those countries have furnished the funds for paying formidable military levies made in them. As we are certainly under no obligation to augment their manufactures or facilitate their exports, we must think ourselves at liberty to decline measures which would have such effects, even though attended with some intermediate advantage to Oudh, if those measures were to interfere with the interests of our own territories."

In the year 1802 the Board of Directors of the East India Company directed that no further advances or other pecuniary encouragement be given to indigo planters. They considered that "from the large profit that has arisen from the produce of that commodity at the Company's sales, the merchants may be able from their own resources to make the necessary advances for carrying it on." Thus for about 22 years (1780—

* Jessore Indigo first cultivated, 1795.

HISTORY.

1802) the East India Company directly supported the indigo industry, and placed India once more in the foremost rank among the indigo-producing countries of the world. When pecuniary aid was no further necessary, the Company retired from direct ownership of indigo-factories. They, however, continued to make large purchases and to trade in indigo. In 1806 they directed a mart to be established in Calcutta for ready-money purchases in indigo, and they consigned to India test samples valued by the home brokers "for the purpose of assisting the judgment of our servants in making their purchases." It was, however, subsequently found necessary to make certain money advances to manufacturers from whom the Company proposed to purchase the dye, so that a certain amount of encouragement was continued, though direct support and supervision were discontinued. It may be of use to quote here the specification of qualities and prices which in 1810 were sent to India:—

		s.	d.
A. Blue	worth per ℔	10	0
B. Purple	do.	9	0
C. Violet	do.	7	6
D. Copper	do.	5	6

The despatch forwarding these valuations gives the results of sales for 1807 to 1809:—

Valuation of Indigo in 1810. 159

		℔		s.	d.
March sale,	1807	2,022,113	@ an average of . . .	8	6
September ,,	,,	3,091,202	@ do. . . .	6	6
March ,,	1808	2,652,428	@ do. . . .	5	6
September ,,	,,	(none arrived in time).			
March ,,	1809	3,995,191	@ an average of . . .	4	7
Ditto Company's ,,		280,502	@ do. . . .	5	6
September sale ,,		371,370	@ do. . . .	4	6
Ditto Company's ,,		98,894	@ do. . . .	5	11

"The general average of all which is 5*s*. 11*d*. per pound; and as the above term comprises a period when the markets were favourable as well as one in which they were adverse, it may be considered as a fair general average." In a "Statement of Advances on the Remittance Plan" the Company is shown to have advanced from 1786 to 1804 close on a million pounds sterling towards their purchases.

Even at this early stage of the new Anglo-Indian Indigo industry, however,—an industry directly fostered by the Government and in which many of its officers traded extensively, or resigned the Company's service to become Indigo planters—difficulties had arisen through the competition of rival concerns. The planters desired that the Company should prohibit the establishment of new factories within certain limited distances of those already in existence. The Directors replied that "since the inconveniences of which they complain may also have contributed to the repeated calls made on the Company to relieve the distresses under which this branch of manufacture laboured, we are the more disposed to wish that new enterprises may be undertaken with prudence, and that the cultivators may avoid collisions by friendly correspondence among themselves, and by mutually restraining their servants from issuing advances to those rayats who may be under previous engagements." The system of advances made by the Company to the owners of factories against annual purchases was thus not only fully established, but that principle was recognised as applicable to the owners of factories in their dealings with the rayats or actual cultivators. At the same time the Company seem to have appreciated the advantages that would likely accrue from the establishment of native manufacturers. In a despatch dated April 1811 we accordingly read:—"It is expedient that your attention should be

Difficulties due to over-production. 160

INDIGOFERA tinctoria.

The Indigo Plant of Commerce.

HISTORY.

directed to the encouragement of the native growers of indigo being proprietors of factories, by issuing advances to them in common with European manufacturers, taking care that the security be sufficiently respectable; as we fully concur in the opinion of our Board of Trade, expressed in their minute of the 28th October 1796, that the cultivation of Indigo cannot be considered as decidedly established in Bengal until the Natives shall chiefly manufacture it of a quality fit for the European market."

"Upon a general view of the present appearance of the state of the indigo trade, we concur in your decision that it is not advisable to establish indigo factories upon our account in the territories under the authority of the Madras Government; for our apprehension that the market in Europe will be liable to be overstocked by the Bengal produce operates as more than a counterpoise to the abstract principle which would otherwise govern our decision, *viz.*, that it is our duty to promote new branches of trade in any of our settlements to which the soil and climate may appear favourable."

In a correspondence between the Board of Directors and the Madras Government, many interesting facts are brought out. For example, in a despatch dated May 1791, the Board approve of the Madras experiments with indigo seed from Bombay. "A new sort of indigo" is referred to and a report on its merit promised. The despatch continues—"The box of **Nerium** indigo by the same conveyance has been inspected by an eminent broker, and the samples valued at from *2s. 9d.* to *7s. 3d.* a pound." Reference is made to European indigo planters in Madras, but the spirit of the correspondence is a recommendation to discourage the expansion of the industry, since the production was viewed as likely sooner or later to far exceed the demand.

Difficulties between the Planters and Cultivators.
161

By 1837 difficulties had arisen in connection with the relation of the zemindars and planters to the actual cultivators. On this subject **Lord Macaulay** wrote a lucid memorandum of the greatest interest, which seems to have had the effect of preventing the Government from interfering, except to organise a greater number of courts so as to obviate the inconvenience necessitated by the distance from each other of those already in existence. **Lord Macaulay** wrote:—"That great evils exist, that great injustice is frequently committed, that many rayats have been brought, partly by the operation of the law and partly by acts committed in defiance of the law, into a state not very far removed from that of partial slavery—is, I fear, too certain." Again, "The regulation, which gave to the Indigo planters who had made advances to a rayat a lien on the Indigo crop, seems to me highly objectionable in principle. But I do not conceive that by rescinding it the Governor General in Council would give any sensible relief to that class of the population whose interests appear to be peculiarly the object of his solicitation. The question appears to be a question between the planter and the zemindar. It is not easy to see how it can be of any consequence to the rayat which of the two may distrain on his crops. I have no reason to believe that the zemindars exercise their power with more justice or humanity than the planters."

engal Indigo Disturbances.
162

Passing over a space of some thirty or forty years, which seem to have been marked only by the continued withdrawal of all official support or encouragement, Indigo next attracts public attention in the troublous times which culminated in the Bengal Indigo Disturbances of 1860-67. These may be briefly summarised as the outcome of the system of land tenure which prevails in Bengal, in which a rent-receiver or zemindar exists between the actual occupant of the land and the Government. In some cases these zemindars temporarily or permanently sold their rights to the owners of indigo factories, thus handing over to the planters a power of

HISTORY

dictating that a certain proportion of each holding should be annually thrown under indigo. The green produce was contracted for at a fixed rate, but while the value of all other crops had steadily improved, the rate paid for the indigo plant practically remained stationary. Where even the shadow of a legal right to enforce indigo cultivation did not exist the system of making advances to the rayat secured compulsory cultivation. Money offered at the season of the cultivator's greatest impecuniosity was a bribe that few could resist, and it was in the interest of the factory owners to secure a maximum amount of indebtedness on the part of the neighbouring rayats. By the law passed in 1830, but repealed in 1835, failure to fulfil contracts could be criminally prosecuted. The effect of that enactment was to give the stronger contracting party the protection of law, while no consideration was shown to the weaker, who might have been forced into contracts the full meaning of which he did not comprehend. But even without special legislation the system pursued placed the cultivators largely at the mercy of the planters if they chose to exercise their power. Complaints of enforced cultivation of an unprofitable crop on the one hand, and of the want of protection against dishonest cultivators on the other, became rife. Expensive supervision and protracted litigation told heavily against the industry and led to unlawful restraint and assault which in certain parts of Bengal culminated in a disturbance that necessitated military protection. A Commission was appointed by Act XI. of 1860 to enquire into the whole subject, and the Report of the Commission revealed the existence of faults on both sides. The Act gave powers to the Magistrate to enforce the completion of contracts to grow indigo where advances had been made; but where such contracts were found to have been obtained by fraud, force, or unlawful intimidation, the complaint was to be dismissed. By Act VIII. of 1868, this special "Indigo Contracts Act" was, however, repealed. His Honour the Lieutenant-Governor of Bengal, in reviewing the Report furnished by the Commissioners, said:—"Setting aside individual cases having no connection, or at least no necessary connection, with the indigo system, that system is fairly chargeable with a very notable portion of those classes of offences the peculiar prevalence of which in Bengal has been from the first a blot in our administration. In my opinion it is rather the system than the planters individually who are to be blamed. It is to the unprofitableness of the cultivation of indigo at the extremely inadequate price given for it, under the system, necessitating either a forced cultivation, or the abandonment of the manufacture from Bengal rayaty plant, that this and every other evil connected with indigo is attributable. An individual manufacturer could not live upon a fair and free system, surrounded on all sides by competitors who get their raw produce without paying nearly its full value. That a whole class did not spontaneously reform itself from within is not surprising. The chief fault was in the defective, and I fear I must say the not impartial, administration of the law, which allowed such a vicious state of things to exist."

Advance-System. **163**

Legislation. **164**

Difficulties also arose in Behar in 1876-77, and an elaborate enquiry was then instituted by the Bengal Government into all the complaints. But the charitable action of the planters to their cultivators during the famine and the kindly way they were then spoken of by the poor in their neighbourhood, satisfied the Government that no cause for interference existed save to put down with a strong hand any violations of the law on either side that might come to notice. The enactments that are still in force regarding indigo cultivation will be found in the Bengal Code, *viz.*, Regulation VI. of 1823, and an amendment to it passed in 1830, as also in the Local Act X. of 1836. These provide that if a rayat by contract takes advances to cultivate a certain prescribed portion of land with indigo, he

Difficulties in Behar. **165**

HISTORY.

can be prosecuted should he fail in his contract, and the person who makes an advance on such a contract is thus recognised as having a lien or interest in the indigo-plant, and is entitled to avail himself of the protection of the court should his interests be threatened. These regulations provide also for the criminal prosecution of persons inducing rayats to break their contracts, and the punishment of persons who damage the indigo-plant. On the other hand they provide for the protection of the rayat against any enforced cultivation, and give him full power to refuse to renew a contract which has just expired. The occupancy rights of rayats in possession of land for twelve or more years is fully recognised. If they have completed their contracts or are prepared to pay the damages awarded by the court for non-fulfilment of contracts, they cannot be compelled to grow indigo; still less can they be deprived of their holdings for not doing so.

Indigo Planters' Association. 166

At several periods the planters have formed associations which have since practically been consolidated into the Indigo Planters' Association. Important reforms have been effected through the planters mutually agreeing to observe certain rules, and the indigo industry has thereby been placed on a more rational basis. But a migration of a pronounced kind took place from Bengal, where the action of the Government was found opposed to certain individual interests, and it may now be said that while the best quality of dye is still produced in Bengal, the chief seats of the industry are in Behar, the North-Western Provinces, and Madras.

Removal of Duty. 167

The action of Government in removing the import and export duties, as also the tax on the green plant, greatly favoured the development of the Indian indigo trade. Previous to the 5th August 1875 there was an export duty of R3 a maund; but from that date it became R3 a maund on manufactured indigo, and R3 a ton on indigo leaves, while on the 25th of February 1880 the duty was entirely removed. Imported indigo was subject to a duty of 7½ per cent., but from the 5th August 1875 it became 5 per cent., and was altogether removed on the 10th March 1882.

Enough has perhaps been said to convey a general idea of the historic facts connected with the growth of the indigo industry. The modern figures of trade, in another paragraph, will complete the sketch. It has been shown that the earliest European accounts of indigo manufacture are associated with Southern and Western India. That the trade then migrated to the West Indies and America, was restored to India through the enlightened, though monopolising policy of the East India Company, but was then established in Bengal. That half a century later it had to leave some of its most productive areas, through a selfish policy of cultivation, and lastly that a new feature arose in the keen competition of the native growers of Madras. An examination of the trade statistics will show how vastly more important Madras Indigo has become within recent years, thus justifying the far-sighted policy of the East India Company in urging the importance of getting the growers to be themselves manufacturers of the dye. If one idea more than another came out pointedly as the result of the enquiry of 1860, it was that the Bengal planters were admittedly paying a very low price to the cultivators for a crop, of which, one of the seasons of cultivation proved irksome to them, since it prevented their devoting themselves to their chief food crops. But the cultivators of Madras and of many parts of Upper India are not only not opposed to growing indigo but do so willingly, since in these regions it not only does not conflict with, but improves the food crop with which it is grown as a valued rotation. In the reports of occupancy manufacture, highly remunerative returns are given. It would appear to richly pay the grower and manufacturer of small holdings, while it may not be remunerative to a private

Objections to Indigo cultivation. 168

HISTORY.

European factory to prepare the dye alone. But the danger anticipated and carefully guarded against by the East India Company, *viz.*, the undue production of inferior native indigo, is in full operation to-day; and were the number of the European factories from any cause to be seriously diminished, it seems just possible that India might once more witness a partial migration of the industry to European colonies. The one saving feature seems to be a wholesale reduction of expenditure in manufacture similar to what has taken place in Tea—thus narrowing the profits but paying on large concerns. The decline in the price paid in England for indigo since 1880—a decline of fully 35 per cent.—may be viewed as, in part, the result of the competition within India itself, a country which enjoys a practical monopoly in the production of the dye; though doubtless, the competition of inferior dye-stuffs, and fluctuations in currency, have also exercised a depressing influence. The competition of Madras, for example, with other parts of India, must have been serious; for within the 25 years ending 1881-82 the cultivation has expanded in South India over 133·8 per cent. and the exports over 116 per cent., while those of Bengal have decreased 3·2 per cent. The exports from Madras in 1855-56 were 2,852,713℔, while last year (1887-88) they were 4,794,944℔ or an increase in 32 years of 68 per cent. From Bengal in the same years they were 10,106,768℔ and 9,781,520℔. The exports from Bengal, however, convey a much less accurate idea of the actual Bengal trade than do those of Madras of the Madras trade. The Bengal exports include the production in the North-Western Provinces, and Oudh. Indigo cultivation has seriously declined in the Lower Provinces, though it has greatly expanded in Behar and the North-Western Provinces. In fact the chief modern feature of the so-called Bengal indigo trade is the rapid expansion of cultivation in tracts under canal irrigation and the great increase of factories owned and worked by natives. Tirhút may be described as the most productive indigo area, as it is also the only tract in which native manufacturers have not as yet succeeded in establishing themselves in competition with Europeans.

Growth of Madras Trade. 169

Expansion in Canal-Irrigated Tracts. 170

CULTIVATION.

CULTIVATION. 171

Soils, Methods, Yield, Cost, and Profit.—The system of cultivation pursued in the various provinces of India is so nearly alike, that it need be necessary only to describe in detail one or two provinces—the notices of the others being more or less restricted to the indication of departures from the general types. Thus, for example, there are three distinct methods pursued in Bengal, characteristic of three important sections of the Presidency, *i.e.*, Lower Bengal, Northern Behar, and Southern Behar. In Lower Bengal a two-fold system arises from the fact that large portions of the Indigo lands are within the area of Gangetic inundation, while in different districts or portions of the same district, higher lands that become annually submerged by the rainfall are also thrown under indigo. In both of these conditions irrigation is consequently quite unnecessary, and, the soil being recent alluvial deposit, a simple system of cultivation with broad-cast sowings is all that is required. In South Behar (*i.e.*, Patna, Gaya, and parts of Chutia Nagpur) irrigation from tanks and wells, and in Shahabad from canals, is found necessary, so that the system of indigo cultivation, there pursued, resembles very much that of the North-Western Provinces and the Panjáb. In Northern Behar (Tirhút, Champarun, Sarun, &c.), a very high class cultivation with drill sowings is practised, and although well and tank irrigation is available, such aid, owing to the fertility of the soil, is very frequently quite unnecessary. To a small extent similar high class cultivation is accom-

CULTIVATION. plished in parts of Lower Bengal on high lands but mostly as spring sowings.

While, therefore, Northern Behar may justly be described as the head-quarters of the indigo industry, some of the districts of Lower Bengal, such as Jessore and Nadiya, still continue to produce the most highly prized qualities. The advance of the industry from Lower Bengal to the Upper Provinces may, in a measure, be due to the loss of a power to enforce cultivation, but the rapidity with which it has spread into the new or greatly enriched tracts, under canal influence, would seem to point to another explanation—direct gain. It is possible also that exhaustion of the soil to indigo, though improvement to rice, may have in some cases necessitated irrigation, which was not in every district easily obtainable. At all events through the aid of canals, large tracts can now be cultivated, during a period when the land was formerly left in fallow, and it seems highly probable that, instead of being exhausted to rice and wheat, the soil is greatly improved by the extra crop. The rayats of Upper India would thus seem to recognise in indigo an additional source of profit without any serious injury to their other crops. *In Lower Bengal, on the other hand, the spring-sown crop has to be attended to at the very time of the principal rice sowings.* It is thus not a favourite crop, since it is less remunerative than rice; but to the planter it yields a very superior quality of dye.

Objection to Indigo cultivation 172

BENGAL. 173

I.—BENGAL.

1st—North Behar.

Digging. 174

Preparing the Land.—Mr. W. M. Reid (in his *Culture and Manufacture of Indigo*) urges the necessity of deep and thorough DIGGING (*tumnee*) for high non-inundated lands such as those of Behar. This should be commenced as soon as last year's manufacturing is completed. "On this work of digging deep down will depend," he says, "in a great measure, the future health and growth of the indigo plant, and it cannot be too carefully supervised." The digging ended, the PLOUGHING next begins. The ploughs used are the ordinary native wooden plough with an iron ploughshare. Usually five or six ploughs work alongside of each other, passing, say, north to south along the field, while a second set work across the furrows from east to west. In this way the surface soil is completely broken up over the deeper digging which was effected by the hand by using the ordinary large hoe or *kodal.* When the ploughing has been completed, the field is next ROLLED by wooden rollers, of smooth cylinders, or constructed with a rough thread like a screw. By this means the clods are broken up, and the field levelled. For the purpose of rolling, the oldest and most worn-out bullocks are employed which, as Mr. Reid explains, "not being smart and active enough for ploughing are put to this slow and arduous work in which dead weight tells more than strength and activity. After rolling down the clods, the ground is again ploughed up, perhaps three or four times, according to the dryness, stubbornness, or clayiness of the soil. The smaller clods which remain are then finally broken by hand by gangs of, from 50 to 100 women and children, boys and girls, in one long row, who keep up a perpetual din, beating time on the clods with thick short sticks amidst clouds of dust until there is not a lump left bigger than an ordinary walnut."

Ploughing. 175

Rolling. 176

Sowing. 177

Seed-drill. 178

Sowing.—The land having been thus thoroughly prepared, sowing commences about February. For this purpose a SEED-DRILL is used—an adaptation from a native implement—specially constructed for indigo purposes. This has been briefly described (in *Gleanings of Science, May 1829*), thus:—"The shares cut the furrow, the wheels of the machine

CULTIVATION. BENGAL.

turn those of the trough, the slanting holes bored in the wheels of the trough, during their passage through the seed, take up each one or more seeds (seldom more than one), and, in the downward part of their revolution, unload themselves with precision into the hoppers, which lead them into the hollow of the ploughshares, which last deposit the seed in the furrow and enclose the seed in an instant." **Mr. Reid** gives a more detailed description of the drill, urging that the wheels and trough must be constructed of *Shisham* wood (**Dalbergia Sissoo**), the latter lined with galvanized iron. Tc ascertain the rate of discharge of seed, the number of revolutions of the seed-wheels in a given distance is tested and the quantity of seed discharged carefully collected in small bags. If too much or too little seed be distributed, the holes are enlarged or diminished until the desired rate of sowing is obtained. The eight-shared drill is that generally employed, but a four-shared to a twenty-shared drill may be used. "After the drill follows an instrument, drawn by bullocks, like a long bamboo ladder, which smooths the earth over the seeds, and then the lands are left. In from four to five days the seeds germinate, generally on or about the third day, and the two first leaves of the plant peep forth."

Manuring.—As the lands become exhausted by constant indigo cultivation, where renovation is not accomplished by annual inundations from the rivers, it becomes essentially necessary to manure. For this purpose the cheapest and best material is the *seet* or refuse plant after the indigo has been extracted. This is dried and carefully stacked, as it is not only of value as a manure for the higher lands not inundated, but it is the fuel used in the succeeding season for the boilers. In Lower Bengal *seet* manuring is not practised, or only rarely so, since the rivers overflow the indigo fields and richly manure them with fertile deposits. In Tirhút, on the contrary, and the upper parts of the province, *seet* manuring is regularly followed. For this purpose, it is carted to the fields and deposited in heaps in regular lines all over the field, and is then trenched into the soil. Manuring. 179

Weeding.—By the time the plant is a few inches above ground, weeds will have grown to such an extent as to necessitate weeding. This is done by a small cutting instrument held in the hand and plunged two inches under ground, so as to cut the weeds off without injuring the indigo. The weeds should then be carried to the margin of the field and not piled in heaps on the field, by which latter careless process many young indigo plants would be killed. Weeding may have to be repeated until the indigo crop rises above the height of being injured. Weeding. 180

Cutting.—Indigo is ripe when the flowers begin to appear. This generally takes place about June—earlier or later according to the nature of the season. Cutting. 181

Irrigation.—In exceptionally dry seasons irrigation may be necessary, and in Bengal this is generally accomplished by a slung basket worked by two men; holding the ropes attached to the basket they throw the water to the higher level. At other times a bucket suspended from a beam, with a mud weight at the further end, is employed. Occasionally a leathern bag drawn from a well by bullocks running down an inclined plane is used. On the bag coming to the top of the well it is swung over the end of the drain and the bullocks made to step backwards. The bag thus coming to rest collapses and the water runs off. China pumps, Persian wheels, or various steam pumps may also be used, but in Bengal, when irrigation is practised (which is but rarely), the two first-mentioned methods are generally pursued to raise water from a tank or blind canal flowing from a river; except in Behar well irrigation is rare. Irrigation. 182

CULTIVATION. BENGAL. Carting. 183

Carting to the factory.—All arrangements for manufacturing having been completed the planter fixes a day for commencing cutting. The lower-lying portions of his estate or plantation are selected for first cuttings. This is necessitated through the fact that the rivers at this season (June) are liable to rise rapidly and overflow their banks, when many acres of ripe crop may be destroyed in a few hours. This remark is, however, more directly applicable to the Lower Provinces than to Behar, where inundation only occasionally occurs. The crop is cut with a sickle and tied into bundles. When cultivated by the rayats and sold to the owner of a factory, it is paid for by the bundle. The size of the bundle varies so greatly, however, in the different parts of the country, that much confusion in estimating the outturn is due to this cause. Most factories use a chain for measuring their bundles. As soon as possible after being cut and bundled, it is conveyed to the factory, either in boats on the river and up the short canal leading to the factory, or by carts. On arrival at the factory it is packed in the STEEPING-VAT when the cultivators' interest in the crop ceases, and the operations of the manufacturer commence.

Steeping-vat. 184

SOUTH BEHAR. 185

2ND—SOUTH BEHAR—PATNA, GAYA, AND SHÁHÁBÁD.

Mr. Reid remarks that the system of cultivation adopted by European factories in the districts of Behar south of the Ganges, is closely allied to that pursued almost universally by the native concerns of Benares and the North-West Provinces. The soil is non-retentive of moisture, and consequently the sowings in Sháhábád and contiguous districts are chiefly carried on during the rainy season in July, August, and September.

Sowing. 186

Seasons of sowing.—The early rainy season sowings are called *Assarhi*, and the crop from these is cut in September-October. The later sowings have no special name; they continue to grow throughout the year and are reaped in July and August, when the crop is known as the *Khunti*, but that term is also applied to a second year's crop from the early sowings. The *Assarhi* crop is a precarious one, and it is only in years of light and favourable rainfall that it yields a good return—consequently no concern in Sháhábád exclusively confines its sowings to June-July, the *Assarhi* crop. **Mr. Reid** proposes the name *Nunda* for the late sowings. That crop is weeded in December, January, and February, and is also pruned; sometimes a further pruning and cleaning is necessary in April or May. It may be designated the backbone of the Sháhábád indigo cultivation, though often largely supplemented by the *Assarhi* crop. On opium fields there is also a certain amount of *Jamana* indigo, that is, a crop sown in March and April under the *assamiwar* system—the rayat taking advances of money from the factory on the understanding that he will sow so much indigo on opium lands after the opium season. These fields are irrigated mostly from wells, the water being raised in a leathern bag through the traction of bullocks made to run down an inclined path. The opening of the Sone Canal added a fourth and very important crop to Sháhábád, which is known as the *Falguni*, from the fact of its being sown in the month of February-March. Lands owned by the factory (*Niz* cultivation) are generally used.

Manufacture. 187

Manufacture.—The operations of the factories in Sháhábád thus commence in July, and the *Nanda* and *Jamana* crops yield the best returns. The second term of factory work generally commences in October, during which time the *Assarhi* and second cutting of the *Khunti* crops are reaped.

The system of cultivation in Patna, Gaya, and Sháhábád is very nearly similar to that which will be described under the paragraph on the

North-West Provinces, and, indeed, the indigo manufactured in this district is in the trade classed with that from the North-West Provinces.

CULTIVATION.

3RD—LOWER BENGAL.

LOWER BENGAL. 188

The method of cultivation pursued in the Lower Provinces is less elaborate than in Northern Behar—the nature of the soil and the conditions of climate alike operating against the necessity for high cultivation. The bulk of the indigo fields are annually inundated, and thus covered with a rich alluvial deposit which, even when dry, does not require a rigorous method of ploughing and clod-breaking.

Sowings. 189

Sowings.—There are two October sowings, and a less extensive though highly valued crop in April. As the water subsides in September-October from the muddy banks and *chars* (islands and shallow promontories in the river bed), the seed is sown BROADCAST. PLOUGHING would be impossible, but by means of a bamboo or raft of plantain stems, the cultivator is enabled to move about over the soft slimy surface and thus scatter the seeds. His supply of seed is carried in a water-tight basket made by winding rattan canes round and round continuously, each coil being fastened hard on the others by ascending ties of split cane. On being plastered with cow-dung these baskets are rendered water-tight, and they are very durable. Floating on the surface of the water or slimy mud the seed supply is thus pushed over the field, and the seed scattered by the hand as uniformly as possible. By its own weight it soon sinks to a depth of two inches, and within a few days, begins to germinate. By this time in some districts the higher *char*-lands have dried sufficiently to allow of the ladder-like harrow being drawn over. thus smoothing the surface and consolidating the soil around the seeds, but such harrowing is in most cases quite unnecessary.

Ploughing. 190

This method of cultivation is in Lower Bengal known as *Chitaní* (meaning to scatter); and so far it is very cheap, but as the plants spring up there appears with them a rampant growth of weeds, which proves a source of very considerable annoyance, and involves a very heavy weeding expenditure. Sometimes cattle are turned on in December and January to browse down the rank vegetation, and in the early state of the October sowings little harm is thereby caused to the crop, while a much valued concession is granted to the rayats. But as the banks of the rivers of Bengal are often densely covered with the scrubby plant, Tamarisk (*Jhâu*), when this extends over the *char*-fields (as it only too frequently does), the greatest possible annoyance is occasioned. The roots penetrate, if anything, deeper than the indigo, and if not uprooted early, the crop may often be thus seriously injured if not entirely destroyed. By the time the hot weather in April and May has come round, the *char* lands have become very dry and hard. They have split up into great squares or brick-like sections—the cracks often gaping two or three inches and penetrating to the tips of the roots of the crop. Further growth is arrested, and the plants wither up to a crop of thin twigs, with a few sickly leaves. They are, however, by no means killed, and when the rains in June begin to soften the soil, the invigorating moisture of the first showers is carried by the cracks straight to the roots, with the result that the sickly-looking twigs at once burst into a fresh vigour and produce an abundant crop of highly valuable dye-yielding leaves.

On more elevated lands the second class of October sowings are made. For this purpose the land is ploughed once or twice and the clods broken by drawing the ladder-like harrow (the *múie*) over the surface. The soil of Bengal being of a more recent alluvial nature than that of Tirhút and Behar generally, a much less elaborate system of cultivation is neces-

CULTIVATION. LOWER BENGAL.

sitated. The soil is easily pulverised, but as with the *char* lands, the seed is sown BROADCAST and not drilled in, which is the method in Tirhút. In these higher *dengali* lands the sowings have to be made expeditiously, for if the soil be allowed completely to harden on the evaporation of the surface moisture, it becomes impossible to plough and harrow sufficiently for indigo sowings until it again becomes softened by rain. Cattle are also sometimes allowed to browse over the *dengali* indigo fields; and in April and May they are carefully weeded.

Mixed Crops. 191

Mixed Crops with Indigo, and Rotation.—Along with the October sowings of Indigo the *rayat* frequently sows winter crops, such as oil-seeds, and thus ekes out an additional return of R2 a bigha. **Mr. F. C. Manson**, in his interesting report on indigo cultivation in Rajmehal, mentions two sowings,—October and April,—and alludes to the advantages of mixed crops with the October sowings. These advantages may be said to be twofold: (*a*) the plants are established in the fields against the time of the vigorous growth with the fall of rain; (*b*) the chief labour of indigo cultivation is accomplished at a season when the rayat has very little else to occupy his time. The rice crop has been harvested, and the ROTATION of indigo is highly advantageous. The roots penetrate to a greater depth than the rice, and thus while directly improving the soil on the principle of a leguminous rotation, the crop tends to break up the soil, and bring a greater depth into bearing, but above all an extra crop of winter produce is obtained by the same operation.

Rotation. 192

Reaping. 193

The reaping takes place in June, July, and August, or if a very full crop, even as late as September. These autumn sowings thus occupy the soil for a period of eight months.

The April or spring sowings take place at the very time the *rayats* may be said to have their busiest season. They are then anxious to prepare their lands for the *ausdhan* (or spring rice), a crop on which it may be said they largely subsist. The interests of the cultivator and the indigo-planter at that period, therefore, seriously conflict, for to *both* delayed sowings (*i. e.*, of rice in the one case and indigo on the other) would mean a complete failure. The indigo obtained from these April sowings is the most profitable of all, and a compulsory April cultivation may be said to have occasioned many of the serious misunderstandings that arose between the *rayats* and the planters. The cultivators, while willing, as a rule, to undertake sowings in October, seriously object to extensive April crops. There is, however, one consideration in favour of indigo that seems to have been overlooked. While the return is far less than with rice, the labour connected with the indigo crop is correspondingly less, nor is the crop exposed to so many risks. The cultivation consists of one or two PLOUGHINGS after showers with light HARROWING, both before and after the BROADCAST sowings. The crop requires very little further supervision, and it is ripe in June and July (*Ram Sunker Sen, Agricultural Statistics of Jessore, 14*).

Profit. 194

Profit of Cultivation; Amount of Seed; Yield of Plant and of Dye.—In Lower Bengal the *bigha* of land is a little more than one-third of an acre, while in most parts of Tirhút it is 4,225 square yards or six-sevenths of an acre. The average amount of seed sown to the acre appears to be from 20 to 30℔ (10 to 15 seers). In the compilation of all available information regarding the cultivation and manufacture of indigo in Bengal, furnished for this work by the Director of Land Records and Agriculture, only one or two isolated passages allude to the yield of plant and dye per acre. Of Midnapur "the average yield of a *bigha* is said to be four bundles; a maund of dye is equivalent to about 250 bundles; and 100 *bighas* yield an average of 3 maunds of dye." But this is obviously incorrect, since by the one statement the yield of dye would be 3·6℔ and

CULTIVATION.

LOWER BENGAL.

by the other 6·7lb an acre. The bundle of green plant is an ever changing quantity, but of Jessore it is said, the yield of dye is 3 to 4, or even 7 maunds per 1,000 bundles. Again, in **Mr. Sheriff's** report it is stated, "The yield of dye is about 6 cwts. per 1,000 bundles (a bundle weighing about 300lb). This would therefore be 1 in 461, and it is probable from this that 5,532lb of plant were produced per acre. In **Mr. Manson's** interesting report on Indigo cultivation in Rajmehal the yield of plant is put down at 30 to 40 bundles per acre and the yield of dye at very nearly 12lb an acre. **Dr. McCann**, in his *Dyes and Tans*, fortunately devoted much careful study to the Records in the Economic Museum relating to the yield of dye per acre. His conclusions have been verified by the writer, with the result that it appears safe to take the yield in the Lower Provinces at 10 to 12lb an acre, and in Behar at nearly 20lb an acre.

Cost of Cultivation. Profit. 195

The cost of cultivation has been so variously stated that it is perhaps scarcely necessary to review all the figures. The differences proceed mainly from the view taken as to the advantages or otherwise of indigo cultivation. Certain items have been debited or credited in one return and omitted from another; and the system of making advances has been claimed as a boon or pronounced a curse according to the prejudices of the writer. But the system of making advances, however objectionable it might be made in individual cases, is a necessary outcome of all production not intended for homestead consumption, and is carried out by Government successfully in the opium industry. Unless a market exists the plant is of no value in itself to the cultivator, and unless plant be available the capitalist could not be expected to erect a factory. The planter is required therefore either to own or to lease the land, or is compelled to hold out such inducements to occupancy tenants as may secure a regular cultivation sufficient to support his factory. The fault lies, not so much in the system of advances, as in the monopoly thereby secured to each concern by which the price paid for the field produce is kept stationary, instead of being allowed to progress with the increasing value of all other produce. That the price paid until recently was too low there seems no doubt; and the thousands of ruined factories now met with in many parts of the country bear testimony to the natural punishment that befell those who either could not, or would not, correct and reform their system of treatment of the rayats. The loss to the cultivator, as matters stand, is hypothetical. It is a loss of the difference in value of the indigo as compared with another crop. The assumption is involved that the land could or would have been required to grow that other crop. It might fairly be contended that the increased value of agricultural produce is the outcome of the multiplicity of crops brought about by foreign demands. The land is required for oilseeds, fibres, dyes, food-stuffs, &c. The surplus not absolutely necessary for food crops would naturally be thrown under the most profitable market crop. In many parts of Bengal, where indigo was formerly grown, jute has proved more profitable; but jute and indigo are only two of the crops by which the foreign markets have enhanced the value of the Indian soil and extended the area of cultivation beyond the good requirements of the people. The subsequent compensating influences of indigo cultivation, which cannot be shown on a balance sheet, have to be borne in mind, and are fully appreciated by the more intelligent rayats. The cash advance is a distinct gain, coming as it does at the time of the cultivator's greatest impecuniosity. Prosperity is not accompanied by increased rent, as is only too frequently the case with the ordinary zemindar. The crop is easily grown and requires little or no attention, thus enabling the rayat with his limited resources to cultivate a greater area. Mixed crops are frequently raised on the same fields, and

Advantages of Indigo cultivation. 196

CULTIVATION.

LOWER BENGAL.

for these (which afford many requisites of life) the same labour and expense would have to be incurred as are necessary for a joint indigo crop. The rotation of a leguminous stock such as indigo is essentially necessary. The roots penetrate to a great depth and draw their moisture from soil which the cereals never touch. They tend thus to bring a greater depth into bearing. The manure given to the indigo (and debited to the indigo balance-sheet) together with the cultivation producing on the surface soil all the effects of a fallow, greatly enrich the soil. It is in fact universally admitted that a better crop of wheat or rice is obtained after indigo than could possibly have been the case after a millet or other minor food crop. While the indigo crop is young, the fields are freely permitted to be used as grazing grounds by the cultivators, and as the cattle eat the weeds but will not touch the indigo, the advantages of this concession are mutual to the rayat and his European neighbour. The presence of a European capitalist in a rural district must also have its advantages. He will at least be a less exacting creditor than the village money-lender, for it is in his interest to keep his cultivators and their cattle in good healthy condition. His factory affords employment to a large community who are food purchasers, and even the *rayat* directly participates in some of the collateral advantages, for he owns cattle and carts and sometimes even boats, and gets paid to convey the produce of his tithe of indigo plant to the factory.

Cost of Cultivation. 197

Mr. Collin reports that in Bengal as a whole, the planter pays from R12 to R15 a bigha for indigo land, including rent of R3. Of that amount R4 to R5 is the actual cost of cultivation, and something is privately paid to the factory servants, leaving a profit to the *rayat* of from R4 to R7. The net profit on a good crop of rice on the same field would be about R8 to R10, but such a crop could not be continuously obtained, and if all land were under rice the profits would seriously decline.

N.-W. P. & OUDH. 198

II.—NORTH-WESTERN PROVINCES AND OUDH.

In the commercial returns of indigo, the produce of these provinces is referred to two sections—Benares and the Doab. The former includes Goruckpore, Ghazipore, Azimgarh, Benares, Jaunpore, and Allahabad, also the small tracts of Oudh and Rohilkhand that have indigo; the latter, all places above Allahabad in the inter-fluvial tract between the Ganges and the Jumna, extending to Meerut and Saharanpur, and south of the Jumna to Jhansi. This distinction is made on account of the difference in the quality of the indigo produced in the two regions, and it may be here pointed out that the former is well and tank, and the latter canal, irrigated. This difference is of considerable importance, for, while the weight of plant obtained by canal irrigation is heightened, the yield and quality of the dye is lessened and deteriorated. The valuable communication on indigo in these provinces received from **Colonel Pitcher**, Director of Land Records and Agriculture, contains the following brief passages regarding the cultivation of the plant:—"The number of PLOUGHINGS in the Doab rarely exceeds two, but in the Benares Division, indigo fields often receive as many as four ploughings. MANURE is very seldom used, but when possible indigo follows sugar-cane or cotton, and it thus derives some benefit from the manures applied to these crops. Experiments in the Government Farm at Cawnpore show that gypsum is a valuable manure for indigo as for other leguminous crops. For indigo sown in spring (*Jamana*) the number of WATERINGS vary from one to three according to the dryness of the weather. Indigo sown at the commencement of the rains (*Asarhi*) receives no watering. As already remarked under "Rotation," "Indigo often follows sugarcane or cotton, and unless kept for seed it is generally followed by wheat or barley the same year." Speaking of the rotation of

Ploughings. 199

Manure. 200

Waterings. 201

CULTIVATION. N.-W. P. & OUDH.

Indigo, Messrs. Duthie and Fuller (*Field and Garden Crops, page 45*) remark:—"The outturn of a *rabi* crop (wheat or barley) will not be above half what it would have been if grown after a fallow in the rains, but it is only with indigo that this much can be obtained unless the land be manured, the outturn of a *rabi* crop grown on unmanured land after millet or maize being extremely small." It is most commonly grown alone, as the period of its growth does not coincide with that of any other crop. Occasionally, however, it is mixed with *jaur* or *arhar*, and is surrounded with a border of castor plant, or of *san* (hemp) more with the idea of one crop insuring the other than with any hope of reaping the produce of both. A loam is preferred, but much of the cultivation is on the lightest possible sand in tracts where copious irrigation is possible from canal." "It is essential that the crop be kept free from weeds, and two WEEDINGS are the least that are given." Weeding. 202

Irrigation. 203

Extended Cultivation in Canal-irrigated Tracts; its Effects.—The Director of Land Records and Agriculture in these Provinces observes:—"A good deal has been written during the past few years about the pernicious effects of the method of indigo cultivation pursued in canal-irrigated districts. There is no doubt that the abundance and cheapness of canal water has enabled the cultivator to grow indigo as an extra crop on lands which, before the advent of the canal, ordinarily lay fallow after the *Kharif* or *Rabi* harvest until the setting in of the next rains. In the old days, when the sugar-cane or cotton crop had been harvested in January or February, the soil had rest for several months. Now an indigo crop is promptly put down, and on being cut in August is followed by sowings of cereals in October or November. The amount of manure received by the land is certainly no greater, and frequently less, than what was given in former times. Many observers have seen in these facts a clear proof of a reckless and soil-exhausting method of cultivation. It may, however, be doubted whether the effects are practically as bad as they at first appear. The long tap-roots of the indigo prevent the plant from trenching on the food-supplies of other crops which follow or precede it, and after the indigo is reaped the decaying roots and stalks when ploughed into the ground constitute a distinct accession of wealth to the surface soil-bed. This seems the only possible and natural explanation of the phenomenon that the sterility predicted in canal-tracts from the cultivation of indigo has not occurred in any material degree."

Seasons. 204

Seasons.—In these Provinces "Indigo may be sown either in the spring (March and April) or at the commencement of the rains. In the first it is called *Jamana* or *Chaiti*, in the second *Asarhi*. The former is ready for cutting in August, the latter in September. The stumps of *Asarhi* indigo are often left in the ground till the following rains, when they spring up again and yield what is known as a *Khunti* or *Ratoon* crop. The *Jamana* system is commonly followed in the Doab, but in the Benares Division much of the indigo is sown in June (*Asarhi*). In some parts of Gorakhpur indigo is sown in October with **Brassica juncea** (Indian Mustard, *lahi*) on alluvial land. The mustard is said to protect the crop in winter, but the area under this mixed crop is limited.

Sowing. 205

Method of Sowing and Quantity of Seed to the Acre.-"Seed is sown broadcast and is mixed up in the soil either by ploughing lightly with the native plough or by passing the long clod-crusher over the field. The quantity usually applied is 15℔ to the acre. It is essential for germination that the land must be thoroughly moist and the seed must not be buried deep."

Reaping. 206

Reaping.—"The plant is fit for reaping just when the flowers appear.

CULTIVATION. N.-W. P. & OUDH.

It is reaped with sickles, generally on contract, the cutters receiving one pice a cwt."

The Benares Division of these Provinces is practically an extension of the indigo tract of Behar—the most productive area. In that division indigo cultivation is conducted entirely with well or tank irrigation, but west of Allahabad (the so-called Doab of commercial returns) indigo follows the path of the canal. Thus in the Agra Division the canal-irrigated districts of Etawah, Etah, Mainpuri, and Farukhabad have much indigo, while Muttra and Agra, which till lately had no canal-irrigation, had very little indigo. So also Cawnpore, a district of Allahabad Division, with canal irrigation has three or four times as much indigo as Fatehpur and the Allahabad Districts.

Profits. 207 *Profit of Cultivation; yield of Plant and of Dye per acre.*—In the report furnished by the Director of Land Records and Agriculture, the following information is given regarding the cost of cultivation:—"The average cost of cultivating an acre of Indigo for plant to be followed by a *rabi* crop amounts to R15-8 in the case of *Jamana* indigo, watered from canal by lift, and R11-3 in the case of *Asarhi*. The average outturn of *Jamana* plant amounts to 80 cwt. and of *Asarhi* to 60 cwt. per acre. The price of plant varies considerably from year to year, and according as the bargain is struck at the time of sowing or at the time of harvest. The former is called *badni*; its price is generally fixed at a favourable rate to the buyer, who has, however, to advance the cultivator about half the stipulated value of his produce to enable him to get on with his cultivation. The latter is called *Khush kharid*, that is to say, the cultivator has the option of selling his produce. The rates paid to the cultivator under the two systems often vary more than R10 per cwt. The former system is dying out in many districts, for, though the factory owner apparently gets the plant at a cheaper rate, yet he often suffers heavy losses by cultivators leaving one village and settling in another whenever they find themselves in debt. In the long run the *badni* costs him no less than *Khush kharid* purchase. Taking the average price at R35 per 100 cwt., an acre of indigo yields a profit of R12-8 in case of *Jamana* and of R9-13 in the case of *Asarhi*." Messrs. **Duthie and Fuller** (in *Field and Garden Crops*) give the yield of plant for *Jamana* 80 maunds per acre and *Asarhi* a little less the first year but equal to *Jamana* the second, and these authors put down the cost of cultivation of *Jamana* land at R15-7. **Sir E. C. Buck** (in *Dyes and Tans of the North-Western Provinces*) says: "The outturn of Indigo varies: as much as 80 to 120 maunds of plant per acre may be cut for the factory: but the cultivator cutting for his own rough manufacture will perhaps cut as little as 50 maunds of plant, leaving the stalks for seed." **Sir Edward** gives the cost of cultivation by well irrigation at R41-2, by canal at R23-2, and, including the value of seed raised and of the subsidiary croos, the acre produce is given as worth R47-2, thus showing a net profit of R6 in the one case and of R24 in the other.

Yield. 208 As to YIELD of dye the Director of Land Records and Agriculture writes —"The yield of dye varies from 15℔ to 30℔ per cwt. of plant. It is highest in years of moderate or light rainfall and lowest in years of heavy rainfall." There would appear to be a clerical error in this statement, as in a further passage the yield is spoken of as three maunds of dye from 1,000 cwt. of plant or 1 in 500. Messrs. **Duthie and Fuller** say the yield of dye varies from 2½ to 4 per 1,000. That would be about 1 in 333 or from the acreage yield of plant which these authors give, 18 to 20℔ of dye would be obtained. In **Sir E. C. Buck's** work the opinions of several planters are given as to the yield of dye. A vat holding 75 maunds gave as a rule 6 to 7 seers of dye, *e.g.*, 14℔ from 6,000℔ or 1 in 428. Another

CULTIVATION. N.-W. P. & OUDH.

writer gives his result as 12 seers of dye from 100 maunds, *e.g.*, 1 in 333: another 100 maunds of plant to 8 seers of dye, and still another 119 maunds to 8 seers. One writer gives the results of careful observations extending over 13 years, in which the mean yield was 1 in 426 for European factories and about 1 in 300 for Native. **Mr. Liotard** speaks of natives obtaining ⅓℔ of dye from 1 cwt. of plant or 1 in 224, while **Mr. Reid** (in his *Culture and Manufacture of Indigo*), who says little of a definite nature regarding the yield of plant or dye in Bengal, gives the average in the North-West Provinces as 300 maunds of plant to one maund of dye. Speaking of the impure indigo or *gád*, so largely prepared in the North-West Provinces and many other parts of India (for local use), **Mr. Reid** remarks, 100 maunds of plant will as a rule yield the planter 2⅓ maunds of *gád*. He speaks also of one *katcha bigha* (about ¼ acre) as yielding 15 to 20 maunds of plant. From these facts it would appear that the acreage yield of dye would be about 18℔.

III.—PANJÁB.

PANJAB. 209

Indigo is said to be grown more or less in all districts except Simla, Kangra, Amritsar, Gurdaspur, Sialkot, Gujranwala, Jhelum, Rawalpindi, Peshawar, and Kohat. The chief seats of the Panjáb indigo industry are, however, in Multan, Muzaffargarh, and Dera Ghazi Khan. It is entirely grown on irrigated land, but a strong prejudice exists against excessive irrigation; hence lands subject to river inundation are considered unsuited. **Mr. Baden Powell** gives much interesting information regarding Panjáb indigo, and mentions that it is produced in two of the districts above (Sialkot and Amritsar), which in recent official reports are shown as not producing it. **Mr. Baden Powell** gives the seed per bigha at 6 seers for the early and 8 seers for the later sowings. The Director of Land Records and Agriculture has furnished a report which embraces the main facts obtained from the three chief indigo districts. On the subject of methods of cultivation he writes:—"The land is usually prepared during the COLD SEASON after the winter rains. It receives about four PLOUGHINGS before the seed is sown and none after. The more labour thus expended the better the crop; but there is no ploughing for the second year's crop obtained off the stems left in the ground from first year's crop.

Ploughing. 210

Sowing. 211

Seasons of sowing.—"The seed is sown from 1st March to 15th May. It is customary to make fresh sowings every year, but sometimes a second or even a third year's crop is obtained off the same plants. The crop takes from three and a half to four months to ripen before it is cut. To sow the crop the field is first flooded and the seeds then scattered broadcast on the water. This practice is followed to ensure the seeds sinking into the ground, and that none may remain exposed to the sun. For a month after sowing the field is irrigated every third day or until the plants are a foot high. Irrigation is then given every eighth or tenth day. When the plants are young water is given only in the evening, and then only sparingly, for fear of their rotting from standing in water heated by the sun. Even when older and stronger, the nicest discrimination is required in regulating the supply of water. An over-supply causes the leaves to turn yellow. Continued heavy falls of rain do much injury and often destroy the crops."

Irrigation. 212

"For the second year's indigo, less irrigation will suffice, the crop being watered every eighth or tenth day; but the watering must take place early, if possible by the 15th April. First year's indigo requires in all from 18 to 20 waterings, second year's only 13 to 14."

Manure. 213

Manure "is not generally used."

Weeding. 214

Weeding "is, however, common either by hand or by causing sheep or goats to pasture on the fields."

CULTIVATION. PANJAB. Reaping. 215 *Reaping*—" Indigo is ready for cutting from 15th July to 15th September. It is in its prime when it has been twelve to fifteen days in flower. If the flowers fade and become yellow before it is cut, the outturn will be small. It is cut in the morning and carried in bundles to the vats where it remains till the afternoon."

SIND, &c. 216

IV.—SIND, RAJPUTANA, CENTRAL INDIA, AND THE CENTRAL PROVINCES.

According to **Dr. Stocks** a portion of the Sind Indigo is obtained from two wild species of **Indigofera**, *viz.*, **I. cordifolia**, *Heyne*, and **I. paucifolia**, *Delisle*; but the cultivated plant, a *rabi* crop, is **I. tinctoria.** He urged the necessity of an effort to have the plant grown by European planters, and seemed to have been confident that Sind might hope to recover its lost industry were this done. In a file of official correspondence conducted in 1851, useful information will be found regarding Sind Indigo. Regarding the Khairpur Territory, **Mr. J. F. Lester** then wrote that the indigo season commences with the *kharif* sowings, about July, and extends over three months. The principal desiderata, he says, are " abundance of water " and " a careful and experienced manufacturer." It is estimated, he continues, that one *kusa* of seed suffices to sow a *bigha* of land. A bigha produces 12 seers of indigo, which is valued at the average rate of R65 per maund or R1-10 per seer. **Mr. George Wood**, Collector of Shikarpur (in the papers alluded to above) furnishes a most instructive report, accompanied with plans and estimates, of indigo cultivation and manufacture. For a thousand bundles of plant **Mr. Wood** says the cultivator receives R125, and that quantity would yield from 6 to 7 maunds of dye. The Commissioner, in reviewing these reports, stated that there could be no doubt as to the capabilities of Sind to produce indigo. In the Gazetteer of Sind indigo is frequently mentioned, but of the Frontier District of Upper Sind it is said— " Indigo was first cultivated in 1859, and may now be considered one of its chief products, as the soil of the Jacobabad Taluka in several parts is said to be admirably adapted for the cultivation of this dye." **Mr. Liotard** writes of Sind as follow :—" During the inundation season, generally in May and June, the land being weeded is moistened. It is then ploughed two or three times, and the seed, after having been steeped in water on the previous night, is sown broadcast. Each acre requires three *kasas* of seed ; after sowing, the land need not be watered for three days or until the seed germinates. Vegetation having begun, the young plants are watered once a week ; and this is continued for ten weeks or till September, when the crop is generally ready to be reaped. A second year's crop is raised after cutting by watering the old plants about thirteen times during the same space of time. In this second year, besides indigo, the seed is also obtained, by leaving at the time of reaping one plant untouched at every four or five feet distance. These solitary plants are watered for two months more, or till they begin to flower." Sind indigo is alluded to as the only kind used in the Nasik District. It is said to be procured from Bombay and to occur in irregular conical cakes, the better specimens of a good blue but most of a hard black, or pale blue (*Nasik Gaz., p. 170*).

RAJPUTANA. 217 Throughout the whole of the records regarding the provinces named above, repeated mention is made of wild indigo being regularly used by the local dyers. The writer made an effort to procure samples of these wild indigoes, and in two instances (from Rajputana) a variety of **Tephrosia pauciflora** or **purpurea** was furnished as wild indigo. At Palanpur in Gujarat the same plant is reported to be extremely abundant, and the

CULTIVATION. SIND.

Political Agent is believed to be presently endeavouring to establish there an indigo industry. In Merwara Mr. Duthie found four species of **Indigofera** growing wild; of **I. argentea** *var.* **cœrulea**, he says that it is on the Aravalli hills known as *níl*. The extent to which indigo cultivation is actually carried on in Rájputana cannot be discovered, but in the Gazetteer it is mentioned in a list of the chief *khárif* crops of Bundi. In a special report on the dye-stuffs of Ajmír, Pandit Bhag Ram says: "The average approximate extent of indigo cultivation for the last five years may be stated at about 500 *bighas* per annum. Wild indigo is more or less spontaneous in all the forest tracts. It is cultivated chiefly in the lands of Srinagar, Rámsar, Gangwara, and Kikri by native dyers, to meet their own requirements. In the hill jungle of Rájgarh it is found to grow spontaneously, which would seem to indicate that the land is fully adapted to the cultivation of indigo; but it is to be regretted that the agricultural classes take little or no interest owing to the difficulties with which the dye is extracted. This circumstance accounts for the scanty supply raised in the districts and the total absence of regular indigo factories." The Pandit then states, however, that the cultivation leaves a "very small margin for profit." He remarks that it is grown on *baráni* land, which is ploughed twice or thrice and then levelled. "The seed is sown broadcast early in the month of June, or as soon as the rains set in. When the young plants are above ground, the grass that grows with them is weeded out by the process of hand-hoeing, which is performed about four times. The reaping season commences in October.

CENTRAL PROVINCES. 218

Of the indigo of the CENTRAL PROVINCES very little has been reported. It is mainly grown in the districts of Sagar, Damoh, Narsinghpur, and Nimar, but it seems probable there are not more than 150 acres annually under the crop. The method of cultivation appears to be similar to that in the North-West Provinces. In Sagar the sowings take place in July or August, and the plants are cut when they attain the height of 2 to 2½ feet. In Damoh and Narsinghpur the fields are ploughed twice in June; the seed is sown broadcast. The plants are cut in October, and dried and stored. The roots are, however, left in the ground and yield a second and a third year crop, after which they are dug out. In the Upper Godavari, the seeds of the wild indigo, which abounds in this district, are sown at the beginning of the monsoon, and the leaves are picked during the month of September. These plants yield dye leaves for three seasons, and thereafter fresh ground is cultivated. The ground is ploughed twice and is not manured. Mr. Liotard, from whose report on Dyes and Tans the above passages regarding the Central Provinces have been compiled, adds that, according to the local officer, "the leaves of the wild indigo, in the neighbourhood of towns and villages, yield a finer dye than the cultivated indigo, because the plants grow up much further apart when left to themselves and are more luxuriant."

The writer has had occasion to point out that such references as the above to wild plants may probably, when carefully investigated, throw much light on the plant, which might be called the original *Nílá* of Sanskrit authors, but there is nothing to justify the opinion that they would prove to be **Indigofera tinctoria.**

V.—BOMBAY.

BOMBAY. 219

In 1787, Dr. Hove described the cultivation of Indigo in Gujarat as follows: "The indigo was partly inter-sown with cotton, and on some plantations with millet and other grains. The lines were divided about 16 inches from each other, in which the cotton shrub stood pretty thick; the above-mentioned grains were scattered without the least regularity. I

INDIGOFERA tinctoria. **The Indigo Plant of Commerce.**

CULTIVATION. BOMBAY.

understood from the planters that they suffer the indigo to grow for two seasons and commonly have three crops a year. The first crop was already removed (November 25th) and on the lower plantations the second was just being cut. The third crop is inferior and is not ready before the hot season sets in." But according to the author of the Broach Gazetteer, even so early as 1820, this Gujarat industry had all but ceased to exist. Of Cambay it is said, "the cultivation of indigo has of late greatly fallen off. Hindu peasants dislike growing it, because in making the dye much insect life is lost, while the Muhammadans, with whom this objection has less force, do not till land enough to raise any large quantity. Sown towards the end of the hot season, indigo is harvested in August before coming to flower." Of Kaira it has been reported: "The small quantity now produced is grown in light or *gorát* soil. At the first fall of rain (June) the field should be ploughed more than once, and if possible manured. The seed is sown in DRILLS from the *tarphan* or drill-plough. After the plants have come up constant weeding is required. The crop reaches maturity in September, and in gathering the leaves great care must be taken that they are not exposed to wet" Of Ahmedabad it is said that, owing to the destruction of insect life and low price obtained for the dye, the cultivation has almost ceased.

In the Deccan, indigo seems at the present day to be mainly grown at Khandesh. "A two-year, and sometimes a three-year crop, is grown to a very small extent, owing to the great expense of preparing it for the market. The seed is sown in July in carefully tilled ground. It can be thrice cut during the rains and lasts two and sometimes three seasons generally without being watered. On account of its mixture with wood ashes, Khandesh indigo classes rather low. The first cutting takes place when the plant is two or three months old: the second year another crop of leaves is cut from the shrub, which is then considered useless and is generally destroyed by ploughing up the land and preparing it for some other crop. Some cultivators let the plant remain in the ground a year longer in order to get a third crop, but the yield is too poor to be remunerative. In the neighbourhood of Faizpur good indigo is raised in considerable quantities by Gujars." "Formerly large quantities were imported from Gujarat. But of late the manufacture of Gujarat indigo has almost entirely ceased, and Khandesh indigo now goes to Surat and other Gujarat markets."

The Director of Land Records and Agriculture, in his *Report for 1885-86 (pp. 36 and 37)*, gives a brief history of the Bombay indigo industry, narrating the decline from the period of Dr. Hove's visit to Gujarat. These facts have already been indicated in the above quotations from the Gazetteers and will be found to have been further elaborated in the paragraph under the heading of "History of Indigo." There are certain features of the Bombay indigo trade regarding which it would have been specially interesting to have obtained definite information, not from a historic point of view only, but with a possible beneficial influence on the efforts to resuscitate the industry. Certain writers allude to **Indigofera tinctoria** being a wild plant in Gujarat. This point has, however, not been definitely determined. At the same time it seems probable that the indigo plant of Western India grown in the fifteenth century and earlier was or may have been a **Tephrosia,** not an **Indigofera,** and that the plant of later date was, in all probability, **Indigofera cœrulea.** Should this supposition prove well founded, it would be desirable to ascertain whether **I. tinctoria** is equally suitable to Bombay as to Bengal, and if not, what other indigo-yielding plant should rather be grown. At the same time it is to be regretted that more precise information as to the methods of cultivation pursued in the districts of Bombay, where indigo is even now grown, has not been forthcoming. In one district, Kaira, it is drill-sown, a practice carried to perfection in Tirhút and only occasionally

CULTIVATION. BOMBAY.

followed in the other indigo localities. There may be other peculiarities still surviving of the early methods of cultivation and manufacture that would prove instructive. It is significant of the migration from Bombay to Bengal, that a similar migration has recently taken place from Bengal to the Upper Provinces. In Tirhút and the Benares Division of the North-West Provinces the greatest yield is obtained, and in these regions indigo cultivation is not only carried on more extensively but on more scientific principles. It would thus appear as if **Indigofera tinctoria** had only just reached the region of India most suited to it—a region where there is no record of its ever having been grown by natives prior to the European modern efforts, still less of its ever having existed as a wild plant. The inference is, therefore, pardonable, that had it been indigenous to India it would most probably have existed from time immemorial in Tirhút and Benares.

Experiments in cultivating indigo have recently been made at Junagad in Kathiawar, and the results seem fairly hopeful. That the Bombay Presidency could produce good indigo seems abundantly proved by the large exports that once took place—and these the earliest recorded exports of the dye as made by the Honourable East India Company—from Surat, Ahmedabad, and Bombay nearly three centuries ago.

VI.—MADRAS.

MADRAS. 220

The Director of Land Records and Agriculture has furnished a series of most instructive reports on the Madras Indigo industry, from which the main facts given below have been derived. In one of these **Mr. Macleane**, Collector of Nellore, writes :—" Indigo cultivation extends along the east coast from Kistna to South Arcot and inland to Kurnool and Cuddapah. Cuddapah is the principal district for indigo : next Nellore. The crop requires a rich friable soil, neither too moist nor too dry. Bright sunshine favours the development of the dye principle, but frequent rains cause a more luxuriant growth. It is generally cultivated as a dry crop. In some parts, it is sown mixed with millet crops, and after the latter are harvested, the indigo shoots up. In dry land, as a rule, one cutting is obtained in October and a second in January. Indigo is also, however, grown on wet land. Here two cuttings are certain and sometimes even a third. It is not unfrequently sown in paddy fields a few days before reaping. On whatever soil sown, the crop, so far from exhausting, is believed to strengthen the soil. The first step in its culture is to plough and manure the ground. A flock of sheep is sometimes enclosed on the ground. At other times the ground is lightly scattered with wood-ashes, cow-dung, &c., or the refuse of the indigo plant from the steeping-vat, or its ashes, are scattered over the fields. The seed is generally sown broadcast. The plant in favourable damp weather, or if lightly-irrigated, makes its appearance above ground in four or five days and ripens in about two months and a half. Great care is taken to weed the plantation thoroughly and to keep the ground moist, but not too wet. When the seedlings attain two or three inches, the fields are weeded. The leaves are cut just before the plant flowers. The first cutting takes place when the crop is three months' old, after which the field is lightly ploughed. The other cuttings follow after intervals of two or three months and are better than the first. The leaves are not gathered after heavy rain, as this washes off from them the farina which constitutes their chief value."

Other two reports give particulars regarding South Arcot, Kistna, Cuddapah, Kurnool, and Vizagapatam. Regarding Cuddapah, **Mr. Sewell** writes :—

Dry cultivation. 221

Dry Land Cultivation.—" After the land has been moistened by

CULTIVATION.

MADRAS.

rain it is ploughed and manured. If the April rain is sufficiently heavy this is done then. If not, in July or August. The ground, loosened by ploughing, is allowed to remain unsown for a few days until the next heavy showers when the seed is sown in rows by drills followed by a roller. It sprouts in about three days. In about a month weeding commences, which is done by labourers with a sort of spud. On dry lands the crop is entirely dependent on rainfall. The first cutting is taken from three to five months after sowing, the interval being short if rain is abundant and frequent, and longer if it is scanty. At intervals of about three months after the first cutting of the leaves a second and third cutting is taken. It is not usual to attempt to take more than three crops, but the plant is allowed after the third cutting to seed."

Wet cultivation. 222

"*Wet Cultivation.*—This is either under tanks or wells. Under wells, the water supply of which is certain, the cultivation is commenced in March or April. The land being constantly under cultivation, if the soil is loose the seed is sown without any previous ploughing, which is, in such cases, thought injurious: otherwise under tanks the land is watered, then ploughed and smoothed by a roller. The land is then manured either with cow-dung or by penning flocks of sheep on it for two or three days. It is then watered, and when it has dried again (*i e.,* in three or four days) the seed is sown. It sprouts in about three days. If it does not do so the land is watered once to make the seeds germinate. The crop is then watered at regular intervals of not less than a week, or more than twenty days. The intervals depend on the rainfall and the consequent state of the soil and atmosphere. Weeding commences a month after sowing, and continues at intervals of a month. The first cutting is taken three or four months after sowing, and the second three months later; when the circumstances permit the crop to be sown early enough, two cuttings are taken and the crop is then ploughed in, and a food-crop, usually paddy, grown. Indigo is not grown continuously on the same land but is alternated with paddy. The refuse of the indigo-vats is much prized as a manure both for indigo and other crops."

Having thus dealt with the methods of cultivation in the two chief indigo-producing districts of Madras little more need be said. To complete the account, however, a few passages may be here given from **Mr. C. Kough's** description of Indigo cultivation in the district of Kurnúl which may be taken to represent the districts in the Presidency where indigo is of less importance than in Cuddapah and Nellore. The plant, **Mr. Kough** says, is a biennial, but is sometimes kept growing for three years. It is sown in black cotton unirrigated soil in July and August after the first showers. "In such lands it grows luxuriantly, and there are often as many as five cuttings lasting into the next season, *viz.*, two in the first year and three in the second. Indigo is also grown on land artificially irrigated. When it is cultivated in paddy land, it is sown in January or February, after the first paddy crop has been harvested, and one or two cuttings are made before the end of the *fasli,* and a third cutting in July and August. After this the stumps are ploughed up for manure and paddy planted; the rayat, moreover, almost invariably gets back the refuse after steeping, and this also he uses as manure." Speaking of the dry land cultivation, **Mr. Kough** further remarks:—"The time of sowing must be selected with care, for the chance of a heavy fall of rain caking the surface of the ground and thus preventing the young germs from coming through is considerable, but this difficulty having been got over, when once a fair start is attained, there is scarcely any crop more hardy and less susceptible of damage by drought. It escapes damage from grazing of cattle, sheep, or goats, as none of these animals will touch it. Thus the soil, climate, and season are the three

CULTIVATION. MADRAS.

factors that affect the prospects of the crop. The indigo grown on dry black cotton soil yields dye of a superior quality. The winter crop yields more prolifically than the summer crop. "Weeding," Mr. Kough remarks, "is sometimes done by the drill, at other times by hand. The crop is cut four months after sowing when the plants begin to blossom. Two acres fair growth yield about three tons in the first cutting, and will give four or five cuttings, of which the first and the last are usually the least heavy." He gives the cost of cultivation at R14, the price paid for the produce as R20, and the profit, accordingly, as R6 per acre.

Cost. 223

Cost of Cultivation: Yield of Plant and of Dye.—In the official reports of the Saidapet Farm a slightly improved process of cultivation is recommended in which the crop is drill-sown after deep ploughing. The danger of the surface soil getting hardened by evaporation after rain, before germination sets in, is strongly urged. If the sowing takes place in November, the crop will be ready for cutting by the beginning of March. An acre at the Farm was found to yield 200 bundles weighing 12,000℔, and to be worth to the cultivator about R30, giving R10-3-0 an acre as net profit. An acre of land yielding 12,000 ℔ of plant was believed to afford 2 maunds of dye worth R90, but the report on Cuddapah quoted above states that a fair yield of dye would be 1½ to 1¾ maunds per acre. Of the Kistna district it is said that the net profit per acre on two years' cultivation of indigo plant is R20-8-0. In Nellore 100 bundles of plant are said to give one maund of dye; and in the Kurnúl report it is said that ½℔ of dye is obtained per cwt of plant. In Bellary, however, the yield of dye is given as only ½ maund per acre. The Madras maund equals 24·68℔ or in mercantile usage 25℔. It seems probable that these estimates of yield of dye to the acre of land are very much overstated even were a deduction of 50 per cent. made for weight of adulteration. The yield for the Presidency obtained by dividing the total production by the ascertained area is only 9·5℔ an acre, and this is probably not far from correct.

VII.—BURMA & ASSAM.

BURMA. 224

In a communication from the Political Agent, Mandalay, dated 1872, it is stated that Indigo cultivation was first commenced in that part of Burma in 1860, when the King of Burma procured manufacturers from Bengal to teach his subjects the art of preparing the dye. Four factories were established. His Majesty required his subjects to sow a third share of their land within the vicinity of the factories when the land in question was found suitable. It is remarked that "Indigo made from the Burmese plant, indigenous to the country, is found not to be equal to the Bengal Indigo." Seed was accordingly annually imported. The Chinese merchants of Mandalay purchased the indigo produced, and exported it to the Shan States and China. It is said to have fetched R100 per 4½ maunds. Dr. Mason (*Burma and Its People*) says that **Indigofera tinctoria**, the *mai-nay* or *shan-mai*, is much less extensively used as a source of blue dye than **Strobilanthes flaccidifolius**, the *mai-gyee*, or than **Marsdenia tinctoria**. An "indigenous shrub," a species of "**Indigofera**," he remarks, is also sometimes used.

In a recent communication on the subject of Indigo cultivation in Burma, it is said to be grown in the Lower Chindwin district. "There are two crops—the wet weather and the dry weather crop. It grows anywhere. The ground is ploughed up and the seeds scattered over it. The wet weather crop is sown in June and collected in July and August, and the dry weather crop is sown in October and collected in December and January." In the Mandalay district it is reported to grow wild "and the plant is never used for extracting the dye. It is employed medicinally

CULTIVATION. BURMA.

by Burmese doctors for the cure of urinary diseases in men and ponies." It may be inferred from this brief notice of Mandalay that the factories started by His Highness the King have since been abandoned. Of Tharawaddy the report states that "Indigo is cultivated to a very small extent. It is planted in July and reaped in October. The soil is first burnt." In Mergui "it grows wild and is used by the Karens for dyeing purposes." In Amherst "the wild plant is used for dyeing to a trifling extent." In the Eastern Division of Upper Burma "it grows wild extensively and is used medicinally only."

It would seem from these brief notices that Burma affords a hopeful prospect of becoming a good field for Indigo cultivation. Large tracts of rich land are uncultivated, and were European planters to open out factories, an emigration of coolies, such as has taken place to Assam, in connection with tea, might be possible, and if so, the cultivation of indigo would prove highly beneficial to the country as well as remunerative to the pioneers of the new industry.

ASSAM. 225

In Assam and Manipur, **Strobilanthes**—the *Rúm*—is regularly cultivated, very much after the same system as in North-West China. The writer took the opportunity, in his *Calcutta International Exhibition Catalogue*, to urge on the Indigo planters the desirability of their experimenting with the cultivation of the *rúm* plant as an additional source of Indigo. He then wrote:—"In many respects it possesses properties eminently suited for a profitable indigo crop, and in China at least the dye is pronounced finer than that obtained from any other plant. It is propagated freely by cuttings, yields prunings twice or three times a year, and is perennial. It would give little or no anxiety to the planter, and if not sufficiently remunerative to take the place of the Bengal Indigo plant, it seems natural to expect that the two plants might with great advantage be cultivated together. The *rúm* would flourish on the higher dry tracts and yield its crop at the seasons when the present indigo factory is silent." Since publishing the above, now some six years ago, the writer has come across many facts to strengthen his convictions, and is satisfied that the greater portion of the Indigo of Assam is the produce of **Strobilanthes flaccidifolius** not of **Indigofera tinctoria.**

CULTIVATION OF SEED.

SEED. 226

Bengal may be said to derive its annual supply of seed from the North-West Provinces. The earliest records of the Bengal Indigo cultivation allude to this fact, so that it seems to have been soon discovered that certain tracts of country favoured the production of leaf while others were better suited for seed cultivation, and in fact that a much superior crop of leaf was obtained from plants raised from specially cultivated seed than by allowing a few of the shrubs in the indigo fields to ripen seed for next year's sowings.

In the report furnished by the Director of Land Records and Agriculture, Bengal, frequent reference is made to the subject of seed production. The following passages may be specially quoted:—

"The seed grown at Kotechandpur, known also as *deshi* (*i.e*, local) seed, is adapted only for high land above the inundation level, requiring five or six ploughings. The crop is harvested late, in Jessor the harvest from Purneah and *deshi* seed going on to September. The Purneah seed, like the *deshi*, is generally used for high land, but it is also adapted for *chur* land. Like the *deshi*, the Purneah seed gives a late crop. The seed from Patna and Cawnpore is adapted for *chur* and *dehra* lands, which receive an annual top-dressing of silt at the time of the overflow of rivers. These lands are either not ploughed at all, or ploughed only two or three times before the Patna or Cawnpore seed is broadcasted. The crop is

CULTIVATION.

SEED.

ready early, *i.e.*, about the middle of June. The Madras seed yields a crop even earlier than that from the Patna and Cawnpore seed. But this is rarely an advantage. The factory operations are dependent on there being plenty of clean water ready at hand, and in May, when the crop from Madras seed is ready for cutting, the rivers are still dry.

"The price of seed varies very much, from R4 to R40 per maund."

The Collector of Midnapore reports:—

"No indigo seed is produced here. The rayats are given 6 seers of seed for each bigha. Last year the seed cost R15 a maund: this year only R5. The high price having induced many to produce it, the supply was greatly increased: hence this decline in price."

The cultivation of seed in the Gya district, and in the Purneah district, is reported on as follows:—

"Seven seers of seed are sown per bigha in the Gya district. In October a portion of the *Asarhi* and *Jamana* are left standing for bearing seed. The plants are kept apart at a distance of 6 inches for this purpose. There are two kinds of indigo seed sown—that grown in the locality known as *dasi*, and that of the North-Western Provinces, known as up-country; the latter being sown more extensively, especially in the chief concerns to the south and south-east and south-west of the station of Purneah, where little or no *dasi* seed is sown. The north and north-west and east of the district have few working factories, and in these the *dasi* is sown in high lands only. For a seed crop it is grown to the west and north-west only; sown in July and August; reaped in December; sells in January; prices vary according to the demand, averaging about R4 to R5 per bazar maund."

North-West Provinces.—The Director reports as follows on the subject of seed cultivation:—"When the plants remain stunted and do not promise to fetch much by their weight they are reserved for seed and cut in December, when they yield about 6 maunds of seed to the acre. The stumps of a *Jamana* crop are also sometimes left in the ground for seed, after the stalks have been cut, and yield about 4 maunds of seed to an acre."

Madras.—**Mr. H. G. Turner**, Collector of Vizagapatam, writes:—"As the lands in which Indigo is cultivated are required almost immediately after for paddy, there are no opportunities of raising a crop for seed purposes." **Mr. C. Kough**, Collector of Kurnúl, reports:—"Seeds are collected in January and February from plants sown in the beginning of the monsoon, and those sown during the previous year. The latter yields more seed. The plant, when the leaves have been cut after the monsoon is over, is allowed to grow for the purpose of yielding seed. With the seed the leaves are also collected and allowed to dry, and when the seed is separated, indigo is extracted from the dry leaf also."

DISEASES AND INJURIES.

DISEASES. 227

In a recent circular, calling for information regarding the Indigo cultivation and manufacture as presently practised (which the writer issued through the Revenue and Agricultural Department), the following replies were received regarding the above subject:—

Bengal.—**Mr. N. G. Mookerji** reported to the Director of Land Records and Agriculture that "the crop suffers from the following causes:— In May and June indigo plants sometimes drop their leaves. This is probably due to droughty weather, but it is only some plants that behave in this way.

(2) A green caterpillar, about 1" long, when of full size, locally called *malpoka*, appears when plants are mature. The appearance of this caterpillar is in fact a sign of the maturity of the leaf for cutting. If, however,

DISEASES.

the harvest is neglected for a few days, the caterpillar does great harm by eating away the leaves.

(3) A big geometridæ caterpillar, 1½" to 2" long, ¼" in diameter when full sized, occasionally appears just before harvest, and does considerable damage. The caterpillar may be just noticed some afternoon for the first time, and the leaf of a whole bigha of flourishing crop may disappear before the next morning.

(4) High winds, hailstorm, constant loading and unloading after harvest, soaking in water, or any other circumstances that spoils the leaf, injures the crop for the dye. To get the best result the largest quantity of leaf must be gathered when mature and taken as quickly as possible to the factory house.

(5) The crop is very liable to blights of different sorts, and is also injured by extremes of either drought or wet. It is also liable to injury from insects at different periods of its growth, to injury from hail, and also to injury from severe storms of wind.

(6) The indigo is a very hardy plant, but owing to the length of time required to bring it to maturity—'a period extending, in the case of the crops sown in October, to from eight to nine months'—it is very liable to injury from drought and from the high winds which are prevalent during the months of April, May, and June. A very severe gale of wind lasting some days, will almost entirely destroy a crop of indigo, stripping it of its leaf, and consequently rendering it useless for manufacturing purposes."

The Collector of Midnapur reports :—"The crop is liable to very few diseases. The spring crop is sometimes eaten by caterpillars and grasshoppers. The autumn crop is too old to be generally attacked by these insects. The quality of the crop, however, is sometimes destroyed by an untimely drought in June, when all the leaves get scorched up after the plant has once started its second growth, or by excessive rain which washes the colour out."

North-West Provinces and Oudh.—The Director of Land Records and Agriculture reports that, "In the early part of its growth it is liable to the attacks of a winged insect called *Gadlie* which feeds on the young leaves. Another source of damage to the plant is continued wet weather, which renders it tall and woody without much foliage, and by a kind of etiolation prevents the proper development of the dye property in the leaves."

Panjáb.—The Director in this province writes :—"As above stated too large a supply of water is fatal. Heavy and continuous rain is also injurious. The plant is also subject to be attacked by certain insects resembling blight and by the grass-hopper. But if rain falls in moderate quantities the insect dies out. Another disease is also common in its earlier stages from the 15th to the 30th day, brought on by excessive heat and hot winds, which cause the plant to dry up and wither away. Wet-fogs too are very injurious."

Madras.—**Mr. L. M. Wynch**, Kistna district, writes :—"The indigo crop is said to be liable to the ravages of several insects, amongst them being locusts, the *gongali purugu*, and *kambali purugu* (caterpillars). The rayats have no means of destroying these insects. The *Buddi tigalu* attacks the crop after it has grown up 6 or 9 inches; if this occurs, the crop is ruined and has to be left till the following year." **Mr. E. J. Sewell**, Collector of Cuddapah, in his reply states that, "At the age of two months the plant is liable to be attacked by diseases known as *Budide-rógame* and *Agui-mandalaputhegulu*. In the former case the leaves turn white, in the latter black, and in both they fall off." **Mr. C. Kough**, Collector of Kurnúl, alludes apparently to these diseases, but he speaks of a third form :

DISEASES.

when subject to the first disease the leaves contract and gradually fall off: when to the second the leaves turn white and the plant ultimately decays: when to the third the leaf is covered with white* matter and the plant dies. "While still young, the plant is subject to the attacks of caterpillars. They attack it when germinating. When the plant attains a height of three inches it is almost free from the insect pest."

YIELD, AND AREA UNDER INDIGO.

AREA & YIELD. 228

With perhaps no other commercial commodity does the inconvenience of a want of uniformity in weights and measures tell more injuriously than with indigo. The *Bígha* of land in Lower Bengal is one-third of an acre; in some parts of Tirhút it is six-sevenths; in the greater part of the North-West Provinces five-eighths; and in many other provinces three-quarters of an acre. The factory maund of dry indigo is 74℔ 10 oz.; the Madras maund 25℔; the Bombay maund 28℔; the Government maund for Bengal, the North-West Provinces, and the Panjáb, 82℔. In this connection it may be remarked that it has been found impossible to discover whether the green indigo plant is purchased in factory maunds or in Government maunds. The calculation of yield of plant to acre has been made on the assumption that the maunds of Upper India given by writers on the subject are 82℔. If this be found hereafter to have been incorrect, an alteration from 82 to 74⅝ will have to be made, but this would not affect the yield of dye, which is estimated on a relation of quantity to a unit. The assumption of the maund being equivalent to a cwt., adopted in certain compilations, has led to the greatest confusion and gross misrepresentation.

The Bigha. 229

The Maund. 230

In working out the indigo statistics given in this article, the attempt has been made to reduce areas to acres and weights to pounds, but the writer is conscious he may have made some mistakes from the difficulty experienced with local returns in determining the system pursued by the original authors. He has, however, corrected these weights and measures with some degree of care, and is hopeful that his results will, on the whole, be found to convey a fairly accurate statement of the indigo industry of India. With regard to yield of dye per acre of crop the very greatest difficulty has been encountered. This has arisen not only from the causes indicated above, but through numerous inaccuracies in the returns, and from a possible desire to avoid definite statements on the part of those most interested. For example, the range of variation of yield of dye has been found to be from 4℔ to 150℔ an acre. Such a discrepancy is obviously due to inaccuracy in compilation or to want of definite information, and not to differences of climate and soil, or method of manufacture. The writer has arrived at the conclusion that a yield of 12℔ an acre for Lower Bengal (and the other non-irrigated tracts, indeed, the greater part of India), and of 18 to 20℔ for Behar, and certain parts of the North-West Provinces and of the Panjáb, would be fairly safe standards.

Dye to Plant. 231

It seems probable that the yield of plant per acre is in direct ratio to the degree of irrigation, but that the quantity and value of the dye obtained is in inverse ratio. This rather complicates matters, but it accounts largely for the greater quantity, though inferior quality, obtained from the canal-irrigated tracts of the North-West Provinces and the Panjáb. A further fact of great importance is that with all Native factories, increased yield is due to two causes: (*a*) the extraction from the plant of a greater amount of organic but non-indigo materials, and (*b*) the

Plant to Acre. 232

* [Probably a species of Periospora.—*Ed.*]

AREA & YIELD.

systematic adulteration that is practised. The greater weight of indigo obtained from an acre of land by the Native as compared with the European methods of manufacture has its explanation in these facts, but its consequence in lessened value. After a consideration of all the facts brought to light by the numerous reports and publications on the subject, it seems safe to infer that the production of indigo by the European factories varies from 1 in 340 to 1 in 640, the average being about 1 of dye from 420 of plant. These extreme variations are due to two causes—the difference in the dye-yielding property of one year's crop as compared with another, and the difference already alluded to between indigo plant from canal-irrigated land and that from non-irrigated land. With native manufacturers the range is 1 in 220 to 1 in 330. Accepting these figures as factors to estimate the yield of crop per acre, we obtain for Lower Bengal 420×12=5,040℔, and in Behar and many parts of the North-West Provinces 420×20=8,400℔. These results are dependent doubtless on test cases (obtained by well-known factories) and may not be applicable absolutely to every case. On turning, however, to the published figures of actual yield of green plant to acreage, it is found that the above estimates correspond remarkably. Thus, the average of all returns gives the yield as ranging from 50 to 80 maunds, the maximum being 120 maunds to the acre. These returns would therefore be 50×82=4,100℔; 80×82=6,560℔, or 120×82=9,840℔ of green plant to the acre of land. The lesser quantity of plant required by the native manufacturers to produce a unit of dye, has been already explained; and it therefore seems preferable to accept the published results of European factories for all estimates of outturn per acre. But the application of this system of calculation can be made with safety only where the percentage of native manufacturers is low, and irrigation not extensively practised. For example, it would be unsafe to assume that an average of 20℔ an acre was produced in the North-West Provinces. By dividing the total recorded production by the area, it would appear in fact as if these provinces yielded only 10·2℔ an acre. Both in Madras and in the North-West Provinces repeated mention is made of the yield being ½℔ of dye to 1 cwt. of plant, or 1 in 224; but such a high return can only be accounted for by an adulteration of close on 50 per cent. of the weight of dye. Indeed, in many parts of India, a large trade is done in preparing and selling the peculiar clays that are mixed with the liquid dye, preparatory to its being made into cakes and dried. A much more correct estimate of yield is 25℔ of dye to 73 cwt. of plant, or 1 in 327, and that is the yield stated by several writers on the indigo industry of the North-West Provinces; but of course it is the yield of good indigo, not of the highly-adulterated stuff.

Area under Indigo in Bengal. 233

Perhaps the earliest attempt to estimate the area of land under indigo in Bengal was made by the late Dr. H. McCann (*Dyes and Tans of Bengal*). His figures were compiled from the reports of local officers furnished in connection with the formation of the Economic Museum. Dr. McCann arrived at the conclusion that in 1877-78 there were 637,956 (or as subsequently quoted in round figures, 700,000) acres of land under indigo. A more recent estimate published in the *Statistical Atlas* gives the area in 1884-85 as having been approximately 1,300,000 acres. If we apply the mode of testing these figures given above, the Bengal production in 1877-78 would have been thus—Behar 297,716 acres, giving 20℔ an acre, and Lower Bengal 340,240 acres, giving 12℔ an acre, or a total of 10,037,200℔. The writer has no means of analysing the second estimate to discover how much was regarded as Behar, and how much Lower Bengal, but assuming that only 12℔ an acre was obtained all over, the

AREA & YIELD.

Bengal.

production in Bengal for 1884-85 would have been 15,600,000℔. According to **Messrs. Thomas and Co.**, the actual production in these years was respectively 3,832,605℔ or 6℔ an acre to **McCann's** area, and 5,351,806℔ or 4·1℔ an acre to the *Statistical Atlas* area. It would thus appear that either the area as given by McCann and the still greater area in the *Statistical Atlas* must have far exceeded the actual area of the years in question, or that the average yield per acre must have been about one-third of what the writer believes it to be. Applying the test of 12℔ an acre for Lower Bengal and 20℔ for Behar, the area under indigo in Bengal, calculated from **Messrs. Thomas and Co.'s** returns, would have been 232,682 acres in 1877-78, 316,914 acres in 1884-85, and 401,302 acres in 1887-88. The average of these acreages is probably much nearer the actual area under the crop than either of the above which have been accepted as correct by most recent writers. It is probable that these accepted areas of Bengal indigo were dependent on some calculation based on the Bengal trade in the dye—a trade which drains 90 per cent. of the North-West, together with a certain amount of the Panjáb, Rajputana, and Central Province indigoes. There being no available agricultural statistics for Bengal, it may serve a useful purpose to test the accuracy of the estimated area of 401,302 acres for the year 1887-88.

According to the agricultural statistics of the provinces of Upper India, the following were the areas under the crop during the year 1887-88:—

	Acres.
Panjáb	75,986
Central Provinces	25
North-West Provinces	298,790
Oudh	18,391
Bombay and Sind	4,160
Berar	319
TOTAL	397,671

The bulk of the indigo produced in cakes is exported, and the Government returns of exports give the quantities and values that left Bengal, Madras, Bombay, and Sind. These will be found tabulated below (Chapter on TRADE IN INDIGO) for the past ten years. To arrive, however, at an approximate estimate of the actual Bengal share in the production of the exports, we may analyse last year's figures thus:—

	℔
From Bengal	9,781,520
„ Bombay	727,328
„ Sind	336,336
TOTAL	10,845,184

The writer believes that, while a production of 20℔ an acre occurs in many parts of the North-West Provinces and possibly throughout Tirhút, for all the provinces (the acreage of which under indigo has been given above), a yield of 12℔ would be found fairly liberal. For the 397,671 acres it may therefore be estimated that 4,772,052℔ were produced. By deducting that amount from the total exports from Bengal, Bombay, and Sind, there would remain the Bengal production, *viz.*, 6,073,132℔, and according to **Thomas and Co.** Lower Bengal and Tirhút produced nearly 800,000℔ more than that amount, but of course the whole production was not exported. Bengal supplied other provinces and consumed locally a large amount. In fact in the table below showing the balance of production over exports it will be seen there remained in Bengal 880,640℔ in

AREA & YIELD.

1887-88. During the past five years the North-West Provinces and Oudh have had an average of 397,557 acres under indigo which have produced the average of 3,609,029℔ of dye or about 10℔ an acre. If the estimated area of Bengal indigo of 400,000 acres (or to guard against any possibility of understating, say 500,000 acres) proves correct, the relatively greater production in Bengal, as compared with the North-West Provinces, shown in the above figures, must be accounted for by the greater acreage yield of Behar, which last year manufactured 5,192,780℔ of indigo.

According to the agricultural statistics of Madras, the Southern Presidency had last year 501,721 acres under indigo, the exported produce of which amounted to 4,794,944℔, which (disregarding local consumption) would be 9·5℔ an acre. Accepting the estimate of 500,000 acres for

Total area for all India.
234

Bengal as correct, the ***total area under the crop for all India*** would, last year, have been 1,399,392 acres. The total foreign exports were 15,640,128℔, which shows a total average production per acre of 11·1℔, to which must be added the Indian consumption probably close on 2,000,000℔ of dye, thus raising the mean acreage production to 12·6℔. But if the calculated acreage and actual yield of Tirhút be excluded, the average yield for the rest of India is only 10·9℔ an acre; so that in round figures one-tenth of the pounds of dye produced is the acreage of the crop.

In conclusion, the writer would add that, although personally satisfied that the above criticism of the published statements regarding yield and acreage will be found far more nearly correct than the opinions hitherto advanced, he is conscious that he has had to be guided by indications more than by direct evidence. Many persons express the yield of dye from plant as one in so many. But applying the factor thus afforded to find the yield of plant per acre, in one case it comes out as 2,000℔, in another as 12,000℔. These are the extremes, but in working through many such statements every intermediate result has been met with. The yield of dye to acre, where given by writers, is almost invariably twice as great as the yield obtained by dividing the total production of the province by the recorded acreage. It is thus probable that the planters who furnished the figures published by Government gave their own results, and that these are far higher than could be admitted for an average provincial production which included Native manufacturers. This error is, however, minimised in such provinces as Tirhút or even Bengal as a whole, where the majority of planters and manufacturers are Europeans, who conduct their operations on a uniform and scientific principle, so that the returns published by the few may be accepted as fairly applicable to all. **Messrs. Thomas & Co.**, in their statement of annual averages, exhibit the relative amounts of Native as compared to European indigo. Last year, while the latter produced in Lower Bengal 20,946 maunds, the former sent into the market only 1,834 maunds. The entire Tirhút production of 69,585 maunds is shown as European. A very different state of affairs prevails in the North-West Provinces. Of the Benares indigo, Europeans produced 6,515 maunds, Natives 5,750 maunds. Of the Doab, the shares were, European 3,120 maunds, Native 24,000 maunds. These facts demonstrate why it may be safe to accept the average of 12℔ for Lower Bengal and 20℔ for Tirhút in estimating area. While some of the European Benares factories produce as good a result as those of Tirhút, a yield of 20℔ for these provinces all over would be incorrect, as may be seen by the fact that it would reduce the necessary indigo area by fully one-half of what is given in the Agricultural Statistical Tables of these Provinces.

LAND TENURE IN INDIGO DISTRICTS.
235

SYSTEMS OF LAND TENURE IN THE INDIGO DISTRICTS.

Prior to the Muhammadan conquests of Upper India the Hindu law

LAND TENURE IN INDIGO DISTRICTS.

and practice was for the ruler (whether king or petty chief) to obtain a share of the produce of the land. This share was by Akbar converted into a tax or rent, based on an actual survey and settlement, but the assessment was fixed for only ten years, and was subject to a revision every decade. In South India (which did not fall under Muhammadan rule till a much later period), the Hindu law prevailed, with the modification that the share to be paid to the ruler was often given in cash fixed at the prevailing price of produce. In Bengal, the Muhammadan rulers introduced the system of farming out the tax collection to certain individuals who were sometimes the descendants of the original owners, at others mere speculators. During the period of anarchy between the death of Aurangzeb and the establishment of the British rule, every power which became predominant taxed the land to the utmost, and the system of farming out the tax collection was naturally found the easiest and most effectual mode of tyranny and persecution. The collector or *zamíndar* obtained the right over a certain district or tract of land from his being the highest bidder, and it became customary to sub-let or even sub-sub-let the *zamíndari* right, until the land had to support not the Government and cultivator only, but two or three sets of collectors, some of whom did absolutely nothing but realise a large personal income. The British early found it necessary to reform this system and to restore more or less the Muhammadan method of direct taxation, based on what is called a "Settlement" of the land revenue. It was part of the policy also to encourage or develop proprietorship. But the varieties of land tenure found in existence gave rise to three distinct systems of settlement, since the general rule had to be accepted that the person who had the highest existing interest in the land should be held responsible for the revenue. In some cases peasant cultivators paid, as they do now, their land rent directly to the State. In others they held their lands from private persons who may briefly, though not accurately, be described as having been the landlords. Through these landlords the State obtained its revenue in such cases. Elsewhere village communities held the lands in common and paid conjointly the revenue.

Bengal, which early came under British rule, was settled under what is now known as the *zamíndari* system. The peasants who cultivated the lands were accepted as having a tenantry right, but the superior holders (the descendants in some cases of ancient families, in others of military chiefs, or only of the farmers of revenue) were recognised as tax collectors in perpetuity on fixed or unchangeable rents. This system prevails not only in Bengal (excluding Orissa) but in the North-Eastern Districts of Madras, in Oudh, and in the Central Provinces. **Zamindari. 236**

In the greater part of Southern and Western India there was found to exist no superior class between the Government and the actual cultivators, and in the village communities certain families or persons had always cultivated certain fields. The settlement was therefore made with the peasants on what is now known as the *Rayatwari* system. **Rayatwari. 237**

A third system of settlement, the *Lambardari*, was made in many parts of the country, especially in the North-West Provinces and in the Panjáb, where village communities, more or less united, were found to hold tracts of land. On the members of these communities jointly the settlement was made, and these communities, now recognised as having proprietary rights, can let the lands they do not wish to cultivate to tenants, but the communities, not the tenants, are responsible for the Government revenue. In the *Rayatwari* provinces village communities own only the fields they cultivate—the waste lands adjoining belong to Government and consequently the revenue may be increased by the cultivation of waste lands or **Lambardari. 238**

LAND TENURE IN INDIGO DISTRICTS.

decreased by the non-cultivation of fields formerly cultivated. In *zamíndari* provinces the village communities are created actual proprietors on a fixed revenue for a tract of land whether cultivated or not, and this often includes large expanses of waste lands which belong to the village.

The above sketch of some of the general features of land tenure has been compiled from **Mr. W. G. Pedder's** memorandum on the subject, and it is hoped that it may convey some idea of the position of land tenure as it affects indigo cultivation. It need only be necessary in fact to give further a few passages from writers who specially deal with the advantages of the one and disadvantages of the other system, in their special bearings on indigo.

BENGAL. 239

Bengal.—**Mr. A. P. MacDonnell**, Magistrate of Durbhanga, in his Administration Report for the year 1876-77, gives a full account of the indigo cultivation in that district, one of the divisions of what is by planters spoken of as Tirhút. **Mr. MacDonnell** wrote of the systems of tenure and conditions of indigo cultivation:—"The cultivation of indigo is either '*Zerat*,' '*Assamiwar*,' or '*Khooshgee*.' It is *Zerat* when the land is in the planter's sole possession, and the rayat employed on it is a mere hired labourer. It is *Assamiwar* when the land is in the rayat's possession, and when he is compelled (being usually the planter's tenant) to grow indigo on it at fixed rates per *bigha*. It is *Khooshgee* when the rayat, under no compulsion, grows the plant as a remunerative crop."

The Commissioner of the Patna Division, in his Administration Report for 1872-73, wrote as follows:—

"In the northern districts of Tirhút, Champarun, and Sarun, the indigo is cultivated, as a rule, in villages let to the planter by the zamindars, and is either *Assamiwar* or *Nij*. Under the former system when the lease is completed the rayats attend the factory and execute agreements to cultivate a specified portion of their uplands with indigo. The common proportion now agreed upon is two to three cottahs per bigha of upland or *bhit*, though in some few factories the proportion demanded is larger, amounting to five or six cottahs, which it appears was the rate prevailing, at least in Champarun and Sarun, before the indigo difficulties in 1867. The agreement is generally for the same term as the lease. At the time of executing it an advance is given, which remains unpaid without interest till the end of the term, and during each year the price agreed on to be paid for the cultivation is given in advance at the beginning of the year.

"The sum paid varies according to whether it includes the rent of the land or not, and also according to the size of the bigha. The average rate in Tirhút, where the bigha is about 4,225 square yards, is from R8-8 to R9, inclusive of rent; and in Sarun, where the bigha is the same as in Tirhút, it is from R7 to R9. In Champarun, where the bigha averages 7,225 square yards, the usual rate is now about R15; but up to the last few years R12 was generally paid. In all cases the lands for indigo are assessed much below the average rent paid for other lands of similar quality.

"In Champarun the rent is generally included in the price paid to the rayat, and the same practice is pursued in Sarun, where lands are taken from the rayats; but in the latter district the majority of planters cultivate their own lands, and consequently the arrangements made with cultivators affect but a comparatively small number. In Tirhút the more usual practice is to write off the rent of the land in the factory books, and to give the rayat R5 to R6-8 per bigha.

"The lands taken from rayats are retained for three to five years by the factory, after which time they are useless for growing indigo, though, as the indigo plant has a long tap-root and draws its nourishment prin-

LAND TENURE IN INDIGO DISTRICTS.

BENGAL.

cipally from the subsoil, they are improved for the growth of cereals and green crops, which subsist upon the upper layer, which has had the advantage of a long fallow and of being manured by the indigo leaves. In lieu of the lands thus given up, a similar area of other lands is taken from the rayats for the rest of the term of the agreement, and in some instances a clause is inserted that these exchanged lands shall be selected by the factory from the best of those in the rayat's holding.

"Speaking generally, the crop may be said in these districts to be sown in February, and the cutting and manufacture to commence early in July. A second cutting of the *khunti* crop generally takes place in September and the land is clear in October, except in the very few instances, and these are mainly on the south of the Ganges, where poppy lands are taken for the growth of a crop of irrigated indigo. No other crop can be grown during the same year: as when the crop is taken off the ground in October, the preparations for fitting the ground for the next year's crop are begun. The soil best fitted for the crop is a rich loam, with a good subsoil, neither too sandy nor too stiff, and old river deposits not liable to inundation give the best yield; but fine crops are also grown in inland villages, or uplands or *bhit*.

"In the districts south of the Ganges the system is totally different from that above described. The cultivation is for the most part *nij*, and is carried on in lands leased by the factory from the zamindars or rayats. The whole expenses of cultivation are paid directly by the planter, who employs his own labourers and bullocks. The seed is sown at the beginning of the rains, and the plant remains in the ground during two years, in each of which it is cut. In strong lands a third year's crop is sometimes taken; but, generally speaking, the land is given up at the end of the second year (when it is said to be eagerly sought after for the growth of green crops) and engagements made for other lands. Many factories have running agreements for two sets of land, one of which is occupied by indigo, and the other remains in the hands of the rayats."

N.-W. P. 240

North-West Provinces and Oudh.—"The tenure of land on which indigo is grown in the North-West Provinces and Oudh does not present any special features. The great majority of the factories belong to Natives and not to Europeans, and of the Europeans engaged in the industry very few grow indigo themselves. The petty cultivator grows indigo just as he does any other crop and sells it to a factory in the neighbourhood. The particular phase which the industry has taken in Bengal, where indigo is cultivated direct by capitalists on lands either owned by them or held on lease from the zamindar, is, though not altogether unknown, rare in these provinces. Here the motive power most frequently employed by the indigo factor is a system of advances to the surrounding cultivators. The Indian rayat can seldom resist a proffered loan, however onerous the obligation which it brings; and once in the factor's debt for indigo contracted for and not delivered, he is often compelled to continue growing the plant even at unremunerative prices. As long as the rate keeps fairly high and seasons are good, the arrangement works well for the factor, though he must expect in the best of years to have to write off some bad debts, and is exposed to losses by rayats fraudulently selling the produce, which they had contracted to deliver to him, in the open market. But if the Calcutta rate for indigo drops and bad seasons intervene, the weakness of the factory system manifests itself. The rayats abscond or are sold up, and the factor is left with a ledger full of debts which there is no hope of ever recovering. The dangers of the 'advances' system are now fortunately beginning to be realised by indigo dealers, and in many districts it is giving way to cash dealings for produce when ready to be

LAND TENURE IN INDIGO DISTRICTS.

N.-W. P.

cut" (*Report by the Director of Land Records and Agriculture*). Mr. Reid (*Culture and Manufacture of Indigo*) devotes a special chapter to the systems of land tenure in these provinces. He speaks of indigo as having "flourished under the liberal Government of the North-West Provinces." He states that his experience is connected with the district of Gorakhpur, which he adds "I emphatically believe to be *the* direction in which the industry of the *blue dye*, forced out of Behar, will, in the near future, compulsorily seek its legitimate development." All disputes, of whatever nature, between landlord and tenant, have to be adjusted before the Collector or Assistant Collector, with appeal, if desired, to the Commissioner, or finally to the Board of Revenue. Mr. Reid reviews briefly the conditions on which land may be held in these provinces, but the reader may consult on this subject for fuller details *The Rent Law Manual* published by Mr. L. W. Teyen. By the provisions of the Rent Act (XII. of 1881) the term "tenant" includes, in addition to the cultivators who inhabit the villages and cultivate the lands pertaining thereto, the sub-lessees and also the sub-sub-lessees. This tenancy, as Mr. Reid puts it, divides itself into four great sections, (*a*) occupancy tenants; (*b*) non-occupancy tenants; (*c*) tenants at fixed rates; and (*d*) ex-proprietary tenants. The *Lambardari* (from "*lambar*" a corruption of the English word "number" and *dar* holding) system, it is contended, gives to the North-West Provinces, a great advantage over Bengal, since every field, plot of grass land, cattle track, and village site has been surveyed and numbered. A tenant by reference to the *patwari* (or village accountant) can obtain an absolute statement of the fields and grazing lands over which his tenancy right extends. On this point Mr. Reid adds, "The numbering of each field secures the tenant against the shifting about or exchange of lands and the confusion consequent upon this which obtains, with such miserable results, in Bengal. An *occupancy tenant* is one who has held unbroken possession of the same lands under cultivation for twelve years, or who inherited the same by legal right. But a gift of land by the landholder confers no occupancy right, nor is an occupancy right transferable. But actual occupation of the land for a time as proprietor, and afterwads continuously as tenant for twelve years, established the rights of an occupancy tenure. *Non-occupancy tenants* are tenants who have not cultivated the land for a term of twelve years. These can be deprived of their tenures at the will of the land-owner, on being served by a written notice of ejectment under the provisions of the Act. *Tenants at fixed rates*: This class of holdings only occurs in the permanently-settled districts, and the term 'fixed rates' implies the grant of land on terms agreed upon at the settlement. This right is inheritable as well as transferable, and the rate cannot be enhanced except on the ground of increased value of holding by alluvial deposits or otherwise. *Ex-proprietary tenants* are those who have lost or parted with their proprietary rights, but who retain an occupancy tenant right on the land, holding the same at a rental of 4 annas in the rupee less than the prevailing rate payable by *tenants-at-will*."

The above sketch, which more immediately concerns the North-West Provinces, with slight modifications, is applicable to a great part of India—the regions where a *rayatwari*, or a village settlement, prevails.

MADRAS. **241**

Madras.—The *Madras Mail*, in an article on indigo cultivation, reviews briefly the difference between the tenure of land in the Southern Presidency and that in Bengal. "In Bengal," the *Mail* remarks, "everything is under the so-called *zamindari* system. The whole country has been parcelled out into blocks, and leased, for sums far below their actual value, to zamindars. These persons do not farm their estates but sub-let them

LAND TENURE IN INDIGO DISTRICTS.

MADRAS.

in blocks or in detail to renters. Under a system of this kind it is of course easy for the European to acquire an estate and grow on it indigo, opium, or what he likes; the proprietor will lease to the man who gives him the best rent and is the most punctual paymaster. Under the *rayatwari* system as it prevails in Madras, this is impossible. Here we have scarcely any zamindars. The great bulk of the land belongs to the Government and is let direct to the rayats, who are petty farmers holding on an average about 10 acres each. To these acres they cling with the utmost tenacity, so that there is no opportunity for the European to step in and inaugurate an enterprise of his own. It is here that the reason why the zamindari system is often called 'the curse of Bengal' is to be found. Though the country is prosperous, the people derive but little benefit therefrom. This goes to the zamindars and their sub-renters for whom the people have to work. In Madras, the rayat is independent; he works for himself and takes all the profits to be derived from his land. This, it is true, is only small in extent, and he pays a higher rent per acre than the Bengali zamindar, but the whole of the proceeds derived from the land is divided between himself and his fellow-rayats, and there are far fewer middlemen to be enriched at his expense."

The Collector of Cuddapah, in his report for this work, says: "There is nothing special about the tenure of land cultivated with indigo; the land-owner and cultivator divide the crop in equal shares." The Collector of Kurnúl says there is no separate assessment for indigo lands: the tenure for indigo cultivation is not distinguished from that for any other crop. The Collector of Vizagapatam says: "We have no particular land tenure here for indigo, as the crop is raised by rayats in common with other crops. The *rayatwari* system is that observed in the Government taluks of this district." But although the Madras system would seem to throw a larger share of the profits into the hands of the actual cultivators than is the case in Bengal, still the same evils from the system of advances would seem, to some extent at least, to prevail in Madras. In the Settlement Report of the Cuddapah district it is, for example, stated that "The poorer rayats will not sow indigo unless they are forced to do so by their creditors—the rich Reddies and Banians,—who always manage to cheat the rayat of a large amount of the profits which is placed to the credit of interest due on loans."

MANUFACTURE OF INDIGO.

MANUFACTURE. 242

The process of preparing Indigo from the plant differs but little in the various provinces of India (at least in the hands of European planters and with larger Native concerns), so that a general description may suffice. Departures from the type will be found discussed in the Chapters on "The INDIGO OF COMMERCE" and "The CHEMISTRY OF INDIGO."

Briefly, it may be said, the necessities of Indigo manufacture are—a manager who is methodical, shrewd, ubiquitous, with a business turn of mind fitting him alike for office routine and the securing of contracts; a liberal supply of good water; an influence, direct or indirect, over the neighbouring cultivators, sufficient to guarantee a regular supply of green plant; a factory, commonly called "The Concern," comprising certain vats, machinery, and appliances; and a staff of loyal employés who can be educated to carry out their various duties tidily and promptly. Space cannot be afforded in the present article to treat in detail of every necessity of indigo manufacture. The respective duties of the manager and of his army of employés must be left untold. A supply of good water and a regular production of indigo plant are necessities without which the Concern cannot exist. In the chapter on CULTIVATION and in that on LAND TENURE,

MANUFACTURE. the production of plant has been fully dealt with, as well as the difficulties which exist in securing a uniform supply. It remains, therefore, to narrate, as briefly as may be possible, the construction of an Indigo Concern, and the manipulation in the various stages of Manufacture.

It may be remarked that there are primarily two processes according as WET (*e.g.*, fresh or green) plant or DRY leaf are employed.

WET PROCESS.

I.—Wet Plant Process.

Water. 243 *Supply of water.*—Proximity to a river or other unfailing source of water is indispensable. In order that the water may be brought on a level with the highest vat in the concern, pumps may be required. The water should be raised to a reservoir, sufficient to contain about 10,000 cubic feet. It should be there retained for some hours, so as to allow the sediment to deposit, before being used. Muddy water would of course impregnate the dye with a quantity of dirt from which it could with difficulty only be again freed. Stagnant and foul water is found to impart to the infusion (the fecula or *mal* as it is called) a property which it is extremely difficult to conquer—namely, of retarding the process of settling and drying.

Vats. 244 *The Vats.*—Pipes are required to connect the supply of water with the *Steeping-vats*, the *Beating-vats*, the *Pulp-boilers*, and the *Engines*. The size of the reservoir as well as the number of vats is dependent on the amount of plant which has to be annually treated. We may assume a concern of twelve *Steeping-vats*. These are built of brick and portland-cement, and are in dimensions 24×18×5 feet. These should be arranged in a row and have in front and below the floor level, a corresponding series of shallower and wider vats (3 to 4 feet deep) known as the *Beating-vats*. The *Steeping-vat* communicates with its *Beating-vat* by a hole at the bottom into which a large wooden plug is driven from the outside. By removing this plug the infusion prepared in the Steeping-vat is drained into the Beating-vat. In a similar way the Beating-vat has a number of holes with plugs arranged in a vertical series, which open into a drain. When the fecula has been precipitated to the bottom of the Beating-vat these plugs are successively removed so as to draw off the useless surface fluid which is allowed to run to waste, or is used as a manure for the fields. The sides of the Beating-vat are made to curve inwards at the top so that the liquid when set in motion may not splash over. When the surface fluid has been carefully drawn off, the thickish deposit is raised by buckets and thrown on a strainer. This removes large mechanical impurities. It then flows by a pipe to a cistern under the boiler pump, sometimes called the Pulp-vat, which has the dimensions of 15×10×3 feet. Before leaving the pipe it passes through a second strainer in the form of a cloth or bag tied over the end. The mouth of the suction pipe of the pump which dips to the bottom of the cistern is closed by a sieve or rose which also assists in the process of straining. The infusion is then raised by the pump and discharged through a further cloth strainer into the boiler.

Boiler. 245 *The Boiler* should be of thin copper in preference to iron and placed in a room instead of in the open air as are the vats. If made of copper the boiler heats more rapidly and uniformly, and the indigo is less liable to get burned: it also lasts longer and is therefore more economical, though the initial cost is greater. The boiler for a factory of the size described should be 25 feet long by 12 feet wide and 4 feet deep; and this will, if required, produce 5 maunds (374℔) of dye. As soon as the fecula has been discharged into the boiler (or after the removal of any surface liquid that may have formed) a certain quantity of clean water is poured into the boiler, a fire is lighted, and the heat gradually raised until it reaches the boiling point. For this purpose steam may be employed in addition to a

MANUFACTURE.

WET PROCESS.

furnace. All the while the pulp is carefully stirred and kept gently boiling for three hours. A fragrant smell and the formation of bubbles on the surface indicate the completion of the boiling process.

Table. 246

The Table or Dripping-vat.—The boiling fecula is now discharged on to a table over which a large damp cloth has been spread. The water percolating through the cloth is pumped up from the surrounding drain and poured over the mass until it passes away in a dark red tint without being charged with indigo. In five or six hours on a wide-spread table (or shallow wooden vat) (40 × 10 × 2 feet), the major portion of the water will have drained away, and the fecula settled on the surface of the cloth. If the straining and re-straining does not readily result in the production of a clear fluid, alum water may be poured over the pulp, but this is rarely necessary. The pulp should now be scraped together in a corner, the cloth folded over, and a weight placed on the top. In this condition it should be left in a few hours to cool and drain still further.

Press. 247

The Press.—This consists of a square wooden box well-fitted and covered with a damp cloth or sheet inside—the framework having on all sides numerous perforations. The inside measurement is 42 inches long, 24½ inches broad, and 12 inches deep, and it is open at top and bottom. This framework or mould is placed on a strong table, and is provided with loose boards for the top and bottom which exactly fit the interior of the mould. It is arranged under a powerful screw which can be turned with long levers. The fecula from the table is placed within the box, the cloth is folded over the top, and the lid adjusted over all when the screw-press is brought into bearing. One turn of the screw is given every now and then for about five hours until the mass of indigo in the press, at first about 8½ inches deep, is compressed to 3 or 3½ inches, or until no more water is seen to be oozing from it. The pressure is then removed evenly and gently, the sides of the press lifted off and the cake (42 × 24½ × 3½ inch) exposed.

Cakes. 248

Cutting the Cake.—It is now marked off into 3 or 3½ inch square blocks, each of which receives the brand of the Factory and the number assigned to the day's manufacture. The cutting up into cakes is accomplished very often by carrying the slab resting on the bottom loose lid of the press to another room and depositing it in a frame of the same dimensions which has slits on its sides; thus allowing of the cake being cut lengthwise and breadthwise, so that by 12 cuts in the former direction and 7 in the latter, it is broken up into 84 cubical cakes, each 3½ inches in size. The cutting is generally accomplished by means of a brass wire stretched between two wooden handles; but a knife may also be employed. The cubes or cakes are now ready for removal to the drying house.

Drying House. 249

Drying or Cake House.—This should be a room well ventilated, with means of preventing severe draughts and blasts of dry wind. It should be about 100 feet long by 50 feet wide and 20 feet high, and be fitted with shelves of a light bamboo framework, sufficiently far apart to allow boys to crawl along the shelves. The cakes are carried from the cutter's hands to the drying room and arranged on the shelves, each cake being turned over from side to side until quite dry.

Sweating Room. 250

The Sweating Room.—The cakes are now conveyed to a closed room in which they are arranged in small walls, each day's manufacture being distinctly marked off from the other. The walls of cakes are next covered over with blankets and dry bran, and the doors of the room secured, so that as little air as possible may be admitted. In about fifteen days time the sweating process will be completed when air should be let in very slowly and the walls of cakes uncovered by degrees, the uncovering process taking four or five days. A sudden exposure would very

THEORY OF MANUFACTURE.

WET PROCESS.

probably crack the cakes. This process of sweating improves the brilliancy and imparts the much appreciated white skin.

Packing. 251 *Packing.*—The cakes being thus thoroughly dried (a process which lasts for about three months), they are each brushed and packed into specially prepared boxes of well-seasoned wood, each box or "Case" being filled, if possible, with the cakes bearing the same mark denoting the day of manufacture. The actual number of cakes in the box should be registered, and if it be necessary to mix the different days' manufacture, the number of cakes placed in it bearing, for example, No 5 and those bearing No 25, &c., &c., should be carefully recorded

Steeping. 252 *Action of Steeping.*—So far we have followed the fluid from the steeping-vat till it is ready for the market, and in doing so we have briefly alluded to the main features of the indigo factory. It is necessary to return to the steeping-vat and discuss the process by which the fecula is extracted and manipulated. The manager, having seen that all the necessary repairs have been made on the factory, that the vats are clean, the reservoir full of clear water, the pumps and machinery in good working order, the carts and oxen, or boats, in good condition, and the cultivators ready to cut the crop, fixes a day for the commencement of operations. The plant is ready, on the flowers having begun to form. The stems are cut within a few inches of the ground, baled into "bundles," and carted off to the factory as soon as possible. The plant is measured or weighed, and stacked in the shade until the afternoon, when the packing of the steeping-vat commences and must be completed before nightfall. Much difference of opinion prevails as to the best method of packing the steeping-vats. The most general way is to arrange the bundles in two layers, one standing on the floor on their stump ends, the other inverted so that the tips of two below embrace the one above, and all the young leaf-bearing parts are thus fixed in the middle of the vat at a uniform depth. Most planters hold, however, that the method of packing is immaterial, so long as two conditions are observed: (*a*) facility for the fluid to drain readily and completely through the tap; (*b*) uniformity horizontally, so that when the beams used for fixing the plant are brought to bear, they may press the plant to the same depth. There should always remain some six inches to allow of the rise of the water above the highest part of the plant. By means of beams and levers acting on the side of the vat the plant is firmly fixed and compressed. When this has been done water is allowed to flow into the vat until the plants are nearly covered. "The night being moderately fair, and the plant good and not inundated before, in nine or ten hours it will generally be ready to draw off. Now see" Mr. Reid remarks, "that your people are ready and have the beaters thoroughly washed out before use. Those who aim at making extraordinarily fine blue, steep a much less time, and generally lose a considerable portion of their produce, and not unfrequently make their indigo of too light a colour for general use. As the fermentation goes on, air bubbles rise upon the surface of the water, until at length the vat has the appearance of a broccoli head; now watch carefully the ebullition of the vat, and the moment it begins to subside and sink down in the steeper, let off the liquid, and you will generally be safe. If it runs out a bright straw-colour, tinged with green, the indigo will be fine; if a strong madeira colour, good; if of a very pale straw-green, violet; and dirty red, bad (coppery). The first indicates good steeping, the second, a little too long, the third not enough, and the last, overdone. This will be your guidance, attending, of course, to the state of the weather, to steep longer or shorter, according to the quality of indigo you wish to make" (***W. M. Reid, Culture and Manufacture of Indigo***). Thus the plant is not only macerated in water but fermentation of a vigorous nature is set up in the steeping-vat, for the fluid rises up from 6 inches to one foot, an expansion that has to be provided for by not completely filling the vat. In the morning the temperature of the water is found to stand at 85° to 90° Fahr. After the fluid is run off into the beating-vat the plants are green but have lost about 12 per cent. of their weight. The indigo refuse (*seet*) is removed to a short distance and dried and stacked. It is of great value as a manure to the fields, and is also invaluable to the planter as fuel for his next year's operations.

THEORY OF MANUFACTURE.

WET PROCESS.

Fermentation in the Steeping-vat. 253

Seet as a manure. 254

Beating and its Effects.—The fluid having been run from the Steeping, to the Beating-vat, various contrivances are adopted for beating it, in all of which a two-fold object seems to be attained: (*a*) the liquid is subjected to the action of the oxygen of the air; (*b*) the particles of the dye-stuff are consolidated or aggregated into sufficiently large particles to facilitate their rapid precipitation. That the fluid obtained from the steeping-vat is not blue-indigo seems very generally accepted, and it is accordingly said that the beating oxidises it, hence several chemical processes have been patented for directly oxidising the fluid and thus lessening or dispensing with the necessity for beating. The fact that oxidising agents do actually facilitate the precipitation of the blue dye-stuff seems to prove that combination with oxygen is necessary before the greenish fluid of the Steeping-vat can be converted into the blue fecula of the Beating-vat. It would, therefore, appear that the steeping and fermentation ultimately result in the formation of the substance known as white-indigo—a compound that contains one atom more hydrogen than blue-indigo. An atom of oxygen combining with two atoms of hydrogen (derived from two proportions) would form water and thereby reduce or reconvert the white soluble indigo into the blue insoluble compound which is the essential property of the indigo of commerce. It has to be admitted, however, that no chemist has actually subjected the materials of the indigo factory to analytical test, and that the above explanation is based, therefore, on the results obtained under similar circumstances with other plants than **Indigofera tinctoria.** If the fluid obtained from the steeping-vat contains, before beating, the substance known as white-indigo, a piece of cloth dipped in it and then exposed to the air should rapidly show that it has been dyed with indigo. In a further chapter, the chemistry of indigo and the explanation of the dyers'-vat, will be discussed; but in concluding this brief allusion to the formation of the blue fecula, it may be said that if the fluid of the steeping-vat does not contain white-indigo, but only finely divided blue-indigo, the beating can exercise no oxidising influence whatsoever, and the patent processes that liberate oxygen within the fluid must in that case have, like the cruder methods of beating, only a mechanical, not a chemical, action.

Oxidation in Beating-vat. 255

White-indigo produced by Fermentation. 256

The simplest form of a Beating-vat is that used by the primitive dye manufacturers. The plants having been steeped in a large earthen pot are removed, and by means of a date-palm leaf or other such substance the fluid is lashed about until the change takes place from a green to a blue precipitate About this time a certain quantity of lime water is added which is supposed to facilitate the precipitation. The contents are set aside to allow of slow evaporation of the water, and when quite thick and almost dry a lump of lime is generally used to close the mouth of the jar until such time as the dye-stuff is to be utilised. This in principle is the process formerly pursued by all European planters, and by most of the smaller factories even at the present day. The fluid having been drawn from the steeping- into the beating- vat, from 10 to 12 men are sent into the fluid (which is sufficiently deep to rise to about their waists) and, standing in

Primitive Beating-vat. 257

Use of Lime. 258

Beating by manual labour. 259

THEORY OF MANUFACTURE.

WET PROCESS.

two rows facing each other, they commence to lash the fluid with wooden oars or staves, or simply by means of their hands. At first slowly and regularly, but as the froth forms and then begins to disappear, the rapidity of beating increases until the liquid is often made to rise in a great wave between the beaters. The higher the fluid is raised into the air the better, so long as irregular beating does not cause it to lash over the sides of the vat and be thus lost. At no time, however, should the beating be violent or spasmodic, but the beaters may slowly move to right and left, thus giving a greater impetus to the movements of the fluid. Violent beating is thought to break the grain of the forming dye too much, and the steadily increased motion is held to prevent coarse grain forming. The exact period of beating cannot be arbitrarily fixed, as it depends on the degree of fermentation that took place in the steeping-vat, and the temperature and the humidity of the atmosphere. In about two to two-and-a-half hours the froth will have disappeared, the vat will have passed from bright to dark green, then purple and ultimately to dark blue. Before the order is given to discontinue the beating, a small quantity of the fluid is taken up by the dipper from near the bottom and run on to a white plate. If the grain readily settles at the bottom forming a sharp edge between the sediment and the fluid, the beating may be considered as satisfactorily accomplished. At this stage it was the habit (and in some factories is so still) to throw into the vat a quantity of lime-water (about 6 *calsis* = 10 to 12 gallons). Instead of using lime-water to complete the precipitation of the dye as it is called, a more favourite and modern process consists in allowing a quantity of pure cold water to be discharged on the surface of the beating-vat from pipes arranged on the walls around the circumference. This ready supply of water is also of advantage in washing out the vat. The explanation may be here offered that the effect of the discharge of fresh water into the beating-vat may be to cause the decomposition of the important soluble substance which has been derived from the plant by simple maceration. In a further chapter it will be explained that by the action of water this substance splits into indigo-blue and sugar. The fermentation employed apparently reduces the indigo-blue to indigo-white, hence necessitating the beating; but should any proportion of the soluble glucoside obtained from the plant remain undecomposed, this would prove highly injurious in the boiler, since by heat, it is known to give origin to browns, reds, &c., that depreciate the value of the dye. The beaters are now told to walk round the vat about a dozen times so as to give a rotatory motion to the contents, afterwards to rapidly jump out. In two or three hours after beating according to the state of the weather, the indigo will have settled to the bottom and the surface fluid may be slowly drained off.

Effects of violent beating. 260

Danger of Heat. 261

Chemical explanation of Beating. 262

It seems to be universally admitted that the beating oxidises the infusion. If the oxygen combines with the indigo instead of with other organic materials, we must assume that the indigo (or a certain proportion of it) exists in the condition of white-indigo. The chemistry of the substance clearly shows that no other indigo compound can be made to combine with oxygen. It was formerly held that carbon was removed by oxidation from the compound that was thus reduced to indigo. This view seems opposed to the chemistry of the substance, and carbon, if liberated, is not derived from the soluble glucoside (obtained by the maceration of the indigo plant), the compound which, by decomposition, forms indigo. Many writers still hold, however, that carbonic acid is liberated; and, if so, it must be viewed as a secondary product of the fermentation, but one not derived from the substance which ultimately affords indigo. The liberation of carbonic acid, it is held, is the cause of the froth observed both in the steeping- and the beating- vats. The use of

Liberation of Carbonic acid. 263

THEORY OF MANUFACTURE. WET PROCESS. Effects of boiling the Fecula. 264

lime, at the close of the beating process, is regarded as destroying the fermentation, which, if still existing in the fecula, would cause destructive changes during the period of subsequent manufacture. The boiling which follows is viewed as liberating volatile materials, expelling air, oxygen, and carbonic acid while at the same time coagulating the albumen and concentrating the pulp.

Antagonistic ideas.

The writer, while endeavouring to criticise the various opinions that are held regarding the reactions and decompositions that take place in the manufacture of indigo, must not be viewed as advancing any peculiar theory of his own. The literature of the subject establishes two entirely antagonistic ideas, the one followed by the actual manufacturer and the other inculcated by the chemist. In the absence of the results of a chemical examination of the products of each stage of manufacture, in each and every modification of the process, it has seemed to the writer that the best course for him would be to exhibit the arguments in favour of, and the defects in, the practical and the chemical methods of indigo manufacture.

PATENTS. 265

Instead of using men to beat the liquid of the oxidising-vat (as it is sometimes called) various patent machines may be employed. For this purpose, for example, the ends of the vat are rounded, and on the bottom along the centre lengthwise is formed an interrupted ridge upon which is adjusted two paddle wheels. These are made to revolve by steam power and a free circulation of the liquid is thus set up, the rapidity of which may be controlled with the utmost nicety. By another process air is forced through the fluid, or by still another method it is conveyed to the particles of white-indigo by centrifugal force. To discuss all the patents that exist would take greater space, however, than the writer has at his disposal. So also, for example, various patents have been granted for chemical inventions in which the liquid both in the steeping- and beating- vats is brought under the influence of direct oxidising agents. Certain properties have been claimed (and planters have been found to substantiate these claims) for patents in which ammonia, alum, nitre, caustic soda, essential oils, &c., are added to the steeping- or beating- vats. If fermentation be a practical necessity of the isolation of indigo from the plant, such patents are probably fully justified, but the chemist holds that fermentation is not a necessity, that simple maceration in cold or hot water slightly acidulated extracts from the plant a soluble glucoside which, by the action of water alone or still more readily in the presence of acids, splits into indigo-blue and a form of sugar. For this decomposition neither oxygen nor alkalies are necessary; indeed, the latter are highly injurious, as they cause, at certain temperatures, a decomposition of the glucoside from which indigo-blue cannot again be prepared.

Machinery used in Beating. 266

Chemical methods.

Fermentatio not necessary. 267

In concluding this brief review of the various stages of manufacture and of the probable reactions which take place, it may be remarked that **Mr. Richard Olpherts** has patented several mechanical improvements that seem of much value. He proposes that the dye-yielding parts of the plant only—the leaves—should be introduced into the Steeping-vat. For this purpose he has large iron-wire cages constructed so as to fit carts and so many of them to accurately fill the vat. These are taken to the fields, charged with leaves as tightly packed as possible, and conveyed to the factory. They can then be lowered into or raised from the vat with great ease, while the leaves are held at a given depth in the water until the desired infusion has been obtained. Each cage may then be raised and placed on a platform over the vat, where it may be washed out by having the leaves trodden on, while clean water is being poured over. The patentee claims several advantages which are, as it would appear, borne out by a chemical study of the substances obtained. A lesser

MANUFACTURE.

PATENTS.

volume of water relative to the bulk of dye yielding material is necessitated, which has its advantages chemically and mechanically, while fermentation is more under control. From the writer's theoretical convictions he would add that the lessened fermentation is probably an immense gain, but the process might be still further perfected by slight improvements having in view the prevention of fermentation entirely. On the cages of leaves being removed, the beaters are sent into the steeping-vat, and after a degree of beating, the surface fluid is drained into the ordinary beating-vat. It is there subjected to a further beating, whereby a proportion of indigo is recovered which, **Mr. Olpherts** affirms, is, by the ordinary process, allowed to run to waste. The sediment from both the upper and lower vats is next carried to the pulping-vat, and for most of its stages of subsequent treatment, **Mr. Olpherts** appears to hold patents for methods of his own. The advantages of **Mr. Olpherts'** system have been commended by **Professor Church**, who appears to have examined his improvements. It cannot but be regretted, however, that a chemist of eminence has not reperformed the analysis of indigo with special reference to the various processes in use for its manufacture, since it has to be admitted that the industry is still in an experimental condition, instead of having matured into one, in which, like soap and candle-making, every stage is dependent on some ascertained chemical fact.

Messrs. Geneste and **Akitt** have obtained a patent for a process of using boric acid in the steeping-vat. They account for the advantages they claim by affirming that two classes of organisms are always present in the water of the steeper and that these set up two kinds of fermentation. The one resulting in the decomposition of Indican into indigo-white and sugar, and the other probably without the production of any indigo-white whatever. They also profess to have discovered that boric acid acts as a selective antiseptic preventing the latter injurious fermentation. But this explanation of the action of boric acid would seem at variance with the established chemistry of Indican which, according to **Schunck**, cannot by any reaction split into indigo-white and sugar, and there would seem nothing to justify the property of a selective antiseptic which the patentees attribute to the acid they employ. Had they recommended the use of an antiseptic acid that would prevent fermentation *entirely*, and thus allow of the steeping-vat liquor being boiled, their process, in the writer's opinion, would have been in keeping with the chemical nature of Indican, but it is possible the native process of using **Eugenia** bark may be found quite as good as the boric acid.

DRY PROCESS. 268

II.—The Dry Leaf Process.

Instead of rapidly conveying the bundles of freshly-cut plant to the factory, as pursued in Bengal, in Madras it was formerly very generally, and to some extent it is still, the habit to dry the leaves before maceration. An inferior quality of dye is said to be produced by this system, but it has the advantage of being suited to the necessities of small growers who are themselves manufacturers. Factories of a primitive kind are built by speculators, who hire these out to the growers. The dry leaf may have in consequence to be carried to a distance, or only one factory existing in a district, each grower may have to wait his turn of being able to engage it. The plant may have to be cut when it is ripe, but by the dry leaf process it can be manufactured at the convenience of the owner.

In the official report furnished for this work by the Madras Government, the following description is given of the dry leaf system: "The dry-leaf process is carried out in the same way as the green-leaf, except that the plant is cut, dried, and threshed to collect the leaves separately,

MANUFACTURE.

DRY PROCESS.

and these are sold to the manufacturers by weight or by measure. When kept dry, the leaves undergo, in the course of four weeks, a material change, their green tint turning into a pale blue grey, previous to which they afford no indigo. The dried leaves are infused in the steeping-vat, with six times their bulk of water, and allowed to steep for two hours, with continual stirring, till all the floating leaves sink. The fine green liquor is then drawn off into the Beating-vat. In this method the fermentation of the fresh leaves, often capricious in its course, is superseded by a much shorter period of simple maceration." The Collector of South Arcot writes: "I may say that the dry-leaf process is more largely resorted to in this district than the fresh-leaf process of manufacture, seeing that the cultivation is carried on chiefly in the dry season, when very little moisture is retained in the leaves; should, however, there be rain at the time of cutting, the fresh-leaf process necessarily takes place." Most of the Madras reports allude to the leaves as being used whether by the dry or wet process. It thus seems probable, that as in **Mr. Olpherts'** process described above, a much less amount of fermentation is involved than in Bengal; indeed, when the leaves are boiled and a decoction prepared, in the presence of an acid (as is often the case), no fermentation whatsoever takes place. In the description of the method pursued in Cuddapah (the Kurpa indigo of commerce), the Collector, **Mr. E. J. Sewell**, speaks of the fresh leaves as employed, but makes no mention of fermentation being set up in the steeping-vat. The fluid is, however, subsequently beaten for three or four hours, the liquor changing from green to deep blue with the grains of indigo settling to the bottom. He then makes an interesting remark that may have a wide significance. "The juice of '*Neradu*' bark, and sometimes a little gingelly oil is added in the steeping-vat when bubbles begin to rise." "The adulteration with blue mud, which has become so common, is done in the boiling process." The *Neradu* is perhaps the bark of **Eugenia Jambolana**, the bark which **Mr. C. Kough** says is used in Kurnúl for the same purpose. That gentleman adds that—"In some parts, before heating the water containing the decomposed *indican*, sweet-oil is mixed in order that the *indican* may sooner be separated. A small quantity of decoction of *Neradi* bark (**Eugenia Jambolana**), or of *Tugarsa* seed (**Cassia**) and lime-water are mixed to improve the quality of the indigo." The writer in another part of this work (see **Cassia Tora**, *Vol. II., p. 225*) has already dealt with the somewhat puzzling use of the seeds of **Cassia Tora** along with indigo. The knowledge of this property is so widespread as to demand more careful consideration than has hitherto been paid to it. It is mentioned by authors writing of the Pánjab, of Bombay, of Madras, of Bengal, and of China. **Mr. Wardle** found these seeds to contain by themselves a useful yellow colour, and it may be assumed that they have something to say, from this property, to the formation (along with the indigo of a species of **Rhamnus**) of at least one of the forms of Chinese green-indigo. But that they are regularly collected and sold to the indigo manufacturers of Madras, and not for any purpose of making green-indigo, is a fact, and a fact that has to be accounted for. They seem to be of special use in the preparation of *Pala* indigo, a dye prepared from the leaves of **Wrightia tinctoria**, though, as **Mr. Kough** informs us, they are also used in the boiling stage in the manufacture of the indigo of commerce. What action oil can serve is difficult to see, but **Mr. Akitt** has patented the use of *essential oils* which he claims, serve as vehicles to convey oxygen to the white-indigo. But demonstrated science, it may be remarked, has done so little for indigo that the dye manufacturers of Madras must be credited with practical and useful discoveries.

No Fermentation. 269

Kurpa Indigo. 270

Oil used. 271

SHORTT'S PROCESS. 272

Dr. Shortt, in his prize essay on Indigo, describes a process of manu-

MANUFACTURE.

SHORTT'S PROCESS. Preparing by Boiling the Sleeping-vat.

273

facture of much interest; but it is somewhat difficult to discover whether that process is only recommended by him as an improvement, or is actually pursued by some of the European manufacturers of Madras. The fresh plant on arrival at the factory is at once placed in large boilers instead of in a steeping-vat. "The boiling should be continued," says **Shortt**, "until the leaf yields its colouring matter to the water; in the meantime, each boiler should have an attendant with a large prong with which the plants should be constantly kept under water, and who should be careful to see that the heat of the fire is reduced the moment ebullition occurs, for if the boiling is carried on longer than necessary, the colouring matter will be destroyed. This requires a great amount of practical knowledge, and no rule can be laid down which alone will be sufficient for the guidance of the inexperienced. By opening the cocks occasionally, and observing the colour of the decoction (which should be somewhat oily in appearance, and have a reddish colour with a peculiar musty odour), we may judge that the period for discontinuing the boiling has arrived." **Dr. Shortt** then directs the decoction to be run into a beating-vat where, by manual labour or steam power, it should be lashed about. He claims a great economy by this decoction process, from the fact that an hour at most will suffice to cause the precipitation of the dye particles. While believing that the decoction is oxidised by the beating, he repudiates any necessity for chemical agencies, stoutly holding that lime or other precipitants are not only useless but even injurious. He recognises that the beating serves also a mechanical purpose by consolidating or agglutinating the fine particles of dye into grains of sufficient weight to cause rapid precipitation. **Dr. Shortt**'s essay was published some years after the date of **Schunck's** analysis of indigo (1855), and had he seen the admirable results obtained by that chemist, it is probable, **Shortt** would not only have been saved from republishing exploded theories but would have recognised that the explanation of the admirable process he was describing centered around the fact that the boiling process dispensed with fermentation. A decoction so obtained, according to **Schunck**, consists mainly of the glucoside *indican* which, in hot water, especially if slightly acidulated (and the juices of most plants are in themselves acid), will split directly into indigo-blue and indigo-sugar. There being no fermentation, reduction to indigo-white does not take place; oxidation is, therefore, unnecessary, and alkalies added to such a decoction would prove highly injurious, as they would in fact form, with *indican*, compounds from which indigo could not be reduced. **Dr. Shortt's** process is so admirably on the lines of the most recent chemical opinions that, although defective in some respects, it is surprising that it has not been extensively adopted.*

Chemical and other Precipitants unnecessary.

274

No fermentation.

275

From the beating-vat the fecula is conveyed to the pulp-boilers by passing through several strainers, and for this purpose, **Shortt** recommends flannel to be used in preference to cotton or flax. After the supernatant brownish fluid has been run off, the fecula should be freely washed with cold clean water, and as rapidly as possible conveyed through the strainers to the pulp-boilers. If this stage of the operation be delayed, fermentation, **Dr. Shortt** says, will be set up and the indigo seriously injured. To avoid the possibility of this he recommends that clean water should be boiled in the boilers, and the fecula discharged at once into the hot boilers and boiling water. The attendant at the pulp-boilers should be furnished with a perforated ladle, and with this he should keep stirring the pulp, the more so as it nears ebullition. As it boils it begins to give out a strong sugary odour in place of its former musty smell. It should now be turned out by

* It is almost identically that patented by **Mr. F. W. Tytler** in May 1888.

MANUFACTURE. SHORTT'S PROCESS.

the tap from the pulp-boilers and run on to the tables in the press-room, where it is made into cakes by the process already described.

Fermentation injurious. 276

It is impossible to avoid the conviction that the process described by **Shortt** not only prevents fermentation, but that **Shortt** recognised such a decomposition as injurious instead of necessary. It is probable, however, that the process would be improved by the introduction into the beating-vat of an acid to assist in the decomposition of the *indican* into indigo-blue and indigo-sugar. It is even probable that through the aid of acids beating might be found to be altogether unnecessary, though its mechanical action of agglutinating the particles of the insoluble blue dye may always be found of practical advantage. The idea of oxidation, however, either by beating or by direct chemical means, must be rejected as inadmissible, since in a cold or hot acid infusion, where fermentation is prevented, indigo-blue is at once formed, and there is nothing, therefore, to oxidise. In the use of an acid to aid in the decomposition into indigo-blue and indigo sugar, a further advantage might be secured, namely, the prevention of the pulp from becoming by any unforeseen cause alkaline, for in the vats and the pulp-boilers serious injury would then be caused through *indican* being decomposed into non-dye-yielding compounds.

Indican decomposed into non-dye-yielding compounds. 277

INDIGO-GREEN.

INDIGO-GREEN. 278

Olpherts' Process. 279

Recently, **Mr. Richard Olpherts** has taken out a patent for the preparation of both indigo-green and indigo-blue on a new principle. By gentle pressure of the fresh dye-yielding parts of the plant, which, baled in compressible cases or sacs and deposited in the steeping-vat, on being subjected, under water, to the necessary pressure, yield the dye-substance. The pressure requires to be sufficient to burst or break the tissue of the plant. For the preparation of indigo-green the fluid is run off before fermentation and for indigo-blue after fermentation has taken place. In both cases the liquor is beaten in the lower vat, as already described, the fecula washed, strained into the pulp-boilers, and after being boiled is made into green or blue cakes by the usual process. Many years ago **M. Charpentier DeCossigny** drew attention to the fact that a green indigo could be produced by a process somewhat similar to the above, one of the features of which consisted in crushing the plant. Indeed, so long ago as 1790 **Mr. Prinsep** sent to England a green indigo which **Bancroft** pronounced a substance analogous to chlorophyl, mixed with a certain proportion of ordinary indigo. The exact nature, however, of the green dye obtained from the plant on being crushed by **Olpherts'** process has not been determined. Indeed, the chemist will have to furnish other facts than have as yet been made known before the simple act of crushing can be accepted as producing so remarkable a change in the dye-yielding juice of the plant.

Crushing the Plant. 280

Olpherts has also taken out a patent for manufacturing coloured indigoes by chemical means from the waste water of the beating vat. The whole subject of the indigo-reds, browns, greens, and yellows, is too imperfectly understood to allow of their being discussed in the limited space at the writer's disposal. The first Aniline dyes (the *mauves* and *magentas*) were obtained from indigo, and although these are now more economically prepared from coal-tar, the subject of their possible production as by-products of indigo is daily gaining in importance. That much useful material has for many years past been run off from the beating-vat as useless is undeniable, and it seems probable that this will not much longer continue so. **Berzelius** obtained indigo-green by adding potash in small portions to an alcoholic solution of an alkaline *hyposulphindigotate*. The precipitate thereby thrown down after washing

By-products of the Indigo Factory. 281

INDIGO-GREEN. with alcohol, was dissolved with oxalic acid, the filtrate (carrying the green in solution) being next treated with carbonate of lime to remove the oxalic acid. By filtration the lime is separated, and by evaporating to dryness the purified green dye is obtained. This substance is soluble in water and may be used direct as a dye. **Chevreul** also obtained a green substance which was found to be a mixture of indigo-brown and indigo-blue. The examination chemically of the green dye said to be obtained by **Mr. Olpherts'** process, and by a similar method by **Mr. Tytler,** should commend itself to the attention of the chemist. If a green dye can be obtained by a process which practically amounts to crushing the plants, its simplicity should secure its universal adoption, and it is probable a large market may yet be found for an INDIGO-GREEN if such can be actually manufactured. *(See para. below on Green Indigo of China.)*

COST & PROFIT. 282

COST OF MANUFACTURE AND PROFIT.

It is impossible to arrive at any very definite conclusions on this subject, but all authors agree in the opinion that, even at present prices, it pays very well to manufacture the dye. Some writers adduce in support of this statement the well-known fact that the planters are often able to pay heavy interest on borrowed capital and still obtain remunerative returns. Under these circumstances it seems enigmatical that indebtedness should exist at all, and it is more probable that, as with tea and other industries, indigo manufacture pays, but does not give the cent per cent returns which have been asserted. It is commonly maintained that the Madras cultivators are better off than those of Bengal, because they are also manufacturers, and that R40 to R60 which they obtain for the maund of dye gives them a handsome return. Their capital invested is probably less proportionately to that of the European, who obtains R200 for the same quantity, seeing that the latter also produces a purer article.

The following statements regarding indigo manufacture are extracted from the report furnished by the Director, Land Records and Agriculture, Bengal:—

In **Mr. Sheriff's** Report the cost of manufacture in Nuddea is said to be about R30 per factory maund (a factory maund is 72lb 10¼oz.) If the crop be a good one, and prices are a fair average, the profit is from R50 to R75 per maund; but if the reverse is the case, there is little or no profit, and very frequently a loss.

"Cost of manufacture in Purneah comes to about R35 per maund, factory weight, equal to 74lb 10 oz. The profits of a factory depend on a favourable season and good prices. The cost of cultivation, including seed, manufacturing, establishment, and all other charges and contingencies, comes to R150 per factory maund on an average."

"The figures of the budget of a factory in Behar cultivating 1,500 acres are stated in the *Indian Agriculturist* of the 23rd February, thus—The rent payable to the zamindar was R69,000, but the amount recovered from the villages was R70,020, so that the rent was more than paid by the villagers, and the profit to be derived from the indigo land was all to the good. Nearly R1,20,000 was required under various items for working expenses, including establishment, cattle, salaries of three European managers, and everything; and the total working charge for the year was in round numbers R4,20,000 [? misprint for R1,20,000—*Ed. Dict.*] The actual yield, however, was 1,150 maunds (of 82lb), which sold for an average price of R200 per maund, thus bringing in R2,30,000; or, in other words, after allowing 10 per cent. on the capital sunk in the factory, and another 10 per cent. as a reserve fund for wear and tear, the factory yielded nearly cent. per cent. on its working charge. These facts, it is added, show

COST & PROFIT.

what enormous profits can be made out of the indigo industry—profits so large that the planter can afford to borrow his capital and pay 22 to 23 per cent. for charges and interest." The writer regards the above balance sheet as absolutely misleading, though it has been confidently published and republished by several writers since it originally appeared. In addition to the obvious misprint, to which attention has been drawn above, parenthetically, the yield of dye is estimated at three times as high as we have any data to believe it could be, even in Tirhút. The factory is said to have cost about R50,000.

Factory Employés. 283

The following particulars from a recent official source may be here given regarding the number of employés and the salaries paid to the same at a Bengal factory:—"The number of persons constantly employed in a factory runs from 25 to 40, but at the manufacturing time there will be from 200 to 300. The native employés are paid from R10 to R30 per month; the more subordinate servants from R4 to R8 per month; carpenters, sawyers, and brick-layers from R5 to R8 per month; coolies from R4 to R6 per month, and at the manufacturing time as high as R8 and R9 per month.

"The following are the employés in a factory at Gya: –

		R	*a.*	
One mukhtar	at	12	0	per mensem.
„ karinda	„	8	0	„
„ jamadar	„	6	0	„
„ mori mate	„	4	8	„
Nil maha mate	„	4	0	„
„ maha	„	4	0	„
Pechawa	„	4	0	„
Chulhakash	„	4	0	„
Ziladar	„	3	0	„
Total		49	8"	

Number of Factories. 284

NUMBER OF FACTORIES AND OF EMPLOYÉS ENGAGED ON INDIGO MANUFACTURE.

In Madras a large number of small native factories are returned simply as "Vats," while in the North-West Provinces a very large number are returned as "Factories," which probably have as little claim to that designation as the vats of Madras. Accepting the returns as they stand, there are in all India 2,762 factories and 6,032 vats, and these give employment to 356,675 persons. The greater proportion of the employés, however, are daily labourers, being engaged during the working season of the factories only. The above estimate of the persons who find employment at Indigo factories does not of course include the rayats who cultivate the plant, although many of these doubtless earn an additional wage from serving as temporary hands at the factories.

Indigo Cases. 285

MATERIALS USED IN THE CONSTRUCTION OF INDIGO CASES.

The principal Wood for this purpose is the mango, but others may also be employed, but of whatever wood made the cases should be tarred. It is urged by planters that the wood employed should be seasoned for at least one year, since the effect of unseasoned boxes is to cause a difference in the weight at the factory, at Calcutta, or at London. For a list of the woods suitable for Indigo chests see under the heading PACKING CASES.

COMMERCIAL INDIGO.

ISOLATION. 286

ITS FORMATION, ISOLATION, VALUATION, AND PRICE.

This substance is chiefly a product of vegetable origin, but it is found also, under certain conditions, in the blood and urine of animals, and even in the milk of cows fed on *saintfoin*.

ISOLATION.

Though obtainable from many plants, in none of them does indigo appear to exist ready formed. According to the accepted chemical history of the substance, it is derived by the decomposition of a soluble glucoside now known as *Indican*. Under certain influences, *Indican* splits into *Indigotin*—blue-indigo—and *Indiglucin*, a saccharine substance, soluble in water and alcohol. When boiled in nitric acid Indiglucin forms oxalic acid. In an aqueous, and still more so in an ammoniacal, solution of nitrate of silver, it reduces the salt to metallic silver, and similarly separates gold from the tri-chloride. It is not fermentable but turns acid by prolonged contact with yeast

Roxburgh (*Trans. Soc. Arts, Vol. XXVIII.*) supposed the indigo plants to contain only the base of the colouring matter, which of itself was green, and that alkalies were necessary to eliminate the indigo-blue. **Giobert** was of opinion that indigo plants contained a colourless substance called by him *Indigogen*, which was soluble in water and possessed more carbon than indigo-blue, into which it was converted by the removal of the excess carbon through combination with the oxygen of the air. This oxidation he regarded as promoted by heat and by the presence of alkalis such as lime, and as arrested by acids, even carbonic acid. He thus accepted the *rationale* of the process as pursued, *viz.*, fermentation and subsequent treatment with lime. **Chevereul** supposed woad and other indigo-yielding plants to contain the soluble or reduced form of indigo known as *white-indigo,*—a substance which has one atom of hydrogen more than blue-indigo. This explanation being extremely simple and in conformity with the treatment of the colour in the hands of the dyer was for a time very generally accepted. On this theory, to some extent at least, is probably based the Anglo-Indian method of manufacturing the dye. The fact was forgotten, however, that white-indigo was soluble in alkaline substances only, and also the further fact was lost sight of, *viz.*, that the leaves of plants in the processes of respiration and assimilation are constantly permeated by, and exhaling, pure oxygen. White-indigo, therefore, could not exist as such in the sap of plants, but must necessarily be converted by oxidation into blue-indigo—a substance which, if present, could be seen as blue grains in the sap. **Schunck**, to whom we owe most, if not all, our knowledge of the chemistry of Indigo, affirms that the juice of most indigo-yielding plants is acid. But **Sachs** has shown that though the sap in the parenchyma and in the vessels of plants are not invariably distinct, yet in most cases the cellular tissue contains chiefly sugar, starch, oil, &c., and also organic acids and acid salts, and are in consequence acid to litmus paper, while the vascular tissues contain mainly albuminoids and give indications of being alkaline. Although the juices of most plants, on being mechanically squeezed out of them, may be acid, or even a decoction obtained by maceration in water may be so, there still remains an important subject of enquiry—the exact location within the tissue of the indigo plant of **Schunck's** *Indican*. The physical property which protoplasm exercises, so to speak, at the necessities of plant organism, in making starch soluble or insoluble, is a potent argument against the advisability of accepting as conclusive chemical laboratory experience. Without desiring to throw doubt on the value of **Schunck's** discoveries in the chemistry of the indigo derived from indigo-yielding plants, it may be remarked that there are certain phenomena witnessed at the indigo factory that suggest the desirability of a thorough examination, both botanically and chemically, of the Indian plant. The results would most probably confirm, and in certain practical directions even amplify, **Schunck's** opinions and conclusions. Indigo, as **Schunck** very properly remarks, is "a substance which is formed sparingly indeed, but in widely distant parts of the organic world.

ISOLATION. **Indigo does not pre-exist in the plants from which it is prepared.** 287

The properties of indigo-blue, which are so peculiar as almost to separate it from all other organic bodies, and to constitute it one *sui generis*, naturally suggest the inquiry, in what form it is contained in the plants and animals from which it is derived. If it exists ready formed in the indigo-bearing plants, how is it that though, when in a free state, insoluble in water, acid, alkalies, alcohol, and most simple menstrua, it should so easily be extracted from those plants by a mere infusion with cold water? If it does not pre-exist in the plant, in what state of combination is it contained therein, and what is the nature of the process by which it is eliminated? The usual method of preparing indigo from the INDIGOFERÆ consists in steeping the plant, especially the leaves, in water, drawing off the infusion, allowing it to undergo fermentation, and then precipitating by means of agitation with air and the addition of lime-water. Now it may be asked, is this process of fermentation, which is often very tedious and difficult to manage, essential to the formation of indigo-blue, or is it merely an accidental phenomenon attending its preparation? If it is essential, at what stage of the process is the formation of the colouring matter to be considered as completed; and is it necessary, as some persons assert, to continue it until actual putrefaction has commenced, or not? These are points which, though perhaps of little consequence to the dyer and consumer of indigo, are of great interest in a chemical point of view, and are of the greatest importance to the manufacturer of indigo. To the latter it must surely be extremely desirable to know the exact nature of the process on which his manufacture depends, and to ascertain whether this process yields into his hands the whole quantity of the product which the material employed is capable of yielding, and also whether the manner of conducting it is in perfect accordance with theoretical requirements" (*Dr. E. Schunck, Philos. Magaz. Aug. 1855*). Although **Schunck** in his interesting articles on the chemistry of indigo does not himself sum up his conclusions and answer the question as to the necessity of fermentation, his results abundantly do so. Some twenty years later **Crookes** remarked:—"What are the chemical phenomena observed during the process of the fermentation of the plant, and under what influence and from what substance present in the plant is indigo derived? Since no researches on the fresh indigo plants have yet been made, and since, moreover, the different phases of the operations whereby indigo is produced on the large scale have never been scientifically examined, we can therefore only reason by analogy or from observation of what happens with other similar plants, and give an explanation which is probably, but not necessarily, correct."

It must be admitted that the salient points raised in the above quotations remain to a large extent unanswered, and that indigo manufacture will continue in the position which, until recently, brewing, distilling, and many other such industries were,—namely, one of experience, of tasting, smelling, and guessing,—until the whole process from the germination of the plant to the production of the cake has been subjected to a critical examination by a committee of botanists and chemists, who alone are ever likely to be able to place it on a platform of scientific accuracy. As it seems to the writer, the feature of importance from the manufacturers' point of view is whether fermentation is a necessity or a consequence of the steeping-vat. If not necessary, whether any advantage would result from the establishment of a system of manufacture where fermentation would be prevented. On this point **Schunck** has elucidated certain facts which seem to have been entirely overlooked, and indigo patents have accordingly been directed towards securing the most complete and rapid oxidation of the decoction in the beating-vat without the idea having apparent-

Is fermentation necessary? 288

INDIGOFERA tinctoria.

Commercial Indigo.

ISOLATION.

ly occurred to any one of attempting the direct isolation of indigo-blue in the steeping-vat. On this point **Schunck's** experiments afford valuable evidence. "**Roxburgh,**" he says, "first directed attention to the fact, that it is possible to obtain indigo by merely treating the plant with hot water, and then agitating the infusion with air, from which it follows that fermentation is not an absolutely essential condition of the formation of indigo-blue." Further on **Schunck** remarks, "The process of decomposition by which colouring matters are formed from other substances are of two kinds. The first consists in the absorption of oxygen and the elimination of hydrogen in the form of water; it is a process of decay, and requires the presence, not only of oxygen, but of some alkali or other base. The second process is one which consists in the splitting up of the original compound into two or more simpler bodies, of which one or more are colouring matters; it is a process of fermentation, and may in general be effected as well by the action of strong acids as by that of ferments. The first process gives rise to colouring matters of a very fugitive nature, such as the colouring matters of logwood and archil. Indeed, in this case the colouring matter, if this name be applied merely to substances endowed with a striking and positive colour, is only one of a long chain of bodies succeeding one another, and is generally not the last product of decomposition. The other process, of which the formation of alizarine is an example, yields colouring matters of a fixed and stable character, which are not further changed by a continuance of the process to which they owe their formation. Now, if indigo-blue be a body which is formed from some colourless substance existing in the plant, we should infer *à priori* that the process by which it is formed is one of fermentation or putrefaction, not requiring the intervention of oxygen or of alkalies." **Schunck** then proceeds to describe the process he adopted for preparing from the powder of dried woad leaves the extract containing the substance which, as stated, he called *Indican.** **Schunck** remarks: "By these and similar simple and easily performed experiments, I was enabled to infer with positive certainty, that **Isatis tinctoria** contains a substance easily soluble in hot and cold water, alcohol and ether, which, by the action of strong mineral acids, yields indigo-blue; that the formation of the colouring matter from it can be effected without the intervention of oxygen or of alkalies, and that the latter, indeed, if allowed to act on it before the application of acid, entirely prevents the formation of colouring matter." The use of lime, a practice now almost abandoned by European manufacturers, would thus appear to be dangerous and often positively injurious. In the steeping-vats in general use, fermentation takes place. But to extract the substance or compound from which, by decomposition, indigo is afterwards prepared, fermentation is not necessary. That compound is soluble in both hot and cold water, and simple maceration is therefore all that is required to remove it from the plant. By fermentation, however, the decomposition is effected into indigo-blue and indigo-sugar, but that decomposition **Schunck** preferred, apparently, to bring about by the alternative process of acids. Fermentation would further reduce indigo-blue to indigo-white, and it seems only natural to suppose that this always takes place more or less, thus necessitating the subsequent oxidation in the beating-vat to reconvert the dye back to the insoluble state. **Gehlen** believed that the beating was necessitated, not so much to oxidise the produce of the steeping-vat, as to aggregate the precipitate into particles, and thus more easily allow of the separation of the dye from the solution.

* He has since prepared it from other plants, and **Michea** has obtained it from **Indigofera tinctoria.**

I. 288

ISOLATION.

Schunck next proceeds to deal with the effect of boiling or heating *Indican*. "There is another very remarkable property, he says, which I have to describe—a property the knowledge of which will probably throw great light on the process of manufacturing indigo. If *indican*, in the form of syrup, as obtained by evaporation of the watery solution, be heated for some time in the water-bath, or if its watery solution be boiled or even moderately heated, it undergoes a complete metamorphosis." When this change has once been accomplished, *indican*, he affirms, "ceases to give the least trace of indigo-blue with acids." It has, however, chemically only taken up the elements of water; but does so more readily if the *indican* be heated in the presence of alkalies. Instead of indigo-blue and its allied red-colouring matter, it now yields substances which possess no tinctorial property. It may, therefore, fairly be assumed that if the temperature of the steeping-vat be raised too high, not only will *indican* be decomposed into indigo-blue, and that substance still further reduced to indigo-white, but that a certain proportion (probably sufficient to materially injure the value of the dye product), will be completely destroyed through its entering into chemical combination with water. And this danger is possibly not confined to the effects of the heat of fermentation in the steeping-vat, but may also take place with the *indican* that may not have been reduced, but carried to the boiler, where the syrupy mass is concentrated. It may, however, be emphasised that these changes can only take place in *indican* as such, and not in the blue indigo obtained by the decomposition of that glucoside. If the fermentation adopted does not at once reduce the whole of the *indican*, and the temperature of the fluid be allowed to rise, the metamorphic change may take place in the vat, or, if undecomposed *indican* passes over to the boiler, it will there undergo a change, producing dark-coloured compounds to the detriment of the indigo with which it will then be mixed.

The enquiry thus seems justified as to whether a process of fermentation in the steeping-vat is the most convenient, economical, and expeditious. In the laboratory **Schunck** found that *indican*, while undergoing decomposition with acids, "splits up immediately into one equivalent of indigo-blue and three equivalents of sugar." It is, however, possible, he says, "that these three equivalents of sugar may not be eliminated all at once." He concludes his first paper on this subject by three inferences drawn from his experiments with woad:—

"1. The **Isatis tinctoria** does not contain indigo-blue ready formed, either in the blue or colourless state.

"2. The formation of the blue-colouring matter in watery extracts of the plant is neither caused nor promoted by the action of oxygen or of alkalies.

"3. Indigo-blue cannot be said to exist in any state of combination in the juice of the plant; it is merely contained in them potentially."

Indigo is thus one of the colouring materials which, in the passage quoted above, **Schunck** affirms, can be derived only by a decomposition caused through fermentation *or by means of acids*. *Indican* is decomposed even by certain dilute acids in the cold, and *more quickly when heated*. This decomposition is, for example, induced by tartaric and oxalic acids, and less easily by acetic acid. **Schunck** gives it the formulæ of $C_{26}H_{31}NO_{17}$, and expresses its decomposition thus:—

$$\underset{\text{Indican}}{C_{26}H_{31}NO_{17}} + \underset{\text{Water}}{2H_2O} = \underset{\text{Indigo-blue}}{C_8H_5NO} + \underset{\text{Indiglucin.}}{3C_6H_{10}O_6}$$

Isolation of Indigo by means of acids. 289

Without experimentally testing the views which a study of the chemistry of this subject induces, it is impossible to suggest practical reforms; but it would seem as if a system of adding dilute acid to a steeping-vat, after a few hours' maceration, so conducted as to prevent fermentation, might

ISOLATION.

result in the immediate decomposition of *Indican* and the formation of Indigo-blue, and thus dispense with the expensive and laborious process of oxidation at present pursued. Indeed, it is probable that the waste product *Indiglucin* might be utilised as the material from which to manufacture the acid necessary. Were such a system found practical, there can be little doubt but that it would result in the production of a purer article, and one freer from adulteration with the Indigo reds, browns, yellows, and greens which so materially lessen the value of the dye as manufactured at the present day. The facts here published seem well worthy the attention of the planter. The old woad manufacturers of Europe extracted their indigo by means of tepid water and lime, a process which rendered fermentation impossible. The use of lime, while precipitating rapidly the indigo-blue when once formed, doubtless acted injuriously, in that it caused the destruction of a large amount of the *indican*. Throughout India, wherever the European planters' influence has not extended, the natives extract their blue dyes by a very similar process to that formerly pursued by the woad manufacturers. Thus the Madras method of boiling, as described by **Dr. Shortt** in the passage quoted above under the Chapter on "Manufacture of Indigo" (p. 436), is practically that experimentally tried by **Roxburgh** and is in principle identical with the old woad method. So in the various provinces of India, a primitive system of manufacture (as it is called) is spoken of as pursued by the Natives, in which fermentation is prevented. Thus, for example, of the Godavery we read, "a fire-place of mud is constructed sufficiently large to hold sixteen *chatties*, and in these *chatties* the plants are placed, and water being poured upon them, the whole is boiled for an hour and a half. The stalks of the plants are then taken out of the *chatties* and the liquor is poured into large jars in which it must be well stirred for two hours, the scum and froth being removed as it rises to the surface. A decoction is made from the green wood of the *Nerídu* tree, and a small quantity of it being poured into the jars containing the indigo liquor, the indigo separates itself from the liquor and falls to the bottom of the jar." Now, by boiling the plant in water a hot solution of *indican* must be obtained, and that too without any fermentation. It is not quite clear what object is served by the subsequent beating of the liquor, but it seems evident that the decoction from the wood which is thereafter added probably contains some acid principle, since it is only at this stage we are told the indigo separates itself from the liquor. By the primitive indigo manufacturers, the juices or gums of certain plants are considered necessary adjuncts in precipitating the dye, some of which would seem well worthy of scientific investigation. It is just possible that, as organic products mostly unknown to the chemists, they may actually possess the property assigned to them. Thus the native manufacturers of the North-West Provinces are said to drain off the surface water from the steeping-vat after which "the soaked plants are trodden out by men for six hours. Some *dhak* **(Butea frondosa)** gum is then added in the proportion of about eight ounces per vat. Water is again run into the vat and the steeping continued for a few more hours, when the water, loaded with indigo particles, is drawn off into another vat, well beaten, and the fecula allowed to settle; when this has taken place the surface water is drained away and the muddy-looking residuum collected in earthen or metal pans, strained, and spread out on cloths resting on a layer of sand. In this way the remaining moisture is absorbed. The indigo is then moulded into cakes, each about the weight of eight ounces, finally dried on ashes and in the sun, and, after about a week, is ready for the market. The imperfect manner in which the leaves and fibres of the plant are

Native method without fermentation. 290

Aids to precipitation. 291

separated from the fecula, and the addition of foreign substances, like gum, to assist fermentation, result in the production of indigo very inferior to that made up in factories so called, however well suited to the moderate requirements of the dyers of the country" (*Report of Muzaffarnagar on Dyes and Tans of North-West Provinces*). It would seem highly improbable that so powerfully astringent a gum as Bengal kino (the gum used in the above process) could assist in fermentation. The process as narrated would, moreover, lead to the supposition that by cold maceration *indican* was extracted from the twigs, and that the **Butea** gum assisted in the decomposition of that substance into its blue indigo and sugar. Mr. Sheriff (*in the Report furnished by the Director of Land Records and Agriculture, Bengal*) says: "Formerly cold water or weak lime-water was added to the water in which the indigo plant was steeped, to hasten the precipitation of the dye, or as a substitute for beating. But this practice is not followed now by the European manufacturers, nor by native manufacturers who prepare the stuff for the European market. The use of lime is said to make the indigo hard and red. It is, however, still employed by village dyers in Bengal, who prepare their own dye. In the manufacture of *kachha* indigo, or the dye prepared for consumption in the country, the gum of **Butea frondosa** is added in some parts of the country to hasten the precipitation. Sometimes the bark of **Eugenia Jambolana** or of **Zyzyphus Jujuba** is also used for the same purpose." The danger of using lime or any other alkali in the steeping-vat, and its utter uselessness for the purpose, causing the decomposition of *indican* into *indigotin*, has already been dealt with, but the habit of throwing a quantity of cold water into the hot extract or fermented decoction seems well worthy of more careful examination. The three plants mentioned above, as used in Bengal to cause the formation of indigo-blue, are, curiously enough, employed for the same purpose in Madras and in the North-West Provinces, or from one end of India to the other. There is a uniformity and correspondence to woad manufacture in the native method of preparing indigo, that either denotes a principle of ascertained merit, which the European planters have neglected to investigate, or must be viewed as an additional proof that the art is an imported and not a naturally evolved one. Thus, in the special report furnished by the Madras Government for this work, the dry-leaf process pursued by petty manufacturers is briefly described:—"Rayats sometimes manufacture, on a small scale, in large earthenware pots about 2¼ feet high; this generally with dried leaf. Such pots are placed in a range of four, six or eight, over one fire and slowly boiled. As soon as the boiling has reached a certain point, the indigo leaves are removed and the liquor is poured into earthen vessels and churned for about half an hour. Then there is added to it a solution of the bark of the *jaumoon* tree (**Eugenia Jambolana**, *Lank.*, MYRTACEÆ) for the purpose of separating the dye from the water. The indigo liquor is then allowed to stand until the product deposits itself at the bottom of the vessels, when the water is generally poured off and the deposit thrown into moulds to harden." It must be allowed that there is a simplicity about this process which, if susceptible of improvement, would very greatly cheapen the price of the dye. The plant by the native process is said to be boiled, but no fermentation takes place. Soluble *indican* is doubtless thereby extracted, which must early begin to split into blue-indigo and sugar. The danger evidently connected with this process is the indiscriminate heating or boiling of the aqueous solution (*cf. p. 60*). Schunck has shown most conclusively that oxygen is in no way necessary to the decomposition of *indican*, nor for the formation of its resultant compounds—*Indiglucin* and *Indigotin*. Oxidising agents in the steeping-vat are therefore not only unnecessary and wasteful, but may be

ISOLATION. Aids to precipitation.

Butea gum. 292

Lime. 293

Eugenia bark. 294

Cold water. 295

Oxygen not necessary. 296

ISOLATION. Aids to precipitation.

absolutely injurious. An alkali added at this stage to the hot decoction would, however, cause its reduction into three or four useless non-dye-yielding compounds. The juice of the plant is naturally acid, however, and this may be thereby prevented. The solution of **Eugenia** bark most probably contains the acid necessary to accomplish the complete decomposition of *Indican* into *Indiglucin* and *Indigotin* (blue-indigo), a decomposition which would take place to a certain extent in water alone. This decomposition had better not be fully started however, until after a short stage of maceration or careful heating so as to remove the *indican* from the plant and not decompose it within the tissues, as might take place if boiled at once in a strong acid fluid. The absurdity of fermenting the decoction in order to reduce *Indigotin* to *Indigogen* (white-indigo) only to have to reconvert *that* back by oxidation, either by means of beaters or patent chemical processes, is abundantly exemplified by the above simple process, which, in the hands of a careful manipulator, would, chemically speaking, produce blue-indigo by the direct decomposition of *indican*.

Eugenia bark could be cheaply enough procured, but its chemical analysis would reveal the nature of the acid which, in a hot decoction, reduces the *indican*. The immense volume of indigo-sugar which, from the factories, is annually poured on the fields as manure, would, however, suggest the desirability of testing the writer's suggestion of the possibility of utilizing that by-product as the source of the reducing acid.

VALUATION. 297

TESTING AND VALUATION OF INDIGO.—As met with in the market the approved form of the dye occurs in small cubic pieces; when of good quality these are so light as to float on water. They are of a violet-blue colour and assume a coppery aspect when rubbed with a hard polished body. They are free from flaws or cavities and are not traversed by veins of white or brown. Good sorts contain from 50 to 60 per cent. of *indigotin*, but in addition to the existence of other products of the decomposition of *indican*, such as red-indigo, brown-indigo or indigo-glutin, mechanical impurities are to a certain extent unavoidably present. The percentage of dirt is, however, greatly reduced by careful and clean manufacture. The inferior sorts of indigo contain a large amount of fraudulently added dirt, such as sand, peculiar muds, starch, powdered lead, &c. The decline of the Bombay indigo manufacture is generally attributed to the adulteration practised. Of Khandesh it is said, for example, that wood-ashes are used in adulteration. In the Manual of the Cuddapah district, Madras, it is reported that there occurs a sort of bluish mud which is used in adulteration of indigo. "The demand for the silt is so great that it has become an article of commerce: it is collected and sold in the bazars and is readily bought up by some native vat-owners during the manufacturing season at 8 annas a maund." A number of similar passages might be quoted from nearly every district of India where native indigo is made. It may have been inferred from what has already been said that the Indian dyers recognise two primary kinds of indigo: (*a*) that made by, or on the principle pursued by, European planters, namely, of boiling down the decoction to a thick syrup and compressing the same into cakes. This is generally spoken of as "boiled" or "English indigo;" (*b*) that made for the local market, in which the decoction is thrown into moulds and allowed to dry slowly at the ordinary temperature of the atmosphere. The latter is generally known as *gád* indigo.

ADULTERATION. 298

PURIFICATION. 299

Various methods exist of preparing pure indigo for laboratory purposes and of estimating the actual percentage of *indigotin* in commercial cakes. It may be purified from most of its foreign matters by treating it successively with dilute sulphuric or hydrochloric acid, with boiling water, and with alcohol. But the most thorough method is to have resort to the

PURIFICATION.

principle of the vat, in making soluble indigo-white, removing the insoluble materials by filtration, and from the filtrate throwing down the insoluble blue-indigo. The weight of the dry indigo-blue to the original weight of the sample is relatively the amount present. The sample should be first dried over the water-bath to ascertain the amount of moisture present. In good samples this should not exceed 3·5 to 6 per cent. The amount of ash may next be determined by calcining a certain amount in a platinum crucible. Good indigo generally yields only 7 to 10 per cent. The exact amount of sand or other mechanical impurities may be discovered by washing out the soluble ingredients of the ash. The detection of the presence of starch is more difficult. If the starch has been coloured blue by iodine the indigo cake is pale coloured, has a great density and friability. The actual presence of starch may be demonstrated by treating the indigo with slightly alkaline water, neutralising the filtered liquid with a few drops of acid and testing with iodine. The characteristic blue reaction will be restored proving the presence of starch.

DECOMPOSITION. 300

Decompositions of Indigo.—Indigo-blue melts and boils when heated in contact with air. At high temperatures it burns with a bright and smoky flame. Dry chlorine does not act on it between 0° and 100°C., but if indigo-blue be stirred up with water in a paste, and chlorine passed through it, the mass becomes first green then yellow. A similar decomposition takes place with bromine and with iodine when heated. When boiled with nitric acid it is converted into *isatin* with the evolution of gas. Concentrated chromic acid immediately decomposes it with violent evolution of carbonic anhydride. Sulphuric acid acts as a solvent forming the so-called Indigo-extract.

PRICE 301

Price of Indigo.—There is perhaps no other article that fluctuates so much as does indigo. The mean annual price varies from year to year according to production and demand, but the outturn of each factory also varies from day to day, depending mainly on the condition in which the plant reached the factory, the nature of the available water, the temperature at the time of steeping, and the success or failure of the manufacturers' instructions as to steeping, fermenting, beating, boiling, drying, cleaning, and packing. One point on which all writers agree is, that the yield and quality of dye is greatly injured if fermentation be set up in the leaves before maceration, and accordingly wet leaves are never collected. It seems quite possible that wet leaves closely packed together would originate a certain amount of decomposition of *indican* within the tissue, the insoluble blue-indigo becoming imprisoned as it were, and the leaves accordingly dyed. Considerable experience is also necessary as to the exact age at which the plants should be cut, for it seems established that there is a period of maximum yield beyond which the leaves give less and less, and before which they also afford an inferior yield and quality of dye. All these dangers and precarious conditions have to be guarded against, and hence, with no chemical apparatus to forewarn him, the manufacturer must often fail and thus produce one day an article of much less value than another. In carefully worked factories each day's manufacture receives a current number, and the various days' productions are never mixed, so that each case of indigo may be depended upon as being of one quality. This has greatly raised the reputation of high merit for Bengal indigo, and enables large consignments to be sold on approved samples. With Native manufacturers, each producing a few maunds, or it may be only pounds, this is quite out of the question, and consequently Native indigo often fetches relatively a far lower price than it merits. The whole traffic in Native Indigo is, as it were, a matter of accident, since it would be too laborious to test each small consignment for the actual amount of dye it contains. One

PRICE.

sample may be sold for relatively more than it is worth; another for considerably less. Fraudulent adulteration is the main cause of the low prices realised for Native indigo. The prejudice against it is in fact likely to survive long after any possible education into the advantages of producing a cleaner article, so that the course of reform is necessarily slow. The small Native manufacturer, obtaining, as he does in Madras, some R30 to R70 for a maund of his adulterated article, is handsomely rewarded for his labour and capital expended, and he is but little interested in the news that his Bengal contemporary can earn R200 a maund more than is obtained for Madras indigo.

The price of the dye will be found dealt with in various places in this brief summary of the main facts regarding Indian indigo. The reader is referred to the chapter on the History of the Dye where the valuations are given of samples of good quality in 1810 as furnished by the Board of Directors of the East India Company. In the chapter on the "Chemistry" and also on the "Trade in Indigo" the prices ruling both in England and India have been discussed. **Mr. J. E. O'Conor** of the Finance and Commerce Department has obligingly furnished the following facts:—

"Price of Indigo (Middling to Good) in Calcutta in the seasons of—

Conf. with prices in 1810, see p. 395.

	R	
1887 November . . .	232½	per Factory maund.
,, December . . .	230	,, ,,
,, January to February .	210	,, ,,
1888 November . . .	210	,, ,,
,, December to February	215	,, ,,
1889 November to February	245	,, ,,

"Average prices of Bengal Indigo in London—

1880	7*s.* 2½*d.*	per lb.
1881	6*s.* 6⅛*d.*	,,
1882	6*s.* 3½*d.*	,,
1883	5*s.* 10⅞*d.*	,,
1884	6*s.* 1½*d.*	,,
1885	5*s.* 3¾*d.*	,,
1886	5*s.* 1¼*d.*	,,
1887	4*s.* 10¾*d.*	,,
1888	4*s.* 9½*d.*	,,
1889	4*s.* 7⅞*d.*	,,"

The fall in the price shown by the above figures is all the more remarkable when it is recollected that India now holds a practical monopoly in the world's supply of the dye. This is doubtless to some extent due to the increased production of Madras indigoes, indeed, of indigo generally as a native industry.

The following table also furnished by **Mr. O'Conor** exhibits the comparative fluctuations in the value since 1850—

Variations in the wholesale prices of Indigo in London, expressing the average of 1845-50 as 100.

1845-50, average of 6 years . =	100	1868 1st ,, ...	154
1851 1st January . . .	128	1869 1st ,, ...	143
1853 1st July . . .	161	1870 1st ,, ...	151
1857 1st ,, ...	121	1871 1st ,, ...	137
1858 1st January . . .	163	1872 1st ,, ...	159
1865 1st ,, ...	137	1873 1st ,, ...	169
1866 1st ,, ...	126	1873 1st July	124
1867 1st ,, ...	145	1874 1st January . . .	123

PRICE.

Variations in the wholesale prices of Indigo in London, expressing the average of 1845-50 as 100—contd.

1874 1st July	156	1882 1st January	195
1875 1st January	157	,, 1st July	188
,, 1st July	150	1883 1st January	190
1876 1st January	130	,, 1st July	144
,, 1st July	137	1884 1st January	151
1877 1st January	173	,, 1st July	167
,, 1st July	147	1885 1st January	157
1878 1st January	169	,, 1st July	133
,, 1st July	164	1886 1st January	153
1879 1st January	167	,, 1st July	130
,, 1st July	169	1887 1st January	131
1880 1st January	205	,, 1st July	131
,, 1st July	206	1888 1st January	129
1881 1st January	197	,, 1st July	129
,, 1st July	189	1889 1st January	125

INDIGO SUBSTITUTES.

SUBSTITUTES. 302

Blues derived from Coal Tar.—Mr. Alfred H. Allen (*President of the Society of Public Analysts*) has done a good service to the indigo industry, by making known the exact position of the chemical investigations which have been conducted with the aim of producing chemical indigo from coal tar. The complete success that attended the artificial production of madder, and the consequent collapse of the European trade in cultivating the plant (**Rubia tinctorium**) and in manufacturing the dye, gave to the enquiry into artificial indigo a seriousness that could not be ignored. The result was awaited anxiously; but, for the present at least, it may be said that the Indigo Planter is left master of the situation. Mr. Allen has pronounced the chemists' endeavours as unsatisfactory from a practical point of view; for, while he can now produce "*Indigotin,*" he cannot do so at a price to compete with the natural product. The following interesting passage may be here given from Mr. Allen's report:—

"If a piece of cotton be introduced into a vat, containing the solution of reduced indigo, the liquid penetrates to the interior of the fine tubes constituting the fibres of cotton, and on subsequently exposing the material to the air, oxidation takes place, the insoluble blue-colouring matter is formed, and, being *in the interior of the fibre*, the goods are dyed a fast blue, incapable of removal by washing, and wholly unacted on by light, soap, or diluted acid or alkaline liquids. By chloride of lime the colour may be bleached, but this has a distinct practical advantage, for a white pattern may be produced on a uniformly blue ground by restricting the treatment with the bleaching agent to those parts of the cloth from which a discharge of the colour is desired. Some years since, by a series of highly complex and ingenious processes, artificial ***indigotin*** was obtained from Coal-tar. Owing, however, to the extremely tedious character of the preparation and the unavoidable loss occurring at certain points of the process, resulting in the formation of a large proportion of valueless products, the production of indigo from Coal-tar has not proved an industrial success, and the manufacture has been almost discontinued, and doubtless will be entirely so when the present stocks are exhausted. The artificial indigo from Coal-tar is identical with the natural product in every respect, and it is merely a question of ***price*** which has prevented it from superseding the colouring matter made in nature's laboratory.

Artificial Indigotin. 303

"But this question of price has been settled, for the present at any rate, in favour of the natural products. The case is not parallel to that of Alizarin, the colouring matter of madder, the synthesis of which from Coal-tar has resulted in the complete abandonment of the cultivation of madder.

SUBSTITUTES.

Alizarin was contained in madder to the extent of about 1 per cent. only against an average of, say, 50 per cent. of colouring matter in natural indigo. Hence, natural indigo-blue is much cheaper than ever alizarin was, in the form of madder, and is obtained from the anthracene of Coal-tar by processes which, if complex, are simpler, fewer, and less wasteful than those by which artificial indigo is obtained. Hence, the conditions of the two problems are not on a par, and no trustworthy deduction can be drawn from the history of artificial Alizarin. But, besides true artificial indigo-blue or *Indigotin*, a number of blue colouring matters are now manufactured from Coal-tar, some of which at least are claimed as more or less formidable rivals of indigo. The most important of these indigo substitutes is unquestionably the colouring matter known as *Alizarin blue.*

Alizarin Blue. 304

This substance resembles indigo in its insolubility in water, and its property of forming a soluble reduction-product which is reconverted into the insoluble blue on exposure to air. Hence, it is capable of being applied in vat-dyeing in a manner precisely similar to that employed with indigo, but is unable to compete in price with the older dye when used in that manner; besides which, its tendency to form insoluble lime-lakes is not in its favour, owing to the inconvenience attending the use of hard water. Even if reduced in price, as it probably will be when the patent expires (about 1895), Alizarin will not produce the same style of print (in Calico-printing) as indigo blue, which is dyed on both sides of the cloth, and the patterns —white, red or orange—got by a discharge process. Hence, so far as cotton-dyeing and calico-printing are concerned, Alizarin-blue is not a serious competitor of indigo. In wool-dyeing, the case is different, for a modified form of Alizarin-blue is obtainable, which, besides being soluble in water, possesses over the insoluble form the advantages of ready applicability and greater fastness to light. In some cases of wool-dyeing the *soluble Alizarin-blue* is pushing indigo somewhat severely. As regards actual price Alizarin-blue may successfully compete with indigo, but the working expenses are considerably greater, and it can only be used for dark shades. Hence in some cases indigo has the advantage, and in others the Coal-tar dye is to be preferred. By dyers or manufacturers, who only use vat-blue occasionally, or who only require it for giving a bottom—that is, a ground colour on which other colours are afterwards dyed —or who want a fast blue for the production of *compound shades*, such as browns, olives, greys, &c., Alizarin-blue will be preferred, as its application is more simple than that of indigo. Wherever dark indigo-blues are a speciality, as for admiralty cloth, there indigo reigns supreme, the chief complaint against Alizarin-blue being that it gives blues that are lacking in body and depth of colour. This objection, however, may be overcome by means known to certain skilful dyers. Alizarin-blue is being tried on a somewhat extensive scale in Germany for dyeing the dark blue army cloth and is said to be preferred, because it is less liable than indigo to rub. The admiralty and officers on the Atlantic liners are now engaged in observing the relative fastness of indigo and Alizarin-blue when exposed to sea-air."

Other Indigoes. 305

Other Indigo-yielding Plants.—Besides the artificial manufacture of indigo there is a feature of the industry that in India at least has been greatly overlooked, *viz.*, the cultivation of other indigo-yielding plants than **Indigofera tinctoria.** Some of those known to yield the dye might be grown on the same plantation and probably with great advantage, since they might be made to afford plant on which the manufacturing operations could be continued during the periods at which the factories, as presently worked, are silent. But there is still another bearing of this question, since some of the indigo-yielding plants might be grown in regions not

SUBSTITUTES.

suited to **Indigofera tinctoria.** Should this idea be taken up, the area of indigo-production might be greatly extended and lessened prices result from an internal rather than an external influence. Brief notices will be found of three indigo-yielding plants which are regularly employed as a source of dye. Thus under the Chapter on "History of Indigo" allusion has been made to samples of **Nerium** indigo (the produce of **Wrightia tinctoria,** the *Pala* indigo of Madras) as having been sent from Madras to the Board of Directors of the Honourable East India Company for examination. This easily-grown tree yields a large supply of leaves annually, and might be made a source of dye at a nominal cost. Numerous papers will be found on this subject in the earlier volumes of the Journals of the Agri-Horticultural Society of India, but **Mr. Fisher's** report, Vol. IV. (Selections), 129—131, may be specially consulted.

In the "Chapter on Cultivation of Indigo" two other indigo-yielding plants have been mentioned. Under the account of Bombay and of Ráj-putana reference has been made to **Tephrosia purpurea.** The Indigoes of Egypt and of the Niger in Africa, are obtained from species of that genus, and it is probable **Tephrosia** might be successfully grown in many parts of India where **Indigofera** would fail. In the account of Indigo cultivation in Assam, a brief passage has been quoted from the writer's account of the *Rum* dye of the Eastern side of India—the produce of **Strobilanthes flaccidifolius.** It does not seem necessary to comment further on this subject, suffice it to add that much of the dye of China is obtained from this source, and that long ago **Fortune** (*see Chinese Indigo by R. Fortune, Jour. Agri.-Hort. Soc. Ind., IX., Sel. 34*) urged the value of this plant. It was apparently first botanically made known to Europe through **Dr. W. Griffith,** who found it in the Mishmi Hills.

Besides the above there are other Indian plants known to yield the dye. The hill tribes on the Himálaya extract indigo from several plants, but more especially from **Marsdenia tinctoria** (the *ryóm* of the Eastern division), a creeping Asclepiad.* In Java **M. parviflora** is cultivated as a source of indigo. The Yoruba Indigo of West Africa is obtained from **Lonchocarpus cyanescens,** *Bth.* **Isatis tinctoria**—the woad—yields the Indigo of Afghánistán and of some parts of China. In Burma two plants not as yet mentioned are said to afford indigo, *viz.*, **Gymnema tingens,** an extensive asclepiad climber, and **Acacia rugata,** a small tree, probably only a variety of **A. concinna,** the soap-nut tree. In Cochin China a composite plant, **Spilanthes tinctoria,** is cultivated as a source of indigo. Many other plants are also reported to yield the dye, though little advantage is taken of these sources of blue-dye. Thus, for example, the leaves of the custard apple, **Anona squamosa,** and of the gram plant, **Cicer arietinum,** yield indigo. In Russia, **Polygonum tinctorium** is grown as a source of the dye, and it is probable several of the Himálayan species might be so used. The fruits of two or three **Polygonums** afford a pulpy substance which dyes the fingers blue on being squeezed, and the young cones of **Abies Webbiana** afford a curious purple dye, which is probably a form of indigo. The tree is known to the hill tribes of the Western Himálaya as the Indigo-fir.

The study of all the indigo-yielding plants should commend itself to the attention of planters.

GREEN INDIGO OF CHINA AND OTHER REPUTED GREEN DYES.

GREEN DYES 306

In the concluding paragraphs of the Chapter on the MANUFACTURE OF INDIGO reference has been made briefly to the processes of making Indigo-green. That substance should not be confused with Chinese

* For further information on the Indigo from **Marsdenia,** see Vol. V.

GREEN DYES.

green-indigo, which, in all probability, is not obtained from the Indigo plant at all; if, indeed, it can be accepted as anything more than a combination of two materials resulting in a green dye. This subject is, however, at present too little known to allow of a succinct review of the literature, and little more therefore can be here done than to furnish the reader with references to the chief works which treat of it. The writer, in his Calcutta International Exhibition Catalogue, referred to four reputed Indian green dyes, *viz.*, **Baccaurea sapida, Gymnema tingens, Hedyotis capitellata,** and **Jatropha glandulifera.** These will be found treated of in their respective places in this work. But the two former are affirmed to be used in conjunction with other substances, so that they may be set on one side. The two latter, on the other hand, are reputed to yield a green dye without the aid of any other dye-stuff. **Gamble,** and also **Schlich,** speak of **Hedyotis** as the Lepcha green dye. "The green leaves are put into water and infused and the cloth to be dyed is steeped in the infusion." **Dr. Thomson,** in the Journal of the Agri-Horticultural Society of India (1862), described the process of preparing the green dye from **Jatropha,** and it is somewhat remarkable that neither of these green dye-yielding plants have been subjected to fresh investigations, still less have they been put to any practical use. If it be the case that **Jatropha glandulifera** yields the dye said to have been prepared from it, an unlimited supply might be obtained in Bengal, since the plant is one of the most prevalent of roadside weeds.

Subsequent to the appearance of the Catalogue cited above, the writer published in the Colonial and Indian Exhibition Catalogue an account of what promises to be another subject of future interesting investigation. A sample of Kampti green dye was procured, the plant from which it was said to have been obtained proving the common pulse **Vigna Catiang,** the *Urohi* of Assam. The leaves and twigs were reported to yield on decoction a fluid which, with the fruits of **Garcinia pedunculata,** produced the green dye. The former may therefore, like **Cicer** (gram), afford a kind of indigo which, with the yellow from the **Garcinia,** would form green. In a recent communication, however, **Mr. H. Z. Darrah** (Director, Land Records and Agriculture, Assam) throws some doubt on the accuracy of the above information. He describes an Assam green prepared from *rúm*-indigo with turmeric and the leaves of *urohi mahorpat* (? **Vigna**). He adds that the leaves of the common plum may be used in place of *urohi*; the plum referred to is probably **Zizyphus,** a plant allied to **Rhamnus.** Major Hannay (Note on the Dye-stuffs of Assam, *Jour. Agri.-Horti. Soc. Ind., VI.* (*Old series*), *p. 69*) describes another Assam green prepared with *rúm*-indigo, the yellow employed being *Mishmi-teet* (**Coptis Teeta**). Throughout India, however, greens are produced with indigo and some vegetable yellow, so that little novelty exists in observations of that nature.

In an account of the Resources of Pegu (*Jour. Agri.-Hort. Soc., IX., Sel. 54,*) it is stated that the leaves of **Photinia serratifolia** (? a form of **P. Notoniana,** *W. & A.*) yield a green dye. The writer can find no confirmation of this statement. **Mason** alludes to the above plant (on the authority of **McClelland**) as the *donk-yat* of the Burmese, but he describes in another page a green dye as prepared by the Burmans from turmeric and the leaves of the soap-nut **Acacia.** In Volume II. (Old Series), p. 251, mention is made by **Mr. A. H. Landers** (*Vegetable and other Products of the Shan Country*) of a green vegetable dye of a fading colour, which, he remarks, would no doubt be lasting with mordants.

Green Indigo of China. 307

Turning now to the information available regarding the GREEN INDIGO OF CHINA, *Lo-kao,* it may practically be only necessary to refer the reader to the interesting information on this subject that exists in the Journals

GREEN DYES.

Green Indigo of China.

of the Agri.-Horticultural Society of India. In Volume VIII., Old Series (1854), 232, **Mr. T. F. Henley** addressed the Secretary of the Society, recommending that **Mr. R. Fortune's** attention (during his contemplated travels in China) should be directed towards investigating the source of the famed green-indigo. **Mr. Henley** referred the Secretary to **Monsieur Persoz's** account of the substance which had then only recently appeared (*in Comptes Rendus de l' Academie des Sciences*, October 1852). A copy of **M. Persoz's** memoir was furnished to **Mr. Fortune** along with a grant of £50 to be spent on plants and seeds including, if possible, those of the green indigo shrub (*Proceedings, clxxxiii.*). In vol. IX., 105, is published **Mr. Fortune's** "Account of the Chinese Green Indigo Plant with specimens of the dye." **Mr. Fortune** in the above paper expresses his satisfaction that the green dye plant forwarded by him had been received in safety by the Society. He then proceeds to say that he found that the seeds of a cultivated along with a wild form of the plant were used in producing the dye. He performed the experiments for himself as described by his Chinese informants. "The extract," **Mr. Fortune** wrote, "from the seeds of the cultivated species which I have evaporated on paper, is yellowish in colour, while that from the wild kind is of a purplish or violet tint, and apparently very beautiful. These mixed together give a green of various shades, according to the proportion of each kind, and this colour is much varied and improved by the addition of alum, sulphate of iron, and such substances used as mordants by dyers. The Chinese invariably informed me the seeds were employed to paper only, and that cotton and silk fabrics were dyed with the bark." **Mr. Fortune** after some trouble found that the chief seat of the green dye (or *Lokao*) manufacture was at Kia-hing-fu situated between Hung-chon-fu and Shanghai, and as he was unable to visit that city he communicated with **Dr. Lockhart**, who entered warmly into the investigation. **Mr. Fortune** then adds that on reaching Shanghai later in the year he found that the **Rev. Mr. Edkins** had, on passing through Kia-hing, procured a bundle of the chips. Speaking of these chips he remarks: "**Dr. Lockhart** had prepared the extract by boiling down a decoction obtained from them." At page 274 a further communication from **Mr. Fortune** is given where he furnishes a passage from a letter received from **Dr. Lockhart** in which the process of preparing the dye is carefully described. The barks of the wild plant called WHITE and of the cultivated called YELLOW are boiled in iron pans.

"The residuum is left for three days, after which it is placed in large earthen-ware vessels, and cotton cloth prepared with lime is dyed with it several times. After five or six immersions the colouring matter is washed from the cloth with water and placed in iron pans to be again boiled. The colouring matter is taken up in cotton yarn several times in succession, and then washed off and sprinkled on thin paper. When half dry, the paper is pasted on light screens, and thoroughly exposed to the sun. The produce is called *Luk-kaon*. In dyeing cotton cloth with it ten parts are mixed with three parts of sub-carbonate of potash in boiling water." The dye obtained "*does not fade with washing, which gives it a superiority over other greens*"

In the same article **Mr. J. McMurray**, Head Gardener to the Society, described the process he had pursued in cultivating the plants and seeds of the Chinese green dye, obtained from **Mr. Fortune**. **Dr. Thomson** (then Superintendent of the Royal Botanic Gardens) identified the plant as a species of **Rhamnus**. Reprinting from the *Pharmaceutical Journal and Transactions* the Agri-Horticultural Society, in their *Selections, page 110*, gives **Mr. Daniel Hanbury's** paper on the notes furnished him by

GREEN DYES.

Green Indigo of China.

Dr. Lockhart. These are in substance the same as briefly indicated above; **Dr. Lockhart** appears thus to have sent the same information to both **Fortune** and **Hanbury**.

In the Proceedings to Volume IX., the Society's Gardener continued to report most favourably of the progress made with **Mr. Fortune's** green dye plants, but the interest in them seems to have died out about the time a translation appeared (by *Mr. H. Cope*) of **Monsieur Natalis Rondot's** most elaborate and instructive paper on "*The Green Dye of China, and Green Dyeing of the Chinese.*" This will be found in Volume X., 275—336, and continued in Volume XI., 139—178. **M. Rondot** affirms that the first European mention of this green indigo occurred in 1845, but that the discovery of the properties of the dye only dates from 1852, *viz.*, with the publication of **Mons. J. Persoz's** paper. It was shown at the Exhibition at Turgot in 1846 both as a paint and as a green dye for silk, but only a passing notice was taken of it. According to the **Rev. Mr. Edkins** no mention is made of the dye in the Chinese and Japanese works earlier than the first decade of the present century. In Volume XI. of the Journals, 171—178, will be found a complete record of all the works that treat of the Green Indigo of China and the Green Dyeing of the Chinese, and although in **D. Hanbury's** *Science Papers* and other modern works repeated mention is made of the dye, our knowledge of the substance has not materially advanced. There seems no doubt, however, but that it is prepared from two species of **Rhamnus** (*viz.*, **R. davuricus**, *Pall.*, and **R. tinctorius**, *Waldst. et Kit.*). Although India possesses some eight indigenous species of Buckthorn none of them are reputed to yield dyes, nor apparently has any person ever thought of trying if they could be made to afford a green dye similar to that of China, a fact probably due to the advance made in the production of green dyes from coal-tar. It is significant, however, that a **Zizyphus** (a genus closely allied to **Rhamnus**) should in Assam be employed as an ingredient in the preparation of a green dye. **Dr. D. Prain** (*A note on Lo-kao*), *Jour. Agri-Hort. Soc, Ind. VIII.* (*New Series*), *278—281*, reviews the present position of the botanical literature of the dye-yielding species of **Rhamnus**. He points out that, in the *Flora of British India* under "**D. dahuricus**" the plants described as **R. globosus**, *Bunge*, and **R. virgatus**, *Roxb.*, have been included as synonyms. If this reduction be confirmed then India possesses one of the Chinese dye-yielding species. **Dr. Prain**, however, is disposed to regard the Roxburghian species as probably quite distinct from **R. davuricus**, *Pall.*, in which case none of the dye-species occur in India. **M. Michel**, at Lyons, experimented with the bark of **Rhamnus Catharticus** and prepared a green dye, which, while stated to be inferior to that obtained from the Chinese species, confirmed the accuracy of the reports that had appeared regarding the *Lo-kao*. **Charvin** (in 1864) chemically prepared the *Lo-kao* from **Rhamnus Catharticus** and other species, and he sold the dye for 37*s*. a ℔, the true Chinese article having previously fetched 7*s*. 6*d*. an ounce. Although a remarkably beautiful dye it is highly problematic if it will ever be able to find a place in the European markets in competition with Aldehyde green, gas green, Hofmann's green, Paris green, and other such cheap colours prepared from coal-tar. The chances of such a trade are dependent, as indeed is the continuance of indigo, upon whether or not it can undersell the Aniline dyes. The Indian species are all temperate plants, and even if found to afford the dye could not be so easily grown on the plains of India as the more tropical Chinese species. **Father Helot** describes the preparation of the green dye thus:—The fresh bark of the species known in China as *hong-pi-lo-chou* is boiled with water and left standing in the liquid for two days: the preparation of the decoc-

GREEN DYES.

Green Indigo of China.

tion from the other species *pé-pi-lo-chou* is continued for ten days. Each of the decoctions is used separately, and on each occasion lime water is added. The fabric is dipped in the former decoction seven to ten times and in the latter three times, but the fabric is allowed to dry between each immersion. The drying is effected on meadows where the cloth is spread out at about night-fall and left until noon of the next day. Only the side of the fabric exposed to the sun becomes coloured, from which circumstance it is inferred that sunlight is indispensable to the formation of *Lo-kao*. The process of preparing *Lo-kao* as a pigment or dye-stuff (described above by **Fortune**) is thus quite different from the process of dyeing with the fresh barks as employed by the Chinese. In the former a fabric is used to remove the dye, the excess being brushed and washed off and then exposed to the sun on paper so as to cause the change necessary to its formation. In the latter the change effected by the sun takes place on the fabric, the exposed surface being dyed. In **M. Michel's** experiments with **Rhamnus Catharticus**, as with the Chinese **Rhamnus** barks, the surface of the fabric exposed became green coloured, but if prepared *Lo-kao* be used as a dye both surfaces of the fabric are dyed, since the tinctorial agent employed has already been acted on by the sun.

The berries of a species of **Rhamnus** are sold in Europe under the name of Persian Berries. These are largely used in calico-printing for steam-orange, olive, green, &c., and this dye exhibits certain actions under the influence of light analogous to the reaction indicated above in the formation of *Lo-kao*. This is especially seen when copper sulphate has been used as a mordant. The olive colour imparted by the berries gradually becomes deeper in colour, but after twelve months' exposure to light the olive green produced becomes as fast to light as vat indigo. These berries are obtained from several species of **Rhamnus**, wild or cultivated, in France, Spain, Italy, Turkey, the Levant, or Persia. In addition to the works named, the reader may consult the following:—*Crookes, Dyeing and Calico-printing, 428—435; Hummel, the Dyeing of Textile Fabrics, 366; Spons' Encyclopædia; Balfour, Cyclopædia; Linn. Soc. Journals, &c*

CHEMISTRY. 308

CHEMISTRY OF INDIGO.

The Explanation of its Tinctorial Reactions.

The Indigo of Commerce varies greatly in quality and purity, and its value, therefore, depends upon the proportion of actual colouring matter contained. Thus good indigo, according to **Mr. A. H. Allen**, ranges from about 4*s*. to 5*s*. 6*d*. per ℔—the proportion of colouring matter being about 50 to 70 per cent. Good Kurpah (Cuddapah) indigo ranges from 2*s*. 9*d*. to 3*s*. 9*d*. per ℔, and usually contains 33 to 52 per cent. of colouring matter. Oudh is generally similar to Kurpah in price and richness. Good Java indigo contains from 60 to 80 per cent. of colouring matter, and ranges in price from 5*s*. to 6*s*. 6*d*. There are many low qualities of indigo as, for example, the so-called "Fig" Indigo, which some times contains only 6 to 9 per cent. of colouring matter. "From a list of the prices and richness of various parcels of indigo sold in London, it appears that the lower qualities usually fetch a higher price per unit of colouring matter contained than is the case with better kinds. On the whole it may be taken that the unit of colouring matter in indigo has a present market value of about 1*d*. per ℔, or perhaps, more strictly, 15—12 of a penny. There are many exceptions, however, as some dyers are willing to pay rather more for a particular quality, which their experience or fancy leads them to prefer. In some cases, certain dyers will consider that Oudh indigo gives the best results, and they will buy Oudh indigo only; while others doing exactly the same kind of work, find that Oudh

CHEMISTRY.

will not answer their purpose, and will use only Bengal indigo, and so on." The dye-colouring matter of the commercial articles has been subjected to the most careful chemical analysis, and, as stated, has received the name of *Indigotin*. The commercial article contains, however, in addition the substances known as indigo-red, indigo brown, indigo-gluten, and a number of brown resinous products which have not as yet been fully determined. Without entering into the history of the development of our knowledge of *Indigotin* further than has already been indicated, it may suffice to state that its chemical composition is now accepted as being represented by the formula C_8H_5NO (or $C_{16}H_{10}N_2O_2$). It is insoluble in water, alcohol. ether, and essential oils, or in dilute acids or alkalies hot or cold. But it is soluble in sulphuric acid forming what is known as EXTRACT OF INDIGO or SULPHATE OF INDIGO. Creosote, phenic acid, alcohol, fixed oils, and aniline, also dissolve small quantities at a boiling heat, but on cooling *Indigotin* is re-deposited : chlorine and hypchlorates destroy its blue colour. Crookes remarks on the insolubility of *Indigotin* as follows :—

"When finely pulverised *indigotin* is mixed with anhydrous acetic acid, and when a single drop of concentrated sulphuric acid is added, there is obtained a beautiful deep blue liquid, from which the *indigotin* is separated unaltered on the addition of water. If the original acetic acid solution is applied to any woven tissue, and immediately afterwards washed in water, the tissue is dyed blue. It should be observed that this process is the only manner hitherto known in which *indigotin*, without being reduced, is reproduced in its primitive state. *Indigotin* is a thoroughly neutral substance, void of taste and smell; its specific gravity is 1·35. When decomposed by dry distillation *indigotin* yields, among other products, aniline. By oxidising agents, such as concentrated solution of chromic acid and chlorine, *indigotin* is converted, when water is simultaneously present, into a new substance, which differs from *indigotin* only by having one atom of oxygen more in its composition, and which is named *Isatin*." By the action of strong and hot nitric acid *indigotin* is converted into *indigotic* and *picric* acids, and in this reaction carbon is eliminated. But the most interesting, as it is the reaction on which the industrial usage of *indigotin* depends, is its property of combining with hydrogen to form a colourless substance, white-indigo, soluble in alkaline earths or alkalis, which is re-convertable into *indigotin* by oxidation through exposure to the air."

METHODS OF DYEING.

METHODS OF DYEING WITH INDIGO.

Cotton vats.

309

COTTON VATS.

The above phenomena, vulgarly called "reduction," requires the hydrogen to be in a nascent state. An agent which, in the presence of an alkali will decompose water, retaining the oxygen and liberating the hydrogen, is that employed. For this purpose ferrous sulphate or copperas and slaked lime are mixed in a large bath or "vat" with the indigo. The indigo is first reduced to a fine powder with water and the contents of the vat are carefully stirred and then allowed to settle. The action of the lime is to form ferrous hydrates; this seizes the oxygen from the water and is thus converted into ferric hydrate, and hydrogen is liberated in the nascent state. The hydrogen unites with the *indigotin* or BLUE-INDIGO to form WHITE-INDIGO and this becomes soluble in the presence of the excess of lime. The liquid of the vat thus gradually gets discoloured to a pale yellow, and sulphate of lime is deposited. These reactions may be expressed thus—

I. $FeSO_4$ Ferrous sulphate + $Ca(OH)_2$ Lime = Ca SO_4 Calcium sulphate + Fe $(OH)_2$ Ferrous hydrate.

METHODS OF DYEING.

Cotton Vats.

II. 2[Fe $(OH)_2$] Ferrous hydrate + 2 H_2O Water. = $Fe_2(OH)_6$ Ferric hydrate + H_2 Hydrogen.

III. 2 (C_8H_5NO) Indigotin or blue-indigo + H_2 Hydrogen=2 (C_8H_6NO) White-indigo.

The order in which the ingredients are generally added to the vat is to fill the vat with water and then to add first the ground indigo and milk of lime, and after these have been stirred about well to add the solution of ferrous sulphate. he mixture is systematically raked about at frequent intervals for 24 hours.

The above process, while perhaps the most rational chemically, is sometimes not followed, owing to the tendency of the fluid to thicken. To avoid this the ferrous sulphate and indigo are first added, and the milk of lime afterwards, and gradually, as required. Lime is preferred to caustic soda, because cotton is found to dye more rapidly and the film of calcium carbonate that forms on the surface of the fluid prevents its oxidation into *indigotin*. The reduction into white-indigo is known to be complete when numerous thick dark-blue veins appear on raking up the liquor and the surface has become covered with a strong blue scum or "flurry." The liquid should then be clear, and of a brownish-amber colour; if greenish, undecomposed indigo is present and more ferrous sulphate should be added. If very dark, more lime is required.

The above is the usual vat preparation, but metal zinc may be employed to liberate the hydrogen:—

$$Zn + H_2O = ZnO = H_2.$$

The presence of lime and indigo are found to facilitate this decomposition, the liberated hydrogen reducing the indigo.

A hydro-sulphite vat may also be used:—

$$H_2SO_3 + Zn = H_2SO_2 + ZnO$$

Sulphurous acid + Zinc = Hypo-sulphurous acid + Zinc oxide.

In practice Professor Hummel (*Dyeing of Textile Fabrics*) says the zinc is usually allowed to act upon concentrated solution of sodium hydrogen sulphite (bisulphite) instead of sulphurous acid, there being produced a solution of sodium hydrogen hyposulphite and zinc sodium sulphite:—

$$Zn + 3\,NaHSO_3 = NaHSO_2 + ZnNa_2(SO_3)2 + H_2O.$$

The reduction of the indigotin to white-indigo by means of acid sodium hyposulphite may be expressed thus:—

$$2(C_8H_5NO) + NaHSO_2 + NaHO =$$

Indigotin Acid Sodium hyposulphite Caustic soda

$$2(C_8H_6NO) + Na_2SO_3$$

White-indigo Disodium sulphite.

The hyposulphite is a somewhat difficult process, but a concentrated solution of reduced indigo may be obtained which can be employed for charging or replenishing a dye-vat. It should be used cold for cotton and hot for wool. It is a useful vat for silk-dyeing.

INDIGO EXTRACT, OR SAXONY BLUE.

Saxony Blue. 310

The above three processes are those generally used with cotton, though, as remarked, the hyposulphite may also be employed with wool or silk. Instead of the vat process, however, animal fibres—wool, silk, &c.—are often dyed with what is called INDIGO-EXTRACT or INDIGO-CARMINE. By the action of strong sulphuric acid *indigotin* is converted into soluble *indigotin-disulphönic* acid. The indigo in this preparation has undergone so complete a change that it cannot be restored. The derivatives of indigo thus obtained have no affinity for cotton, though used by bleachers for

METHODS OF DYEING.

Saxony Blue.

tinting. The colours are fugitive, but their ease of application is a great recommendation, and animal fibres may be dyed on being simply steeped in the slightly acidulated hot extract solution. The colour is much brighter than can be obtained from a vat-dye, but it does not stand milling. Silk to be dyed with the extract is sometimes mordanted with alum, in order, on the addition of cochineal, logwood, or orchil, to produce reddish shades—purples or violets. Indigo-carmine is produced by the addition of sodium sulphate.

Fermentation Vats for Woollens and Silks. 311

FERMENTATION VATS FOR WOOLLEN AND SILK GOODS.

Indigotin is also applied to woollen goods by means of a vat containing a special preparation of white-indigo. These vats are prepared or "set" in various ways, but in all, the indigo is reduced in the presence of some organic substance such as woad, madder, molasses, bran, &c. The organic material undergoes fermentation when macerated in water with the application of heat. The hydrogen evolved by the fermentation transforms the *indigotin*, and this is dissolved in the vat by the addition of some alkali or alkaline carbonate.

Woad Vat. 312

The Woad Vat.—The woad plant itself, formerly employed as a source of blue, is now used entirely as a fermentation agent in the woollen vat, but it is probable that cabbage leaves might be equally serviceable. The vat is first partly filled with water, and crushed woad is then added, the whole being well stirred and the temperature raised to 50° or 60° C. This temperature is maintained from 24 to 30 hours; bran and madder are now added and half the total required quantity of lime. After being well stirred, the vat is covered over and left for 12 to 24 hours. By this time fermentation has been established, the surface of the fluid becomes covered with a coppery-blue scum, and on being stirred the liquor is seen to be of a greenish yellow colour interspersed with blue veins of restored indigo. If the bottom of the vat be disturbed a froth will appear on the surface, and the sediment, if brought up, will smell sour. A piece of wool dipped in the bath and then exposed to the air is coloured blue. All these indications denote the favourable progression of fermentation, and it now becomes necessary to keep the further stages under control. The remainder of the lime is added by degrees, vigorously stirring the contents of the vat all the while. In the course of 24 hours if the fermentation continues favourably, the vat will be ready for use. Too rapid fermentation is checked by the addition of lime, or if too slow it is accelerated by the addition of bran. The dyeing power of the vat is maintained by the daily addition of lime and bran and every other day of indigo. After three or four months, or when the vat sediment becomes bulky, no further additions are made, but the vat is used for lighter shades until its indigo is exhausted, when it has to be emptied out and a fresh vat started.

The above account of a fermentation-vat has been compiled from **Professor Hummel's** excellent work on the subject, and for fuller details and quantities of the ingredients the reader is referred to the original text.

Potash Vat. 313

The Potash Vat, which contains indigo, madder, bran, and carbonate of potash, is not so liable to get out of order owing to its not containing so highly nitrogenous a substance as woad. It also dyes more rapidly and gives deeper though duller shades of blue. The colour does not come off so much on milling with soap and weak alkalis. It is, therefore, best adapted for very dark shades of navy blue and is better suited for silk-dyeing than a vat containing lime. The best silk vat is, however, one in which indigo is reduced by powdered zinc and ammonia.

Soda Vat. 314

The Soda, or German Vat contains indigo, bran (or treacle), carbonate of soda, and slaked lime.

METHODS OF DYEING.

Urine Vat. 315

The Urine Vat is used by dyers who have to resort to vat dyeing only occasionally. It contains stale urine, and common salt, heated for several hours up to 50° or 60° C. Madder is then added and indigo, fermentation being allowed to proceed as before until white-indigo is produced.

Native Vats. 316

NATIVE VATS, OR THE PROCESSES OF REDUCING INDIGO PURSUED BY THE PEOPLE OF INDIA.

Though crude in comparison with the vats prepared in Europe, the native dyers of India are fully aware of most of the processes of rendering indigo soluble, and in fixing it by making it again insoluble within the tissue of the fabric. The vat most generally preferred, however, is a fermentation-vat. A pound of powdered indigo is soaked with three pounds of lime and four pounds of impure carbonate of soda in water; four ounces of sugar are added. If fermentation does not commence in seven or eight hours, more lime and sugar are added, and if the weather be cold the vat is heated to encourage the fermentation. When the reduction has taken place the vat is used in the ordinary manner (*Buck, Dyes and Tans of North-Western Provinces*).

In the *Nasik Gazetteer* an interesting account of indigo-dyeing, as practised in Bombay, will be found. "To prepare the solution of indigo the dyers have two vats—a salt vat, *khára pip,* for dyeing cotton, in which poor indigo, and a sweet vat, *mitha pip,* for silk, in which good indigo, is used." The vat employed is an open-topped wooden barrel or earthen vessel sunk in the ground, and large enough to hold 300 gallons of water.

Salt Vat. 317

The Salt Vat.—About 150 gallons of water are thrown into the vat. To this is added 8℔ impure carbonate of soda (*sáji khár*), and 4℔ of lime. The mixture is stirred well and allowed to stand over-night, while 5℔ of indigo are left soaking in a separate dish of water. Next morning the indigo is reduced to a paste by being worked in a stone trough, the operator's hand being covered by a cloth. The indigo is then thrown into the vat and well stirred for two or three hours, the stirring being repeated at night-fall when other 5℔ of indigo paste is thrown into the vat. On the third day a copper pot of about ten gallons capacity is filled with sediment from an old vat and the sediment thrown into the new vat, the mixture being again stirred and the vat securely closed. If there is no old sediment, two pounds of lime, two pounds of dates, and ten pounds of water are boiled till the mixture becomes yellow. This hot preparation is thrown into the vat as it is being stirred in place of sediment from an old vat. On the fourth day the vat mixture is yellow, and, when stirred, gives off foam. The colour of this foam indicates whether or not the vat has been successfully prepared. It is collected, made into balls, and dried, and proves useful to colour patches where the dye has not taken uniformly. If the foam be reddish the vat has been correctly prepared; if white 3℔ of carbonate of soda should be added; if it irritates the skin, clots, or is oily, about 4℔ of dates should be added. The vat should be ready for use on the fifth day.

Sweet Vat. 318

The Sweet Vat.—For this purpose 120 gallons of water should be placed in the barrel, 4℔ carbonate of soda added, and the vat kept covered for three days. On the third day 4℔ carbonate of soda and 2℔ of lime should be added, and the mixture stirred three times a day. Next morning, the fourth day, 4℔ of old brown sugar are dissolved in cold water, the solution is thrown into the vat and the mixture stirred occasionally for three days more. When fermentation sets in the mixture begins to crackle. At this stage the foam is examined, and, if it is reddish, the vat is ready. If white carbonate of soda and lime have to be added, the proportions of lime and sugar requiring the nicest adjustment. If this

METHODS OF DYEING.

Sweet Vat.

be not attended to the indigo rots, has an offensive smell, and is unfit for dyeing (*Dr. Náráyan Daji, Dyeing in Western India, 23*).

In Bengal sugar is used in only one or two districts as the fermentation agent. In Darjeeling the refuse water obtained in washing sheep is employed in the vat. In Rajshahi district the twigs of **Morinda tinctoria** are added to the vat, apparently much after the same manner as madder is employed in Europe. In Malda the seeds of **Cassia Tora** are added to the vat. This subject has been alluded to above in connection with the manufacture of Indigo in South India and is apparently of much interest, since the employment of these seeds is so widely known. For a similar purpose apparently the seeds of a plant known as *ichi-bichi* are used in Chittagong (*Dr. McCann, Dyes and Tans of Bengal*).

Indigo-Dyeing and Calico-Printing.

319

INDIGO-DYEING AND CALICO-PRINTING.

The principle of indigo-dyeing has been indicated by the foregoing remarks regarding the chemistry of the drug, and the formation of dye-vats. The fabric or yarn is immersed in the reduced indigo solution for the required number of times until the texture of the substance has been impregnated with a sufficient amount of white-indigo. Care must be taken not to bring the fabric or yarn above the surface of the vat when "hawking" it in the fluid, nor of course should it be allowed to touch the sediment. If raised above the surface, it will be at once oxidised and the portions so acted upon will receive a different colour from the rest. Cloth dyes best in a vat in which moderate fermentation is preserved. After the necessary degree of immersion the fabric is hung up to drain, and is thereby exposed to the air, when the oxygen combines with the excess hydrogen to form water, and this evaporates away, leaving the insoluble particles of *indigotin* in the interior of each filament of the texture. If the exact shade has not been obtained the process is repeated. The greatest difficulty is in regulating the fermentation bath, and considerable experience is required to know when it is right. Woollen materials should be first boiled well in water, then plunged into tepid water, and squeezed before being entered in the indigo vat. They should never be allowed to lie in heaps after boiling, since this would cause unequal dyeing. The boiling serves the double purpose of allowing the damp fabric to dye evenly and prevents the admission into the vat of too much air which would oxidise the indigo-white. Before beginning to dye, the blue scum should be taken off the surface of the liquor of the vat; otherwise spots will be given to the parts of the texture coming first in contact. After dyeing, the fabric should be rinsed in acidulated water and washed thoroughly. In order to remove every trace of loosely adhering indigo, a good milling with soap and fuller's earth is given, so that the finished piece will not colour a white handkerchief when rubbed on its surface. If it be desired to make the colour extra fast the fabric should be boiled in a solution of alum or bichromate of potash and tartaric acid. Some dyers also regard a boiling of the dyed goods in barwood, sanderswood, or camwood as necessary to completely fix the colour and to secure its being fast in light. Steaming makes the colour a little violet, but faster against light.

Calicoes, after leaving the vat, are first rinsed in cold water to remove the loose lime and indigo adhering mechanically, and then in dilute sulphuric acid to dissolve off the calcium carbonate. They are afterwards washed in water and dried, and thus a uniform blue colour of any required shade may be produced. Cotton fabrics are also dyed with indigo by printing or pencilling white-indigo upon the surface. The alkali employed to hold the dye in solution is washed out as soon as the *indigogen* has been absorbed by the tissue and the oxidation produced by washing transforms the dye into its insoluble form, thus exhibiting the patterns printed. "Fast-blue" prints are obtained by precipitating from a paste of indigo and a strong reducing metallic oxide (commonly hydrated protoxide of tin), a too rapid oxidation of the white-indigo being prevented. The paste is printed on the fabric which is then passed through an alkali bath of lime-water or soda solution. The alkali replaces the oxide of tin, forming a soluble combination of white-indigo, which penetrates and is fixed by oxidation. "China-blue" prints are somewhat similar. Indigo previously pulverised is printed on the tissue and the fabric placed under the influence of reducing agents; when absorbed the white-indigo is oxidised. But parti-coloured goods may be produced having a blue ground colour by either of two processes technically known as *resist* and *discharge*. In the former the fabric, prior to immersion in the indigo vat, is printed with a substance which, either chemically or mechanically, prevents the *indigogen* penetrating into the fabric. After being dyed blue on the exposed parts the *resist* is washed off and the white patches printed with any other colours or patterns. In the latter process (*discharge*) the *indigotin* of the uniformly dyed fabric is removed by chemical agents which are printed in the desired pattern. The mechanical *resists*, generally used, are wax or pipe-clay (much after the manner in which the *Palampúr* calicoes of South India are dyed) and the chemical *resists* are powerfully oxidising agents, such as salts of copper or bichloride of mercury, which convert the *indigogen* of the vat, coming in contact with the printed portions, into *indigotin* which is deposited mechanically on the surface, instead of being absorbed into the substance of the fabric. *Resists* printed on a blue ground and the fabric again plunged into the vat results in a pattern in shades of blue. The *discharges* are in many respects inconvenient, as they do not give sharp prints and very strong vats are required to prevent running. Chromic acid is the best discharge, but chromate of lead is often used. Red prussiate of potash and caustic soda also form a useful discharge to indigo.

METHODS OF DYEING. Indigo-dyeing and Calico-Printing.

Resists. 320

Discharges. 321

Defects.—All fermentation-vats are subject to derangements by which they become more or less useless. Deficiency of lime is the most common cause, the fermentation being thus unchecked and allowed to proceed until the liquor is useless. This may be recognised by the scum disappearing, the liquor becoming muddy, and the escape of a disagreeable odour. The only remedy is to heat the vat to 90°C. and add lime. If this does not check the fermentation the preparation is useless. Another danger, according to **Professor Hummel**, is too rapid working of the vat. The colours become faint and the liquor gives off an ammoniacal odour. The third defect is too much lime. The indigo-white is precipitated and the liquor becomes of a dark-brown colour, while the healthy odour and the scum disappear. If noticed in time this may be corrected by the addition of a little ferrous sulphate or dilute sulphuric acid to precipitate the lime in an insoluble condition.

Defects. 322

TRADE IN INDIGO.

TRADE IN INDIGO. 323

It is perhaps unnecessary to endeavour to carry the modern trade in Indigo further back than one hundred years. The total imports into Great Britain were, in 1782, 495,100℔, of which the States of America furnished 161,216; the West Indies 64,309; British Continental Colonies 128,640; Flanders (Austrian) 78,070; Portugal 27,308; Asia 25,535 (including, of course, India); Ireland 6,373, and Spain 200℔. In the year following, the imports were 1,284,565℔, Asia contributing 93,047; the States of America 518,980; the West Indies 204,645, and 117,235℔ from Holland. From that year the British imports steadily increased

TRADE.

till 1788, they attained the magnitude of 2,096,911℔, of which Asia furnished 622,691℔; the States of America 1,060,164; Spain 204,461, and the West Indies 94,550℔. In 1795 they amounted to 4,368,027℔, of which Bengal furnished 2,955,862℔. In a Despatch, dated 1792, the Board of Directors congratulated the Indian Government, that, as the British imports of Bengal Indigo increased, those from the Spanish and French colonies declined, while at the same time a large export trade from Great Britain to the Continent had been established. In 1790 the re-export trade amounted to close on a million pounds of the dye.

Restoration of Indigo Trade to India. 324

It may thus be said that in less than 20 years, from the date of their first effort, the East India Company, by the enlightened action they took, completely restored to India its indigo industry. It is also noteworthy that whereas the total imports of indigo into Great Britain were in 1782 less than half a million pounds, the production of a superior quality at a cheap rate, in Bengal, had so increased the demand that ten years later Great Britain alone received over four million pounds. This brilliant result is fully borne out by the statistics of the present indigo trade. The French and Spanish manufacturers have practically ceased to exist. In 1888 the United States of America furnished Great Britain with 123 cwt.; Central America with 6,769 cwt., and Holland gave to England a small amount, derived most probably from Java mainly. With the single Dutch exception, however, the other European nations that are annually shown as participating in the British supply may be viewed as re-exporting Indian Indigo. Practically nine-tenths of the Indigo trade is now Indian. Thus, for example, out of a total of 78,128 cwt. (8,750,336℔) of indigo, imported by Great Britain last year (1888), India furnished 69,416 cwt. The Philippine Islands produce a small amount, but their trade does not appear to expand. It is doubtful how far any other country (except perhaps China) could ever compete with India. We possess a complete monopoly, and prices are not governed by competition with other producing countries, but by the caprice of demand or fashion and the uncertainty of the Indian supply. It is very generally admitted that few crops are more precarious. Too much or too little rain is equally destructive, and it is impossible to guarantee the quality that can be produced, still less to secure uniformity and continuity of a particular kind from a given factory. It accordingly follows that prices are governed to a large extent by internal rather than external circumstances, and they may be said to fluctuate annually by supply and demand, and from day to day by the quality brought into the markets. A certain percentage of inferior native-made indigo or of the adulterated article has come to be a recognised factor in governing prices; but, apart from such considerations, the dye property of the produce of even the best and most carefullv superintended factories is by no means constant. This fact renders it often extremely difficult to draw comparisons between the returns of Indigo trade year by year, or month by month. At the same time other difficulties also exist. Until comparatively recent times it was customary to give the trade returns of Indigo in pounds The foreign trade is at the present time expressed in cwts., the railway, road, and river traffic in maunds of 82℔, mostly gross weight, and the commercial statistics in factory maunds of 74℔ 10 oz., net weight of indigo. In the following tables an effort has been made to facilitate comparison between the official and commercial returns by expressing quantities in pounds, but a constant error in all such statistics, when dealing with one year as compared with another, exists in the fact that the official year extends from 1st April to 31st March, the commercial Indigo year from 1st October to 30th September. This is the more serious, since the indigo sales, in Bengal, last only for four

TRADE.

months, *viz.*, November to February; though it is lessened by the fact that the whole of the indigo intended for the foreign markets has left India by the end of March. The foreign transactions of each official year would, therefore, nearly correspond to the commercial returns, but the internal trade of the official year would seriously overlap the commercial. The most important error, however, lies in the fact of Government figures being the gross weight of cases containing indigo, while the commercial are for net weights of actual indigo. It would, in fact, be a safer mode of comparison, therefore, to take the Government gross maund of 82lb as corresponding to the commercial net weight in factory maunds of 74⅝lb—the difference being the weight of packages, &c., but merchants put down the difference in the two returns as equal to 25 per cent. While bearing in mind these sources of error, an average of several years would give a fairly trustworthy mode of estimation, more so at all events than the actual figures for the indigo trade of any one year.

In 1813 **Milburn** stated that the European demand for indigo amounted to three million pounds, but that, if required, Bengal alone might furnish five times that amount. It is somewhat significant that the Indian production is today a little over its prophesied possibilities. The average annual exports of indigo from all India during the past ten years have been 134,798 cwt. (15,097,376lb). According to **Messrs. Thomas & Co.'s** annual statement of "Indigo Averages" the production (during the past ten years) for Lower Bengal, Behar (Tirhút, &c.), Benares, and Doab (North-Western Provinces), has been 9,881,163lb, or during the past three years 9,813,187lb. In the official returns of Bengal a certain amount of indigo is recorded as brought from the Panjáb, Rájputana, the Central Provinces, Assam and Madras, which apparently Messrs. Thomas & Co. do not take into consideration. These sources of supply increase slightly the returns for Bengal, but there exists a sufficient correspondence in the commercial and official statistics to confirm the accuracy of both. According to the official returns of Bengal the production *plus* the imports have averaged 9,857,745lb, during the past three years. At the same time all exports, whether by land or sea, have averaged 9,460,878lb, thus leaving a balance of 396,857lb in stock or available for local consumption. It seems probable that the Bengal local consumption averages from 250,000lb to 400,000lb annually. A similar analysis might be worked out for Madras, but it is believed the figures of trade in the four tables here given will convey a sufficiently comprehensive idea of the Indian indigo trade. It may, however, be remarked that the growth of the Madras indigo trade is the most characteristic modern feature of the Indian industry. The Bengal production has fluctuated within more or less fixed limits, but that of Madras has steadily improved for the past thirty or forty years, and has attained a proportion which could scarcely have been anticipated by the pioneers of the trade who, at the beginning of the present century, laboured untiringly to establish an industry which is now annually worth 2½ million pounds sterling to the Lower Provinces; Madras adding its quota of close on another million sterling.

(*Conf.* with remarks under the Chapter on "Yield and Area under Indigo.")

TRADE.

TABLE No. I.

I. 324 *The following table gives the production of Indigo for the past ten years as recorded by the Calcutta Merchants (compiled from Messrs. Thomas & Co.'s Annual Statement of "Averages"). The figures are those of net weights in maunds of* $74\frac{5}{8}$ ℔.

YEAR.	Lower Bengal.	Behar, *i.e.*, Tirhút, Champaran. Chuprah, &c.	Benares, *i.e.*, Goruckpore, Shahabad, Jaunpore, Allahabad.	Doab.	Total in Factory Maunds of 74℔ 10 oz.	Total in ℔.
1879	14,535	28,387	8,696	21,510	73,128	5,457,177
1880	23,490	66,390	10,830	35,490	136,200	10,163,925
1881	17,425	58,062	14,484	45,434	135,405	10,104,598
1882	18,957	58,569	15,710	57,042	150,278	11,214,495
1883	17,206	58,748	18,648	64,786	159,388	11,894,329
1884	19,668	62,038	25,141	59,660	166,507	12,425,584
1885	19,829	51,887	13,450	23,526	108,692	8,111,140
1886	23,353	66,080	11,574	30,254	131,261	9,795,352
1887	21,700	67,800	14,750	27,250	131,500	9,813,187
1888	22,780	69,585	12,265	27,120	131,750	9,831,843
Average production for each of above years. { Mds. .	19,894	58,754	14,554	39,207	132,409	9,881,021
{ ℔ .	1,484,590	4,384,517	1,086,092	2,925,822	...	
					Averages for all the Provinces here dealt with.	Average per ℔.
Highest average prices for past 10 years per factory maund . R	289-12	263-14	233-2	227-2	253-7	3-5-9
Lowest average price for past 10 years per factory maund . R	197-2	192-8	173-3	180-4	185-12	2-7-0
Average of all prices for each Province per factory maund . R	239-5	234-2	214-0	195-0	220-0	2-15-0

TRADE.

TABLE NO. II.

Analysis of Bengal Trade in Indigo during the past three years derived from all available Official Returns—the maunds being mostly (though not invariably) gross weight of 82 ℔.

Routes.		Imports and Production. 1885-86.	1886-87.	1887-88.	Exports and Consumption 1885-86.	1886-87.	1887-88.
1. By East Indian Railway	Mds.	90,833	1,13,767	1,13,638	855	196	447
	℔	7,448,306	9,328,894	9,318,316	70,110	16,072	36,654
2. By Eastern Bengal State Railway	Mds.	9,642	8,958	11,646	21	177	338
	℔	790,644	734,556	954,972	1,722	14,514	27,716
3. By Inland Steamer	Mds.	...	461	515	...	16	16
	℔	...	37,802	42,230	...	1,312	1,312
4. By Boat	Mds.	3,173	1,878	3,900	138	232	90
	℔	260,186	153,996	319,800	11,316	19,024	7,380
5. By Road	Mds.	21	...	...	...	...	...
	℔	1,722	...	...	...	...	...
6. By Sea—							
(a) Coastwise	Cwt.	258	401	839	135	25	22
	℔	28,596	44,912	93,968	15,120	2,800	2,464
(b) Foreign	Cwt.	8	45	75	76,109	87,941	87,335
	℔	896	5,040	8,400	8,524,208	9,849,392	9,781,520
7.	Total ℔	8,530,350	10,305,200	10,737,686	8,622,476	9,903,114	9,857,046
8. Balance over exports left in Bengal		...	402,086 ℔	880,640 ℔	Exports exceeded the imports by 92,126 ℔ thus drawing on old stocks.	...	...

INDIGOFERA tinctoria.

Commercial Indigo.

TRADE.

In the Bengal Report of 1887-88 on the Internal Trade (River, Rail, and Road) it is stated that the following were the sources from which Calcutta derived its indigo:—

	1885-86. Maunds.	1886-87. Maunds.	1887-88. Maunds.
Behar	59,481	73,798	71,457
North-West Provinces and Oudh.	30,580	39,194	40,522
Bengal . . .	13,423	11,871	16,664
Other places . .	546	808	2,296
TOTAL .	1,04,030	1,25,671	1,30,939

"The increase in the imports noticed in the last report was fully maintained, and still further extended during the year 1887-88. According to the customs return the price of indigo averaged R231 per maund, against R219 per maund in the previous year. It is stated that higher prices were obtained in Calcutta in consequence of stocks in London having been reduced by increased consumption in Europe and America." "With regard to the North-West, the cultivation was considerably more than last year, but the produce is said to have been very poor. As regards Bengal, the only districts where the outturn is shown to have been appreciably more than in the preceding year were Jessore and Rajshahye. In Jessore the quantity manufactured was 2,762 maunds, against 1,339 maunds in the previous year. The price varied from R190 to R222 per maund, while in the previous year it varied from R190 to R207. The outturn in the Rajshahye District amounted to 1,165 maunds, against 848 maunds in 1886-87." "The Indigo industry is said to be increasing and doing very well in the Patna Division; but in the Bhagulpore Division the season's crop suffered to a certain extent by bad distribution of rain."

The Director of Land Records and Agriculture of the North-West Provinces and Oudh furnishes the following information regarding the indigo produced in these provinces during the past five years:—

YEAR.	Area in acres.	Total exports in maunds of 74⅗lb.	Total exports in lb.	Exports to Calcutta in maunds of 74⅗lb.	Exports to Calcutta in lb.
1884-85	565,733	78,474	5,856,122	71,861	5,362,627
1885-86	478,319	37,068	2,766,199	30,580	2,282,032
1886-87	290,070	42,947	3,204,920	39,193	2,924,777
1887-88	317,179	43,576	3,251,859	40,522	3,023,954
1888-89	336,586	39,746	2,966,045	37,199	2,775,975
Average for past five years	397,557	48,362	3,609,029	43,871	3,273,873

In the following table the exports from India are given, Bengal indigo thus including the above quantities from the North-West Provinces as well as smaller amounts from the Panjáb, Central Provinces, and Assam. Taking the returns of Foreign exports as approximately representing the Bengal trade, on an average Bengal itself contributes 6,500,000lb, and the North-West Provinces 3,200,000lb, the balance being derived from the other provinces named above.

TRADE.

Table No. III.—*Foreign Exports of Indigo from India during the past ten years.*

		1878-79.*	1879-80.*	1880-81.	1881-82.	1882-83.	1883-84.	1884-85.	1885-86.	1886-87.	1887-88.	Average of ten years.
Bengal .	R	2,30,88,814	1,88,04,499	2,88,45,753	3,15,12,595	3,02,35,399	3,37,31,810	3,02,33,957	2,55,96,913	2,62,43,170	2,74,56,889	2,75,74,980
	Cwt.	74,747	47,928	88,111	91,898	99,715	110,015	106,069	76,109	87,941	87,335	86,987
	℔	8,371,664	5,367,936	9,868,432	10,292,576	11,168,080	12,321,680	11,879,728	8,524,208	9,849,392	9,781,520	9,742,521
Madras .	R	59,19,466	94,33,527	62,35,270	1,22,78,358	76,30,967	1,15,35,134	94,35,208	1,03,61,412	86,40,177	97,39,620	91,20,914
	Cwt.	26,111	43,899	25,295	51,088	33,474	51,724	42,001	45,828	38,314	42,812	40,054
	℔	2,924,432	4,916,688	2,833,040	5,721,856	3,749,088	5,793,088	4,704,112	5,132,736	4,291,168	4,794,944	4,486,115
Bombay .	R	3,45,583	8,12,788	4,15,981	8,53,312	7,71,503	7,62,110	6,34,295	15,75,039	15,88,247	13,58,984	9,11,784
	Cwt.	2,229	5,331	1,954	4,507	4,504	4,006	3,376	8,015	8,319	6,494	4,874
	℔	249,648	597,072	218,848	504,784	504,448	448,672	378,112	897,680	931,728	727,328	545,832
Sind .	R	2,50,762	4,21,451	2,18,810	4,46,537	4,92,101	3,80,818	3,85,536	2,98,237	4,45,175	3,51,001	3,69,043
	Cwt.	1,964	3,765	1,510	2,870	3,348	2,845	3,183	2,543	3,822	3,003	2,885
	℔	219,968	421,680	169,120	321,440	374,976	318,640	356,496	284,816	428,064	336,336	323,154
Burma .	R	...	...	...	...	...	34	...	...	...	...	3
	Cwt.	..	...	...	...	...	...	...	...	...	...	...
	℔	...	...	...	...	...	...	...	...	...	...	...
Total .	R	2,96,04,625	2,94,72,265	3,57,15,814	4,50,90,802	3,91,29,9,0	4,64,09,906	4,06,88,996	3,78,31,601	3,69,16,769	3,89,06,494	3,79,76,724
	Cwt.	105,051	100,923	116,870	150,363	141,041	168,590	154,629	132,495	138,396	139,644	134,800
	℔	11,765,712	11,303,376	13,089,440	16,840,656	15,796,592	18,882,080	17,318,448	14,839,440	15,500,352	15,640,128	15,097,622

* These were famine years.

Commercial Indigo.

TRADE.

TABLE No. IV.

The following analysis of the Foreign Exports of Indigo from India in 1887-88 exhibits the countries to which consigned, and in the previous table will be found the share of these exports taken by each Presidency:—

Countries to which exported.	Cwt.	Value in Rupees.
United Kingdom	56,986	1,54,34,982
Austria	11,780	35,70,817
Belgium	208	69,610
France	17,406	49,76,119
Germany	6,392	20,24,291
Greece	145	38,857
Italy	1,533	4,57,013
Malta	205	45,831
Russia	1,668	5,13,964
Turkey in Europe	222	50,770
Egypt	13,154	29,49,421
United States	21,350	71,33,493
Aden	18	2,487
Arabia	404	52,234
China—Hong-kong	73	21,861
Persia	5,229	7,98,723
Straits Settlements	5	1,625
Turkey in Asia	2,841	7,58,833
Australia	13	3,500
Other Countries	12	2,063
TOTAL	139,644	3,89,06,494

TABLE No. V.

The imports into India of Foreign Indigo are somewhat remarkable. The following may be given as an analysis of the transactions in 1887-88:—

Exported from	Cwt.	Value in Rupees.	Imported by	Cwt.	Value in Rupees.
United Kingdom	143	27,438	Bengal	75	17,310
China—Hong-kong	72	16,785	Bombay	76	12,718
Straits Settlements	1,259	12,756	Madras	68	14,800
Other Countries	4	580	Burma	1,259	12,731
TOTAL	1,478	57,559		1,478	57,559

The remarkably low-priced indigo of the Straits seems to find a ready market in Burma, where it is largely used by the Shans for dyeing their blue clothes, and doubtless also by the Chinese residents.

Inter-Provincial. 325

INTER-PROVINCIAL AND TRANS-FRONTIER TRADE IN INDIGO.

In Table No. II, given above, will be found a balance sheet for the Indigo trade of Calcutta—the Indigo Emporium. The coastwise and foreign marine trade is there given, as also the internal trade carried by railways, and on the roads and rivers. Of the general inter-provincial coast-wise trade it may be said that the total IMPORTS were, in 1887-88, 9,516 cwt., valued at R13,47,292. Out of that amount Bombay received

TRADE.

Inter-provincial.

8,483 cwt., and Bengal 839 cwt. The coatwise EXPORTS during the same period, were 9,584 cwt., valued at R12,69,305. Of these exports Sind contributed 8,477 cwt., and Madras 1,038, the bulk of the former going to Bombay and of the latter to Bengal. The railway-borne trade showed inter-provincial transactions of an even much larger amount than the coast-wise. The total imports were 43,513 cwt. and the exports 46,681 cwt. Sind (including Karachi) received from the Panjáb, for example, 5,703 cwt., and Bengal drew from the North-Western Provinces 27,235 cwt. The railway returns of indigo conveyed from district to district within each province was also very considerable, amounting to 99,235 cwt. A very large trade is also conducted by road and river, and perhaps a considerable annual production is consumed by the local dyers of which nothing can be learned in official statistics, since consumed where produced.

Trans-frontier. 326

The trans-frontier trade is of some importance: the exports average in value close on five lakhs of rupees, and the imports five or six thousand rupees. Of the exports across the frontier, the Panjáb last year sent 2,916 maunds, valued at R3,08,275, of which Kabul alone received more than half that amount, Kashmír and Bajaur each taking about R 40,000 worth of the dye.

It is however, practically impossible to arrive at an absolutely definite statement of the production and consumption of indigo in India. A very large amount is probably produced by primitive manufacturers and used up by local dyers of which nothing can be learned. The movements of the dye from province to province both by sea and land overlap each other and are thus returned time after time. Thus a given quantity of indigo comes into Calcutta from Madras and is recorded in the coast-wise statistics. How much of that re-appears in the railway exports from Calcutta to the provinces of Upper India, or in the trans-frontier exports, or still further again and again in the transactions between district and district cannot be discovered, but each transaction is marked by the entries in marine, rail, road, and river returns. The large rail-borne quantities shown as conveyed from the districts in Bengal to Calcutta and also from the Provinces of Upper India to Calcutta figure again as the foreign exports from that port. The only constant factor therefore, from which a definite conception can be drawn as to the condition of the trade is furnished by the total foreign imports and exports. The latter represents the surplus over-consumption for all India but is by no means the surplus over local production.

Inga dulcis, see **Pithecolobium dulce**, *Benth.*; Vol. V.

INK AND MARKING NUTS.

327

Various substances are used by the Natives of India in making INK, the usual process being to mix some astringent principle such as galls or myrobalans with one of the iron salts or oxides. The charcoal of rice with lac and gum arabic is, however, employed in Madras, and the Muhammadans generally prepare the ink they use from lamp-black, gum arabic, and the juice of the aloe. The following are the plants specially mentioned as adjuncts in the formation of inks:—

(1) **Alnus nepalensis**, *D. Don.* Bark forms an ingredient in native red inks.

(2) **Cordia Myxa**, *Linn.* Stewart says the unripe fruit is used as a marking nut, though its colour is less enduring than that from **Semecarpus.**

(3) **Phyllanthus Emblica**, *Linn.* Fruits largely employed in making black ink.

INK. (4) **Semecarpus Anacardium,** *Linn. f.* The MARKING NUT is used by all the Indian washermen to mark the linen given them to wash. The juice of the unripe pericarp with lime forms the ink, but as it is apt to cause severe inflammation it has to be used with caution. Without the addition of lime it is often employed as ordinary writing ink.

(5) **Terminalia belerica** and **Chebula,** the unripe fruits of either species, or indeed of any **Terminalia,** is combined with iron in making ink.

(6) In schools, and even by the village merchant, black boards are employed, the ink being a white soft clay mixed with water and applied with a reed.

328 INSECTS, SPIDERS, &c.

Insects. With the exception of the silk-worm, the bee, the cochineal, and lac, and the gall-forming insects, the INSECTA are not of very great economic importance. They belong to the great sub-kingdom of Arthropoda or Articulata, which by modern zoologists is divided into two sections—water breathing and air breathing. The former corresponds to the class Crustacea, while the latter comprises the classes Arachnida (Spiders, Scorpions, and Mites), Myriapoda (Centipedes and Millipedes), and Insecta (insects proper or articulated animals with three pairs of legs, borne on the thorax). The INSECTA are referred to three divisions—Ametabola, Hemimetabola, and Holometabola, according as they attain the adult condition without passing through a metamorphosis, or have an incomplete or a complete metamorphosis. The insects of the first division are parasites on man and animals. The second division comprises three orders—the Hemiptera, Orthoptera, and Neuroptera. The Hemiptera are parasitic on plants and animals such as the Plant-lice (**Aphides**), Field-bug (**Pentatoma**), the Cochineal, Lac, and Wax-forming Insects (**Cocci**), &c. To this order belong some of the gall-producing insects such as those on the lime leaves and the petioles of the poplar. Chinese galls of commerce are stated to be produced by **Aphis chinensis** (on **Rhus semialata**) and the **Pistacia** galls are known to be formed by Hemipterous insects. The Orthoptera includes the Crickets, Grasshoppers, Locusts, Cockroaches, &c. Although some of these are eaten by man, interest in them is more connected with the destruction they often effect to crops. The Migratory Locusts (**Acrydium migratorium**) of Africa and Southern Asia causes destruction against which human ingenuity has practically proved powerless. The Neuroptera are the Dragon-flies, Caddis-flies, May-flies, the Antlion, and Termites. The last mentioned, the white-ants, are, in most parts of India, highly destructive to timber. They live in communities like the Bee and the true Ant (see Hymenoptera), but the workers are individuals of no fully developed sex, whereas among the ants they are undeveloped females; on emergence from the egg the young termite does not pass through a quiescent stage.

The third division Holometabola includes all the insects that pass through a complete metamorphosis, *i.e.,* they exist in a larva stage between the egg and the perfect condition. The orders embraced by this division are Aphaniptera—the Flea and the Chigoe, Diptera (one of the largest orders of INSECTA embracing the House-flies and Flesh-flies, Gnats, Forest-flies, Crane-flies, Gad-flies, &c.; Lepidoptera—the Butterflies and Moths, the most beautiful and, as embracing the silk-worm, the most useful of all insects; Hymenoptera—the Bees, Wasps, Ants, Ichneumons, Saw-flies, &c.; Strepsiptera, an order of small parasitic insects found on bees and other Hymenoptera; and lastly, the Coleoptera or Beetles.

INSECTS.

Many of the Hymenoptera are of considerable use to man such as the bees and the gall-forming species, of which **Cynips gallæ-tinctoria,** forms the Oak-galls of commerce.

The above sketch of the classification of Insects may help to bring out the natural affinities that should exist in an arrangement of the Economic Products derived from the insect world. But the following popular and commercial classification may be given, as it will serve to direct the reader's attention to the positions in this work where fuller details may be found :—

COMMERCIAL CLASSIFICATION. 329

(1) **Resin, White-wax, Bees-wax, &c.**—These products are formed by Hemipterous and Hymenopterous insects See **Lac (also Coccus), Wax, Honey, Oils, &c.**

(2) **Dyes and Tans.**—These are afforded by Hemipterous, Dypterous, Hymenopterous, and Coleopterous Insects. See **Coccus, Galls, &c.**

(3) **Fibres.**—The most important insect fibre is, of course, SILK (which see), but as a curiosity it may be added that some of the spiders spin a web of such strength as to be possible of utilisation as a textile.

(4) **Medicine.**—The beetle known as the Spanish-fly and the Cochineal are the chief medicinal insects. See **Cantharis** and for Indian substitutes see also **Mylabris.** The distillation of ants yields Formic acid. Galls are also employed medicinally, and the beetle **Meloe trianthema** is sometimes collected for its medicinal oil.

(5) **Food.**—The most important food substance afforded by insects is, of course, Honey (which see, as also **Bees**). Grasshoppers, Locusts, and Termites are eaten, and according to the **Rev. A. Campbell,** the red-ant is by the Santals viewed as edible. (See ANT-GREASE.) Many of the aphides also cause the excretion of Manna (which see). In Burma a grub found in palms is considered a great luxury by the Burmans and Karens.

(6) **Ornament.**—The wing cases of beetles, but especially of **Buprestis viltata,** are employed for ornamental purposes, and a considerable trade is done in these. They are used to ornament *khas-khas* fans, baskets, lace, muslin, &c. The brightly coloured back of an Ichneumon is often placed on the foreheads of married Hindu women.

(7) **Pests.**—A large number of Insects belong to this section, for a description of which the reader is referred to the article **Pests** in another volume, and to the remarks under Insect diseases in the articles on Coffee, Cotton, Indigo, Rice, Tea, Wheat, &c.

(8) **Destructive Insects.**—In this section has to be mentioned the white-ants, weevils, wood borers, &c. Although these mostly attack dead wood or grain, they have been described along with those that infest living structures ; see the article on **Pests.**

(9) **Scavengers.**—Excreta injurious to man are buried deep in the earth by the useful dung-beetles which belong to the genera **Coprida** and **Dynastida.** The **Necrophaga** are carrion feeders. The scavenger beetle of Burma is not a pellet roller; it selects for its abode a bed of ordure which it excavates and carries to a chamber two or three feet below the surface.

(10) **Sacred Insects.**—The Hindus do not reverence any insect, though many other animals are held sacred together with a large list of plants. Among some of the aboriginal tribes, however, beetles are viewed with religious feelings. The Mâls of the Rajmahal hills bury in the ground small vessels of water into which they place the large black scavenger and other beetles.

INTRODUCED ECONOMIC PLANTS. 330

To give an exhaustive list of all the introduced plants found in India would occupy many pages and, perhaps, after all, serve a very unimportant purpose—at least from an economic point of view. Of the more important may be mentioned the following (taken mainly from **DeCandolle's** *Origin of Cultivated Plants*) but arranged in two sections :—

331 A.—Indigenous to other parts of Asia, or to Africa and Europe.

The country mentioned after each plant indicates the region from which it is presumed to have been originally derived.

1. **Allium Ampeloprasum,** *var.* **Porrum.** LEEK. Europe.
2. **A. Cepa.** ONION. Persia, Afghanistan, &c.
3. **Areca Catechu.** ARECA NUT. Malay.
4. **Artocarpus incisa.** BREAD FRUIT. Sanda Islands.
5. **Avena sativa.** OATS. Eastern Temperate Europe.
6. **Beta vulgaris.** BEET. Canaries: Mediterranean basin: Western Temperate Europe.
7. **Brassica Napus.** RAPE. Europe: Western Siberia.
8. **B. oleracea.** CABBAGE. Europe.
9. **B. Rapa.** TURNIP. Europe: Western Siberia.
10. **Cajanus indicus.** DAL.
11. **Cannabis sativa.** HEMP. Dahuria: Siberia.
12. **Carthamus tinctorius.** SAFFLOWER. Arabia ? India. [Syria.
13. **Ceratonia Siliqua.** CAROB TREE. Southern Coast of Anatolia:
14. **Cicer arietinum.** CHICK PEA or GRAM. South of the Caucasus and of the Caspian.
15. **Cichorium Intybus.** CHICORY. Europe: Northern Africa: Western Temperate Asia.
16. **Citrullus vulgaris.** WATER-MELON. Tropical Africa.
17. **Citrus Aurantium.** ORANGE. China and Cochin-China.
18. **C. decumana.** POMELO. Pacific Islands to east of Java.
19. **Cocos nucifera.** COCOA NUT. Malay and Polynesia. [sinia.
20. **Coffea arabica.** COFFEE. Tropical Africa: Mozambique: Abys-
21. **Corchorus capsularis.** JUTE. ? Java.
22. **Cucurbita maxima.** SPANISH GOURD. Guinea.
23. **Cydonia vulgaris.** QUINCE. Northern Persia.
24. **Cynara Cardunculus** } ARTICHOKE. Southern Europe: North **C. Scolymus** } Africa: Canaries: Madeira.
25. **Eriobotrya japonica.** LOQUAT. Japan.
26. **Faba vulgaris.** BEAN. South of the Caspian. [Siberia.
27. **Fagopyrum esculentum.** BUCKWHEAT. Mantschuria: Central
28. **Ficus Carica.** FIG. Mediterranean basin, from Syria to the Canaries.
29. **Fragaria vesca.** STRAWBERRY. Europe: the plant is a native of Temperate India but was never cultivated by the Natives.
30. **Glycine hispida.** SOY BEAN. Cochin-China: Japan: Java.
31. **Gossypium arboreum.** TREE COTTON. ? Upper Egypt and Africa, but by some authors is believed to be a native of India. The true **G. herbaceum** does not exist in India, the indigenous cotton being mainly **G. Wightianum.** A large series of Exotic or American cottons are also grown, more especially **G. hirsutum,** New Orleans or Saw-ginned Darwar.
32. **Hibiscus esculentus.** OCHRO. Tropical Africa.
33. **Hordeum distichon, vulgare,** and **hexastichon.** FORMS OF BARLEY. Western Temperate Asia.
34. **Lactuca Scariola.** LETTUCE. Southern Europe: Northern Africa: Western Asia.
35. **Lathyrus sativus.** CHICKLING VETCH. South of the Caucasus.

36. **Lawsonia alba.** HENNA. Western Tropical Asia : Nubia.
37. **Linum usitatissimum.** FLAX. Western Asia.
38. **Myristica moschata.** NUTMEG. Moluccas.
39. **Nephelium Litchi.** LITCHI. Southern China : Cochin-China.
40. **Panicum miliaceum.** CHENA-MILLET. Egypt : Arabia.
41. **Papaver somniferum.** POPPY. Mediterranean basin.
42. **Petroselinum sativum.** PARSLEY. Southern Europe.
43. **Phœnix dactylifera.** DATE PALM. Western Asia and Africa from the Euphrates to the Canaries.
44. **Piper Betle.** BETEL PEPPER. Malay.
45. **Pistacia vera.** PISTACHIO. Syria.
46. **Pisum arvense.** FIELD PEA. Italy.
47. **Prunus armeniaca.** APRICOT. China.
48. **Prunus domestica.** PLUM. Anatolia, South of the Caucasus : North Persia.
49. **P. insititia.** PLUM. Southern Europe : Armenia : South of the Caucasus.
50. **P. persica.** PEACH. China.
51. **Pyrus Malus.** APPLE. Europe.
52. **Ricinus communis.** CASTOR OIL. Abyssinia : Sennaar : Kordofan.
53. **Saccharum officinarum.** SUGAR-CANE. Cochin-China.
54. **Sesamum indicum.** TÉL or GINGELLY. Sunda Islands.
55. **Setaria italica.** ITALIAN MILLET. China: Japan : Indian Archipelago.
56. **Sorghum saccharatum.** SUGAR-CANE. Tropical Africa.
57. **S. vulgare.** JUAR MILLET. Tropical Africa.
58. **Spinacia oleracea.** SPINACH. Persia.
59. **Triticum vulgare.** WHEAT. ? Region of the Euphrates.
60. **Zizyphus vulgaris.** COMMON JUJUBE. China.

B.—Indigenous to North or South America. 332

1. **Achras Sapota.** SAPODILLA PLUM. Campeachy, Isthmus Panama, Venezuela.
2. **Agave americana.** AMERICAN ALOE. Mexico.
3. **Anacardium occidentale.** CASHEW-NUT. Tropical America.
4. **Ananas sativa.** PINE-APPLE. Mexico : Central America : Panama, &c.
5. **Anona muricata.** SOUR SOP. West India Islands.
6. **A. reticulata.** BULLOCK'S HEART. West India Islands : New Granada.
7. **A. squamosa** CUSTARD APPLE or SWEET SOP. West India Islands.
8. **Arachis hypogæa.** EARTH NUTS. Brazil.
9. **Bixa Orellana.** ARNOTTO. Tropical America.
10. **Erythroxylon Coca.** COCA. Peru : Bolivia.
11. **Capsicum annuum.** CAPSICUM. Brazil
12. **C. frutescens.** CAPSICUM. Peru to Bahia.
13. **Carica Papaya.** PAPAYA. Central America and West Indies.
14. **Chenopodium Quinoa.** QUINOA. Peru : New Granada.
15. **Cinchona Calisaya.** QUININE. Bolivia, south of Peru.
16. **C. officinalis.** Ecuador.
17. **C. succirubra.** Ecuador.
18. **Cucurbita Pepo.** / **C. Melopepo.** } PUMPKIN. Temperate North America.
19. **Gossypium barbadense.** BARBADOES COTTON, SEA ISLAND, &c., &c. New Granada : Mexico, West Indies (see remark above).
20. **Helianthus tuberosus.** JERUSALEM ARTICHOKE. North America.
21. **Ipomœa Batatas.** SWEET POTATO. Tropical America.

22. **Lycopersicum esculentum.** TOMATO. Peru.
23. **Manihot utilissima.** MANIOC. Brazil.
24. **Maranta arundinacea.** ARROW-ROOT. Tropical America.
25. **Nicotiana rustica** / **N. Tabacum** } TOBACCO. From Mexico and Ecuador.
26. **Opuntia**, various species. PRICKLY PEAR. Mexico.
27. **Psidium Guayava.** GUAVA. Tropical continental America.
28. **Physalis peruviana.** CAPE GOOSEBERRY. Tropical America.
29. **Solanum tuberosum.** POTATO. Chili: Peru.
30. **Theobroma Cacao.** CACAO. Amazon and Orinoco Valley.
31. **Zea Mays.** MAIZE or INDIAN CORN. New Granada.

The above enumeration by no means embraces all the introduced plants now met with in India, nor indeed does it exhaust those of economic value. It will suffice, however, to convey some idea of India's indebtedness to other countries and more especially to America. The kindly way American plants have taken to India is one of the most remarkable features of the present vegetation, and throughout India vast tracts may be found literally covered, to the extermination of indigenous types, with useless American weeds. An examination of the above list will reveal the fact that India has produced remarkably few products to give in exchange for those it has received. Rice is, perhaps, the most remarkable food-stuff that would appear to be indigenous and it is followed in importance by tea and indigo (two doubtfully indigenous products) and by a few millets, pulses, and fruits, chief amongst which is the mango. The early Hindu invaders of India and the Arabian and Persian traders, brought to it many of the long list of Asiatic products, but it is probable that the later Muhammadan conquerors (through the European settlers) added a far larger number, including most of the American plants, which now constitute so much of the important features of Indian agriculture. Amongst the African plants may be specially mentioned **Sorghum saccharatum, Sorghum vulgare** (the juar), **Cajanus indicus** (the dal), **Citrullus vulgaris** (the water-melon), and **Hibiscus esculentus** (the ochro).

INULA, *Linn.; Gen. Pl., II., 330.*

In India some 20 species of **Inula** occur, many of them extremely abundant plants, as, for example, **I. Cappa**, *DC.*, a shrub met with on the Temperate Himálaya from Kumáon to Bhután, chiefly between 4,000 and 6,000 feet, but in some localities ascending even to 10,000 feet; also distributed to the Khásia Hills, Manipur, Burma, China, and Java. No distinct record exists, however, of any of these being used as substitutes for the Elecampane, though by a mistake in synonymy, apparently, some writers regard the species below as the Indian form of that drug. In Kashmír the root of **I. Royleana,** *DC.*, is used to adulterate *kút* (see **Saussurea Lappa**).

333 **Inula racemosa,** *Hook. f.; Fl. Br. Ind., III., 292;* COMPOSITÆ.

Syn.—I. HELENIUM, *Herb. Ind. Or. H. f. & T. not of Linn.*

I. HELENIUM, *Linn.*, is the ELECAMPANE, *Eng.;* AUNEE, AULNEE, *Fr.;* ALANT, BRUSTALANT, *Ger.;* ENULA CAMPANA, *It. & Sp.;* DEVIASIL, *Russian.*

Vern.—*Rásan*, ARAB.; *Zanjabil-i-shámi*, PERS.

References.—*Voigt, Hort. Sub. Cal., 410; Ainslie, Mat. Ind., I., 119; O'Shaughnessy, Beng. Dispens., 419; Flück. & Hanb., Pharmacog., 380; U. S. Dispens., 15th Ed., 799; Bent. & Trim., Med. Pl., 150; Year Book Pharm., 1874, 626; 1875, 238; 1881, 119.*

Habitat.—Western Himálaya, on the borders of fields, &c.; Kashmir, altitude 5,000 to 7,000 feet; Piti, altitude 9,000 to 10,000 feet.

MEDICINE. Root. 334 **Medicine.**—The ROOT of the Elecampane is now mainly used in veterinary medicine. When dry the roots have a weak, aromatic odour, resembling Orris and Camphor; in flavour they are slightly bitter, and their action is as a mild tonic. They also possess diaphoretic, diuretic, expectorant, and emmenagogue properties. By the ancients Elecampane was much employed in the complaints peculiar to females, and in America the drug is still resorted to in the treatment of amenorrhœa, while it is found to be sometimes beneficial in chronic diseases of the lungs when complications of general debility or want of tone in the digestive organs exist. In France and Switzerland it is used in the distillation of Absinthe; see **Artemisia Absinthium,** Vol. I., 323. MEDICINE.

Iodine, see **Barilla,** Vol. I., p. 399, also **Fucus, & Gracilaria,** Vol. III., 451, and IV., 174.

IONIDIUM, *Vent.; Gen. Pl., I., 117, 970.*

Ionidium suffruticosum, *Ging.; Fl. Br. Ind., I., 185; Wight, Ic., t. 308;* VIOLACEÆ. 335

Syn.—VIOLA SUFFRUTICOSA and ENNEASPERMA, *Roxb.;* V. FRUTESCENS and ERECTA, *Roth.;* IONIDIUM ENNEASPERMUM, *DC.;* I. HETEROPHYLLUM and ERECTUM, *DC.;* I. HEXASPERMUM, *Dalz.*

Vern.—*Ratan-purus,* HIND.; *Núnbora,* BENG.; *Tandi sol, bír suraj mukhi,* SANTAL; *Ratan-purus,* DEC.; *Orilai hámarai,* TAM.; *Suryákánti, nila kobari, purusha ratnam,* TEL.; *Chárati* (according to **Ainslie**), SANS.

References.—*Roxb., Fl. Ind., Ed. C.B.C., 218; Voigt, Hort. Sub. Cal., 77; Dalz. & Gibs., Bomb. Fl., 12; Rev. A. Campbell, Econ. Prod. Chutia Nagpur, Nos. 7517 and 8171; Rheede, Hort. Mal., IX., 60; Elliot, Fl. Andhr., 131, 160, 171; Ainslie, Mat. Ind., II., 267; O'Shaughnessy, Beng. Dispens., 209; Dymock, Mat. Med. W. Ind., 2nd Ed., 66; U. S. Dispens., 15th Ed., 817; Moodeen Sheriff, Mat Med. South India, 32; Pharmacographia Indica, I., 139; Trans. Agri.-Hort. Soc., VII., 71; Gazetteers:—Bombay (Kanara), XV., 427; N.-W. P. (Bundelkhand), I., 79; (Agra), IV., lxxii.; (Himalayan Districts), X., 305; Mysore and Coorg, I., 57; Nellore Man., 137.*

Habitat.—A small herbaceous plant abundant throughout the warmer parts of India, from Agra to Bengal, Orissa, and the Southern Konkan (common at Belgaum and Kánara).

Medicine.—According to **Ainslie** the LEAVES and tender STALKS are used by the natives of Southern India as a demulcent. They are employed in decoction and electuary in conjunction with some mild oil, in preparing a cool liniment for the head. Of the decoction, he adds, about an ounce and a half is given twice daily. **Dr. Moodeen Sheriff** (in his forthcoming work) says it is demulcent and refrigerant and is useful in some cases of gonorrhœa and of scalding of urine. According to the **Rev. A. Campbell** the Santals employ the ROOT in bowel complaints of children accompanied with black excreta. In the *United States Dispensatory* it is stated that the root of a species of **Ionidium** has attracted some attention in the treatment of elephantiasis. MEDICINE. Leaves. 336 Stalks. 337 Root. 338

Chemistry.—According to the *Pharmacographia Indica,* "The root contains an alkaloid soluble in ether, and alcohol, not easily crystallized; its solution in the form of a salt, which it readily forms with the mineral and vegetable acids, is precipitated by potassio-mercuric iodide, iodine in potassium iodide, tannin and the alkalies. It also contains quercitrin, allied to the viola-quercitrin of Mandelin; and another colouring matter soluble in water, but insoluble in amylic alcohol, an acid resin, and a quantity of mucilage and oxalates." CHEMISTRY. 339

Sir W. O'Shaughnessy failed to detect the least trace of the active principles emetine or violine in samples of the plant procured from the Royal Botanic Gardens, Calcutta.

The Kulmi Sag.

MEDICINE. SPECIAL OPINIONS.—§ "An infusion of the root is used as a diuretic and given in gonorrhæa. The leaves are demulcent. Indigenous in Belgaum" (*Surgeon-Major C. T. Peters, South Afghánistan*).

Ipecacuanha, see **Cephælis Ipecacuanha**, *Rich.*; VOL. II., p. 247.

IPHIGENIA, *Kunth.*; *Gen. Pl., III., 824.*

[LILIACEÆ.

340 **Iphigenia indica**, *Kunth.*; *Linn. Soc. Jour., Vol. XVII., 450*;

Syn.—MELANTHIUM INDICUM, *Linn.*; ANGUILLARIA INDICA, *R. Br.*

Vern.—*Chutia chandbol*, SANTAL.

References.—*Rev. A. Campbell, Report on Econ. Prod. Chutia Nagpur, No. 8703; Kánara Gazetteer, XV., 444.*

Habitat.—A small herb appearing during the rainy season in tropical and sub-tropical India: ascends the Himálaya to altitudes of 6,000 to 7,000 feet. Is frequent from Chutia Nágpur to Kánara, also common in Burma and Ceylon.

DYE. 341 **Dye.**—The **Rev. A. Campbell** mentions that the Santals obtain a red dye from the flowers.

IPOMÆA, *Linn.*; *Gen. Pl., II., 870.*

342 **Ipomæa angustifolia**, *Jacq.*; *Fl. Br. Ind., IV., 205*; CONVOLVULACEÆ.

Syn.—I. DENTICULATA, *Br.*; I. FILICAULIS, *Blume*; CONVOLVULUS LINIFOLIUS, *Wall., Cat., 1389*; C. MEDIUM, *Roxb.*; C. FILICAULIS, *Vahl.*

Vern.—*Tala-neli* (according to **Rheede**), TAM.; *Konda, sita savaram* (according to **Elliot**, who remarks that *savaram* and *lanjasavaram* are names given by **Heyne** to this plant, as also the Sanskrit synonym *Prasáran*, which more correctly should be given to **Pæderia fœtida**), TEL.

References.—*Roxb., Fl. Ind., Ed. C.B.C., 159; Dalz. & Gibs., Bomb. Fl., 165; Sir W. Elliot, Fl. Andhr., 14, 62, 97, 106, 169; Rheede, Hort. Mal., XI., t. 55; Gazetteers:—Bombay (Kanara), XV., 439; N.-W. P. (Agra), IV., lxxv.; Mysore and Coorg, I., 56.*

Habitat.—A diffuse twining biennial, found on the hills of the Deccan Peninsula and distributed to Bundelkhand, the Khásia Hills, Malacca, Ceylon, &c.

343 **I. aquatica**, *Forsk.*; *Fl. Br. Ind., IV., 210.*

Syn.—CONVOLVULUS REPENS and REPTANS, *Roxb.*; I. REPTANS, *Poir.*; I. REPENS, *Roth.*; I. SUBDENTATA, *Miq.*

Vern.—*Kalmi-sák*, BENG.; *Kulum sag* (PARBUTIAH), NEPAL; *Kalmi-ság, nári*, N.-W. P.; *Ganthian, nárí, náli*, PB.; *Naro*, SIND; *Nálichi baji*, BOMB.; *Sarkarei valli, koilangu*, TAM.; *Túti-kura*, TEL.; *Kan-kun*, SING.; *Kalambi*, SANS.

References.—*Roxb., Fl. Ind., Ed. C.B.C., 162; Voigt, Hort. Sub. Cal., 355; Dalz. & Gibs., Bomb. Fl., 164; Stewart, Pb. Pl., 150; Elliot, Fl. Andhr., 185; Rheede, Hort. Mal., XI., t. 52; Stocks, Account of Sind; Rev. A. Campbell, Report Econ. Prod. Chutia Nagpur, No. 8133; O'Shaughnessy, Beng. Dispens., 506; U. C. Dutt, Mat. Med. Hind., 302; Dymock, Mat. Med. W. Ind., 889; Taylor, Medical Topography of Dacca, 48; Atkinson, Econ. Prod. N.-W. P., V., 96; Gazetteers:—Bombay (Kánara), XV., 439; N.-W. P. (Bundelkhand), I., 82; (Agra), IV., lxxv.; Indian Forester, III., 237; XII., App. 17; XIV., 391; Agri.-Hort. Soc. of Ind. Trans., IV., 103; Journal (Old Series), X., 20.*

Habitat.—An aquatic species common throughout India, and especially abundant on the surface of tanks in Bengal. In Madras and Ceylon it is regularly cultivated as a vegetable.

MEDICINE. Juice. 344 **Medicine.**—In Burma the JUICE is said to be employed as an emetic in cases of arsenical or opium poisoning. **O'Shaughnessy** says that the juice when dried is nearly equal to scammony in purgative efficacy.

The Kulmi Sag. (*G. Watt.*)

IPOMÆA aquatica.

FOOD. Shoots. 345 Leaves. 346 Roots. 347

Food.—The young SHOOTS, LEAVES, and ROOTS of this floating aquatic species are universally eaten as a vegetable in the region in which it is found. In Lower Bengal, where a superabundance of stagnant water occurs, in the form of tanks, it is to be found throughout the year, but in other parts of the country it is more or less cultivated. At times bunds are made across the outlets of marshes to prevent their being drained, and cuttings of the plant are then placed in the mud all over the swamp. In this semi-cultivated condition crops are obtained, where otherwise the plant could scarcely exist. In Madras and Ceylon, however, it is regularly cultivated and trained to exist on fields not inundated. In a circular letter addressed to Local Governments, in which information was called for on the Economic Products described in this volume, full particulars were recorded regarding this vegetable. The following abstract of these replies may be here given:—

Bengal.—Found wild in tanks: is regularly eaten as a vegetable.

Assam.—An aquatic weed which grows in almost all tanks and *jeels* or low-lying tracts covered with water. The weed generally begins to grow with the commencement of the rains and remains till January or February when it begins to wither on the top of the water. In *jeels* and tanks where the water never dries up, this weed is found all the year round. The leaves and stems are gathered, cut to pieces, boiled in acid, alone or with fish, and eaten by the Assamese with rice. They are supposed to have a very cooling effect if boiled on the previous night and eaten the first thing in the morning. The weed is considered to be very wholesome for females who suffer from nervous and general debility.

Madras.—In the Madura District alone the extent of cultivation is said to be about 600 acres. The kind produced is to a large extent that with white roots. It is sown in August and harvested from December to March. The creeper is cultivated in *Padugai, Senval,* and *Karisal* soils. The land is ploughed three times, manured, and ploughed again, then plots are formed and watered. Thereafter the creepers are planted in small pieces of six inches at a span's length from each other. Ten loads of the creeper are required to plant one acre. The creepers are first grown amidst *ragi*. Weeding is done after one month. The cost of cultivation is R20. The roots are sweet and after being boiled largely eaten.

Burma.—In the Pegu District it is found growing wild as a creeper floating on tanks, fields, streams, &c. The Burmese eat it, and it is so much a favourite that many persons are engaged collecting and selling it. The price of a bundle is one or two pice, and being very plentiful only a small profit of a few rupees can be made on it.

In the Akyab District the area of cultivation is about 30 acres. It is first a spontaneous growth in tanks, swamps, &c. Slips are broken off and about June are planted out. It is fit for use in three weeks and continues growing nearly throughout the year so long as there is water. Cost of cultivation about R10 or R15 an acre. Profit about R60 or R70 per acre. Portions of land are embanked by which the water is kept from escaping and the cuttings are then put down. It is used as a vegetable.

In the Mandalay District it is found in abundance in tanks and marshy localities, and is extensively used as one of the most favourite vegetable diets.

In the Thongwa District it grows wild and is largely eaten by the people. In the Maubin Sub-division the annual produce is about 1,000 baskets.

In the Bhamo District it is commonly met with and is much used by the natives in making curries for which the leaves and tender tips are employed. Similar information has been communicated regarding the Upper

Chindwin, Kyaukpyu, Bassein, Amherst, Eastern division of Upper Burma, Toungoo, and Ruby Mines Districts.

SPECIAL NOTE.—§ "*Kalmi* is largely cultivated in shallow ponds near the villages of Bengal on account of its leaves which are used as a culinary vegetable. They are said to be cooling" (*Assistant Surgeon Shib Chunder Bhattacharji, Chanda, Central Provinces*).

Fodder.—See Vol. III., p. 415.

348 **Ipomæa Batatas,** *Lamk.; Fl. Br. Ind., IV., 202.*

SWEET POTATO.

Syn.—BATATAS EDULIS, *Chois;* CONVOLVULUS BATATAS, *Linn.;* C. EDULIS, *Thunb.;* C. ESCULENTUS, *Salisb.*

Vern.—*Mita-alú* (sweet potato), *shakarkand*, HIND.; *Ranga-alú, lal-alú* or *lal-shakarkand-alú* (= the red form), *chine álu* (the white), BENG.; *Sakar-kenda*, SANTAL; *Goria alú* (white form), *boga* or *ranga* (red), ASSAM; *Fukum* (Newari), *Kerkalú* (Parbutiah), NEPAL; *Shakarkand*, PB.; *Gajar lahori* (= Lahore Carrot), SIND; *Ratalú, shakar-kandu*, BOMB.; *Ratáli*, MAR.; *Sakaria*, GUZ.; *Vallikilángu, sakkarei-vellei-kelangu*, TAM.; *Chelagada*, TEL.; *Genasu*, KAN.; *Kapa-kalenga*, MALAY.; *Kaswan, myouk-ní*, BURM.; *Batala*, SING.; *Lardak Lahori* (= Lahore Carrot), PERS.

References.—*Roxb., Fl. Ind., Ed. C.B.C., 162; Stewart, Pb. Pl., 150; Watt, in Sel. Rec. Govt. of India, 1888-89, 147-154; Mason, Burma & Its People, 455, 783; Rev. A. Campbell, Econ. Prod. Chutia Nagpur, No. 8163; Linschoten, Voyage to the East Indies (1596), Vol. II., 42, 279; Stocks, Account of Sind; Rheede (Kuppa-kelengu), Hort. Mal., VII., 95, t. 50; Rumph., Herb. Amb., V., t. 130; DeCandolle, Orig. Cult. Pl., 53; Baden Powell, Pb. Pr., 259; Drury, U. Pl., 72; Lisboa, U. Pl. Bomb., 166; Agri.-Hort. Soc. Ind., Trans., II., 1838; (App.), 306; III. (Proc.), 229; IV., 103; Journals (Old Series), I., 128, II. (Proc.) 311; III., (Proc.) 40; IV., (Sel.) 34; V., (Sel.) 29, (Proc.) 45; IX (Sel.), 58, 75; X., 20; (New Series,) I.; (Sel.) 9, 16, 17; V., 35, 43; VII., (Proc.) cxvi; Kew Off. Guide to the Mus. of Ec. Bot., 99; Kew Off. Guide to Bot. Gardens and Arboretum, 76; Home Dept. Cor. regarding Pharm. Ind., 282; Bengal—First Ann. Rept. Director of Agriculture (1886), p. lxvii.; Madras—Board of Rev. Proc., 1889, No. 266, App. A.; Man. Trichinopoly, 74; Desc. Acc. of the Godavery, 8; Man. Coimbatore, 230; Exper. Farm Rept., 1879, 111; Bombay—Man. Rev. Accounts, p. 101; Gaz., VII., 40, XII., 170; Crop Experiments, 1884-85, pp. 16, 81; Central Provinces—Rev. Settle. of Jubbulpore, 86; Nursingpore, 53; Mysore & Coorg, Gaz., II., 11.*

HISTORY. 349

Habitat and History.—There are two forms of this plant met with under cultivation in India, distinguished by the colour of their tubers, *viz.*, red and white. These are apparently two races derived from a common stock which was originally a native of tropical South America, but it seems probable that the latter was a later development than the former. It is, at all events, customary in India, to hear the red form spoken of as the hardier and country sweet potato, while the white is the more delicate, superior quality introduced by the English. The reader is referred to the discussion given under **Dioscorea** *(Vol. III., 115—122)* regarding the Yams and Sweet-potatoes of India. It will be there seen that the author is disposed to regard several of the Yams, now cultivated, as Indian in their origin, and as having been once upon a time known by some of the specific names since transferred to the sweet-potato. That all the forms of the sweet-potato found in India have been introduced there can be no doubt, and since the **Batatas** of **Linschoten,** which he described as found in India three hundred years ago, was a form of **Dioscorea** (Yam) and not the sweet-potato, the date of introduction of the latter must be fixed at a more recent date. This idea is confirmed by the universality, in the provinces of India, of descriptive names for the sweet-potato that either compare it to indigenous tubers, or allude to its sugary property—of these names being ancient.

Medicine.—The ROOTS of this plant have a laxative property. Baden Powell says that in the Panjáb the natives regard the sweet-potato, "as hot, useful to strengthen the brain and in special diseases." MEDICINE. Roots. 350

SPECIAL OPINIONS.—§ "The laxative properties of the roots must be very slight, as they are eaten in large quantities as a vegetable without any apparent effect" (*Brigade Surgeon G. A. Watson, Allahabad*). "Sweet potato is indigestive" (*V. Ummegudien, Mettapollium, Madras*).

Food.—The sweet potato is eaten by all classes of the natives, either in curry, or simply roasted, or after having been cut in half, lengthwise, and fried. Another way of preparing the sweet-potato is to boil it, cut it in slices, and add rasped cocoanut milk and sugar. In this way it becomes a fairly good dessert. It is also boiled, mashed, and made into pudding in the usual European style with sugar, egg, and milk (*L. Liotard*). FOOD. 351

Fodder.—The TOPS of the sweet-potato are used as cattle fodder, and when young are eaten as a pot-herb. See Vol. III., p. 415. FODDER. Tops. 352

Cultivation.—*Bengal.*—There are two varieties cultivated: one with purple tubers and purple flowers which is known as *Sakarkanda*, and the other with white tubers and flowers—the *Chine álu*. Sweet potato requires a sandy soil and is generally raised on lands that will scarcely grow any other crop. Five to seven ploughings are given and the land then raised in ridges often leaving a narrow belt between the trenches and ridges. It is propagated by cuttings, to supply which creepers are grown near the homesteads in a damp locality. Before planting, these creepers are pulled up and cut into pieces, each containing two knots, the portions close to the roots being rejected. The cuttings are planted in the ridges, one of the knots being buried in the ground and the other left above. The planting time is from August to September. When the cuttings have taken root, the ridges are properly made, and later on the plants are earthed up. The harvest time extends from October to November. The yield is from 20 to 30 maunds per bigha (=⅓rd acre). *Chine álu* sells from eight or ten annas a maund and *Sakarkand* from twelve to fourteen. A kind of thread-like worm sometimes grows inside the roots and does much injury to the plant (*Annual Report, Director of Land Records and Agriculture, 1886, p. lxvii*). CULTIVATION. Bengal. 353

In a special report furnished for this work the Director of Land Records and Agriculture writes: "*Sakarkand* —In the Bhagulpore Division the fields are ploughed and harrowed two or three times. The seeds are then scattered in June-July, and dug out in January-February. It is grown to a small extent in Chota Nagpur for home consumption. There are two varieties: one is known as *desi* (country) or red, and the other *belati* (English) or white. The latter is fast driving the former out of cultivation. The area under *aluas* and *sulknis* (species of sweet potatoes or yams) is said to be considerable in the Mozufferpore districts, *viz.*, 6 per cent. of the whole cultivation. This article of diet is one of the chief foods of the poorer classes in Mozufferpore, and lasts from October till March, being dug up as required."

Assam.—In a special report the Director of Land Records and Agriculture stated: "The sweet-potato is grown in the villages in or around the homesteads. It is sown in September or October and taken up in January." Assam. 354

North-West Provinces.—Mr. E. T. Atkinson (*Economic Products, Part V., 19*) says the sweet potato or *Shakrkand, rakt-alu*, or *pind-alu* is cultivated all over these provinces. There are two principal varieties, one with white and the other with red tubers, both of which are highly N.-W. Provinces. 355

CULTIVATION. N.-W. Provinces.

esteemed as palatable and nutritious articles of food when boiled or roasted. The young leaves and tender shoots are eaten in curries and used as greens, and the leaves themselves are considered an excellent fodder for cattle."

Central Provinces. 356

Central Provinces.—**Mr. J. B. Fuller** reports: "This is commonly cultivated during the cold weather in the beds of streams and on the sloping banks of large rivers in company with *Brinjal* or *Bhata* (**Solanum melongenum**). Full details of area are not available. In the Saugor, Seoni, and Hoshangabad districts it is returned as covering over 500 acres, and it is grown on over 700 acres in Jubbulpur."

Panjab. 357

Panjáb.—**Stewart** wrote: "the sweet potato is commonly cultivated (in the cold weather) in the eastern part of the Panjáb plains; the root is eaten as a vegetable."

Bombay. 358

Bombay.—In a special report furnished for this work the Director of Land Records and Agriculture writes:—"Sweet potato is grown everywhere in the Presidency. In 1887-88 it occupied 10,666 acres. It is irrigated and manured. There are two varieties, the white and the red. The latter is smaller and sweeter. In Gujarat it is grown in the cold weather, but in the Deccan and the Konkan any time in the rains or cold weather. It takes about four to six months to mature. The land is ploughed, harrowed, weeded, and dressed with manure and sown from a last year's bed which has been watered and kept for the purpose. The potato rots in the ground and sends out small shoots which consist of a knot or eye, with a single leaf attached. The shoots are planted in the prepared beds and irrigated. About 4,000 to 5,000 shoots go to an acre, and produce from $1\frac{1}{2}$ to 5 tons. The price varies from R$\frac{3}{4}$ to R2 per Indian maund.

The young shoots and leaves are eaten as a pot-herb. The tubers are eaten both boiled and roasted. They are largely used on fast days. Dried and ground into flour they are also made into cakes which the higher classes eat on fast days. During the rainy season, when the grain stock of poorer cultivators runs short, sweet-potato not uncommonly forms part of their bread stuff. The mature vine is an excellent cattle food. The average yearly outturn may be estimated at 8,90,400 maunds."

In the report of the Crop Experiments mention is made of two kinds (1) *Deshi* or *Mulki*, with heart-shaped leaves only slightly indented and root rather small but sweet, often red coloured: (2) *Viláyiti* introduced some fifteen years ago, leaves deeply lobed, roots large and never red coloured. The cultivation of the crop is regarded as good for the land. Out of a field of sweet-potatoes three rows were pulled and the produce weighed, *viz.*, 947℔. **Mr. Heaton**, who reports on the experiment, remarks that if the southern part of the field had given as good a return as the northern, the outturn would have been 13,489℔ per acre. The yield was by no means a first-rate one. The roots were of medium size, not big. I estimate, he adds, that a 16-anna crop would yield from 18,000 to 20,000℔ an acre. **Mr. Price**, of the Southern Maratha Survey, experimented on sweet-potatoes many years ago and obtained as the result a yield of 6,200℔ per acre.

Madras. 359

Madras.—In the Coimbatore Manual it is stated "white and red varieties are grown in gardens, the cuttings being planted from September to December, and the roots dug out from January to March. The land is well ploughed, manured, and ridged. The cuttings, which are from creepers planted in a nursery a month previously, and are a foot long, are planted on the sides of the ridges. Watering is done every three or four days, but weeding only once or so, as the creepers cover the ground. It is said that manure is not applied, as the crop always follows heavily manured *ragi* (**Eleusine Corocana**). The outturn is valued at from R70 to 100, and the expenses of cultivation at from R30 to 40, of which the watering is a

CULTIVATION.

Madras.

nominal half. A garden land of 1·35 acres has been known to let to a tenant for R70 per annum, one of the crops being sweet potatoes. The value of the outturn of sweet potatoes was perhaps R130. The Government assessment on 1·35 acres would be R2 and water rate R4-8."

The following special reports, contributed for this work, may be here given. **Mr. H. Sewell**, Collector of Cuddapah, wrote :—"The creepers are first planted in June and July, and when multiplied are transplanted in October. The crop of potatoes is dug out in January and February. The soil required is red sand. The cost of cultivation is R15 and profit R25 a cawndy. The potatoes are sold at 2 annas a tûk."

Mr. H. Willock, Collector of Trichinopoly, stated :—"The area under sweet potato (*valli kalangu*) cultivation in this district is only about 150 acres. The outturn is estimated at about R10 worth per acre per annum. The cost of cultivation is about R6. This tuber is cultivated to a very small extent in patches in *olapperi* lands under well and channel irrigation. It is generally grown on red loam or sandy soil." **Mr. Goodrich**, Collector of Bellary, reported :—"The area devoted to sweet potato in three taluks is 112 acres, but the extent in other taluks cannot be estimated. The time of sowing varies according to locality from June to November, and that of reaping from December to March. The average cost of cultivation per acre is R29, the highest being R45, and the lowest R12. Red and mixed soils are those best suited to this crop. The ground should be ploughed and harrowed twice, and irrigated once, in eight days. It is eaten as a vegetable, and is also sometimes made into dishes. The annual yield of this crop is about 50,700 maunds, and the cost price varies from 8 annas a maund to 10 maunds for a rupee."

Burma.
360

Burma.—The following special reports have been received regarding sweet-potato cultivation in Burma :—"In the Upper Chindwin district it is grown on soft soil. The ground is broken and small bits of stems planted. No watering is required. It is planted in September and October, and the potatoes are collected in February and March. It is grown largely, and sold at R3 per 100 viss. But the export is not much."

"In the Lower Chindwin district it is very plentiful. The grounds selected are usually on the islands in the river. It is boiled, cooked, or fried and then eaten."

"In the Prome district, the area under cultivation is 21·50 acres. It is sown in July and reaped in March. The cost of cultivation is R3-8 for a quarter of an acre, yielding potatoes to the amount of 150 viss, which being sold for R9 leaves a profit of R5-8. Slips are taken and planted in sandy soil in a hole of a span's depth. It is cooked and eaten. The annual production is estimated at 13,000 viss, and the cost at the place of production is given as R5-8 or R6 per 100 viss."

"In the Pegu district it is cultivated in moderate quantities by Shan and Chinese gardeners, The ground is divided into ridges about a foot high and on them the tubers are planted. When ready for market the tubers are dug out and sold at 10 to 12 annas per 10 viss. It is generally grown on sandy loam near streams, as it does not form large tubers in the shallow hard soil of the hills. It is planted at the beginning of the rains, and the tubers are taken up in the cold weather."

"In the Akyab district, the area cultivated is about 50 acres. It is sown about October or November and dug up in about three months. Cost of cultivation about R30 per acre. The soil has to be ploughed, manured, and made into ridges on which the cuttings are planted. It thrives better when cowdung is used as manure. It is used as food, but only as a delicacy and eaten in small quantities."

The Sweet Potato.

CULTIVATION. Burma.

"In the Kyoukpyu district it is grown all over, as both the leaves and the tubers are eaten by the natives."

"In the Eastern Division, Upper Burma, it is cultivated to some extent in two varieties—the yellow and white."

"In the Toungoo district it is largely sown by Shans in the river silt, at the end of the rains, and at other times where loose, friable, and sandy soil is obtainable."

Similar notices are given for the other districts of Burma, thus showing a very extensive cultivation.

ALCOHOL. 361

Alcohol prepared from Sweet Potatoes.—The utilisation of the sweet-potato as a source of alcohol has recently been discussed in several newspapers. **M. Ralu**, it would appear, has obtained patents for the distillation of a specially prepared flour from these tubers. A Company has also been formed at the Azores, and the alcohol it produces is said to be readily purchased by Lisbon merchants to be used in the fortification of wines. "The alcohol, **M. Ralu** reports, of which we have samples, is superior in quality to that of the best marks of France. The distillery obtains 12 per cent. (*i.e.*, 12 litres per 100 kilogrammes of sweet-potato) of alcohol. We have experimented with the sweet-potatoes of Algeria. They give 13·4 litres of alcohol per 100 kilog. The sweet-potato of Martinique and Brazil has given 15 litres. There is here therefore a very rich material for distillation. Ordinary potatoes yield only 3 litres of alcohol per 100 kilog." As compared with these figures it may here be added that with the exception of Indian-corn sweet-potatoes are cheaper than any of the substances used in distillation, while they yield the highest recorded percentage of alcohol. Thus, for example, wheat yields from 28 to 30 litres per 100 kilog.; barley from 22 to 23; oats from 20 to 21; buck-wheat from 24 to 25; Indian-corn from 28 to 30; rice from 32 to 33; and sweet-potato flour from 38 to 39 litres per 100 kilog. As compared with the preparation of alcohol from maize it has been stated there is a great saving of time and combustibles when distilling from the flour of sweet-potato. Alcohol from maize costs 10*f*. per hectolitre more to make, and when made, sells from 8*f*. to 10*f*. less than the alcohol from sweet-potato. The writer of the article in the *European Mail*, from which publication the statements given above have been derived, very rightly remarks that it remains to be seen whether it would pay to cultivate the sweet-potato on a large scale as a source of alcohol. Doubtless the price of the tuber would rise considerably, and other difficulties may be found to exist which the chemist could not have anticipated. It seems desirable, however, that the fact should be made widely known that these tubers could be used as a source of alcohol.

362 **Ipomæa biloba,** *Forsk.; Fl. Br. Ind., IV., 212.*

Syn.—I. MARITIMA, *R. Br.; Blume;* I. PES-CAPRŒ, *Roth.; Chois; DC., Prod.; Dalz. and Gibs., Bomb. Fl;* CONVOLVULUS PES-CAPRŒ, *Linn., Roxb., Fl. Ind;* C. BILOBATUS, *Roxb.;* BATATAS MARITIMA, *Bojer.*

Vern.—*Dopati-latá,* HIND.; *Chhágulkúrí,* BENG.; *Kansárinata,* ORISSA; *Marjádvel, marayadavel, marjavel,* BOMB.; *Bálabándi tige*(=hare creeper), *chevulapilli tige,* TEL.; *Pin-lai-ka-zum,* BURM.; *Múdú-bintambúrú,* SING.

References.—*Roxb., Fl. Ind., Ed. C.B.C., 163; Voigt, Hort. Sub. Cal. 356; Dalz. & Gibs., Bomb. Fl, 164; Trimen, Cat. Ceylon Pl., 60; Rheede, Hort. Mal., XI., t. 57; Rumph., Herb. Amb., V., t. 159, fig. 1; Sir W. Elliot, Fl. Andhr., 20, 24, 37; Mason, Burma and Its People, 437, 783; Bidie, Report on Madras Drugs Calc. Inter. Exh., 1883-84; Dymock, Mat. Med. W. Ind., 2nd Ed., 563; Lisboa, U. Pl. Bomb., 285; Indian Forester, IX., 238; XII., 329; Hunter, Gazetteer Orissa, II., 178;*

Madras Manual of Admin., II., 27; Journal, Agri.-Hort. Soc. Ind., IX., 176, 177; X., 20.

Habitat.—A common plant near the sea, especially on the western coast.

Medicine.—The parenchima of the ROOT contains starch and large conglomerate raphides. The whole plant is very mucilaginous. The LEAVES are applied externally in rheumatism and colic, and the JUICE is given as a diuretic in dropsy, and at the same time the bruised leaves are applied to the dropsical part (*Mat. Med. West. Ind.*). MEDICINE. Root. 363 Leaves. 364 Juice. 365

Fodder.—**Roxburgh** remarks that it is most useful in binding loose sand, thereby preparing the way for the growth of other plants upon previously barren shifting sands. The Madras Manual pronounces it as one of the most important sand-binding plants. Goats, horses, hares, tame rabbits, and rats are also said to be fond of it. Cows fed on it yield tainted milk. From the summit of the thick, fleshy, tap root spread many succulent red branches, bearing bi-lobed leaves on long petioles, very much resembling a Bauhinia leaf. FODDER. 366

Sacred.—Lisboa alludes to a reputed practice of twining the creeper around the cot of a Hindu mother, apparently from the idea that as it binds the sand on the sea-shore, so it will secure the child to the mother against the goddess of destiny. SACRED. 367

Ipomæa bona-nox, *Linn.; Fl. Br. Ind., IV., 197; Wight, Ic., t. 1361.* 368

THE MOON-FLOWER, a name derived from the fact that the large, white, and sweetly-scented flowers open only at night.

Syn.—CALONYCTION SPECIOSUM, *Chois.;* IPOMŒA GRANDIFLORA, *Roxb.*

This should not be confused with **Lettsomia bona-nox,** *Roxb.* (=**Rivea hypocrateriformis,** *Chois*).

Vern.—*Dúdiya-kulmi* (**I. grandiflora,** *Roxb.*), *ilál kalmi,* BENG.; *Gulchandni,* BOMB.; *Naga-múghatei,* TAM.; *Nagara-múkuttykai,* TEL.; *Munda-valli,* MALAY; *Nway-ka-zún-a-phyú, nweka-zumbyi,* BURM.; *Alanga,* SING.; *Pathmapu-todami* (or *patmapu*), SANS.

References.—*Roxb., Fl. Ind., Ed. C.B.C., 167; Kurz, For. Fl. Burm., II., 217; Dalz. & Gibs., Bomb. Fl., 164; Griff. Notul, IV., 286; Sir W. Jones, Asiatic Researches, IV., 257; Mason, Burma and Its People, 437, 783; Ainslie, Mat. Ind., II., 219; O'Shaughnessy, Beng. Dispens., 506; Dymock, Mat. Med. W. Ind., 2nd Ed., 561; Journal Agri.-Hort. Soc. Ind., VI., 47; X., 20;* (**Rev. J. Long** *gives* **Roxburgh's** *information regarding* **Ipomæa Tarpethum,** *Br., as applying to this species; according to* **Roxburgh** *these species both bear the name Dud* (*or Dudiya*) *Kulmi*).

Habitat.—The typical form of this plant is a native of America, early introduced into India, and in some parts of the country now become quite naturalised. The form which **Roxburgh** called **I. grandiflora** is, by the *Flora of British India*, viewed as an indigenous variety of the species, with a less tendency to have lobed leaves. It is found in most parts of tropical India, and from Assam and Bengal to Tenasserim, Malabar, &c. The introduced form has, however, become so thoroughly acclimatised that it is often difficult to distinguish the one from the other. The **Rev. J. Long** says that it flowers in the cold season in the Circars and in the rains in Bengal.

Medicine.—Dried, the CAPSULES and SEEDS, as well as the FLOWERS, LEAVES, and ROOTS are included amongst the medicines supposed to have some merit as remedies against snake-bite (*Ainslie*). In the *Materia Medica of Western India* it is stated that the capsules are reported to be used in the Konkan for the same purpose as described above in connection with Madras. MEDICINE. Capsules. 369 Seeds. 370 Flowers. 371 Leaves. 372 Roots 373

SPECIAL OPINIONS.—§ ' It is sometimes used as food, and is supposed to have mild alterative and diuretic properties" (*Civil Surgeon D. Basu, Faridpore, Bengal*).

FOOD. Seeds. 374 **Food.**—Ainslie says the SEEDS of this species are eaten when young.

875 **Ipomæa cærulea,** *Koen.;* Syn. for **I. hederaceæ,** *Jacq.*

376 **I. cymosa,** *R. & S.; Fl. Br. Ind., IV., 211.*

Vern.—*Lál danah?,* HIND.; *Karmbi arak,* SANTAL.; *Cháta kattu-tiva, viru malle, kappa-tiva,* TEL.; *Kiri-madu.* SING.

References.—*Roxb., Fl. Ind. Ed. C.B.C., 163; Elliot, Fl. Andhr., 35, 192; Rev. A. Campbell, Report Ec. Prod. Chutia Nagpur, No. 9253; Trimen, Cat. Ceylon Pl., 60.*

Habitat.—A scandent plant frequent throughout India, except in the drier areas, ascending to 4,000 feet. Seeds covered with brownish black hairs.

MEDICINE. Seeds. 377 **Medicine.**—According to the *Pharmacopœia of India* this may be one of the species that yields the medicinal SEEDS known as *Shapussundo,* see Ipomæa sp.?

FOOD. Leaves. 378 **Food.**—The LEAVES are eaten by the Santals as a pot-herb.

379 **I. digitata,** *Linn.; Fl. Br. Ind., IV., 202.*

Syn.—I. PANICULATA, *R. Br.;* IPOMÆA GOSSYPIFOLIA, *Willd.;* CONVOLVULUS PANICULATUS, *Linn.; Roxb, Fl. Ind.;* BATATAS PANICULATA, *Chois.*

Vern.—*Bilái kand,* HIND.; *Bilai-kand, bhúmi* or *bhúi-kumrá,* BENG.; *Bhui-kohala,* BOMB.; *Matta-pal-tiga,* TEL.; *Bhumichekri-gadde,* KAN.; *Phal-modecc* MALAY; *Vidári, bhumikashmánda,* SANS.

References.—*Roxb., Fl. Ind., Ed. C.B.C., 160; Dalz. & Gibs., Bomb. Fl., 167; Rheede, Hort. Mal., II., 101—102, t., 49; Ainslie, Mat. Ind., II., 307; U. C. Dutt, Mat. Med. Hind., 205, 294, 323; Dymock, Mat. Med. W. Ind., 2nd Ed., 564; S. Arjun, Bomb. Drugs, 92; Year Book Pharm., 1880, 249; Birdwood, Bomb. Pr., 57; Kew Off. Guide to Bot. Gardens and Arboretum, 27; Bombay (Kanara), XV., 439; Journal Agri.-Hort. Soc. Ind. (Old Series), X., 20.*

Habitat.—A native of tropical India from Bengal and Assam to Ceylon, but found in the drier Western portion.

Largely cultivated on account of its pink purple flowers, which appear in the rains.

MEDICINE. Roots. 380 **Medicine.**—The large tuberous ROOTS are very much used in native medicine, being regarded as tonic, alterative, aphrodisiac, demulcent, and lactagogue. The powdered root-stock is given with wine, for the purpose of increasing secretion of milk, and for the emaciation of children with debility and want of digestive power, the early Sanskrit physicians prescribed a diet of equal parts of *vidári,* wheat-flour, and barley (*Dutt, l.c.*). **Rheede** refers to the use of the powdered sun-dried root, boiled in sugar and butter to promote obesity and moderate the menstrual discharge. **Dymock** says that the tuberous roots are in considerable demand, being extensively used in Bombay, and that the young tapering roots, of a small variety, are known in the herbalists' shops as *asgand.* The name *asgand* is also, however, given to **Withania coagulans,** *Dunal* (see Vol. VI.), the roots of which are twisted, white, covered with a thick soft bark and contain a good deal of red-colouring matter, which is visible through the greyish epidermis. The roots of the **Ipomæa** (*asgand*) are only half the length of those of **Withania,** or about 6 to 8 inches long; they are light yellowish brown externally, white internally, brittle, with a short and starchy fracture. They are mucilaginous and bitter in taste. According to the *Makhzan-el-Adwiya* they are tonic and alterative, and by **Chakradatta** are recommended to be given to weakly children. The **Rev. J. Long** says the powdered root is used in spleen disease; it is purgative in its action. In Davies' Trade and

Resources of the North-West Frontier frequent mention is made of *Asgand*.

SPECIAL OPINIONS.—§ "Much used as a lactagogue" (*Assistant Surgeon Shib Chunder Bhattacharji, Chanda, Central Provinces*). "Cholagogue, useful in liver complaints" (*J. N. Dey, Jeypore*).

Fodder.—Stems and leaves eaten by cattle. FODDER. 381

Ipomæa eriocarpa, *Br.; Fl. Br. Ind., IV., 204; Wight., Ic., t. 169.* 382

Syn.—I. SESSILIFLORA, *Roth.*; CONVOLVULUS SPHŒROCEPHALUS, *Roxb.*; C. HISPIDUS, *Wall.*

Vern.—*Kalman*, ASSAM; *Haran-khúri, hara* (BIJNOR), *bhánwar*, N.-W. P.; *Bhánwar*, PB.; *Pwiti tige*, TEL.

References.—*Roxb., Fl. Ind., Ed. C.B.C., 158; Dalz. & Gibs., Bomb. Fl., 166; Elliot, Fl. Andhr., 160; Rev. A. Campbell, Report Econ. Prod., Chutia Nagpur, No. 8119; Atkinson, Econ. Prod. N.-W. P., V., 95; Gazetteers:—Mysore & Coorg, I., 63; N.-W. P. (Bundelkhand), I., 82; (Agra), IV., lxxv.; Himalayan Districts, X., 314; Journal Agri.-Hort. Soc. Ind. (Old Series), XIII. (Sel.), 63.*

Habitat.—Found throughout India up to altitudes of 4,000 feet; common in Bengal and distributed to Ceylon, Afghánistan, &c.

Food.—According to most writers this plant is eaten in times of famine. **Atkinson**, however, says it "occurs wild in the plains and lower hills and is sometimes sown with wheat. The leaves and stems are eaten as a vegetable, and the leaves are used in medicine." FOOD. 383

I. hederacea, *Jacq.; Fl. Br. Ind., IV., 199.* 384

Syn.—I. CŒRULEA, *Koen.*; *Roxb., Fl. Ind.*; I. PUNCTATA, *Pers.*; I. NIL and I. BARBATA, *Roth.*; CONVOLVULUS NIL, *Linn.*; C. HEDERACEUS, *Linn.*; C. DILLENII, *Lamk.*; PHARBITIS HEDERACEA and P. NIL, *Chois.*; P. DIVERSIFOLIA, *Lindl.*; P. VARIIFOLIA, *Dcne.*

Vern.—The seeds are officinal and are known as *Kálá-dánah, mirchai*, HIND., BENG., and BOMB.; *Nil-kalmi*, BENG.; *Baunra*, N.-W. P.; *Hub-ul-nil*, KASHMIR; *Bildi, ker, kirpáwa, phaprú sag, ishpecha*, PB.; *Hub-ul-nil*, SIND; *Káli-zirki, zirki*, DEC.; *Kálá-dáná*, GUZ.; *Kodi, kákkatán-virai, jirki-virai*, TAM.; *Kolli-vittulu*, TEL.; *Ganribija*, KAN.; *Hab-un-nil*, ARAB.; *Tukm-i-nil*, PERS.

Dr. **Moodeen Sheriff** says the Deccan name *Kali-zirki* should be exclusively applied to the seeds of this plant, and that great ambiguity has been caused through *Kálá dánah*, &c., being applied to the seed of **Clitora Ternatea,** *Linn.* (which see). Dr. **Dymock** thinks it doubtful whether the Hindustani name *Zirki* is properly applied to **I. hederacea.** For *Kálá-zira* see **Nigella sativa.**

References.—*Roxb., Fl. Ind., Ed. C.B.C., 168; Stewart, Pb. Pl., 151; Mason, Burma and Its People, 783; Elliot, Fl. Andhr., 93; Pharm. Ind., 155; Waring in Jour. Pharm. Soc., VII., series 2, 498; Waring, Bazar Med., 75; O'Shaughnessy, Beng. Dispens., 505; Moodeen Sheriff, Supp. Pharm. Ind., 196; U. C. Dutt, Mat. Med. Hind., 205; Dymock, Mat. Med. W. Ind., 2nd Ed., 558; Fluck. & Hanb., Pharmacog., 448; Bent. & Trim., Med. Pl., 185; S. Arjun, Bomb. Drugs, 95; Murray, Pl. and Drugs Sind, 166; Med. Top. Ajmir, 144; Irvine, Mat. Med. Patna, 51; Baden Powell, Pb. Pr., 367; Atkinson, Him. Dist., 314, 745; Drury, U. Pl., 336; Lisboa, U. Pl. Bomb., 255; Birdwood, Bomb. Pr., 57; Home Dept. Official Corresp. regarding proposed new Pharmacopœia, 227, 282, 319; Manual Cuddapah District, Madras, 200; Gazetteers:—Orissa I., 63; II., 159, 181; Mysore and Coorg, I., 63; Bundelkhand, 82; Agra, lxxiv.; Meerut, 82.*

Habitat.—A common plant, widely cultivated in India, but also apparently found wild.

There are two varieties:—

α **integrifolia**: leaves entire.

Syn.—I. CŒRULESCENS, *Roxb., Fl Ind., Ed. C.B.C., 168.*

IPOMÆA hederacea.

The Kala-danah.

β **himalaica**: leaves and flowers large, seeds densely closely villous. Found at altitudes of 4,000 to 5,000 feet.

MEDICINE. 385 **Medicine.**—This drug does not appear to have been known to the
early Sanskrit physicians, and is not described by **Ainslie**. The author
of the *Makhzan-ul-Adwiya* says that it is a drastic purgative, useful in
the treatment of bilious and phlegmatic humours, and that it acts also as
Seeds. 386 an anthelmintic. **Roxburgh** was apparently the first to make these SEEDS
known to European physicians, and it may be said they now hold an im-
Extract. 387 portant position as a useful and cheap substitute for jalap. They were
made officinal in the *Pharm. India* in 1868, where directions will be found
Tincture. 388 to make the preparations in which the drug is now administered, namely,
in the form of a TINCTURE, an EXTRACT, a compound POWDER, or a resin,
Powder. 389 thus supplying the place of the corresponding preparations of Jalap. The
resin (introduced by **Dr. Bidie**) appears to be the most satisfactory form
Resin. 390 of administering the medicine; the dose is 4 to 8 grains. This substance
is known as Pharbitisin. It has a nauseous acrid taste and an unpleasant
odour, especially when heated.

Dr. Dymock says of the extract "that no directions for separating the albumen and mucilage are given, consequently the result of the operation is an enormous bulk of almost inert extract, which in a short time becomes putrid. Five to ten grains of this extract have no perceptible effect as a purgative." **O'Shaughnessy** says that "the powdered seeds in doses of 30 to 40 grains act as a quick, safe, and pleasant cathartic." Out of 100 cases tried, **Sir William O'Shaughnessy** says that it proved an effective purgative in 94, occasioned vomiting in 5, but in 15 it produced on an average five stools within 2½ hours and the action generally commenced in an hour's time. **O'Shaughnessy** adds that in his time the seeds sold at 4 seers per rupee. **Dr. Kirkpatrick** regards *kálá dána* as intermediate in strength between rhubarb and jalap. According to **Irvine** the natives roast the seeds before reducing them to a powder.

SPECIAL OPINIONS.—§ "A certain and efficacious purgative. Dose ʒ½ to ʒi of the powdered seed mixed with powdered ginger" (*Assistant Surgeon Shib Chundra Bhattacharji, Chanda, Central Provinces*). "This has been only lately used as a purgative. It acts fairly well, producing four to five copious motions" (*Apothecary J. G. Ashworth, Kumbakonam, Madras*). "The powdered seeds are in daily use as a substitute for jalap, and given in doses of 20 to 60 grains are very efficient, acting as a diuretic purgative" (*Surgeon-Major S. H. Browne, M.D., Hoshangabad, Central Provinces*). "I have found the extract of *Káládána* a very useful purgative especially in doses of 5 grains either alone or in combination with a grain or two of calomel. I once gave a ten-grain dose alone on an empty stomach and it caused vomiting, but I never knew a five-grain dose to do so. I have formed a favourable opinion of the value of the extract in cases of constipation connected with a torpid liver" (*Surgeon-Major H. W. E. Catham, M.D., M.R.C.P., Lond., Bombay Army, Ahmednagore*). "*Káládána*, used as a purgative, instead of jalap in ʒ*ss.* to ʒi *ss.* doses" (*Assistant Surgeon Nehal Sing, Shaharunpur*). "Used here as a purgative, and can be obtained in the bazárs at Umballa" (*Brigade-Surgeon R. Bateson, I.M.D., Umballa City*). "I was the first to extract and give an account of Pharbitisin as a medicinal agent" (*Surgeon-General and Sanitary Commissioner G. Bidie, C.I.E., Madras*). "Pharbitis powdered is a first-rate substitute for jalap" (*Surgeon-Major L. Beech, Cocanada*). "The seed usually known in the bazár as *Káládána* is a very useful and safe purgative and is much employed in dispensary practice. Its action resembles very much that of jalap, the dose being usually ʒi" (*Civil Surgeon R. Macleod, Sarun*). "Several native

MEDICINE.

practitioners have told me of the value of '*Káládána seeds*' as an active purgative, causing copious liquid stools, and as being very safe and efficacious, but inadmissible in inflammatory states of the alimentary canal" (*Surgeon J. Ffrench Mullen, M.D., I.M.S., Saidpore*). "Quite equal to jalap and in many respects superior to it. Very largely used in charitable dispensaries. The powdered seeds given in drachm doses" (*Civil Surgeon S. M. Shircore, Murshidabad*). "Much used as an ordinary purgative in dispensary practice, the usual dose of the powdered seeds being half a drachm" (*Surgeon G. Price, Shahabad*). "A cheap and certain purgative. I constantly use it in jails" (*Surgeon R. D. Murray, M.B., Burdwan*). "An excellent and sure purgative, operating quickly. The seeds reduced to a powder are used as a substitute for jalap. Dose ʒi with 5 grains of *pulvis zingiberis* and a little sugar" (*Civil Surgeon C. M. Russell, Sarun*). "The seeds of **Ipomæa hederacea** or the *Kálá-zirkí-kebínj* of Madras and *Káládána* of Calcutta. They are undoubtedly one of the few good and cheap cathartics India possesses, and also one of those purgatives which act very efficiently and satisfactorily when used alone. In this respect *Káládána* is preferable to jalap. The simple powder is prepared by drying the seeds well and reducing them to fine powder in the ordinary way. Dose, from 45 grains to one drachm. The compound powder should be prepared with cream of tartar in equal proportions as follows:—Take of *Káládána* and cream of tartar, in powder, each seven ounces; ginger, in powder, one ounce; rub them well together and pass the compound powder through a fine sieve. Dose from one drachm to a drachm and a half" (*Honorary Surgeon Moodeen Sheriff, Khan Bahadur, G.M.M.C.*, Triplicane, Madras).

CHEMISTRY. 391

Chemistry.—For an account of the chemical nature of these seeds see *Flückiger and Hanbury's Pharmacographia* (reproduced in *Dymock's Materia Medica, Western India, 560*).

TRADE. 392

Trade.—In Bombay the seeds of **I. muricata,** *Jacq.* (imported from Persia), are sold much more commonly than those of the true *kálá dána* (*Dymock*). Their action appears to be identical (see the account of that species below). Irvine remarks that in Patna the *kálá dána* seeds sold at 2 annas a pound during his time.

Ipomæa muricata, *Jacq.; Fl. Br. Ind., IV., 197.* 393

Syn.—I. BONA-NOX, *Bot. Reg.*; CONVOLVULUS MURICATUS, *Linn.*; CALONYCTION MURICATUM, *G. Don.*; C. BONA-NOX, *Chois., var.* MURICATA, *Chois. in DC. Prod.*

Vern.—*Bárik bhauri* (= lesser *bhauri*), KONKAN; *Gariya*, BOMB.; *Kaludlanga*, SING.; *Tukin-i-nil*, imported into Bombay from PERSIA.

References.—*Roxb., Fl. Ind., Ed. C.B.C., 167; Trimen, Cat. Ceylon Pl., 59; Grah., List Bomb. Pl., No. 972; Dymock, Mat. Med. W. Ind., 2nd Ed., 561, 562; Bent. & Trim., Med. Pl., 185; S. Arjun, Bomb. Drugs, 219.*

Habitat.—Common in India, cultivated, and also apparently wild. In some of its forms it approaches so close as to be almost indistinguishable from certain conditions of **I. hederacea,** *Jacq.*; it is quite distinct from **I. bona-nox.** It is said to occur on the Himálaya at altitudes of 1,000 to 5,000 feet, extending from Kangra to Sikkim.

MEDICINE. Seeds. 394

Medicine.—The SEEDS are used chiefly as a substitute for those of the preceding species, from which they can be distinguished by their larger size, lighter colour, and thick testa. Dr. Dymock was the first to establish that the *Tukm-i-níl* imported from Persia was identical with this species. When exhausted, he writes, with rectified spirit, the Persian seeds yield a brown resin and yellowish oil, having the same appearance as the resin and oil of the genuine article. The medicinal properties of the *Tukm-i-níl*

MEDICINE. Juice. 395 seem to be the same as that of the *Káládána*, but **Dymock** adds "accurate observations are required." The JUICE of the plant is used to destroy bugs. The value of the imported seed is said to be R5 per maund of 37½lb in Bombay.

No information is available as to whether or not the seeds of the indigenous stock of **I. muricata** are utilised by the people of India.

FOOD. 396 **Food.**—**Dymock** refers to this species as one of the plants eaten in times of famine.

397 **Ipomæa obscura,** *Ker.; Fl. Br. Ind., IV., 207.*

Syn.—I. INSUAVIS, *Bl.;* I. OCULARIS, *Bartl.;* CONVOLVULUS OBSCURUS, *Linn.*

Vern.—*Sirutali,* TAM.; *Tsinnataliaku,* also *nalla kokkita,* TEL.; *Mahamádu,* SING.

References.—*Roxb., Fl Ind., Ed. C.B.C., 158; Dalz. & Gibs., Bomb. Fl., 166; Wight, in Madras Journ., V., 6, t. 12; Ainslie, Mat. Indica, II., 394; Trimen, Cat. Ceylon Pl.. 60; Drury, U. Pl., 259; Mason, Burma and Its People, 784; Elliot, Fl. Andhr., 125; Gazetteers:—Bundelkhand, 82; Mysore & Coorg, I., 56; Bombay (Kanara), XV., 439.*

Habitat.—Fairly common throughout India, ascending to altitudes of 3,000 feet. The variety known as **gemella** is frequent in Bengal and Burma.

MEDICINE. Leaves. 398 **Medicine.**—**Ainslie** appears to be the only writer who describes the properties of this plant. He says "the LEAVES of this twining plant have a pleasant smell and mucilaginous taste; when toasted, powdered, and boiled with a certain portion of *ghí*, they are considered as a valuable application in aphthous affections."

399 **I. pes-tigridis,** *Linn.; Fl. Br. Ind., IV., 204; Wight, Ic., t. 836.*

TIGER-FOOTED IPOMÆA.

Syn.—CONVOLVULUS PES-TIGRIDI, *Spreng.;* C. PALMATA, *Mœnch.*

Vern.—*Languli-latá,* BENG.; *Mekamu aduga,* TEL.; *Divi-adiya, divipahuru,* SING.

References.—*Roxb., Fl. Ind., Ed. C.B.C., 169: Elliot, Fl. Andhr., 114; Mason, Burma & Its People, 437, 783; Gazetteers:—Bundelkhand, 82; Agra, lxxv.; Himálayan Districts, 314; Mysore & Coorg, I., 56; Indian Forester, XII., App. 17; Trimen, Cat. Ceylon Pl., 59; Rheede, Hort. Mal., XI., 121, t. 59.*

Habitat.—A very common species, throughout India from the Panjáb to Malacca and Ceylon. There are two varieties:—

α **hepaticifolia**: leaves 3-lobed or angular. This is the most abundant species in the Deccan Peninsula.

β **capitellata**: leaves ovate, cordate, acute, entire. A common **Ipomœa** from the Deccan to Behar and Hindústan proper.

MEDICINE. 400 **Medicine.**—Said to be used medicinally.

401 **I. purga,** *Hayne.*

JALAP.

Syn.—EXOGNONIUM PURGA, *Benth.*

References.—*Year-Book, Pharm., 1874, 85; Kew Report, 1881, 49; Kew Off. Guide to the Mus. of Ec. Bot., 99.*

Habitat.—A native of the Mexican Andes, occurring at altitudes between 5,000 and 8,000 feet above the sea. In the localities where it occurs, rain falls almost daily, and the diurnal temperature varies from 60° to 75° Fh. It flourishes in shady woods in a deep rich vegetable soil.

It is now being cultivated in Europe and India; in the south of England it is said to grow freely if planted in sheltered borders, but its flowers are produced so late in autumn, that they rarely expand, and the tubers, which develop in some abundance, are liable to be destroyed in winter unless protected from frost (*Pharmacographia*).

CULTIVATION.

CULTIVATION. N.-W. P. 402

N.-W. PROVINCES.—About the beginning of the present century, **Ainslie** urged that the Government of India should endeavour to cultivate the jalap plant. In 1836, the same recommendation was made by **Dr. Royle** (*Trans. Agri.-Hort. Soc. Ind., III., 40*), but apparently the first attempt at its cultivation was not made until in 1854. **Dr. Jameson**, Superintendent of the Botanic Gardens, Saharanpur, in that year reported:—" Through the kindness of **Dr. Royle** I have been able to introduce jalap into the Himálaya, where it is now growing with great vigour. In a few years there will be a sufficient quantity of roots to supply this valuable medicine to the public service. It grows well at altitudes between 4,000 to 4,500 feet." Commenting on that report **Mr. J. F Duthie** informs the author that "this experiment did not lead to any practical result, as no further mention was made by **Dr. Jameson** regarding the cultivation of the plant. In 1884, 300 small tubers weighing 5 to 6℔ were received from Ootacamund for cultivation in the Arnigadh garden, and from these the crop has increased to 640℔. The plant grows with great vigour at Arnigadh; but the root growth, **Mr Gollan** thinks, would be all the better if the vine growth were less vigorous. As the Bengal medical depôts require in all 2,000℔ of the dried root annually, and only 640℔ have been produced during four years, I fear it will take some time before the garden is in a position to supply the full amount. **Mr. Gollan** tells me that the tubers lose five-sixths of their weight in drying, so that in order to prepare 2,000℔ of dried tubers no less than 12,000℔ of green tubers would be required; and to keep up a steady annual supply of 2,000℔, 36,000℔ of green tubers should be in hand before commencing to supply the drug."

Madras. 403

MADRAS.—In a foot-note to page 443 of the *Pharmacographia*, **Flückiger** and **Hanbury** say that on the 15th January 1870 they received from **Mr. Broughton**, of the Nilghiri Hills, a cluster of tubers of jalap weighing 9℔. It would thus appear that jalap cultivation was commenced in the Madras Presidency prior to that date. **Mr. Jamieson**, however, in his Annual Report of the Botanic Gardens and Park of the Nilghiris for 1879-80, states:—" The propagation of jalap was started in the Ootacamund gardens in 1877. Our stock then consisted of 100 plants that had been grown in the gardens for ornament. Although the tubers increase to a large size and the stems grow luxuriantly and flower profusely, the flowers do not fertilize freely, and produce but few seeds in the climate of Ootacamund, consequently the increase of the plant has been confined almost entirely to propagation from root and stem cuttings. This work has been carried on steadily until there are now upwards of 25,000 plants and root cuttings permanently planted out. The tubers were at first planted 4×4 feet apart, but it was found that they did not exhaust the land, and thrive equally well when planted closer. They are, therefore, now planted in rows three feet apart and the distance between each plant in the row does not exceed two feet, so that 25,000 plants give a total area planted of five acres. The cost of cultivation, weeding, &c., is also much reduced by close planting."

Method of cultivation.—Having no data to guide me regarding the climate and soil which are natural to the jalap plant, I tried it in a variety of situations, and now find that it thrives best in a tolerably rich, dry, and friable, loamy soil; in fact, conditions of soil that are indispensable to the production of good potatoes seem equally necessary to the growth of jalap; good grass land is preferable to open rich forest land; when

IPOMÆA purga.

The Jalap Plant.

CULTIVATION.
Madras.

planted in the latter it has a greater tendency to produce a mass of succulent roots than to form tubers. In opening land for the cultivation of jalap several acres of well drained grass land sheltered from the south-west winds and with a stream of water on or near it should be laid out in terraces 10 feet wide. The terracing should be completed by the end of January and the ground dug over to the depth of two feet, and left exposed to the action of the sun until the beginning of April, when it should be drilled, manured, and planted with potatoes (an early ripening variety of kidney would be the best variety to plant). The potatoes should be lifted in June, the land cleared of weeds, and forked over in order that the manure (not taken up by the potatoes) may be thoroughly incorporated with the soil. The ground is now ready to receive the jalap plants, which should be planted when the tubers are about the size of pigeon eggs in rows (across the terraces) on ridges a few inches higher than the general level of the ground, in order that they may be raised sufficiently high to prevent water from lodging immediately around them. If the weather be dry, the plants should be watered occasionally until they have begun to grow; when once established ordinary garden culture as to weeding, &c., is all that is necessary.

Jalap is a herbaceous plant throwing out twining stems, which should be supported by stakes or wire trellises in the same manner as ordinary garden peas are. The stems die down annually and the tubers remain dormant for two or three months. In addition to the ærial stems, jalap throws out a mass of underground shoots, which emit roots and form tubers at intervals of from six to nine inches. It is from these underground shoots that the greater proportion of our plants have been raised; when cut about three inches long and planted, they root freely and gradually enlarge into tubers; by this means the plants can be multiplied to any extent.

One acre of land planted, as I have described, should at the end of three years produce 5,000℔ of green tubers, which will yield, when thoroughly dried, 1,000℔ of jalap powder; the cost of cultivation, collection, and drying of the root for the same period will not exceed R300. I, therefore, estimate that dried jalap tuber can be produced in Ootacamund at a cost of 4 annas and 10 pies per ℔.

In 1881-82 **Mr. Jamieson** reported: "The cultivation of this plant has passed beyond the experimental stage, and it is now an established fact that jalap can be grown successfully in Ootacamund and will pay a fair return on the outlay even at the price allowed by the Medical Stores. The outturn of dry root was not so large as I anticipated, or what it would have been had the tubers planted been allowed to come to maturity. But so long as I have to lift them to supply the public, it will be some years before I have a sufficient area brought under cultivation to meet the demand of even the local medical department, the annual requirements of which **Surgeon-General Cornish** puts at 1,200℔."

In 1882-83 **Mr. Jamieson** remarked: "The total number of tubers sold and distributed gratis during the year was 9,051, which is 3,741 more than was sent out in 1881-82. From the above figures it will be apparent that the cultivation of this drug is being taken up by private individuals. To those who purchased a large number of tubers a considerable reduction in the price (R5 per 100) was made. This plan I consider preferable to supplying tubers to any and every applicant. As directed the area under this crop has not been extended, the present stock being sufficient to meet a considerable increase on last year's demands for tubers and leave a quantity available for the Medical Stores."

Mr. M. A. Lawson, Government Botanist and Director of Chinchona Plantations, Nilghiris, has favoured the author with the following special report, which brings the facts of Madras cultivation down to the present date:—"The true jalap (**Ipomæa purga**) grows extremely well on these hills, especially in peaty soil. The tubers in four or five years' time, weigh, when first taken up, from five to ten pounds, and upon analysis are found to be as rich in the purgative resins, as the best kinds imported from South America. I only know of one person who has attempted to grow them on any large scale, and he has not been very successful. The Government of Madras ordered the re-cultivation of the jalap in these gardens two years ago, and about one acre has been planted with the tubers, which in another two years' time will yield probably some thousands of pounds of the drug, and if the cultivation of them is found profitable it is hoped that planters will be induced to give it another trial." CULTIVATION. Madras

Bengal.—The Superintendent of the Royal Botanic Gardens, Calcutta, reports that the cultivation of the jalap plant in Bengal was not successful. Bengal. 404

Ipomæa Quamoclit, *Linn.; Fl. Br. Ind., IV., 199.* 405

RED JASMINE.

Syn.—CONVOLVULUS PENNATUS, *Lamk.;* C. QUAMOCLIT, *Spreng.;* QUAMOCLIT VULGARIS, *Chois.*

Vern.—*Taru latá, lál* or *swétá, kámlatá,* or *lál* or *swéta turú latá,* BENG.; *Vishnukrant,* MAR.; *Kámlatá,* HIND.; *Tsjuria-cranti,* MAL. (In *Rheede, Hort. Mal., XI., 123, t. 60); Myatlaœ-ni,* BURM.; *Kámalata, tarúlatá* (Cupid's flower), SANS.

References.—*Roxb., Fl. Ind., Ed. C.B.C., 169; Sir W. Jones, V., 88, No. 20; Dymock, Mat. Med. W. Ind., 2nd Ed., 561; U. C. Dutt, Mat. Med. Hind., 302; Mason, Burma and Its People, 438, 783; Jour. Agri.-Hort. Soc. Ind. (New Series), IV., 94; Rumph., Amb., V., 155, t. 2.*

Habitat.—Roxburgh regarded this species as a native of various parts of India, but the *Flora of British India* gives it as "common throughout India, in gardens and as a denizen; native of tropical America." There are two varieties of the plant met with in gardens—red and white flowered; the former only occurs wild or ? acclimatised.

Medicine.—The Hindús consider it to have cooling properties. The pounded leaves are, for example, applied to bleeding piles, while a preparation of the juice with hot ghí is administered internally. Dr. **Dymock** says that in Bombay the leaves are used as a *lép* for carbuncles (*Kálpúli*). MEDICINE. 406

I. reniformis, *Chois.; Fl. Br. Ind., IV., 206.* 407

Syn.—CONVOLVULUS RENIFORMIS, *Roxb.;* EVOLVULUS EMARGINATUS, *Burm.;* IPOMÆA CYMBALARIA, *Fengl.*

Vern.—*Undirkáni,* BOMB.; *Perretay kiray,* TAM.; *Toinuatali,* TEL.

References.—*Roxb., Fl. Ind., Ed. C.B.C., 161; List of Drugs contributed Calcutta Inter. Exhib. by Baroda Durbar, No. 118; Dymock, Mat. Med. W. Ind., 2nd Ed., 566; S. Arjun, Bomb. Drugs, 94; Lisboa, U. Pl. Bomb., 202; Indian Forester, III., 237; Gazetteer, Banda, 82.*

Habitat.—A procumbent plant with reniform leaves and small yellow flowers; common in the hotter parts of India in damp places, extending northwards to Behar and Rajputána and westwards to the Deccan Peninsula. It ascends the hills to altitudes of 3,000 feet above the sea.

Medicine.—The plant is described as deobstruent and diuretic (*Sakharam Arjun*). **Dymock** remarks that the Hindus administer the juice in rat-bite, and drop it into the ear to cure sores in that organ. MEDICINE. Juice. 408

Food.—Used as a pot-herb. FOOD. 409

I. sepiaria, *Kœn.; Fl. Br. Ind., IV., 209.* 410

Syn.—IPOMÆA STRIATA, *Roth.;* I. HEYNII, *Wall.;* CONVOLVULUS MAXIMUS AND STRIATUS, *Vahl.;* C. MARGINATUS, *Lamk.;* C. SEPIARIUS and INCRASSATUS, *Wall.*

Vern.—*Ban kalmi*, BENG.; *Thali kirai*, TAM.; *Metta túti, puriti tige*, TEL.; *Tiru-tali*, MAL.; *Rasa-tel-kola*, SING.

References.—*Roxb., Fl. Ind., Ed. C.B.C., 168; Rev. J. Long in Agri.-Hort. Soc. Ind. Journ., X., 21; Dalz. & Gibs., Bomb. Fl., 166; Elliot, Fl. Andhr., 115; Gazetteers:—Bombay, III., 203; Mysore & Coorg, I., 63; Banda, 82; Indian Forester, III., 237; Trimen, Cat. Ceylon Pl., 60; Lisboa, U. Pl. Bomb., 202.*

Habitat.—A species frequently met with throughout India to Malacca, Ceylon, &c. The variety- **stipulacea—(=Convolvulus stipulaceus,** ***Roxb.***), occurs in Chittagong.

MEDICINE. 411 **Medicine.**—According to the *Pharmacopœia of India*, this is presumed to be one of the species that yield the seeds known as *Shapussundo*: see **Ipomœa sp. ?** It seems more than likely. however, that *Lál danah* is not, as generally affirmed, synonymous with *Shapussundo*.

FOOD & FODDER. 412 **Food and Fodder.**—Eaten as a pot-herb and given as fodder to cattle.

413

Ipomæa sp. ?

Vern.—*Shapussundo*, and by some also *Lal danah.*

References.—*K. L. De, Indigenous Drugs of India, 86; Pharm. Ind., 157; Drury, U. Pl. Ind., 337; Home Department Official Correspondence regarding a New Edition of Pharm. Ind.; Memo. by K. L. De, No. 14.*

Habitat.—Although dried specimens of this plant do not appear to have been submitted to any botanical authority for determination, **Dr. De** affirms that it "grows abundantly in the Upper Provinces," and he adds: "When cultivated in Bengal, it grows equally well. The seeds are sold in large quantities in Patna and other markets of the Upper Provinces. Each capsule contains three seeds of a brownish-red colour, and studded with minute hairs. When soaked in water, they swell and yield a mucilage. The seeds are to be dried in the sun previous to being powdered."

The above passage practically contains all we know regarding the identification of these seeds. **Waring**, in the *Pharmacopœia*, referred them to **Ipomæa**, apparently on the assumption that they were identical with the *lal-dana* of **Mr. T. C. Lahory**, but he does not inform us whether

Lal-Dana. 414 or not he personally examined either or both of these seeds. If the plant that yields the so-called *shapussundo* be as plentiful as **Dr. De** would lead us to suppose, it is remarkable that it has never been botanically determined. Numerous writers refer to it, but perhaps repeat in other sentences only the statements originally published by **De**, so that during the past twenty years nothing further has been made known than is contained in De's *Indigenous Drugs of India*.

The remark that the capsule contained only three seeds would (assuming that it is an **Ipomæa**, of which the writer has some doubts) assign it to the sub-genus **Pharbitis**, whereas **Waring** presumed *shapussundo* to be the seeds of **Ipomæa cymosa**, *R. & S*, and **I. sepiaria**, *Kœn.*, "which have their seeds covered with short brown hairs," but these species do not belong to **Pharbitis.** Turning to works on Economic Botany, however, to discover references to *shapussundo*, **Roxburgh** (*Fl. Ind., Ed. C.B.C., 607*) is found to assign the name *shah-pusund* to **Amberboa (Centaurea) moschata,** *DC., Prod., VI., 560.* and **Voigt** (*Hort. Sub., Cal, 424*) repeats this as *sapusund* (the Sweet Sultan). **Baden Powell** (*Panjáb Products, 355 & 386*) adds to *shahpasand* the name of *lál dána* as the vernacular of **Centaurea moschata,** and refers these to the same position with **Centaurea Behen,** *Linn* (*DC., Prod., VI., 567*), which he describes as a mild purgative root imported into India from Kabul. **Stewart** gives the name *shápiang* to **Withania coagulans** and *asgand* to **W. somnifera.** **Dymock,** while discussing the properties of **Centaurea Behen,** remarks "**Ainslie** confounds it with *asgand*,* and also with the root of **Physalis (Withania) somnifera**" (*Conf. with Mat. Ind, II., 14*). The writer has read through Ainslie's account, and is of opinion that he is there describing, as he affirms, **Withania somnifera**, which **Dymock** himself admits is the true *asgand*, though the tubers of an **Ipomæa** imported into Bombay from Persia are also sold in the bazárs as *asgand*. (See the information given under **I. digitata above**.)

It will thus be observed that much confusion exists regarding *shapussundo*, and were it not that the seeds of neither **Amberboa** nor **Centaurea** would answer to **De's** description, the writer would be much more disposed to transfer that drug to either of these genera than to leave it in **Ipomæa**. It is curious, however, that *shapussundo*, *asgand* (**Ipomæa**), *asgand* (**Withania**), and **Centaurea Behen**—mild aperient drugs—should be so much confused one with the other, and that **Amberboa moschata** should at the same time be the only plant of the lot that can be said to be met with on the plains of India, though of course under cultivation only, still further that **Withania** should be the only indigenous Indian plant of the series. **Irvine**, in his *Materia Medica of Patna*, does not allude to *shapussundo*, though **De** specially mentions that city as the market where it is chiefly obtained.

Ipomæa Turpethum, *Br.; Fl. Br. Ind., IV., 212.* 415

TURPETH ROOT OR INDIAN JALAP.

Syn.—CONVOLVULUS TURPETHUM, *Linn.; Roxb.; Wight; Hassk.;* C. ANCEPS, *Linn.;* C. TRIQUETER, *Vahl.;* I. ANCEPS, *R. & S.; Bl.; Chois.;* I. TRIQUETRA, *R. & S.;* SPIRANTHERA TURPETHUM, *Bojer;* OPERCULINA TURPETHUM, *Manso.*

Vern.—*Nisoth, tarbud, nukpatar, pitohri,* HIND.; *Teori, dhud kalmí* (or *dudiya-kalmi*), BENG.; *Bana etka,* SANTAL; *Chita bánsa* (the root in bazárs, *turbud, nisot*), PB.; *Nishotar, phútkari,* BOMB.; *Nishottara,* MAR.; *Nashotar,* or *nahotara,* GUZ; *Tikuri,* DEC.; *Shivadai,* TAM.; *Tella tégada, tégada,* TEL.; *Bilitigadu,* KAN.; *Trasta-wálu,* SING.; *Trivrit, triputá,* SANS.; *Turbund,* ARAB.

References.—*Roxb., Fl. Ind., Ed. C.B.C., 160; Kurz, For. Fl. Burm., II., 218; Thwaites, En. Ceylon Pl., 212; Dalz. & Gibs., Bomb. Fl., 165; Stewart, Pb. Pl., 150; Trimen, Cat. Ceylon Pl., 60; Mason, Burma and Its People, 784; Elliot, Fl. Andhr., 174, 179; Rev. A. Campbell, Report Econ. Prod. Chutia Nagpur, No. 9242; Pharm. Ind, 156; Ainslie, Mat. Ind., II., 382; O'Shaughnessy, Beng. Dispens., 504; Moodeen Sheriff, Supp. Pharm. Ind., 346; U. C. Dutt, Mat. Med. Hind., 203, 321; Dymock, Mat. Med. W. Ind., 2nd Ed., 556; De, Indigenous Drugs India, 64; S. Arjun, Bomb. Drugs, 94; Year-Book, Pharm., 1880, 249; Irvine, Mat. Med. Patna, 116; Home Dept. Official Corresp. regarding proposed New Pharm. Ind., 238; Baden Powell, Pb. Pr., 367; Atkinson, Him. Dist., 314; Drury, U. Pl., 259; Lisboa, U. Pl. Bomb., 255, 353; Birdwood, Bomb. Pr., 58; Davies, Trade and Resources of N.-W. Frontier (exports from Nurpur & Kullu to Leh & Yarkand), App. ccxi; Gazetteers:—Orissa, II., 160; Mysore and Coorg, I., 56; Kanara, 439; Banda, 82; Agra, lxxv.*

Habitat.—Found throughout India, ascending to altitudes of 3,000 feet. Distributed to Ceylon, South-East Asia, Malaya, Australia, &c.

Cultivated occasionally in gardens as an ornamental plant; it grows well, for example, in the Botanic Gardens, Calcutta. The supply for medical purposes is, however derived entirely from the wild plant. The Surgeon General, Bombay, reports that the plant is extremely cheap, selling at R7 a cwt.

Medicine.—Dutt (*Hindu Materia Medica*) says of this drug: "Two varieties of *trivrit* are described by most writers, namely, *sveta* or white, MEDICINE. 416

* He was probably misled by the *Tuhfat* which gives *asgand* as Hindi for "white Behen."

IPOMÆA Turpethum. **The Turpeth Root or Indian Jalap.**

MEDICINE.

and *krishna* or black. The white variety is preferred for medical use as a moderate or mild cathartic. The black variety is said to be a powerful drastic and to cause vomiting, faintness, and giddiness. *Trivrit* has been used as a purgative from time immemorial and is still used as such by native practitioners, alone, as well as in various combinations. In fact this medicine is the ordinary cathartic in use amongst natives, just as Jalap is among Europeans. The usual mode of administering it is as follows. About two scruples of the root are rubbed into a pulp with water and taken with the addition of rock salt and ginger or sugar and black pepper." **Roxburgh** wrote in his *Flora Indica* of this drug that " the bark of the root is by natives employed as a purgative, which they use it fresh, rubbed up with milk. About six inches in length of a root as thick as the little finger they reckon a common dose." In a separate and more recent note on this subject he says, quoting **Dr. Gordon**: " The drug which this plant yields is so excellent a substitute for jalap, and deserves so much the attention of practitioners, that I doubt not the following account will prove acceptable. It is a native of all parts of continental and probably of insular India also, as it is said to be found in the Society and Friendly Isles and the New Hebrides. It thrives best in moist shady places on the sides of ditches, sending forth long, climbing, quadrangular stems, which in the rains are covered with abundance of large white, bell-shaped flowers. Both root and stem are perennial. The roots are long, branchy, somewhat fleshy, and when fresh contain a milky juice which quickly hardens into a resinous substance, altogether soluble in spirits of wine. The milk has a taste at first sweetish, afterwards slightly acid; the dried root has scarcely any perceptible taste or smell. It abounds in woody fibres, which, however, separate from the more resinous substance in pounding, and ought to be removed before the trituration is completed. It is, in fact, in the bark of the root that all the purgative matter exists. The older the plant the more woody is the bark of the root; and if attention be not paid in trituration to the removal of the woody fibres, the quality of the powder obtained must vary in strength accordingly. It is probably from this circumstance that its character for uncertainty of operation has arisen, which has occasioned its disuse in Europe. An extract, which may be obtained in the proportion of one ounce to a pound of the dried root, would not be liable to that objection. Both are given in rather larger proportion than jalap. Like it, the power and certainty of its operation are very much aided by the addition of cream of tartar to the powder, or of calomel to the extract. I have found the powder in this form to operate with a very small degree of tenesmus and very freely, producing three or four motions within two to four hours. It is considered by the natives as possessing peculiar hydragogue virtues, but I have used it also with decided advantage in the first stages of febrile affections."

About the same time that **Roxburgh** was examining *turbud*, **Ainslie** was engaged writing his *Materia Indica*, and in addition to much interesting information derived from other sources, he republished the further facts given in **Carey** and **Wallich**'s edition of *Roxburgh's Flora Indica*, in which **Wallich, Gordon**, and **Glass** pronounce the root " a medicine of very considerable value as a cathartic." **Ainslie** urges that it is in the root bark, not the entire root that the medicinal property resides, so that the drug may be depreciated by the use of the whole root. **Ainslie** remarks: "In the valuable Sanskrit Dictionary, the *Amara Kosha*, and also in the *Bhávaprakása* and *Rájanighantu*,* may be found many synonyms for this

* These **Mr Colebroke** mentions as amongst the best writings of the Hindus on Materia Medica.

plant; in the last of these the root in question, *teorí* (BENG.), is recommended as of use in removing worms and phlegm." "Our present article," he continues, "had long a place in the British Materia Medica (**Convolvulus indicus, Alatus maximus**), but of late years it has fallen into disuse. I find it mentioned by **Avicenna** (264) under the name تربد *turbad;* but the first amongst the Arabs who prescribed it was **Mesue**, who gave the root, in powder, to the extent of from ʒi to ʒii, and of the decoction, from ʒii to ʒiv. **Alston**, in his *Materia Medica* (Vol. II., 530), speaks of *turbith* as a strong and resinous cathartic, and recommended in his days in gout, dropsy, and leprosy. The plant is known to the modern Greeks by the name of *τουρπεθ*." "**Virey** speaks of the root of the **Convolvulus Turpethum** as more drastic than the common Jalap." Muhammadan physicians recognise, like the Hindus, two forms of *turbad*, a white and a black, and recommend that the black should be avoided on account of its poisonous properties. **MEDICINE.**

In India the usage of this drug, apparently, fell off after the appearance of *O'Shaughnessy's Bengal Dispensatory*, where he wrote, while commenting on the facts published in *Roxburgh's Flora of India*, "we have also subjected its properties to careful clinical experiment, and we feel warranted in asserting that the action of the medicine is so extremely uncertain that it does not deserve a place in our Pharmacopœia." That verdict was reproduced in *Waring's Pharmacopœia of India*, and European interest in the drug, from the date of that work, may be said to have died out. It will, however, be found (from the selection of special opinions quoted below) that native practitioners continue strongly in favour of it, and that **Dr. Moodeen Sheriff** and others think that the adverse opinion held by European authorities is largely due to the whole root having been used instead of the root-bark. In this opinion **Merat** and **DeLens** concur, recommending that the roots should be chosen in which the bark is intact, "as most of the activity of the root resides in that part."

Chemistry.—The bark is without odour and has little taste. "Examined by **M. Bontron Charland**, it was found to contain resin, a fatty substance, volatile oil, albumen, starch, a yellow colouring matter, lignin, salts, and oxide of iron. The root contains 10 per cent. of resin (*M. Andouard*). According to **M. Spirgatis**, this resin is a glucoside, *turpethin*, C_{36} H_{56} O_{16}, like that of other CONVOLVULACEÆ, insoluble in ether, but soluble in alcohol, to which it imparts a brown colour not removable by animal charcoal. To obtain it pure the alcoholic solution is concentrated; the resin precipitated by water, boiled with this vehicle, then dried, reduced to powder, digested with ether, and finally redissolved by absolute alcohol, and thrown down by ether." "It is inflammable, burning with a smoky flame, and emitting irritant vapours" (*United States Dispensatory, 15th Ed., 1770*). Accordingly **Spirgatis turpethin** is only present in the root to the extent of 4 per cent. Under the action of alkaline bases it is transformed into turpethic acid (C_{34} H_{60} O_{18}), and by hydrochloric and other dilute acids it is decomposed into *turpetholic* acid (C_{16} H_{32} O_4) and glucose. (*Watts, Chemistry, V, 925-926*). According to **Dr. Dymock** the FLOWERS are in Western India applied to the head in hemicrania. **Chemistry.** 417

SPECIAL OPINIONS.—§ "My experience of the root of **Ipomæa Turpethum** is much greater now than nineteen years ago, when I spoke of it in the *Supplement to the Pharm. of India*, pages 346—348. There are two varieties of it which are well known in the bazar as *sufed-turbud*, and *kala-turbud*, or, the white and the black *turpeth* root, respectively. The drug consists of pieces of root and root-bark, and the latter being the best and not generally sold separately in the bazár, requires to be picked **Flowers.** 418

MEDICINE. out. The root-bark occurs in cylindrical pieces, varying in length from 2 to 4 inches, and in thickness from one quarter to one inch; quilled or curved; smooth, but often wrinkled longitudinally; taste slightly acrid, and odour agreeable if new. The colour of the white variety is grey or reddish-grey; and of the black, brown. The cut ends of the piece, particularly in the white variety, are often covered with some resinous substance. The thickness of the bark itself is about a line or two, and the bark of the white variety is generally thicker than that of the black.

The *turpeth* root, notably the white variety of it, is quite equal to *jalap* and superior to *rhubarb* in its action, and preferable to both for having no nauseous smell or taste, and for being a very efficient and satisfactory purgative when used alone. Its dose is somewhat larger than that of jalap, but this is no disadvantage, as long as it is safe and free from nauseous taste and smell. The dose is larger only by 10 or 15 grains. As a cathartic and laxative, the *turpeth* root is useful in all the affections in which either jalap or rhubarb is indicated. The best way of administering it is in simple powder; but it may also be employed in combination with cream of tartar in euqal proportion, and with or without a few grains of ginger in each drachm of the compound powder. Dose of the simple powder is, from fifty to seventy grains, and of the compound powder from a drachm to nintey grains" (*Honorary Surgeon Moodeen Sheriff, Khan Bahadur, G. M. M. C., Triplicane, Madras*). "What is sold as black turband has the same structure as the white, but it is of smaller size and darker colour" (*Surgeon-Major Dymock, Bombay*). "Used as a purgative by native practitioners" (*Civil Surgeon W. H. Morgan, Cochin*). "Is a purgative which is very frequently used with rock salt and long pepper" (*Surgeon-Major Robb, Civil Surgeon, Ahmedabad*). "The root powdered in doses of a drachm and a half is used as a purgative, but is less efficient than káládáná" (*1st Class Hospital Asst. Lal Mahomed, Main Dispensary, Hoshangabad, Central Provinces*).

FODDER. 419 **Fodder.**—Cattle will not eat this plant, though they greedily devour many of the other species of **Ipomæa.**

SACRED. 420 **Sacred.**—The Hindus offer the flowers to Shiva.

421 **Ipomæa vitifolia,** *Sweet.; Fl. Br. Ind., IV., 213.*

Syn.—CONVOLVULUS VITIFOLIUS, *Linn.;* C. ANGULARIS, *Linn.;* IPOMÆA ANGULARIS, *Chois.*

Vern.—*Nawal,* BOMB.; *Kya-hin-ka-lae-nway,* BURM.

References.—*Roxb., Fl. Ind., Ed. C. B. C., 160; Dalz. and Gibs., Bomb. Fl., 165; Kurz, For. Fl. Burm., II., 219; Thwaites, En. Ceylon Pl., 426; Voigt, Hort. Sub. Cal., 361; Indian Forester, II., 26; Dymock, Mat. Med. W. Ind., 566; Journ. Agri.-Hort. Soc. Ind. (Old Series), VI., 147; Gazetteers:—Kanara, 439; Him. Dist., 314.*

Habitat.—Met with throughout India (except the dry North West) and from Sikkim to Assam, Chittagong, and Burma. Distributed to Ceylon, Malacca, Malay Islands, &c.

MEDICINE. Juice. 422 **Medicine.**—Dymock says the JUICE is in the Koncan considered cooling and is given with milk and sugar. A *lép* is prepared; consisting of the juice, with lime juice, opium, and **Coptis Teeta,** which is applied around the orbit of the eye in inflammation.

IRIS, *Linn.; Gen. Pl., III., 686.*

The economic information regarding the various species of Indian Iris is very obscure. It seems probable that a re-arrangement of the facts here published may have to take place when the bazar products have been more definitely referred to the species to which they belong. This enumeration may, therefore, be accepted more as giving the names of the Indian wild and cultivated species than as placing on record a satisfactory statement of each individual. There would appear to be some six or seven indigenous species: within Quetta three additional forms, and in Afghánistan some four or five more. In Davies' Trade and Resources of the North-West Frontier *Iris* and *Kut* (**Saussurea Lappa**) are confused together.

Iris ensata, *Thunb.; Jour. Linn. Soc., XVI., 139;* IRIDEÆ. 423

Syn.—I. MOORCROFTIANA, *Wall.*

Vern.—*Irisa, sosun,* HIND.; *Tesma,* BHOTE; *Krishún, ánarjal, marjal,* KASHMIR; *Begbunuphsha,* PERS.

References.—*Stewart, Pb. Pl., 240; Boissier, Fl. Orient., V., 127; Med. Top. Ajmir, 129.*

Habitat.—Common on the temperate North-West Himálaya and Kashmír, in damp places, often grown in gardens.

Fodder.—§ "The LEAVES are largely collected and employed as fodder and bedding for cattle, for thatching, matting, and basket-work in Ladak" (*Surgeon-Major J. E. T. Aitchison, Simla*). FODDER. Leaves. 424

I. florentina, *Linn.; Jour. Soc. Linn., XVI., 146.* 425

THE IRIS, ORRIS ROOT.

Vern.—*Irsa, sosun,* HIND.; *Bekh sosan, ersó,* KASHMIR; *Irisa,* PB.; *Bekh-i-banfsa,* PERS.

References.—*Stewart, Pb. Pl., 240; Boissier, Fl. Or., V., 137; Ainslie, Mat. Ind., I., 182, 284; Bent. & Trim., Med. Pl., 273; S. Arjun, Bomb. Drugs, 142; Murray, Pl. and Drugs, Sind, 19; Irvine, Mat. Med., Patna, 10; Med. Top. Ajmir, 129; Year-Book, Pharm., 1873, 114; Birdwood, Bomb. Pr., 89.*

Habitat.—This is the European plant so much used in the preparation of the sweetly-scented Otto of Orris. It is said to be sometimes met with in Indian gardens. In Kashmír this, as also several other species, are cultivated, and either wild or only acclimatised are frequently seen on the margins of fields near water. The writer collected what appears to be this species in Pangí. It is fairly plentiful in the Simla district and in Kullu, but always under circumstances that would lead to the suspicion of its being cultivated or at most only an escape from cultivation. This may probably be the **Iris longifolia** mentioned by **Adams** (*Wanderings of a Naturalist, p. 196*) as seen by him on the Northern Pinjal, though it would seem that **I. pallida** is in many parts of the Himálaya also acclimatised and may possibly be the *shrúl* or *skecho* of *Stewart's Panjáb Plants.*

Perfumery.—It is not known whether the natives of India employ the tubers of any **Iris** in perfumery. But it is significant that while in the *Ain-i-Akbari* a special chapter is devoted to Perfumes, no mention should be made of Orris. Presumably the perfume prepared from these rhizomes in Europe was not known to the Muhammadans of India 300 years ago. (*Conf. with Piesse, Art of Perfumery, 172-174.*) PERFUMERY. 426

Medicine.—There seems considerable doubt as to the identification of the *írisa* root; in fact, it is not by any means proved to be obtained from this species. **Stewart** says *írisa* is used externally in the treatment of rheumatism. **Irvine** mentions that it comes to India from **Kabul**. But **I. germanica** is cultivated in Kashmír and is more probably the source of the drug mentioned by **Stewart** than is **I. florentina.** MEDICINE. 427

I. germanica, *Linn.; Jour. Linn. Soc., XVI., 146.* 428

Syn.—I. DEFLEXA, *Knowles and Westc.;* I. NEPALENSIS, *Wall., in Lindl. Bot. Reg., t. 818, von Don;* I. VIOLACEA, *Savi.*

Vern.—*Bikh-i-banafshah,* PERS.; *Irsá,* ARAB.; *Keore-ká-múl,* HIND. & BOMB.

References.—*Dymock, Mat. Med. W. Ind., 2nd Ed., 793-794; Boissier. Fl. Or., V., 139.*

IRIS; Morea. **The Nil Kamal.**

MEDICINE. Root. 429 **Medicine.**—"Orris root is not mentioned by the old Hindu physicians. Muhammadan writers mention several kinds of **Iris** under the names of *Irsá* and *Susán*, but their descriptions are not sufficiently accurate to enable us to identify them with any certainty. From **Mir Muhammad Husain's** account (*Conf. Makhzan*, article 'Súsán'), I conclude that both the **Iris fœtidissma,** *L.*, the *ζυρις* of **Dioscorides** and **I. pseudoacorus,** *L.*, the *νερόκρινος* of the modern Greeks, are used medicinally in Persia. The correct Persian for **I. germanica** appears to be *Súsán-i-ásmánjúní*. The word *Súsán* is said to be derived from the Syrian *Súsání*. *Irsá* is evidently a corruption of the Greek Iris."

"Iris root is considered by Muhammadan hakims to be deobstruent, aperient, diuretic, especially useful in removing bilious obstructions. It is also used externally as an application to small sores and pimples. From the large number of diseases in which this drug is recommended, it would appear to be regarded as useful in most diseases."

"Indian Orris root differs from the European drug, inasmuch as the bark of the rhizome has not been removed; it is also small and badly preserved." "Bombay is supplied with Orris root from Persia and Kashmír" (*Dymock*).

430 **Iris kumaonensis,** *Wall.; Jour. Linn. Soc., XVI., 144.*

Syn.—I. LONGIFOLIA, *Royle, Ill., 372, t. 91.*

Vern.—*Piáz, karkar, tezma,* PB.

Habitat.—An exceedingly plentiful plant on the grassy slopes of the temperate and alpine North-West Himálaya, especially in Lahoul and Ladak, and in the Chenab and Ravi basins generally.

MEDICINE. Root. 431 Leaves. 432 **Medicine.**—Dr. **Stewart** states that in Chumba the ROOT and the LEAVES are given in fever. In parts of Ladak the leaves appear to be given as fodder.

It seems probable that this or the next species is the source of the *Lakrí-pashánbed* or *pokhánbed*, which **Dymock** says is obtained from Tibet. Speaking of the drug **Dymock** says:—"When soaked in glycerine it stains it of a rich reddish-brown. The taste is acrid and astringent: the odour musky and aromatic. The minute structure of the rhizome has a general resemblance to that of Orris root, but numerous stellate raphides are present. The drug is chiefly used as a diuretic. It is also believed to be an antidote for opium (dose, grs. 15).

FODDER. Leaves. 433 **Fodder.**—In Ladak the leaves are said to be used as fodder.

434 **I. nepalensis,** *D. Don., non Wall.; Jour. Linn. Soc., XVI., 143.*

Syn.—I. DECORA, *Wall.;* I. SULCATA, *Wall.*

Vern.—*Chalnúndar, sosan, shoti, chilúchí,* PB. and N.-W. P. HIMÁLAYAN NAMES. **Madden**, in his account of the Himálayan CONIFERÆ, says that this species is known in Kumaon as *Nîl Kamál* (blue Lotus) and **Crinum toxicarium** as *Chunder Kamal* (white Lotus): both he says are found near temples. **I. nepalensis** is the *Nîl Kanwal* in *Atkinson's Himálayan Districts.*

Habitat.—A very plentiful plant in Nepál, extending west to the Sutlej and the Beas, and met with in cultivation in Kashmír, being the broad-leaved form often seen in grave-yards in the Panjáb.

435 **I. (Morea) Sisyrinchium,** *Linn.; Journ. Linn. Soc., XVI.;* [*Boiss., Fl. Or., V., 120.*

Habitat.—Baluchistán (Quetta), Afghánistan.

FOOD. 436 **Food.**—Mr. **J. F. Duthie** reports that the bulbs of this species are eaten in times of scarcity.

Iris pseudacorus, *Linn.; Journn. Linn. Soc., XVI., 140; Boiss., [Fl. Or., V., 127.* 437

Habitat.—Europe, Africa, Persia, &c.

Stewart mentions this plant as seen in and near Kashmír, but is doubtful of its having been correctly determined. The plant he refers to is known to the natives as ***Kríshún ímarjal, marjal,*** and was found at altitudes of 5,000 feet above the sea. That species is not known, however, to be indigenous to Kashmír, so that **Stewart's** plant was much more likely to have been **I. ensata.** (*J. Murray.*)

IRON, *Ball, Man. Geology of India, III., 335-420.* 438

Iron exists in nature in three great classes of ores—(1), as SULPHIDE, or Iron Pyrites, with the chemical composition of FeS_2; (2), as CARBONATE; (3), as OXIDE. .The FIRST class is not directly worked for the purpose of obtaining the metal; the second and third are consequently those of commercia importance as sources of iron. The SECOND class of ore, in which iron exists as the carbonate, $Fe\ CO_3$, consists essentially of two kinds, namely, those in which the salt is crystalline and little admixed with earthy matter, and those in which a larger or smaller amount of clay is intimately intermixed with the ferrous carbonate. The former class is termed ***spathic iron ore*** (***siderite spathose, sparry ore***), the latter generally bears the name of ***argillaceous ore*** or ***clay-ironstone. Clay-ironstone*** exists in large deposits in many coal measures, and in this situation is known as ***black-band.*** The THIRD class of ores in which iron exists as an oxide, may be divided into three sub-classes, namely, (1) that in which the metal essentially occurs as anhydrous ferric oxide (Fe_2O_3), (2) that in which it occurs as hydrated ferric oxide (Fe_2O_3, H_2O), and (3) that in which the iron is in the form of ferrous and ferric oxide combined, of which sub-class magnetic oxide of iron, Fe_3O_4 is the type.

The ores of the first sub-class are chiefly known as ***red hæmatite*** and ***specular ores. Red hæmatite*** is generally soft and friable, when it is known as ***red ochre*** or ***puddler's ore,*** but it also occurs in hard crystalline masses called ***kidney ore. Specular ore*** is a hard well-crystallized form, of a greyish or black colour, and brilliant lustre.

The ores of the second sub-class most frequently occur mixed with a large amount of earthy matter, together with considerable quantities of sulphur and phosphorus. They are known as ***brown hæmatite, bog-iron ore,*** or ***limonite.*** In texture these ores differ considerably from red hæmatite, being cindry or earthy in character, and rarely massive. A pure, definitely crystallized hydrate, called ***göthite,*** is, however, occasionally found.

The ores of the third sub-class are known as ***magnetic iron ore,*** or ***magnetite.*** The purest varieties are strongly magnetic, and often exhibit polarity, when they are known as ***loadstone.*** The magnetic iron ores may be easily distinguished from the other oxides by the fact that they give a black streak when drawn over a white surface.

Iron exists in great abundance in many parts of India, occurring in one or more of the forms briefly detailed above, and has long been known and worked by the Natives. The prevailing ore is the magnetic oxide, but members of the other classes are also met with abundantly. For the following remarks regarding the mode of occurrence and distribution of iron in this country, the writer is indebted to the admirable and exhaustive account of the metal published by **Ball** in his ***Economic Geology,*** from which the facts here given have been mainly compiled.

GENERAL REMARKS. 439

GENERAL REMARKS.—In the peninsular area magnetite occurs in beds or veins, in most of the regions where metamorphic rocks prevail. In some places, as in the Salem District of the Madras Presidency, it

GENERAL REMARKS.

is found in immense quantities, whole hills and ranges being formed of the purest oxide. Similarly, in the Chanda District, Central Provinces, enormous deposits of specular ore are found with which magnetite is also present. The abundance and wide distribution of these ores in the oldest rocks no doubt explains the frequent recurrence of deposits, and the general dissemination of ferruginous matter which more or less characterises the sedimentary rocks of all subsequent periods.

In the submetamorphic or transition rocks, magnetite is known to occur in some localities. Also, along faults and fractures both in the metamorphic and submetamorphic rocks, very considerable veins of limonite are to be found in many parts of the country, such as Cuddapah, Karnúl, Manbhum, Jubbulpur, &c. The rich ores of Central India principally consist of hæmatites and occur in the Bijawar or lower transition series of rocks.

In the Vindhyan formation ferruginous matter is commonly disseminated through immense thicknesses of beds, tinting the rocks it pervades; but in all localities in which the ores are sufficiently concentrated to be workable, the deposits are found in veins, not in beds.

In certain of the conglomerates, sandstones, and shales of the Gondwana system, concretionary masses of limonite are abundant, while in some of the coal-fields siderite, and clay ironstone, altered more or less into limonite, occur imbedded in sufficient quantity to be of considerable importance. It is believed that this limonite is used by Native smelters in all the coal-fields, but the unaltered carbonates are rarely if ever utilised in India, the Native furnaces being apparently unsuited for their reduction.

The next group, or so-called ironstone-shale, is only represented in the Damuda valley, and is largely utilised as a source of iron ore in the Raniganj coal-field, where there is an inexhaustible supply of readily accessible ore. This ore, which originally existed in the form of black-band or siderite, has been to some extent altered into limonite on the surface, but deeper it probably exists in its original condition.

In the next succeeding group, the Raniganj-Kamthi, ferruginous matter is distributed somewhat unequally, in certain parts being scarce or wholly absent, in others sufficient to tinge the whole rock. Segregated masses and thin layers of ore also are found here and there, and are occasionally utilised. Ore occurs in a similar way in the following groups of the Gondwana series.

The cretaceous rocks of Southern India, in the Trichinopoly District, contain nodules of iron ore in some abundance, which were formerly smelted to a considerable extent by the inhabitants. The Deccan trap includes a large amount of iron ore, disseminated for the most part in minute crystals of magnetite, but occasionally met with in nests of hæmatite, more rarely in layers which pass into ferruginous earth or bole. The beds of rivers which traverse this trap not unfrequently contain magnetic iron-sand, which might be collected in sufficient quantity for the requirements of Native furnaces, but rarely is so, owing to the brown hæmatite described below being generally more readily obtainable in the same localities.

Segregated bands frequently occur towards the bases of beds of laterite, which are often prolific sources of an easily worked brown hæmatite, and sometimes contain a high percentage of metal. These lateritic ores have been worked by native smelters in various localities, and at Bepur in Malabar, and Muhammad Bazar in Birbhum, they have been utilised by British Companies.

Lastly, detrital ores of sub-recent age occur in many localities, derived from the breaking up of the above deposits and from superficial accumulations of ferruginous matter. These being in general easily obtained and more or less soft and decomposed, are frequently preferred by Natives to the harder and more refractory ores *in situ*, which are more difficult both to mine and reduce.

In the extra-peninsular area the principal sources of iron ores are the tertiary formations; but in the North-West Himálaya, and also frequently in Afghánistan and Burma, there are considerable deposits in the older metamorphic rocks.

Iron, *Mallet, Geology of India, IV. (Mineralogy), 28, 49, 50, 52, 59.* 440

FER, *Fr.*; EISEN, *Ger.*; FERRO, *Ital.*; HIERRO, *Sp.*

Vern.—*Lohá, lohah, lohá chúr*=iron filings, *kheri*=steel, *giri*=bloom or ball of iron, *nar*=iron furnace, *khit*=slag from iron furnace, *karguha*=iron furnace tongs, *sansri*=iron tongs, HIND.; *Lohá, láhá*, BENG.; *Lokhanda*, MAR.; *Lévu*, GUZ.; *Irumbu*, TAM.; *Inumu*, TEL.; *Kabina*, KAN.; *Irumba*, MALAY.; *Dán, thán*, BURM.; *Yakada*, SING.; *Ayam, lóham, lauha*, SANS.; *Hadíd*, ARAB.; *Áhan*, PERS.

References.—*Baden Powell, Pb. Prod., 1-10, 53, 66; Atkinson, Him. Dist., 262-290; Economic Geology, Hill Districts, N.-W. P., Allahabad, 1877; Mason, Burma and its People, 565, 729; Oldham, Yule's Misson to Court of Ava (1855), 346; Balfour, Cyclop., II., 371; Rep. on Iron Ores, &c., of Madras Presidency; Hunter, Statistical Acct. of Assam (1879), I., 21, 231, 260, 299, 380; II., 210, 235; Hooker, Himalayan Journals, II., 310; Ain-i-Akbari, Blochmann's Trans., 40, 113; Gladwin's Trans., II., 42, 126, 139, 170; Pharm. Ind., 364; Ainslie, Mat. Ind., I., 522-527; Moodeen Sheriff, Supp. Pharm. Ind., 141; U. C. Dutt, Mat. Med. Hind., 46; Fleming, Med. Pl. & Drugs as in As. Res., XI.; Royle, Prod. Res., 3; Hope, Note on the Iron Industry in India; Govt. of India, Dept. of Revenue and Commerce, No. 6, Sept. 1873, Dept. of Finance and Commerce, No. 2899, August 4th, 1882; Molesworth, Note on Iron Manufacture by Private Enterprise, August 15th, 1882; R. C. Von Schwarz, Rep. on Bengal Iron Works, Oct. 1881; Trade and Nav. Reps. Madras; Trade & Navigation Statistics of British India; "Pioneer," Jan. 25th, 1883; Gazetteers:—Mysore & Coorg, I., 429; Orissa, II., 68, 75, App. III., Bombay, IV., 130; V., 19, 124; VI., 11; VIII, 90, 91; X., 30; XI., 134; XV., 19; Panjab:—Amritsar, 44; Gurgaon, 13; Gurdaspur, 60; Hoshiarpur, 113; Shahpur, 113; Peshawar, 24; N.-W. P., I., 58; Madras Manual of Administration, II., 36, 37, 38, 115; Bengal Adm. Rep. (1860), 125; Forbes, Settlement Rep. Calcutta (1872); Central Prov. Adm. Rep. (1864-65), 5; (1865-66), 67; (1866-67), 78; District Manuals:—Madras;—Cuddapah 45; Trichinopoly, 68; Coimbatore, 22, 23; Settlement Reports:—Central Provinces;—Chanda, 4, 105, 113, 114; Damoh, 88; Upper Godavery, 41; Jubbulpore, 87; Nagpur (Suppl.), 276; Raipur, 7; Saugor, 3; Panjab;—Dera Ghazi Khan, 7; Dera Ismail Khan, 7; Hazara, 9; Hoshungabad, 16; N.-W. P. & Oudh;—Sabtpore, 12; also many passages in the publications of the Geological Survey and in the Journals of the Asiatic Society of Bengal.*

OCCURRENCE, &c.—Mr. F. R. Mallet, Superintendent of the Geological Survey of India, has kindly furnished the following account of the regions in which Iron is found and of the manufacture of the metal in India:— OCCURRENCE. 441

Iron has not been found native except in meteorites, and, very rarely in volcanic rocks. It forms, however, an essential constituent of over a hundred minerals, although only a few of these are utilised for the extraction of the metal. The most important ores are as follows, the percentage of iron they contain when pure being indicated by the figures:—

Magnetite, or *magnetic iron*, protosesquioxide; 72·4. The black sand which is smelted in many parts of India consists of magnetite, with, very frequently, a varying proportion of titaniferous iron. *Hæmatite*, or *red hæmatite*, sesquioxide; 70·0. *Limonite*, or *brown hæmatite*, hydrous sesquioxide; 59·9. Ordinary laterite is an impure earthy variety of this ore. *Siderite*, or *spathic iron*, carbonate, 48·3: *clay-ironstone*, or *argillaceous iron ore*, is an impure variety of the carbonate. *Black-band* is a carbon- Magnetite. 442 Hæmatite. 443 Limonite. 444 Siderite. 445 Black-band. 446

OCCURRENCE Ilmenite. 447

aceous clay-ironstone. *Ilmenite,* or *titaniferous iron,* an ore allied to hæmatite, but having a portion of the iron replaced by titanium, may also be mentioned, although it is of comparatively little importance.

MANUFACTURE.

MANUFACTURE. 1. Native methods of—Iron. 448

NATIVE METHODS—IRON.—In former times the manufacture of iron by native methods was widely practised throughout the various provinces of India, but within the last few decades, and more especially during the last quarter of a century, since the opening of railway communications, the competition of English iron has done much to curtail, and in many cases has entirely extinguished, the native outturn. A long list of places might, however, be given, in which the manufacture survives.

According to the method most usually adopted, the furnace, which is built of clay, is some three or four feet in height, more or less, although the dimensions vary very considerably in different parts of the country. Both the exterior and interior are more or less conical in form, the circular cross section at the bottom being greater than that at the top. Near the bottom in front is an orifice which is stopped with clay while the furnace is in blast, and through which the bloom is removed at the end of the operation. Clay tuyeres are inserted near the base which convey the blast from two skin bellows, worked alternately, so as to keep up a continuous stream of air. The fuel used is charcoal, and no flux is added with the ore. After the furnace has been brought to a sufficiently high temperature, the ore, sometimes in the form of natural sand, but more commonly after having been pounded to small fragments or coarse powder, is sprinkled in at the top, in small quantities at frequent intervals, and alternately with a sufficiency of charcoal to keep the charge nearly level with the top of the furnace. From time to time during the operation (which lasts several hours) the slag is removed through a hole which is then stopped with clay. When such a quantity of ore has been added as will produce a bloom of iron of the size proper to the furnace used, the blast is stopped, the orifice opened, and the bloom removed. It is a pasty mass of *malleable* iron containing a good deal of intermixed slag, which is removed as far as practicable by immediate hammering. But before use the iron needs further refining by re-heating and hammering, a process which is sometimes carried out more than once, and which serves at the same time to bring the iron into more usable shapes.

In some cases the refining is done by the smelters, but very commonly they cut the bloom almost in two halves, to show the quality of the mass to the purchaser, and sell it in that state, the refining being done by others. The expenditure of charcoal is very great in proportion to the result obtained (according to **Mr. Ball**, as much, sometimes, as 14 tons of fuel to 1 ton of finished iron), and a large quantity of metal remains in the slag, so that the process is an extremely wasteful one (thus it is stated by **Mr. Heath** that in the Salem district magnetite, containing 72 per cent. of metal, yields only 15 per cent. of its weight of bar iron). At the same time the iron obtained is generally of very good quality, and were it not on account of the convenient forms in which English iron can be obtained, the Indian product would have less difficulty in maintaining its position amongst the native smiths. On account of this competition, and also from increasing scarcity of fuel, "one finds" (as remarked by **Mr. Ball**) "the iron smelters in many regions the hardest worked, but poorest, amongst the population. The iron is sold at a high price, but the bulk of the profit goes to the traders through whose hands the metal passes. The amount of iron produced bears but a miserable proportion to the labour, time, and material expended."

MANUFACTURE. Native Methods of—Iron.

Although the method most commonly adopted approximates, more or less closely, to that described above, wide variations from it are in vogue in some regions. Thus, in Manipur the tuyeres are inserted at the back of the furnace, while in front, and opposite to them, is a roughly semi-circular aperture some nine inches across. The products of combustion pass out through this, the chimney above being merely used for feeding in the ore and fuel, so that the furnace is practically an open hearth (*R. D. Oldham, Mem. G. S. I., XIX., 240*). At Puppadoung, in Upper Burma, the furnaces are worked with a natural draught only, no bellows being used (*W. T. Blanford, Journ. A. S. B., XXXI., 219*). In Kathiawar the furnace, according to a not very clear description, is rectangular in section, and horizontal instead of vertical (*Jacob, Selec. Rec. Govt. Bomb., XXXVII., 467*). In Birbhum, as observed by Dr. Oldham, in 1852, the furnaces were of unusual capacity, and produced *pig* iron, which was reduced to a malleable condition by subsequent refining (*Selec. Rec. Govt. Bengal, VIII.*). Contrary, again, to the practice in India, the smelters of the Waziri hills add limestone as a flux (*A. Verchère, Jour. A. S. B., XXXVI., pt. II., 21*).

The bellows also vary greatly in their form and size. For the smaller furnaces, a pair, each made from a single goatskin; are most frequently used, while those employed for the larger are sometimes made of bullock-hides. Some forms are worked by hand, while in others one or two persons stand with one foot on each bellows, and transfer their weight alternately from one to the other.

Steel. 449

STEEL.—The process by which the Indian steel, known as *wootz*, is produced, has been described by several writers. The Salem district in Madras used to be one of the most important centres of production, and Mr. Heath has written a very clear account of the method employed (*Journ. Roy. As. Soc., V., 390*). The iron, which is smelted from magnetite in the usual way, is refined by repeated heatings and hammerings, and eventually formed into bars measuring about 12 inches $\times 1\frac{1}{2} \times \frac{1}{2}$. These are cut into small pieces, a number of which, aggregating from half a pound to two pounds in weight, are packed closely in a crucible, together with a tenth part by weight of dried wood chopped small; the whole is covered over with one or two green leaves, and the mouth of the crucible filled up with tempered clay, rammed in close. As soon as the clay is dry, from 20 to 24 crucibles are built up, in the form of an arch, in a small furnace, which is lighted, and the blast kept up for about $2\frac{1}{2}$ hours; the crucibles are then removed, allowed to cool, broken, and the cakes of steel, which have the form of the bottom of the crucible, taken out. When the fusion has been perfect, the top of the case is covered with striæ radiating from the centre. The cakes are subsequently heated for several hours at a temperature just below their melting point, and turned over in the current of air from the bellows. The object of this treatment, in Mr. Heath's opinion, is the elimination, by oxidation, of the excess of carbon which the cakes contain, and to which they owe their comparatively low fusing point. When this operation is completed, the cakes are hammered out into short stout bars, in which state the steel is sold.

The crucibles are formed of a refractory "red loam," mixed with a large proportion of charred rice-husk. The wood used is that of the **Cassia auriculata**, and the leaves those of the **Asclepias gigantea**, or, if such are not procurable, those of the **Convolvulus laurifolia** [**Lettsomia elliptica**, *Wight*].

It appears that in some cases the blooms produced in the ordinary iron furnace, after refining in the usual way, are sufficiently steely for employment in the fabrication of edged tools, which are tempered by plunging them while hot into water. A steely iron thus made in the Jabalpur district

IRON. **Manufacture of Iron.**

MANUFACTURE.

is produced from a highly manganiferous hæmatite (*Rec. G. S I., XVI., 101*; *P. N. Bose, Ibid., XXI., 88*). But in the district of Darjeeling similar iron is smelted from magnetite containing no manganese (*Mem. G. S. I., XI., 67*).

2. English Method of— 450

ENGLISH METHODS.—Numerous attempts have been made to manufacture iron on the English system in India, but nearly all of these have been unsuccessful and have been long since abandoned; one of the chief causes of failure being the difficulty in keeping large furnaces supplied with charcoal. In 1833 a company was formed, with works at Porto Novo, in South Arcot, in the Salem district; and at Beypur, in Malabar. Pig-iron was smelted from the Salem ore, part of which was sent to England, and commanded a good price for conversion into steel; a large quantity of it was used in the construction of the Britannia and Menai bridges. Excellent steel was also produced at the Indian works, and iron was smelted at Beypur; the company, however, appears never to have paid a dividend, and the concern was a steadily losing one. About 1855, the Birbhum Iron-works Company commenced operations, and produced pig-iron of good quality. But from various causes the works were financially a failure. The Kumaun Iron-works were started in 1857, and after many vicissitudes ended like the preceding ventures. About 1861 works were commenced at Barwai, in the State of Indore; but, when the preparations for smelting were nearly complete, it was decided to abandon the project. In 1880, or thereabouts, works were established at Náhun, in the Punjab State of Sirmur. The enterprise, however, has not hitherto proved a financial success, owing, chiefly, to the long distance and bad roads over which the ore has to be conveyed from the mines.

In all the attempts noticed above, charcoal was the fuel used, or proposed for use. In 1875, however, an experimental trial was made, to determine whether the rich iron ores of Chánda could be smelted on a commercial scale with the coal raised in the same district. The experiment was not successful, the failure being attributed mainly to the inferior character of the fuel, and its unsuitability to the purpose in hand. About the same time (1874) the Bengal Iron Company commenced operations at Barakar, in the Raniganj coal-field. The ore employed, which was obtained from the surface in the neighbourhood of the works, occurs abundantly in the ironstone shale group of the Damuda (coal-bearing) series. In these rocks the ore forms bands and nodules of clay-ironstone, but near the surface this has been altered into an earthy peroxide, partly hydrated and partly hematitic. It was this altered form which was smelted, the fuel used being mainly coke made from coal raised near the works and in the Karharbari collieries. For flux, limestone obtained at Lachete hill, and at Hánsapathar, a few miles further off, was employed. Altogether the company produced 12,700 tons of pig-iron, but it was commercially unsuccessful, and stopped work in 1879. Three years later the concern was taken over by Government, and placed under the management of Ritter von Schwartz. One blast furnace was re-started on the 1st January 1884, and is still in active operation, the outturn of pig-iron up to the end of 1888 being 30,616 tons. A second blast furnace, on an improved type, has lately been erected to meet the increasing demand for pig-iron, and arrangements have been made to smelt about 15,000 tons during the official year 1889-90, and 20,000 in the year following. The annual consumption of pig-iron in the foundry is at present about 2,000 tons, the castings being chiefly pipes, sleepers, bridge-piles, ornamental work, railway axle-boxes, and agricultural implements. To Ritter von Schwartz belongs the credit of having been the first to show that iron *can* be successfully

INDIAN IRON ORES.

smelted in India on European principles." The Barákar Iron Works were again sold to a Company by Government in 1889.

IRON ORES OF INDIA.

MADRAS.

Madras. 451

Travancore.—Iron is, or used to be, smelted in this district, the ore apparently being obtained chiefly from the laterite. The outturn, however, was not very great even more than forty years ago, the taluk of Shenkotta being the only one in which any considerable quantity was produced.

Black magnetite sand is cast upon the beach, at frequent intervals, in enormous quantities, along the coast of South Travancore, the source of which is at present unknown.

Tinnevelli.—Lateritic ore and magnetite are both found. The ore that has been used in the native furnaces is magnetic ironsand, from which also steel has been produced in the village of Vanga-colum.

Madura.—Lateritic ore occurs in abundance, and there are traces of considerable smelting having been carried on formerly, but the industry is now quite extinct.

Pudukattai.- Magnetite has been noticed near Mallampatti, 19 miles N.-W. by N. from the chief town of the State.

Trichinopoli.—Ferruginous nodules from the cretaceous rocks were utilised at some former period, but no iron is made in the district now-a-days.

Coimbatore,—Iron is, or used to be, smelted from "black sand."

Nilgiri Hills.—Hematite and magnetite are tolerably abundant, forming short, irregular bands or masses in the gneiss, the former being far the more common at the surface; but this may be due to atmospheric alteration of magnetite. The comparative scarcity of fuel renders it improbable that these ores could be utilised.

Malabar.—Iron ores are abundant, the principal being magnetite and laterite. The former occurs in the form of bands through the metamorphic strata, and of black sand; here, as elsewhere, the sand being produced by the degradation of the banded form of ore. Iron has been produced on a considerable scale, the ore being obtained, and also smelted, in the taluks of Kurmenaad, Shernaad, Walluwanaad, Ernaad, and Temelpuram. In 1854 there were over 100 furnaces in operation, which were of a larger size than those commonly used in India, some being as much as ten feet in height. Nearly a ton of ore is said to have been the charge, from which, in round numbers, from 250 to 400℔ of metal was obtained. At the date mentioned, the quantity of iron produced yearly in the district was estimated at about 475 tons, but these figures seem very small, considering the number of furnaces in blast. The Beypur works, previously alluded to, were situated in the Malabar district.

South Kanara.—Iron ore is said to be abundant.

Salem.—Ore exists in inexhaustible quantities. It occurs in the form of numerous beds (in the metamorphic rocks), some of which are 50 to 100 feet thick and may be traced for miles. The ore consists of magnetite interlaminated with quartz, the relative proportion of the two minerals varying greatly; in the richest parts as much as seven eighths is magnetite From the degradation of such beds magnetic ironsand has been formed in abundance. The beds which are most important in extent and richness are those belonging to the four following groups :—

1st—The Godumullay group, E and N.-E. of Salem.
2nd—The Tullamullay-Kolymullay group.
3rd—The Singiputty group.
4th—The Tirtamullay group.

IRON. **Iron Ores of India.**

IRON ORES. Madras.

Excellent iron is produced in the Salem district, but, as in so many other parts of India, the outturn is decreasing, partly from the growing scarcity of charcoal, and partly from the influx of English iron. The manufacture of steel has been described on p. 503. The ore smelted at the Porto Novo Iron-works was brought from Salem.

South Arcot.—Ore is abundant, and has been smelted in a very large number of villages, more especially in the taluks of Trinomalai and Kallakurchi. The manufacture of steel has also been carried on.

North Arcot.—About the year 1855 there were 86 villages where smelting was conducted. The ore principally used was "black sand," but solid ore (magnetite?) was raised in many places. It is said that ore was mined "in every taluk."

Chingalpat.—Magnetite is stated to exist in considerable quantity in the hills near the town of Chingalpat.

Nellore.—Numerous beds of magnetite occur in the schistose gneiss near the town of Ongole, and in the lower part of the Gundlakamma valley. Hæmatite schist has been noticed in the Chundi hills. Iron used to be smelted in the district, but the industry appears to be now extinct.

Kadapah and Karnúl.—Iron ore is pretty generally distributed in the rocks of the two formations named after these districts, but that which is worked is in the Kadapah series, where it occurs in layers among the beds, or in veins, strings, and nests. The finest occurrence is in the Gunnygull ridge, south of Karnúl, which is seamed with great veins of very pure specular iron. Rich deposits also occur south-east of Ramulkota; but there are numerous other localities in both districts where mining is carried on, the ore being generally hæmatite.

Anantipur.—Iron is said to be abundant, the ore being, in part, at least, of the same character as that in Bellary.

Bellary.—In the Saṇdur and Copper hills to the west of the chief town of the district "the supply of splendid hæmatite ore is absolutely unlimited." It occurs in a series of bands interstratified with schist and contemporaneous trap belonging to the Dhárwar series. The old iron industry is, however, nearly extinct, owing mainly to the scarcity of fuel.

Similar ore also occurs in profusion in the "Penner-Haggari band" of the same rocks, which passes, in a south-east to north-west direction, through the districts of Bellary, Lingsugur (in Haiderabad), and Kaladgi.

Kistna.—Iron has been worked in the old collectorates of Guntur and Masulipatam, now amalgamated into the district of Kistna. The principal localities are in the Guntur taluks of Datchapali, Timmarkatta, and Narsaranpett, and the taluks of Jaggiapetta, Tirwar, Gállapali, and Pentapand in Masulipatam.

Godávari.—Iron ore (limonite and hæmatite) occurs in large quantity, distributed through the sandstones of Gollapili, Tripati, and Rájamahendri; but it is only worked in a few localities, and on a small scale.

Vizagapatam.—Iron has been manufactured in numerous places, but little definite information appears to be available respecting the ores. The zamindári of Jaipur has been one of the principal seats of the industry.

Ganjam.—Although iron ore is believed to exist in this district, no smelting operations are carried on there at present.

Mysore. 452

MYSORE.

Ashtagram division.—Ores, which are in large part magnetite, are abundant in many places, but the number of mines is not very great.

Bangalore division.—Both iron and steel are produced, the ore principally used being "black sand." Chinapatam has long been celebrated for

IRON ORES. Mysore.

the manufacture of steel wire, which has been sent to remote parts of India for the strings of musical instruments.

Nagar division.—Here also iron and steel are manufactured, "black sand" being the ore chiefly employed. In the Shimoga district iron ores are worked in some parts. In Kadur, they are largely obtained and smelted along the hills east of the Baba Budan, and those round Obrani. Ore, probably hæmatite, is largely developed in the hills near Chitaldrug.

In Mysore, as in so many other parts of India, the outturn of iron has greatly diminished during the last few decades.

Haiderabad. 453

HAIDERABAD.

Hæmatite occurs in profusion in the "Pennar-Haggari band" of Dhárwar rocks, which traverses the *Lingsugur district*, as noticed under Bellary. Fairly rich beds of magnetic iron have been found nine miles to the south-east of the town of Khamamet, and a very large and important band of the same ore occurs close to the Singareni coal-field, in the same district (*Khamamet*). Hæmatite is obtainable about four miles north of the field. The co-existence of such rich ores, coal, and limestone in the same neighbourhood naturally suggests Singareni as one of the localities in India most favourably circumstanced for the manufacture of iron on European principles.

Smelting has been carried on extensively, in native furnaces, in the parganas of Kallur and Anantagiri. Hæmatite, titaniferous ironsand, and yellow and red ochre are stated to occur in *Warungul.*

Steel-making has been carried on at several villages in *Yelgandal*, but one of the most celebrated places for its manufacture is Konasamundram in *Indor*, 12 miles south of the Godávari and 25 from the town of Nirmal. Accounts written some fifty years ago state that most of the outturn was bought by dealers from Persia who travelled to Konasamundram for the purpose, and that the steel was used in making the celebrated Damascus sword blades. Two kinds of iron were used in its production—one from Mitpalli smelted from "ironsand," or from Dimdurti, reduced from magnetite occurring disseminate through gneiss and mica schist, and the other from Kondapur smelted from an ore "found amongst the iron clay" (laterite?): three parts of the former were employed to two of the latter.

A boundless supply of the above ores is said to be obtainable.

Central Provinces. 454

CENTRAL PROVINCES.

Bastar.—Iron ore is said to occur, in immense quantities, on the Bela Dila, and in the valley of the Jorivág river; also less plentifully towards the North-Western boundary. The quality is good, but it is little worked, as the demand is insignificant.

Sambalpur.—"Iron ore is found in nearly all the zamindáris and Garhját States. It is most plentiful and of the best description in Rairakhol." The metal has been smelted in some quantity at Kudderbuga, 20 miles north-east of Sambalpur, from altered magnetite. Ores obtained from the Barákar group, and, less frequently, from the Upper (Hingir) sandstone are used in other places.

Biláspur.—Ores are to be found in many places, although they are not worked in any considerable scale.

Raipur.—Very rich ores occur abundantly in parts, especially in the Dáundi Lohára zamindári, and the feudatory state of Khairágar, where a considerable number of furnaces are at work. The zamindáris of Gandái, Thakurtola, and Worárband, and the State of Nándgáon, also contain ores which are worked in some places. Those principally used are hæmatite and laterite.

IRON. **Iron Ores of India.**

IRON ORES.

Central Provinces.

Chánda.—This is one of the most remarkable districts in India, containing "enormous and splendid accumulations of ore" which occur principally in the metamorphic rocks. Hæmatite is the most abundant, and furnishes all the supplies for the native furnaces; but magnetite and limonite also occur. The most noted localities are: Lohára, where hæmatite of extreme purity forms an entire hill ⅝ths of a mile long, 200 yards broad, and 120 feet high; Pipalgaon, near which there is an enormous amount of similar ore; Dewalgaon, in the vicinity of which there is a hill 250 feet high, composed in large part of hæmatite; Ratnapur, where a very rich lode of brown iron ore, 40 and 50 feet thick in places, has been found; and Dissi, the lode near which is of hæmatite and magnetite. Immense deposits also exist at Gunjwáhi, Ogalpet, Metapur, Bhánapur, and Mendki. The unsuccessful attempt to smelt the Chánda ores with the coal of the same district has been already noticed.

Bálaghát.—The iron industry is very near extinction, although a few furnaces still struggle for existence in the Baila, Kini, and Bhánpur zamindáris. Ore from the lateritic rocks is that chiefly used.

Bhandára.—The ore commonly smelted in this district is lateritic, and the metal produced from the mines at Agri and Ambághari in the pargana of Chandpur is said to be of excellent quality. About the middle of the century 160 furnaces were at work.

Nágpur.—Ore of good quality is reported to exist near Mansar.

Mandla.—Lateritic ore is used, the best iron being, it is said, produced from that obtained near Ramgarh and Mowai in the Raigarh Bichhia tract.

Seoni.—Iron is stated to be produced in Juni and Katangi.

Chhindwára.—Lateritic ore is known to exist.

Nimár.—Some ores of poor quality have been noticed to the south-east of Barwai and near Punása.

Hoshangabád.—In the area occupied by transition rocks to the north-west of Harda and south of the Narbada, hæmatite has been worked in several places.

Narsinghpur.—Tendukhera has long attracted notice owing to the excellent quality of the iron manufactured there, which is said to command a higher price than any other iron made in the Narbada valley. In 1830 a suspension bridge over the Bias river, in Sagar, was opened, the iron for which had all been smelted in native furnaces at Tendukhera. The ore is an earthy manganiferous hæmatite and limonite, and the good quality of the iron has been attributed to the ore being somewhat calcareous (the gangue being partly limestone), which produces the same effect as would a purposely added flux. The quality may also be due, in part, to the manganiferous character of the ore. It appears that part of the iron that is produced is converted into steel, by a method different from that practised in Madras.

Jabalpur.—Iron exists in immense quantities in this district, both in the transition rocks and in the laterite. The ore in the older series is mainly hæmatite, in part manganiferous, while that in the lateritic formation is limonite. The transition ores are worked in several places, and iron is still smelted on a somewhat considerable scale in the district. Gangai is perhaps the only spot where magnetite is raised. At Agaria and Partabpur there are whole hills of schistose and micaceous hæmatite, containing only traces of phosphorus and sulphur. The micaceous variety is mined in preference to the other from being softer. The Janli mine, however, is the most extensively worked in the district, the ore, being an ochreous hæmatite, is easily extracted, and probably has been found by experience to be easy of reduction. "Olpherts' metallic paint" is made from the Janli ore by grinding it to an impalpable powder.

IRON ORES. Central Provinces.

At Mangeli, Mogála, Gogra, and Danwai in the Lora ridge, a manganiferous hæmatite is worked, the iron produced therefrom being a hard steely kind used for edged tools. The iron smelted from the Agaria, Partabpur, and Janli ore is much softer. Manganiferous hæmatite also occurs in large quantity at Gosalpur, where limonite is also found.

The laterite ores occur in beds near the base of the formation, and consist of rich pisolitic limonite. They are obtainable over a wide area to the east, north, and south-west of Marwára, the Kanhwára hills being especially rich in such. The laterite ore used to be smelted on a considerable scale, but of late years the mines have been abandoned.

The opening out of the Umaria Colliery has recently directed attention to the advantages north-eastern Jubalpur possesses for the manufacture of iron on European principles. Magnificent ores, pure limestone, and coal are all comprised within a comparatively small area.

Central India.

Central India. 455

Rewah.—The broad band of transition rocks which traverses this State parallel to, and south of, the river Son, is known to contain iron ore, which is also believed to exist in the coal-measure strata.

Bandelkand.—The transition rocks, which occupy a large area in the State of Bijawar, in some places contain abundance of rich hæmatite. It is worked on a considerable scale, Herapur being one of the chief centres of production.

Gwalior.—Iron ores occur most profusely in the transition rocks which extend over a wide tract to the southward of Gwalior Fort. The mines at Santan, Maesora, Gokalpur, Dharoli, and Banwari are described as extremely rich, the ore being schistose and more or less ochreous hæmatite, which is manganiferous at Santan. Hæmatite is also worked at Raypur, Pár hill, and Mangor, while magnetite is obtained at Gokalpur and Girwai. All the above mines are within 10 miles of Gwalior, except those at Pár hill and Mangor which are at a somewhat greater distance. The ores are said to contain only traces of sulphur, and no phosphorus. Limonite occurs at Binaori, Baroda, Imilia, Gunjari, and Baron, villages between 45 and 80 miles from Gwalior. But the ore is inferior to the anhydrous kinds.

About 50 miles to the north-west of Gwalior there is a forest, which it is estimated would, without replanting, supply fuel for an outturn of 12 tons of bar iron a day for a period of 900 years.

The mines of Bágh (about 60 miles west-south-west of Indore) have long been celebrated. They are in the transition rocks, the ore being hæmatite.

Indore, Dhár, and *Chándgarh.*—The same rocks are exposed near Barwai, and cover large areas in the *Dhár* forest and north of Chándgarh. Rich deposits of hæmatite occur in connection with them, some of which are found along fault lines, while others are surface accumulations derived from the older rocks.

Ali Rajpur.—Some ore has been observed in the metamorphic rocks.

Bombay.

Bombay. 456

North Kanara.—Ore (apparently lateritic) occurs in the Sahyádri range.

Dhárwar.—Formerly, when fuel was plentiful, much iron was smelted in the Kappatgudd hills.

Kaladgi.—The hæmatite, occurring in the *Dhárwar* series, has already been noticed in connection with the Bellary district.

Belgaum.—Iron was formerly smelted in several places, but the manufacture is now extinct.

IRON ORES. Bombay.

Goa.—Iron ore is found at Bága, Satára, Pernem, and especially in the District of Zambaulin.

Sawantwári and Kholapur.—Iron is (or used to be) produced from (apparently lateritic) ore.

Ratnagiri.—The metamorphic rocks are said to contain magnetite or hæmatite near Málwán. Lateritic ore is also found in the district, and formerly was largely smelted.

Sátára.—Lateritic ore has been utilised here also, but, apparently, on only a small scale.

Surat.—Iron ore is said to be found in the Balsár and Párdi Subdivisions.

Rewa Kántha.—Iron seems to have been once worked on an extensive scale along the western limits of the district. Near Jámbughoda, Limodra, and Ladkesar large heaps of slag still remain. Near Mohan, in Chhota Odaipur, a bed of hæmatite, of limited extent, occurs in the Nimar sandstone.

Panch Mahále.—At Pálanpur, in Godhra, ore of considerable richness is found.

Kaira and Ahmadabad.—Heaps of slag, in certain places, show that iron was formerly produced.

Káthiawár.—Even fifty years ago the manufacture of iron was in a moribund condition, and to-day there is not a single furnace in blast. In the west, the laterite yielded very rich ore, while the ironstone bands, near the top of the Omia (jurassic) beds, were utilised in the northern part of the peninsula.

Cutch.—Here also the industry is extinct. Much of the ore formerly used was derived from the hæmatitic laterite of the sub-nummulitic group.

Rajputana.
457

RAJPUTANA.

Sind.—Almost the only ore hitherto discovered, in sufficient quantity for iron-making, is in the passage beds between the Kirthar and Ranikot groups, north-west of Kotri. Masses of magnetite, and bands of red and brown hæmatite, more or less pure, occur, however, in considerable quantity, in many places. The scarcity of fuel in Sind renders the rarity of iron ore of little importance. Large deposits of iron ore occur, in several places, in the Arvali (transition) rocks. Some of these are situated in ***Marwar, Ajmere, Bundi, Kota,*** and ***Bhartpur.***

Meywár.—A promising bed of limonite exists near Gangar, which is worked to some extent.

Jaipur.—Large quantities of ore have been raised at Karwar, near Hindaun, but the workings are now abandoned.

Alwar.—The mines at Bhángarh still produce large quantities of ore, and are now the only source of supply for the numerous furnaces in the Alwar territory. The ore is a mixture of limonite, magnetite, and manganese oxide. The old and extensive mines near Rájgarh are no longer worked, but the deposit extends in a regular belt for a distance of over 1½ miles, and has an average width (it is said) of 500 feet; the ore consisting chiefly of rich hæmatite and limonite.

Panjab.
458

PANJÁB.

Bannu.—The hills 25 or 30 miles south-east of Bannu are reported to contain hæmatite in abundance. The iron produced from it is in great demand at Kálabágh.

Peshávar.—It is said that iron is smelted in Pesháwar, and other neighbouring places, from ore (magnetic ironsand?) obtained from Bajaur.

Jhelam.—Hæmatite is abundant in the Kot Keráana Hills.

IRON ORES. Panjab.

Kángra.—Magnetite, which occurs in the metamorphic strata, "is available in any quantity," and the iron produced from it is of great strength and tenacity. Magnetic ironsand, derived from the above, and hæmatite, are also found.

Mandi.—There are a few villages where smelting is carried on from the same ore as in Kángra.

Simla Hill States.—Some iron is produced in Bashahr, Jabbal, Dhámi, and perhaps in other States, the ore commonly used being magnetic ironsand. There are mines of magnetic ore in Kot Khai, and at Chaita, in Sirmur. The last named supply the ore for the Rájá's smelting works at Náhan.

Gurgaon.—Firozpur, in the extreme south, once possessed considerable smelting works, the ore used being hæmatite. It would appear that some iron is still produced there.

KASHMÍR.

Kashmir. 459

Limonite has been extensively mined on the Punch river, in the outer hills. Iron is also worked in the neighbourhood of the Dragar mountain to the north of Pansir, in the Riási district, at the village of Soap, or Sufahán, situated on the Bimwár river, at the south-eastern end of the Kashmír valley, at Arwan, near Sopur, and Shar, near Pámpur. In Ladákh, at the village of Wánla, nearly due south of Khalsi on the Indus there are very extensive iron works.

NORTH-WEST PROVINCES.

N.-W. Provinces. 460

Kumaun.—In several parts of Kumaun deposits of iron ore occur some of which are rich and abundant. Amongst the localities which have attracted most notice is Rámgarh, near which several mines (Pahli, Loshgiáni, Natna Khán, Parwára, &c.) have been worked in micaceous hæmatite and limonite. Hæmatite and micaceous iron also exist near Khairna, but the deposits are not of much importance. Very large quantities of ore, which is earthy hæmatite and limonite, occur in beds in the Siválik strata, in the neighbourhood of Káladhungi and Dechauri. The Dechauri ore is of better average quality than the other.

Lalitpur.—At Salda, in the Maraura pargana, a considerable amount of iron is smelted from hæmatite, while a hard steely kind is made at Pura.

Bánda.—In pargana Kalyángarh iron is rather extensively worked at several points, especially at Gobarhái. There are also mines at Deori and Khiráni.

Mirzapur.—The metamorphic rocks in the southern part of the district contain some small bands of magnetite, which have been worked on a trivial scale.

BENGAL.

Bengal. 461

Orissa.—Iron of excellent quality has been produced in Lálcher, the ores used being chiefly derived from the sandstones of the coal-measure and upper groups. Smelting is (or used to be) also carried on in Dhenkánal, and probably other tributary states as well as in Katták.

Burdwán.—In the Ranigánj coal-field throughout the ironstone shale group, but more especially in the upper part of the group, bands and nodules of clay ironstone are common, and constitute an inexhaustible supply of ore. When unaltered, the ore is either clay-ironstone or black band (a carbonaceous variety of the same), but near the surface it has been converted into earthy hæmatite or limonite, or a mixture of both. It is this altered form that is used at the Barákar Iron-works, which have been briefly noticed in the preceding part of this article.

IRON ORES. Bengal.

Birbhum.—Iron used to be made on rather a large scale, the *kacha*, or crude, metal unlike that produced in other parts of India, resembling pig-iron, from which *pakka*, or refined, iron was produced by a sort of puddling process. The ore used in the Birbhum Iron-works Company's furnaces was obtained from beds near the base of the laterite.

Bhágalpur and *Monghyr.*—Some smelting has been carried on in both districts.

Gyá.—A rather large deposit of magnetite has been found in the Barákar hills, which has not been worked, although iron has been made from other ores.

Hazáribágh.—Furnaces are numerous in the vicinity of the Bokárs and Karanpura coal-fields, where ore from the ironstone shale group is used. Magnetite from the metamorphic rocks is also smelted in places.

Mánbhum.—A portion of the Ràniganj coal-field, noticed under Bardwán, is situated in Mánbhum. Magnetite exists in both the metamorphic and transition rocks, but not in very great quantity. Ilmenite occurs also, and the excellent quality of some of the native-made iron has been attributed to the presence of this mineral in the magnetic iron-sand used in the furnaces. In places along the line of disturbed junction between the rocks just mentioned, there are veins or lodes of red and brown hæmatite, the latter sometimes in great abundance. Other deposits, again, are in close relationship to the Dalma trap.

Singhbhum.—As in Mánbhum, nests of very pure hæmatite occur in the Dalma trap dyke. The ores smelted are chiefly from ferruginous schists, and from the laterite. The most promising ores, however, are to be found in a number of lodes or veins in the transition rocks near, and west of, Chaibasa. Some of these are manganiferous.

Lohárdaga.—There is a remarkable abundance of ores in the Palamau subdivision, which includes magnetite in the metamorphic rocks; black-band, limonite and hæmatite in a well-defined zone of ferruginous shales in the Barákar group of the Aurunga coal-field, and to the north of Balunagar; and, thirdly, red and brown hæmatite in the laterite. The magnetic ores are not very important, and the lateritic ores, although plentiful and of excellent quality, are found on the tops of lofty plateaux, where they are practically almost inaccessible. The Barákar ores, however, are both favourably placed and abundant, and excellent crystalline limestone occurs in the immediate vicinity of the coal-field. There is some doubt about the quality of the coal, but should railway communication with the district be established the position will probably attract attention as a site for iron-works.

Tributary States.—In the Tributary States of Chutia Nagpur iron smelting is carried on to some extent.

Dárjiling.—A valuable bed of extremely pure magnetite, with some micaceous hæmatite, occurs in the metamorphic rocks near Sikbhar, about 5 miles south-east of Kalingpung. The magnetite has been smelted on a trifling scale, and is said to yield a steely iron suitable for making *kukris* and *bans.* There is a strong ferruginous band in the tertiary sandstones of Lohárgarh hill, from which an immense supply of ore might be obtained; but it is of very indifferent quality.

Assam. 462

ASSAM.

Khási-Jaintia Hills.—Iron used to be made in the Khási-Jaintia hills, but the manufacture has completely died out. The ore employed was a titaniferous magnetite, occurring in the form of minute grains disseminated through decomposed granite. The soft friable rock was raked into

IRON ORES.

Assam.

a narrow channel with a rapid current of water running down it, and the heavy ironsand caught by a small dam, while the lighter particles were washed away. After further purification by re-washing, the ore was smelted, and yielded an iron which, after refining more than once by re-heating and hammering, was of excellent quality.

Nága Hills.—In former times nodules of clay-ironstone, occurring in the tertiary coal-measures of the Nága hills, were smelted for iron, and during the most flourishing period the outturn must have been very considerable. As long ago as 1841 the manufacture was all but extinct, and has, for many years, been wholly so. In some localities impure limonite is very plentiful in the Siválik strata, but the ore is of poor quality.

Manipur.—In more than one locality in the valley of Manipur, a layer in the alluvium, occurring a few feet below the surface, is utilised as a source of iron. The clay, which contains small pisolitic nodules of limonite, is washed to separate the latter, which are afterwards pounded and smelted. Titaniferous iron ore is said to be obtained in some of the streams. The peculiar furnace used in Manipur has been alluded to previously.

BURMA.

BURMA. 463

Upper Burma.—A very large proportion of the iron used is smelted in the country around Puppádoung, an extinct volcano about 30 miles south-east of Pagán, ferruginous concretions, occurring in the tertiary strata, are employed as ore, and the iron produced is of excellent quality. The furnaces are peculiar in being worked by a natural draught only, no bellows being used. Iron is also smelted at Maldu, a good distance north of Shue-bo-myo. To the west of Sagain, for miles of the Irawadi, rich hæmatite is said to abound.

Iron ore is reported to be very plntiful in the Shan States, and within the last year, the occurrence of an enormous deposit of hæmatite near Sengaung in the vicinity of Myit-nge, in the Shan hills, has been brought to notice.

Pegu.—Limonitic concretions in the newer tertiary strata in eastern Prome furnish excellent ore, which was smelted on a considerable scale prior to the British occupation of the province.

Tenasserim.—East of the Sittaung the ore usually met with is magnetite, which often occurs in thick beds or lodes; but specular iron is also found. The lower ranges of hills to the east of Shue-gyin station are said to contain abundance of ore.

Between Maulmain and Tavoy seventeen localities have been noted where iron ore occurs in the hills of tertiary strata. The best locality, however, with respect to quantity, quality, and position, is one near Tavoy. On the right bank of the river, opposite the town, runs a range of low hills at a distance from the river varying from $1\frac{1}{2}$ to 3 miles, and extending to a distance of 5 or 6. These are said to be formed almost entirely of magnetite, the supply of ore being inexhaustible.

Some 10 miles south-west of the town of Mergui, there is an island, comprising a hill about 200 feet high which is said to be formed of iron ore. A similar island is reported to exist 4 miles to the southward of the first mentioned. Magnetite and hæmatite have been noticed on an islet between King and Tawlang islands. Limonite is stated to occur in some abundance at Podan-or.

ANDAMAN ISLANDS.

Andaman Islands. 464

Within the last few years hæmatite has been discovered at Rang-u-Cháng, a place some miles south of Port Blair; but it is not very abundant, and is mixed with too much quartz and pyrites to be of any value.

PROSPECTS of the IRON INDUSTRY. 465

GENERAL SUMMARY OF PROSPECTS OF IRON MANUFACTURE.

The following summary by **Sir G. Molesworth**, written in 1882, may be here quoted in entirety, giving, as it does, the unbiassed opinions of a careful and competent expert:—

"It appears to me to be doubtful whether iron manufactured from charcoal will be produced at rates sufficiently low to compete with English manufacture; but if practicable, it must be combined with a very perfect and well-organised system of forest conservancy in a *district capable of producing timber of a quality eminently suited* for the manufacture of charcoal fit for smelting operations, at a low price; and the timber must be in a position near the iron works, as charcoal is bulky and consequently expensive in carriage, and the expense of carrying ore and flux to any great distance would be very great. Abstracting the information which I have given above as bearing on the economic manufacture of iron, I arrive at the following conclusions:—

In the Panjáb and Sind the scarcity of fuel is an insuperable obstacle. In the North-West, after a patient trial, iron manufacture has not been successful, and **Mr. Bauerman** reports against the probability of success.

In the Central Provinces, Jubbulpore and Chanda afford very fair promise of success.

In Central India the Gwalior iron-field may possibly be successful; but the absence of coal and the distance of the forests militate against success. At Berwai the failure has been without a fair trial, but the late enhancement of the price of fuel renders its success doubtful.

At Rewah there appear to exist elements for success; but works in this district would naturally be combined with the Jubbulpore works.

In Rájputana and Bombay Presidency the scarcity of fuel and competition with England appear to preclude the establishment of iron-works on a large scale.

In Bengal the Bengal Iron Works appear to hold out promise of success, but other localities in that district, so far as they have been examined, do not appear to hold out any such promise.

In Orissa the coal is too poor to hope for success.

In Assam the ore where abundant is not sufficiently rich, and in the coal measures where it is rich, it is not in sufficient abundance.

In the Nizam's Dominions the ore and fuel are not in juxta-position so far as is known at present.

In Madras the scarcity of fuel prevents the utilisation of most valuable ore.

In Burma there appear to be elements for the production of iron, but the market will doubtless be limited.

Taking all these circumstances into consideration, we can only, in the present state of our knowledge, assume three centres as the *probable* fields for iron industry on a large scale by the agency of English capital, *viz.*:—

1st—The Bengal Iron Works,
2nd—The Jubbulpore District,
3rd—The Chanda coal and iron fields."

IRON WORKS. 466

IRON WORKS AND FOUNDRIES.—**Mr. O'Conor**, in his Statistical Tables for British India, 1888-89, gives a list of forty-seven iron and brass foundries, of which eight are in Bombay, three in Sind, five in Madras, seventeen in Bengal, one in Lower Burma, two in the North-West Provinces and Oudh, and eleven in the Panjáb. Of these the most important are the Barrakar Iron Works, and Burn and Co.'s Iron Works in Bengal, the Byculla Iron Works, Bombay, and the Iron Foundry and Ship-building Yard, Rangoon. The total number of persons employed is stated to be between 12,000 and 13,000.

IRON ORES. Uses. 467

USES.—It is needless to enter into a consideration of the multitudinous methods of application of this invaluable metal which do not in any noteworthy manner differ in India from those followed in other countries. The methods of utilisation of the chief ores and salts will be found detailed below.

TRADE.

TRADE. Internal. 468

The trade in iron and steel may be considered under two headings—I, Internal, II, External.

I. INTERNAL.—The total rail and river transactions in iron during the past year amounted to 49,37,135 maunds, valued at R2,22,60,226. The largest exporters were Bombay with 17,65,925 maunds; Calcutta with 13,57,136 maunds; Madras seaports with 5,67,774 maunds; and Karachi with 5,02,873 maunds; Bengal with 1,64,779 maunds; and Bombay Presidency with 1,47,551 maunds. The largest importing centres were Bombay Presidency with 10,42,159 maunds; the Panjáb with 7,08,147 maunds; Bengal with 6,90,450 maunds; the North-West Provinces and Oudh with 6,16,118 maunds; Madras with 4,47,911 maunds; Rájputana and Central India with 2,61,185 maunds; and the Central Provinces with 2,38,117 maunds.

The coasting trade returns for the same year differentiate Indian and Foreign iron. The imports of the former amounted to 16,400 cwt., valued at R99,561. The largest importer was Bombay with 8,701 cwt., value R44,718; followed by Madras with 3,744 cwt., Bengal with 2,143 cwt., Burma with 921 cwt., and Sind with 891 cwt. Of the Bombay imports, 6,553 cwt. was received from British Ports within the Presidency, 1,115 cwt. from Goa, and about 1,000 cwt. from Kathiawar, Kach, and the Gaekwar's Territory. Madras received its supply almost entirely from Bengal and Bombay; Bengal was supplied altogether by other British Ports within the Presidency; Sind received nearly the whole of its supply from Bombay, and Burma from Bengal. The exports coastwise for the same year were returned as 32,634 cwt., or nearly double the imports, from which it may be assumed that the period of most active trade nearly corresponds to the time of returning the trade figures. Of these, Bombay is shewn as having exported 30,915 cwt., 11,879 of which was destined for Sind, and must have been on its way at the time of compilation of the import figures.

The coastwise trade in foreign iron and steel during the same year was much larger. The imports of iron amounted to 262,455 cwt., value R15,46,255. The largest item of this amount was the import into Madras 94,270 cwt. Bombay followed with 74,414 cwt., then Sind with 39,855, Burma with 31,641, and Bengal with 22,275 cwt. Of the Madras imports 60,128 cwt. was derived from Bombay, 21,729 cwt. from British Ports within the Presidency, and smaller quantities from Bengal and Burma. Nearly the whole of the Bombay imports were derived from other British Ports within the Presidency. Those of Sind with the exception of a few cwt. came from Bombay, those of Burma from Bengal, Bombay, and other British Ports within the Presidency, and those of Bengal from Madras, Bombay, Sind, Burma, and British Ports within the Presidency.

The export table for the same year again, probably for the same reason, shews a considerably larger trade, amounting to 332,094 cwt., valued at R19,20,168. Of this a very large proportion, *viz.*, 244,114 cwt. was shipped from Bombay, chiefly to British Ports within the Presidency, Madras, Kathiawar, and Sind.

The coasting trade in steel is unimportant, the imports in 1888-89

IRON. **Trade in Iron.**

TRADE. Internal.

amounting to only 13,050 cwt., valued at R1,05,053. The share of each of the Presidencies and Seaboard Provinces in this trade was as follows:—Bombay, 4,419 cwt., chiefly derived from British Ports within the Presidency; Madras, 4,439 cwt., obtained nearly entirely from Bombay; Sind, 3,374 cwt. from Bombay; Burma, 661 cwt. from British Ports within the Province, Bengal, and Bombay; and Bengal with 157 cwt., most of which came from other ports within the Province.

The exports are returned as 19,839 cwt., value R1,57,524, of which 17,563 cwt., value R1,37,761, was shipped from Bombay.

The trans-frontier trade in the metal shews a marked increase during the past three years, as will be seen from the following table of the exports from British India during that period:—

	Cwt.	R
1886-87	94,626	11,17,951
1887-88	112,379	12,95,822
1888-89	154,043	18,75,306

The chief importers, in the year under consideration, were Burma with 71,741 cwt., the country supplied by the Sind-Pishin Railway with 54,044 cwt., Nepál with 12,769 cwt., Kashmír with 8,322, and Kábul with 5,726 cwt.

The trans-frontier imports into British India are small, unimportant, and shew no tendency to increase. They amounted in 1888-89 to 2,818 cwt., of which 1,590 came from Nepál, 677 from Upper Burma, 509 by the Sind-Pishin Railway, and a few cwt. from Bajaur, Thibet, and Bhútan.

External. 469

2. EXTERNAL.—*Imports.*—The import trade in iron and steel, manufactured and unmanufactured, is one that has naturally undergone a rapid and steady increase with the development of spinning and weaving mills, and of railways in India. It is impossible, from the statistics available, to give an exact representation of the actual amount imported, since, under the headings "Railway Plant and Rolling Stock," "Machines and Machinery," and "Hardware and Cutlery," other metals and manufactures of wood are included. But figures shewing the development of these, as well as of the imports of the metal alone, may be given, since they constitute a large proportion of the iron and steel imported into the country.

The increase in the import trade may be best shewn by the following tabular statements in which the quinquennial averages for the years from 1871-72 onwards are given:—

Quinquennial Averages of the Imports of Iron, Steel, and their Manufactures, excluding Government Stores.

	Cwt.	R
IRON—		
1871-72 to 1875-76*	...	1,01,23,056
1876-77 to 1880-81	2,366,196	1,43,73,815
1881-82 to 1885-86	3,248,719	1,87,50,330
1886-87	3,280,389	1,78,29,940
1887-88	4,321,587	2,44,73,951
1888-89	4,002,793	2,51,51,791
STEEL—	Cwt.	R
1871-72 to 1875-76	36,480	8,14,854
1876-77 to 1880-81	89,311	8,35,869
1881-82 to 1885-86	241,966	16,74,036
1886-87	349,317	20,98,619
1887-88	420,927	25,88,107
1888-89	542,388	34,45,892

* Years 1871-72 to 1874-75 include cutlery and hardware, thereafter excluded.

TRADE.
External.

HARDWARE AND CUTTLERY—	R
1875-76	47,53,383
1876-77 to 1880-81	45,96,710
1881-82 to 1885-86	77,06,770
1886-87	86,53,973
1887-88	1,09,39,396
1888-89	1,10,22,046
MACHINERY AND MILLWORK—	**R**
1850-51 to 1854-55	4,81,100
1855-56 to 1859-60	52,08,990
1860-61 to 1864-65	61,40,630
1865-66 to 1869-70	72,32,230
1871-72 to 1875-76	90,06,214
1876-77 to 1880-81	78,59,004
1881-82 to 1885-86	1,36,56,374
1886-87	1,37,14,591
1887-88	1,80,02,178
1888-89	2,31,68,714
RAILWAY PLANT AND ROLLING STOCK—	**R**
1871-72 to 1875-76	48,44,664
1876-77 to 1880-81	95,93,112
1881-82 to 1885-86	1,48,79,824
1886-87	1,43,51,244
1887-88	2,57,76,029
1888-89	2,49,32,389

To the above figures must be added the raw and manufactured metal imported by Government, which represents a considerable sum and has largely increased during the same period. In the year under consideration these imports were as follows:—

	Cwt.	R
Iron	288,611	18,30,488
Steel	20,464	1,53,448
Railway plant, &c.	...	74,91,809
Machinery, &c.	...	5,61,664
Hardware and Cutlery	...	7,72,267

A review of the above figures reveals the fact that during the past eighteen years the imports of iron have doubled, both in quantity and value, while those of steel have increased more than fifteen times in quantity, though only four times in value. The imports of Hardware and Cutlery have increased in value more than twice in the same period, while those of Railway plant and rolling stock have increased five times. The development of the imports of machinery from about 5 lakhs in 1850-51 to nearly 2⅓ crores in 1888-89 is a startling testimony to the enormous advance made by steam power in India during a period of less than forty years. Though the above facts point to increased mercantile and industrial activity and prosperity, they at the same time accentuate the fact that no practical development in iron-making has as yet been made in India. As already shown, in the Barakar works and elsewhere, an important industry might be created, but up to this time the production of Indian metal is limited to trifling quantities of cast iron,—and that, notwithstanding the fact the Government and the public require and import close on a quarter of a million tons of iron and steel yearly, without including the enormous amount imported as machinery, railway plant, &c. In connection with this subject it may be noted that the increase in cost of all machinery, railway plant, and other manufactured iron from freight is very large. This is well shewn in a series of tables recently compiled by Mr. E. H. Stone, Assistant Secretary, Railway Branch, Public Works Department, shewing the cost of the material for one mile of single track, of the several

TRADE. External.

forms of permanent way in India. One of these may be taken as an example. The cost, free on board at the port in England, of the material required for one mile of single track of a permanent way 5 feet 6 inches gauge, with flat-footed steel rails 80℔ per yard, on cast iron bowl sleepers, is £2,020-8-11. The charges for freight and insurance on this amount to £318-1-4, and the landing charges in India to R655, making the total cost delivered in India, converted to Indian currency with exchange 1*s.* 6*d.* to the rupee, R31,835. Further, the cost of carriage of the material by railway for every 100 miles is shewn to vary from R900 to R1,858 according to the railway on which carried. These figures plainly shew the possibilities of large profit, supposing that as good rails as those manufactured in England could be made in India at anything like the same price. On this subject **Mr. Stone** has kindly furnished the following note:—

"Assuming that the cost of production would be the same, the inducement to the capitalist would be that he could sell rails at the same profit as his English competitor, plus an enormous extra profit represented by the cost of carrying the English-made material to India. In other words, if the quality were in both cases the same, Railways would buy of the Indian firm if they could get rails delivered at a lower rate than they could get them delivered by the English firm. Hence the Indian firm could charge anything they liked, provided their charges were something less than English charges plus carriage to India. In the English charges there is of course a good profit to begin with,—the profit to be made by saving cost of carriage is extra profit." From these notes it would appear that, with good and experienced management and a possibility of obtaining good fuel, the iron industry in India ought to be capable of enormous development.

Analysis of Imports. 470

ANALYSIS OF IMPORTS.—The total imports of iron during the year under consideration were made up as follows:—

DESCRIPTION.	Cwt.	R
Old iron for re-manufacture	7,729	16,573
Cast (pig)	138,542	3,47,742
Wrought—		
Bar	1,235,814	63,43,693
Angle, bolt and rod	442,160	21,67,543
Sheets and plates (including tinned plates)	486,655	30,74,635
Galvanised (other than wire)	360,603	36,46,909
Hoop	111,589	6,45,030
Anchors, cables, and kentledge	25,957	2,89,594
Nails, screws, and rivets	144,064	14,44,944
Wire	11,616	1,20,692
Beams, pillars, and bridgework	455,768	33,40,271
Pipes and tubes	444,760	23,04,691
Rice bowls	61,848	4,99,911
Other manufactures of wrought or cast Iron	76,108	9,09,583

Out of the total, 3,715,081 cwt. came from the United Kingdom, 262,436 cwt. from Belgium, and smaller quantities from Germany, Ceylon, Sweden, the Straits, and Austria, named in order of importance.

The old iron was obtained almost entirely from Ceylon, the cast iron altogether from the United Kingdom, the other descriptions almost entirely from the United Kingdom and Belgium.

Of the total Bombay imported 1,630,152 cwt.; Bengal, 1,472,934 cwt.; Madras 444,088 cwt.; Burma, 270,827 cwt.; and Sind, 184,792 cwt. Per-

TRADE.
External.

haps the greatest increase has taken place in galvanised iron, and piping and tubing. The former is now largely utilised for roofing sheds and workshops in the plains, and residences in the hills. The increase in the latter has been brought about by the constant progress of waterworks and drainage schemes in municipal areas.

Of the total imports of steel, 137,108 cwt. consisted of "hoop" which is now largely replacing iron hoops for all purposes, markedly so for binding bales of cotton, &c. Of the remainder 32,201 cwt. consisted of "cast steel," and 373,070 was returned under the miscellaneous heading of "other sorts." Here, again, as in iron, the United Kingdom was the chief source of supply, contributing 456,550 cwt. out of the total. Belgium came next with 77,976 cwt., followed by Germany with 7,063, and the United States, France, and other countries with small and unimportant quantities.

The chief importer of steel was Sind, with 212,595, followed by Bombay with 165,476 cwt., Bengal with 136,160 cwt., Madras with 22,065, and Burma with 6,092 cwt.

The hardware and cutlery came almost entirely from the United Kingdom, which furnished an amount valued at R99,23,931, out of the total of 110 lakhs. Austria, Belgium, Germany, and the Straits Settlements contributed to the value of R3,26,873, R2,45,318, R1,55,408, and R1,12,404, respectively. Bombay imported to the value of R52,97,815; Bengal, R31,59,111; Burma, R12,05,248; Madras, R9,03,773; and Sind, R4,56,099.

The machinery and millwork came almost entirely from the United Kingdom. Bombay imported R1,12,05,163 of steam machinery and R81,222 of other sorts; Bengal, R10,89,867 of the former and R75,41,341 of other sorts (coffee, tea, sugar, agricultural, &c., &c., machinery); Madras, Burma, and Sind smaller amounts.

Practically the whole of the railway plant was supplied by the United Kingdom. Bombay imported the greatest proportion, *viz.*, locomotive engines and tenders, to the value of R38,60,878; carriages and trucks to R60,23,427; materials for construction, &c., to R97,08,678. Bengal imported locomotives, &c., value R40,244; carriages and trucks, value R,264,671; and materials for construction, &c., value R28,75,378. The imports into Madras, Sind, and Burma under this heading consisted almost entirely of materials for construction, and were unimportant.

The re-export trade is inconsiderable. In the year under consideration the re-exports were as follows:—

Iron, 306,041 cwt., value R8,19,505; steel, 2,753 cwt., value R28,398; hardware and cutlery to the value of R4,48,799; machinery and millwork to the value of R1,06,161.

The largest importer of re-exported iron was the United States with 102,660 cwt., followed by Italy with 80,560, the United Kingdom with 40,138, China with 36,654, Persia with 14,601, and other countries with small amounts; 257,566 cwt. was exported from Bombay; 45,437 from Bengal; and small quantities from Sind, Madras, and Burma.

Of the re-exported steel, 1,038 cwt. went to Arabia, 751 cwt. to Aden, 431 cwt., to Persia, 222 cwt. to Turkey in Asia, 188 cwt. to other countries (unenumerated), and 101 cwt. to the United Kingdom. Nearly the whole, *viz.*, 2,719 cwt., was shipped from Bombay.

The re-exports of machinery went almost entirely to the United Kingdom and Australia, but small quantities were also despatched to the Straits Settlements, Ceylon, Arabia, and other countries. Of this, Bombay exported to the value of R57,311, Bengal, R31,658, Burma, Madras, and Sind small quantities.

The comparatively large re-export of hardware and cutlery took place

IRON Oxides.

Trade in Iron.

TRADE. External. 471 almost entirely from Bombay to the United Kingdom, the East Coast of Africa, the Straits Settlements, Persia, and Turkey in Asia.

EXPORTS.—The exports of Indian iron and ironwork are naturally, from the very backward state of the iron industry in the country, very small and unimportant.

In the past five years they have been as follows:—

	Iron.		Hardware and Cutlery.	Machinery and Millwork.
	Cwt.	R	R	R
1884-85	5,712	50,440	42,707	2,107
1885-86	6,480	59,213	70,877	1,443
1886-87	7,329	61,901	65,955	3,137
1887-88	5,375	47,818	67,086	878
1888-89	6,406	62,170	93,609	1,757

There is no export of steel.

The above figures have been enumerated, not so much for the value of the trade, which, indeed, is almost unworthy of anything but a passing mention, as for the sake of comparison with the large figures of the import trade. It is unnecessary to go into a detailed account of the ports exporting and receiving these small quantities, but it may be mentioned that Bombay is the largest exporter, and that the trade is almost entirely with small Eastern ports. In the case of hardware and cutlery, however, the United Kingdom in 1888-89 imported goods to the value of R36,014.

472 **Iron Oxides**, *Ball, Geology of India, III., 416.*

Vern.—Rust or impure red oxide of iron=*Lóhéka-zang, lóhé-ka-gú, mandór*, HIND.; *Lohár-gú, lohár-jhangár*, BENG.; *Lokhan-dhácha-katai*, MAR.; *Lohánu-zang*, GUZ.; *Ayach-chenduram, irumbu-chittam*, TAM.; *Inapa-chittam, aya-shindúramu*, TEL.; *Khabbanada-kittá*, KAN.; *Irumbuk-kítam*, MALAY.; *Sánpiyá, támbiyá, sánkhí, tánkhí*, BURM.; *Yakkada-kittam, mallokodá*, SING.; *Mandúram*, SANS.; *Khabsul-hadíd, záfará-nul-hadíd, zanjárul-hadíd*, ARAB.; *Zange-áhan, chirke-áhan, rime-áhan, zangáre-áhan*, PERS.

Red ochre=*géru, hirmji*, HIND.

Yellow ochre=*Rámraj, haldimáti*, HIND.

Magnetic oxide of Iron, or loadstone=*Chamak-ká-patthar, chamak*, HIND.; *Miqnátis, mighnátis, hajrulmighnátis*, ARAB.; *Sange-áhanrubá, sangí-chamak*, PERS.

References.—*Pharm. Ind., 366, 367; Ainslie, Mat. Ind., 522; Moodeen Sheriff, Supp. Pharm. Ind., 140, 141; U. C. Dutt, Mat. Med. Hind., 46-54; McCann, Dyes and Tans, Beng., 32, 33, 68, 91; Buck, Dyes and Tans, N.-W. P., 42; Balfour, Cyclop. II., 375.*

LOCALITIES. 473 **Occurrences.**—Iron ochre, consisting of the earthy varieties of the hæmatites, occurs abundantly in India, as a necessary consequence of the wide distribution of iron ores and laterite. The following localities, however, are mentioned by Ball as the sources from which it is principally obtained:—Madras,—Trivandpuram in Trichinopoly, various localities in the Nilghiris: Bengal,—the Rájmahal and Kharakpur hills: Central Provinces,—Madnapur and Thakurtola in Raipur, the Salitikri hills in Balaghat, Jauli in Jubbulpore, and generally in the Chanda District: Bombay,—in several parts of Cutch: Sikkim: Burma,—the banks of the Tavoy and East Tenasserim river.

Magnetite is also very generally distributed, as will have been observed from the account of the occurrences of iron. Iron rust is collected and employed in medicine, and an artificial combination of oxides is manufactured for the same purpose under the name of powdered iron.

Iron Oxides. (*J. Murray.*)

IRON Oxides.

DYE & PIGMENT. 474

Dye and Pigment.—The ochres, both yellow and red, are largely employed in many localities for the adornment of the walls of huts and houses, by Hindus for painting caste-marks on the forehead, for dyeing the clothes of certain castes, and for other similar purposes. **Sir E. C. Buck** writes in his *Dyes and Tans of the North-Western Provinces*, "This substance (*geru*) is found in some parts of these Provinces, and is largely imported from Gwalior. It is much used by *fakirs* and ascetics for dyeing their clothes of a dull orange colour, but plays a more or less important part in the hands of the dyers in the production of several well known colours. The earth is simply pounded and mixed with water, into which the cloth is dipped. Red ochre of a lighter colour than *geru* is known as *hirmji*, while yellow ochre is occasionally used as a dye under the name of *rámraj*." He then gives the following examples of dyes produced by *geru*:—In Allahabad a buff colour known as *Geru*, is produced by red ochre and alum; in Cawnpore a dove grey called *Faktái*, is produced by the same oxide with myrobalan, sulphate of iron and alum; and in Allahabad a purple called *Nafarmáni*, is produced by red ochre, safflower, and indigo. *Geru* is also used in calico-printing.

At Jauti or Jauli in Jabalpur large mines of massive hæmatite have been leased by **Mr. W. G. Olpherts** for the purpose of manufacturing a mineral paint, prepared by grinding the ores to an impalpable powder between stones worked by water-power. **Ball** states that, at the time he wrote, the value per ton in London was £9-10—retail £13. "This mineral paint has proved to be the cheapest in the Indian market, it lies smoothly on wood or iron, and has been successfully used against damp on porous tiles, bricks, and plaster. It is now used by the principal Rail and Steamship Companies, and has stood a practical test on the metal work of the principal bridges in India; it has been found most useful on the inside of boilers. In the preparation of one cwt. of paint, ready for use, the following components are required: Dry oxide of iron, 65¾ lb, linseed oil 5¼ gallons. The addition to the oil of one-fourth of its weight of common bazár resin renders the colour more brilliant and lasting. Three pounds for a first coat and two pounds for a second are sufficient to cover 100 superficial feet on an average" (*Ball*).

MEDICINE. Yellow Ochre. 475

Medicine.—A yellow impure OCHRE is believed in Sikkim to be a cure for goitre.

Magnetic Oxide. 476

The MAGNETIC OXIDE is used in doses of from five to ten grains as a tonic and hæmatinic. It is highly esteemed as a medicine in various localities, *e.g.*, Mysore, where a drinking cup of magnetic oxide was recommended to the late Raja by his native physicians. They held that by drinking out of it he would prolong his life. In the same district it is believed that milk, if boiled in such a cup, cannot flow over.

Hydrated peroxide. 477

HYDRATED PEROXIDE of iron is employed in doses of five to thirty grains or more as tonic, emmenagogue and anthelmintic, and mixed with water is a valuable antidote in cases of arsenical poisoning. In Hindu medicine a mixture of the proto- and per- oxides is largely administered in doses of from six to twelve grains in many diseases, under the idea that it is in reality powdered and refined iron. The following account is given by **U. C. Dutt**:—"Three varieties of iron are used in Hindu medicine, namely, *kanta lauha* or cast iron, *mandura* or iron rust, and *lauhasara* or salts of iron, produced by iron being kept in contact with vegetable acids. The form of cast-iron used in the manufacture of pans for boiling milk is considered superior to all others for medicinal use. The small particles of iron which are scattered around when hot iron is beaten on the anvil, are called *mandura*. They are allowed to remain in contact with the earth till they become very rusty and brittle, when they are considered fit for

IRON Pyrites.

Iron Oxides and Pyrites.

MEDICINE.

use. The properties of *mandura* are said to be analogous to those of cast-iron." "Cast-iron is purified by beating it into thin plates, heating the plates in fire, and sprinkling them with cow's urine, sour *congi* oil, and a decoction of the pulse **Dolichos biflorus** (*kulattha*) seven times in succession. The plates are reduced to powder by pounding them in an iron mortar, rubbing them with cow's urine, and roasting the mixture in a covered crucible repeatedly till it is reduced to a fine impalpable powder, that will float on water and will not irritate the eyes when applied to them. It is usual to rub the iron with cow's urine and roast it about a hundred times in succession. In some cases it is recommended that iron should be thus roasted separately for a thousand times. *Mandura* is purified and prepared for use in the same way. Prepared iron is a fine impalpable powder of a reddish-grey or brick-dust colour." This preparation of iron oxides was believed by the practitioners of Sanskrit medicine to increase strength, vigour, and longevity, to cure all sorts of diseases, and to be the best of tonics. **Dutt** writes, "When gold and silver are not available, iron is substituted for them. It is used in painful dyspepsia, chronic fever, phthisis, anasarca, piles, enlarged spleen and liver, anæmia, obesity, urinary diseases, diseases of the nervous system, skin-diseases, &c. When iron is administered, the following articles of diet should be avoided, namely, *kushmanda* (fruit of **Benincasa cerifera,** *Savi*), sesamum oil, *kulattha* (pulse of **Dolichos biflorus,** *Linn.*), mustard, wines, and acids."

The above described preparation of iron oxides enters into many complex preparations, in which it is combined with the other metals, spices, aromatics, and other vegetable drugs. Each preparation is supposed to have specific virtues, and is specially employed for certain diseases. It is needless in such an article as the present to enter into a detailed description of these, for which the reader may be referred to the exhaustive account of the utilisation of iron in Hindu medicine by **U. C. Dutt.**

478 **Iron Pyrites,** *Ball, Geol. of India, III., 418.*

SULPHIDE OR SULPHURET OF IRON; FeS_2.

Vern.—*Rúpamakhi,* HIND.; *Kangsmúki,* BENG.; *Swarna mukhi,* MAR.; *Súrna múki, sonmakki* (more properly applied to copper pyrites), MADRAS; *Svarnnamákshika, táramákshika,* SANS.

Reference.—*U. C. Dutt, Mat. Med. Hind., 56.*

LOCALITIES. 479

Occurrences.—This ore is widely distributed through formations of various ages. Thus it occurs in quartz reefs associated with the ores of other metals, and often with metallic gold, and is also to be found in slates, limestones, coal, and in the extra-peninsular tertiary rocks, *e.g.*, the nummulitic formations of Sind. Though it is possible that this ore may occur in considerable abundance, and, indeed, it is stated that it does so in the Bhima limestones of the Deccan, still it appears probable that the supplies of Indian pyrites are not easily accessible, and are unlikely to become of commercial importance. At the present time the chief supply is imported into Calcutta from Arabia, where it is obtained on the surface and in the beds of rivers.

In the Calcutta market, the wholesale value is said to be R6 to R10 a maund. It is used almost entirely as a drug (*Ball*).

MEDICINE. 480

Medicine.—Dutt writes, "Iron pyrites has been used in medicine from a very remote period. It occurs in two forms, namely, in dark yellow nodules with a golden lustre, and in silvery radiating crystals. The former is called *Svarnamákshika,* and the latter *Táramákshika.* The ancients supposed that they contained gold and silver respectively, in combination with other ingredients, and possessed in part the properties of those precious metals. Iron pyrites is purified by being boiled in lemon juice with one-third its weight of rock-salt in an iron vessel, till the pot becomes red-hot. It is reduced to powder by being rubbed with oil or goat's urine, and then roasted in a closed crucible. Thus prepared it has a sweetish bitter taste. It is considered tonic, alterative, and useful in anæmia, urinary diseases, ascites, anasarca, prurigo, eye-diseases, &c. As an alterative tonic it is generally used in combination with other and in the medicines of its class such as iron, talc, mercury &c." MEDICINE.

Iron Sulphate, *Ball, Geol. of India, 419.* 481

GREEN VITRIOL, GREEN COPPERAS, FERRIC SULPHATE, $FeSO_4$, [H_2O.

Vern.—*Kasis, hirá-kasís, káhi* (*káhi-máti*=sulphate of iron earth, *káhi-sabz*=impure green vitriol, *káhi-saféd*=white anhydrous iron sulphate, *kahi-siyá*=black iron sulphate; *kahi-zard*=yellow variety of anhydrous sulphate), HIND.; *Hirá-kos, hirá-kosís,* BENG.; *Kashish, híra-kashish,* BOMB.; *Híra-kasís,* GUZ.; *Anna-bédi,* TAM.; *Anna-bhédi,* TEL., MALAY., KAN.; *Kasisa* (*dhátukásis*=green variety, *phushpa kásis*=yellowish variety), SANS.; *Záje-asfar,* ARAB.; *Záke-zard,* PERS.

References.—*Mason, Burma & Its People, 568, 730; Ainslie, Mat. Ind., I., 529; O'Shaughnessy, Beng. Dispens., 352; Moodeen Sheriff, Supp. Pharm. Ind., 141; U. C. Dutt, Mat. Med. Hind., 55; McCann, Dyes and Tans, Beng., 3, 17, 42, 132, 134, 135, 138, 140, 142, 144, 149, 150, 151, 162, 169; Buck, Dyes and Tans, N.-W. P., 41; Liotard, Dyes, 116, 127, 128, 129, 130, 134; Balfour, Cyclop., II., 375.*

Occurrences.—Ferrous sulphate occurs in the form of green crystals soluble in water. It is formed abundantly by natural oxidation of iron pyrites, and is apt to undergo a further alteration into the red sulphate of the sesquioxide, or ferric sulphate. For use in the arts it is made on a large scale by exposing moistened iron-pyrites to atmospheric influence, when the same changes which occasion its natural formation occur. Natural green copperas is to be found in many parts of India, and is largely employed by natives for dyeing. LOCALITIES. 482

Dye.—This salt is used extensively in dyeing partly as a colour producer, partly as a mordant. It is generally employed mixed with organic substances to obtain various shades of black, brown, grey, purple, violet, and green, and is also much used in calico-printing. In certain localities the natural copperas is employed, in others it is artificially prepared by placing clean bars of iron in a tub containing a solution of coarse sugar and other substances. **Sir E. C. Buck** states that the weight of iron used is about four times that of the sugar dissolved, that when the solution has become of a deep black colour it is ready for dyeing purposes, and that myrobalans are occasionally added to clear the colour. DYE. 483

A large trade is said to be carried on in the preparation of the salt for the dyer's use in Lucknow.

Medicine.—Native sulphate of iron has been known to Sanskrit medicine from a very early age, but its chemical nature does not appear to have been recognised by those who recommended its use. It was only rarely employed internally, but was much used externally in skin diseases as an astringent, and was recommended as a strengthening application to glandular structures. In European medicine it is valued as a powerful chalybeate tonic, astringent, emmenagogue, antiperiodic, and anthelmintic. In large doses it is an irritant poison. Locally applied it is used as an astringent and stimulant. MEDICINE. 484

The dried sulphate is employed for the same purposes and also as a styptic.

Other salts of iron, especially the persulphate, perchloride, phosphate, carbonate, pernitrate, citrate, acetate, arseniate, and iodide, are largely employed in European medicine, and are all manufactured from the metal directly or from some of its salts. In Indian medicine, however, the mix-

MEDICINE. ture of the oxides already described is almost the only form of iron used, and it would be superfluous in this place to give an account of the well-known properties to the above enumerated artificially-prepared salts.

(*G. Watt.*)

ISATIS, *Linn.; Gen. Pl., I., 94.*

485 **Isatis tinctoria,** *Linn.; Fl. Br. Ind., I., 163;* CRUCIFERÆ.

THE WOAD OR DEVIL'S WEED.

An erect, herbaceous plant, like a large cabbage, common in Western Tibet, Afghánistan, &c., occurring both wild and cultivated. Also largely grown in certain regions in China.

INDIGO. 486 Yields the indigo of China. **Dr. Aitchison,** in his report upon the Kuram Valley, informs us that it is used for this purpose in Afghánistan. For further particulars see the article on **Indigofera tinctoria,** *pp. 387-469.* Also *Jour. Agri.-Hort. Soc. Ind. (Old Series), VIII. (Proc.), 180. XI., 157, 158.*

ISEILEMA, *Anders.* (Anthistiria, *Linn.*) ; *Gen. Pl., III., 1136.*

487 **Iseilema laxum,** *Hack.; Duthie, Fodder Grasses, N. Ind., 43.*

Syn.—I. PROSTRATA, *Anders.*

Vern.—*Champ* (SIMLA HILLS), *luinji* (KANGRA), *Chhat* (RAWALPINDI), *gándi* (HISSAR), PB.; *Karar-gandhel-dungarko* (JEYPUR); *Musel, musiál, machauri* (LALITPUR), BUNDELKHAND; *Masán, manchi-malwa, malwajari* (CHANDA), *ghorayal* (SEONI), C. P.

Habitat.—Common on the plains of Northern India on low-lying land.

FODDER. 488 **Fodder.**—See Vol. III., 423.

489 **I. Wightii,** *Anders.*

Syn.—ANTHISTIRIA WIGHTII, *Nees.*; A. BLADHII, *Wight*; A. PROSTRATA, *Trin.*

Vern.—*Gandel* (ALIGARH), N.-W. P.; *Ganni* (GUJRANWALA, SHAHPUR, and LAHORE), PB.; *Ghor-masan, musán, pulsu-malwa-gadi* (CHANDA), *musan* (BALAGHAT), C. P.

Habitat.—Associated with the preceding.

FODDER. 490 **Fodder.**—See Vol. III., 423.

ISINGLASS, GELATIN, GLUE, AND GELOSE.

Animal tissues contain an adhesive substance known as "Histose." This, on pieces of flesh, skin, bone, horn, &c., being boiled, is changed into Gelatine, which is soluble in hot water and is thus separated from the other substances present. By simple evaporation of the water the gelatine forms into a dry, hard material, which receives in commerce different names according to the nature of the substances from which it was extracted. These may be briefly indicated :—

491 **(a) Isinglass.**

COLLE DE POISSON, *Fr.*; HAUSENBLASE, *Germ.*

Icthyocolla Greek (from ἰχθυς a fish, and κόλλα, glue). The English name is a corruption from *icing* and glass or iceglass, and vulgarly the name isinglass is often applied to Mica.

As implied by the Greek name for this substance, it is a gelatine derived from fish. Isinglass of European Commerce is chiefly prepared from the air-bladders of **Acipenser Sturio** and other Sturgeons, found in the Caspian Sea, the Black Sea, and the Arctic Ocean. The Brazilian isinglass (obtained from Brazil and Guiana) is prepared from one or two species of **Silurus,** chiefly **S. Parkerii.** The isinglass of India is derived from a large number of fish and sharks, &c., but only recently has India come to be viewed as a source of isinglass for Europe, the trade having been previously with China in the edible substances known as FISH MAWS and SHARKS' FINS. Many writers, however, affirm that all that is necessary to expand an Indian trade in Isinglass is to instruct the fishermen to bestow greater care in selection and preparation of the substance. ISINGLASS.

The uses of Isinglass are in the preparation of jellies and confections and as a clarifying or filtering medium for wine, beer, coffee, and other liquids. It is also employed along with gum as a dressing to give lustre to ribbons and other silk articles.

The reader is referred to the article under **Sharks' Fins** and also to that on **Fish.**

(*b*) **Gelatine** (or **Gelatin**) and (*c*) **Glue.** 492

These substances are prepared from animal flesh, bones, skins, hoofs, and horns. The former is a purer article than the latter, all the stages in its manufacture being carried out with scrupulous care and cleanliness. Glue may be defined as a glutine manufactured from the tan-yard refuse. But there is no chemical difference between gelatine, glue, and isinglass. The purer forms of transparent gelatine are used for culinary purposes (calves'-feet jelly, milk-white blance-mange, usually containing gelatine instead of isinglass) but for certain purposes gelatine like glue is used as an adhesive medium and hence glues of every quality passing into gelatine are met with in commerce.

The Edible Swallows' Nest may be described as an Indian edible gelatine (see Vol. II., pp. 505-509). In some parts of India a glue is said to be prepared from fish bones, but ordinary glue is also made as, for example, in Lucknow. The article chiefly used is apparently, however, entirely imported (see *Hoey, Trade and Manuf. N. Ind., 176*).

(*d*) **Gelose** (a name given by **Payen**).

This is a gelatinous substance prepared from the various **Algae** designated in India as *Agar-Agar*, and in European commerce as CHINA MOSS. The best known substances of this nature are the so-called Ceylon Moss (see **Gracilaria**) and the Japanese Isinglass. The last mentioned is derived from one or two species of sea weed (according to **Mr. D. Hanbury**), but more especially from **Gelidium corneum.** 493

Gelose differs from gelatine in not being precipitated by tannic acid: from the starch jellies in not being rendered blue by iodine, and from gum, by its insolubility in cold water, and its greater gelatinising power. In this last respect it forms ten times as much jelly as an equal weight of isinglass, and may, therefore, be economically substituted. It contains no nitrogen, and is therefore not nutritious; but as isinglass and gelatine are most frequently used in culinary purposes as vehicles for other nutritious materials, this need not be a serious objection to the employment of gelose.

Gelose contains 42·770 per cent. of carbon, 5·775 hydrogen, and 51·455 oxygen. It, therefore, belongs to the class of proximate principles which possess a larger proportion of oxygen than is required to form water with the hydrogen. It swells up in cold water and dissolves almost wholly in boiling water and on cooling forms a jelly of 500 times the weight of water to the quantity of gelose employed. It differs, however, from jelly with isinglass in the important character that it requires a high temperature for its fusion when once formed. Hence culinary articles prepared with this substance do not dissolve in the mouth, and in consequence the jellies known as seaweed are exported from China without deterioration. One of the chief forms of gelose appears to have been shown at the Paris Exhibition of 1878 under the name of Thao or Ta-o. If the hot solution of that substance be stirred until cool, it forms a liquid instead of a jelly, and that preparation gives body and gloss to fabrics without stiffening

GELOSE. them—the defect of dextrine and starch. The properties of Ta-o in this respect are further heightened by combination with glycerine. It may be mixed also with talc, and even while hot it will intermix with gum, starch, dextrine, gelatine, or gum tragacanth, thus forming *size* or *dressing* that may be employed with great advantage to silk goods. Since this dressing is only again softened at high temperatures, the fabrics are not deprived of it by rain or when exposed to damp atmospheres. Gelose may also be employed in double-dyeing by being combined with sulphate of copper and the chlorides of aniline and potassium. For sizing paper it is of great value; indeed, its chief obstacle to an extended application in the arts is its high price. It is somewhat surprising that so valuable a substance should not have been made better known and more carefully examined. **Watt**, for example, remarks that since no definite compound has been formed with it, its atomic weight and rational formula has not been determined. Its chief properties seem, however, to ally it with the gums (**H. Morin**) and according to **M. Porumbaru** the formula for pure gelose is $C_6H_{10}O_5$ thus analogous to lichenin, inulin, and tunisin.

Dr. Dymock remarks of China Moss: "There is, however, no reason why a similar substance should not be made from our common native seaweeds, of which **Gelideum corneum** and **Gracilaria confervoidea** approach most nearly in character the Algæ from which Thao is made." This statement would justify the recommendation made by **Royle** nearly half a century ago that the indigenous Algæ and other sources of isinglass on the seashores of India should be carefully examined with the view to discovering whether any of them might find a place in European commerce. At present the *Agar-agar* used in India is entirely imported.

The reader is referred to Iceland Moss (**Cetraria**), to **Fucus**, and to Irish or Carrageen Moss, &c., for descriptions of other allied substances; but is cautioned against accepting all the known Indian products that form jellies as of necessity belonging to the Isinglass, Gelatine, and Gelose group. For trade statistics regarding Indian Isinglass the article on **Sharks' Fins & Fish Maws** may be consulted.

ISCHŒMUM, *Linn.; Gen. Pl., III., Vol. III.*

494 **Ischœmum angustifolium,** *Hack.; DC. Monogr. Phan., Vol. VI., 241; Duthie, Fodder Grasses of Northern India, 27* (under **Pollinia**).

Syn.—ANDROPOGON BINATUS, *Retz.*; A. NOTOPOGON, *Steud.*; A. OBVALLATUS, *Steud.*; A. INVOLUTUS, *Steud.*; SPODIOPOGON ANGUSTIFOLIUS, *Trin.*; S. LANIGER, *Nees.*; S. NOTOPOGON, *Nees.*; POLLINIA ERIOPODA, *Hance.*

Vern.—*Bhábar*, HIND.; *Baboi, babui, saba*, BENG.; *Sabai* (HIND.), CHUTIA NAGPUR; *Bachkron*, SANTAL; *Bankas, ban-kush, baib, bamoth*, N.-W. P. and OUDH; *Pan-babiyo* (ALMORA), KUMAON; *Bhabar, babbar, munji, baggar*, PB.; *Nulka-gadi, som, moya*, C. P.

References.—*Stewart, Pb. Pl., 249; Hooker, Icones Plant. Pl., 1773; Royle, Ill. Him. Bot., 415, 416; Report Royal Bot. Gard., Calcutta (in part), 1877-78; 1882-83; Kew Bulletin, July 1888, pp. 157-160; Journ. Linn. Soc. (in part), Vol. XX., 409; Trans. Agri.-Hort. Soc of India (in part), VIII., 272; Jour. (Old Series), XII., 332; XIII., 293 (New Series), V., (Proc.), 21; VIII., 98-106.*

Habitat.—A perennial grass with strong wiry stems, clothed at their bases with woolly pubescence. The leaves are long and narrow and with involute edges. Each stem bears two to four racemes, composed of numerous spikelets which are densely clothed with brownish or golden coloured silky hairs. This species is plentiful in the Sub-Himálayan tract as well as in the hilly parts of Bundelkhand and Central India. It is frequently found associated with **Eriophorum comosum** (which see, *Volume III., 266*), and hence from **Wallich** and **Royle** proceeded an error, only recently corrected, of viewing both as one and the same. **Eriophorum** is a sedge and **Ischæmum** a grass, but both are no doubt employed for the same purposes.

Fibre.—[This grass is used in paper-making and in the construction of strings, ropes, and mats. It seems doubtful whether the natives of India separately distinguished it from the sedge above alluded to, and these two plants accordingly appear to bear the same vernacular names. The passages that would appear more especially to allude to the sedge allude to it as employed in the construction of rope-bridges, but as a source of paper there is probably no doubt that the bulk of the material so employed is the grass here dealt with. It is perhaps unnecessary to repeat the facts published in the *Kew Bulletin* (*l. c.*) showing what led to the confusion regarding these two plants, nor those which ultimately cleared the matter up. It would seem that had **Dr. Stewart's** remarks republished below (made in 1863) received more careful consideration, the ambiguity that for long existed would never have existed. The following passages regarding the grass as a source of paper, &c., are, however, of practical interest and may be said to convey all that is known on this subject:— FIBRE. 495

Mr. R. W. Bingham, writing of the Sasseram District of Bengal in 1862, remarked:—

"*Buggaie* is a wild grass which is ready for cutting in October and November. Several hundred thousand maunds are cut annually in the forest of the Kymore range, and it is sold for roofing and other purposes; when made into a coarse twine it answers for tying thatch and bamboos, as well as for bottoming the cots, or charpoys of the lower orders, as well as a more expensive article. Much of it finds its way to the river ports, but I fancy it is all consumed in this country, and would be but of little use in the manufactures of Europe" (*Agri.-Hort. Soc. Journ., Vol. XII., p. 332*).

Dr. J. L. Stewart wrote of this grass in Bijnor (September 1863) as follows:—"This grass, which is abundant in this part of the Himálaya and occasional on the skirts of the Siwaliks, appears to furnish almost all the material called *bhabar* so largely used for string in these parts. Botanists, from **Wallich** and **Royle** downwards, have stated this to be the produce of **Eriophorum comosum,** of which, however, apparently a very small proportion of that brought to the plains consists. **Dr. Brandis** first drew my attention to the probability of the ordinary belief being erroneous, and subsequent enquiry has shown that the case is as above stated.

"The string is very coarse but strong, and although there is great waste in the manufacture, exceedingly cheap. It is well adapted for boat ropes, the rope-work of bedsteads, and other ordinary purposes. Possibly the *bhabar* may come into play as a paper-material, at least it is worth the trial, and probably larger quantities of the raw article could be got than of any other fibre that I know of in this part of the Himálaya." **Dr. Stewart** says, the price of the raw material is annas 8 per maund and of the string R2 per maund (*Agri.-Hort. Soc. Journ., Vol. XIII., 293*).

Dr. King, in his report of the Royal Botanic Gardens, Calcutta, for 1882-83, re-established the independence of the two forms of the *bhabar*.

"In several former reports I have referred to the leaves known by the vernacular name *bhabur* as the produce of **Eriophorum comosum.** I have now satisfied myself that the bulk of the *bhabur* used by natives for rope-making is not derived from **Eriophorum,** as I had supposed, but from **Andropogon involutus.** This grass, I find from enquiry locally made,

FIBRE.

abounds in the hill parts of Behar and Chutia Nagpore, where it is known as *Sabai*. From these regions it can be obtained in quite considerable enough quantity to make its utilization as a paper material a feasible project, and the people who actually collect it, sell it at a reasonable enough rate. But in order to get it brought to Calcutta in sufficient quantity for local manufacture, or for shipment to Europe, middlemen have to be employed, whose ideas of profit are pitched so high that, until they become modified, the utilisation of *bhabur* must remain in abeyance. This is only in accord with the common experience in the mofussil, that competition in trade is not sufficiently keen to have much effect in keeping down prices, but that, on the contrary, traders still form guilds banded together to enhance prices, even at the risk of choking off demand."

In 1883 **Mr. C E. Edwards**, Manager of the Lucknow Paper Mills, furnished the following report regarding *bhabur* grass :—

"This grass we have used here, but not to any great extent, owing to the price being too high; besides the outturn is not so great as with jute. I found it not to yield more than about 35 per cent. of paper. This is to a great extent owing to the top part of the plant being somewhat perished, I presume owing to the tops being more exposed to the atmosphere, as this part appears to get ripe much earlier than the bottom.

"In the process of boiling the perished or top part gets destroyed before the bottom part of the stems gets sufficiently reduced to a pulp. This accounts for the great loss in the manufacture, but this could be obviated by having the top parts cut off before despatching it to the paper mills, and if it could be had at the same price with the tops off, I have no hesitation in saying it would be a good and cheap enough fibre for paper-making purposes."

In the same year the late **Mr. T. Routledge** furnished the following facts regarding *bhabur* as a source of paper :—"I believe it will make a fair sheet of paper, much the same as fine esparto, but does not contain so much glutinous and amylaceous matters, nor so much silica. The sample sent, you will remark, was cut, and not pulled from the roots as esparto is. Like esparto, *in situ* it is worth very little, and is used for similar purposes—roping, matting, baskets, &c. The cost of esparto consists in collection, carriage to port of shipment, and, latterly, baling charges, freight to England, &c. Whether from India, with long and probably costly inland carriage, with heavy freight added, it can come into competition with esparto, is doubtful, and I do not think it would pay to convert it into stock."

The **Rev. H. P. Boerresen** of Rampore Hát gives the following particulars regarding the grass :—"(1). The *Sabai* or *Babui* grass yields two crops in the year—one in September and the other at the end of October, or early in November, without any irrigation as the rainy season is then prevalent. It might yield a third cutting if irrigated, but I cannot say anything on this head, never having made the experiment, nor have I seen it attempted by others.

"(2). I believe it will grow anywhere, as we have transplanted it from here to all our other out-stations in the Santal Parganas, and it thrives in them all. The Santal Christians have also taken some of it to our Christian Colony in Guma Duar, Assam, where it also grows well."

"(3). I have never attempted to propagate it by seed, but always by roots; when a clump or tuft is dug out, it may be divided into as many small divisions of roots as one pleases, and these are put down again in rows, about three feet from one another, and the same interval between each root planted. It will yield a very trifling return the first two years, but by the third or fourth year, when the roots have spread and multiplied

FIBRE.

it gives a *good* crop. The plot on which it is planted must be kept free from other grass. When it is seven or eight years old, the roots should be beaten down with wooden mallets, or a plough should be run through them in every direction, and fresh earth thrown over the whole; this will increase the yield. If not treated in this way, it will cease yielding any crop. When grown too old, it must be taken up entirely, re-divided in small bunches of roots, and transplanted to a fresh locality.

"(4). We bought the grass always in local *háts* for roofing purposes (as rope) before we grew our own, and nowhere in our neighbourhood am I aware of its being cultivated in any but very small patches by a solitary man here or there. It is not cultivated as a source of income or trade, so that I am unable to say where the roots may be bought, or at what price. We got a small quantity of the roots originally from a Hindu village, but by fostering and spreading their cultivation we have now a considerable quantity. It should be planted in a *dry* spot where no water lodges, as experience has shown in one of our stations, where the water oozed up from below and rotted the roots, that it would not grow there. A sloping site is probably the best."

"When we first started the Mission here we had to pay R4 a maund in the *háts* for the grass, in order to twist it into rope or string, and it was the having to pay so much that led me to try and cultivate it ourselves. The grass runs to seed in the hot months, shortly before the rainy season, but these must be cut off and removed, or the crop will deteriorate."

It is understood that the Bally Paper Mills, Calcutta, still continues to use a considerable amount of the grass, drawing its supplies from Chutia Nagpur and the Nepál Tarai. In the *Colonial and Indian Exhibition Catalogue* the writer stated regarding this paper material :—

"In a country teeming with fibres, it is a surprising fact that the question of a paper-fibre should still be under consideration. European authors seem always to forget, however, the immense size of the continent of India, and that a few hundred miles are of little consideration in most Indian questions; but, when this assumes the form of railway freight upon a bulky article, it becomes prohibitive even to industries that can afford to pay more than paper-making. To cultivate fibre, especially for paper, has many drawbacks, and one only need be mentioned—the cost of land near commercial centres is so high as to almost preclude the idea of the cultivation of paperfibres. But the paper fibre must, like esparto, be fit almost for immediate immersion in the vats, for paper-making can never pay for cultivation and separation of fibre, even where land can be got at a merely nominal rent. No fibre known to commerce can compete with jute in point of cheapness, yet the paper-maker can afford to purchase jute waste and jute cuttings only, so that but for the demand to meet an altogether different purpose, the paper-maker could never procure jute. The two most important indigenous paper grass-fibres in India are *munj* grass and *bhabar* grass. These are now being used by our Indian paper mills; the supply at a remunerative price is the chief obstacle to an extended employment. The roots and lower stems of rice have been suggested as a paper material, and if it be shown that these are worth the trouble of collecting, the supply might be practically unlimited; but it is doubtful whether the paper-maker could pay sufficient to cover freight and expense of collection. It has been demonstrated that bamboo affords excellent paper, but practical difficulties exist which have, for the present at least, dispelled the hopes once entertained of the immense tracts of bamboo forest becoming of commercial value as supplies of paper material.

The most valuable paper materials in India are, after all, old rags, waste gunny bags, and old *sunn* ropes."

FIBRE. The results of the Exhibition abundantly justified the above opinions, and the gentlemen who were present at the conference on INDIAN PAPER MATERIALS agreed that "while the *bhabar* and *munj* grasses were very good for paper, it would never pay to import them into Europe in that condition, the more so since esparto was obtainable at a price far below what would be charged in freight alone from India. Unless these grasses could profitably be shipped in the form of half stuff they were out of the question." Thus, so far as a foreign trade in these grasses is concerned, it would seem that no hope need be entertained as against esparto, and it is probable that next to the want of a uniform and large supply, the question of railway freight within India itself is likely to operate unfavourably against the development of a large Indian paper trade in *bhabar* grass. The Agri.-Horticultural Society of India, in order to ascertain the extent of production, prices, and uses of *bhabar* grass, and at the instance of the Chamber of Commerce issued to its members in 1886-87 a series of questions. The replies are published in the Journal (Vol. VIII.). These may be briefly summarised here. **Mr. Dear** of Monghyr reported that 10,000 to 15,000 maunds might be obtained in a season from the Kharrackpore hills at about R1-14 a maund. **Dr. Hill**, Parulia, stated that it was cultivated in some parts of Manbhoom, but to a larger extent in Singhbhoom (Chutia Nagpur). The people make strings and ropes of it which are exported; the raw material is never exported. Large quantities of the grass can be obtained during February and up to May; it sells at from 12 annas to R1 per maund. **Mr. T. M. Gibbon, C.I.E.**, replied that "the *sabai* grass is grown largely in the northern portion of the district, but I believe only in the north, in the jungle lands of Ramnuggur and Rajepore Sohureah. It is also grown in the Nepál jungles bordering the above lands." "Last year the grass was sent direct to the Bally Mills at, I believe, R2-8 per maund." **Syed Ali, Nawab**, of Jainugger, Durbhunga, gave the exports of the grass from the Nepál Tarai from April 1883 to February 15th, 1887, as amounting to 3,003 maunds. **Mr. W. Claxton Péppé** of Gorakhpore reported that "This grass is called here *bunkas*, and grows on the first range of hills in Nepál: that growing at the bottom of the hills is not so good as that growing higher up, though both are actually the same grass. The grass does not grow in Gorakhpore, but on the first range of hills in Nepál. It is cut by the Nepálese and sold by them to Binjaras and others at Bahadurgunj, about 3 miles south of the hills and about 25 miles north-west from the frontier here. By these men it is taken to Uska Bazár, the terminus of the new branch of the Bengal-North-Western Railway." The price of the raw material varies considerably as shown by **Mr. Péppé**, *viz.*, from 66½℔ per rupee to 114℔ per rupee. It is chiefly sold in the manufactured state as string, and regarding the trade in string he adds: "In February the new crop comes in. The grass made into string I buy at 31℔ per rupee. During the last year I bought 2,563℔ and paid for it R83. These purchases are made at my door from Binjaras and other carriers." Other reports were procured by the Society from Manbhoom, Hazaribagh, &c., but the above, it is believed, convey the important facts which need be here republished. *G. Watt, Ed., Dict. Econ. Prod.*]

Fodder. 496 **Fodder.**—See Vol. III., 423.

497 **Ischœmum ciliare,** *Retz.; Duthie, Fodder Grasses of N. India, 30.*

Syn.—I. ARISTATUM, *Willd.*

Vern.—*Bara-toriya-gadi, pyana-koru-gadi, paba, irkor, guhera,* CENTRAL PROVS.

Reference.—*Hackel in DC. Monogr. Phan., Vol. VI., p. 225.*

Habitat.—Abundant over the greater part of India, from the Himálaya to Ceylon, growing in wet marshy ground.

This species is divided by Hackel into the following varieties and sub-varieties.

Var. α—genuium.

Sub-var. 1.—PROREPENS. **Syn.**—I. TENELLUM, *Roxb., Fl. Ind., Ed., C.B.C., 108.* Nepál to South India.

Sub-var. 2.—SCROBICULATUM. **Syn.**—I. SCROBICULATUM, *W. & A.* Ceylon.

Sub-var. 3.—MALACOPHYLLUM. **Syn.**—SPODIOPOGON OBLIQUŒVALVIS, *Nees;* ANDROPOGON MALACOPHYLLUS, *Hochst.;* I. ARISTATUM, *Roxb., Fl. Ind., Ed. C.B.C., 107.* Terai to South India, and in Burma.

Sub-var. 4.—VILLOSUM. **Syn.**—SPODIOPOGON VILLOSUS, *Nees;* S. OBLIQUŒVALVIS, *var.* VILLOSUS, *Benth.* Central Provinces.

Var. β—Wallichii. Sylhet.

Var. γ—longipilum. Ceylon.

Fodder.—Occasionally used for feeding cattle. FODDER. 498

Ischœmum laxum, *R. Br.; Duthie, Fodder Grasses of N. India, 31.* 499

Syn.—ANDROPOGON STRIATUS, *Klein;* A. NERVOSUS, *Rottl.;* A. BROWNEI, *Kunth.;* A. MACROSTACHYUS, *Anders.;* POLLINIA STRIATA, *Spreng.;* HOLOGAMIUM NERVOSUM, *Nees;* ISCHÆMUM NERVOSUM, *Thw.;* I. MACROSTACHYUM, *A. Rich.;* SEHIMA MACROSTACHYUM, *Hochst.*

Vern.—*Sairan, hirn,* RAJ.; *Sein, seind, sarun,* C. INDIA; *Sira, sedwa, pona,* C. P.; *Sainad,* BERAR.

References.—*Hackel, in DC. Monogr. Phan., Vol. VI., 243; Indian Forester, XII., App. 24.*

Habitat.—Hilly parts of Rájputana, Bundelkhand, and the Central Provinces. It forms a large portion of the undergrowth in the Nímar forests, and seedling trees find beneficial shelter in the spaces between the clumps.

Fodder.—It is considered as one of the best of the Central India fodder grasses. FODDER. 500

I. pilosum, *Hack.; Duthie, Fodder Grasses of N. India, 31.* 501

Syn.—ANDROPOGON PILOSUS, *Klein;* SPODIOPOGON PILOSUS, *Nees.*

Vern.—*Khuri,* CENTRAL INDIA; *Khunda, kunda,* C. P.; *Nattu,* TEL.

References.—*Wight, in Mad. Journ. Lit. & Sc., II., 139; Elliot, Fl. Andhr, 130; Hackel, in DC. Monogr. Phan., Vol. VI., 240.*

Habitat.—A tall, rather coarse, glaucous grass, with thick deeply penetrating roots. The long spikes, usually in pairs, are densely clothed with white silky hairs. A characteristic black soil species and very common in the Central Provinces. Elliot mentions that "it infests the *regáda,* or black cotton soil, to the great detriment of cultivation. It is called *Kunduru nattu,* or the 'grievous weed,' to distinguish it from *Garaka nattu* or 'grass, weed,' which is **Cynodon Dactylon.**"

Fodder.—Considered to be a good fodder grass in the Nimar district. **Captain Masters,** writing from Gúna, reports this grass as being of indifferent quality. FODDER. 502

I. rugosum, *Salisb.; Duthie, Fodder Grasses of N. India.* 503

Syn.—ANDROPOGON RUGOSUS, *Stend.*

Vern.—*Mehat, munmuna,* PB.; *Jalgundya, toli,* RAJ.; *Dhanua maror,* N. W. P. & OUDH; *Amarkarh, maggru-gadi, murdi,* C. P.; *Tudi,* BERAR *Marudi,* SANTAL.

References.—*Roxb., Fl. Ind., Ed. C.B.C., 107; Indian Forester, XII., App. 25; Rev. A. Campbell, Report Chutia Nagpur; Hackel, in DC. Monogr. Phan., Vol. VI., 206.*

Habitat.—A common grass of wet marshy ground in the plains, and at low elevations on the Himálaya. **Roxburgh** observes that it is generally found growing amongst rice, and is so much like it that until the plants have come into flower they are with difficulty distinguished from it. The outer glumes of the florets are hard and transversely wrinkled, hence the specific name.

FOOD & FODDER. 504 **Food and Fodder.**—Cattle and horses eat this grass when it is young. In some parts of the Central Provinces the grain is used as food.

505 **Ischœmum sulcatum,** *Hack. in DC., Monogr. Phan., Vol. VI., 248.*

Vern.—*Pownia,* C. INDIA; *Pona,* C. P.

Habitat.—Abundant along the edges of cultivated land in the black soil of Central India. Allied to **I. laxum**, but more slender, and with the culms branching. The leaf blades are rough with minute prickles directed upwards.

FODDER. 506 **Fodder.**—A very nutritious grass, and largely used as fodder wherever it occurs in abundance.

507 ## ISONANDRA, *Wight; Gen. Pl., II., 657.*

The species of this genus, as it has now been restricted, are of little or no economic value. They are trees which doubtless yield timbers of some use, but the plants of interest formerly referred to this place have been transferred to **Dichopsis** (*which see, Vol. III., pp. 101-109*).

ITEA, *Linn.; Gen. Pl., I., 647.*

A genus of shrubs or small trees comprising some five species, of which three occur on the Himálaya and Khásia mountains.

The timber, though small, is considered useful by the hill tribes, and in Chamba the bark is employed medicinally.

508 **Itea chinensis,** *Hook. & Arn.; Fl. Br. Ind., II., 408;* SAXIFRAGACEÆ.

Habitat.—A small tree of the Khásia hills between 4,000 and 6,000 feet.

509 **I. macrophylla,** *Wall.; Fl. Br. Ind., II., 408.*

Vern.—*Teturldum,* LEPCHA.

Habitat.—A small tree in the Eastern Himálaya from Sikkim and Bhután to the Khásia hills; altitude 2,000 to 4,000 feet. Found by the author in Manipur at altitudes of 4,000 feet.

510 **I. nutans,** *Royle; Ill., 226; Fl. Br. Ind., II., 408.*

Vern.—*Lelar,* KAGHAN; *Garkath,* KUMAON.

Habitat.—A small tree found on the North-West Himálaya from the Indus to Nepál. **Stewart** specially mentions it in his account of Hazara and Kaghan.

IVORY.

For an account of this substance see under **Elephas indicus**, Vol. III., 226.

IXORA, *Linn.; Gen. Pl., II., 113.*

The various species of **Ixora**, though extensively cultivated in the gardens of the hotter parts of India, are not, as a rule, of much economic value. For an account of the cultivated species and varieties the reader is referred to **Firminger**, *Man. Gardening for India, pp. 580 to 585.* Several species such as **I. nigricans** and **I. spectabilis**, are large bushes or small trees which yield timbers of some value. Of the latter, **Kurz** remarks that it is yellowish-white, heavy, close-grained, hard, and brittle; on exposure turning pale-coloured and blackish-streaked.

511 **Ixora acuminata,** *Roxb.; Fl. Br. Ind., III., 137;* RUBIACEÆ.

Vern.— *Churipat,* NEPAL; *Thekera,* ASSAM.

Reference.—*Roxb., Fl. Ind., Ed. C.B.C., 128.*

Habitat.—A robust, glabrous shrub distributed from Sikkim 3,000 feet to Bhután, Assam, Khásia hills, and Chittagong.

Dye.—In a note on the Dyes of Assam, the Director of Land Records and Agriculture mentions this plant as used as a mordant along with Arnotto (**Bixa Orellana**). DYE. 512

Ixora coccinea, *Linn.; Fl. Br. Ind., III., 145.* 513

The FLAME TREE OF THE WOODS.

Syn.—IXORA GRANDIFLORA, *Br.*; I. PROPINQUA, *Br.*; I. INCARNATA, *DC.*; I. OBOVATA, *Heyne*; I. BANDHUCA, *Roxb.*; PAVETTA COCCINEA, P. INCARNATA, *Blume*; P. BANDHUCA, *Miq.*

Vern.—*Rangan, rajana*, BENG.; *Pankul*, MAR.; *Pán-sá-yeik*, BURM.; *Bandhúka, raktaka, bandhújivaka*, SANS.

References.—*Roxb., Fl. Ind., Ed. C.B.C., 126; Brandis, For. Fl., 275; Kurz, For. Fl. Burm., II., 26; Rept. Pegu, App. cxi.; Beddome, Man. cxxxiv., 7; Gamble, Man. Timb., 230; Dalz. & Gibs., Bomb. Fl., 112; Sir William Jones, V., 80; Mason, Burma and Its People, 415, 786; Dymock, Mat. Med. W. Ind., 2nd Ed., 410; Lisboa, U. Pl. Bomb., 344; Balfour, Cyclop., II., 391; Gazetteers:—Orissa, II., 178; Mysore & Coorg, I., 70; Kanara, 71; Banda, 81; Journ. Agri.-Hort. Soc. Ind. (Old Series), II., 356; X., 11; (New Series), IV., 118.*

Habitat.—A small shrub extensively cultivated in gardens throughout the hotter parts of India. According to **Roxburgh** it is a native of China and the Moluccas, but **Kurz** regards it as a native of Lower Burma also.

Medicine.—" In dysentery 2 *tolás* of the flowers, fried in *ghí*, are rubbed down with 4 *gunjás* each of cummin and *nágkessur*, and made into a bolus with butter and sugarcandy and administered twice a day " (*Dymock*). MEDICINE. 514

SPECIAL OPINIONS.—§ " The root of the plant is used in the treatment of acute dysentery by **Assistant Surgeon U. Lal Deb** attached to the Howrah General Hospital who first introduced the remedy. The mode of preparing the drug is given as follows:—The root is ground to a pulp with a little water on a stone with a small proportion of long pepper, and about 30 to 40 grains of the fresh root should be taken in this way every three or four hours. A tincture made with 4 oz. of the dried root to one pint of proof spirit has been also found efficacious though less so than the fresh root. The remedy has been largely used in the form of tincture in the Howrah General Hospital in both the European and Native Departments. It is a valuable remedy, but not as certain and effective as Ipecacuanha. The remedy is not disagreeable to the taste, nor emetic or poisonous" (*Civil Surgeon J. G. Pilcher, Howrah*).

Sacred.—The flowers are held sacred to Siva and Vishnu.

I. parviflora, *Vahl.; Fl. Br. Ind., III., 142.* 515

THE TORCH TREE.

Syn.—IXORA ARBOREA, *Roxb. MSS. ex Smith*; I. DECIPIENS, *DC.*; I. PAVETTA, *Andr.*

Vern.—*Kota gandhal* (*loha jangia*, in Chutia Nagpur), HIND.; *Rangan*, BENG.; *Pété*, KOL.; *Merom met'*, SANTAL; *Tellu, kurwan*, URIYA; *Disti, kori*, GOND.; *Kurat, lokandi, narkurat, raikura*, BOMB.; *Kúrat, lokandi, khura*, MAR.; *Kura*, KONKAN; *Shulundu kora*, TAM.; *Korimi pála, korivi pála, putta pála* (Elliot), *karipal, kachipadél, tadda pallu* (Gamble), TEL.; *Gorivi*, (COORG), *korgi, hennugorvi*, KAN.; *Pán sá yeip* (**Ixora** generically), BURM.; *Maha-ratambalá* (*Karankuttai, punki rai*, TAM. in Ceylon), SING. NOTE.—The Telegu, Kanarese, and other South Indian names denote the use of the green twigs as torches.

References.—*Roxb., Fl. Ind., Ed. C.B.C., 128; Brandis, For. Fl., 275; Kurz, For. Fl. Burm., II., 21; Pegu Report, lxxiii.; Beddome, Fl. Sylv., t. 222; Gamble, Man. Timb., 230; Dalz. & Gibs., Bomb. Fl., 113; Elliot, Fl. Andhr., 93, 98, 112, 161; Rev. A. Campbell, Rept. Chutia Nagpur,*

No. 8411; Mason, Burma and Its People, 786; Trim., Cat. Ceylon Pl., 44; Lisboa, U. Pl. Bomb., 88, 344; Indian Forester, III., 203; Gazetteers:—Mysore and Coorg, I., 60, 70; Thana, 25; Kanara, 71; Settle. Rept., Chanda, C. P., App. VI.

Habitat.—An evergreen shrub or small tree of Western Bengal, Behar, Burma, Western, Central, and South India: from the Satpura range southwards. Distributed through Chittagong, and Pegu to Ceylon.

MEDICINE. Root. 516 Fruit. 517

Medicine.—**Rev. A. Campbell** states that the Santals employ the ROOT or FRUIT as a medicine to be given to females when the urine is high coloured.

FOOD & FODDER. 518

Food and Fodder.—The Santals eat the ripe FRUIT. In the Central Provinces buffaloes are said to eat the leaves.

Structure of the Wood.—Light brown, smooth, very hard, close-grained. Well suited for turning and might do for engraving. **Beddome** remarks that it is used for furniture and building purposes. **Kurz** says that it takes a good polish, and **Lisboa,** that it is used for fuel. Weight 57 to 66℔ per cubic foot.

DOMESTIC. 519

Domestic.—The branches are employed for torches and are used for that purpose, by letter carriers (*Brandis*).

JACK-FRUIT.

Jack-fruit, see **Artocarpus integrifolia**, *Linn.*; Vol. I., 330.

JADE.

Jade, *Ball, in Man. Geol. of India, III., 516.* 1

Under the name Jade several different minerals are included, which are not always easily ditsinguishable. Scientifically, however, the name is restricted to a definite substance called NEPHRITE. This term, derived from *νεφρός*, the kidney, refers to the reputed value of the mineral in renal diseases, whence also it was formerly known as *Lapis nephriticus* by the Romans. The word Jade is probably a corruption of the Spanish "*hijada,*" from the old name "*piedra de hijada*" applied by the Spaniards in Mexico and Peru to the mineral.

True jade or Nephrite is a native silicate of calcium and magnesium, and may be regarded as a crypto-crystalline variety of horneblende. It has a specific gravity of from 2·91 to 3·06. Most specimens can be scratched by flint or quartz, but the mineral is said to become harder after exposure of its broken surfaces. It is compact, non-crystalline, and extremely tough, and varies in colour, from every shade of green to yellowish grey or white.

JADEITE. 2

JADEITE is one of the substances most commonly confused with true jade, from which it differs in its more complicated structure, containing as it does a number of bases combined with silica. Much of the Chinese so-called jade is in reality Jadeite. This substance was employed for making axes, and for ornamental purposes, by the inhabitants of the ancient Swiss Lake-dwellings, by the Egyptians, and by the Mexicans, &c.

A considerable proportion of the Chinese "Jade" also consists of another mineral, Prehnite (a silicate of lime and alumina). Amongst other substances passed off as Jade are serpentine, and clever imitations in glass, which may be seen even in the bazárs of Burma and Central Asia.

Vern.—*Yashm*, HIND; *Sang-i-yashab, changtaw, sang-i-kas*, PB.; *Yashm, sang-i-yashm*, PERS.; *Sútashi, kashtashi*, TURKI.

References.—*Mallet, Geol. of India, IV. (Mineralogy), 85, 94; Balfour, Cyclop., II., 395; Forbes Watson, Ind. Survey, I., 413; Baden Powell, Pb. Man., 148, 202; Cat. Panjáb Pro. Cal. Exhib., 28; District Manual, Coimbatore, Madras, 72; Burm. Admin. Rep.:—1874-75, 86, 89; 1882-83, 79, 80, 84, 86, 89, 96, app. cxx., cxxii., cxxiv., cxxvi., cxxviii., cxxx.; 1883-84, 43, 44, 48, app. cxii., cxiv., cxvi., cxviii., cxx., cxxii.; 1884-85, 35, 36; Rep. Yarkhand Mission, 1873, 75; Anderson, Expedition to Yunan, 66, 827; Mallet, Rec. G. S. I., V., 22; Stoliczka, Rec. G.S.I., VII., 51; Hannay & Prinsep, J. A. S. B., VI., 265; Yule, "Marco Polo," I., 177; Warry, Rep. Jade-mines of Mogaung, 1888; Rev. Dept. No. 522-1m., 1887, Correspondence on Minerals of Burma; Indian Agri., Nov. 27, 1886; May 19, 1888.*

SOURCES.

SOURCES. Bengal. 3

Begnal.—The horneblende rock west of Dumrahur and Urjhut, in South Mirzapur, passes into a finely granular to nearly compact tremolite forming a coarse Jade. In this state it is met with, more especially between Kotamowa and Bhamni, and at the top of Kurea Ghát. An olive-green Jade occurs north-west of Kisari, and the mineral is also found associated with the Corundum at Pipra (*Ball; Mallet*).

Turkistan. 4

Turkistan.—For many centuries Jade mines have been worked by the Chinese at Karakásh on the Kuenlun range. These mines were visited and described by **Dr. Stoliczka** in 1874, from whose account it appears that the rocks in which the Jade-bearing veins occur are syenitic gneisses with micaceous and horneblendic schists. The mineral varies from pale

SOURCES.

green to a dark green, but some of a white colour also occurs. Certain of the veins observed by **Dr. Stoliczka** were 10 feet thick. Associated with the Jade in these veins a coarsely crystalline white dolomite was found. Besides these mines at Karakásh, Jade is found in most of the streams issuing from the Kuenlun, and south of Kotan other mines of great antiquity exist, which are mentioned as producing Jade by Chinese authors of 2,000 years ago (*Ball; Mallett*).

Upper Burma. 5

Upper Burma.—The Mogaung District has long been known to the Chinese as a source of a very valuable white or greenish "Jade." **Mallet** states that all the specimens of Mogaung so-called jade which he has had an opportunity of examining, have proved by their specific gravity, fusibility, texture, hardness, and colour to be Jadeite, a mineral which the Chinese hold in higher value than true Jade or Nephrite. The Mogaung mines were first described by **Dr. Anderson** in his Report on the Expedition to Yunan, 1871, but since the British occupation, they have attracted much attention, and have formed the subject of an exhaustive and interesting report by **W. Warry, Esq.**, Political Officer, Bhamo. From this it appears that the discovery of the mineral was accidentally made by a small Yunanese trader in the thirteenth century. A large party was organised for the purpose of returning to procure more from the same place, but they appear to have been unsuccessful. Another attempt was made by the Yunan Government in the fourteenth century, but all the members of the expedition are said to have perished by malaria or at the hands of hostile hill tribes. No further exploration of the Jade country seems to have been made by the Chinese till 1784, when, on the termination of hostilities between China and Burma, a regular trade opened out between the two countries. Adventurous bands of Chinese soon discovered that the Jade-producing districts lay on the right bank of the Uru river, and a small but regular supply of the stone was now conveyed every year to Yunan. Impracticable roads, a malarious climate, and an unsettled country, however, prevented the expansion of the trade, and year after year numbers of Chinese merchants perished in their attempts to procure the stone.

Early in the present century the Burmese Kings seem to have become aware of the importance of the Jade trade and of the revenue which it might be made to yield. In 1806 a Burmese Collectorate was established at the site of what is now the town of Mogaung, and a guard was regularly stationed at the mines to protect the trade and to maintain order. Regarding the duty levied, and the mine regulations **Mr. Warry** writes:—

"Mogaung now became the head-quarters of the Jade trade in Burma. Comparatively few Chinese actually went up to the mines; the Kachins themselves brought down most of the stone to Shuitunchun, a sandbank opposite Mogaung, where a large bazár was held during the season. The Burmese Collector imposed no tax upon the stone until it was ready to leave Mogaung, when he levied an *ad valorem* duty of 33 per cent. and issued a permit which was examined by his deputy at Tapaw, one day's journey from Mogaung by river. After this the stone passed freely anywhere in Burma without further charge or inspection. The value of Jade was determined for purposes of taxation by an official appraiser. This officer, however, by private arrangement with the traders and the Collector, estimated all stone at about one-third of its real value. The actual duty paid was therefore small and business proceeded smoothly, cases of friction between the traders and the customs officers being of very rare occurrence. All payments were made in bar silver. The metal used was at first fairly pure, but it was soon debased by a large admixture of lead. Rupees did not come into general use until 1874."

Besides the duty levied at Mogaung, the stone had to bear certain

SOURCES. Upper Burma.

minor charges at the mines and at Namiakyaukseik, distant one day's journey from the former.

"Under the system just described the Jade trade continued to flourish for many years. The period of its greatest prosperity is comprised within the years 1831-40, during which time at least 800 Chinese and 600 Shans were annually engaged in business or labour at the mines. All the stone was sent by one of the above mentioned routes to Yunnan-fu, at this time the great emporium of the trade. The business there was mainly in the hands of Cantonese merchants, who bought the rough stone in large quantities and carried it back to be cut and polished at Canton.

"In 1841 war broke out between Great Britain and China. Hostilities first commenced at Canton, and the effect on the Jade trade was not long in making itself felt. Cantonese merchants no longer came to buy stone at Yunan-fu. Stocks accumulated, and Yunan traders ceased to go up to the mines. The Kachins, suffering from this stoppage of business, made urgent representations to the Burmese at Mogaung; and in 1842 a Burmese officer proceeded from Mogaung to Momein to inquire if any offence had been given to Chinese traders that they did not come as usual to the mines.

"There was a partial revival of the trade for a few years commencing with 1846, but the disturbed state of Southern China, consequent upon the Taiping rebellion of 1850, prevented a complete recovery; and with the outbreak of the Panthay rebellion in 1857 the roads leading to Yunnan-fu were blocked and all business in Jade came to a standstill for several years.

"During the early part of the period just passed in review the Chinese estimate that the average amount of duty collected each year did not exceed R6,000, the output of Jade being small and the official appraisers venal. About the year 1836, when the trade was most flourishing, R21,000 was the probable amount of the annual collection. After 1840 the duty fell to R3,000 or less, and then it dwindled away to nothing. The above estimates are probably below the mark, as the Chinese would, for obvious reasons, be inclined to understate the real amount.

"The year 1861 witnessed a great improvement in the Jade trade. From that date until now the bulk of the stone has been carried by sea to Canton. In 1861 the first Cantonese merchant arrived in Mandalay. He bought up all the old stocks of Jade and conveyed them to China by sea, realising a large fortune on this single venture. His example was quickly followed by other Cantonese, and once more the trade in Jade revived, and numerous Yunanese went up to the mines. The principal quarries were now at Sanka. Stone had been discovered there many years before but had been pronounced poor in quality and scarcely worth the trouble of working. Now, however, upon a second trial, it proved to be equal or superior to that from the earlier mines, the colour having, as the Kachins alleged, matured and deepened in the interval. The annual duty collected at this time probably amounted to at least R27,000.

"Hitherto the collection of the duty had been in the hands of an official who had paid a very high price at Ava for his appointment and who was in the habit of remitting to the capital only as much as he thought fit—usually about one-fifth—of the actual receipts. In 1866 the tax was farmed out for the first time. The price obtained was R60,000 for a three years' lease. At the expiration of this term the King, dissatisfied with the amount of the Jade revenue, determined to buy all the stone from the Kachins himself, and appointed a high official to act as his agent at the mines. For a whole season Chinese and other dealers in Jade were excluded from the mines. The stone, as it was dug up, was purchased by

Jade and Jadeite.

SOURCES. Upper Burma.

the King's agent, carried to Mogaung, and there retailed to the traders. This arrangement was of course highly unsatisfactory to the Kachins, who first protested against the exclusion of other purchasers, and then, finding their protest of no avail, resorted to the much more effectual method of curtailing the supply of stone and producing only pieces of indifferent quality. For this reason the King's experiment was a failure, and the total revenue he secured did not equal the proceeds derived from the sale of the monopoly in the preceding year. The Chinese explain the failure on other grounds. The experiment, they say, was doomed from the outset 'owing to the inherent impropriety of a sovereign descending into the arena of trade and taking the bread out of the mouths of his own subjects.'

"During the years 1870, 1871, and 1872 the King obtained an annual remittance of R12,000 from the Collector at Mogaung on account of the Jade duty. In the following year new deposits of fine Jade were discovered at Mantiemmo, and the King again determined to become the sole purchaser from the Kachins. On this occasion, too, the revenue he realised fell far below the average of former years.

"In 1874 the old system was reverted to and the collection amounted to R6,000. Once more, in 1875, the King undertook to buy the stone himself from the Kachins, and again the experiment failed, though not so badly as on the two previous attempts. About this time the Iku quarry was discovered, and the output being very good the right of collecting the duty was sold in 1876 for three years for the sum of R60,000. In 1880 Wu Chi, the son of a Canton Chinaman by a Burmese mother, obtained a three-years' lease of the monopoly at the rate of R50,000 a year. In the second year of his term the Tomo quarries were opened and he made an immense fortune.

"In the autumn of 1883 Mogaung was sacked by the Kachins, and during the ensuing winter and spring there was no trade in Jade. In June 1884, order having been partially restored, a Chinese syndicate, represented by Li Te-su, took the monopoly for three years, agreeing to pay R10,000 the first year, R15,000 the second, and R20,000 the third.

The up-country was still unsettled, and the lessees, by arrangement with the traders, were permitted to collect duty at Bhamo instead of as heretofore at Mogaung. During the first two years of their term, owing to the disturbances connected with the adventure of Hsiao Chin and the British occupation of Upper Burma, they collected little or no duty; but the proceeds of the third year left them with a margin of R20,000 over and above their total expenses for the three years."

The tax (at the same rate of 33 per cent. *ad valorem*) was then farmed out by the British Government to a Cantonese lessee for about R50,000, who appears to have been very strict in exacting his rights, levied new and unauthorised taxes, and made every endeavour not only to collect the duty, but to get the management of the mines into his own hands. As a result he became extremely unpopular, and in 1887 was assaulted and outraged at Mogaung.

Method of Mining. 6

METHOD OF MINING.—The method of mining followed in the deeper quarries is described as follows by **Mr. Warry**:—

"In important matters, such as the discovery or the opening of a Jade mine, the action of the Kachin tribes is entirely determined by superstitious considerations. In their search for stone they are guided by indications furnished by burning bamboos; when it is discovered, favourable omens are anxiously awaited before the discovery is announced to the Kachin community. A meeting is then convened by the Chief Sawbwa, and again sacrifice and other methods of divination are resorted to in

SOURCES.

Upper Burma.

order to ascertain if the mine should be worked at once or be allowed to remain undisturbed for a period of years until the colour—such is the Kachin belief—is sufficiently matured. If the indications are favourable to the immediate opening of the mine, the land at and around the out-cropping stone is marked out by ropes into small plots a few feet square which are then apportioned among all the Kachins present. No Kachin belonging to the same family is refused a share, no matter how far away he may live

"The ground thus parcelled out, traders are invited to the mine, and after an elaborate ceremonial it is declared open and the digging commences. A similar ceremonial is held at the opening of each successive season. This year (1888) the sacrifices were on an unusually large scale, an abundant output being desired in order to meet expected orders on behalf of the Emperor of China who is to be married shortly. On the occasion of the Emperor Tungchih's marriage in 1872 it is said that a sum amounting to four lakhs of rupees was expended at Canton in buying Jade for use at the ceremony, and a great impulse was thereby given to the trade in Burma.

"The Kachins have always claimed the exclusive right of digging at the mines. They have, however, from time to time allowed Shans to assist them, and in the early days Chinese were permitted to work certain quarries temporarily abandoned by the Kachins. The Chinese, however, found the labour severe and the results unsatisfactory, and they have now for many years contented themselves with buying stone brought to the surface by Kachins.

"The season for Jade operations begins in November and lasts until May, when the unhealthiness of the climate compels all traders to leave and the flooding of the mines suspends further operations on the part of the Kachins.

"This flooding of the deepest and most productive quarries is the greatest difficulty with which the Kachins have to contend, and they have spent much labour and money in devising expedients, with indifferent success, to meet it. There were at the time of our visit elaborate bamboo structures over some of the largest quarries for the purpose of baling out the water. When the floor of the pit can be kept dry for a few hours,—and this is as a rule possible only in February and March,—immense fires are lighted at the base of the stone. A careful watch must then be kept, in a tremendous heat, in order to detect the first signs of splitting. When these occur the Kachins immediately attack the stone with pickaxes and hammers, or detach portions by hauling on levers inserted in the crack. All this must be done when the stone is at its highest temperature, and the Kachins protect themselves from the fierce heat by fastening layers of plantain leaves round the exposed parts of their persons. The labour is described as severe in the extreme and such as only a Kachin would undertake for any consideration. The heat is insupportable even for onlookers at the top of the mine, and the mortality among the actual workers is very considerable each season. The Chinese take a malicious pleasure in reminding the Kachins that in the early days when quarrying was easy the right of digging was jealously withheld from outsiders; and they assure them that under present conditions they need not be apprehensive of an infringement of their monopoly."

In summing up the present condition of the Jade-mining industry, and the trade in the stone in Burma, Mr. Warry writes :—

"The demand for Jade is universal throughout China, and the price of the best stone shows no tendency to fall. Burma is practically the only source of the supply, and there seems no reason to think that the supply is likely to fall short of the demand. Considering the large area over

JADE. **Jade and Jadeite.**

SOURCES. Upper Burma.

which the Jade stone has at one time or another been discovered, the impracticable nature of the country, covered for the most part with thick jungle, and the rough character of the prospecting, which consists merely in examining large and obvious outcropping stones, it is probable that the jade hitherto discovered bears a very small proportion to that still concealed. It is likely, therefore, that in the Jade country our Government possess a source of revenue capable of considerable development. Putting out of sight the probability of future discoveries of jade, there is no doubt that the revenue derived from the present mines might be much improved if free access could be obtained to the country. The introduction of European appliances, which should supersede the present injurious method of working the quarries, would add considerable value to the output, a good part of which is now calcined by the action of heat. And the smuggling of stone overland to China would at the same time be effectually prevented."

The rate of taxation and the method of levying it is at present under the consideration of Government.

TRADE. 7

Trade.—For a statement of the exports from Burma during the years from 1880 to 1886 the reader is referred to the article **Cornelain,** *Vol. II., 168.* In continuation of the table there given, it may be stated that the exports in the last year for which statistics are available (1888-89) were 4,898 cwt., value R6,54,040.

It would thus appear that the Jade trade from Burma which had seemingly fallen off from 1881-82 onwards has again begun to revive, and hopes may be entertained that it will continue to increase as the country becomes more settled.

A considerable amount of a false Jade, in reality serpentine, is imported into the Panjáb from Afghánistan. It is said to be found near Kandahar, and to be brought down the Indus on rafts floated with inflated skins to Attock, whence it is carried to Bhera in the Shahpur district to be used in the manufacture of knife-handles.

MEDICINE. 8

Medicine.—Liquor drunk from a Jade or Agate cup is supposed to allay palpitation of the heart, and the cup itself is believed to act as a protection against poisoning.

DOMESTIC. 9

Domestic.—Cups, vessels, and ornaments of Jade are highly valued by the Chinese, and also by certain classes in India. Not only are its hardness, weight, sonoriety and peculiar sombre tint admired, but wearing the stone is supposed to impart to the wearer many good qualities and virtues. Easy to work when first extracted, Jade acquires a considerable degree of hardness after exposure, a property which together with its toughness renders it very useful as a material for ornaments. In Burma, it is principally employed in the manufacture of finger- and ear-rings, bracelets, chains, &c. The chief seat of the manufacture of these ornaments is at Momein, where the stone, according to **Dr. Anderson,** is cut by means of circular discs of copper and rotating conical tipped cylinders, charged with siliceous mud and what appears to be ruby dust. At Momein a pair of bracelets of the finest Jade cost about R100. The most highly valued Jade or Jadeite is of an intense bright green colour resembling emerald, but red and pale pinkish varieties are also prized.

Exceptionally beautiful specimens of Indian Jade work date from the times of Jehangír and Shah-Jehan, both of whom appear to have taken much pleasure in cups and ornaments made of this stone. At the present time, the stone is not so much employed as it was formerly in Indian art work, still figures, cups, and other ornaments are occasionally made from it, especially in Southern India. As already mentioned, articles of spurious Jade, glass, &c., are frequently met with, and are passed off by merchants as the true article. Perhaps the most commonly employed of these, in India, is the Afghán Jade-like serpentine, which is largely used in Bhera (Shahpur district) for making hilts of *peshquabs* (or Afghán knives), and of the hunting or dinner knife. Baden Powell describes the method of cutting this stone as follows:—"The pale green plasma called '*sang-i-yesham*' is cut by means of an iron saw, and water mixed with red sand and pounded '*kurand*'" (corundum). "It is polished by application to the '*sán*'" (polishing wheel), "wetted with water only, then by being kept wet with water, and rubbed with a piece of '*wáti*' (a smooth fragment of stoneware, pottery or potshera), and lastly by rubbing very finely pounded burnt *sang-i-yesham* on it. This last process must be done very thoroughly. The other stones used for handles come from the Salt Range and the hills near Attock." DOMESTIC.

Jaggery, see **Borassus flabelliformis,** *Linn.* (Vol. I., 495); **Caryota urens** *Linn.* (Vol. II., 208), **Cocos nucifera,** *Linn.* (Vol. II., 449), also **Phœnix sylvestris,** *Roxb.*; Vol. V. Consult also the article, **Narcotics,** Vol. V.

Jalap, see **Ipomœa purga,** *Hayne*; see p. 488.

JASMINUM, *Linn.*; *Gen. Pl., II.*, 674.

A genus of scandent or erect shrubs, belonging to the Natural Order OLEACEÆ, and comprising 90 species distributed over the tropics or warm temperate parts of the Old World. Of these, some 43 are natives of India.

Jasminum angustifolium, *Vahl.*; *Fl. Br. Ind., III.*, 598; *Wight, Ic.*, *t.* 698; OLEACEÆ. 10

Syn.—JASMINUM VIMINEUM, *Willd.*; J. TRIFLORUM, *Pers.*; NYCTANTHES ANGUSTIFOLIA, *Linn.*; N. VIMINEA, *Retz.*; N. TRIFLORA, *Burm.*; MOGORIUM VIMINEUM, *Lamk.*; M. TRIFLORUM, *Lamk.*

Vern.—*Mwari, ban-mallica,* HIND.; *Chattu mallika, caat-mallica,* TAM; *Chiri malle, adevie-mallie,* TEL.; *Katu-pitsjegam,* MALAY.; *Walsuman pichcha,* SING.; *Kanana mullika, asphota, vana malli,* SANS.

References.—*Roxb., Fl. Ind., Ed. C.B.C., 32; Elliot, Fl. Andhr., 42; Ainslie, Mat. Ind., II., 52; Drury, U. Pl., 267; Balfour, Cyclop., II., 419; Treasury of Bot., I., 636; Gazetteer, N.-W. P., I., 82.*

Habitat.—A common shrub on the lower hills of the Deccan and Ceylon.

Medicine.—Ainslie writes: "The bitter ROOT, ground small and mixed with powdered *vassumboo* (root of **Acorus Calamus**) and lime-juice, is considered a valuable external application in cases of ringworm." MEDICINE. Root. 11

Domestic.—Roxburgh says: "It grows easily in every soil and situation, is constantly covered with leaves, the bright, shining, deep colour of which renders it always beautiful, and particularly well adapted for screening windows, covering arbours, &c." DOMESTIC. 12

J. arborescens, *Roxb.*; *Fl. Br. Ind., III.*, 594; *Wight, Ic.*, *t.* 699. 13

Syn.—JASMINUM MONTANUM, *Roth*; J. ARBOREUM, *Rœm.*

VAR. **latifolia,** *Roxb.*; *Wight, Ic., t.* 703.

VAR. **montana,** *Roth*; J. PUNCTATUM, *Wall.*; J. GLABELLUM, *Wall.*

Vern.—*Saptala, nava-mallika, bela, muta-bela,* HIND.; *Bura-kúnda, nuva-mullika,* BENG.; *Gada hund baha,* SANTAL; *Kussar,* BOMB.; *Kusara,* MAR.; *Adavi malle,* TEL.; *Saptala, nava-mallika, madhvi,* SANS.

References.—*Roxb., Fl. Ind., Ed. C.B.C., 32; Brandis, For. Fl., 311; Dalz. & Gibs., Bomb. Fl., 138; Works of Sir W. Jones, V., 73; Rev. A. Campbell, Ec. Prod., Chutia Nagpur, No. 9277; Elliot, Fl. Andhr., 11; Dymock, Mat. Med. W. Ind., 2nd Ed., 475; Atkinson, Him. Dist., 313; Lisboa, U. Pl. Bomb., 223; Balfour, Cyclop., II., 420; Indian Forester, VI., 321; XI., 273; Gazetteers:—Bombay (XV., 71), 437; N.-W. P., IV., 74.*

Habitat.—A large shrub, or scrubby tree, native of the Tropical North-West Himálaya, also of the hot lower hills of Southern India and Ceylon at altitudes from 500 to 3,000 feet. *Var.* **latifolia** occurs along the base of the Himálaya from Kumáon to Bengal, while *var.* **montana** is a native of the Deccan.

OIL. 14 **Oil.**—**Lisboa** mentions this species in his list of oil-yielding plants, and states that a volatile oil may be obtained from it by distillation.

MEDICINE. 15 **Medicine.**—The **Rev. A. Campbell** states that a preparation of the plant is given by the Santals in certain menstrual complaints. **Dymock** describing *var.* **latifolia** writes: "The JUICE OF THE LEAVES is used with pepper, garlic, and other stimulants as an emetic in obstruction of the bronchial tubes by viscid phlegm. Seven leaves will furnish sufficient juice for a dose. For young children the juice of half a leaf and of four leaves of *agásta* (**Sesbania grandiflora**) may be mixed with two grains of black pepper, and two grains of dried borax and given in honey." **S. Arjun**

Leaves. 16 (*Cat. Bomb. Drugs, 89*) states that the LEAVES of the same variety are slightly bitter and astringent, and might be used as a tonic and stomachic.

FOOD. 17 **Food.**—The seeds of *var.* **latifolia** were eaten at Poona during the famine of 1877-78.

18 **Jasminum grandiflorum,** *Linn.; Fl. Br. Ind., III., 603; Wight, Ic., t. 1257.*

THE SPANISH JASMINE.

Syn.—JASMINUM AUREUM, *Don.*

Vern.—*Chámeli, játi,* HIND.; *Játi,* BENG.; *Jáhi,* N.-W.P.; *Chamba, chambeli, jati,* PB.; *Chambeli,* BOMB.; *Ghambeli,* GUZ.; *Jaji,* TEL.; *Myatlæ,* BURM.; *Játi,* SANS.

References.—*Roxb., Fl. Ind., Ed. C.B.C., 34; Brandis, For. Fl., 313; Kurz, For. Fl. Burm., II., 150; Gamble, Man. Timb., 255; Stewart, Pb. Pl., 141; Sir William Jones, V., 74; Elliot, Fl. Andh., 72; Mason, Burma and Its People, 409, 802; U. C. Dutt, Mat. Med. Hind., 190, 300; Dymock, Mat. Med. W. Ind., 2nd Ed., 474; Baden Powell, Pb. Pr., 359; Atkinson, Him. Dist., 740; Piesse, Art of Perfumery, 133-137; Balfour, Cyclop., II., 420; Treasury of Bot., I., 626; Agri.-Hort. Soc. of India, Jour. X. (Old Series), 15; IV. (New Series), 118; Gazetteers:—Mysore and Coorg, II., 7; N.-W. P., IV., 74; I., 82; X., 313.*

Habitat.—A large, scandent, glabrous shrub, found wild in the sub-tropical North-West Himálaya, at altitudes from 2,000 to 5,000 feet, often cultivated throughout India.

OIL. Flowers. 19 **Oil.**—The fragrant FLOWERS are used in India for preparing a scented oil, which is considered cooling and is much prized by the richer classes of Natives. In certain localities such as Kumáon it is also employed for making perfumed waters. In Europe, the flower of the Spanish Jasmine is very highly esteemed by the perfumer, owing to the delicacy and sweetness of its odour, and to the fact that it is impossible to imitate it by artificial combinations. **Piesse** states that the chief source of Jasmine oil is near Cannes in the South of France, where the plant is largely cultivated. An acre of land yields about 500fb of blossom, which is said to be bought by manufacturing perfumers at the rate of 1*s.* 6*d.* to 2*s.* 6*d.* the pound. In the perfumer's laboratory the method of obtaining the odour is by absorption or "*enfleurage.*" In this process the flowers are sprinkled over a layer of lard and beef suet, the operation being repeated every few days with fresh flowers as long as the plants remain in blossom. The grease absorbs the odour, and is finally melted at as low a temperature as possible and strained. Oils are similarly prepared, a cotton cloth steeped in olive oil being substituted for the layer of lard and suet. The extract of jasmine is prepared by adding one quart of rectified spirit to two pounds of jasmine fat or oil, and leaving the mixture to stand for a fortnight at summer heat. The extract is then strained off, and the still-scented residue of fat or oil is sold as pomatum or hair oil. The extract is highly valued, and enters into the composition of a great many of the handkerchief perfumes of Europe. From Jasmine oil a crystalline camphor has been separated, which is inodorous, fusible at 125°, hardly soluble in water and easily soluble in spirit and ether. It has been suggested by Gmelin that this so-called camphor may be a fatty acid. OIL.

Medicine.—The plant has long been known and recommended by Sanskrit and Muhammadan writers. The former believe the oil prepared from the FLOWERS to be cooling when applied externally, the LEAVES to be of value internally, in cases of skin diseases, aphthæ, otorrhœa, &c., and the FRESH JUICE OF THE LEAVES to be a valuable application for corns between the toes. According to Muhammadan writers the plant is deobstruent, anthelmintic, diuretic, and emmenagogue. The flowers made into a plaster are also supposed to act as an aphrodisiac (*U. C. Dutt, Dymock, &c.*) Atkinson states that in the North-Western Provinces "the flowers and their essence are used as an application in skin diseases, headache, and weak eyes; the leaves are used in toothache." MEDICINE. Flowers. 20 Leaves. 21 Juice. 22

SPECIAL OPINIONS.—§ "The leaves when chewed are found useful in ulceration of the mucous membrane of the mouth. Boiled in ghí they are likewise frequently used with benefit for the same affection" (*Civil Surgeon C. J. W. Meadows, Barrisal*). "The oil is largely employed as an external application to the healthy skin to protect it from the effects of dry and cold winds" (*Surgeon-Major A. S. G. Jayakar, Muskat*).

Structure of the Wood.—Similar to that of **J. officinale.** In Catalonia and Turkey the plant is cultivated for its long straight stems, which are made into slender pipe-tubes, much prized by the Moors and Turks. TIMBER. 23

Jasminum humile, *Linn.; Fl. Br. Ind., III., 602; Wight, Ic., t. 1258.* 24

Syn.—JASMINUM CHRYSANTHEMUM, *Roxb.*; J. REVOLUTUM, *Sims.*; J. BIGNONIACEUM, *Wall.*; J. WALLICHIANUM, *Lindl.*; J. INODORUM, *Jacquem.*

Vern.—*Malto, pitmalti,* HIND.; *Svarnajui,* BENG.; *Sonajahi,* KUMAON; *Chamba, juari, tsonu, summun, jái, kuja, sím, ri, shing, púring, márti, naugei,* PB.; *Pachcha adavi molla,* TEL.; *Svarnajuthiká, hemapushpiká,* SAN

References.—*Roxb., Fl. Ind., Ed. C.B.C., 33; Brandis, For. Fl., 313; Gamble, Man. Timb., 255; Stewart, Pb. Pl., 141; Elliot, Fl. Andhr., 69, 139; Aitchison, Fl. Kuram Valley, 147; Stewart, Bot. Tour in Hazara and Khágán, in Jour. Agri.-Hort. Soc. of India, XIV. (Old Series), (App. 15; U. C. Dutt, Mat. Med. Hind., 299, 319; Baden Powell, Pb. Pr., 359; Atkinson, Him. Dist., 313; Drury, U Pl., 267; McCann, Dyes and Tans, Beng., 154; Balfour, Cyclop., II., 420; Indian Forester, II., 25, 26; V., 184; XII., App. 16 and 28; Gazetteer, N.-W. P., IV., 74.*

Habitat.—A small, erect, rigid shrub, native of the sub-tropical Himálaya from Kashmír to Nepal, at altitudes from 2,000 to 5,000 feet; found also in South India and Ceylon, from 2,000 to 6,000 feet; widely cultivated in gardens throughout India.

Dye.—Aitchison states that in the Kuram Valley a yellow dye is extracted from the ROOTS. It is curious that this fact should apparently be unknown to the hill tribes in India where the plant is equally abundant. A dye-stuff much used in Chittagong under the name *júrí* may, however, possibly be derived from this jasmine (*Dr. G. Watt*). DYE. Roots. 25

Oil.—Like many other jasmines, this species bears FLOWERS which yield an aromatic essential oil, used in Native perfumery. OIL. Flowers. 26

Medicine.—The ROOT is said to be useful in ringworm (*Baden Powell*). MEDICINE. Root. 27

SPECIAL OPINIONS.—§ "The milky JUICE which exudes on an incision in the bark of this plant, is alleged to have the power of destroying the Juice. 28

MEDICINE. 29 unhealthy lining walls of chronic sinuses and fistulas" (*Surgeon-Major Bankabehari Gupta, M.B., Pooree*).

Jasminum officinale, *Linn.; Fl. Br. Ind., III., 603.*

Vern.—*Chamba,* HIND.; *Chambeli,* KUMAON; *Chamba, chirichog, kiri,* KASHMIR; *Bansu, kwer, dúmni, dassi, suni, somun, samsem, sim, ri, shing, púring, márti, naugri, jái,* PB.

References.—*Brandis, For. Fl., 313; Stewart, Pb. Pl., 141; Boiss., Fl. Orient., IV., 43; O'Shaughnessy, Beng. Dispens., 51; Balfour, Cyclop., II., 420; Smith. Dic., 227; Treasury of Bot., I., 636.*

Habitat.—A climber of the Salt Range and Kashmír, at altitudes from 3,000 to 9,000 feet. Often cultivated in India, China, and other countries; distributed through Cabul and Persia.

OIL. Flowers. 30 **Oil.**—The FLOWERS of this species yield a fragrant oil, similar to that already described under J. **grandiflora.**

MEDICINE. Root. 31 **Medicine.**—The ROOT has been found useful in ringworm (*Stewart*).

32 **J. pubescens,** *Willd.; Fl. Br. Ind., III., 592; Wight, Ic., t 702, 1248.*

Syn.—JASMINUM HIRSUTUM, *Willd.;* J. MULTIFLORUM, *Roth;* J. CONGESTUM, *Wall.;* J. BRACTEATUM (by error "FRACTIATUM"), *Wight;* NYCTANTHES PUBESCENS, *Retz.;* N. MULTIFLORA, *Burm.;* MOGORIUM PUBESCENS, *Lamk.*

VAR. **bracteata.** *Roxb., Hort. Beng.*

Vern.—*Kúnd phul, kúnda, kúndo,* HIND.; *Kúnd phul, kúnda,* BENG.; *Mógra,* MAR.; *Katu-tsjiregam-mulla,* MALAY.; *Kúnda,* SANS.

References.—*Roxb., Fl. Ind., Ed. C.B.C., 31, 32; Brandis, For. Fl., 312; Kurz, For. Fl. Burm., II., 154; Dals. & Gibs., Bomb. Fl., 138; U. C. Dutt, Mat. Med. Hind., 306; S. Arjun, Bomb. Drugs, 89; Atkinson, Him. Dist., 313; Balfour, Cyclop., II., 420; Gazetteers:—N.-W. P., IV., lxxiv; Bombay, XV., Pt. I., 437.*

Habitat.—A large tomentose shrub, common throughout India up to an altitude of 3,000 feet; distributed through Burma to China. It is also frequently cultivated in gardens.

MEDICINE Leaves 33 **Medicine.**—S. Arjun states that "the dried LEAVES soaked in water and made into a poultice are used in indolent ulcers to generate a healthy action." The ROOT is said to be an efficient antidote in cases of snake-

Root. 34 bite.

35 **J. Sambac,** *Ait.; Fl. Br. Ind., III., 591; Wight, Ic., t. 704.*

THE ARABIAN JASMINE.

Syn.—JASMINUM FRAGRANS, *Salisb.;* J. UNDULATUM, *Willd.;* J. ZAMBAC, *Roxb.;* J. QUINQUEFLORUM, *Heyne;* J. PUBESCENS, *Wall., not of Willd.;* NYCTANTHES SAMBAC *and* UNDULATUM, *Linn.;* MOGORIUM SAMBAC, *Lamk.;* M. UNDULATUM, *Lamk.*

VAR. 1.—**Sambac proper**; corolla tube not twice the length of the calyx.

VAR. 2.—**Heyneana,** corolla tube 2 to 5 times as long as the calyx teeth.

Vern.—*Motia, bel, banmalliká, mogra,* HIND.; *Bel, banmalliká, nabamalliká, mallik, mogra,* BENG.; *Chamba, chambeli, múgra,* PB.; *Mogri,* BOMB.; *Mográ,* MAR.; *Mogro,* GUZ.; *Mallippu, malligaip-pú,* TAM.; *Boddu malle, nava málika, malle,* TEL.; *Mallige,* KAN.; *Pun mulla, mulláchcha-pú, mullappú,* MALAY.; *Thæmbaumali, sabay, mali,* BURM.; *Pich-chi-mal* (=flowers), SING.; *Navamáliká, malliká, vanamalliká, várshiki, asphota,* SANS.; *Saman, suman, yásaman, vardeabyaz,* ARAB.; *Zambak, gule-supéd,* PERS.

References.—*Roxb., Fl. Ind., Ed. C.B.C., 30; Brandis, For. Fl., 311; Kurz, For. Fl. Burm., II., 153; Dalz. & Gibs., Bomb. Fl., 137; Stewart, Pb. Pl., 141; Elliot, Fl. Andhr., 29, 110, 130; Mason, Burma, and Its People, 409, 802; Works of Sir W. Jones, V., 73, 74; Pharm. Ind., 136; Moodeen Sheriff Supp. Pharm. Ind., 162; U. C. Dutt, Mat. Med. Hind., 190, 308, 311, 322, 323; Dymock, Mat. Med. W. Ind., 393; Atkinson, Him. Dist., 313; Drury, U. Pl., 267; Lisboa, U. Pl. Bomb., 223; Cat. Baroda Durbar, C. I. Exhib., No. 119; Balfour, Cyclop., II., 420;*

Smith, Dic., 227; Treasury of Bot., I., 636; Kew Off. Guide to the Mus. of Ec. Bot., 94; For. Ad. Rep., Chutia Nagpur, 1885, 32; Gazetteers:—Mysore & Coorg, II., 7; Bombay, VII., 40; VIII., 184; Orissa, II., 179 (App. VI.); N.-W. P., I., 82; IV., lxxiv; Madras Manual of Administration, I., 362.

Habitat.—A fragrant climbing shrub, common throughout India, Burma, and Ceylon, at altitudes up to 2,000 feet; extensively cultivated in the tropics of both hemispheres.

OIL, Flowers. 36

Oil.—This species, together with **J. officinale** and **J. grandiflorum**, is extensively cultivated for its fragrant FLOWERS, from which a considerable proportion of the oil of Jasmine used in India is derived (see **J. grandiflorum**). The method of *enfleurage* is resorted to in India, as in Europe, for extracting the odorous principle, but instead of fat or oil, crushed sesamum seeds are employed. **Dr. Dymock** states that in Persia almonds are similarly used.

MEDICINE. Flowers. 37

Medicine.—The FLOWERS are, in Madras, generally believed to possess considerable powers as a lactifuge. For this purpose about two to three handfuls of the flowers are bruised, and unmoistened are applied to each breast, and renewed once or twice a day. The secretion is sometimes arrested in 24 hours, though the period is more frequently two to three days. **Dr. W G. King** states in a special opinion quoted below that the action of the fresh flowers is quicker than that of belladonna. The dried LEAVES soaked in water and made into a poultice are used in indolent ulcers. "The ROOT of the wild variety is used as an emmenagogue in Goa" (*Dymock*). **Baden Powell** states that this plant is used in the Panjáb as a remedy in cases of insanity, in weakness of sight, and affections of the mouth.

Leaves. 38
Root. 39

SPECIAL OPINIONS.—§ "Largely used by the natives as a lactifuge, the flowers being applied externally to the mammae" (*Surgeon W. F. Thomas, 33rd Madras Infantry, Mangalore*). "The flowers applied over the mammae suppress secretion of milk" (*Surgeon-Major A. F. Dobson, M.B., Bangalore*). "Stomachic—useful in heart-burn" (*Surgeon-Major J. M. Houston, Travancore*). "The flowers used as a poultice or wash next to the breast, suppress the secretion of milk" (*Surgeon-Major L. Beech, Cocanada*). "The fresh flowers when applied to the mammary gland direct, act more quickly as a lactifuge than belladonna" (*Surgeon W. G. King, M.B., Madras*). "The flowers applied to the breast act as a very efficient lactifuge" (*Surgeon-Major J. North, Bangalore*).

DOMESTIC & SACRED. Flowers. 40

Domestic and Sacred.—The FLOWERS are highly esteemed in the East on account of their fragrance, and are frequently referred to by Arabian, Persian, and Sanskrit poets. Those of one of the double varieties are held sacred to Vishnu, and are used as votive offerings in certain Hindu religious ceremonies. They are also supposed to form one of the darts of Kama Deva, the Hindu God of Love.

JATROPHA, *Linn.; Gen. Pl., III., 290.*

A genus of woody plants belonging to the Natural Order EUPHORBIACEÆ, and comprising about 70 species, chiefly natives of America. Of these, four are indigenous to India, and three others are naturalised or cultivated.

Jatropha Curcas, *Linn.; Fl. Br. Ind. V., 383;* EUPHORBIACEÆ. 41

Syn.—J. MOLUCCANA, *Herb. Russ.*

Vern.—*Bagberenda, safedind, páhári erand, jangli-arandi, bhernda, bag-bherenda,* HIND.; *Bágbherendá, bon-bhérandá, erandá-gáchh, safed ind, pahari erand,* BENG.; *Kulejera, totka bendi,* KOL.; *Bhernda,* SANTAL.; *Baigab,* ORISSA; *Kadam,* NEPAL; *Rattanjot, japhrota, jablota, pún,* seeds=*jamálgota,* PB.; *Mogalieranda, kurukarlu, maraharalu,* BOMB.; *Mogali eranda, rána-yerandi,* MAR.; *Ratanjota, jamalgota, jangli-araudi,* GUZ.; *Erundi, ehandrajot, jangli-yarandi,* DEC.;

Jatropha Oil.

Kattámanakku, koat amunak, kaat-amunck, caar-noochie, TAM.; *Népálam, adivieamida, pépálam,* TEL.; *Maranarulle, maraharalu, bettada-haralu,* KAN.; *Kaak-avenako. káttá-vanakka,* MALAY.; *Thinbau-kyeksu, thinbaw-kyetsu, thinban-kyeksu, késu-gi,* BURM.; *Vel-endaru, erandu,* SING.; *Kanana eranda, kanana-kerundum, nepala, paravata-yeranda,* SANS.; *Dande-nahri, dande-barri,* ARAB. & PERS.

References.—*Roxb., Fl. Ind., Ed. C.B.C., 689; **Brandis**, For. Fl., 442; Kurz, For. Fl. Burm., II., 403; Gamble, Man. Timb., 365; Dalz. & Gibs., Bomb. Fl., 77; Stewart, Pb. Pl., 192, 196; Rev. A. Campbell, Ec. Prod., Chutta Nagpur, No. 8426; Mason, Burma and Its People, 510, 762; Elliot, Fl. Andhr., 133; Pharm. Ind., 203; Ainslie, Mat. Ind., II, 45; O'Shaughnessy, Beng. Dispens., 558; Moodeen Sheriff, Supp. Pharm. Ind., 163; U. C. Dutt, Mat. Med. Hind., 302; Dymock, Mat. Med. W. Ind., 2nd Ed., 694; Fleming, Med. Pl. and Drugs, as in As. Res., Vol. XI., 169; U. S. Dispens., 15th Ed., 1582; Cat. of Med. Pl., Baroda Durbar, Col. Ind. Exhib., No. 120; Baden Powell, Pb. Pr., 583; Atkinson, Him. Dist., 740; Drury, U. Pl., 268; Lisboa, U. Pl. Bomb, 221, 248, 255, 269; Birdwood Bomb. Pr., 77, 289; Christy, Com. Pl. and Drugs, VIII, 4, 64; Liotard, Dyes (App.) VIII.; Cooke, Oils and Oilseeds, 49; Watson's Rep., 33; Balfour, Cyclop., II, 425; Smith, Dic., 319; Kew Off. Guide to the Mus. of Ec. Bot., 75; For. Ad. Rep. Ch. Nagpur, 1885, 34; Ind. For., 1885, V., 211; VI., 240; XI., 5; Agri.-Hort. Soc. of India, Trans. II., 168; Journ. (Old Series), VII., Sel. 56; X., 36; Gazetteers:—Orissa, II., 181; Mysore and Coorg, I., 50; Bombay, XII., 153; XV., Pt. I., 443; N.-W. P., IV., lxxvii; Hoshiarpur (Panjáb), 11.*

Habitat.—An evergreen shrub, indigenous to America, but cultivated in most parts of India, especially on the Coromandel Coast and in Travancore.

RESIN. Juice. 42 **Resin.**—**Dymock** writes: "The JUICE when dried in the sun forms a bright reddish-brown, brittle, substance like shell-lac, which may yet be put to some useful technical purpose." **Dr. Warden** reports that an acrid resin has been isolated by **Saubeiran** from the plant. **Drury** states that "the juice is of a very tenacious nature, and if blown forms large bubbles, probably owing to the presence of caoutchouc."

DYE. Juice. 43 **Dye.**—"The JUICE dyes linen black" (*Liotard*).

OIL. Seeds. 44 **Oil.**—The SEEDS yield about 30 per cent. of an oil of a pale yellow colour and acrid taste, which is employed by the poorer classes of natives for illuminating purposes. In England the oil obtained from African seeds at one time obtained a reputation as a lubricant, and is said to be used in the manufacture of certain transparent soaps. Of late years, it has also been recommended as a substitute for olive oil in dressing woollen cloths, and as a good drying oil. The Chinese are said to form a varnish by boiling the oil with oxide of iron.

CHEMISTRY. 45 **Chemical Composition.**—**Dymock** gives the following: "The kernels of the seeds were found by **Arnandon** and **Ubaldini** to contain 7·2 per cent. of water, 37·5 of oil, 55·3 of sugar, starch, albumin, casein, and inorganic matters. The kernels yielded 4·8 per cent. ash, and 4·2 per cent. nitrogen; the kernels and husks together 6 per cent. ash, and 2·9 per cent. nitrogen. The oil yielded by saponification, glycerin and an acid, which, as well as the unsaponified oil, produced octylic alcohol by distillation with hydrate of potassium." Ricinollic acid, a substance which also occurs in castor oil, is said to be a constituent of the seeds. This oil may be distinguished from castor oil by its very slight solubility in alcohol.

MEDICINE. Seeds. 46 **Medicine.**—The SEEDS, like those of the allied castor oil and croton oil plants, are purgative, more drastic than the former, and milder than the latter. They are employed for this purpose by the more indigent classes of natives, but the uncertainty of their action, and the frequently violent symptoms following their use, render them unsafe, and consequently little esteemed.
Oil. 47 The OIL possesses similar properties, but labours under the same disadvantage of uncertainty of action. The acrid, emetic, and drastic principle

MEDICINE.

appears to reside chiefly in the embryo. It has been stated that if the embryo be wholly removed, four or five of the seeds may be used as a gentle and safe purgative. Numerous cases of poisoning from eating the entire seeds are, however, on record. The symptoms appear to be, burning in the mouth and faces, a feeling of distension and pain in the abdomen, nausea, vomiting, violent purging, heat and congestion of the extremities, delirium, and insensibility, followed by depression before recovery. The best antidote is said to be lime-juice. Owing to the uncertainty of action in the case of both the seeds and oil, it appears very improbable that either will ever occupy a place as an internal remedy in European medicine. Externally applied, however, the oil is held in high esteem as a remedy for itch, herpes, chronic rheumatism, and as a cleaning application for wounds and ulcers. The LEAVES are rubefacient, and a decoction externally applied is said to excite the secretion of milk. According to **Roxburgh** the leaves warmed and rubbed with castor-oil are used by the Natives as a suppurative. The viscid JUICE which flows from the stem on incision has a high reputation as an application to wounds and foul ulcers. It checks bleeding and promotes healing by forming an air-tight film like that produced by collodion. It has been tried in European practice as a hæmostatic apparently with success. Thus, in a report of an interesting case published by **Dr. Evers** in the *Ind. Med. Gazette*, March 1875, it is stated that a drachm was injected into a varicose aneurism, with the effect of almost immediately stopping pulsation, while no ill-effects resulted from the operation. The ROOT-BARK is applied externally for rheumatism in Goa, and the same part of the plant, mixed with asafœtida and butter-milk, s, in the Konkan, prescribed in cases of dyspepsia and diarrhœa (*Dymock*).

Leaves. 48

Juice. 49

Root-bark. 50

SPECIAL OPINIONS.—§ "Milky juice valuable as a styptic. For further particulars, *vide Ind. Med. Gazette* for March 1875, page 66" (*Civil Surgeon B. Evers, M.D., Wardha*). "The viscid juice is used like collodion for closing cuts or abrasions" (*Brigade-Surgeon W. Dymock, Bombay*). "The viscid juice from the tree is applied to inflamed gums, and the sticks are used as tooth-brushes" (*T. Ruthman, Moodelliar, Chingleput, Madras Presidency*). "The juice is stated to strengthen the gums" (*Assistant Surgeon T. N. Ghose, Meerut*). "The milky juice is useful in toothache. The tender branch is used by the natives to clean the teeth with the view of strengthening the gums. A decoction of the leaves forms an excellent wash in eczema and indolent ulcers" (*Surgeon-Major J. M. Houston, Travancore, and Medical Store-keeper, J. Gomes, Travandrum*). "Oil is purgative, but not safe as it is apt to cause vomiting and over-purgation" (*Apothecary T. Ward, Madanápalle, Allahabad*). "The juice of twigs after rubbing between the palms of the hand has been frequently used by me as a good detergent" (*Assistant Surgeon Nanda Lall Ghose, Bankipore*). "The leaves applied over chronic and indolent ulcers have the effect of producing granulation in a few days" (*Surgeon W. Wilson, Bogra*). "The juice is a useful application for old unhealthy sores and sinuses, and (with a little salt) in cases of toothache and gumboils. The twigs used as tooth-brushes are believed to cure spongy gums and gumboils" (*Civil Surgeon J. H. Thornton, B.A., M.B., Monghyr*). "The milky juice is a powerful styptic. I have found it most serviceable in checking hæmorrhage from small vessels and unhealthy ulcers, &c." (*U. C. Dutt, Civil Medical Officer, Serampore*). "The fresh juice of this plant seems to have a considerable power of arresting hæmorrhage" (*Surgeon-Major Bankabehari Gupta, M.B., Pooree*). "The juice is said to be an astringent made into lather, it is applied to indolent ulcers and sinuses. The fresh twigs are used as tooth-brushes in cases of spongy and relaxed gums" (*Surgeon A. Crombie, Dacca*).

DOMESTIC. 51 **Domestic Uses.**—This plant is readily propagated by cuttings and is much employed in making hedges. The young twigs are used as detergent tooth-cleaners and are supposed to strengthen the gums. They are also said to be employed in making pipe stems. **Lisboa** states that the young leaves and branches are used in Damaun as a manure for cocoanut trees. **Captain Jenkins** writes that if castor oil leaves are not procurable, the leaves of this plant, with others, are used in Assam to feed silk-worms (*Trans. Agri -Hort. Soc. of Ind., II., 168*).

52 **Jatropha glandulifera,** *Roxb.; Fl. Br. Ind., V., 382.*

Syn.—JATROPHA GLAUCA, *Vahl.*

Vern.—*Lál-bherenda,* BENG.; *Verendi,* KOL.; *Undarbibi, jangli-erandi,* BOMB.; *Addalai, udalai,* TAM.; *Dundigapu, káti ámudapu, néla ámudamu,* TEL.; *Nikumba,* SANS.; *Abab* (**Ainslie**, according to **Moodeen Sheriff** this name is more properly applied to a plant of the Natural Order SOLANACEÆ, ARAB.

References.—*Roxb., Fl. Ind., Ed. C.B.C., 689; Kurz, For. Fl. Burm., II., 403; Thwaites, En. Ceylon Pl., 277; Dalz. & Gibs., Bomb. Fl., 229; Elliot, Fl. Andhr., 48, 89; Pharm Ind., 204; Ainslie, Mat. Ind., II., 5; O'Shaughnessy, Beng. Dispens., 558; Dymock, Mat. Med. W. Ind., 2nd Ed., 697; Liotard, Dyes (App.), II.; Cooke, Oils and Oilseeds, 50; Wardle, Rep. on Dyes and Tans. of Ind. (1887), 11, 35; For. Ad. Rep., Ch. Nagpur, 1885, 34; Agri.-Hort. Soc. Ind. Journ. (Old Series). XII., Pro. 25, 26, 29, 30; Gazetteers:—Mysore and Coorg, I., 65; Bombay, XV., Pt. I., 443.*

Habitat.—A shrub or (?) small evergreen tree, common near villages in Bengal, Burma, the Northern Circars, and the Deccan, rare in Oudh and the Panjáb. One of the most abundant of hedge plants in the lower provinces but also prevalent on village waste lands. Distributed to tropical Africa.

RESIN. 53 **Resin.**—**Kurz** states that this species yields a resin, which appears to have been unnoticed by other writers, and is probably not used.

DYE. **Dye.**—One of the most interesting economic features of this plant is the fact that it is one of those reputed to yield a green dye. The first person to bring this to notice was **Dr. Thompson**, Civil Surgeon of Malwa, who in 1862 submitted to the Agri.-Hort. Soc. of India, specimens of cotton, silk,

Leaves. 54 and wool, which were said to have been dyed green by the juice of the LEAVES. From recent investigations, however, it would seem extremely improbable that the beautiful green colour then shown, was obtained without the aid of indigo or some other blue dye. In 1877 samples of the dye were forwarded by the Government of India, to **Mr. Wardle**. Careful experiments were made by many processes with silk and cotton, but the colours obtained were all yellow, drab, fawn or brown, with, by one or two of the processes, a faint greenish tint. **Mr. Wardle** reported as follows:—"In the report of **Surgeon-Major G. Bidie, M.B.**, it is stated that these leaves yield a green dye of great beauty; but none of the methods which the limited quantity of dye-stuff at my disposal has enabled me to apply, have produced any such result, nor do I think it likely that they will produce a green without the aid of indigo or other bluish dye."

OIL. Seeds. 55 **Oil.**—The SEEDS yield a light straw-coloured fluid oil, similar to castor oil in appearance. According to **Lepine** it has a sp gr. of 0·963, and solidifies at 5°C. "For preparing this oil, the seeds should be collected as the capsule begins to split or change colour from green to brown; the latter should then be thrown down on a mat, and covered with another mat, and on a few hours exposure to a bright sun the seeds will have separated from the shell; for if allowed to remain on the shrub till quite ripe, the capsule bursts and the seeds are scattered and lost" (*Thompson*). The oil is then obtained in the ordinary way by expression, or "by roasting the seeds in a perforated earthen vessel, fitted upon another vessel, into which when the whole apparatus is heated in a pit filled with burning cow-dung the oil drops" (*Dymock*). It is chiefly used medicinally as an external application. OIL.

Medicine.—The OIL has long been used as an external stimulating application in cases of chronic rheumatism and paralytic affections. Like that obtained from **J. Curcas** it is purgative, but is little employed internally. It is also used as an application to sinuses, ulcers, foul wounds, and ringworm. Dymock states that the ROOT brayed with water is given to children suffering from abdominal enlargements. It purges, and is said to reduce glandular swellings. The JUICE of the plant is used in various parts of the country as an escharotic to remove films from the eyes. MEDICINE. Oil. 56 Root. 57 Juice. 58

Jatropha Manihot, *Willd.*, see **Manihot utilissima,** *Pohl.*; EUPHORBIACEÆ.

J. nana, *Dalz. & Gibs. Bomb. Fl. 229; Fl. Br. Ind., V., 382.* 59

Vern.—*Kirkundi,* MAR.

References.—*Dymock, Mat. Med. West. Ind., 2nd Ed., 699.*

Habitat.—A small sparingly branched shrub found rarely in the Konkan, in stony and waste places near Poona, Bombay, &c.

Medicine.—The JUICE is employed as a counter-irritant in ophthalmia (*Dalz. & Gibs.*). MEDICINE. Juice. 60

Jinjili oil, see **Sesamum indicum,** *Linn.*; PEDALINEÆ; Vol. VI.

Job's Tears, see **Coix lachryma,** *Linn.*; Vol. II, 492.

Jonesia Asoka, *Roxb.*; see **Saraca indica,** *Linn.*; LEGUMINOSÆ; Vol. VI.

JUGLANS, *Linn.*; *Gen. Pl., III., 398.*

The typical genus of the Natural Order JUGLANDEÆ, comprising three or four species, natives of Asia and North America. Of these one is Indian. All the species are large, handsome, deciduous, and monœcious trees. The generic name is derived from the old Latin appellation of the walnut *Jovis glans* or the nut of Jupiter.

Juglans regia, *Linn.*; *Fl. Br. Ind., V., 595*; JUGLANDEÆ. 61

THE WALNUT TREE.

Syn.—J. REGIA, var. KUMAONICA, *Cas DC.*; ? J. ARGUTA, *Wall.*

Vern.—*Akhrót, ákrót,* HIND.; *Akhrót, ákrút,* BENG.; *Tagashing,* BHUTIA; *Kabsing,* ASSAM; *Kowal,* LEPCHA; *Ak, ákhor, ákhrot, kharat,* N.-W. P.; *Akhor, kharot,* KUMAON; *Akhor, krot, dún,* KASHMIR; *Akhrót, dún, chármaghz, than thán, khór, ká, darga, akhóri, krot, ka-botang, starga, ughz, waghz, thanka,* bark=*dindása,* PB.; *Ughz, waghz,* AFG.; *'Akróda,* MAR.; *'Akhrot,* GUZ.; *Akróttu,* TAM.; *Akrótu,* TEL.; *Akródu,* KAN.; *Siskhyá-si, tikyá-si, thitkya,* BURM.; *Yang-gulk,* TURKI; *Akshota áks-chóda, ákhóda, akhóta,* SANS.; *Jouz,* ARAB.; *Girdagán, chár-maghz, chahár-maghz, jaoz,* PERS.

References.—*Roxb., Fl. Ind., Ed. C.B.C., 670; Brandis, For. Fl., 497; Kurz, For. Fl. Burm., II, 490; Gamble, Man. Timb., 392; Stewart, Pb. Pl., 201; Jour. of a Bot. Tour in Hazara, &c., in Jour. Agri.-Hort. Soc. of Ind., XIV. (Old Series), 12, 45; DC., Origin Cult. Pl., 425; Aitchison, Fl. Kuram Valley, in Jour. Linn. Soc., XVIII., 95; Bot. of Afghan Del. Com. in Trans. Linn. Soc., III., Pt. I. (1888), 110; Voyage of J. H. van Linschoten (1596), Ed. Burnell, Tiele. and Yule, II., 279; Mason, Burma and Its People, 460, 777; Ain-i-Akbari, Gladwin's Trans., I., 83, 85; II., 143, 169; Ainslie, Mat. Ind., I., 463; O'Shaughnessy, Beng. Dispens, 605; Moodeen Sheriff, Supp. Pharm. Ind., 163; U. C. Dutt, Mat. Med. Hind., 290; Murray, Pl. and Drugs, Sind, 36; Year-Book of Pharm., 1874, 529; Baden Powell, Pb. Pr., 267, 384, 385, 583; Atkinson, Him. Dist., 317, 716, 740; Ec. Prod., N.-W. P., Pt. V., 88, 89; Cooke, Oils and Oilseeds, 51; Spons' Encyclop., 1413; Balfour, Cyclop., II.,*

451; Smith, Dic., 433; Treasury of Bot., I., 641; Kew Off. Guide to the Mus. of Ec. Bot., 140; Agri.-Hort. Soc. Ind. Trans, I., 66; IV., 105, 147; VI., 248; VII. (Pro.) 83; Journals (Old Series), II. (Sel.), 298; IV. (Pro.), 34; V., 186 (Sel.), 175-179 (Pro.), 15, 55; XIII., 371, 384, 385; Indian Forester, I., 92, 306; VIII., 404; X., 57; XI., 5, 284; XIII., 66, 127; Gazetteers:—Panjáb, Peshawar, 28; Bannu, 23; Dera Ismail Khan, 19; Gurdaspur, 55; Hazara, 14; Rawal-Pindi, 15; Simla, 10; Settlement Report, Kohat, 29; Madras Man. of Administr. II., 27.

Habitat.—A large deciduous tree found wild and cultivated in the Temperate Himálaya and Western Tibet, from Kashmír and Nubra eastwards at altitudes from 3,000 to 10,000 feet, also in the Manipur and Ava Hills; distributed to Northern Persia, the Caucasus and Armenia (*Fl. Br. Ind.*). **Aitchison** states that it also occurs wild in the ravines of Shéndtoi and Darban in Afghánistán, and **DeCandolle** considers that it has a wider range of distribution than that described above. According to the latter learned author it has been proved that this tree also occurs wild in Japan, in the mountains of Greece, and in the mountains of Banat, giving a range extending from Eastern Temperate Europe to Japan. This opinion is confirmed by the existence of palæontological evidence of its having at one time extended as far as Provence.

CULTIVATION. 62

Cultivation.—The use of the walnut and its consequent cultivation date from very remote times; so remote, indeed, that it is impossible to say at what period the tree was first cultivated in India. For many centuries the nuts have formed an important article of export from the hills to the plains. In the *Aín-i-Akbarí* mention is made of Kashmír walnuts, which, even in the sixteenth century, had a wide reputation, and the inferior "Tartarian" nuts are also alluded to. At the present day, the tree is extensively cultivated all over the Temperate Himálaya, especially in Afghánistán and Kashmír. It requires a climate with neither extremes of heat nor cold, good soil, and, in the young state, careful weeding. **Brandis** mentions three varieties:—**α tenera,** (thin shelled) **β beluchistana, γ kamaonia.** At Chumba and Kashmír a superior quality is grown, which is characterised by having a soft shell.

DYE. Bark. 63 Rind. 64

Dye.—The BARK and RIND of the fruit are used in dyeing and tanning.

OIL. Kernel. 65

Oil.—The albuminous KERNEL affords, by expression, about 50 per cent. of a clear sweet oil, largely used in the hills for culinary purposes and illumination, but rarely seen in the plains. **Stewart** states that a large proportion of the hill walnut oil is prepared by simply bruising the kernel between stones. It is one of the most important vegetable fixed oils of Europe, and is said to form one-third of the total oil manufactured in France. In Spain and Italy also, outside the olive-growing regions, it is largely expressed. The following description of the mode of preparation, composition, and qualities of the oil may be quoted from *Spons' Encyclopædia*:—"The oil should not be extracted from the nuts until two to three months after they have been gathered. This delay is absolutely necessary to secure an abundant yield, as the fresh kernel contains only a sort of emulsive milk, and the oil continues to form after the harvest has taken place; if too long a period elapse, the oil will be less sweet, and perhaps even rancid. The kernels are carefully freed from shell and skin, and crushed into a paste, which is put into bags and submitted to a press; the first oil which escapes is termed 'virgin' and is reserved for feeding purposes. The cake is then rubbed down in boiling water, and pressed anew; the second oil called 'fire-drawn' is applied to industrial uses. The exhausted cake forms good cattle food.

"The virgin oil, recently extracted, is fluid, almost colourless, with a

feeble odour, and not disagreeable flavour. Its sp. gr. is 0·926 at 15°C. OIL.
and 0·871 at 94°C.; it thickens to a butter-like consistence at—15°C., and
solidifies to a white mass at 27½°C. In the fresh state, it is largely used
in Nassau, Switzerland, and other countries as a substitute for olive oil in
salads, &c, but is scarcely to be considered a first class alimentary oil.
The fire-drawn oil is greenish, caustic, and siccative, surpassing linseed oil
in the last respect, and exhibiting the property more strongly as it becomes more rancid. On this account it is preferred by many artists before
all other oils. It affords a brilliant light, and may be used in the manufacture of soft soaps."

Medicine.—The BARK is used as an anthelmintic and detergent; the MEDICINE. Bark. 66
LEAVES are astringent and tonic, in decoction are supposed to be specific
in strumous sores, and to be anthelmintic; the FRUIT is also believed to
have an alterative effect in rheumatism. The fruit was believed by the Leaves. 67
ancients to be alexipharmic; thus the famous antidote of Mithridates
was composed of two walnuts, two figs, and twenty leaves of rice, rubbed Fruits. 68
together with a grain of salt. The same belief was held by the Arabians.

Food and Fodder.—The FRUIT or "walnut" ripens in July-September,
and consists of two seed-lobes (the edible part) crumpled up within a hard
shell. Several varieties are met with in the Himálaya. The wild tree FOOD. Fruit.
bears a nut with a thick shell and small kernel, which is rarely eaten. Of 69
the cultivated varieties the best is a thin-shelled form known as *kaghazi-akhrot*. FODDER. 70
The walnut forms an important article of diet in Kashmír and
the North-West Himálaya, and, as already stated, is exported in considerable quantities from these localities, also from Afghánistán and Persia, Twigs. 71
to the plains. The TWIGS and LEAVES are used as fodder and are regularly lopped for this purpose. Leaves. 72
Stewart states that the inhabitants of the
Sutlej Valley believe that if the trees get a rest in this respect every fourth Oil. 73
year, the quality of the fruit is not deteriorated. As already mentioned
the OIL is edible, and the OIL-CAKE a good cattle food. Oil-cake. 74

Structure of the Wood.—Heartwood greyish-brown with darker TIMBER. 75
streaks, often beautifully mottled, moderately hard, even-grained, seasons
and polishes well; weight—European 40 to 48℔, North-West Himálaya 41℔, Sikkim 33℔ per cubic foot (*Gamble*). A valuable timber, not
only on account of its useful properties, but also owing to the large size at
which it can be obtained. Thus Brandis describes trees up to 28 feet in
girth and 100 feet in height from the North-West Himálaya, and in Sikkim
the tree often reaches a height of 100 to 120 feet with a girth of 12 feet
or more. Walnut is at present being grown entirely for the sake of its timber in large plantations at Rangbúl and other localities near Darjíling. It
is largely used throughout the Himálaya for making furniture, for ornamental, carved, and inlaid work, and as a timber for indoor house-work in
European buildings. The wood takes a long time to season, but when
thoroughly dry does not shrink nor warp by heat or moisture. This property, together with the fact that it does not corrode iron, renders it most
valuable for gun-stocks, in the manufacture of which a large amount of
walnut is yearly employed. It is occasionally used for shingles by the
Bhútias but is not so good for this purpose as chestnut. At one time a
European Company is said to have been formed with the object of purchasing the right to remove the knots from the stems of the walnut trees
in Kashmír It is believed this concession was, however, cancelled owing
to the reckless injury done to the trees.

Domestic Uses.—"The BARK is largely exported to the plains to be sold DOMESTIC. Bark.
under the name of *dindása* for women's tooth-sticks, or for chewing to give 76
a red colour to the lips; it is said to prevent the formation of tartar"
(*Stewart*). Honigberger states that a twig is recommended to be kept in

DOMESTIC. a room to dispel flies. **Abul Fazl**, in his description of the Súbah of Kashmír, mentions a curious custom which appears to have prevailed in his time. He writes: "In the village of Zinabúl is a spring and basin, into which people throw walnuts, to know what will be the issue of any affair; the walnuts floating is a good omen, and on the contrary if they sink, it is a sign of bad luck."

77 JUNCUS, *Linn.; Gen. Pl., III., 867.*

The typical genus of the Natural Order JUNCACEÆ; it comprises many Indian species. In India none appear to be utilised, but in Europe the common rush, **Juncus effusus**, *Linn.*, is employed for making mats, baskets, and chair bottoms, and its pith is used for the wicks of rush-lights. In Spain it is said to be also used in the preparation of a textile fibre. Two other species are utilised for paper-making in Australia, and in Italy one affords a cordage fibre. It is extremely probable that certain of our Indian species might be similarly utilised. Thus **Royle** has suggested that **J. glaucus**, *Ehrb.*, a common Himálayan rush which is closely allied to **J. effusus**, *Linn.*, might be employed for all the uses to which the latter is put in Europe.

JUNIPERUS, *Linn.; Gen. Pl., III., 427.*

A genus of evergreen monœcious trees or shrubs belonging to the Natural Order CONIFERÆ, and comprising some 25 species, natives of the temperate and cold regions of the Northern Hemisphere. Of these four are Indian.

78 Juniperus communis, *Linn.; Fl. Br. Ind., V., 646;* CONIFERÆ.

THE JUNIPER.

Syn.—J. NANA, *Willd.*

Vern.—*Aaraar*, HIND.; *Chichia*, KUMAON; *Núch, páma, pethra, bentha, betar*, KASHMIR; *Petthri, petthar, betthal, wetyar, páma, giáshúk, lassar, núch, chúch, betar, dhúp, lewar, langshúr, thélu, gúgil, chúi, shúpa*, fruit=*haulber, abhúl*, PB.; *Langshúr thélu, lewar*, KUNAWAR; *Chúni, shuha*, PITI; *Sbama*, LAHOUL; *Abhal*, DEC.; Fruit=*Abhal, habbul-aaraar, samratul-arraar*, ARAB.

References.—*Brandis, For. Fl., 535; Gamble, Man. Timb., 411; Stewart, Pb. Pl., 222-3; Pharm. Ind., 223; Ainslie, Mat. Ind., I., 379; O'Shaughnessy, Beng. Dispens., 619; Moodeen Sheriff, Supp. Pharm. Ind. 164; Dymock, Mat. Med. W. Ind., 2nd Ed., 762; Flück. & Hanb., Pharmacog., 624; Bent. & Trim., Med. Pl., 255; S. Arjun, Bomb. Drugs, 134; Murray, Pl. and Drugs, Sind, 26; Year Book Pharm., 1874, 14, 65, 629; Irvine, Mat. Med. Patna, 35; Baden Powell, Pb. Pr., 378; Atkinson, Him. Dist., 318, 842; Birdwood, Bomb. Pr., 85; Piesse, Art of Perfumery 363; Spons, Encyclop., 1422, 1624, 1681; Balfour, Cyclop., II, 454; Smith, Dic., 100, 230; Treasury of Bot., I., 642; Kew Reports, 131; Ind. For., III., 25, 32; VIII., 409; XI., 5; XIII., 62, 70; Agri.-Hort. Soc. Ind., Jour. (Old Series), IV., Sel. 119, 257; VII., 131, 149, 150, 153, 154.*

Habitat.—A large shrub of the North-West Himálaya from Kumáon westwards, at altitudes from 5,500 to 14,000 feet; distributed through Asia to Temperate and Sub-arctic Europe, North Africa, and North America.

RESIN. Wood. 79 Fruit. 80

Resin.—The WOOD is highly resinous, and the ripe FRUIT also contains a resin—neither are of special economic value. The name of "juniper-gum" or "resin" has been erroneously applied to Sandarach, a product of **Callitris quadrivalvis.** See Vol. II., 28.

OIL. Fruit. 81

Oil.—A volatile oil is obtained, to the extent of from 0·4 to 1·2 per cent., by distilling the FRUIT with water. This oil is most abundant in the fully-grown green fruit, since on ripening a certain amount is transformed into resin. It is colourless or pale yellow, has a strong odour of the fruit, a sp. gr. of 0·847 to 0·870, and is slightly soluble in alcohol. Chemically it consists of a mixture of levogyre oils, one of which has the composition of $C_{10} H_{16}$ and boils at 155°C., the other and prevailing portion of the oil consists of hydrocarbons, polymeric with terpene, $C_{10} H_{16}$ and boils at 200° C. (*Flückiger and Hanbury*). Juniper-oil is considerably used in medicine. The WOOD and YOUNG TWIGS also yield a volatile oil which is frequently distilled abroad. It is sometimes substituted for the officinal oil in Europe and America, but though cheaper, it more closely resembles Turpentine in its composition and properties (*Bentley and Trimen*). **Piesse** states that it is much employed on the continent in the manufacture of Juniper Tar Soap, made by dissolving the oil in a fixed vegetable oil, such as almond or olive oil, or in fine tallow, and adding a weak soda ley. The result is a moderately firm and clear soap, which possesses the advantage of ready mixibility with water. OIL. Wood. 82 Young Twigs. 83

Medicine.—The FRUIT and the OIL above described are officinal in the Pharmacopœia of India and Great Britain. The former, commonly called a berry, is in reality a galbulus, and is round, about the size of a pea, and of a deep purplish black colour, covered by a glaucous bloom. In addition to the oil it contains a large quantity of glucose, 33 per cent. (**Trommsdorff**), 41·9 per cent. (**Donath**), 16 per cent. in the dried fruit (**Rielhausen**), 1877—also small quantities of albumenoids, formic, acetic, and malic acids, resin, and inorganic substances. In addition to these **Donath** in 1873 eliminated ·37 per cent. of a bitter principle which he termed *Juniperine*. MEDICINE. Fruit. 84 Oil. 85

When bruised the fruit has an agreeable, aromatic odour, and a warm, somewhat spicy, sweetish, slightly terebinthinate taste. The medicinal properties of the fruit and oil are similar, both possessing carminative, stimulant, and diuretic qualities, their physiological action closely resembling that of turpentine. The remote local stimulant action of Juniper on the kidney, however, is more marked, and it further possesses the advantage of being neither disagreeable nor dangerously powerful. But it also must be administered with caution, as in large doses it causes strangury and renal inflammation.

In European practice the oil is employed in doses of 1 to 4 minims, as a diuretic in dropsy not dependent on acute renal disease, *i.e.*, in hepatic and cardiac anasarca, and in some cases of chronic Bright's disease. It has also been employed in mucous discharges, such as gonorrhœa, gleet, and leucorrhœa, and in some cutaneous diseases.

The Juniper Tar Soap above described has been considerably used on the continent of Europe as a remedy for skin diseases, its ready mixibility with water, and the cleanly nature of the application, giving it considerable advantages over terebinthinous ointments.

The NUTS are sold in the bazárs of Northern India for medicine, and are similarly prescribed as diuretic and stimulant. **Irvine** mentions that they are imported into Patna from Nepál, and are used in the treatment of gonorrhœa. Nuts. 86

Structure of the Wood.—White, moderately hard, fragrant, highly resinous, with a small mass of darker wood near the centre. Weight 33 to 34℔ per cubic foot. Used for fuel, and burnt as incense (*dhúp*). TIMBER. 87

Domestic Uses.—The WOOD, TWIGS, and LEAVES, are used as incense in the Panjáb under the name of *dhúp* (*Stewart*). **Madden** states that the BERRIES are used with barley meal in distilling a spirit, the former being probably only added for the purpose of flavouring the liquor. In Europe as is well known they are extensively employed for the same purpose in the case of gin. DOMESTIC. Wood. 88 Twigs. 89 Leaves. 90 Berries. 91

92 **Juniperus macropoda,** *Boiss.; Fl. Br. Ind., V., 647.*

THE HIMALAYAN PENCIL CEDAR.

Syn.—JUNIPERUS EXCELSA, *Brand.; also Wall., Cat., 6041, (not of Bieb);* J. GOSSAINTHANEANA, *Loddig.*

Vern.—*Dhúpi, dhúpri, chandan, shúkpa,* NEPAL; *Dhúp, padam, padmak surgi,* N.-W. P.; *Súrgi, lewar, newar, dupri, chundun,* KUMAON; *Chalai, shúkpá, shúr, shúrgu, lewar, luir,* PB.; *Shúrbúta, shúrgú, shúkpa,* TIBET; *Charái,* HAZARA; *Apúrs,* BELUCHISTAN.

References.—*Brandis, For. Fl., 538, t. 68; Gamble, Man. Timb., 412; Stewart, Pb. Pl., 223, 224; Jour. Bot. Tour in Hazara, &c, in Agri.-Hort. Soc. Ind. Jour. (Old Series), XIV., 55; Aitchison, Bo.. of Afgh. Del. Com. in Linn. Soc. Trans., III., Pt. I., 1888, 113; Baden Powell, Pb. Pr., 583; Atkinson, Him. Dist., 318, 843; Balfour, Cyclop., II., 455; Agri.-Hort. Soc. of Ind., Jour. (Old Series), IV., 246, 250, 252, 256, 257, 267; VII., 117, 129, 137-150; XIII., 383, XIV.. 267, 269, 271, 272; Settlement Reports: Panjab Kangra, 160; Hazara, 12; Ind. For., III., 25; V., 183, 184, 185, 186; VII., 127; VIII, 114; IX, 180; XIII, 70, 278; XIV., 365, 371; Gazetteers: Panjab, Simla Dist., 10; Hazara Dist., 14.*

Habitat.—A tree attaining 50 feet in height, native of the arid tract of the Inner Himálaya, from Nepál westwards, and of Western Tibet, at altitudes from 5,000 to 14,000 feet; distributed to Afghánistán, Beluchistán, Persia, and Arabia. Most of the Indian information regarding this tree has been published under the synonym **J. excelsa,** *Bieb.*, a synonym which is rejected in the *Flora of British India,* though **Hooker** remarks, "I doubt its proving distinct from **J. excelsa,** *Bieb.*"

Mr. E. E. Fernandez (*Notes for a Manual of Sylviculture, in the Indian Forester, VIII, 114*) alludes to the injury done to this tree by the parasitic action of a species of **Arceuthobium. Mr. Fernandez** writes:—"Thus the **Arceuthobium oxycedri,** as far as is known, grows in India only on **Juniperus excelsa,** gradually over-spreading the plant on which it has once taken root, often killing the branch or the entire tree." In the *Flora of British India* the discovery of a new species (the only Indian one) is announced, *viz.*, **A. minutissimum,** *Hook. f.*, collected by **Mr. J. F. Duthie** on **Pinus excelsa.** It seems probable that both these statements refer to one and the same parasite, and that a mistake has been made as to the plant on which it is found.

RESIN. Wood. 93

Resin.—The WOOD, like that of the preceding species, is highly resiniferous.

MEDICINE. Fruit. 94 Branches. 95

Medicine.—The FRUIT is used medicinally, and appears to have similar properties to that of **J. communis.** The smaller BRANCHES when burnt are supposed to exercise a deodorising and cleansing influence, and, in Khágán, they are believed to act as a remedy for the delirium of fever.

TIMBER. 96

Structure of the Wood.—Sapwood white, heartwood red, very fragrant, often with a purplish tinge; growth slow, weight from 25 to 37℔ per cubic foot, the average being 32℔. It has the same agreeable odour as the wood from which pencils are made, is light, and though not strong, withstands, to a remarkable degree, the action of moisture. Owing to this property it is much used in the regions of its growth for making water channels, house-posts, beams alternating with stone in the exterior walls of houses, and for vessels for water, milk, &c. It is also, owing to its semi-sacred character, in Tibet and Lahoul, employed in the construction of temples. It forms excellent fuel, and is said to yield good charcoal when burned.

The wood of this species would probably form an excellent substitute for the American cedar at present employed in pencil-making, but the price of the latter is too low to admit of the export of the former from the Himálaya at profitable rates.

Domestic and Sacred.—In the Buddhist regions of the Inner Himálaya this Juniper is regarded as sacred. It is frequently planted near temples; the WOOD and TWIGS form part of the incense called *dhúp;* the FRUIT is also burned as incense, and the young twigs are used as votive offerings. In Ladák small BRANCHES are frequently seen placed on cairns, &c. A certain quantity of the wood is said to be exported to the plains for use as incense. DOMESTIC & SACRED. Wood. 97 Twigs. 98 Fruit. 99 Branches. 100

Juniperus pseudo-sabina, *Fisch. & Mey.; Fl. Br. Ind., V., 646.* 101

Syn.—JUNIPERUS EXCELSA, *Wall., Cat., 6041* (not of Bieb); J. WALLICHIANA, *Hook. f. & Thoms.;* J. SABINA? *Herb. Ind. Or.;* J. INDICA, *Bertoloni.*

Vern.—*Bhil,* HIND.; *Tchokpo,* SIKKIM; *Shirchin,* BYANS; *Poh,* TIBET.

References.—*Brandis, For. Fl., 537; Gamble, Man. Timb., 412; Aitchison Fl. Kuram Valley in Linn. Soc. Jour., XVIII., 97; Balfour, Cyclop., II., 455; Ind. For., VIII., 410; XI., 5.*

Habitat.—A variable species, occurring on the Temperate Himálaya, from Kashmír to Bhutan and in Western Tibet, at altitudes of 9,000 to 15,000 feet. In the North-West Himálaya it is found as a bush, while in Sikkim it is a tall tree attaining 60 feet in height, with a stout trunk and black thick ramification and foliage.

Structure of the Wood.—Similar to that of **J. macropoda.** TIMBER. 102

Domestic Uses.—**Aitchison** writes: "The BARK exfoliates in long fibrous strips, which are collected and employed by the natives for making pads for carrying their water-jars on, and for other similar purposes." DOMESTIC. Bark. 103

J. recurva, *Ham.; Fl. Br. Ind., V., 647.* 104

THE WEEPING BLUE JUNIPER.

Syn.—JUNIPERUS SQUAMOSA, *Ham.*

VAR. **squamata,** *Parlat.,* differs from the typical form in being a decumbent or prostrate bush; J. SQUAMATA, *Ham.;* J. DENSA, *Gord.;* J. LAMBERTIANA and RIGIDA, *Wall.;* J. RELIGIOSA, *Royle.*

Vern.—*Tupi,* NEPAL; *Deschú, chakbu,* SIKKIM; *Bettir, bhedára, bidelganj, thelu, phulu, jhora, gúggal, bil, úrún, agáni,* N.-W. P.; *Wetyar, bettar, chúch, thelu phulu,* PB.; *Pama,* TIBET.

References.—*Brandis, For. Fl., 536; Gamble, Man. Timb., 412; Atkinson, Him. Dist., 318, 842; Royle, Ill. Him. Bot., 350; Balfour, Cyclop., II., 455; Agri.-Hort. Soc. of Ind., Jour. (Old Series), IV., (Sel.), 257, 258; VII., 137, 139, 146-155; Gazetteers:—Mysore and Coorg, I., 66; Panjab, Simla Dist., 10; Ind. For., I., 97; III., 25; V., 466; VII., 127; VIII., 409; IX., 321.*

Habitat.—A very variable plant of the Temperate and Alpine Himalaya, from 7,500 to 15,000 feet, which as a tree attains a height of 30 feet with a straight trunk and pendulous branches, at higher elevations becomes stunted, in alpine or exposed situations even passing into the condition and described as *var.* **squamata**: it is distributed to Afghánistán.

Resin.—The WOOD contains a large quantity of resin similar to that of **J. communis.** RESIN. Wood. 105

Medicine.—**Aitchison** reports that the SMOKE from the GREEN WOOD is known in Kashmír as a powerful emetic, producing long-continued vomiting. MEDICINE. Smoke. 106

Structure of the Wood.—Sapwood white, heartwood light red, very fragrant; similar to that of **J. excelsa,** except that the short broad medullary rays are wanting; weight 38 to 42lb per cubic foot. Used for fuel. TIMBER. 107

Domestic Uses.—This species, like those already described, is one of the chief sources of incense, the WOOD, LEAVES and TWIGS being exported from Sikkim for this purpose. **Atkinson** describes an interesting use of the leaves and twigs in the *North-West Provinces Gazetteer, Vol. X.,* in which he writes: "This bush is used in the manufacture of the yeast called *balma,* DOMESTIC. Wood. 108 Leaves. 109 Twigs. 110

DOMESTIC. which forms an adjunct in the preparation of spirits from rice. The yeast is made by moistening coarse barley flour, which is formed into a ball and covered all round with leaves and twigs of juniper. The whole is then closely wrapped up in blankets, kept in a warm place, and allowed to ferment, which it usually does in three or four days."

JURINEA, *Cass.; Gen. Pl., II., 473.*

111 **Jurinea macrocephala,** *Benth.; Fl. Br. Ind., III., 378;* COMPOSITÆ.

Syn.—DOLOMÆA MACROCEPHALA, *DC.;* SERRATULA MACROCEPHALA, *Wall.*

Vern.—*Dhúp,* KASHMIR; *Dhúp, dhúpa, gúgal, zhangar,* root=*dhúp, pokharmúl?* PB.

These terms are applied not only to the root of this plant but also to many other fragrant products used as incense. (See species of **Juniper** above.)

References.—*Stewart, Pb. Pl., 125; Baden Powell, Pb. Pr., Vern. Index, xlix.; Atkinson, Him. Dist., 312; Christy, Com. Pl. and Drugs, VI., 93.*

Habitat.—A herb, with perennial root, common in the Western Himálaya, from Kashmír to Kumáon, at altitudes from 11,000 to 14,000 feet.

MEDICINE. Root. 112 **Medicine.**—"The bruised ROOT is applied to eruptions, and a decoction of it is taken in colic, &c. **Aitchison** also remarks that some part of the plant is used medicinally. The root is from many places exported to the plains, sometimes after being pounded and made up into cakes with its own juice, as I was informed in Khágán. And it appears to be officinal in the Panjáb bazárs under both of the above names. It is considered cordial, and is given in puerperal fever, &c. This is probably the *jarí* (root) *dhúp,* of which nearly seven maunds from Bissahir were exposed for sale at the Rampúr fair in 1867, according to the Official Report" (*Stewart*).

In confirmation of the above statement of **Stewart's** it may be mentioned that **Dr. G. Watt** saw the preparation *dhúp* being made from the roots of this plant in Rampúr, and also higher up on the neighbouring hills bordering on Kulu witnessed the plant being collected for export to Rampur.

DOMESTIC & SACRED. 113 **Domestic and Sacred Uses.**—**Madden** states that the odorous root is used as incense offered at shrines and to rajás. The flowers also are placed in temples on the Sutlej. (*Cf. dhúp* under **Juniperus**) A considerable quantity of the root is believed to be exported from Kashmír to Tibet for use as incense.

JUSSIÆA, *Linn.; Gen. Pl., I., 788.*

114 **Jussiæa suffruticosa,** *Linn.; Fl. Br. Ind., II., 587;* ONAGRACEÆ.

Syn.—J. EXALTATA, *Roxb.;* J. VILLOSA, *Lamk.;* J. FRUTICOSA, *DC.;* J. SCABRA, *Willd.;* J. BURMANNI and OCTOPHILA, *DC.;* J. LONGIPES, *Griff.;* J. DECUMBENS, *Wall.;* J ANGUSTIFOLIA, *Lamk.;* EPILOBIUM FRUTICOSUM, *Lour.*

Vern.—*Lal-bunlunga, bun-lung,* BENG.; *Petra da, dak ichak,* SANTAL.; *Panalavanga,* MAR.; *Niru agni véndra páku?* (ELLIOT), TEL; *Carambu,* MALAY.; *Hæmarago,* SING.; *Bhalava anga,* SANS.

References.—*Roxb., Fl. Ind., Ed. C.B.C., 371; Kurz in Jour. As. Soc., 1877, Pt. II., 90; Thwaites, En. Ceylon Pl., 123; Dalz. & Gibs., Bomb. Fl., 98; Rheede, Hort. Mal., II., t. 50; W. & A., Prodr., 336; Campbell, Ec. Prod. Chutia Nagpur, Nos. 8720, 9211; Elliot, Fl. Andhr, 135; Ainslie, Mat. Ind., II., 66; Dymock, Mat. Med. W. Ind., 2nd Ed., 325; Drury, U. Pl., 269; Gazetteers:—Bombay, XV., 434; N.-W. P., I., 80; IV., lxxii.*

Habitat.—A perennial under shrub, found all over India, in moist places, which are overgrown with small jungle except in the western desert regions; also a native of Ceylon; distributed through the warmer moist parts of the whole world.

Medicine.—**Ainslie,** quoting **Rheede,** states that the plant, ground fine, and steeped in butter-milk, is considered good in dysentery, and that a decoction is said to dissipate flatulency, increase urine, and to act as a purgative and anthelmintic. MEDICINE. 115

JUSTICIA, *Linn.; Gen. Pl., II., 1108.*

Justicia Adhatoda, *Linn.*; see **Adhatoda Vasica,** *Nees.* Vol. I., 109.

J. bicalyculata, *Wight.*; see **Peristrophe bicalyculata,** *Nees.*

[*468*; ACANTHACEÆ.

J. Gendarussa, *Linn. f.; Fl. Br. Ind., IV., 532; Wight, Ic, t. 468.* 116

Syn.—GENDARUSSA VULGARIS, *Nees.*

Vern.—*Udi-sanbhálú, nili-nargandi,* HIND.; *Jagat-madan, jogmodon,* BENG.; *Teo, kala-adulsa,* BOMB.; *Kálishanbáli,* DEC.; *Karu-noch-chi, karuppu-noch-chi,* TAM.; *Néla-vávili, nalla-noch-chili, nalla-vávili,* TEL.; *Karelakki-gidá,* KAN.; *Karun-noch-chi, váták-koti, vátan-golli,* MALAY.; *Bavanet,* BURM.; *Kalu-varaniá,* SING.; *Níla-nirgundi, krishtna-surasa,* SANS.; *Aslak-ásvad,* ARAB.; *Banj-angashte-siyáh,* PERS.

References.—*Roxb., Fl. Ind., Ed. C.B.C., 43; Kurz, For. Fl. Burm., II., 247; Gamble, Man. Timb., 280; Dalz. & Gibs., Bomb. Fl. Supp., 71; Rheede, Hort. Mal., IX., t. 42; Pharm. Ind., 162; Ainslie, Mat. Ind., II., 67; O'Shaughnessy, Beng. Dispens., 483; Moodeen Sheriff, Supp. Pharm. Ind., 164; Dymock, Mat. Med. W. Ind., 2nd Ed., 592; S. Arjun, Bomb. Drugs, 199; Bidie, Cat. Raw Prod., S. India, 37; Rumph., Herb. Amb., IV., t., 28.*

Habitat.—An evergreen, dense shrub, 2 to 4 feet high; found in tropical forests throughout India, from Bengal to Ceylon and Malacca, often an escape from cultivation. **C. B. Clarke** writes in the *Flora of British India,* "Distribution, Malaya and China to the Phillipines (?wild). . . . . This commonly cultivated plant is considered by **Nees** and **T. Anderson** to be wild in various parts of India, but the rarity of the seed renders this doubtful. **Colonel Beddome** says 'Wild on Mooleyit in Tenasserim.'"

Medicine.—The LEAVES and TENDER SHOOTS when rubbed emit a strong but not unpleasant odour; used in decoction in chronic rheumatism (*Ainslie*). The Malays employ the PLANT as a febrifuge. In Java it is considered to have emetic properties (*Pharm Ind.*). **Dr. Dymock,** in his *Materia Medica,* says that the medicinal properties attributed to this plant by **Ainslie** and in the *Indian Pharmacopœia* do not correspond with those which it actually possesses, and conjectures that it has been confounded with some species of **Vitex.** MEDICINE. Leaves. 117 Tender Shoots. 118 Plants. 119

SPECIAL OPINIONS.—§"An oil prepared from the leaves when applied locally is said to be useful in eczema and an infusion of the leaves is given internally in cephalalgia, hemiplegia, and facial paralysis" (*Surgeon-Major J. M. Houston, Travancore*). "The juice of the fresh leaves is dropped into the ear for earache, and into the corresponding nostril on the side of the head affected with hemicrania" (*Honorary Surgeon P. Kinsley, Chicacole, Ganjam, Madras*).

J. procumbens, *Linn.; Fl. Br. Ind, IV., 539.* 120

Syn.—JUSTICIA MICRANTHA, *Wall*; J. HIRTELLA, *Wall.*; ROSTELLULARIA PROCUMBENS, *Nees.*; R. ADENOSTACHYA, *Nees.*

VAR.—**latispica,** *Clarke*; SYN.—ROSTELLULARIA PROCUMBENS, *Wight, Ic., t. 1539*; R. MOLLISSIMA, *Nees.*

Vern.—*Ghâti-pitpápra, pitpápada,* BOMB.

References.—*Roxb., Fl. Ind., Ed. C.B.C., 45; Dalz. & Gibs., Bomb. Fl., 193; Dymock, Mat. Med. W. Ind, 2nd Ed., 591.*

Habitat.—A herb of South-Western India and Ceylon, extending as far north as the South Konkan; distributed to Malaya and Australia.

MEDICINE. Plant. 121

Medicine.—**Dymock** states that the whole PLANT, gathered when in flower, is dried and used as a substitute for **Fumaria**, the true *Pit-pápṛá*, which it resembles in having a faintly bitter, disagreeable taste. (See **Fumaria**, *Vol. III., 454*).

FOOD. 122

Food.—It is used more or less as an article of food in certain parts of Bombay.

123

JUTE.

The reader should consult the article **Corchorus**, *Vol. II., 534-562*. Since the date on which that article was written certain new and interesting facts have been brought to light, the most important of which is perhaps, the exhaustive chemical study to which the fibre has been subjected by **Messrs. Cross** and **Bevan**. These chemists published, in the *Journal of the Chemical Society* (*1889*), the results of their detailed analysis both of the formation of the fibre and its chemical nature. Their paper is too elaborate and technical for the present work, but doubtless, as their discoveries become better known, practical developments of an industrial nature will be brought to light that may extend the uses of the fibre. The following brief *résumé* may, however, be here given:—

CHEMISTRY. 124

Chemistry.—Jute, the best representative of the group of lignified celluloses, in the chemical investigations contained in the paper, is treated as the type. Although from its nature jute fibre is an aggregate, it exhibits, notwithstanding, a constancy in composition and properties which denote a chemical individual. The simplest expression of its elementary composition is the empirical formula, $C_{12} H_{18} O_9$ (C=47·0, H=6·0, O=47·0) whilst its proximate resolution into *cellulose* (78-80 per cent.) and *non-cellulose* (20-22 per cent.) may be represented by the formula $3 C_6 H_{10} O_5$, $C_6 H_6 O_3$. This is for the time advanced more as a statistical than a molecular expression. The authors, however, proceed to show that the more oxidisable constituents of the jute-fibre (non-cellulose) are compounds, or a compound, characterised by an atomic ratio approximating to $C_6 : H_6 : O$ (more exactly $C_{76} H_{80} O_{37}$, and associated with cellulose in chemical rather than mechanical union. The intimate nature of this union is shewn, by—

1st—the resistance of the fibres to the action of all simple hydrolytic agents;

2nd—the fact that in those reactions of the fibre substance which depend on its alcoholic characteristics (OH groups), its molecular homogeneity is equally manifest.

This second proposition is proved by a series of very elaborate experiments, for an account of which the reader is referred to the pamphlet in question.

The authors further state that, from the earliest appearance of the fibre bundles, for example in sections cut a few centimetres below the growing point, they have all the chemical characteristics of the mature fibre. "In the isolated fibre," they write, "we find no sensible variation in the proportion of cellulose to non-cellulose, nor in their mode of combination, throughout the entire length of 2 to 3 metres. This evidence appears to carry with it the suggestion that both cellulose and non-cellulose may have a common and simultaneous origin in a parent substance, and that lignification is the result of progressive modifications and differentiations of this original complex molecule."

The authors, in conclusion, briefly summarise the main points which they have endeavoured to establish, as follows:—

1. The fibre is a compound of cellulose and non-cellulose empirically $C_{12} H_{18} O_9$, of the general chemical features of the cellulose, resisting hydrolysis and yielding explosive nitrates, of which the highest is the tetranitrate.

Trade in Jute. (*J. Murray.*) **JUTE.**

CHEMISTRY.

2. The non-cellulose is a complex molecule, from the products of resolution of which, by chlorination and hydrolysis, it may be inferred that it contains the following groups of molecules:—

(*a*) $C_{18} H_{18} O_{10}$, ketone transitional to a quinone, chlorinated directly to form mairogallol: (*b*) $C_5 H_4 O_2$, furfaral in combination, by condensation with (*a*), and with (*c*), an acetic residue. These are combined in the approximate molecular proportion (*a*) 2, (*b*) 6, (*c*) 5, $= C_{76} H_{80} O_{37}$.

(3) The ligno-celluloses in the earlier stages of growth are constituted on this type. The true woods, on the other hand, appear to represent a more condensed type. (*Conf.* with the remarks under **Corchorus,** *Vol. II*, *551*.)

TRADE. 125

Trade.—It is, perhaps, unnecessary to re-write the statistical portions of the article **Corchorus** (*Vol. II., 557-561*), since the fluctuations in the Jute trade are not of a spasmodic or extraordinary nature. A review of any recent period of years would naturally convey the main features of the trade, and render it alone necessary to add annually a bare statement of the chief transactions. With this view the exports of raw jute from India up to last year may be here given in continuation of the table at *p. 557* of the article **Corchorus.** The average exports for the five years ending 1887-88 was 8,223,859 cwt, but the returns for 1888-89 are the largest on record, exceeding those of 1882-83 by 204,234 cwt., the actual figures having been 10,553,143 cwt., valued at R7,89,71,539.

At page 559 of the article **Corchorus** particulars are given of the Indian consumption of jute and the Indian manufactures. Through the kindness of **Messrs. Barry & Co.,** it has been possible to considerably amplify the information on these subjects. The following table gives the manufacture of bags down to last year:—

PORTS.	1885.	1886.	1887.	1888.	1889.
	Bags.	Bags.	Bags.	Bags.	Bags.
Burma . .	10,579,300	12,193,300	12,981,400	14,726,700	13,163,000
Straits . .	7,748,000	7,433,000	11,413,300	13,880,600	10,536,400
Bombay . .	22,494,700	17,373,000	19,578,500	16,926,400	15,761,400
Coast . .	8,346,300	7,329,100	6,029,400	7,624,500	8,699,300
Up-country and Local .	33,801,900	36,434,500	34,658,700	32,078,600	28,057,500
Home Consumption, Total . . .	82,970,200	80,762,900	84,661,300	85,236,800	76,217,600
Australia . .	12,813,800	6,191,800	16,373,300	18,278,000	13,620,900
New Zealand .	3,278,500	2,611,400	3,972,300	3,928,100	6,584,100
Cape . .	1,519,000	2,095,300	2,582,900	2,636,100	4,284,500
Egypt . .	4,726,500	3,788,000	2,694,600	3,135,800	4,189,400
New York . .	7,714,500	6,964,900	8,736,800	6,621,400	24,244,600
San Francisco .	15,821,200	31,592,700	24,736,100	34,484,900	24,547,700
Europe . .	9,150,200	8,865,200	4,939,100	10,886,200	31,631,000
Foreign Exports—Total	55,023,700	62,109,300	64,035,100	79,970,500	109,102,200
GRAND TOTAL OF BAGS MADE IN INDIA .	137,993,900	142,872,200	148,696,400	165,207,300	185,319,800

There are now in Bengal (according to Commercial returns) 23 mills at work furnished with 157,175 spindles and 7,673 looms. The probable

JUTE. **Trade in Jute.**

TRADE. number of operatives employed has been estimated at 62,000 hands. (*Conf. with Vol. II., 558*). In the article **Corchorus** the probable consumption of Jute by the Bengal mills was assumed to be about 8,571,428 cwt. The commercial returns, which the writer has now been able to consult, give the Bengal consumption at 5,000,000 maunds and the probable value of their productions at R300 lakhs. **Mr. J. E. O'Conor** (*Statistical Tables of British India*) shows 24 mills at work in Bengal, one in Madras, and one at Cawnpore, and he gives the amount of Jute worked up by these mills during 1888-89 as 3,796,944 cwt. But in comparing commercial with official returns, it has to be borne in mind that the years do not correspond. It, however, seems probable that the estimate of total production of jute (*p. 558* of the article **Corchorus**) of 15,000,000 cwt. per annum is not far from correct, being an under, rather than over, estimate.

KÆMPFERIA, *Linn.; Gen. Pl., III., 641.*

A number of species belonging to this genus are found in Indian gardens, but those of economic value are mainly **K. Galanga** and **K. rotunda.** Other species are referred to by writers on Indian botany such as **K. candida** (the *Pán-u-phyú* of Pegu) and **K. parviflora** (the *Ká-mong-ni* of Pegu). **K. pandurata,** *Roxb.*, is said by Rumphius to be cultivated for culinary and medicinal purposes, and Rheede notices it as a medicine used in dysentery. This is very doubtfully a native of India, but it appears to belong to Ceylon at all events. Thwaites remarks that the roots are used medicinally.

[SCITAMINEÆ.

Kæmpferia angustifolia, *Roxb., Fl. Ind., Ed. C.B.C., 6;* 1

Vern.—*Kanján-búra, mudú-nirbisha,* HIND., BENG.

Habitat.—A native of Bengal.

Medicine.—Roxburgh remarks that the people of Bengal use the ROOTS of this plant as a medicine for their cattle. MEDICINE. Roots. 2

K. Galanga, *Linn.* 3

Syn.—ALPINIA SESSILIS, *Kön.*

Vern.—*Chandra múla,* HIND.; *Chandú múlá, humúla,* BENG.; *Katsjulum,* MAL.; *Katsjolum,* TAM.; *Kachoram,* TEL.; *Kha-mung,* BURM.; *Chandra-mulika,* SANS. It is probable some of the above names refer to **Curcuma Zedoaria,** *Roscoe;* see Vol. II., 669.

References.—*Roxb., Fl. Ind., Ed. C.B.C., 5; Voigt, Hort. Sub. Cal., 566; Kurz, Pegu Rep., App. C., xx; Thwaites, En. Ceylon Pl., 316; Grah., Cat. Bomb. Pl., 208; Elliot, Fl. Andhr., 75; Mason, Burma and Its People, 501, 804; Jour. Agri.-Hort. Soc. Ind., X., 341; Pharm. Ind., 232; O'Shaughnessy, Beng. Dispens., 649; Moodeen Sheriff, Supp. Pharm. Ind., 165; Drury, U. Pl., 271.*

Habitat.—A fairly abundant plant, met with in the hotter parts of India; much cultivated in gardens.

Perfumery.—Roxburgh remarks that the TUBERS are used in perfumery. Mason says that the ROOTS may be often seen attached to the necklaces of Karen females, for the sake of their perfume. They also place them in their clothes for the same purpose. PERFUMERY. Tubers. 4 Roots. 5

Medicine.—It is probable that, as implied in the Pharmacopœia of India, the TUBERS of this, and the next species, are used indiscriminately in Hindu medicine. They are agreeably fragrant, and of a warm, bitterish, aromatic taste. O'Shaughnesssy justly remarks: "Notwithstanding its specific name, it is not the source of the true *Galanga* root of the drug-shops." Drury quoting from Rheede says that the tubers reduced to powder and mixed with honey are given in coughs and pectoral affections. Boiled in oil it is externally applied to stoppages of the nasal organs. MEDICINE. Tubers. 6

Food.—The TUBERS are said to be eaten as an ingredient of *pán* (betel leaf, &c.), the supply being obtained mainly from Chittagong. Thwaites alludes to its use as a masticatory in Ceylon. FOOD. Tubers. 7

K. rotunda, *Linn.; Wight, Ic., t. 2029.* 8

Vern.—*Bhui-champa,* HIND.; *Bhuichámpá,* BENG; *Bhui champo,* GUJ.; *Konda kalava,* TEL.; *Malan-kua* (or mountain ginger), MALAY.; *Myae-ban-touk* (or *myae-pá-douk*), BURM.; *Yawakenda, loukenda,* SING.; *Bhúmichampa, bhúchampaca,* SANS.; *Kúntshi,* JAVA; *Nagai mio,* COCHIN CHINA.

References.—*Roxb., Fl. Ind., Ed. C.B.C., 6; Asiatick Reser., XI., 327; Voigt, Hort. Sub. Cal., 566; Kurz, Pegu Rep., App. C., xx.; Thwaites, En. Ceylon Pl., 316; Trimen, Cat. Ceylon Pl., 91; Bot. Mag., t. 920; Grah., Cat. Bomb. Pl., 208; Rheede, Hort. Mal., XI., t. 9; Elliot, Fl. Andhr., 95; Mason, Burma and Its People, 428, 804; Sir W.*

Jones, V., p. 69; Ainslie, Mat. Ind., I., 489, 491; O'Shaughnessy, Beng. Dispens., 650; Moodeen Sheriff, Supp. Pharm. Ind., 165; U. C. Dutt, Mat. Med. Hind., 294; Dymock, Mat. Med. W. Ind., 2nd Ed., 785; U. S. Dispens., 15th Ed., 1782; Year Book Pharm., 1880, 251; Drury, U. Pl., 272; Gazetteer, Bomb. (Gujrat), VI., 15; Jour. Agri.-Hort. Soc. Ind. (Old Series), X., 342.

Habitat.—An elegant stemless plant with large rotund leaves: native of the damp hot regions of India and Burma. Distributed to Ceylon, Java, and Cochin China. Often met with in cultivation both on account of its graceful leaves and its sweetly-scented flowers, which appear in the hot season when the plant is leafless; the perfume of the flowers, as the Sanskrit name implies, recalls those of the *champaka* (**Michelia**).

MEDICINE. 9 **Medicine.**—**Roxburgh** points out that **Woodville** was in error when he supposed this plant to be the source of the Round Zedoary (see **Curcuma Zedoaria,** *Roxb.;* Vol. II., 669), but this mistake continued to be made by many subsequent writers. **Ainslie** remarks: "It is the **Zedoaria rotunda** of **Bauhin** and has been well described by **Sir William Jones** in the fourth Volume of the *Asiatic Researches.* On the Malabar Coast it is termed *Malanqua*; and **Rheede** informs us (*Hort. Mal., II., 18*) that the whole PLANT,

Plant. 10 when reduced to powder, and used in the form of an ointment, has wonderful efficacy in healing fresh wounds, and that, taken internally, it removes any coagulated blood or purulent matter that may be within the body; he

Root. 11 adds that the ROOT is a useful medicine in anasarcous swellings." **Dr.** **Dymock** writes that in Bombay a powder of the TUBERS "is used as a

Tubers. 12 popular local application in mumps (*Galgand*), but as they are generally combined with more active remedies, such as Croton seed, Aconite, and Nux-vomica, it is probable that they do not contribute much to the cure." "The substance of the rhizomes and tubers is of a pale straw-colour, has a bitter, pungent, camphoraceous taste, much like that of true Zedoary; the whole plant is aromatic." In the Gazetteer of the Rewa-Kanta District it is stated that the roots are stomachic and are also applied to swellings. **Thwaites** remarks that in Ceylon the root is employed medicinally, but he does not state for what purpose. The almost universal belief (from one side of India to the other) that the rhizomes are useful in reducing swellings, would suggest the desirability of this subject being more carefully investigated in the future.

SPECIAL OPINION.—§ "According to Sanskrit writers the root, used in the form of a poultice, promotes suppuration" (*U. C. Dutt, Civil Medical Officer, Serampore.*)

13 **Kaing-grass,** a name used in Burma for certain grasses which, according to **Sir D. Brandis,** consist chiefly of the following:—**Arundo sp., Phragmites Roxburghii, Saccharum procerum,** and **S. spontaneum.** On this subject a considerable amount of official correspondence took place in 1881, between the Chief Commissioner of Burma and the Government of India. For further information see the respective species of grass named above.

Kaladana, see **Ipomœa hederacea,** *Jacq.;* CONVOLVULACEÆ, p. 485.

Kalamander or **Calamander Wood,** see **Diospyros quæsita,** *Thwaites;* EBENACEÆ; Vol. III., 154.

KALANCHOE, *Adans.; Gen. Pl., I., 659.*

14 **Kalanchoe laciniata,** *DC.; Fl. Br. Ind., II., 415; Wight, Ic., t. 1158;* CRASSULACEÆ.

Syn.—K. TERETIFOLIA, *Haw.;* COTYLEDON LACINIATA, *Roxb.*

Vern.—*Hemsagar*, HIND., BENG.; *Zakhmhyát, parna-bij*, BOMB.; *Malakulli*, TAM.; *Hémaságara* (=sea of gold), SANS.

References.—*Roxb., Fl. Ind., Ed. C.B.C., 388; Voigt, Hort. Sub. Cal., 268; Kurz in Jour. As. Soc., 1876, Pt. II., 309; Thwaites, En. Ceylon Pl., 129, 417; Dalz. & Gibs., Bomb. Fl., 105; Grah., Cat. Bomb. Pl., 81; W. & A., Prod., 360; Mason, Burma and Its People, 773; Dymock, Mat. Med. W. Ind., 2nd Ed., 360; Pharm. Ind., I., 590-591; Gazetteer, Mysore & Coorg, I., 61; Honigberger, Thirty-five years in the East, II., 263.*

Habitat.—An erect, stout, perennial herb, common throughout the tropical regions of India, especially the Deccan Peninsula, and from Bengal (Behar-Dacca) to Burma, Yunan, Malacca, Java, and Tropical Africa.

MEDICINE.

Leaves. 15

Medicine.—**Baillon** (*Natural History of Plants, III., 317*) remarks of the CRASSULACEÆ that the succulent species owe their cooling properties as topical applications to the quantity of water contained in their fleshy organs. It seems probable that most of the large-leaved species might be so used, as, for example, the house-leeks of Europe, but at the same time it is probable that certain of the genera and even species possess special peculiarities. It is on this account that the writer does not follow the example, recently given, by the authors of the *Pharmacographia Indica*, in treating **Kálanchoe laciniata, K. spathulata,** and **Bryophyllum calycinum** conjointly. **Dr. Dymock**, in his *Materia Medica of Western India*, gave a detailed account of **Bryophyllum,** and quoted under **Kalanchoe** a brief sentence said to be derived from **Ainslie.** It would thus seem that no very good object has been served by assigning the properties of **Bryophyllum** to the species of **Kalanchoe.** There is a historic interest, however, in this subject which it seems desirable to point out When **Roxburgh** wrote his *Flora Indica (1820)*, **Bryophyllum** was a garden curiosity, which he states had been procured from the Moluccas. At the present day, thousands of square miles of Bengal possess that plant as their most prevalent weed. Every lane around the Botanic Gardens, Calcutta, possesses a rampant growth of **Bryophyllum,** and throughout the greater part of Bengal this is the case, the plant becoming less abundant as the drier areas are reached, though it is also met with as far north as the Sutlej valley in the Panjáb (20 miles from Simla), also in Bombay, Madras, and Burma. We are told it was brought to India by Lady Clive in 1799 (see **Rev. J. Long**, *Journal Agri.-Hort. Soc. Ind. (Old Series), X., 7*), but whether that date is absolutely correct or not, botanists agree in viewing it as an introduced plant. Even supposing it existed in Bengal prior to a foreign introduction, such as that attributed to Lady Clive, it must have been an extremely rare plant indeed, to have escaped **Roxburgh's** scrutiny of the Flora of the Lower Provinces, and even so recent a writer as **Stewart** apparently did not know of its existence in the Panjáb. This rapidity of distribution within the past sixty years practically proves, therefore, its exotic nature. **U. C Dutt**, in his *Hindu Materia Medica*, does not allude to either **Bryophyllum** or **Kalanchoe,** and, indeed, everything points to an opinion adverse to the belief that **Bryophyllum** could have been known to even the most modern of Sanskrit writers. In the *Pharmacographia Indica*, however, we are told of **Kalanchoe** and **Bryophyllum** that they "are called in Sanskrit *Asthibhaksha* and *Parna-vija* or 'leaf-seed,' because their leaves, when placed upon moist ground, take root and produce young plants." Now this application of these Sanskrit names could only be made of **Bryophyllum,** since none of the **Kalanchoes** possess the remarkable property exhibited by **Bryophyllum** of producing budules in the crenatures of the leaves. **Dr. Moodeen Sheriff** gives a series of vernacular names under **Kalanchoe laciniata** which, in the present work (Vol. I., 543), have been transferred almost entirely to **Bryophyllum.** It seemed to

MEDICINE. the writer at that time, that these names, together with the others given in the passage cited, were, by modern usage, assigned to **Bryophyllum**, but it may be added that **Kalanchoe laciniata** being not only the most abundant and best known species, but the one with the strongest claims to the possession of medicinal properties, is in all probability the plant alluded to by all Muhammadan and Hindu writers on the Materia Medica of India. On the Himálaya **K. spathulata** is a fairly plentiful species, but it has the reputation of being poisonous, and is, therefore, said to be rarely used as a cooling external application, though special preparations from it are so employed.

The medicinal properties given in the *Pharmacographia Indica* will be found in the first volume of this work under **Bryophyllum** (*l. c.*), but the sentence said to be derived from **Ainslie** regarding the leaves of **K. laciniata** may be here given :—" I can myself speak of their good effects in cleaning ulcers and allaying inflammation." The authors of the *Pharmacographia* add :—"We have seen decidedly beneficial effects follow their application to contused wounds; swelling and discoloration were prevented, and union of the cut parts took place more rapidly than it does under ordinary treatment. The juice of the leaves is administered in doses of ¼ to 1 tola (45 to 180 grains) with double the quantity of melted butter in diarrhœa, dysentery, and cholera; it is also considered beneficial in lithiasis." The above passage has been quoted as it seems of very considerable interest, but the writer is not quite certain whether or not it applies to **K. laciniata** or to **Bryophyllum calycinum.**

Juice. 16 SPECIAL OPINION.—§ " The JUICE is used externally in bruises, sprains, and burns; also to cure superficial ulcers. As a styptic it is used on fresh cuts and abrasions." (*Civil Surgeon J. H. Thornton, B.A., M.B., Monghyr*).

Kalanchoe pinnata, *Pers.*, also **Cotyledon rhizophylla**, *Roxb.*, and **Bryophyllum pinnatum**, *Kurz.*; see **Bryophyllum calycinum**, *Salisb.*, Vol. I, 543.

17 **K. spathulata**, *DC.*; *Fl. Br. Ind., II., 414.*

Syn.—K. VARIANS, *Haw.*; K. NUDICAULIS, *Ham.*; K. CRENATA, *Oliv.*; K. ACUTIFLORA, *Kurz.*

Vern.—*Tatára, rungrú, haiza-ka-patta*, PB., HIND.; *Pát kudri, bakal patta*, KUMAON.

References.—*Voigt, Hort. Sub. Cal., 268; Kurz in Journ. As. Soc., 1876, Pt. II., p. 309; Haw. in Phil. Mag. Lond., N. S., VI., 302, 303; W. & A., Prodr., 360; Oliv., Fl. Trop. Afr., II., 394; H. f. & T. in Journ. Linn. Soc., II., 91; Atkinson, Him. Dist., 350, 476.*

Habitat.—A succulent perennial of the tropical and subtropical Himálaya, from Kashmír to Bhután. Found generally between 1,000 and 4,000 feet above the sea, but near Simla it ascends to 6,000 feet; common also in Burma. Distributed to the warm tracts of China and Java.

MEDICINE. Leaves. 18 **Medicine.**—In Lahore this is reckoned a specific in cholera, and in Kangra the burned LEAVES are applied to abscesses.

Juice. 19 SPECIAL OPINION.—§" The expressed JUICE of the bitter variety of this species is used in enlarged spleen. It acts as an antiperiodic, tonic, and drastic purgative." (*Civil Surgeon J. H. Thornton, B.A., M.B., Monghyr.*)

FODDER. 20 **Fodder.**—It is poisonous to goats and is not eaten by cattle.

Kalar, see **Reh.** Vol. VI.

Kale, a name applied to certain loose-leaved forms of the cabbage; see **Brassica (oleracea) acephala**; CRUCIFERÆ, Vol. I., p. 534.

Kale, Indian, a name sometimes given to edible Aroids in those parts of the country where the leaves are eaten; see **Colocasia antiquorum,** AROIDEÆ, Vol. II., p. 510.

Kale, Sea, a name given to **Crambe maritima**: see the Indian form **Crambe cordifolia,** CRUCIFERÆ, Vol. II., p. 582.

Kamela, see **Mallotus philippinensis,** *Müll.*; EUPHORBIACEÆ. Vol. V.

KANDELIA, *Wight et Arn.; Gen. Pl., I., 679.*

Kandelia Rheedii, *W. & A., Fl. Br. Ind., II., 437; Wight, Ill., [I., t. 89;* RHIZOPHOREÆ. 21

Syn.—RHIZOPHORA CANDEL, *Linn.*

Vern.—*Guria,* BENG.; *Rasunia, rasuria,* URIYA; *Tsjeron-kandel,* MALAY.

References.—*Voigt, Hort. Sub. Cal., 41; Brandis, For. Fl., 218; Kurz, For. Fl. Burm., I., 449; Beddome, Fl. Sylv. Anal. Gen., 100, Pl. XIII., fig. 6; Gamble, Man. Timb., 176; W. & A., Prodr., I., 310; Hook., Ic. Pl., t. 362; Grah., Cat. Bomb. Pl., 68; Rheede, Hort. Mal., VI., t. 35; Drury, U. Pl., 272; Gazetteer, Orissa, II., App. VI., 176; Indian Forester, VI., 124.*

Habitat.—An evergreen shrub, or small tree, found on the muddy shores in the tidal creeks of Bengal, Burma, and the Western Coast. Distributed to Ceylon and the Malay Islands.

Dye and Tan.—The BARK is used in Tavoy in dyeing red, but probably as a mordant. It is said to be employed in Cochin as a tanning material. DYE & TAN. Bark. 22

Medicine.—According to **Rheede** the BARK, mixed with dried ginger or long pepper and rose-water, is said to be a cure for diabetes. MEDICINE. Bark. 23

Structure of the Wood.—Soft, close-grained: weight 38lb per cubic foot; used only for firewood. TIMBER. 24

Kankar.—The concretionary carbonate of lime, which usually occurs in nodules on alluvial deposits, see Vol. II., 147.

KAOLIN; *Mallet, Man. Geology, Part IV., 129.* 25

Mr. H. B. Medlicott, late Director of the Geological Survey of India, has kindly furnished the following note regarding Kaolin:—

"According to **Von Richthopen** the most famous porcelain clay of China is derived from a fine, hard, greenish slate (white when powdered), quarried in the mountains of King-te-chin, and hence called Kao-ling ('high ridge'). We are more familiar with crude Kaolin in the form of 'Cornish-stone', a highly felspathic granite rock in an advanced state of decomposition, from which the clay is obtained by elutriation. This process is sometimes effected naturally by denudation, when the clay is found ready made. It is only the purest varieties that can rank as Kaolin, the inferior kinds being available for crucibles, bricks, or common pottery, according to the greater or lesser trace of fluxing or of colouring matter. Where crystalline rocks occur, as they do in parts of India, this material is probably not wanting. The quality of such kaolin can only be tested by adequate practical trial. Promising material has been noticed in Trichinopoly, North Arcot, Mangalore, Gya, Delhi, Darjeeling, and Dalhousie."

It is understood that a very superior quality of **Kaolin** has recently been discovered in the Central Provinces, and that **Messrs. Burn & Co.** have obtained a concession to work the same. There are many articles made of **Kaolin** which have to be imported into India, such, for example, as electric insulators. If these can be made of Central Provinces **Kaolin** it is probable that a large industry will rapidly be developed.

The reader is referred for further information to the article **Clay**, Vol. II., 364-367.

Kapok, see **Eriodendron anfractuosum**, *DC.* Vol. III., 258-264.

Kapur-kachri, see **Hedychium spicatum**, *Ham.;* p. 207; also **Curcuma** [**Zedoaria**, *Roscoe;* Vol. II., 671.

Karyat, kiryat, creyat, see **Andrographis paniculata**, *Nees;* Vol. I., [240.

Katira Gum, see **Sterculia urens**, *Roxb.;* Vol. VI.: also **Cochlospermum** [**Gossypium**, *DC.;* Vol. II., 412-414.

Kath, or **Catechu**, see **Acacia Catechu**, *Willd.;* Vol. I., 27-44.

Kauri Pine, see **Dammara australis**, *Lamb.;* Vol. III., 18.

Keersal, see **Acacia Catechu**, *Willd.;* Vol. I., 35.

Kelp, see **Barilla**, Vol. I., 399; also **Fucus**, Vol. III., 451.

Kelu, kilan, &c., see **Cedrus Libani** *var.* **Deodara**, Vol. II., 235.

Kerosene, see **Petroleum**, Vol. V.

26 **Khaki**, an earthy or clay colour, now largely used to dye the uniforms of soldiers. "Khaki" is the name given to a sect of Vaishnava Hindus founded by **Kil**, a disciple of **Krishna Das.** They apply ashes of cowdung to their dress and persons, hence the name of *khaki* as given to them. There are numerous processes for producing this colour—some with dyes, others with pigments. The following may be mentioned:—

Allahabad Khaki.—This is produced by boiling myrabolans, gall-nuts, and sulphate of iron together.

Oudh Khaki.—This is prepared from the barks of **Acacia arabica** and **Butea frondosa** with the extract, Cutch.

Palanpur (Bombay) Khaki.—Obtained from clay in combination with sulphate of iron. In Palanpur a good khaki is also prepared with **Terminalia belerica.**

In many parts of the country, such as in Manipur, a natural earth is used. The *laynung* earth of Manipur seems capable of much development.

It is needless to attempt an enumeration of all the Khaki dyes of India, and the reader is, therefore, referred to the article **Dyes** (Appendix to this work) for further information; also to **Pigments**, Vol. V.

Kharif, see **Crops**, Vol. II., 594.

Kidney-bean, see **Phaseolus vulgaris**, Vol. V.

Kinka Oil, see **Vernonia anthelmintica**, Vol. VI.

Kino, see **Butea frondosa**, *Roxb.;* Vol. I., 548, and **Pterocarpus Marsupium**, [*Roxb.;* Vol. VI.

KLEINHOVIA, *Linn.; Gen. Pl., I., 219.*

27 **Kleinhovia Hospita**, *Linn.; Fl. Br. Ind., I., 364.*

References.—*Roxb., Fl. Ind., Ed. C.B.C., 505; Beddome, Fl. Sylv. Anal. Gen., t. 4; Gamble, Man. Timb., 45; Dalz. & Gibs., Bomb. Fl., 23; W. & A., Prodr., I., 64; Lisboa, U. Pl. Bomb., 22; Agri.-Hort. Soc. Ind., Trans., VII., 48, 81; Journ., VI., 40.*

Habitat.—A handsome tree met with in the Eastern and Western Peninsulas of India and distributed to Malacca, Singapur, Ceylon, Java, the Philippine Islands, and East Tropical Africa. Frequently grown as an avenue tree, especially in Calcutta, Poona, &c.

Structure of the Wood.—The old timber is stated to be much valued in Java. No information is available regarding the Indian-grown timber. TIMBER. 28

Knol-kohl, Kohl-rabi, or Ole kole. 29

This is the turnip-stemmed cabbage which is every year being more extensively cultivated in India. There are two varieties—a purple and a green. Firminger says the best seed is obtainable from the Cape of Good Hope. The plants take about six weeks or two months to be ready for the table. They should be transplanted when they have about three or four leaves and on ridges 20 inches apart, the plants being 14 to 15 inches asunder.

The reader is referred to *Firminger's Man. Gard. for India, 137-138; Jour. Agri.-Hort. Soc. Ind. (New Series), V., 41; Rept. Govt. Bot. Gard., Saharanpur, 1884, 5, &c.*, also to Vol. I., 534.

KOCHIA, *Roth.; Gen. Pl., III., 60.*

Kochia indica, *Fl. Br. Ind., V., 11; Wight, Ic., t. 1791;* CHENOPODIACEÆ. 30

Syn.—K. GRIFFITHII, *Bunge;* PANDERIA PILOSA, *Herb. Ind. Or., H. f. & T. in part.*

Vern.—*Kaura ro, búi* (bazar name *búi chhoti*), PB.

Reference.—*Stewart, Pb. Pl., 179.*

Habitat.—A herbaceous plant found in North-West India from Delhi to the Indus (very abundant), also in the Deccan Peninsula and distributed to Afghánistán.

Medicine.—According to **Stewart** this PLANT is employed medicinally in the Panjáb. MEDICINE. Plant. 31

SPECIAL OPINIONS.—§ "Used as a vascular (cardiac) stimulant in cases of weak and irregular heart, especially when following on fevers." (*Civil Surgeon F. F. Perry, Jullunder City, Punjab*).

Fodder.—It is used as a camel fodder, see Vol. II., 60. FODDER. 32

KŒLERIA, *Pers.; Gen. Pl., III., 1183.*

Kœleria cristata, *Pers.;* GRAMINEÆ. 33

Syn.—ARIA CRISTATA, *Linn.*

Fodder.—See Vol. III., 436.

K. phleoides, *Pers.; Duthie, Fodder Grasses of Northern India, 61.* 34

References.—*Stewart, Botanic Tour in Hazara (Jour. Agri.-Hort. Soc. Ind., XIV. 6).*

Fodder.—See Vol. II., 423.

KOKOONA, *Thw.; Gen. Pl., I., 362.*

Kokoona zeylanica, *Thwaites; Fl. Br. Ind., I., 616;* CELASTRINEÆ. 35

References.—*Beddome, Fl. Sylv., t. 146; Anal. Gen., lxx; Thwaites, En. Ceylon Pl., 52; Kew Reports, 31.*

Habitat.—A tree with pale-coloured bark found on the banks of streams in the Western Peninsula and Ceylon.

Oil.—Thwaites says that an oil is expressed from the SEEDS, which is used for burning in lamps. OIL. Seeds. 36

Medicine.—The inner yellow BARK is employed medicinally. It is also made into a kind of snuff which excites copious secretion, and is considered beneficial in headache. MEDICINE. Bark. 37

Kola Nut, see **Cola acuminata,** *R. Br.*, Vol. II 500.

Koot or **Kut**, see **Saussurea Lappa**, *C.B.C.*; Vol. VI.

Kousso, see **Brayera anthelmintica**, *Kunth.*; Vol. I., 534.

KRAMERIA, *Linn.*; *Gen. Pl.*, *I.*, *140.*

38 **Krameria triandra**, *Ruiz et Pavon*; POLYGALEÆ.

KRAMERIA; RHATANY.

Habitat.—A native of Peru and Bolivia.

MEDICINE. Root. 39 **Medicine.**—Rhatany ROOT is powerfully astringent and tonic. It has been successfully employed in chronic diarrhœa, in passive or atonic hæmorrhages, and locally in leucorrhœa, ophthalmia, &c. (*Pharm. Ind.*). The powder may be used as a dentifrice when mixed with prepared chalk or myrrh. The drug is entirely obtained from foreign sources.

40 **Kúlú.**—In the Report on Madras fibres (published in the Selections of the Records of the Madras Government, 1856) mention is made of a strong

FIBRE. 41 and useful fibre known as *Kúlú* which is said to be prepared by retting. This may be the fibre of **Sterculia urens**, *Roxb.*, which see, Vol. VI.

KYDIA, *Roxb.*; *Gen. Pl.*, *I.*, *203.*

[MALVACEÆ.

42 **Kydia calycina**, *Roxb.*; *Fl. Br. Ind.*, *I.*, *348*; *Wight*, *Ic.*, *t. 879-80-81*;

Syn.—K. FRATERNA, *Roxb.*; K. ROXBURGHIANA, *Wight*; K. PULVERULENTA, *Ham.*

Vern.—*Pola, púla, puli patha, potari,* HIND.; *Pattrá, paláo,* BIJNOUR; *Varanga, várangada, warung,* BOMB.; *Baranga, bhoti,* C. P.; *Bittia, gonyer, pata dhamin,* KOL.; *Poska olat',* SANTAL; *Derki,* KHARWAR; *Puta, puttiya,* N.-W. P.; *Kubindé,* NEPAL; *Sedangtaglar,* LEPCHA; *Mahow, moshungon,* MECHI; *Boldobak,* GARO; *Kopásia,* URIYA; *Púlí, púlá, polá,* PB.; *Potari, pandiki, peddapotri, pedda kunji,* TEL.; *Buruk, bosha, kunji,* GOND.; *Bendi, bende-naru, bellaka,* KAN.; *Warung, iliya,* MAR.; *Dwabote, bokemaiza,* BURM.; *Dwabók,* SHAN.

References.—*Roxb., Fl. Ind., Ed. C.B.C., 521; Voigt, Hort. Sub. Cal., 108; Kurz, For. Fl. Burm., 124; Beddome, Fl. Sylv., t. 3; Gamble, Man. Timb., 43; Dalz. & Gibs., Bomb. Fl., 24; Stewart, Pb. Pl., 25; Rev. A. Campbell, Rept. Econ. Prod. Chutia Nagpur, No. 9213; Elliot, Fl. Andhr., 144, 156; Mason, Burma and Its People, 536, 755; O'Shaughnessy, Beng. Dispens., 227; Pharmacog. Indica, 228; Atkinson, Him. Dist., 306, 740, 791; Lisboa, U. Pl. Bomb., 16, 229; Royle, Prod. Res., 230; Fib. Pl., 266; Report on Fibres of India by Cross, Bevan, King & Watt, 10; For. Ad. Rep. Chutia Nagpur, 1885, 28; Gazetteers:—Burm., I., 139; Bomb., XV., 71; XI., 24; Mysore & Coorg I., 68; Indian Forester, I., 84, 275; II., 18; III., 200; IV., 323; VIII., 119, 417; X., 222, 325; XI., 381; XIII., 119; XIV., 269, 298; Agri.-Hort. Soc. Ind., Trans., VII., 81; Jour. (Old Series), VIII., (Sel.) 177; IX., (Sel.) 40; XIII., 308, (Sel.) 60.*

Habitat.—A small tree, or large bush common in the forests of the sub-tropical regions of India and Burma except the arid tracts. It ascends the Himálaya to altitudes of about 2,000 feet.

FIBRE. Bark. 43 **Fibre.**—The inner BARK yields a bast fibre used for coarse ropes, &c. It was chemically analysed by **Cross, Bevan** and **King**, and found to possess moisture 10·9; ash 2·5. The decomposition (hydrolysis), by alkalis, gave the following results: loss after boiling for five minutes in solution of 1 per cent. Na_2O, 13·2, after boiling for one hour, 25·2. Amount of cellulose 70·2. Effect of mercerising [*i.e.*, subjection to action of strong caustic soda (33 per cent. Na_2O) in cold for one hour] loss 7·2. It will thus be seen that in point of cellulose, and in power of resistance to hydrolysis, **Kydia** fibre is fairly useful, being about twentieth in order of merit of a list of some 300 fibres met with in India.

Medicine.—The Rev. A. Campbell tells us that among the Santals the LEAVES are pounded and made into a paste and applied to the body for pains. They are also chewed when there is a deficiency of saliva. The chemistry of this plant has not been carefully worked out, but it seems probable (from the properties assigned to it) that it much resembles the mallows. MEDICINE. Leaves. 44

Food.—The BARK is mucilaginous, and is used to clarify sugar. The knowledge of this application of the bark seems very general on the lower Himálaya. **Stewart** in his admirable report on Bijnour says: "In the manufacture of sugar, the juice of the cane is heated to a less degree than when *goor* is to be made, and by the addition of a cold infusion of *chukha* (the bark of **Kydia calycina**, and *sajji*, or impure carbonate of soda) it becomes *ráb* (syrup). Into this is put a quantity of *sarwáli* (a **Potamageton**), also more *sajji*, and by degrees *khánd*, coarse sugar, to be afterwards clarified by boiling into *boora*, *chíní*, *misrí*, &c., crystallizes, and *shírá* passes off." FOOD. Bark. 45

Structure of the Wood.—White, soft, no heart-wood. Weight 40 to 45lb a cubic foot. Used for house-building, ploughs, and oars, and for carving. In the Records of the Government—No. IX., A Report on the principal trees found in the forests of Pegu—it is stated that the saplings are used, from their great strength and elasticity, by the natives, for making banghy sticks, but it is large enough to afford timber of three or four feet girth. Dr. Stewart, in his Report on the Forests of Bijnour, says: "The wood is little used, but it is occasionally employed for making ploughs and spoons." TIMBER. 46

KYLLINGA (or KYLLINYIA), *Rottb.*; *Gen. Pl.*, *III.*, *1045*.

Several species belonging to this genus are mentioned by various writers as constituting a feature of the pasturage of certain tracts of country. The following is the most important species, since, in addition to being eaten by cattle, its roots are medicinal.

Kyllinga monocephala, *Linn.*; CYPERACEÆ. 47

Vern.—*Shwet-gothúbi, nirbishi*, BENG. and HIND.; *Nirvishá*, SANS.

References.—*Roxb., Fl. Ind., Ed. C.B.C., 61; Voigt, Hort. Sub. Cal., 724; Wight, Contrib., 91; Dymock, Mat. Med. W. Ind., 2nd Ed., 853; Dalz. & Gibs., Bomb. Fl., 285; U.C. Dutt, Hindu Mat. Med., 311; Jour. Agri.-Hort. Soc. Ind. (Old Series), X., 356.*

Habitat.—A creeping sedge met with throughout the Peninsula of India in low shady pasture ground.

Medicine.—This is believed by many authors to be the *Nirvishá* of certain Sanskrit medical writers, who recommend the use of the ROOTS as an antidote to poisons. The roots are fragrant and aromatic. (Conf. with Vol. III., 66, 68.) MEDICINE. Roots. 48

LAC.

The name "lac" is applied to the resinous incrustation formed on the bark of twigs of certain trees, by the action of the lac-insect, **Coccus lacca**. For a description of the method of production, of the trees on which the insect feeds and produces the incrustation, and of the different forms of commercial lac, the reader is referred to the article under the name of the insect, Vol. II., 409—412. In this place the distribution of lac in India, its utilisation in Native and European industries, and the extent of the lac-trade, will be considered.

1 **Lac.**

Vern.—*Lákh*, HIND.; *Gálá*, BENG.; rough lac=*Khám lákh*, seed-lac=*lákh dáná*, shell-lac=*chapra lákh*, lac-dye=*kirmai*, PB.; *Lák*, GUZ.; *Komburruki*, TAM.; *Kommolaka*, TEL.; *Arakku*, *ambalu*, MALAY.; *Khejijk*, BURM.; *Lakada*, SING.; *Lákshá*, SANS.

References.—*Stewart, Pb. Pl., 41, 43, 60, 213, 214; J. H. van Linschoten, Voyage to the E. Indies, ed. Burnell, Tiele, and Yule, 1885, II., 88; Aín-i-Akbari, Blochmann's Trans., 226; U. C. Dutt, Mat. Med. Hind., 276; Fleming, Med. Pl. and Drugs, as in As. Res., Vol. XI., 195; Honigberger, Thirty-five years in the East, II., 296; Baden Powell, Pb. Pr., 190, 194; Pb. Manuf., 242; McCann, Dyes and Tans, Beng., 50-65; Buck, Dyes and Tans, N.-W. P., 24, 25; Liotard, Dyes, 33—41; Royle, Prod. Res., 4; O'Conor, Note on Lac, 1876; Annual Reviews of the Trade of India; Trade Statistics of Br. Ind.; J. L. Kipling, Note on Japanese and Indian Processes of lacquering; Balfour, Cyclop., II., 246; Ure, Dic. Indus., Arts and Manuf., II., 785; Indian Forester:—I., 269; II., 292; VIII., 81; IX., 163; Agri.-Hort. Soc. of Ind., Trans:—II., 200; III., 61, 67; VI., 47-52, 127, 157; V. (Pro.), 34, 40, 49; VI., 157; VII. (Pro.), 190; VIII. (Pro.), 345, 406; Journals (Old Series):—I., 102; II., 230-233, 252; VII., 263; X., 50-53, (Pro.) 108, 115, 148; XI., 45-51 (Sel.), 37-45; XII., 110; (New Series):—I., 179; VI. (Pro.), 6; Bomb. Admin. Rep., 1871-72, 378; Gazetteers:—Panjáb, Peshâwar, 152, 153; Hoshiarpur, 116; Gurdáspur, 60; Lahore, 101; Dera Ghází Khan, 90; Delhi, 129; Montgomery, 140; Bombay, XVI., 180; Mysore and Coorg, I., 436; C. P. (1870), 118; Madras Man. of Admin., I., 314; Settle. Rep.:—C. P., Upper Godaveri Dist., 40; Chanda Dist., 111.*

DISTRIBUTION. 2 **Distribution.**—The following summary of the occurrence of lac in India is given by **Mr. J. E. O'Conor** in the note above cited: "In India lac occurs in Bengal and Assam (abundantly), the North-West Provinces and Oudh (sparingly), the Central Provinces (abundantly), the Panjáb, Bombay, Sind, and Madras (more or less sparingly), and Burma (abundantly in some places). Lac is also found in some other countries of Southern Asia, *viz*., Siam, Ceylon, some of the islands of the Eastern Archipelago, and China, Siamese lac being held in high estimation. In India the best lac is obtained from Assam and Burma." Lac from the last mentioned locality would appear to have been long known and esteemed. Thus in the sixteenth century it was noticed and described by the traveller **Linschoten**, who wrote, "Lacke by the Malabares, Bengalens, and Decaniins is called, *assii*, by the Moors *lac*. The men of Pegu (where the best is found, and most trafiqued withall) doe call it *trick*, and deale much therewith by carrying it into the island of Sumatra, and there they exchange it for Pepper, and from thence it is carried to the Redde Sea, to Persia, and Arabia, whereupon the Arabians, Persians, and Turkes call it Lac Sumutri, because it is brought from thence into their countries."

The largest amount of lac, annually exported from India, is obtained from the Central Provinces, less from Bengal and Assam, and a comparatively small quantity from Burma. That produced in the other districts, above enumerated, is chiefly consumed in local manufactures.

Bengal and Assam. 3 The following more detailed account of the distribution of lac in the various Provinces is condensed from **Mr. O'Conor's** valuable report:—

Bengal and Assam.—In Bengal, lac is produced abundantly in the jungle tracts of Bírbhúm, Chútia Nagpur, and Orissa. In various places in the

DISTRIBUTION. Bengal and Assam.

forests of Assam, it is also found in large quantities and forms a regular article of trade, a portion of the production being manufactured at Dacca, and the rest sent to Calcutta. In 1867 the Deputy Commissioner of Púrúlia in Chútia Nagpur reported that the smallest average yearly supply from the district was 15,000 maunds, though the actual yield was, he believed, considerably more, while it was capable of great extension. From Singhbhúm, at the same period, about 1,250 maunds of lac were exported annually. In the Gya district the supply was estimated at 12,000 maunds; in Kamrup (Assam) about 5,000 maunds, with great capacity for development; in Hazaribagh at 2,000 maunds. **Mr. O'Conor**, however, remarks that these figures probably do not by any means approximate to the actual yield of the districts named.

In Bengal, lac is gathered twice a year, *viz.*, from about the middle of October to the end of January, and from the middle of May to the middle of July.

N.-W. Provinces. 4

The North-West Provinces.—Lac is obtained in some quantities from the Garhwal forests, and is said to have been largely exported to the plains some years ago. **Mr. O'Conor**, however, considers it probable that most of the substance thus brought down from Garhwal is consumed in the Province, very little, if any, being sent to Calcutta.

Oudh. 5

Oudh.—Lac is gathered in the more wooded parts of the south-eastern districts, and is exported to the Mirzapore factories and elsewhere.

Panjáb. 6

The Panjáb.—**Baden Powell** states that the production of lac is universal in this Province. According to **Mr. O'Conor**, however, Panjáb lac is inferior in quality, and is probably not exported, the whole production being consumed locally.

Sind. 7

Sind.—Lac is, in Sind, found only in the forests about Hyderabad, 12 miles north and south of the town. It occurs on **Acacia arabica**, a tree which, in moister regions, appears to be exempt from the attacks of the insect. The substance thus obtained is largely used in the manufacture of the well known lacquered ware of Hyderabad.

Central Provinces. 8

Central Provinces.—Large quantities of lac are found in all the districts of these Provinces, but particularly in the more Eastern parts. It has been stated that the Central Provinces could readily supply some 25,000 tons of stick-lac annually. A considerable amount is consumed locally for the manufacture of bracelets and other articles, but most districts also export to a greater or less extent. The incrustation is collected by jungle-tribes,—Bahelyas, Rajhors, Bhirijas, Kurkus, Dhanuks, Nahils, Bhois, and some classes of Muhammadans—who sell the produce in small quantities to Patwas, who again retail it in larger quantities to the regular dealers.

Mysore. 9

Mysore.—In this Province lac is produced in all three districts, but chiefly in Nundidrug. It is, however, not known how much might be procurable annually, the produce being collected in many places by several petty contractors, but the supply is certainly large, and probably capable of increase. At present nearly the whole production appears to be consumed locally, since the exports from Madras and Mysore are very small.

Burma. 10

Burma.—As already stated the vast forests of Burma are capable of producing an almost unlimited quantity of lac. The chief sources of commercial Burmese lac are the Shan States and Upper Burma, "stick-lac" from these places being imported in considerable quantity into Calcutta, where it is remanufactured into "shell-lac," for export.

Endeavours have been made in various parts of India to increase the quantity and improve the quality of lac by artificially rearing the insect on the most suitable trees. For a complete *résumé* of the information regarding these experiments the reader is referred to **Mr. O'Conor's** exhaustive note on the subject.

PESTS. 11

Pests.—Besides the damage brought about by fire, drought, and frost, the lac-insect is subject to the attacks of various enemies, which frequently cause great havoc and materially decrease the lac crop. These are fully dealt with in the *Indian Forester, Vol. I., 281,* by **Mr. McKee,** who writes:—"The ant, both large and small, attends the female cells for the purpose of licking up the sweet excrement; they do not appear to hurt the insect beyond biting off the ends of the white filaments, and thus bringing many an occupant of the cells to a premature end by cutting off the supply of breathing air, which the filaments serve to convey through the holes in the lac. Where ants are seen about the lac, it never appears healthy, and many cells are found with the insect dead inside them. The lac whilst on the tree is also attacked by the larva of a moth, which appears to be a species of **Galleria,** belonging to the ninth section of the **Nocturnæ,** named **Tinetes** by **Latreille,** one of which is famous for eating the honeycomb of bees, living on their larvæ, and destroying their wax. A second species was also detected, which appeared to belong to the Genus **Tinea.** The ravages of these insects destroy the colouring matter contained in the females, and a brood of young is never obtained from the cells visited by them.

"At present there seems to be no way of protecting the lac from the depredations of these larvæ. The ants, however, may be circumvented in two ways—either by surrounding the trees with wood-ashes, or by something sufficiently attractive to draw their attention away from the encrustations."

CHEMISTRY. 12

Chemical Composition.—The following analyses are given by **Mr. Hatchett**:—

Stick-lac.—Resin, 68; colouring extract, 10; wax, 6; gluten, 5·5; extraneous substances, 6·5; loss, 4 per cent.

Seed-lac.—Resin, 88·5; colouring extract, 2·5; wax, 4·5; gluten, 2 per cent.

Shell-lac.—Resin, 90·9; colouring extract, 0·5; wax, 4; nitrogenous matter, 2·8 per cent.

In *Ure's Dictionary of Arts, Manufactures, &c.,* the following more elaborate analysis by **Dr. John** is given:—

Stick-lac in 120 parts contains—An odorous common resin, 80; a resin insoluble in ether, 20; colouring matter analogous to that of cochineal, 4·5; bitter balsamic matter, 3; dun-yellow extract, 0·5; acid (*laccic acid*), 0·75; fatty matter like wax, 3; skins of insects and colouring matter, 2·5; salts, 1·25; earths, 0·75; loss, 3·75. The writer of the same article continues,—"The resin may be obtained pure by treating shell-lac with cold alcohol, and filtering the solution in order to separate a yellow-grey pulverulent matter. When the alcohol is again distilled off, a brown, translucent, hard and brittle resin of specific gravity 1·139 remains. It melts into a viscid mass with heat, and diffuses an aromatic odour. Anhydrous alcohol dissolves it in all proportions. According to **John** it consists of two resins, one of which dissolves readily in alcohol, ether, the volatile and fatty oils, while the other is little soluble in cold alcohol and is insoluble in ether and the volatile oils. **Unverdorben,** however, has detected no less than four different resins and some other substances in shell-lac. Shell-lac dissolves with ease in dilute muriatic and acetic acids, but not in concentrated sulphuric acid. The resin of shell-lac has a great tendency to combine with salifiable bases, as with caustic potash, which it deprives of its alkaline taste.

This solution, which is of a dark-red colour, dries into a brilliant, transparent, reddish-brown mass, which may be re-dissolved in both water and alcohol. By passing chlorine in excess through the dark-coloured alkaline solution, the lac-resin is precipitated in a colourless state. When this pre-

CHEMISTRY.

cipitate is washed and dried, it forms, with alcohol, an excellent pale-yellow varnish, especially with the addition of a little turpentine and mastic. With the acid of heat, shell-lac dissolves readily in a solution of borax. The substances which **Unverdorben** found in shell-lac are the following:—1, a resin soluble in alcohol and ether; 2, a resin soluble in alcohol, insoluble in ether; 3, a resinous body, little soluble in cold alcohol; 4, a crystallisable resin; 5, a resin soluble in alcohol and ether, but insoluble in petroleum, and uncrystallisable; 6, the unsaponified fat of the coccus insect, as well as oleic and margaric acids; 7, wax; 8, the *laccine* of **Dr. John**; 9, an extractive colouring matter" (*p. 786*). White shell-lac changes rapidly on exposure to the air, becoming, within a fortnight of its preparation, incapable of solution, and is generally kept in water for this reason.

MANUFACTURE. 13

Manufacture.—The method ordinarily employed has been already described (*Vol. II., 411*), but the following interesting account by **Mr. O'Conor** of the improved process adopted at Cossipore may be quoted as likely to be of value:—

"The lac is first separated from the twigs by the action of rollers, worked by steam. Of these rollers there are three sets, each consisting of an upper and under roller with a sieve attached. Between these the stick-lac passes from a feeder, and the lac is, by the turn of the roller, separated from the wood and broken up, falling on to a sieve, while the twigs are thrown off aside in a heap. If the lac has not been sufficiently broken up by the first roller to pass through the sieve, some of the twigs not having been separated, it passes on to the second roller, and goes through the same process, passing again, if still not fine enough, to the third, whence the lac is dropped, as the sieve is filled, into a series of small troughs arranged on an endless chain working with the machine, and is projected thence as the chain moves into a heap on the floor. The twigs are thrown off on to a platform on the other side. These are afterwards again examined by women, and all the remaining lac separated by hand, and, as far as it may be worth while, used in manufacture. The refuse is bought by natives for the manufacture of *choories* and other articles made of lac. The sticks are used for fuel in the furnace of the steam-engine.

"The lac is now placed in a horizontal cylinder furnished internally with arms, arranged on a bar passing through the cylinder from end to end. These arms are worked by steam power, and their action, combined with water with which the cylinder is filled, breaks up the lac into very small pieces, and separates the colouring matter which forms lac-dye. Lime is frequently employed to assist in the precipitation of the dye when the water is not naturally impregnated with lime. In the liquid thus obtained the lac is left to soak for twenty-four hours in a large vat, the liquid being then drawn off, by the removal of plugs, into a vat on a lower level, and there left to settle in the same way as indigo, the colouring matter being precipitated to the bottom. The clear water at top is drawn off, and the sediment, after having been passed through a strainer, much of the same nature as that used by paper-makers for the straining of pulp, is finally allowed to settle and consolidate, when it is pressed in frames into cakes, which are afterwards dried in the sun. These cakes are the lac-dye of commerce.

"The lac, now called 'seed-lac,' after maceration, is thoroughly melted in a close vessel heated by steam, and thence conducted into open shallow troughs, also heated by steam, where the melting continues Some resin is here mixed with the lac, to act as a flux and to prevent the lac from burning and adhering to the vessel. The resin, which is probably useful for this purpose, flies off, at least in great part, during the process of ebullition. But I may remark here that great complaints have been made in Calcutta of the adulteration of shell-lac with resin by some unscrupulous

MANUFACTURE.

native manufacturers. Undoubtedly the high prices offered in 1874 encouraged adulteration. It is said that what was then offered as lac often consisted of resin to the extent of 50 and even 60 per cent., and it may be that the practice still exists. Lac thus adulterated may be detected by its smell when broken.

"Ranged round the troughs are a series of zinc columns, inclined outwards at an angle of 45°. These columns are hollow, and being supplied by pipes with tepid water, are maintained at a certain temperature. They must never become too hot, or the fluid lac would not consolidate; nor must they become too cool, for then the lac would harden at once, and break up into small fragments which would adhere to the surface of the column. A quantity of the melted lac is now taken up by a workman in the concavity of a piece of plantain-bark, this being the material best adapted to the purpose, and dexterously flung on to one of the columns. Here the liquid mass is spread evenly and thinly over the surface by a man who makes use, for the purpose, of a leaf of the pine-apple plant or some other tough fibrous material. The leaf being held in both hands, its edge is drawn over the liquid until the mass is properly spread over the surface of the column to the required degree of fineness. It begins to consolidate at once, and becomes of a pliable, leathery texture. As soon as the lac is thoroughly consolidated it is taken off by a workman, while still so hot that it would burn the fingers of any person not accustomed to the work, a considerable section of the upper portion of the sheet of lac being torn off, because it is thicker there than in the rest of the sheet, and thrown back into the trough to be melted again. The sheet is placed on a rod held in readiness by a woman, each extremity of the sheet hanging down like a towel on a rack, and the whole is hung up to dry in a large drying-shed, the rods supporting the lac being ranged on supports running across the shed from side to side, just like a tobacco drying-house. The next day it is fit for despatch, and it is then packed in boxes and sent away.

"The points in which **Messrs. Angelo's** machinery differs from the ordinary processes are—(1) the triple system of rollers and sieves, and the endless chain of receivers; (2) the strainer for lac-dye, which is made of metal; (3) the system of melting the seed-lac by the action of steam; and (4) the hollow zinc columns with the arrangement for keeping them at an equable temperature. The natives use stems of plantain-trees, which do not last of course for more than a day, and the European improvement of porcelain and stone columns is inferior to **Mr. Angelo's** invention. This factory turns out on an average 100 maunds of shell-lac a day, from 200 maunds of stick-lac, for six months in the year, working from October to April; and with a sufficiency of raw material it could turn out 150 maunds a day during the whole year."

USES.

USES. Dye. 14

Dye.—For a description of lac-dye, the reader is referred to Volume II., 412. In addition to the information there detailed the following account of its utilisation as a cosmetic in India may be quoted from **McCann**:—"Lac-dye seems to be employed in various parts of Bengal by native women as a cosmetic for dyeing the soles of the feet and the palms of the hands or tips of the fingers, taking the place of *mehndi* or *henna* **(Lawsonia alba)** which is almost universally employed for that purpose. To prepare this cosmetic pieces of stick-lac are bruised in water, and cakes made either of cotton (Murshidábád) or of the similar floss covering the seeds of the *mudar* **(Calotropis gigantea)** are steeped in the water so that the fibres may attract the dye. These are the cakes used as cosmetics, either by wetting them on the hands and feet, or else by soaking them in water, and applying the water to the skin. The cakes are called *alta*."

Medicine.—A decoction of shell-lac is much used in Hindu medicine for preparing several medicinal oils, such as *Lákshádi taila, Chandanádi taila, Angáraka taila, &c.* (*U. C. Dutt*). **Honigberger** states that the substance was officinal at Lahore during the Sikh rule, being used in enlargement of the liver, dropsy, ulcers, &c., and also, in the form of a varnish, as an application to wounds. MEDICINE. 15

Domestic, &c.—Lac is largely used in India for the manufacture of bracelets (*chúris*), rings, beads, and other trinkets worn as ornaments by women of the poorer classes, and also in the production of turned wood-lacquer work. It appears to have been employed from very remote times for the latter purpose, perhaps the earliest description of the process being that of **Linschoten** in the sixteenth century, who, in describing the lac of Pegu, wrote as follows:—"Being refined and cleansed they make it" (the lac) "of all the colours of India. Then they dresse their bedsteds withall, that is to say, in turning of the woode, they take a peece of lac of what colour they will, and as they turne it, when it commeth to his fashion, they spread the lac upon the whole peece of woode which presently, with the heat of the turning (melteth the waxe, so that it entereth into the crestes, &c.), cleaveth unto it about the thickness of a man's naile: they then burnish it (over) with a broad straw or dry rushes so (cunningly) that all the woode is covered withall, and it shineth like glasse, most pleasant to behold, and continueth as long as the woode being well looked unto. In this sorte they cover all kinds of household stuffe in India, as bedsteddes, chaires, stooles, &c., and all their turned wood-work, which is wonderful common and much used throughout all India." This description agrees almost exactly with the process now commonly employed, excepting that, instead of the grass and rushes referred to as used for polishing, a dry palm leaf stalk, cut chisel-wise, and an oiled piece of rough muslin, are generally substituted. Lac-turnery of this description reaches a high development in certain parts of the Panjáb and Sind, perhaps the most esteemed being that of Pák Pattán in the Montgomery district. The following description of the method employed may be quoted from a note by **Mr. Kipling**:—"The most notable industry of the Montgomery district is the lac-turnery of Pák, Pattán. There are several families who send out a variety of toys, boxes spring wheels, charpoy legs, &c., to all parts of the Panjáb. The wood used is chiefly *bhán*, locally *obhán* (**Populus euphratica**)—the black or Lombardy poplar,—a soft, light, easily worked wood, containing no resin, and not liable to the attacks of insects, all of which are essential points. Nothing could be simpler in principle than the craft of the *Kharadi* (lacquer-worker), while his lathe is a perfect example of the many Indian contrivances which produce wonderful results with the most elementary and apparently inadequate means. The varnish which is produced by pressing what is virtually a stick of coloured sealing-wax, against a rapidly revolving wooden object, has been found by the experience of generations to resist damp, dust, and excessive heat and dryness, better than any known paint, and it is employed for all articles of domestic use which can be turned on the lathe. If this fine coating could be as cheaply applied to flat surfaces it would be of immense use. But this essentially simple art is capable of infinite variations. Though there are few towns in which it is not wrought in some fashion, there are some which, like Pák Pattán, enjoy a special reputation. The work from this town, though strongly resembling that of Sind, with which province the south-west of the Panjáb has some noticeable affinities, may be recognised by the use of a rich, mottled or purple, alternating with bands of black, on which delicate floral borders and designs appear to be painted in red and green. This ornament is, however, produced in a manner analogous to the Sgraffito of Italian architectural decoration. DOMESTIC. 16

LAC. **Domestic Uses of Lac.**

DOMESTIC.

Coats of different colours are super-imposed on the surface, and the pattern is produced by scratching through these with a sharp stylus. Thus a red flower is made by scratching through the black and green films; for the leaves the black only is cut away exposing the green; and for a white line all these are cut through to the white wood. This is obviously work requiring great delicacy of hand and long practice. The articles made at Pák Pattán, besides objects for native use, are tea-poys, toys, flower-stands, plateaux, chessmen, work-boxes, &c. The workmen are Muhammadans."

Work of a similar nature but less pleasing, owing to the brilliancy of the aniline colours which of late years have been largely employed to colour the lac, is made in the Hoshiarpur district. In the work of the same district transparent lac colour is also used, being applied over a ground of tin foil. Lines are then scratched in the layer, and filled in with lac of another colour. Various coloured lac is also considerably employed for colouring metal work.

The lac-ware of India is thus entirely different from the true lacquer of Japan, which is made by applying a varnish made of the sap of **Rhus vernicifera**, to the article on which, the reader is referred for a description of true lacquer-ware, and the possibility of its manufacture in India.

Lac is also used as a varnish in India, for which purpose the dye is generally left in the resin to secure a deep colour. In Burma it is employed to fix the blades of knives and similar instruments in their handles. In Bombay and elsewhere it is employed in manufacturing grindstones, for which purpose it is mixed, after being ground, with fine river sand in the proportion of one of the former to three of the latter, and moulded to the desired form.

In Europe lac is chiefly used in the preparation of varnishes, and by hatters, who stiffen the silk hats in common use by an application of a mixture of shell lac, sandarach, mastic, and other resins, dissolved in alcohol or naphtha. It is also largely employed in the manufacture of sealing-wax, which is prepared from a mixture of shell lac, Venice turpentine, colophony, and colouring matter. Lac is also used for making lithographic ink, and in the manufacture of the pigment "lake," which last utilisation **Mr. O'Conor** recommends to the consideration of manufacturers of lac-dye in India. Lacquer, prepared for giving a golden colour to brass and other metals, at the same time preserving their lustre, is made by adding gamboge. saffron, or some other transparent yellow, to an alcoholic solution of shell-lac.

TRADE.
Lac.
17

Trade.—LAC.—A consideration of the foreign trade in lac—shell, button, stick, and other kinds, not including lac-dye—shews that while the quantity exported during the last ten years has on the whole increased, the price of the article, and the consequent relative value of the total exports has considerably decreased. During the ten years ending 1875-76 the quantity increased more than 120 per cent. The last year of that decade shewed a special increase, the total exported being 92,915 cwt, value R72,91,751. In the five years ending 1883-84 the average total export was 99,723 cwt., value R57,48,415, while during the five years ending 1888-89 it amounted to 139,263 cwt., value R52,08,973. The average value of 1 cwt. of lac has thus fallen from R78 in 1875-76 to R37 in 1888-89. For many years shell and button-lac were subject to an export duty, which, however, was remitted, together with that on most other articles of merchandise, in February 1880. Lac is exported chiefly in the form of shell-lac, and, to a smaller extent, as button-lac. The unmanufactured form, stick-lac, is exported only to a very small extent.

The countries which constitute the principal markets for lac are the United Kingdom and the United States; smaller quantities are also import-

TRADE. Lac.

ed by France, Austria, Germany, Italy, Australia, Belgium, China, the Straits Settlements, Spain, and Holland. Nearly the whole quantity goes from Bengal.

During the five years ending 1883-84 an average of 1,641 cwt. of lac, valued R47,573, is shewn as imported from foreign countries; in the four years ending 1888-89 the amount increased to 2,194 cwt., value R49,808. Nearly the whole of this consists of stick lac, imported into Calcutta from the Straits Settlements to be manufactured into shell-lac and re-exported. The inland trade in lac for the year is valued at 101 lakhs of rupees. Calcutta received 17 lakhs from Bengal, 32 lakhs from the North-West Provinces and Oudh, and 11 lakhs from Assam. The North-Western Provinces collect the produce of the Central Provinces, the Panjáb, and Rewah for exportation, and a portion of the consignments from Bengal to Calcutta consists of re-exports of arrivals from Assam.

Lac-dye. 18

LAC-DYE.—The export trade in this substance was once a very important and profitable one, but of late years it has dwindled away to almost nothing. As early as 1875 Mr. O'Conor wrote:—

"Commencing in 1865 with a valuation twice as high as that of shell-lac, we see lac-dye fall gradually from 1872, until during the last two years the market value has been for the better sorts barely half the price of shell-lac, while the inferior kinds are hardly saleable at any price. Lac-dye in fact is now of very minor importance, both in the eyes of manufacturers and shippers, as compared with shell-lac. It has always had competitors in cochineal and other dyes, but lately the competition of mineral dyes has become very formidable. These aniline dyes are produced so cheaply and are worked so easily, that they threaten to supersede the use of most vegetable dyes, and it is probable that the prospects of Indian dyes will before long require much consideration from the State and all interested in them.

"When in 1871 a tariff value of R45 per cwt. was placed on lac-dye, the article occupied a place relatively to shell-lac which it no longer occupies, and to which it will probably never again attain. The tariff valuation was far above the real value of the inferior kinds, and the duty therefore pressed heavily on them, representing indeed a real levy of six times 4 per cent and more. Taking these facts into consideration, and having regard to the small importance of the revenue derived from the duty (about R28,000 a year on the average) the Government of India, by Notification dated 27th November 1874, removed lac-dye altogether from the list of dutiable exports. The remission of duty has no doubt so far been a relief to manufacturers, but it has not had any substantial effect in reviving the trade."

At the present time lac-dye is manufactured only because it is a necessary by-product in the preparation of shell-lac, but being almost unsaleable in Europe, it is an export of very little value. During the five years ending 1883-84, the average export was 6,010 cwt., value R1,01,803, while in the past five years ending 1888-89, the average export was only 684 cwt., value R63,335. In 1887-88 the total quantity was only 279 cwt., in 1888-89, 334 cwt. The average price per cwt. has fallen from approximately R57 in 1865 to R24 in 1888-89, and while the former average includes the inferior kinds once saleable, the latter represents the price obtained for the very best quality, which is the only form now saleable at any price. The foreign market is the United Kingdom.

LACTUCA, *Linn.; Gen. Pl., VII., 524.*

A genus of the Natural Order COMPOSITÆ, which comprises some 60 species, of which 22 are known to be natives of India. The name is derived from the milky juice contained in the stems and leaves, &c.

19 **Lactuca remotiflora,** *DC.; Fl. Br. Ind., III., 403;* COMPOSITÆ.

Syn.—LACTUCA SCHIMPERI, *Jaub. & Spach.;* BRACHYRAMPHUS SONCHIFOLIUS, *DC.;* CACALIA SONCHIFOLIA, *Wall.*

Vern.—*Undira-cha-kan* ("rat's ear"), MAR.; *Taraxaco,* PORTUG. IN GOA.

References.—*Dalz. & Gibs., Bomb. Fl., 132; Dymock, Mat. Med. W. Ind., 2nd Ed., 462.*

Habitat.—A small delicate herb, native of Banda and Sind; distributed to Arabia.

MEDICINE. 20 **Medicine.**—The whole plant is used at Goa as a substitute for Taraxacum (*Dymock*).

21 **L. Scariola,** *Linn., Fl. Br. Ind., III., 404.*

THE LETTUCE.

Var. sativa.—SYN.—LACTUCA SATIVA, *Linn.;* L. BRACTEATA and SATIVA, *Wall.*

Vern.—*Káhú, salád, khas,* HIND.; *Sálád, káhú,* BENG.; *Káhú,* PB.; *Káhú,* SIND; *Shalláttu,* TAM.; *Kávu,* TEL.; *Salada,* SING.; *Khas,* ARAB.; *Káhú,* PERS.

References.—*Roxb., Fl. Ind., Ed C.B.C., 593; Stewart, Pb. Pl., 127; DC., Origin Cult. Pl., 95; Mason, Burma & Its People, 473, 789; Aitchison, Bot. Afgh. Del. Comm. in Trans. Linn. Soc., III., Pt. I., 83; Ainslie, Mat. Ind., I., xxiii; O'Shaughnessy, Beng. Dispens., 406; Moodeen Sheriff, Supp. Pharm. Ind., 167; Dymock, Mat. Med. W. Ind., 2nd Ed., 459; Fluck. & Hanb., Pharmacog., 396; Bent. & Trim., Med. Pl., 161; S. Arjun, Bomb. Drugs, 79; Murray, Pl. and Drugs, Sind, 187; Irvine, Mat. Med Patna, 57, 114; Baden Powell, Pb. Pr., 355; Atkinson, Him. Dist., 703; Ec. Prod. N.-W. P., Pt. V., 18; Lisboa, U. Pl. Bomb., 163; Birdwood, Bomb. Pr., 49; Cooke, Oils and Oilseeds, 51; Spons' Encyclop., 1414; Smith, Dic., 243; Kew Reports, 88; Gazetteers:—Mysore and Coorg, I., 62; Bombay, VIII., 184; N.-W. P., I., 82.*

Habitat.—A large, somewhat prickly herb of the Western Himálaya from Marrí to Kunawar, at altitudes of 6,000 to 11,000 feet, found also in Western Tibet, at altitudes of 9,000 to 12,000 feet; distributed to Siberia and westwards to the British Isles and the Canaries. **Var. sativa** (the common garden lettuce) is more succulent, much smaller and quite smooth, and is cultivated throughout India, as a cold season garden vegetable. The economic information given in this article deals entirely with the cultivated variety.

OIL. Seeds. 22 **Oil.**—The SEEDS yield a clear, sweet, transparent oil, a sample of which from Lahore was shown at the Panjáb Exhibition in 1864. No definite information appears to exist regarding its method of preparation, cost, or probable economic value.

MEDICINE. 23 **Medicine.**—The lettuce has always been held in high estimation in the East on account of its cooling and refreshing properties. "The seeds are one of the four lesser cold seeds of old writers, and as such still retain their position in the Materia Medica of the East. **Mir Muhammad Hussain,** in his *Makhzan,* mentions several kinds of lettuce, and also lettuce opium, but he acknowledges the superiority of the lettuces raised from English seed in India over that of Persia, and enlarges upon the cooling and purifying action of the herb upon the blood." (*Dymock*). The plant appears, however, to be unknown to Hindu medicine. To the ancient Greeks and other European nations, the soporific action of the plant (θρίδαξ of the Greeks) was well known. Lactucarium. 24 LACTUCARIUM, the concrete juice obtained by incision and spontaneous evaporation of the juice, is officinal in the United States Dispensatory, but though formerly recognised in the London, Edinburgh, and Dublin Pharmacopœias, it is no longer officinal in the Pharmacopœia of the United Kingdom, nor in that of India. In these publications its place is supplied by the inspissated juice of **Lactuca virosa,** *Linn.* In addition to these two sources, Lactucarium may also be obtained from **L. altissima**, *Biel.*, a native of the Caucasus, now cultivated in Auvergne for that purpose. MEDICINE.

CHEMICAL COMPOSITION.—Lactucarium is a mixture of various organic substances, together with 8 to 10 per cent. of inorganic matter. It is not completely taken up by any solvent, and when heated softens, but does not melt. The principal constituents are *lactucon* or *lactucerin*, *lactucin*, and *lactucic acid*. Lactucon when pure occurs in the form of colourless needles, which are without odour or taste, neutral, and insoluble in water, though readily soluble in alcohol or ether. It resembles mannite, a peculiar variety of sugar contained in manna (see *Vol. III.*, *442*), and is also analogous to euphorbon, echicerin, taraxacerin, and cynanchol. Latucin forms white pearly scales, readily soluble in acetic acid, insoluble in ether, and with a strongly bitter taste. Lactucic acid when first obtained is a light yellow, amorphous mass, but after standing for some time, it assumes a crystalline appearance. CHEMISTRY. 25

ACTION AND USES.—Lactucarium appears to be a mild hypnotic, which, however, owing to its extreme uncertainty, is now very rarely prescribed. It may be administered in doses of from two to ten grains, as a mild soporific in cases not suited for the exhibition of opium. It has the reputation of being sedative, anodyne, purgative, diuretic, diaphoretic, and antispasmodic, and useful in the treatment of phthisis, bronchitis, asthma, and pertussis. Most of these properties, however, are probably imaginary, since numerous experiments have failed to show that lactucarium possesses more than very slight sedative properties, if, indeed, it is not absolutely inert. The anodyne property probably exists more largely in the wild than in the cultivated lettuce. Action. 26

SPECIAL OPINIONS.—§ "The seeds are given boiled, or made into a confection, in cases of bronchitis, especially chronic ones" (*Surgeon-Major C. W. Calthrop, M.D., Morar*). "Lettuce poultice acts as a soothing application to painful and irritable ulcers" (*S. M. Shircore, Civil Surgeon, Murshedábád*).

Food.—The lettuce appears to have been employed for food from very remote times; indeed, **Herodotus** informs us that it was served at the tables of Persian Kings more than 400 years before the Christian Era. This being the case it is somewhat remarkable that all knowledge of its cultivation for food appears to have died out in India, until, comparatively recently, when it was re-introduced by Europeans. At the present time it is largely grown all over the country, from October to February as a cold season crop, but is rarely if ever eaten by the Natives, being cultivated almost entirely for the European population. FOOD. 27

[*III.*, *406*.

Lactuca tatarica, *C. A. Meyer*, **var. tibetica**, *Clarke*; *Fl. Br. Ind.*, 28

Vern.—*Khâwe*, LADAK.

References.—*Clarke, Comp. Ind., 267; Stewart, Pb. Pl., 128; Atkinson, Him. Dist., 312.*

Habitat.—A common herb in Western Tibet, at altitudes of 12,000 to 16,000 feet; distributed to Central Asia and Siberia.

Fodder.—"It is occasionally browsed by sheep, but is said at times to produce bad effects" (*Stewart*). FODDER. 29

L. virosa, *Linn.*

THE STRONG-SCENTED LETTUCE.

This species is a native of Europe, imported into India for the preparation of the extract, which is officinal in the Pharmacopœia. Its properties are similar to those already detailed under Lactucarium, in the description of **L. Scariola**, but as the latter is frequently prepared from the cultivated and more inert variety, **sativa**, the extract from **L. virosa**, is, perhaps, more constant in its properties, and consequently more trustworthy.

LAGENARIA, *Ser.; Gen. Pl., I., 823.*

30 **Lagenaria vulgaris,** *Seringe; Fl. Br. Ind., II., 613; Wight, Ill., [t. 105;* CUCURBITACEÆ.

Syn.—LAGENARIA VITTATA, HISPIDA, and IDOLATRICA, *Seringe;* CUCURBITA LAGENARIA, *Linn.*

Vern.—Cultivated form=*Kaddú, al-kaddú, gol kaddu, lauki, lauká, láu, kashiphal, mithi tumbi,* wild form=*tumri, tita láu,* HIND.; cultivated form=*Kodu, láu,* wild form=*tikta láu,* BENG.; *Kadu,* SANTAL; *Me-kuri,* NAGA; *Boga lao,* ASSAM; *Phusi, konkra,* NEPAL; *Kaddu, láu, lauki,* N.-W. P.; cultivated form=*Kauka,* wild form=*tumri,* KUMAON; *Gol-kaddu,* BIJNOR; *alava,* SANAK; *Keddú, kábulí kaddú, gol-kaddú, lauki, tumba,* PB.; *Kaddú, iráo, hurrea-kadu,* SIND; *Kadú bhopalá, dudhya, bhopla bija,* MAR.; *Tumadá,* GUZ.; *Soriai-kai, shora-kai,* TAM.; *Kundánuga, ánapa chettu, nélánuga, alá-buvu, ánuga kaya, ánapa káya, sora káya, gubba kaya,* TEL.; *Bella-schora,* MALAY.; *Bú-sin-swai,* BURM.; *Diya-laba,* SING.; cultivated form=*Alábu,* wild form=*katutumbi,* SANS.; *Kaddú,* PERS.

References.—*Roxb., Fl. Ind., Ed. C.B.C., 700; Kurz in Jour. As. Soc., 1877, Pt. ii, 100; Stewart, Pb. Pl., 98; DC., Origin Cult. Pl., 245—249; Elliot, Flora Andhr., 13, 15, 16, 37, 63, 103, 131; Mason, Burma and Its People, 470, 747; O'Shaughnessy, Beng. Dispens., 343; U. C. Dutt, Mat. Med. Hind., 290, 304; Dymock, Mat. Med. W. Ind., 2nd Ed., 349; S. Arjun, Bomb. Drugs, 59, 203; Med. Top. Ajm., 142; Baden Powell, Pb. Pr., 347; Atkinson, Him. Dist., 700; Econom. Prod., N.-W. P., Pt. V., 5; Duthie & Fuller, Field and Garden Crops, II., 48, Pl. xlviii; Lisboa, U. Pl. Bomb., 158, 265; Stocks, Note on Sind; Balfour, Cyclop., II., 652; Kew Reports, 70; Indian Forester, IX., 202; Agri.-Hort. Soc. of Ind., Transactions, I., 41; III., 196; IV., 104; VII., 64, 66; Journals (Old Series), IV., 202; VII., 69; IX., Sel., 58; X., 3; XIII., Sel., 52, 53; Gazetteers:—Mysore and Coorg, I., 61; Bombay, V., 29; Orissa, II., 180; N.-W. P., I., 81; IV., lxxii; Settlement Reps.:—N.-W. P., Kumaon App., 33; Lahore, 14.*

Habitat.—A climbing plant found wild in India, the Moluccas, and Abyssinia; extensively cultivated in America, Australia, and China, and many parts of India. The fruit assumes many different forms, the result of cultivation. Perhaps the most remarkable of these are the pilgrim's gourd (in the form of a bottle), the long-necked gourd, the trumpet gourd (shapes which are indicated by their names), and the calabash, which is generally large, rounded, and devoid of a neck. The bottle gourd sometimes reaches as much as 6 feet in length. The flesh of the cultivated forms differs from that of the wild, the former being sweet and edible, the latter bitter, unpalatable, and sometimes even purgative. The species may, notwithstanding the number of its forms, be popularly recognised by its white flowers, and of the hardness the outer rind of the fruit.

CULTIVATION. 31 **Cultivation.**—As already stated, this gourd is largely cultivated throughout India. The following information regarding the best methods, given by Mr. **Gollan** in the *Indian Forester, IX., 202,* may, however, be quoted as generally applicable:—

"It can be sown as early as February and as late as July. However, for rainy season use, two sowings should be made, the first in April and the second in June. The first sowing will be ready for use in the beginning of the rains. The second will come in about the middle, and keep up the supply until the cold season. It can be sown in nurseries and transplanted, or sown at once where intended to be grown. The latter mode is preferable, but if an empty plot be not available when the sowing season arrives, it is better to adopt the first named than let the sowing season slip past. It succeeds best in heavily manured sandy soil, but will thrive ordinarily well

CULTIVATION.

in any. When sown or transplanted, the seeds or plants should be inserted in patches 6 feet apart. No supports are required, as it prefers to trail along the ground. It should be weeded when necessary, until the patches interlace and cover the ground Afterwards it will not require to be touched, as the dense network of branches will keep down the weeds."

As an example of the extent to which this gourd is cultivated all over India, the following figures relating to the area under *kaddú* in certain districts of the North-West Provinces, during the rainy season of 1880, may be given on the authority of Messrs. **Duthie & Fuller**:—' Allahabad, 202 acres; Meerut, 140; Mainpuri, 76; Shajahanpur, 61; Bulandshahr, 54; Bijnor, 37; Jalaun, 28; giving a total in these districts of 598 acres.

OIL. Seeds. 32

Oil.—The SEEDS yield a clear limpid oil, similar to that derived from the seeds of the cucumber. Beyond the fact that it is employed medicinally in certain localities, nothing is known of its properties.

MEDICINE. Seeds. 33

Medicine.—The SEEDS were originally one of the four cold cucurbitaceous seeds of the ancients, but pumpkin seeds are now generally substituted for them (*Dymock*). They are, however, still considered cooling and are given internally as a remedy for headache. The OIL above described is [Oil, 34] also employed for headache, being applied externally. The PULP of the [Pulp. 35] wild form *tumri* is purgative, sometimes excessively so. Thus **Lindley** states that certain sailors were poisoned by beer which had been standing in a hollowed bottle gourd, the symptoms produced being similar to those attending cholera. It is said to be largely used, by native farriers in the Panjáb, as a purgative for horses. The pulp of the cultivated forms is occasionally employed as an adjunct to purgatives, and is also considered cool, diuretic, antibilious, useful in coughs, and an antidote to certain poisons. It is applied externally as a poultice, and as a cooling application to the shaved head in delirium. The LEAVES are purgative, and are recom- [Leaves. 36] mended by Hindu physicians to be taken in the form of decoction for jaundice.

Juice. 37

SPECIAL OPINION.—§ "The JUICE of the leaf is given for children's diarrhœa" (*V. Ummegudien, Mettapolliam, Madras*).

FOOD. Fruit. 38

Food.—The cultivated forms are eaten both by Europeans and Natives. By the former the FRUIT is boiled when young and used as vegetable marrow. **Firminger** says that "cut up into slices in the manner of French Beans, it affords a palatable, but rather insipid dish about the beginning of the cold season." By the latter it is sliced and cooked in curry, or the pulp is eaten with vinegar, or mixed with rice. If hung in a free current of air it will keep well for three or four months, a property which ought to render it of [Shoots. 39] value as a vegetable for sea-voyages. The YOUNG SHOOTS and LEAVES [Leaves. 40] are in India also eaten by all classes.

DOMESTIC & SACRED. 41

Domestic and Sacred.—The dried shell of the fruit of the bottle-shaped gourd is used as a bottle for water, and by the Nagas for holding their *zú*, or beer. The small wild form *tumri* is used for making the stringed instrument, *sitar*, and the wind instrument called *bín*. The latter, an instrument principally used by snake-charmers, consists of a double flageolet, fitted to a hollow and narrow-necked gourd.

In the Deccan and other localities the hollowed-out gourd is used as a float for crossing rivers, four or five being considered sufficient to support a man with a burden on his head.

LAGERSTRŒMIA, *Linn.; Gen. Pl., I., 783.*

A genus of trees or shrubs which belongs to the Natural Order LYTHRACEÆ, and comprises 18 species, natives of South-east Asia, extending to Australia. Burma is the centre of the genus. All the species are highly ornamental, and may be either propagated by seed, or by cuttings in garden soil.

42 **Lagerstrœmia Flos-Reginæ,** *Retz.; Fl. Br. Ind., II., 577; Wight, Ic., t. 413;* LYTHRACEÆ.

Syn.—L. REGINÆ, *Roxb.;* L. MACROCARPA, *Wall.;* ADAMBEA GLABRA, *Lamk.*

Var angusta, *Wall.*

Vern.—*Arjuna, jarúl,* HIND.; *Jarúl,* BENG.; *Gara saikre,* KOL.; *Sekra,* SANTAL; *Ajhar, jarul,* ASSAM; *Bolashari,* GARO.; *Taman, bondara-* BOMB.; *Bondara, mota-bondara,* KONKAN; *Taman, tamana, mota bon, dara,* MAR.; *Kadali,* TAM.; *Chennangi,* TEL.; *Challá, holedásál, maruva,* KAN.; *Adamboe,* MALAY; *'yengma, kone-pyinma, pyinma,* BURM.; *Kamaung,* MAGH.; *Murute, múrúta-gass,* SING.; *Arjuna,* SANS.

References.—*Roxb., Fl. Ind., Ed. C B.C., 404; Brandis, For. Fl., 240; Kurz, For. Fl. Burm., I., 524; Beddome, Fl. Sylv., t. 29; Gamble, Man. Timb., 202. 203; Dalz. & Gibs, Bomb. Fl., 98; Mason. Burma and Its People, 406, 537, 538, 758; Elliot, Flora Andh., 36; Works of Sir W. Jones, V., 147; Lisboa, U. Pl. Bomb., 80; Birdwood, Bomb. Pr., 330; Smith, Dic., 55; Kew Reports, 69; Ind. For.:—I., 112, 363; II., 19; III., 23; IV., 47, 101; V., 190, 497; VII., 42, 196; VIII, 387, 402, 414; IX., 358; X., 33, 134, 532; XI, 258, 288, 320, 321, 374, 375; XIII., 127, 376, 553; XIV., 119, 339; Agri.-Hort. Soc. of Ind., Journals, (Old Series):—IV., 128, 134, 208; VI., 41; VIII., Sel., 177; IX., 252, 423; XI., 446; XIII., 336; Gazetteers:—Mysore and Coorg, I., 47, 61; Bombay, XIII., 24; XV., 37.*

Habitat.—A large deciduous tree of Eastern Bengal, Assam, Burma, and the West Coast, extending north to Ratnagiri; cultivated as far north as Lahore.

RESIN. 43 **Resin**—"Exudes a resin" (*Kurz*).

MEDICINE. Root. 44 Bark. 45 Leaves. 46 Flowers. 47 Seeds. 48 **Medicine.**—The ROOT is prescribed as an astringent. "Its root, BARK, LEAVES, and FLOWERS are used medicinally by the Natives" (*Beddome*). The **Rev. J. Long**, in an article on the Indigenous Plants of Bengal, states that the SEEDS are narcotic, the bark and leaves purgative (*Jour. Agri-Hort. Soc. of Ind.* (*Old Series*), *IX., 423*). **Dr. Thomson** reports, that the fruit of the *Pyenma* is used in the Andamans as a local application for aphthæ of the mouth (*Jour. Agri.-Hort. Soc. of Ind.* (*Old Series*), *XI., 446*).

SPECIAL OPINION.—§ "The bark of this and of **L. indica,** *Linn*, is considered stimulant and febrifuge" (*Surgeon-Major W. D. Stewart, Cuttack*).

TIMBER. 49 **Structure of the Wood.**—"Shining, light red, hard; annual rings marked by a belt of large pores, weight about 40lb per cubic foot. This is the most valuable timber of Sylhet, Cachar, and Chittagong, and in Burma the next in value after teak. It is used in ship-building, and for boats and canoes, all kinds of construction, timber and carts. The Ordnance Department use it for many parts of their gun-carriages. In South India it is used for building, and in Ceylon for casks" (*Gamble*). **Beddome** states that it is very durable under water, though it soon decays under ground, also that, in his time, it was employed at the Madras Gun-carriage manufactory for light and heavy field checks, felloes and cart naves, framing and boards of wagons, limbers, ammunition box boards, and platform carts.

At a conference held on Timbers, at the Colonial and Indian Exhibition in 1886, it was recommended to carriage-makers in England, and the suggestion offered that it might be exported from Calcutta, Chittagong, Rangoon, and Madras. The chief supply comes from the forests of Assam and Cachar, and the timber would, therefore, be procurable most readily from Calcutta.

50 **L. hypoleuca,** *Kurz.; Fl. Br. Ind, II., 577.*

Vern.—*Pymmah,* BURM.; *Pábdá,* AND.

References.—*Kurz, For. Fl., I., 523; In Jour. As. Soc., 1872, Pt. II., 307.*

Habitat.—A tall tree of the Andaman Islands, attaining a height of 60 to 70 feet.

Structure of the Wood.—Greyish-brown, narrow-streaked, close-grained and heavy, weighing 35 to 45lb per cubic foot. It is largely used in the Andamans for building, shingles, and for constructive work generally. TIMBER. 51

Lagerstrœmia indica, *Linn.; Fl. Br. Ind., II., 575; Wt., Ill., t. 86.* 52

Syn.—LAGERSTRŒMIA ELEGANS, *Wall.*; VELAGA GLOBOSA, *Gærtn.*

Vern.—*Phurush, telinga-china,* HIND., BENG.; *Dháyti,* BOMB.; *Telanga-china,* TAM.

References.—*Roxb., Fl. Ind., Ed. C.B.C., 404; Kurz, For. Fl., I., 521; Mason, Burma and its People, 407; Indian Forester:—V., 79—81; VI., 125; VIII., 303, 438; Gazetteers:—Mysore and Coorg, I., 61; Bombay, XVII., 24; Orissa, II., 178; Agri-Hort. Soc. of Ind. Journals:—II., 356; IV., 208; VI., 41; (New Series), IV., 95; VIII., 8.*

Habitat.—A handsome flowering shrub, possibly wild on the Eastern boundary of India, largely cultivated in gardens throughout the country on account of its beautiful lilac-coloured flowers.

L. lanceolata, *Beddome; Fl. Br. Ind., II., 576; Wight, Ic., t. 109.* 53

Syn.—L. MICROCARPA, *Wight.*

Vern.—*Boda, bondaga,* HIND.; *Nana, sokutia,* MAR.; *Sokutia,* GUZ.; *Bentaek, venteak,* TAM.; *Bandára, nandi, bolundúr,* KAN.

References.—*Brandis, For. Fl., 240; Beddome, Fl. Sylv., t. 30; Gamble, Man. Timb., 201; Dalz. & Gibs., Bomb. Fl., 98; Lisboa, U. Pl. Bomb., 79; Balfour, Cyclop., II., 653; Indian Forester:—II., 19; III., 23, 357; Gazetteers:—Mysore and Coorg, I., 48, 61; Bombay, XV., Pt. I., 33; Manual of the Coimbatore District, 407.*

Habitat.—A large tree met with in the forests of the Malabar coast from Bombay to Travancore.

Structure of the Wood.—Red, moderately hard, straight and elastic, weight from 41 to 48lb per cubic foot; much used in construction, for ship-building, in making coffee cases and furniture, also in the Konkan for fish stakes, and firewood. If left exposed in the forest it soon rots, and is rapidly attacked by white ants. TIMBER. 54

L. parviflora, *Roxb.; Fl. Br. Ind., II., 575; Wight, Ic., t. 69.* 55

Syn.—LAGERSTRŒMIA FATIOA, *Blume*; FATIOA NEPAULENSIS, *Wall.*

Var. majuscula=L. LANCEOLATA, *Bedd.* not of *Wall.*

Var. benghalensis.

Vern.—*Bákli, kat, dhaura, lendya, seina, sida, asid,* HIND.; *Sida,* BENG.; *Salora,* URIYA; *Saikre,* KOL.; *Sidha,* BEHAR; *Sekrec,* SANTAL; *Sida,* ASSAM; *Borderi, bordengri,* NEPAL; *Kanhil,* LEPCHA; *Shida,* GARO; *Shej,* BANDA; *Sida,* MICHI; *Seji,* BIJERAGGOARH; *Sina, nelli, leria,* GOND; *Chekerey,* KURKU; *Dhaura, shej,* N.-W. P.; *Bákli, dháu, dhaura,* PB.; *Sahine,* CHANDA; *Lendya,* BAITUL; *Kulia sija, lendi,* BHUNDARA; *Tendiya,* SEONI; *Nana, bondára, nandi, belli-nandi, sina, lendi,* MAR.; *Kakria,* GUZ.; *Chungi, pilúgu,* HYDERABAD; *Katcha-catta-marum,* TAM.; *Chenangi,* TEL.; *Ventaku, cheninge, channangi, bandára,* KAN.; *Tsam-belai,* BURM.

References.—*Roxb., Fl. Ind., Ed. C.B.C., 401; Brandis, For. Fl., 239; Kurz, For. Fl. Burm., I., 521; Beddome, Fl. Sylv., t. 31, 32; Gamble, Man. Timb., 200; Dalz. & Gibs., Bomb. Fl, 98; Stewart, Pb. Pl., 90; Rev. A. Campbell, Ec. Prod., Chutia Nagpur, No. 8406; Mason, Burma and Its People, 407; Atkinson, Him. Dist., 310; Lisboa, U. Pl. Bomb., 79, 245; Liotard, Dyes, 160, 169; McCann, Dyes & Tans, Beng., 160, 169; Balfour, Cyclop., II., 653; Kew Reports, 69; For. Ad. Rep., Chutia Nagpur, 1885, 6, 31; Agri-Hort. Soc. of India, Journals (Old Series):—VIII., Sel., 177; XIII., 309; (New Series):—V., 70; VII., 134; Indian Forester:—I., 87, 88, 99, 275; III., 189, 202; IV., 322, 366; V., 212; VIII., 270, 412, 414; IX., 438; X., 325, 326; XI., 252, 288; XII., 311, 313; XIII., 121; XIV., 147, 151; Gazetteers:—Mysore & Coorg, I.,*

50, 52, 61; II., 8; III., 22; N.-W. P., I., 81; IV., lxxii; Bombay, VII., 32, 36; XII., 71; XIII., 26; XVII., 25; XVIII., 49; Settlement Reports, C. P.:—Nimar, 306; Seoní, 10; Chindwara, 110; Bhundara, 19; Baitúl, 127; District Manual:—Cuddapah, 285.

Habitat.—A large deciduous tree met with in the sub-Himálayan tract from the Jumna eastwards, also in Oudh, Bengal, Assam, Central and South India. VAR. **majuscula** is common on the north-eastern edge of the Deccan plateau, in Chutia Nagpur, and in Bengal; while VAR. **benghalensis** occurs in Nepál, Sikkim, Assam, and Burma.

GUM. Bark. 56 **Gum.**—Lisboa writes, "The gum which exudes from the BARK is said to be sweet and edible."

FIBRE. Bark. 57 **Fibre.**—**Campbell** reports that the BARK yields a fibre, which is employed in Chutia Nagpur for the manufacture of ropes.

DYE & TAN. Bark. 58 Leaves. 59 **Dye and Tan.**—The BARK, along with that of *ashna* (**Terminalia tomentosa**), is employed in Midnapur for dyeing skins black (*McCann*). Lisboa states that the bark and LEAVES are used in Bombay for tanning, and the **Rev. A. Campbell** reports a similar utilisation of the bark in Chutia Nagpur.

TIMBER. 60 **Structure of the Wood.**—Very hard, grey or greyish-brown, often with a reddish tinge, darker near the centre, weight from 40 to 60℔ per cubic foot. It is tough, elastic, seasons well, works freely, and is fairly durable. Ten sleepers, laid down on the Oudh and Rohilkhand Railway in 1870, were reported, on being examined in 1875, to be quite sound. A number of sleepers have also been tried on the Northern Bengal State Railway, but the results have not been published. The timber is largely employed for ploughs and other agricultural implements, for construction, for boats, buggy shafts and axe handles, and is said to yield good charcoal (*Gamble; Beddome, &c.*).

DOMESTIC. 61 **Domestic.**—This tree is one of those on which the *tasar* silkworm is fed.

62 **Lagerstrœmia tomentosa,** *Presl.; Fl. Br. Ind., II., 578.*

Syn.—L. PUBESCENS, *Wall.*

Vern.—*Laieya*, BURM.

References.—*Kurz, For. Fl. Burm., I., 523; Gamble, Man. Timb., 204; Indian Forester:—VIII., 414; IX., 216; XI., 230; Gazetteer, Burma, I., 130.*

Habitat.—A large deciduous tree of Burma, frequent in Pegu and Martaban.

RESIN. 63 **Resin.**—It exudes a red resin (*Kurz*).

TIMBER. 64 **Structure of the Wood.**—Grey or greyish-brown, close-grained, moderately hard, weight from 46 to 53 ℔ per cubic foot; valued for bows and spear handles, also used for canoes and cart wheels.

LAGGERA, *Sch.-Bip.; Gen. Pl., II., 290.*

65 **Laggera aurita,** *Schultz-Bip.; Fl. Br. Ind., III., 271;* COMPOSITÆ.

Syn.—BLUMEA AURITA, *DC.*; B. GUINEENSIS, *DC.*; CONYZA AURITA, *Linn.*

References.—*Roxb., Fl. Ind., Ed. C.B.C., 602; Moodeen Sheriff, Suppl., Pharm. Ind., 73.*

Habitat.—A common herbaceous weed in waste places from the Panjáb and Sind to Chittagong, Burma, and South India.

MEDICINE. Plant. 66 **Medicine.**—The whole PLANT has an odour of turpentine but does not appear to have any known economic property. Of late years, however, it has been confused with some other compositous plant of medicinal value, regarding which the following remarks by **Dr. G. Watt** in the Calcutta Exhibition Catalogue may be quoted: "**Moodeen Sheriff** has, in his Supplement to the *Pharmacopœia Indica*, gone into a long discussion as to the correct scientific name of the *kakrondá* (HIND.) or *kamáfútis* (ARAB.). He describes the flowers as white and about the size of a large pea. This precludes it from being **Blumea aurita,** *DC.,* which has pink flowers. The colour of the capitula of this great natural order is a character of much more importance than almost in any other family, and we are therefore almost justified in saying that the white-flowered plant alluded to cannot be either a **Blumea** or a **Laggera.**" MEDICINE.

LALLEMANTIA, *Fisch. et Mey.; Gen. Pl., II., 1200.*

Lallemantia Royleana, *Benth.; Fl. Br. Ind., IV., 667;* LABIATÆ. 67

Syn —DRACOCEPHALUM ROYLEANUM, *Benth.;* NEPETA ERODIIFOLIA, *Boiss.*

Vern.—Seeds=*Gharei kashmálú, tukhm-lealanga,* HIND.; seeds=*Gharei kashmálú, tukhm-malanga, tukhm-bálangú,* PB.; seeds=*Tukhm-i-bálangú,* BOMB.; seeds=*Tukhm-i-bálangú,* PERS.

References.—*Stewart, Pb. Pl., 168; Aitchison, Bot. Afghan Del. Com., in Trans. Linn. Soc., 2nd Ser., III., Pt. I., 97; Dymock, Mat. Med. W. Ind., 2nd Ed. 614; S. Arjun, Bomb. Drugs, 100; Irvine, Mat. Med. Patna, 111; Honigberger, Thirty-five years in the East, II., 272; Atkinson, Him. Dist., 315; Birdwood, Bomb. Pr., 62; Settlement Rep., Lahore, 9.*

Habitat.—An annual erect herb, found doubtfully wild but usually cultivated in the hills and plains of the Panjáb.

Medicine.—The plant is largely grown on account of its mucilaginous SEEDS, which are considered cooling and sedative, and are extensively employed in the preparation of a mucilaginous beverage. MEDICINE. Seeds. 68

LAMINARIA, *Lam., Baillon, Traité de Botanique Médicale Cryptogamique, 278.*

A genus of seaweeds, distinguished by largely developed, coriaceous, entire, or palmate ribless fronds, which are supported on a foot, generally elongated, simple, cylindrical and terminating in a strong scutiform base by which the alga is attached to rocks. The only species which are known to exist on the Indian coasts are **L. digitata,** *Lam.*, **L. bulbosa,** *Ag.*, and **L. saccharina,** *Lam.*, of which the last alone is of economic importance.

Laminaria saccharina, *Lam.; Baill., Bot. Med. Crypt., loc. cit.;* THE SWEET TANGLE. [*Turn., Fuce, t. 163;* ALGÆ. 69

Vern.—*Gélarpatr,* HIND.

References.—*Stewart, Pb. Pl., 269; Dymock, Mat. Med. W. Ind., 2nd Ed., 874; Murray, Pl. and Drugs, Sind, 3; Baden Powell, Pb. Pr., 384; Balfour, Cyclop., II., 664.*

Habitat—A greenish-red seaweed, with a slight odour and sweetish taste, found in all deep seas, and frequently cast up on Indian coasts. MEDICINE. 70

Medicine—After having been washed in fresh water and dried, the thallus becomes covered with a white sweet efflorescence composed to a great extent of mannite. The plant also contains much iodine and is employed as an alterative for the cure of goitre, scrofula, and syphilitic affections. **Murray** states that for the latter disease it is a favourite remedy in Sind, being exhibited in the form of a syrup, in conjunction with a decoction of quince seeds.

The Sweet Tangle forms a regular article of commerce through Kashmír to India, and is to be found in most bazárs of the Panjáb and Sind. **Cayley** mentions that 16 seers were imported from Yárkhand to Leh in 1867. **Honigberger** states that in his time the plant was officinal at Lahore and Kashmír, and that the fronds were procured from Tibet, where, according to native report, they grew in a salt lake. In all probability, however, they may have been obtained from the Caspian, or may have come from the sea *via* China.

The same writer describes it as useful in cases of anorexia, dyspepsia, fever, scrofulous ulcers, buccal ulceration (syphilitic ?), and tenesmus.

Another species, **L. Cloustoni**, is employed medicinally in Europe in the manufacture of "tents" for expanding fistulæ, the os uteri, &c. The foot stalk is dried, pressed into as small cylinders as possible, cut to the desired size, &c.; by its expansion when placed in contact with moisture it exercises a powerful effect as a dilator. When washed in fresh water the thallus of this species sheds a large quantity of mucilage. When dried in the air, like the others it produces a copious white efflorescence, which, however, is quite distinct chemically, being bitter and saline, consisting for the most part of sulphate of soda. No record exists of this species having been found in Indian seas.

INDUSTRIAL. 71 **Industrial.**—All the Indian species of **Laminaria** are remarkably rich in iodine, and if obtainable in any quantity might be utilised for the manufacture of that substance. **L. digitata**, which, according to **Murray**, is found thrown up on the Clifton and Manora beaches, is said to yield a valuable charcoal, which, when purified, dried, and finely powdered, possesses more deodorizing and decolourising properties than animal charcoal itself.

LAMPRACHÆNIUM, *Benth.; Gen. Pl., II., 225.*

72 **Lamprachænium microcephalum,** *Benth.; Fl. Br. Ind., III., 229;* COMPOSITÆ.

Syn.—DECANEURON MICROCEPHALUM, *Dalz.*

Vern.—*Brahma-dandi*, BOMB.

References.—*Dalz. & Gibs., Bomb. Fl., 122; Clarke, Comp. Ind., 5; Dymock, Mat. Med. W. Ind., 2nd Ed., 423.*

Habitat—A much-branched annual, having an odour of chamomile; common in the Konkan.

MEDICINE. 73 **Medicine.**—"Used medicinally as an aromatic bitter" (*Dymock*).

74

LANTANA, *Linn.; Gen. Pl., II., 1142.*

A genus of rambling shrubs belonging to the Natural Order VERBENACEÆ, and comprising 40 species, chiefly natives of Tropical and Sub-tropical America. Of these three are natives of India, but none are of economic importance. One species, however, **Lantana Camara**, *Linn.*, a native of America, has run wild from cultivation in many parts of the country, especially along the Western Coast and in Ceylon, and is of considerable interest, owing to the damage, which it has already done, and which is likely to increase with its extension.

This prickly plant grows so densely as to become perfectly impenetrable to men and cattle, and consequently may completely prevent the utilisation of large areas. Owing to its extraordinary vitality it is extremely difficult to exterminate. The only method of getting rid of the growth is said to be by digging it up and burning it, root and branch, reappearing twigs being similarly destroyed as soon as they are seen. A few writers, however, have held that under certain conditions the **Lantana** growth is by no means an unmixed evil, since, owing to its shade-giving properties and density, it exercises a markedly renovating effect on the soil. Thus Mr. **R. Thompson**, writing in the *Indian Forester* (Volume VI.), states that in his opinion it is highly useful in tracts which have been deforested. The plant, according to him, by rapidly overgrowing such land, serves to retain the humus in the soil. He further states that in the event of the land again coming under forest, the light-loving **Lantana** quickly dies out.

In cultivated and pasture lands, on the other hand, there is no doubt that the shrub is a serious evil, and one against which it is difficult to combat. Thus in the *Gazetteer of Mysore & Coorg* it is stated that it has spread over the whole of Coorg and, covering as it does, many tracts which might be much more profitably occupied, has rendered its extermination a matter of imperative necessity.

Lapidary. For an account of the stones principally used in India in the lapidary's art, see the articles on **Carnelian, &c.,** Volume II., 167; also **Jade,** p. 535; and **Precious Stones,** Vol. VI.

LAPIS LAZULI, *Ball in Man. Geology of Ind., III., 528-530.*

A mineral of somewhat complex composition, containing silica, alumina, soda, lime, iron, sulphuric acid, sulphur, and chlorine. Its value has considerably decreased in modern times, owing to the discovery, by a French chemist, in 1828, of an efficient artificial substitute, the composition of which is identical or nearly so with that of the natural stone. The colour of both, which varies from a rich Berlin to an azure blue, is supposed to be due to sodium sulphide.

Lapis Lazuli, *Geol. of Ind., IV. (Mineralogy, Mallet), 99.* 75

Vern.—*Lajwárd,* HIND.; *Lajburud,* BEHAR; *Rájávaral,* GUZ.

References.—*Baden Powell, Pb. Prod., 64; Irvine, Mat. Med. of Patna, 60; Ain-i-Akbari, Gladwin's Trans., II., 183; Balfour, Cyclop., II., 679; Spons, Cyclop., 1042, 1548; Bombay Gazetteer, VI., 201.*

Occurrence.—This mineral is found in Persia, Bucharia, China, Siberia, and Chili, and is also said to occur in Afghánistán and Biluchistán. Though not known with certainty to occur in India, it is imported into the country, where it is employed for several purposes. LOCALITY. 76

Pigment.—True ultra-marine, composed of finely powdered **Lapis Lazuli,** was at one time largely used in India for house-decoration and book-illuminating. It is said to have cost, when purified, from R80 to R100 a seer, but now, as in Europe, its place has been taken by the artificially-made substitute, which is sold in the bazárs under the same name of *lajward* for some R4 a seer. In Europe, the refuse in the manufacture is calcined and affords delicate gray pigments, known as ultra-marine ash (*Ball*). PIGMENT. 77

Medicine.—It is employed as an astringent and refrigerant by native practitioners (*Irvine*); it is also said to be mixed with jalap and other purgatives, and to be used as an external application to ulcers. MEDICINE. 78

LAPORTEA, *Gaud.; Gen. Pl., III., 383.*

Laportea crenulata, *Gaud.; Fl. Br. Ind., V., 550; Wight, Ic., t. 668;* URTICACEÆ. 79

THE FEVER OR DEVIL NETTLE.

Syn.—L. GIGANTEA and LATIFOLIA, *Gaud.*; URTICA CRENULATA, *Roxb.*; U. GIGANTEA, *Poir.*; U. SINUATA, *Blume*; U. CHURTA, *Ham.*; URERA JAVENSIS and GIGANTEA, *Gaud.*; U. CRENULATA and COMMERSONIANA, *Wedd.*; DENDROCNIDE CRENULATA, *Miq.*

Vern.—*Chorpattu, surat,* BENG.; seed=*Utigun ka bij,* BEHAR; *Moringi,* NEP; *Sir-nat,* ASSAM; *Mealum-ma, sunkrong,* LEPCHA; *Pheytakyee,* BURM.; *Maúsa,* SINGH.

References.—*Roxb., Fl. Ind., Ed. C.B.C., 657; Brandis, For. Fl., 404; Kurz, For. Fl. Burm., II., 421; Beddome, Fl. Sylv., t. 306; Gamble, Man. Timb., 323; Thwaites, En. Ceylon Pl., 259; Hooker, Him. Journ. II., 188; Med. Top. Ajm., 148; Irvine, Mat. Med. Patna, 79; Royle, Fib. Pl., 366; Cross, Bevan and King, Rep. on Ind. Fibres, 10, 54; Agri-Hort. Soc. of Ind.:—Journ. (Old Series), IV., 216; VII., 222;*

VIII., 89; X., Sel., 26; Balfour, Cyclop., II., 280; Indian Forester, II., 21; XIV., 269, 273.

Habitat.—An evergreen arborescent shrub, met with in the Tropical Himálaya from Sikkim eastwards, also in Assam, the Khásia mountains, and Perak, distributed to Ceylon and Java. It has powerfully stinging hairs, contact with which causes excessive burning pains which last for several days and are greatly intensified by the application of water. **Beddome** remarks that with natives the sting often brings on fever, hence the name by which the plant is known to coffee planters and other English residents. **Hooker** states that the plant possesses this property of stinging only in the autumn, a fact which appears not to have been noticed by most Indian writers.

FIBRE. 80 **Fibre.**—This plant yields a strong useful fibre, used by the hill tribes of Assam for cordage, and in the manufacture of a coarse cloth. **Major Hannay**, who was one of the first to bring the fibre to the notice of Europeans, stated that it was fine, white, apparently of no great strength, and by report, not very lasting. **Messrs. Cross, Bevan, and King**, however, in their recent report on Indian fibres, appear to hold a more favourable opinion, stating that the fibre "is good, is more or less allied to rhea, and, like that fibre, is very strong." On comparing the results of the chemical analyses of these two fibres, however, by the same observers, it will be found that while rhea fibre contains 80·3 per cent. of cellulose, that of **Laportea** has only 53·4 and is also very largely acted on by hydrolysis. The small percentage of cellulose would appear to indicate a very great inferiority in the latter, and this fact, together with the heavy loss it sustains by hydrolysis, stamps the fibre as unlikely ever to become of commercial importance.

Medicine. Seeds. 81 **Medicine.**—**Irvine** states that the SEEDS, in doses of ½ drachm to ½ oz., are used in Patna in the same way as coriander, also that they are imported from Nepal, and are sold for 8 annas per ℔.

Lard, see **Hog**, p 253.

LARIX, *Miller; Gen. Pl., III., 442.*

A genus of trees belonging to the Natural Order CONIFERÆ, and comprising eight species, natives of North Temperate and Arctic regions. Of these only one is indigenous to India. The European Larch, **L. europæa**, is extensively cultivated for its timber, and for ornamental purposes, in the hilly regions of Europe. It yields VENICE TURPENTINE, a substance used to a small extent medicinally.

82 **Larix Griffithii**, *Hook. f. & Thoms.; Fl. Br. Ind., V., 655;* CONIFERÆ.

THE HIMALAYAN LARCH.

Syn.—LARIX GRIFFITHIANA, *Gord.;* ABIES GRIFFITHIANA, *Lindl. & Gord.;* PINUS GRIFFITHII, *Parlat.*

Vern.—*Boargasella,* NEPAL; *Sah, saar,* SIKKIM.

References.—*Brandis, For. Fl., 531; Gamble, Man. Timb., 410; Hooker, Him. Jour., II., 44; Balfour, Cyclop., II., 681; Kew Off. Guide to the Mus. of Ec. Bot., 147.*

Habitat.—A small deciduous tree found in Eastern Nepal, Sikkim, and Bhutan, between altitudes of 8,000 and 12,000 feet.

TIMBER. 83 **Structure of the Wood.**—Heart-wood red, inner zone of each annual ring soft and spongy, outer zone narrow, firm, and shining; resinous ducts scanty, large; weight 32℔ per cubic foot (*Gamble*). **Hooker** states that he never saw the wood red, as above described, but always white and soft, a statement, however, which **Gamble** holds to be incorrect. The wood splits well, and is considered one of the most durable of the coniferous timbers exported from Sikkim into Tibet.

LASIA, *Lour.; Gen. Pl., III., 995.*

Lasia spinosa, *Thwaites; DC., Monographiæ Phanerogamarum, II., [273; Wight, Ic., t. 777;* AROIDEÆ. 84

Syn.—DRACONTIUM SPINOSUM, *Linn.;* POTHOS LASIA, *Roxb.;* P. SPINOSUS, *Ham.;* P. HETEROPHYLLA, *Roxb.;* LASIA ROXBURGHII, *Griff.;* L. HETEROPHYLLA, *Schott;* L. ZOLLINGERI, *Schott;* L. JENKINSII, *Schott;* L. DEHISCENS, *Schott.*

Var. Hermanni; L. HERMANNI, *Schott.*

Vern.—*Kanta-katchú,* BENG.; *Kanta saru,* SANTALI; *Múlasari, kanta kachóramu,* TEL.; *Kohilla, mahakshilla,* SING.; *Záyap,* BURM.

References.—*Roxb., Fl. Ind., Ed. C.B.C., 147; Kurz, Prelim. For. Rep. on Pegu, App. C., xix.; Thwaites, En. Ceylon Pl., 336; Elliot, Fl. Andhr., 82, 118; Campbell, Ec. Prod. Chutia Nagpur, No. 8421.*

Habitat.—Native of India, occurring in Sylhet, Bengal, and Assam, distributed to Java, Sumatra, and Ceylon.

Medicine.—**Campbell** reports that the ROOT is highly esteemed by the Santals as a remedy for affections of the throat. The LEAVES and roots are also used medicinally by the Singalese. **MEDICINE. Roots. 85 Leaves. 86**

Food.—Thwaites states that the LEAVES and ROOTS are employed as a vegetable in curries. **FOOD. Leaves. 87 Roots. 88**

LASIOSIPHON, *Fresen.; Gen. Pl, III., 197.*

Lasiosiphon eriocephalus, *Dcne.; Fl. Br. Ind., V., 197; Wight., [Ic., tt. 1859, 1860;* THYMELACEÆ. 89

Syn.—LASIOSIPHON SPECIOSUS, *Dcne.;* L. SISPARENSIS, HUGELII?, and INSULARIS, *Meissn;* L. METZIANUS, *Miq.;* DAPHNE ERIOCEPHALA, *Wall;* LANCHNÆA ERIOCEPHALA, *Heyne;* GNIDIA ERIOCEPHALA, *Meissn.;* G. SISPARENSIS, *Gardn.;* G. INSULARIS, *Gardn.;* G. MONTICOLA, *Miq.*

Vern.—*Ramatta, rametha,* BOMB., *Rámita, rámetta,* MAR.; *Rámi,* KAN.; *Naha,* SING.

References.—*Beddome, For. Man., 179, t. 25, f. 2; Gamble, Man. Timb., 314; Thwaites, En. Ceylon Pl., 250; Dalz. & Gibs., Bomb. Fl., 221; Dymock, Mat. Med. W. Ind., 2nd Ed., 674; Lisboa, U. Pl. Bomb., 113, 268; Mad. Bot. Gard. Rep. 1883-84, 10; Gazetteers:—Bomb., X., 403; XV., 72.*

Habitat.—A small tree or large shrub met with in the Deccan; in Southern India, ascending to 7,000 feet on the Nilghiris; and in Ceylon, where it reaches an altitude of 4,000 feet.

Fibre.—The BARK yields a fibre which has been recommended as a paper-making material. **FIBRE. Bark. 90**

Medicine.—The BARK is a powerful vesicant (*Dymock*). **MEDICINE. Bark. 91**

Domestic.—The LEAVES and BARK are acrid and poisonous, and are frequently used to poison fish. In the *Kanara Gazetteer* it is stated that the wood or its ash destroys the teeth, and that the natives are most careful not to use it. **DOMESTIC. Leaves. 92 Bark. 93**

Lastrea Filix-mas, THE MALE-FERN, see Ferns, Vol. III., 323.

LATERITE.

Laterite, *Ball., Man. Geology of India, III., 549.* 94

Vern.—*Kabúk,* SING.

The following note has been kindly furnished by **H. B. Medlicott, Esq., F.R.S.**, late Director of the Geological Survey:—

"An essentially Indian rock, to which this name (deriv. *later*, a brick) was given in the first instance by **Dr Buchanan**, from its generally brown or reddish-brown colour, and perhaps from the easy way in which it can be cut or chipped into blocks of all sizes on its first quarrying. The largest and best-known laterite regions are those of the Deccan, the West Coast, from between Bombay and Ratnagiri, nearly down to Cape Comorin, **LOCALITY. 95**

LOCALITY. and the eastern coast of the Madras Presidency. This rock also extends up into Orissa and Midnapur, and is found in smaller and detached areas all over the Peninsula, even capping the summits of some of the highest hills. In Ceylon a similar rock goes by the name of *Cabook* or *Kabúk*.

"Wherever it occurs it is used as a building or road material, for it is easily dressed or broken up. When first quarried, it is generally soft and of a bright red colour, but on exposure it hardens and turns a dark brown colour. This peculiar property of hardening on exposure is best developed in the drier areas; it being difficult, along the west coast, for example, to procure a stone which is sufficiently ferruginous in its composition to harden quickly under that moist climate." For further information see "**Iron**," pp. 500, 505, 507, 509, 510; also **Stones, Building**, Vol. VI.

LATHYRUS, *Linn.; Gen. Pl., I., 526.*

A genus of annual or perennial herbs, belonging to the Natural Order LEGUMINOSÆ, which comprises some 170 species, of which 7 are natives of India.

96 **Lathyrus Aphaca,** *Linn.; Fl. Br. Ind., II., 179;* LEGUMINOSÆ.

Vern.—*Jangli-matar, musúr-chúna,* BENG.; *Kaibú,* NEPAL.; *Rawan, rawari,* PB.

References.—*Roxb., Fl. Ind. Ed. C. B. C., 566; Voigt, Hort. Sub. Cal., 227; Stewart, Pb. Pl., 70; Jour. Bot. Tour in Hazara, &c., in Jour. Agri-Hort. Soc. Ind., XIV. (Old Series), 119; Aitchison, Bot. of Afgh. Del. Com., in Trans. Linn. Soc., 2nd ser., III., Pt. I., 59; Atkinson, Him. Dist., 308; Agri-Hort. Soc of Ind., Trans., IV., 82, 102; Journ. (Old Series), IX., 416; Gazetteers:—N.-W. P., I., 80; IV., lxxxi.*

Habitat.—A much-branched field weed, found throughout the plains of Bengal, the North-West Provinces, Oudh, the Panjáb, Hazara, Kumaon, and Kashmír.

FODDER. 97 **Fodder.**—It is used in many localities for cattle fodder, and is said to be cultivated for that purpose near the Jumna. According to **Voigt** the ripe seeds are narcotic, their effects being noticeable when eaten abundantly, but when young they are perfectly harmless.

98 **L. imphalensis,** *Watt, ms.*

Habitat.—A new species, named as above by **Dr. G. Watt**, was collected by him at Myang-khong in Munipur at an altitude of 4,000 feet.

FODDER. 99 **Fodder.**—The plant is largely used as fodder in Munipur, where it is allowed to cover the fields like a weed, after removal of the crops.

100 **L. sativus,** *Linn.; Fl. Br. Ind., II., 179.*

JAROSSE OR GESSE.

Vern.—*Khesári, kassúr, kasári, kassar-tiuri, latri,* HIND.; *Khesári, teora, kassúr,* BENG.; *Kesari,* SANTAL; *Kesari,* NEPAL: *Kisára, churál chapa, mattar, kása, latri, tiura,* N.-W.P.; *Kisári, mattar, churál, karas, karil,* PB.; *Mattar,* SIND.; *Lákh,* MAR.; *Láng,* GUZ.; *Triputi,* SANS.; *Masang,* PERS.

References.—*Roxb., Fl. Ind., Ed. C.B.C., 566; Voigt, Hort. Sub. Cal., 227; Stewart, Pb. Pl., 71; DC., Origin Cult. Pl., 110; Campbell, Ec. Prod. Chutia Nagpur, No. 9214; Pharm. Ind., 80; Pharmacog. Indica. I., 489, 490; O'Shaughnessy, Beng. Dispens., 317; U. C. Dutt, Mat. Med. Hind., 321; Murray, Pl. and Drugs, Sind, 121; Baden Powell, Pb. Pr., 242; Atkinson, Him. Dist., 308, 694; Duthie & Fuller, Field and Garden Crops, Pl. II., 16, Pl. xxxii; Lisboa, U. Pl Bomb., 139, 140, 152, 264; Birdwood, Bomb. Pr., 120; Royle, Ill. Him. Bot., 200; Church, Food Grains of India, 132; Smith, Veterinary Hygiene, 156; Balfour, Cyclop., II., 686; Agri-Hort. Soc. of Ind.:—Transactions, I.,*

41; IV., 102; VI., 77; Journals (Old Series), IV., sel., 150; XIII., sel., 51, 52; Gazetteers:—Bombay, II., 63; N.-W. P., I., 80; IV., lxxi; N.-W. P., Selections, 1866, II., 265-295; Settlement Reports:—Panjáb, Jhang Dist., 90; Lahore, 10; Montgomery, 107; N.-W. P. and Oudh, Azamgarh, 121, 123; C. P., Chanda, 81; Stocks, Rep. on Sind.

Habitat.—An annual herb, indigenous, according to **DeCandolle,** in the region extending from the south of the Caucasus, to the north of India. It has spread as a weed of cultivation from its original home. It is cultivated all over India from the northern indigenous area, through the North-West Provinces to the plains of Bengal, and the Southern and Western Presidencies.

CULTIVATION. 101

Cultivation.—It is grown as a cold weather crop on land which will raise no other kind of pulse. Regarding the method pursued in the North-West Provinces, **Messrs. Duthie & Fuller** write: "Its cultivation is commonest on very heavy clay soils, and it is frequently sown on land submerged in the rainy season, which hardens during the cold weather almost to the consistency of a stone, splitting up into long deep fissures. After prolonged floods it occasionally offers a means of raising a cold weather crop from land which would otherwise be unculturable, since it can be sown broadcast on wiry ground, and is not so injuriously affected by the subsequent hardening of the surface as would be the case with any other *rabi* crop. For similar reasons it is occasionally sown in rice fields even before the rice is cut, springing up between the rice stalks, and yielding a crop in the spring while the rice stubble is still standing.

"Its cultivation is commonest in the Eastern Districts, and is of considerable extent in that part of the Allahabad District which lies south of the Jumna. It is also much grown under the name of *latri* in the Azamgarh Districts."

The method employed in the Panjáb is very similar to that above detailed. New alluvial lands or the beds of *nallahs* are generally chosen, and the seed is scattered broadcast, and left to take its chance. The crop is sown in September-October, and reaped in March. Occasionally the land receives one or two ploughings, and the seed is in that case sowed by drill. When other food-grains are obtainable the crop is valued as fodder only; it is rarely threshed except for seed, though the young pods are occasionally used as a green vegetable.

The writer of the Settlement Report of the Jhang District states that nearly 8,000 acres were under this crop at the time at which he wrote; that the best crop he had seen was one raised on silt only a year old in Shorkot, and that the plant occasionally suffered severely from the ravages of caterpillars which attacked the pods.

Area. 102

AREA UNDER THE CROP.—No trustworthy statistics are available of the area under **Lathyrus** in the Panjáb, Bengal, Oudh, Madras, Assam, Burma, Ajmere, or Coorg. The figures quoted below regarding the North-West Provinces are unavoidably incomplete, but those regarding Bombay, Sind, the Central Provinces, and Berar may be received as accurate.

N.-W. Provinces. 103

The North-West Provinces.—The only available statistics (which, as above stated, are incomplete) give the area under this crop in 1887-88 as having been 56,100 acres, or more than double the average of the same statistics for the three years immediately preceding, which amounted to only 24,820 acres.

Bombay. 104

Bombay.—The returns of the Director of Land Records and Agriculture shew the crop to be an unimportant one, and to have varied during the past few years between 32,783 acres in 1888-89 and a maximum of 57,661 acres in 1886-87. The only locality in which it is cultivated to any extent is Broach, which in 1886-87 had 48,000 out of the total of 57,661 acres.

LATHYRUS sativus.

The Khesári Pulse.

CULTIVATION.

Sind. 105 *Sind.*—In 1888-89, the only year for which statistics are available, 72,876 acres were under the crop.

Central Provinces. 106 *Central Provinces.*—These Provinces appear to possess the largest area under **Lathyrus** in all India. Thus in 1887-88 the total area under the crop was 357,883 acres, an increase of more than 100,000 acres on the average area for the preceding four years, which was approximately 250,000 acres.

Berar. 107 *Berar.*—The area under the crop in this region appears to be remarkably constant. Thus, during the past five years, it has only varied some 9,000 acres from 46,500 to 53,900, the latter figure being returned for the year 1888-89.

OIL. 108 **Oil.**—**O'Shaughnessy** states that the oil expressed from the seeds is a powerful and dangerous cathartic. **Church** gives 0·9 per cent. as the proportion of oil contained in the seed.

FOOD & FODDER. 109 **Food and Fodder.**—As already stated, this pea is cultivated principally as a fodder, but being very cheap and easily grown it is considerably used as food by the poorer classes, largely so in times of scarcity. It is also used to a considerable extent to adulterate *dál* from which it can scarcely be distinguished. The following chemical composition is given by **Church**: water, 10·1; albuminoids, 31·9; starch and fibre, 53·9; oil, 0·9; ash, 3·2. The nutrient ratio is about 1 : 1·75, while the nutrient value is nearly 8·7.

Poisonous property. A recent analysis by **Astier** has revealed the presence in the grain of a volatile liquid alkaloid, probably produced by some proteid ferment, which exhibits the toxic effects of the seeds, and the action of which is destroyed by heat. The evil effects of habitual consumption of the seeds have long been known, and though the subject is one which has been much discussed, there appears to be no doubt that continued use of this article of diet has a tendency to produce paralysis. In one district of Bengal alone, according to **Irving**, nearly 4 per cent. of the population suffered from its toxic effects in 1860. That observer went into the subject most extensively, and found that if used occasionally and in small quantity, the results were constipation, colic, or some other form of indigestion. But, on the other hand, if freely employed, and especially without admixture with other sorts of grain, he found palsy to be a frequent sequel. **Dr. Irving's** results further showed that ill-effects were more apt to occur in the rainy season, and that the great majority of sufferers were males, the proportion in the cases which came under his observation being 6·11 males to 0·59 females.

During the years from 1829 to 1834 the grain formed, by a series of accidents, the chief food of some of the eastern villages of Oudh, and, apparently, as a direct result many cases of sudden paralysis of the lower extremities ensued. The circumstances which gave rise to this condition and the characteristic features of the disease were carefully described by **Colonel Sleeman**, from whose account the following may be quoted:—

"In 1829 the wheat and other spring crops in this and the surrounding villages were destroyed by a severe hail-storm; in 1830 they were deficient from the want of seasonable rains, and in 1831 they were destroyed by blight. During these three years the *kasári* which, though not sown of itself, is left carelessly to grow among the wheat and other grain, and given in the green and dry state to cattle, remained uninjured, and thrived with great luxuriance. In 1831 they reaped a rich crop of it from the blighted wheat fields, and subsisted upon its grain during that and the following year, giving the stalks and leaves only to their cattle. In 1833 the sad effects of this food began to manifest themselves. The younger part of the population of this and the surrounding villages, from the age of thirty downwards, began to be deprived of the use of their limbs below the waist by paralytic strokes, in all cases sudden, but in some more

FOOD & FODDER.

severe than in others. About half the youth of this village of both sexes became affected during the years 1833 and 1834; and many of them have lost the use of their lower limbs entirely, and are unable to move. The youth of the surrounding villages, in which *kasári*, from the same causes, formed the chief article of food during the years 1831 and 1832, have suffered in an equal degree. Since the year 1834 no new case has occurred, but no person once attacked had been found to recover the use of the limbs affected, and my tent was surrounded by great numbers of the youth in different stages of the disease, imploring my advice and assistance under this dreadful visitation. Some of them were very fine-looking young men of good caste and respectable families, and all stated that their pains and infirmities were confined entirely to the joints below the waist. They described the attack as coming on suddenly, often while the person was asleep, and without any warning symptoms whatever, and stated that a greater proportion of the young men were attacked than of the young women It is the prevailing opinion of the natives throughout the country that both horses and bullocks which have been much fed upon *kasári* are liable to lose the use of their limbs, but if the poisonous qualities abound more in the grain than in the stalk or the leaves, man, who eats nothing but the grain, must be more liable to suffer from the use of this food than beasts, which eat it merely as they eat grass or hay."

Again, **Lisboa**, commenting on this disease, writes: "The subject was taken up by **Dr. Kinloch Kirk** in Upper Sind. A villager had brought him his wife, about thirty years old, who was suffering from paralysis of the lower extremities. When questioned as to what he thought the cause to be, the man replied: 'It is from *kasári*; we are very poor, and she was obliged to eat it for five months on end.' **Dr. Kirk** hereupon instituted enquiries into the subject, which confirmed the statement; and he adds that, 'the natives know this *dal* is poison, but they eat it because it is cheap, thinking that they can stop in time to save themselves from its consequences.'"

This condition, which has lately received the name of "lathyrismus," has formed the subject of investigations by **Dr. B. Suchard**, by whom it has been found that the chief effect produced on the human subject is upon the muscles of the lower extremities, especially on those below the knee. In horses, paralysis of the hinder extremities also takes place, but an affection of the larynx resulting in asphyxia and death has been observed, a complication which has not been recorded in man. **Cantarri** of Naples has published a number of cases in which he has carefully observed the conditions after death. The muscles of the face, neck, and trunk were found not to be affected; those of the lower extremities, especially the abductors, were found to have undergone a fatty degeneration, the transverse striæ being diminished, and the ultimate fibres containing little drops of oil. No affection of the spinal cord was discovered.

Until lately it was extremely doubtful as to what could be the cause of this poisonous action of the pulse. Various ideas were entertained by different writers, some holding that the noxious properties were due to the large percentage of albuminous material contained in the seed. The recent isolation of an alkaloid which possesses the toxic characters of the grain would seem, however, definitely to settle the point. The importance of this discovery lies in the fact that the alkaloid is volatile. It may not be present in properly cooked preparations of the grain, such as thin pressed cakes made at a high temperature. On the other hand, if made at low temperatures or imperfectly cooked, these cakes, or preparations, such as curries, *dál*, pasteballs, &c., in all probability retain sufficient of the principle to produce poisonous effects, if eaten during a prolonged period.

FOOD & FODDER.

It seems probable that the volatile property of the poison may afford an explanation of the apparent capriciousness of the effects of the pulse on persons who habitually consume it. In any case, it is highly desirable that experiments should be made for the purpose of definitely ascertaining the presence or otherwise of the alkaloid in the ordinary **Lathyrus** diet, *chapatis, dál,* curries, paste-balls, &c., consumed as food by the natives of India. Should it be possible, by simple increase of heat and greater care in cooking, to render this avowedly noxious food-grain, a wholesome and nutritious article of food, one of the cheapest and most easily obtainable pulses of India might take a much higher rank than it now holds, and consequently become much more widely utilised.

As already stated, the effects of the pulse when given as a food to cattle is similar to those observed in man. Thus, from the following extract from **Smith's** *Veterinary Hygiene,* it would appear to have been used with deleterious results in Europe :—"The disease, arising from the use of **Lathyrus sativus** (*kisári dál*) has been described by **Messrs. Leather,** and **Professors McCall** and **Williams** as affecting horses in England and Scotland. The grain is brought home to this country as ballast, and its action on horses is to produce the most intense dyspnœa and roaring when put to work; the appetite is not affected, and when in the stable the animal appears in perfect health. Several fatal cases are reported by these observers whose articles in the *Veterinary Journal* and *Veterinarian,* April 1885, and *Veterinarian,* November 1886, may be consulted with great advantage." (*Conf. with* **Cicer arietinum,** *Vol. II., 279.*)

Don writes regarding its effects on other animals: "Swine fattened on this meal lose the use of their limbs, but grow very fat lying on the ground. Kine are reported to grow lean on it, but sheep not to be affected. Pigeons, especially when young, lose the power of walking by feeding on the seeds. Poultry will not readily touch it, but geese eat it without any apparent damage. In some parts of Switzerland cattle feed on the herb without any apparent harm."

SPECIAL OPINIONS.—§ "I have seen many cases of paralysis while a Civil Surgeon in the Panjáb, which the patients themselves and their family all believed to be due to the use of *khesari dál,* and I have seen the specimens of the seeds and of the bread made from them" (*Surgeon-Major C. W. Calthrop, M.D., Morar*). "The occasional use of the *dál* does not bring on paraplegia, but many poor people are obliged to live almost entirely on it. They eat the green undressed plant, cook it, make *dál* of the seeds, and *chapáti* of the flour. It is people of this description that suffer from paralysis of the lower extremities" (*Bolly Chund Sen, Teacher of Medicine*).

LAUNÆA, *Cass.; Gen. Pl., II., 529.*

110 **Launæa aspleniifolia,** *DC.; Fl. Br. Ind., III., 415;* COMPOSITÆ.

Syn.—MICRORHYNCHUS ASPLENIIFOLIUS, *DC.*; PRENANTHES DICHOTOMA, *Wall.*; P. ASPLENIIFOLIA, *Willd.*

Vern.—*Tík-chana,* BENG.; *Birmalla,* SANTALI.

References.—*Roxb., Fl. Ind., Ed. C.B.C., 594; Campbell, Ec. Prod. Chutia Nagpur, No. 9263.*

Habitat.—A perennial herb met with in the plains of India, from the Panjáb to Assam, and southwards to the Sunderbands, Circars, Andamans, Burma, and Tenasserim.

MEDICINE. Root. 111

Medicine.—"The ROOT of this plant, along with that of *uttri dudhi,* pounded and boiled in mustard oil, is given as a lactagogue" (*Campbell*).

Launæa nudicaulis, *Less.; Fl. Br. Ind., III., 416.* 112

Syn.—MICRORHYNCHUS NUDICAULIS, *Less.;* M. FALLAX, *Jaub. & Spach;* ZOLLIKOFERIA NUDICAULIS, *Boiss.;* CHONDRILLA NUDICAULIS, *Linn.;* LACTUCA NUDICAULIS, *Murray;* L. OBTUSA, *Clarke;* PRENANTHES PATENS and P. DICHOTOMA, *Wall.;* P. OBTUSA, *Ham.;* P. PROCUMBENS, *Roxb.;* BRACHYRAMPUS OBTUSUS, *DC.*

Vern.—*Batthal, dúdhlak, tarizha, spúdukei, Pb.;*

References.—*Roxb., Fl. Ind., Ed. C.B.C., 593; Stewart, Pb. Pl., 127; Indian Forester, XII., App., 15.*

Habitat.—A common weed throughout the plains, from Bengal and Behar to the Panjáb, ascending the Western Himálaya (in Kumaon) to 8,000 feet; found also in Sind and in the Deccan.

Medicine.—In the Southern Panjáb the plant is used medicinally in *sharbat* (*Stewart*). MEDICINE. 113

L. pinnatifida, *Cass.; Fl. Br. Ind., II., 416; Wight, Ill., t. 133.* 114

Syn—MICRORHYNCHUS SARMENTOSUS, *DC.;* PRENANTHES SARMENTOSA, *Willd.;* LACTUCA SARMENTOSA, *DC.*

Vern.—*Ban-kahu,* JUICE = *khi-khowa,* SIND; *Pathri,* BOMB.; *Almirao,* GOA.

References.—*Dymock, Mat Med. W. Ind., 462; Murray, Pl. and Drugs, Sind, 188; Lisboa, U. Pl. Bomb., 163; Bombay Gazetteer, XV., 436; Madras Man. of Admin., II., 27.*

Habitat.—A common plant of the sandy coasts of India from Bengal to Ceylon, Madras, and Malabar; very plentiful at Karáchi and Goa.

Medicine.—"In Bombay *pathri* is given to buffaloes to promote the secretion of milk" (*Dymock*). According to **Murray** the JUICE is used as a soporific for children in doses of half a *massa*, and is also externally applied in rheumatic affections, mixed with the oil of **Pongamia glabra,** or the juice of the leaves of **Vitex leucoxylon.** MEDICINE. Juice. 115

Agricultural Use.—In the Madras Manual of Administration it is stated that this is one of the most important sand-binding plants on the coast. AGRICULTURAL USE. 116

LAURUS, *Linn.; Gen. Pl., III., 163.*

A genus of the Natural Order LAURINEÆ of which none are natives of India. The berries of one species, however, **L. nobilis,** *Linn.*, the "sweet barg," or "laurel" of the poets, is employed in medicine in India, under the name of *Zafui*, or *habul-ghar.*

Laurus Camphora, *Linn.*, see **Cinnamomum Camphora,** *Nees.*; Vol. II., 317.

LAVANDULA, *Linn.; Gen. Pl., II., 1179.*

A genus of herbs or shrubs belonging to the Natural Order LABIATÆ, which comprises some 20 species, mostly Mediterranean. Of these, two—**L. Gibsoni,** *Grah.*, and **L. Burmanni,** *Benth.*, are natives of India. Neither of these is known to be of economic value, but **Lisboa,** in his *Useful Plants,* has affirmed that the latter, a highly aromatic species, might be utilised in the manufacture of Oil of Lavender, a substance at present entirely imported from Europe.

Lavandula Stœchas, *Linn.; DC. Prodr., XII., 144;* LABIATÆ. 117

Vern.—*Dhárú,* HIND.; *Ustúkhúdús, alphajan,* BOMB.; *Ustúkhúdús,* ARAB.

References.—*O'Shaughnessy, Beng. Dispens., 489; Dymock, Mat. Med. W. Ind, 2nd Ed., 618; Flück. & Hanb., Pharmacog., 479; S. Arjun. Bomb. Drugs, 100; Year Book Pharm., 1873, 85; 1875, 259; Irvine, Mat. Med. Patna, 79; Birdwood, Bomb. Pr., 63; Spons, Encyclop., 1423; Balfour, Cyclop., II., 687.*

The true Lavender.

Habitat.—A fragrant herb found in the Canaries, Portugal, and Eastward throughout the Mediterranean region to Constantinople and Asia Minor.

OIL. 118 **Oil.**—It appears doubtful whether an oil is, or ever has been, distilled from this species. Certain authors, however, affirm that it is the source of the true oil of spike, used by painters on porcelain, and for artists' varnishes. At the present day, however, most if not all the oil of spike of commerce is derived from **L. spica,** *DC.*, another European species.

MEDICINE. Dried Plant. 119 Flowers. 120 **Medicine.**—The DRIED PLANT and the FLOWERS have long been employed medicinally. **Dioscorides** mentions the plant, and states that it is called Stœchas, from the fact that it is found in the Stœchades, a group of islands on the south coast of Gaul. According to **Sprengel** it is the *'ἴφυον* of **Theophrastus.** In early European medicine the flowers were known as Flores Stœchados, or Stœchas arabica, and held a place in the London Pharmacopœia down to 1746.

In the East the herb has also long been prized, and is frequently mentioned by ancient writers on medicine. At the present day it is a regular article of import into Bombay from the Persian Gulf. **Dymock** writes, "It is much used by native practitioners, who consider it to be resolvent, deobstruent, and carminative, and prescribe it in chest affections; they also think that it assists in expelling bilious and phlegmatic humours." **Irvine** states that it is used in Patna for the preparation of scents, and in doses of grs. 40 to ½ oz. in infusion as a refrigerant. According to **Dymock** the price in Bombay is R8 per maund of 37½ lb.

121 **Lavandula vera,** *DC., Prodr., XII., 145.*

References.—*Pharm. Ind., 165; O'Shaughnessy, Beng. Dispens., 488; Flück. & Hanb., Pharmacog., 476; Bent. & Trim., Med. Pl., 199; Year Book Pharm., 1879, 466; Spons, Encycl., 1423; Balfour, Cyclop., II., 687; Smith, Dic., 241; Kew Reports, 104.*

Habitat.—A native of South Europe and the Mediterranean shores, extending into Western Africa; introduced into England, where it is now extensively cultivated for the spikes of flowers from which a volatile oil is distilled.

OIL. 122 **Oil.**—From the herb an oil is obtained by distillation which is pale yellow with grateful odour, and bitter, aromatic taste. The characteristic properties of the plant have been developed to an enormous extent by English cultivation and soil. But they are not equally developed, for, indeed, there are only two districts that can be said to suit the plant: these are Mitcham in Surrey and Hitchin in Hertfordshire. In the latter locality over 50 acres are annually under cultivation. The bushes are grown in rows 4 feet apart, the plants being 3 feet removed from each other; an acre so planted contains about 3,547 bushes. When about four years old they yield the best otto, and are improved by keeping back the flowering. An acre should yield about six to seven quarts of otto. Essence of lavender is best prepared by distilling a mixture of essential oil of lavender with rectified spirit, in the proportions of 4 oz. of the former to 5 pints of the latter (60 over proof), with 1 pint rose water. (*Piesse on Perfumery.*)

The cultivation of lavender does not appear to have been attempted in India except to a small extent in the Nilghiri Botanic Gardens, but there seems to be no reason why it should not do well on the Himálaya.

MEDICINE. Oil. 123 Spirit. 124 Compound Tincture. 125 **Medicine.**—In the Pharmacopœia of India three preparations are described as officinal—an OIL, a SPIRIT, and a COMPOUND TINCTURE. Of these the oil is rarely administered internally in its simple form, and is employed almost entirely for disguising the unpleasant odour of ointments and other preparations. The spirit and tincture (diluted preparations of the oil), are stimulant, carminative, and antispasmodic, and are considerably used in nervous and hysterical cases, incipient syncope, flatulence, and flatulent colic. The dose of the spirit is 30 to 60 drops, of the compound tincture 30 drops to 2 drachms.

LAWSONIA, *Linn.; Gen. Pl., I., 782.*

[LYTHRACEÆ.

Lawsonia alba, *Lam.; Fl. Br. Ind., II., 573; Wight, Ill., t. 87;* 126

THE HENNA PLANT, CAMPHIRE, CYPRESS SHRUB, OF EGYPTIAN PRIVET.

Syn.—L. INERMIS, *Roxb.;* L. SPINOSA, *Linn.;* L. PURPUREA, *Lamk.*

Vern.—*Mehndi, mhindí, héna,* HIND.; *Shudi, méhédi, mendi,* BENG.; *Manghati, muljuyáti* URIYA; *Mihndi,* SANTAL; *Simrú,* BHOTE; *Mehndi, hinna, nakrize, panwár* PB.; *Mehendi,* RAJ; *Mendi,* SIND; *Mhéndi, ménhdí,* C.P; *Méndi, méndhi,* MAR.; *Médi, méndi,* GUZ.; *Marithondi, marutónri, aivanam,* TAM.; *Góranta, pachcha pedda góranta, iveni,* TEL.; *Gorantlu, górante,* KAN.; *Mayilánchi, marutónni, ponta letshi, daunlacca,* MALAY.; *Dan, danbin,* BURM.; *Maritondi,* SING.; *Mendhi, kuravaka, mendiká, sakachara,* SANS.; *Hinná, yoranná,* ARAB.; *Hiná, panna,* PERS.

References.—*Roxb., Fl., Ind., Ed. C.B.C., 325; Brandis, For. Fl., 238; Kurz, For. Fl. Burm., I., 519; Beddome, Fl. Sylv., anal. Gen., xiv., f. 6; Gamble, Man. Timb., 200; Dalz. & Gibs., Bomb. Fl, 97; Stewart, Pb., Pl., 90; DC. Origin Cult. Pl., 138; Griffith, Ic. Pl. Asiat., t. 580; Elliot, Fl. Andhr.; 62, 140; Pharm. Ind., 87; Ainslie, Mat. Ind., II., 190; O'Shaughnessy, Beng Dispens., 331; Moodeen Sheriff, Supp. Pharm. Ind., 167; U. C. Dutt, Mat. Med. Hind., 309; Dymock, Mat. Med. W. Ind., 2nd Ed., 307; Fluck. & Hanb., Pharmacog., 305; S. Arjun, Bomb. Drugs, 53; Murray, Pl. and Drugs, Sind, 144; Year Book Pharm., 1875, 207; 1879, 213; Cat Baroda Durbar, Col. Ind. Exhib., Nos. 121, 122; Baden Powell, Pb. Pr. 348, 451; Atkinson, Him. Dist., 310; Drury, U. Pl., 275; Lisboa, U. Pl. Bomb., 245; Birdwood, Bomb. Pr., 298, 344; Christy, Com. Pl. and Drugs, VIII., 50; McCann, Dyes and Tans, Beng., 56, 90; Buck, Dyes and Tans, N.-W. P., 29; Liotard, Dyes, 47; Crookes, Dyeing, 512; Wardle, Rep. on Indian Dyes, 5, 30; Campbell, Ec. Prod. Chutia Nagpur, No. 8408; Mason, Burma and its People, 407, 758; Honigberger, Thirty-five years in the East, I., 4, 167, II., 297; Ain-i-Akbari, Gladwin's Trans., I., 96, 100; II., 41; Stocks, Rep. on Sind; Spons, Encyclop., 858, 1424; Balfour, Cyclop, II, 690; Smith, Dic., 81; Kew, Reports, 68; Simmonds, Trop. Agri., 390; Piesse, Perfumery, 402; Indian Forester, XII., App., 4; Agri.-Hort. Soc. of Ind.,—Transactions, VI., 240; Journals, (Old Series), II., 357; IV., sel., 154; VI., 41; IX., 422; XI., 268; Gazetteers:—Orissa, App. VI., 179; Mysore and Coorg, I., 61; Bombay, V., 25; N.-W. P., I., 81; IV., lxii; Panjáb, Shahpur, 66; Settlement Reports:—Panjáb, Delhi, App. xxv., 271.*

Habitat.—A small, elegant, and sweetly-scented bush, cultivated commonly throughout India, perhaps wild in the Western regions, and according to **Roxburgh** certainly so on the Coromandel Coast. **Stocks** found it indigenous in Balúchistán, and **Brandis** states that it is perhaps wild in Central India. **De Candolle** writes that, owing to the shrub being at the present day more or less wild in all the warm regions of Western Asia and Africa to the north of the equator, he finds it difficult to ascertain its original area. From a consideration of the vernacular names in several languages, however, and particularly owing to the wide dispersion of the Persian word *panna,* he is inclined to the opinion that the plant first grew on the borders of Persia, and that its use, as well as its cultivation, probably spread from that region west into Africa and eastward to India.

This opinion would appear to be supported by the fact that the use of the dye as a cosmetic, in India at least, is evidently of Muhammadan or Western origin.

Cultivation of the Henna Plant.

CULTIVATION. 127

Cultivation.—The henna plant is commonly cultivated in all the provinces of India, partly for the sake of its dye and fragrant flowers, partly as a hedge plant. The following account by **Major Davies** of its cultivation in Shahpúr, may be quoted from **Baden Powell's** *Panjáb Products*, as descriptive of the method which is likely to produce the greatest amount of dye:—

Panjab. 128

"The soil is prepared by repeated ploughings, not less than sixteen, and heavy manuring. Before sowing, the seed is allowed to soak in water for twenty-five days. It is then spread on cloth and allowed to dry partially. The plot of land in which it is proposed to grow the *mendhi* is then formed into small beds, and some days before sowing these are kept flooded. The seed is scattered on the surface of the water, and with it sinks into the ground. For the first three days after sowing, water is given regularly night and morning, after that only once a day. The young plant first shows above ground on the fifteenth day, after which water is only given every other day for a month, when it is supplied at intervals of three days, and this is continued for another month, by which time the plants have become nearly two feet high. They are now fit for transplanting. The mode of conducting this operation is as follows:—The young plant on being taken out of the ground is reduced by nipping off about six inches from the centre shoot. After having been subjected to this treatment, the young plants are singly put into holes, previously dug for them, at distances of about a foot from each other. They are then watered daily until they have recovered the shock of transplanting, and afterwards, as they may require it. The fields are weeded regularly once a month. The first year nothing is taken from the plants, but after that they yield, for years without intermission, a double crop. At each cutting, about nine inches is taken from the top shoots of the plants. The two crops are got in *Baisakh* (April and May) and *Katak* (October and November) of each year.

"The labourers employed in planting out the *mendhi* are liberally fed as long as the operation lasts, and a distribution of sweetmeats takes place when it is over. The season for sowing is during the month of *Baisakh*, that of transplanting, *Sawan* (July and August). A year's produce of an acre of well-grown *mendhi* is twenty maunds of dry leaves, of which about six maunds are gathered in the spring, and the rest during the autumn months, and the same plants continue to yield for twenty or twenty-five years.

"The selling price of the leaves averages a rupee for twelve seers, so that the value of the crops per acre is about R66. After the first year the expenses of cultivation do not much exceed those of other crops. The produce of the *mendhi* grown in this district is nearly all carried across the Jhelum and sold in the northern districts, none of it finds its way to the south."

Sind. 129

Sind.—Reports from district officers inform us that in the Surat and Hyderabad districts it is grown in gardens as a hedge, the leaves being used for the dye. In the Shikarpur district it is also cultivated, the profit accruing therefrom being estimated at 50 per cent. On every 100 square feet of soil two seers of seed are sown in February. The seed germinates in about ten or twelve days; when a foot high, the plants are transplanted and in four months the crop is fit to yield its leaves for dyeing. The average quantity of dye produced in Shikarpur is said to be 5,000 maunds, of which 1,000 maunds are consumed locally, and the rest is exported to Sukkur and other places.

Bengal and N.-W. Provinces. 130

Bengal and the *North-West Provinces.*—Henna is quite as common in these provinces as elsewhere, and, according to **Gamble**, is occasionally found in cultivation in the Terai.

CULTIVATION. Madras and Bombay. 131

Madras and *Bombay.*—Henna is extremely common in these provinces as a hedge plant, and also occurs in gardens, but no statistics are available regarding the area under cultivation nor the degree of utilisation of the leaves.

DYE. Leaves. 132

Dye.—A decoction of the LEAVES is occasionally used in dyeing cloth, the colour produced being a shade of yellowish or reddish-brown known as *malagiri.* It is also employed in combination with other dyes in Rajputana for dyeing handkerchiefs of an almond colour. Samples were sent for report by the Government of India to **Mr. Wardle,** who wrote as follows: —

" These leaves, said to produce a red colour, only produce a grey colour on silk by my methods after repeated trials; it is probable that the sample sent may not have been a good one. They possess no tannin as shown by no blackening of the colour with an iron salt. There seems to be more colour produced on unbleached Indian *tussar* without a mordant than on any of the others where a mordant was used. What there is of the colour seems to be very artistic." **Wardle's** statement that " they possess no tannin " is at variance with the published results of chemical analysis, a fact that would tend to confirm his own suspicion that " the sample sent may not have been a good one."

Dr. G. Watt (*Selections from the Records from the Government of India, Revenue and Agricultural Department., I., 56*) mentions that an Indigo-planter recently informed him that he was experimenting, on a large scale, with the leaves of this plant, in the preparation of a reddish dye, isolated in the same way as indigo. This remark is of great interest in connection with the red dye described by **Hove** as similarly prepared from **Bixa Orellana,** and which at one time formed an important article of export from India to China. No evidence, however, appears to exist that the natives of India are, or have ever at any tmie been, acquainted with a cake-dye similar to Indigo, prepared from **Lawsonia alba.**

Much the most important use of *Henna* in India, as it has been all over the East for many centuries, is as an article of the toilet, the leaves being used for staining the fingers, nails, hands, and feet, and for dyeing the hair. The utilisation of henna for the former purpose dates from the earliest times, as is shown by ancient mummies and paintings. In India it is universal among Muhammadan women, and is also adopted to a greater or lesser extent by Hindus. To dye the nails the freshly-gathered leaves are pounded with catechu or lime, and applied to the fingers over-night; for staining the hands and feet the leaves are simply rubbed over these parts. For dyeing the hair the following are the recipes principally in use, as given by **Honigberger**: " Of **Lawsonia alba** and **Indigofera Anil** take each two parts, and of dried myrtle leaves and emblic myrobolans each one part; these are made into powder and mixed with water to the consistency of a soft poultice, which is applied to the hair. It is stated that this composition stimulates the growth of the hair, blackens that which is grey, and prevents its splitting. The second recipe is one which I have myself used in earlier years in India and in Persia, and it is generally recognised by the orientals (who are extremely fond of possessing fine long black hair) as the best preparation for the purpose. The powdered **Lawsonia alba** is made by water into a soft pap, and applied in that state to the hair, taking care that all the hair is completely overspread to the roots. It is then covered and fastened up with leaves, or by a piece of wax cloth or oil skin, and having been suffered to remain for from half an hour to an hour, the preparation is then washed off. The effect of this is to dye the hair a bright red colour; which colour by the next process is converted into a beautiful black.

" The second application is a paste prepared with water from the powder

DYE. of the indigo plant before mentioned; it is used in a similar manner to the first but must be allowed to remain on the hair for three hours; being then washed off, the operation of dyeing is completed, and the hair is rubbed with oil or pomatum, to give it lustre and suppleness. The only inconvenience of the processes I have described, which are so cheap in the East, is that they require to be repeated about once a week, for, as the hair grows from the roots, it would otherwise, after a few days, show at that part the natural colour of the hair, and consequently present a very unsightly appearance. These operations are generally performed at noon, a time when every one is at home either for rest or indoor occupation."

By certain classes of Muhammadans the process is stopped at the first stage, leaving the hair and beard of a brilliant red colour, and in Persia, Arabia, and Northern India, the manes and tails of horses are frequently dyed red by a similar process.

OIL. **Oil.**—The SEEDS yield an oil of which little is known; the FLOWERS are
Seeds. used in perfumery and embalming, and a fragrant otto (*mehndí*) is distilled
133 from them in Benares, Lucknow, and other localities. In *Spons' Encyclopædia*
Flowers. it is stated that this otto is "remarkably and delightfully fragrant."

134 **Olivier** states that a fragrant water distilled from the flowers was known to the Jews, and was employed in baths, and as a perfume in religious ceremonies, such as marriage, circumcision, and the feast of Courban-Bieram. The ancient Egyptians made a similar use of the flowers for the purpose of perfuming the oils and ointments with which they anointed the body, and for embalming.

MEDICINE. **Medicine.**—*Henna* has been employed and esteemed as a medicine
135 from the remotest times. It was used by the Egyptians as an astringent; **Dioscorides** mentions it under the name of *κυπρος*. **Pliny** says that the most esteemed cypress of Egypt comes from Canope, on the banks of the Nile; the second from Ascalon in Judæa; and the third from the island of Cyprus. **Prosper Alpinus** also makes reference to the drug, stating that invalids procure ease by inhaling the perfume of the flowers, and by applying them to the forehead. **Avicenna** compares the properties of *henna* with those of dragon's blood, and states that its leaves possess the same property of curing ulcers, and that a decoction of them is employed in cases of inflammation and burns, and as a remedy against ulcers of the mouth.

Leaves. In India *henna* is much esteemed by Muhammadans, indeed, a tradi-
136 tion exists that their prophet spoke of it as *syyadu riáhín*=the best of herbs. Arabic and Persian writers describe the LEAVES as a valuable external application in headache, combined with oil so as to form a paste, to which resin is sometimes added. "They are also applied to the soles of the feet in small-pox, and are supposed to prevent the eyes being affected by the disease. Applied to the hair and nails, as already described, they have the reputation of promoting healthy growth. An ointment made from the leaves is said to have valuable healing properties, and a decoction is used
Bark. as an astringent gargle. The BARK is given in jaundice and enlarge-
137 ment of the spleen, also in calculous affections, and as an alterative in leprosy and obstinate skin diseases; in decoction it is applied to burns,
Seeds. scalds, &c. The SEEDS, with honey and tragacanth, are described as
138 cephalic. An infusion of the FLOWERS is said to cure headache and to be
Flowers. a good application to bruises; a pillow stuffed with them has the reputa-
139 tion of acting as a soporific. An ointment is also used as an external application to bruises" (*Dymock*). **Ainslie** states that in Southen India the *Vytians* "prepare a kind of extract from the pale-yellow, strong-smelling flowers, as also from the leaves and tender shoots, which they consider as a valuable remedy in cases of lepra and other depraved habits of the

MEDICINE.

body; prescribing it in the quantity of half a tea-spoonful twice in the twenty-four hours." **Dymock** remarks that "in the Konkan the leaf juice mixed with water is given as a remedy for spermatorrhœa, and with milk in the condition popularly known as 'hot and cold fits.'" *Henna* occupies a place in the non-officinal list of the Pharmacopœia of India, where attention is specially drawn to its use in "that obscure affection termed 'burning of the feet' often met with in India." **Dr. Grierson** is quoted as mentioning an obstinate case benefited by the use of the leaves, and it is stated that "the editor, when in Burma, witnessed, in some cases, a great amount of temporary relief to the distressing sensation of burning, from this remedy, when numerous other means had previously failed. The fresh leaves beaten up into a paste with vinegar and applied as a poultice to the soles of the feet, was the common mode of application, but some patients obtained greater relief from using strong frictions with the bruised leaves over the part. Like all other remedies, however, they not unfrequently fail to afford more than temporary relief."

It may be of interest to note that **Honigberger** mentions a use of this plant not dealt with by other writers. Remarking on a severe attack of rheumatism from which he suffered in Greece, he writes, "Amongst the many remedies which I applied, **Lawsonia inermis** proved the most effectual. Every evening before going to bed, I applied to the affected part the pulverized herb, with as much water as was sufficient to make a soft poultice. In the morning I washed it off, but the place continued red; it is the same herb with which the fair sex in the East stain their hands and feet."

Chemistry. 140

CHEMICAL COMPOSITION.—The only chemical substance of medicinal value known to be contained in *Henna* leaves is an astringent principle. **Berthelot** supposed them to contain gallic acid; later investigations by **M. Abd-el-Aziz Herraory** have shown that the colouring matter consists of a sort of tannin to which he gave the name of *Hennotannic acid.* **Dymock**, quoting from *Journal de Pharmacie, January 1863*, writes: "This principle is brown, of a resinoid appearance, and soluble in boiling water. It possesses the properties of tannin, such as blackening the sesqui-salts of iron, and precipitating gelatine. It reduces oxide of copper in Trommer's test, and heat decomposes it, with the production of crystalline needles which reduce nitrate of silver."

SPECIAL OPINIONS.—§ "A decoction of *mendhi har* and *kath* is used for gonorrhœa as an injection (*Surgeon-Major C. W. Calthrop, M.D., Morar*). "The fresh leaves, beaten into a paste with vinegar or lime juice, are said to be useful in cases of 'burning of the feet'" (*Civil Surgeon J. H. Thornton, B.A., M.B, Monghyr*). "A poultice of the fresh leaves has a soothing effect when applied to swollen and painful parts; it is also used to harden the skin of newly healed wounds" (*Civil Surgeon J. Anderson, M.B., Bijnor, North-Western Provinces*). "An infusion made from the leaves has been found particularly useful in cases of scurvy and ulcers of all kinds. In cases of ulceration of the mouth, it is a valuable remedy, used as a gargle" (*Civil Surgeon F F. Perry, Jullundur City, Panjáb*). "In Belgaum the seeds are called *isbon* and are considered useful in delirium. They are given internally" (*C. T. Peters, M B., Zandra, South Afghánistan*). "The leaves, both dry and green mashed with water and applied to the body, relieve prickly heat." (*Surgeon-Major A.S.G. Jayakar, Muskat*).

TIMBER. 141

Structure of the Wood.—Grey, hard, close-grained, used for tool handles, tent pegs, &c.

DOMESTIC. 142

Domestic.—This handsome shrub has always been a favourite garden plant in the East from the times of the ancient Egyptians to the present

DOMESTIC. day. Thus Solomon writes, "My beloved is unto me as a cluster of camphire in the vineyards of En-gedi." The use of the otto by the Jews has been already referred to, and they also had a custom of sprinkling the flowers on the garments of the newly-married. In India it is much cultivated as a garden plant and for hedges, and is held in particularly high esteem by Muhammadans, who consider its flowers amongst the most suitable with which to compose a posy to be presented in token of esteem and honour. The utilisation of the otto and dye for domestic purposes has already been described.

LEAD, *Ball, in Man. Geology of Ind., III., 281-311.*

This metal rarely occurs anywhere in the metallic or native state, and appears never to be found in this condition in India. Its commonest ore is the sulphide, or galena. Many of the other ores are of little interest save to the mineralogist, but bournonite or antimonial lead, cerussite, or the carbonate, pyromorphite or phosphate, mimetite or arsenate, and crocoisite or chromate, are all of economic importance. Many compounds of lead which do not occur abundantly in nature, but which are extensively used in the arts for pigments, dyeing, glass-making, medicine, &c., are artificially prepared.

143 **Lead,** *Mallet, Man. of Geol. of India, IV. (Mineralogy), 6, 15.*

Vern.—The Metal :—*Sisa, sisu, shish,* HIND.; *Sisá, búndi,* BENG.; *Sisha, surb,* PB.; *Shísh,* DEC.; *Ikam, eiam,* TAM.; *Shíshum, sheshumu,* TEL.; *Temaétam,* MALAY.; *Khai-pok, khai-ma-pok,* BURM.; *Sísaka,* SANS.; *Anúk, rassás,* ARAB.; *Surb,* PERS.

The Oers and Salts :—

I. GALENA.—*Surmá,* krishna surmá,* HIND.; *Anjana, sanviránjana,* SANS.

II. RED OXIDE OF LEAD.—*Sandhur, ingur,* HIND. & PB.; *Séndúr,* DEC.; *Segapú, senduérum, ium sindúrum,* TAM.; *Yérra sindúram,* TEL.; *Temaméra, sadda langam, galanggam,* MALAY.; *H'sang,* BURM.; *Sindúra,* SANS.; *Isrenj,* ARAB.

III. LITHARGE.—*Murdasang,* HIND. & PB.; *Marudar singhie,* TAM.

IV. CARBONATE OF LEAD.—*Saféda,* HIND. & PB.; *Vullay, múthú vullay,* TAM.; *Sibaydú,* TEL.; *Tima-putih,* MALAY.; *Asfidáj,* ARAB.; *Saféda,* PERS.

V. CHROMATE OF LEAD.—*Peori-wilayti,* HIND.

References.—*Forbes Watson, Indust. Survey, II., 406, 407; Baden Powell, Pb. Pr., 11, 54, 63, 102, 114, 115; Mason, Burma and Its People, 565, 729; Abul Fazl, Ain-i-Akbari (Blochmann's Trans.), 40; Atkinson, Him. Dist., 290; Pharm. Ind., 360-364; Ainslie, Mat. Ind., I., 532-537; U. C. Dutt, Mat. Med. Hind., 72, 73; Irvine, Mat. Med. Patna, 98; Crooke, Dyeing and Calico-Printing, 152, 154, 541; Gazetteers:—Panjáb, Shahpur, 13; Settlement Reports:—C. P., Raipore, 7; Panjáb, Dera Ghazi Khan, 7; Dist. Man.:—Madras, Cuddapah, 26, 27; Madras, Man. of Admin., II., App. vi., 35; Bomb. Admin. Rep., 1872-73, 366.*

LOCALITY. 144 **Occurrence.**—Although at the present time lead ores are not largely worked in India, Ball states that there is probably no metal of which the ores have been worked to so large an extent in ancient times, excepting those of iron. This is testified to by the number of old mines in Southern India, Rájputana, Balúchistán, and Afghánistán It is possible that in some at least of these cases the galena may have been chiefly worked for the silver which it contained. Ball states that ores of lead are found in Peninsular India, in the rocks of metamorphic, submetamorphic, and lower Vindhyan ages only; but that in the extra-peninsular area they occur in, at least, one younger formation, namely, in the cretaceous rocks of Balúchistan.

(* Confused with antimony.)

LOCALITY.

As the working of lead ores in India is at present exceedingly unimportant and precarious, it does not appear to be necessary to enter into a full account of the localities in which they are to be found. For such information the reader is referred to the long and exhaustive account by Ball (*loc. cit.*). The following short description of the principal localities kindly furnished by **Mr. Medlicott** (*late Director of the Geological Survey*) may suffice for the purposes of this work:—

"At Jungamrazpilly or Baswapur, in the Cuddapah district of Madras, argentiferous galena occurs and was formerly very extensively worked, but the workings have been long deserted. Gazalpully, or Baswapur, in the Karnúl district, is also another locality for similar ores. Deposits of galena containing silver occur in the Bhaugulpur district, and have been more or less worked at various times. There are some galena deposits, rich in silver, at Dhadka in the Manbhúm district. In the Central Provinces, lead ores have been found in Sambalpur at Talpuchia, Zhunan, and Padampur, and in Raipur at Chicholi. In Rájputana, galena occurs and has been mined for in Ajmir, Alwar, and Udaipur. In Balúchistán there are ancient lead mines at Sekran of considerable extent, but long deserted. In the Panjáb argentiferous galena occurs at Manikaru in the Kulu district. Lead mines and works were started a few years ago near Subathu in the Simla district, and mines have been for long worked by natives in Sirmur."

DYE & PIGMENT. Red Lead. 145

Dye and Pigment.—The RED OXIDE OF LEAD or red lead is largely used as a pigment. It is made in certain parts of the country by exposing melting lead to a stream of air, and is also imported. It is not only largely employed for painting and decorating, but is also used for religious purposes by the Hindus, who make marks with it on their idols, and on their money at certain seasons for good luck. Boulders may frequently be seen in the Himálaya marked with a patch of red lead, which is supposed to convert them into idols or objects of worship. All married Hindu women employ red lead as a paint for the forehead.

Chromate. 146

The yellow CHROMATE OF LEAD or *peori-wilayti*, so called from its resemblance to the Hardwari *peori*, is imported from Europe. It is used in dyeing and as a pigment for all shades of yellow, from the palest primrose to deep orange chrome; and is also considerably employed in calico-printing, the colouring material being formed in the cloth. The process employed is described by **Crookes** as follows: "24 ozs of white sugar of lead (acetate) are dissolved in each gallon of water and thickened with 4℔ of gum. The solution is printed on the cloth which is then dried and passed for two minutes through a cold bath of bichromate of potash and common salt, containing 6 ozs. of bichromate of potash and 24 ozs. of salt to the gallon. The cloth is finally well washed and dried." By this means yellow chromate of lead is precipitated in the fibres of the cloth. By slightly varying the chemical process different shades are obtained, and the process is applicable to yarns as well as to cloth.

Acetate. 147

The ACETATE and SUBACETATE are occasionally employed as mordants and for the production of lakes, but the colours obtained are inferior, and are apt to blacken in the presence of sulphur. A lead mordant is, however, in common use for fixing aniline purple and mauve.

Nitrate. 148
Sulphate. 149

NITRATE OF LEAD is employed in mixing murexide purple, and it, as well as the SULPHATE, may be used instead of the acetate for producing chromate of lead in cotton goods.

Plumbite. 150

PLUMBITE and PLUMBATE OF SODA are also occasionally similarly employed.

White Lead. 151

WHITE LEAD (the CARBONATE) is employed largely as a ground colour, and as a white paint.

LEAD. **Various uses of Lead.**

MEDICINE. 152 **Medicine.**—The monoxide or litharge (PbO.), the carbonate, acetate, subacetate, nitrate, and iodide are officinal in the Indian Pharmacopœia. All these salts, except the acetate, are employed externally only, as sedative and cooling astringents, in various skin diseases. The acetate is similarly used, but is also administered internally as an astringent and hæmostatic. The properties and applications of the officinal salts are too well known to require further description.

Preparations of lead have long been similarly employed, and valued in Indian medicine. Thus **U. C. Dutt** states that the Sanskrit name for Galena —*anjana*—literally signifies "*the* collyrium," and was applied to the substance, owing to the fact that it was considered the best medicine for the eyes. It is largely used at the present day as an application to the eyes, and is supposed to strengthen the sight, improve the appearance of the eyes, and preserve them from disease. In ophthalmia, blepharitis, and allied diseases of the conjunctiva it is prescribed in various preparations. **U. C. Dutt** describes two of these as follows:—(1) Galena is heated over a fire, cooled in a decoction of the three myrobolans for seven times in succession, then rubbed up with human milk. (2) To one part of purified and melted lead, add an equal proportion of mercury and two parts of galena, rub them all together and reduce to powder. Now add camphor, equal in weight to one-tenth part of the mass, and mix intimately. (*Mat. Med., Hindus.*)

Metallic lead purified, by pouring the molten metal into the milky juice of **Calotropis gigantea,** is also employed in Sanskrit medicine. It is reduced to powder by again melting it in an iron cup, adding to it one-fourth part of its weight of *yavakshára* (impure carbonate of potash) and powdered tamarind shells, and agitating with an iron rod till the mass is reduced to fine powder. The preparation thus produced consists of an impure carbonate of lead. It is supposed to be a valuable remedy in various forms of urinary disease. A lead probe is preferred for the application of collyria to the eyelids.

Litharge is unknown in Sanskrit medicine, but is said by **Ainslie** to be used by the "Tamools" for the preparation of certain *kálímbús* or plasters, and to be placed by the Arabs amongst their anodynes.

Red lead is described in Sanskrit medicine as useful for external application in cases of eruptive skin diseases. It is applied in the form of an oil, or liniment, and mixed with black pepper and butter as an ointment.

DOMESTIC & SACRED. 153 **Domestic and Sacred.**—The ordinary industrial uses of lead and its salts require no comment. The metal is comparatively little employed by the natives of India owing to its scarceness. Red lead is, as already stated, used to a considerable extent as a pigment for several religious purposes. Galena is employed not only as a medicinal application to the eyes, but also as a cosmetic in the place of antimony. It is also utilised to a considerable extent for glazing pottery.

TRADE. **Trade.**—No statistics can be given regarding the inland trade in lead, since it is not sufficiently important to be returned separately. The import trade from foreign countries is, however, considerable, having amounted, in

Imports. 154 the year under consideration, to 113,130 cwt., valued at R15,88,631. In addition Government imported 49,285 cwt., valued at R4,62,535. The trade has also increased very considerably during the past ten years. Thus, during the five years ending 1883-84, the imports (private) averaged 70,758 cwt., valued at R10,05,196, while in a similar period ending 1888-89 they averaged 100,256 cwt., valued at R12,63,241. This increase of about 30,000 cwt. is believed to have been almost entirely caused by a development of the trade in lead sheeting for tea-chests, the imports of which have increased from 39,735 cwt. in 1883-84, to 66,415 cwt. in 1888-89. The imports in the other descriptions enumerated in the trade returns, *viz.*, pig, ore, and various kinds of wrought lead other than sheeting for tea-chests, have remained comparatively steady. **TRADE. Imports.**

The imports on account of Government have increased from an average of 6,854 cwt., valued at R60,769 in the quinquennial period ending 1883-84, to 17,099 cwt., value R1,98,220 in that ending 1888-89 (the figures of weight are defective, those of value accurate). The amount of the metal imported, however, fluctuates very erratically year by year, and consists almost entirely of pig lead for ordnance purposes.

The greater part of the general imports comes from the United Kingdom. Thus in the year under consideration the sources of supply were as follows:—The United Kingdom, 94,262 cwt.; France, 15,788 cwt.; Italy, 2,111 cwt.; Belgium, 825 cwt.; Ceylon, 97 cwt.; and other countries 47 cwt. The shares of the seaboard presidencies and provinces in the imports were—Bengal, 76,407 cwt.; Bombay, 30,023 cwt.; Sind, 3,415 cwt.; Madras, 2,339 cwt., and Burma, 946 cwt.

RE-EXPORTS.—With the increase in imports a corresponding increase has taken place in the re-export trade, which is, however, very unimportant. **Re-exports. 155**
In the five years ending 1883-84 it averaged 2,658 cwt., value R29,707, while, in a similar period ending 1888-89, it averaged 3,962 cwt., value R39,136. Nearly the entire trade is in pig-lead. In 1888-89, the whole re-export amounted to 5,095 cwt., value R56,330, of which 1,612 cwt. went to Arabia; 1,272 cwt. to Turkey in Asia; 899 cwt. to China; 690 to Persia, and unimportant quantities to other countries. Nearly the entire export was from Bombay.

LEATHER. 156

Leather consists of hides and skins of certain animals, separated from fleshy and fatty matters and prepared by means of chemical agents in such a way that they resist the influences to which they are naturally subject. Skins in their fresh state are tough, flexible, and apparently well suited for clothing, &c.; but on drying they become hard, horny, pervious to water, and finally putrid. These changes are prevented by the processes of the leather manufacturer, and at the same time the skin is rendered stronger, more impermeable to water, more supple and less likely to be affected by wear and tear. The final result of the processes ordinarily employed is a chemical combination of certain constituents of the skin with tannin. Two other processes are, however, occasionally followed, by one of which "tawed leather" is produced, through the action of mineral salts, by the other "shamoyed leather" is formed, in which the skins are combined with oils or fatty substances.

Vern.—*Charmo, chamra,* HIND.; *Tol,* TAM.; *Kulit, balulang,* MALAY.; *Charm,* PERS.

References.—*Sel. from Rec. Govt. of Ind., R. & A. Dept., I., 83-111; Baden Powell, Pb. Prod., 156; Pb. Manuf., 121-136; Balfour, Cyclop. II., 693; Ure, Dict. Indus., Arts and Manu., II., 836; Gazetteers:—Bombay, IV., 139; Peshawar, 149; Hoshiarpur, 113; Gurdaspur, 61; Shahpur, 78; Settlement Rep., Kohat, 127.*

The Natives of India have long been acquainted with the methods used in tanning, though it is very doubtful if they have ever made leather of superior quality. They employ for this purpose the hides of the buffalo, bullock, and cow, also the skins of goats and sheep. For an account of these the reader is referred to the article on **Hides**, pp. 248-251; also to that on **Skins**, Vol. VI.

The following note by **Captain John Stewart, R.A.**, on the methods pursued at the Government Tannery, Cawnpore, may be reprinted from the Records of the Government of India, Revenue and Agricultural Department, as it gives a full and interesting account of the processes most likely to give good results with Indian hides.

TANNING.

TANNING.

Buildings. BUILDINGS.—" The TANNERY consists of a large building, in the floor of
157 which are masonry pits plastered with chunam. The floor is on two levels.
The Tannery. The lower contains the BEAM-HOUSE and LIME-PITS, where the hair and flesh
158 of hides are removed by the action of milk of lime; also the BARK-TAPS and
Lime pits. 'SPENDERS' where the bark is infused. The latter are large masonry pits
159 with false bottoms of wood, through which the infusion drains off by plug-
Bark taps. holes into a well adjoining, where a pump is fixed, and the liquors are
160 raised and carried into the tan-pits on the higher level; these are the pits
Spenders. in which the hides are "tanned" and they drain off into the spenders and
161 taps on the lower level.

Tan pits. " The CURRIER'S SHOP is a long two-storied building: the lower story is
162 furnished with currier's beams, and scouring tables of stone, where the
The Currier's shop. hides are "shaved" and scoured preparatory to being oiled and "dubbed." The upper story is for finishing the "currying" process. It is fur-
163 nished with wooden tables for setting out and "dubbing" the hides, which are hung up to dry on battens suspended from the roof. There are arrangements for hanging the hides on both stories.

HIDES. Raw material. " HIDES.—The hides and skins tanned and curried are—buffalo, bullock, and cow, and goat and sheep skin.

164 " BUFFALO HIDES are obtained from the slaughter markets of Cawnpore
Buffalo Hides. and adjacent towns and cities. They are either green, direct from the
165 butcher, or dry-salted. They are best suited for tanning in the former condition, as the salt cure of the North-Western Provinces is inferior and imperfect, absorbing so much moisture that in the damp heat hides are apt to rot; while the dry heat of the climate so hardens and contracts the fibres of the skin that much labour is required to loosen the pores to receive the lime and tan. Great precaution, therefore, is necessary in the selection of dry-salted hides, especially as native dealers lay on the *kharee* or salt very thick to gain weight. If the hides are fresh slaughtered and have been lightly cured, they soak down to a natural state in about two days, but if they are stale, that is, have been cured some months and are besmeared with *kharee*, the tanner should reject them, for they will very likely decompose in the soak before becoming soft. Buffalo is the only available hide that will produce leather thick enough for harness work in this country; but there is no doubt that much of the inferiority of country leather arises from the poorness of the skin of that beast. It is poorly fed, not generally cared for, and usually killed when too old to breed or give milk.

The hide of the male buffalo is too coarse, and it gets such bad treatment in the plough or the cart that it is generally full of sores and goad-marks. In large towns there is a market for buffalo beef for the low caste and poorer Mussalman population, and also for grease, but the younger and better cattle are rarely slaughtered; it is from these that the local tanners select their hides for the finer uses of harness, saddlery, and accoutrements. Many good hides are ruined by the butchers in "flaying," from inefficient arrangements in the slaughter-houses and from injudicious use of their tools.

Bullock and Cow Hides. " BULLOCK AND COW HIDES are also procured green and dry-salted,
166 but the same care is necessary in selection. The animals slaughtered by the Commissariat Department are usually the best, but they are small and unfit for any thick work. This class of hides is much exported to England and extensively used there for boot and shoe upper leather, for which it is much esteemed. In the English market there is great objection to North-

TANNING.

West cured hides. Patna, Dacca, and Durbungah cures, though the hides are not in themselves superior, are far preferred. The finest hides of this description are those killed at Agra, Delhi, and Meerut. Much damage is done to the hides by branding on the butts and shoulders.

Goat and Sheep Skins. 167

"GOAT AND SHEEP SKINS are always obtained green from the local market; goat skins are generally very good; sheep are poor and small.

MATERIALS USED IN TANNING, &c.

MATERIALS USED IN TANNING. 168

White or stone lime. 169

"Lime is brought into Cawnpore, chiefly from the Banda District, and is used for loosening the hair and flesh of hides and killing the grease. It should be taken in lump, unslaked.

Babul Bark. 170

"*Babul* bark is obtained locally, the wood being extensively used for firing. The bark season extends from January to June, that is, the spring of the year, when the sap is upon the trees; from six to ten-year old trees are the best for bark. The bark should be peeled from the trees immediately after they are cut down. The natives are rather careless in this particular, and greatly injure the bark, for, to get it separated from the wood, they beat it with wooden mallets, and gash it about to get it loose; they then peel it off by hand; each gash is a wound in the bark through which the tannin escapes. In England a peeling iron is used, and long strips of bark are taken off without any beating, but of course this must be done before the sap has dried. The tannin is contained in the white or inner stratum of the bark. The tannin is stronger in babul than in oak bark, but the quality is not considered so good.

"Experiments tried lately in England by **Professor Abel**, the Chemist to the War Department, proved *babul* to have keeping qualities quite equal to oak bark, if not superior; it is thus a valuable tanning agent. But it has more colouring matter, that is, it gives a reddish liquor, and this is somewhat against it, but a great advantage is found from mixing with it, *hurr*, or *bahaira* (the myrobalan of commerce—the dried fruit of **Terminalia Chebula**,), which is plentiful in the market, and is used extensively as a dye. The liquor from the *hurr*, or *hurra*, is a powerful tan, and though it is not reported to be of a quality that would make good leather of itself, it is highly esteemed in England to mix with other tanning agents, owing to the bright colour it imparts, and herein is its usefulness in combination with *babul*. The *hurr* is a product of the forest, and is very common, but the natives of Cawnpore use it only as a dye. *Sumach* is another tanning substance, which has been used in small quantities at this tannery, but only for finishing and imparting colour. It is imported from England and is therefore very expensive.

Myrobalan. 171

Sumach. 172

Terra Japonica or Kutch. 173

"Cutch is a very powerful astringent and rich in tannin. It is the inspissated juice of **Acacia Catechu** (the *khair* of the forests), and is used by the natives as a dye and is also eaten with *pán*; its tannin is three or four times stronger than that of oak, but of poor quality; the leather made of it is of dark colour and does not last well, but tans so quickly and therefore so cheaply, that it is used extensively in England: such leather is believed to be unsuited for wear in this climate. At Cawnpore the Cutch is employed only for darkening colour when that is required.

Divi-Divi. 174

"*Divi-Divi* is the pod of a shrub that is a native of South America, but grows well in the Madras and Bombay Presidencies; it is called the **Cæsalpinia coriaria.** The pod is exported from Bombay to England, but in small quantities. The tannin is strong and considered good for mixing with barks. The shrubs were grown at Cawnpore from seed and a plantation of about 6,000 or 7,000 trees is doing well. The barks of *Sál*, *Asaina*, and *Amultas* are known to contain tannin, and experiments are being carried out with them. (For a complete *résumé* of available information

Other tanning barks. 175

TANNING.

regarding Indian tans the reader is referred to the article on '**Tanning materials**, Vol. VI.)

Cod-oil. 176 "The only oil used for currying is cod-oil; it is obtained from England, and is the best known for the purpose. Indian fish-oil would answer, if more carefully extracted. Mutton or goat tallow is used either alone or mixed with cod-oil; in the latter state, it is called "dubbing," which is applied to all leather intended for harness straps, or like pliable purposes.

Tallow dubbing. 177

OPERATIONS OF THE TANNERY.

Labour. 178 "Native tanners and curriers are all of the *Chamar* caste; the former are *Runghias*, a trade class of that caste; such labour is very plentiful about Cawnpore. The wages vary from R5 to R9 per mensem according to skill.

Liming. 179 The hide when received green is only washed and put into a pit of milk of lime, but if it has been salted and dried out, it has to be soaked in soft water before it can go into lime. It is often so hard dried that it will not soak down, and heavy fulling stocks are required to break it down and soften it ready for the liming process.

Unhairing and fleshing. 180 "The hides are first put into weak lime and then into stronger, until the hair is loosened and the skin 'plumped' up; they are then placed over tanner's beams made of cast iron and convex, so that a two-handled, blunt, concave knife can be worked over them to push off the hair. When this has been accomplished, the hide is turned over on the same beam, and the flesh and fat which was left on and has now become loosened by the action of the lime, is shaved off by means of a two-handled, concave, fleshing knife, which has sharpened edge, and takes off a greater or less shaving as required. Some dexterity is needed in this work, as a slip of the knife may shave too deep and cut into the hide. The most skilled tanners are employed in this, and called 'beamsmen.'

Use of Lime. 181 "There is great diversity of opinion in the trade as to the use of lime. It is generally allowed that it is an evil but a necessary one; it is not good for leather, but the hair and flesh must be removed, and there is no safer way of doing so. In the French and in some of the American tanneries the hair is removed by sweating the hides and producing partial decomposition, but this would be too dangerous in a climate like that of India.

Graining or Bating. 182 "Lime being an evil, it becomes necessary, after it has done its work, to rapidly obliterate all traces of it in the texture of the hide, and this forms the second process of the tanning and is called 'bating' Sole leather does not require this process, but for all harness and dress leather and for every soft purpose it is essential. The hides after being unhaired, fleshed,

Bran grainer. 183 and washed, are thrown into a pit called a 'grainer.' in which bran and water have been allowed to ferment; the acid thus produced removes the lime from the texture of the skin and loosens and distends the pores, so that they are cleaned of every foreign substance and brought into a state highly sensitive to the action of tannic acid.

"Much caution is needed in India in the management of grainers, for if the hides are left a few minutes too long, the acid dissolves the gelatinous materials, the action being quicker than in England owing to the higher temperature.

Pigeon or hen dung grainer. 184 "Bran being sometimes expensive, the ammonia from pigeon or hen dung has been tried with some success; this is a usual grainer in England, but there have been difficulties in collecting sufficient quantities of it at Cawnpore, and even when procured, it is often so full of extraneous substances that bran is preferable.

Stale liquor grainer. 185 "The native tanners of Cawnpore use a grainer of stale fermenting tan liquor, and this is so inexpensive that it is used for all heavy leather in the Government tannery. It is notable that the French use stale rotten liquors

TANNING.

to bring down or soften their skins, and they are most successful in their soft pliable leathers.

Sulphuric acid grainers. 186

"Weak solutions of sulphuric acid have been tried; to plump up the skins; but the acid is expensive, and all the above mentioned grainers give more or less the same result. An experienced tanner can tell by feeling the hides whether they have been brought down enough, and are fit for the next process, which is the tanning proper.

"After the process of bating has been gone through and before being taken to the tan pits the hides are once more put on the beam, and all dirt, short hairs, &c., removed. A new and recently patented process of bating is being tried in the Harness and Saddlery Factory at Cawnpore. The process consists simply of using carbonic acid gas to swell the hides and get rid of lime and greases by chemical action. The gas is generated in the usual way from whiting and sulphuric acid; and is introduced into a pit where the hides and skins are in water. After an hour or an hour and a half, the chemical action separates and deposits the lime, and the greases come out in the form of soap; and by scudding the hide over a beam, all the impurities are dissipated, and the hide is far better prepared for the tan yard than by any of the old processes.

Bark grinding. 187

"Babul bark is broken and ground in a bark mill with breakers. The latter breaks the fibres of the bark, and the mill crushes and grinds it, thus rendering it easy of infusion in cold water and facilitating the extraction of tannin. Three mills with breakers are used at Cawnpore for this purpose, which are driven by a powerful steam engine.

"Myrobalans are ground and crushed in a disintegrator which is specially made for the purpose.

Tanning proper 188

"First infusions are made in the bark-taps above described, with fresh cold water. Soft water is of course the best both for this and for soaking hides in during the first stages in the beam house, but water should be free of lime even for the tanning process.

Infusing bark. 189

"English tanners who produce cheap articles infuse with hot water; but this is not considered good as it extracts too much of the colouring and other earthy matters, and hastens the development of gallic acid which is injurious to the liquors.

Spending bark. 190

"After a pit of bark has given off a liquor, the half-spent material is cast over into one of the spenders, where afterwards another half-spent liquor is put on it, the mixture being freshened up and so on till the bark is quite spent, when it is thrown out. The liquors are continually worked over and over again to spend the bark.

Handling. 191

"The hides are first put into the oldest or weakest liquor. In the state in which they come from the grainer they should not be subjected to the too sudden action of tannin; they are therefore continually handled or taken in and out of the pits. The pits are called 'handlers.' For the first month the hides are handled daily, then gradually allowed to remain longer and longer till they are fit for the next stage, which is called dusting. The pits are called 'dusters.' The hides are put into a stronger liquor with some finely ground bark thrown in between each layer of four or five hides, to keep the strength of the liquor up as the hides suck in the tannin.

Dusting. 192

"In the dusters, the hides are taken up weekly, and then fortnightly, when new liquors are supplied and fresh bark added.

Laying away. 193

"The dusting goes on for three or four months, and then the hides are ready for laying away. The pits are now called 'layers,' and the hides are put into a new strong liquor, with quantities of fine bark between each. In these pits they lie for a month or two at a time absorbing the tannin. The half spent liquor from layers, dusters, and handlers are always drained off on to the top of the spenders, and these after being freshened up are

TANNING.

brought up again by means of the pump; thus they are kept moving, which is very important, as it retards the formation of gallic acid which is injurious to the tannin, and which is but too apt to form in hot weather.

"Babul bark and myrobalans are infused together, and the addition of the latter adds great strength, and takes away from the deep colour of the liquors.

Time taken in tanning.
194

"The hides are kept three or four months in layers, and by that time in general the tannin has penetrated right through them, and they are tanned. Buffalo hides take from nine months to a year from first to last according to weight and substance; cow hides are tanned in from four to six months; sheep and goat skins are done in tubs, and take from a month to two months.

"The climate of India is favourable to the penetration and combination of tannin with the gelatine of hides. The chief danger to be guarded against is the fretting of liquor and production of gallic acid, a compound more readily produced in Indian Tanning owing to the atmospheric heat.

"After the hides are tanned, they can be dried and stored, but in this country they are apt to dry so hard, and to darken so much by the action of light and air that it is thought best to curry them at once, especially for harness work or dress leather. For sole leather or crop hides no currying is required, but they are struck, that is, pressed down and rolled, and then stored.

NATIVE PROCESS OF TANNING.

Native tanning.
195

"Native tanners carry out the liming process very much like that described; they, however, rather overlime the hides according to English ideas. Their grainer is very effectual as far as softening the skins; they then rinse out all moisture, and, while the pores are yet opened, they work the hides in liquor, and afterwards sew the hide up into a bag having one end open. This bag is filled with finely pounded bark and hung up over a pit from which liquor is ladled into the bag, and there a very strong infusion is formed which percolates through the pores of the hide. This tannin therefore goes right through the hide, but it has not time to make the chemical combination, which is considered so essential for the leather. Native tanning does not occupy more than a week.

CURRYING.

Currying.
196

"For harness and all dressing purposes, the currying is a most necessary and important process. By tanning, the hide has become leather, but it is not fit for use without currying. This consists of a number of manipulations, *e.g.*, stoning, shaving, scouring, oiling, reshaving or flattening, setting, resetting, dubbing and finishing.

Stoning.
197

"For ordinary harness leather a buffalo hide, as it comes from the tan pits, is hung up till it is half dry, when it is stoned out with a rough stone on a flat table, to straighten it as much as possible.

Shaving.
198

"It is then thrown over a currier's beam, an upright thick plank faced with **Lignum vitæ**. This is the beam board for shaving on. The hide is smoothed down the board, and the currier, with a two-handled shaving knife, having a turned edge, takes off the outer flesh and inequalities from the flesh side, in fact, levels the hide as nearly as he can do so, having regard to economy in not wasting leather. The object of this shaving is to get the hide to lie flat on a table, so that the scouring stones and sleepers may touch each part of the surface with equal pressure and thus thoroughly clean it, as well as press out all dirt and bloom, or earthy deposit from the bark that remains in the tissue.

Preparing the shaving knife.
199

"The preparation and use of the currier's shaving knife requires skilled men, each man sharpens and turns the edge of his own knife. He first

CURRYING.

rubs it and clears a perfect straight edge, then places his knife, edge upwards, before him on the floor, holding it between his knees, and resting it against the wall; in this position he takes in both hands a heavy steel and presses it along the straight edge, with more or less pressure as required to turn the edge. This edge catches the lumps and inequalities of the flesh side and shaves them off. When shaved the hide is laid flat on a stone table, and undergoes a series of scourings and pressings out on both sides, to remove bloom and dirt and take out crease. After this it is steeped for two days in sumach or myrobalans to give it a finishing colour. These infusions are made with boiling water.

Scouring. 200

"The hide is then taken to another table, and sleeked out on both sides with a steel sleeker; then while still wet, cod-oil is rubbed lightly on the flesh, and more heavily on the grain. It is then hung up to allow the oil to be drawn into the pores of the skin as the water evaporates.

Steeping in Sumach water. 201

Oiling. 202

"When about three parts dry it is passed through a stoning machine and pendulum roller; this is done to make the leather firm and to get out the stretch, and, then, for a second time, it is placed on a currier's beam to be lightly re-shaved on the flesh side, that is, the roughness is taken off without sacrificing much of the leather. This process is called flattening.

Flattening. 203

"After this it is hung up again to dry a little more. It is afterwards stoned out on a table. Then the table or sufficient portion of a table for the hide to lie on is rubbed over with dubbing (a mixture of oil and tallow) The hide is laid on the dubbed table, grain upwards, so that the flesh adheres to the table; in this state it is well set out, that is, the hide is stretched out as much as it will go by pressure with a hard setting stone with a smooth edge. To do this the hide is damped down in any place in which it may have dried too much.

Setting on dub. 204

"When fairly set, it is allowed to dry a little, then stoned out on the flesh side, and afterwards reset on the grain side, to prepare for the dubbing which is now laid on with a brush, rubbed in and smoothed down with the palm of the hand.

Resetting. 205

"Care must be taken that the hide is equally damped all over before it is 'dubbed,' otherwise the dry parts will remain dark in colour. In this dry climate it is very necessary to watch that hides do not get too dry in the processes of 'currying.' When dubbed, the hides are hung up to dry, and when the grease has gone well in, leaving a white coating of refuse-tallow on the outside, they can be taken down and finished off, which is done on a table. First, the flesh side is smoothed with a pebble or glass, then the grain is 'sleeked' with a fine 'sleeker' to remove the coating of tallow, then sleeked with the pebble and glass to make smooth and bring up a gloss; the hide is then dried out. After all this, the hide is ready for the saddler and harnĕss maker. It is prepared for various other purposes with more or less dubbing, as the leather may be required for soft or for hard purposes.

Dubbing or stuffing. 206

Finishing. 207

"Cow hides do not require so much setting as buffalo, but they have oiling and dubbing in proportion to their thickness. For very soft uses the cow hides are softened by hand labour with a crippling board which loosens the grain.

Currying Cow Hides. 208

"Sheep and goat skins are oiled and then set, and dubbed, hung up, and dried out, after which they are softened.

Sheep and Goat skins. 209

"The following tools are used in tanning:—tongs for lifting hides out of limepits; unhairing, fleshing, and rounding knives; tanner's hooks for lifting hides out of tanpits; scudding knives; striking pens.

Tools for Tanning 210

"The following tools are used in currying:—currier's shaving knives; sharpening and turning steels; scouring, pumice and setting stones; scour-

and for Currying, 211

CURRYING.

ing, finishing, and glass sleekers; pebble; stuffing brush; crippling board, rub-stone and clearing-stones for sharpening knives.

NATIVE CURRYING.

Native currying. 212

"The natives neglect the currying process of their leather. There is no trade corresponding to the currier's in this country. The best harness-makers curry their leather before cutting it up into straps, in fact they curry it in strips of a foot wide, and apply tallow only; very often they apply nothing but buttermilk, which gives the leather a very bright appearance, but the moisture very soon evaporates, and the leather becomes hard and brittle."

Native Leather.

BEST KINDS OF NATIVE LEATHER,

As above stated the usual product of the Indian tanner is a hard, imperfect, thick leather which is very pervious to water, and is not at all durable. In certain localities, however, fancy leathers are produced of a very good quality. Perhaps the best of these is the soft wash-leather of the Kangra and Hushiarpur districts of the Panjáb, and the red skins of Núrpur. The former is of a very soft texture, of a greenish buff colour, and so thin and pliable as to be made into gloves, or the breeches which are constantly worn by the Kangra hill-men in the colder months. The method of preparation is described by **Baden Powell** as follows:—"First, the skin is wetted, and then steeped in a *matka* or earthen pot full of lime and water, the mixture to be sufficient in quantity to cover the skin completely. Every day the earthen pan, which is kept carefully closed over, is opened and the skin turned and shaken in the lime water; if the skin is thin, twenty days of this treatment suffice; if thick, one month. After this the skin is washed in clean water, and then well rubbed with the dried and powdered leaves of the *dhao* tree (**Anogeissus latifolia**, *Wall.*) for two hours successively. After the rubbing is over the leaves and skin are put together in a vessel and water added. Next day the skin is tied on a stick and wrung out, (as a *dhobí* does clothes). Again it is steeped in a fresh solution of the leaves, and this process of steeping and wringing out is repeated for four days consecutively; fresh leaves are to be used each time. The skin is then sewed up with a *munj* string, into the form of a hollow sack, filled with *dhao* leaves, and hung up. On being taken down it is reversed and hung up again by the other end; this ensures both ends being well impregnated with the solution—the process occupies two days. The skin is then opened out, dried, well rubbed with oil, and eventually washed. When dry it is scraped with an iron implement called *rambi*, and is thereafter again rubbed with oil. After three or four days it is washed in cold water and dried. It is then rubbed with a mixture of curds and water and again dried, the process being thus completed. The skin chiefly used is that of the *sábur* or *sambar* deer, but others, such as those of the goat, buffalo, &c., may be employed." A similar method is followed in the preparation of skins with the hair on, with excellent results, care being taken to avoid injuring the fur.

The coloured leather of Núrpur is, according to **Baden Powell**, prepared as follows:—"The leather is of two kinds, thick and thin. The thickest is that prepared from buffalo hides, next that from the hides of cows and bulls. The thin leather is made from the skins of goats, sheep, rams, and deer. The hides of horses, asses, and camels, are also occasionally utilised. The thick kinds are made of two colours, black and red; and the thin of four—black, yellow, red, and green; the red or crimson is the commonest. The method of preparing the small skins is as follows:—The skins are rubbed when fresh with two chittacks of salt, and put in the sun; in one day they dry. They are then washed, rubbed with wood ashes, and the hair, thus loosened, is scraped off with a piece of wood. The skins

NATIVE LEATHER.

are then put into water, rubbed with pieces of rough potsherd, which completely removes the hair, &c., and again washed. When forty skins have been prepared thus far they are together put into a great earthen cauldron or *nánd*, along with barley-meal 6 seers, salt 2½ seers, and water is poured on. They are left to soak in this for four days, then 2 seers of *bankath* (coarse catechu) are added. The soaking is continued for four or five days longer, after which the skins are taken out and again scraped with an iron *khurpa* or scraper. They are again soaked in a *nánd* with 7 seers of barley-meal and 3 seers of salt dissolved in fresh water, for a period of three to four days. The process of dyeing is now performed by soaking the skins for three or four days in solutions prepared as follows :—Four seers of lac are boiled in water with 2 chittacks of *sajji* (coarse soda) and two chittacks of the bark of the *lodar* tree (**Symplocos ? cratægoides,** *Ham.*). From this solution the skins are removed to an infusion of the bark of *amaltas* (**Cassia Fistula**) and allowed to soak for three or four days; they are then removed and the colour fixed by rubbing the skins with powdered salt The cost of preparing 40 red skins is R12-9-6." The leather thus prepared is employed in the manufacture of belts, ornamental native shoes, saddlery, &c., for an account of which the reader is referred to works on Indian Manufactures and Art Industries.

A soft leather of very fine quality is prepared to a small extent in Gujrat, for the purpose of covering the easy chairs known as " Gujrat " chairs. It is probably manufactured by a more or less satisfactory imitation of the European process of softening, dyeing, and polishing leather.

English leather, manufactured and unmanfactured, is largely imported, as are Russia leather and several ornamental leathers from Afghánistán.

TRADE.

Internal. 213

Trade.—The following passage occurs in the Review of the Inland Trade of India for 1888-89 :—" Having regard to the consumption of so large a population as that of India, the inland trade in leather manufactures, valued at 83 lakhs of rupees in 1888-89, seems altogether disproportionately small. The apparent discrepancy is, however, readily accounted for by the communal institutions of the people. Each village has its own workers in leather, who are also, to a large extent, their own tanners; and it is part of their recognised duties to keep their patrons in boots, and to cure and make up the hides required for the leathern buckets made for irrigation. The needs of the agricultural population being thus locally supplied, without the intervention of traders, the traffic is practically restricted to the demands of the city classes and the requirements of the army.

"There are several large towns which enjoy a special reputation in the leather goods trade. Delhi, Agra, Lucknow, Benares, Patna, Poona, &c., are manufacturing, collecting, and distributing centres for native shoes. Cawnpore, the seat of a large Government leather factory, is noted for its harness, saddlery, and European boots. Calcutta supplies the cities of Bengal with shoes in the European style, which have supplanted those of old fashion in the Eastern Provinces." Mr. O'Conor (*Statistical Tables for British India, 1888-89*) states that 48 tanneries are returned in all, of which 30 are in Madras, 11 in Sind, 3 in Calcutta, 2 in Cawnpore, and 2 in Bangalore. The estimated value of the outturn, as far as known, from these concerns in the same period was R24,09,895. The 15 principal tanneries employed an average daily number of 2,224 men, out of which one Boot Factory in Cawnpore employed 1,388. A review of the Rail and River-borne Provincial Trade Returns for the same year shows that the total quantity of manufactured leather thus transported amounted to 71,768 maunds, valued at R82,72,625. Of this amount Calcutta exported 14,137 maunds, of the considerable value of R29,97,044; the North-West Provinces exported a much larger quantity, 20,237 maunds, but of markedly less proportionate

LEATHER. **Coasting and Foreign Trade.**

TRADE. Internal.

value, R19,22,515. Similarly, the exports from the Panjáb are seen to be large, 17,928 maunds, and again of considerably less proportionate value than those from Calcutta, amounting only to R10,03,968. Following on these were Bombay Town, Bengal, Sind, Madras Seaports, Bombay, Karáchi, Madras, Mysore, Rájputana and Central India, the Central Provinces, the Nizam's Territory, Berar, and Assam, named in order of importance.

The largest importing provinces was Bengal, with 24,537 maunds, valued at R36,25,089, followed by Calcutta town with 11,703 maunds, value R10,86,074. Next in order of importance were the North-West Provinces and Oudh with 8,390 maunds, value R7,85,971, the Panjáb witr nearly the same amount, then Sind, Bombay, Rájputana and Central India, the Central Provinces, Madras Seaports, Bombay Town, the Nizam's Territory, Assam, Berar, Madras, Karáchi, and Mysore with much smaller quantities.

Coasting. 214

COASTING TRADE.—The trade in foreign leather is not recorded, apparently because it is too unimportant to be worthy of notice. In Indian leather, the average imports value somewhat less than a lakh of rupees, and consist chiefly of unmanufactured leather, transhipped from Bengal and Madras to Burma. The average imports of boots and shoes (Indian) are 1¾ lakhs of rupees, and chiefly comprise supplies from Bengal to Burma.

Foreign Imports. 215

FOREIGN TRADE.—During the past ten years the imports of Foreign Leather show a marked increase, notwithstanding the development in the home manufacture of articles for European use from country leather. Thus during the five years ending 1879-80 and that ending 1888-89 the imports were—

	1875-1880.		1884-1889.	
	Cwt.	R	Cwt.	R
Unwrought Leather .	740	1,01,865	1,514	2,46,873
Leather Manufactures	...	8,01,110	...	12,09,173
TOTAL .	...	9,02,975	...	14,56,046

The principal market from which both unwrought and manufactured leather is imported is the United Kingdom. For example, in the last year for which statistics are available, 1888-89, out of a total value of 16½ lakhs of rupees, over 15 lakhs represented the imports from the United Kingdom. Small quantities are also obtained from the Straits Settlements, France, Australia, Austria, Germany, Belgium, Ceylon, and other countries. A large proportion of the leather manufactures (6 lakhs of rupees in value in 1888-89) consists of Saddlery and Harness. The distribution of the imports in 8881-89 was as follows :—

		Unwrought.	Saddlery and Harness.	Manufactures, other sorts.
Bengal	Lakhs R	1'19	2'49	'84
Bombay	,,	1'97	2'32	5'17
Sind	,,	'06	'32	'21
Madras	,,	—	'36	'65
Burma	,,	'48	'44	'19
	Total Lakhs R	3'7	5'93	7'06

The imports in 1889-90 (details not given) were valued at 18·48 lakhs of rupees, against 16·69 in 1888-89. The figures of trade here quoted are exclusive of those of boots and shoes, the imports of which average about 12 lakhs rupees. The United Kingdom is the largest supplier of these, and Bengal and Bombay take by far the greatest share between them. **TRADE. Imports.**

FRONTIER TRADE.—The imports from across the frontier average 1·20 lakhs of rupees. The average exports are about R2¼ lakhs (chiefly manufactured leather): the trade is mainly between Kábul and the Panjáb. **Frontier. 216**

The RE-EXPORT TRADE is small and unimportant. The average for the five years ending 1883-84 was 25 cwt. of unwrought leather, valued at R14,566, and manufactured leather, value R30,610. During the past five years the average in both cases was smaller, *viz.*, 21 cwt. of unwrought leather, valued at R2,721, and manufactured, valued at R19,023. The re-export trade is principally between Bombay and the United Kingdom. **Re-export. 217**

EXPORTS OF INDIAN LEATHER.—The exports of unwrought leather have slightly increased in quantity but diminished in value during the period under consideration; the exports of manufactures have slightly increased in value. Thus, in the five years ending 1883-84, the average was 184 cwt. of unwrought leather, value, R14,566, and manufactures to the value of R22,295, while during a similar period ending 1888-89, the average was 291 cwt., value R14,382, of unwrought leather, and the manufactured article to the value of R26,368. In the above statistics a small quantity of saddlery and harness, annually exported as Government stores to Aden, have not been included. In 1888-89 the value of this amounted to only R119. In conclusion it may be noticed that the above figures show a steady increase in the amount of leather imported from Europe, notwithstanding the concurrent development of the Indian tanning industry. Also, that the amount of unwrought leather imported has increased in a higher ratio than the amount of leather manufactures, thus shewing a development in the demand for European leather, presumably intended to be manufactured in this country. **Exports—Indian. 218**

LEBIDIEROPSIS, *Müell. Arg.; Gen. Pl., III., 268.*

A genus of EUPHORBIACEÆ reduced in the *Genera Plantarum* to **Cleistanthus,** *Hook. f.*

Lebidieropsis orbicularis, *Müell.-Arg.;* EUPHORBIACEÆ. 219

This species is reduced to **Cleistanthus collinus,** *Benth.*, in the *Flora of British India, V., 274.* **Sir J. D. Hooker,** however, remarks that "the genus **Lebidieropsis** should probably be restored; the globose seeds differ wholly from those of any other **Cleistanthus.**" When the description of the genus **Cleistanthus** was written in Vol. II. of this work, this species was inadvertently omitted, all the economic information regarding the plant in India being described by **Beddome, Gamble, Brandis,** and others under **Lebidieropsis.** The description may, therefore, for the purposes of this work, be given here.

Syn.—BRIDELIA COLLINA, *Hook. f. & Arn.;* AMANOA COLLINA, *Baill.;* LEBIDIEROPSIS COLLINA, *Muell.-Arg.;* CLUYTIA COLLINA, *Roxb.;* C. RETUSA and PATULA, *Wight.;* ANDRACHNE ORBICULATA, *Roth.;* A. CADISHAN, *Roxb.;* EMBLICA PALASIS, *Ham.*

Vern.—*Garrar, garári,* HIND.; *Karada,* URIYA; *Parasu, pas, pasu, larchuter,* KOL.; *Kargalli,* SANTAL; *Ghara,* BERAR; *Kergali,* KHARWAR; *Ganari,* C. P.; *Garari,* MAR.; *Wodayu, waddan, wúdacha-marum,* TAM.; *Korsi, wodesha, kadishen, korse, koricha,* TEL.; *Madara,* SING.

References.—*Roxb., Fl. Ind., Ed. C.B.C., 704; Brandis, For. Fl., 450; Beddome, For. Man., 203, t. 26, f. 5; Gamble, Man. Timb., 358; Thwaites,*

En. Ceylon Pl., 280; Drury, U. Pl., 276; Campbell, Ec. Prod. Chutia Nagpur, No. 7543; For. Ad. Rep. Ch. Nagpur, 1885, 6, 34.

Habitat.—A small deciduous tree, met with in the dry hills of various parts of India; extending from Simla eastwards to Behar, and southward to Central India and the Deccan; rare in Ceylon.

MEDICINE. Bark. 220 Fruit. 221 Leaves. 222

Medicine.—Roxburgh writes, "The BARK or outer crust of the capsule is reported to be exceedingly poisonous." The Rev. Mr. Campbell reports that the FRUIT and bark are employed in Chutia Nágpur to poison fish, and that the latter is also considered a useful application in cutaneous diseases. For severe headache, the head and upper part of the body are bathed in water in which the LEAVES have been steeped.

TIMBER. 223

Structure of the Wood.—Red, tough, moderately hard to hard, close and even grained, warps in seasoning; heart-wood dark red, sapwood small, lighter coloured; weight 54 to 55lb per cubic foot. It resembles the wood of **Eugenia operculata**, and is almost identical with that of **Flacourtia Ramontchi** (*Gamble*). Beddome states that it is much used in India, and since it coppices well and is procurable in great abundance, it may be specially recommended where fuel is required in the more tropical parts of the country. It has also been described as a useful wood for turning.

Ledebouria hyacinthoides, *Roth.*; see **Scilla hyacinthoides,** *Linn.*; [*Vol. VI.*

LEEA, *Linn.; Gen. Pl., I., 388, 999.*

A genus of small trees, shrub, or herbs, belonging to the Natural Order AMPELIDEÆ, which comprises about 25 species; most abundant in the tropics of Asia and Africa. Of these 15 are recognised by the *Flora of British India* as indigenous to this country.

224 **Leea aspera,** *Wall.; Fl. Br. Ind., I., 665;* AMPELIDEÆ.

Syn.—? L. HERBACEA, *Wall. Cat., 6829 and 6824 E & G (in part) under* L. STAPHYLEA.

Vern.—*Kumála,* N.-W. P.; *Kumáli, kúrmáli,* KUMAON; *Holma,* PB.; *Thakya-nway-than,* BURM.

References.—*Brandis, For. Fl., 102; Kurz, For. Fl. Burm., I., 281; Gamble, Man. Timb., 93; Stewart, Pb. Pl., 35; Atkinson, Him. Dist., 307; Ec. Prod., N.-W. P., Pt. V., 93; Indian Forester, X., 325; Agri.-Hort. Soc. of Ind.:—Journ. (Old Series), XIII., 309; Gazetteers:—Mysore & Coorg, I., 58; Bombay, XV., 431; N.-W. P., IV., lxx.*

Habitat.—A herb or small shrub of the Western Tropical Himálaya from Jamu to Nepál, ascending to 6,000 feet; also met with in Oudh and Western India from Khandésh to the Konkan. According to Kurz, it is common in the mixed forests all over Pegu, up to 2,000 feet.

FOOD. Fruit. 225

Food.—Produces a small, black, succulent FRUIT, about the size of a black currant, which is pleasant to the taste, and is eaten by the Natives.

226 **L. crispa,** *Willd.; Fl. Br. Ind., I., 665.*

Syn.—LEEA PINNATA, *Andr.*

Vern.—*Ban-chálitá,* BENG.; *Nalugu,* MALAY.; *Kaletthein, kaphet-theing,* BURM.

References.—*Roxb., Fl. Ind., Ed. C.B.C., 220; Kurz, For. Fl. Burm., I., 280; For. Rep. Pegu, App. C., iv; Gamble, Man. Timb., 93; Dymock, Mat. Med. W. Ind., 2nd Ed., 187; Pharmacog. Indica, I., 365; Gazetteer, Bomb., XV., Pt. I., 430; Agri.-Hort. Soc. of Ind.:—Transactions, VII., 53; Journals (Old Series), VI., 36; IX., 411.*

Habitat.—A tall shrubby perennial, met with in the Sikkim Himálaya and Khásia mountains, from 1,000 to 4,000 feet; also in Sylhet, Assam, Chittagong, the Konkan, Pegu, and Martaban.

MEDICINE Leaves. 227

Medicine.—The Rev. J. Long states (*Agri.-Hort. Soc. of Ind. Journal, IX.*), that the LEAVES when bruised are employed in Bengal as an application to wounds. Dymock, in his description of **L. macrophylla**, *Roxb.*, writes, "The TUBERS of **L. crispa**, *Willd.*, are also used as a remedy for guinea-worm, and are said to be more efficient than those of **L. macrophylla.**" MEDICINE. Tubers. 228

Leea hirta, *Roxb.; Fl. Br. Ind., I., 668.* 229

Syn.—L. SCABRA, *Steud.*

Vern.—*Kákjanghá,* HIND., BENG.; *Sura pádi,* TEL.; *Nagamauk, nágámouk,* BURM.; *Kákajanghá, kákángah,* SANS.

References.—*Roxb., Fl. Ind., Ed. C.B C., 220; Kurz, For. Fl. Burm., I., 281; For. Rep. Pegu, App. C., iv; Gamble, Man. Timb., 93; Elliot, Fl. Andhr., 170; U. C. Dutt, Hind. Mat. Med., 301; Pharmacog. Indica, I., 365; Campbell, Ec. Prod. Chutia Nagpur, No. 8136.*

Habitat.—A shrubby perennial, met with in the Sikkim Himálaya, Assam, Sylhet, the Khásia mountains, Eastern Bengal, Chittagong, Pegu, and the Andaman Islands. The **Rev. A. Campbell** reports having found it also on Parisnath Hill in Behar.

Medicine.—The authors of the *Pharmacographia Indica* write, "**Leea hirta,** *Roxb.* (*Kákájangha*) is also used medicinally," but they neither give an account of its properties, nor state in what part of the country it is utilised. No information on the subject appears to be obtainable from the works of other Indian writers; but it may be safe to assume that in this, as in other species of **Leea**, the TUBERS and STEMS are probably astringent and mucilaginous. MEDICINE. Tubers. 230 Stems. 231

L. macrophylla, *Roxb.; Fl. Br. Ind., I., 664; Wight, Ic., t. 1154.* 232

Syn.—LEEA LATIFOLIA, *Griff.*; L. SIMPLICIFOLIA, *Griff.*

Vern.—*Samudraca, dholsamudra,* HIND.; *Dholsamudra,* BENG.; *Hathan,* SANTAL; *Dinda,* BOMB.; *Dinda,* MAR.; *Kya-bet-gyi, kyah-phet-kyi,* BURM.; *Dholasamudriká,* SANS.

References.—*Roxb., Fl. Ind., Ed. C.B.C., 219; Griff., Ic. Pl. Asiat., 645, f. 1; Brandis, For. Fl., 102; Kurz, For. Fl. Burm., I., 278; For. Rep. Pegu, App. C., iv.; Gamble, Man. Timb., 93; Dalz. & Gibs, Bomb. Fl., 41; Pharmacog. Indica, I., 364; U. C. Dutt, Mat. Med. Hind., 297; Dymock, Mat. Med. W. Ind., 2nd. Ed., 187; Drury, U. Pl., 276; Lisboa, U. Pl. Bomb., 151, 291; Campbell, Ec. Prod. Chutia Nagpur, No. 8770; Mason, Burma and its People, 503, 742; Works of Sir W. Jones, V., 92; Agri-Hort. Soc. of Ind., Transactions, VII., 53; Gazetteers:—Bombay, XV., Pt. I., 430; N.-W. P., IV., lxx.*

Habitat.—A large herbaceous perennial found throughout the hotter parts of India, from the Tropical Himálaya as far west as Kumaon, to Bengal, Assam, and Tenasserim; also met with in Western India.

Dye.—**Roxburgh** writes, "The ROOT promises to yield a colour fit for dyeing," but he does not state whether that remark is the result of personal observation, nor for what reasons he was led to make it. The tubers are of a deep red colour and very mucilaginous, but no record exists of their ever having been utilised for dye-producing purposes. DYE. Root. 233

Medicine.—**Roxburgh** says that the ROOT is astringent and a reputed remedy for ring-worm, and **Dymock** adds that it is employed in the cure of guinea-worm, and when pounded is applied to obstinate sores to promote cicatrization. The **Rev. A. Campbell** reports that in Chutia Nagpúr, it is supposed to have anodyne properties, being applied externally to allay pain. **Mason** mentions the plant, and states that it is sometimes cultivated in Burma for the astringent properties of its roots, the Burmans using it as an application to wounds to stop the effusion of blood. MEDICINE. Root. 234

Food—The LEAVES were used as an article of food in the Násik district during the famine of 1877-78. **Lisboa** states that, cooked as a vegetable, they are regularly eaten in the Konkan. FOOD. Leaves. 235

SACRED. Leaves. 236 **Sacred Uses.**—**Lisboa** writes, "The large LEAVES of this plant are used as platters for food every Monday during the month of *Shrávan*."

237 **Leea robusta,** *Roxb.; Fl. Br. Ind., I., 667.*

Syn.—LEEA ASPERA, *Wall. Cat., 6825 in part;* L. COMPACTIFLORA, *Kurz.*

Vern.—*Haramada, hatkan,* SANTAL; *Galeni,* NEPAL; *Pantom,* LEPCHA; *Gino,* GOA; root=*ratanhia,* PORT. AT GOA.

References.—*Roxb., Fl. Ind., Ed. C.B.C., 220; Kurz, in Jour. As. Soc. Beng., 1873, II., 65; Gamble, Man. Timb., 93; Dymock, Mat. Med. W. Ind., 2nd Ed., 186; Atkinson, Him. Dist., 307; Campbell, Ec. Prod. Chutia Nagpur, No. 7553.*

Habitat.—A large robust shrub, native of the Sikkim Himálaya and Khásia mountains, at altitudes of 1,000 to 5,000 feet; also Martaban, the Northern Circars, and Western India.

MEDICINE. Root. 238 **Medicine.**—**Dymock** writes, "The ROOT of this **Leea,** which is a common undershrub in jungles in the Goa territory, is called *ratanhia* by the Portuguese, and is much used in diarrhœa and chronic dysentery." In the more recent *Pharmacographia Indica,* however, the same information is given regarding **L. sambucina,** *Willd.,* and no mention is made of **L. robusta.** It may, therefore, be presumed that the above remarks by **Dr. Dymock** were inadvertently applied by him in his earlier work to the wrong species.

Campbell says that in Chutia Nagpur the soft and fleshy root is applied externally as an anodyne, and is also given to cattle for diarrhœa.

TIMBER. 239 **Structure of the Wood.**—Moderately hard; the stems are used for fencing, for stakes, and for building temporary huts (*Gamble*).

DOMESTIC. Stems. 240 **Domestic.**—The dry STEMS are sometimes used as torches.

241 **L. sambucina,** *Willd.; Fl. Br. Ind., I., 666; Wight, Ic., 78; Ill., [I., 58.*

Syn.—LEEA ÆQUATA, *Wall. (not of DC.);* L. ACUMINATA, *Wall.;* L. GIGANTEA, *Griff.;* L. LÆTA, *Wall.;* L. OTTILIS, *DC.;* L.? VAR. SANGUINEA, *Wall.;* L. STAPHYLEA, *Roxb.;* AQUILICIA SAMBUCINA, *Linn.;* A. OTTILIS, *Gærtn.;* GILBERTIA NALUGU, *DC.;* GUSTONIA NALUGU, *Spreng.;* OTTILIS ZEYLANICA, *Gærtn.;* STAPHYLEA INDICA, *Burm.*

Vern.—*Kurkur-jihwa,* HIND., BENG.; *Dino,* GOA; *Karkani,* MAR.; *Ankados, ankadósa,* TEL.; *Nalúgu,* MALAY.; *Kalet, nágá-mouk,* BURM.; *Búrúla-gass,* SING.

References.—*Roxb., Fl. Ind., Ed. C.B.C., 221; Griff., Notul., IV., 697; Brandis, For. Fl., 102; Kurz, For. Fl. Burm., I., 279; in Jour. As. Soc. Beng., 1873, II., 65, 66; Gamble, Man. Timb., 93; Thwaites, En. Ceylon Pl., 64; Dalz. & Gibs., Bomb. Fl., 41; Rheede, Hort. Mal., II., 43, t. 26; Elliot, Flora Andhr., 15; Pharmacographia Indica, I., 363; Atkinson, Him. Dist., 307; Mason, Burma and Its People, 742; Agri.-Hort. Soc. of Ind.:—Transactions, VII., 53; Journals (Old Series), VI., 36; Gazetteers:—N.-W. P., IV., lxx.; Bombay, XV., Pt. I., 430.*

Habitat.—A glabrous shrub with stout soft-wooded stems; widely spread throughout the hotter parts of India, from the Himálaya (as far west as Kumaon), to Burma, and southwards to Ceylon and Malacca.

MEDICINE. Root. 242 **Medicine.**—**Rheede** calls the plant *Nalugu* and gives *Dino* as its Brahminic name. He states that a decoction of the ROOT is given in colic, and that it is cooling and relieves thirst. Under this species, the authors of the *Pharmacographia Indica* give the information quoted from **Dymock** above, under **L. robusta.** They, however, add, "The roasted LEAVES are applied

Leaves. 243 to the head in vertigo; the juice of the young leaves is digestive. . . . In Réunion the root is called *Bois de Sureau,* and is said to be used as a sudorific."

TIMBER. 244 **Structure of the Wood.**—"Rather heavy, close-grained, soft, pale-brown, turning darker, with a silvery lustre, the pith medullary, brown, small; soon attacked by xylophages" (*Kurz*).

LEECHES.

The LEECHES employed in medicine belong to the family GNATHOBDELLIDÆ of the Natural Order HIRUDINEÆ, of which they form the typical genus, HIRUDO.

Leeches. 245

Vern.—*Jók, jónk,* HIND.; *Jonk,* BENG.; *Drik,* KASHMIR; *Jala,* GUJ.; *Attai,* TAM.; *Jelagalu, attalu, jeriku,* TEL.; *Jigani,* KAN.; *Atta,* MALAY.; *Miyon, minyon,* BURM.; *Kudallu, púndal,* SING.; *Jalúkaha,* SANS.; *Aalaq, zaló,* ARAB.; *Zaló, zalók,* PERS.

Medicine.—Many distinct species of leech appear to be met with in India, but they have not been scientifically determined. Several are used medicinally and have long been known and valued by Sanskrit writers. **Susruta** gives a detailed account of the different forms and of their habits, modes of application, &c. He describes twelve kinds, six of which, he says, are venomous, six useful. The former live near putrid fish and animals, in foul stagnant pools, and are to be avoided; the latter inhabit clear and deep pools which contain water-lilies and other sweet-smelling plants. Of these, he states, that the middle-sized are the best. He directs that they should be kept in a new water-pot, in which some clay, grass, or fresh leaves of aquatic plants and water have been put, and that they must be fed on the roots of water-lilies and dry flesh. When the leech is sluggish, he recommends that a drop or two of milk should be applied to the part on which it is required to fix, or that a slight scarification should be made. The leech is induced to drop off by sprinkling salt on its head, and is cleaned by stripping. (*U. C. Dutt, Mat. Med. Hindus, 275.*) MEDICINE. 246

The use of the leech was probably adopted from the Hindus by the Muhammadans, with whom it is now greatly in vogue. At the present time, the leeches used in Bengal are chiefly caught in tanks in the neighbourhood of Baraset, by persons who collect them by entering the water and allowing the leeches to catch hold of their legs. They are chiefly gathered in May as the tanks begin to dry up. But the best leeches are said to be obtained from the North-Western Provinces, chiefly from Shekoabad. Large numbers are, however, also caught at Patiala in the Panjáb.

SPECIAL OPINIONS.—§ "Leeches boiled with honey are used as a local application for tonsilitis, and dried and rubbed with aloes as a local remedy for hæmorrhiods" (*Dr. Emerson*). "Boiled and ground up, they are employed as an application to make the hair grow (*Surgeon-Major C. W. Calthrop, M.D., Morar*).

LEERSIA, *Swartz.; Gen. Pl., III, 1117.*

A genus of aquatic grasses comprising five species, of which one occurs in India. They are similar in habit to **Oryza,** but the spikelets are smaller, the glumes (of which there are only two) are also thinner, and there is no pale.

Leersia hexandra, *Swartz.; Duthie, Fodder Grasses of N. India, 21;* [GRAMINEÆ. 247

Syn.—L. AUSTRALIS, *R. Br.*

Vern.—*Layú,* SING.

References—*Symonds, Grasses of the Indian Peninsula, 58, t. 49; Thw., En. Pl. Zeylanicæ, 356; Campbell, Ec. Prod. Chutia Nagpur, No. 8169; Von Mueller, Select Extra-Trop. Plants, 219; Maiden, Useful Pl. of Australia, 93.*

Habitat.—Found occasionally bordering tanks and on wet ground throughout India and Ceylon; distributed to Africa and the warmer parts of America and Australia.

FODDER.
248 **Fodder.—Symonds** states that horses and cattle are fond of this grass, a remark confirmed by Australian writers, who assert that it is one of the most esteemed of the aquatic grasses of Eastern Australia. **Baron von Mueller** mentions that it is regularly cultivated for fodder in the Philippines, in the same way as rice.

249 **Leingang,** a sort of earth found everywhere in Manipur and used for imparting a *khaki* colour to cloth (see **Pigments**); also employed as an article of food (see **Clays,** Vol. II., 361).

250 **LEMNA,** *Linn.; Gen. Pl., III., 1001.*

A genus of water plants, popularly known as "duck-weeds," which belongs to the Natural Order LEMNACEÆ, and comprises seven species, natives of fresh waters of temperate and tropical regions. Of these five are (according to *Kurz, Linn. Soc. Jour., IX., 264*) natives of India, namely, **L. gibba,** *Kurz,* **L. minor,** *Linn.*, **L. oligorrhiza,** *Kurz,* **L. polyrrhiza,** *Linn.*, and **L. trisulca,** *Linn.*

Certain of these species, particularly **L. minor,** frequently invade flooded rice fields, and by their thick growth on the surface of the water are supposed by Natives to do much damage to the crop. **Dr. G. Watt,** in a recent paper on **Adhatoda Vasica,** has shown that the leaves and twigs of that plant are applied to rice land, not only as a green manure, but also for the purpose of killing the aquatic weeds, which would otherwise greatly injure the crop. He writes, "The natives pointed to fields that had been treated in this way; and there could be no mistake that these were clean or free from the green scum caused through floating duck-weeds (**Lemnas**) and submerged **Charas**" (*Select. from the Rec. of the Govt. of Ind., Rev. and Agric. Dept. (June 22nd, 1887), I., 69*). **Dr. Watt,** however, suggested that this antiseptic property might be extended to the destruction of the pests on other crops, such as sugar-cane, tea and coffee, besides the weeds in the rice fields.

The leaf of **Lemna** has acrid properties and is employed medicinally in China, being deemed cooling, diuretic, antiscorbutic, astringent and alterative. It is used in cutaneous diseases, and as a wash for ophthalmia, carbuncles, and syphilitic sores. In India the duck-weeds appear to be put to no economic use, excepting with other water plants in refining sugar (See **Hydrilla,** p. 311; also **Saccharum officinarum**). It is, however, noteworthy that the members of this genus are very rich in inorganic substances, and have, accordingly, been recommended for use as manure. In a specimen of **L. minor, Herve Mangon** found:—Nitrogen, 3·6; silica, 6·7; lime, 8·2; phosphoric acid, 1·1; and other mineral constituents, 19·3 per cent. (*Watts, Dic. Chem., III., 570*).

Lemon, see **Citrus,** *Linn.;* Vol. II., 333.

Lemon-grass and **Lemon-grass oil,** see **Andropogon citratus,** *DC.;* [Vol. I., 242.

251 **Lendí-piplí,** *Dymock, Mat Med. W. Ind., 2nd Ed., 728.*

Dymock writes regarding an undetermined species of the Natural Order PIPERACEÆ employed medicinally in Bombay under the above name: "Globular catkins of a species of pepper occasionally found in the Bombay market, said to come from Singapore. They are of the size of the pellets of sheep's dung, hence the name. The taste is very hot and acrid. The individual fruits are nearly as large as cardamom seeds, the whole catkin having much the appearance of a small blackberry."

LENS, *Gren. et Godr.; Gen. Pl., I., 525.*

Lens esculenta, *Mœnch.; DC., Prodr., II., 366;* LEGUMINOSÆ. 252

THE LENTIL.

Syn.—ERVUM LENS, *Linn.;* CICER LENS, *Willd.*

Vern.—*Masúr,* HIND.; *Masúri, buro-mussúr,* BENG.; *Masur moha,* ASSAM; *Masúri,* N.-W. P.; *Mohr, masúr, masar, mohri, manhri, chanching, kerse, maúri,* PB.; *Adah, adas,* AFG.; *Masuri dál,* GUZ.; *Misurpurpur,* TAM.; *Misur-pappu, chiri sanagalu,* TEL.; *Massur, chanangi,* KAN.; *Masura,* SANS.; *Adas,* ARAB.; *Mirjumak, adas,* PERS.

References.—*Roxb., Fl. Ind., Ed. C.B.C., 567; Stewart, Pb., Pl., 68; DC., Origin Cult. Pl., 107, 321; Elliot, Fl. Andhr., 42; Aitchison, Bot. Afgh. Del. Comm., in Trans. Linn. Soc., 2nd Ser., III., Pt. I., 59; U. C. Dutt, Mat. Med. Hind., 309; Bent. & Trim., Med. Pl., 76; Murray, Pl. and Drugs, Sind, 120; Watts Dict. of Chemistry, III., 570; Baden Powell, Pb. Pr., 241; Atkinson, Him. Dist., 308, 694; Duthie & Fuller, Field and Garden Crops, II., 13, Pl. xxx.; Birdwood, Bomb. Pr., 119; Royle, Ill. Him. Bot., 200; Church, Food Grains of India, 138, f. 25; Balfour, Cyclop., II., 700; Smith, Dic., 243; Agri.-Hort. Soc. of Ind.:—Transactions, I., 41; IV., 101; V., 64; Journals (Old Series), V., Sel., 30; IX., 416; Reports of Agric. Dept., Bombay, 1885-86, 1886-87, 1887-88, App. viii.; Bengal, 1886, App. xxvii., lxxii.; C. P., 1885, Statement I.; Quarterly Journ. of Agric., V., 63; Special Reports from J. B. Fuller, Esq., C.S., Cent. Prov.; Dir. Land Rec. & Agric., Bengal; Dir. Land Rec. & Agric., Assam; Govt. of Burma; Gazetteers:—Bombay, XVI., 100; XVII., 269; N.-W. P., I., 80; IV., lxxi.; Oudh, I., 528; Panjáb, Ludhiana, 140; Gurgaon, 84; Hoshiarpur, 91; Simla, 57; Muzaffargarh, 95; Kurnal, 172; Settlement Reports:—C. P., Nimar, 197; Baitúl, 62; Chanda, 81; Narsinghpur, 52; Panjáb, Simla, App. II., xxxix.; Hazara, 88; Kangra, 24; Jhang, 84, 90; Montgomery, 107; Lahore, 9; Madras, Man. of Administration, I., 289.*

Habitat.—A valuable pulse, grown as a winter crop all over India. **DeCandolle** writes: "The lentil appears to have existed in western temperate Asia, in Greece, and in Italy, where its cultivation was first undertaken in very early prehistoric times, when it was introduced into Egypt. Its cultivation appears to have extended at a less remote epoch, but still hardly in historic time, both east and west, that is, into Europe and India." The lentil has been cultivated in the East, in the Mediterranean basin, and in Switzerland, from prehistoric times. According to **Herodotus, Theophrastus,** and others, the Egyptians used it largely. The ancient Jews also employed it to a considerable extent as an article of food, the hulled pulse having probably formed Esau's mess of red pottage. **Aristophanes** mentions it as an article of food of the poor among the Greeks, by whom it was known as "φάκος." By the Latins it was known and used under the name of "lens," the derivation of which is doubtful, but the dissimilarity of the Latin and Greek words show that the species probably existed wild in each country before it was cultivated. Another proof of its ancient existence in Europe is the fact that lentils have been found in the lake-dwellings of St. Peter's Island, Lake of Bienne, which are of the bronze age (*DeCandolle*).

From a consideration of the Sanskrit, Persian, and Arabic names of the lentil, and the evident connection of most local Indian names with these, **DeCandollé** arrives at the opinion that the plant was not known in India before the invasion of the Sanskrit-speaking races.

Cultivation.—The lentil is at the present day cultivated in all parts of India, especially in the Central Provinces and Madras. For convenience CULTIVATION. 253

LENS esculenta.

Cultivation of the Lentil.

CULTIVATION.

of description the methods pursued, amount of seed used, area under the crop, &c., may be detailed under the headings of the various Provinces from which trustworthy information is available :—

Panjab. 254

The Panjáb.—It is commonly cultivated all over this Province, and is said to be grown as high as 5,500 feet on the Chenab, and in parts of Ladak at 11,500 feet. The following remarks, extracted from the Settlement Report of the Jhang District, may be taken as typical of the methods in general use throughout the Province :—

"Area under cultivation 2,404 acres. *Massur* is a *sailaba* (inundation) crop, and is never sown on other soils. Either new alluvial soils, or light lands that are not good enough for wheat, are selected. *Massur* is often the first crop sown on new *sailab* lands, or it may follow *mattar* (**Lathyrus sativa**). The land is ploughed once or twice, and the seed is sown broadcast. One ploughing takes place after the seed is sown. *Maggan-Poh* (December-January) are the months for sowing, from 1 to 1½ *panapis* per *kanal* (or 30 to 45℔ of seed per acre) being the amount used. The crop ripens in the end of *Chait* and the beginning of *Baisákh* (March-April). It is reaped, not pulled. The yield is light, and the crop is subject to much the same injuries as gram (**Cicer arietinum**), the pods being attacked by caterpillars. Rain, wind, and thunder are hurtful when the plant is in flower." During the past five years the acreage under the pulse has been returned as follows :—1884-85, 170,975 acres; 1885-86, 100,820; 1886-87, 142,121; 1887-88, 146,931; and 1888-89, 159,461 acres.

N.-W. Provinces. 255

North-West Provinces.—The following account of the method pursued in these Provinces is given by Messrs. Duthie & Fuller :—"Lentils are grown as a cold-weather crop under much the same conditions as peas. In the Tarai district the area under them constitutes nearly 7 per cent. of the total *rabi* cropped area; in the Rohilkhand Division it amounts to 2 per cent., and in the districts of the Benares Division to 1·3 per cent. In no other division does it reach so high a proportion as 1 per cent., being next largest in the Meerut and Allahabad Divisions (0·8 to 0·7 per cent.), and smallest in the Agra and Jhansi Divisions (0·1 to 0·3 per cent.), which together comprise the driest tract in the Provinces. Taking the thirty temporarily-settled districts as a whole, *masúr* is grown on almost exactly 1 per cent. of the total *rabi* cropped area.

It is grown on all kinds of soils, but chiefly in low-lying land. It is comparatively seldom grown after an autumn fallow, but most commonly follows early rice. It is often sown while the rice stalks are standing, and allowed to grow up amongst them. Three ploughings are as a rule sufficient.

The quantity of seed per acre varies with the condition of the ground, but is commonly about 1 maund. The average produce from unirrigated land is from 6⅓ to 8 maunds of grain, but with irrigation from 10 to 12 maunds would not be an excessive outturn. The average area under *masúr* in the thirty temporarily-settled districts as deduced from the agricultural returns of 1878-79-80 was 15,961 acres irrigated, and 98,264 acres, dry, giving a total of 114,225 acres, of which much the largest proportion, a total of 42,388 acres, was in Rohilkhand." The acreage during the past five years has been as follows :—1884-85, 137,597; 1885-86, 140,446; 1886-87, 169,832; 1887-88, 169,949; 1888-89, not received.

Central Provinces. 256

Central Provinces.—J. B. Fuller, Esq., C.S., reports as follows :—"**Ervum Lens** is grown throughout the Provinces as a *rabi* or cold-weather crop. Its cultivation is largest in the Narbadda Valley, the Satpura District, and Chhattisgarh. In the Nagpúr country it is less popular. The

CULTIVATION.

area under *masúr* in each division during the last two years is shown below:—

	1885-86. Acres.	1886-87. Acres.
Jabalpur Division	70,984	79,840
Narbadda	65,864	69,669
Nagpúr	7,090	7,080
Chhattisgarh	82,869	108,113
TOTAL	226,807	264,702

(In the last two years the crop has been considerably smaller, in 1887-88 the total acreage was only 203,417; and in 1888-89, 186,460.)

The crop is generally grown on the best black soil; indeed, the growth of *masúr* on black soil may be accepted as an indication that the land is of first quality. A curious fact connected with its cultivation in Chhattisgarh is that the Santami *chamars*, who form an important section of the agricultural community of that tract, decline to have anything to do with it on the ground that in its red colour it resembles flesh."

Berar. 257

Berar.—The crop is grown to a small extent in this province, the area during the past five years having averaged between 20,000 and 25,000 acres.

Bengal. 258

Bengal.—The following is extracted from the Report of the Director of Land Records and Agriculture:—"There are two varieties of **Ervum Lens** grown in Dacca; one is the Patna variety, and the other is met with in Lower Bengal. The plants of the latter are bushy and give a better outturn. It does best in clay soil, as in very light soil the plants wither away. It comes after *neali* or *kele kartiksal* paddy, and is followed by paddy. Three or four ploughings suffice. It should be sown when the land is fairly dry, owing to the fact that, if too much moisture be present, the plants shoot up quickly, but afterwards make no progress.

Five seers of seed are sown per *bigha*, from the middle of October up to the end of November, and barley is sometimes sown mixed with it. It is harvested in February-March, and yields from 3 to 4 maunds.

In the Bhagulpore Division one or two ploughings and one or two harrowings are considered sufficient; the seed is sown in October-November, and the crop reaped in March-April. Lentils are also grown to a very limited extent in Chutia Nagpúr, and in the Behar Districts to the extent of about 2 per cent. of the winter-sowings."

Bombay. 259

Bombay.—In Bombay the lentil crop is a very unimportant one compared with other pulses. As in other districts, it is sown in October and harvested in February, and a small quantity is annually sent to Bombay from the area in which it is chiefly grown, namely, Násik. From the Annual Reports of the Director of Land Records and Agriculture for 1885-86 to 1887-88, it would appear that the average area under the crop, during these three years, was 18,323 acres. In 1888-89 it was only 18,002 acres. By far the largest proportion of this was in Násik, which had an average area of 9,562 acres or more than half, Belgaum had 5,139 acres, Poona 1,527, and Sátára 721 acres.

No statistics are available regarding the amount of seed used, outturn, or profit per acre.

Madras. 260

Madras.—No accurate information is available regarding this Province. In 1886-87 101,473 acres were returned as under lentils, but since that date **Lens esculenta** has not figured in the returns.

Assam and Burma. 261

Assam and Burma.—The lentil is said to be grown in these Provinces, but no accurate information is available regarding the area under it, nor of the extent to which it forms a part of the food of the people.

LENS esculenta.

Medicinal and Other Uses of Lentils.

CULTIVATION.

A consideration of the preceding reports and extracts shows that the crop is universally a cold weather one; that it is especially suited to rich, moist, low-lying soils; that with a fair amount of seed (about 1 maund) it yields from 6 to 8 maunds an acre, and if irrigated 10 to 12 maunds; and that the soil requires little preliminary working, and the crop little after-attention.

Church remarks that the yield might be increased if more pains were taken in the selection of seed for sowing, since there are some varieties of the lentil, which produce seed weighing twice as much as the small common sort, and which, notwithstanding this, do not make a proportionately increased demand upon the resources of the soil. The total area under the crop in the districts from which statistics are forthcoming averages between 600,000 and 750,000 acres, but these figures do not include Bengal, Oudh, Burma, Assam, Central India, and the Native States.

MEDICINE. 262

Medicine.—Lentils have long held the reputation of being useful medicinally in cases of constipation and other intestinal affections. It is probable, however, that they have no special value in such cases. In the form of a paste or poultice, the pulse is used as an application to ulcers occurring after small-pox.

SPECIAL OPINIONS.—§ "I have found lentils of great use in the treatment of chronic and obstinate constipation" (*Surgeon-Major C. W. Calthrop, M.D., Morar*). "A paste prepared from the seeds is a useful cleansing application in cases of foul and indolent ulcers" (*Assistant Surgeon T. N. Ghose, Meerut*).

FOOD. 263

Food.—The lentil, as already indicated, is a food of the greatest antiquity, and has for many centuries been much prized by the natives of most Eastern countries. In Egypt, Syria, and other Eastern countries it is parched and eaten at the present day, and is considered one of the best foods for those who undertake long journeys. In India it is eaten as *dál*, flavoured with various aromatics and carminatives, also as a component part of the dish called *kichri*, and is considered the most nutritious of the pulses. If too freely indulged in, however, it is supposed to be heating and apt to give rise to eruptions. The young pod is also occasionally eaten as a vegetable, and the dry leaves and stalks are employed as a fodder for cattle.

The ground pulse mixed with a little flour of barley or some other cereal, and common salt, is sold in Europe, under the names of Ervalenta or Revalenta, as a food for invalids.

CHEMISTRY. 264

CHEMICAL COMPOSITION.—**Church** gives the following as the average composition :—

	IN 100 PARTS. Husked.	With husk.
Water	11·8	11·7
Albumenoids	25·1	24·9
Starch	58·4	56·0
Oil	1·3	1·5
Fibre	1·2	3·6
Ash	2·2	2·3

The same authority writes: "The nutrient ratio is 1 : 2·5, and the nutrient value 87. The lentil is generally regarded as a pulse of the second class, inferior to *múng* (**Phaseolus Mungo**), but equal to *urhur*, the pigeon-pea. It is highly nutritious, but somewhat heating; it should be carefully freed from the husk or coat. The bitter substance which occurs in lentils may be removed by soaking them for a short time in water in which a little carbonate of soda (common washing soda) has been dissolved. The meal of lentils, deprived of their coat, is of great richness, containing generally more albumenoid or flesh-forming matter than bean or pea-flour."

CHEMISTRY.

Other chemists have obtained analyses differing to some degree, from that given by **Church**, especially in the percentage of starch. Thus according to **Krocken** the air-dried seed contains from 34 to 35 per cent. of starch; **Fresenius** found 35·5 per cent., and 7 per cent. of gum, and **Poggiale** 44 per cent. of the two combined. **Boussingault** found the air-dried straw to contain 27 per cent. of water, 57 per cent. of woody fibre, and 3·9 per cent. of ash, which was found to contain a large percentage of lime, potash, silica, and phosphoric acid (*Watts, Dictionary of Chemistry*).

Trade.—A considerable trade is carried on in the lentil, but since the returns do not specify the various pulses separately, it is impossible to give separate figures regarding this species. TRADE. 265

LEONOTIS, *Br.; Gen. Pl., II., 1214.*

(*t. 867*; LABIATÆ. 266

Leonotis nepetæfolia, *Br.; Fl. Br. Ind., IV., 691; Wight, Ic.,*

Syn.—PHLOMIS NEPETÆFOLIA, *Linn.*; LEONURUS GLOBOSUS, *Mœnch.*

Vern.—*Hejurchei,* BENG.; *Dare dhompo, janum dhompo,* SANTAL; *Mátijer, mátisúl,* BOMB.; *Dípmal,* MAR.; *Rana bhéri, béri, mulu golimidi, hanumanta bira,* TEL.; *Maha-yakwan-assa,* SING.

References.—*Roxb., Fl. Ind., Ed. C.B.C., 461; Thwaites, En. Ceylon Pl., 241; Dalz. & Gibs., Bomb. Fl., 211; Grah., Cat. Bomb. Pl., 153; Elliot, Fl. Andhr., 25, 26, 68, 119, 163; Campbell, Ec. Prod., Chutia Nagpur, No. 8487; Dymock, Mat. Med. W. Ind., 611; Year Book of Pharmacy, 1880, 250; Atkinson, Him. Dist., 316; Indian Forester, XII. (app.), 19; Gazetteers:—N.-W. P., I., 83; IV., lxxvi.*

Habitat.—A tall annual shrub, found cultivated and naturalized, but doubtfully indigenous, throughout the hotter parts of India, from the Panjáb to Travancore and Singapore, also in the drier regions of Ceylon.

Medicine.—**Campbell** states that in Chutia Nagpur the ash produced by burning the FLOWER-BUDS is applied to burns and scalds, and **Dymock** writes that in Bombay the ash of the FLOWER-HEADS mixed with curds is applied to ringworm and other itchy diseases of the skin. MEDICINE. Flower-buds. 267 Flower-heads. 268

(COMPOSITÆ.

Leontodon Taraxacum, *Linn.*; see **Taraxacum officinale,** *Wigg.*;

LEONURUS, *Linn.; Gen. Pl., II., 1210.*

Leonurus sibiricus, *Linn.; Fl. Br. Ind, IV., 678;* LABIATÆ. 269

Syn.—L. TARTARICUS, *Burm.*; L. HETEROPHYLLUS, *Sweet*; STACHYS ARTEMESIÆ, *Lour.*

Vern.—*Gúma,* PATNA.

References.—*Roxb., Fl. Ind., Ed. C.B.C., 461; Burm. Fl. Ind., 127; Irvine, Mat. Med. Patna, 129.*

Habitat.—An annual, erect, leafy stout herb, occurring on the plains of India from Bengal and Sylhet to Coorg, doubtfully indigenous.

Medicine.—**Irvine** writes: "The ROOT, LEAVES, and JUICE are bitter, and used as a febrifuge. Dose 2 drachms to 2 ounces in infusion; price 1 anna per pound." MEDICINE. Root. 270 Leaves. 271 Juice. 272

Leopards, see **Tigers**; Vol. VI.

LEPIDAGATHIS, *Willd.; Gen. Pl., II., 1101.*

[ACANTHACEÆ.

Lepidagathis cristata, *Willd.; Fl. Br. Ind., IV., 516;* 273

Syn.—L. SHUTERI, *T. Anders.*; L. SP., *n. 18*; *Herb. Ind. Or. H.f. & T.*

Var. rupestris, with pubescent elliptic leaves and stems; L. RUPESTRIS, *Nees.*

Vern.—*Ot dhompo,* SANTAL; *Koli-che-chútar,* BOMB.; *Bhui terada,* MAR.

References.—*Roxb., Fl. Ind., Ed. C.B.C., 476; Dalz., & Gibs., Bomb. Fl., 191; Dymock, Mat. Med. W. Ind., 2nd Ed., 594; Campbell, Ec. Prod., Chutia Nagpur, No. 7862; Gazetteers:—Bombay, V., 27; N.-W. P., I., 83; IV., lxxvi.*

Habitat.—A common plant on dry, elevated barren ground, in Coromandel.

MEDICINE. Ash. 274 Plant. 275 **Medicine.**—Campbell states that the ASH of the burned plant is employed in Chutia Nagpúr as an application to sores. **Dymock** remarks that it is applied to cure itchy affections of the skin, and that the PLANT is considered a good bitter for administration in fevers.

SPECIAL OPINION.—§ "The plant is considered to be a bitter tonic" (*Surgeon-Major C. T. Peters, South Afghánistán*).

LEPIDIUM, *Linn.; Gen. Pl., I., 87.*

276 **Lepidium Draba,** *Linn.; Fl. Br. Ind., I., 160;* CRUCIFERÆ.

Syn.—L. CHALEPENSE, *Schrenk.;* L. REPENS, *Boiss.*

Vern.—*Bijindak,* AFGH.; *Búsk,* QUETTA.

References.—*Boiss, Fl. Orient., I., 356; Aitchison, Bot. of Afgh. Del. Com., in Trans. Linn. Soc., 2nd Series, III., Pt. I., 36.*

Habitat.—A common weed of cultivation in the Panjáb, distributed westwards, through Afghánistán and Baluchistán to Europe.

FOOD & FODDER. Leaves. 277 **Food and Fodder.**—The LEAVES are largely collected and eaten as a pot-herb by the inhabitants of Afghánistán and Baluchistán. It is one of the principal spring fodder plants of Quetta for cattle and horses (see **Fodder,** Vol. III., 415).

278 **L. Iberis,** *Linn.; DC., Prodr., I., 207.*

THE PEPPER GRASS or PEPPERWORT.

References.—*Dymock, Mat. Med. W. Ind., 55; Pharmacog. Indica, I., 118.*

Habitat.—A herb of which the indigenous area extends from Southern Europe to Siberia.

MEDICINE. Seeds. 279 **Medicine.**—Dr. **Dymock** and the authors of the ***Pharmacographia Indica*** describe the medicinal SEEDS, called *towdri,* as products of this plant. The origin of these seeds, which are imported into Bombay from Persia, has for long been a subject of discussion. By some English writers they have been attributed to **Malva sylvestris,** by others to **Cheiranthus Cheiri,** but **Dymock** has pointed out that, owing to the nature of the fruit, neither of these suppositions can be correct. Again, **Stewart** states that in the Panjáb and Sind **Matthiola incana,** *R. Br.,* is valued for its seeds, which constitute one of the several kinds of *todri.* The authors of the *Pharmacographia Indica* commenting on this write: "In short this Persian name appears to have much the same meaning as the λευκόϊον of the Greeks, being applied loosely to several Spring flowers." **Dr. Dymock,** however, has arrived at the conclusion that at least "some of the *towdrí* seed is doubtless the produce of **L. Iberis,** *Linn.*"

Mir Muhammad Husain described the seeds and the plant which was supposed to yield them, and believed them to be "hot in the second degree, and moist in the first; some say dry. Properties aphrodisiac, fattening the body and purifying the blood." Commenting on this **Dymock** writes: "The drug is in general use for the above-mentioned properties, which are attributed by the natives to most of the cruciferous seeds" Again, in describing the properties of **L. Iberis,** he writes: "This plant was known to the ancients and employed as a rubefacient in rheumatism; the seeds taken internally were prescribed in rheumatism and dropsy. According to **Pliny** they were first used by **Damocrates.** **Corre** and

Lejanne state that the plant is called 'Cresson de Savane' in the Antilles, and is considered to have all the properties of water-cress." MEDICINE.

Towdrí seeds as imported into Bombay are of three kinds—"pale red" (called white); "red;" and "dark red" or "brown" (sometimes called black *towdri*). When soaked in water they become thickly coated with mucilage.

It appears from the descriptions in the works above quoted that the Persian "*towdri*" seeds have not been examined chemically. Such an examination might in all probability result in the exact determination of their origin, or at least would prove the correctness or otherwise of Dr. Dymock's opinion. The flowering tops and seeds of **L. Iberis** were analysed by **Leroux** in 1837, and found to yield a characteristic amorphous bitter principle which that chemist named "Lepidin" along with a sulphuretted volatile oil (*Pharmacographia Indica*). Should lepidin be proved to exist in *towdrí*, Dr. Dymock's opinion would be materially confirmed.

Lepidium latifolium, *Linn.; Fl. Br. Ind., I., 160.* 280

Vern.—*Gonyúch*, LADAK.

References—*Stewart, Pb. Plants, List of Omissions; Aitchison, Bot. Afgh. Del. Com., in Trans Linn. Soc., 2nd Series, III., Pt. I., 36; Year Book of Pharmacy, 1874, 622.*

Habitat.—An erect, leafy herb, abundant in Western Tibet, distributed to Europe, North and Western Asia, and North Africa.

Medicine.—In Europe this plant is popularly believed to be a useful antiscorbutic. MEDICINE. 281

Fodder.—**Stewart** states that it is browsed by sheep and goats in Ladak, but is little eaten by *yáks*. FODDER. 282

L. sativum, *Linn; Fl. Br. Ind., I., 159.* 283

THE CRESS.

Vern.—*Hálim, chansaur, hurf*, HIND.; *Halím, aleverie*, BENG.; *Hálim, halang*, N.-W. P.; *Shargundei, tezak, hálim*, PB.; *Ahreo*, SIND; *Asáliya, ahaliva*, BOMB.; *Ahliva*, MAR.; *Asálio, hálim*, GUZ.; *Halím*, DEC.; *Ali-verai*, TAM.; *Adit-yalu, adala-vitala, ádeli, ádiyalu*, TEL.; *Allibija*, KAN.; *Mong-nyin, sa-mung-nee*, BURM.; *Chandrasura, chandrika*, SANS.; *Jarjir, hurf, half, reschad*, ARAB.; *Turrah-tizkah, háleh, towdri, seeds=tukhme-turrahtézak*, PERS.

References.—*Stewart, Pb. Pl., 14; DC. Origin Cult. Pl., 86; Elliot, Fl. Andhr., 10; Mason, Burma and Its People, 469, 749; Ainslie, Mat. Ind., I., 95; II., 12; O'Shaughnessy, Beng. Dispens., 188; Honigberger, Thirty-five years in the East, II., 298; Moodeen Sheriff, Supp. Pharm. Ind., 168; Mat. Med. Madras, 24; U. C. Dutt, Mat. Med. Hind., 115, 295; Dymock, Mat. Med. W. Ind., 2nd Ed., 57; Pharmacog. Indica, I., 120; S. Arjun, Bomb. Drugs, 11; Murray, Pl. and Drugs, Sind, 47; Irvine, Mat. Med. Patna, 127; Cat. Baroda Durbar at Col. Ind. Exhib., No. 123; Watts, Dic. of Chem., VIII., Pt. I., 584; Atkinson, Him. Dist., 305, 702, 708; Ec. Prod., N.-W. P., Pt. V., 13, 39; Lisboa, U. Pl. Bomb., 145; Birdwood, Bomb. Pr., 7, 138; Cooke, Oils and Oilseeds, 52; Stocks, Rep. on Sind; Smith, Dic., 142; Agri Hort. Soc. of Ind., Transactions, III., 195; V., 64; VII., 69; Journals (Old Series), XIII., 390; (New Series), IV., 23; V., 33, 40, Sel., 5; Pro., 21; Gazetteers:—Orissa, II., App. VI., 181; Mysore and Coorg, I., 57; Bomb., VIII., 183; N.-W. P., I., 79; IV., lxvii.; Settlement Reports, Kumaon, App., 23; Lahore, 11.*

Habitat.—A small annual herb, cultivated throughout India and Western Tibet, also in other parts of Eastern Asia, Europe, and North Africa. Its origin is extremely obscure. **DeCandolle** writes: "The cultivation of this species must date from ancient times and be widely diffused, for very different names exist." He then proceeds to discuss the various terms applied to the cress in Arabic, Persian, Albanian, &c., and states that the plant has no known name in Sanskrit. In later Sanskrit works, however, the medicinal uses of the seeds are described under the name of *Chandra sura*, a name probably of recent origin, since it does not appear to occur in older Sanskrit works on medicine. **DeCandolle** enumerates the localities in which the plant has been found, and concludes his remarks as follows :—"I am led to believe, by this assemblage of more or less doubtful facts, that the plant is of Persian origin, whence it may have spread, after the Sanskrit epoch, into the gardens of India, Syria, Greece, and Egypt, and even as far as Abyssinia."

CULTIVATION. 284

Cultivation.—No statistics are available regarding the exact extent to which the cress is cultivated in India. It is, however, largely grown all over the country, not only for European consumption, but also as a pot-herb for native use. The following account of the best method to pursue in its cultivation, quoted from *Firminger's Gardening for India*, may be found useful :—"Cress seed may be sown in the open ground when the rains cease in October. It is best to sow only a small quantity at a time, and to keep up a succession of sowings, at short intervals, during the cold season. As the *malís* rarely cut it for use till it is three or four inches high, it is as well to sow it broadcast, and rather thinly.

"If, however, it be required to be eaten, as is usual in England, when little more than the seed leaves are formed, it may be raised at nearly all times in the year. It is best in that case to make the sowings in large shallow pans, filled with good light soil. The soil should be well watered and the seed then scattered thickly over its surface. Over a pan a covering should be placed till the seed germinates, and then be removed. In a few days the cress will be fit for cutting. In order to save seed, in the early part of the cold season, plants in the open ground at about six inches apart should be reserved. These by the commencement of the hot season will yield a plentiful supply of seed, which should be carefully stored away for future use."

OIL. Seed. 285

Oil.—The SEED yields an oil, said to be very similar to that obtained from mustard seed. The only record of its preparation in India is by **Lieutenant Hawkes** (*Report on Oils of Southern India, 37*). It is described by **Schübler** as brown-yellow, with a sp. gr. of 0·924, turning thick and turbid at—6° (21°F.) and solid at—15° (5°F.), and drying slowly. The following remarks by **Lieutenant Pogson**, extracted from a letter read before the Agri.-Hort. Soc. of India, may be quoted, as they show that the oil at one time received a certain amount of attention in India. He writes :—"I would strongly recommend European landed proprietors to cultivate the common cress, or *halim* of India, for the sake of its oil. One hundred pounds of the seed on being pressed will yield no less than 57lb of oil, whilst that of the sunflower gives but 15lb. [*Journal* (*New Series*), *V*., *Pro., xxi*]. No writer appears to have recorded the useful properties of cress-oil, but if the amount obtainable is even approximately that mentioned by **Lieutenant Pogson**, it would appear to be worthy of greater attention than it has up to this time received.

MEDICINE. Seeds. 286

Medicine.—The SEEDS are described by later writers on Sanskrit medicine as tonic, alterative, useful in hiccup, diarrhœa, and skin diseases arising from disordered blood (*U. C. Dutt*). By Muhammadan writers the cress is identified with the καρδαμον of the Greeks, and the seeds are considered hot and dry in the third degree, aphrodisiac, and diuretic. They are recommended for the dispersion of certain chronic enlargements of the spleen, &c., and as an alterative in various conditions supposed to be produced by cold humours (*Dymock*). **Ainslie** states that the Arabians place the seed amongst their *mokerchat* (vesicatoria); that the *hakíms* are in the habit of prescribing it as a stomachic and gentle stimulant. He remarks,

MEDICINE.

however, that the latter doubt whether it does not sometimes, if imprudently given, bring about abortion. The *Vytians* esteem it, when bruised and mixed with lime-juice, as a valuable repellent in cases of local inflammation. **O'Shaughnessy** quotes these remarks, and adds that he found the drug answer as a gentle and warm aperient.

By modern practitioners of Indian medicine the seeds are considered useful in dysentery and dysenteric diarrhœa, and also in febrile and catarrhal affections. **Moodeen Sheriff**, in his *Mat. Med. of Madras*, writes: "Externally it is of great service in all the diseases in which the mustard is resorted to. The thick and very gummy mucilage of the seeds acts as a mechanical antidote in cases of poisoning by irritant substances, enveloping the poisonous particles and sheathing the mucous membrane of the stomach and intestine." The same writer regards the seed as a more satisfactory rubefacient than those of mustard-seed prepared in India, and asserts that, in his opinion, if as carefully and finely ground as English mustard-flour, it would probably be superior to that also. He also states that the mucilage obtainable from the seeds may be used as a substitute for imported tragacanth and gum-arabic.

Honigberger remarks that the seeds were officinal in the Panjáb in the early part of this century, that the PLANT itself was administered in cases of asthma, cough with expectoration, and bleeding piles, and that the ROOT was used in secondary syphilis and tenesmus. According to **Bellew** the seeds are also considered to be lactagogue in the Panjáb, and are administered, after being boiled with milk, to cause abortion.

Plant. 287
Root. 288

Doses.—"Of the seeds, from one to two and a half drachms; and of the decoction from one to three fluid ounces; three or four times in the twenty-four hours." (*Moodeen Sheriff.*)

CHEMICAL COMPOSITION.—"The herb and seeds distilled with steam yield a volatile aromatic oil which does not separate spontaneously from the watery distillate, but may be extracted therefrom by agitation with benzene. Three-fourths of the crude product boiled at 226·5° exhibited the composition of pure α toluonitril phenyl-acetonitril, or phenyl-methyl cyanide, $C_6H_5CH_2CN$, and when heated to 207° for a short time with hydrochloric acid, yielded phenyl-acetic acid" (*Watts, Dictionary of Chemistry, VIII., Pt. I., quoting from A. W. Hofmann*).

Chemistry. 289

SPECIAL OPINIONS.—§"The decoction of the seeds is thus prepared: Take of the seeds, bruised, six drachms; liquorice root, cut into small pieces and bruised, one drachm; water, one pint and a quarter; boil for 10 minutes in a covered vessel, and strain when cool. The dose of the seeds is from one to two drachms and a half; and of the decoction, from one to three fluid ounces; three or four times in the twenty-four hours. The best medicinal property of this drug, according to my own experience, is its usefulness in dysentery and dysenteric diarrhœa. The coarse powder and the thick and very gummy mucilage of the seeds appear well-suited to allay the irritation of the mucous coat of the intestines in those diseases, and they thus relieve or check their symptoms to a considerable extent. The coarse powder or mucilage of course is not always sufficient to complete a cure by itself, but, like many other remedies in dysentery, requires the assistance of some other medicines according to the circumstances of each individual case. The entire seeds swallowed with water have also some control over the above-named diseases, in consequence of their property of becoming soft and mucilaginous when moistened, but are much less efficacious in this respect than their powder or mucilage" (*Honorary Surgeon Moodeen Sheriff, Khan Bahadur, Triplicane, Madras*).

FOOD & FODDER. Leaves. 290

Food and Fodder.—The LEAVES are used as an article of food, by Natives, either raw as a relish, or boiled as an ingredient of vegetable curries.

By Europeans, cress is chiefly consumed as a salad, and is also used for garnishing dishes. **Stocks** states that in Sind it is employed as a fodder for horses, camels, oxen, &c.

TRADE. 291 **Trade.**—The seeds are largely imported from Persia into Bombay and are to be found in all bazárs. **Dymock** states that their value is R3-4 per maund of 37½lb. **Moodeen Sheriff** writes that the wholesale price is R3 per maund; retail, annas 3 per pound.

LEPTADENIA, *Br.; Gen. Pl., II., 778.*

[Ic., t. 350; ASCLEPIADEÆ.

292 **Leptadenia reticulata,** *Wight & Arn.; Fl. Br. Ind., IV., 63; Wight,*

Syn.—L. APPENDICULATA, *Dcne.;* L. IMBERBE, *Wight;* L. BREVIPES, *Wight;* CYNANCHUM RETICULATUM, *Retz.;* C. ASTHMATICUM, *Ham.;* (?) C. OVATUM, *Thunb.;* SECAMONE CANESCENS, *Sm.;* ASCLEPIAS SUBEROSA, *Roxb.;* GYMNEMA AURANTIACA, *Wall.*

Vern.—*Rayadodi, shinguti, dodhi, pala-kuda,* BOMB.; *Palakúdé,* TAM.; *Pála tige, kalasa,* TEL.

References.—*Roxb., Fl. Ind., Ed. C.B.C., 254; Dalz. & Gibs., Bomb. Fl., 152; Elliot, Flora Andhrica, 143; Lisboa, U. Pl. Bomb., 165, 201; Gazetteers:—Mysore and Coorg, I., 62; N.-W. P., IV., lxxiv.*

Habitat.—A twining shrub with corky bark, found in the Eastern Panjáb, from Banda southwards through the Deccan Peninsula, also in Burma, Singapore, and Ceylon.

FOOD. Leaves. 293 Shoots. 294 Follicles. 295 **Food.**—The LEAVES and tender SHOOTS are used in some parts of India as a vegetable, especially during times of scarcity, and the FOLLICLES, known as *shinguti* or *dodhi*, are also sometimes employed in the same way (*Lisboa*).

296 **L. Spartium,** *Wight; Fl. Br. Ind., IV., 64.*

Syn.—LEPTADENIA PYROTECHNICA, *Dcne.;* L. JACQUEMONTIA and GRACILIS, *Dcne.;* CYNANCHUM PYROTECHNICUM, *Forsk.;* SARCOSTEMMA PYROTECHNICA, *Br.;* MICROLOMA PYROTECHNICA, *Spreng.;* M. ANGUSTIFOLIA, *Ham.*

Vern.—*Kip,* SIND.

References—*Royle, Fib. Plants, 306; Stocks, Rep. on Sind; Gazetteers:—Bomb., V., 27; Sind, 559; N.-W. P., IV., lxxiv.*

Habitat.—A glabrous shrub of the Panjáb and Sind, extending eastwards to the Jumna, and distributed to Balúchistán, Arabia, Egypt, and Senegambia.

FIBRE. 297 **Fibre.**—Dr. **Stocks** states that the fibre of this species is "much used in Sind for making ropes to bring up water from wells, as water does not rot it." Dr. **Royle** adds on the authority of the same observer that it is frequently combined for this purpose with the fibre of **Periploca aphylla,** *Dcne.*

FODDER. 298 **Fodder.**—In the *Sind Gazetteer* this plant is described as "one of the most important fodder plants in the province for horses, cattle, and camels." Mr. **J. H. Lace** informs the writer that it is largely used as a camel fodder in Sibi.

DOMESTIC. 299 **Domestic.**—The plant is said by Mr. **Lace** to be extensively employed in Sibi for thatching purposes.

L. viminea, *Bth.,* see **Orthanthera viminea,** *W. & A.;* Vol. V.

LEPTOCHLOA, *Beauv.; Gen. Pl., III., 1172.*

300 **Leptochloa chinensis,** *Nees.; Duthie, Fodder Grasses of Northern [India, 59, Pl. lxxi.;* GRAMINEÆ.

Syn.—L. TENERRIMA, *R. & S.;* POA CHINENSIS, *Kœn.;* P. DECIPIENS, *R. Br.;* ERAGROSTIS DECIPIENS, *Steud.*

Vern.—*Chánhel, chipa-chi-magadi, jhira, phulkia,* N.-W. P.

References.—*Symonds, Grasses of the Ind. Penin., 29; Gaz., N.-W. P., IV., lxxx.*

Habitat.—A common grass on the sides of water-courses, and in moist places, throughout the plains of India.

Fodder.—See **Fodder**, Vol. III., 423.

LEPTORHABDOS, *Schrenk.; Gen. Pl., II., 971.*

[SCROPHULARINEÆ.

Leptorhabdos Benthamiana, *Walp.; Fl. Br. Ind., IV., 303;* 301

Syn.—GERARDIA PARVIFLORA, *Wall.*; LEPTORHABDOS VIRGATA, *Benth.*; L. PARVIFLORA, *Benth.*; DARGERIA PINNATIFIDA, *Dcne.*

Vern.—*Siri lasht,* BALUCH.

Habitat.—An erect slender herb, found in the Western Temperate Himálava, from Kashmír to Kumaon, at altitudes from 5,000 to 11,000 feet, also in Western Tibet; distributed westwards to Afghánistán, Balúchistán, and Persia.

FODDER. 302

Fodder.—Mr. **J. H. Lace** informs the writer that this plant is one of the principal summer fodders for sheep and goats throughout the Juniper Forests of Balúchistán.

Lepus, see **Hares**, p. 202.

LESPEDEZA, *Mich.; Gen. Pl., I., 524.* 303

A genus of herbs or undershrubs belonging to the Natural Order LEGUMINOSÆ, which comprises some twenty to twenty-five species, of which nine or ten are natives of India. None of the Indian species appear to be of economic value, but it is possible that some, if not all, might be utilised as fodder. **L. stricta**, the "Japan Clover," a Chinese or Japanese plant, has of late years attracted much attention, in America, as a fodder plant. It is not known how the plant was originally introduced into that country, but it is supposed that it spread from a few seeds obtained from a cargo of Japan tea, and started gowth near Charleston about 50 years ago. The plant has spread over large tracts and appears to possess many very valuable properties. The *Indian Agriculturist* of January 1st, 1887, contains the following:—"The plant is an annual but comes up year after year, self-sown, and grows so vigorously, that it smothers many other grasses with which it comes in contact. One writer is of opinion that it will never be of real value as a crop to be cultivated like clover, and a New York agricultural paper states that, although on rich, damp, soil, it will grow as high as 18 inches, the leaves are small, and the stems too hard and woody to make good hay. On the other hand, a Virginia farmer gives his experience as entirely favourable to the plant. Ten years ago he procured some seed and sowed it on an acre in the middle of a 60-acre field, being informed that it would spread all over the field and afford excellent grazing. This turned out to be true, Japan clover and Bermuda grass being the only fodders that kept green during the drought of 1885. Japan clover, he adds, starts early and by July or August, when other fodders are failing, it forms a heavy dense sod, and affords excellent grazing till killed by sharp frost, but it starts again in the following spring, and spreads year by year. It will grow and form a sod on the thinnest land, and even in roads and other bare places where nothing else will grow. In Louisiana it is extensively used for hay." In another passage the following occurs:—"Dr. **Schomburgh**, quoting from a Californian paper, writes, 'It grows on every kind of soil—rich or poor, clay or sandy, dry or wet—and is deep rooted and improving to the soil. It is wonderfully fattening and contains:—nitrogenous matter 16·6; fat 1·1; ash 5·92 per cent., the latter consisting of lime, magnesia, potash, soda, phospheric and sulphuric acids.'"

From these accounts it would appear that the plant is remarkably hardy, and has great powers of resistance to drought,—qualities which ought to render it an extremely useful fodder plant for India.

LETTSOMIA, *Roxb.; Gen. Pl., II., 870.*

304 **Lettsomia elliptica,** *Wight; Fl. Br. Ind., IV., 192;* CONVOLVULACEÆ.

Syn.—CONVOLVULUS ELLIPTICUS, *Spreng.;* C. LAURIFOLIUS, *Roxb.;* IPOMÆA ELLIPTICA, *Roth.;* I. LAURIFOLIA, *Sweet;* ARGYREIA ELLIPTICA, *Chois.*

Vern.—*Vonangadi,* KAN.

References.—*Roxb., Fl. Ind., Ed. C.B.C., 158; Dalz. & Gibs., Bomb. Fl., 169.*

Habitat.—A scandent shrub of the Deccan Peninsula and Ceylon.

INDUSTRIAL. Leaves. **Industrial Use.**—The fresh LEAVES are employed in the manufacture of steel in Mysore. (See **Iron**, p. 503.)

305 **Lettuce,** see **Lactuca Scariola,** *Linn.,* **var. sativa**; p. 578.

LEUCÆNA, *Benth.; Gen. Pl., I., 594.*

306 **Leucæna glauca,** *Benth.; Fl. Br. Ind., II., 290;* LEGUMINOSÆ.

Syn.—ACACIA FRONDOSA, *Willd.;* A. CARINGA, *Ham.;* A. GLAUCA, BICEPS, and LEUCOCEPHALA, *DC.*

Vern.—*Tagarai-virai,* TAM.

References.—*Brandis, For. Fl., 172; Gamble, Man. of Timb., 145; Bidie, Cat. Raw. Prod., S. India, 120.*

Habitat.—A shrub or small tree, met with throughout India, and other parts of Tropical Asia and Africa, but probably indigenous only to Tropical America.

TIMBER. 307 **Structure of the Wood.**—White, hard.

DOMESTIC. Seeds. 308 **Domestic.**—Dr. **Bidie** states that the SEEDS are employed in Madras for making fancy bags, baskets, and ornaments.

LEUCAS, *Br.; Gen. Pl., II., 1213.*

A genus of woolly or villous, rarely glabrate, herbs or undershrubs, which belongs to the Natural Order LABIATÆ and comprises some 50 species, natives of Asia and Africa. Of these 38 are indigenous to India.

309 **Leucas aspera,** *Spreng.; Fl. Br. Ind., IV., 690;* LABIATÆ.

Syn.—L. PLUKENETII, *Benth.;* L. DIMIDIATA, *Benth.;* PHLOMIS ASPERA, *Willd.;* P. PLUKENETII, *Roth.;* P. ESCULENTA, *Roxb.;* P. OBLIQUA, *Ham.*

Vern.—*Chota-hal-kúsa,* HIND., BENG.; *Thurdúri baji,* DEC.; *Tamba,* BOMB.; *Tumbai-chedi, thombay-keeray,* TAM.; *Tumma-chettu, tummi-kura,* TEL.

References.—*Roxb., Fl. Ind., Ed. C.B.C., 461; Dalz. & Gibs., Bomb. Fl., 211; Grah., Cat. Bomb. Pl., 153; Pharm. Ind., 169; Dymock, Mat. Med. W. Ind., 2nd Ed., 889; Bidie, Cat. Raw Pr., Paris Exh., 38; Lisboa, U. Pl Bom., 203; Agri.-Hort. Soc. of Ind.;—Journals (Old Series) IV., 15; X., 23; Gazetteers:—N. W. P., I., 83; IV, lxxvi; Ind. Forester, II, 25; III., 237; District Manual, Coimbatore, 247.*

Habitat.—A small very variable herb of the plains of India, extending from Sikkim and Behar westwards to the Panjáb, and southward to Cape Comorin; distributed to Mauritius, Java, and the Philippine Islands.

MEDICINE. Juice. 310 **Medicine.**—Reputed, apparently without reason, to be an antidote for snake-bite. The JUICE of the leaves, according to Dr. **J. Shortt**, is an application of some value in cases of psoriasis and other chronic skin diseases.

MEDICINE.

SPECIAL OPINIONS.—§ "The leaves are said to be useful in chronic rheumatism" (*Civil Surgeon C. J. W. Meadows, Burrisal*). "In North Bengal the flowers of **Leucas aspera, L. Cephalotes,** and **L. linifolia** are given warmed in a little honey for coughs and colds to children. The juice of the leaves is applied to disperse painful swellings, and answers the same purpose as tincture of iodine. Also given in skin diseases" (*C. T. Peters, M.B., Zandra, South Afghánistán*).

FOOD. Leaves & Shoots. 311

Food.—The LEAVES and TENDER SHOOTS are used as a pot-herb even in ordinary seasons, and were largely utilised in the Deccan Famine of 1877-78. [*l.* 337.

Leucas Cephalotes, *Spreng.; Fl. Br. Ind., IV., 689; Wight, Ic.,* 312

Syn.—LEUCAS CAPITATA, *Desf.;* PHLOMIS CEPHALOTES, *Roth.*

Vern.—*Dhurpi sag,* HIND.; *Bara halkasá,* BENG.; *Andia dhurup arak,* SANTALI; *Phúman, sísaliús, maldoda, guldoda, chatra,* PB.; *Kubi,* GUZ.; *Tumba,* MAR.; *Pedda tumni, tumni,* TEL.; *Kedári,* KHANDESH.

References.—*Roxb., Fl. Ind., Ed. C.B.C., 461; Grah., Cat., Bomb. Pl., 153; Dals. & Gibs., Bomb. Fl., 211; Stewart, Pb. Pl., 168; Elliot, Flora Andhrica, 150, 184; Campbell, Ec. Prod. Chutia Nagpur, No. 9106; Dymock, Mat. Med. W. Ind., 2nd Ed., 617, 888; S. Arjun, Bomb. Drugs, 101; Cat. Baroda Durbar, Col. Ind. Exhib., No. 124; Atkinson, Him. Dist., 316; Agri.-Hort. Soc. of Ind.:—Jour. (Old Series), X., 23; Gazetteers:—N.-W. P., IV., lxxvi; Bombay, V., 28; Ind. Forester, XII., App., 19.*

Habitat.—A common herb on the Himálaya at altitudes of 2,000 to 6,000 feet, from Simla to Bhután; found also in the plains from Chittagong and Assam westwards to the Panjáb, and southwards through the Deccan; distributed to Afghánistán.

OIL. Seeds. 313

Oil.—The SEEDS yield an oil, which is used medicinally in Chutia Nagpúr (*Campbell*) and by the Manipuris as an adjunct in dyeing with **Rubia** (*Watt*). See **Rubia sikkimensis,** *Kurz.*

MEDICINE. Plant. 314 Oil. 315 Juice. 316 Flowers. 317

Medicine.—The PLANT is considered a mild stimulant and diaphoretic and is considerably used in fevers and coughs. The OIL, as above stated, is also said to be employed medicinally by the Santals, but Mr. Campbell does not mention its therapeutic action. The fresh JUICE is used in certain localities as an external application in scabies. The FLOWERS are administered in the form of a syrup as a domestic remedy for coughs and colds.

FOOD. Leaves. 318

Food.—The LEAVES are eaten as a pot-herb, in some parts of the country, and were largely used as an article of food in the Khandesh District during the famine of 1877-78.

DOMESTIC & SACRED. Leaves. 319 Flowers. 320

Domestic and Sacred.—Stewart states that the LEAVES are stirred up with milk in the Panjáb for the sake of the odour they impart. The FLOWERS are sacred to Siva, and are frequently used as an offering.

L. Clarkei, *Hook. f.; Fl. Br. Ind., IV., 688.* 321

Vern.—*Enga dhurup, 'arak,* SANTALI; *Dhurpi sag,* HINDI, CHUTIA NAGPUR.

Reference.—*Campbell, Ec. Prod., Chutia Nagpur, No. 9105.*

Habitat.—A common weed of cultivation in Chutia Nagpúr and Behar.

FOOD. Leaves. 322

Food.—"The LEAVES are eaten as a pot-herb" (*Campbell*).

L. linifolia, *Spreng.; Fl. Br. Ind., IV., 690.* 323

Syn.—LEUCAS LAVANDULÆFOLIA, *Sm.;* PHLOMIS LINIFOLIA, *Roth.;* P. ZEYLANICA, *Roxb.;* LEONURUS INDICUS, *Burm.;* **var. decipiens,** *Sm.;* L. ASPERA, *Pl. Ind. Or. Hohenack.*

Vern.—*Gumá, hal-khusa,* HIND.; *Halkasá,* BENG.; *Dron,* ASSAM; *Gumi, kumbha,* GOND; *Goma,* DEC.; *Púla tummi,* TEL.; *Dronapushpi,* SANS.

References.—*Roxb., Fl. Ind., Ed. C.B.C., 461; Grah., Cat. Bomb. Pl., 153; Dalz. & Gibs., Bomb. Fl., 211; Elliot, Flora Andhr., 158; U. C. Dutt, Mat. Med. Hind., 297; Dymock, Mat. Med. W. Ind., 2nd Ed., 617; Drury, U. Pl., 277; A Note on the Condition of the People of Assam, 1888; Agri.-Hort. Soc. of Ind.:—Jour. (Old Series), X., 23; Gazetteer, Bombay, VI., 15.*

Habitat.—A stout herb met with in the plains, from Assam, Bengal, and Sylhet to Singapore, also in the Deccan, and on the West Coast from the Konkan to Travancore.

MEDICINE. Leaves. 324 **Medicine.**—Mr. J. F. Duthie informs the writer that the natives of Central India believe that the LEAVES, when roasted and eaten with salt, have febrifugal properties. Mr. Long writes, in his account of the indigenous plants of Bengal, "The Ceylonese attribute almost miraculous curative powers to this plant. The leaves are bruised and a tea-spoonful of juice given, which is snuffed up the nostrils, and used by natives in the North-West, as a remedy against snake-bite; the mungoose also eats it. The fresh juice is employed as a remedy against headaches and colds." In the foregoing statement it is difficult to understand how much applies to the utilisation of the plant in Ceylon, how much appertains to Bengal,—a point of some importance since, according to the *Flora of British India*, **L. linifolia** is said not to occur in Ceylon. It appears possible that certain of the properties attributed to this species in the above description should properly be referred to **L. zeylanica,** *Br.*, and that the confusion may have arisen from the fact that **Phlomis zeylanica,** *Roxb.*, is identical with this species (**L. linifolia,** *Spreng.*), while **Phlomis zeylanica,** *Linn.*, is the true Ceylon plant, **L. zeylanica,** *Br.*

It may be mentioned, however, that in the *Gujarát Gazetteer* the statement is repeated.

FOOD. Tops. 325 **Food.**—Darrah writes (*Note on the Condition of the People of Assam*), "The young TOPS usually fried in oil. They are not eaten alone, as the flavour is bitter, but are generally mixed in small quantities with *khuturia*."

DOMESTIC & SACRED. Flowers. 326 **Domestic and Sacred.**—Roxburgh states that the FLOWERS are used by the Brahmins to decorate their idols.

327 **Leucas mollissima,** *Wall.; Fl. Br. Ind., IV., 682.*

Syn.—LEUCAS PILOSA, VAR. PUBESCENS, *Benth.*

Var.—**scaberula,** *Clarke.*

,, **angustifolia,** Z. ANGUSTIFOLIA, *Wall.*

,, **strigosa,** Z STRIGOSA, *Benth.*

Vern.—*Gitil arak,* SANTALI.

References.—*Campbell, Ec. Prod. Chutia Nagpur, Nos. 9406 & 9491; Stewart, Jour. of a Bot. tour in Hazara, &c., in Jour. Agri.-Hort. Soc. Ind. (Old Series), XIV., 27, 42; Gazetteers, N.-W. P., I., 83; IV., lxxvi; X., 315.*

Habitat.—A slender much-branched herb, met with in the sub-tropical Himálaya, the Khásia mountains, the Konkan, Central India, and Ceylon.

FOOD. Leaves. 328 **Food.**—Campbell writes, "A very common plant, exceedingly abundant in the Tundi hills. The LEAVES are eaten as a pot-herb by the Santals."

329 **L. zeylanica,** *Br.; Fl. Br. Ind., IV., 689; Wight, Ill., t. 176.*

Syn.—LEUCAS INVOLUCRATA, *Benth.*; PHLOMIS ZEYLANICA, *Linn.*; LEONURUS MARRUBIASTRUM, *Burm.*

Vern.—*Gatta-túmba,* SING.

Reference.—*Thwaites, En. Cal. Pb., 240.*

Habitat.—A herb of Assam, Kachar, and Chittagong, the Malay Peninsula, and Ceylon.

Medicine.—**Thwaites** states that the natives of Ceylon use the LEAVES as a medicine. It is possible that the information given by **Long** and quoted by **Drury** under **L. linifolia** may refer to this species. MEDICINE. Leaves. 330

Food.—The Singalese are said to employ the LEAVES as a vegetable in their curries. FOOD. Leaves. 331

LICHENS, *Baill., Traité de Botanique Médicale Cryptogamique, 56.* 332

Lichens may be defined as cellular perennial cryptogamous plants, furnished with a vegetative system containing gonidia, and a reproductive system consisting of female apothecia and male spermogonia. Many botanists have adopted a theory, originally promulgated by **Schwendener**, according to which Lichens are compound structures, composed of a fungus (the hyphæ) and certain elementary algae (the gonidial elements), the former of which produces the reproductive bodies and is nourished by the latter. For the purposes of this article it is convenient, however, to follow the older system, and to retain the classification and terminology of **Nylander**.

Vern.—*Chharilá, pathar-ka-phúl,* used as a dye=*chulcherila,* HIND.; *Chalchalira, charcharila, aúsneh, hiúnsew,* PB.; *Mota dágada phúl, barik dágada phúl,* BOMB.; *Kulpasi,* TAM.; *Ráti páchi, ratha pu,* TEL.; *Sailaja, saileya,* SANS.; *Ushirah,* ARAB.; *Dowálah,* PERS.

References.—*Acharius, Lichenographiæ; Nylander, Lichenes Scandinaviæ; Krempelhuber, Lichenologie; Lindsay, British Lichens; Crombie, Lin. Soc.; Stewart, Pb. Pl., 269; Elliot, Fl. Andhr., 163; Stirton, on the Lichenology of India, 1879; Encyclop. Britannica, XIV., 552; Pharm. Ind., 258; Ainslie, Mat. Ind., II., 170; O'Shaughnessy, Beng. Dispens., 671; U. C. Dutt, Mat. Med. Hind., 315; Dymock, Mat. Med W. Ind., 2nd Ed., 864; Flück. & Hanb., Pharmacog., 737; Bent. & Trim., Med. Pl., 301, 302; Honigberger, Thirty-five years in the East, II., 299; Royle, Prod. Res., 232; Watts, Dic. Chem , VIII., Pt. II., 1759; Smith, Dic., 144, 244; Ure, Dic. Indus., Arts, and Manu., III., 313; Agri.-Hort. Soc. of Ind.:—Transactions, V., 87, 94, pro., 87, 100; VII., pro., 123, Journals (Old Series), II., 208, sel., 307, pro., 402; IX., 297-300; X., 363; Indian Forester, XIV., 209, 318.*

Habitat and Distribution.—Lichens are chiefly found on the barks of trees, rocks, the ground, mosses, and more rarely perennial leaves. It is to be regretted that the distribution in India of the numerous species of Lichens of economic interest is not accurately known. With the exception of the collection made by **Sir J. D. Hooker** on the Himálaya, and by a few observers in the same region, on the Nilghiris, and in the neighbourhood of a few of the larger towns in the plains (such as those made near Calcutta by **Dr. Watt**), little or nothing has been done towards the extension of a knowledge of Indian lichenology. DISTRIBUTION. 333

A consideration, however, of the very varying climatic and geological conditions and the great range of altitude to be found in India, leads to the conclusion that the numbers of species must be very great, and in all probability most of the forms which possess economic interest are met with in many parts of India. In the following article, therefore, the description will not be confined to forms known to be Indian, but others of economic interest will also be referred to.

Chemical Composition.—Before entering into a consideration of the economic uses of Lichens, it may be advisable first of all to give a brief *résumé* of the known facts regarding their chemical composition. The principal substance occurring in lichens, especially in such as are foliaceous and fruiticulose, is *lichenine*—a special kind of carbohydrate peculiar to these plants. This substance is intermediate in character between dextrin and starch, having the formula $C_6H_{10}O_5$. It is obtained as a jelly-like deposit from an aqueous extract of many lichens, and perhaps exists in greatest abundance in **Cetraria islandica.** By boiling with sulphuric or hydrochloric CHEMISTRY. 334

CHEMISTRY. acid it is converted into sugar. Starch also occurs though rarely in large lenticular grains scattered through the tissues. Chlorophyll exists in small quantities only as do the following substances enumerated by **Nylander,** (1) phosphate of lime, sea-salt, manganese, iron; (2) picrolichenine, variolarine, orceine, cetrarine, inuline, erythrine, rocceline, picro-erythrine; (3) gyrophoric acid, parellic acid, usneic acid, orceic acid, erythrinic acid; (4) sugar (not crystallisable), oil, waxy matter and resinous matter. The ash of lichens is said to constitute from 8 to 10 per cent. of the whole, and consists of various alkalies, alkaline earths and metals, such as potash, soda, lime, magnesia, alumina, silica, manganese, and iron, in combination with certain acids, the principal of which are carbonic, phosphoric, sulphuric, and hydrochloric.

The chemical re-actions effected by several re-agents such as iodine, ammonia, &c., on the thallus of lichens are of great interest, and of considerable importance in determining the systematic position of species, but are not of sufficient economic importance to require description in this work.

GUM. 335 **Gum.**—At one time several species were employed in the arts for the manufacture of a gum as a substitute for gum-arabic. The author of the article "Lichens" in the *Encyclopædia Britannica* states that the chief of these were **Ramalina fraxinea, Evernia prunastri,** and **Parmelia physodes,** all of which contain a considerable proportion of gummy matter and were employed in calico-printing, and for making parchment and card board. Of these the first and last have been found on the North-West Himálaya by **Dr. Watt,** and several species of **Evernia** are also to be met with in India, though **prunastri** does not appear to have been recorded.

DYE. Litmus. 336 Orchil. 337 Cudbear. 338 **Dye.**—The most important utilisation of lichens in the acts is in the preparation of various dyes, the principal of which are known commerce as LITMUS, ORCHIL, OR ARCHIL, and CUDBEAR. Orchil proper is derived from various species of **Roccella,** of which the chief is **Roccella tinctoria.** The value of this lichen varies greatly according to the locality from which it is obtained, the best being derived from the Canaries and Cape de Verde. A certain amount is obtained from Ceylon, where it grows on the stems of certain palms; and in 1886 was said to fetch from £40 to £50 a ton. Samples of the true orchil, **R. tinctoria,** were also sent to the Colonial and Indian Exhibition, as a drug, from Madras, where, however, no knowledge appears to exist of its value as a dye; nor does any account of the amount obtainable, and the locality from which derived, appear to exist. **Lindsay** alludes to **R. fuciformis** as found on the mango tree at Pondicherry.

Orchil differs in appearance according to its mode of preparation, two kinds being distinguished in commerce under the names of red and blue orchil. Both are purplish-red, thick liquids, with an ammoniacal odour, and are prepared by the action of ammonia on the acids contained in the lichen structure. The ammoniacal liquor may be prepared in the ordinary way from sal-ammoniac, lime, and water, or may consist of a mixture of stale urine and lime. The preparation of red orchil differs from that of blue in the fact that heat is used. The addition of potash or soda to the ammoniacal liquor results in the production of a blue colouring matter, LITMUS, a substance largely employed in chemistry, pharmacy, medicine and the arts, as a test for alkalinity or acidity.

Perelle. 339 Another dye product of considerable value is PERELLE prepared from **Lecanora parella** and much used in the preparation of a red or crimson dye. CUDBEAR, a dye stuff of inferior quality to either of the above, is derived from **Lecanora tartarea** and was formerly extensively employed by the peasantry of Northern Europe for giving a scarlet or purple colour to woollen cloths. Neither of these species have been recorded from India, but several other members of the genus do occur, though apparently they are not known to possess any economic value.

DYE.

Amongst other lichens which afford dyes may he mentioned **Parmelia saxatilis, P. omphalodes, P. Acetabulum, P. centrifuga, P. conspersa, P. olivacea, P. physodes,** and **P. stygia,** all of which yield brown dyes. Of these the first, sixth, and seventh have been recorded from the North-West Himálaya; and another species, **P. kamtschadalis,** found in the same region, is largely employed in the north of India in calico-printing, with the object of imparting a peculiar perfume, and a very pale rose tinge to the cloth, **Umblicaria pustulata,** several species of **Gyrophora,** and **Urceolaria scruposa,** have also been employed to yield red or brownish dyes in Europe. Yellow dyes are obtained from **Chlorea vulpina, Platysma juniperinum, Parmelia caperata,** and **P. conspersa, Physcia flavicans, P. parietina, P. lychnea, P. pulverulenta,** and **P. candelaria.** Certain of these are to be found in India, others are represented by different species of the same genus.

Of the numerous dye-yielding lichens above mentioned the true "Orchils" are alone of commercial importance. It therefore appears very desirable that the question of the occurrence of **Roccella** in India should be definitely settled, and that the amount obtainable and the localities where it exists should, as far as possible, be accurately determined.

MEDICINE. 340

Medicine.—Several species of the genera **Cladonia, Usnea, Ramalina, Evernia, Cetraria, Sticta, Parmelia, Physcia, Peltigera, Partusaria,** and **Peltidea** were at one time extensively used as medicine in Europe, most of which enjoyed a great reputation as demulcents, febrifuges, astringents, tonics, purgatives, and anthelmintics. **Peltigera canina** was long regarded as a specific for hydrophobia; **Platysma juniperinum** was supposed to be a sovereign remedy for jaundice, and other species had attributed to them equally imaginary properties.

At the present day only one species is regarded as really of remedial value, namely, **Cetraria islandica,** or the Iceland moss. This, as above mentioned, contains a very large proportion of lichenine, and is undeniably a valuable demulcent in various dyspeptic and pulmonary complaints. Several species of the same genus occur in India, and it is not unlikely that investigation might prove certain of these to possess equally valuable properties. The only lichens which are described as in general use as medicines in India are **Parmelia kamtschadalis, P. perlata,** and **P. perforata.** The first has been well known to Muhammadan writers for many centuries. The author of the *Makhzan-el-Adwiya* describes it as astringent, resolvent, and aperient, and states that when burnt the smoke relieves head-ache, and that the powder is a good cephalic snuff.

Honigberger describes the same species as officinal at Lahore, and writes, "The *Hakíms* administer it in disorders of the stomach, dyspepsia, vomiting, pain in the liver and womb, induration of the uterus, amenorrhœa, calculi, and nocturnal spermatic discharges." He recommends its administration in general pain, salivation, soreness of the throat, and toothache.

Ainslie mentions **Parmelia perlata** and **P. perforata** under the name of **Lichen rotundatus.** He describes it as "a dry, pale-coloured rock-moss, which the *Vytians* suppose to possess a peculiar cooling quality and prepare with it a liniment for the head." In the first half of the present century this lichen attracted attention as a diuretic, for which purpose it was first boiled in water, then beaten into a pulp or bruised in a mortar, and placed as a poultice over the renal and lumbar regions. Owing to its former reputation it has been referred to in the Pharmacopœia of India, but **Waring** remarks that in all probability the continuous application of warmth and moisture by a poultice of linseed or rice would be found equally efficacious.

FOOD & FODDER. 341

Food and Fodder.—Many lichens form valuable articles of food and fodder in the arctic regions and arid steppes of Europe, where food of a bet-

LICHENS.

Economic Properties of Lichens.

FOOD.

ter kind is often scarce or altogether wanting. The principal of these are **Cetraria islandica** (the Iceland moss), **Cladina rangiferina,** and **C. sylvatica** (the Reindeer moss), and several species of **Gyrophora** which constitute the "Tripe de Roche." This substance when boiled is often eaten by the Indians of North America and by Canadian hunters when pressed by hunger. Another is the "Manna Lichen" **(Lecanora esculenta),** which has been described by **Pallas** as occurring on the steppes of Russia in little globular masses, as large as a walnut. **Baillon** writes: "It is believed that these masses, torn from the rocks, can be carried by the wind and deposited so as to form thick beds, which have not unnaturally given rise to the belief in showers of manna. This must to all appearance be the manna of the Hebrews, of which **Parrot** saw in several localities in Persia in 1828, beds of over a foot in thickness. These facts agree with what has been observed in Algeria, in which country men and animals can, in case of necessity, subsist on this lichen. It is very rich in oxalate of lime, and contains nearly a quarter of its weight of lichenine, along with a certain amount of saccharine material, and a small quantity of a nitrogenous matter."

The writer can find only one record of a lichen being employed as food in India. In August 1886, **Mr. Brougham,** Assistant Conservator of Forests, Bellary, drew attention to a lichen called in Telegu *ratha pu*, or "rock flower," which he described as used as a food by the inhabitants of that district, and considered to be a delicacy. It was submitted for identification to Kew, and was pronounced a species of **Parmelia** near **perforata.** This fact was published in the *Indian Forester* for May 1888, with the result that in July of the same year the following further information was communicated. "The *rathi puvu* found in Bellary is probably identical with that found on the large rocks on Horseley konda and Tettu hill in the Cuddapah District. Its use is unknown in the district, but some merchants from Anantapur export a few bandy-loads annually. They pay a seigniorage of R2 per maund of 25lb, and sell the lichen in Anantapur at R10 or R12 per maund. In Anantapur it is eaten with curry, and is also used medicinally. The lichen is collected during the hot weather, *i.e.*, in April and May.

The business is very profitable, and the seigniorage rate might easily be doubled without causing any hardship to the merchants. Collection by departmental agency is being tried at present, and it is possible that this may induce the Anantapur merchants to make higher bids for the right of collection. The cost of collection and export and the profit obtained are as follows:—

	R	*a.*	*p.*
Seigniorage on 40 maunds	80	0	0
Collection of do.	20	0	0
Cartage to Anantapur, 80 miles	8	0	0
Total .	108	0	0
Sale of 40 maunds at R10	400	0	0
Profit .	292	0	0"

DOMESTIC.
342

Domestic.—In the seventeenth century certain species of **Usnea, Ramalina, Evernia,** and **Cladina** were employed in the art of perfumery in Europe. From their supposed aptitude for imbibing and retaining perfumes, these lichens, after being powdered, formed the base of various perfumes such as the "Poudre de Cypre" of hairdressers, but their employment in this way has long died out (*Encyclop. Brit.*). **Cetraria islandica** and **C.**

rangiferina are employed in Sweden as sources of alcohol. They are treated with sulphuric or hydrochloric acid, by which means glucose is formed from the lichenine, which is then allowed to ferment. DOMESTIC.

In India **Endocarpon Moulinsii** is used as a substitute for cork, for lining entomological collecting boxes, &c., for which its large, tough thallus renders it peculiarly suitable.

Líchí, see **Nephelium Litchi,** *Camb.*, Vol. V.; also list of **Fruits,** Vol. III., [448.

LICUALA, *Thunb.; Gen. Pl., III., 928.*

[*ccxxii.*

Licuala peltata, *Roxb.; Griff., Palms of Brit. Ind., 120, Pl.* 343

Vern.—*aPtti, chatta-pat, kuruchipat,* ASSAM; *Kurud, kurkuti,* BENG.; *Sa-lu, salúben,* BURM.; *Kápadah,* AND.

References.—*Roxb., Fl. Ind., Ed. C. B. C., 299; Kurz, For. Fl. Burm., II., 527; Gamble, Man. Timb., 418; Royle, Fib. Pl., 97; Balfour, Cyclop, II., 709; Indian Forester, I., 366.*

Habitat.—A peltate-leaved palm of the forests of Assam, Eastern Bengal, Chittagong, Burma, and the Andamans, said to extend westwards as far as Sikkim.

Domestic.—Griffith writes, "**Major Jenkins** informs me, 'the leaves of the *chatta-pat* are used for the same purposes as those of the *toko*'" (**Livistonia Jenkinsiana**), "'but are much coarser and only made use of by the lower orders. The demand for them is very great, scarcely a single ploughman, cow-keeper, or cooly but has his *jhapee* or *chattah*'" (umbrella-hat or umbrella) "'made of *chattah-pat.*'" The leaves are largely used for thatching, in the Andamans and the hill tracts of Chittagong, and when grass is scarce, are largely exported to the plains for the same purpose from the latter district. In certain parts of Burma they are employed in the manufacture of torches from *In* oil (see **Dipterocarpus tuberculatus,** *Roxb.;* Vol. III., 161). DOMESTIC. 344

LIGNITE.

345

Lignite.—Is a fossil wood, somewhat carbonised, but still displaying its woody structure. It is intermediate between peat and coal, and comprises jet, moor coal, bovey coal, brown coal, and basaltic coal. The following note regarding its occurrence in India has been furnished by **Mr. H. B. Medlicott**:—"Much of the substance called 'coal' and included under that head in the compilation on that subject, particularly that of cretaceous age, might be called lignite. True lignite occurs on the West Coast, near Warakilli in the Travancore State, and again in parts of the Malabar District. Very thick beds of a carbonaceous mud, which, however, can hardly be called lignite, have also been found in the alluvium between Pondicherry and Cuddalore." OCCURRENCE. 346

In addition to the above-mentioned localities, the following are alluded to by various writers. **Atkinson** states that in the North-West Provinces indications of lignite appear near Ránibágh close to Haldwáni in the Barakheri Pass near Bhamauri, and in the streams of the Sub-Himálaya near Najibabad in the Bijnor District (*Him. Dist., 291*). Again, in the Panjáb, traces of **Lignite** are said to be found in the Shekh Budín range (*Dera Ismail Khan Settl. Rep., 34*) and in many parts of the Salt Range, more particularly in the hills between Kálábágh and the Chicháli Pass in Isa Khel. That from the latter source was tried as fuel in the Indus Flotilla steamers for several years before they ceased running up to Makhad, but was found to be deficient in heat-giving power, to give off an immense quantity of smoke, and to be in every way inferior to coal (*Bannu Settl. Rep., 91*). Lignite has also been recorded from Chittagong, Burma, and Assam.

Lignum-aloes, see **Aquillaria Agallocha**, *Roxb.*; Vol. I., 279.

Lignum colubrinum, see **Strychnos colubrina**, *Linn.*; Vol. VI.

Ligusticum Ajawain, *Fleming*, see **Carum copticum**, *Benth.*; Vol. III., [198.

LIGUSTRUM, *Linn.*; *Gen. Pl.*, *II.*, 679.

A genus of shrubs or trees, which belongs to the Natural Order OLEACEÆ, and comprises 25 species, of which 11 are natives of India. **L. vulgare**, the privet, is a common hedge plant in Europe.

[*Ic.*, *t.* 1243; OLEACEÆ.

347 **Ligustrum neilgherrense**, *Wight*; *Fl. Br. Ind.*, *III.*, 615; *Wight*,

Syn.—L. CANDOLLEANUM, *Dcne.*, *not of Blume*; OLEA ROXBURGHII, *Wall.*

References.—*Dalz. & Gibs.*, *Bomb. Fl.*, 159; *Lisboa*, *U. Pl. Bomb.*, 98.

Habitat.—A large shrub or small tree, commonly met with on the hills of Southern India.

TIMBER. 348 **Structure of the Wood.**—"Light-brown, rather close-grained, and durable; generally used at Mahableshwar in the construction of huts and for fuel" (*Lisboa*).

349 **L. Roxburghii**, *Clarke*; *Fl. Br. Ind.*, *III.*, 615; *Wight*, *Ic.*, *t.* 1242.

Syn.—LIGUSTRUM CANDOLLEANUM, *Blume*, *not of Dcne*; L. ROBUSTUM, *Bedd*, *not of Blume*; L. COMPACTUM, *Dcne.*, *not of H. f. & T.*; OLEA ROXBURGHII, *Wall.*, *not of Spreng. nor of Wight*; VISIANIA ROBUSTA, *Wight*; PHILLYREA TERMINALIS, *Roxb.*

Considerable confusion exists in Indian economic literature regarding this plant. Most writers appear to have followed **Brandis** in giving the economic information of this species under **L. robustum**, *H. f. & T.* (**L. robustum**, *Blume*); and the same author has still more added to the confusion by giving two North-West Provinces vernacular names for that species, which in all probability ought to be applied to **L. compactum**, *H. f. & T.* or **L. nepalense**, *Wall.* In any case, they can belong neither to **L. robustum**, *Blume*, a Bengal, nor to **L. Roxburghii**, *Clarke*, a West Indian, plant.

References.—*Brandis*, *For. Fl.*, 310; *Beddome*, *For. Man.*, 154; *Gamble*, *Man. Timb.*, 258; *Lisboa*, *U. Pl. Bomb.*, 211; *Ind. Forester*, *II.*, 25; *VIII.*, 410.

Habitat.—A small tree, common on the mountains of Western India, at altitudes of 4,000 to 7,000 feet.

DOMESTIC. Bark. 350 **Domestic.**—In South India the BARK is put into the toddy of **Caryota urens** to cause rapid fermentation.

LILIUM, *Linn.*; *Gen. Pl.*, *III.*, 816.

351 **Lilium giganteum**, *Wall.*; *Baker*, *Tulipeæ*, *in Jour. Linn. Soc.*, [*XIV.*, 227; LILIACEÆ.

Syn.—L. CORDIFOLIUM, *Don.*

Habitat.—A large lily of the Temperate Himálaya, from Kumáon and Garhwál to Khásia and Sikkim, at altitudes of 5,000 to 10,000 feet.

MEDICINE. Leaves. 352 **Medicine.**—The LEAVES are employed as an external cooling application to alleviate the pain of wounds and bruises.

[*XIV.*, 228.

353 **L. Wallichianum**, *Schultes fil.*; *Baker*, *Tulipeæ*, *in Jour. Linn. Soc.*,

Syn.—LILIUM BATISUA, *Ham.*; L. JAPONICUM, *D. Don*; L. LONGIFLORUM, *Wall.*

Vern.—*Findora*, HIND.

References.—*Atkinson, Him. Dist., 319, 740; Indian Forester, II., 26, 27.*

Habitat.—A herb of the Central Sub-temperate Himálaya, at altitudes of from 3,000 to 4,000 feet.

Medicine.—The dried bulb SCALES possess demulcent properties and are used like salep in pectoral complaints (*Atkinson*). MEDICINE. Scales. 354

Limbolee oil, see **Murraya Kœnigii,** *Spr.;* Vol. V.

Lime (Mineral), see **Carbonate of Lime,** Vol. II., 142—152.

Lime (Fruit), see **Citrus Medica,** *Linn.;* **var. 3, acida,** and **var. 4, Limetta,** Vol. II., 355, 357.

LIMNANTHEMUM, *S.P. Gmel.; Gen. Pl., II, 819.*

Limnanthemum cristatum, *Griseb; Fl. Br. Ind., IV., 131;* [*Wight, Ill., t. 157, bis. fig. 4;* GENTIANACEÆ. 355

Syn.—MENYANTHES CRISTATA, *Roxb.;* VILLARSIA CRISTATA, *Spreng.;* V. INDICA, *Wall.*

Vern.—*Tagarmul, cumuda, cairava, ghain-chú,* HIND.; *Tagarmul, shiulichhop, chúli,* BENG.; *Khatara, kumuda,* BOMB.; *Kolaréchikal,* MAR.; *Antara tamara,* TEL.; *Kálanusárivá, sitali, kúmúdwuti,* SANS.

References.—*Roxb., Fl. Ind., Ed. C.B.C., 154; Rheede, Hort. Mal., XI., t. 29; Thwaites, En. Ceylon Pl., 205; Dalz. & Gibs., Bomb. Fl., 158; Sir W. Jones, Treatise Pl. Ind., V., 35; U. C. Dutt, Mat. Med. Hind., 302, 318; Lisboa, U. Pl. Bomb., 165, 201; Gazetteers:—Bombay, XV., 438; N.-W P., IV., lxxiv.; Ind. Forester, XIV, 392.*

Habitat.—A common aquatic herb throughout India.

Food and Fodder.—The STEMS, FRUIT, and LEAVES are eaten in certain localities at all times, curried or boiled; in others they are employed only in times of scarcity. FOOD & FODDER. Stems. 356 Fruit. 357 Leaves. 358

L. nymphæoides, *Link; Fl. Br. Ind., IV., 131.* 359

Syn.—MENYANTHES NYMPHÆOIDES, *Linn;* VILLARSIA NYMPHÆOIDES, *Vent.*

Vern.—*Kúrú, khair posh, gul jafari purnka,* PB.

References.—*Stewart, Pb. Pl., 148; Honigberger, Thirty-five years in the East, II., 364.*

Habitat.—A common aquatic herb in the lakes of Kashmír, at altitudes of 6,000 to 9,000 feet.

Medicine.—Honigberger states that the fresh LEAVES are useful in cases of "periodic headache." MEDICINE. Leaves. 360

Fodder.—Largely used as fodder in Kashmír. Honigberger remarks that it is supposed by the natives to increase the flow of milk in cows. FODDER. 361

LIMONIA, *Linn.; Gen. Pl., I., 303.*

Limonia acidissima, *Linn.; Fl. Br. Ind., I., 507;* RUTACEÆ. 362

Syn.—L. CRENULATA, *Roxb.;* HESPERETHUSA ACIDISSIMA, CRENULATA AMBIGUA, *Rœm.*

Vern.—*Beli,* HIND.; *Belsian,* CHUTIA NAGPUR; *Bhenta,* URIYA; *Bali,* N.-W. P.; *Keiri, kara,* MERWARA; *Rán-limbú, naringi,* BOMB.; *Kawat, nai-bél,* MAR.; *Tór-elaga,* TEL.; *Naibela, nai-bél,* KAN.; *Tsjeru, caabnaregam,* MALAY.; *Thihay-asa,* BURM.

References.—*Roxb., Fl. Ind., Ed. C.B.C., 364; Brandis, For. Fl., 47; Kurz, For. Fl. Burm., I., 192; Beddome, Fl. Sylv., Anal. Gen., xlv.; Gamble, Man. Timb., 59; Grah. Cat. Bomb. Pl., 23; Dalz. & Gibs., Bomb. Fl., 29; Elliot, Fl. Andhr., 183; Dymock, Mat. Med. W. Ind., 2nd Ed., 32; Pharmacog. Ind., I., 267; Atkinson, Him. Dist., 307, 740; Ec. Prod., N.-W. P., Pt. V., 49; Lisboa, U. Pl. Bomb., 33, 274; Indian Forester, III., 200; XII., App., 9; Gazetteer, N.-W. P., lxix.*

Habitat.—A spinous glabrous shrub or small tree, native of dry hills in various parts of India. It has been recorded from the North-West Himálaya in Simla and Kumaon, ascending to 4,000 feet, from the Monghyr hills in Behar, from Assam, and from the Bombay ghâts and Coromandel southwards. It flowers in April-May, and fruits in the rainy season.

MEDICINE. Leave. 363 Root. 364 Fruit. 365 **Medicine.**—**Rheede** noticed the medicinal virtues of this plant, writing, "The LEAVES are supposed to be a remedy for epilepsy; the ROOT is purgative, sudorific, and employed for the cure of colic and cardialgia. The dried FRUIT is tonic, diminishes intestinal fermentation, has the power of resisting the contagion of small-pox, malignant and pestilent fevers, and is also considered an excellent antidote to various poisons, on which account it is much sought for, and forms an article of commerce with Arab and other merchants." **Graham, Drury,** and others quote the above, but give no record of actual observation of the uses and value of the plant as a medicine. **Lisboa** states that the berry is much used as a tonic in Malabar, where it forms an article of commerce, and that its red coloured mucilage is considered to be an antidote against snake-bite and the poisons of other venomous animals.

FOOD. Fruit. 366 **Food.**—The FRUIT, a round berry, is very acid, but is occasionally employed as an article of food by Natives. The authors of the *Pharmacographia Indica* state that it is exported to the Arabian coasts, where it is used as a condiment with fish, meat, &c., being powdered along with the spices commonly used in cooking.

TIMBER. 367 **Structure of the Wood.**—Yellow, hard, and close-grained, weight 59℔ per cubic foot; used for axles of oil-presses, rice-pounders, &c., and adapted for the lathe. (*Lisboa*).

DOMESTIC. Fruit. 368 **Domestic.**—**Drury** remarks that the FRUIT is used in Java instead of soap, a use which the authors of the *Pharmacographia Indica* state to be also known in India, being indicated by the Marathi name, which signifies "Barber's Bael-fruit."

LINARIA, *Pers.; Gen. Pl., II., 932.*

Linaria glauca, *Spreng.;* see **Schweinfurthia sphærocarpa,** *A. Braun.;* Vol. VI.

369 **L. ramosissima,** *Wall.; Fl. Br. Ind., IV., 251; Wight, Ill. t., 165.*

Syn.—LINARIA ROYLEI, *Chavannes.*

Var. pubescens, *Stocks ms.*

References.—*Dalz. & Gibs., Bomb. Fl., 176; Stewart, Jour. of a Bot. Tour in Hazara, &c., in Agri-Hort. Soc. of Ind., Jour. (Old Series), XIV., 6, 33; Murray, Pl. and Drugs of Sind., 179; Gazetteers:—Bomb., V., 27; N.-W. P., I., 83; IV., lxxv; X., 314.*

Habitat.—A spreading, branched herb, met with on rocks and stony places throughout India, from the Panjáb and Sind to Chittagong and Ceylon, ascending to an altitude of 5,000 feet; distributed westwards to Afghánistán and eastwards to Ava.

The variety **pubescens** has been recorded by **Stocks** from Sind, and may probably be the **L. cirrhosa** described by **Murray.**

MEDICINE. 370 **Medicine.**—This species (including the variety, if the above supposition be correct,) is valued in Sind as a remedy for diabetes (*Murray*).

LINDENBERGIA, *Lehm.; Gen. Pl., II., 948.*

[SCROPHULARINEÆ.

371 **Lindenbergia urticæfolia,** *Lehm.; Fl. Br. Ind., IV., 262;*

Syn.—STEMODIA RUDERALIS, *Vahl.;* S. MURALIS, *Roxb.;* BRACHYCORIS PARVIFLORA, *Schrad.*

Vern.—*Dhol,* MAR.

References.—*Roxb., Fl. Ind., Ed. C.B.C., 490; Dalz. & Gibs., Bomb. Fl., 176; Dymock, Mat. Med. W. Ind., 581; Indian Forester, XII. (App.), 18; Gazetteers:—Bombay, V., 27; N.-W. P., I, 83; X., 314.*

Habitat.—A diffuse, downy annual, common on walls and banks throughout India, from Jamu to the Nilghiris and Tenasserim, ascending to 6,000 feet on the Himálaya; distributed westwards to Afghánistán.

Medicine.—**Dymock** writes, "The JUICE is given in chronic bronchitis, and, mixed with that of the Coriander, is applied to skin eruptions. The plant has a faint aromatic odour, and a slightly bitter taste." MEDICINE. Juice. 372

LINDERA, *Thunb.; Gen. Pl., III., 163.*

Lindera Neesiana, *Benth.; Fl. Br. Ind., V., 186;* LAURINEÆ. 373

Syn.—APERULA NEESIANA, *Blume;* BENZOIN NEESIANUM, *Nees;* TETRANTHERA NEESIANA, *Wall.;* T. PRUNIFOLIA, *Wall;* LAURUS MACROPHYLLA, *Don.*

Reference.—*Kurz. For., Fl., II., 309.*

Habitat.—An aromatic tree met with in the Temperate Himálaya from Nepál to Sikkim, at altitudes of 6,000 to 8,000 feet, also in the drier hill forests of Burma, from Martaban down to Tenasserim—between 4,000 and 5,000 feet.

Medicine.—**Kurz** remarks, "Yields excellent sassafras." It seems possible that this plant may produce part of the "Sassafras of Nepál," supposed to be principally derived from **Cinnamomum glanduliferum,** *Meissn.* (See Vol. II., 317.) MEDICINE. 374

L. pulcherrima, *Benth.; Fl. Br. Ind., V., 185.* 375

Syn.—DAPHNIDIUM PULCHERRIMUM, *Nees.*

Vern.—*Dadia,* HIND.; *Dingpingwai,* KHASIA; *Sisi,* NEP.; *Nupsor,* LEPCHA.

References.—*Brandis. For. Fl., 383; Kurz., For. Fl. Burm., II., 306; Gamble, Man. of Timb., 312; List of Trees, &c., of Darjiling, 65.*

Habitat.—A large tree with hoary branchlets met with on the Temperate Himálaya from Kumaon, at altitudes of 5,000 to 8,000 feet, to Sikkim and Bhután at altitudes of 4,000 to 9,000 feet; also on the Khásia mountains between 5,000 and 6,000 feet, and on hills in Martaban ascending to 6,000 feet.

Structure of the Wood.—"Reddish-white, moderately hard, even-grained, weight 33 to 40lb per cubic foot, used for building, cattle yokes, and occasionally for tea-boxes" (*Gamble*). TIMBER. 376

LINOCIERA, *Swartz.; Gen. Pl., II., 678.*

Linociera intermedia, *Wight; Fl. Br. Ind., III., 609; Wight, Ic., tt. 735, 1245;* OLEACEÆ. 377

Syn.—CHURIANTHUS INTERMEDIA, *Bedd.*

Var. Roxburghii, OLEA PANICULATA, *Roxb.;* O. ROXBURGHII, *Spreng.;* O. ROXBURGHIANA, *Koem. et Sch.*

References.—*Dalz. & Gibs., Bomb. Fl., 159; Bedd., Fl. Sylv., t. 239; For. Man., 153; Lisboa, U. Pl. Bomb., 97.*

Habitat.—A small or middle-sized glabrous tree met with on the Nilghiris, from 1,000 to 6,000 feet, also on the Anamallays, 5,000 feet. The variety **Roxburghii** is recorded from Orissa and the Circars, Chutia Nagpúr, the Western Ghâts, and the Siwaliks.

Structure of the Wood.—**Beddome** states that the typical species is valued as a timber by the Natives. **Lisboa** writes of the variety, "Wood pale-brown, hard, close-grained, durable, used for agricultural implements and turning." TIMBER 378

the United States
asters